Managing Design

Managing Design

Conversations, Project Controls, and
Best Practices for Commercial Design
and Construction Projects

Michael Alan LeFevre

WILEY

Registered Office
John Wiley & Sons, Inc., 111 River Street, Hoboken, NJ 07030, USA

Editorial Office
111 River Street, Hoboken, NJ 07030, USA

For details of our global editorial offices, customer services, and more information about Wiley products visit us at www.wiley.com.

Wiley also publishes its books in a variety of electronic formats and by print-on-demand. Some content that appears in standard print versions of this book may not be available in other formats.

Library of Congress Cataloging-in-Publication Data is Available:

ISBN 9781119561767 (Hardback)
ISBN 9781119562009 (ePDF)
ISBN 9781119561972 (ePub)

Cover design: Wiley

Set in 9.5/12.5pt and Life LT Std by SPi Global, Chennai, India

V10010203_051119

Contents

Preface

It's Time

Managing design is an oxymoron but can be done – by those who want to and know how. This is a book for readers who seek to fulfill those conditions – one that explores the hearts and souls, minds and processes of design and construction collaborators. What core issues create this conundrum? Why is it so hard for designers to grapple with the constructs of budget and schedule control? How can team members understand one another better to mutually intensify the value of their teamwork? These decades-old questions have resurfaced with the convergence of pressures such as more participants, growing complexity, and hyper-track schedules.

In the thirty years since Chuck Thomsen's *Managing Brainpower* books were published, we've seen significant change in the design and construction industry. In the span of these decades, technological change such as the advent of CADD and BIM, and new contracting approaches like Integrated Project Delivery (IPD), have had dramatic impacts on design practice and the ways projects are delivered. Who could have imagined virtual meetings, Web-based access, remote computing, and the schedule compression and new collaboration forms we've witnessed? In the last few years, Randy Deutsch has addressed this change and looked to the future of practice. In the foreword I've asked both to link past to present to frame current industry challenges.

What challenges are inherent in managing design? To begin with, centuries-old cultural norms – the very belief systems of the design and construction professions – cling to individual prowess and ego as primary values. A predominant white, male culture lingers. The number of players on project teams is ever increasing, as is the complexity of designing and building, and doing it sustainably. Uncertainty in global sociopolitical and economic arenas compounds tight schedules and budgets. For the past few decades, an industry typically slow to change has been forced to evolve and assimilate automation at record speeds with dizzying results. Finally, economic downturns and recessions have impacted the profession's talent pool. If change is in the wind, a new understanding of these forces, those who use them, and their minds and processes is needed.

This book is an "exposé." It exposes issues and poses solutions to longstanding industry ills. The design-construction landscape is shifting in response to "chaos theory" intervening events – forces that threaten our stability. As we redesign ourselves to adapt, talk and action are needed. In this book, I give you something you can't get anywhere else: an *insider's* look at the mysteries of managing design for yourself, team, firm, and future, a self-help book for design and construction.

Managing Design is written to help you find your place in today's design management continuum. What you do when you find that place, and how you advance it for future projects and generations, is up to you.

Michael Alan LeFevre

This book is for all who have faced the challenge of trying to take a circular, messy, seemingly random, creative process and force it into a neat, linear, objective process with deadlines.

This is my dad going to work. He draws buildings.

FIGURE 0.1   This Is My Dad Going to Work, drawing courtesy Danielle LeFevre.

Foreword

Charles Thomsen

FAIA, FCMAA, PAST CHAIRMAN 3D/I, AUTHOR, MANAGING BRAINPOWER, 1989

The Question

The question Michael poses so eloquently is: *Can design be managed? If so, how?*

I know of no more important issue in the construction industry. Design shapes our environment, the quality of our buildings, and the productivity of construction – the nation's second largest industry, and the world's largest. In the 20th century, the number of subcontractors on projects went from very few to a great many. Less work was done in the field. Since project teams were populated with companies that hadn't worked together, complexity escalated. Management became as important as craftsmanship and knowing how to put a building together. In response, colleges and universities added construction management curricula. "Construction management" became a profession, a delivery strategy, and a contract form.

Similar changes happened in design. Subconsultants increased. Global organizations replaced local sources. Industrialization offered an infinite number of new products and technologies and buildings became more sophisticated. By the 21st century, most of the knowledge of construction technology and cost resided with subcontractors and manufacturers, not architects and engineers. Designers changed from being experts in construction to understanding complicated user needs, evaluating technology developed by others, and integrating it all. CM, Bridging, Design Assist, and IPD emerged to bring more collaboration to extended project teams that included contractors and subcontractors.

The changes in design and construction are similar in both design and construction organizations. But while the construction industry has developed management tools, little attention has been given to developing tools for managing design – no new curricula in colleges and universities, no new profession. The unwritten assumption is that construction managers will manage project delivery of design and construction. Not a bad idea, but design – so different from construction – requires different management concepts.

Managing Design Is Different from Managing Construction

Construction involves a series of definable tasks in predictable sequences using calculable resources. A construction manager can plan a fixed number of masons, a finite number of days, and a specified number of bricks to build a wall of a given size. It is subject to analysis and modification: if the number of masons is increased, the days can be decreased.

But the ideas and decisions that led to the design of that wall, its material choices, their relation to other building systems, and the cost and the functional and aesthetic requirements weren't governed by

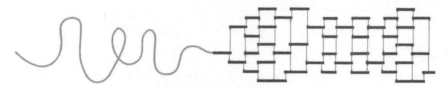

FIGURE 0.2 Design versus Construction Process, Chuck Thomsen.

a finite plan. While construction of a brick wall is complete when the last brick is in place, there's no objective measure of when a design is "done." Builders want decisions. Designers want time. An effort to improve has no definable end. Design deadlines, while necessary, become arbitrary. As projects progress, designers understand them better, have epiphanies, discard work, double back, and redesign. Builders, if they want to stay in business, don't do that. In construction, excellence in craftsmanship, schedule, and logistics is clear. In design, functional and aesthetic excellence lack objective yardsticks. Construction demands sequential discipline: place rebar before pouring concrete; build foundations before walls. Not true with design. An architect can think about cabinet details and the site plan on the same day.

Construction teams focus on a clear goal: build what is specified, efficiently, profitably. Design teams are divided. Some may be production and profit oriented, others interested in aesthetics or functional optimization. Some will be inclined to ask for more time to study the problem in search of a better solution. But "better" may be defined in personal terms and invite disagreement on what that means. A generation of construction managers appeared at the end of the 20th century, educated to train contractors and use tools appropriate for construction. When those same tools don't work with design, CMs feel designers are "flakey" and "unbusinesslike." Designers mutter about crass contractors who only care about cost and schedule. Stereotyping raises its ugly head. Collaboration suffers.

Construction is resource and production intensive. It deals with materials and defined results. The sequence of work can be mapped and managed. Design is information and decision intensive. One must conceive the processes to acquire information, organize ideas, prioritize issues, and formulate decisions. It's a lot fuzzier than pouring concrete and laying brick. Design deals with ideas and inexplicit issues among diverse people with varied legal relationships. The path and result are difficult to map.

So How Do You Manage Design?

Design shapes our world. Managing it is crucial, yet little guidance exists for how to do it. But do it we must. Projects can't be run like artist colonies. Clients want to know what their projects will look like, what they will cost, and when they'll be done. Design controls the answers to those questions. It must be managed. Michael LeFevre has attacked these questions by interviewing thought leaders. The conversations offer a variety of opinions. Michael's discussions include how to connect with clients, education, and the designer's mentality. These interchanges offer fuel for thought for design and construction teammates in the future. Most important, he discusses the growing expectation that construction managers and trade contractors must help shape designs with constructability, cost, and schedule input. Like the design process he describes, readers—present or future design managers—are presented with an opportunity to digest the data, grasp the ideas, and plot a course. Good luck!

Chuck Thomsen

Randy Deutsch, AIA, LEED AP
<small>Author, Superusers: Design Technology Specialists and the Future of Practice, 2019</small>

Design Isn't What It Used to Be

What we call *design* continues to evolve at an ever-increasing clip. With data-driven design we design by manipulating data, not form. With generative design we design leveraging algorithms and parametric modeling with predetermined constraints. Design is changing due to the introduction of new computational tools, including algorithms. Soon, AI-enabled design will be informed and improved in pre-design by post-occupancy evaluations that take place before the project is even designed. Design professionals today use visual programming tools to automate and complete work in hours that might otherwise take days. The cloud enables data visualizations to be a real-time product of the design act, something that designers working with their managers used to undertake as a separate activity. No more. Many activities designers do today can be transformed into data, and many design process tasks have been automated.

Indeed, much of design management has been outsourced – not to countries or people, but to software – and will increasingly be in the years ahead. Will design management still be needed as we further redesign design? The answer depends on whom you ask.

Why a Book of Conversations?

In the spirit of MIT professor and author Sherry Turkle, Michael LeFevre does his part in this book to reclaim conversation at a time witnessing the rise of technology and all its negative effects. In this time of technology and machine solutions, people – with their attendant conversation, connecting and collaborating – become more important. We are finally learning that quicker and more convenient communication is not necessarily better communication; where reading interviews – reliving conversations that have already taken place – slows down time, providing the reader with the opportunity to digest and question.

The heart of the book is not a collection of isolated interviews but a series of connected conversations. What kind? Conversations that argue for a larger role for the architect, a larger outlook and understanding of their role in the construction process and industry; conversations on the ambiguities and uncertainties of managing design in an industry ripe for transformation; conversations that, as Michael asserts, take the pulse of the profession, conducted live, in one sitting, recorded and transcribed nearly verbatim. Having participated, I can attest they were indeed intimate, candid, and substantial. Reading this book is like eavesdropping on a stimulating conversation among industry stalwarts at a dinner party.

Why Now?

Today, as we more and more design with the end in mind, design and construction are becoming increasingly indistinguishable. Written at a time of interdisciplinary collaboration when project phases are merging, disciplines blurring, roles blending, and tools converging; when architects are increasingly moving into means and methods, and builders are increasingly providing design services. It's a time when more firms and emerging professionals are exploring vertical integration and entering the entire project pipeline. One can imagine in the not-too-distant-future "design'" no longer as a stand-alone phase, a world where design technologists become just technologists. Does it still make sense to address the management of design as a separate subject? As Philip Bernstein FAIA, whose conversation appears in these pages, has suggested elsewhere, in the near future "there will be a stronger connection between what's designed, built, and how it operates." And with that stronger connection comes a transformation of design management. But to *what*, exactly?

One Well-Connected Author

What makes this book unique and worth reading? In two words: Michael LeFevre. Michael is particularly well-qualified to write a cross-industry book on managing design as he has been *living* it for five decades. His unique perspective, stemming from having practiced as both architect and construction manager – and their connector – offers much.

What also makes this book unique and worth reading is Michael's incomparable network of industry luminaries. I don't know of anyone else in the industry with such a wide network of industry experts, or who else could have written this, a book of conversations that form a foundation for industry change. Only Michael. While this book has but one author, as a collaborative effort it comprises the collective intelligence of many voices. Take Chuck Thomsen. Chuck is an architect, construction manager, corporate executive, and educator. So too is Michael. And now, with this book – like Chuck – he is now also an author. Just as I reached for Chuck Thomsen's *Managing Brainpower* in 1989 (the original small box set of three books is still here next to me as I write this), if I were starting off in the industry today, I would undoubtedly reach for Michael LeFevre's *Managing Design*.

The book may be called *Managing Design,* but the theme is unmistakably that of *change*. A fount of institutional knowledge himself, Michael conducted interviews with more than 40 industry leaders, each, as Michael says, with *a passionate agenda for change*. So much has changed in the years that separate the publication of these books. But one thing hasn't: *conversation*.

These conversations represent multiple voices and points of view from all core project parties – owners, architects, engineers, contractors – as well as the many industry advancers critical to a thorough understanding of managing design. Listening in to these varied perspectives will help you think like others on project teams, making you a more effective communicator and, importantly, an empathetic team member.

By understanding what's important to others, we can more effectively shape our message. Together, what they share offers a valuable look at the state of design and construction management in the U.S. Just as a rising tide lifts all boats, LeFevre talks to industry leaders to elevate the rest of us. Like an architectural Robin Hood, he takes from the experience-rich to reward those hungry for knowledge and hard-won wisdom.

The Never Futile and Always Sisyphean Task of Managing Design

What Michael LeFevre has attempted to accomplish in this book is to capture the current knowledge, thinking, and insights of an industry in a moment of transition, before it is lost. Michael's observations from leading industry change agents and thought leaders led him to, as he puts it, "a series of deeper conversations – to help us understand why our siloed professions persist – and what we can do about it." The book does not merely bemoan the many familiar inefficiencies, obstacles, and challenges to project success; it poses solutions: insights, strategies, and propositions that readers can "test the wheels of" and apply to their own organizations and practices.

Architects may be comfortable with ambiguity. Many others aren't. Owners want certainty – in a project there's too much risk, too much unknown at stake. Managing the design process takes a design from a state of uncertainty and ambiguity to one of certainty and clarity – something every owner wants and can appreciate. Similarly, through a litany of new tech tools and collaborative workflows, we can move from a complex, time-intensive design to one that is instantaneous and simple.

Recent research shows that managers are overconfident about their skills. And, per the Dunning-Kruger effect, the worse they are, the better they think they are. This book is a necessary antidote to this illusory bias and will stop readers from confusing confidence with competence. Good design management is about creating good experiences. You may not be able to pinpoint what design management is, but you always recognize an enjoyable project experience when its design is managed well. Good design management is about creating good project experiences.

Will Reading These Conversations and Ideas Help Us Work Together?

I believe so. But reading isn't enough. It is up to the reader to take the advice presented here and apply it. At this watershed technological moment, change is no longer on the horizon. We're in the midst of it. War soldiers famously carried with them to the front commonplace books containing snippets of wisdom serving to keep their spirits up, lift their morale in troubled times, and keep them focused on what's important. You can think of this book of conversations and principles as a commonplace book for our profession and industry in this time.

Dip in, find a useful nugget of first-hand industry experience, meditate on it, then apply it, and move on to another. Arm yourself with all the current knowledge you can find, none of which incidentally can be found on the Internet: book knowledge combined with the most expert, useful, and actionable insights available. That's why a book like this is a great resource, helping you navigate the best of both worlds. Herein, you'll find decisive direction that will help you and your organization not only manage, but lead, change. As Michael says of these conversations in the introduction, "via shared history and future visioning, they posit a professional GIS system using the wisdom of many." Consider this book your professional GIS system.

What could possibly be of more use in this time of uncertainty?

Randy Deutsch

Introduction

Premise

Managing design is difficult. To tackle the task, this book speaks across traditional industry silos. For designers: "Why are contractors so impatient? Why don't they understand design?" For contractors: "What motivates architects and engineers?" For owners: "Why can't they get along? Why is this so difficult?" For students and teachers: "What do we need for future practice? We've proven we need each other, but can we *understand* one another better to collaborate at higher levels?" For all of us: "What does the future hold and can we shape it?"

> *"Opportunity dances with those on the dance floor."*
>
> – Anonymous

The following interviews with industry leaders—owners, architects, contractors, and academics—help us answer these questions and presents a passionate agenda for change. Dialogues with diverse, experienced leaders have the power to expand our understanding. What makes these conversations on the ambiguities of managing design valuable? Honesty, awareness, and empathy. The interviewees' openness on the issues and solutions around managing design shows that at the heart of our industry and its future is humanism.

> *"Yes, risk-taking is inherently failure-prone. Otherwise, it would be called 'sure-thing-taking.'"*
>
> – Jim McMahon

How better to understand our teammates than to hear them speak? Discussions on the "dark side" of architect, contractor, and owner relationships are uncomfortable. They are also a prerequisite to navigating between creativity and discipline. The inherent conflicts between these groups make our work challenging, rewarding and infinitely human. Maybe by making ourselves aware of our team members' issues, we will learn to appreciate their point of view. In project debriefings and discussions with design partners and clients, the answer ultimately comes down to one thing: *people*. Team trust and communication is part of that dialogue. In a profession fraught with methodological obstacles, egos, complex programs, and evolving toolsets, the issue always comes back to *people*. I talked to dozens of them, experts all. They offer a valuable look inside the state of design and construction. And a way forward. As a management model, the Project Design Controls framework in Part 2 does too. As you contemplate your next project or career move, ask yourself, do you care—*about others?* Do you really have a team? In a team you have a chance to be a multiplier—part of something larger than yourself. Do you have the fundamentals to manage design?

"Necessity is the mother of taking chances."

– Mark Twain

Mission

In 1997, after a thirty-year career as a practicing architect, I felt the need for industry change. Witnessing the emerging digital revolution and complexities, and recurring overbudget, late designs, I was compelled to create an opportunity for contractors and designers to work better together. I had lived it myself. After too many budget-busting nightmares I had the opportunity to work with Atlanta-based Holder Construction Company on two projects: Zoo Atlanta's Action Research Conservation Center and WXIA-TV's Newsroom Studio. In both instances we collaborated to resolve the challenges and build successful projects. We created a new position—a role focused on connecting designers and contractors. Holder's Planning & Design Support Services group was born. I changed my life for this purpose: I switched careers to build this bridge. While architectural colleagues accused me of moving to the "dark side," my newfound construction associates believed I had "seen the light." Whatever the illumination, I have spent the past twenty years working with more than eighty design firms throughout the country, in the penumbra between design and construction, enabling and managing design on projects.

In that time, my colleagues and I have continued to respond to on-project challenges in managing design. In each case, we applied tools developed in response to project needs to enable teamwork. We called the process "design management." Like the slogan from chemical company BASF, design managers "don't *make* the designs or the buildings, they make the collaboration *better*." With company support, I compiled these practices to share with a wider audience. This book is the result. Use it to get better projects and more collaborative teams.

Perspective

My years under a hard hat deepened my appreciation of design leadership. Leading a design team while conquering fee, schedule, and budget bogeymen is not easy. Teams need design managers, but not to apply rote management practices. Teams need leaders who "get" design. Designers do too. Working in one of the country's best construction management companies, I have found ways to bridge design and management differences. That cross-industry perspective drove me to share my insights.

"Literary style is nothing less than an ethical strategy—it's always an attempt to get the reader to care about people who are not the same as he or she is."

– Zadie Smith

Surveying my personal career asymptote made me abandon the status quo. It was time to buck inertia – and my comfort zone – to take on a larger cause. This is the book I *had* to write. In it you'll find issues and opinions, problems and solutions. One thing you won't find: apathy. People care deeply about designing and building. It's profound. When you talk to people who make buildings for a living, you will not find nonchalance, you will find passion. That was a joy to affirm and gives me hope. Look at the interviews. None of them say things are fine, or that they don't care. If you're reading this, you feel the same way.

Owners, designers and builders share another trait. They have a common desire to create *new* realities. Facing obstacles like conflicting information and limited resources, the best of us stay positive. What we need now is a new reality for how we work together – positive thinking about "managing design."

While many recognize and discuss these problems, few are moved to action. In the decision to redirect my career, I decided to do something about them. That choice changed my life and broadened my reach. I am trying to cultivate others to do the same.

"Watch out – he's a dual agent!"

– Antoine Predock, FAIA, AIA Gold Medalist

"Dual Agent"

In 2002, my employer, Atlanta-based Holder Construction, was working on the Flint RiverQuarium in Albany, Georgia. Holder served as construction manager and worked on the project with Antoine Predock Architects. Their project architect was Sam Sterling. As Sam and I got to know each other in the early days, I proudly described my role: *"As someone who has practiced on both sides, I can speak two languages! I can translate design intent into constructible form."* Sam looked at me and said, *"You know Mike, I mentioned that to Antoine. He said, 'You know what that means. It means: watch out, he's a dual agent!'"*

It gave us a good laugh, even though the sentiment is emblematic of the mistrust that lurks in the weeds of our professions. We've got some weeding and fertilizing to do.

"Without you guys pulling us out of the budget inferno, we wouldn't have had *a project."*

– Sam Sterling, AIA

A few years later, I ran into Antoine at the American Institute of Architects' national convention in Los Angeles, moments after he had been awarded the AIA's Gold Medal before thousands in the plenary session. After I congratulated him, he was quick to remember "how great it was to work with Holder." We'd succeeded in getting our project back in budget and realizing its vision. (See Case Study 3 in Chapter 24.) His. Ours. A nice closure. His colleague Sam Sterling would later say, "Without you guys pulling us out of the budget inferno we wouldn't have *had* a project."

Creating more of those kinds of memories is what this book is about.

Methods

I interviewed more than forty people for this book. Many I know from working together and sharing a passion for the subject. Others represent an important position on one or more issues. The group's initial composition held some of the familiar, closed-culture thinking I sought to expel. At the suggestion of Rebecca W. E. Edmunds, the demographic expanded to include broader perspectives, more women and ethnicities, millennials and younger contributors – voices I hadn't yet sought. The collection and conversations got richer. Questions were tailored to each respondent's background and sent in advance. Interviews were conducted live, recorded, and transcribed with modest editing for brevity. The expanded reach needed a convergence: principles to bring it together to apply on projects. You will find that in Part 2, "Project Design Controls."

Evocation, Provocation, and Paths

Can talking with people through interviews evoke enough emotion to change the order of things? Can it invoke thought or provoke action? I hope so. It would be rewarding if it could cause even a small positive movement in one person, project, or firm. Designing and building can be filled with the joys of serving, collaborating, and making, but need a little attention, redirection, and inspiration. Like those of other technical professions becoming more complex, design and construction graduates are being forced to choose a specialized path or create a niche. For architects, the traditional vectors are still available: design, technical prowess, communication and people skills, marketing, and more. But now, add energy, environmental skills, design management, facility management, and digital wizardry.

Odyssey: Me to We

Like most design projects, this book began with an idea and a step. Then another. It gained focus and took shape. It began as an introspective look at the nature of design – a bridging, psychological inquiry into the minds of diverse team members to draw conclusions on the amorphous, immensely challenging nature of managing design. But I quickly learned that before drawing conclusions, the analysis had to move beyond "me" to "we." It grew to have an extroverted focus, polling experts from a broad landscape to recognize problems and to find possible solutions.

Terms

In this book, the terms "architect" and "designer" are used as generic shortcuts for architects, engineers, interior designers, landscape architects, graphic designers, and the cornucopia of design professionals and consultants who contribute to projects. Gender terms (e.g. he, she, we, and they) are used interchangeably.

The term "owner" is used for owners, users, owner's representatives, program managers, developers, and others whose job it is to manage and shape design, and direct teams.

"Contractor" refers broadly to construction managers, general contractors, trade contractors, manufacturers, suppliers, and vendors who implement, build, and install designers' visions.

"Team" refers to the breadth of participants needed to design and build.

The "industry" spoken of refers broadly to planning, design, construction, ownership, and operation of commercial and institutional building projects.

Countless others who offer support are consciously excluded, including the legions of code officials, regulators, financiers, and others who react to and support the work of designers, owners, and contractors rather than manage it.

Focus

If you seek a comprehensive design methods overview, you have chosen the wrong book. J. Chris Jones wrote that book in 1970. Its title is *Design Methods*. His pioneering book surveys more than thirty-five design methods in an academic research context. Readers seeking instruction on how to run a design firm should look to Chuck Thomsen's *Managing Brainpower* and Art Gensler's *Art's Principles*. Those wondering whether to become an architect should consider Roger K. Lewis's approachable classic *Architect? A Candid Guide to the Profession*. If you're looking for the one-size-fits-all answer on how to manage design, you won't find it here. What you will find in Part 2 is a conceptual model of fundamental principles useable to manage a team during design and construction of a project. They are called Project Design Controls.

"The positive thinker sees the invisible, feels the intangible, and achieves the impossible."
– Winston Churchill

Precedents: Anecdotal Dearth

Shockingly, despite centuries of practice, the art and science of managing design are still new subjects. Scant literature exists to inform us. Little research or applied sciences exist to bridge the disparate cultures of designers, contractors, and owners. A few good books have been written on managing design. Most approach the subject from a pure management point of view, as if design were an objective, measurable set of tasks. In my experience, that is far from the truth. Working beside notable professionals has given me the sense of how architects think. As designer, manager and principal, I did what they did.

Managing Uncertainty

At the 2018 AIA National Conference in New York, I reconnected with a study called *Managing Uncertainty and Expectations in Design and Construction*. This McGraw-Hill Smart Market Report by Clark Davis, Steve Jones, and Carol Wedge, with the AIA's Large Firm Roundtable, studied issues we will discuss. Presenter Steve Jones's observations included:

"In designing and building we're trying to build one-of-a-kind assets out in the weather. In this study, we're trying to find out what drives them into the ditch. This data has never existed in our industry. With it, we can have adult conversations about these issues, not act like a bunch of mercenaries out to protect ourselves."

Industry best practice calls for planning and teaming first. But many projects do not start that way.

"What's the root cause of these dysfunctions? It's that the pace of change is exceeding the pace of construction."
– Carol Wedge, 23rd June 2018

Why don't owners correct this? Because their corporate incentives and jobs can discourage it. For one-time, nonserial builders, projects do not follow patterns. They are risky. As business people busy with their jobs, these owners may push design and construction onto the backs of their hired professionals.

"Owners don't want to own the fact that – with their project team – they've created a startup."
– Stephen Jones, 23rd June 2018.

Consistent with this book's findings and framework, this research by Dodge Data & Analytics offers hard data and useful tools. Among them is a free project contingency calculator, useful in managing design risk. To download this report, see www.construction.com/toolkit/reports/project-planning-guide-owners-project-teams.

Outreach: Provoking Change

Standing on the shoulders of Tom Peters and Bob Waterman's 1982 *In Search of Excellence: Lessons from America's Best-Run Companies*, and Jim Collins and Jerry Porras' *Built to Last: Successful Habits of Visionary Companies*, I reached out to leaders from the country's best firms. This semi-scientific approach would be more credible than my own experience. How were things working for *them*? What were *they* experiencing? What could we *change* to be better partners? Unlike Peters and Collins, I did not study firms. Instead, I asked questions to get perspectives on issues facing the profession to understand the nature of the design-management continuum, and ways forward.

At the same time, I gave presentations to some of their leadership groups, speaking the truth, planting seeds to shift perspectives and provoke change. I modeled these talks after a campaign I led from 2005 to 2012 called BIM Education Awareness Momentum and Use by Partners (BEAMUP). That effort contributed to changing behavior within the industry.

Audience

What kind of readers can benefit from *Managing Design?* Everyone! Owners, architects, contractors, and emerging professionals. Whether you are an engineer, interior designer, trade contractor, student, teacher, experienced pro, or early career aspirant, the issues and opportunities in the introduction, conversations in Part 1, and Project Design Controls and actions in Part 2 can open your eyes. This book is for:

- Architects wrestling with the age-old demons of designing to budget, fending off contractor advances, or musing their future
- Owner searching for ways to lead design and construction teams in a complex world – with unfamiliar tools and processes
- Contractors and trade contractors faced with incomplete, uncoordinated drawings and untenable schedules, and mystified about design thinking and process
- Teachers, students, engineers, software gurus and technologists

The notion that architects, contractors, and owners are the only ones suffering from these issues is limiting. The design–management continuum applies to any creative endeavor – software, cooking, advertising, graphic design, software coders – you name it. The discussion looks at alternate thinking to bring these thought modes together for a common cause.

In reading this book, experienced professionals may find it tells them what they already know for their own discipline, but not for the disciplines of others. Blending disparate cultures is challenging. Designers, contractors, and owners have different educations and motivations. They think and speak in different languages. By understanding one another better, we can better align to serve our clients and one another.

Project-First Attitude

What is design management? Activities that achieve projects whose program, design, and scope stay within their budget, delivered on time, with good collaboration, design, and documents. Achieving this takes many forms. Tools and techniques help. Focus, responsibility, and assigning team member responsibilities do too. On good teams it is not about contracts or who's in charge. No one says it better than Holder Construction's Michael Kenig: "It doesn't matter who's doing what, as long as what needs to be getting done is getting done." His advice advocates a project-first attitude – the hallmark of a successful collaborator.

How to Use This Book

Managing Design is organized to accommodate a variety of reader types. Those who prefer a linear approach can absorb it front to back, building conclusions. Others may prefer a selective route. If you are interested in one voice type or theme (e.g. owner leadership) or are familiar with one or more interviewee, skip around. Digest it as you like. Solution-oriented readers can move ahead to Part 2 to find a skeleton whose experiential bones offer design control best practices. Learn something new or affirm something you believe. Design your own path.

In the book's margins, in addition to interviewee quotes, you'll find empowering nudges from famous artists, musicians, improvisors, and rulebreakers, including unexpected people like John Prine, Jimmy Carter, Miles Davis, T.S. Eliot, and B.B. King. Why? Because design involves having the freedom to venture without fear of failure. Risk taking. Managing design does too. It's an art *and* a science; one that demands that its practitioners go freeform and play off their bandmates.

Play on.

Applied, Not Basic

This book is about understanding design process to keep it in some semblance of organized, rational behavior, time, space, and financial parameters that support construction. These are foreign constructs to many architects—things foisted upon them. Maybe they had one management class in school. This study avoids management basics: how to set objectives, fee budgets, proposals, and workplans.

The book does not offer legal advice. Readers must apply its findings in their own way at their own risk. The focus is on what's different from conventional management – design anomalies: how to apply thinking, resources and tools to manage design directly – and with constructors and owners.

The book's tone and interviews cast doubt on whether design can be managed at all. Certainly, there's no one way to try. The process seems more a question of how to ameliorate differences. What you'll find is a trove of perspectives from industry leaders and a conceptual grid for design control thinking you can apply in a way that suits you best.

Questions

Three questions can be drawn from the book's title: *Are* you managing design? C*an* design be managed? And w*hat* does the future hold? This book will answer all three, through dialogue. Do we need to manage design and change processes? Keeping quiet, or simply talking and not doing anything differently has not worked.

*"Qui tacet consentire videtur."
(He who is silent is taken to agree.)"*

This expression reflects the "silence procedure" or "tacit acceptance" in law. As you will discover, neither the people interviewed in book, nor its author, are being silent. There is no tacit acceptance. Designers who work in isolation, disengaged owners, and builders who shun design process ownership perpetuate the problems.

The active verb/gerund title form "managing design" is used more than the passive/noun form (i.e. "management") because managing design is an ongoing pursuit. Starting with intention and desire, attempts to manage design are often followed by quick reaction and adjustment. Experienced practitioners do it more slowly, with wisdom, self-leveling feedback mechanisms, and teamwork.

Goals, Data, People

The goal of this book is to expose issues and help like-minded people change them. Being iconoclastic was never a goal. Attacking cherished beliefs and long-held traditions offers no understanding or solutions. There's a lack of clear thinking and new directions. It's time for a new S-curve in architecture and building.

There is ample momentum – even hype – for using computers to cure our ills. Some interviewed in this book are that movement's staunchest advocates. But data driven design doesn't foretell the solution. Acts of design – and management – *start* with the numbers. Data are the basis that inform your thought process, but they don't *tell* you what to do. You can ignore, question, or refine them, or project them into the future, but without them what do you have? Nothing. How can you practice that way? You can't. That's why architects are "losing": they're losing at data.[1] Contractors and owners wield cost data like clubs. In this book, data is one point of *beginning* for a new way to collaborate, not the be all and end all. *People* offer that.

Lost Horizon

In researching this book, I looked for precedents and visions to guide the writing. I didn't find many, and then it struck me. In the 1970s, during a quiet, graduate school summer in Ann Arbor, I learned of James Hilton's classic 1933 book, *Lost Horizon*. I devoured it – carefully, judiciously, appreciatively. The Zen of the experience was a marked contrast to my college-student excess. I sat quietly on my balcony absorbing this new thinking rather than washing it down with a cold one.

What did I discover in my reading? The valley of Shangri-La, a mythical place where peace and brotherhood are the norm, and no one ever gets old. Peace, serenity, and utopia in a hidden valley in the Himalayas. A place where monks and citizens live in harmony under one rule: Be kind. Wouldn't that be a fine precedent for how to design and build projects instead of fighting and burning out? A utopian fantasy? Maybe. But what if there's something to learn from Shangri-La about managing design? A little moderation? Perhaps one of you will be the next Robert Conway, the chosen one who ascends to be the new leader – to show us the way. Conway was a fighter, diplomat, and leader. Skeptical at first, he ultimately recognized his calling. Will you?

In Frank Capra's 1937 film adaptation of the book, one of Conway's traveling companions articulates our traditional approach: "If you can't get it with smooth talk, you send your army." The default

---

[1] At an Autodesk executive breakfast briefing at the AIA Conference in New York in June 2018, Phil Bernstein shared this anecdote. The design team happily presented their design options to the client, complete with renderings, models, and images. The software client's response: "But where did these come from? Where's the data, the decision tree, the substantiation?"

to conflict serves no one. Another says, "It's not knowing where you're going that's the problem, it's wondering what's going to be there when you get there." Fear of the unknown is something we all face in our work. Conway looks to the future without fear. He believes in a new horizon. In design and construction, when it comes to collaboration, it seems we've lost sight of our horizon. Maybe *you* can help us find it again.

Time to Act

Is now the right time for this book? The "change or perish" mantra has become a cause célèbre in the building industry. Is this book merely one more jalopy in a decades-long, slow-moving traffic jam? Like most design efforts the book was an exploration. Start and see where it goes. Now complete, its synthesis has multiple forms: a unity of minds with change as bond – a divergence of possible directions to foster that change, and a framework for design control. I accept them all. Traveling this path, I found interesting people, each with a unique perspective. Each with a passion for change. There were no passive observers, only experts willing to share their convictions and act on them.

How will it end? Will some deus ex machina resolve the situation? Thomas Friedman[2] talks about America becoming a dictatorship for one day, to demand that utility companies work cooperatively to share and fix America's energy grid. Maybe an omniscient benevolent government overlord can figure out how to fix the building industry's malaise. Until then, we're all we've got: enlightened owners, change-ready contractors, and fed-up architects, students, and teachers determined to go about things in a smarter way, together, starting with figuring out how to manage ourselves.

Issues

Managing Design is a book about the people and problems that drive our industry. Its perspectives expose core issues we face as teams: What contractual, educational, and economic roadblocks constrain us? How do our motivations differ? What themes reappear? Expert viewpoints bring currency to these issues and provoke investigation into solutions. Their questions are familiar:

> *"Everything's already been said, but since nobody was listening, we have to start again."*
> – André Gide

- *Why can't architects and designers design to budget?*
- *How can owners better understand and manage their teams?*
- *How can contractors relate to their design partners to collaborate better?*
- *What are the root causes of design and construction team dysfunction?*
- *Why does our industry's productivity lag others?*
- *Is technology the savior?*

---

[2] Friedman, Thomas: *Hot, Flat and Crowded*, Picador, 2008, pp. 430–431: "If only America could be China for a day - just one day. Just one day."

- *What actions can each of the parties take to improve, both individually and collectively?*
- *What cultural legacies constrain and enable teams?*
- *How must our educational institutions change to serve the industry?*
- *How do our partners think and speak? What are their motivations?*
- *Are new entrepreneurial approaches the solution?*
- *Can design be managed? If so, how and by whom?*
- *What unspoken character traits do designers possess that mystify partners?*

We are all witnesses to these outdated industry excuses and practices. Racial and gender inequity, misogynistic behaviors, and other longstanding problems have smoldered in design and construction. In seeking to manage design, we must correct centuries-old, closed-culture problems. For many in practice, self-indulgence, design excess, and mismanagement are endemic. Many firms have been reluctant to change out of fear or inertia. How will old guard purveyors react to the specter of change? Will they let it make them stronger, focused and better or allow it to baffle and confound them.

Problems

What problems are we trying to solve? Every few years, the industry conducts surveys to identify the concerns that demand "design management." Management consultants produce whitepapers that restate the obvious. Invariably, owners and contractor surveys expose these issues:

- Late, uncoordinated, incomplete drawings
- Scope creep
- Over budget designs
- Uncoordinated consultants
- Challenged relationships, collaboration, and trust
- Inefficient processes
- Conflicting mindsets and objectives – not listening to client goals

Designers' issues are similar but with a different slant:

- Not enough time
- The change rate's impact on traditional design process
- Low fees and profits, commoditized services
- Role encroachment
- No respect from others for the value of design and architecture

The alarming consistency in these every-few-year polls results reveals two things:

1. Our ability to manage design has challenges.
2. We must do something about them.

Oxymorons and Continua

What are the oxymorons related to managing design? Opposed pairs like design scheduling, designing to budget, design work planning, design risk management, design follow-up, and design communication. I assert these phrases to be opposites for several reasons. Having worked in architecture for 30 years I can tell you, the typical designer is not inclined to:

- Get done on time (more studies can *always* be done).
- Overcommunicate (most would rather *do* the work than talk about it).
- Listen to clients (clients do not *know* the language of design).
- Manage *money* (it is the currency of *others*).
- Spend time *managing or following up* on action items (it cuts into creative time).

This will be helpful information for all who have ever been frustrated by their designers' inability to do these things. On a personal level, anyone who has wondered why their interior designer has not called back, was late with color samples, or blew the budget will understand these behaviors more if they know:

- Design types are wired differently. Their DNA has selected them to create. They're genetically predisposed to be late and over budget (or they would have been managers).
- Their education has been honed to reinforce their exploratory skills.
- Culture and colleagues have supported their migration to creation, away from rule following.
- Their practice and experience have conditioned a focus on esoteric artistic pursuits.
- They have been rewarded for talents like drawing, creativity, abstraction, individualism, and ingenuity over conformance and compliance.
- Many have chosen (or been trained) to value beauty, art, and architecture over client service, profit, and traditional business virtues.

History shows designers aren't good at these things. Enigmatic as they are, they're true. What can you do about such a fate? Read this book. See Part 2. What *are* the skillsets of designers? Designing. Creating. Exploring. Combining. Synthesizing. Making. These are things they love and will work late and fight for, even over profit or personal gain.

"You don't think your way to creative work. You work your way to creative thinking."
                                                                      – George Nelson

The Collaborative Quest

Design and construction leaders have survived by delivering great projects to clients, even as we acknowledge our flaws. On every project, owners, designers, and contractors work together. We rely on one another. This has worked surprisingly well, despite cultural differences and process-driven obstacles. Still, after many years and billions of dollars, inefficiencies, budget overages, and schedule challenges persist – all stemming from our collective inability to *manage* design.

A growing disquiet remains among designers and builders. Architects who left during the Great Recession of 2008 and 2009 are reluctant to return. Contractors battle a scarcity of skilled onsite trade workers. What's behind all this? How can we handle these issues and the different behaviors of project players? Can we fix what is wrong?

Absurdities

The absurdities of inefficiency, lack of work/life balance, and waste remain inherent in design and construction. Demoralized teams, unhappy owners, and frenzied contractors are forced to redesign and build before design is complete. Yet we do it again and again. Albert Camus, in his essay "The Myth of Sisyphus," offers a fitting reaction:

"I draw from the absurd three consequences, which are my revolt, my freedom, and my passion."[3]

I'm with Camus – it's time for revolt. And through this book, I exercise freedom and share my passion to reverse the absurdity. Join me.

Willing Suspension of Disbelief

As we ruminate on the state of things, as you evaluate the veracity of these claims, ask yourself: Do you consider yourself a victim? A poor soul laboring inside a lost profession? Facing burnout and losing cache'? Or are you one of its champions, exemplifying its value through actions and work? What motivates you? How you think and what you do can remake who you are. Speakers in this book (including its author) rely on generalizations in making their claims. As best we can, we'll break them down and get specific. Until then, I ask for your trust on this anecdotal journey to seek answers. While this book probes expert wisdom to gather data and confirm hypotheses, it also offers design management principles to cement understanding and use.

A Symposium on the Future of Practice

In October 2017, I attended "AEC Entrepreneurship: Creating a High-Tech Building Economy – A Symposium on the Future of Practice" at the Georgia Institute of Technology (Georgia Tech) in Atlanta. Issues and themes discussed included:

- The shift from "service" to "product design" in the design professions
- New organizational forms
- Supply chain integration and management
- Change influences external to the design profession (e.g. manufacturing and prefabrication)

Here's what conference presenters shared to frame the issues:

"When I founded the AEC Integration Lab at Georgia Tech in 2005, it was a problem. The contractors wanted to talk about problems onsite this week, [while] the research students wanted to talk about theoretical, leading edge work that would come to fruition in 10 years."
– Chuck Eastman, 10/6/17, Founder, Georgia Tech Digital Building Lab

---

[3] Albert Camus, "The Myth of Sisyphus."

"At Columbia, I had a studio addressing anxiety about moving from service to product. As architects, we value the one-off nature of what we do. I'm trying to elevate this for even more impact at Georgia Tech, to create an educational system to prepare the next generation. I see six future practice drivers:

- *Integration (we have to work with others)*
- *Specialization*
- *Automation*
- *Research*
- *Organization*
- *Value*
- *Transformation is happening. And it's coming from outside our industry.*

"Design firms are getting paid more for change management services than for traditional design services per Jim Cramer and the Design Futures Council."

"We're setting up vertical studios. Teams define a topic, choose an interdisciplinary partner, then look for funding. But the traditional architectural educational system hasn't supported that kind of interdisciplinary approach. And the accreditation board reinforces the old norm. It's a problem."

– Scott Marble, Program Chair, Georgia Tech, School of Architecture

"It's rare to have the trust and respect for each other's minds where you can listen, hear, and see something deeper in what they say – that's collaboration."

– Daniel Kahneman

"We got tired of trying to convince people to do things a different way. So, we decided to ignore what the rest of the world has done and do it ourselves."

"When you think of building design like product design, the full promise of the feedback loop is realized. We engage with what's actually happening. We start with a research team, pose questions, do interviews, collect data, and use that information to inform design. . . It's the difference between siloed disciplines and a guild approach. More like deployment on a mission-based task. When we only had 15–20 people there was no bureaucracy, just direct communication. Little teams, like startups, that built autonomously. Each had capabilities for fabrication, surveying, construction management, sourcing, logistics, and a supply chain. We're doubling every year. We did 6 million SF in 2017, we'll do 12 million next year. We're redesigning how we work every year. We have to."

"We hold the risk, so we can do whatever we want. Uncoordinated drawings become hedged assessments. No time is spent talking about fault. It's a global strategy with local tactics, so we allow for deviation. We give authorship, authority and control at the team level."

– Federico Negro, WeWork

"I'm interested in the intersection of technology and project delivery—the exchange of information and risk. That affects how you get to widespread change. The building industry is suboptimized to the point of failure. How do the players relationships change as a result? The BIM problem is largely done. As we've seen in the conference today, AECOM, Katerra, WeWork, all are horizontal attempts to eliminate transactions. Now, architects make construction documents and "dare" contractors to build from them. The problem is the supply chain structural challenges.

"We need to isolate the pieces and optimize performance. Then, the exchanges will evolve on a different set of transactional principles. We need to control outcomes by controlling the supply chain."

– Phil Bernstein, Technology Consultant, Assistant Dean,
Yale University College of Architecture

These leading change agents and thought leaders opened the door to a series of deeper conversations that help us understand why our siloed professions persist, and what we can do about it. Let's open our minds and listen to what they say.

The Emperor

It has become clear that design management is the unclothed emperor in the room. We do not talk about it, it lingers lonely in the background. We have conditioned ourselves to expect it to be delinquent. As a designer, you know management is needed, but those tasks are not what you signed up for. Yet clients and contractors care deeply about management. In the often-conflicted reality of practice, managing design is a fundamental question and has been for as long I have been involved. Likely for centuries. Just read *Brunelleschi's Dome*[4]. Author Ross King recounts the architect Brunelleschi doing cost estimates, being over budget, and needing value analysis cost reduction, all after competing for the work. That was 500 years ago. How can these issues be 500 years old and still not understood? One notable difference: Brunelleschi was intimately involved in deciding (and designing) how his creation would be built. He had to confront reality by devising ingenious cranes, gears, and mechanisms to bring his creation – the dome – to fruition. Today, we pass these things off as the contractor's means and methods. Clearly, the "master builder" is a bygone concept. Rightly so. Things are too complex for any one entity to manage everything. What to do?

Teams and Delivery

Today, complexity dictates that projects are realized in teams. A multiplicity of skills is needed. No one entity has them all. This book's discussions assume a "team-approach" project delivery of one sort or another. Construction Management-at-Risk (CMAR), Design-Build (DB), or Integrated Project

---

[4] Ross King, *Brunelleschi's Dome: How a Renaissance Genius Reinvented Architecture* (Penguin Books, 2000).

Delivery (IPD) are the contractual scenarios assumed by most interviewees, but readers using classic delivery methods such as Design/Bid/Build (D/B/B) can also benefit, despite their different incentives.

Context

How did things get this way? What socioeconomic factors and industrial changes led to today's maladapted industry? Can context explain the challenges? While far from a comprehensive history of the last hundred years in design and construction, significant events have shaped current practice. Here are a few likely contributors, set to a U.S.-biased, by-the-decade timeline:

1900s Industrialization begins, enabling factory production, elevators, HVAC, steel, glazing, new materials, and division of labor.

1910s Modernism and means of production challenges Beaux Arts traditional practice and design education.

1920s WWI industrialization, economic depression.

1930s Recovery, many current systems evolve.

1940s Post-WW2 management practices, speed, and urgency give birth to construction management, fast track, and other project-delivery modes.

1950s Mid-century modernism, suburban and economic growth in U.S.

1960s Civil rights and environmental issues influence design processes.

1970s Overlay drafting, CADD drafting migrate to AEC industry, leading to widespread adoption.

1980s Personal computing, the internet, software's rise.

1990s Internet age, www emerges, mobile devices appear, sustainability reemerges.

2000s Integrated project delivery, recession depletes design and construction talent pool.

2010s Terrorism, global uncertainty, cloud computing, artificial intelligence, inclusion, diversity.

In addition to the above time-specific developments, other issues have evolved steadily over these decades to reshape contemporary practice:

- Design education, contracts and incentives retain decades-old foundational beliefs – losing pace with changing practice.
- Ongoing schedule compression reduces quality of design documents to precipitously new lows
- Mobile computing and practice complexity distribute expertise globally (integration and communication changes conceptually, requiring machine assistance).
- New infrastructures require capitalization, funding and maintenance to enable new processes. Current/past AE firm management lacks necessary skills.
- Design and construction specialization increase, adding complexity, requiring more management.

In the face of such change, and amid still-held practices such as reverse incentive contracts, manual drawing, data hoarding, re-creation, bias, and rework, it is little wonder teams face challenges.

Punctuated Equilibrium

As we look to necessary adaptation, most of us have a limited understanding of evolution. We imagine a straight, ever-inclining line. Experts tell us that things remain in stasis until an external force exerts pressure on a system or species. It might be the Ice Age or a forest fire, the invention of computers, or the internet. Whatever it is forces necessary adaptation. These evolutionary "cusps" generate stepped, not gradual change. Evolutionists call it "punctuated equilibrium." Said another way, most of us don't change until we have to – or want to. This adaptation distinguishes survivors. Contrary to popular belief, Darwin didn't coin "survival of the fittest." He said:

"It's that organism most readily adaptable to change that survives."

– Charles Darwin

There's a difference between being the strongest, smartest, or fastest at a single point in time and being the best *adapter* over the long haul – Darwin's "fitness." Which are you? Evolution selects based on survival traits. If we don't change, we face extinction. The data suggesting necessary changes exist. Let's use them. Longevity is a good success metric. Adapters survive.

Shifting Expectations, Value and Capitalization: Vicious Cycles

The equilibrium-punctuating events over these decades have had an unanticipated consequence. Beyond making it harder to cope with change they have shifted owners' expectations. Now that chaos theory and Moore's Law have given us software, hardware, the internet, and automation, owners expect teams to deliver buildings at close to the same rate these same owners can do a Google search. "Google can give it to me in seconds. Amazon can ship it to me tomorrow. Why can't a good team design and build my project in months?" They can, with a prepared owner and the right processes. Without those things, they don't stand a chance, but many are forced to try.

To get to the bottom of why these old ways persist let's look from several angles.

From a design lens, the average practitioner was not trained to think as an entrepreneur. Design culture has not attracted or rewarded professionals with finance or technology backgrounds. It celebrates award-winning designers – so-called "starchitects." The commoditization of the profession and its fees prevents most design firms from funding recapitalization. (See the Bernstein and Kanner interviews in Chapters 6 and 13.) Finally, the industry doesn't generate the profit levels required to significantly invest in new technologies. Why? Architects' fees are stuck at the same level they were 20 years ago.

"There is scarcely any passion without struggle."

– Camus, The Myth of Sisyphus, p. 73

Contractors build wooden formwork by hand, pour concrete, then tear the formwork down – an archaic method, but still cost effective due to available, inexpensive migrant labor. Optimists draw parallels to the air travel industry whose profitability languished for years until it could finally raise prices and reinvest in better planes and airports. The AEC industry is still searching for its value points and the epochal timing that will let it do the same. Until we change it, the cycle continues.

Landscape

In surveying the sometimes apocalyptic design and construction landscape, little can be gained from posing a Mad Max future scenario of contractors, owners, and architects roaming the planet, killing one another off to survive. Based on leading practitioners' feedback, I borrow insight to pose recommendations for a positive future state. Far from being a searing indictment of the professions of architecture or construction, or their decline, the book posits challenges:

- *Architects*: Get with it! Learn to be servant leaders from industry leaders who have conquered management demons and figured out how to design and manage concurrently. Get outside of your traditional self-indulgence or comfort zone. Lead by example. Prove your value and learn to be teammates.
- *Owners*: Lead and engage with your teams. You do not have to know it all to lead designers well.
- *Contractors*: Practice random acts of kindness. Help your partners manage the unmanageable. Look out for *everyone's* interest.
- *Trade Contractors and Manufacturers*: Insert yourself into teams. Add nascent value.
- *Students and Educators*: Become agile. Let go of historic idiosyncrasies. Prepare practitioners for our current reality. Integrate future thinking. Shape a new culture, value model, attitude, and skill set.

Together, we can have a profound impact on creating more potent relationships and a richer collaboration landscape.

Themes

The book suggests alternative worldviews for how teams might work together. With past as prologue, retrospective and current state, it portends an optimistic outlook for collaborators. An explosion of themes emerged from its interviews, including:

- Educational change, studio legacies, and apprenticeship
- Self-image, self-esteem, and attitude
- Technology's impact
- Lean thinking and processes
- Entrepreneurial thinking, profitability, and value propositions
- Increasing complexity and risk aversion
- Contracts, incentives, and project delivery
- Owner leadership and client intimacy
- Engineers' mentalities: "leading from behind"
- Contractor awareness, design sensitivity, and empathy
- Relationships and trust
- Planning, project definition, and process

- Trade contractor issues, perceptions, and realities
- Strategic alliances, horizontal integration, and supply chain transformation
- Vertical integration and consolidation
- Scale: small/local versus large/global versus mid/regional
- Specialization versus generalization
- The "Silver Tsunami" – the mass exodus of baby boomer firm leadership
- Debunking the designer as "lone wolf" myth[5]
- Gender and racial equity, diversity, and inclusion

> *"Is the anger a reaction against their own insecurity, impatience and irritation?"*
> – Ursula K. Le Guin, No Time to Spare: Thinking About What Matters

Two themes persist: self-respect and self-image. Some designers have hidden behind the excuse of turf encroachment. In their eyes, construction, program, and design managers pose threats. This is misdirected.

"I see the migration of architects to working within CM firms, and the involvement of others in architecture to support design as an expansion of the profession, not encroachment."

Nondesign teammates put it simply: "Architects must end their risk-averse behavior and stop blaming others. No risk, no reward. Get some skin in the game." Colleague Ennis Parker reframed the issue:

Anger and resentment against the encroachers also came up in the interviews. Anger is a good start. It recognizes and accepts the problems. But it's only productive if it motivates change. What psychology fuels such sentiments? Jealousy of contractors' profitability? Defense mechanisms? Are architects *glad* to not be like results-oriented contractors or clients too busy to care about aesthetics? That's how many client and contractors have been coached. These themes weave a tapestry. Via shared history and future visions, they posit a professional GIS system using the wisdom of many. This data set lets readers draw their own roadmaps. Together they form an anecdotal foundation to inform an inquiry into the nature of design – new thinking to connect disparate players in design and construction.

Self-Imposed Disservice

A recurring theme in many interviews is the architect's profitability. Fixed percentage, lump sum, or hourly fees constrain profit and limit innovation. Owners and AEs pine for a better way but are low on change tolerance and motivation. What we are doing is not working. Let's coach owners and project mates to set fees, schedules, budgets, and incentives that reward performance and value, not treat our

---

[5] In her opinion essay in ARCHITECT Newswire, 8/3/2018, "To End Abuse in Architecture, Start with the Lone Wolf Myth," Esther Sperber refers to stories of "great white men who work – and excel – on their own. Nothing can be further from the truth. . .. The romantic image of the creative genius . . . [is] an aggregate of cultural beliefs and biases that restricts our understanding of how innovation actually happens."

design work as a subpar commodity. Can we agree on that? Fee reduction in the face of competition is a false economy in the long run.

Stereotyping

Stereotypes in any form present frightening challenges to overcome. They persist in our industry, but we do not have to resign ourselves to typecasting. Its very nature compels us to genuine change. Client and partner feedback and metrics, along with anecdotal surveys and experience, suggest inefficient teams are still too common. Recurring stereotypes plague us, including these:

Owners "are" faced with business uncertainty; behind schedule on activating their projects; challenged to make decisions; unable to read drawings, use BIM, or understand current design practices; too busy to manage teams; reliant on contractors and architects to get along and deliver their projects.

Architects "are" often behind schedule; designing over budget; less adapt in business currencies (time, money, risk); culturally isolated; ceding consultant control; lagging in digital technology; fee and profit poor; time pressed; challenged by code, consultant, regulatory, and material complexities; pressured to shape a new generation of leaders.

Contractors "are" insensitive to, or ignorant of, design processes; schedule and cost driven; without regard for design; claiming the architect's territory.

Can we change? Can we prove "them" wrong?

Overlooking Things

I get reflective as I gaze back on the profession I have been a part of for 50 years. How are we doing these days? I ask.

"Fine, if you overlook a few things."[6]

In that time, the universe we knew became a multiverse of tools, experts, digital processes, and communication modes. Some of us are having trouble adapting to it all and wondering what we can do. Including me. But it's hard to overlook things anymore. Design and construction are mature markets, yet after several decades are still in the early stages of adapting to the new economy. It's now clear that new forms of technology, process, and collaboration are required to embrace that shift. This book talks about how.

Answers are predictable when you ask practitioners how *they* feel about the state of things. Complaints ensue. A few optimists cheerily share half-full outlooks. But, after the venting ends, what you discover from the great majority is complacency. When asked how things are, the predominance of players agree with John Prine's assessment from his song "Pretty Good":

"Pretty good, not bad, I can't complain. But actually everything's about the same."

But they do complain. They have for years. The *"same"* condition described is one of declining profitability and increasing complexity. Not enough has happened to react or adapt.

---

[6] A conversation between David Kundtz and his mother, in M.J. Ryan, *"365 Health and Happiness Boosters,"* (Conari Press, 2000), p. 215.

Motivation

The average architect is not "pretty good" at business or managing design. Why? Because they are not taught it in school. With the emphasis squarely on the art and science of design, business is a distant third in most curricula. After graduation, most practitioners find business precepts and self-management inaccessible. Such matters are handled by a few top-level, business-savvy principals and project managers. Few architects care that they are not being exposed to business in the workplace. This skill, seldom valued during school, continues to be relegated to the back burner during most architects' careers. They don't want it. They're not good at it. They're *glad* someone else is dealing with it. Most are not rewarded or penalized for it by their superiors who hold similar laissez faire attitudes. "Business" isn't why they entered the profession. This needs to change to sustain the design profession.

Servant Leadership

Designers are misunderstood. In their hearts they are interested in long-term value creation, not a quick fix or fast buck. But they need help with delivering and getting paid from the rest of us. Because without them what do we have? Most likely an on-time, on-budget, God-awful mess where nobody would want to live or work. Are you with me? This book is about leadership and servant leadership, understanding and empathy, compassion and support. Managing design is about creating a shared sense of purpose: designers and builders who work together better. If that is not what you are here for, you may want to move on. If it is, let's go.

> *"Life is hard. It's harder when you're stupid."*
>
> – John Wayne

Fundamental Flaws and Changed Conditions

In the theoretical "good old days," AIA contracts, practice and processes were linear and assumed little change. Owner programs formed the basis for design. Contractors built based on largely complete, coordinated document sets. All this in a reasonable time frame. Architects used experience, judgment, and intuition. No machines, and limited data informed them as they worked alone or in small groups with little architect/contractor interaction. Architects got paid fixed or hourly fees absent financial incentives. Under this "fixed" worldview and slower moving era, owners who changed their minds, and contractors who discovered errors or omissions in the documents grappled with change orders and delays.

We do not live in that world anymore. While most design firms still labor under commoditized fees, they now have interest in surviving in a profession some say is "three recessions away from extinction." Yet they show minimal ability to redesign their processes or value propositions. Those who want to revalue their services must reengineer themselves offline, on projects, or in radical new ways.

New Realities

In today's projects, things are different. Concurrent, not linear, phased processes are the norm. Change is constant. Contemporary teams know:

- Programs are often nonexistent, insufficient, self-created, and ever-evolving.
- Everything is changing and there is lots of data.

- A wealth of powerful, complex, largely-untapped machines, software, and infrastructures exist.
- It takes lots of people and mandatory interaction to cope with it all in much less time.

No wonder practitioners (and the people in this book) are talking. Borrowing Robert Venturi's analytical method, Tables 1, 2, 3 and 4 below contrast the new contexts, design conditions, and drivers that shape current versus past practice, the "complexity and contradiction" of modern projects.

It is clear we need new processes for designing, building, and collaborating. For architects, this includes demonstrating value and devising new ways of getting paid.[7] These methods must accommodate and embrace change, machines, data, and people, not ignore them. For all of us, it means new ways to manage design processes to help creative teammates shore up their historic shortcomings.

"Quality," Construction Administration, and the Afterlife of Design

Clients are most concerned about the quality of the final product. Long after the schedule, budget, design and experience are over, the building remains. Design, the construction it drives, and any design management discussion should serve the full project life cycle. Attention has traditionally focused on three areas: time, money and drawings – during design. The Project Design Controls framework speaks to starting projects, design scheduling, designing to budget, and the quality of design documents. But what about *finishing* them and *building* the building?

> "Everybody is a mad scientist, and life is their lab. We're all trying to experiment to find a way to live, to solve problems, to fend off madness and chaos."
>
> – David Cronenberg

TABLE 1 Design Conditions and Contexts

Design Conditions and Contexts		
Aspect	**Past Practice ("Old")**	**Current Practice ("New")**
Program	Provided/Fixed	Team Developed/Changing
Participants	Few	Many
Speed	Slow	Fast
Data	Minimal	Maximal
Change Basis	Little/None	Significant/Ongoing
Process	Linear/Sequenced	Networked/Concurrent
Decision Basis	Intuitive/Internal	Quantitative/Visible
Stability	Static	Dynamic
Delivery	Program/Design/Bid/Build	CM-AR/DB/IPD
Conceived By	One	Many

---

[7] On the opening day of the 2018 AIA Conference on Architecture in New York, ARCHITECT magazine's headline noted Katerra's acquisition of my former firm, Lord, Aeck & Sargent Architects, an example of a vertically integrated company bringing design in-house to create new value propositions to change the industry.

TABLE 2   Team (OAC Relationship)

Team (Owner-Architect-Contractor Relationship)	
Past Practice ("Old")	*Current Practice ("New")*
Owner	Ditto Left and add:
Architect	Construction Manager
Engineers/Consultants	Funding
Contractor	Program/Development Manager
Subcontractors	Users
	Facilities Planners and Managers
	Real Estate
	Health/Environment/Safety
	Sustainability Director
	Communication Director
	MWBE/EBO/Diversity Officer

"If . . . self-respect springs from "the willingness to accept responsibility for one's own life" then rising to anger upon feeling slighted by another is a maladaptive abdication of that responsibility."
— Maria Popova, Brain Pickings, 12/10/20171

TABLE 3   Design Drivers and Influences

Past Practice ("Old")	*Current Practice ("New")*
Goals/Objectives	Goals/Objectives
Program/Function	Program/Function
Site/Context	Site/Context
Structure	Structure
Form	Form
Systems	Systems
Skin	Skin
Cost	Cost
	Branding/Marketing
	Customer/User Experience
	Collaboration/Interaction
	Adaptability/Repurposing
	Energy/Performative Design
	Workplace Equity and Diversity

Past Practice ("Old")	Current Practice ("New")
	Safety/Risk Management
	Cost/Value/Options Analysis
	Purchasing Sequence
	Construction Field Need
	Sustainability/Resilience
	Market Forces/Geopolitical Events
	Constructability/CM Input
	Computer Generated Form/Fab
	Community Input
	Decision Tracking
	Life Cycle Cost Analysis
	Regulatory Issues
	Prefabrication/Modularity
	Healthy Materials
	Orchestrating Complexity

TABLE 4 Design Disciplines and Consultants

Design Disciplines and Consultants	
Past Practice ("Old")	**Current Practice ("New")**
Architect	Architect
Civil Engineer	Civil Engineer
Structural Engineer	Structural Engineer
Mechanical Engineer	Mechanical Engineer
Electrical Engineer	Electrical Engineer
Plumbing Engineer	Plumbing Engineer
	Fire Protection Engineer
	BIM/VDC Specialist
	Audio Visual/Security
	Energy Analysts
	Info Tech/Lo Vo/WiFi/DAS
	Interior Designer/Graphic Designer
	Life Safety/Code Consultant
	Food Service

What about the construction administration phase? As a twenty-year employee of a national construction management company, I can assure you: the lion's share of our employee's time is spent in the field, and, design and documents continue throughout construction. Why? Because design teams are not given enough time to do their work – or, collectively, we have not been able to figure out how to retool or rethink things to deliver projects fast enough.

Preconstruction and other early groups manage design during the design phase. But after we have done our best to use collaborative contracting approaches, set up our project, and employ all the Project Design Controls we can, the building still needs to be built. And inevitably, despite our best efforts, design is rarely done at that point. It never really is. After months of design and documents comes years of construction – and *more* design. What now?

Having spent time in the field and witnessed the hard work and commitment of those involved – the rest of us can't do enough during the design phase to ensure that construction, construction administration – and the poor souls forced to build our planning, design, and purchasing efforts – go as smoothly as they can. The alternative is more requests for information (RFIs) and longer hours. Heed the experts' advice: make it easier on those who come behind design to build. They bear the burdens set in place by the designs we shape. God love them.

Movement

Transformative Practices

Are there any shining lights? In the face of pressures to "change or perish," some *have* changed. Firms such as Adrian Smith+Gordon Gill Architecture are using technology, new processes and a culture of integration to produce world-class, performative work. Kieran Timberlake's research enables informed, valuable, sustainable, prefabricated work. (See the Cheng interview in Chapter 4.) Millennial professionals are designing new career paths with technology and information at their core. Leading academics and seers such as Phil Bernstein are challenging early-career design students to devise new compensation models. My questions for you are: Are you looking? Listening? Changing your tools, strategy, structure, or culture? If not, one can only assume your projects, processes and people are happy, efficient, and content. Are they?

Construction Management and Field Technology: Leading Adopters

Contractors are leading the way in technology adoption to cope with changing conditions, declining documents, and change during construction. For good reason. construction managers (CMs) and general contractors (GCs) have had higher profit margins than designers. They carry greater risk and revenues (roughly 10–20 times design revenues, i.e. the full value of construction cost: 100 percent versus 5 to 10 percent historic design fees). These higher stakes, greater ability to manage risk, and greater profitability have converged to put construction in the fore of industry retooling. Their faster BIM technology adoption rates prove it. Autodesk and software companies figured this out in the early BIM-adoption days and redirected marketing efforts to contractors in addition to designers. Contractor digital reality capture, field survey, and layout boomed with technology. Mobile computing and digital models for data, coordination, sequencing, quantities, purchasing and materials tracking became the norm in a mere ten years. These innovations are contractor coping strategies to "manage" design or recover from its impacts, the result of not having figured out how to manage design. Save for the practices shared in Part 2's Project Design Controls framework, there are few alternatives.

Owning the Problem

At a recent presentation to a design firm's emerging leaders, I challenged them to change behaviors. A young architect raised his hand and innocently asked, "Maybe *we* should be managing design ourselves?" "Great thought!" I said, not wanting to embarrass him or be impolite to my hosts. "You should!" The opportunity to own the problem has existed for hundreds of years. With only occasional exceptions, the profession has failed to respond. To a disturbing degree, architects have lost the trust and respect of owners and contractors. Behavior must change to earn it back. We must do things differently, starting with managing time and money (our own, and the owner's). These are the basic currencies of all other businesses but are foreign to most designers. Otherwise, others manage us: owners, contractors, program managers all become the architect's de facto babysitters.

"Where do new ideas come from? The answer is simple: differences. Creativity comes from unlikely juxtapositions."

– Nicholas Negroponte

Change Precedents

Several recent change movements offer hope we can move beyond inertia:

The Digital Revolution: It has been twenty years since building information modeling (BIM) revolutionized our business. Data from organizations such as McGraw-Hill Smart Market reports show that BIM use and impact has accelerated dramatically. The BIMForum and Building Smart initiatives unite their followers and increase data sharing and new work processes. Visionaries, such as Kimon Onuma, manage, migrate, and share design data while generating thousands of square feet of building models collaboratively, instantly, in "BIMStorms" using the cloud. Technology is dramatically changing our processes.

Sustainability: Organizations such as the United States Green Building Council (USGBC) and Well Building are making great strides for energy and resource efficient design. The American Institute of Architects (AIA) and others have joined Vision 2030 to set a high bar for measurable change.

Collaboration: In March 2017, more than 100 leaders gathered in Atlanta at the Design Futures Council's inaugural conference on "Collaboration Across the Design Continuum." These leaders framed problems and posed possibilities for better collaboration. Luminaries such as Barbara White Bryson, Renee' Cheng, Howard Ashcraft, Ray Daddazio, and Phil Harrison spoke about the industry's needs in this area. Healthcare clients such as Banner Health, Sutter Health, and Universal Health are leading the way with new collaborative contract forms. Government bodies are adopting new contract types, such as CMAR, Design-Build and IPD, in contrast to their history of Design-Bid-Build delivery.

In the current mainstream these examples are viewed as leading-edge. We emulate their work and try to convince our teams to follow their leads. In the meantime, life goes on. In equal measures, our challenge is to follow these leaders, lead ourselves, and use the wisdom in this book to ameliorate our work together. Show up, keep the faith, and get better. Until the next big leap happens, or we cause it, what else can we do?

"There is no design without discipline. There is no discipline without intelligence."
— Massimo Vignelli

Swing Hard

In the 1990s, the Atlanta Braves began an unprecedented decade-and-a-half long run of division titles. After a big hit in a big game, a reporter asked the team's diminutive shortstop, Rafael Belliard, to describe his approach. His reply was: "Swing hard, you just might hit it."

So, this is my big pitch to help fix what's wrong: *Swing hard*. Accept the challenge. Determine what needs to be fixed and make changes. Create a better future for collaboration. Let's listen closely to the experts here and come out swinging.[8]

Can Design Be Managed?

Some may consider managing design a futile endeavor. Who can hope to bring order to the chaos that is design? As lovers of architecture who depend upon it for our livelihood, we have a vested interest in design's well-being. The more we can be in synch with our design partners and teams, the better our lives will be. I have not fully figured out how yet, but a lot of us are thinking about it. What can we possibly expect to accomplish? Plenty, if we listen.

Managing Design

PART 1

PERSPECTIVES

"If you risk nothing, then you risk everything."

— Geena Davis

"One does not discover new lands without consenting to lose sight of the shore for a very long time."

— André Gide

"I'll play it first, and tell you what it is later."

— Miles Davis

CHAPTER 1

The Interviews

Thought Leaders: Current Realities and Future States

To get a feel for the state of things, I surveyed industry leaders on all sides of the owner-architect-contractor triangle, and a few outside it. For those not skilled in statistical analysis or data mining, "research" is little more than talking to people. I adopted that model and interviewed industry leaders. What they gave back was astonishing; a series of outlooks and candid admissions of issues they face and optimistic future visions. While the introduction unearthed the issues broadly, Part 1 dives deep to explore individual views.

> *"Talent wins games, but teamwork and intelligence wins championships."*
>
> – Michael Jordan

Topics

Interviewees are grouped in chapters by topic. To achieve a panoply of perspectives, I built a Noah's Ark of vantage points – two or more of almost every point of view: owners and architects, engineers and technologists, contractors and trade contractors, students and teachers, strategists and manufacturers. Having boarded this vessel, they share their world views for survival amid a heavy downpour of industry turmoil. With their shared goal to stop the rain, their words give us hope for sunny days and dry land. While many could be in more than one category, this topic-based organization offers a roughly chrono-logical, subject-based logic. Moving from old to new, and clients first, past to present and future, it ends with new voices of advocacy from emerging leaders. Our mission is to build a cohort of understanding, an army for action. For readers looking for a place to start, here are synopses of what you will find.

> *"We talk a lot about hope, helping, and teamwork. Our whole message is that we are more powerful together."*
>
> – Victoria Osteen

3

Chapter 2 Client Empathy: Listening, Collaboration, and Expertise

Chuck Thomsen, author of *Managing Brainpower,* and former chairman of 3D/I, has sustained design and construction relationships over decades. He opens the discussions with reminders of the importance of client intimacy, developing expertise, and the marked differences between design and construction processes. His comments on design philosophies frame the challenge: on one hand, you have a design firm, process, and building type that is predictable, rational, functionally driven by client needs, and straightforward. At the other extreme may be a design team diametrically opposed to rational process. Something may have to give. Assess your team's tendencies to find out.

Beverly Willis, a seven-decade practitioner, gives perspective on the evolution of practice from simpler "post-and-beam" times to technology-enabled, transdisciplinary experts.

Chapter 3 Owner Leadership: Programs, Users, and Talking

Barbara White Bryson challenges owners to lead and engage with their teams, making the case that *owners* have the biggest opportunity and responsibility to make a difference in our teams. Bryson's outlook is bold and optimistic, but not enough owners feel the way she does. How do we educate and activate more of them?

John Moebes's' distinction between serial "program" owners and one-time owners reveals that leadership and forethought can develop policies and procedures for repeat use. There is not enough time to develop them on a single project.

Arthur Frazier's candor about the spectrum of owner responsibilities – leaks, diverse, needy unknowing users, and the perils of design micromanagement are revealing.

Together, Bryson, Moebes, and Frazier's insights into the owner's world are enlightening in their call for owner leadership and project-specific approaches, particularly by program, serial builders.

Chapter 4 Building Learning Organizations: Knowledge, and Research

Jim Cramer shares the history of developing the country's leading design think tank using the strategies of "learning from the best" and having "courage for the future."

Renee Cheng's primer on knowledge management and research give value-adding levers to design practices. Her diverse cadre of students leads the charge.

Randy Deutsch invites design graduates to wear multiple hats as they deploy new tool sets and integrated thinking.

Chapter 5 Firm Culture: Management and Attitudes

Scott Simpson's integrated view of design and its management represents a studied approach from someone who has led several of the world's leading firms.

Thom Penney discusses firm culture and the viability of a regional practice model.

John Busby reflects on the evolution of AIA issues such as continuing education, international practice, the architect's education, cost awareness, and the many meanings and misunderstandings of design.

Agatha Kessler's analogies for empathy, leadership, and tough love suggest a necessary approach to leading designers. She uses different keys to open different doors.

As current and past firm and national AIA presidents, this group's perspectives offer a broad over-look of the state of the architectural profession, firm culture, and practice.

Chapter 6 Strategy: Early Questions, Planning Horizons, and Socialization

Bob Carnegie and *Marc L'Italien* offer practitioners' perspectives on the importance of collaborative practice.

Matthew Dumich gives us a look inside a global practice that enlists his firm's mastery of state-of-the-art tools to create performative buildings.

Chapter 7 Process: Lean Scheduling – Agile and Efficient

Jeff Paine and *Peter Styx* speak openly about architect-contractor relationships, trust, budgeting reli-ance, schedules, expectations, and showing empathy to satisfy client and team needs.

Chapter 8 Collaborators: Performative Design (Better Together)

Phil Freelon and *Allison Grace Williams* expound on the designer's mentality. They share a passion for client-focused, site-specific talent and team-led, high-design solutions that express the art of architecture. Designers asking tough questions to challenge assumptions is at the core. Phil's coachlike admonitions urge readers to push through the rain and pain. Allison's determined championing of design's value suggests we ask tough questions and challenge givens to transcend mere building.

Chapter 9 Design and Budgets: Architect/Contractor Collaboration and Trust

Dan Nall's honest assessment of the range of design subconsultant service types may give pause to architects, owners and contractors who procure such services. Which do you need? What have you bought: industry-leading synthesis and synergy, or a repeat implementation of a standard formula? Nall's elucidation of the need for "catalog engineering" makes sense.

Kurt Swensson, in his role as firm founder, owner, and practice thought leader, offers keen insight into the subconsultant's survival mindset. How to lead from behind? Can we anticipate outcomes, inno-vate, and manage at the point of attack? Serve situationally? Teams face these everyday realities in managing themselves and their clients. In a wide-ranging discussion he opens up about frontline decision making, trust, integrated teams, and the myths they can carry – realistic expectations for shaping the market.

Chapter 10 Strategy: Early Questions, Planning Horizons, and Socialization

Good owners challenge their architects to ask them the right questions. There is a market opportunity to serve those who manage and use buildings, not just those who design and build them. Why not move the conversation upstream to affect strategic business thinking?

Margaret Serrato's candid conversation about strategic questions reveals flaws in current approaches and the need for early analysis. While this may not be currently thought of as traditional design service, there is a need for it. Her contrarian, data-driven questions provoke clients to reconsider conventional thinking.

Phil Bernstein frames futures for practice: technology-driven change, identifying firms' reason for being, value propositions, forging alliances, and planning horizons.

David Gilmore illuminates early team-building best practices, leadership versus management, and the coming exodus of design firm leadership. His data and experience base draws from the world's best firms. He shares his business acumen to guide them, and us, in team socialization and design leadership.

Chapter 11 Engineers and the Consultant's Mindset: Leading From Behind

Bruce Cousins' application of Lean methodologies offers new rigors to contrast traditional design thinking. Readers encountering these processes for the first time may surprise themselves.

Chad Roberson, through his firm's simple, democratic, self-determining Lean work planning system, offers a palatable way to manage design: by the piece.

Chapter 12 Contractors: Risk and Design Assist Expertise

John Rapaport and his colleagues speak honestly about the untapped potential of trade contractor expertise to inform design and construction and the failure to use valuable project history and feedback. BIM's untapped potential and the need for capital investment in research and data reuse are discussed.

Jeff Giglio and *Don Davidson* discuss timing of trade contractor involvement, trust, and the collective challenge of keeping up with new materials and systems.

Wayne Wadsworth offers an eye-opening analysis of the importance of the participant supply chain in designing and building, including the industry trade contractor network in assessing and managing higher risk levels stemming from innovative design, and translating it into reality.

Jon Lewis shares a boots-on-the-ground constructor's point of view. As a general superintendent responsible for building major projects, he is a leading construction technology adopter and a sympathetic design thinker, manager, and collaborator who copes with and builds unmanaged design.

Chapter 13 Technology: Leveraging Data

Arol Wolford's legacy of data management and reuse in the design construction industries offer a clear opportunity. His investment in gaming engines shows continued leadership to enable new ways of seeing, thinking and collaborating.

Casey Robb's perspective and decades of experience as a manufacturer's representative reveals gaps in technical product data, knowledge sharing, and respect for relationships.

Josh Kanner's entrepreneurship in technology-enabled data reuse has resulted in industry-leading applications that have transformed workflows. Arol Wolford and Josh Kanner are serial entrepreneurs: builders, acquirers, and sellers of multiple enterprises that employ data to inform design and building.

Chapter 14 Entrepreneurship: Vertical Integration and Value Propositions

Scott Marble posits the need and opportunity for entrepreneurial attitudes and new alliances in redesigning the business of design.

David Fano and his firm WeWork are changing design and construction via vertical integration. Rather than complain about inefficiencies, they reinvented the market. Their model to design, develop, automate, and control their employee and customer experiences forms a new paradigm, reminiscent of Apple, Google, and the like.

Chapter 15 Change Agents: Advocacy, Equity, and Sustainability

As midcareer professionals, and firm and national leaders, this group points to larger, industry-wide issues with fresh voices and growing momentum.

Simon Clopton paints a picture of future practice for current graduates. His nontraditional career aspirations suggest a myriad of paths for millennial practitioners.

Emily Grandstaff-Rice frames a position of advocacy, holistic design, and the architect's responsibility to have a voice – to go beyond traditional practice. She is active in raising awareness for gender and racial equity in design and construction through Equity by Design.

Based on experience, facts, and personal passion, these leaders' experiences tell stories designed to help readers imagine alternative futures.

Let's get to them.

"Science fiction is the great opportunity to speculate on what could happen. It does give me, as a futurist, scenarios."

– Ray Kurzweil

CHAPTER 2

Client Empathy:
Listening, Collaboration, and Expertise

INTERVIEW

Chuck Thomsen, FAIA, FCMAA

Past Chairman, 3D/I International

4th August 2017

Chuck Thomsen, FAIA, FCMAA, is former chairman of 3D/I International, Houston, Texas, a global design and construction firm with more than 500 employees. He has practiced as architect, construction manager, and program manager, and was the author of *Managing Brainpower* in 1989. He was the first to become a Fellow in both the American Institute of Architects and Construction Management Association of America. He shares universal truths about process differences, expertise, and interpersonal interaction.

Client Intimacy and the Creative Continuum or "Snow Cards, Squatter Sessions, and Goody-Goody Talk"

You remain an inspiration at 86. You're still posting articles, consulting, taking time to talk to guys like me; you and your wife just took your new boat from Boston to Houston. How was it?

Thomsen: We had a grand time, from Boston to Houston, in four legs starting in July. Loved it. The secret to old age is to try new things.

"The secret to old age is to try new things."

The *Harvard Medical Journal* and my doctor say if you're good at crosswords, that's good, but you gotta do *new* things to keep your brain active: learn to navigate a boat or learn to play a piano, things like that.

So, my first reaction to your book's topic is that design is *really* hard to manage. . .

It's an oxymoron, virtually impossible.

> "My first reaction is that design is really hard to manage."

Thomsen: We were at a meeting at 3D/I years ago, and our lead designer got up and drew a Gantt chart–looking thing with bars on the whiteboard and said, "You guys in the construction management world, you've got a dance card with bars and precedent relationships on it, and you know it takes so many days to build a brick wall and that you can increase the number of masons and shorten the days." It's engineering, calculable, that sort of thing. Design isn't like that. Then he drew a squiggly line that doubled back on itself. [See the diagram in the foreword.]

In design, there's nothing you can't spend more time on and make better. And you can have an epiphany and go back to start it all over again. I can work on a door detail and the site plan in the same day or discover an aspect – or develop a concept – that will change it all. I can switch back and forth. The process isn't the same at all. Maybe there's some logic to it, but I don't know what it is. The best designers care a lot about their work. You can always spend more time on something to make it better. That's why so many architects work all night before a deadline to do that.

If you're really looking for creative design, it's a great mistake to overmanage it. Or to try to manage design as you would a construction process: with steps along the way progressing in a logical sequence. Designers, as they work on the problem, become more acquainted with it and are inclined to go back and make changes. It's easy to do on paper, hard to do with concrete in the field.

True. Other interviewees describe that aspect.

> "It's a great mistake to overmanage it, if you're really looking for creative design."

Thomsen: You can build a continuum, with one extreme being the design that finds its inspiration in the requirements of the program, owner, and project realities. The concept comes from the needs of an owner. At the other end is the architect who develops an aesthetic signature – a style that he or she is known for. With the latter, the form is a given and the function fits within to make a notable piece of architecture (or perhaps a flop).

I was on Ellerbe's board for a while, and of course, 18 years at CRS, and 24 at 3D/I. Those companies owed their remarkable success to their ability to learn the realities of a building type and their clients. Tom Ellerbe took over his dad's firm when he was, I think, about 28. He had a project with a country doctor named Charles Mayo. I met Tom when he was 96. I asked him what he did to keep a client for 75 years. He said: "Well, we never would have become what we did without the Mayos. We never stopped paying attention to them." He had to figure out what a hospital was. And it turned Ellerbe into one of the best hospital firms in America.

Do you know the story of the Squatter Sessions at CRS?

Yes, I love them, but don't know the history.

Thomsen: Bill Caudill and John Rowlett were on the faculty at Texas A&M in College Station, Texas. They opened a firm over a grocery store. Bill had done his Master's thesis at MIT on school design, and the young firm won a school project in Blackwell, Oklahoma. They designed a scheme for the owner, who didn't like it. They designed it again and the owner didn't like that one either. So, Bill said, "We're going to go out of business if we don't get this right. So, here's what we're gonna do. We're gonna take our drawing boards and T-squares and drive to Blackwell and work in their conference room. We're gonna meet with them every two hours until we get it right." They had a student, Wally Scott. Wally had a car, so he was a necessary part of the team. (And that's where Caudill Rowlett Scott, at one point the largest AE company in the country, got its name.) But the lesson is that they got close to their client – like Tom and the Mayos. Since the project was in Oklahoma, they called the process of working in the client's office a "squatter" and they did it for every project. Their clients really educated them on school design.

"It was the client intimacy that made them."

When CRS started doing schools in the late 1950s, the baby boomers were reaching school age. CRS's practice boomed with that demographic surge. They figured out how to collaborate with their clients. Client intimacy that made them the country's leading school firm.

Great expression.

Thomsen: That's not mine, it's Tom Peters's. He fell in love with CRS's Squatter process.

Tom Peters exemplifies how the design profession needs to get out of its comfort zone. You said it at the start: try new things!

Thomsen: That concept, developing a trusting relationship with a client, made CRS, Ellerbe, and 3D/I. Ellerbe developed it with the Mayos and it made Tom's firm. CRS did it with the Squatters design process. Soon after Vic Neuhaus started his firm, Neuhaus and Taylor (3D/I's predecessor), a friend, Gerry Hines, said he wanted to get in the development business and build warehouses. Vic designed it and everything else Gerry did for years. They figured out what a good office building was. Their collaboration was critical to 3D/I's future and their dominance of office building design in the '70s.

Part of managing design is finding a way to truly understand a client's business, like Ellerbe, CRS, and 3D/I. To do it is difficult. You look at a market and figure out how you're gonna learn it. Recognizing demographic shifts and getting to know your client are key.

Those ideas are still so valid. We're doing that at Holder with data centers – a building type developed recently – and we're leading the way in learning that business and how it's different for each client. Someone else will come along with the next shift, building, or client type, and blaze a trail. Maybe it's repurposing buildings or shopping malls. . .

Thomsen: When Lean construction emerged, those ideas were already embedded in CRS's processes. Remember the A3 sheets? A Lean construction concept is that every issue and solution should be condensed to one A3 sheet. Make it simple. We used "Snow Cards." During a squatter session, the client or someone else would say something important, and instead of burying it in a report it was written on a 5x7 card with a magic marker and pinned on the wall. The joke was you "snowed" the client by making them feel like you understood their need by feeding back what they said. But the reality was writing it down made the design team understand. It was one way that Bill Caudill managed design.

"Writing it down made the design team understand."

It showed you heard them.

Thomsen: Right. The squatter sessions made sense because the process always had a deadline. You could always work on something longer, but it set a limit. I wrote a report – more like a diary – on a week-long squatter session once. They always started on Monday morning, but by Wednesday, it was a complete breakdown – a train wreck. On Monday, we started out with a lovely meeting, everybody's friends, we set goals. On Tuesday, we're underway. By Wednesday night, a design concept emerged, and we're way over budget! By Thursday, somebody was saying we need to put off the presentation. But instead, we work all night and meet the deadline, and everybody is in love again. Both the client and the team feel like veterans from the same war.

I've done "wallpaper jobs" – filling the walls with information – and used the cards. George Heery called those sessions Pre-Design Project Analyses. I still preach them and do them whenever I can. As CMs, sometimes we think a good kickoff meeting is: "Hi, I'm Chad your project manager. We need to buy precast next month, so we need your drawings next week." What happened to: Who are you? Our common goals? Our mission? How to communicate and work together? A plan?

> "That sounds 'goody-goody,' but it's true."

Thomsen: You're so right. My son is a contractor, a grandson is in real estate development, another a contractor. Contractors see their work as manageable, "time-able." And design isn't that at all. But design has to support it. When Total Quality Management emerged, everybody said quality was doing everything once. Well, that's bullshit! Design is looking at 15 variations. You have to know you can say, "All those are rotten. Let's do 15 more."

> "Projects that have a good project definition cost 17% less than the average. Those that don't cost 20% more."

I like to be the contrarian and say we need to do things "many times 'wrong,'" adding certainty as we know more.

Thomsen: You have to know every process is different and help each other as best you can. Ed Merrow's firm, Independent Project Analysis in Vienna, Virginia, maintains a large database. Ed told me projects that have a good project definition cost 17% less than the average. Those that don't cost 20% more. I repeated that every time I made a presentation to construction professionals. I asked: "Do you believe it?" People nodded. CII, the industry's leading research organization, created the Project Definition Rating Index (PDRI.) Clear understanding before you start is important to get good results. Although the PDRI should be redesigned for each project. The concept is good.

> "We need to do things 'many times, wrong' with increasing certainty, until we know more."

I've pushed something I call "ScopeDoc" to define project scope. PDRI covers everything: services, roles, responsibilities, and project. ScopeDoc focuses only on scope of work. I was at CIFE in Stanford a few years back. An industrial-systems engineer described design as "isolating the variables and using repeat processes." I said, "Design's not like that." He argued, "Every building has a roof, doesn't it? Footings? Floors? Walls?" He was trying to get us to use checklists. I agreed. Designers do too much intuitively. We get sets of drawings with no mention of half the scope or systems. "We're only in SD or DD," they say. "Fine, but if you can't tell me what kind of roof you're thinking about, how can I estimate it?" I'm a proponent of industry checklists, like CSI, not to constrain creativity, but to augment our brains for the rote parts, and add rigor.

Thomsen: Yes. Suddenly, there are so many people. Literally thousands that create little bits of information or service – users, owners, subconsultants. We had 35 subconsultants on one project and the contractor had 75 subcontractors. You've got to get the goals clear, so everybody knows what to do, and contribute at the right time. And guess what? As the future unfolds, the process is going to get even bigger and more complicated!

When I presented at the AIA's "Future of Practice" conference in Washington D.C., the prevailing mantra was: "We architects must return to being the master builders!" I stood up and said: "That's crazy. No one knows that much. It takes teams. We don't care who's in charge, or who works for whom." Despite the grousing and griping, no one's willing to change. If you started a project or firm tomorrow, what would you change to make it a success?

Thomsen: The "soft stuff" of management is important. We saw the extreme in new project delivery strategies going from Design/Bid/Build to CM, Design/Build to IPD, and Qualifications-based. All these are trying to create new relationships between architects and contractors. That's the soft stuff. It's more a social construct. I'd work to build those relationships in a culture where no one can be allowed to fail. That sounds "goody-goody," but it's true.

"If I had it all to do over again, I'd spend more time on that."

Ann Lamott's book Hallelujah Anyway: Rediscovering Mercy *is about letting your guard down, thanking and being nice to others, maybe even when they don't deserve it, because you probably had moments when someone did that for you. We've tried the adversarial thing. It hasn't worked.*

Thomsen: Right! Believe it or not, we had a really hard time knowing if we were doing a good job. Here we were, the CEOs, COOs, CFOs, the C-suite, the top of the company, and we couldn't figure out if we were doing a good job. Our clients didn't tell us we weren't performing. They didn't want to alienate our people they were working with. Our project team wasn't inclined to tell us if we were screwing up either. We couldn't measure the quality of our performance accurately. I finally figured out how: I'd simply ask our project manager if he liked our client. Feelings are so reciprocal. If our PM says: "Our client is difficult, I can't get him to respond," or something else critical, we knew we were in trouble. If our PM says something positive like, "He's a good person," I know I'll get a similar response from our client. It's amazingly consistent.

If you create a collaborative relationship, where it's a cultural standard to befriend and help everyone, and be trustworthy, you've created the atmosphere that produces successful projects. But who talks about these things? Who teaches those skills? If I had it all to do over again, I'd spend more time on that.

Most of us didn't get into design and construction because we were psychologists, but it's much of what we do. Your perspective is invaluable. The principles you share were applicable 30 years ago and still are. I think they will be 30 years from now.

Thomsen: I'm delighted to hear that. I'm happy to talk to you anytime.

Chuck Thomsen Key Points:

1. Managing design is not aligning a linear process: it's messy.
2. Focus on clients. Spend time with them. Become expert at what they do. Learn from them.
3. If you seek true creativity, manage design with a loose tether.
4. Define project scope early.
5. Build and sustain trusting friendships with the extended project team.

INTERVIEW
Beverly Willis, FAIA

Beverly Willis Architects Inc.   21st September 2017

Beverly Willis, FAIA, has played a major role in the development of architectural concepts and practices that influenced the design of American cities and architecture. She is the co-founder of the National Building Museum in Washington, D.C., and the Beverly Willis Architecture Foundation, a nonprofit organization working to change the culture for women in the building industry through research and education.[1] During a seven-decade career, Willis has been a technology pioneer and a force for industry change. In 2009, she served on the AIA's National Technology in Architectural Practice BIM Honor Award Jury with the author. She is still active at 90.

Specialization and Generalization or "Orchestrating the Post and Beam Crowd"

We're talking about extremes. On one hand there's the iterative, messy, process that erodes confidence in designers meeting schedules and budgets. At the other extreme are those who embrace rigor, and things like BIM. What has your approach to managing design been?

Willis: I started my career long before BIM. My firm was the very first to use the computer to do in-house programming, creating software we called CARLA, Computerized Approach to Residential Land Analysis. That was 1971, before the desktop computer, Microsoft, Apple, or Google. In the '70s, as part of our programming efforts, we had to become knowledgeable about systems analysis, which was new, and preceded coding. Coding was a more European approach. But what systems did for us, all stages of our work, was allow us to analyze the process and create matrix forms that could be filled out. This was useful for cost estimating, schedule, controls, and that sort of thing. But on a bigger level, since even before I started practice in 1966, the design, engineering, and construction world has changed dramatically.

"BIM, from my standpoint, is a bit old fashioned."

When I started, projects were relatively small. A 12-story building was considered a high-rise. The architect did design, cost estimating, construction supervision, and construction management – the old-fashioned idea of the architect as the "master builder." Over the years, my projects have grown increasingly until today, they're mammoth, in terms of size and the teams – the architects, engineers, and constructors it takes to deal with them. In that process, the architectural community has become increasingly specialized, giving tasks to specialists in cost estimating, environmental analysis, and other specialties. The architect's role, as I see it, has increasingly become one of organization, planning, and conceptual design, because even some large, well-known, so-called "expert" firms can do schematic

---
[1] Wikipedia

design, but not the detailed drawings. That's usually farmed out to a young, high-tech firm that can apply all sorts of computer analysis.

> *"To be successful you need a wide range of knowledge."*

BIM, from my standpoint, is a bit old-fashioned, because the prevailing software today is essentially using the intranet, where the global firms can work 24/7, day and night, simultaneously, and the consultants can too, simultaneously. That model developed on the intranet can drive production in the factory, cutting steel or stone, so the accuracy is incredible. That's where we are in meeting the large project challenges that take a variety of talents and firms to execute them.

You suggest an increasing need for management in the architect's role, if nothing more than in organization, choreography, and conducting larger, more complex teams.

Willis: I want to emphasize the planning. Somebody has to develop a program and lay it out. That *is* the work of the architect. The first step on any project. Know the objectives, the people they serve, and how that flows together. That's a challenge if you're trying to do it without a leader. From there, you have the plan and can assess the budget and make management decisions. When you have a relationship with a client and can say you meet their objectives, then you can begin to do design.

Early feasibility testing, and project definition as design management starting points. What advice would you give aspiring architects entering the profession today?

> *"We have to deal with the results of that research by a whole range of disciplines. Architects making their mark today are the ones who have knowledge about the specialties, the new materials . . ."*
>
> *"We were really sort of a 'post-and-beam' crowd. It's a different world today."*

Willis: To be successful, you need a wide range of knowledge. More than what you'll get in a typical design education. You have to understand business, development, and different aspects. In my case, because we pioneered a system for environmental assessment, we had to learn about biology, flora, fauna, and a different range of issues not normally considered architectural design. For example, a firm came up with the idea that instead of spending millions to clean up the Hudson River, just toss in a lot of oysters, and they'll clean it up. The relationship between biology and architecture came up with a better – a cheaper – solution for environmental work. We're at a point in history where the research of all sorts of disciplines is coming to a single point, and we deal with the results of that whole range.

Architects making their mark today know the specialties, new materials, how to use structure in a different way, how to use the natural environment, and weave it into architecture. That's where it's going. As a young person in school today, you need to understand how biology, architecture, and chemically based materials interweave. It's an expansion of knowledge in architecture and engineering we didn't see in my day. We were really sort of a "post-and-beam" crowd. It's a different world today.

Fewer things can be done intuitively, based on what an individual can know. You're saying there's room for those who go beyond boundaries and create value and expertise, and for organizers, planners, and generalists. Designers can't do it all themselves anymore; they need specialists.

Willis: It's like the conductor leading the orchestra. In my day, it was still the conductor leading the orchestra, but today the conductor has a far more complex music to conduct.

And a larger orchestra, with more tools, complexity, pressure – and less-prepared clients, and always less time.

Willis: Exactly.

Beverly Willis Key Points:

1. Future design managers will need an expanded skillset, beyond the provinces of traditional education and practice. Technology is a given.
2. Intuition and experience are no longer enough to inform design.
3. Rigor, systems thinking, and processes are now required.
4. Test and balance design, budget, and program early to manage design.
5. New processes are required as project scale and complexity have grown.
6. Future design managers will include expert specialists and expert generalists as conductors, choreographers, organizers, and planners with greater technical proficiency to manage.

CHAPTER 3

Owner Leadership: Programs, Users, and Talking

INTERVIEW
Barbara White Bryson, Ed.D., FAIA

Associate Dean for Research and Academic Affairs, University of Arizona, College of Architecture, Planning and Landscape Architecture   1st September 2017

Barbara White Bryson, Ed.D. FAIA, is an architect, owner, and educator. She is currently Associate Dean for Research and Academic Affairs at the University of Arizona's College of Architecture, Planning and Landscape Architecture, and Managing Principal of the DesignIntelligence/ Design Futures Council DI/Strategic Advisors consultancy. Her professional design experience includes work with Rice University and Spillis Candela (now AECOM). She is co-author of *The Owner's Dilemma: Driving Success and Innovation in the Design and Construction Industry.* Her challenges to owners, architects, and educators are timely.

Collaboration and Communication or "Teaching Masochists"

Let's start with a reflection. Since your book challenges owners to lead, what are the issues around collaboration? Are challenges being met? How are owners feeling?

"We don't teach collaboration."

Bryson: Different types of architects are doing different kinds of practice. So, in managing design, there's no one size fits all, not now or in the future, and industry disruption is coming, much more quickly than people imagine. In 2017, I wrote an article published in DesignIntelligence: "The Future of

Architects: Distinction or Irrelevance?" I wrote it provocatively on purpose, because I believe architects are working their way towards being irrelevant in the design conversation – especially if they do not make a few simple adjustments in their approach to the profession. One point I wrote about was building foundational knowledge, which I also mentioned in *The Owner's Dilemma*. How difficult it is to stand in front of corporate boards, legislatures, and town halls without proof that one delivery method is better than another? Or. . .that we understand building costs or the value of design? We should have research to prove to support our work. The second point I made in the article, which you note in your question, is the need for architects to practice collaboration. We often do not teach collaboration in school. We talk it, but do not teach this very specific set of skills. Collaboration isn't easy; it is hard. Business schools start teaching collaboration day one. You cannot walk into a business school orientation without hearing the words "Form, Storm, Norm and Perform."

In architecture school, there's little discussion about why teams work or why they don't. It's frustrating. It seems students are sometimes taught that the loudest, most egotistical, most critical voice wins, because that's the environment they're thrown into. Another reason for potential irrelevance is that we architects often speak and write to each other in this private, insular, secret language we love and enjoy. We neglect to remember how important it is to speak clearly to the rest of the world, articulating design value and demystifying our processes, so others can appreciate the value we provide.

Is it our closed culture? Tradition? The vestiges of the guild or studio? If so, somebody has to acknowledge that and change it.

Bryson: The culture challenges may go back to a lack of rigorous or scientific research within the discipline. Many architects say, "We do research," but it's often theoretical, experimental, or esoteric, not based on scientific method, which reinforces separateness from other disciplines. Not that design theory research is not important, but it is narrow. If we created an environment in architectural education where students could learn basic methodology during a research-based experience, it would profoundly change our profession. Renee Cheng and Laura Lee have developed tremendous research around collaboration – it changes how students think. Students and stakeholders realize knowledge adds value to what's delivered to the owner. Kieran Timberlake is a fabulous example of a firm that includes a research component in their practice. It adds extraordinary value to owners. I know from personal experience their design delivery is as excellent as any in the world.

"How to communicate that to the stakeholders so they become part of the process."

I want to add one additional point regarding the relevance of architects: architects have been risk-avoiders for decades, which means we have stepped back from responsibility. We've listened to lawyers and insurers preaching risk avoidance rather than risk management or reduction for far too long.

AIA documents haven't served us well. These documents say, "You shall not. . ." and that gets translated into professional practice courses in universities. We have to understand that risk is about what you cannot know or what you cannot control. Let's learn how to create processes and environments to help us know more and control more. We need to teach students how to think about the business of architecture.

It's avoidance and contraction mode. I see examples in OAC teams. A roof substitution is suggested, but the risk of evaluating it is thrown back to the designer. Maybe the team should share the risk?

Bryson: Whether you believe it or not, teams that work in silos ARE sharing risk. . .very, very high risk. The greater the knowledge sharing and the greater the collaborative problem solving, the lower the shared risk. It is wonderful when we craft contracts that reflect collaborative behavior and lower shared risk.

That's how we're working, but sometimes at the end, we still push the responsibility back to the architect. Maybe it's tradition, or not having an IPD contract?

Bryson: We all assist each other in a collaborative process and risk goes down dramatically. The architect helps the contractor reduce risk on his or her decisions, and in turn the contractor helps the owner or architect reduce risk on his or hers. It's a lower risk environment even if contracts are old school.

Yes, we're all "designing." But our old contract forms aren't set up to address that.

Bryson: But we can't make "perfect" the enemy of "good." That's why I embraced IPD-ish delivery. Howard Ashcraft and Will Lichtig have done some great work. I quote them in my book. I was not in the position to change the contracts, but I could absolutely change the processes, the teams, and the behavior. Will says, "Change the people or change the people."

There's one last point related to architects becoming irrelevant: U.S. architects, even if they work internationally, aren't appreciating what's going on internationally. The world has changed significantly. Cities appear almost overnight; a high-rise building can be assembled in days with no U.S. firms in sight. Super-professionals build more square feet in a year than U.S. firms do in a decade. The world has needs we aren't aware of. American professionals are not sufficiently prepared for technology disruption. If we continue to practice the same old way, there's a good chance, in a decade, the rest of the world will have figured out things we haven't. Then we'll be competing for owners impatient with cost increases and inefficiency. Amy Edmundson and Susan Reynolds wrote that 75% of activities occurring on projects sites add no value. The *Economist* in 2017 said 90% of infrastructure projects are over budget or schedule, and 60% of building projects in the UK are not meeting schedule goals. When the economy gets good, we pump up prices. When owners get impatient and frustrated with poor schedule performance and high prices, they'll find other ways to buy buildings like prefabrication and robotic construction. When that happens, architects may or may not be involved.

I believe we should teach our emerging leaders both basic business leadership skills and research skills. Most young professionals look at you with a blank stare when you talk about business planning. We have to target emerging firm leaders, to give them basic business concepts and an understanding of the disruptors coming.

Firms aren't going to be able to take on every one of these disruptions: algorithmic design, hyper-prefabrication, hyper-collaboration, robotic design, swarm intelligence, modular design, mini-robotics. It's like 25 freight trains. A firm can't make sense of them all. But if they take on one, complementary to their strengths, they can seize an opportunity. For example, HKS does it in sports design. They say there's a lot that's

"'You shall not'. . . We can no longer practice in that manner."

formulaic in stadium design so, it's a natural fit. I've read that they use algorithmic design to produce 75% of it, independent of the individual project, efficiently.

The research piece is important because architectural students don't get a research experience – most never had that light bulb moment. They think a case study is simply documenting the project experience. They don't understand how knowledge is built within other disciplines. We need to give them that experience, so they can build a research culture in their firms and understand the value of rigorously built knowledge.

We've talked about research, business planning, and collaboration. The fourth issue you mentioned is value proposition. In the future, not many firms will survive using the old mantra "We're a four-person firm, we can design anything." They'll need something more compelling.

Bryson: A ton of small firms will be doing things that will surprise you. They're going to be defining problems, working with artists and furniture manufacturers, doing design thinking with downtown organizations, leading in areas where, for years, the AIA's been telling them they can't go. I still believe in small firms if they learn how to manage risk, not avoid risk.

I can draw a parallel to apps. There are apps for everything, to connect, translate, enable. More, not less. There's an app for everything, so there's a firm for everything.

Bryson: Young firms are going to be masters, designing hardware, new kinds of bathtubs, intellectual property, and they're not going to look anything like the small firm in our day that just does houses and restaurants. They're taking off the handcuffs to do wild, wonderful problem-solving jobs for all kinds of people in all kinds of places. They will also be part of redefining our relationship with the business plan.

It's an extrapolation – or "atomization" – of the profession. Maybe it's because we've been a profession of many small firms; 90 percent of firms are 3 people or less. With technology, small, nimble practices can produce a lot and be profitable. But it could also be risk management, because they'll be focused and good at what they do.

Bryson: I suppose that's true. Architects have to stop believing their only revenue source is owners' fees. That's broken. Architects are so dependent on fees that when owners stop building, they have nothing to fall back on. Architects have developed intellectual property for years and done nothing with it. They've been redesigning and redesigning. Very little – materials, products, stairs, details, windows, hardware, furniture – has been translated into additional revenue.

I think the reasons are: a) architects don't think about making money, don't like the business side, and aren't enterpreneurial; and b) not capturing intellectual capital for reuse gives us a another chance to do what we love: design things from scratch, not reuse them, play in our favorite sandbox. Business isn't in our DNA. A self-defeating pattern that needs to change, correct?

"Most young professionals look at you with a blank stare when you talk about business planning."

Bryson: Absolutely, because we're taught through our education to be somewhat masochistic. Here's the first rule of effective collaboration: every collaborator must ask for what they need to be successful. Architects aren't taught to do that. If you don't ask for what you need, the entire team fails.

Good rule. One ask has to be a way to make money. Architects have to figure that out.

Bryson: Yes! We need to stop teaching students to work all night and take so much negative criticism. Constructive criticism absolutely. Life is tough. You're have to learn how to fail and pick yourself up. But also to ask for what you need to be successful. And how to function in collaborative teams: "If I don't get my share of the materials, I'm not going to be able to deliver what I need to for us to be a successful team." It's a simple set of concepts. Architects need to think, "I'm going to deliver a building. Maybe I should be paid, not based on hours to design it, but on how the building performs." Like a commercial that runs again and again, maybe I should get a percentage each time it's used. Maybe we can develop an agreement: "If this building performs well – here's the criteria – and we keep it running well, I get paid royalties over time."

Intriguing concept. If young architects are paving the way for change, and contractors can manage risk, what about owners? How do we fix percentage fees, and commoditized services? Can owners bring change?

Bryson: That's a multipart problem. I was recently talking to the Texas Owners at College Station. "Collaboration in a Risky Business" was the title; it's based on a story about my uncle who dove unknowingly into the shallow end of a pool. He could not see below the sparkling surface. That's a metaphor for what owners do when they choose Design-Bid-Build. They dive in without being able to see beyond the surface of the water.

As architects, we're not good at articulating the value of collaborative processes, because we talk about "best practices," which makes board members' eyes roll. We should instead talk about risk. In Design/Bid/Build, risk is higher. A board member's responsibility in D/B/B doesn't stop at lowest first cost. Their risk for total cost extends to the project's end. But, here's the catch: the industry doesn't have foundational knowledge – it can't prove one delivery process is better than another. Renée Cheng's research on GSA's collaborative processes is getting us there, but more is needed.

It's gaining momentum, but only among a small group preaching to the choir at conferences. The "rest of us" are saddled with archaic processes. If we want the job, we take it under the owner's terms.

Bryson: That's right. We must communicate effectively and talk about risk. Boards can understand discussions about specific issues that cannot known or controlled. When we sign a D/B/B contract, we cannot control or know anything after bid day, and it's like diving into a pool without knowing whats below the shining surface. You cannot hold the team accountable for results after that. The other thing we have to stop saying is in collaborative environments that there's still high risk. That's bull. When we get great people, tools, and processes risk is dramatically reduced. Yes, Hurricane Harvey can come in – but that's another deal. Beyond that, when you communicate effectively you deliver a great project.

"Leading in areas where, for years, the AIA's been telling them they can't go."

You're right – when a great team works for a common purpose, we're successful.

Bryson: We've got to change our language because too many still believe even in a collaborative environment there's still risk. Because they haven't done the work. I believe windfalls and overruns are symptoms of the same problem set: not monitoring your budget every week, and having an uncollaborative team.

You asked earlier what's going on in my world. The Association of University Architects is a peer group. Recently a question came from a colleague that asked, "When do you get board funding approval? We get it too early and it causes problems." I blogged about it. I said, "That's the wrong question." Why don't we set price based on needs? They're a key part of the business plan. If you bring in all the players and go through scoping and predesign, get that approved and it meets their needs, that ought to be fine. You ought to manage it with gusto: it's your mission! Some say if you set it too early, you'll get what you get. I'm here to tell you that's also what's gonna make us irrelevant: not being dedicated to the success of the businesses and stakeholders we serve.

How can we be more efficient? Figure it out. Prefab bathrooms and deliver them 25% more efficiently to last longer. Be creative. Think differently about what you're delivering. Meet your customer's needs.

Maybe that needs to be a criteria in an AE RFP? Look for people who come in with a prefab mindset or capability? Focused on a systems approach?

"You ought to manage it with gusto: It's your mission!"

Bryson: At Rice we had Michael Hopkins, a phenomenal designer. We had the opportunity to do prefab bathrooms. They not only embraced doing them, they designed them elegantly and had beautiful fiberglass forms. Prefabricated doesn't have to mean lack of design.

Barbara Bryson Key Points:

1. Architecture schools must teach specific collaboration skills and defeat the culture of masochism.
2. Architects must learn to communicate in owner terms.
3. To collaborate openly, ask for what you need to be successful.
4. To change the architect's value proposition, change the profession's historic mantra: "You shall not. . ." and risk avoidance attitude.
5. To manage design, their practices, and futures, architects must adopt a business mindset and learn business planning.

INTERVIEW

John Moebes, AIA

Senior Construction Director, Crate & Barrel   15th August 2017

John Moebes, AIA, is Senior Construction Director with Crate & Barrel. He has used building information modeling (BIM) to transform the design and construction delivery process. He is a frequent presenter at industry conferences and an avid change agent. Through owner leadership as a serial builder, he has demonstrated metrics and results with new design management processes. His insight into owner "program" versus "project" responsibilities is timely.

Programs versus Projects or "Savvy Owners"

You have done great things as an owner using BIM with architects and contractors, and shared them with the industry. What have you done lately to solve the problems of managing design?

Moebes: Well, I'm Crate & Barrel's Director of Construction, so my work is 85% construction manager and 15% design. I share a large fiduciary responsibility to manage the company's capital investment in real estate. We have a group of 8 architects and 4 construction managers. We produce a design intent package, and then look for engineers and architects-of-record to augment and localize our content.

You're doing bridging?

Moebes: A form of it. We do something called "augmentation and localization." With the exception of structural and MEP, we create the design intent around the aesthetics and operations and give it to outside engineers and architects to bridge the gaps. It is one-part bridging and one-part reconfiguring of the AIA's traditional SD, DD, CD packages. In terms of technology, we're deep into Revit, but not like a traditional design firm. We don't use it to generate CDs, just for design content – more a "light-duty" DD/CD package.

Repeat projects and technology give you "reach," with tight margins in a retail environment.

Moebes: Our biggest success is the tri-party relationship we developed with structural steel, BIM, and IPD to make our steel systems more efficient. It took us three years to develop it and we tested several partners to find the right team. In the end, we chose one of the largest steel fabricators in the U.S. and a very small, but sophisticated, engineering firm to partner with us. We gained efficiency by leveraging the engineer's models to go right to final detailing and fabrication.

"I'm seeing BIM retrograde these days."

We had a mutually beneficial relationship among the three parties. It started with technology but evolved to be more about trust, teamwork, and respect. That's been our biggest success. We do other things I take for granted, like using the model to do material estimating and ordering. It was revolutionary 10 years ago, but now it's old hat. Not to drift, but I'm seeing "BIM retrograde" these days. The millennial generation is drifting back to producing the same 2D documents we produced 15–20 years ago.

Why in the world is that happening? Because the marketplace demands it, or that's the way leaders are pushing them?

Moebes: A little of both. BIM passion has subsided to a large degree. The adoption level is high, but the passion to solve problems has subsided. You don't have firm leaders saying, "I want to plow a bunch of money into this to revolutionize my business."

We've hit a flat spot. At BIMForum meetings, the same 300 people preach to each other and get lathered up. We've pushed and pulled each other up, but where's the other 99%? They're not in the room. They don't even know what we're doing. You're facing that trend within your own organization, and you're one of the leaders!

Moebes: Not to detract from the millennials, but we leaders need to formally train the younger generation to be well-rounded designers. BIM software is a great design platform, but its mastery saps a lot of time out of the first three years of a graduate's career that was once used to master the nuts and bolts of practicing architecture.

We've closed one door and opened another. That takes us to your process. There aren't many "good" owners yet. New processes introduce risk. What did you do to change your leadership's minds? Data? Pilots? Do we need to set up an owner's training academy, and will you be its dean?

Moebes: Ha! Maybe professor emeritus. . . The difference between an architect who does healthcare and one who does retail is small. The design processes are similar, the documents too. But the difference in those two types of owners is enormous – they have nothing in common. For an owner like us, it's easy to make the BIM business case.

Lots of owners are making excellent use of BIM but don't talk about it. At first, I led Crate & Barrel's work as a consultant from the outside. Then I joined the company and mandated that all our consultants use BIM. My early adoption predecessors showed it could help construction and I took it to the next level.

What is the secret to improving collaboration? More time to plan and develop processes as teams? Or should architects just get over themselves and stop whining?

Moebes: I wish I would've stopped to consider that question every Friday for the past 20 years. Owners who build a lot or do expensive, complicated buildings see benefit from technology, and they're the savviest owners with respect to BIM.

Different owners and building types have different design processes. One is open-ended, in search of formal ideas, the other is refinement of a box. We just finished Mercedes-Benz Stadium, where everything was first-of-kind. Very different than refinement.

Moebes: Are the processes that different? Both buildings are there to make money.

It's just harder. As a process person, I get mad when we, or owners, don't allow time to plan. We do it to ourselves – we come in and just start. No goals. No testing. No getting to know you. You never get that time back. Why don't experienced owners know better?

Moebes: As a company, we're doing it. I'm running a building program. Others are too. Projects like the stadium are just a single project. No ongoing learning. Singular projects cause the most stress. I'm creating a program to manage multiple building projects. When you talk about BIM protocols, that's fundamentally not part of a project. It's part of a program. BIM protocols, travel policies, contracts, in my mind, are all part of a program.

Implicit in what you're saying is that processes take time. As the owner, you have to call the shots and develop them in advance to be ready to use them – you have to invest the energy, because you'll never get the time on a project. So, there's great power and potential in programs. That's a great observation. That's how you get the time.

Moebes: One service the AIA provides that's terribly underrecognized, and they do it for everybody, is standardized forms. That whole body of contracts and forms is their "gift" to us to jumpstart program management.

On one hand, if you want creativity, you can't overdirect design. On the other, working for a serial builder, I'm an enabler. We use lots of checklists. Like you, we rewrite most of them. Despite their value, the standard AIA forms don't apply to all situations. Even when we're all at the table and use our best tricks to manage, we still go over budget and owners can't believe it: "That's what I hired you for!" How do we regain respect?

Moebes: The profession is going to change radically. It started with the last recession. The frustrated people left. The well-adjusted ones remain. I see fewer malcontents. What I'm seeing in the U.S. is different than globally. Globally, architects have more respect on the AEC team than in the U.S. I'm not sure why. A lot is self-inflicted. Whenever I meet people at parties and on vacation, I never get sneered at for being an architect. They always say, "Wow, architecture, that's really cool!" When I'm talking with other owners, I don't bash architects anymore. Sure, we make mistakes, but that's not the profession. My advice would be just "walk away" from that stigma. Be cool again, know your game, and foster respect.

"That's fundamentally not part of a project. It's part of a program."

When I look at our company now, I'm impressed at how much instruction we give to young retailers, merchants, and buyers. So much institutional knowledge is passed to new employees. AEC in general, needs to do that.

John Moebes Key Points:

1. Repeat "program" owners have more time, opportunity, and feedback loops to develop or project protocol, and reap the benefits.
2. In some markets, "BIM retrograde flat spots" are occurring. Clients have to slow down to align.
3. BIM requires more collaboration and communication. That comes with a cost, which the need must justify.
4. If the project need is to execute a commodity, it may be okay to "just say no" to collaborative, customized services.
5. Owner's may need technical help from architects and contractors. Current contracts treat owners like "idiots."
6. Architects must focus on their value proposition to reassert their confidence and stature.

INTERVIEW
Arthur E. Frazier III, AIA

Director, Facilities Management and Services, Spelman College   1st February 2018

Arthur E. Frazier, III, AIA, is Director, Facilities Management and Services at Spelman College, a private black liberal arts college for women in Atlanta, Georgia. He is a registered architect with 30 years experience at national design firms, including Lord, Aeck & Sargent, Cooper Carry, and the Philadelphia Architects Workshop. He holds a Bachelor of Arts and a Master's in Architecture from the University of Pennsylvania. As owner's representative and construction manager with H.J. Russell and Emory University, he has championed sustainable design and LEED certification for many projects. In 2003, he joined Clement and Wynn Program Managers. In 2005, he joined Spelman College to oversee design, construction, sustainability, and facilities management. In this interview he reveals the pros and cons of owner micromanagement.

Paying Utility Bills, Fixing Leaks, and Dictating Design, or "How to Get Less"

You've had a wide-ranging career on all sides of the team. How did you get there?

Frazier: When I was with Lord, Aeck & Sargent, designing the Aaron Diamond Aids Research Center in New York City in the '80s, the building operations engineers made more money than I did. It made me ask, "What's wrong with this picture?" So, I began a journey of crossing over to the owner's side. I stayed out of politics and bureaucracy after I found I was an amateur at finger-pointing.

How has that shaped your perspective?

Frazier: Design, and design management, bring awareness of the impact of design decisions and issues to maintenance, operations, and sustainability. You're more attuned to the issues of others. On the owner side, we pay the utility bills, so we see the benefits of sustainable design and LEED. Very different from developer work. He's going to sell the building, so he doesn't need a long-term solution. You care about all of it. You've probably seen the same sensitivity to construction issues having crossed over.

> *"A distinction I make is whether our project managers are 'managing' the design process or 'directing' it."*

I was on a design call with Studio Gang last week for one of our projects when a pipe burst. It was raining water from the attic down three floors all over one of our historic buildings. I had to change uniforms and go into maintenance mode. That's the gamut. There's always something going on.

As an owner, you know the users' refrain: "I want, I want, I want." How do you cope with the constant conflict of ever-limited resources?

Frazier: Users bring a challenge in that they don't understand our process. Some do better than others. Some can read drawings, some can't. We do this all the time and they don't, so helping them is what we do. They want everything. They need everything. Getting them to understand their own priorities is key, sorting "must haves" versus "luxuries" is a challenge. Faculty are particularly difficult, let's say "challenging." Other users, such as doctors and researchers, are more straightforward.

When experts live in their silos, each can think their discipline more important than the others. We got into the design and construction business because we love creating and making, but our greatest challenge is dealing with people. Did we learn how to do that in school? No! We are doing a job we weren't taught. All we have is on-the-job training.

Frazier: I face that on many levels. In managing designers, but also employees. Leading and coordinating design teams is one thing. Program management – managing managers – is different. I said to my boss, "I need to understand the people!" On our Arts and Innovation Center project, we have many different users. One is an engineer. He comes to meetings with an organized, informed approach to sharing his needs – precise, with lists, Venn diagrams, square footage information, and facts. The other group from the drama department is theatrical. Totally different. They're loose, descriptive, hard to pin down. They deal in emotions and feelings, words and experiences. Hard to nail down exactly what they want: it's a different script every time we meet. Understanding your users and how they communicate is where it starts.

Working with architects is similar. They have different styles and approaches. A distinction I make is whether our project managers are "managing" or "directing" design process. Some architects revolt when they're directed. They feel their creativity is taken away when they're directed to do specific things, versus being managed or steered.

Another interviewee talked about "leading" versus "managing" design. I'd say, all of us, regardless of role, would rather be led, than managed or directed. Designers in particular.

> *"The users bring a challenge in that they don't understand our process."*

Frazier: I like your description of it as leading. You're hiring the architect for their creativity and problem solving. When you start overmanaging, you lose that. You put handcuffs on them. When they're led, or can lead, they come up with more creative solutions.

Managing people is central, but how does technology play into your process?

Frazier: We've struggled with it due to our size. We haven't taken BIM to the next level – after project completion – for Facilities Management. We have some models, but none of us use them. We're exploring how but haven't taken advantage of it. We're behind some other campuses. We want to get there, but our maintenance staff is older and slower to adapt to new technology. As we bring in younger blood, we'll get to the highest and best use of technology.

How does your project development unfold? The project we're working on together now has taken the predictable path. It gets designed, goes over budget, we beat it back down, rescope it, and start anew. Are you doing anything to avoid that in the future?

Frazier: Several things led to where we are: a new administration, multiple user groups, the building type. The existing building dates to 1964. The users have had no resources for a long time. They ask for, and need, a lot. I talk to other campuses. We're not doing anything differently than they are. Most have external design review boards. We don't. Teams like working with us because we have a "short decision tree": just our team and the business and finance office.

> *"Teams like working with us because we have a short decision tree."*

That's a benefit of being a small, private organization – you control your destiny. Knowing that, what would you do differently to break the vicious cycle of overbudget projects and the stigma of unmanaged design? None of us got into this business to do value analysis. What could we do as a lean process if we all got in a room together and dreamed?

Frazier: That's what we plan to do. We've got the president on board. Nothing more can be added. She'll say no. We'll say no to her. That's what led to the scope creep. The design team, Studio Gang, designed a very complex building. We're going to work with you as the contractor to build a cost model and work to it. Years ago, we did a project at Lord, Aeck & Sargent that way. The wild cards were the structure, skin, and HVAC system. We set component targets and hit them. It worked. The way we bought it ticked off the contractors, though, because we went directly to the submarket and bought trades. I don't want to do that again, because I got tired of hearing: "Your subcontractor did this!" "That scope was left out!" "Those were not coordinated or included."

You discovered that doing it yourself, there was value and service in scoping and coordinating trades by the CM after all? Those pronouns "yours" versus "ours" are telling. There was a strategy. And it worked. Teammates get browbeaten and lapse into complacency and abject civility: "Just tell me what to do." They give up whatever semblance of value they have earned after all these years.

Frazier: That's what happens when you direct architects and take away their creative freedom: you get less. The Winship Cancer Institute at Emory went well from a collaboration perspective, and it had nothing to do with design or construction, it had to do with purpose. The Cancer Center representative stood up at the partnering meeting and said, "Everybody in this room has been touched by cancer." It gave us a common purpose and theme for the entire project. We may not have been able to cure cancer personally, but we could do our best to give those who were trying the best building we could. It wasn't just a common purpose, it was a higher purpose. We repeated it when tradespeople came in, and at the topping out, we had cancer survivors sign the steel beam. It was inspiring. That was 1998, and I remember it like it was yesterday.

> *"What would you do differently to break the vicious cycle of overbudget projects – the stigma of 'unmanaged design'?"*

Anybody lucky enough to be a part of great project with a great team remembers it. It's life-changing. But we miss the opportunity to look for that common goal, find that purpose, and get to know one another. Two good things happened in your example: one, a common, higher purpose; two, an engaged, motivating owner. Can we tolerate a "design only" mentality, letting others control cost after the fact?

> "When you direct architects and take away their creative freedom, you get less."

Frazier: There has to be balance. I've worked on projects where design was all that mattered. Everything else was a distant second. When that happens, you better shore up the team. On this project, time is the issue. It's a conflicting agenda. Designers say they'll do it, then it's rushed and we all suffer. Some firms were so intent on design, their CDs suffered. I knew we were in trouble when the superintendent showed up in the OAC meeting. It meant he didn't have dimensions to build by. It starts with communication and common goals. We don't need people serving their own agendas, who don't listen to what we want. It's interaction between people.

Owners could help themselves by setting up project planning phases. Too often we jump in because there's "no time"; we never recover. If we don't do it before we start, we are forced to do it on the fly, and it costs more. Do it upfront and you have something.

Frazier: Excellent point. I've said to my boss on multiple occasions recently: "We need to figure what we want before we go to an architect and contractor." Part of the problem is our own clarity of need. It would make things go smoother. It affects project delivery. We did a project using George Heery's Bridging method, a two-part approach. Our challenge was we weren't clear on what we wanted. We paid more after we were underway.

I like half that method: defining project scope before design and implementation begins. I don't agree with bringing in a new team and handing it off. The team has just gotten to know one another and the project. Continue the momentum; don't lose it in a handoff.

Frazier: On a residence hall renovation last summer, we used a flexible scope approach. The contractor came on early. We didn't know what the scope was. We knew we wanted to spend $3 million to make improvements, that's it. We identified $10 million of options. Those that couldn't be done over the summer were dropped. We made scope choices, trading watts for windows, and it worked. The paint was still wet when the kids arrived for school in the fall, but it worked.

Arthur Frazier Key Points:

1. Owners have responsibility for the entire gamut of planning, design, construction, and operation. Partners who understand that can serve better.
2. Designers and constructors who employ cross-discipline perspective and empathy (owner-architect-contractor) will understand owner objectives and offer more value.
3. Owners who lead their teams (or let them lead) versus directing or managing them will be the beneficiaries of highly motivated, more creative professional service. Owners who direct or dictate solutions will be doomed to the opposite.
4. Owners who can define their needs, demand planning phase time, and articulate common, higher purposes for projects will be rewarded.

CHAPTER 4

Building Learning Organizations: Knowledge and Research

INTERVIEW

James P. Cramer, Hon. AIA

Chairman Emeritus, Design Futures Council   6th September 2017

Jim Cramer, Hon. AIA, Hon IIDA, CAE, began his career with the Minnesota Society of Architects in 1978. He was named president of CACE in 1982 and served on the National AIA Board from 1982 to 1994; the last six years as its EVP/CEO. In 1994, with Dr. Jonas Salk, he cofounded the Design Futures Council and its journal, DesignIntelligence. He is chairman emeritus of the DFC; a Richard Upjohn Fellow of the AIA, and senior fellow of the DFC. He has served as strategic advisor and board member to several dozen of the world's leading AEC organizations. He is author of six books, including *Design Plus Enterprise: Seeking a New Reality in Architecture*. Today he teaches courses on the business of architecture and on leadership development. In this reflection, he refocuses us on learning from the best.

Building Design Futures: Courage For The Future Or "Moving The Laggards"

In creating the Design Futures Council, you're one of the first to capture knowledge, create content, and distribute it across the profession. Looking to the future of your organization and the profession, are we facing a different world?

"My focus has been on courage for the future."

Cramer: We've always focused on what's next. Believing that so much of what we know comes from conformance, risk reduction, and self-preservation, we fear what's next. So, my focus has been on courage for the future.

It's taken me 30 years following in your footsteps to catch up. I'm hoping to do the same.

Cramer: The paradox is well-managed organizations can be so focused on improving management – getting the details so perfect, they lose their leadership edge. Or they take on an arrogance, where they have a big blindside. Management and leadership go hand in hand, with a lot of personal self-esteem added in. I've watched great organizations lose their edge. It starts with the leaders. Firms can transform. You go to one strategic inflection point, react, and go on. Rather than succumbing to the life cycle, which is birth, growth, prime, then decline and death, you challenge it strategically and recognize growth points. We see so many in construction and architecture in their 60s, 70s, 80s, and even 90s reinventing themselves – being more relevant than they were in their 40s.

> *"You can move from being a laggard into that top group."*

I'm not seeing the effects of much strategy on average. On jobsites, we face people with neither of these qualities – architects who have lost credibility and authority. I'm trying to get them to recognize this and evolve.

Cramer: We need that kind of book.

Recent books talk about cutting-edge BIM use and tech change – moving away from drawings. But we're still seeing drawings. Maybe it's a Robin Hood approach – robbing the rich to help the poor – for the rest of us. Maybe we can get on the leaders' jet stream and benefit from their momentum? What can the other 90% of us do to get going?

Cramer: I'd say it's 80 percent of us. In my teaching, lecturing, and leading conferences, we target the 20 percent – the top quintile. If you look at the top performers and their metrics, you see what real talent is doing, motivating their people, using technologies and communicating with their audiences. That's where the edges are. We know what their compensation is, what their bonuses are, their equity. Case studies show us. We know who they meet with in their free time and how they interact with their audiences. This top quintile group is exciting to study. This is where the future is. And you can move from being a laggard into that top group. That's the other thing we've learned. This isn't fixed. You can change. That's the good news.

Even if you pick up just one tip, learn to do a strategic plan, or tighten your value proposition, as nimble as some firms are, they can change quickly. Almost overnight. Those are the people who can benefit from this book – those on the cusp of leadership who want to move into the top group.

You were one of the first to embrace the importance of research and knowledge sharing. How did you start?

Cramer: Jonas Salk, founder of the Salk Institute, was a huge architecture patron. I met him as he received the AIA Twenty-five Year Award for the Salk Institute at the Kennedy Center, and he became a lifetime friend. We founded DFC to get richer data. Jonas would say, "To achieve success, you need to study success. So, for over 20 years we've studied success. Salk loved this industry. He loved design and architecture and credits the great architecture he worked and lived in for his inspiration.

So, an inspired owner was the genesis of the DFC. Let's stay with owners. They're lagging too. They have their own business issues to cope with. They want the designers and contractors they hire to figure it all out. Are they driving the project train to set up project delivery?

Cramer: Great clients have values. They know what they want, and they've never stopped creating their organization. They're not just running it, they're continuing to create it. They have synergy with their architects and designers. They make decisions, but they are nurturing organizations. That's why they have great buildings in cities all over the country. Not just because of their vision and what they do, but because they connect their values to projects and inspire their teams.

The human side, nurturing, and the ongoing creation of their organization never goes away. But how are they evolving with technology, infrastructure, and process?

Cramer: We sometimes get so focused on product we miss process. It can evolve. Value imagination and the time to get process right. Don't rationalize that things can't be fixed, fix them.

Are you seeing game-changing leadership in process evolution?

Cramer: A model in *Design Plus Enterprise*[1] uses "corner" factors – professional services, operations, human resources, and technology, marketing, and finance services. In the center are leadership and foresight. Just like in designing a building, an organization's process must be designed to operate better, be responsive, and deliver value.

I tell clients they can be equally motivated to design their next great process as they are their next building. The model serves as a dashboard to value imagination over rationalization, and connection with people over financials, to achieve mutual satisfaction. People say, "That's too good to be true," but I see it unfold time and time again: people can fall in love with the profession all over again! I've done it with major firms, and it's been very satisfying.

I'm also doing a study called "Ten Under Ten" – ten firms in America with under ten employees achieving at levels that match large firms – and having fun. DesignIntelligence and AIA are co-sponsors. Sixty firms were nominated and ten selected. They avoid the excuse: "We're only a small firm." These are small firms that think big. Some of their success can't be replicated by a huge organization.

I love the expression "fall in love with the profession all over again." Sometimes we use that as an excuse for designer shortcomings: "We love what we do." But it's part of the problem – the singular-focus, opaque culture, and disregarding the change around us. You've reframed it in a hopeful, optimistic way. If we can help designers reinvent their process, it's a better outlook.

Cramer: I'm teaching a class at Georgia Tech called "Fees and Profits: Making Money in Architecture." When students walk into a conference or reception room of a thousand people, I remind them to connect with the people they can learn the most from. It's okay to be a little selfish about that. Norman Foster says the two most important things that contributed to his success were the people he chose to spend time with and the books he chose to read. This is what inspired him, gave him fuel to bring out the best – and his unique abilities to succeed. When he was at Yale, the naysayers said, "The best days of this profession are behind us." He challenged that, as many great leaders have. There's still plenty of

---

[1] James P. Cramer, *Design Plus Enterprise: Seeking a New Reality in Architecture* (New York: Greenway Communications, 2002).

"Fall in love with the profession all over again!"

room for imagination, reinvention, and entrepreneurial success. Foster has even started a not-for-profit foundation to improve global lifestyle quality. It supports the ghettos around some of their big airport projects with drone ports and a housing initiative for impoverished communities. Foster has also been smart about evolving through the down times with capital infusions. I admire these kinds of leaders.

Let's go to the Design Futures Council. How has it evolved since you turned over the reins? As you pass the baton, what parallels can we make to firms looking to the future?

Cramer: My game plan has been to turn it over. The DFC can thrive into the future. Sometimes owners unintentionally stifle things; I wanted to get out of the way. They have a much more sophisticated research arm than I ever did. They're digging into the economics of the AEC industry. They've already started a new publication called Design Intelligence Advantage. They have an economist panel. They believe clients want more. I was narrow and deep; they want to go broad and see themselves evolving faster. They see themselves as having breakthroughs. It's riskier, because they have to deliver, but they see more value, purpose, and promise than in the past.

"Plenty of room for entrepreneurial success."

Perfect. It's broader, different, sliced into more parts, and maybe riskier, but less dependent on any one person as a singular source. That's applicable as founders pass their firms on to next generation leaders.

Cramer: The one word for that is agility.

And a great thought to end with.

Jim Cramer Key Points:

1. Focus on the top quintile performers. Get their data, emulate their behavior, or modify and adapt it for your use. Migrate your firm from being a laggard into the top-performing group.
2. Think small, act big. Agility is key for future firms.
3. Have the courage to change. Fall in love with the profession all over again.

INTERVIEW
Renee Cheng, FAIA

Dean, College of Built Environments, University of Washington   20th September 2017

*(Interview conducted while at University of Minnesota.)

Renée Cheng, FAIA, is a nationally renowned professor. Formerly she was Associate Dean for Research and Engagement at the School of Architecture at the University of Minnesota. She has pioneered research in design, emerging technologies, and industry IPD, BIM, and Lean adoption. In 2017, she became the inaugural chair of the Lean Construction Institute's Research Committee, advising the LCI Board and advancing research for collaborative practice.

Cheng has a Master of Architecture degree from Harvard University's Graduate School of Design, and a Bachelor's degree in Psychology and Social Relations from Harvard College. She has worked for Pei Cobb Freed, Richard Meier & Associates, and in her own practice. She was a contributing author to the AIA's "Report on Integrated Practice," 2006; and "Goat Rodeo: Practicing Built Environments," 2016. In 2016, she was elected to the AIA's College of Fellows. Her primer on knowledge versus research is valuable.

Knowledge Management, Design Culture, Research and Credibility, or "Going It on Your Own"

You chose to leave practice to focus on research and teaching. Research has been a gap in our profession for 40-plus years. How has that shaped your work?

Cheng: I liked working on large, complex projects, but found it too much to do both practice and teaching. I miss practice, and my many years there greatly inform my teaching. Now, I'm doing case studies on those large projects I love, documenting them in other ways.

"There's confusion around what research means."

You are one of the leading educators in architectural research, while having "the designer's mind" background. A gap exists in our profession in knowledge and research. I don't have a clue what you do – I didn't learn it in school – other than a few case studies. I'm not alone. Can you talk about what you're doing?

Cheng: There's confusion around what "research" means. I've been learning from my collaborators, and my background in psychology and statistics has helped. To start, I'd draw a difference between knowledge and research. There's knowledge management, which is a field unto itself, that the profession needs to do better with. There are promising pockets of research around managing the knowledge we already have.

For a firm, that's internal processes – keyword searches and expertise – repeatable processes so you don't keep making the same mistakes. Those can form a strong practice basis. This goes to our credibility as experts, ability to serve clients, work with collaborators, and put our fingers on information. Knowledge management is mechanical. The strategy is: do it. Build it into the firm's daily habits.

Our profession has a history of the individual designer starting with a blank page. Many practices have particular detailing or signature aspects. Even if you start from a blank page, you work from a family of details, similar specs, materials, or strategy. This can be technical or programmatic. Having knowledge of the many pieces that go into making a building is difficult for any one person.

Even on a small project we need a strategy for knowledge management. The more that's folded into daily practice, the more it becomes integral to what you do and doesn't feel onerous to document afterward. Knowledge management works when it becomes almost invisible. You can grab what you need. You're posting, getting feedback, using it to learn for the next project, and thinking of projects as more than one-off designs. Knowledge management gives us a handle on what's needed – some firms are further along getting started.

The next step is knowledge management sharing across firms. Not just being an expert within your own firm and proprietarily marketing that expertise, but finding things larger than what any one firm can do. You share your knowledge management and resources with partners, builders, consultants, owners, and other architects, to honestly talk about what happens, what's most efficient, how we measure these things. That starts to segue into research.

When you share things across the industry you need to vet that information. Not just anecdotal knowledge or individual experts' opinions based on hard-won knowledge over years of practice, but can we vet this intelligence over more objective sources? Either what's been done historically, or what might be cutting edge? Or, what have we heard from manufacturers that might be best practices, or what new things our client might want to try? Then you start to get into research.

What starts you might be an area you don't know much about, or know quite a bit about, but are asked to apply in a different way. In the latter case, you perform the classic research hallmarks: literature review, sharing your information, a methodology. Not just design explorations. Not saying, "Oh, I have a method – I'm just gonna start with schematic design and develop the idea," but a methodology saying I believe we can optimize this curtainwall for performance and view, or something along those lines. What's been done before? Is it a physical thing, a screen? How does it fit where people sit and their view range, or what am I doing to figure this out? The first question is: "Has someone else figured this out?"

This process is literature review. Someone has done something similar. It might be manufacturer testing, or something perceptual. This work might not be directly relatable but was important. Literature review lets you see who else has asked similar questions and what methodology they used to test hypotheses against what we think is going to happen. And can this process gather evidence or metrics of success?

With metrics, we define the question and move towards an answer. Potentially across multiple projects with partners who might also provide expertise. Then we act more like a research team. A lot less like a bunch of designers doing what we think works based on what we've done in the past.

"We need to consider a new education system that shifts our values."

Why do most of us believe we can do better on our own? A systems guy tried to introduce us to systematic decision making once. I hit him with the old: "Well, design doesn't work like that. It's iterative, messy, explorative..." His reply was, "Every building has a roof, doesn't it? It's subject to the laws of gravity, correct?" More repeat processes than we acknowledge or want to accommodate. Now I try to use more checklists. But why do designers insist on reinventing the wheel?

Cheng: That tendency goes back to our educational principles and the Beaux Arts. It's strong still, despite being centuries old. That sentiment was based on raw talent, and, in the Beaux Arts days, on who could render the best and had the best compositional eye. Students were assigned the parti prix. You might get the dumbbell scheme, and I might get the courtyard, and whoever could execute their scheme better was rewarded. So, it's not even the design idea, but the execution, with a lot of beautiful proportions, orchestration, and sequence, and how well we render and depict it. We've put premiums on raw talent. We have people facile at formal composition who rely on others to deal with construction and materials. In the Beaux Arts, we mean monolithic buildings, mostly stone, and basic systems for thermal mass, lag, shading, and orientation. If you think about architectural education's roots, almost everything is traceable to the Beaux Arts or Bauhaus. The Bauhaus challenged the Beaux Arts' formal approach. It proposed: "What if we talk about the systems of production?" How can we capture industrial production and be inspired by airplanes, ships, and manufacturing products? Bauhaus was still formal and romantic and about who could evoke those things best.

We need to consider a new education system that shifts values. While we appreciate formal composition and people that make spaces to touch our hearts, we also need people who can speak the language of owners, engineers, partners, and contractors – for buildability and affordability, structure and performance. How do we make that part of our education? Not just in training, but in our core education and values. It requires partnership. Individuals can't hold all the design knowledge alone. You can't be expert in all those things; you have to orchestrate in a different way. The process still relies on design form but requires more now – it has to be objectively measurable.

> *"No one wants to be in an industry that has consistently lost productivity since 1964."*

We can no longer get away with being singular. I interviewed an engineer, and he must have used the phrases "figure it out" and "solve the problem" ten times. Individual problem solving is so built in to the mindset. That way of thinking – challenging the status quo to believe we can solve problems alone, better than anyone has before – call it "designer's chutzpah" – it's our culture. We were trained: whatever the problem, we can solve it. In the old days, that was generally by ourselves.

Cheng: Problem solving versus problem seeking has been a longtime tension within architectural education. We want people to be able to solve problems. On the other hand, to "solve a problem" is a very engineering approach. Whereas the architect begins with: "What is the problem here? Can we articulate it?" The solution may not be a structure or a building. Maybe we end up rethinking the process. The client ends up delegating resources, reusing space they have, or changing the process completely.

In the sixties and seventies, systems thinking and sustainable design were common discussions. We're in a demanding era now, with complex systems no individual can know on their own. We have increasingly data-driven clients. They're no longer willing to simply trust we know what we're doing based on instinctual feeling, intuition, or past similar project experience.

They want evidence we have that expertise. It doesn't have to be hard data; it might be patterns we've seen in a specific building type, such as that cost overruns could be from x, y, or z. Even in residential architecture, we should know more of what real estate agents know. When we say to a client, "Come to an architect to do your addition instead of going to Home Depot," and the client asks, "Will I have higher resale value?" we can reply, "In your neighborhood, we've found an architect can increase resale value by 20% or more with a faster sale." A real estate agent knows this, but architects haven't tracked it.

"It requires more now. It has to be objectively measurable."

Not "Trust me, I'm a genius artist?"

Cheng: Or "Trust me, you're gonna like the way it looks. Form a relationship with me, I'm going to be your creative partner and help envision your dreams." That's important information, and we don't want to give it up, but emotions and human connection aren't enough today.

It's a challenge to our educational heritage, our cultural baggage, and the culture of owners. We all need to change and lead the process.

Cheng: Owners must establish goals and values in evidence-based design and research. Architects need to be able to show how that value plays out for owners. Owners must say, "We're willing to invest in this." Even if it doesn't help our immediate project, it may help the next one or the industry – which is experiencing enormous issues with one-third efficiency lost to mistakes, inefficiencies, and delays. No one wants to be in an industry that has lost productivity since 1964. We have to solve this together. Yes, owners can lead in their way. Either they can demand or support it, but we must show we can do it and work with them to determine where value lies. It's not in shortening design time – it might extend the time needed to see how design strategies play out in post occupancy.

We've already surpassed the laws of physics in reducing design time. We need data to show what that time compression means.

"Owners can lead in their way. Either they can demand it or support it, but we also must show we can do it and begin to work with them to determine where the value lies."

Cheng: We need to show why shortening design time is problematic, especially with the benefits proven in Lean construction. There's frustration around why designers can't work in a "lean" manner. The foundational issue with Lean, which borrows from manufacturing because construction is a kind of manufacturing process, is that one can gain those efficiencies. But when those techniques are applied to design, they fail. Because we're not manufacturing things or producing widgets. We need a circular, spiraling, iterative process, and that looks like waste and wasted time. When we talk about increasing value, we need a basis to say, "If you gave your designer more time to study options and measure and predict what will happen with information from subcontractors and manufacturers, you'll end up with fewer changes, a better, easier to maintain building, more aligned with your users."

That argues for designers having more time upfront, but saving time, wasted effort, rework, and dissatisfaction later. But we can't say that yet. That's my ideal: to be able to say that. Right now, we're far from proving: "Look where this extra design time has gotten us!" That would be my goal: research data that backs up our process claims credibly to establish the value of what we do. Then, it's no longer about a service industry based on hours spent and billed.

That "billable hours" basis is the norm for many firms to manage themselves and their fees. They're going about it wrong. Not value, data, or client-based results, just: "We're running out of hours." One solution is demonstrating the paltry amount of an architectural fee versus the overall project budget, including facility life cycle cost. Most design firms aren't good at it.

Back to time, once the project train has left the station, few of us can convince the owner to plan or slow down. It's impossible once the owner calls and says, "I'm late, make it happen." Only serial, program builders can find the time, unless the group is willing to pause midcourse.

Cheng: On IPD projects we've seen teams go extraordinarily below target schedule and cost – sometimes reaching 20 percent savings. Then you say, "If they had that kind of savings (often during construction), can we plan for it, knowing planning will save cost?" If we look at the overall project timeline, it's harder to make the case to allow two more months, or 20 percent more design time. But the team can make the case they can deliver more performance and hold the schedule. It depends on owner priorities. Speed to market? China proved a high-rise hotel can be built in 15 days.

But that planning occurred well before construction began. The Lean Construction Institute and others are collecting that data. If owners are too busy, we can tell them how to reallocate work up front. In managing schedule and budget, contractors have earned that trust. Architects must get better at it.

Cheng: Architects have to revise their value proposition from being hours-based. We've been squeezing to be faster based on tools. But the framework should be value. How when we think thorough problems, they save on building performance. Healthcare has done this. Roger Ulrich's work shows a gall bladder patient's recovery can be shorter based on having windows. Even if it's only one day shorter, the insurance companies' savings pay for the windows, let alone the millions of dollars of impacts annually. This type of thinking allows us to have justification. Not just because they're beautiful or won design awards, but because of patient outcomes. When we talk as architects about reducing first-cost, and our fees as part of that first-cost, that discussion is dwarfed when we think about personnel retention costs. Can we translate that into how indoor air quality equals fewer sick days? Retail and rental are the most powerful drivers for a developer. Patient outcomes are the powerful metric for healthcare. Productivity is for office and commercial projects. Student outcomes are for schools. We're behind in showing a correlation between things we design, let alone being able to prove causation that a good design causes positive outcomes.

Common to every issue you raise is a need to change from an inward to an outward focus, from self-focus to results-oriented client focus. A radical mindset shift for our culture.

Cheng: But you can do that and still have a beautiful building people like to be in.

Let's finish the knowledge management versus research discussion.

Cheng: Knowledge management is not research. Knowledge management is important and can be effective in improving the quality of the outcomes and establishing expertise. It has value on its own. Research is more challenging for firms to invest in. It crosses more boundaries for sharing and dissemination, being open to critique, and borrowing critiques from social and other sciences. It's easier to sell KM as something we should all do, because we can close the door and market it. Research must bridge outside a firm. Not only in a literature search, but rarely are we inventing a new method. We're probably using and sharing results that either replicates a method or shows one, so others can assess your outcomes. Was there a large enough sample size? Was it documented properly? We have to be willing to open ourselves up to that level of critique, which is not natural for many firms. Increasingly, firms are seeing research as a differentiator. They market it. Although much of it is simply iterative design that results in case studies – that's not research. Some firms are looser on how they define research. The industry would be better if we were more rigorous in how we talked about research.

Only a handful of firms are doing that, but if you're seeing a movement that's a good sign.

Cheng: More people are marketing it, but I'm not sure they have content for what they're selling. It's like the early days of sustainable design. Firms would say, "100% of our staff is LEED accredited, including the receptionist." Well, so what? Don't get me wrong, LEED was good in the end. It changed the industry to do more measurement. Now we see more firms talking about research. Whether they really do it and share it is another thing.

Other than the emergence of research as a differentiator, what might you advise firms trying to find a better process or value proposition? Can we help them have an optimistic view of managing design? What process would you lay out to put all you've talked about to work?

"That would be my goal, where research data backs up our ability to make claims about process in a credible way and are able to establish the value of what we do. Then, it's no longer about a service industry based on hours spent and billed."

Cheng: In projects with conventional design, budget and schedule goals, IPD might not be worth considering. In those cases, use BIM and Lean and document some success metrics. Somewhere in your Conditions of Satisfaction with the owner, you can measure outcomes. Not just "We want to be on time and on budget," but rather, "We want increased sales," or "fewer errors in our warehouses," or "a higher retention rate" or "lower electricity bills." Something the owner wants to do the team can rally around and offer ideas geared to those goals. Projects should establish goals that relate to the owner's business objectives and measure. If more projects, even ones not aspirational for design, technology or program, set even modest goals, beyond budget and schedule, and shared those with the industry, we could answer the question: "Did we improve?" A lot can be done in the majority curve you talk about in the book. Those who think the crazy stuff is too far out there can take manageable steps towards thinking about design and giving people information for future use.

That's a great approach for those whose process is set or may be underway with modest steps. A place to start with little cost and few obstacles.

"That's my advice: Let's get clearer about real project objectives, beyond: 'on time/on budget.'"

Cheng: It has a profound effect on the team if they understand the goals. It can genuinely help the owner. I've seen it on GSA projects for The American Recovery and Reinvestment Act. That was an amazingly powerful goal about getting America back to work. It helped unify teams. Other extraordinary programs include the first LEED and Petal building in Canada or a cancer research center saving lives. Those things can change the tenor of a team and reduce animosity in the Design/Bid/Build world. Spending time with owners to articulate business objectives and how the building advances them is worthwhile and generates good things. Track those objectives and note the associated meaning. How can I help as the architect or trade partner? That's my advice: Let's get clearer about real project objectives, beyond: "on time/on budget."

What a concise sound bite. As projects get complex, and owners less able to cope, they're not doing programs. They say, "You tell me what my objectives are!" To your point: "Let's get clearer," even if it's done on the fly, we can bring more rigor and re-learn a problem definition skillset. Architects, contractors and trades are programming and goal setting these days. It's how we're working. Any final thoughts?

Cheng: We're crossing many roles; that's why IPD is great – it encourages crossover. But it's not without issues. You were at the Design Futures Council session where Howard Ashcraft described "IPD-ish" as problematic; roles can get blurred and be riskier. The work I'm doing – the Master of Science and Research Practices – prepares students to lead in research-focused practices. We give them research methods to identify key industry problems – specific collaboration and leadership skills, across architect/contractor, gender or racial differences. We're preparing a generation of students – our cohort is 80 percent women of color. It's a small program, and our students succeed when they graduate, because they're prepared to address strategic firm goals as they relate to projects. We're generating a new graduate type, so I hope the industry receives and supports them, so they can flourish. They come from diverse backgrounds, international, educated in the U.S., dreamers. As they accelerate into leadership, we prepare them with the background and training; they bring their gender and diversity, which brings benefits and different perspectives – this is the right thing.

Renee Cheng Key Points:

1. Knowledge management and research are separate, distinct endeavors and disciplines. One within a firm, the other shared across projects, firms, and the industry using a higher level of rigor and methodology.
2. Both can add value in applying management discipline to design/construction practices, and shifting to client, result, and performance-based focuses.
3. Clearly defined objectives, programs, and goals make a difference in results.
4. Projects underway, with modest goals, benefit from clear articulation of simple business objectives beyond "on time, on budget."
5. Teams rally around shared aspirational goals to achieve better results.
6. Research requires a mindset shift, and a conscious investment by firms, owners, educators, and industry.

INTERVIEW
Randy Deutsch, AIA

Associate Director, Graduate Studies, University of Illinois   4th October 2017

Randy Deutsch AIA, LEED AP is Associate Director for Graduate Studies and Clinical Associate Professor at the University of Illinois Urbana-Champaign teaching design, professional practice, building technology, and digital technology. He is an author, architect, educator, and social media influencer in creating and illuminating the convergence of BIM and architectural practice. A leading design architect and technology strategist, he has been an international keynote speaker, Exec Ed program leader at Harvard GSD, and responsible for design of over 100 large, sustainable projects.

He has written for DesignIntelligence, been featured in *ARCHITECT* magazine and *Architectural Record*, and is the author of four books: *Convergence: The Redesign of Design* (AD, 2017); *Data-Driven Design and Construction: Strategies for Capturing, Analyzing and Applying Building Data* (Wiley, 2015), *BIM and Integrated Design: Strategies for Architectural Practice* (Wiley, 2011), and *Superusers: Design Technology Specialists and the Future of Practice* (Routledge, 2019.) In his pursuit of future practice and emerging practitioners, Randy Deutsch continues as a leading industry voice.

Integrated Education and the Role of the Academy or "Wearing Three Hats"

Deutsch: I'd love to open with a question first. When you use the two words "design management," what do they mean to you?

In my world, it applies to commercial and institutional design and construction, as practiced in the U.S. I'm appealing to a wide audience – architects, engineers, contractors, owners, and academics –

> **"It's the essence of being an architect."**

who wrestle with the inability to keep design process "under control." In other words, all of us who work in collaborative teams. I started this as a personal inquiry, because I haven't seen anyone delve into the minds of designers or those who work with them in collaborative capacities. I'm interested in the player's mindsets and the nature of the process. Not managing a firm, or hours budgets, but collaboration from my viewpoint as a practitioner who's seen both sides as architect and contractor. The book is an attempt to help us understand one another so we can work together better. It may be an unanswerable question.

Deutsch: That's the *way* you want it. That makes a good book. And you are the living, walking example of somebody who has been on both sides. Your viewpoint is a great value of this book.

I appreciate your thoughts. Some interviewees have been analytical. They need things defined.

Deutsch: In my experience, if I'm interviewing a convergent thinker, who needs decisions and definitions, then I tend to be the *associative* divergent person in the conversation. And vice versa. So, I tend to take off from where the person is. You and I both may be associative, open minded, so this may be a mess.

The book's focus is managing design. As a director of a graduate program in architecture, can you talk about your curricula in that area? If you have them, are they characterized as professional practice, design, or methods classes?

Deutsch: Without critiquing our program, I've taught at a university where the philosophy flipped every decade or so. From theory and paper architecture, then switching back to being practical. Illinois has always been practice-oriented. We originally produced architects meant to end up in the Chicago market. Now, in the global economy, we're developing leaders to practice effectively wherever in the world they want to. So, there are three ways I can answer.

First, our curricula have a tradition of integrated architecture, not just coordinating with disciplines, but a whole student mindset that engages design with practice-friendly classes. Even performance courses like HVAC, energy, and structures. We bring in consultants, trades, firm members, and an alumni network that stays close. Students go to their offices and they come in to critique projects.

We have a comprehensive, NAAB accredited Capstone project, but many of our classes are versions of that. We have a saying here, like "what happens in Vegas stays in Vegas," only for us it's "what happens in one class doesn't stay in that class." Take technology: we treat design as something that is indistinguishable from technology. With that mindset, you don't get to cherry pick your focus. Why? Because practitioners don't get to cherry pick on real buildings. You build a repertoire of applied knowledge and theory.

The second point is, I lead and teach the Professional Practice course. It's the final semester of Graduate school, the segue between school and practice, a "flipped" class where we meet face-to-face, but content is also available online anytime. Students can watch a video, write a response, take a quiz, ask a question. The content is the NAAB stuff – design management, risk management, financial management, having the wherewithal to take a project from predesign through construction and beyond. All that is online. Then we meet face to face 3 hours a week, and that's the more exciting part of the class, where we cover an important component of design management: the soft skills. From industry, statistics, insurers' or attorneys' perspectives, we talk about personal habits and communication styles, how you might approach your design presentation to lower risk and be more financially responsible.

With the success of our graduates in both soft and hard skills, we want them to not just be critical thinkers. Every university says that. We want them to be strategic thinkers, to think about the audience they're presenting to, the processes, value engineering, working with municipal groups, and negotiating, that every decision isn't just for the owner or client, but others too. The public, neighbors, users – the architect is the only one representing all those groups. That's not just an overlay or something on a checklist, it's the essence of being an architect.

I'm lucky to get the guy leading the effort. When I think of the history of the Chicago School – at Michigan we were Chicago School "derivative" – we emphasized function and systems integration, and less on professional practice. It may have been there, but I was too focused on design. I'm glad you're teaching it. Too many design partners lack strategy and design risk management, forget about managing their fee budgets and construction costs.

How is the subject of design management received by faculty and students? What's the current mentality towards managing or being more disciplined in design?

Deutsch: When students come here, I interview them. They know what they're getting. They're not surprised. But to your other question, it's not perfect. Every faculty at every university is diverse. There are realities such as "academic freedom" and "tenure," where one can think and teach the way they want. It's difficult to get anyone to change their mind. Within an academic culture, there are both those who've practiced and those who've been in academia their entire career. There's a healthy, diverse series of viewpoints students get exposed to, and it's up to them to decide where they want to focus.

Almost daily, I hear colleagues say, "Software will never save you" or make disparaging comments about practice. They say, "We're not a vocational school." Or "You'll have plenty of time to pick this up in practice." It's difficult when colleagues say, "It's inappropriate for students to learn technology in school. They can get it online." But clients and firm leaders don't want interns to learn on their projects. I sometimes find myself between a rock and a hard place.

I was a faculty member at Michigan years ago. Anyone suggesting practical or hands-on experience was dismissed or chastised by some camps: "We're not a trade school!" That centuries-old baggage of the architectural academic culture – the limitations of the Beaux Arts era – is frustrating, but still happening. We need theory and reality.

Deutsch: You have to be resilient and let it roll off your back. Given that, in the ivory tower of academia, I've learned that practice will always be 10 to 20 years ahead of school. Yes, there is prize-winning innovative research, but generally speaking, we're not rewarded for looking into the crystal ball and projecting into the future, or preparing students for it, so what you see is faculty groups banding together to lock into an approach or technique, and that's what they'll cover.

It goes to the core of university culture – an inclusive, universal place with many viewpoints that's traditionally conveyed slowly changing knowledge – not the fastest to change or future looking.

> "That's why I teach project designers who start on smaller projects: you're going to play all three of those roles."

Deutsch: Even in a great economic market, where 100% of our students get jobs, many are fearful, nervous, anxious. They're not looking far ahead. If they graduate in May, they're worried about June. I make sure, even for design-oriented students, that they can take on the project architect's and even the project manager's mindset.

It's great practice to make them walk in the other guy's shoes.

Deutsch: A lot of graduates aren't thinking about how you shape information delivery graphically and verbally to your audience to get the outcome you expect. That's strategic thinking and design risk management. They're just hoping to get invited to the meeting, not beyond. That's why I teach designers who start on smaller projects: you're going to play all three of those roles, so you have to be able to play all three. You need to cover all three – design, management and technology – with every answer you give.

You're a clear advocate for technology. For all its advantages, it also brings other management complexities. How are you teaching students to cope? In my experience, it can be a foreign skillset to manage technology the way design is done today. Twenty-five years ago, the first time I set up a sheet in CAD, I was an hour into it, frustrated and frazzled, and I hadn't yet had a single idea, drawn a line, or added value in any way. Those things are simpler now, but other complexities have crept in. For all its automation and added intelligence, technology brings other issues. Can managing the technological aspects of design be taught?

Deutsch: You're talking about empathy. When I was in school, I learned Fortran, where we'd model building using 500 punch cards. When I got out of school, I was heartbroken to learn that while I was in school working hard to learn Fortran, there was this thing called a Macintosh, which had been around a year. It confirmed that school was behind, holding on to old technology. I knew when I'd come back to teach later in my career, I'd make sure students didn't get the same whiplash experience. To empathize. I have to face this comment every week: "We're not teaching technology, BIM, Revit." You have this stalemate where schools say, "Learn it in practice" and practice says: "You really ought to learn it in school,"

> *"That's an emerging professional with the right mindset to manage technology – and design – and they are connected."*

so now there's third party trainers, consultants, and YouTube. So, I just don't tell anybody and teach it in all my courses, so students have a working knowledge of things like artificial intelligence and reality capture. They don't necessarily need to learn to write code but do need to understand the technology ecosystem and how firms are leveraging it.

On technology management, the attitude I take is that proficiency in any one tool isn't enough. When students go to a career fair, the first question after "What is your name?" is: "Do you know Revit?" You probably won't get an interview if you answer "no." But that's still not enough. You have to build a repertoire of tools to solve whatever intractable wicked problem you're asked to solve. Once you have that toolbox filled, you also include things like Moleskines, metaphors, and Excel spreadsheets. Because, sometimes technology tools aren't the best way to answer a challenge. I want to make sure students understand that.

Your word mindset is on target. I want students to get away from specific technological proficiency and develop the mindset, confidence, willingness, and wherewithal to approach any employer or client request. To look for opportunities to automate and innovate. Take software home at night. Even if you don't have an IT background. Maybe your roommate did. If you have that gumption to say, "I'm not familiar with that tool, but I'll take it home and come back with an opinion on whether it's a good tool for our team," that's a great thing to offer. That's an emerging professional with the right mindset to manage technology – and design – and they are connected.

It's interesting you have to sneak technology in the back door. Many of us in big organizations face the same issue. Large firms aren't likely to have the full-on passion we do. And you can't teach it all because, as you said, technology will change as soon as students get in the real world. Nothing prepares them to be the technology leader at a large firm, so they need what you're giving: a design technology leader's, or entrepreneur's mentality. That's design thinking, an ability to empathize, communicate, develop vision, persuade, translate languages, and learn.

While we've crusaded for industry change for over a decade, we might be approaching the subject at different ends of the professional spectrum. You're talking to and teaching future leaders and emerging professionals; I'm trying to help the rest of us. The laggards. I include myself and some of the country's best design firms and owners in the "laggard" category. On jobsites, we still face plenty of 2D drawings, obsolete practices, and resistance to change. While I love where you're taking us, I feel we – the laggards – are the majority curve. Any advice for the late bloomers, other than read your books and get with it? How can we bring up the rear guard?

Deutsch: First, if you walked away from my books and think we're on opposite ends of the spectrum, then I've failed. I've dedicated the last 17 years to make technology consumable: user-friendly, understandable, approachable and nonthreatening. I've taken my presentations and talks to outlying areas and met with local and regional AIA groups. I haven't always been successful. At the end, there'll be clapping; then we end up talking about the reality happening right around the corner.

Oh no, you've succeeded. It's just a difference in perspective. You're operating on the right half of the continuum, and I'm on the left. With contractors and owners who don't get it and aren't reading your books. That's the "reality around the corner."

Deutsch: As an anecdote, in my second book it was important to include the word "construction" in the title. Not because contractors read books, or CMs. I was assured by the publisher they don't. It's a cultural thing. Every contractor I spoke to was completely open to talk, but not one would sign the publisher's agreement, so I couldn't use any of it. Easy, helpful, insightful, and understood it all, but they wouldn't sign. It comes down to motivation, and they didn't have it. Their mantra is getting in and getting out. Making money. What on earth did their signing away the publishing rights to their words have to do with their work? Nothing.

Interesting that cultural and motivational differences come out in that way.

"Design problems are different than other problem types. There is no right or wrong answer."

Deutsch: Part of my charge, and my job, is to continue to reach out to contractors and subs where they see things not as separate, but where we're in cahoots with each other. We're in it together. I mean that literally, where we go right from design to fabrication with technology; and I mean it figuratively, in the sense that owners don't want finger pointing. They want teams.

You and I met when I gave a talk at the KA Connect Conference, and I spoke about the AIA loving emerging professionals – and late-career architects and fellows. What about the middle? That thirty-five- to forty-five-year professional – the middle of the Oreo cookie – gets forgotten. They're on their own to find a way through their career and to cope. I see a lot of that. When they get retirement in their sights they say, "If I can get through the next 10 or 15 years without another downturn in the economy, maybe I don't have to learn the new flavor of technology or practice methods running through the office." They have that "I'm good" attitude. Those are my people – the middle is my audience. That's also where I am in my career.

I don't think we ever stop learning. I think everyone should have a main career and a second, or backup, career. Every seven years, I've changed my side career to keep fresh. In academia, we're not encouraged to run our ideas past the real world to test them. When you're in mid-career, you have an opportunity to take something on to keep your mind sharp and be of more use to emerging professionals.

When I wrote my first book, I thought we'd have digital natives sitting with experienced professionals. As I visited firms they said, "Look around. Do you see any of that in our firm?" I didn't. So, I changed my approach to teach students how to put a building together. That still leaves a place for experienced mid-career architects. Because buildings are more complex and no one person can do it all.

The 45-year-old architect may not need to learn generative design or coding, but they need to know what they are, how they affect their role, and how to work with them.

One has to manage the design of one's career. We only change when it becomes too painful to continue as we are. In the contractor's case, things are good. There's work and profits. But inherent cultural differences and inefficient processes are burning us out. Contractors, owners and architects. We're working too hard. That's why we're evangelists. I'm also targeting the mid-career group with a vested interest in change. I believe in a multiple-career approach. It's worked for you, and I've had five career iterations as an architect and five more as a contractor. Maybe it's the quest for the new.

Management seems externally applied to design. A supervisor, a set of rules, limits, or self-discipline. Activities such as planning and organizing, evaluating, measuring, reporting, acting – that's management – making decisions, resolving conflict, and making change to return the process back within limits. With our cultural and academic heritage, and the nature of design, how do we get better at managing?

Deutsch: I'll begin with motivation. A lot of students and design professionals think of management as something other people do to them. It's external. It comes from the outside. It's structured, restrictive, controlling. It limits freedom.

I approach management differently. From an empathetic standpoint. I try to get every student to think of an outcome they wanted but couldn't achieve. One recent student was having a hard time keeping up with all the technology and deliverables at the quick pace we set. They came in asking about the tenets of time management. They're going to go far! They felt a pain point and are taking action. They're dealing with it. They desire an outcome. Maybe it's just an A in the course or to lower their stress level. But there is one. A lot of times we're pushing, but first we have to get them to a point where there's a gap. Something's missing. There's a need. They want it. It's not a "nice to have."

Given that, management isn't externally applied, it's a qualifier for internal ways of selecting strategies to get the outcomes you're looking for. Financial, time, design, and other management types. That requires you to know an outcome you want.

When we try to train contractors on technology, we have to put it terms of what they're already doing, or they think they don't have time to learn it. It's not a "must have." We need to get everyone to understand something is a "must have" before they'll learn it. Otherwise they'll forget it as soon as class is over. Or, professionally they'll have "training erosion" because they didn't apply it. A lot of it is having the right mindset.

I'd be remiss if I didn't bring this up: Design problems are different than other problem types. There's no right or wrong answer. That's the heart of design management to me. You can design with checklists, as a lot of firms do. But there's still the design culture, and bad behavior. Like the day before you go in for permit, you check all the boxes without vetting it. That's unprofessional. Those people aren't going to let others restrict their role as a designer. Other examples of bad designer behavior are when a principal designer is in a meeting and the topic of technology comes up, and they look at their cell phone or check email. They're disinterested because they don't own that aspect of their process.

"I try to discourage those things and get people to empathize . . . management is a tool they can overlay . . . and achieve that outcome they're looking for. It frees them up to design more! . . . If they're going to automate to get more design time . . . that's the motivation . . . tools as a management strategy."

Another one is when a lead designer sets the management tone. Let's say they leave the team for a period and leave someone in charge. Many times, they'll come back and modify the design in some way to put their imprint on it, their ownership, because they can't imagine it – the design, the project – surviving without them. That's bad behavior. I try to discourage those actions and get people to empathize and understand that management is a tool they can overlay on their process to achieve the outcome they're looking for. If nothing else, it frees them up to design more! That's what motivates them. If they're going to automate to get more design time, that's the motivation, using tools as a management strategy.

If they don't want it intrinsically, then the management isn't likely to work. They'll resist or work around it. Someone can always game the system. Your example of the design leader swooping in to undermine their team – is not management or leadership – it's destructive, demotivating, and a horrible example. But in a culture where design is king, it happens too often.

I want to end on a happier note: Thanks for your insight and hard work to shape the next generations of architects who are well versed, motivated, and equipped to manage design.

Randy Deutsch Key Points:

1. To be effective, managing design must be intrinsic, internally driven, not an external overlay. The designer must want it to achieve a desired outcome. Find a pain point and do something about it.
2. Resistance to, or ignorance of, the need to change is futile.
3. Design problems are different from other kinds: there is no right or wrong answer.
4. Good design managers must be able to empathize and think with multiple minds – to play all roles, wear different hats.
5. Design leaders who charge others with a project's design shouldn't modify the resulting design solely to put their imprint on it. They're exhibiting bad, demotivating behavior.

CHAPTER 5

Firm Culture: Management and Attitudes

INTERVIEW

Scott Simpson, FAIA, NCARB, LEED AAP

Former President & CEO, The Stubbins Associates; Senior Fellow, Design
Futures Council   21st November 2017

Scott Simpson, FAIA, NCARB, LEED AP, is an award-winning architect and former CEO of The Stubbins Associates in Cambridge, MA. He is an AIA Richard Upjohn Fellow of the AIA, and a Senior Fellow and Board Member of the Design Futures Council. He has lectured at the Harvard Business School; Urban Land Institute; National Press Club; architecture schools at Yale, Harvard, Rice, University of Illinois, and University of Wisconsin; AIA, as well as multiple international conferences. He has published more than 180 articles on innovation in the A/E/C industry and co-authored three books: *How Firms Succeed, The Next Architect: A New Twist on the Future of Design,* and *Lessons from the Future.* Over a 40+ year career, Simpson has been a principal in five firms including president & CEO of Flad & Associates. He was awarded the Distinguished Achievement Award by the Connecticut Construction Institute and the "Hall of Fame" award from the Massachusetts Building Congress. An acknowledged expert on BIM and Integrated Project Delivery (IPD) and LEED, he holds degrees from Yale, Harvard Graduate School of Design, and has held registration in 22 states.

An Integrated Approach: Avoiding Voluntary Misfortune or "Doomed to Be Successful"

Where did you get such a broad perspective? Design? Collaboration? Leadership? What drove you initially, on the design–management spectrum, and how has that changed?

Simpson: At first, I wanted to be a designer, like 99 percent of my classmates, the next Frank Lloyd Wright. Everybody starts with high ambitions. I really enjoyed the design process, conceptualizing and translating ideas into three dimensions. But as I started my career, it became clear that design is more than lines on paper and compelling graphics. A lot goes into making a building. It's multidimensional. You need to be a good designer, but also good at business, marketing, financial management and team leadership, blending those skills in a way that optimizes the outcome. I didn't lose my interest in design but took on more dimensions as my career advanced – I began to play different instruments. That shifted me toward the management side. I realized architecture and design are a team effort. You need great talent, technical skill, business and financial acumen. Leadership is where you can make the most difference. That's how I migrated to the leadership side.

> "A lot of architects engage in what I'll call 'voluntary misfortune'. They do it to themselves."

I remember as a young designer sitting at my desk, thinking the design was emanating from me as I drew lines. It didn't take long to see that other things drove design: cost, schedule, program, information, relationships, technical input. So, I shared a similar migration to yours – to a broader involvement.

Simpson: Clients don't build buildings to spend money, they build buildings to make money. So, we have to approach architecture with a value-added perspective. Looking at design as a way to increase owner value opens windows of opportunity. These issues aren't discussed in design school. There, it's form, space, texture, aesthetics, etc. There's so much more to design. Effective design professionals embrace all aspects.

But there are still those who retain their love for being primarily a designer.

> "Clients don't build buildings to spend money, they build buildings to make money."

Simpson: There are many examples. Bob Stern is one. Michael Graves was one. Rafael Vinoly. Frank Gehry. Plenty of others are primarily form givers. That's where they hang their hats. But when you look at their practices and how they produce work, they're the visible tip of the iceberg in their firms. They're surrounded by all kinds of others with professional skills that make the work possible.

You moderated a Design Futures Council's meeting in New York: "The Business of Design" with Robert A.M. Stern. He was presented their Lifetime Achievement Award. Bob confessed to being the "Last of the Mohicans" when it comes to technology – still printing emails and relegating technology to a subservient role – behind design, architecture, humanity, client service and history.

Simpson: Bob Stern is very smart, and he doesn't want to get bogged down with the business or marketing sides, although he's very good at both. He wants to deal with design and drive the firm on that basis. But he knows that without others in the firm who focus on those things, he wouldn't be nearly as successful as he's been. He knows it. He's done a great job of positioning RAMSA as a design-driven firm. It's a very strong brand.

His comment was: "Some people play golf or network, I think about buildings." Having led five large, successful firms over your career, what were the prevailing philosophies about managing design? Which came first – management or design? Were any characterized by a division between design and business? Or did you have an integral approach?

Simpson: The Stewart Design Group was my first principal position as VP in Charge of Design. It was an offshoot of Perry Dean Rogers in Boston. In the 1980s, it focused on application of computer technology to architecture. We had our own PDP911 computer. We wrote our own software and used computers in design process. Then I moved to Cannon Design, with a healthcare and higher ed focus. They did some great work but were more process-focused than design-focused. I went to Flad as president and CEO to do knowledge-based research buildings. We tried to change the firm character from mostly healthcare to a broader range of work. Then, I went to Stubbins. Hugh Stubbins was very design focused, a great designer. The Citigroup building in New York is still one of the great buildings of the 20th century. But although he was known as a designer, he was also a very smart business guy, a developer and clever with money. There is no black versus white in terms of design versus process management—they're two sides of the same coin. Hugh embodied that in an underappreciated way.

We morphed into Kling Stubbins to become an A/E firm and work internationally, then got bought by Jacobs. It's a gigantic, publicly traded company, more process focused and concerned with building function and processes—timesheets, budgets and so on. At Stubbins, design always came first, as a strong ethic. But we were smart business people. It's a mutually supporting system if you conceptualize it correctly. This is never taught in design school. Nobody talks about marketing, management or business, but they're the essential underpinning of a successful design practice.

I learned that the hard way. We drew sumptuous drawings. We didn't focus on anything else. I learned from those negative experiences: There's got to be a better way.

Simpson: A lot of people make the mistake of thinking it's design versus business, but it's really design and business – two strands of the same rope. If you weave them together, they're stronger than they are separately. As we got smarter about business, we got better at design. We got smarter tools, better processes, business results, talent and people, all founded on sound business practices.

How about firms where design and management are not balanced – where one side is dominant. Do you have experience with those?

Simpson: Absolutely. Jacobs, due to its size and scope, is more process-oriented. There, it's more about utilization and a bit less about design quality. That's not to say they don't care about design, but design doesn't drive the business. Some firms are more service or business-focused

"Design always came first. It was a very strong ethic. At the same time, we were smart business people. You realized it is a mutually supporting system if you conceptualize it correctly. This is never taught in design school – as you well know."

than design focused, and that's perfectly okay. Markets have a broad spectrum of clients with different goals and strategies. Some clients understand the value of design. They seek firms that are clever, creative and interesting to deal with because they know they're going to get more value that way. Others have a different focus. There's nothing wrong with this; they say: "We're primarily concerned about budgets and schedules and getting things done predictably, and don't mess up my life with all this other stuff called design." Firms will seek their rightful place in the market, based on what they're good at and inclined to do. There will always be a place for high-end, design-oriented firms, and for service-oriented firms. Clients ultimately control that.

As contemporaries in school, we were both likely taught that plans and a structural grid were a good way to begin design. Generally, a functionalist approach. Most of our work was started with sketches, then migrated into hard lined solutions.

Simpson: Yes. Everyone has a different point of view on how to start to think about design. Some begin with structure. Mies did. Some of the best architects I know are finely attuned to the business aspects, such as meeting budget first. At Pickard Chilton, Jon Pickard came out of Cesar Pelli's office. He and Bill Chilton have a very interesting and powerful marketing message. They do a lot of corporate high-rise headquarters. They tell their clients: "We will create buildings that are distinctive from a design point of view, but our focus is also going to be to increase your real estate value." They've kept records and have the numbers. They'll prove to their clients that the buildings they design are worth more in the marketplace than comparable office buildings. That's an extremely powerful message because it reinforces the business value of good design. They do wonderful work. As another example, I was involved in a Boston Seaport building for Price Waterhouse Coopers. We pushed hard to use high-tech triple glazing and sustainable design strategies to make it distinctive, arguing it would stand out in the marketplace and command higher rents. As it turned out, before the building was completed, it had sold for the highest price ever paid for a building in Boston. There's an example of taking design, technical and business perspectives and blending them to get great results. The building won awards, but as far the client was concerned, they spent $400 million to build it and cashed in for around $1.2 billion in 9 months.

Let's talk about time and money and the architect's reputation for designing over budget and schedule. How can they overcome these perceptions? What has to change to deliver and measure better value from designers?

"Designing over and over doesn't necessarily improve the outcome."

Simpson: A lot of architects engage in what I'll call "voluntary misfortune." Designing over and over doesn't necessarily improve the outcome. If you design without knowing construction cost, all you're doing is guaranteeing change orders and coordination problems later. It doesn't lead to a better result. This is self-punishment in an odd kind of way. People are beginning to realize design has business value. And vice versa. Look at Apple. Apple is a design company that created a product called the iPhone. It's the most profitable product in the history of mankind. Apple's fierce attention on design value is what sets it apart. People are realizing there's genuine business value in this thing we call design. It's difficult to define, but it's real. More and more people are thinking in those terms. We see the convergence of business and design thinking for enterprise values. I wouldn't be surprised to see Chief Design Officers in most corporations soon. It will amplify architects' value and make them more potent, adding design and business value together.

Have you seen firms that have transformed how design is being done? A new secret sauce? Where other forces – technology, management integration, customer service, or research – have changed design thinking? Or is design still being generated via hand sketches then given to production teams to be digitized, with management as a separate overlay?

Simpson: Peter Beck bought a design firm and folded it in to his construction firm. They developed their own software. You can pump in parameters for lot size, circulation, or whatever and algorithms generate options with linked costs. The program spits out solutions in seconds. It's pretty cool. SHoP, Kieran Timberlake, and Norman Foster's firm all have radically advanced processes. Paul Doherty at The Digit Group is doing things with gesture driven software. You can have coffee with a client and create 3D models at the same time. Before lunch is over, the model is ready. Going from an idea into production mode is happening now.

I'm not sure the profession takes advantage of the collaborative options available. Most projects are still delivered conventionally, with plenty of owner changes and RFIs.

Simpson: Look at Apple, Google, Uber, and Tesla. They're changing the process of design to execution. Going direct from idea to production. Tesla's stock is now valued more than General Motors. They've automated the process for cars and changed the driver experience. Technology is giv-

> *"Change is going to come from the client world."*

ing us leverage to connect thinking with doing. The ability to model, make things, 3D printing . . . I can see using gaming engines instead of construction documents. You see it in manufacturing, the movie and publishing industries. It's tough for design school faculties to keep up with the changes. They're behind the curve. Kids coming out of school are having to relearn design skills to align with the market. It's going to be a new profession in 5 to 7 years.

Arol Wolford is already involved with gaming engines. The leaders are taking us there. What can the rest of us do? Where can emerging leaders make change?

Simpson: I've been consulting with a prestigious major East Coast university, about to build billions of dollars in projects. Their average project cost overrun was 44 percent – it's an issue. Change is going to come from the client world. Even with those metrics, change is hard for humans. BIM arrived in 2003. Phil Bernstein was an early evangelist at Autodesk, hired to wave that flag around the world. He was very effective, but it took the industry a decade to embrace BIM. I was CEO of Stubbins then. We flipped to 100 percent BIM in 2003. That changed the nature of our practice. We were ahead of the curve for a couple years, and got business

> *"It's really tough for design school faculties to keep up, the kids coming out of school are having to relearn."*

because of it, then everybody else caught up. That's the nature of change – humans are reluctant, afraid to embrace change. Most clients are conservative. I was Barbara Bryson's consultant ten years ago, when she put her toe in the water for a version of IPD on a lab at Rice University. We tried it with huge success, but even after that, it took her a while to embrace process innovation on the next few projects. Now she's a huge proponent. Clients across the country now embrace it.

We start too late, without time to set up a more effective process. We miss the opportunity to plan. If we sat down to create a mutually beneficial, "lean" process, we wouldn't come up with one that had 1000's of drawings reissued, followed by thousands of RFIs. But that's the way we build jobs.

Simpson: At CIFE at Stanford, John Kunz said any building of any size can be built in 9 months. Period. That's all it should take as implementation. But this trillion-dollar industry and its processes are disintermediated. Look at the auto industry in 1910. There were lots of players. Now there are only 2 U.S. companies that drive the industry. It's the same in construction. Primarily lots of small firms, guys with pickup trucks. But the larger leaders will shape the future and move the needle.

If you had a magic wand, what would you say to an aspiring design firm to carry them to a prosperous future? What should they do at the intersection of design and management?

"The profession you're about to enter doesn't exist . . . [It] needs to be reinvented . . . I have great faith good things will happen . . . It's inevitable: we're doomed to be successful!"

Simpson: Great question. I gave that lecture at Harvard and Yale. I told students, "The profession you're about to enter doesn't exist. It's clear the AEC Industry has a problem complying with budget and schedule. The old guard isn't embracing change. As you emerge from school, you have to create your own industry and future." If it's a trillion-dollar industry and we're 30 percent inefficient, we've got a $300 billion-dollar value gap to redeploy. That's enough funding to make a difference. The problem is attitude, not aptitude. The profession needs to be reinvented. I have great faith good things will happen, probably sooner than we expect. In a New Yorker cartoon, two caterpillars are crawling along a branch. They see a butterfly flying overhead. One caterpillar says to the other, "You'll never get me up in one of those things." But of course, it's inevitable – we all become butterflies someday. The next generation will have the tools we didn't have when we came out of school. Let's start with that. It's inevitable: we're doomed to be successful!

Scott Simpson Key Points:

1. Design and management are two strands of the same rope, not opposing forces.
2. "Design-first" firms can thrive but need a strong supporting cast with supplementary skillsets.
3. Design schools are challenged to keep up with the pace of change in tools and processes. The burden will fall to students to create their own futures.
4. Design curricula should emphasize client value as a design objective, differentiated career path and added value for designers, versus traditional form and function focus.

INTERVIEW
Thompson Penney, FAIA

CEO, LS3P   17th August 2017

Thompson Penney, FAIA, is CEO of LS3P Architects, a 55-year-old design firm of 340+ professionals with offices in Charleston, Charlotte, Columbia, Wilmington, Raleigh, Greenville, Myrtle Beach, and Savannah. He is active in the AIA/AGC Joint Committee and was the 79th National President of the American Institute of Architects in 2003. He was a contributor to the Managing Uncertainty study in 2014. His thoughts on firm culture and attitude are manifest in his firm's success.

The Regional Model or "Sittin' Around the Table"

LS3P has been around a long time and grown to have multiple offices. You must be doing something right. In my experience, a firm's culture and attitude come top-down from its leader. What's your personal philosophy and attitude towards managing design? Big focus? Necessary evil? Impossible? Are there subcultures within your firm – management versus design? And are you succeeding in pushing leadership's philosophy down to the firm?

Penney: To me, culture trumps just about anything. We've been in business 55 years and have developed a strong culture. I wouldn't describe it as top-down process. It's established by setting examples. It's difficult to say, "This is what it is, and you need to follow it." We have distributed leadership – 77 shareholders. I became LS3P's president in '89, and we changed from firm names to initials to be able shine a spotlight on others. We did this for two reasons – to let team members develop and to invest in them so they'd feel a part of the firm and stay with us. If you've got five names on the door, clients just want to work with those people. It allowed our people to grow and the firm to grow. In 1999, we opened up ownership because our people were the ones making the growth happen. We're inclusive. With each merger the firm's complexion changes. We have an open view toward culture. Our core values are unchanged, but our culture evolves because our newest offices have good ideas too. Nothing remains the same. We have names for our studios and groups. Our top leadership group is called Evolution. Culture is most important.

> *"The first thing that has to happen is that architects have to value themselves. If we don't value ourselves, why should anybody else?"*

The next thing is very close – measurement. Do you have open information? People want to succeed, and they will if you give them the information. It's important to show we're open in the firm's business aspects. If you don't know what's not working, you can't fix it. We're open about metrics. We measure a lot of things. Even when someone isn't succeeding, or we find a problem with a system, we let team members know the facts and figures, so they can help fix it for everyone.

I've talked to other small boutique firms that shun management. Yet you devote attention to trying to manage yourselves with typical fee breakdowns, and manhour budgets?

Penney: A big part of my focus from '89 is the firm's business infrastructure. We use Deltek Vision. It's designed, in my opinion, for very numbers-oriented, linear thinkers. So, we've written programs ourselves. Every person has a dashboard. It gives varying levels of information depending on your position in the firm. When you go up the chain, PA's and PM's see all their project intel. Studio leaders see all their people's data. So, we slice and dice it every which way. The Board sees everything. Everybody knows where we are and how we're doing. We are, I'm proud to say, a very high-performing firm, per metrics of the Large Firm Roundtable. If you give people the right information and tools, they'll be successful! We believe in delegation. Most people know their work better than I do, so I give them the tools and measurements and get out of the way. That part of our culture came from Frank Lucas, our founder. He was one of the best delegators I've ever seen. He has an amazing level of judgment, about anything. He wouldn't tell you what to do, but he got it. Judgment is one of the most valuable skills a leader can have. You can have all kinds of rules and regulations, but you end up breaking them. Because, you're going to run across that client you've been working with for 20 years, who needs you to do things differently. We need that judgment by people with boots on the ground who have the relationship. You have to trust them. Our systems are set up for people to succeed on their own, without us telling them what to do. It's a delicate balance between systems, delegation and trust - one of my favorite things.

Do you work on training your firm's attitude and culture for a management and service focus?

Penney: Yes, but I'd add design focus as well. The book *Good to Great* has a saying about "the tyranny of the or" and "the genius of the and." I firmly believe in this. We're not a service firm or a design firm – I think you can be both. Whether we're successful or not, only others can tell, but I think we're doing a good job balancing those two things. Serving the client, doing good design, and managing the business. And those doing the hard work, with the skills and talent, get rewarded.

Years ago, I had lunch with Dr. Eugene Odum, who wrote Fundamentals of Ecology. *He was the godfather of Ecology. I asked how he reconciled economic pressures with sustainability. He referenced the Greek root of the word ekos, or "house." Ecology is "knowledge" of house, economy is "management" of house. They're integral, compatible, he argued. If we don't manage the house, it falls. It's a duality, not an either/or.*

Penney: I tell all new employees about our "Paradoxical Balance." It includes:

LS3P: Our Paradoxical Balance

1. ENR top 20 "A" firm/eight local offices with deep community involvement.
2. We believe in the value of specialized expertise/while keeping a well-balanced generalized portfolio.
3. We are a design focused practice/and understand the value of being a well-run business.
4. We value being a high-performing firm/that balances a commitment to work with a commitment to family.
5. We are both a recognized design firm/and a valued service firm to our clients.
6. We excel by individual initiative/and by collaboration as a team.
7. We have a great respect for and celebrate our past/but we are clearly focused on the future.

Can I assume you're hiring "balanced" people? Not "one-sided" ones who are going to grouse in the back room about how bad things are . . . "Gripers." I still see it – the moaning and grieving. It lingers.

Penney: Yes, it drives me crazy. We're our own worst enemies sometimes. The first thing is that architects have to value themselves. If we don't value ourselves, why should anybody else? I get so tired of hearing that self-fulfilling prophecy: "We're not gonna make any money. Woe is us." I've learned so much from contractors.

Architects can do just as well. We've got leverage and deal with millions of dollars entrusted to us. If we're complaining, why do we expect anybody else to value us? It seems as simple as that.

Penney: Sometimes I find myself getting on my soap box about the negatives, but you have to recognize your deficiencies before you can do anything about them. There are so many positive things: the way we're educated and think. We're problem solvers, but one of the biggest faults is not valuing what things cost. I have two sayings: *If you don't know how much something costs, you don't care.* And, *how can I know if I like it if I don't know how much it costs?* If there are no prices on a restaurant's menu, I don't know what I want to order.

I did a project as a young designer, and the contractor looked at the owner and said, "This Bradley electric water cooler is $1,200. I can get you one for $300." I felt like a fool. As an architect, I'd just picked out the coolest one. It was beautiful, but I didn't have a clue how much it cost. Before everything was online, I restricted catalogs in the office to those that had prices in them. Without prices, catalogs tempt architects to ignore cost and only look from an aesthetic or performance standpoint. Technology will get us there.

Recently CBRE merged with Kahua to access a "Software as a Service" global online database to "change the practice of design, construction and real estate." We have the data, but don't use it.

Penney: I met with a Kohler rep yesterday, and she pulled up a fixture online and it had all the cost data right there. It is, or will be, integrated into the model. A running tickertape that will have implications for contractors as well.

What are the new generations showing you? Are they pushing to affect your process?

The old guard complains most about their work-life balance: *"What do you mean you're leaving at 5:00? I'm staying here all night trying to get the work done and they're ready to leave?"* It's a short-term problem, but good in the long run. It will drive efficiencies in a very positive way. One pet peeve: I learned no respect for time in school. I worked all the time. Millennials are more attuned to balance. We have to adapt.

They're already smarter than we were! Years ago, an architect-friend's wife, who was in finance, said: "What the hell's wrong with you guys? Why do you have to stay until 7 every night? Why don't you get your work done and go home?" I've never forgotten it. There'll be a new balanced way.

Penney: Another interesting thing I see is the level of confidence. I'm on a bank board, and we have young graduates. They're confident and know what they're doing. Sometimes, young designers don't feel as empowered or as confident. Maybe our business is so complex it takes longer to grasp it.

In construction, we get entry level people in front of clients and expose them. The old architectural culture puts young professionals slaving away at a desk on details and stairs. We have a responsibility to expose them to reality, not shelter them, so they can do better than we did.

Penney: We do a great job of getting young professionals out in the field. We used to say we don't need 5 people at a meeting, but it's an investment to let them see what goes on. An educational experience.

Looking 5-10-15 years out, do you see anything radical or different?

Penney: Our model is different from most firms our size. We have 340-some-odd people, 8 offices, in 3 states, all within driving distance. All those offices are community ingrained and participatory. That's different from other firms our size in Tier 1 cities, where you have to fly between offices. I'm bullish on our model. I liken it to a regional law firm that has the roots and relationships. They know their clients. They have the expertise of a Big Four, with tax experts and such, but are grounded in their community.

Everything goes in pendulum swings. We've seen over past decades the rise of expert firms. When I got out of school, I worked on a healthcare project with radiology. I didn't know about lead shielding in walls; I figured it out as I went. Those days are over. With knowledge sharing, and firms like ours becoming experts in healthcare, senior living or high-end facilities, we combine the best of both worlds: but our local architects are sitting around a table at the Chamber of Commerce with community leaders and having trust. I think the pendulum will swing back to local experts rather than flying experts in from out of town. I know local owners who regret choosing an out-of-town firm. Our model is to sleep in your own bed more often than not. The excitement of travel soon wears off. I don't know why other firms aren't working this way, but I'm prejudiced.

I'm also excited about technology. It has the power to take the drudgery out of our work, so we can focus on what we do best: leadership, design, and value-based services, not just being a communication tool between designer and contractor. A client recently said "We've got a major problem here! Our architect is not doing complete plans and specs," but the client was doing repeat buildings, so why draw everything again if the team knows what it is? It was eye-opening. There are many ways to success. In our model we have diverse markets, so every office has a different way of approaching work. You have to give them leeway and communicate.

You have been so optimistic. Anything else you want to share?

Penney: In terms of collaboration bright spots, I'd like to plug the AIA/AGC Joint Committee, where there's so much empathy and collaborative spirit between architects and contractors. Your company taught us how to put BIM on iPads. I'm bullish on collaboration increasing between architects and contractors. We need to look at starting it in the academy. That's where the bad habits start. The good ones too. If we can get colleges to start collaborative supportive thinking there, where we're not *at* each other, that'll mean something. Architecture and construction need each other. I am so positive about our shared future.

Thom Penney Key Points:

1. Architects must first return to valuing themselves and restoring their confidence.
2. Collaborations between architects and contractors will be essential in future practice.
3. A regional, mid-size firm with diverse expertise and depth can thrive.

INTERVIEW

John Busby, FAIA

John Busby, FAIA, is founder, and principal of Jova Daniels Busby, a leading architectural firm in Atlanta from 1966 until 2010. He was the national president of the American Institute of Architects in 1986, president of the National Architectural Accreditation Board in 1994 and 1995, chancellor of the AIA's College of Fellows in 1993, and a graduate of the Georgia Institute of Technology in 1959. In his tenure as national AIA leader, he played a groundbreaking role in developing many policies used by architects practicing today. The demise of his once-thriving firm offers a warning.

AIA Reflections or "Breaking Boundaries"

Historically, many of the great three-principal firms used a classical three-part organization to manage their firms. Did you follow that model?

Busby: Yes. Henry Jova was focused on the design, Stanley Daniels led the administrative and management duties, and I focused on technical matters, documents, materials and such, although we all wanted to stay involved in projects and made efforts to enable design and professional growth for others. At its peak, our firm had 75 people, so we were still able to oversee it all. But the firm is no more; we closed the doors in 2010.

It was a leading firm for several decades. Architects have a reputation for being bad at managing design. That is, schedules, budgets, and themselves. Many owners and contractors don't understand the process. They ask, "What's the problem? Why does it take so long? Why can't you get in budget? Let's go from a to b and get it built." These are the core questions I'm after. How did your firm go about managing design?

Busby: Like any firm, we'd sit with the three principals and senior associates for meetings every two weeks to discuss policies and approaches. To be honest, we got to the point we all agreed, "Let's do that." But I'd find out later, we had no continuity or follow up. We didn't track morale, status, continuing education, firm development, counseling employees. We didn't have any "internal management" except as principals and associates, and sometimes even we failed. We'd come together in projects and generally see how we were doing on schedules and budgets, but *that's it.*

> *"We'd come together and generally see how we were doing on schedules and budgets, but that's it."*

Those processes are quite different in a construction firm. We do projections, multiple management levels, risk management, contingency planning and precise reporting – weekly, monthly, quarterly – to track performance and profitability, by project and company. Processes are clear, visible and managed as company goals. Quite different from your periodic discussions where design and the work were at the fore and managing them was more of an "oh by the way."

Busby: Very much. For cost control, we'd talk to contractors about CM versus bid. At first, we did a lot of bid work. Then, to control costs we asked clients to bring in capable contractor partners, and we'd interview them and bring them on board. That worked well. Around preliminaries, DD, we'd begin to initiate cost containment and take their advice. However, the last project I worked on the costs fluctuated too much. We didn't get the right advice.

That issue remains. We're still doing something wrong.

Busby: Architectural schools are beginning to diversify. When I was accrediting schools, the architecture school was stand alone. Right next to it was usually an engineering or construction school that offered classes in construction, lighting, cost control, you name it. When I asked if students could cross over and take those classes most of them said, "Oh, no." How about if the professors came over here and taught? "Oh, no, no." It was forbidden. They have a budget for their faculty and we have one for ours. Departmental budget limits, quotas, or territorial boundaries ruled.

"Management," budgets, structure, and paradigms were getting in the way?

Busby: Very much so. Isolated schools – still focused on design – had cost programs, but it was out of Means or Dodge, some other book, or their imagination. Academics just

> *"Being aware of how a cost estimate was prepared was unheard of in the education of an architect."*

didn't get into details for costs or how contractors prepare them. Architects' estimates either worked with a contractor, or were done on average, similar costs by building type – school, lab, etc. – on square foot costs. Maybe add a 10 percent contingency. Being aware of how a cost estimate was prepared was unheard of in the education of an architect. Most of it concentrated on design, design, design.

Most schools have an engineer on the faculty, who would advise fourth- or fifth-year students. Keep in mind, back then, there was a limited range of discipline and curricula in architecture. Today with CAD, BIM, energy, LEED, and so many other things, its explosive. They can't learn it all. Now you can come up with a dynamic, distorted façade that can only be worked out by a computer, but not know how a contractor estimates it. Contractors previously unfamiliar with new materials are working with manufacturers. They're beginning to understand innovative, complex forms and materials, but it takes more energy to figure out, versus the simple forms and buildings of the past.

You taught, mentored, and juried at Southern Polytech, now Kennesaw State.

Busby: I had an exercise where I asked students to calculate the volume of concrete they needed for a slab. Well, you'd be surprised. They couldn't do it!

One of my beliefs, in looking for causes, is the architectural profession's closed culture. We can't cross boundaries to take classes. We stay locked in a "secret studio" working all night. Not enough socialization. Did you begin to see curricula change to fix that closed culture? Are student or faculty mindsets moving that direction? Any collaboration classes? We didn't have any.

Busby: In one class, architecture and construction faculty tried to come together to teach a course collaboratively, on codes. We talked about being compatible; we worked together on the lesson plan. There was also an introductory course in architectural practice that we team taught and that students engaged in. I did see, in juries, courses introduced in work planning, bar chart scheduling, even construction scheduling and work sequencing.

So, some management, scheduling, and planning skills, but not learning to collaborate?

Busby: Right. Some were in the *opposite* direction from collaboration. I did see difficulty, because of the independent thinking. We would challenge something – and they would resist. Even when we would come up with a clear reason why it didn't work, they would insist. So, I said, "Well, if you *insist* on that approach, even if after I've given you something which, in a jury, might bring criticism, then go ahead." Sure enough, a lot of them came back and said, "Well, I wish I'd have taken your advice." And I said, "That's ok. You learned something."

That's a great example of what architects have taught and rewarded in our culture: independent thinking, ego-driven intuition, stubbornness. It doesn't work in today's profession. We need teams.

Busby: Yes. But the public does not understand design. They understand the design of their kitchen cabinets and patio area, but when you look at the totality of a project, it gets very complicated. We're having to understand new systems we never thought about. Now we have IT systems, graphic design, and so many new disciplines.

> "But the public does not understand design."

It's always been the case that clients didn't understand design or couldn't read drawings. But now that things are more complex, not even experienced professionals understand what design is anymore. We've got our work cut out for us. So many kinds of – and meanings for – design, all complex. How do we create new ways of working with a collaborative mindset?

Busby: I ran into some critical opposition on that, 5 or 6 years ago in the AIA. They all said, "Design is important." I said, "Yes, but what the hell does it mean?" What about understanding materials, or project delivery, or land disturbance and codes? Preliminary design begins with a sketch on paper or on a digital pad and moves rapidly into a contract. Great. You have a 20,000 SF floorplan, but no definition of materials or scope. How do we know what it costs? What are the finishes? Beginning to

> "Design is important. I said, 'Yes, but what the hell does it mean?'"

understand design is important, but controlling costs is part of design. And presenting owners with options is part of design. Do you want a 5-year or 30-year roof? Maintenance costs? Life cycle costs? Those are part of design. A lot of folks don't understand that. Architects sometimes miss it too. Early contractor involvement and new methods of teaming shape design.

What were some of the biggest issues or challenges you faced as AIA president?

Busby: Liability insurance was big. Prices went up, and the profession screamed, "What the hell?" Another was other countries providing architectural services across the world. Creating accords compatible for registration across the border to bring in a Canadian firm in, or vice versa. We set up accords with Canada and Mexico, then Great Britain, Australia, Japan, and the Pacific Rim. We began to see international practice come into play in Saudi Arabia for the new universities, campuses and big housing projects.

You opened the doors for many of the large global practices of today: Gensler, HOK. That was pioneering work.

Busby: And specialization. For liability insurance, everybody wanted million-dollar policies, when sometimes the entire project wasn't worth that much. So, we looked into pooled, contingent insurance by the owner rather than having individual policies. We dealt with arbitration and negotiation. We approved an updated set of AIA Documents. From '85 to '95 we began to see alternative delivery agreements in the industry. We had to get a lot of architects to understand we were *part* of the construction industry, not the *entire* industry. That we provided services to it and needed to get more knowledgeable about what we were designing and why.

We still face these questions. Delegated responsibilities, and how they blur in teams.

> "We had to get a lot of the architects to understand that we were part of the construction industry, not the entire industry. That we provided services to it."

Busby: We even sent a team to Lloyds of London, the documents and insurance committees. Lloyds was the overwriter for Schinnerer and others covering construction. We had to explain how much coverage we needed because the entire project didn't usually fail, just some aspect, and the owner needed to pick up some of that risk. Getting the industry to understand that was difficult.

Groundbreaking work. Those days saw the emergence of CM, fast track, and the erosion of the architect's role as "master builder." Thirty or forty years later, it's back. To survive, we need to embrace construction partners, specialists and owners in new ways. It's far from being "all about the architect," as we were trained.

Busby: Yes, even a house is getting complicated. If the architect is going to take on construction administration duties, he has to know what's going in that house, rough-in, framing systems, digital systems, etc. Smart buildings . . . what is that?

We need specialists, but somebody's still got to coordinate it. There's a place for generalists, but they must be smarter. As we look forward, what would you tell the aspiring future architect? What should they do that you and I might not have?

Busby: Be conscious of what you think you know, and what you don't know, depending on the kind of practice you want. If you want to be a generalist, you can – there will still be small practices in cities across the country. But that doesn't relieve you of the responsibility to understand what else is out there. I advocate for continuing education. Never assume you know everything – there's so much innovation. Be ready to engage in a dialogue to better understand and advise your client and others so you know where to attach yourself and dial in to the process.

I show a slide of Frank Lloyd Wright with his colleagues gathered around in rapt attention. He knew all he needed to know. Now, we can't know it all. Nor are most of us practicing in the same room. We have a more inclusive process that takes management and dialogue with others.

Busby: We had a big debate on continuing education at the Boston Convention. I said: "If we don't pass this, we're not gonna play in the game. NCARB, or the public is going to dictate what we learn. If we want to be in the game, we've got to do this." We didn't for LEED and lost. Keep in mind, the architect is designing the future, and the contractor is implementing it. We must understand systems, fire codes, and so on. Architecture integrates everything. That's my difficulty with defining "design." Does it mean the façade? Or how its connected to the structure? What about the curtainwall subcontractor's expertise?

You advocated for many important issues for the profession – liability insurance, globalization, continuing education – and our part in the larger industry. I appreciate your leadership.

Busby: In the '80s, the Institute contemplated the future. We said, "We need to look 10 years out." In the '90s, we said, "No, we need to change it to 5." Then we said 3. I said, "The institute needs continuity of programs and message." I wanted to deemphasize the convention and get to issues the profession needs for the future. Some people will listen. Some ignore it. But the institute has to be aware of the changes.

There was a schism between the profession and architectural education. Professors and administrators complained they didn't have time to explore broader curricula. But we knew we needed to encourage different courses in architecture. In 1966 or '67, Harvard or Penn recommended expanding. They said, "We can't do it all, we're gonna add another year." That's when we went from a 5-year degree to the 4 plus 2-year program. We understood what a graduate student moving into the profession would require. Now, that graduate student takes the ARE before their experience is fulfilled. That's profound, with the extended practical experience they'll need. That's the progress we need to make. We can no longer plan for 10 years, because change is too rapid. To this day, I *still* don't know what design is. I know what the design *process* is, but I don't think we've defined or delivered what design is.

There are so many meanings for it, not just one.

Busby: Some things the contractor can answer, some the architect can. We have to think about it in all those ways, with all those inputs to get the totality of design process. And we wonder why design costs so much? Well, why does medicine cost so much? Do you want the doctor to leave something out? Even in construction, we must get owners to understand the complexity and process, and understand it's a process of *discovery*, because we're defining it as we go. It's something you need; we recommend you put it in. If you don't want it, don't blame us, or say we didn't fulfill the contract. We've got to communicate that to the owner's project manager. We're moving into that phase now.

> *"To this day, I still don't know what design is. I know what the design process is, but I don't think we've defined, or delivered, what design is."*

The movement to teaming and new communication tools are taking us there.

John Busby Key Points:

1. Design firms retaining a casual approach to managing would benefit from following other businesses, like construction, which measure and manage actively.
2. Architects need to be taught about construction cost estimating and take ownership for designing to budgets.
3. "Design" has many meanings. (e.g. appearance, systems, performance, maintenance) – a source of confusion in trying to understand and manage.
4. Owners and the public don't understand "design" or design process.
5. AEs are part of the construction industry and can't operate in isolation.
6. Continuing education and an expanded expertise range will be required for future practice.

INTERVIEW
Agatha Kessler, Assoc. AIA

Chairman, Fentress Architects   21st February 2018

Agatha Kessler has worked as an executive in finance and technology, building international businesses in emerging products with VISA and Hewlett-Packard. Energized by the intersection of technology, business, and design, in 2007 Agatha joined Fentress Architects as CEO. She holds an MBA and has lived in many cities around the world. As chairman of Fentress, Agatha serves on multiple boards, including Opera Colorado and the Design Futures Council. She is cofounder of the Aerial Futures international think tank https://www.aerialfutures.org and is pursuing a PhD in aviation at Embry-Riddle Aeronautical University.

Her tenure at Fentress, a global practice with large, diverse teams, offers insights on inclusivity and diversity in managing design. Her people-centric focus advances strategic vision through continuous learning and high-performing teams. Leading with her heart, she brings energy, spirit, and laughter to the workplace, for exceptional performance.

The Known Unknown: Keys to Change Readiness or "Weeding and Nurturing"

Your personal page on your company website opens with this quote: "What you don't know you cannot manage. Acquire the fundamentals to drive revolutionary change." That's a core issue of the book. Design manages the unknown. Coming from business to lead a global design firm for the past ten years, how have you approached the challenge?

> *"If you can find the known you can manage the variances."*

Kessler: I want to praise you first. Your questions are very thoughtful. They show that you listened and did your homework. The book is very well organized, and your questions are connected to who I am. You don't often find all that in one person. I appreciate that.

Thank you very much. That's nice to hear. It's new ground for me, listening, interviewing, the book.

Kessler: Maybe doing the book will make you a better partner?

I hope so.

Kessler: I really believe what that quote says. Because without it, you're just throwing darts in the dark. To know what you know is important because we often don't know what we don't know. But I'd say design is *not* always unknown. If you can find the known you can manage the variances. Some are fixed continents, others are moving islands. Not all the parts are moving. So, one of the first things I did was listen, watch, observe and learn, like a baby. I was thinking about Aikido, the Japanese martial art. Aikido is a way of combining forces. It's the principle of blending with your opponent's movement to minimize your efforts. You go with the flow, but you use that force to do what you, or your company needs, to get to where you envision. By understanding that rhythm and intent, we can find optimal solutions.

The second thing is to treat people with respect and empathy. I call it different keys to open different doors. When I worked for VISA and Hewlett Packard, my mentors had high expectations. They showed me what I needed to learn. One of the first things they said was necessary to be a good GM was to acquire a repertoire of skills. I can't be the same person for everybody, so I talk to each individual in their unique style and place to unleash their talents.

Great analogies. I agree, design is not all unknown.

Kessler: I was also thinking about the technology world. In technology we have just as much creativity as in design. We manage even more unknowns. Think of the folks at Apple who invented the iPhone. They had nothing. A clean sheet of paper. They had to know more than technology. There were many unknowns including inventing a market for their product. When we design buildings, we know the client, the site, the budget. We know many things. A lot more than in technology. Everything is relative.

> *"You need different keys to open different doors."*

The more I get into it, the more I'm learning that this book is about empathy and inclusion. Like design – and many stories – you don't entirely know where it's going when you start.

Kessler: That must feel good.

It does. You commented that compared to other industries' willingness to change and adopt new technologies, architecture is a "dinosaur." There's a "gap" – can you explain?

Kessler: I call architecture a dinosaur because in relative terms it hasn't undergone much change. Take a butterfly and a parrot. Some butterflies live only one week. Some parrots live 50 years. So, for a butterfly, a day is a very long time. Coming from the technology world, we did version updates every 2 months, so we made lots of decisions quickly. In architecture, being slower to change and not making decisions seems the norm. Both for clients and the design team. Architecture has changed relatively little in the last few hundred years. Look at how we teach architecture. We teach it by batch processing – our educational system is a batch processor for the convenience of the processor, not the learners.

In architecture, changes have been more about technology, not culture or radical changes in process. But looking to the future, William Gibson said, "The future is already here, it's just not evenly distributed." Machine design can already use algorithms to do much of what architects do. Artificial intelligence has already composed plenty of music via algorithm. I'm talking about the future. Should we be very afraid? I don't think so – designers will just do more valuable stuff.

Take digital photography. Mass access to digital cameras and iPhone cameras haven't made photography disappear as a discipline, they've expanded it. We're at an inflection point where those who move with the current will thrive and be more valuable. That current will only get stronger. Those who don't will not be around. The cycle naturally screens out some practitioners. We won the national AIA Award for BIM a few years ago. We started using BIM at least 10 years ago – when we started doing paperless projects. We are mindful to evolve with the technology current.

You describe yourself as a global citizen. How can your global background and acumen in building new business opportunities be applied to changing the building industry?

> "Should we be very afraid? I don't think so – designers can just do more valuable stuff."

Kessler: I look at this from two perspectives. One, global practice versus the U.S. The building industry is structured differently outside the U.S. In Asia, and much of Europe, it's almost all design-build. Vertically integrated. In the U.S., it's moving that way. The DBIA would say so. What's similar? Stronger companies can ride volatility because they're prepared and have big projects that span multiple years. For me, it strengthens the idea of discipline. Being disciplined doesn't mean you aren't flexible. When you need to change you can. Working from a position of strength allows agility. Discipline and agility go together. That's what I mean by understanding the fundamentals. Like a good ballet dancer, if you don't know your basics, you can't go beyond them to be creative.

Then there's volatility. The design industry is still like Dallas and Denver were in the 1980s. Houses sold for $10 because they were one-industry towns. Architecture is similar. Some companies don't diversify and are caught by being inflexible. We have learned that lesson. We diversified. One should do something anti-cycle – take work that spans multiple years. Diversify your portfolio and hone your expertise.

Shifting to the human side, you talked about your favorite questions in the studio being: "Are you having fun? And are you learning?" and if the answers are yes, things are good. Teams are happy, productive, the firm does well, clients are happy. What can firms do to ensure the answers are yes?

Kessler: Two things. First, nurture. Second, create a safe, stable environment. We have Fentress University. We do a lot of classes. If you aren't careful, you can be in Fentress University full time. Once you do that, you've created a safe environment. The rest of the time you give staff more rope and coach them. Our young designers – most have master's degrees – get licensed within 2–3 years because we create that accelerated learning and safe, nurturing environment. If it's not safe it's difficult to learn and perform.

On the business side, as leaders, our job is to ensure we're financially strong - so our people can focus on their design. We're not a hire and fire company. If we need to borrow money to keep good people, we do. Otherwise, if they are insecure, how can they give you their trust, talent and effort?

Let's look at the designer's bias. Coming in as an "outsider," you've had success. Your firm was shrewd to bring you in, and you showed courage to take the helm of a design organization. What lessons from that fresh perspective can we apply to a perennial problem set: closed culture, myopia, ego-driven, lack of inclusivity, and a history of silos?

Kessler: Mr. Fentress is very secure. He says, "I'm a designer. If I were a fisherman, I should be fishing." So, he looked for someone else to do the other things. I complement him. He was brave to bring me in – it wasn't courage on my part. This is not a hobby. It's a business. Running a badly managed firm is being irresponsible to the people who dedicate their talents. We focus on strong finances and culture. Because of that we've been successful.

I learned quickly that architects are nice conflict-avoidance people. They don't understand if we don't agree, it's okay. If we try to understand each other's perspectives the outcome is likely to be stronger. To begin change, you have to weed, and the architects couldn't do it because they were so nice. They think being nice is most important, but you have to be strong. If your people aren't committed, or don't care, that's not what you want. Build an environment that rewards performers that thrive in our culture first and help those who don't fit find happiness somewhere else. I'm not a hard-hearted person. I'm a people person. I foster people. But being indiscriminate is unfair to the good people.

So, the first thing is weed, then nurture, then hire the right people. Conflict avoidance? I've seen it. The perception is: "architects" are "architects," but they're all different. They do different things. Some design, some manage, some detail. They're individuals who migrate between roles and evolve. I've jokingly said I got into design because it didn't involve conflict. No life or death situations – inanimate objects. I couldn't work in an emergency room or be a lawyer. But surprise! Design and construction have conflict. Because they attract conflict-avoiders, we aren't pre-disposed to seek it and resolve it. We're content to do what we love. The job of managers and leaders is to seek and address conflict. Designers need tough love to ensure they can thrive. Can you share your future visions of the air travel industry? What can firms look for?

Kessler: Signals for change are not easy to spot. If they were, everyone would see them coming. Industry convergence is the basis for co-founding the Aerial Futures because Artificial Intelligence, machine learning, and automation mean things will be different. Then, we have demographic changes. When those elements come together, the convergence will be amazing. In a few years, millennials will be three-quarters of the work force! Air travel democratization means more people are flying. With driverless cars, drop-off points will be different. Fewer parking spaces. Hyperloops

"I learned quickly that architects are nice people— conflict-avoidance people."

will be part of our mobility system. Commercial supersonic flights are coming back. The changes will be exciting. I love the analogy about horses at the turn of the century. People said, "Why do we need these gas-belching cars? All we need is faster horses." Myopia happens all the time. Michelin was a tire company. Why did they assign stars to restaurants? To encourage people to drive more. A hundred years later, people are still change resistant. To tie that to the building industry, I see artificial intelligence and machine learning influencing the industry in a big way. Little quakes will occur on the periphery, but a tsunami is coming. Machines can design it. Architects will do higher-level things. You'll still need people to build buildings, but drones can do more – as well as prefabrication. The point is, we need to be ready for change.

Agatha Kessler Key Points:

1. To manage design, separate known from unknown and manage both.
2. Go with the flow. Match your opponent's [partner, client, teammates] force and direction to minimize effort. [Aikido]
3. Use context, empathy, listening, and relational skills to build strong working relationships: "different keys unlock different doors."
4. Learn from technology to embrace change faster. Hone decision-making skills.
5. As leaders, create safe nurturing environments, which may include some weeding and diversification to build stability and agility.
6. Supplement "nice" people, a.k.a. "conflict avoiders," with conflict resolvers for the good of the team and to create environments where everyone can thrive.
7. Be ready for change. The future is here.

CHAPTER 6

Strategy: Early Questions, Planning Horizons, and Socialization

INTERVIEW
Phil Bernstein, FAIA

Associate Dean and Senior Lecturer, Yale School of Architecture   25th August 2017

Phil Bernstein, FAIA, is the Associate Dean and Senior Lecturer at the Yale School of Architecture, co-author of *Building (in) The Future: Recasting Labor in Architecture*, contributor to multiple books and essays on technology's impact on design practice, and a frequent presenter at national conferences. He was formerly Vice President of Strategic Industry Relations, and continues as an Autodesk Fellow and a consultant to AEC firms helping them with technology strategy. He practiced most recently as a principal in the office of Pelli Clarke Pelli, Architects. He was recently called "The Prophet" in *Architect* magazine. His new book, *Architecture | Design | Data: Practice Competency in the Era of Computation*, was published in 2018.

Post-Pasture Value Propositions or "Three Horizons?"

You've been a practitioner, strategist, and consultant, focused on the future. Let's look at a typical architecture firm. Most are led by guys like me – fifty or sixty – overwhelmed by technology, the CM's fast-track processes, and compressed schedules. Those who want to change are undercapitalized or unaware of how to start. Can we speculate on an approach to firm survival?

> "Firms that haven't answered these questions about why they exist, and their value propositions don't deserve to continue."

Bernstein: One good indicator is whether they have a succession plan in place. They need a view toward the future – it's a philosophical discussion about a leader's view of their firm. There are really two types of strategies, either explicit or implicit. The explicit strategies are: Strategy A, you pass the firm on to the next generation. In B, you get acquired. The implicit strategy I see all the time is C, you let the firm fritter away on its own and shut the doors. It just peters out. The next generation sees no one is pitching the ball to them, so they move on.

Let's talk about Strategy A.

Bernstein: These questions aren't disassociated from the larger questions about what the firm's fundamental value proposition is, and why they exist and why that existence needs to be perpetuated. Firms that haven't answered these questions don't deserve to continue.

Let's say you have an older baby boomer like you and me. They've been thinking about passing the firm on, and they've done so by having a conversation they started 10 years ago: "You're going to take this firm over, and we're going to spend the next 5 to 8 years transitioning you into a leadership role. So, we're going to identify what this firm is really good at and make sure everybody understands that, and we're going to transfer the equity from me to you. So, when I finally step down, you'll take the reins of this firm that we both agree is doing a great job with x, y, and z."

Part of that analysis has to be understanding the firm's use of technology and how it connects to their value proposition. You can't skip over any of those steps. Otherwise you get into this weird prescriptive thing: "If you don't learn how to use some latest software, your firm's going to die." That's one absurd end of the argument. The other is: "If you're 65 and still don't get it, that's okay. You've had a good career. Pass the torch and get out of the way."

If you're saying you're coming across people – leaders – who are 50, and you're seeing the same problems, that's a bigger issue. Then you have to say, "Really? You're 50 years old and came up through the prime growth years of your career and you don't understand technology? Did you go to school somewhere where they didn't have electricity?"

Consider the slope of two curves. One curve is the generational curve. The early baby boomers are gone. We're the late boomers. At this stage of the game, it's probably too late for even late boomers to undergo some kind of existential transformation where they suddenly decide that new, technology enabled ways of working are somehow interesting and they're going to change their business to embrace them. That's just not happening. If you don't get it by now and you're our age, you need to be put out to pasture. If you're not willing to be put out to pasture for a variety of demographic reasons related to the global financial crisis and better healthcare – some of the great architects are working into their 90s – and you're still working, then you better have people who understand these things. Because the next generation is going to push change. The other curve . . . the slope of change is pretty slow.

> "They're having the same experience most owners are: somewhere between 15 and 90 percent of their projects miss their budgets and schedules."

Let's talk about the 40-year-old architect – an emerging leader and firm owner with time to do something about these issues. They have a vested interest and a leadership foothold. The 25-year-olds are using and pushing technology, but they're busy on projects, not leading their firm's futures. How can their bosses – these emerging leaders – approach it? How do they start?

Bernstein: Let's look at the spheres of influence that we as individuals within firms have the power to change, versus the endemic problems of the building industry. At some level, you'd think companies with technological vision would be interested in the most advanced possible use of technology in construction, but when they when they get to the design, construction and operational end, most of them operate like everybody else. Why is that? Why do they not see the endemic problems? Why are they interested in continuing and repeating the same issues?

That's the essence of this inquiry.

Bernstein: Okay, a company is operating, and they've been conditioned to a certain set of circumstances and market conditions. Those mostly have to do with meeting budgets and schedules; we'll set quality aside. They're having the same experience most owners are: somewhere between 15 and 90 percent of their projects miss their budgets and schedules, depending on which studies you read. So, as an owner, they see no opportunity for innovation in the industry. It's a self-reinforcing system. There are a million reasons for these attitudes, which are well-studied. That road was paved in 2004 with the CURT Whitepaper.[1]

The real question is why clients continue this? I'll contrast this with how I spent my day yesterday. I was up at Brown University, who are beginning their 6th IPD project. Brown has turned into one of the country's leading IPD owners, for 3 reasons: better results, saving money – millions – and IPD jobs are more fun. Everybody enjoys working on them.

One premise you present is: "Embrace technology as a firm leader – do it or die." Another is owners leading. For owners who have too much to do: "Engage and lead, embrace IDP."

Bernstein: The 40-year-old practitioner has to first decide what the essential value of her firm actually is, and then work out the value proposition relative to the overall systems through which the firm's buildings are delivered. Then, we can talk about the implication of technology.

That 40-year-old firm owner needs to decide in 3 planning horizons – short term, mid-term, and long term – about her attitude about design methodology and its relationship to technology. Right now, it's not a question of whether a 40-year-old with a firm of more than 3 or 4 people *can* compete in the marketplace without technology. The efficiencies that BIM has brought to the U.S. architectural profession – unless a firm has some kind of magical, mystical sauce – make competing without BIM impossible, because that firm simply can't produce at the same rate as everybody else. I point this fact out in several places: net revenues of U.S. architecture firms are the same as they were before the 2008 recession, but there are 20 percent fewer people in the profession doing the same amount of work. What's the explanation for that?

> *"Owners don't want to look at drawings because they don't understand drawings, they don't care about them, and they don't see the world through them . . . they see the world through a lens of 3D video games, their iPad, and high-resolution, IMAX Batman movies."*

---

[1] Construction Users Roundtable (CURT) Whitepaper 1202, "Collaboration Integrated Information and the Project Life Cycle in Building Design, Construction and Operation," August 2004.

From 2008 to 2015, BIM adoption tripled in the U.S. That's a correlation not a causation – but it seems like there's an obvious relationship. What else could have caused that increase in efficiency?

The other side is that the pasture-ready and the greener-grass groups left the profession. And the two factors worked together.

Bernstein: Let's look at those three horizons, short, medium, and long – what we call in the software business, "H1, H2, and H3." H1 is the next 24 months, the stuff – systems and market forces – you're executing against right now. H2 is 3 to 7 years, the market forces coming soon to affect H1, but not quite here yet. H3 is the great beyond you need to think about but can't plan explicitly for. So, in H1, you've got the BIM productivity question and high-res visualization to support clients today who demand higher fidelity output since those folks grew up playing video games.

"This goes back to the H2 question: can a modern, thriving architecture firm tackle the functional, labor and cost challenges of construction without sustained relationships with construction firms."

Owners don't want to look at drawings because they don't understand them, don't care about them, and don't see the world through them. Instead, they see the world through a lens of 3D video games, their iPad, and high-resolution, IMAX Batman movies. So, you're not going to communicate with those people though traditional artifacts: plans, sections, and elevations.

Contractors are struggling with a labor shortage. Assuming the economy doesn't collapse, this shortage is going to get dire. This person's children – that 40-year-old future firm leader doesn't aspire for her kids to be construction workers, and, with our current uncertainty around immigration, the labor supply is going to dry up. Thirty to forty percent of the U.S. workforce is immigrants, many undocumented. That flow is slowing. But we still have to build stuff, and there's more competitive pressure, so what's going to happen? We'll see more automated machinery on jobsites, and pressure toward offsite prefabrication. Why not start thinking about how all these issues are going to work? Those ideas might not come from a design firm, since they relate to means and methods. A design firm typically doesn't invent a systematized way of construction or a unitized subsystem – it's up to guys like you at Holder to figure out how to get things built. It isn't going to happen any other way. This goes back to the H2 question: whether or not a modern, thriving architecture firm can tackle the functional, labor and cost challenges of construction without sustained relationships with construction firms. If I were a 40-year-old right now, I'd be playing golf with, going to the school play with, having picnics with a bunch of people who build stuff, so I build a network where these things become possible. These are all H2 questions. Design firms must get smarter about the means and methods of construction being developed today.

At Yale, students in one studio have to build a house the students designed and had a competition around. This year we had problems. We didn't take possession of the site until three weeks ago. So, the students have been in a warehouse, prefabricating the house, modularly, and trucking it to the site. Now they're assembling it. They were forced to go prefab – they didn't have a choice – they couldn't finish without it. They accommodated a means and methods challenge. Firms need to think about that same H2 question.

Many clients have a data-driven perspective because they have the computation power to collect and reason about information. In the next 3–5 years firms will have to figure out what data means for them. How do you prove your design results? How do you demonstrate to a client that all you're telling them *will* happen? These involve H2 strategies for projecting future construction and measuring performance.

The H3 stuff has to do with the disaggregation or "de-skilling" of the profession. Ten years from now, many of the routinized tasks of design: code analysis, spatial configuration, schedule generation – the rote work – is going to be automated. Not just by scripts, but by expert systems that learn how to do things. It happened to textile workers first, now to lower level knowledge workers, and it's coming to an AE firm near you in the next 10 to 15 years. The H3 question is: What does your firm look like then? What's the right way for the profession to design itself for that future? Someone has to make the systems to perform these functions. Perhaps the professions should create those, so we control our meta-destiny.

"The 60-year-olds don't have to worry about this . . . But today's 40-year-olds better keep an eye on it and plan for it soon."

Phil Bernstein Key Points:

1. Firms unable to articulate why they exist and their unique value proposition should hang it up.
2. Those who don't "get" technology should find someone who does.
3. Plan in three planning horizons," H1, H2, H3 (short, medium, long term).

INTERVIEW
Margaret Gilchrist Serrato, PHD, MBA, AIA, ASID, LEED AP

Workplace Foresight Architect, Herman Miller   14th August 2017

With over 30 years of research and design experience, Margaret Gilchrist Serrato focuses on creating high-performance working and learning environments. Her education in architecture, interior design, and environmental psychology, MBA, and PhD in architecture, culture, and behavior, enable comprehensive and dynamic strategic planning for a wide range of clients.

Margaret's professional design experience includes 10 years at Lord, Aeck & Sargent, as Senior Architect & Interior Designer, and 5 years at Tvsdesign as Associate Principal in the corporate workplace studio. She is co-author of "Post Occupancy Evaluation" in *The Architect's Handbook of Professional Practice*, and "Supporting Work Team Effectiveness" in *Facility Design for High-Performance Teams*. Dr. Margaret is a frequent presenter and provocateur.

Strategic Business Planning or "Asking the Right Questions"

The premise of this investigation is: Our process isn't working. You're in a new role, doing something very different to help owners redefine their needs early in the process. They can set the wheels in motion but may not know how. Tell us about it.

> *"You take a process which is difficult at best – design – and make it almost unmanageable."*

Serrato: The reason I started doing what I'm doing is I became frustrated with the workplace design process and the workplaces that resulted. Part of the problem was the culture of the design firms themselves. Culture drives process. At some firms, people share and are eager to explore new ideas. At others, they're protective and territorial. You take a process – design, which is difficult at best – and make it almost unmanageable.

Herman Miller contacted me, and we invented my current role, free of traditional fixed processes. I said, "What would I be doing?" They said, "We don't really know. We're going to call it workplace strategy. Do whatever you think needs to be done to help our clients thing about workspace and processes differently." I knew the approach couldn't be about furniture, because let's face it, a chair is a chair. So, I went out and talked to clients all over the country. Everyone said the same things: reduce real estate costs, boost innovation, support employee engagement. The same buttons, just different mixes.

It's funny to see how recurring these issues are.

Serrato: Yes, the Einstein quote . . .

Doing the same thing repeatedly and expecting a different result?

Serrato: Exactly! It's insanity! It took me a while to realize the thing that was broken was the process. People talked about doing something different, then fell back into doing the process to get try to get there. What I've done is go back and create a new programming process. It's very different than the traditional one. It allows a new process by eliminating some of the a priori baggage.

I usually start by giving clients a brief history of the workplace, so they have a better idea of how it evolved to today. For instance, some clients may not know their current workplace was influenced by the layering of military style management methods over the workplace in postwar America. Before 1945, there was little notion that if you were a junior executive, your office was smaller and, if you were a senior executive, you got a bigger office or a corner office with a window on a higher floor. The idea of status equated to space had a powerful effect on workplace design. For example, we recently worked with a military equipment manufacturer and discovered they had 12 different office and workstations types tied to people's paygrades. I asked what that had to do with how people worked, and they couldn't really explain.

People understand that the way work happens is changing. We talk about all the things that have created different ways of working. We're not trying to tell them to change the way they work, we just recognize people are already working differently and help them figure it out the best way to support it, which is a relief to them, because they thought, "Here come the furniture people to tell us we need open office." Well, I'm not here to do that.

I have a simple process, some steps I've created to help clients realize a new workplace that is very different than what they were expecting. When I step through it, what usually happens after about 45 minutes, is that people *physically react*. They say, "Oh no, we're doing this wrong! We've got to start over!" The CEO gets up, they pace, sweat, hold their head in their hands. They stand up and push the drawings away. It's amazing.

> *"The thing that plagues us . . . is time for planning."*

The reason they put their head in their hands is they had no idea. No one told them. You provide this translating, eye-opening service, which is about managing design in an impactful way. Most of the time someone like you isn't at the table, so we follow traditional patterns. I do too. I'm not going to design the building, I'm there to translate and connect. But the thing that plagues us, and I'm curious to hear your reaction, is time for planning. Usually we, or the design team are brought in early. Clients face uncertainty, then, suddenly, it's: We need our new building in 8 months! GO! So, we have the kickoff meeting, set the footprint, buy trades, and there's NO time to discuss goals, process, innovation . . . Just GO! No wonder we get the same result and keep using our messy ways: accelerated processes, no planning. No team building means fractured processes.

Serrato: You think *you* come in late? Think about us in the furniture world! We're usually at the end. Often clients come into our showroom and say they want a new workplace to support new ways of working, but they are really there just to kick the tires on the furniture. Our salespeople are trying to get us involved earlier in the process. No chairs for you yet! But, if they client already has a completed set of drawings, I'm out. It's too late. With little boxes on the drawings that read Cubicle A, it's just about price. If I do try to enter the process at that point, I'm so disruptive I have no positive effect, only negative. I have to be there at the very beginning and introduce a new process to be effective.

An important advantage of this new process is that it's short and to the point. In the traditional programming process, the designer interviews all the department heads and user groups, creates exhaustive documents of unique program requirements and headcount projections for each department, assigns space types and sizes based on an org chart or other existing standards then tries fit everything into the space. When everything doesn't fit, someone inevitably will suggest eliminating some private offices or more usually, making workstations smaller. The result is a diminished workplace where people complain about noise, privacy, distraction, lack of storage, and general feelings of loss. When I see this happening, I respond, "Stop, that's crazy. Let's use a different process based not on headcount or status, but based on how often people are at work, and how much time they spend working alone or together." Then, we help them figure that out *scientifically,* not by just asking them. At Herman Miller, we measure seat utilization with motion sensors attached to individual and group seats. We find that although most people tell us they are in their seats all the time, their actual utilization is probably closer to thirty-six percent, which is average for the U.S. Office seats utilization is even lower.

When I show clients their data and point out the folly in that old way of thinking, they usually get it right away. With their minds open to new possibilities, we can go on to create a workplace that responds to what people actually do at work. Based on their data, and with the goals of utilization, innovation and engagement, most of the clients I work with are moving to free address, activity-based workplace, with few or no assigned offices or workstations, but with a rich variety of open and closed, individual and group settings. People have robust mobile technology and are free to work when and where they want. I'm not saying that shifting to this new process and new workplace doesn't come with challenges sometimes. After all, we're dealing with 75 years of doing it the old way. But in the end, the new process is shorter, easier, less risky, and cheaper, so clients are happy with it.

There aren't many like you, working outside the mainstream. We need this service.

Serrato: Many people want to do this type of workplace strategic consulting, across many industries. But to be successful, it's important to balance the real estate, numbers-driven approach, with the design and the human psychology approach. It can't be just about the numbers.

That gives us hope, but there's inertia for the old way.

"'Oh my God, we're doing this wrong! We've got to start over!' The CEO gets up, they pace, sweat, hold their head in their hands. They stand up and push the drawings away. It's amazing."

Serrato: We can only chip away at the iceberg. Sometimes I work with colleagues who want a truly collaborative partnership. Sometimes I've worked in less collaborative relationships, and that is more difficult. "Culture matters" can't be lip service. It starts with the leadership. Their ability to let go of control, honor people they work with . . . it's there or not. Same thing really for the success of these new workplaces. That is, the leadership has to trust their people.

A subtext exists beyond cultures between contractors, architects and owners: cultures within firms. We have biases there too, which we may never fix.

Serrato: You don't think you can teach people?

I do. We work on culture. We teach classes on it. We promote it and reward it. But I have to be careful when I come in. People want to know, "Who is this guy? Is he going to try to design the building?" I have to earn their trust by proving I'm there to help – by my actions. People don't like you to take their power or come in with a language they don't know.

> "We're dealing with 75 years of doing it the old way."

Serrato: It happens to me too. I'm telling them that they might have problems, especially with their process. But I don't want to tell them what to do. I want to provide the context and suggest some different ways of approaching their process and let them come to their own conclusions. When I'm with design firms, I say, "Here's what we're seeing, I'll bet you are too." They usually agree, and I ask if I can explain it to their client. I find that the key to unlocking the process is the data. Data can be powerful. It's a language that clients understand. In some ways, my approach is clinical, not just furniture, colors and fluff. I try to get them to not move incrementally but jump from the Red Ocean into the Blue Ocean, to get rid of the things holding their organizations back.

It takes boldness. We can agree to jump, but let's set parameters.

Serrato: I manage it by collecting the data and running scenarios. We can show multiple measurable potential outcomes around utilization, innovation and engagement.

We used to be able to practice intuitively. Now we have to be quantitative and run the numbers.

Serrato: There's still a lot to be said for experience and intuition. People say, "How do you apply the data?" I say: right here – with my brain and 30 years' experience. The robots won't take over.

So, people, human judgment, and experience still have a place. Machines can do a lot, but someone's got to ask, "Should we?"

Serrato: The next question is always: How to mitigate risk? How do we start? What do we do next? Well, I was a Girl Scout. We had these structured lists for earning badges, broken into categories and steps. My process is similarly methodical, then there's an evaluation. The next question is: "How do we sell it up to higher levels of decision making and convince the CEO?" I did a slide deck based on the Seven Slide Solution. It poses the problem then offers a solution. And we do a pilot test, which is good, because everyone knows it's just a pilot, so the risk is managed.

Try one and establish a format for "lessons learned?"

Serrato: Yes. For example, people often ask for more collaboration, but they might not know what it is. Why do you want more? How will you measure it? How will you know you were successful? We call this "challenger sales" or "rational drowning." I use it as way to organize the metrics and lessons learned.

To let them come to the idea on their own?

Serrato: The best way to shift their thinking is to disrupt their thinking, then change their experience. That's why pilot studies work. People's perceptions are based on their experience. Their beliefs are based on what they know. It's a long chain. People are amazingly predictable in their responses.

> "The robots won't take over."

The parallels between what you're doing and the ideas behind this book are strong. I see you coping in similar ways, with toolsets and checklists. Just the right amount of structure.

Serrato: Sometimes I help clients come to the structure or process themselves, but other times I have clients who are looking for someone to give them the structure, so I let them know that it can be simple. There is a process. We're can do these things, we're going to try it with a pilot and you'll be happy. What I'm *really* doing is making *them* look good – the facilities staff, the designers, everybody. In the end, we sell furniture, but I'm there to differentiate Herman Miller and help my customers.

That works all day long. The marketplace needs more of what you do: strategic planning, programming, questioning, great client-focused service, and, importantly, a better process. And to make the time to get in before the train has left the station.

"So, open your arms, let go, amazing things happen, join in their mistakes and you may find they embrace your expertise."

Serrato: The traditional, linear programming process was driven by engineers and architects; it's taken the life out of the process and its obsolete. I use all sorts of digital technology to visualize and evaluate project information, and that's powerful. Much of what we used to do by hand can be done more effectively by digital technology.

Few are using digital modeling to its potential – it's as much art and craft as hand drawing ever was. Even more so, because you have to decide what to include – just like always – but you have to structure the data. When we bring 50 people into an expensive meeting and people don't bring data, it's nuts. I have coping tools, checklists to help us remember.

Serrato: Coping is a good word. Even if I'm just having a dialogue, I weave it into a story that includes their data, even their own mission statement, a great coping tool to get clients back to center on their goals. That makes it easier to topple them a bit by pointing out they might be working on the wrong things.

Margaret Serrato Key Points:

1. Teaching clients to ask the right questions early is a competitive advantage.
2. Planning and programming time is rare.
3. Use research data to add design credibility, manage it and make decisions.
4. Include specialty expertise in early consulting.
5. A results-oriented, problem seeking process is a huge need, versus traditional design approaches.

INTERVIEW
David Gilmore

President, CEO, DesignIntelligence   30th November 2017

David Gilmore is President and CEO of DesignIntelligence, a family of companies that includes the Design Futures Council, and DI Media, the publishing arm of DI Quarterly, and DI Advantage, a geopolitical and economic impact publication for the AEC industry. Dave also serves as Managing Director of DI Strategic Advisors, a consulting firm to leading design and construction firms.

With 30 years of experience as a strategy executive, management consultant, and private equity investment leader, he has led strategic initiatives across an array of industries. He sits on multiple boards offering expertise in decision-support, strategic growth planning & execution, mergers & acquisitions, and intellectual property productization. With exposure to the country's leading design firms, he offers a keen perspective on AEC industry firm best practices and future outlooks.

Team Building and Project Planning or "Last Exit"

Design Intelligence and the Design Futures Council have had a mission to inform and study design excellence for years. Can you share your motivation to lead those organizations? Tell us about your background and your vision, personally and organizationally.

Gilmore: Years ago, in Silicon Valley, I was involved with a series of companies. All our work was related to design. Designing software, hardware, systems – we were thinking constantly about design. Much of it was in high-tech mergers and acquisition. The design thinking dynamic applied. Roger Martin contextualized this in his book *The Design of Business,* which captured the essence of what we're talking about. Design has never been just the architect's purview. We used design-based principles to arrange mergers and acquisitions. Everything is design. How you structure a deal. Financing and assimilation of assets. Everything I did was related to design to the point where I became curious about formalizing thinking around the premise.

> *"My question is: have large firms become fat, dumb and happy? Or are they on that knife's edge of clarity to drive their business forward?"*

I began attending the American Institute of Architects annual national convention almost 20 years ago and have attended nearly every year since because of what I was exposed to in lectures and workshops. I saw what architects thought about and how they identified problems – what I call "problem spotting" and "problem solving." It intrigued me, and I began to incorporate these perspectives into our management consulting work. That's what got me into design – recognizing architecture as a wonderful mix of the artistic with the disciplined.

Gilmore: I became good friends with the entrepreneur Arol Wolford; he connected me with the Design Futures Council. That's how I began to understand what was going on at the council to the point where we invested in and eventually acquired it. I did that because I believed that the AEC industry – architecture, engineering, *and* construction – is dramatically underserved from a business, strategy, and technology perspective.

It's one thing to state there's a problem, and another to do something about it. So much of my career – and the folks who work with me – has been in corporate strategy. To help organizations understand their values, vision, and how they can restructure to go to market in a new way. We thought that was applicable to the AEC space. We influence through DesignIntelligence publications, convene Design Futures Council thought leading gatherings, performing deep-dive research projects, and, when applicable, leveraging the expertise of DI Strategic Advisors. That's what gets us up every morning and motivates us: we believe we can be a significant transforming force in this industry.

"A massive exodus – mature, seasoned leaders will exit the industry in the next 10 years. This enormous class of mid-fifties to sixties baby boomers are about to retire. The "male, pale, stale" dynamic will move on. The question is: what's coming behind this generation? Are there true leaders in that next demographic?"

While you haven't been a design practitioner per se, you've seen and worked with thousands. How have your efforts to study practice been received by practitioners? What mentalities are you seeing? You deal with the leaders. You noted most of them are "pale, male, and stale" – the old guard. Now, through the "Ten Firms under Ten," study you're targeting the small, younger, fresh firms. Do you embrace the middle?

Gilmore: The middle is an important part of what we do. Small firms are what we've known least about. Data suggests that the large majority of architecture firms are 10 people or fewer – there hasn't been many services or help for them. The "Ten Under Ten" initiative looks at those firms. I've talked about the importance of not forgetting your roots, and of the value from a single client to small firms. In those cases – a single act with a single client – is win or lose everything. So, how small practitioners serve clients is important. Small firms exist on a "win all/lose all continuum." They've got to meet cash flow. They need laser focus, and to communicate that to clients. That's powerful.

My question is, have large firms become "dumb, fat, and happy?" "We win some, lose some." Or are they on that knife's edge of clarity to drive their business forward? What are some exemplars pressing forward? We want to bring that out because I fear large firms are forgetting their roots. We're looking at 300 firms in a DI research study. We're realizing a massive exodus - mature, seasoned leaders will exit the industry in the next 10 years. This enormous class of mid-fifties to sixties baby boomers are about to retire. The "male, pale, stale" dynamic will move on. The question is: what's coming behind this generation? Are there true leaders in that next demographic? Are they investing to ensure practice sustainability? Is there a new level of resilience and thinking about the practice? Because this isn't your father's Oldsmobile. Thank God the baby boomers are leaving, because they won't survive what's coming! But God forbid we forget the wisdom of their experience, lest we make even larger mistakes.

They and the mid-career folks are this book's target audience. Everyone else is too inexperienced or on their way out.

Gilmore: Architecture leadership over the next 20 years will come from the Gen Xers, those now 38 to 50. The question is, will they behave the way their bosses did? Or will they say, "I don't want what was done to me to be done to the next generation. I'm going to invest in the growth, intelligence, and maturity of millennials to create management teams in their 20s to 40s." If you invest in people you can get that. The most important asset in professional services is people, but they're often the least cared for. We put more attention on financial assets than people assets. If we changed that emphasis, we'd get exponential output improvements and drive the industry forward.

One of the features I introduced to DI and the DFC since my coming in is to open the doors wide to engineers and contractors, because we are the *AEC* industry. I believe in the Arup philosophy: total design, total integration.

That's matches my own beliefs, education, and career. An inclusive direction.

Gilmore: Ecclesiastes said, "There is no new thing under the sun." There's truth to it. We're seeing this amazing cry for collaboration, to bring intuitive and nonintuitive thinkers together. My mom used to keep all her old dresses, saying they'll come around again. Twenty years later they came back into fashion. The collaboration case is the same: it's being driven not by it being fashionable to do total design, but because it's the only way to meet market demands in a logical, rational way with outcomes that sustain our relevance. We must move into cross-functional collaboration in everything we do. That's why we look at AEC broadly and why I see everything as design. It's not a new theme. It's a renewed theme. And we – including you – are pushing it.

> *"It's not a new theme. It's a renewed theme."*

I've seen a similar example in the several cycles of environmental reawakening over my career. Now, everyone gets sustainability, not just the advocates. It's mainstream thinking. Your comments about investing in people – that idea has been around since Peter Drucker. My experience is that kind of investment is rare in design firms because it requires resources, leadership vision, and risk tolerance to take those leaps. Are you finding differently?

Gilmore: Absolutely. They don't talk about it because it's competitive. But you have people at the table, saying, "Really? You're doing that? It's going to lower distributions and bonuses, and so on."

I was with a firm yesterday and I said, "You're spending *too much* on training and *none* on orientation and education." The training was about butts in seats, listen to the lecture, answer a few questions, and check the box. We both know firms have to move beyond that to knowledge and insight capture through orientation of people: who the firm is, why it exists, and how you as an individual one fits into the overall firm design and strategy.

It's almost an immersion or baptism into company culture, not one time. It goes on to ensure the intelligence sticks. Education, or training, is the second aspect. The best education used to be through apprenticeship – this was an apprentice-based profession. Not anymore. Today you get a degree, get put on stair details and grunt work for a couple years and then get your license. Hopefully you get put on something more exciting and learn from the person next to you, or you don't. But there's rarely a direct assignment of a master-apprentice relationship. The question is: "How do we achieve that kind of education without the overhead and expense?" We've talked about it at length because it has material impact on the profit/loss statement.

I sense many older firm leaders still share the old paradigm: "New hires can wait until 'they learn how to put a building together,'" as the saying goes. Maybe the New Guard is ponying up.

Gilmore: I'm seeing it. The big firms are steadily investing in people. The Big H firms are HOK, HKS, HDR, HGA, and others. Dialog, a wonderful Canadian firm is doing it. At Arup, it's in the water system. People are made part of research teams and constantly rotated in and out, so the opportunity is there to assimilate knowledge. Investment is being made by the baby boomers. I'm excited about it but worried they won't have the conviction – when the downturn occurs – to sustain the investment. Conviction depends on what leaders think is critical. They invest in that.

Technology has been a major force on practice. Some firms have made 7-figure investments. For all its advantages, technology brings a host of other management complexities – unintended consequences. How are you coaching firms to cope with those? It seems a never-ending money pit.

Gilmore: I'm convinced there's a massive gap between what Revit, as an example, offers and what people are taking advantage of. The products are potentially overengineered and have too much complexity and functionality: "It's all in there." I wish the products were layered based on the way they get used – like "gaming" where you master "levels" before you can move to the next feature/functionality. Had software designers done that, the results could have been unbelievable.

Autodesk software innovation hasn't been as effective as it could have been had they invested in visioning use and users. As a result, they opened the door to other software products that enable, translate and supplement the base software, like Rhino, Grasshopper, and others. There's a litany of programs created as workarounds, extractions from, or additions to the core applications. If they'd stratified the base product like gaming, they'd have a strategic defense in place.

The problem is this glut of software options creates such an overhead for companies to constantly manage. Now we're talking to firms about how to "step back" from BIM and ask, "what problem am I trying to solve?" It's like when you're drowning, you're just flailing – in overhead, cost, detailing, fees, client data, and inability to share models. You don't get time to step back and say, "Wait a minute! We've waded into this thing, and now we're in way over our head." What if we reversed, went back to shore and said, "We're going to do this differently, applying all we've learned. How would we approach BIM's pervasiveness in our firm going forward?" That exercise could yield unbelievable benefits. Some things might be left behind, while other, better things get picked up along the way.

"This was an apprentice-based profession. Not anymore."

I interviewed a leading BIM-advocate-client, John Moebes. He also talked about reevaluating BIM business use cases. I'm glad you and a few leading firms are taking the time, now that BIM has entered relative maturity, to reflect on what the software and its users want to be or do next. This is a key question for firms and tech providers. You're studying and talking to leaders and successful firms. I'm trying to help the rest of us cope. On jobsites, we still face plenty of old-fashioned drawings, flawed practices and change resistance; it's the reality. How do late bloomers and day-to-day strugglers get with it? Owners don't seem to care – they say, "That's why I hired you." They don't open models or allow time for cooperative process, or BIM execution plans. Any advice?

"Take 5 to plan."   **Gilmore:** A major outcome of our Atlanta collaboration conference was the value of assembling the team – anyone who touches the project. The discussion didn't go far enough, but we do this with software. If we were going to build a software system, we bring everybody together. We literally spend time not talking about the project. We spent time getting to know one another. What turns you on? What ticks you off? What things are most important to you? What are your motives? What moves you and prompts you? We go through that dynamic because once we break the ice of interpersonal knowledge, business discussions go unbelievably smoothly. Then, we can begin to talk about the "desired state." Not just the end product, but the process to achieve it. What would it look like? How do we approach it? We spend at least ten days in facilitated meetings, painting our journey to realizing the software. But only after everyone is united. By the way, if some participants are stuck in the mud and antagonistic, we exit them: "You don't get to play. You can't be on this project."

You "exit" them?

Gilmore: Yes. "You're not gonna work. This isn't gonna work." You'd have a subcontractor come in and say, "No, this is the way I do my job, thank you very much." And we'd say, "Wrong." We'd work with them, and if they wouldn't change we'd say, "You're out."

We care about collaboration first, *then* expertise. When we handle the process that way, we always finish ahead of time, under budget, as a unified group. It is sad when projects are over. Nobody wants to go home. We've created a fellowship. This is where AEC has to head, instead of oddball groupings, mixes and matches of participants each time. This time it's one mix, next time another. Why aren't firms forming alliances? For example, you could say, "We work best with these three firms. We know each other's process, have relationships, talk the same, have figured out how to mesh our technologies, and these four engineering firms fit this need." When we go to owners to do a deal, we say, "Here's my team. We're an alliance. You're going to save time, money, effort, and frustration if you go with us." As opposed to the fractured bidding process, we go through diving to the bottom line.

> *"If we were going to build a software system, we brought everybody together. 'Bring all the liars in the room and lock the door,' I said. We would literally spend time not talking about the project. We would spend time getting to know one another . . . once we broke the ice of interpersonal knowledge, business discussions went unbelievably smoothly."*

Industry data show team building and project analysis are the top guarantors of success. Too often we jump into the work and forget the people and process, wondering why it's so difficult. We know better. My sound bite for owners and teams who think they're too busy to get to know one another, or plan is: "Whatever time you have, take five percent of it to plan."

How expensive and what was the scale of planning for an enterprise like software building?

Gilmore: We used it to merge two mortgage operations in Australia. They needed all new software and agreed to a new workflow to be used across 1,500 people, processing thousands of mortgages. We conceptualized at a high level how it would work, the owner said exactly what they wanted, and we created a team. The software project was $53 million alone. That didn't include hardware, network, and so on. The project total was north of $300M. The software team alone was 200+ people – from Moscow; Riga, Latvia; San Francisco, and Australia. We brought the leaders in and spent 30 days together. End users too. Forty folks in Adelaide, Australia, for 30 days, were we socialized and learned how to connect. The owner said, "If you think this is the right thing to do, will you put money into it?" So, we did a risk/ reward contract that said if we hit major milestones, there were incentives and bonuses. It was a 3-year project completed 6 months early. An investment in 40 people for 30 days of "nonproductive" work accelerated the project dramatically. Because now I had a relationship with this person I'm talking to in Moscow. We've breathed the same air. Broken bread together. Now he's going to bend over backward to help me if I need him, because we have that relationship.

We do lots of $200 to 300 million projects. To convince people to come together and spend even one day getting to know each other can be like pulling teeth. I guess it's because the leads are business-focused, linear thinkers, who say, "We're late starting. Let's get the task done."

"An investment in 40 people for 30 days of 'nonproductive' work accelerated the project dramatically. Because now I had a relationship with this person I'm talking to in Moscow. We've breathed the same air . . . broken bread together. Now he's going to bend over backward to help me if I need him, because we have that relationship."

Gilmore: It's even more myopic: there are business managers in each AEC silo who say, "We pay you X thousand dollars a year, and we've gotta have X return on your labor for our investment. So, why in the *world* would I allow you to go into a room for 30 days doing *absolutely nothing for this project?!*" That's how they see it. Metric-driven utilization managers rarely understand how construction really works. There is a massive disconnect between the business and the process of design realization.

People who don't understand design process make decisions based on business-only logic. They've never done design or construction and don't know how to relate.

But everything is not simply a transaction. In seeing, studying and consulting with 54 of the world's best firms this year, have you seen examples of firms managing the design process in innovative ways?

Gilmore: A large East Coast construction company is asking me, "What do we need to not just be another GC/CM?" That they're even asking the question shows humility and is so different. Many organizations think they've figured it out. They brought in an EVP of Innovation from outside the industry to shake up the norm. It's counterintuitive. Others from MIT were brought in to do deep-dive analytics to drive positioning. This is smart stuff by smart people. By the way, these people aren't billable. This company is doing it because they believe acting smarter in front of owners will yield benefits, profits and revenues. They're one exemplar. In the architectural space, HKS and HDR are doing great things. In engineering, Walter P. Moore is working to rethink their business.

In researching the book, I've uncovered firms rethinking their approach. Many architects are challenged to be profitable. Can we find ways to get them to take more risk and derive more long-term value? How can we embolden them to try when their own bread on the table is at stake? Are architects risk averse or just complacent?

Gilmore: The root of all evil has been aberrant institutional contractual documentation. They're not just risk averse, they're paranoid. What that mentality generates is an "all about me" consciousness. All that kind of contract does is cover butts and force people to think about themselves. That's why I'm a big advocate of Howard Ashcraft and his work. He's focused on integrated project delivery, so you have to take that into consideration. But there's truth in his idea: "Let's get away from fault-based contracts." Let's think about win/win for everybody. He's coming at it from a legal basis to drive behavior, because no one trusts anybody.

They've lived and been taught that for hundreds of years. If architects are genetically predisposed, or environmentally conditioned to be risk averse, maybe the collaboration rebirth will team talented designers with risk managers to maximize team outcomes and help those who can't help themselves – in this case architects – for risk and business.

Gilmore: We might do a Transformative Think Tank (T3) on this to convene on the condition of AEC contracts, behaviors they drive, and what could be different. Lawyers make money on contracts, and even more when people sue each other. They have to change their value proposition too. I want to dive in deeply to do something about it.

Let's look at those who lead versus manage. Great leaders were ready to fall on their swords because they believed so strongly in their mission. Mahatma Gandhi, General MacArthur, Martin Luther King, Jr. . . . What are the traits of great design leaders?

Gilmore: The first thing that comes to mind is the authenticity, or litmus, test: is anyone following? Leading means followers. The question is, what causes people to follow? We typically associate great leadership with inspiration and vision. In their absence, people get confused and make up their own stories. Cohesive vision draws people to follow, and, more significantly, to follow together. In successful organizations, leadership leads with vision and inspires people to draw together, so they're all following the same thing. Organizations whose leaders aren't cohesive, unified and inspiring draw both conflicting and complementary followers, but not *unified* followers. What ultimately happens is division, conflict, strife, and disunity. "Armed camps" form. "My vision is better than your vision, and yours is wrong." These judgments come from these conflicted visions. They ask, "How did we ever get to this?" It comes down to leaders having inspiration from a vision we can all get into together and lead.

Some exemplar firms have done this. And people can point to nuances as exceptions. Being human, we'll always vary from purity, but we look at organizations as a whole. Are they inspired? Are people following in the same direction? That's how the Bjarke Ingles Group, BIG, is. They have a unified vision; their partners hang out, laugh, argue, and have fun together. They work ridiculously hard together. Make no mistake about it, they are *together*. Their vision is exciting! You can see their outcomes in design. They've become a talent magnet – contagious in their exceptional design – a great example of an organization with unified vision and unity in design. Another example is Gensler, which has always had a unified vision around their work. It started with Art, who had a clear vision his leaders could get behind and march to. It was exciting to watch as they grew to one of the largest global firms with thousands of talented people.

Let's turn to your point: managing design is more about leadership than management. Designers, probably all of us, would rather be led than managed. To have the freedom to decide, choose tools, and work in ways that suit us. Sometimes lead designers are willing to sacrifice personal financial gain to uphold design principles – what behavior does that drive?

Gilmore: "Out-in-front" leaders lead with dynamism. Oh my gosh! We're caught up in charisma! Wonderful. If three or more leaders do it, it's magic. That type of leader leads alone. Humans love heroes. We create a personality cult around the out-front flag waver yelling, "Follow me over the cliff!" A better leadership model is shared. Effective leaders lead from the middle of the pack – servant leadership. I'm not sure how valid that is in design practice. An illustration is the pebble in the pond. It drops in the center and the water ripples in concentric rings. Great leaders do this with influence radiating from the center. A collaborative team holds them accountable. Balance is built in – when you feel bad, someone else feels good. When you're pessimistic, someone else is optimistic. When you act like a jerk, somebody calls you out. That's collaborative balance. If a team acts that way from the center, the ripples impact the next ring of leadership, management, subject matter experts, and staff positively – a collaborative splash. Teams like that are rare. More often, leaders are placed by expedience rather than strategy – roles are empty, and we need to fill them. As opposed to: "What do we as an organization need to accomplish? Will we go through the hard work of finding the right fit?" Rarely.

Is culture behind that? Because historically the heroic, lone genius has been rewarded.

Gilmore: A firm in San Antonio leads "from-the-middle" – a partnership of equals. Another point is to defer to one another instead of competing. Trust. Unless I trust you, I'll always second-guess you. I'll project the potential negative outcomes of your agenda against mine. When we trust in a collaborative leadership team, magic happens. That dynamic in design firm leadership is rare.

Design firm leaders haven't been willing to invest the time to develop that common, consistent leadership message. Within Holder, on the construction side, we invest heavily in it. It's a core tenet: we're one company with a unified message.

Gilmore: There's "Heads Down" versus "Heads Up" leadership. Heads Up people are fully aware and prepared to execute the inspiration – always cognizant, but not so caught up in it they can't get work done. They trust their leader and translate their message into execution. "Heads Down" is self-protection against leadership disillusionment. "I'm gonna survive, endure, just make it through." We all know those kinds of folk. There's no life around the work. It's self-protective or defensive – a posture of closure. Heads Up leaders have an open posture. Both can get work done, but do people follow the Heads Down leader? The Heads Up leader is effective with inspiration and vision and believes it. The Heads Down leader inoculates themselves against inspiring others.

"What ultimately happens is division, conflict, strife, and disunity. 'Armed camps' form. "My vision is better than our vision, and yours is wrong." These judgments come from these conflicted visions.

It comes down to leaders . . . inspiration . . . vision we can all get into together. . ."

As a lifelong designer, I relate to not wanting to be managed. One alternative is to transform yourself, or ally with someone, so management is integral to design. I've always respected the difference between doing things right (management) versus doing the right things (leadership). If designers are averse to being managed, leading is what we need to work on. In design, management seems externally applied by outside forces – a supervisor, rules, or self-set criteria. How do we get better at building it in?

Gilmore: Good questions. We need to continue the conversation.

David Gilmore Key Points:

1. Design thinking has never been the sole purview of architects. It informs software, products, and other businesses with good effect. And vice versa.
2. A massive exodus of design firm leadership will occur in the next ten years. Is the next generation of leaders prepared? Are investments in human capital and corporate knowledge being made?
3. Firms would be well served to step back from technology and do a strategic business process assessment for BIM. What do we really need?
4. Time spent in team socialization, relationships, and early project definition is a drastically underused technique despite proven, significant, measurable payback.
5. A disconnect exists between business and design realization. Smart teams bridge this divide.
6. The industry needs contracts focused on team versus individual gain and new risk/reward bases, particularly for architects.
7. Change starts locally, at home, within your own organization.
8. Design firm leadership needs shared vision, inspiration, and trust, at all levels.
9. "Heads Up managers" who share a vision are preferable to defensive "Heads Down managers."

CHAPTER 7

Process:
Lean Scheduling – Agile and Efficient

INTERVIEW

Bruce Cousins, AIA

Founder, Sword Inc., San Francisco, Denver, Santa Fe   14th September 2017

Bruce Cousins, AIA, founded Sword Inc., a lean, continuous improvement and design management firm offering consulting services, in 2014. Previously, he was Regional Continuous Improvement (BIM/Lean) Director for Turner Construction for Northern California Prior to this he was National VDC/BIM Manager with Weitz Construction, and owner of a full-service architectural firm for over 35 years. An early adopter and advocate of Building Information Modeling, Lean and systems thinking, he received his Bachelor's in Urban Studies from the University of Denver and Master Architecture degree from the University of California at Berkeley. He is a Registered Architect in Colorado. He shares Lean processes in design management.

Standardized Methods or "Leaning in, Softly"

The book's target audience includes owners, architects, contractors, and the idea of "team" in the broadest sense. Those who swoop in from other disciplines and think they're going to apply traditional management practices to architectural practice are usually misguided. You've been applying Lean principles – share what that means.

Cousins: I have a lot of ideas on how I'd like to see design managed both by owners, design services providers and builders in Design/Build projects. Much of it is contained in a paper I just wrote focused

"*Designers are often challenged with that approach. There's no standardized method. In fact, it's frowned upon. Architecture uses a more independent approach. They're leading their brand.*"

on the Lean Agile Kanban process[1]. In the past ten years I participated on the construction side to establish Integrated Project Delivery (IPD) Big Rooms and coached conventional project teams in the Last Planner System®, Target Value Design and Production Planning using Takt Planning methods.

With a dual view of design and construction, how did you start?

Cousins: I studied industrial design and construction in college with Richard Bender at UC Berkeley. His book *Crack in the Rearview Mirror* was an in depth look at industrialized construction and "systems building." Then I went into custom resort and residential design in Colorado. There, we always had a limited window to build due to the weather, so we worked closely with contractors. It was mostly Design/Bid/Build, but we created foundation packages, because we had to get foundations in the ground before November and the projects came to us in July. In the spring, we'd do a framing package after uncovering the foundations. This focused us on deliverables and schedules. After getting into construction, I saw how schedules managed the supply chain. How to keep continuous flow without stopping and starting, eliminate waste, and be efficient? As construction became more compressed, we had to focus on managing the front end of design. Designers are challenged by that. In general, there has been no standardized method to manage the design phase. In fact, it's frowned upon. Architects tend to create an individualized approach to manage early design phases. They're leading their brand. Design is iterative, "nonsequential." For the right reasons, not because we're disorganized, but because we're solving problems. Problems take you in different directions.

Good design is iterative. But good designers also have developed processes that have served them to both to create good design and meet the client's project objectives. But a standard process for managing design has really only been evolving over the past ten years as projects became more complex and schedules became more compressed.

Over the years, I've tried to figure out what design management means. My first presentation on it was at Build Boston in 2006. Design management, in my mind, became strongly linked to BIM. We adopted BIM in my office in 1993 and used it to recreate a 2D approach. We evolved to a 3D approach to manage design because BIM did the drawings for us. The software was better at realizing a complete coordinated set under compressed schedules. When we had to do foundation permit sets in a few months, it had to be efficient. That was the beginning.

Then Weitz hired me as national Lean manager. I had a services group. We got onboard with BIM early. From the construction view, it was more about model management than authorship. Our modus operandi was to not change the model, because it was the official construction documents. We built our own federated construction model. As I got into Lean, the work became about process improvements and the underlying conceptual structure of Lean. At Turner, we had 29 offices in the U.S. and 6 or 7 of us were creating the company's national standard for implementing Lean. We put together policy and procedure including BIM, IPD and Lean as an item on a scorecard. Let's call it "alternative design man-

---

[1] In Lean methodology, a Kanban is a card containing information required to be done on a product at each stage on its path to completion and parts needed at subsequent processes. Cards are used to control work-in-progress (W.I.P.), production, and inventory flow. A Kanban system allows a company to use Just-in-Time (J.I.T) Production and Ordering Systems that minimize inventories while satisfying customer demands. (Source: Beyond LEAN.com)

agement." It evolved into creating "great practices" not "best practices," because processes were evolving, and we wanted to get better. Regional offices could score Lean and BIM participation and evaluate cost and savings in time, or RFI reduction. Big Rooms[2] and contracts were tracked. On design/build projects, did you have shared savings or incentives? The Levi's Stadium did. The first iteration of a large search engine company campus was an IPD Joint venture with Turner and DPR. In each case, we had different financial arrangements and design challenges.

Even though we used IPD the design team still managed their work in the old way – in the design phase. The architect, NBBJ, would go to Seattle to meet with the company's principals and come back and tell the Big Room, "Here's what we decided." Not much input from the IPD team. So, even though we had a triparty agreement, design management was ineffective – it would have benefited from pre-thinking to help the team. Over the last few years I've been working with both owners and builders on improving design process outcomes with a product planning approach. Taking a page from product development thought leaders has guided me in developing my approach to design management that is a collaborative and integrated approach. (See *The Principles of Product Development Flow: Second Generation Lean Product Development*, Donald G. Reinertsen, ISBN-13:978-1-935401-00-1 ©2009.)

There is a building boom in data centers around the world. Many have adopted Lean construction approaches that are now more integrated during the design phase. Owners historically just signed a design/bid/build contract without much integration of designer and builder. They realized they weren't getting the desired quality, or costs escalated because they didn't have a handle on costs upfront. From a Lean standpoint, we have coached owners the benefits of this approach and so many are now implementing a collaborative approach that starts before a line is on paper.

Intel was focused on Target Value Design, but nobody knew what that meant. *That* was an interesting experience. We helped figure out that Target Value Design meant smaller batches of work, rapid prototyping with the contractor collaborating early in the design phase. Now, the focus is on "pull planning"[3] and establishing a commitment-and-relationship-based work environment between design service providers and builders. We coached the designer who was also producing shop drawings an Agile/Kanban technique and using a cloud-based Kanban software application in conjunction with the Last Planner System®. Kanban is a signal, a card, not rocket science. When you finish something, you pull a card. It started with Toyota's production system. Finish a task, pull a card, and it tells you materials needed for the next stage to keep inventory low. (See David J. Anderson Kanban, *Successful Evolutionary Change for Your Technology Business*, Blue Hole Press, 2010.)

Do you mean a visual control, like the copier tray that tells you to reload paper?

Cousins: Same idea. To manage production, you want metrics. Kanban relates to queuing theory: how many items go through a system in a period of time, like cars across a bridge. In design management, we have ideas, or knowledge. A design team can list them on a card and call them backlog. Say we need to do the underground power vault; we talk

> "Problems take you in different directions. The nature of good design, quality design, is that it's iterative . . . non-sequential."

---

[2] "Big Room" refers to a large team work area in which digital information (i.e. BIM, et al.) can be shared live, onscreen, collaboratively among all parties to enhance efficiency.

[3] Pull planning refers to a Lean work planning process in which participants develop schedules, by "pulling" and sharing information, for example: What do you need to start your work? I need Bill to have the ceilings framed. Then, how long do you need to finish? And so on, until a complete schedule sequence is built with real information and ownership.

to the power company and identify where it comes into the site. Those decisions can be written, on stickies or digitally. That becomes "Work in Progress" (WIP). The team takes the WIP, assigns it, and it goes through the system – through "gates." For example, an approval gate. It's the combination of the Kanban board and perhaps replacing Schematic Design with a BIM Level of Development (LOD) approach. The process adds clarity and removes ambiguity.

You create cards and move them through the system, so the team knows where each individual is and has a visual of where they are as a team. With so many players and complexity, the process is broken into clusters: creating swim lanes for site, enclosure, interiors, permits. This visual organization is an efficient, responsive way to manage design. Most architects do this anyway, but their process is less organized. This system lets one know how many people (resources), and easily adjusts to new under-standing and knowledge. when something needs to be done. It's a visual way to see what's been accom-plished and creates a look ahead of a team's productivity and an ability to balance the work load if necessary and project staffing assignments daily and weekly. It's more objective and efficient than most waterfall approaches that encourage people to fill in the estimated time to completion.

In my experience, the opportunity to do what you've laid out has been limited because those applying it aren't in tune with the realities of commercial design process. Other constraints are time, not knowing what you don't know, and owners not having someone like you. They throw everyone into the hop-per and say "Go." The train has left the station. The process, in all its messiness, is set. How can you implement Lean once you're underway? The clock's ticking. It's time consuming to change process mid-stream. If you're not working offline to establish a program like this, you've missed the chance, right?

Cousins: No. The process can start anywhere. I agree contractors think of design as wasting time. They say, "We can pick colors in 10 minutes, why can't you?" On the academic side, Lean processes are defined. But I've been implementing this process for 10 years. Yes, there are roadblocks. Some archi-tects are against it: "It won't work. It's gonna cost a million dollars." On the flip side, I've done it with open minded firms willing to try it and propel change. It *requires* change. We have new technology, shorter schedules. Some people don't want to change. In that case, we're not interested. It takes con-scious effort to change. There's a great YouTube video called "The Backwards Bicycle" about learning to ride a bike reengineered for the wheel to move opposite of the handlebars – backwards. This video highlights how your brain struggles to change when you have to learn in a new way. It's a "soft way" to bring continuous improvement to organizations.

The process becomes less threatening.

Cousins: There's a book, *The Art of Doing Twice the Work in Half the Time*, by Jeff Southerland – one of the original authors of Agile and Scrum software and processes. In the Lean Construction Institute, we're doing Target Value Design and Target Value Project Delivery. I'm working on a "model-centric" approach to design schedule management' with the premise we should use a BIM Execution Plan (BxP) – the "Project Playbook." It's about exchanges, standards, and who needs what from whom. We're using the BxP as the core of a management approach. Everybody's an asset on a daily basis.

The question we always got in the early BIM adoption days was "How do we start?" What are your recommendations for newcomers to these processes?

Cousins: Start where you are. It doesn't have to be the beginning. Don't let perfection be the enemy of goodness. Map your processes: how you're doing it now. Understand current process, strengths and weaknesses. Make the analysis transparent. Get people in a room. Then set goals for where you want to be. Use the programs available. It's about deciding you *want* to. Value Stream Mapping visualizes current workflow and identifies issues. One consistent challenge is planning to do too much. So, we limit batch sizes and subdivide processes into smaller groups with cross-functional teams, not silos. Then, a general group brings it together. Make rules explicit. Make workflow visual digitally or with stickies. Then check it: Plan, Do, Check, Adjust – the original W.E. Deming sequence. Start without a focused schedule, set milestones. See how things are going, resources used, how decisions are made. Create a constraint log and do daily huddles.

> *"Start where you are. It doesn't have to be the beginning. Don't let perfection be the enemy of goodness."*

Combining the Last Planner System® and Kanban is different than how most designers work now. But the complexity of projects has increased, schedules have become more compressed and technology (BIM and cloud-based) have both evolved to support a collaborative real time work environment that breaks down silos and improves design solutions that don't need to be reworked later. You can see what activities are being accomplished by whom. As a design project manager, you can plan manpower and adjust commitments and needs. That's a benefit. I've seen work planning make a huge difference for firms. The Last Planner System® methodology is very beneficial on projects as a production planning and Value Stream Mapping Tool to align goals and objectives and get everyone marching to the same drummer.

Through Kanban and Lean you've presented a very different perspective to potential design managers. Some will embrace it with open arms, some will think it's Beelzebub speaking.

Bruce Cousins Key Points:

1. To manage themselves and add value to their teams, designers should bolster one-off invention with the study of LEAN process mapping.
2. Knowing key inputs, exchanges, and planning for outcomes can offer efficiency.
3. LEAN Process change can start where you are.
4. LEAN processes may take planning time and discipline currently unfamiliar to AECO teams.

INTERVIEW
Chad Roberson, AIA, LEED AP BD+C

Principal, Clark Nexsen, Asheville, N.C.   6th February 2018

Chad Roberson AIA, LEED AP, BD+C is a principal and licensed contractor and leads Clark Nexsen's Asheville, North Carolina, office. His experience encompasses design and management of public and private projects, college and university buildings, performing arts facilities, and commercial development. Complex building programs, dense urban or campus contexts, and diverse user groups characterize his projects.

Chad led design and construction for the Health and Human Sciences Building on the Millennial Campus of Western Carolina University, a LEED gold-certified project that received a 2015 international Rethinking the Future Sustainability Award. He leads the design team for the Joint Medical Education Facility Cancer Center for Pardee Hospital, Wingate University's Pharmacy School, and Blue Ridge Community College's nursing program. He has a Bachelor of Environmental Design in Architecture from North Carolina State University and a Master of Architecture from Columbia University. His firm's use of Lean work planning shows potential.

Self-Determined Lean Work Planning and Career Paths or "Playing Cards"

As a degreed architect and licensed contractor, you have a dual perspective. What drove you to that?

> **"I definitely did not learn collaboration in school."**

Roberson: It's been an interesting path. I remember helping my dad. We always had something going on. My first job was as the cut man for a framing crew at 12 or 13. I probably shouldn't have been out there legally. In college, I had my own framing crew. I learned a lot. Not only technically, but in dealing with people. Talking to graders and carpenters changed my perspective and gave me respect for what they do. I've always had a strong affinity for design. I worked for professors doing high-end design. That led me here. I admit it's a little schizophrenic sometimes, but it works.

You're the poster child for contractors who want architects to spend time in the field. What's your philosophy on managing design? Yours, and the firm's?

Roberson: There's no clear answer, but maybe there's a story. We just do it. We have a very young office. Me and one other fellow are the old guys at 47. Our average age is 32. They're super talented, and we look for that in hiring, regardless of where they are in the country. We have a network to find them. When they arrive, they're amazed at our approach. In the past they may have been associated with a principal who gave them a sketch or parti and then stayed on them. Here they have more design input. It becomes their project and I act as critic, guide, and managing principal for budget, schedule, and technical soundness. I'm involved in programming and design. Less in design development, and less during CDs unless there's a problem. We try to instill an entrepreneurial attitude, so we can take them in a different direction, where they're responsible for it all. Some don't know how to respond to that. Some aren't scared by a white piece of paper, some are. In the last 10 years we've won God knows how many design awards and none of the projects look the same. Such creative solutions for each. I act as a guide, not a force from the top. They do things I could never do on my own.

Your project scale promotes that. Your firm works in a wide range of building types and across the AE spectrum – from vertical construction to civil and infrastructure work. Do project types affect how you manage design teams?

Roberson: $15–20 M is our sweet spot for project size, but the approach is the same regardless. Our arrangement here is different than the firm's overall. We have 17 architects here, but 280 people in Virginia. I want to stay involved. A technique we use to manage growth is that everyone has multiple projects going on. It's a juggling act, but it's made us profitable and nimble. We move between projects without momentum loss.

Are you still using design skills?

Roberson: Absolutely. I've been in design mode for the last 2 weeks pursuing a project. Developing the story, ideas, sketches, and doing a cash flow projection at the same time.

Your role seems as much coach or collaboration conductor as anything. Did your education prepare you with collaboration skills or are they self-learned? Are newer graduates better at it, or do we need a curricula shift – away from the lone designer model?

Roberson: I definitely did *not* learn it school. More like on the baseball field, football field, and on projects. The system is set up to create individual designers. It's changing though, to project-based learning and collaborative approaches.

As a mid-career professional, you fit the demographic of an important target audience: those with a vested interest in shaping the future of the profession and the experience and stature in their firms to do it. You mentioned your office – and the company – is going through a LEAN exercise to bring rigor to your design process. Some have embraced it, some less so. Can you share that story?

Roberson: We won an award. While in Savannah to receive it, I went to a class in Lean. It wasn't really Lean, it was just being smart and strategic. We became interested at the firm board level when we began looking for a new president/CEO and greater profitability. We investigated it and rolled it out on a few projects.

We have a 5-minute morning meeting where you put cards on the table and move them around. We break work down into 4-hour increments. You do a task, get it done, then move to the next task, and pick another card. It helps you manage your time because you have short-term deadlines. You don't draw 10 iterations because you know the deadline. It helps you make decisions and keeps you on task. When you're stuck, it's just one task – it won't eat the entire fee. It breaks the process down into bite size chunks. Especially if staff is younger and doesn't have a broader perspective yet. They can manage it. Anybody can make a card. Anybody can work on a card. You have to know you have the skill to do it, or consciously push your skills. We coach them. It's helped them grow personally, professionally, and helps us grow. You can determine your own destiny.

The other part is QC. When you finish a task, you get someone to check it. It doesn't have a be a superior – it can be an equal. It keeps the project moving forward. We don't miss big things or have major rework. It streamlines us and avoids "head down" overfocus and big busts. That allows time to do high-quality design. Rather than 15 percent of the effort for schematics, we can shift to 20 or 25 percent, with better solutions. It doesn't take us 35 percent of the fee to do CDs. We're bringing it along equally, with less rework.

How do you deal with resistors?

> "You pick a card . . . do a task, get it done, then move to the next task . . . It breaks the process down into bite-size chunks."

Roberson: We don't have any. Somebody will say, "Hey, just be quiet and go pick another card." It defuses it and removes the emotion, and it's not long term. They get so excited and motivated by being able to control their own destiny. In the past people got bored. Not now. Now, they're constantly challenged.

You're using self-determination, motivation, and a democratized, open work planning process to allow personal freedom, growth and teamwork – in a very flat, visible way.

Roberson: It's a tool, but it works.

Any advice for young professionals considering a career in design or construction?

> "They get so excited and so motivated by being able to control their own destiny."

Roberson: Public speaking and writing. I don't know that I do either one well, but we are constantly striving to improve. The last four years we've made it a point how to improve we speak and communicate with clients. We brought in a consultant. Among other things, she taught us choreography and how to calm ourselves. It wound up being a great team building exercise. We still have sayings from that that we share like: "having dragon tails." (They hold you down . . .) Now, someone just has to say "dragon tails" and the person stops their negative behavior instantly.

One more thing, sharing an office over the years has helped me learn how to be a better collaborator. People can't believe I choose to do it, but it's been so valuable to constantly bounce ideas off and confide in one another. We just moved in to new space and we continued it. We tore down the wall between us. We sit in a glass box in the middle of everybody. There's so much dialogue. Are we communicating this properly? Should we pursue this job? So many things. It's changed the team's perspectives – and my own – for how to collaborate. Very different.

Chad Roberson Key Points:

1. Breaking down design work into short-term manageable tasks – a form of Lean pull scheduling minimizes risk and manages design.
2. Allowing team members to self-select and self-determine tasks (and the resulting familiar, productivity versus less efficient personal and professional growth) yields high satisfaction and motivation. A support network ensures quality.
3. A visible, shared work planning process builds trust versus "assignment."
4. Long-term retention and outcomes are enhanced by higher than normal levels of designer dependency, motivation, and love for their projects.

CHAPTER 8

Collaborators:
Performative Design (Better Together)

INTERVIEW
Marc L'Italien, FAIA

Principal, Associate Vice President, HGA   28th July 2017

Marc L'Italien, FAIA, is Principal/Associate Vice President with HGA, San Francisco. His work focuses on museums, science centers, aquariums and zoos, and higher-education learning environments, civic buildings, and community engagement projects. His innovative approach to planning, design, and sustainability has had a lasting impact on many communities. As a key leader of HGA's nationally recognized Arts, Community, and Education (ACE) Practice Group, Marc provides strategic planning, project leadership, and design expertise to further strengthen the firm's growing West Coast presence. His firm's practice is built on a collaborative culture.

On Collaboration or "We're Better Together"

Let's talk about flaws in our industry's process: Projects are over budget; we have cultural differences; and owners don't seem to manage their teams anymore. What issues do you see around process and are you doing anything different to cope?

L'Italien: Process is a great place to start. It's at the epicenter of everything we do. It's a different world from when I went to school, and the expectations of people coming out of school are too. I recently left a firm of 80 people in one studio to join one that's ten times as big, with ten offices. So, we have a deeper bench and more resources and experts. Technology helps that. We meet every week. If it wasn't

> "Re: over budget: We've had success with contractors on board as our partner from the beginning . . . At the end of the day we all have to think like we're on the same team."

for technology, we couldn't have that smooth communication. The meeting technology – not even drawing technology and programs – enables that. And BIM has been a real game changer.

My work is more front-end, client relationships, setting up expectations and a process that works with the client, and determining early concepts. I still draw by hand a lot. But many of the graduates coming out of school don't have hand drawing skills. I've tried to work with them, but success has been mixed. I'm a big believer in bringing technologies together: Overlay a computer drawing, hand sketch, then sweeten it in Photoshop. We're starting to see more Revit models at early stages as staff becomes more facile – put in the parameters and let it evolve. It's interesting to see how staff combine software systems. That's going full tilt today.

Physical modeling is still in play, but not as much as it should be. We have a Digital Practice Group and do a digital profile for each project. I think a physical model is valid at every step. That's met with resistance: "Why? We have the BIM?" But I've yet to witness a client, after they see a physical model, not have a positive reaction, like: "This is fabulous! Or, Wow, I like it, but I had no idea it was going to do *this*." It's fascinating to see how they react to physical models. Digital only tells part of the story; it goes hand in hand with a physical model.

For all its promise, pure BIM doesn't seem to provide as much client comfort. The interfaces aren't quite there yet.

L'Italien: We're doing virtual reality work. We just won a research award for the visual impacts of modeling eye diseases. What's to be determined is how that affects the spaces designed around those conditions. Before we didn't have that user perspective, now we do. It's going to be groundbreaking.

What about the architect's value proposition? At Yale, Phil Bernstein is challenging students to invent a firm whose value proposition isn't based on fixed fee or hourly services. They have to propose another way to be compensated and derive value. Are you thinking about that?

L'Italien: I wish I had a succinct answer. It's a major problem. Architects don't have much skin in the game anymore. No more master builders. With the separation of architecture and contracting, things went in different directions. You have firms doing interesting work that's expensive to build and prone to more liability; it's a mess. It doesn't always align with client needs, budgets or schedules. That fringe is trying to create value, but it's risky.

How is your firm dealing with the perennial problem of designs being over budget? Do you have a strategy for changing that?

L'Italien: We've had success with contractors on board as partners from the beginning. Sometimes we have two parallel estimates because there's too much at risk. If you can keep a project vision in line, it's money in the bank later. Design-build now presents more opportunities for architects and contractors to come together. I wish it could be set up equally for architects to share in the risk and the profits. We've got to think like we're all on the same team. It benefits everybody. The adversarial relationship doesn't work. We struggle with it. We have a good reputation for keeping our projects in budget. It's difficult to balance everything. Having great partners makes a difference. When difficult people work only for their faction's benefit, it breaks down.

Maybe if we just keep trying, in 20, 30 years we'll be better collaborators. I'm teaching classes at both ends: design sensitivity training for contractors and what contractors need from architects.

L'Italien: Students need to be exposed to that. When I was in school, any class in management or construction wasn't interesting. We saw our time in school as the opportunity to learn the craft. Now, people are discovering if you don't come out with those chops, you're at a disadvantage.

"We're looking for those kinds of people."

I want to come back to something you said: we might have a great relationship with the architect and contractor, but the challenge is with the clients! They often want it all, but don't have the budget. They're not always good at deciphering conflicting needs within their own organization and giving proper direction; it often falls on the architect. Sometimes there's stubbornness: "You're doing something wrong! It's gotta be cheaper! We're gonna *beat* this out of you!" We're developing sophisticated buildings to last 50–70 years. They have to recognize how complicated this is and be better equipped to collaborate with us on streamlining decisions about what they're going to give up. If not, we're all at risk and not a good reflection on anybody.

It doesn't take long for architects and contractors to agree that owners are a common problem. We're trying to do something about it. We're using Owner Decision Tracking lists. So many people do nothing but complain. We're taking ownership and action.

L'Italien: Thank God people like you are doing this because they're good at it. They get us the kind of answers we need so we can do our job better. We live in a complicated age. Technology, the internet, more stuff to respond to. As we've rebranded and looked at our internal design process, we've realized the solo-hero-designer model is gone. We're looking for facile people, good listeners, leaders, and collaborators who can rally teams collectively. A lot of designers have that need – a comfort zone – to be able go off in a corner and "create greatness." They get offended when things don't go well, or design gets eroded. We want to empower a team collective. We see better results, satisfaction, and buy in. The more people you train to collaborate and work with different personalities – the more it helps. We're looking for *those* kinds of people.

It's a sign you've recognized it and are doing something about it. Get rid of the Howard Roark[1] model. We've earned this (bad) place we're in; we've got to do something different.

L'Italien: It's exciting. Its gonna take time to turn the ship, though. Leadership is embracing it. In the future, collaboration will define the firms that survive versus the ones that wither on the vine.

It brings to mind the books on practice in the '80s and '90s by Gutman, Cuff, and Kostof, on firm evolution. Big firms and boutique firms survive with the mid-sized ones focusing and the ones afraid to change going away. I'm heartened by what you're talking about.

---

[1]   Howard Roark, protagonist in Ayn Rand's novel *The Fountainhead*, the quintessential ego-driven architect driven to blow up his building rather than see its design integrity compromised by committee mediocrity – a longstanding architectural role model.

L'Italien: The collective is better than the individual. A lesser performance by a huge firm with expert resources can have more significant environmental impacts than one by a boutique firm.

Mediocrity can creep in on the big projects and big firms too . . . Maybe you command a bigger fee if you can deliver more, and find an owner who will listen? I don't know how we got into this commodity position. CM's are better negotiators. We need to learn.

"The collective is better than the individual."

L'Italien: Yes, clients want more service without paying for it. Because technology can streamline our processes, they assume it equates to less time and we can credit them these costs. Then architects undercut each other up competing because they want it so bad. I love talking about this stuff. It's great you're doing this.

Marc L'Italien Key Points:

1. The design over budget conundrum can be combatted by architect/contractor partnerships.
2. Good firms are looking for young talent who can break the "lone designer," ego-based model and work collaboratively.
3. The collective is better than the individual.

INTERVIEW
Bob Carnegie, AIA

Director of Architecture, HOK Houston   6th August 2017

Bob Carnegie, AIA, is Director of Architecture, HOK Houston. He has 30 years of experience as an architect, owner and developer, in commercial real estate and project management working for Trammell Crow Company, Dienna Nelson Augustine, and now HOK, a global design, architecture, engineering, and planning firm. He is principal and senior project manager with responsibilities in design management, architecture, engineering, and planning of commercial facilities. He talks about design and management process, sharing and trust.

Sharing Expertise, Trust, and Owner Engagement or "Assume This"

You began your career in the real estate industry, working with Trammell Crow.

Carnegie: Yes, most of my experience is as an owner, but still in project management. These are my answers, not necessarily HOK's.

Tell me about your design process.

Carnegie: I've been doing this long enough that I've seen and used traditional and nontraditional design delivery methods. Most of the time, we're hired by the client to study or design something they can price for feasibility. If the seed germinates, we're asked to proceed with design and construction.

> *"We now spend more time (and labor) in DD than CD."*

   For architecture, at HOK, the process is organized and managed. We have dedicated project managers, responsible for initiation (in some cases even pursuit and capture), work planning (with project architect input), negotiating agreements, managing the team (including morale), and project financial success. Project designers are responsible for design, and project architects are responsible for technical delivery. "Dedicated" mean's that's their project role. They may have multiple projects.

How have BIM/VDC and digital tools impacted your design process? What systems effects, or unintended consequences good or bad, are you seeing?

Carnegie: The level of detail demanded by clients and collaborators has grown with technological advances. Unfortunately, fees have lagged – architects do a lot of work for free. The other trap is the capability to quickly show options; we spend too much time looking at every possible solution. We should

"Project designers are responsible for design, and project architects are responsible for project delivery."

stick to what we've been asked to provide: 3–4 options in enough detail to be decision-ready. BIM and other technologies have changed the way we design and document. The impact is most evident with the level of detail documents can provide today. It's created a labor shift from CD's to DD. We now spend more time and labor in DD than CD.

Consequences include the expectation that some work traditionally done by the contractor and their subcontractors is now expected to be done by us and isn't accompanied by a transfer of fee. Today, tremendous additional detail is included in a set of CDs, including some not required to construct the work. While it's an amazing tool, Revit can lead architects to spend more time in production.

Are concepts still generated via sketching, or direct to Revit models or other software?

"The best experiences are on projects with people who really know what they're doing and are willing to share experiences and best practices. This creates a positive environment and enables broad sharing team members can use for the rest of their careers."

Carnegie: "Old school" was hand sketching, which evolved into hard-lined drafting. Today, we start with a hardline drafted (Revit or Auto-CAD) plan or massing diagram. Then comes sketching, by hand, or with software (depending on who's doing it). There are a lot of new tools – software and hardware – that give us options. It comes down to what the designer's comfortable with, so the medium doesn't get in the way of the art.

This doesn't necessarily serve the transition into production well. If the design concept is created and finalized by hand, or in specialty software, it's got to be regenerated into production format; 95 percent of the time that's Revit. If it started and evolved in Revit, that would be better for production, but can limit the concept's potential. It depends on the users. Today we design in whatever medium allows the most flexibility and creativity, then transfer it to a production format.

If we had the best of all possible worlds, designers and technical architects agree on a medium, design is born, and progresses to completion in a single format. That's not happening in our firm yet; the best design tools differ from the technical delivery tools.

How has your process changed over the last 10 or 20 years?

Carnegie: BIM is the major change in how we deliver design. Before that, CADD. The big question is: *what's next?* We're nearly to the end of the cycle and should see the "next big thing" on the horizon. Virtual reality (VR) is big, but focused on design and visualization, not technical delivery.

Are you still doing physical models?

Carnegie: They continue to be a great way to visualize design. Model methods have changed and continue to evolve. We use several, depending on the assignment, and what we're trying to communicate. Modeling's not just a good way to visualize a design. In the next decade, we'll have component design in our models used by manufacturers to fabricate building components. What was previously thought of as "custom" will become the norm. That will put more demands on A/E's – more detail required in CD's, more time and cost.

Any other innovative ways of parallel processing or automating your processes to cope with schedule and fee demands?

Carnegie: Small innovations are happening. Sometimes it's simply discovering something the software already does. On some projects, a partnership is formed between the architect, engineers, and GC around BIM and the detail the design team includes for the construction team's benefit. This collaboration doesn't always form, but when it does it can be effective.

As you contemplate your design process "reinvention," or "evolution," what are you focusing on to drive out waste, add value, work faster, and keep people happier?

Carnegie: One I've struggled with: the new generation of architects has a different paradigm of architecture. Everything processes faster and is more complex due to technology, code requirements, sustainability, and other priorities. As technology evolves, not just for design tools, but for construction means, methods, and materials, architects will be challenged to come up with different ways to document. Keeping people happy is tough. The youngest generation of architects seems to have different priorities than older generations. They're not as willing to work long hours; their work space is important, and amenities are as important as income.

What's the 5 or 10-year future vision for your firm to ensure survival? How might you be practicing differently than today?

Carnegie: Diversification. Reduce work in the corporate sector and increase other sectors like commercial, healthcare, science and technology, aviation and transportation, and government. If we can't open our markets, the next cycle may be difficult.

What trends or behaviors, specifically from owner and contractor partners, are giving you pause?

Carnegie: We've recently had some contractors target design as an excuse for poor management or bad estimating. They blame imperfect design for their delays or cost overruns. It's more than the distraction of frivolous RFIs and can be dangerous and disheartening.

What changes would improve your ability to practice and collaborate? (i.e. "We hate it when . . .")

Carnegie: My biggest "stop-doing" relates to trust, or lack thereof. Projects are smoother and more successful when team trust is built. Without trust the team struggles and the project suffers. Trust has to be earned, so the best projects are usually with repeat clients and contractors. Second is the attitude that owner, architect, and contractor relationships are adversarial. We see this with an owner who thinks of the project as purely a commercial venture – cut expenses, maximize profits. I've worked with owners who believe the project is successful when the architect and contractor barely break even or *lose* money. They feel like they did their job controlling expenses. The reality is, if they allowed the architect and contractor to focus on the project rather than control their costs, projects would be more successful. It's about the project, not individual success. I wish everyone got that.

It's not just a transaction. Are there are emerging best practice behaviors from owners and contractors that should be continued? Or start doing (i.e. "We love it when . . .")

Carnegie: The best project experiences are with people who know what they're doing and are willing to share experiences and best practices. This creates a positive environment and broad sharing team members can use for the rest of their careers. Personal commitments are formed between team members – some last well beyond the project.

That's how I got half the tools I use!

Knowing architects are challenged to be profitable, would you entertain alternative incentives or compensation models if added value resulted?

Carnegie: I'd absolutely be open to alternative compensation if it makes sense and is achievable. With my development background, we've looked at many compensation models. As an architect, there are few choices: percentage of hard cost or cost per square foot, which are translated to hourly (with a cap, always with a cap) or lump sum fees.

How do you develop design schedules? Who does it? What do you use? And does it work?

Carnegie: Project managers create work plans. They work with the project architect to develop a team and plan labor for each phase. This is communicated to the team, with the deliverables, so everyone knows their objective and how much time they have to reach it. I use a quick Excel spreadsheet to rough out the plan and align the proposed fee and labor model. Formal work plans are entered into Deltek and blended with accounting, but there are other platforms. It works inconsistently. Plans need to be updated constantly, either due to a schedule shift, change order, or over or under working the plan. The good news is even a loose plan helps with staffing and lets us see where help is needed and where folks have availability. It helps fee forecasting.

What are you doing to combat the perennial overbudget condition that plagues projects, causes value analysis and redesign, and demoralizes and bankrupts teams? How can architects change the rampant owner/contractor perception that they spend other-people's money, not manage it?

> *"Many times, the architect (yes, on my teams too) just assumes their design is what the project needs."*

Carnegie: We remind owners that design is a process and they're a participant. I remind my teams there should always be options to reduce cost. While we have the habit of pushing for the best possible solution, often the most expensive one, we're doing it for project benefit, with owner input and buy-in throughout. To change, we need to be more budget sensitive and communicate cost issues to clients. If we conceive a way to improve the design, we need to approach our client early to warn of the potential budget impact, and give them the chance to say no. Many times, the architect (yes, on my teams too) just assumes their design is what the project needs. While they're not ignoring cost, they don't highlight it, and probably don't do enough to find less expensive worthy alternatives.

Are CM, Design-Build, and IPD project delivery having an impact on your practice? Do you behave differently under these collaborative modes versus a Design/Bid/Build delivery? Why?

Carnegie: We do behave differently – there's usually an incentive to collaborate as a single team!

That tells us something.

Bob Carnegie Key Points:

1. Due to practice evolution, BIM, front-loaded data, and decisions, teams are spending more time in design development than in CDs.
2. Memorable project experiences come from knowledgeable team members willing to share expertise.
3. Owners focused on long-term value versus short-term fee and cost cutting can benefit.
4. Owner engagement and architect's communication efforts in budget control are key.
5. Architects must discontinue the practice of assuming they know what owners need.

INTERVIEW
Matthew Dumich, FAIA

Senior Project Manager, Adrian Smith + Gordon Gill Architecture   1st February 2018

Matthew Dumich, FAIA, is a Senior Project Manager with Adrian Smith + Gordon Gill Architecture, with offices in Chicago and Beijing. He is deeply devoted to getting emerging professionals engaged with each other, with the AIA (having created innovative ways to reach them), and with translating these connections into project work and practice. His work on large international projects uses mobile and virtual technology to connect voices from different practice levels and parts of the world and inform building design. He was behind bringing Black Spectacles online learning modules for ARE training to the national level. In his role at AS+GG he is responsible for oversight of project teams, management, consultants, and clients on large global projects.

Performative Design or "Welcome to the Machine Age"

Let's start with process. Your firm is a leader in technology use to inform complex, high-rise, high-performing buildings. Your intention is buildings that respond to environmental and urban issues. That's a more integrated approach than we were taught in school decades ago, using sketches, physical models, and only our brains and intuition. How do technology and other forces shape your design process? How do you manage the many experts you must need to execute such integrated work? Is it different from practice decades ago? Do "management" activities weigh in? Testing budgets, manhour limits, deadlines? That's several questions, but let's start there.

Dumich: There's a misconception that we do only high-rise work. I've done very few tall buildings with AS+GG. We do highly complex, large-scale master planning and architecture projects all over the world, including lots of arts and cultural projects. Recently we've had the privilege to shape two different world expos or world's fairs – Expo 2017 in Astana, Kazakhstan, and the Dubai Expo 2020.

   We couldn't do what we do without leveraging technology – in every way. We have consultants in 9 time zones on four different continents on a current project. We are on videoconferences, WhatsApp, texting back and forth, Skype, or FaceTime, communicating in multiple ways every day. Our projects are often fast track with advanced package delivery to start construction during the design process. We need a detailed work plan and robust design software to coordinate complex, high performance buildings. Early in the design process, we're working in all formats – physical and digital. On a recent project, the first "sketch" was a physical "model": two balloons and some string, tacked to a site plan. That became the building. It was a tent with structural elements, represented by this initial sketch model.

> "We couldn't do what we do without leveraging technology – in every way."

The balloon represented a membrane structure?

Dumich: Yes. The idea was that it was responsive to the environment, with openings in the membrane that could respond to temperature but shade the space inside. This was the first hour of the project, after reading the brief and doing some precedent research. The balloon idea came from the project being in a harsh desert climate. It was a tent. The tent metaphor was also a practical decision that the roof canopy could be built first to shade the construction workers completing the rest of the building below.

Constructability, environmental context, worker comfort, and safety integrated into the design concept from day one – hour one, even. Not something you encounter every day.

Dumich: That's one example of a kernel of an idea. We work in hand sketches. I'm sitting in a field of foam models right now. We do a lot of crude massing studies in master planning, stacking blocks to find the first group of big ideas. We're sketching in the computer, usually Rhino with plugins, for computational design to develop algorithms to manipulate forms and surfaces. You come up with this idea, then apply rules to the surface to rationalize the shape into panels or modules. They give an underlying rationale to an otherwise complicated thing. You hear about architects using the computer to enable wild form making – but the computer can rationalize complex forms into a constructible building. This might be optimizing mullion spacing or panelization of a sculpturally complex surface into repetitive panel modules. At Expo 2017, we created a wild, undulating interior wall with cast glass hexagonal panels but made it efficient with limited panel types.

The two-way nature of that never struck me before. The computer allows generation of complex form, but also allows translation back into rationalized, buildable pieces.

Dumich: How are we working differently from 20 years ago? Although we typically have a contractual requirement to produce 2D drawing sheets, we don't necessarily need to do thousands of drawings. We just exchange digital files with the contractors. Some things can't be communicated in 2D in a useful way. You can take a million section slices through a complex thing, but that doesn't tell the whole story. Sophisticated partners prefer the digital model files. We start modeling in Rhino and have scripts and workflows to bring Rhino models into Revit. Our projects are all documented in Revit. We feel strongly about using the best technology. Most people in the office, particularly the younger ones, don't know Autocad, or life at a hand-drafting table.

Throughout the design process we make iterations of physical models. We have a fantastic model shop. I'd put it up against almost anybody's. For everything from building massing to detailed full scale mock-ups, we make tons of physical models at all scales. Sometimes it's a ¼ scale model of an exterior wall, or a detail. Some are presentation models for clients and some are study models to inform our design process. We also have a 4-axis router that cuts metal, wood, or plastic into complex forms, 3D printers, and a vacuum form to mold plastic and silkscreen on. We can create wild shapes and make professional models in-house. Looking around the office now . . . we have tower models, façade details, a custom doorknob, and furniture. We study projects at all scales. Digital doesn't tell the whole story. We want to touch, feel, kick, and experiment with it.

Our design is rooted in context. We often work in emerging cities that don't always have what you'd traditionally think of as context – neighboring buildings. So, our context is cultural and environmental. Solar positioning and prevailing winds are drivers that shapes the design. All our projects are designed to be performative in passive and active ways. Extending exterior surfaces to maximize areas exposed to solar radiation for photovoltaics, canting walls to reduce heat gain and glare. We shape buildings to optimize wind impacts, with ribbed elements to disrupt wind, in some cases, scoop and collect it, integrating wind turbines to generate energy. The goal is to have the buildings be power plants. We aim for net zero or positive energy. That's an exciting thing about our work: we have ambitious design and performance goals, and forward-thinking clients who support those goals. That's our brand, if we have one: advancing performative design.

It's good to see you using such a vast toolset – for communication, design, documentation, analysis, and representation. All methods, hand and automated, are still employed. You won a national AIA Technology in Architectural Practice award recently, right? It enabled what you're talking about.

Dumich: Absolutely.

Let's overlay management attitudes. Design is iterative. To orchestrate such complex, integrated, global designs needs minute-by-minute, networked schedules showing inputs, exchanges, decisions, and outputs. In an integrated process like yours I could see design as all-encompassing and separate "swooping in" to beat schedules and budgets into shape. How is it really done?

"The irony is we have all these complex design tools, but we don't really have great management tools, say for deliverables tracking or manhours."

Dumich: Yes, we do all that. The irony is we have all these complex design tools, but don't really have great management tools. A little about the firm structure. We have 3 partners and 10 directors, structured in the classic three parts: design, technical, and management. I'm one of three project managers that cover all our projects, all over the world. Each project has a design, technical, and management lead. My style is to put an organizational framework on a project and have a regular rhythm of meetings. But I stay flexible. A great Chicago architect, Dan Wheeler, told me once I was overplanning an AIA initiative. He said, "Don't overthink it. Just make it, break it and fix it." So, you develop a plan, implement, and reassess as you move forward. We use MS Project for scheduling and other Excel management tools we created. We set up project schedule and staffing plans, then monitor them. We do weekly internal and consultant team meetings, monthly in person workshops with clients with weekly videoconferences. International clients often want us to be local, onsite, but we insist that all design work is done in Chicago. As a result, we need to be available for extended work hours. I have a matrix that shows each time zone and workhours. There's no time when they all overlap for all consultants and the client. As a team, we're working a 36-hour workday. We also travel a lot. It's exhausting.

Your description is thorough. Being at the top of your game, you must command market level or above fees to support the tools you need. You've changed the game for the architect's value proposition using automation to add client value - and make more money for your firm. True?

Dumich: AS+GG has an amazing team of leaders with a proven track record and reputation for executing iconic, world class projects. Our fees reflect the value of our ideas and high level of service. These projects are difficult to benchmark due to scale, complexity and accelerated delivery. We tailor our services to the project and work with clients to establish an appropriate fee range. Success relies on a strong team. We have excellent people and recruit the best, all over the world. Diversity makes design dialogue richer and helps us interpret different cultures.

Having just wrapped up two first-of-kind mega projects, I'm sensitive to meeting budgets. Both had designs and budgets set day one. How do you deal with budget control on integrated, first-of-kind, global projects, with little precedent and predictability?

"We take budget very seriously. If you're over budget you don't have a job. You have to fix it."

Dumich: It's a daily conversation. We take budget very seriously. If you're over budget you don't have a job. You have to address it. Budgets are set in a variety of ways, with and without our input. We are often required to have a cost consultant, not just for milestone estimates, but almost real time cost tracking. We submitted a 100-page cost estimate in a recent competition submittal that was later adopted as the project budget. The cost consultant does market testing to validate systems. On complex assemblies we rely on mock-ups and specialist contractor involvement early in the design process. For Al Wasl Plaza in Dubai, we designed a 200-foot-high domed canopy structure over a civic plaza that is 400' diameter at the base that we call the "Trellis." It's not just a shade, it's an immersive, 360-degree projection surface, like an outdoor IMAX theater. How do you estimate that – the world's largest domed, IMAX trellis structure? The client brought on a design-assist specialist contractor to work with us as the design developed. Should steel members be cast, welded, bolted? Visual mock-ups? Constructability? We completed our documents for competitive bidding and we held our breath hoping the bids would align with the cost planning and market testing. Through this process of due diligence, the bids were on budget. When you look at value analysis on performative buildings with integrated design strategies, you can't just cross off line-items. This can protect design, for better or worse, so you better be in budget.

At times in my career this issue drove us to a "nonintegrated" design approach where we could trade out systems to meet budgets. On your projects, you can't just change the skin to a standard storefront, or EIFS. You have to pay more attention. Your process sounds thoughtful, integrated. What's the catch? Any dirty laundry you can share? It sounds like such a well-oiled machine.

Dumich: It's not. It's high pressure. Our projects represent national or civic pride. To execute iconic architecture is a huge responsibility. You better have the right team and a client willing to take risks for things never done before. What can go wrong? Lots. In the case of this Trellis dome, our process was right. But, because it's so custom, we're at the whim of the bidders. We're doing fast-track delivery – designing while building. That speed inherently brings potential for mistakes. Everyone needs to understand the process and risks. Coordination will take time and drawings will need to be updated and reissued. Great projects are never easy. We always have to fight to maintain design and performance quality through many challenges for the benefit of the project.

Few of us want to be managed. How often do you set up a planning phase?

Dumich: I'm big on culture. We work hard. The office environment needs to be mutually beneficial – where people learn and grow. Bigger projects *never* have enough time. We *absolutely* should be doing some team bonding with the client and consultants to get to know each other, but we *never* have time for it. Worse, the client interface often starts with contract negotiation. It's the worst way to start a relationship. In some ways it would better to have lawyers handle this, but we know the project and what's important to us as architects, so we lead the negotiations ourselves. We miss that early "get to know you" part. We try to build rapport, to get outside the board room to build trust so we can go to war together. But the speed doesn't afford time to do it right.

Well, it took me some digging, but I finally found a chink in your armor. One place for improvement. At least you're aware and working on it, but it may be owner driven. We suffer from the same thing, but that's why people come to us. We get it done. I just wish we could design a happier, smarter process – to get home to see our kids more. Have you always been on the management side?

"We absolutely should be doing some team bonding with the client and consultants to get to know each other, but we never have time for it."

Dumich: I've grown into a management role. I was trained as a generalist. As a young architect, I had a hotel project where I coordinated the complex structure and picked the drapes. I saw all sides of the project from the inside-out. As I gained more experience I gravitated to larger, complex projects that required dedicated management. I enjoy the personal interaction, I'm client-facing. I'm an orchestrator and communicator, organizing and leading multidisciplinary teams. While I'm not drawing or modeling anymore, I'm still part of the design dialogue. Because of the breadth of my experience, I am able to take calls from the client, defend the design and solve problems on the spot. So, I've played many roles, but gravitated to team leader and team therapist.

We share that migration. It probably comes as no surprise to you, we need people like you. Connectors and people skilled at being good listeners.

"I believe design leadership is more about being the master collaborator or ringleader than being the lone genius."

Dumich: As architects we are well trained in graphic communication, but verbal and written communication is fundamental to success. We believe in honest communication and dealing with issues head on. That goes a long way with clients and contractors. Leave the ego aside. Demonstrate a commitment to mediate problems, not fight, but to roll up your sleeves, work together, and solve problems for the benefit of the project.

That comes up so often – our cultural heritage: ego-driven, lone heroes. As a member of the design profession's future leadership demographic, I'm glad to hear you say that.

Dumich: I believe design leadership is about being the master collaborator or ringleader rather than a lone genius, with a cape. We are facilitators and connectors solving complex problems through design. The other thing I wanted to touch on is the AIA. It's important to foster healthy and inclusive professional culture. I've always tried to do that. I've been lucky to have had great mentors that have guided my career. I've always tried to be a mentor, seek leadership opportunities that advance the profession, and create tools for others to succeed. I co-founded the AIA Chicago Bridge mentoring program that has been replicated as a national model. I also helped launch a partnership with AIA and the Black

Spectacles online ARE training as a resource for all architecture school graduates. I've been fortunate in my career and I want to equip the next generation to lead. I have a passion for sharing my experiences and supporting the pipeline or talent and leadership to ensure our profession continues to have a strong voice in the future.

Matthew Dumich Key Points:

1. Technology is essential for high-performing firms doing high-performing buildings – for communication, design, documentation, analysis, and representation.
2. Management thinking, with technology and design, is an integral part of the three-legged stool, not something to be ignored or layered on externally.
3. Designing to budget is a requirement of responsible firms and takes diligence. Hope, luck, and risk remain unavoidable due to market forces and custom designs.
4. Great design requires architects as master collaborators not lone artist geniuses, and like-minded clients and teams willing to share risk.
5. Owners should set aside time and invest in project definition and team building at project starts to prepare teams to "go to war" together to deliver a successful project.

CHAPTER 9

Design and Budgets: Architect/Contractor Collaboration and Trust

INTERVIEW
Jeffrey Paine, FAIA

Founding Principal, Duda|Paine Architects   16th January 2018

Jeffrey Paine sees architectural process as an opportunity to expand the power and function of design. He is a facile leader and thinker, with a "long view" of the design's impact on communities and economic growth. As cofounder of Duda|Paine Architects with Turan Duda, he places design thinking at the forefront of the architectural process. He leads technical execution of every project, focusing on contributions of clients, design team members, and other experts. As advocate for smarter clients and teams, he advances communication and information sharing systems in the practice. His inclusion of diverse perspectives fuels iconic, transformational buildings for public institutions and private entities throughout the US and Mexico. He is a frequent speaker at graduate programs in architecture, business, and professional forums for the Urban Land Institute, Duke University, NC State, NC Chapel Hill, and others. In 2016, he completed research on the history, implications, and evolution of public spaces as visiting scholar at the American Academy in Rome. He was elevated to AIA Fellowship in 2019.

"They believed
in a somewhat
adversarial
contractor-architect
working relation-
ship. They weren't
unusual in that
regard."

Pressing Schedules or "Are We Done Yet?"

Please share your background including career influences, and their impact on your thinking about an approach to practice. Do you see management as integral to design or as an overlay – a necessary evil?

Paine: Management of design varies from firm to firm. Early in my career I worked for Kevin Roche John Dinkeloo and Associates. At the time, they were very departmentalized, with a similarly departmental management style, and an ad hoc approach to design, documentation and getting things built. In the field, they believed in a somewhat adversarial contractor-architect relationship, which wasn't unusual then – most architects saw their role as protecting the client from the contractor and change orders. Design-build and "construction management at risk" didn't exist. An architect produced 100 percent construction documents, put them out to bid, helped the owner select the winning bid, and argued all the way through the project on behalf of the client.

In 1982, I started working for Cesar Pelli and, at 28, was "the gray beard." Back then, Pelli's firm mostly did design only, working with an architect of record, which produced the CDs. I became liaison between two large firms in New York and Toronto on the World Financial Center in New York City to ensure the documents and construction met our design intent. I'd redline their drawings over and over. It started out adversarial, but we evolved into a collaborative relationship. That spirit transferred into how we worked with contractors and subcontractors. This was long before any mention of "construction manager at risk" – that came later in the mid- to-late 1990s, but was something Pelli's firm, and then Turan and my firm, were suited to – engaging with a construction manager as a member of the design team. Some CMs were good at preconstruction services, cost estimating, constructability, and detailing assistance. Some aren't. But we've always had this mentality of seeing everyone we work with – engineers, consultants, contractors, and subcontractors – as collaborators.

In terms of design management, Kevin Roche and Cesar Pelli are very different. Kevin's approach was mysterious in that he would not involve the client much in the design process, then he would overwhelm them with presentation models and drawings. The firm had a warehouse devoted to building models large enough to poke your head inside to get a sense of its spaces. Kevin's philosophy was you hired an expert, so you should listen to him.

Cesar Pelli was different. He taught us to bring three or four potential design solutions to a meeting, from early concepts through schematics. Clients were involved and engaged. Even when talking about materials, we showed options. You'd think that Kevin Roche's design management process would be more efficient, but Pelli's was, because the client bought in quickly and on schedule. Along the way we built a belief we were creating the right solution. The design process wasn't haphazard, but about engaging clients, so they felt involved and believed in the solution.

One of the most fulfilling moments for Turan and me is on the site, when our building is completed, standing next to someone unfamiliar with the design, and he or she explains why things look the way they do or why certain things were done. We're teaching clients the value of design and a well-managed architectural process. We're demystifying design and engaging them. By doing so, we're making them believers in what we do, why we do it, what the best design solution is, and why it was done.

Your use "oxymoron" to describe the conflict between design and management. We've never been interested in the idea of a master builder or designer. We don't engage clients artificially. We like being challenged with budget, context, usability, and functionality issues. It makes for a better building, and we don't have problems after the fact because our client has seen and been a part of the process. Engaging everyone is more interesting, more collaborative, and provokes us to produce better work.

Working for commercial and higher-education clients, do you tailor your design approach?

Paine: They're different. Not to generalize, but in commercial projects, especially build-to-suit, you're often designing for an organization's key leaders. That might be one person or a small leadership group. Academic clients typically have more layers of approval.

> *"We're pulling back the curtain."*

Building consensus in an academic environment involves hard work and buy-in. When you're designing a project such as a student union, the process can be complicated. Forty or fifty different users might be involved. Leading a consensus building process takes skill and patience. It requires you to be more than a good designer – you often need to build consensus among a conflicted group. It's not speculative; you are designing for the people using the building, so you become engaged with their goals and aspirations, the work they do and how they intend to use the building. For example, for the Duke University School of Medicine, we learned how Duke's Medical Center aspires to change the way future medical education is conducted.

How has your design process evolved with virtual and digital tools? Are there unintended consequences? Are concepts still generated by sketching and physical modeling?

> *"The client thinks it's what we're going to build. And we're scratching our heads saying, 'Is it really what we want? Are we really there?'"*

Paine: We use the latest 3D modeling technology, which allows us to work simultaneously with our teammates using the same digital model. However, with digital tools it's sometimes easy to jump over a step or two, and short circuit the design process. That's our biggest concern. Because there's a market expectation now, even when we're competing for a project, that we'll produce 3D photorealistic renderings of the exterior and interior spaces. This pushes the project forward before we have a sense of what is needed or wanted. It often doesn't give us enough time to consider whether what we are presenting is this the best alternative.

The process between your brain and the computer screen is different than between your brain and a sketch pad. With a computer, you pick a point on a screen, pick another point and connect them. When you're drawing on paper, you're not sure where the line's going. It's a different way of thinking and talking. We encourage people to communicate design ideas in sketch form. So, it's not just a matter of the tool a person uses, it's a matter of thinking and having the capacity to generate ideas and talk about them.

The visual sophistication of digital technology can also make a rendering look too finished, and clients can believe we're ready to start construction documents and build the project. In an iterative design, digital tools add efficiency, but you have to make sure they don't short circuit.

One interviewee said, "You can always keep designing." You have to set limits, or it goes on and on. While a change can always be made, it isn't always necessary. With your early client involvement, as you get closer to a finished project, are changes more manageable?

"We're leading a consensus building process that takes work, talent and patience. It requires you to be more than simply a good architect, you need to be a good communicator who can build consensus among a sometimes-conflicted group."

Paine: Trust is an enormous part of any project, especially a large, complicated, expensive one. Clients want to trust they're hiring a firm that will spend their budget dollars wisely while creating a great building. Because we engage clients from the beginning, and they are invoiced in the process if changes are made – whatever the reason – they generally understand. We don't abuse the trust they put in us, but part of managing design is rethinking design solutions when situations require.

Do you develop design schedules? Who does it? What tools do you use? Since design can be a messy, uncontrollable process, how do you reduce the unpredictability?

Paine: The first thing is to constructively argue with the client and CM to allow the time necessary to design and document the project. They're chomping at the bit to put a price to our drawings, and we're usually pushing for a little more time. B.I.M. forces more work to be done in schematic design, and expectation for a design resolution in schematics is greater. We used to give a 50 percent CD package to a contractor for pricing and create a guaranteed maximum price, or GMP. Now we're seeing owners asking the CM for a GMP at 50 percent DD, or earlier. Budgets are becoming hard numbers before the design is adequately resolved.

"They're chomping at the bit to put a price to our drawings. And we're always asking for a little more time."

The pressure to produce design work and release documents earlier is a risk for the entire industry, not just for architects. Buildings are being commoditized – design work is banged out and documented to make it cheaper and faster. We have to start asking, "To what end?" What does this push mean to future of the built environment?

How can architects change the owner/contractor perception that architects are adept at spending money and not so good at managing it?

Paine: In the 1970s, architecture stepped away from confirming and documenting construction costs. The result? Contractors became construction managers and got involved earlier. We welcome that, but it comes with baggage. We've become reliant on the CM to provide accurate information early in design, before being able to give subcontractors fully developed drawings to estimate materials. Thus, the CM has to forecast construction costs and fill in blanks to come up with estimates before documents are done. Some CMs are much more adept than others at this, and we sometimes have to change the design to stay on budget. We need CMs to be responsible team members and advise us accurately. This is one reason we like to work with a competent CM like Holder – having experience working together is a plus.

The desire for accurate estimates early reflects an expectation to produce more information in less time. It's a push-pull, but we need time to work back and forth the CM, and we need client involvement. The CM can help us teach clients about the entire project process and warn them when decisions are negatively impacting the budget or schedule. It's not us trying to spend the client's money and the conractor trying to save it – staying on budget is a shared responsibility.

Do you work with a CM on a repeat basis or do you start relationship building fresh?

Paine: We're often asked to help the client select a CM. We'll create a set of drawings illustrating the basic project parameters and schedule, and then interview three or four CMs with our client to hear their initial thoughts and how they'd approach cost estimating and construction. We meet their team to see how we'd work together. We check references. It helps to be part of that process, because we're picking teammates with whom we and the client can work well. Again, it's easier when we've worked together before.

If you haven't worked with a CM before, is your role to drive the relationship-building process?

"It's not a matter of us trying to spend the client's money and the contractor trying to save the client's money – maintaining the budget must be a shared responsibility."

Paine: Relationship building is part of our work. We have an excellent reputation as people who listen, who don't dictate or try to control the conversation. We're willing to learn something from a contractor or subcontractor. We want to be locked arm-in-arm with the CM, strategizing and looking out for the owner's interests together from the beginning. Then, risks become more apparent and we both protect the owner.

Do you use any BIM costing technology?

Paine: The scale, technology and complexity of our projects typically prevent this. And we rely on our CMs for cost estimating. However, I do believe in "tandem cost estimating." We encourage clients to let us add a qualified cost estimator to the team. Not to question or push back on the CM's pricing, but to create a system of checks and balances, with two sets of eyes looking at the details and costing them.

What's bugging you about the practice of architecture? What can your owner-contractor partners do to make things better? You've mentioned time and spoken about contractors getting engaged and looking out for the team as opposed to their own interests.

Paine: We're finishing a corporate campus with a CM at risk. Although we had some frustrations throughout the process related to budget and schedule, the project ended both on time and on budget. However, a few things were taken out that didn't need to go. We can't put those back when we find out later we can afford them. It's frustrating when design features are lost in this process.

Having a strong working relationship where we can ask each other tough questions during design and documentation is welcome. It comes down to the people involved. I remind clients, when talking about any architect, consultant, or contractor, a good design process is not just whose name is on the door, but who's doing the work? Because we've had excellent and terrible experiences with banner firms. It's always about relationships. To be successful, we need trust and respect between all parties. And friendly accountability where people are willing to say, "Why are you proposing that when it doesn't fit the budget?" or "We think *this* should be in the budget."

What issues does your firm face in managing design around staffing, skills, risk/reward, training, financial, recruiting, short-term and long-term perspectives? You've talked about young staff, and the need to give them a range of experiences.

Paine: I've been thinking about this. Our 20th anniversary was in 2017, and Turan and I are excited about where we are and what we have accomplished. We've built a body of work and are working with the next generation to continue what we've begun. We spend more time co-laboring in the design studio and onsite to make certain the work meets our expectations. Our typical employee is just out of school, and we want them to learn the unique way we work. It's not the most efficient way, but it can be if everyone learns the routine of studying different options, using three-dimensional models, sketching, and thinking about what we're designing and why. When we hire people, we look as much at attitude as aptitude. We've become more selective. We look for good judgment, not just design, and how a person talks with others: in the office, with clients, consultants, and contractors, and how someone builds relationships to accomplish what needs to be done. We depend on our teams to elevate the quality of our design work. That helps us push design boundaries and investigate new solutions and builds trust with our clients.

Anyone can step up if they have basic life skills, good judgment, and know how to treat people with respect. They also need people backing them up. Many firms have a synthetic mentoring process. Not here – one of my biggest joys is going from desk to desk to ask individuals, "How are things going?" It gives me opportunity to connect one-on-one, and opens them to say, "Well, I'm struggling with this issue." Then, I can pull up a stool and offer advice. It may not be about a design; it may be a personal issue, or a problem they're having with a client, consultant, or contractor. Them knowing they can ask questions and get advice is important.

What trends or behaviors from owner and contractor partners are giving you pause? Things that, if you changed them, would improve your ability to practice and collaborate, i.e. "We hate it when . . ." Conversely, any positive emerging behaviors from owners and contractors, "We love it when . . ."?

"What does this push to get it faster and cheaper mean to future of the built environment? Will it be commoditized, or simply value-engineered to be efficient? What are we losing in that transfer?"

Paine: It's easy to typecast architects as having their heads in the clouds and being clueless about the impact they may cause on a client's budget or schedule. And it's easy to forget we too are in business to make money. We look to our contractor partners to help us expedite construction review and approval. For example, if the standard closeout procedures aren't being followed, and we have to go back multiple times to see that the work is being complete, we lose any profit we may have in the project.

Beyond conscientious cost estimating, I'd ask our CM partners to focus on constructability, and schedule review to ensure our time for design and on site is used efficiently. For instance, lately we're seeing submittals only date-stamped; that takes more time to review. We expect the CM to review all submittals before sending them to us. This means we must spend extra time in review, and time is money.

On the owner side, being an advocate for design, realizing once the design has been presented and approved, that we need to protect it and not change our minds. Things like paying invoices on time, treating the architect and consultants with respect, and listening to their ideas make a great client. Architects do their best work for great clients.

In just date-stamping shop drawings, is the contractor is trying to avoid the cost of reviewing the submittals themselves?

Paine: I think it's often a lack of training. We're seeing younger people on site representing contractors. They could say the same about us, but I'd argue our people in the field are supported daily by senior people. And we don't send shop drawings back that were only reviewed by an intern. They'll do an initial review, and then sit with our construction administration director to make sure they're marked up correctly.

Can you share a 5- or 10-year future state for your firm? Will you be practicing differently than today? Any new compensation models or value propositions to increase the architect's value?

Paine: Our business model is simple: a design process that's inquisitive, led by experts who are generalists. One that engages our clients but is not a free-for-all. Turan and I are involved in all the work, and our experience brings direction to the design process. We've created a firm purposefully not made up of studio groups. Instead, we've created thinkers – people who don't take things for granted, who want to look at any design problem in a fresh way.

> *"There has to be a degree of a level of trust and respect between all parties. And friendly accountability."*

A firm of thinkers can take a design problem, whether it's 2 million square feet or a lifeguard station on the beach, give the client what they need on time and on budget, and make two plus two equal five – something no one anticipated. In Cesar's forward to our book *Individual to Collective*, he writes, "There's a place for the mega firms in the future, but there's still going to be a place where a firm like Duda|Paine led by people who clients want to be engaged with, who bring a more personal approach to design and engaging the client in the design process."

Jeffrey Paine Key Points:

1. Design firms rely on CM's as partners in budget, schedule, and constructability expertise.
2. The insatiable curiosity of designers is immutable – a core value.
3. Qualities such as communication, consensus building, and trust are essential leadership traits for future architects.
4. Contractors are valued, needed partners and must respect designers' time, values, and process.
5. The drive for faster and cheaper has surpassed the laws of physics. Design and CM teams can't keep up and produce quality work in a humane manner. Clients who desire commoditization may get what they wish for.

INTERVIEW
Peter Styx, AIA

Director of Architecture, AECOM, Minneapolis   2nd August 2017

Peter Styx, AIA, is a Director of Architecture for AECOM, a global design and construction firm, with more than 87,000 employees. With more than 40 years' experience as an architect, designer, contractor, and project manager, his projects are varied in aesthetics and geography, including projects in the United States, South America, and Bahrain. Peter maintains licenses in 7 states and specializes in sustainable design, transit and transportation, corporate offices, and data centers. He speaks to communication, value analysis, cost control, and selling.

Meeting Budgets or "How to Work with Contractors"

Let's start with process.

Styx: I just reread a memo I wrote 16 years ago about how to work with contractors – and the things I said all hold true. I wouldn't change a thing that was said. Process is number one. For all the talk of design and construction processes changing and getting better, I'm struggling to see any real differences in the way we collaborate with contractors. What's different is that we're technology driven. We prefer to start in SketchUp, then switch to Revit to flush things out. I feel the same today about technology as I always have: it's wonderful when it works; miserable when it doesn't. The learning curve is long, the intricacies are amazing.

> **"Email is the scourge of our industry because it decreases or disjoints information."**

Communication is still king. The inclination is to "throw something over the wall." Even more so today, communication has to be reinforced. It's so easy to rely on an email. It's hard work getting people on the phone, getting face to face, bringing people together, but if you don't, you're buried in a myriad of emails that are worthless on their own. One becomes a hundred by the time you get all the people involved. Email is the scourge of our industry because it decreases or disjoints information. The amount of time to reread them and sort priorities is massive. Unbelievable.

I joke that despite the promise of BIM, we're really building our buildings by email, crude, marked-up pdfs, and RFI's.

Styx: You've characterized it well. We've dumbed it down, but at least it's visual, so people can understand it.

Has becoming a part of AECOM impacted your process?

Styx: Not yet, because I haven't interacted with our construction arm, which will be a big change. I like that estimating is internal now and can directly support the design effort.

That prompts the perennial overbudget question. If you were to work together, conceivably you could solve it. Even when we pull out all our best tricks, our projects still spike over, and we have to bring them back. Are you doing anything to break the cycle?

Styx: We're looking at project budgets instead of only construction budgets – that's one thing we're doing differently. We want clients to look at permits, financing fees, FFE, and all pieces of a project. We push them to ensure they have a total project cost. When they forget something, it doesn't work because it typically erodes money available for construction. We also work hard to set big volume square-foot budgets and get in the right cost range. We use published cost database services and work with clients to establish our place in the range. When we start macro, we don't have the massive Value Engineering or Value Analysis that comes later.

You're changing the scale by using total project budgets and dealing in comparable ROM costs vs. getting into the weeds. We've taken the opposite approach to get into detailed scope and CSI checklists early. You've gone high end, and we've gone low. Anyone who can get us out of VA is a saint.

Styx: We could go on forever on that, but both of us work to get the total picture, not just the construction costs.

You get to the point you can't take it anymore. Same people, same remarks, and you say: Oh God, are we doing this dance again?

Styx: Absolutely. An intern works directly with me now, and I work on what to impart to her, but the biggest thing is to think like an owner. We've been driven by budgets and methods to think, "Just tell us what you want, and we'll execute it." Now I'm coaching people to think as if the building is for them! How comfortable do you want it? Traffic, acoustics, etc. It's difficult to get people to think that way. Then, after we're told what to do, if it needs to change, we have to say, "It's at your expense, not ours," and then we get into these value engineering (VE) wars and we're no longer offering professional services, we're simply executing.

And then the team is demoralized. They think, "Just tell me what to do." That's not why we got into this industry. Firms say, "I'm afraid to do anything really good and interesting, it'll just be VE'ed out."

Styx: I've been lucky to work on a variety of projects. It takes the right attitude – and more research. We're having to think about markets, business sectors, tourism. Who would have thought architects would do that?

That's outside the normal limits. It's client and business focused. I only hope there's more fee and value in it, which would give more room to be creative, add value, design . . .

> "Oh God, are we doing this dance again?"

Styx: For years using the word "sell" was a horrible thing in architecture, but we do it every day. We sell to our clients, we sell our ideas to other disciplines, yet we're not really taught to sell. We're taught *not* to. Young professionals must be encouraged to accept sales as their work, to get out of the belief that I'm "just" somebody who designs a building. We're not waiting, we're finding those kinds of clients and employees.

When I see the higher profitability of contractors vs. architects it makes me wonder. How did architecture become a commodity? Because we let it? Why is the superintendent's truck paid for and not the architectural principal's? It's a continuum. On one end, you have risk management smarts and negotiation ability. On the other, you have sales and persuasion. Believing both are your job is a new mindset for architects – or needs to be.

Styx: My best "worst" story is when I asked a colleague to go to lunch with me and a client. "Why do you need me?" he asked. "I don't personally know that client. I don't know what to say to them." That was so indicative of how people have to get out of their technician mindsets. You want clients as colleagues and friends – more than just clients, so you're more than a stilted professional who says, "Oh no, I can't do that."

What can we as contractors do to help?

Styx: Quit sending us those damn RFI's! No, I can honestly say that for 20 years I've worked with very good contractors. It's a compliment. It shows they get it; they're being client driven, helping architects, not being adversarial.

If architects haven't caught on as readily, what are you doing to catch us up? Are you hiring differently to create the breed of architect we'll need in ten years?

Styx: We're looking for younger people, and Revit is a required proficiency. That's quite different from just 5 years ago, but we also don't want them just to be chained to the screen. We're exposing them to clients, and I demand they truly participate if they attend a meeting. You also need colleagues as "partners," not just employees. Also, outside your own employees, you should decide who is going to be supportive and go back to them. This group is broader than it used to be, including CMs, subs, and vendors.

> *"For years using the word 'sell' was a horrible thing in architecture, but we do it every day."*

Yes, you were our "go to" partner. Sometimes owners call us "cronies," but it's their money and risk. If they want to go to the low, risky people, fine. We'll all see what that's like.

Styx: I think it's really cool you're doing this – it's a great initiative.

Peter Styx Key Points:

1. Current value analysis processes are painful. Smart teams break that pattern and avoid the dance of redesign.
2. Email is not adequate for many communication needs.
3. Focus on client needs and cost control (versus the architect's agenda): keys to a new value proposition.
4. Schools and firms need to teach architects selling, communicating and empathy skills. "Selling" is not a bad thing. It's essential.
5. Wise firms find long term partners to help manage design.

CHAPTER 10

Art and Architecture: Design Leadership and Conviction

INTERVIEW

Phil Freelon, FAIA

Design Director, Perkins+Will

21st February 2018

Phil Freelon, FAIA, LEED AP BD+C is Design Director of Perkins+Will's North Carolina practice. He leads design studios in Durham and Charlotte, is a member of the board of directors, and a key leader for the firm's cultural and civic practice. Freelon's design achievements include cultural, civic, and academic projects for America's most respected cultural institutions. He led the design team for the $500 Million Smithsonian Institution's National Museum of African American History and Culture in Washington, D.C., and the National Center for Civil and Human Rights in Atlanta, Georgia. In 2011, then President Barack Obama appointed him to the U.S. Commission of Fine Arts, where he served from 2012 to 2016. Freelon is a Fellow of the American Institute of Architects and a recipient of the Thomas Jefferson Award for Public Architecture. In 2017, Fast Company named him one of the world's 100 most creative people in business, and "America's Humanitarian Architect."

"On smaller projects you can't compartmentalize management and design and technical details."

Cultural Understanding, Design Tools and Ideas or "We're Still in Charge"

Can we talk about your migration into design? What took you there?

Freelon: It started with my architectural education and being a standout in studio – I was named top designer in my class. Years later, evaluating new hires, I always want to see their portfolio. Those who show promise during their academic careers can expect to do well in the profession. Conversely, if students haven't distinguished themselves while in design in school, it's hard to imagine they'll come into the professional realm and suddenly perform at the top levels.

You've been responsible for many high-profile, public projects, many focused on civil rights and the African American experience. Fast Company *magazine labeled you America's Humanitarian Architect. How has practicing as one of America's leading minority architects on these building types informed your work? How did this practice focus come to be?*

Freelon: I worked for firms in Boston, Houston, North Carolina, and other places for fourteen years before starting my own firm. So, I had quite a bit of experience in other practices, and with public buildings. I did higher ed, K-12, office buildings, institutional work, and more. When I started my firm as one, then two, then three people, growing organically, opportunities came for small projects in the cultural realm – community centers, libraries, galleries – but on a small scale. The choice is obvious for clients looking for an architect who understands the context of the building they're trying to create. For instance, in the case of NC State University, the African American Cultural Center was one of our early projects in the cultural realm. NC State was looking for someone who understood the school and the culture. Well, I'm an NC State graduate and I'm African American. There's a commonality and a congruence of thought. Clients sought me out at first, then I began to pursue projects based on my expertise. Over the years, building that expertise on buildings around the country positioned us to compete for projects like the Smithsonian National Museum of African American History and Culture. The process was anything but overnight. The Smithsonian Institution came along at a time when we were experienced, mature and had relevant work experience in San Francisco, Atlanta and other cities. We gradually became known for work of this type. It was a long time in the making.

What's your attitude toward management? How do management forces shape your design work? i.e. Keep it in budget, on schedule, profitable? Are they opposing forces? Things done integrally, or best done by those with different, complementary skills? As a director of design, your focus is clear. How do you accomplish the rest? Do you see design and management as alternating perspectives?

Freelon: No. My career started in small firms. On smaller projects you can't compartmentalize management, design and technical details. You engage others who are stronger in those areas. In bigger firms, sometimes it's necessary to parse out aspects because projects are so large. You can't expect one person to manage the project and lead the design. But even in that scenario, you expect team members to respect each other's roles, so everyone is pulling in the same direction, with buy in and mutual understanding. We're all required to deliver a successful project. I prefer the integrated method. Even though you're pursuing design at the highest level, you need all aspects covered. I'd argue that if a project is just aesthetics, it's not complete. Similarly, if there's no regard to budget, schedule or technical quality, it's incomplete.

Perkins+Will is one of the oldest and largest firms in the U.S. Has their acquisition of Freelon changed your practice and daily tasks?

Freelon: It's a bigger firm, so management is more critical. The stakes are higher. Our CEO, Phil Harrison, earned his Master of Architecture degree from the Harvard Graduate School of Design. Naturally, design excellence is important to him. He doesn't design much anymore, but it shapes his thinking and approach to firm leadership. He champions it. Some of the other large firms are headed by lawyers, MBA's, or engineers who are much less design-focused. Perkins+Will and the Freelon Group had similar design-focused cultures. That why we decided to join forces. I still do what I'm good at, and my role as a design leader is evolving.

> *"It's the ideas that count, not whether they're put down on paper or in a computer . . . We're still in charge. We hold the mouse."*

How does technology shape your approach to design? How do you feel about machines? Do they make it easier to get the best out of people or add undue complexities?

Freelon: I've seen it all. When I started architecture school I had a T square and manual instruments, drawing on vellum. And then on through pens and pin bars, precursors to the computer. I've embraced it all. Not that I can do it all. I still draw by hand primarily because that's how I was brought up and I am more comfortable with the manual tools. I also draw using my computer. Some people rely too heavily on the computer to inform their design solutions. I embrace it. I admire it. But I also recognize that design is all about ideas. Computers can't imagine the same way humans can. We're still in charge. We hold the mouse. If you can't get what you imagine with one tool, use a different one. But I'm not going to criticize those who use digital tools. Some great work has resulted from people who start the ideation process on the computer. I happen to do it the other way because that's how I was trained. I'm good with all of it. Ultimately, it's the ideas that count, not whether they're put down on paper or in a computer. I'm annoyed by architects who talk about the "good old days." It's like sports – everyone wants to romanticize the period they came from: "The players were better back then." I don't buy it. I love what's happening now, how we're using BIM, Rhino, and the rest. It's exciting. Let's embrace it.

One analogy is: I have a manual screwdriver and a power one. I use them both. Valuable tools for different jobs. Although I do admit to joining the reminiscers sometimes.

Freelon: That's a good analogy.

The book's title, core questions, and themes imply conflict and portray an ever-tougher profession. Everybody wants faster, better, cheaper. Are you feeling that? How do your teams cope? Are there elegant solutions to what seems an age-old, culturally entrenched, complex problem set that can pave the way for a kinder, gentler, more collaborative profession?

Freelon: Absolutely. I feel it. Faster, better, quicker, more accurate. That's the world we live in today. There's a broader conversation about what architects do, and how we've lost control, or ground, to program managers, CMs and others. But it's like complaining about the rain. Well, it's raining on everybody. You can complain about it or do something. Open an umbrella. Deal with it. No one's going to slow down because architects feel like they don't have enough time to do their best work. Can we be more effective in how we engage our clients and explain the value we add? Can we be persuasive enough? That's on us. If you're doing something on the National Mall that's never been seen before, can you convince the Smithsonian Institution, National Parks Service, Secret Service, and the National Fine Commission on Fine Arts that it can be done, and *should* be done? It's incumbent upon us.

Did you get any formal training in communication, presenting, persuasion, strategy, or politics in architecture school?

Freelon: You know the answer: certainly not. What we do as architects is only partially addressed in school. I think that's okay. Look, you have precious few semesters to focus on design, theory and learning how to think creatively and critically. I'm not in favor of taking time from important lessons in the studio environment to teach students to draw in Revit or learn to be a project manager. We have an internship period for that. In school, let's focus on what you can't get outside of academia. Your own life experience is part of it. My dad was in sales and marketing. I got to see and absorb those skills by being around people who were excellent at it. My folks didn't know any architects, but we knew what a profession was. I'm not saying we shouldn't teach it. We all had professional practice class in school. I taught it at NC State and MIT, but from a designer's perspective. I taught that factors like cost, schedule, and fees can influence the design approach. I'm an optimist. Let's solve it, not whine about it. Use what you have. I played football very briefly, and I remember complaining to the coach, "It's cold, it's wet," and how tough it was. My coach told me, "That's happening on both sides of the field. Don't talk to me about that. Deal with it."

The design profession is experiencing some angst about the "rain" and about other professions encroaching. Have you forged any new processes or secret sauce in response?

Freelon: It comes down to talent. In design, to start with, you have to have ability, then hone and work hard at it. Not everybody can be an excellent designer. You have to find, attract, retain and nurture talent. Give them opportunities to grow and work on interesting buildings. In a larger firm, that gets more difficult. At Perkins+Will, we have the Design Leadership Council. Internal reviews inform and evaluate our design product, so we're not focused only on financial factors. That's unusual for a big firm. It starts at the top. Phil Harrison understands the value of design. It's not cheap to fly people around to do design reviews and weigh in on projects during the conceptual phase. But we do. We're big on stakeholder involvement. A lot of our work is civic and public, so we involve users early. Great ideas come out of it. It's not necessarily a secret sauce, but it's important.

Chuck Thomsen recalled the effectiveness of client empathy and involvement as far back as in the '60s - getting to know clients, valuing their work, becoming more expert and valued as you do.

Freelon: I used to work at 3D/I. Chuck transitioned it from a design firm to a Construction Management firm. When I was Loeb Fellow at Harvard, our paths crossed again. Yes, those things still work. They're what we do every day. As a design director, I take a hands-off approach. People need to be given leeway to be their best selves. You want to be there as a backstop but let them spread their wings. It's been a challenge to delegate design. But our firm wouldn't have grown beyond 15 people if I did all the design work. I'm trying to be the coach instead of the quarterback. I try to guide and influence more projects that way.

It's a common designer's malady, because we love what we do and are so hands on. It's a classic hurdle to let go of the "doing." What you refer to as "talent," in design, technical issues, or other proficiency, has historically been highly valued in the old individualized practice mode. For some, that implies disdaining management. Doing and managing are different skills. Historically a hard leap for many designers. I'm not sure our owner and contractor friends understand that or know how to help us cope. Their world is more about managing than doing.

> *[Re: civil rights] "Architects are most distinguished by your thunderous silence and complete irrelevance."*
>
> – Whitney M. Young, 1968

Freelon: I want to come back to where you started. Our profession, and the world in general, suffers from a lack of diversity. It's horrible that only 2 percent of architects are African American, and only 13 percent are women. It's appalling and embarrassing that the statistics are the same as 40 years ago, when I entered in the profession. It's not just a lost opportunity for individual development, it's also a lost opportunity to leverage the creative thinking that comes from having diverse conversations. If we're all the same, we miss it. Where is the diversity? It's a huge design problem. If you're not writing about that in this book, you're missing a very important point.

I am. I didn't set out to, but I'm learning. Diversity is becoming a big part of it. Not just racial and gender diversity, but including and embracing all perspectives: race, gender, architects, contractors, owners. More voices, understanding, richer teams, better ideas.

Phil Freelon Key Points:

1. Ideas are important; tools merely enable them.
2. Humans are still in charge.
3. The broader range of persuasive and managerial architectural skills – communication, sales, and the like – can be learned outside the studio.
4. No complaining. Deal with it. Take charge.
5. Diversity yields richer ideas.

INTERVIEW
Allison Grace Williams, FAIA

Principal Provocateur, AGWms_studio   6th February 2018

Allison Grace Williams FAIA, NCARB, LEED AP, is an architect, urban designer, and artist. Her international portfolio of civic, cultural, and research buildings and places is an inventive, inspiring narrative that connects culture, technology, and environment to convey traditions of audience and place. In her 35 years in practice, Allison has led design studios as Senior Associate Partner with SOM, Design Principal and Director of Design at Perkins+Will, and AECOM's Western Region. Her career includes design awards from Progressive Architecture, AIA, and GSA Design Excellence Commissions. In 2017 she established AGWms studio, a consulting practice at the intersection of architecture and art. The studio leverages design instincts to engage clients collaboratively to discover provocative, ideas and embed authentic, artful narrative in the built environment. Her clients include developers, institutions, and other architects and designers at all stages and scales.

She is an Adjunct Lecturer at Stanford, studio critic, speaker, and juror. Awarded a Harvard Loeb Fellowship, she is the current Chair for the Harvard GSD Visiting Committee. She holds a Master of Architecture (Distinguished Alumnus, 2015), B.A. in the Practice of Art from University of California, Berkeley. In 2018, she received the Norma Sklarek Award, conferred by the AIA California Council Board of Directors to on an architect or architecturally oriented organization in recognition of their social responsibility. An avid sketcher as her medium for design investigation, Williams's resume and portfolio are available at: www.agwms.com.

> "Tangled up in all that is instinct, gut, invention, nuance, and things hard to quantify, measure, or predict. It's still very much the way I think about design in practice, and the integration of beauty still sponsors an authenticity and fundamental design excellence."

Art and Beauty, Architecture and Building, or "Instinct, Innovation, and Respect: Managing Ourselves"

Considering your distinguished and widely recognized body of work, I'd be interested to know what drove you to design and architecture?

Williams: In art school I gravitated toward printmaking . . . zinc plate etchings . . . a very process-driven medium. In etching, one can test how intentional deviation disrupts the outcome. But art was quite solitary, too private an audience, disconnected from meaningful engagement and contribution, with no way to know if a work was successful. Success was too subjective. I was drawn to architecture as a process, a process of design, a tool for discovery, and an opportunity for an artful approach to the built environment. Environmental and social concerns were (and are) central responsibilities of this profession as a social art and, for me, have always been integral. But art exploration as a tool for projects in the public realm was my primary motivator.

Today we talk about social justice in a braver, up-front way, because we see, and can demonstrate, that design impacts outcomes indirectly and directly. That buildings create environments and a resonance (or discord) with people and places was always important. We are not designing in a vacuum. There was always a human factor or face, a balancing of specific aspects with the purpose and need, with a specific site, and with the desire to establish timeless importance and collective value.

I've mostly worked on large projects and design proposals in urban places. My training was at SOM – a true extension of formal education, a methodology around a timeless aesthetic. It was about the rigor, the iterative process of exploration, a reductive and frequently-technical interdisciplinary approach that distinguished SOM's learning environment. Tangled up in all that is instinct, gut, invention, nuance, and things hard to quantify, measure, or predict. It's still the way I think about design in practice: the integration of beauty still sponsors an authenticity and fundamental design excellence.

It's remarkable you've been able to work primarily on design-focused commissions. Have any been driven by commercial objectives?

Williams: I've been privileged to have captured, led, and collaborated on numerous great commissions, many won by competition. Notably, during the heyday of the GSA (Government Services Administration) Design Excellence Program. There, the intention was to position the design conversation, and design leadership, on major public and civic commissions very early through the selection of a lead designer. This allowed the design process to impact the program, sometimes the site selection, and, in theory, the coherency, and a whole, innovative, performative approach to the work. Though shortlisted for many more than I won, those pursuits influenced how I worked. That is, landing on a strong conceptual diagram, and following a holistic approach often defined, re-defined or re-positioned a project with specificity – with genuine responsive expression as an outcome. Moving upstream sets a framework for a big idea that tethers values and ties parts together. It informs how things evolve and frequently even shapes the details. I've always been most impactful earlier in the design process.

I've worked with developers where the pro forma, site plan, massing, and more might already be cast. Even then, I advocate (and rather enjoy) unpacking things, as far back as the project and client can stand it, to challenge or confirm the assumptions or approach and broaden the conversation beyond just the building. The hope in doing this is that a developer-led team might discover (together) some enhanced potential, some fundamental benefit to the bottom line, some meaningful programmatic contribution that they assess to be worth taking a calculated risk.

"A successful design leader doesn't always make friends but gains and promotes respect within the team. Respect for one another is crucial. Going forward as a team in unison around an idea of respect is powerful as it evolves. It infects every aspect of how a project develops, strengthening and enriching it continuously."

How have the "new" forces, schedule compression, technology, and increasing complexity, affected your approach?

Williams: The profession has changed in that the industry and the art are often not aligned. Clearly technology gives us the conceptual and analytical tools to study, iterate, analyze, and discover what we could not even imagine. It also pushes the process of design, documentation, and building faster to meet the speed with which everything needs to happen in this aggressive global economy. But is something inherently less valued when the priority is speed and cost? I have no regrets, but I sometimes imagine, what if I'd come along now – as a freshly minted architect – as opposed to 30 some years ago. Because things move faster, and collaboration and trans-disciplinary thinking is becoming more prevalent, more valuable. Beyond interdisciplinary collaboration between civil, structural engineers and architecture, crossing into disciplines such as medicine, the sciences, and, and, and. That's extremely interesting to me right now as an evolving way to deploy our design training.

"I don't accept that designers are unable to manage themselves!"

The rigor, the rationalization, and concept development – it's the notion of thinking outside the status quo, to think of ourselves as scientists as much as artists. I've always enjoyed coming at opportunities from all sides. Notably, in my experience, scientists are the most engaging clients when there is a shared sense of curiosity. Even when I'm just playing devil's advocate with myself, the best way to take the top off something is to look at it from a bunch of different angles. I've always been more interested in looking at what it *could* be, or what happens if we tweak a key aspect of the proposition.

What does it take to direct design successfully in a team setting, where goals of getting done on time may conflict with studying one more better idea?

Williams: What's required to lead is the ability, after putting ideas out there and taking everything in, to move the team in a clear direction. A successful design leader doesn't always make friends but gains and promotes respect within the team. Respect is crucial. Going forward as a team in unison around an idea with respect is powerful as it evolves. It infects every aspect of how a project develops, strengthening and enriching it continuously.

"When architecture becomes the industry of production of buildings and the creative/artful/ beauty conversations and roles are off the table, it's not much fun as a profession anymore. More of a ball and chain."

Going to your big firm days, you mentioned the 3-legged stool of skills at SOM: design, technical prowess and management. Most are strong in one area. Some have two, but it's rare to find someone strong in all three? How do you solve that?

Williams: Thinking singularly is unproductive. I'm always looking at intersections rather than silos.

Much of the challenge of managing design stems from misunderstanding, bias, ignorance, or siloed views. The knocks against designers are: inability to manage themselves, listen to clients, stay in budget and on schedule. How can we beat those stigmas?

Williams: I don't accept that designers are unable to manage themselves! I'd suggest that design schools are not teaching advocacy, and there is discussion on whether they should or not. Architects are not taught to advance ideas or articulate the value of an approach in convincing ways. We need to get better at stating and demonstrating value propositions, with metrics baked into the approach/proposition, as part of the evolution of the approach and decisions. It ultimately goes to whether we advance and benefit the outcome from the patron's perspective. What do we bring to the table? If we're playing out the predictable, responsible formula, nobody is unhappy, but we haven't advocated for ideas or their power to impact the formula. The goal isn't just to do something different for its own sake. Isn't it more about testing or furthering the value proposition or some other measurable metric? Why does this approach make a difference or matter? Is its impact measurable, on a personal or communal level, or environmentally? We have certain responsibilities as architects, beyond buildings not leaking or falling down. That was "101" in school. You always know where the sun is, you care who and what is next to you – the context. It's fundamental. But we don't always advocate and communicate the metrics that go with decisions. Now, many fresh, talented designers and impactful practices are figuring out how. I'm intrigued by that. It helps the profession restore relevance and necessity.

The pressure to balance creativity with project constraints seems to increase daily. How do we free ourselves? Should we walk away when objectives are misaligned?

Williams: It's a heavy order to suggest, from a *business* perspective, that architects walk away from projects. Most can't. But some do because it's the commitment they made to their practices – the principles for which they stand. We're probably reading and seeing the work of only a small percentage of the firms published in our magazines, pushing the edge, taking creative risks – excluding private residences. Lots of firms do respectable work, but so many buildings become the same, generic, formulaic response. Relatively few take even baby steps toward a dialogue that's reaching harder and keeping architecture-as-art in their conversation. When architecture becomes the industry of production of buildings and the creative/artful/beauty conversations and roles are off the table, it's not much fun as a profession anymore. More of a ball and chain.

Those lucky to have contributed to great projects want to continue high level collaboration and pursue great design. But if those are the goals, we have to meet them all. Your recounting of an 11th-hour design concept reversal brought back memories. We've all done it – then had to pay the price. Such a late-stage shift might draw teammates ire but must be done to reach the higher-level answer. Owners and teams who share such goals need to know this. How have you engaged management discipline?

Williams: Projects need leaders as passionate about management as other players are about design exploration and technical innovation. A good manager manages on behalf of the client, serving them in the evolution of a building. They are also hugely responsible to create a fertile environment in which a design process can flourish.

Allison Williams Key Points:

1. Art and beauty have always been the province of architecture and must continue.
2. Pure production of buildings may be a separate pursuit.
3. Architects can and must manage themselves.
4. Gut, intuition, nuance, instinct, invention, and unmeasurable factors remain, and join new metrics as indicators of good design.

CHAPTER 11

Engineers and The Consultant's Mindset: Leading From Behind

INTERVIEW

Daniel Nall, FAIA, FASHRAE, LEED Fellow, BEMP, HBDP, CPHC

Formerly, Regional Director, Syska & Hennessy SH Group, New York   11th August 2017

Daniel H. Nall, PE, FAIA, FASHRAE, LEED Fellow, BEMP, HBDP, CPHC, was formerly regional director, High Performance Solutions, and vice president with SH Group, Inc. (Syska & Hennessy) in New York. As a practicing architect, engineer, firm owner, and consultant, he was an early pioneer of energy modeling, beginning his investigations in 1981. Nall now focuses on sustainability and high-performance engineering design and innovation. He was chosen as one of ENR's Top 25 Professionals of the year in 2008.

Aligning Objectives and Optimizing Systems or "Catalog Engineering"

You have an incredible resume, range, and career, from engineering to architecture to sustainability. And you had to be one of the first energy analysts in the country.

Nall: Yes, I was teaching at Princeton – a split role in engineering and architecture in the 1970's. Then I got recruited to join Heery Energy Company in 1981.

Let's focus on the engineer's perspective. You're a rare bird who wears multiple hats and have a more valuable perspective than just the engineer's view. How are overcoming issues like siloed design, commoditization and meeting budgets these days?

> **"In that circumstance . . . I don't really give a shit what the vision for the project is."**

Nall: Engineers need to be at the table early. But not all engineers' mindsets allow them to contribute early. Often, they don't get it. They're at the table to help realize a project vision. Part of that is the owner's and part the architect's. The MEP engineer is there to help them both achieve their visions. Sometimes they conflict and the engineer gets caught in the middle. Back before the energy stuff, when I was practicing in pure design with Jones Nall Davis, I didn't want to *hear* from the architect until he had my floor plan backgrounds ready.

For example, we were the engineers for a spec office building in Atlanta. Our fee for base building design and construction administration was $0.27/SF. At 300,000 SF, $80,000. And we would've made money on that project, but we got called in to investigate a vibration they blamed on my cooling tower and I spent days there. Turns out it was the elevator.

You pitched in to serve the owner, but another team member you had no contractual relationship with caused you to not make a profit. You were a good team player, and you got screwed.

> **"Someone needs to add that dose of reality."**

Nall: That's right. In that circumstance, a standard commodity solution – and level of service – I don't really give a shit what the vision for the project is. You, Mr. Architect, have done this many times. So, just give me a machine room and let's get it done. Just shoot me backgrounds and I'll throw in some units, size some ducts and power, and we'll be done. God willing and the creek don't rise, I might even make a penny on it.

Then you have the other world – a vision where an enlightened owner can benefit from someone like you. Are you seeing any new ways of working, or new value propositions?

Nall: One is BIM. Not just the 3D aspect, but populated databases with attributes. Our energy models now start with BIMs created in Schematics. We can translate into energy modeling software and evaluate different design decisions, window wall ratios, shading, daylighting . . .

When I first saw some of the energy visualization tools I thought even I, as a design-focused architect, could use this. It's intuitive. I can't deal with printouts and tons of data.

Nall: Yes. The next thing is developing procedures with responsibilities for encoding as-built data for facility management, with performance parameters and bar coding in the field. You scan the conduit and it tells you what's inside.

We've been doing that for years. We wrote facility management (FM) software and owners are loving it and willing to pay for it. That data has a long life and great payback.

Nall: FM has responsibilities for subcontractors, commissioning agents and the design team. Liabilities too, so, it needs to be clear and well-articulated, so the owner gets a fully functioning FM tool. And we can repeat energy performance modeling to detect malfunctioning components.

I tell people: owners don't build buildings to optimize engineering systems. They build them for other reasons. Owners have to realize they need to cooperate to achieve their objectives. Yes, you've got to lookout for your skin, but it's also possible, even desirable, to optimize your system to help everybody meet *their* goals. In any group endeavor, if one person leaves the field smug and another disgruntled, the project hasn't been optimized. I mentor people to take their satisfaction from the overall project. Did your work contribute to the whole? Or just your small area of expertise? A holistic attitude is a prerequisite to an integrated design concept.

If we're trying to break down silos and cultural stereotypes, from the perspective of one of the country's largest, oldest, most sophisticated engineering firms, are you looking to new models? Or do we simply subject ourselves to how the owner sets up the project?

Nall: Helping an owner achieve their vision as a sophisticated consumer of design and construction services is a good goal. Unfortunately, I'm not convinced current purveyors of that kind of assistance – owner and tenant reps – are serving owners well.

We deal with it too. Those entities, and the owners they serve, seem behind in current leadership skills. They aren't up on technology, or how we work. They're old-school – not only do they not know how to motivate modern teams, they have no sense of our processes or tools.

Nall: If services were delivered in a back-and-forth way, with an architect, a contractor, or an owner's rep, to help clients determine their goals, it would have incredible value. We're starting a project where the owner dictated a decision that's impossible – anyone with half a brain knows these things are totally incompatible. But no one is in a role to explain it, and, even though it won't work, it's a program requirement! Someone needs to add that dose of reality.

They'd probably not get paid for it – they might even reduce their own fee. I argue for programs, but they're no good if they're wrong or impossible. We need the flexibility to be able to change on the fly. Some projects are not set up to reward that.

Nall: A firm that's doing this in limited ways is probably Gensler. They have a group that works hand-in-hand with clients to set goals. What are the implications of those goals in physical form?

There's big value in that early work when can it can make a difference.

Nall: There's something to helping owners become intelligent building consumers. More than budget and schedule. Helping them understand the implications of their goals in physical configurations. Not many professionals offer that service. It takes sophistication to understand the client's business and particularize your advice to their situation. Then you have to convert it into physical terms using the client's process for design and acquisition. You need different expertise. had a client that was a many-headed monster of different constituents. They had a project manager, but he was too weak because of corporate jockeying. It was like a government bureaucracy. They had the arrogance of a church building committee: "You should be doing this almost for free because you're working for a higher authority." The client from hell. They gave us strong direction but had convenient memory lapses when that direction turned out to be bad.

"There is something to helping owners become intelligent consumers of buildings. More than budget and schedule."

We face the same thing. This is our plight. Owners are so busy running their own stuff they're not engaged in the project. They say, "That's what I hired you for!"

Nall: Then you have the other extreme, where a team is incredibly successful implementing a vision, but it was the wrong one.

What do the rest of us need to know to work with engineers better?

Nall: A sense of trust needs to exist among the entire team. Granted, it has to be earned. I try to render opinions or make recommendations based on understanding the whole project. I don't own stock in a particular vendor. The self-interest I express reflects me *and* the team, cost, the goals of the entire project, and protection from risk. We had an owner ask each of us at the kickoff meeting to write one thing we were going to do on the project that had never been done before. I wrote, "We *aren't* going to do any one thing that has never been done before. We're going to take things with a track record of use and put them together in innovative ways that don't involve performing outside their envelope to achieve results that have never been achieved." It minimizes risk by configuring proven things in a clever way and achieves superior performance. Everyone else put down new stuff. Risky stuff, not in the appetite of the great preponderance of owners.

Innovation shouldn't introduce trouble or put equipment outside it's demonstrated performance domain. We can get huge energy efficiencies without pushing any one piece past it's limits. It's just a new way to stack things. That's where we're innovative. We are, after all – and I hate to use this term – "catalog engineers." We're not designing new jet engines, we're arranging systems in different ways that don't void anybody's warranty.

Daniel Nall Key Points:

1. Design teams can bring leadership to owners to help set project objectives, placing a project on the commodity versus first-of-kind risk continuum, to introduce a dose of reality.
2. Most teams are constrained to dealing with standard components and systems in clever assemblies and configurations.
3. Risk can be managed in this way while still allowing innovative design.

INTERVIEW
Kurt Swensson, PhD, PE, LEED AP

Founding Principal, KSi Engineers   14th September 2017

Kurt Swensson, PhD, PE, LEED AP, is founding principal and president of KSi Structural Engineers Inc. He was formerly a principal with Stanley D. Lindsey Associates, Ltd. After beginning his experience in the AEC business at his father's architecture firm (Earl Swensson, FAIA, Nashville, TN) as an office intern and construction laborer in the 1970s, he has seen the evolution of the design industry over multiple decades. Dr. Swensson has been an active thought leader in the American Institute of Steel Construction (AISC), Building Seismic Safety Council, ASCE, Structuring Engineering Institute, and other groups. He is an executive board member and president of the Structural Engineers Association of Georgia. A frequenter presenter and lecturer, he has published more than 40 industry papers, authored a design guide for the National Council of Structural Engineers Association (NCSEA), and been responsible for hundreds of notable projects across the U.S.

"Somebody wise once told me: If you can't control the outcome of a situation that affects you, you have to be able to predict the outcome and prepare accordingly."

Managing at the Point of Attack: Anticipating Outcomes or "The Waiter and the Old Man and the Sea"

Let's start with perspective – yours as a consulting structural engineer. I have my own preconceptions having worked with you; can you share your version?

Swensson: Perspective is a great start. I grew up with an architect who brought everything home, so I've been exposed to design process since the late 1970s. I worked in my father's office in the 1970s, worked construction and then at Stan Lindsey's starting in 1981. The experience goes back more than thirty-five years. My wife and I founded KSi out of my basement in June of 1999. Man, I'm old.

For the past twenty years owning my firm of twenty to forty people, I have been in a niche market. Everything we talk about is from that small to medium size firm point of view. It's where we reside – from $300M projects to very small ones. Most of our work has been $10 to 50 M in higher education, healthcare, science and technology and commercial – that's our sweet spot. The vast majority have been good experiences. Out of several thousand projects, only a few have been train wrecks. So, we come from a generally positive outlook.

People want to hear about what works.

Swensson: The building types and projects we focus on generally lead to good life experiences. Most of the time we put up with the way things are because things aren't too bad. Good is the enemy of better. Every job has its own set of financial, technical and people challenges. We work with the situation and do the best we can. That's the overview of where we're coming from.

Smart. You've chosen a niche and size that suits you and building types generally risk-averse and immune to economic cycles.

You're faced with managing design at multiple levels: on every project, you've got an owner, architect, contractor, and your own firm to manage. What's your philosophy and approach? Do you take things as they come? Or control your fate? Are you resigned to being a victim?

Swensson: Somebody wise once told me: If you can't control the outcome of a situation that affects you, you've got to be able to *predict* the outcome and prepare accordingly.

My philosophy is there are very few things I can control, and they mainly deal with what we can do in our own house. I've tried to be good at reading situations and project direction before issues happen. For example, if I hear people talk about pricing pressure in meetings. Okay, value analysis is coming, which will affect the deadline. Who are the players and decision makers? Then drill down to see what is happening and what is needed from our team. Then I can direct my team to be most efficient – redesign or pause – while the budget is resolved. That's where I can exert influence.

This industry is about people. I've been in the Atlanta market twenty-eight years. You get an idea of how people work; they know how you work. I can exert influence in some situations. I've been told I'm more aggressive, in terms of management, than some consultants. I get involved and ask questions up the line on pricing, construction, permitting, schedule. To stay focused. If we're doing a $300-million project, a lot of people are at the table running different parts and pieces.

The challenge for us is the small office building project that no one's managing – the "I need a 15,000 SF office building now" projects. There's no schedule; no agenda, limited pricing. The client says, "Let's figure out what we're going to do next week." I'm the one who says, "Wait a minute! Where's the plan? Where's the schedule? Permitting? Pricing? How's that gonna work?" They look at me funny because I'm "just" a consultant, but I've got to know how things are going to work and take care of my people. Sometimes the project lead does not listen, and sometimes my approach is welcome.

"You asked about the mindset of consultants. Our mindset is survival. It's that simple. It's all I'm trying to do."

I've been told, "That's not your deal." So, I influence where I can. Our firm is subject to the needs and schedules of others on the team from civil to subs to architects, owners and contractors. When one team member does something, it affects all of us. Schedule, finance, design. All those moving parts touch everyone. That's a problem we run into: so many moving parts not controlled or managed from a unified source. My coping strategies are: engagement, commitment, and a level of aggressiveness. Do I take the offensive in solving a problem, or be more defensive and protect the firm from poor project management?

You're saying that's harder in the commodity projects, where you may be less well suited because you have experienced, thinking, principal involvement?

Swensson: Right. You figure out, "Where's this train running, and am I on board?" A bigger issue is representing the firm. The firm is successful when a project is managed well. If a project is not being managed, what happens when I take a leadership role? How does the project team treat me as we move forward? Am I ignored, buffeted, celebrated? If you get your hand slapped, or realize no one cares, or it's not going to influence the project, you say, "I'll back off." Best case scenario is you bring everything you've got to the table and it adds value. Worst case is you see there's no leadership, you can't tell where things are going, so you can't work ahead to add value or coordinate. At that point you can only react and protect your interests as best you can while still serving the client.

You're at the high end of having an ability to read your clients and training the firm to do the same. But at your size you're able to touch most projects.

Swensson: Well, I've worked in architectural firms, on jobsites, a broad range. Being around a while you say, "I can add more value with a well-placed phone call than I can with reams of paper and structural analysis, or calculations." It's still about the people and working it out. As you noted, in many cases I'm not driving it. But sometimes you can say no. "No, there is no time or money to look at that option." "No, that idea's not going work." I don't do that often. You have to beat me with a stick to get me to say no. But in the industry, many firms just say no. They want to do the simple 30 x 30 bay because they know it works and their risk is not going to be rewarded, then, why do anything else?' That's the "go to" position for some consultants, in my opinion. Personally, I want more for me and my firm.

That's the old-school model of our profession: the principal is adding value, making reads. But the seers are foretelling of disaggregation, that we won't be able to have you there all the time, we'll commoditize and automate more of it.

Swensson: As a subconsultant with limited control, we need experience and trust. Does the project team trust me? I remember you being honest with me about a deadline because you trusted my ability to deliver on time. You gave others a different deadline because they had shown they would not deliver – that's the people part you can't put into software. Automation misses that completely.

A consultant leader needs to be able to discern what information the team needs for different building types and delivery processes, like IPD. It is a custom unique situation. Those will be run by the most influential and powerful person in the room, not necessarily the person with the best answers. Not necessarily who has the *right* answer, but who can get people to do what they want them to do. Automation misses that and cannot control it.

> *"The idea of increasing the amount of money a company makes from any particular project is so disassociated from the teamwork that happens when a decision is made. I don't think fees or profitability influence designers, very honestly."*

Because they are in charge, are the owner, the development manager, the dominant personality, or have the power or money?

Swensson: Or because they're the team member who's got the decision maker's ear. Larry Lord, Founder of Lord, Aeck & Sargent Architects, taught me this: "The right person needs to make the right decision at the right time," to get the best project and team performance. Unfortunately, the right person to make that call is not always the most influential person in the room.

In my experience, the goal of maximizing the client's return on investment on a project can become disassociated from the design team decision making process. In many cases, design firm profitability doesn't influence designers, honestly. It may influence some executive somewhere, but they're not in the meeting when the design decision is made. The executives may review it afterward and discipline or reward accordingly, but that doesn't change the behavior day to day.

Profitability can be disconnected from what happens around the team table. The typical frontline team is driven by knowledge, passion, emotion, commitment – things other than management or business – at the decision moment. Managing risk may not be top of mind for design types in the heat of the moment, or at all.

Swensson: Absolutely.

What are we doing to make you crazy? Everybody else: owners, architects, CMs . . .

Swensson: To use a sports analogy, we need somebody to call a play. A good play. We all need to understand the play and do what we're supposed to within the larger game plan. If you look at a football team, the consultants could be like the linemen, or receivers. I used to play center. They have specific skills and duties to perform to support the team. They know what the quarterback's going to do, so they block a certain way. All of a sudden, the quarterback decides to do something different, and the linemen don't know how to block successfully, so the play doesn't work. If the quarterback keeps going rogue the lineman lose faith and find it hard to "leave it all out on the field." You lose trust and the ability to predict. Trust is huge. When you start, have a plan. You can call audibles. Any plan can work, just communicate. Design/Bid/Build will work when done right. CM-at-Risk, GMP, and Design/Build will too. But when people aren't committed, knowledgeable, or don't follow the game plan, things go sideways. It's human nature. Sometimes people just decide to do what they want or think is best for them rather than following the play.

Much of our discussion comes back to people and their ability to read situations. That's how I've worked most of my career, so I struggle with the idea of disaggregation and atomization of professional judgment.

> "You asked what it feels like? Servant, expert, trusted advisor, salesman, gatekeeper, teacher, translator, and mind reader. Some people treat me as a supplier, some treat me as a partner."

Swensson: In *From Good to Great*, Jim Collins talks about getting the right people on the bus. I think of a team in a boat. Once we figure out where we're rowing, I'm in. Even when the storm comes, I'm still in if it's a good team. Project delays, budget cuts, material or labor shortages can be tough, but we can survive. We can work together and adjust to changing conditions. As soon as people start doing what's good for their firm instead of the team, things start to unravel. Trust is a big part of it and not being able to predict where people are going.

You asked about the mindset of consultants. I would say a consultant's mindset is get the job done well and within their budget. It's that simple. That's how a firm survives.

Your approach makes sense. Your role can't be leadership in normal ways, because you're not driving. You're riding.

Swensson: Right. I've got to read what's going on. Figure out where the project and team are going so our firm can succeed. My role is not as a normal leader. It feels more like servant, expert, trusted advisor, salesman, gatekeeper, teacher, translator, and mind reader. Some people treat me as a supplier, some treat me as a partner. It is a dance, a give and take. To be accepted as a partner you have to be a good partner. In general, people respond to how they're treated. Our idea is, read the situation, determine what's going to dictate the situation, and figure out how to work our way through to the betterment of the project and project team. It's like a being a great waiter. A great waiter listens to and understands her guests, makes good recommendations, delivers great service and is so graceful in the delivery that the guests will only remember the great experience. If I can add value to your experience, that's a great thing. But I also have to read your situation and what you're trying to accomplish. It's all about expectations and performance.

I'm doing something similar now as a servant and enabler to design and construction teams, rather than being in the lead design role. You once said that designers have a "genetic deformity" in their love for, and the need to do, design. It's almost like a drug habit or a co-dependency. You want to do well, so you're asked back. That objective is stronger than making money. I'm in this business to make things, not because I'm a business guru.

Swensson: We're in this to create a lifestyle. It's different for each one of us. What motivates each of us is different. On any project, the "rubber hits the road" when the individual team member sitting at their desk decides what they're going to do about the project issues each hour of each day. That's what drives these projects. That's what everybody misses! Technically, do they take their time to refine the design? Do they make that one more design run or go home? Coordinate one more thing or decide someone else will catch it? When they're sitting in the meeting, do they speak up or not? Those decisions are made based on each team member's mood that day, whether they trust the people in the room, what "floats their boat" and gives them satisfaction at that point in time, at the point of attack – it has nothing to do with the bonus they got 3 months ago.

We have a wide range of choices as professionals. We're managing ourselves as designers every second, and we're trusted to do so.

Swensson: True. It's like my golf game – I know where the errors are, but execution is another matter. As a profession we're working on that. I love one quote on the process being broken: "The present systems are perfectly designed to give us the results we have." So, it is what it is, because of the people involved. We are human beings, not perfect, so the systems cannot be perfect.

How are machines changing your business?

Swensson: Speed of delivery is number one. I use the term "quickness." Speed is one thing. Quickness is another. We can change steel framing elements a lot quicker than before. It's also about perception of our

"I know where the errors are, but execution is another matter."

"The present systems are perfectly designed to give us the results we have."

profession's ability to react: Owners and leaders make decisions quicker, and don't think about the implications on workload, production, coordination, so it's brought another set of issues. I believe our machines have out stripped our human teams' ability to make good decisions and track them. We used to pick up "red lines" on drawings after coordination sessions. We still have to do the same thing in the computer model to resolve conflicts found in collision detection sessions. It may take less time, but we are doing it over and over again for each iteration. So, the result is very similar. Leaders don't feel like they have to make good educated decisions from the beginning because the team with the software can "work it out."

Computer technology has brought that expectation. Yes, we can design a hundred individual beams faster than before. But before we did not have to design 100 beams, we designed the typical conditions for a more rational design. The computer has given us complexity we never saw before, even when it does not add value to the project Materials, geometry, details, delivery methods – it's complex now. I always say, "If we built cars the way we do buildings, we'd still be riding horses." The computer has allowed geometries and material selection to create unforeseen project conditions so complex they defy description – with very little if any added value to the project. Computers also allow amazing things we couldn't do before. Technology has had an amazing impact, good and bad, on the industry.

New technologies and economic and social models are fueling discussions on how the design profession is subdividing for survival – the disaggregation I mentioned. Are you seeing and planning for this? Or aggregation? What's your view on the mega-firm model?

Swensson: Aggregation has a place, but that's not where we're going. We've had opportunities to join mega-firms; routinely the answer is no. We're going to make our own way unless the right situation is presented. I see it more as integration of disciplines, where we have firms with architecture and engineering, all together. My experience dealing with some of these multidisciplinary mega-firms, is that integrated design is a myth. Whether in house or out of house. It has to do with commitment and people. My experience with larger firms is that they have expertise and experience, but it is not focused on any individual project. The management considers the greater good of their firm, not the good of an individual project. Incentives and systems are set up to produce those results. It has been my experience that their experts are spread thin across the firm's backlog. The professionals who know what they're doing are expected to fly around the country, if not the world. The people doing the actual work are less experienced with less expertise. Their compensation is based on hitting a number of hours a project manager budgeted.

Principals can't be everywhere, so the B and C teams do the work on the larger projects.

> "Integrated design is a myth. Whether in house or out of house. It has to do with commitment."

Swensson: Yes, and the principals and experts can't get their work done because they're too busy flying somewhere to land a job or put out a fire. I've sat across from these individuals in meetings. They do not have an intuitive deep knowledge of the project or project team. Someone has briefed them, but they haven't lived it. Integrated mega-firm teams work for some mega projects where the effort scales well. But an integrated mega-firm team on custom work on mid to even large regional projects? There's no advantage.

There's a lot of talk about prefabrication to solve workforce issues, but you have to come at it with a special mentality. To some firms, that would be anathema.

Swensson: There's still the need and demand for custom-programed, or artist-driven architecture. And with the purely business-focused folks, it's the need to "get the best deal." To be able to tell their boss, they bought it for less than anybody else. They have to be able to show competition. To show that you used the lowest bidder to confirm "value." It is hard to do that now with pre-fabricated systems because they are not to the point that they can provide the custom solution. Owners seem to be reticent to buy the system up front. So, they do a custom design, then a supplier tries to apply the prefabricated solution. Most of the time it is not a good fit. This will change as suppliers become more sophisticated in their design and sales efforts. As they begin to affect project decisions at initiation not after design is complete, the prefabrication movement will begin to gain steam.

We do this work because we love it. Others are in it for the deal. Are you doing anything to get yourself a better deal, such as a new value proposition or new services? Are there alternative ways of making deals, if the essential nature of designers is a given?

Swensson: Being a consultant and serving clients every day, I'm not looking at radically different ideas. Just good clients and projects. The decision is: how big an organization can a hands-on service approach drive? 20 people, 40 people, or what? That's a personal choice. I look at work supporting life, not the other way around. I work for personal relationships, professional challenge, and profit. Its more about supporting a lifestyle and sending kids to college and being able to retire someday. For me, it's all about the people. we can take on any technical challenge and compete because we have smart people. That is enough. Futurists predict bolder things, but plenty of work is going on the way you're describing. I think it will be like this for some time. But I recognize people still need to hear their business model is okay. They need change *their* way, not somebody else's.

Change will only happen when the status quo becomes so painful we can't put up with it. The real problems I've had – just a few over 30 years – aren't worth reinventing the wheel over. The problems were due to the people not the process. The process relies on getting good people working smart, together Many people reinvent our industry on a micro level every day. I'm more organic: put me in a situation, let's see what it is, figure out how to solve it and get it done today with the people we have. That's my message: you have the freedom to solve a problem however you need to. Project management systems need be organized to *allow* us to solve the problems at hand, not get in the way. One analogy is Hemingway's *The Old Man and the Sea*. The fisherman lands a huge fish and he tries to bring it to shore, but the sharks and barracudas eat it up. Some projects are like that. We, as consultants, go out and try to land a big project. Once we land the project, the "management system" kicks in. A myriad of project team members, working in their own best interests, and not committed to success of the team, all take their bite. When we get back home at the end of the project, any benefits have been stripped away.

Maybe our individual version of the designer's experience is what makes life exciting – the act of design itself. Making micro decisions. Maybe we love design because of that management struggle. But it's also creative thinking, coping with context, and working with people to solve a problem.

> *"Change will only happen when the status quo becomes so painful we can't put up with it."*

"When I think about who I'd go to battle with, it's the commitment to each other and to solving the problem. If I've got those two things, I have it all."

Swensson: Growing up, my dad talked about who he would "go to battle with" by his side: When I think about who I'd go to battle with, it's the commitment to each other and solving the problem. If I've got those two things I have it all. I know the team will be successful. The experience will be exciting and rewarding.

You don't want the enemy to be yourself. You want to be on a good team, including architect, owner and contractor. The other part is the quest for human connection, collaboration, meaning, and the great feeling that comes from being on winning teams. It confirms the human condition.

Kurt Swensson Key Points:

1. In small to medium firms, many consultant's mindsets are survival, not changing the industry.
2. Leading from below can be a challenge. Reading situations is an essential to serving the team.
3. Great value can be added by those with the ability – if they are allowed, or if they push it.
4. Anticipating and predicting outcomes offers value for design professionals and all team members. Projects need plans, not winging it.
5. On-the-fly judgments about project goals, level of service, and level of effort are key to consultant self-management and survival.
6. Regardless of project delivery or contract, commitment, people, and project knowledge make the difference. Integrated design is a myth if teammates aren't knowledgeable, committed and focused on the individual project.
7. Design professionals at the workplace make thousands of self-managed, micro-decisions daily – innovating and applying creative thinking as they work.

CHAPTER 12

Contractors:
Risk and Design Assist Expertise

INTERVIEW

John Rapaport, with John Lord, David Scognamiglio, and Jeremy Moskowitz

Component Assembly Systems, Inc./Component West   25th October 2017

John Rapaport is Chief Contracting Officer and General Counsel, Component Assembly Systems, a 54-year-old national specialty contracting firm for interiors, carpentry, and other trades. His father, Lewis Rapaport, founded the company in 1964, completing their first project at the New York World's Fair. He is active in the National Institute of Building Sciences (NIBS) Building Smart Alliance, actively pursuing standardized data for interior systems and wall types. He is a frequent contributor to ENR and author of industry papers. Joining John Rapaport in the interview were John Lord, Chief Technology Officer; David Scognamiglio, Sr. Vice President, West Coast Operations; and Jeremy Moskowitz, General Manager. Together, through on-project work such as major new corporate office campuses such as One World Trade Center and ongoing work in data reuse and prefabrication, they champion design and construction industry transformation and trade contractor involvement.

Trade Contractor Expertise or, "My Friend the Architect"

Rapaport: First, what you're doing is important. What's going on doesn't get out there – the reality of it. The sub's perspective. The fact that you care and are writing about it matters.

Can we start with a little company background?

> "To us 'design' means: how are we going to accomplish that with the systems we're installing?"

Rapaport: We are specialty contractors: walls, ceilings, carpenters, tapers, and related subtrades. David was a carpenter in the field and now runs the West Coast. So, he's seen it and can describe what we do better than anybody. Component Assembly Systems is a 54-year-old company. We started doing carpentry then changed the name in the late 1960s to reflect a broader scope. Now it's come full circle. Dad is still the CEO. He loves technology and we have seven offices in the U.S. We're working nationally. John Lord came in the early 1990s and started to digitize our processes. We took estimating onscreen, did our own software and acquired a company called C/F Data Systems for accounting. We got into it, and it helped us drive revenues and clarity. David can expand.

Scognamiglio: We're a carpentry trade contractor. Not strictly interiors. We do exterior framing as well. In different areas of the country we do different scopes of work. In New York we do millwork and acoustic ceilings in addition to framing, drywall and finishing. On the West Coast, we're doing primarily interior carpentry and exterior skin. As a 54-year-old company, the trick is: How do you teach an old dog new tricks? That's what we face as a carpentry-trade subcontractor. A lot of our workforce is mature, so they're used to traditional building methods. Not just for our company, but for many we deal with. You have the traditional folks and this technology wave. That's the biggest challenge we face, internally and externally, with other trade and general contractors.

I want to get a perspective of life from your point of view: a subcontractor trying to manage design. As an architect, I learned to rely on people like you because I didn't know what I was doing. I needed your expertise.

Scognamiglio: Right. But we have to ask: what do we mean when we say "design"? As an architect, you're going to come up with an aesthetic, a shape, or an effect you're looking for. But to us, "design" means: how are we going to accomplish that with the systems we're installing? How can we meet your objectives? So, for us, "design" is the fire rating, the acoustical systems, the details, implementation – all the things we're installing to meet your objectives.

A recurring theme. Design has many meanings to different audiences.

> "Design professionals aren't exploring all the available opportunities . . . we're the Rodney Dangerfield of the industry."

Lord: Design professionals aren't exploring all the available opportunities. We get the job. We come up with our means and methods to put in place what the architect has envisioned. Why doesn't the architect come to us sooner? We have tremendous experience. No architect on the planet has what we have. We've seen it all, over five decades on major projects. Yet, we're brought on board after design is done; we can offer curved treatments they really want, where they've squared things off.

We see opportunities for architects to leverage our experience in ways that aren't being done in today's delivery methods. We're using machines to prefab drywall into unique shapes, multifaceted surfaces. The average architectural draftsman doesn't know where to go. They look at canned books from manufacturers. Yet, if they'd lean on us to say, "I'm trying to achieve this," we have the expertise. We can push boundaries with them, but it takes coming in earlier, rather than saying, "Hey, we need the bid on Friday." We're not considered as a partner. Anything but. We're the Rodney Dangerfield of the industry, at the bottom of the food chain. We get beat up even by our own suppliers: "You're gonna use Revit, Tekla, etc." Creating our own ERP software was a matter of desperation. It helps us with trend analysis. Architects don't lean on us.

"Why architects don't ask for that?" is a good question. What's your answer?

Lord: You may be in a better position to answer. Delivery methods dictate it at times. We're working on prefab opportunities, panelization and standardizing wall types. Walls are being drawn from scratch, when they could be pulled from a standard library. Then we can focus estimating on means and methods rather than relearning new types and recounting walls embedded into a BIM.

It also starts with the AIA Documents, which are one-sided. Ten percent is held as retention *and* a bond is held. Our profits don't come close to that. So, you're double insured. Our reputation doesn't even come into play, that we've completed every project in 53 years. There should be a bond rating applied so we're not commoditized.

Rapaport: Mike, what do you think is reason architects aren't using our expertise?

I agree with what John said, contracts and delivery methods dictate some of it, no contractual privity or direct access. But more than that, it's cultural. In the architect's education and practice, we're taught by implication, that contractors are second-class citizens. Extrapolating, that must mean subcontractors are third-class citizens. This is never stated as an intention; it's implicit in the education. The architect, we're taught in our self-contained culture, is the only one qualified to perform the "high act" of building design. We're supposed to be the experts. We weren't taught to rely on others, except for our consultants, which is another group considered a tier below. Contractually, they are, but God knows we need them.

This is changing, but the past is precedent. When we get cutting-edge projects we need help. If we're lucky enough to have expert partners like you, more of us are taking advantage of it. Architects who don't get that you are professionals too are missing the boat. My sense is you'd welcome a direct call for advice from an architect whether you had the job or not: "Can I pick your brain?" It might help you get it later. At the worst, you'd develop a relationship, or do a favor.

> "We're not considered as a partner. Anything but."

Scognamiglio: Yes. What we're seeing on the West Coast is the beginning of partnerships between architects and subcontractors. But as carpentry trade contractors, we're not part of it yet. We're relegated to the group that comes in later. Now, it's the ones deemed "most valuable" to the job: the mechanicals, the electricians.

On a job we're working on right now, they were brought in 2 years in advance of our contract being issued. This was a fully modeled BIM job – every stud, not just king studs and conflicts – including full shop drawings. So, you have a situation where all the MEP work was fully coordinated in BIM. Then we were brought on and began our modeling process, and we realize we can't put the stud there. The duct has to move. There's no other way to resolve the conflict. You have a fully coordinated mechanical model that now *has to* move and change because we were brought in so late in the design process. Frankly, it's wreaked havoc with field changes and things that could have been avoided bringing us in as a partner at the same time.

> "Mostly, scheduling is a myth. There's no data from subs that drives the work dynamically. It's just done top down, and often abandoned. So, schedules don't matter. They're just people, targets, or hopeful things that don't pan out."

We've seen an expertise food chain: people at the perceived bottom of the list are masonry and drywall – they don't have any moving parts. But why solve the puzzle with parts missing?

Scognamiglio: Any money you could have saved by hard bidding or possible buyout, you could have saved two-fold by coordinating with the other trades and using us earlier for our expertise, BIM coordination, value engineering, and tweaking the design. You didn't have to spend that money in the first place. Some owners get it, but the contractual and lingering, historical misconceptions persist, like hard bid being a good delivery method.

Let's talk about leverage and impact. Where are you seeing the best leverage? Is it people? Or is technology your big survival strategy?

Lord: A long time ago, when we were blaming everyone for everything as "whiners," we found we couldn't rely on anything. Not even "trust but verify." We had to do things ourselves. We took it upon ourselves to get into BIM, doing our own models to highlight clashes and areas we need to go in for an early pass, so we don't get closed out by the GC. A little historical perspective: HVAC contractors brought BIM to the construction world. They were vertical in BIM long before we knew what the term was, because they had a manufacturing component they wanted to tie to the field.

Because they're doing heat calcs, the GC leans on them for design. They're in the best position to get on the job first. We had a meeting with a mechanical contractor and a GC in Boston a few years ago to talk about how delivery is working. By default, the HVAC contractor ends up running the clash detection meetings. Maybe they throw in a schedule. They want to pump out product and aren't doing it floor by floor. That's not efficient. They want to stamp out 1,000 pieces of the same thing. Anyone who understands manufacturing understands that, because you have to change your tooling for the next thing. That lends itself to vertical HVAC construction, and we wonder why we're going back so many times.

That's a generalization, but we have to be nimble with the schedule and do our own. We know the real sequences are nothing like what the schedule shows. Schedules are mostly static. At the tail end of a job, these Excel spreadsheets don't tie to any schedule, they're just expedient: get this floor done, so it looks good.

We lean on technology a lot for the back-office stuff you don't see and to solve the GC's problems, so they don't come back on us. If we don't, we're talking about money, the scraps we're all chasing. So, technology is a big part of our business to manage design, in the field – the Total Stations – and in manufacturing.

If the design firm doesn't have a technology, prefabrication or manufacturing mindset when they come to the table as customer design objectives, or do their part in making costs, schedules and construction more efficient, they miss the chance. They won't benefit from our expertise or improve the process. If they start design hanging on to the mindset that "it's just drywall and studs, we can rip it out or change it," then that's what we'll do. I only know a few architectural firms in the country that even remotely think about those things.

I agree, if you're not there early, the ship has sailed. It's got to be their philosophy. At conferences, people are talking about sharing models and right of reliance, but the majority aren't sharing. I understand why you're having to rework your own stuff.

Rapaport: At the Georgia Tech Digital Building Lab we're the only subcontractor present. They've had success with Tekla, steel and precast, but we're a late award trade. When we come in, duct is already in the air – forget the model – we have to get top track around the duct. It repeats because they buy us late. Many subs are late awarded, price-driven, unknown entities, with no consideration of performance.

One of your questions reminded me: there's little feedback to subs. Not just every day, "We hate you/we love you," that's nonsense. But: "You're doing this well." Are we meeting schedules? Mostly, scheduling is a myth. No sub data drives work dynamically. It's done top down, and often abandoned. So, schedules don't matter. They're just targets or hopeful things that don't pan out. At the end of the job there's not enough feedback on specifics, such as: "You didn't do well giving us proposals on time, you did great BIM work, great job meeting the schedule." There's little feedback to make us better, or vice versa. Most everything's a one-off

> *"At the end of the job there's not enough feedback on specifics . . . to make us better, or vice versa. Everything's a one-off throwaway."*

throwaway, like: "See you on the next job." We need more feedback. We're willing to learn. Not many people approach it as creating a relationship. It's largely a commodity environment, especially with walls and ceilings. I just sent you a magazine cover from our industry that illustrates what we're talking about. It's a voodoo doll with stickpins in it. It will give you a laugh. That's how we feel about our ability to be part of managing design. Unless that changes, I don't see how you can reach out to us ahead of time if you haven't talked to us at the end.

Lord: Project financing is predicated on viability. You have rental rates, costs substantiated by low bids, but if people turned in the honest numbers it might not be a "go." That's how it's initiated. You're in for a penny, in for a pound. You've got to finish it. That could explain why it's still a low bid world. The original funding group has offloaded it and moved on, so there's no accountability. We see BIM's potential to cost and schedule load as offering more meaningful data and feedback than static, disconnected Primavera schedules and PDF printouts.

The original price is seldom the final price. Owners get seduced but pay later when the difficulties come in. One solution is less low bid, price-driven work. The other thing you're talking about is feedback and data. Renee Cheng's interview talked about knowledge management and research to get that feedback and share it. But Josh Kanner found, in contrast to other business sectors, that design and construction leaders see investments in technology as additional problems rather than solutions. We have work to do.

Designer and contractor mindsets are still to figure things out for ourselves for the first time and adapt and overcome. Science, technology, and rigor are not defaults. We generally don't look to see who's done it before and move that forward. Those are things we could stop and start doing. Any other new things you're having success with?

Scognamiglio: We're trying to bridge the gap. Technology's available to a handful of people in the office doing BIM. But how is that leveraged in the field where the rubber meets the road? We've put BIM computers, PlanGrid, Bluebeam technology in the hands of the field guys. We're doing shop drawings from BIM, putting data vaults in the field with Wi-Fi to get it literally in the hands of the people doing construction. We can integrate RFIs on the fly. For each building geography, we have an RFI folder. The minute the contractor uploads an RFI, it's available in the field. When we go to an area to build a room or a pod, we look at the model first, then the plans and shop drawings for RFIs that would affect us. Jeremy and my push has been to get technology out of the office and into the hands of the guys building. We're also using it for layout.

I've had a theory for those using BIM in the field, that the only ones really touching it are at a management level. If I'm a guy hanging wallboard, a supervisor is telling me what to do there. Am I correct in that assumption?

Scognamiglio: Not necessarily. For us it stops or slows down at the framing level. Every stud is modeled. It stops at the board. I need my framing guys to be able to go to a community data center. I have a gang box, a computer, a printer and the model set up at every leading edge of the project. They don't have to depend on the framing foreman, who may be in one of ten different locations. It's like a library: they go to a community desk, pop open the model, and are trained and empowered to use that information, so they don't have to wait. They don't have to bottleneck at the foreman. Each lead person gets that information the minute they need it.

It's not hard to teach someone how to access design information on an iPad. The distinction that it stops at the framing helps – you're not modeling the board.

Scognamiglio: No, but for the drywaller, taper or bead installer, a lot of jobs for the high-end design firms are intricate designs. What you knew on a standard office building no longer applies. Even late tasks need model access: how does this bead work or meet the exposed steel column detail? That design information has to be available to everyone – even if you think they might not need it.

That's managing design implementation in the field by giving open access to current design data.

> *"I like to call it moving from 'design-assist' to 'project-assist.' That's how we subs should be thought of."*

Rapaport: Yes. And more data is getting created. Productivity data. 5D BIM, the most basic data. How much does it cost to put this in? How many man-hours? Managing design, in our terms, is not during its creation, but in implementing it and performance. Owners don't have historical data because they haven't required it. They should have a right to it. It's an industry debate. We're starting to see owners require that subs provide labor coding and task durations. The estimate is ours to keep. It's proprietary. The GC tracks how many carpenters are on the job, where they are and what they're doing, but not in a format useful on future projects. It's all one off.

Yes, there are a lot of one-off designs, but there are also repeat projects it would be useful for, like schools, and simple office buildings, stock designs where that history would help.

> *"It's one of the most remarkable things I've seen in 30 years: they're not collecting historical data."*

We don't know of anyone except the subs who are holding this historical information. The GCs get that. They do conceptual estimating, but they don't know what it's going to cost broken down in meaningful ways. It's one of the most remarkable things I've seen in 30 years: they're not collecting historical data.

Where do you, as a GC, get your historical cost information to begin with? The R.S. Means databases aren't based on actual data. Teams have a right to know it. It would help us get budgets in line with designs. Some are trying to collect it, but they're not organizing it in useful ways. An ENR article[1] talks about how we can't get better if we don't know what we did.

Until that changes, how do you manage design? It'll be over budget and late. We had a major highly visible project – it was over budget – and yet every stick was modeled in BIM. It should be in the AIA contract:

---

[1] "Unsticking Vital Data," Engineering News-Record, April 13, 2015.

you must provide us the codes we tell you. Then we could roll it up, and figure out how much time it took to build that stairwell, how much framing, hours? Then, designs can reflect it, and we can work together to control costs. I like to call it going from design-assist to project-assist. That's how we subs should be thought of.

You mean adding value project wide, not just on your trade?

Rapaport: Yes. Getting in early, starting with design-assist and advice, doing the things we've talked about, then providing expertise and data that benefits the whole project. You don't have to hold all our money because you know we're building on the 5th floor. Held money in this industry shows no one trusts each other. Subs have to bill aggressively because you're holding 10 percent, but we have to pay for labor, materials, and our subs. We don't get paid for sixty-five days, plus you still hold 10 percent. That's gotta change. We'll let you come into our systems and see our data, how we're performing, then you can free up money. Change the values. Hold two percent, maybe, not ten.

An IPD flaw is the idea of full, open-book accounting. We want our profits and risk to be ours. We own it. We're not fee based, but we'll share process data, hours, and history. How we use that for future bids is ours. That will be a future game changer. You'll know what it takes to do things, you can have better conceptual estimating and start learning from each other. Now, everything's hidden. That's why you don't see reaching out: Hey, why don't we work together? Aside from "Gimme this answer now, go faster, we need more people . . ." Well, you didn't pay for more people. It was low bid, done our way.

That's what it will take to make a collaborative team – not just somebody plugging Primavera schedules with no input.

Lord: Think how GC superintendents are graded. In my mind, a perfect one conducts the symphony to perfection with high quality. Everyone works together. How do you get feedback on that? Were the subs productive? No. In fact, superintendents are graded on how hard they beat the subs into submission. Another GC told me they're trained to do that. How is that mentality helping our industry? Let's say there's an issue on the job. We work with GCs long term. Many owners are one-shotters. If there's an issue, who do you think the fall guy is? It's the designer. I understand the architect's point of view. They feel targets on their back. If you understand delivery you know why. When there's a job issue, there may be an attempt to make it look like a design issue, when it's really a GC or project issue. Subs want to get paid for GC-directed inefficiencies, so, there's designer blowback.

> *"Think about the extended services architects could provide. They start the ball rolling, but then get into the 'defensive legal fetal position' once things develop."*

I'm trying to get designers to see that. Contractors are better at business and getting paid for inefficiencies caused by others. Architects aren't. They need to shake up their process and defend themselves.

Lord: Think about the extended services architects could provide. They start the ball rolling, but then get into the "defensive legal fetal position" once things develop. When that happens, it's a lost project asset. They should be camped out on that job. I'd love to see them turn around an RFI in one day, not 45. They should be tied to overall job performance, and mediating payments fairly to get us closer to zero retainage. When we miss the chance to learn from projects it's a shame.

What may drive that retracting attitude is our collective commoditized position. I lobby to owners to pay architects more and fund onsite people. They need our help in the field.

> "Coming in after the fact turning BIM into an as-built, wastes value, defeats the purpose and misses the intent. If that message could get through to owners and architects, quality and cost would be so improved."

Moskowitz: Our architect is onsite every Wednesday. On most projects, we're not using BIM as it was intended. The fail on these projects is late buyout. The intent is to model every stud, dimension everything. But when you buy BIM so late, it defeats the reasons BIM was designed: to make construction efficient, improve quality, and create efficiencies. Coming in after the fact and turning BIM into an as-built, wastes value, defeats the purpose and misses the intent. If that message could get through to owners and architects, quality and cost would be so improved.

We've got to share that blame for late starts. Owners, GCs, architects. We know better but don't change. If we all got on board earlier, it would be better. But owner problems prevent that.

Scognamiglio: They're missing the opportunity on multiple levels. From a GC standpoint, you're working for leaner fees. So, the more coordinated things are, the less supervision, management or rework you need. We're looking at another phase on a recently completed project. We asked: when are you going to buy out the drywall? They said: probably the same as last phase, which equated to 3 months before work starts. We threw up our arms and said: didn't we learn anything from the 1,000 conversations we had on late BIM starts, missed expertise and full coordination? The importance of getting someone involved earlier? But it was just the same mindset.

We've always said we'd have to build the same project twice to get that learning and those efficiencies. So, when you finally do get that chance and still miss it, you want to pull your hair out, if you have any left.

Scognamiglio: We had a situation with a 4" pipe in a 6" wall. Our structural rule said, "No penetration could exceed 5/4 of the track width." It was never flagged as an issue. But nobody realized the floor core was 6", which violated the rule. Because the job was so compressed, nobody saw it until we got in the field. A lot of the benefit of timely BIM was shot. We've done a lot of BIM on the West Coast. What usually happens is that the BIM effort is behind construction – the issue Jeremy mentioned about how BIM serves as a source for not much more than a nice set of as-builts. It could have done so much more to help the project be efficient.

Moskowitz: The other thing we notice in a lot of architect's drawing sets, is that details seem to be pulled from standard, outdated details. After we're awarded, we end up being having to explain more current technology, products, assemblies, that work more efficiently and improve schedules such as top-of-wall products that turn 3-pass operations into 1-pass. Coming in late, it's difficult to get those into the job. In early design, we can share that expertise, the STC (sound transmission class), the fire rating, and end up with solution that streamlines the schedule for everyone. GCs love that.

Scognamiglio: It's got to become part of the process. Some walls have stringent requirements. There's a common misconception among those not intimately involved in drywall, that "a wall is a wall" –just studs and drywall with a bottom track and a top track. That could NOT be farther from the truth. A wall is a dynamic assembly that moves in many different directions under different conditions to meet different criteria. In some cases, a fire rating, a sound rating, a smoke rating. There's a matrix of products and assemblies for *every* wall we build. An architect needs a wall with a certain STC rating. That's created in a lab with strict standards. In a lab environment, it meets them. But in designs and the field, things change, and we have to react. So, a wall is *not* just a wall. It's a dynamic assembly that has to meet a set of dynamic criteria. We have to educate the architect and owner: one of the biggest parts of our Preconstruction efforts and "our" design is how to design walls to meet specified criteria.

I've been on the other side of the table, valuing your help, and it's absolutely a design exercise, focused on technical implementation, but design nonetheless. Without it, and you, the likelihood of achieving a design vision that meets specified criteria is low. What do you see for the future?

Lord: You touched on IPD. We've done some. Insurance companies aren't wild about it. Here's an anecdote about architects sharing BIM models: We met an architect at a conference who said, "Sure, we'll share." We called him Monday and he said, "I'm sorry, I can't." There's innovation in the West, with pods, prefab, not your grandfather's prefab, but higher quality. We also see material handling differently – like exoskeletal material handling that alleviates worker injuries. Technology can help that.

Rapaport: We are leading a NIBS BuildingSmart BIM initiative for Partition Exchange types called PARTie. It was originally called WALLie. We're standardizing wall types and developing digital materials ordering. Now, it's still paper based. Type 1 and Type 2 walls will be the same through all architect's designs. Now, on every job you have to rethink and reunderstand design intent. If they're standardized, you can work at a higher level. Spec writers are hidden. People don't understand their connection. A flawed issue can repeat and cause problems. We should get the quantities. Why are we having to recreate those? No one wants to recount them. In Europe they get them digitally. A better future would be to give quantities to subs. Why have subs spend the time? So many uncompensated hours. We get 2D drawings, have to count them and make mistakes. Specs aren't linked, or intelligent. They're PDF-ed and dumbed down for us. That's what you get when you dumb down information and buy at low price. That's got to change. With less held money, there's money for technology investment. Subs don't have money for servers because things are so tight. Maybe it comes back to self-respect. In the future, the IPD concept is: we're equal partners. If we do well, come with me to the next job. Then, good owners will say, "Let's get that team together again." The future should be clarity, less held money, and more service: working together. Then, others will have to up their game, so it's not just about low price.

> "Details seem to be pulled from standard, outdated details. We end up . . . in the position of trying to explain more current technology, products, assemblies, that work way more efficiently and help improve schedules."

Your work towards standards presents a positive view of the future. We've got work to do though, fueling change through technology, but plenty of hurdles left.

John Rapaport, John Lord, David Scognamiglio, and Jeremy Moskowitz Key Points:

1. Architects are not capitalizing on subcontractor expertise early to inform design, reduce costs and schedules, enhance constructability, and streamline processes.
2. Subcontractors face technology adoption challenges with a mature, change-resistant workforce.
3. Design, to subcontractors, means implementation, means, and methods in executing the architect's concept – an after-the-fact, make-work approach when not done in unison.
4. Subcontractors brought in late miss the opportunity to add value as part of the team.
5. Early BIM coordination with only MEP and structure is wasted if interior systems BIM is added later. This new puzzle piece necessitates recoordinating earlier systems or tearing out work.
6. To streamline processes, design firms should embrace fabrication and installation as part of their design philosophy.
7. Scheduling is a myth. CMs, GCs, owners and architects should embrace Lean team-based pull scheduling with input from all trades, versus top-down dictated schedules.
8. Teams should value on-project and after-project performance feedback and share actual productivity data as knowledge sharing to create feedback loops based on history.
9. To enhance value as partners, subcontractors should move from "design-assist" to "project-assist" beyond their trade to project-wide expertise, schedule optimization, and design management.
10. Owners, CMs and teams should embrace expanded services for architects to ensure their continued expertise and presence during the construction phase, when "design" is made real.

INTERVIEW
Don Davidson and Jeff Giglio

CEO and Chairman, Inglett & Stubbs   7th February 2018

Don Davidson joined Inglett & Stubbs in 1977. Before his current position as CEO, he served 10 years as Vice President of Construction Services, responsible for commercial, institutional, and healthcare estimating, value analysis, and project management. He has current company oversight for safety, quality, and company resources. Don has 33 years of experience in the electrical industry and is a graduate of Kennesaw State University, 1977, with a BS in Architectural Engineering.

Jeff Giglio is chairman of Inglett & Stubbs. He joined the company in 1979. Previously, he served 10 years as president/CEO responsible for overall company oversight. He oversees development and coordination of Inglett & Stubbs' other business entities and has a BS in Electrical Engineering from Georgia Tech, 1977.

Inglett & Stubbs LLC is a leader in the electrical construction industry, consistently ranked in the top 30 electrical contractors in the U.S. by *Electrical Construction & Maintenance* magazine. With a remarkable half century's history of successful projects, the company's tradition of accomplishment and sustained growth has secured a strong market position and promising future for partners and customers.

Planning and Trade Contractor Design-Assist Mindsets or "We Need You Onsite Tomorrow"

Let's talk about "Design." Although you're an electrical contracting firm, you do significant amounts of "design" – finishing, or adapting engineer's designs to reality or market conditions, detailing design, or doing design-build work. What does "design" mean in your terms?

Davidson: To be true design it's from scratch. On a lot of jobs, we partner in design development. We coordinate with end users to create what they are looking for to avoid rework. We don't do a lot of true design, we do a lot of refinement. Clients count on us to have their backs. One of the advantages of a solid MEP team is we get to know that owner's best interest, needs, and wants. They rely on us for that. That's not always the design engineer of record. In many cases we deal with the same engineers over and over. We've earned a good reputation, so they're open to discussion and that we may know more about the owner's need or operation than they do. New people may be reluctant to modify their designs, but when we deal with repeat teams it's easier.

"Manage the project. As opposed to the project managing you."

I'm curious how you approach that, not being the lead. Some call it "leading from behind." You provide design assist services to some degree on almost all projects, right?

Davidson: Design assist is a good term. Leading from behind is too. It's an art. So many are interested in their agenda – not all designers, but many want to put their stamp on something the owner may not want. Give clients what they want without upsetting your design teammates.

Conversely, what about the projects where the owner wants to raise the bar, and builds a process with a stellar team and brings everybody on board early to establish synergy?

Davidson: We like those jobs. You can't do the best design without a good construction team to keep it in budget and on schedule. You have to be cognizant that all these vendors want to push new technology, but the owner is wary. Balance. Solving those puzzles are fun.

How do you get keep on course when technology is pulling you one way and design interests another? With maybe unknowing owners in a third direction?

"All I wanted was two switches."

Davidson: Communication is a big part of what we do. The end user is the key. At one extreme are the lighting designers and interior designers. We try to respect what they're trying to do but have to communicate and go to the end user – the maintenance guys – and they say, "That's not what I want at all." Not long ago we had a job with fancy lighting controls, zones, massive, very complex, very expensive. When we turned it over to the owner's facilities engineer, he said, "All I wanted was two switches."

"When CM clients tell us, 'We can't tell you if you got the job until Friday,' and 'If you get it, we need you onsite Monday,' it just doesn't work."

But what they want may be too simple and not meet code. Some energy codes no longer allow "simple." You have to walk the line between energy codes, fire codes, the janitor and lighting designer.

How do you find the mix of people skills, persuasion, psychologist and tech savvy to keep up? That's not something the average electrician has.

Davidson: We're training for it! We've brought in people who know about these things, but technology changes so fast. Even if you're the factory rep, it changes faster than you can keep up with. We expend effort to stay current as we bid and install. We're doing it, so it works. But lighting control is nearly impossible to keep up with no matter where you fall in the chain.

How do you dispel the lingering mistrust and old cultural biases among teams?

Davidson: That's where history and trust comes in. You can't lead someone down a bad path and keep their trust.

We just finished the new Mercedes-Benz Stadium in Atlanta together. With the end date set, it was a tough process. From your perspective, what could we have done better?

Davidson: That was a design management and design completion challenge. Most would agree if we'd spent another year in design, it would have allowed construction to go smoother. For reasons beyond our control the date was set, and the cart was before the horse. We were brought in to help with Pre-construction and design assist. That level of innovation needed it.

How would you change the design management process in an ideal setting?

Davidson: Planning! Not just about scope, but manpower, management, and cash flow. We don't allow enough time to plan and manage all phases. Calling and saying, "We're starting next week; we need your preconstruction people here" doesn't give us enough time to develop a team. We constantly see owners not allowing enough time at the front end, so we can avoid redesign. It affects all phases. When clients tell us, "We can't tell you if you got the job until Friday," and "If you get it, we need you onsite Monday," it just doesn't work.

That's perhaps the biggest issue everyone has raised. Industry data and personal experience shows that when we do plan, it pays off. Why can't we get owners, and sometimes ourselves to do it?

Davidson: It's owner driven. It pushes the same people under the bus every time. Even the GCs. We've grown accustomed to waiting until last minute. Because we can. That doesn't mean it's the best way, it's just become expected – a default bad behavior. If people feel they have until the last minute to decide, they do. If they'd commit earlier things would be better

Giglio: The State Farm Office complex we did with you is a good example: we were brought on way early. We had a big hole to dig first. We had almost 100 percent drawings before we started, so we had time to estimate, schedule, and determine labor needs. We planned the heck out of it, and it went amazingly well because we had time to do installation layouts and BIM coordination. Instead of just keeping our head above water, we had 95 percent of most of it weeks in advance. All we had to do then was execute the plan and react to daily crises. It showed. It had challenges but wasn't an everyday fight. Our catchphrase is "Manage the project." As opposed to "The project manages you."

What new tools and processes are you using to get better results? What can traditional practitioners apply to our industry?

Davidson: A lot of new tools have come into play to make it easier. BIM, LEAN, and so many others. If we do them, our work is supposed to be better, faster and save money. I haven't seen that yet, but it helps manage conflicts. If you don't do BIM, prefab, and other things today, you can't keep up. It's part of today's means and methods.

Giglio: BIM is a good example. A great tool, great return, but we have to be careful how far we take it. It has diminishing returns. It doesn't all have to be done down to the last locknut. Having the last 10 percent in BIM costs a fortune and doesn't serve anyone. The Trimble units we use now to layout work are great. We can't function without them. They're a standard. Speed-wise, prefabbing is big. It helps with safety, quality control, waste, and logistics. The problem with prefabbing is it requires pre-job planning. You can't make changes the day of, or it's costly and what's on the skiff must be tossed. So, prefab brings risk too.

Logistics is key. How we think about materials and people. Where they park, working conditions. Our ability to manage design projects flows to operations. When people think we don't respect them, they stop caring. They go from being professionals to just laborers. We need to start moving the pendulum, because logistics is a motivator, and it's driven by our ability to manage design process.

How can you work with designers better?

Davidson: When people drafted by hand we could do redlines. Now, they're designing and documenting final product simultaneously. Looking over their shoulder is harder, but necessary. If they let us have access to their systems, we can have input. Some resist because they don't *want* us looking over shoulders. Their fees are low . . . they don't have time. The way contracts are set up, all they have to sell is time. Some still don't think they need us.

Maybe they also resist because they're in the middle of the creative process? Giving access and transparency can be an issue.

Giglio: We need to look at design as they're doing it, versus their doing it, issuing it, sending it, and weeks passing, and then us coming back with feedback. That doesn't keep up or add value. Our work needs to be done in real time.

Davidson: It gets back to being brought on early. When we are, we're there to help ensure a smarter creative process. When we come in late, we can't. Right now, the entire industry is facing an issue of not having enough skilled people – office and field people. For us, it hurts doubly. It's not just getting the right people, but our supply chain is impacted too. We're on the job day one and the last day too, the longest of any subcontractors, because of the nature of our work. We are so dictated by our predecessor's work: if they're not on schedule, it doesn't matter. If not, it's impossible to stay on schedule. We have to plan and resource-level it, so design rework and mismanaged design kill us too.

Giglio: Going back to prefabbing, the more we can preplan and prefab the better. We might only have an hour in a prefab wall onsite versus eight hours for site-built.

Prefabbing is a solution, but it needs two things: motivation and time. First, you need an interested owner and design team. Then, you need upfront planning time to pull it off. Teams who want to do it have a different mentality and goals. We're not seeing that much. We have to drive it more.

Giglio: A downside is prefab removes the owner's flexibility to make changes, because we've spent half the cost in the prefab process. So, there are risks.

Don Davidson and Jeff Giglio Key Points:

1. Today's industry relies on construction expertise for design input and management.
2. Trade contractors offer expert value in interpreting and translating new technologies into design, owner, market, and construction contexts.
3. Even experts and vendors are challenged to keep up with the pace of technological change.
4. Trade contractors can balance and inform often conflicting loftier design objectives with the simpler needs of owners' facilities operations and maintenance staff.
5. Preplanning phases allow smoother processes.
6. Owners and GC/CMs have come to expect last minute decision making, but this comes at a cost.
7. Prefabrication is a solution to worker shortages and compressed schedules but needs motivated teams and upfront time to execute. It has offsetting risks: limited flexibility and committed cost.

INTERVIEW
Wayne Wadsworth, DBIA, LEED AP

Executive Vice President, Holder Construction Company   21st March 2018

Wayne Wadsworth is a graduate of the University of Florida Rinker School of Construction. In 2011 he was selected as their Distinguished Alumni of the Year, in recognition of his ongoing service to the university. In his 30-year career with Holder, he has been responsible for Preconstruction, Planning & Design Support Services, MEP Services, Building Information Modeling and Interiors, and championed company culture. As part of the leadership team directing company growth from annual revenues of $175 million in 1990 to more than $3 billion in 2018, he has been responsible for many of the company's most significant clients and projects.

From 2013 until 2108 he has been Principal-in-Charge/Project Executive for the Mercedes-Benz Stadium in Atlanta, the new home of the NFL Atlanta Falcons and MLS Atlanta United Soccer Team. This $1.5 billion-dollar facility is a first-of-kind, operable roof, LEED Platinum facility completed via CM-at-Risk project delivery. For his exemplary role, he was presented with the company's first ever Chairman's Leadership Award in 2018.

"Eyes-Wide-Open" Leadership and Design Ownership or "Stretching the Market: The Chain"

You won the company's first ever Chairman's Leadership Award for your role on the stadium. And the project has just won the AGC's national Grand Award for Best Project of the Year, in 2017. To a person, everyone involved admits it was the toughest project they faced in their careers. You were a leader, confidant, strategist, spiritual guru, and counselor to all of it to owners, contractors, designers, trades – you name it. How did you do it?

> *"You simply cannot get caught up in the exposure of the dollars, the schedule, your own firm's and the personal challenges and emotions. I stayed focused on accepting the reality."*

Wadsworth: I'm honored by the awards, but as I said when I received them, I did so on behalf of every individual who contributed. Thousands. I had a role in orchestrating the talent. We had an extraordinary team and did a good job bringing the right people to the right situation at the right times. We tried to create a platform where they could do what they do. It's a simple formula.

As I think back, one of the things I tried to stay focused on was not getting into the noise – the level of stress present in every meeting. You simply cannot get caught up in the exposure of the dollars, the schedule, your own firm's and the personal challenges and emotions. I stayed focused on accepting the reality of what is, and then, consistent with Holder's culture, how do we treat people and solve problems for everybody? Because you have to understand their individual problem sets and business needs. I just read something you wrote – that we were "building while we were designing." Doing that brings issues. With such an unbelievable design, we had to keep reconciling the reality of where we needed to be to finish, while we were caring for that design. That was an ever-changing battle. You can't get frustrated. I don't know how many times plans changed. New issues came up and walloped us. There's just no choice. For the sake of Arthur Blank [owner of the Falcons], the city, the state, our own company – you're in this and you're committed. You've got to move forward. You can't get down. You can't fret.

That was the greatest lesson I'll take away: you better be careful what you ask for – especially when you ask for things that have never been done. What it took to get it done is remarkable.

The lesson for a "design-driven," "first ever" activity, by God, is: "Fear it, respect it." Face it together. It will take the best – and even they may have never done it before. It becomes an exercise in managing risk, and the unknown.

Wadsworth: Yes. My preconstruction background was critical, because so many of the problems were "forced reality checks" of going back to square zero and assessing: how do you find a path that still leads you to success? All the fundamental project controls – how do you balance budget, schedule issues? How do you get documents to support it, scope coordinated among trades with commitment and understanding? Then figure out the processes and people to execute it – and it changed a million times.

"Not getting into the noise."

Most readers, having never done it, don't know what "preconstruction" entails. It's not simply getting plans, doing a quantity takeoff and an estimate, and awarding the trade contract. We define it as something very different, much broader.

"So many of these problems were 'forced reality checks.' How do you find a path that still leads you to success?"

Wadsworth: It's more holistic. Taking ownership for the entire process, owner, developer, designers, everybody coming together to plan, set goals. Then, how do we achieve them?

Practiced at its highest level, preconstruction is really an exercise in design management, beginning with setting expectations – often based on an already started design.

Wadsworth: For us as builders, it's about translating it all into construction. We've got to build something. Ultimately, subcontractors have to do shop drawings and fabricate it. It's about understanding design, expressing the goals, and translating it all – in a balanced way – with cost, schedule and quality.

How do we share those lessons? I'm a curator of such things. Being so engaged on a project like the stadium, did you keep a diary or lessons learned log?

Wadsworth: I have become more thoughtful. Interestingly – architects and designers will appreciate this – my notes are becoming much more graphic in nature with more diagrams than words, as I make connections. Words are less efficient. My thoughts about cause-and-effect relationships are becoming clear. I've found diagrams more effective. I have thousands of lessons I intend to share. It'll just take time. I have a difficult time doing them when I'm in the moment on a project like the stadium, when I'm processing as quickly as I need to. But I have so many thoughts I'm dying to pour out, not unlike you are. Beliefs, thoughts, principles that got me through it.

I know in the moment, it's all you can do to sleep and try to clear your head. Diagrams have always been a part of my toolset. The McKinsey 7-S Organization model is a great one. It shows a networked web of 7 organizational systems that must connect and align:[2] Strategy, Structure, Systems, Style, Staff, Skills, and Shared Values. But I add an 8th "S" for the Supply Network. (See Figure 12.1 below.)

Wadsworth: Yes. It's very good.

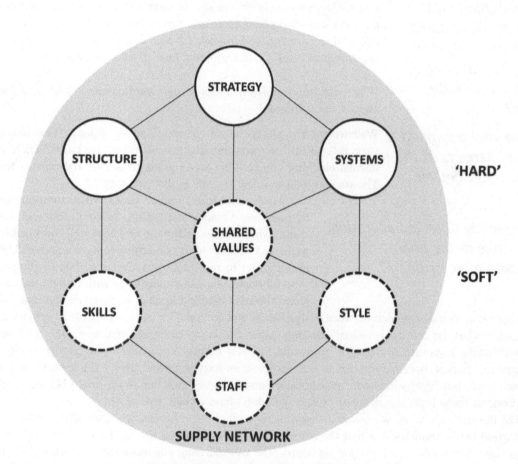

FIGURE 12.1   7-S +1 diagram, M. LeFevre after Peters, Waterman, McKinsey & Co.

You are very "together": confident, capable, positive, a great leader, an uncanny teammate. We've shared in stressful times, I go home late at night and have a beer to unwind, but as an ex–track star and Iron Man athlete, you go home and work out. Did this project threaten your resolve? It did mine. It was one of the more complex buildings in recent years. Did it change your behavior – how you managed yourself?

[2] Peters, Thomas J. "A Brief History of the 7-S ("McKinsey 7-S") Model," January 9, 2011, https://tompeters.com/2011/03/a-brief-history-of-the-7-s-mckinsey-7-s-model/.

Wadsworth: Certainly, my time was modified. Virtually every waking moment was claimed. I had to dedicate myself fully, to be all in. I love to work out and train on that side of my life. The goals dropped off, but the activity didn't. I still worked out, generally twice a day. Before work and after. And I even took to doing it *during* other activities. While brushing my teeth or shaving, grab a chair, stand on one foot, or do squats. You can do it if you're committed. A concentrated effort like that does provide a mental, physical and psychological relief. It's part of who I am. I try to share it with others. There's no excuse. I try not to accept excuses – in any part of my life. There's always a way. You just have a find a way to adapt.

"You always have to be watching and asking: What about an issue is 'one off,' not your familiar, bread-and-butter path?"

I love your no-excuses mentality and how you rose to the challenge with an equal reaction to keep your equilibrium. I'll share – at the other end of the spectrum – after I left the jobsite for the last year, I was usually on the elliptical trainer during our weekly 7 a.m. Friday leadership calls, doing two things at once – with a different level of intensity, I'm sure.

What else did you learn about design management, and how has it shaped your outlook?

"How well equipped are all aspects of the industry to deliver it?"

Wadsworth: You always have to be watching and asking: What about an issue is "one-off," not your familiar, bread-and-butter path? Where your normal blocking and tackling aren't going to deliver the normal result. The stadium had so many systems in that category.

"Eyes-wide-open collaboration, and links in the chain become more critical."

You've said this a million times, Mike, the importance of planning. Lessons learned are to stress-test and understand: what are you asking for, and how well equipped is the industry to deliver it? It's one thing to have wonderful ideas, and another to recognize how they translate to the people who fabricate, ship, and coordinate with trades, and understand the risks inherent in all of it. Some design issues offer a challenge, even a disconnect, between design intent and what the industry is capable of constructing. No matter what the process is called: planning, preconstruction, or design management – it's figuring out and having a grounded execution approach. For new things, it's also how you *change* reality and *improve* the market. Because you *can* do that. You can set lofty goals and stretch the industry to places it hasn't been, but "eyes-wide-open" collaboration and knowing each link in the chain become critical. If just one of those links is missing or broken, you don't have a chain that works.

On the stadium, when we saw and experienced each of those breaks, we made shifts and changes not normal in the industry, but that the industry is capable of, given the need and motivation, to the extent your owner and developer are on board. The transparency, collaboration, and having the right people are exponentially more important as you stretch boundaries. Not just people who *want* to, but who *can*, and are in the right *position* to. There's a big difference.

The 7-S diagram applies. Not only does every link in the chain have to be aligned and working, but there's an eighth component – outside your internal organization – it's your network. In this case, the industry supply chain. You can't operate without it. It takes design management to a new scale. Not only do teams have to manage designers, owners and builders – they have to do it in the context of the industry – in some cases changing that industry.

Wadsworth: Design management isn't just about lines on the paper, or making sure geometry, aesthetics and functions work. It's how that design relates to our ability to deliver it. Constructability, not just means and methods, but how the entire industry comes together to deliver all of it, is key. If you don't understand all that you're fooling yourself.

Let's shift to look internally. Twenty years ago, I left architectural practice to join Holder to create a new role in managing design. You were a big part of the leadership team that brought me on. Did you have a vision for what my role could be? How has it played out?

Wadsworth: It was the right decision then, and I'm a thousand percent all in that it still is. We *are in* the design business – whether we're the CM, the GC, the Design-Builder, or doing delegated design. I say this universally: you can't be a contractor in the way we are and not realize we are in the design business.

To clarify, we are not the designer-of-record, nor we do design in house. We bring on expert partners to do that. You're talking about owning and being affected by design process.

Wadsworth: Yes, I'm speaking philosophically. Relationally, not contractually. But that planning and design success dictates our ability to succeed and execute as a contractor. Putting energy in setting projects up for success and collaborating with AEs and trade partners is essential. My vision for doing that is to make life better, easier. Not for the sake of getting credit, or any other purpose. The more we understand design, the better we'll be. In the traditional education of a contractor, there's just not enough time to learn that. In the risk we take every day as a contractor – to guarantee costs, schedule, quality, and other things – we're 100 percent dependent on design. So, the better job we do managing design . . . you get the point. I still believe today having individuals within the company with those skills makes things so much more efficient. So, I'm drinking the Kool-Aid and I'm going to keep espousing it.

From starting with one architect – me – and a germ of an idea twenty years ago, we've grown to five registered architects – dedicated to transferring design skills and spreading that gospel across the company. Who does design management involve? Clients? AEs? What's your philosophy?

Wadsworth: It's not just managing architects. It starts with an initial conversation and an owner having a need. At that point you're beginning to manage design. It might not be lines on paper yet, but there are concepts, options, ideas, forces that shape the project's birth. What are they? Is there a site, a budget, a program, a need date? You're starting to balance the equation of expectations and drive it down into every aspect. You balance them in a nuanced way. That equation is always running in the background of my head – in conjunction with our own needs as Holder, our core goals and business.

Let's look at each of the OAC perspectives. Since owners drive things, what do our clients need to know about managing design?

Wadsworth: A healthy appreciation for the chain we talked about earlier. Anything they want, has a chain and its links. The best owners don't commoditize *any* link in that chain.

That theme has come up often! As professional service providers, nobody wants to be commoditized. Maybe there's a market for a commodity approach, but we're not in it.

> *"I say this universally: you can't be a contractor in the way we are and not realize we are in the design business."*

Wadsworth: If you respect each link of that specific product and think about what's unique, you'll get it. Maybe it's not even the design. Maybe it's the market. Right now, labor shortages limit manufacturing capacity, or steel tariffs may impact or change the design. How might they potentially dictate a different material or approach? Rather than saying "that's the contractor's or designer's problem," good owners embrace that risk and pull it forward. This is a detail-oriented business. Anytime you take something for granted has the chance to come back and bite you in the butt.

When owners say, "That's what we hired you for," they distance themselves. It sends the wrong message. Not one of enlightenment, trust, togetherness, and risk sharing. In many cases, the design team works for the owner, and the market is a collective team risk.

Wadsworth: Our saying has always been: "We only have team problems and we only have team solutions." Focus on staying together as a team and bring your different skillsets and experience together to solve problems. We share the role to prevent and solve problems.

You touched earlier on my pet peeve: failure to plan and align objectives. We find out mid-project we're trying to do different things. Owners first, but us too. We have a voice.

Wadsworth: We all – owners, architects, and contractors – tend to underestimate what it takes. Particularly on pioneering projects. It's a puzzling issue. We should advise owners to have the fortitude to speak up. Some of our brethren overpromise trying to win work.

Speaking of brethren, what would you like to tell our design partners? Plenty of design firms, good ones, well respected ones, seem to be distancing themselves from what you're talking about.

Wadsworth: That *we* are in the design business. And *they* are in the construction business. Even though we're not designing it, and they're not building it, we share the outcome. Good architects take complete responsibility for helping their contractor execute and vice versa. In the same way we're trying – spiritually – for design. That goes back to education. How much did you learn in architecture school about how to be a contractor? Not much. And motivation. You can't underestimate the extraordinary design talent out there. We want and need it, but somebody has to translate it.

In 20 years we've made a difference, helped teams, and trained many apostles, but there's still much to do. What should contractors know about managing design?

Wadsworth: The opportunity to evolve and take ownership of everything we've talked about doing is still out there. That's not a criticism, it's a celebration. How can we aspire *as teams* to return to being master builders? There's so much expertise. On some projects you can feel the flow, the beautiful music. You have teammates who get it. Somehow the team is aligned. Everybody knows what they're supposed to do and it's fun to be a part of that dynamic. It makes life better. That's what we all have to aspire to.

Wayne Wadsworth Key Points:

1. Leaders need the ability to "tune out the noise."
2. Cutting-edge design requires an eyes-wide-open, shared risk approach.
3. Design and construction are mutually interdependent. In team approach projects, each must share ownership for the other. The industry supply chain network is integral to great design and construction.
4. Design management is most impactful early in planning and pre-project stages.

INTERVIEW

Jon Lewis

General Superintendent, Holder Construction Company   31st July 2018

Jon Lewis is General Superintendent at Holder Construction Company. He a holds a Bachelor's in Civil Engineering degree from South Alabama University. Over his 20-year career he has been lead superintendent for 10 significant projects and overseen 30 others. Notable projects include the 45-story high-rise $670M tower for Devon Energy in Oklahoma City, a $620M, multiple-phase office campus project in Dunwoody, GA, the Business School at Technology Square at Georgia Tech, the Dulles Discovery Office Campus in Ashburn, VA, and the Central Park East Office Tower in Phoenix.

An early technology advocate, in 2005, he was the first company superintendent to use building information modeling on a jobsite leading to significant payback of the then-emerging digital toolset. He has since championed drones, laser scanning, mobile model and document access on iPads, and onsite digital job boxes. A four-time winner of Holder's President's Award of Excellence for Superior Quality, he championed Lean processes in 2006, led the scheduling intuitive, is a winner of the company's Andy Rogero People Development Award for training, and workforce development champion. He shares the field perspective.

Contracts, Collaboration, Construction, and "Chasing Design"' or "Fear the Unknown"

We're talking about managing design as it affects construction. As a boots-on-the-ground guy, you get limited exposure to early design stages but then have to make it all real. What are your issues? I'm giving you the soapbox.

Lewis: The big issue is the industry wants 100 percent collaboration and input but isn't set up with contracts that allow us to behave like that. We're restricted on how free-flowing and open information can be because of contract risks. An example would be, if everybody wants a share of the schedule risk and utmost collaboration, then everybody should be willing to take part of that risk. But in every project in my career the schedule risk is completely ours. Unfortunately, that causes every conversation to

> **"For 100 percent collaboration, everybody should be willing to take part of that risk."**

be led with the premise of "This is going to affect the schedule." I get it that owners need a commitment to move in and run their businesses. But if we want to get to 100 percent collaboration and share in profit, accolades, and rewards, the industry is going to have to figure that out.

Great observation. I don't lead with that because I'm not a lawyer, but it's true. Even when we don't get those contracts, our culture still has us trying to collaborate. Those who know understand contracts are where the leverage is. Not to be defensive, just to be just good risk managers, we have to be aware and honest about how the actions of others – like designers who may not be under contract to us – affect us.

Lewis: As you know, we teach and train our people to go the extra step, never write a cold RFI, never have a conversation that hasn't been vetted. No surprises. And when we don't get what we need we're going to shoulder the load and carry the rest of the team to the finish line. That's who we are. But it gets difficult when you have a developer or owner's rep that doesn't' live in that cooperative world and thrives on conflict – that puts the team in a defensive position. That puts us in turmoil.

People who haven't worked with us or any collaborative group . . . if they only knew how much risk we assume – that we don't need to.

Lewis: Right. I don't know if we'll ever get there. The other issue is time. Owners don't give design teams enough time to complete contract documents before we start. We're always chasing a completed design. It shows in submittals. We'll do them based on the CDs and they'll be sent back with red ink saying "see forthcoming bulletin" or "submit RFI to get that detail" for things that aren't on the drawings. Design is always still happening. That's why the first issue will be hard to achieve: drawings aren't done enough to get construction started. We're holding the candle for the schedule, and we might seem combative, when the opposite is true. We have to constantly force completed design to meet the schedule. 100 percent CD's are issued and when you open them, you can't chase details through. They're missing, incomplete, uncoordinated.

"We're always chasing a completed design."

Some are working on contracts and shared incentives, but until the industry gets there, this book is to educate people on these impacts and cope in the meantime. Even experienced owners don't understand the trouble they cause by failing to allow enough time for planning and design. That's why they need us: we solve their problem. But it demands working until 2 a.m., burning people out, and costing more money. They could get better, smoother projects if they knew this: allow more time, pay architects more, set up collaborative cultures and get better service and buildings. What other issues come to mind?

Lewis: The modeling process. When we don't have complete designs it's hard to coordinate models when we should to build on time. If the documents were complete, we could do shop drawings and trade contractor models. They can't be completed when they need to be because design isn't done. Modeling's in the wrong sequence. Today, it's after we award contracts, and that's where it needs to be. The problem is, we award trade contracts and go to work the same week. The schedule dictates that. We're forced to by the way contracts are set up. Once we're forced into that scenario, we're behind the eight ball in modeling and trade coordination.

We learned that in the early days of BIM, but we haven't made it happen enough – getting coordinated BIMs 45 or 90 days before work gets fabricated.

Lewis: Trade contractors need to model and coordinate as soon as the GC/CM gets awarded. We need to bring them onboard earlier – while protecting costs, competition, and risk – or extend the schedules.

Very clear articulation of three major issues.

Lewis: In my role I have to look at details and also step back to reflect on big issues. When you do that it's always these three issues. I appreciate your comment, but it's what I live with every day.

I'm trying to build empathy and bilateral understanding. First in your direction. How many times have you moved your family for the company to build projects?

Lewis: Nine times. From Bloomington, Illinois, to Atlanta, then Winston Salem, Richmond, Charleston, Phoenix, Oklahoma City, Ashburn, Virginia, and Atlanta again.

What design-driven issues do you face in the field? What keeps you up at night?

Lewis: In my role, I fear the unknown. I don't know what the owner and design team are thinking as it relates to change. We pride ourselves on being flexible and always having a Plan B. But you have your eye on the ball and then suddenly: curveball. You're still asked to hold the schedule. I just call it the unknown. Sometimes those unknowns can derail a project.

Design management is about making the unknown known, and helping our partners achieve interim steps in a more predictable controllable way. Do you understand the architect's process? Do you get enough time and engagement to affect it or are you too busy or committed elsewhere?

Lewis: If design partners would involve us early enough, we could help them march down a less disruptive path. I understand what they have to do, but not how long it takes to do it all and issue it. I don't know the manhours to do a significant bulletin. I'd be willing to spend more personal time and effort with them if I knew they'd design to means and methods, available products and standard industry practices. It is difficult to find that time. You've got your own ship on the water you're trying to keep from sinking. So, you have to justify the time you'd spend with them versus the reward or benefit. If I was sitting next to the architect while he worked, and he'd ask, "Jon, can you build this?" I could answer questions on the spot. We'd save so much time and eliminate so much back and forth. He'd know what he puts on the drawings is affordable and buildable.

It goes back to contracts. Can we require or incentivize that time and expertise? Every time we collaborate it proves that it works. We do it to afterward to get out of problems, but not to prevent them. What can we change to get better drawings to build by?

Lewis: More input on the front end from the guys going to build it to help design go down a track that's affordable and meets the team's objectives.

How can architects take advantage of your knowhow? What would you change about the process?

Lewis: More time together. A more appropriate question might be: what would I have time to do that would improve the process? It's balancing my own project risk and responsibilities versus helping others. We schedule repeat meetings. We have a standing structure coordination meeting every other Wednesday at 9 a.m. People bring design, coordination, or constructability issues. We resolve them, then issue drawings and RFIs prior to issues showing up in the field. It can be that easy, done early enough, when we talk first. But more needs to take place on the front end.

"In my role, I fear the unknown."

Talk about some of the technology you're using in the field to cope.

Lewis: Federated models are the biggest thing. Our coping device with incomplete uncoordinated design is the model put together by subcontractors. We might default to that model more than CDs. That's a risk. But we can hang our hat on the model being right because we've already worked the issues out. Those six other uncoordinated drawings, not as much.

How do you do a constructability review?

Lewis: Every time a significant bulletin or new CD is released, we immediately look through them individually. After 3 days everyone brings their reviews together and we talk about issues. We use BlueBeam overlays to point out changes. Then we ask the AE, developers and owners reps to resolve the comments. In almost every case that's done within a few weeks. If we're building the structure, we focus on those issues first. There isn't time to do it all at once at detailed levels. We do it incrementally to support the sequence.

I think people don't know how much we need to prioritize those reviews due to limited time. What are the most screwed-up processes and how can we fix them?

Lewis: One more issue has to be dealt with moving forward. Every time the owner hires a third-party consultant they issue field reports and make *suggestions*. We don't work from suggestions. Somebody has to read that report and make decisions. Today, the only person who does that is the CM. What about the 14 optional issues? The design team or owner need to fix and own that. That exists to the nth degree with waterproofing consultants. Design details in the CDs are usually generic. But once you hire a CM and a roofer you have to abide by the rules of a specific manufacturer. Changes need to be issued as a Bulletin or solicited by us via an RFI. We're held responsible for fulfilling items in the field report. But we didn't because they're not in the construction documents. When we raise this open-endedness to the owner they look to their designers and ask: Didn't we issue a drawing? Didn't we tell the contractor what to build? Why not? Owners should tell the design team to digest the reports and recommend what to do. This detail no. This detail yes. This one you rejected. That's what we need.

Can you share any good stories of AE/CM collaboration or adversity overcome?

Lewis: I could go on all night on that. So many good things have come from collaboration. One that stands out is the top of the structure – the crown details – at the Devon Energy Tower. The building had a varying radius. There, the structural engineer, erectors, and glass designers and installers spent a year discussing the details and challenges before we ever started construction. In my career, that stands out. It was remarkable, at the end we could all say we fabricated and installed the complex roof crown structure with very few issues.

Jon Lewis Key Points:

1. We need new contract forms to elicit collaboration.
2. Owners should allow time for planning and design completion before construction start.
3. Designers and builders need more time together during design.
4. Award trades earlier to support trade model coordination before construction.
5. Owners and designers must assume responsible for timing and decisions of supplemental late stage consultants and integrate them into design documents for direction.

CHAPTER 13

Technology: Leveraging Data

INTERVIEW

Arol Wolford, Hon. AIA

Owner, VIMaec, Building Systems Design   27th October 2017

Arol Wolford, Hon. AIA, is a serial entrepreneur, visionary, and software developer instrumental in pioneering new technologies to inform and manage design. In the 1980s, he was one of the first to bring data mining to the AEC workspace by founding Construction Market Data, an industry reporting media firm with hundreds of thousands of subscribers and products users – CMD, Architects First Source, and R.S. Means – all sold to Reed Elsevier in 2002. He serves on the board of DesignIntelligence, and the Design Futures Council. He is currently owner of VIMaec, a 3D visualization and data optimization tool using the Unity gaming engine to link 3D models and data to create virtual information models (VIMs) from building information models (BIMs). He recently acquired Building Systems Design (BSD), a leader in linked, intelligent building specifications and cost information since the 1980s.

Through his leadership in applying technology to the AEC professions, he has connected design and construction through managed information for decades. A frequent speaker, presenter, and juror, he was recognized by the American Institute of Architects as Honorary AIA in 1997.

Mining the Data, Counting Your Blessings, and Seeing Clearly or "Life Is Just a Game"

You've had a vision of technology's power to manage design information for almost 40 years. What is the source of your various visions for informing designs?

> **"**"He saw a huge inefficiency in how projects were bid and how quantity surveys were done."

Wolford: Three blessed events shaped it. First, Dad was a mechanical engineer who had gone to Berkeley. He saw an inefficiency in how projects were bid, and quantity surveys were done. As a mechanical engineer working for Carrier, he realized we were all looking at the same set of plans counting things. So, he started a company in 1957 to get back to the real pursuits: helping engineers configure the right equipment in the design phase and making sure that the specified equipment was sold to the Mechanical contractor. Those were the fun parts of the job. But he would spend 25–30 percent of his time on plans, counting and measuring things. He recognized the English and German concept of AEs doing that work up front was better.

Here's a little-known fact – I did the research: in 1929, the AIA, AGC, and SME agreed for efficiency, the best thing would be to have architects and engineers create the quantity surveys. They'd be paid 2 percent for commercial projects and 1 percent for residential ones. Then the depression came, and they didn't do it. Today, it's still inefficient and we haven't fixed it, even though Revit and other software can count objects and do quantity surveys. AEs won't do it because they're wondering how they get paid. The inefficiency continues. Out of $700 billion of commercial construction in the U.S., $70 billion is still spent counting and measuring things. Total AE design fees are only 50 billion. This is a travesty I'm still passionate about solving. With objects, we can link Revit to R.S. Means cost data and do quantities, but AEs aren't interested in doing it because they don't get paid and it's not the fun stuff. But it's necessary, and an impediment to our industry.

And there's the historical risk aversion, or "That's' the contractor's job."

Wolford: Bill McDonough, the environmental champion told me, "We need that same quantity survey data for the environmental analysis! I want to know how many SF of curtainwall there are." I'm still on that pursuit. But the start came from my dad. He started the business and was quite successful. He fell ill and couldn't expand it, so I took over.

My second blessing was growing up in Silicon Valley in the 1960s. I was a "boring hippie." I liked the '60s spirit and idealism, but not the drugs. Silicon Valley was a hotbed. It grew so fast, we had the 6th-highest GDP in the world. That idealism included the notion it was okay to make a mistake. To *not* try to fulfill your dreams was the biggest mistake. The message was, "If you had talents, try them." That supported entrepreneurs. It led to me meeting my wife, who became my partner, and our idea was to set up branches around the U.S. doing quantity surveys. So, we did that, at 22, in 1975.

As we opened these places, we needed good plan rooms. Dodge and Builder's Exchange let us but cut us off in 1981. They forced us to compete with them. That was ridiculous, because Dodge was a $150 million, profitable business with no competitors. But I was too young and dumb to know better. We started using low-end, less expensive IBM computers and selling face to face. We did things in a different way.

In 1982, we started Construction Market Data. By the 2000s, we had moved ahead of Dodge. We had 1,200 employees. In 2000, we sold that business for $300 million to Reed Elsevier. I was on Revit's board at the time. Leonid Raiz, Revit's Founder, used to say, "We need to the virtual design, the virtual construction, then fly the building. Autodesk has done a great job with the design part, but we can do better on virtual construction." John Hirschtick was also on the board. He created Solidworks, a leading CAD manufacturer. I was on the board, and Dave Lamont, a brilliant guy from Apple. I think he came up with the name BIM. When Autodesk bought it in 2002, Phil Bernstein evangelized BIM and did an awesome job.

Leonid said, "Arol, you've gotta keep making objects because we're making the building, and it's important to have the virtual chillers and windows." They were an outgrowth of CAD blocks. When Revit and ArchiCAD came, they had to have virtual objects. Leonid said, "You're putting the "I" in BIM, and without that, BIM is just BM – worthless." We made a wide library of generic objects. These guys were leaders, an awesome team. But it needed a large company like Autodesk to take it to the next place by virtually building it and, to your focus, adding collaboration. Having a guy with a manufacturing background was key. Dr. Jungreis was a brilliant software guy. They were idealistic and believed what they were creating would change the world. We knew the industry was inefficient and Revit would help that. People don't know we had a deal to have Reed-Elsevier buy Revit too. I wanted to give Revit to AEs to exchange information to the construction community without having to use the phone. Revit was set up on our own server back then, on a cloud, accessed through the internet. I thought, I'll give it away free. We'll get our data and turn our reporters into trainers. Then Reed's magazine business tanked, and they pulled back at the last minute. A year later Autodesk bought us. Carl Bass said, "Arol, I just bought the gun. I want to make sure you'll keep making the bullets."

He was smarter than me. I said, "If we keep making the objects, will you let me have the quantity surveys?" He said, "Yes, with limits." I still wanted to give the quantity surveys to the marketplace to help efficiency, but they didn't want the responsibility. But I continue on!

You've consistently led the industry by managing and reusing information and connecting design and construction. Others followed; catch us up on those.

Wolford: My other blessing was my wife, Jane Paradise Wolford. After our daughters grew up, she went back to school and got a PhD in architecture. Her theory was called F.A.C.E., for Function, Aesthetics, Cost, and Environment, advocating architecture's need to balance those factors. Now, the concept of holistic architecture is obvious, but it wasn't then. We need a software and process for managing design that integrates those factors. How can you do one without the others? Function and aesthetics have been served, but cost and environment are open opportunities.

That's when Chuck Eastman stepped in. He was one of the founders of BIM – its godfather. His colleague, Jim Glymph was working with Gehry early on, doing progressive models. They had a great database. He said, "I know you want to bring Revit and objects to the people, but when it goes onto 2D plans, no one gets to see those objects." The manufacturer's reps were frustrated because they never see it. He said, "If you'd take a Revit model and put it in the Unity Gaming engine and bring in the data, people would use it." $7 million and 4 years later, we can do it.

"'To design great buildings' was always [student's] answer. Now, their answer is: 'To save the world.' They are interested in climatic change, and bigger issues."

Anybody can use this gaming engine and fly the building. You touch a wall and it shows the data. Manufacturers like that. At the start, we charged $25K to do it. We knew that was high, so we made a "factory" to do it efficiently. Now, we can do it for $100 to $500. That's what VIMaec does. It can be a great collaboration tool. We can translate Revit, ArchiCAD and others into Unity's gaming engine. When we distribute it for collaboration, it's much less expensive. Owners, architects, everyone can afford it; it's almost free.

Your theme has been connecting people and information. You've never practiced as a designer, but you've brought them together. Did you see resistance or support?

"We need tools to help collaboration."

Wolford: I'm a genetics major so I was driven to have good data. As a science guy, I believe good design needs good data. In 1981, I got criticized for the company name Construction Market Data. They wanted to call it "Information" or "Wisdom," but data was the basic unit. I was driven by that and presented it visually – I'm an architectural groupie. I could serve because I loved architecture. I was lucky to be on a panel for the AIA's 150th celebration. We went to Windsor Castle with Prince Charles and a panel of famous architects. Michael Brown told us we had the key to the environmental future: the lion's share of energy is used by buildings. We were shocked. We came back and did a survey. Architects didn't know either. Forty percent thought cars were the culprit. We'd uncovered an opportunity. For 15 years, I've been on the Georgia Tech Board. I ask students, "Why do you want to go into Architecture?" "To design great buildings," was always their answer. Now, it's "To save the world." They're interested in climate change, bigger issues. That's gratifying.

You and I know systems thinking like this is key. But we need tools to help collaboration. We're starting to get them. I've seen projects virtually designing and building together. They run option analyses and get data to show impacts and outcomes. Cost impact, environmental. Now, we can run through Revit into a virtual information model with cost data. That's exciting. But changing minds takes tenacity and idealism. After we sold Revit to Autodesk, it took them 5 years to go from $1M to $5M in sales. But once they broke through, it took off. The 2008 Recession made it, but many architects lost their jobs. To respond, firms started using Revit, and it went from $5 to $500 million in five years. People called us stupid selling Revit for what we did. Now it's worth billions. But I don't regret it, because it needed a large organization like Autodesk, with marketing breadth, to advance the technology. Even with these tools, we have to change minds and overcome silos.

"The way forward is gamification and the principles of gamification. It includes the pure joy of creating collaboration."

"It's hard to change people's minds. It takes tenacity and idealism."

Having always worked in the power curve to leverage information and technology, you've been successful building and selling companies. But architects aren't. Our culture is strong at designing from scratch. How can we change?

Wolford: Now is an exciting time for architects. Younger ones are gaining comfort around virtualization, computers, Rhino, Revit, collaboration. Beyond the Unity gaming engine, I'm excited about the "Unity mindset." The way forward is gamification and its principles including the pure joy of creating collaboration with Minecraft! You talk to a 5-year-old, a 12-year-old kid. Girls, boys, collaborating, using Lego blocks to create. We've got a generation coming who have been using Minecraft, an amazing software collaboration tool. It collaborates well, but it's not straightforward software. But kids help each other; they share things. That spirit and model – what kids are doing with Minecraft, the passion, excitement, creativity, collaboration – is what *we* need.

We need to learn from those kids. They understand those factors. When you say Unity mindset, you don't just mean Unity software, you mean working collaboratively too, correct?

Wolford: Both. They chose me as a partner because a lot of people are using visualization, but I was the first to bring in the data.

I can think of a third meaning. Unity software, the unity of connecting people, bringing information together, and the unity mindset in the collaboration sense – three ways to look at it. Maybe this "triple unity mindset" is a way to leverage architect's talents through collaboration – through the wisdom and power of crowds.

Wolford: You just recapped the truth that will let us break through: the concept of a unified trinity.

Arol Wolford Key Points:

1. The design and construction industries continue to re-count quantities without sharing. Data is a valued unit of exchange in design. Tools that let us see and have low cost data can clear the path.
2. Collaboration tools enable connections and knowledge transfer. More use is needed to realize their full potential. Beyond tools, changing attitudes, tenacity, idealism, and purpose are necessary.
3. Our current generation is socialized using gaming collaboration tools such as Minecraft. Professionals can learn from these emerging attitudes and use the power of knowledge sharing to manage design of balanced, holistic projects, leverage and enhance the architect's value proposition, and advance their long-term resiliency as professional service providers.

INTERVIEW

Casey Robb, FCSI, CDT, CCPR, LEED AP

CF Robb Consulting Services, LLC   1st August 2018

Casey Robb is a fellow in the Construction Specifications Institute and Certified Construction Product Representative (CCPR) through CSI. In his 40-year career he has been a construction field engineer and contract administrator for Owens Corning Fiberglass Contracting Division and manufacturer's product representative for Thoro Systems Products (now BASF), Dow Corning Corporation, Dupont Building Innovations, and Kingspan Insulated Panels, advising architects, engineers, and building enclosure consultants on technical performance and specifications. He is former national president of CSI, and recipient of multiple sales achievement awards, including the Chairman's Award with Thoro Systems and Dow Corning.

Since 2017, he has consulted with Building Product Manufacturers to leverage product data in a consultative approach with architects, product promotion, and spec development. Having witnessed the industry evolution from paper catalogs and specification sheets to digital, cloud-based, data-rich BIM objects he offers perspective on accessing product knowledge through new and old ways – and the continued importance of building relationships.

Manufacturers, Knowledge and Building Relationships or "Can the Internet Buy You Lunch?"

You've been in the industry for decades. As a designer, I needed people like you to keep me straight. I couldn't possibly know what you knew to keep informed. What issues are shaping product knowledge exchange from your perspective?

> "Can the internet buy you lunch?"

Robb: Can the internet buy you lunch? Can it share experience and knowledge? I don't think it can replace the knowledge of a trusted advisor. Architects have more information at their fingertips than ever but it's often incomplete or lacking application knowledge. Design and construction professionals are working 24x7 to compete and meet time constraints. They don't have time to collect the information they need to produce quality construction documents. Data they might have gotten in the past from McGraw-Hill SWEETS®, desk references or product catalogs now comes from the internet. They used to rely on professional architectural product representatives. They still need people.

I'll share an analogy. If I'm looking at a Ford, I can go on the internet, choose a model, custom build one with my colors and features, and price it with lots of information – an amazing digital capability available in minutes. But if I need a replacement part, or to know if they've had trouble with it, I call the dealer's service rep, or better yet, go in and talk to the mechanic. That's how I really know it works.

BIM is a big change. Concepts still need to be analyzed for constructability and coordinated before rushing into final CDs and bidding. Contractors are forced by schedule pressures and incomplete documents to say, "We can do this better, cheaper, faster." This dilutes design professionals' value. Seeking knowledge from trusted resources like professional architectural product representatives can help in this transition.

The firmwide specifier role of the past has all but vanished or been outsourced. When I started my career as a product rep, the typical architectural firm had a traditional "specifier" as the gatekeeper or knowledge broker. This technical resource typically had the architectural background and constructability knowledge to coach others. Firms are now building their own technical teams around specialties like high-performance building enclosures and sustainability. They also use independent specifiers – people who worked at these firms in the past but now share their talents with multiple firms. As more product decisions are made by younger professionals there's a knowledge gap. This is where experienced architectural product representatives can help.

What plagues you or drives you crazy?

Robb: Are you ready? Just don't call me "vendor," please! I'm not selling hot dogs on a street corner or offering you trinkets from the inside of my sport coat. We exist to educate architects and provide solutions. We offer technical expertise on specialty products and new concepts for architects to consider. Experienced reps are conceptual and strategic in their approach with designers. There's a turnover with the "Golden Rep" generation. Many of the "legends" I once admired early in my career are retiring. Newer product reps have limited training options. This is where I hope to help in my consulting role. My initiative is to teach and mentor the next generation of architectural reps.

Architects should remember: architectural product reps want to earn your respect for the knowledge they bring. Many are certified as CDT (Contract Document Technologist) or CCPR (Certified Construction Product Representative) through CSI (Construction Specifications Institute). Look for those credentials to ensure your reps have been educated on construction documents, formats, standards, and project delivery. This is their career. They've worked in the field training contractors and solving technical issues. They provide technical knowledge and guidance. That can be invaluable in ways that raw internet data can't. Experienced reps reduce your risk and liability, help establish a quality level, a basis of design, and defend design decisions. A few reps could give us all a bad name. If you've had a bad experience, try another one. Seek those who can help and who you can trust. Please don't call me "vendor." Respect what we do.

They don't appreciate your role? Is the architect's insensitivity to supply chain issues a culture, or attitude problem?

Robb: Yes. Many long-term specifier-rep relationships are no longer in place. Younger project managers do internet research and don't rely on the rep. "I don't need a rep – I can find it online." Thousands of us want to serve you and be part of your team to bring you knowledge. We have a vested interest to make your project succeed by using products correctly. I'm working on a network of tools and training to reestablish relationships by training a new generation to provide architects what they want, when they want it. Manufacturers need this training for those new to the business or moving from pure sales roles into architectural promotion. That's is a big step. It needs a different skillset. I've built my business around helping manufacturers with business development strategies, tactical promotion and education. I hope this will be my legacy.

In my day we relied on "trusted advisor" manufacturer's representatives'. Not just to update the catalogs, or conduct a lunch-and-learn, but answer project-specific questions. Has that changed? What do architects, engineers, specifiers, owners, and builders need to know about your world?

Robb: Offering "lunch-and-learns" used to be for product or project-specific discussions, not just generic features or CEUs. They helped design professionals understand product categories, solutions, design considerations, advantages and limitations. With the growth of the AIA-CES program and continuing education requirements, "lunch-and-learns" have become an expectation – your ticket to get into a firm – a free-food entitlement rather than an education forum. Professionals can find CEUs on the internet – just-in-time learning at their fingertips. Many firms do internal CEUs on their own. So why do so many architects require lunch-and-learns just to get in to see them? I don't think we ask the right questions in advance. The dynamics must change.

Lunch-and-learns transfer knowledge and train new generations of architects. Design Principals and manufacturers have to encourage respect for the education, not just expect a free lunch. I've seen promising signs with some firms having rules for lunch-and-learn education programs, such as attendance policies and sharing. We respect the need for CEUs. All we ask is you respect your reps. Welcome them into your firm and connect them with the right people. That's "outside collaboration" – your network of "trusted advisor" representatives.

Are technology and information-rich 3D objects reshaping designing and specifying? Because of the differences between generic model objects and proprietary, manufacturer-specific features, this aspect hasn't taken off as predicted. Do you agree?

Robb: It depends on the firm and their BIM adoption level. Not all have fully adopted BIM, and some products aren't as well suited to BIM. Flashing and waterproofing, for example, aren't easy to model, and are hard to describe and detail at transitions. Every product doesn't need a parametric model for download. It's different across the product matrix and depends on the Level of Development and project stage. If you're in final design and have furnishing or hardware, then yes, you may want to invest in BIMs others can import. If you promote through-wall flashing and are asked to develop expensive BIM objects for architects, maybe less so. BIM object creation has been slow because manufacturers are being solicited to spend huge sums of cash to develop product-specific models that may not be preferred by designers. In the discussions I've had with manufacturers, not all are on board with BIM objects.

Budgeting products has always been as issue. I see this differently now that I've worked with a contractor for 20 years. Manufacturers can only speak to part of the cost. Your thoughts?

Robb: It's still a challenge. Architects ask for budget pricing at the concept level. Reps can talk ballpark pricing all day long. When architects get specific on cost you can see the design-build or value engineering influence. It's better to collaborate and refine budgets along the way. For budget pricing, reps typically have broad-scope discussions with architects and go to reliable subcontractors for detailed input. The best way is in teams with trades involved.

On a recent visit with an architect, she said, "Contractors tell me basis of design doesn't matter." Well, what's happened to level of quality and design intent? Many times, this is "vulture engineering" – attempts to lower the quality for another party's benefit. If you want to drop the quality level to every product under the sun, you'll end up comparing "apples to oranges." Basis of design needs to mean something with architects and GCs. Professional arch-reps can defend design intent and specifications they helped develop.

Are you seeing transformational things in firms? New tools? Processes? As specialization and complexity increase, how are you behaving differently to keep up and be of value?

Robb: I'm seeing new tools from new apps to parametric detailing models. You understand this better than most. AE's are in a unique position to manage information. But to turn the architect's model into a parametric model ready for construction, the GC needs to add interference detection, logistics and constructability. Designers who embrace BIM tell me they can enter 95 percent of the design data into models. But that last 5 percent causes the trouble. Models have design information, but without the knowledge to execute in the field, building them can be a nightmare.

Has your role changed? Where will your kind be in 20 years?

Robb: Yes. Staying relevant is key. In twenty years? I'm not sure. A lot will happen in the next 5 to 10 years. The industry is ripe for change. Early adopters want it to happen quickly and we see it with innovative designers and manufacturers. But, there's not enough critical mass to drive change or long-term investment. That's where leadership helps. Boomers know change is difficult and we struggle with knowledge transfer to younger counterparts. It's a two-way street. We can learn from each other if we listen.

If you're a product rep planning to call on a firm to secure a project, you better have a compelling reason. Architects don't seem to have time for reps or see their value to their daily work. They resort to doing product research on their own and are reluctant to call you back. In six months, when they need an answer, can they find a quality rep when they need one? With all the consolidations, will that trusted rep I had yesterday be there in six months? What are we doing about it? We follow the architects' lead and are a resource when needed (or a pest when not), or wait for them to reach out. If more firms reach out, we can help. Maybe having an inter-rep or trusted advisor network is the answer. We'll see.

We see too many projects with all kinds of design challenges – structural complexities, substrate conditions and cladding types. The architect is rushed to get the project off their desk and onto the street for bidding. They don't see the constructability issues or have time to coordinate construction documents. Bidders are frustrated. "How can I bid this project?" is the repeated call. What can we do in a few days to meet the bid day next week? You can tell there will be substitutions, RFIs, change orders, and maybe legal issues. Having a trusted resource at their side can avoid this. In 20 years I hope we see improved CDs coordinated in a virtual model to get designer's visions to contractor with less chaos.

The architects do it to themselves by introducing complexity and not managing their design?

Robb: In some cases, yes. When design firms deliver poor documents on compressed schedules it opens them up to risk. It's a negative influence on the value they bring. Some firms are leading design delivery in positive directions. Some have internal enclosure specialists and LEED® and LBC experts on-staff with online databases of sustainable products and performance history. That's the dichotomy: they're getting better internally, building knowledge, but they still need trusted external resources. Especially with new project delivery forms.

What is consolidation doing to your business? Are firms adapting to survive?

Robb: Consolidation and acquisition is affecting the entire AEC and manufacturer arena. As companies merge with related products and segments, they have more solutions, relationships, and service ranges, like sustainability, building enclosure commissioning, LEED® and energy efficient design. Some are racing to provide an "all-in-one" approach that needs less skilled labor. Some are looking at partnering, modular thinking, integration of assemblies, and prefabrication. It's a race to differentiate and survive. Many reps are fed up with the status quo and want to see change. Some will be become independent. If you're a paint or hardware rep for one of the big guys, you have a quiver of solutions. There's value when product reps offer a wider range of solutions. In the future, how likely will you be to get an audience with a busy architect with only one building component?

What do you wish the rest of us knew about what you do?

Robb: Architects need to know good product reps are still working for them, even though it might seem we've disappeared. We need you to include us, and we need your honesty. Tell us what you want. We make mistakes too, but we're here to help and do the right thing, not work against you. We're here to share technical expertise and knowledge. I may even call my competitor if it helps you.

The good news – relationships are not dead!

Casey Robb Key Points:

1. Design professionals are challenged to cope with the universe of technical information under compressed schedules. Acknowledge, support, and manage this risk.
2. Good manufacturer's representatives are career professionals who support design and construction professionals and add value, not peddle products.
3. The design community should educate current generations to appreciate and partner with technical product representatives in addition to researching and downloading digital data.
4. Common human courtesy, professional respect, and information sharing is a two-way street.

INTERVIEW

Josh Kanner

Founder, SmartVidio   19th October 2017

Josh Kanner is a software entrepreneur, responsible for creating leading software products in the design and construction industry. Since 2000, he has championed technology and data reuse to manage design and construction information. In 2004, he was co-founder of Vela Systems, a suite of modules for construction administration with punch list, facility management, and quality, before selling it to Autodesk. The program is now R-branded as Autodesk BIM360 Field, with thousands of licenses sold. Based in Boston, he is now president of SmartVidio, a database management system for optimizing and accessing construction photos and videos. His history of helping designers with technology adoption is instructive.

Reusing Data or "The Technology Problem"

You don't have an architecture, engineering or construction degree. How did you get into this field?

Kanner: I learned on the fly. The cofounder of Vela was Adam Omansky. He was an owner's rep at the Four Seasons and had a master's in architecture. I learned a lot from him, from our customers, and from being in the field. The original Vela idea came from work he did the Harvard Graduate School of Design and the MIT Center for Real Estate. He thought it would be interesting if the industry could use RFID tags to communicate with mobile devices in real time what was wrong in a room. Instead of using painter's tape on the walls, he'd use RFID tags that would "wake up" and notify a mobile device. We started working together in 2004.

> *"I learned on the fly . . . the need to have technology in the field with the software products to manage it."*

My background was in enterprise software, working with companies around their software needs. We learned no one needed RFID tags. They were more excited about having a mobile device with plans on it, so they could do workflow in the field. That was the beginning of Vela: combining the need to have technology in the field with the software to manage it.

That was a novel idea at the time; were you one of the first companies with mobile files?

Kanner: Others were trying to do it, from different perspectives, primarily existing players like Prolog, Primavera, Constructware. They tried to make mobile apps as add-ons to their desktop solutions. But no one was thinking, *what do field personnel need?* A superintendent? An architect? We came at it that way. It was a different approach.

Interesting formula: one guy with an architecture background and another with an MBA and experience in building an enterprise software company. An alliance of two different skillsets around a technology vision to manage design information. How did that happen?

Kanner: I had experience building software products. I graduated from business school in 2000. I had run products working directly with the engineering team and customers. We grew from 10 people with 1 customer in 2000, to 250 people with 100 customers in 2005 when I left to start Vela. I'd been through the process of helping design and grow a product with customers in that mode. When Adam and I started Vela, I had an idea what we needed to do – muscle memory around growing a product and a company.

"The way design bridges from concept to reality is in a bill of materials and a way to manufacture and assemble it."

"Architects and owners are losing control of the data, because it's not theirs anymore. It's up to the contractors, trades and specialty experts. Others are creating the data to get the project built."

I brought two disciplines to bear. One was quality management. An original use case was punch list management. Once we started working with contractors, it was obvious that software quality management and the concept of watching for bugs was similar to construction punch lists. Root causes and related practices were important. I thought quality management was a discipline that could translate from software to construction. The other was that the first software I spent five years building was for manufacturing. At its core was the idea of a bill of materials. The way design bridges from concept to reality is in a bill of materials and a way to manufacture and assemble it. We focused on managing data from the beginning, because we wanted to bring the best concepts of manufacturing into construction.

Even though you hadn't built buildings, you built a company with parallel aspects. The manufacturing-to-construction shift was as much adaptation as invention. The promise of manufacturing processes informing design and construction is still untapped.

Most of your work has focused on the "back end" of design: CDs, construction and closeout. Not the creative process. A core question is how to get designers to think systematically – the way you do – to reuse data and work in a business context. How have your efforts to automate and manage design information been received by practitioners? What mentalities are you seeing towards managing, being disciplined or informed in design?

Kanner: There are a couple dimensions to this. One is the same continuum every new technology is on: Geoffrey Moore's "Crossing the Chasm": early adopters, early majority, and so on. User distribution is always an issue. In AEC, another dimension is: where are you in the project delivery chain? What's your relationship to value flow? Are you the owner, architect, engineer consultant, contractor, trade? Value flow changes how you think about technology and risk. Through those lenses, the people who get excited are the early adopters. They also care passionately about the problem. When you're solving problems for field execution of construction documents, the Venn diagram of folks who care is largest in the contracting community. By the time you get to delivery and execution, most folks on the design side are down to the last bit, the CA portion of their fee, and there's not a lot of risk or financial gain in it. They're abrogating to the CM. In those two diagrams, we found greater reception in construction firms than in design firms. There were exceptions. Architectonica did their CA on Vela's systems. They wanted to get a handle on their complex, forward design ideas as they were executed.

There's plenty of risk for design firms during construction, but I think the issue is more that architects love the early process of design more than they do the messy reality of construction.

Kanner: Another interesting concept at the intersection of technology in the design-to-construction flow is managing data versus constructible data. You go through greater specificity, from SD to DD to CD. Because of tools like Navisworks, Revit for contractors, trades modeling steel, and MEP, more modeling is being done later, and detail, by necessity, added later in the process. As a result, architects and owners are losing control of the data, because it's not theirs anymore. It's the contractors, trades and specialty experts. They're creating the data to get the project built. Many times, as you know, that requires rebuilding the model.

"Where are you in the overall project delivery chain? Your relationship to the flow of value? . . . The flow of value really changes the way you think about technology and manage risk. Through both of those lenses, the segment of people who get really excited is the early adopters."

For those who embrace it, the movement to use data goes on, but as a construction company we deal with the laggards. The older leaders at the sunset of their careers still have the old mindsets; they don't embrace it early enough to make a difference. Architects' and owners' reluctance is real; any advice for late bloomers?

Kanner: First, it's never too late to improve your craft. Next, these are tools. Do you want to use tools that get you the best quality and improve productivity? You do. You're going to get better quality, go home earlier and be prouder of what you deliver. I'd deal with those first: you're not too old to learn, and these are just tools.

You're a technology advocate. We get it intellectually, but once we start doing it, we need new hardware, software, training, and infrastructure. For all its advantages, technology also invokes a host of other management complexities. How are you coaching customers to cope?

Kanner: It's time for the industry to think of itself as a series of best practices that can improve outcomes. The industry doesn't believe in the concept of technology ROI, because each project floats on its own bottom, runs on its own P&L. There's a cultural blind spot compared to other industries. The software I built before Vela was for manufacturing companies. They said, "The electronic component industry is a tough industry. Increasingly commoditized. New competition is coming. Margins are squeezed by price pressures. We *need* technology – the electronic component manufacturers are saying this – to help us improve our margins and differentiate ourselves!"

Then fast forward two years. I'm talking to senior executives in multi-million-dollar AEC firms and I hear the same three inputs: "Our industry is tough. It's commoditized. There's new entrants all the time and our margins are increasingly squeezed, and I *don't* have time or money for technology." That blew me away. This cultural mindset doesn't see technology as a way out of the problems, it sees it as an *additional* problem. Until that mindset changes, and it is in some companies, we face an interesting problem.

"Cultural blind spot" is apt. It shows the contrast between industries. Manufacturing is more capitalized than AEC, as productivity graphs show. You've been able to attract money, but most architects don't wake up to do that. They don't have the skills, inclination or motivation. They don't know how look for entrepreneurial opportunities. That's the essential conflict. Any advice from a guy who's had a few successes in a row?

Kanner: A couple questions are posed here. One is: where can architects go for funding? The other is: how can AEC folks get started, even before funding? How can they start thinking about a technology or startups they want to do? Now is the best time in the AEC industry's history to do it, for many reasons. Component technologies are accessible. You can develop a new Amazon app and start working with little investment, almost free. More products are becoming open. You can use their Application Programming Interface and build on components already there, not rebuild a bunch of stuff. So much is open source. Getting an idea off the ground is not a high bar anymore. Folks in the design space are doing this. Thornton Tomasetti had so many ideas they've figured out a way to spin them out as little companies. Others are entrepreneurial, like Ian Keough. He thinks we need a new class of architect: a coder. There's never been a better time to get funding in A&E. In just the last few months several initiatives launched.

> *"This is why I really love the AEC industry. What's been easier than I expected is finding passionate early adopter users, excited to try new tools and be a part of trying to change the industry."*

You need the right mindset and the will to try. For the not-so technology-inclined, maybe there are "nontech" entrepreneurial ideas that can improve their process and value. What have been some of the hardest issues you've faced?

Kanner: Necessity is the mother of invention. People were tired of doing "bloogie" things to translate data or perpetuate dumb processes, so they built software. There are all kinds of risk-building companies. Technology risk: can you build it? Team risk: can we execute? And market risk: is the target market ready for it? At Vela in 2005, we could never have planned that tablet PCs were all that existed for mobility. They were expensive – $4,000 – and only had a 2-hour battery life. They weighed over 5 pounds and required a special operating system and stylus pen. They generated excitement, but we were held back because they were clunky and expensive. Part of my advice is to persevere. The iPad was a turning point. In retrospect, it was a no-brainer, but it wasn't an easy decision. We moved our entire mobile client – what field guys use – to the iPad as soon as it became available. That unlocked our growth. But we had to persevere when it was hard to get field users on board.

Beyond the risks you talked about, there's luck. But you were shrewd enough to see it – you had to be looking – agile enough to pivot, and courageous enough to put your eggs in a new basket. It worked. But if one of those goes wrong it's a different story. Any easy ones along the way?

> *"AEC has a cultural blind spot compared to other industries."*

Kanner: This is why I love the AEC industry. What's been easier than expected is finding passionate early adopters, excited to try new tools and be part of changing the industry. A certain group is passionate. We show up at the conferences. We meet again. There is a bond – we're a crew.

The choir. That's fascinating, your take on the past and the future. Intentional or not, you made some great connections to the book's premise, and "designer's mind" questions.

Josh Kanner Key Points:

1. AECO executives and leaders must change their view of technology as a problem, and embrace it through funding and process change, to join the rest of capitalized, tool-using business and society to improve AECO productivity. AECO has a technology "cultural blind spot" compared to other industries.
2. AEC professionals have a passion for what they do and love the new. Sell that.
3. Manufacturing offers untapped potential as an AEC efficiency model. For example, producing bills of materials via extracted data.
4. Your place in the AEC project delivery food chain may influence your outlook toward process change, technology, and new value propositions. Know it. Do something about it.
5. Architects and owners are losing control of the data, ceding it to contractors to get things built. No wonder why power is being lost – it's gone with the data.

CHAPTER 14

Entrepreneurship: Vertical Integration and Value Propositions

INTERVIEW

Scott Marble, AIA

William H. Harrison Chair, Professor, School of Architecture, Georgia Tech   11th October 2017

Scott Marble, AIA, is an architect, educator, and author focused on technology in architecture and design process. He is the William H. Harrison Chair and Professor in the School of Architecture at Georgia Tech. Previously, he was adjunct professor at Columbia University in New York. He is a founding partner of Marble Fairbanks Architects in New York. A frequent lecturer in digital technologies, he recently completed the book *Digital Workflows in Architecture: Design, Assembly, Industry*, published by Birkhauser. His firm has received over 40 local, national, and international design awards.

Rethinking Relationships, Delivering Value or "Giving Them the Business"

I enjoyed your and Dennis Shelden's conference at Georgia Tech, the Digital Building Lab Symposium: AEC Entrepreneurship and the Future of Practice. Great people, synergy, and momentum. This book focuses on inherent conflicts between the art of design and the discipline of managing it. As chair of a program in architecture, can you talk about your experience, curriculum, and interests in that area?

Marble: That's familiar territory. The topic is part of what I was doing at Columbia, how I've run my practice for 25 years, and something I'm trying to bring to Georgia Tech. I've been thinking about a different way to frame the issue set. Not just managing design, but "designing" design. Part of what I'm advocating is the current state of design in our industry – the way design is positioned – is flawed in many ways. I'm pushing for a rethinking, not only of design, but of the overall relationships and dynamics in our industry.

> "Not just managing design, but "designing" design . . . the way design is positioned in our industry is flawed in many ways. I'm pushing for a rethinking, not only of design, but of the overall relationships and dynamics that exist in our industry."

The minute you called it "designing" design, instead of "managing" design, I immediately got excited, because it sounded new, exploratory – design-like. It exposed my bias toward design and away from business. It's a professional sweet spot, or weak spot.

Marble: From an architect's perspective, my main goal is making sure, as technology restructures the way we all act, that *design* advocacy remains part of the discussion. It's easy, in the rush to solve technical, logistical, legal, contractual, and economic problems, to push design out of the equation. As an industry, we need to be cognizant that design is a central part. A way to integrate design in the discussions, as opposed to something that creates process tensions and inefficiencies.

In our practice I focus on process with clients as well as design. I think about design of processes and buildings. I try to introduce new, technology-driven ways to make projects better. At Georgia Tech, because of its legacy of tech-based design research, there's an opportunity to take it to another level. Not only to train the next generation of architects, but to do high-level, basic and foundational research on the technology that enables it.

> "As technology starts to restructure the way we all act, design advocacy remains a part of the discussion. Because it's very easy, in the rush to solve the technical, logistical, legal, contractual, and economic problems, to push design out of the equation. As an industry, we need to be cognizant design is a central part."

As complexity and other forces creep in, design can get left behind. So, yes, somebody has to advocate for design. When I was in school we focused on the art and science, not business. Technology hadn't arrived. I'm now tackling the business agenda, and technology has become a fourth area. Is technology changing design process? Most firms still use a traditional process. They program it, develop some sketches or ideas, and throw it over the wall to the BIM geeks. The creative part is unchanged.

Marble: I think that's right. The point on business comes up a lot and has started to surface as an urgent issue in schools. Historically, it wasn't part of the discussion. That partly explains why architects have a reputation for not being successful in business aspects. It's part of what the conference was about. I reconceive the term "business" as "entrepreneurship." The way I'd distinguish between those two things is that business implies the way things are done *now*. In emphasizing "entrepreneurship," I want people to think about it from a point of view of innovation. Not only innovation in design, construction, and common aspects, but from a business perspective. In other words, think of the *business* aspects of this new direction as something that can be innovative, entrepreneurial and change old business models. That's why at Georgia Tech, with Dennis Shelden, we're trying to create an entrepreneurship overlay in everything we do, so it's not after-the-fact or something architects don't deal with. They see business and design innovation as hand-in-hand.

I use the term "business," but it's not really the book's focus, as in: how to manage a firm, do hours budgets and billings. It's more the attitude: Why are we bad at managing ourselves to meet schedules, design to budget, and drive to outcomes? The thought process of bringing discipline and higher value to design – and __can__ we? How do mid-career professionals (owners, architects, contractors) get the courage to and change? How can design regain its rightful place?

Marble: My work at Columbia was a five-year, design-studio-based research project to address the issues in the book. Topics included how to collaborate in creative ways, the one-off nature of design, and how to set up technology for "design reuse." Where you could design parametric building assemblies and components to be used by other designers and shared, as opposed to individual, one-off, project-specific processes. We also addressed specialization in design sectors. So many subsets of design now question the core competency of what an architect does.

We tried to introduce the idea that the architect's historic role – generalist for many building aspects – might be up for reconsideration. That architects could work in innovative, specialized areas to reimagine their career path, was embedded in the curriculum. At Tech, the opportunities are greater. The curriculum ranges from introduction in design thinking and entrepreneurship for undergrads, to research labs developing new software and tools in partnership with industry.

When you were talking about design reuse and assemblies, I thought about the ubiquitous efforts in AE firms decades ago to develop "detail libraries" and "standard details" for reuse. You're talking about a more sophisticated version of that?

Marble: That impulse was right, but maybe firms resisted because architects have this real, valuable interest in constantly reinventing things. Even if you've done it before, you still start from scratch. It's a fundamental aspect of how we've practiced for centuries. Today, the pressure to economize process and not start from scratch is positive, but there's a balance between repeating a design, or aspects of it, and understanding that creating anew is important to the profession. We don't want all buildings to look the same. What makes cities great is their diversity. It's a basic human need and architects create it.

"We need to figure out a way where design is more integrated in the discussions, as opposed to something that creates tensions and inefficiencies in the process."

The last thing we want is to move toward a repeatability model that's good from the process and economic side but is a disaster from the cultural and social side. The challenge is to balance those two tendencies. The repeatability question has to creatively reuse design without resorting to mindless repeatability. Research in technology, parametric tools, and logic-based design process expands the designer's capacity to be creative. We can use this economy of means and still maintain creativity and open-ended aspects.

So, in 10 or 20 years, maybe students entering the workforce will have the mindset for it. Are you facing resistance or momentum from the design cadre?

Marble: Momentum. There are so many voices among the faculty, all valuable. Maybe some discomfort. With unfamiliar technology or change in general, they can shy away because it's different from how they think. That's unnecessary. One can embrace the positive aspects of technology by understanding what it's doing without having to be experts in every software. There's general interest in this topic and a sense it's going to happen, just when and how. People are realizing if this *is* going to happen, I need to get on board and be at the table when the conversation happens. Tech has a mandate to get deeper on the application side.

> "Think of the business aspects of this new direction as something that can be innovative, entrepreneurial and change old business models."

The technology train brings its own issues. We're transformed, then we discover, "My God, now what have we gotten ourselves into?" Are you teaching technology management?

Marble: Georgia Tech has been at it for years. Chuck Eastman, who started the Digital Building Lab, has been one of the leaders.

They are clear, global leaders in interoperability and standards. That reduces the need to manage certain things, because they flow.

Marble: We can avoid the lost data, interoperability and many technology management problems. It's well under way.

How do the rest us, who don't have PhD's in interoperability, catch up? How do we bring the "slow set" along?

Marble: That's tricky. We're a small practice. All tech-savvy. No matter how sophisticated the client is with technology, I'll always explain the benefits of certain approaches, how it benefits *them*. We educate clients on this. Sometimes I get them to test something. If you say to a client, "This technology makes design efficient," they'll say, "That's great. I expect you to have it and use it. That's why we hired you."' It's only valuable if it benefits them and their concerns. You have to make that case. Clients assume you keep up with it for no extra cost. If they see no benefit to them, they don't care whether you draw by hand or use the world's most advanced parametric software. There has to be a payoff on their end.

> "That impulse was right, but the reason some firms have resisted is architects have this real, and very valuable, inter-est in constantly reinventing things. Even if you've done it before, you're still starting from scratch."

Make it be about the customer. But the average architect isn't good at making the client benefit case. Articulating it or selling it persuasively. Effective design management involves looking for, recognizing and creating value – sales, and client empathy. Are you teaching in those areas?

Marble: One challenge of being an architect is what we do has so many dimensions. Learning it all would take a 15-year degree program. Topics like you mention have to be covered in professional practice and elective courses.

I keep coming back to this entrepreneurial theme . . . architects understanding a few basic principles and concepts around what it means to innovate through their work. If I'm selling a service to a client, if I have an entrepreneur's mindset, I understand what they're looking for and their tolerance for the new. I'm going to respond not by just giving them exactly what they want, but by giving them something they didn't even *know* they wanted. That's entrepreneurship. Listening to their needs and going beyond.

We're taught to do it in design. It's another thing to do it with services, recognize you're adding value and get paid for it. We've always aspired to overdeliver on the design side and make clients say "wow." It's why we get in trouble on budgets, schedules and fees. That's not where the problem lies: money is.

Marble: Great architecture can get clients to pay more because they're seduced by it. It's something they want. I'm not such an advocate of the model where a client comes to design team with a budget, program and a schedule, and the architect's job is to stick exactly to that. Because so often the program and presumptions are incomplete or wrong. It's rare when a client comes to a team with a well-conceived program, budget and schedule. The team's job is to vet them, so the baseline is balanced. That might involve telling them to do less. It's not always about expanding scope, it's working with them to accomplish their core goals.

If it's only about complying with program, budget and schedule, design can be boring, with no added value. That's not what we got into this profession for. Other careers are based on transactions and complying. Some clients would be thrilled with that, and it's our job to know when that's the case. But there's an inherent, "elephant in the room" issue for architects: by "design," we try to go beyond, consciously, intentionally, not manage within the lines. That's our mystery.

Marble: I distinguish between "conditions" and "assumed constraints," which might be artificial and unnecessary. They may be given to the design team but not needed to accomplish the end. Sometimes our job is to question.

In our old way of working, the assumption was: program and budget are right. The way we work now, much of it isn't there, or isn't right. So, the process gains value from both designer and contractor, which is why we practice in teams. How do we find that balance, increasing value while managing the tendency to overdo?

Marble: There's not even an adjective to describe how messed up fee structure is. On projects in my office right now, fee structures range from a percent of construction, which we all know is a conflict of interest, to a fixed fee generated with incomplete project information, to open ended hourly rates, not really good for a client, to hourly with a cap. All are short sighted. Phil makes the point that what a design team gets paid is a miniscule part of the overall project, and the deal structures are so bad they incentivize teams to do a bad job. Where do we go? Say you design a great office building that yields a doubled rent because of its design or performance. The architect doesn't get any of that in the current model. It never gets fed back into design fees unless you own the building. Some do that to capture the full potential, but there's no incentive for architects. It's a deeply structured, inherent problem. Different models exist. Getting owners to buy in would help – it might be tricky. It's got to happen. If architects keep giving their value away, it'll never change. We can complain all we want, but until we stop shooting ourselves in the foot by giving it away, owners will just go to the next guy who will. Let's figure out a way to capture the value we create. It's a cultural challenge. How to get owners to pay for more value?

It takes more than one party, plus courage. Architects to ask, owners to try. Entrepreneurship summarizes all we're talking about. We need to recognize our value, articulate and sell it, and not be locked in our historically myopic design-zone, then going away and grumbling. That's the issue.

"Let's figure out a way to capture the value we create."

Marble: It is. It captures a lot of these things. It's not something schools are teaching. It's not just business, it's innovation, understanding the value of what we do as architects and engraining that into a curriculum to make it part of the architect's sensibility. In a pure VC/Startup world, it's about identifying a product, finding a customer and making it profitable: "commercializing your idea." It's the VC world's core underpinning. To bring that mentality into architecture is interesting, but architects have different values. Economic, but also cultural, social, and other non-monetary aspects. When we talk about value proposition, we can't compromise other aspects of our work.

For most of us, being an entrepreneur won't mean developing an app and selling it to Autodesk. It will mean delivering value and selling it to the customer. A broader meaning.

Marble: There's a lot to take on. Enormous opportunities. This is an exciting time in architecture and AEC. Much of the promise and potential we've talked about for 30 years is materializing. Owners need to join the discussion to get us there.

Scott Marble Key Points:

1. Undertake the "design" of design. Develop a process in which design is an integral valued part, not an element that introduces inefficiency and tension.
2. Architects must address design reuse and repeatability without succumbing to mindless sameness. Not all endeavors need to be started from scratch.
3. Architects must capture the value we create through entrepreneurship.
4. Educate clients and partners on technology's benefits by showing its value to *them*.
5. Owners must engage in the conversation and be willing to take calculated risks with trusted partners to bring about change, including new fee and incentive models.

INTERVIEW
David Fano

Chief Growth Officer, WeWork, New York   15th September 2017

In 2008, David Fano founded CASE Inc., a technology innovation, BIM consultancy, and building technology strategy firm supporting innovative architectural practice, with Federico Negro and Steve Sanderson. In 2015, CASE Inc. was acquired by WeWork, a global network of workspaces providing more than 268,000 members with space, community, and services because of their expertise in driving innovation. An early adopter and leader of building information modeling, freeware, shareware, and digital tools, David received his Master of Architecture degree with honors from Columbia University. WeWork, valued at $20 billion in 2017, is enabling further transformation in design and real estate processes – and the industry. This interview took place in September of 2017; however, the figures mentioned in the interview were updated in October 2018.

Integrating Vertically, Changing the Market or "We Do Different Things"

Your founding of CASE Inc. broke molds for traditional architectural practice. It was a new business model using technology. I jokingly call your creation of algorithms and free shareware the "drug pusher's model": you gave out free samples (shareware on the internet), got people hooked, and eventually upsold them to a consulting gig or something that pays – they had to have it. I'm assuming you made money, enough to make you attractive to WeWork?

> "*Putting space into production is expensive and takes time, so we had to professionalize how we do design and construction.*"

Fano: We weren't acquired for our revenue, rather for our ability to drive technology innovation.

But you were at least as profitable as a typical AE firm?

Fano: Yes, we were a pretty typical service firm. By all measures of revenue and margins, nothing too crazy, but we made money.

Enough to support your doing what you liked to do and were good at. Which is what most of us strive for on the design side. Most of us didn't enter the profession because we were business people.

Now, at WeWork, you have a more ambitious set of goals: to transform the way design, construction and leasing space works. I picked out some keywords from your website: "a global network of creators . . . movement toward humanizing work." How would you describe what you're doing?

Fano: WeWork is on a mission to help people focus time and effort on doing the things they're passionate about. Our mission is to create a world where people work to make a life, not just a living. We really believe that. When people come together, they can accomplish much more than when they do something on their own. That community is this unmeasurable, enabling thing about being human that we value and gain from: the ability to connect with people.

Part of how we do that is creating environments in which people can come together. That's not the only way, and it's not the only way we'll do it in the long term, but the way this company started to pursue bringing people together was to design and build environments in which people work and challenge the traditional ways people get space to work. That put us squarely in the design and construction industry at first. I don't think Amazon ever had any vision of being purely a book company, that's just where they started.

"It's a new paradigm. Vertically integrated. It's as if Boeing opened an airline. It's more akin to what Apple does with iPhones. They control the full experience. They make the hardware, software, services, and sell and manage the full customer experience end to end."

We had to develop a proficiency around how we design and deliver workspace, because we want to create as many of these environments as fast as we can around the world to help people come together. Putting space into production is expensive and takes time, so we had to professionalize how we do design and construction. The environments and design are tailored to communal experience. The more we can strip away the things that aren't core to their mission, the more likely they'll achieve their mission.

"Professionalize" it? Isn't that what we've been trying to do all these years?

Fano: A lot of companies don't have that in their mission: to create space. It's: "index the internet . . . make the cheapest products on the planet . . . connect people through telecommunications," and that's what they *should* focus on. We want to strip away the noncore things and make it easy for them, so they can focus on their high value activities.

That's a basic need: connecting with other humans, the power of groups, but you're going about it in a tech-enabled, communal way. It supports the book's premise: designers do architecture and engineering because they love it. I joined a company that had experts at management and execution, so I could focus on my high value activities.

In executing your mission, you find yourself in the design and construction space – a world where traditional processes are broken. How do you go about it?

Fano: We do a lot in house. The only thing we *don't* do is sign and seal drawings. We start with the initial real estate sourcing, layout, and schematic design, then 3D scan/reality capture and BIM model every location to 100 percent DD, then work with local architects for code review and local conditions. We do a lot of our own estimating, procurement and logistics, and manage owner furnished items. A lot of volume, so we need to save every penny as we scale the business. This year we'll do 10 million SF of construction, and next year we'll do around 18 million SF.[1] We work closely with select GCs, but we're trying to get to where we're hiring them for labor, and we provide, because of economies of scale, a lot of the materials and equipment. Then we run the space ourselves. We have our own facilities and hospitality teams, repair and maintenance – all of that is done in house.

---

[1] Figures current as of October 2018.

Let's touch on the contracting approach. How are you bringing on Construction partners? CM? Bid?

Fano: All the above. In certain markets, we've got preferred vendors we like to work with. In others, we bid it out, like new versus existing markets. We've tried different contract structures, but across the board, all contracted partners use WeWork processes and procedures.

You've gone beyond the boundaries of traditional design firm revenue sources. You've got a toe in all kinds of ponds, many of which are historically profitable – based on the number of real estate entities, design and construction firms competing in those arenas.

Fano: There are all sorts of arenas to work in. It's not just the space. We help companies get insurance – other services. We do all sorts of things.

And speed to market or occupancy is an advantage. Lower carrying costs? Self-financing?

Fano: Yes, that's why we've raised as much capital as we have. A lot of that is venture capital.

Do you still have the freedom to provide the old, externally focused CASE Inc. oriented strategic technology services, or are you just focused on your own business now?

Fano: CASE Inc. has gone away. We focus on internal efforts. Building 18 million SF next year,[2] our internal team does all that work. My role on the Executive Team is to oversee design and construction globally. In effect, I run a 1,000-person design-build firm.[3] I'm also responsible for sales and marketing. I tell people I "build 'em and fill 'em." I'm our Chief Growth Officer.

With CASE Inc. gone, you've opened the door for others to fill the void following the model you helped build: to consult with architectural firms about technology. That's an incredible story.

Fano: I met you early, you helped shape a lot of the ideas. I remember your presentations at the University of Florida and BIMForum.

What advice do you have for those who want to change their process? Do you have competitors? Maybe you don't want to help create any, but what can you share with next-generation leaders?

Fano: It's a new paradigm. Vertically integrated. It's as if Boeing opened an airline. More akin to what Apple does with iPhones. They control the full experience. They make the hardware, software, services, and sell and manage the full customer experience end-to-end. In that sense, we're a bit of a new paradigm. In the real estate industry, when people look for space, there are many options from signing their own lease to coming to a company like us, but in terms of our offerings, we don't have a feature-for-feature competitor. We're trying to stay innovative, do new things, stay competitive and deliver an exceptional experience for the people in our buildings. And stay true to our mission.

The closest competitors are real estate entities. Others have offered cooperative work spaces, but they are more corporate. You're smart to control the entire experience in a different way.

---

[2] Figure current as of October 2018.
[3] Figure current as of October 2018.

"The one-liner is: challenge every-thing. Take nothing as a given. Force yourself to reevaluate every process and every way we've always done things in this industry, because they can all be better."

Fano: In this industry, if you want to innovate, you can't think in the context of how it's always been done. The one liner is: *challenge everything.* Take nothing as a given. Force yourself to reevaluate every process and every way we've always done things because they can all be better.

So many are stuck in their comfort zone or are afraid to take a risk. You did. Are you part of the establishment now?

Fano: Nah. We're doing different things.

From the WeWork webpage:

"When we started WeWork in 2010, we wanted to build more than beautiful, shared office spaces. We wanted to build a community. A place you join as an individual, 'me', but where you become part of a greater 'we.' A place where we're redefining success measured by personal fulfillment, not just the bottom line. Community is our catalyst."

"The nature of work is changing. Recruitment, retention, innovation, and productivity now require not just coffee, but also yoga, not just printers, but also art installations. WeWork offers companies of all sizes the opportunity to reimagine employees' days through refreshing design, engaging community, and benefits for all."

"As one of the very few companies that acts as an owner, operator, and client organization, WeWork is taking advantage of this unique position by looking at the way we build, design, operate, and engage in a holistic way. Our teams work cross-functionally with stakeholders in real estate, community, technology, and culture because as a vertically integrated company we can focus on developing products that connect members, improve member experience, and cultivate community through an agile and iterative process."

David Fano Key Points:

1. Firms with new organizational models and paradigms are changing the industry.
2. A vertically integrated solution controlling the entire customer experience is a bigger risk than a traditional service organization but offers advantages.
3. Technology is an enabler in new delivery models re: data, speed, integration.
4. Firms that can tap into new, nontraditional revenue streams have the upper hand, more capital, and fewer – or no – competitors.
5. Challenge everything. Take nothing as a given. All processes can be better.

CHAPTER 15

Change Agents: Advocacy, Equity, and Sustainability

INTERVIEW

Simon Joaquin Clopton, MS

31st July 2018

Simon Clopton is a recent graduate with a Master of Science degree in Building Construction/Program Management from Georgia Tech, and an earlier Bachelors Degree from UC San Diego in fine arts and digital media. He was project manager for construction of the Georgia Tech School of Architecture's Hinman Pavilion, a Living Building Equity Champion for the Kendeda Building, and selected as a Georgia Tech Climate Change Fellow and Graduate Researcher in the CONECTech Lab. Before his graduate degree he was managing director of Allotrope Mexico, a division of Allotrope Partners, a venture capital firm investing in clean technologies and renewable energies. Previously, he was co-owner and construction manager at Clopton Construction, a family-owned residential contractor in the San Francisco Bay Area and founder and director of the Centro de Educación Ambiental de la Peninsula Yucateca in Mexico. On the cusp of the Gen X/Millennial generations, he represents the sentiment and zeitgeist of emerging graduates and future practitioners.

> *"It'll take a transition, but if we're doing it right, doing good and doing well should become joined at the hip."*

Sustainable Practice: Tools and Data, Proof & Persuasion or "Doing Right, Good and Well"

As a current graduate looking to enter the industry, you're an interesting case. You've had experience in other industries – residential construction, venture capital, and sustainability consulting and education, but you're exploring returning to construction. Some are going the other way. What's the prevailing sentiment for you and your peers? What's the outlook? The attitude?

Clopton: As a current graduate looking to reenter the industry, my perspective may vary from my peers, but both are positive. Having grown up working in residential construction, my shift back to design and construction is as much a happy homecoming as it is a new and exciting phase in my professional life. I consider myself lucky to be entering at a time when we can finally build the proverbial better mouse-trap. And, we can back up our assertions with demonstrable results. The AEC industry is at a cross-roads. Materials science, advances in design, data science and environmental urgency have the potential to bring about great change. Today we can build with net positive impacts. As we refine processes, costs will drop, and efficiency will rise. All this is wonderful in theory, but with a bit of data crunching we can show the value proposition too.

Back in 2000, LEED seemed ambitious and restrictive, now it's commonplace. I look forward to the evolution of unicorn projects like Apple's new campus, and Georgia Tech's Kendeda Building. That level of sustainability will become commonplace. Energy and resiliency modeling in programming and predesign will be the norm. The economic benefits for financing and lifetime cost projections will be part of builders winning bids – with supply chain sustainability as a selection factor like lowest total cost. It'll take a transition, but if we're doing it right, doing good and doing well should become joined at the hip.

I'm a realist. Pushing the envelope can come across as crazy idealism. Some approaches fit better in one application than another. For example, modeled on simple economic metrics alone, renewables and energy storage perform better in certain utility markets over others. Being able to show what works where under real-world conditions improves project pipelines. On the construction side, the best laid VDC, 5D BIM and lean planning can be challenging to implement, but that doesn't mean we shouldn't try. The shift towards a smarter, better design and construction sector is happening. It's an exciting time to enter the industry.

Are you sensing a trend toward nontraditional careers?

Clopton: Yes and no. The AEC industry is shifting to new technologies. Proficiency in multiple design software, financial modeling, and desktop-to-cloud mobile platforms would have been crazy talk 10–15 years ago, now they're common resume bullet points. Tools like drones and augmented reality have the potential to increase AEC productivity. Whether someone considers themselves an analyst, data scientist or builder – if they get the job done better, they bring huge value. As far as attracting and retaining a younger workforce goes, I suspect we'll see an uptick in recruiting and retention, based on the above.

Having heard some of the issues discussed in this book and perhaps experiencing them through your own work, what's your take on managing design? Can it be done? How?

Clopton: Managing the design phase can and must be done. We need a solid workflow from preprogramming through handoff and O&M. Done right it's a value add to everyone involved – designers included. As part of my Spring 2018 semester at Georgia Tech (GT) I took an architecture special problems course. Architecture students had been designing an addition to the Hinman Building courtyard and the time had come to build it. Without going into minutiae, the design was beautiful, but was nowhere near ready to build. Constructability, compliance with design criteria, paperwork, site conditions, delivery method, and other details hadn't been accounted for. A lot of our work went to maintaining the schedule. The late-phase heavy lifting could have been streamlined had it started day one. That would have taken pressure off the designers. We had an overall design vision and goals at the start, but they weren't tied to a schedule and the nondesign requirements. Involving the management team and trades at the start would have been a game changer. Design management done right shouldn't be a restrictive slog. It should help designers focus and move from vision to an operational project built as intended.

So, you had a design management baptism by fire?

Clopton: Yes. Though I think challenging projects are great learning opportunities. As we moved into construction we got into a balanced, collaborative workflow. I'm a proponent of real-world projects being integrated into AEC higher-education and would love to see them as part of regular curricula. Given the project's positive outcome, I'm confident had holistic scheduling, collaboration, and early stakeholder involvement been done day one, things would have been easier.

Share your personal vision of an ideal career opportunity. Why? What are you doing in 20 years?

Clopton: I had the opportunity to go back to school to get my MS in building construction/program management with the goal of increasing sustainability and value in the AEC industry. There has to be value on both sides of the equation. If not, we're doing it wrong and the shift won't stick. My ideal career would involve problem solving, continued learning, and the chance to make positive impacts. Team, culture, and stability are key. Ideally, all that would combine to provide opportunity at multiple project stages.

"There has to be value on both sides of the equation. If not, we're doing it wrong and the shift won't stick."

What drives you?

Doing something I enjoy. Interesting, meaningful work, and the chance to interact with and learn from great people to support my family all drive me. The ability to make a positive impact does too. I have a do-gooder streak and want to make a difference, but I'm pragmatic. These can be balanced for better business and more fulfilling work.

In the book I challenge emerging leaders to become change agents. How do you feel about that? Ready? Prepared? Wondering why we didn't fix things for you over the past few generations?

I'm eager for the challenge of becoming an industry change agent. Ready – yes, prepared – yes. I'll continue to learn more and rely on a team to complement my skillset. As far as fixing things over the past few generations goes, a lot's been done. Our industry will always be at some stage of Tuckman's evolutionary cycle. That's good. Great strides have been made in project delivery, sustainability, and technology – and statistics show the AEC industry will improve equity and diversity between now and 2050. The changes we'll see over the coming decades will be astounding. A huge opportunity.

Simon Joaquin Clopton Key Points:

1. The design and construction industries are ripe for change. The need for such change and the convergence and availability of graduates with matching skillsets and value is serendipitous.
2. Many new graduates embrace their leadership opportunity and are prepared for it.
3. A cross-industry, cross-discipline mindset, and digital proficiencies and skillsets have become the norm – a minimum expectation for current graduates.
4. Financial acumen and the ability to articulate value propositions are expected, if not within an individual, then within a team.

INTERVIEW

Emily Grandstaff-Rice, FAIA, LEED AP BD+C, ID+C, WELL AP, NCARB, NCIDQ

Senior Associate, Arrowstreet, Boston, 2018 AIA Director-At-Large   9th February 2018

Emily Grandstaff-Rice, FAIA, LEED AP BD+C, ID+C, WELL AP, NCARB, NCIDQ, is a graduate of Rensselaer Polytechnic Institute and Harvard University, and a senior associate at Arrowstreet in Boston. Her leadership includes serving as president of the Boston Society of Architects in 2014, and their board of directors from 2009 to 2015. She has chaired the AIA's Equity and Future of Architecture Committee, Equity in Architecture Commission, and Continuing Education Committee. In 2017, she was elected as Director-at-Large to the AIA's National Board. Her innovative design work reinforces that a building is more than its shell – it is an experience. An architect with 19 years' experience on a broad project range, she is a frequent speaker and writer on the future of practice, addressing technology, the social economy, and environmental urgency. At the forefront of the equity conversation, she uses "hackathon" innovation to foster change in the profession. Representing all voices in the AIA, she focuses on process improvements for public projects, AIA, and peer professional organization action.

The Advocate or "To Be Continued"

You are the one of most active political voices among the interviewee group, active in many initiatives including the Equity by Design movement. How did you get there? What are your issues? How does advocacy affect your design approach?

Grandstaff-Rice: I come to the table with an understanding that politics goes several ways. Having grown up in St. Louis, I understand political actions are locally based and recognize it's important to receive people's ideas in context. Well-meaning may come from different sides, but we're working toward the same goal. That's how I see architects moving conversations forward. It's not about winning or losing, but the larger goal of how to improve the environment, health, and wellness. Sometimes happiness is a goal, and that's great! How can I use architecture as a tool to have a better impact in our day-to-day lives? Creating buildings manifests itself in different ways. Not just today, but 30–50–70 years from now. The urban realm is important. How do buildings present themselves to the public? Are they welcoming? How do they meet the ground?

> *"I'm an architect and . . ."*

How did that history influence you as a practitioner and leading voice for equity in architecture?

Grandstaff-Rice: Equity issues can have multiple meanings: equal access to buildings, services, areas, but also within the profession. I had to work through what it means to be a female architect. I didn't come in wanting to be a great *female* architect, I just wanted to be an architect. That's it. I dealt with things others didn't. So, I want to make it better for the next person. That goes to informing them about legal rights and being a leader, so I can be an example.

I took a strength finder test a couple years ago. Its insight said I always want to make things better, which drives me to advocacy, being a citizen and an architect. I tell the story of Sir Ken Robinson, the educational researcher and advocate who spoke at an AIA conference a couple of years ago. I was taken by his speech. It was about how schools kill creativity. How we all have these creative senses, but the framework inhibits them. He advocated that we all have many jobs. I could say I'm an architect, but I'm more than that. I'm a mother, teacher, citizen, resident, volunteer. I was in line waiting for him to sign my book. "Emily!" he said, "What do you do?" "I'm an architect!" I said. "That's *it*?" he replied. I sat through his whole speech and still fell into a trap. I think architecture would be a better profession if we stood up to that question, when people ask, "What do you do?" What he wanted, and what I've learned since, is for me to say, "I'm an architect *and* I advocate for better schools," or, ". . .*and* I create welcoming lobbies that allow the public and staff to engage," or "I'm an architect *and* I care about the environment and energy efficiency *and* I speak about it in my community." It's the "*and*" that's important. Part of being an architect lies in our ability to be explicit about what we do. So, "I'm an architect *and*. . ."

My first job was after my senior year in high school. I had a major conflict after my 3rd year and needed a break. I did a co-op in New York at Perkins Eastman, then studied abroad in India, so I had a gap in my formal schooling. That shaped my understanding of what it means to be an architect. Then I worked with EYP in Albany. After graduation I moved to Boston and worked with a small practitioner, Adolfo Perez, doing high-end residential, then to Cambridge Seven for 13 years. They were large shapers of my career.

I remember Arrowstreet being known for user and community engagement, public advocacy, and inclusiveness. These were emerging notions in the anti-establishment 1960s, in a firm that studied its process.

Grandstaff-Rice: You're on the right track. The firm has been around for 50 plus years. Yes, it's had a responsibility to community and research in practice. We want to be at the forefront of research, but grounded. Within budget, yet inspirational. Adam Grant, a professor at Penn, talks about innovation. Sometimes those who get there first create, and sometimes those who follow do it better. My point is, we innovate with a grounding in architecture, environment and practice.

A reality-based, approach. You've seen the dawn of computers in practice. How does technology shape your work? Do machines get the best out of people or add complexities? Love 'em or hate 'em?

Grandstaff-Rice: My first job was in 1994. It was a 3-person office. On the first day, there was a new tool called HOK Draw. They wanted to try it and hired me to do that but still do hand drawing too. Lettering! That entire summer I forced myself. Technology! I feel like I'm the last generation to see the other side, before laptops. When I got out in 2000, the shift had been made. I see myself as a bridge. I feel lucky to have seen it. That goes back to the education-communication metaphor. I want people to understand I lived through that. The empathy it gave me has helped.

How do you manage designers?

Grandstaff-Rice: My background is in education. I could have been a teacher rather than an architect, but I'm where I need to be. Education fascinates me. It's a form of communication. It lets us learn about the world and those we share it with. Education is that language, or way of connecting. My undergraduate thesis was about metaphors. I'm fascinated by the idea you can reflect a concept in different ways. Creating a common thread, but letting it work its way through other things. Architecture is a metaphor for ideas. It has its practical purposes: shelter, and thermal conditioning, but also a vehicle that inspires. It brings feeling, instincts or notions into different expressions. I'm always chasing that idea: ways we can use something to do something else.

After I got my professional degree, I got a Master's in educational technology. I worked with cultural organizations to communicate about informal learning environments about pedagogy – adult learning. My thesis was about mandatory continuing education – yes, the driest thesis ever to come out of Harvard – and levels of expertise. There's the baseline, interns, and levels of mastery. Brian Lawson and Kees Dorst's book *Design Expertise* talks about this. You move to project architect, then senior designer. There's proficiency, advanced experts, and mastery. The master knew *everything*. Very few get to mastery. I operate on the proficiency line, then pick something new to become expert at. People kid me because I'm still adding initials after my name, but it's the joy of what I do: learning.

We all have strengths and weaknesses. We don't take enough credit for things we're expert at. When I manage a project, I recognize people's different stages. Some are in the Mihaly Csikszentmihalyi "flow." I try to find the one thing that drives this person. If they're motivated by it, then go for it. Versus someone who's already an expert. I advise them to find someone to mentor. To pull from the entire array makes us richer. I fall in the middle: generalist to specialist.

Managing a team is a design process itself. My job is to let you take the risks and be your safety net. That lets us customize solutions with checks and balances. Talk to me, and we'll find a solution. Know I'll be there to back you up if it doesn't work. I want people to show up with a sense of "I'm making a difference." Personally, or for the community.

Your bio talks about your interests in social economy and environmental urgency. How do they affect your work?

We must recognize we're designing buildings for someone else to use. Part of that is knowing our work will change movement patterns, financial models, communities. The idea that architecture changes things. Phil Bernstein would say, "Our time and skill have such an impact on the GDP, yet we don't get any benefit from it. We get a small slice, yet the building lives on." We talk about that with clients and in the office. More than just the money, it's about that what we do affects others. Environmentally, should we do buildings that use fossil fuel? That's an ethical question; we can do better and move the needle.

To a degree, we've earned our stereotypes as self-serving "Howard Roarks," doing buildings for ourselves. As builders, we're looking for good teammates. If we're not listening to our customers, we won't have much to design. What's your mindset on the management-creativity continuum?

Grandstaff-Rice: Someone told me: "You're not the typical project manager." I said, "Good!" We're all designers, we just design in different realms. The split between management and design did us a disservice. We all have input on form, message, and economics. We're integrated, even though we do different tasks. We have different roles, responsibilities and input on how buildings get shaped. Why? Because we're all trained as architects, but because we're all users of buildings. Some clients come with preconceived notions or development plans. If we can't bring them along in the journey, we do them a disservice.

Matthew Dumich's office has an incredible mastery of tools. It enables their performative work but takes specialists and someone to orchestrate them all. Beverly Willis talked about that too. Arrowstreet is among the few current firms doing practice related research. Most of it seems technical. Can it inform the questions and themes in this book: How to manage design?

"The split between management and design did us a disservice."

Grandstaff-Rice: My graduate degree in educational technology at Harvard gave me a language and facility to talk about research outside the realm of architecture. In science, formulating a thesis and testing it is outcome based. Trial and error, looking for causation. In education, research is observation and data based. The analysis comes from your ability to evaluate in impartial, curatorial ways. Social science. Having an academic basis is a benefit. I did data collection but had to justify it to the Harvard Review Board, because I dealt with human subjects. I learned questions always have an element of bias, even if you work consciously to strip it away. You can't ask leading questions, you must ask clarifying questions.

What I consider research is the data – scientific testing of architecture systems. But research is also understanding my client, site issues, context, imagery and how elements work together. What do I want to learn, and how can I test that, so the next time somebody does it, they can learn from it too? That's giving back. How can I advance the larger practice of architecture? To me, it's all research. It's about understanding. And does it help a larger conversation: are we doing the right thing?

How do we make time to be empathetic? For the last 20 years, I've been a stranger in a strange land: an architect working within a construction company. When the pressure's on, we go with what we know, and trouble comes in. We take short cuts and go to our defaults and biases. We stop listening and rush. Conflict and errors result. How can we be flexible and "in the moment enough" to see it happening and fix it, given not enough time?

"To be continued . . ."

Grandstaff-Rice: Resilience makes things stronger. But you can't just fix one segment – it's got to be fixed for everybody. That means financial, weather, and community resilience. Boston's Chief Resilience Officer used to end our conversations by saying, "We'll keep in touch because we'll be *intentional* about it." Many times, I end conversations with: "To be continued." It's my catch phrase to continue with people.

We revert to bad habits because there isn't bandwidth for it all. We balance by being intentional. When we see default behaviors, we pause, then act. It's not contrived, it's intentional. You can't do everything, but you can decide what's important.

All the players need to get better at being resilient and intentional. I just learned about a phenomenon called the Zeigarnik effect: the psychological tendency to remember an uncompleted task more than a completed one. It's why beach books leave you hanging at the end of each chapter, to keep you turning pages – and how serial TV shows work. We love having something to look forward to. Our attention is heightened – that unfinished thing won't leave our mind.

Grandstaff-Rice: As architects, that's what we do. We create buildings with open-ended futures. We don't know how they'll hold up or evolve over 50 years. Time will take its toll. That's uneasy for many of us. We want to innovate, but don't have to innovate everything on every front. Pick the idea that makes a difference, that motivates you. That's how you focus. Have check-ins. Set goals. Reassess. Track progress. Notice it. That's a first step. Then you can address it.

Those principles are a great place to end: positive, forward looking, open ended.

Grandstaff-Rice: One of the reasons I'm an architect is to be fulfilled. I need a product, a building. I couldn't be an accountant. I need closure, even though the story goes on after I touch it. I'm not sure we answered your questions, but I'd rather be a conundrum than a one-liner. After an

"We'll be intentional about it."

AIA presentation I made, Randy Deutsch tweeted one person's feedback: "Messy but brilliant." I like it that way. I want you to draw your own conclusions. With me, it's not always a scripted dance, but we always learn something. So, "to be continued."

Emily Grandstaff-Rice Key Points:

1. Architects should adopt advocacy for a larger purpose: "I'm an architect and . . ."
2. Managing designers is best done by finding and supporting what they're passionate about.
3. Issues such as equity and environmental and social responsibility should be recognized and supported by research: giving back to those who follow.
4. Empathy comes from intentionality.

PART 2

PROJECT DESIGN CONTROLS: A FRAMEWORK FOR BALANCE, CHANGE, AND ACTION

"Test fast, fail fast, adjust fast."

– Tom Peters

"Twenty years from now you will be more disappointed by the things you didn't do than by the ones you did. So throw off the bowlines, sail away from the safe harbor, catch the trade winds in your sails. Explore. Dream. Discover."

– Mark Twain

"Nothing will ever be attempted, if all possible objections must be first overcome."

– Samuel Johnson

CHAPTER 16

Project Design Controls:
A Framework for Balance,
Change, and Action

Let's get right to it. Managing design is a difficult proposition with problems and opportunities galore. But you've got a project to plan, design, and build. We have little time to waste. In Part 1, the voices of over forty industry experts came together to discuss perennial issues – and offer suggestions. Their conversations were reflective, to frame issues and catalyze action. Enough talking. Time to start doing. We've listened and done our research; let's get to work.

How can we bring these ideas to bear? We need a means of organizing them to make sense of such an amalgam. In Part 2, we explore "Project Design Controls: A Framework for Balance, Change, and Action." Think of it as an armature with which you can "roll up your sleeves and get to work." With an overview and detailed look at each level's focal points, Part 2 offers principles for collaborating, understanding design processes, and using the model. We give you the concepts, tell you why they are important, and give you advice on how to use them.

"It don't make no sense that common sense don't make no sense no more."

– John Prine

Origins: Looking, Seeing, Borrowing, and Common Sense

Where did this tool rack of design management hammers and levers originate? They came from the best firms and projects in the industry. Many were adapted. I saw a recurring problem, extracted its parts, and organized those parts to be actionable on projects. Some, like classic aphorisms, are simply chestnuts – nuggets of wisdom that have stood the test of time, so simple they are likely the wisdom of many.

Others were adapted from related fields: "That's a good idea in airplane manufacturing or psychology, so why don't we apply it in construction? Software just automated that process, we should too." Colleagues, architects, builders, and thinkers – from the Harvard Business School, to W. Edwards Deming, to Tom Peters, Bill Gates,[1] and me – had no bounds to our pilfering. Why? Because these

---

[1] Hat tip to Tom Peters, an unabashed stealer and cross-industry idea-kidnapper. His books share stories of buying magazines in airport newsstands that have nothing to do with his industry. Then he raids them for ideas to pilfer and reapply creatively. When looking for process improvement, why limit ourselves to things used only in our design and construction arena? Let's borrow from the best and aspire to their levels! How do Google, Microsoft, Ritz-Carlton, and the New England Patriots do it?

practices did not involve proprietary knowledge, secret sauce, or magical powers – they were common sense. I've appropriated them, tailored them to connect design and construction teams, and given them to you. Now they are "our" best practices to guide successful teams. They are an assemblage of design management thinking – to ensure that teams speak the same language and work together. They work. Only a fool would disregard them. Let's explore.

Project Design Controls

Project Design Controls demand the attention of well-managed teams. This section organizes and provides structure for their use. While the generic term "project controls" may have alternative meanings in construction (e.g. surveying and field layout benchmarks, quality metrics, or schedule milestones) or finance (e.g. cash flow and other indicators) a more specific term is used here in the sense of controlling and managing design. Project Design Controls include:

- Level 0: Subsurface "Contractual/Forming" (Contracts and Team Assembly; Parties; Performance Period; Termination; Services & Deliverables; Compensation; Conditions)
- Level 1: Foundation "Planning/Organizing" (Project Analysis Kickoff; Programming & Research; Goals and Objectives; Roles and Responsibilities; Communication Protocols)
- Level 2: Structure "Measuring/Baseline" (Budget: Developing Budgets and Controlling Project Cost; Scope: Understanding What's in Our Project; Schedule: Scheduling Design; Documents: Planning, Managing, and Reviewing Design Documents)
- Level 3: Systems "Relating/Collaboration" (Relationships; Trust; People; Collaboration)
- Level 4: Enclosure "Leading/Strategic" (Change Management; Issue Tracking and Completion; Consultant Coordination; Option and Value Analysis)

Some of these elements are objective and easier to measure. Others are less tangible and relationship-based. Together they form a network of interrelated factors that can be used to manage design. They are called Project Design Controls because managing design needs limits. Control involves planning and intervention, monitoring and action. Without tangible guidelines and baselines to set direction, build guardrails, and keep us within necessary boundaries, teams are directionless. They cast their collective fates to the winds of wherever design process takes them.

To make them easy to understand and use, the Project Design Controls are organized into "Levels" and structured into a "Framework."

Framework: Intent, Form and Function

The Project Design Control (PDC) model is both simple and complex. Conceived of as a stable, constrained box, it strives to make order from chaos. This architectonic/building-morphic form is designed to help users construct, remember, and employ a cognitive map to manage design. The associative metaphorical Level names – "Subsurface," "Foundations," "Structure," "Systems," and "Enclosure" – were chosen for easy recall. A memory grid for users, the PDC Framework's form is practically purposed so you can remember and use it.[2]

---

[2] In Joshua Foer's book *Moonwalking with Einstein: The Art and Science of Remembering Everything*, Ed Cooke offers a memory-enhancement trick: "Change whatever thing is being inputted into your memory into something so colorful, so exciting and so different from anything you've seen before that you can't possibly forget it." Foer describes techniques such as "chunking," "elaborative encoding," and building a "memory palace" – a mental model of a familiar room or building into which objects to be remembered can be stored along a route to be easily recalled. In the Project Design Controls Framework, I have borrowed them all.

New Ideas?

Many of these ideas are familiar. Chuck Thomsen discussed many of them in his book *Managing Brainpower* almost 40 years ago. Some may soon be replaced by technology. New automation emerges by the second that eliminates steps in manually processing, translating and exchanging information. The principles remain. I celebrate being able to post central files online in cloud-based project portals. It saves time, ensures access to current information and promotes shared intelligence. But we will always need to translate, interpret and evolve that information. That's where we professionals add value, judgment, and experience. Who's got the steering wheel? What information should we leave out? Who decides which tools to use? We will always have to talk to one another to make those decisions.

Not Here. Here.

In Part 2 you will not find techniques that deal with the core subject matter areas of owning, designing, or building projects. As the provinces of their own professions, those subjects are left for professional business, architecture and engineering schools, construction management programs, and their respective subject matter experts. Do not expect to learn how to become an architect or engineer, to create building forms or size beams. Dispel your hopes to learn about cranes, hoisting and site logistics, purchasing and procurement, jobsite safety and cost estimating, construction scheduling, or pure contractor duties. You will not find a lesson on how to get funding for your project or how to run your business.

What you will find is a comprehensive way of thinking that connects and enables designing and building – control touchpoints to manage design. Decades of design and construction practice have taught us that we have to be connected. The advice in this section explains how to better make those connections. With your help, we can.

Navigation and Adoption: Internalization and Sharing

Readers with specific needs are urged to move directly to the Project Design Controls they need. Need a design schedule? Turn to design scheduling to find rudiments for planning design processes. Building a team? Move ahead to Level 0 or 1 to assemble, organize, structure, and plan your team, and to get to know the talented, passionate individuals who compose it. For scoping, budgeting or managing design documents, Level 2 offers explanations to help you cope.

None of this is rocket science, merely proven practices accumulated and shared over time. It is likely in digesting these concepts, you will experience déjà vu and brow-smacking realizations of how simple some of the ideas are. For example, develop a list of drawings you plan to produce. Good! Do it.

"Creativity is nothing more than combining two things that haven't been together before."

Anon.

Use this section as a best-practices checklist. Share those practices with a colleague, even if they don't work for your firm. They might appreciate it. They might even have one of their own that they prefer, or which works better for their team. They might have to use a different one by contract – or because their owner or boss said so. No problem. The important thing is that the job gets done. Read, share, adapt. Combine, simplify, and automate these principles as best you can. Make them work for you, not the other way around. If any are already obsolete, have been automated by software, or have been accommodated in another way, move on. Advance to your next management touchstone. These precepts are a means to an end.

Applying Advice

This conversation is between us. You have heard the voices of others in this book. Now I will help you apply their advice. Our conversation may continue in interactive form. Media, blogs, papers, digital chats, face-to-face discussions, presentations, or workshops may enable discourse. For now, imagine we are talking, and I have heard your issues, because they are probably like those countless others have faced. These experience-based, cross-industry principles can help you manage design. Adopt and adapt these lessons. Shape them for your own use. The approach is simple: a do-it-yourself, some-assembly-required method, open to interpretation, adaptation, and personalization. Would you have it any other way?

Design Management Prerequisites: Experience, Desperation, and Motivation

Not everyone is ready to manage design. Before you can get it right, it helps to get it wrong. Everything. Be late with your drawings. Change the design concept at the 11th hour. Go wildly over budget, 30, 40, 50 percent – multiple times. I have. Without facing your responsibility for such things, and feeling its pain, you will not have reached the point where you want to manage design. And you can't know how. The motivation to manage design comes from only one place: desperation. If you think you can do it easily, you are either wrong, or not practicing with the risk and invention valued by leading firms.

Your motivation must be fueled by fear, survival, and a desire to succeed at least as great as the desire you had to design, build, or manage a design team in the first place. Such desire comes only from the depths of your soul. Those who believe design can be managed easily or linearly do not understand it. They have not experienced it at the levels required to know how it feels. Start with failure, then decide to do something. Fix it.

Before you read the rest of this book or put its teachings into practice, promise this: design or build a few projects if you haven't already done so. Experience the mistakes and issues discussed in the interviews. Then you will be ready. When you have experienced failure first hand and are as passionate about fixing what is wrong as you are for designing and building, read on. The thoughts shared will help – but only if you are ready to hear them.

Flexible Framework

The original outline of this book considered offering templates and tools for quick, easy use. "Use these top twenty forms, check the boxes and you will have it." That approach is easy in concept but misguided in practice. Now, we offer a framework for how to think about managing design, called Project Design Controls. If you can grasp the fundamentals of design process and thinking you can manage them. I leave it to you to apply these ideas to the unique way of working with your team, project, or firm. Given the basics, you can adopt this framework, modify it, or design and build your own.

Other Theories

Similar management frameworks have been put forward over the last 50 years, including George Heery's book *Time, Cost and Architecture*. Published in 1975, it's one of the first books to propose time and cost as a management basis for architecture, engineering, and construction. Few principles for managing design come from advanced fields like rocket science, neuroscience, or quantum physics. Even fewer principles reveal proprietary magic wands or super tools. Rather, most techniques involve ordinary ingredients often lacking in design processes: common sense, experience, analysis, and subdivision of processes into manageable steps.

Toolmaking: What Gets Measured Gets Done

The best designers and builders make their own tools or customize those adopted from others. They're good at design, reinterpretation and masterful at using the tools of their trade. In time, many of this book's principles will be automated or simplified via software, artificial intelligence, or integration. Some may need to be customized. Their final form is not important; rather, it matters that their guiding functions – and the thinking behind them – get done.

An adage often attributed to Peter Drucker reminds us, "What gets measured gets done." This advice still applies – even to design. Despite design's iterative nature, experience tells us even the most creative teams need goals and thrive on, limits. If we don't set them, how do we know where we are going or if we have arrived? All teams need rules of engagement. In business settings, with multiple players and complex teams, those who wish to manage themselves must write these rules down, and share, monitor, and track their progress against the plan. Don't have a schedule? Then, how can you know how long you have to finish? No budget? No baseline Program of Requirements? Good luck. Then how do you know what problem you are trying to solve? How big is your building supposed to be? What are the sustainability goals? Design managers must be familiar with green metrics, analysis tools, pricing, and scheduling. The interviews in Part 1 with Dumich, Clopton, et al., cover these topics.

"Good artists borrow. Great artists steal."

Picasso

Boundaries, Limits, and Constraints: Enemies or Friends?

Design needs boundaries. Without them, there is no conflict, no boundary to break, no box to get out of: no fun for the people who thrive on the dopamine fueled by creative challenges. Without constraints, no enemy exists to rally against or outwit with design genius or hard work. Project Design Controls offer such limits.

Control with Target Value Design

Can you really "control" these design elements? It may be challenging, but you must try. As a start, set a target value limit for each, apply the three-question litmus test below, then commit to meet these limits as a team. When you miss one, or creep outside the lines, it is the manager's job to direct the team and make change to restore balance. Figure 16.1 shows such an intervention, made possible by having first set target values that allow assessment and correction.

Target value design is a proven principle in design management circles. It is a simple idea: every element to be designed needs a target. Only with an explicit target in place can teams assess whether it has been met. Apply this concept to budgets, schedules, program and performance criteria, sustainability metrics, safety, and others. This is what you need to know to do it: set a target, track it, and meet it. If you miss something, change it to restore balance or adjust another factor.

In managing design, target values and limits offer vision and guidance that is sometimes not absolute. In sharp contrast to traditional metrics as may be seen in finance, engineering, and purely objective pursuits, the team can sometimes decide to ignore or reset their limits in favor of other "discovered," "reprioritized" values. They shift targets midstream, allowing deviation in process.

Project Design Controls Overview

Project Design Controls are a framework within which the infinitely flexible, ultimately mysterious dance we know as design can be constrained. Each control provides guidelines to work within, test, or push beyond. Flexible, adaptable and scalable, this framework and its control touchpoints gives order to related moving parts. Although many design professionals aren't thrilled with having to have limits, the smart ones realize they're necessary. Good designers embrace and use these limits for inspiration and guidance. They use limits to manage themselves. Project Design Controls set up, track, and manage

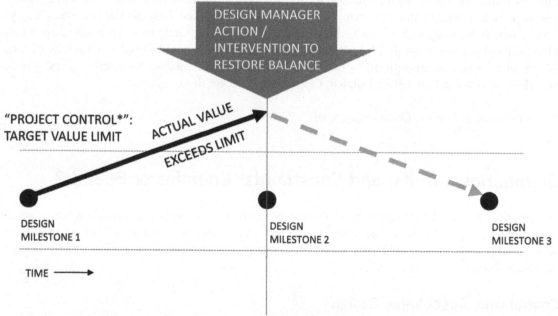

*PROJECT CONTROL: BUDGET, SCHEDULE, PROGRAM, SCOPE, DOCUMENTS, ETC.

FIGURE 16.1 Using target value limits.

design. Kept in balance, they deliver successful projects. Left unchecked they can destabilize projects. Chapter 23, "Understanding and Using the Framework," shares advice for how to cope with such imbalance.

As seen in Figure 16.2, the Project Design Controls Framework has "levels." Configured in a way familiar to those who design and build, they are organized in the logic and sequence of building. Construct a mental model of this structure and use each layer to manage change and project balance. When things inevitably change, adjust them to regain balance and keep on track. To understand the framework, we will begin with an overview of each level, share a litmus test to test project health, then explore each in detail. The level names and their functional purposes (in parentheses) are:

Level 0: "Subsurface" (Contractual/Forming)
Level 1: "Foundation" (Planning/Organizing)
Level 2: "Structure" (Measuring/Baseline)
Level 3: "Systems" (Relating/Collaboration)
Level 4: "Enclosure" (Leading/Strategic)

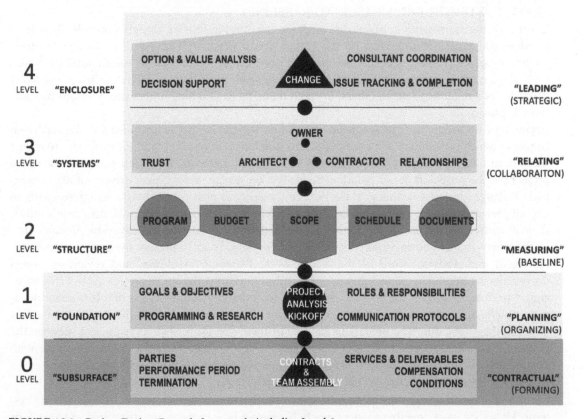

FIGURE 16.2   Project Design Controls framework, including Level 0.

LEVEL 0: "SUBSURFACE" (CONTRACTUAL/FORMING)

Level 0 of the framework contains Project Design Controls for forming teams and establishing team-member duties. Level 0 is the "Subsurface" level, the bedrock for the project's "Foundation," "Structure," "Systems," and "Enclosure" levels which rest upon it. Some design managers have the luxury of influencing the project delivery method, contract forms, and participants that will comprise their team. Others must accept the contract and team as assigned. There is no greater opportunity to influence a project's collaborative outcome than to get the "right people on the bus," be clear about what they are to do, and set common incentives.

LEVEL 1: "FOUNDATION" (PLANNING/ORGANIZING)

Level 1 of the framework contains the Project Design Goals that establish the rules of engagement, that supporting planning, organizing and defining tasks. This level forms the "Foundation" atop the Subsurface Level to support creating a team. These Project Design Controls include Goals and Objectives, Roles and Responsibilities, Programming and Research, Communication Protocols, and the Project Analysis Kickoff Meeting. Once set, teams rely on these controls to record common purpose and method, and advance to set limits in tangible, measurable categories – traditional management metrics – to "structure" their project and establish a baseline to manage to. Good design managers don't miss the chance to plan their projects. They lay a strong foundation upon which to build.

LEVEL 2: "STRUCTURE" (MEASURING/BASELINE)

Level 2 of the framework contains the tangible, measurable Project Design Controls. This level provides the Structure that rests on the Foundation beneath it. These Project Design Controls include program, budget, scope, schedule, and documents. In contrast to softer, subjective controls, these controls can be set, written, tracked, managed, and adjusted. A change in any one likely necessitates a related change in others. Without them, there's nothing to manage. Without balance, there is chaos.

Beginning with program, or project need, these limits include budget as a requisite. Budget is driven by scope, which is the intent and extent of design solutions that respond to program needs. To reflect those needs, scope describes how much and what kind of materials, labor and equipment is in the project per the proposed design solution. Scope is a key design management driver. Another perennial challenge, design scheduling, is the next core metric, to plan detailed tasks integrated with an overall project schedule. As the manifestation of the first four structural controls, documents reflect them in written and graphic form. As the traditional designer-to-contractor deliverable, documents – in all forms – models, construction documents, or exchanged data, represent the governing contractual basis and written record of the team's work. Planning, managing, and reviewing design documents is essential to managing design, but only if done over the entire project cycle. Little value comes from waiting until the project end to deal with design documents to complain about document quality or content, as if no discussion had ever taken place about what they needed to contain or when they were due. The precarious balance of these five elements is ever-present. Change the program? This could mean more scope and budget overruns and delay the documents. Late with documents? A construction delay is likely. A well-managed team mitigates such ripple effects.

LEVELS 3: "SYSTEMS" (RELATING/COLLABORATION)

Level 3 contains the "Systems" Design Controls. In dynamic project contexts, people are the systems. How they work together matters much. No matter how projects are structured at Level 3, or how many tangible metrics may have been recorded and managed, they are not worth much if the team does not trust them. Who cares if some project manager sets a budget? If no one understands, believes, or buys in to it, it won't govern anyone's design behavior. Like their real-life counterparts (e.g. HVAC, and other complex systems with moving parts, thermostats, gauges, valves and controls mechanisms), these human "systems" set and monitor the structure. Communication, trust, common language, interoperability, and teamwork are the relational human systems through which all other control metrics flow in design management.

LEVEL 4: "ENCLOSURE" (LEADING/STRATEGIC)

Level 4 of the framework, "Enclosure," addresses less predictable, strategic factors. I call it this because it acts as the building's skin to protect its occupants from the elements. Used to sense and admit new information, controls at this level form the project's "brains." How do we understand, manage, control, and lead the myriad of participants, consultants, owners, and experts, with external forces to find consensus, coordination, completion, and closure? Level 4 controls help teams make decisions, coordinate work, and manage change. This layer is the "envelope," "roof" or "umbrella" under which projects operate. It requires mastery. Teams that allow raised-arm-excuse-making, (e.g. "I can't get answers") are their own worst enemies. It is their job to anticipate and manage unresolved issues to keep the storms away. Rather than being surprised at rough project weather and changing conditions them should anticipate and respond to these factors.

Setting and adjusting these Project Design Controls is the art and science of managing design. While change is inevitable, balancing it is required and even possible, if we can tell how we are doing. The litmus test shows us how.

The Litmus Test: Project Design Controls

As a quick test of project design health, a quick exercise is to ask three questions for each Project Design Control. This simple 3-part, yes/no litmus test offers quick answers about project balance. These simple queries can reveal much. Questions for each factor are:

1. Do we have this kind of project control set up (i.e. a program, a budget, etc.)?
2. What is the control's status or value? (Does its current value match its targeted value?)
3. Is it in balance with the other project controls?

   If the answer to any of these questions is no for any factor, design managers must act. Why? What's the problem? What can be done? Who must be told? How? When? What adjustments can we make to restore balance. If the answer is no, congratulations, you have looked at your project controls and determined something is out of whack. Now you can act to correct it using the PDCs.

   If the answer is yes, congratulations, you are in balance. You are collaborating and managing design. Now you can focus on what might happen next to get you out of balance and anticipate and prepare for it.

If the answer is "I don't know," you may need more information, or to reframe the questions. If the answer is "I don't care," there is little hope for you. You are not trying. To manage design, constantly ask these three questions for each project control.

Design Management

Design management is the series of actions, processes, and tools that result in a design that matches its program, within budget, for the agreed scope and schedule, as documented. With the framework and litmus test in hand, let's look at each level and Project Design Control in detail.

CHAPTER 17

Level 0: Subsurface (Contractual/Forming)

There is no more important act in managing design than contracting for it and assembling the right team. That means qualifying, selecting, scoping, and assembling design team members into a team with parties, duties, terms, compensation, performance period, and termination conditions. Agreeing on the right partners and service scope sets the stage and establishes the rules of design engagement. While these bodies of knowledge are the subjects of other books, we will lay a foundation for them here – groundwork to support the rest of the framework above.

In this context, setting and using Project Design Controls begins with convening the team and writing contracts to give the collective the best conditions under which to work. Some teams may be provided a team and contract – conditions dictated by others and assigned to the team. Fear not, the rest of the PDC model still applies.

To begin the Project Design Controls–as–building analogy, we stand on a subsurface level (Level 0), the geotechnical soils, or bedrock into which the foundations (Level 1) of project planning are built. Refer to the Figure 17.1 to view this "forming" Project Control Level 0 (Contractual.) (Note: If your project delivery approach and contacts are established and your team is already formed, move to "Level 1: Foundation [Planning/Organizing]" to organize and start your project.)

Project Design Controls

In the following sections, the Project Design Controls are indicated in ALL CAPS.

TEAM ASSEMBLY (Selecting the Right Team)

Selecting and assembling the team is paramount. Teams can be put together in many ways. Most owners use Requests for Proposals (RFPs), Requests for Qualifications (RFQs), interviews, competitions, and other forms. With collaboration as an objective, the bias is not to award design and construction based on price alone. Rather, select teammates who will work closely to provide professional services – as designers and contractors do, based on experience, skills, culture, chemistry and collaboration to deliver value. Do not underestimate the importance of selecting professionals who share your objectives, culture and values.

| 0 LEVEL | "SUBSURFACE" | PARTIES PERFORMANCE PERIOD TERMINATION | CONTRACTS & TEAM ASSEMBLY | SERVICES & DELIVERABLES COMPENSATION CONDITIONS | "CONTRACTUAL" (FORMING) |

FIGURE 17.1   Level 0: "Subsurface" (Contractual/Forming).

CONTRACTS (Project Delivery Forms)

New guard leaders of the legal, financial and project delivery movements argue that project delivery mechanisms, contracts, and incentives hold the key to industry change. Data and momentum show they are right. Without reason or requirement to change, most of us carry on as we always have. A few innovators and change agents try to effect new processes, but without contract leverage, they face an uphill climb. This condition is understandable; it is backed up by data from projects across the country. While this book's focus is not on project delivery or contracts, it offers the uninitiated a brief overview of common methods. Borrowing heavily from the work of M. Kenig's *Project Delivery Systems* (AGC of America, 2011) and H. Ashcraft's *Integrating Project Delivery*, Wiley, 2017), four prevalent common project delivery methods and their features are offered in simplified form to support this book's theses. Depending on the method selected, owners and teams contracting using these methods must articulate:

- Parties, services, deliverables
- Performance period
- Compensation
- Conditions
- Project delivery

PARTIES, SERVICES, AND DELIVERABLES

Design managers must ensure all necessary disciplines are included in contracts. In the pressure to execute agreements quickly, players can be forgotten. Use checklists as reminders of the many participants and responsibilities. In team arrangements, tasks can be done by various parties. Be clear on who is doing what.

PERFORMANCE PERIOD

Projects that continue beyond anticipated schedules can deplete design fees and exhaust teams. The impacts of working with teammates who have lost their profit margin, potentially due to no fault of their own may be felt across the entire owner, architect, contractor (OAC) team. In contracting, allow time and fee contingencies or incentives to compensate.

COMPENSATION

Design fees have historically been based on percent of construction costs, lump sum or hourly bases. Based on the current state of practice all seem inadequate. Teams contracting for design should derive new value-based compensation models that reward providers for service and value. (See Bernstein, Marble, and Bryson interviews.) Compensation that rewards performance that achieves team metrics or other goals is overdue. To understand AE fees as a relative percentage of life cycle cost over a thirty- to fifty-year facility life, consider Table 17.1

TABLE 17.1   Relative Facility Life Cycle Cost

Relative Facility Life Cycle Cost	%
Planning	.02
Design	.09
Construction	8.7
Activation	3.2
Operation	87.0
Total	100.0

R. Tietjen, AIA, RLA, U.S. Department of Veterans Affairs.

Since their work represents only 1.1% of a building's life-cycle cost, design teams should demonstrate that their value contributes to leveraging these other factors. If they could offer such value, even as much as *doubling* current AE fee levels would be cost effective for OAC teams and offer immense payback, in view of design's potential to pay dividends on the other 98.9 percent of a project's total ownership cost or building life cycle cost.

In managing design, reach an understanding of appropriate value-and-service-based scope and fees that pay for what the team needs, rather than basing contracts on current commodity fee levels.

Conditions

Contracts must include clauses for breach, termination, assignment, errors and omissions, key personnel, and other key legal aspects. Where IPD or team arrangements are used, clarify all responsibilities. They will be blurred and harder to distinguish when the team collaborates on what have historically been clear, singular responsibilities.

Project Delivery

Table 17.2 compares the abilities of various project delivery methods to enable design management. It explores differences under each delivery method to show how design management's "when" and "how" changes. While most tools and processes are the same, how and when they're used, and their effectiveness varies.

TABLE 17.2   Project Delivery Methods: Design Management Advantages

Project Delivery/ Contract Type	Single Responsible Entity?	Shared Risk, Incentives, Value Add Possible?	Speed Possible?	Flexible/ Adaptable to Changing Conditions?	Common Process and Tool Benefits?	External Ability to Manage Design?
Design/Bid/Build (DBB)	No	No	Slow	No	No	No
CM-at-Risk (CMAR)	No	Partial	Fast	Yes	Partial	Partial
Design-Build (DB)	Yes	Yes	Fast	Yes	Yes	Yes
IPD (IPD)	Yes	Yes	Fast	Yes	Yes	Yes

Supporting Collaboration

Contract incentives can be a major reason for success in design management, coordination, and meeting budgets and schedules. They carry more legal, contractual, and financial weight than other important factors such as culture and processes. To capitalize and remove the adversarial incentives inherent in old contract forms, get rid of them! Don't miss the chance to use new collaborative contracts and incentives. To reveal the benefits, the Project Delivery Timelines diagram in Figure 17.2 shows the advantages of various delivery approaches and contract forms: how and when Project Design Controls can be set under each. The implication? Contract methods that apply PDCs early offer greater impact, influence, and value in managing design.

Design-Bid-Build

In Design-Bid-Build delivery there is no ability to select and form the team until design is complete and budget and schedule are developed. Opportunities for contractor-and-team-based budget, schedule, and constructability input are lost. No ability to form team-based Level 0, 1, 2, 3 or 4 PDCs exists.

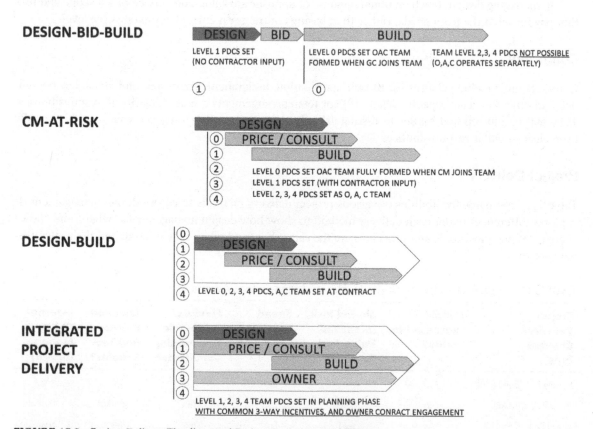

FIGURE 17.2 Project Delivery Timelines and Project Design Control Timing.

CM-at-Risk

In CM-at-Risk delivery, Level 0 PDCs can be formed the moment all parties have joined the team. Levels 1–4 PDCs can be established early to set the stage for teamwork while design is in progress.

Design-Build

An alternative contracting approach to D-B-B or CMAR is Design-Build. In this method, the owner relies on a single entity to design and build their project. Whether performed within a single integrated firm or via subcontracted designer under the builder, (or builder under designer) is of secondary owner concern. In this approach, the owner relies on a single responsible party but faces other unintended consequences. Contractors accustomed to complaining about designers under contract to the owner must change their attitude: contractually, they are the designer! Let design management and understanding begin. Design-Build Institute of America research shows a rapid rise in this delivery method. By 2020, forty-four percent of construction work is anticipated to be delivered under Design-Build delivery.[1]

As in CM-at Risk delivery, in Design-Build, all PDC Levels can be set when the design-builder is brought on board. The difference is that the owner holds a single contract. PDCs can be established early and managed to set the stage for teamwork as design begins.

IPD (Integrated Project Delivery, Integrated Practice, and Team Integration)

In response to issues of dysfunction and contradictory incentives, leaders have developed new contracting forms to solve these problems. Approaches such as Integrated Project Delivery (IPD) have seen success in forming tri-party agreements in which owners, designers and contractors share project responsibility. In contrast to the siloed Design-Bid-Build approach, IPD not only offers – but dictates – a "painshare/gainshare" mentality. At a time when IPD is used more often, this approach, and its "IPD-ish" and "IPD lite" variations are worth exploring for those who wish to manage design collaboratively. IPD offers the advantage shared incentives, from the outset. In this arrangement, the highest potential exists for Project Design Control, incentivized by all PDC Levels including the contract.

Other Resources

Further project delivery analysis is beyond our purpose. For more on project delivery and contract approaches, see M. Allison, H. Ashcraft, R. Cheng, et al., *Integrated Project Delivery: An Action Guide for Leaders* (Charles Pankow Foundation, 2018); M. Kenig, *Project Delivery Systems for Construction* (Associated General Contractors of America, 2011); and M. Fischer, H. Ashcraft, et al., *Integrating Project Delivery* (Wiley, 2017).

This book's intent is to remain "project delivery neutral." Regardless of which contract or project delivery method you're operating under, use that method's conditions to collaborate as best you can. Define your project and get your team before you start design and construction in earnest. Owners and teams savvy enough to conduct a project definition phase to test team empathy and project parameters will be better informed about their project scope and resource needs and can select the right architect, contractor, and contract type for their given situation as they begin design and construction.

---

[1] DBIA 2018 data, per Construction Dive, Kathleen Brown, July 9, 2018.

CHAPTER 18

Level 1:
Foundation (Planning/Organizing)

The best way to manage design in a team environment is to start well. (See Figure 18.1.) Countless teams fail because they forget to prepare or think they don't have time to plan. Make it a point to create a tightly knit working group. Set the following controls in place first, or as soon as you can.

"If you don't know where you're going – you'll end up somewhere else."

– Yogi Berra

| **1** | | GOALS & OBJECTIVES | PROJECT ANALYSIS KICKOFF | ROLES& RESPONSIBILITIES | |
| LEVEL | "FOUNDATION" | PROGRAMMING & RESEARCH | | COMMUNICATION PROTOCOLS | "PLANNING" (ORGANIZING) |

FIGURE 18.1 Level 1: "Foundation" (Planning/Organizing).

Goals and Objectives

Set goals: Setting goals before beginning a project seems obvious. Unfortunately, it is surprising to see how often team fail to set goals. William Pena, in his classic book *Problem Seeking*, talked about the value of breaking goals down into subject-based categories: form, function, time, economy, and quality.

Breakdown: This breakdown method is valuable in grouping goals into categories useful in design and construction. Too many projects are launched with "lazy" goals such as: "Be on time, on budget, with high quality." These clichéd statements do little to define project-specific needs. Strong teams articulate *real* goals, categorize them and commit to them.

Shared goals: The next goal setting principle is that all team members should agree to the goals. Within client organizations, multiple constituencies may have conflicting goals. The construction director nearing retirement may want to implement a safe, easy design, and an easy transition. The visionary CEO may want her new building to set new standards in innovation, energy, or worker satisfaction.

The aging facilities manager may want simple, old-school systems; the environmentally conscious, newly hired advocate may pine for state-of-the-art, complex systems.

Beyond client goals, other team members may want radically different things. The engineer wants efficiency, or to recover profits lost on the last complex design job. The architect could want to execute swiftly based on past formulas, with little innovation, to keep his fee. The contractor, amid a firm reinvention, may wish to revisit the process of designing and building collaboratively, using all the technology, process, and social techniques available. The owner may wish to experiment with new IPD contracts. Without sharing goals and objectives, and strategies to accomplish them, this team is headed for trouble. When they don't resolve their differences, they march off into months of conflict. Good teams make goals explicit and agree on them before they start.

> ## ✳ Tip: Goal Writing and Sharing
>
> Equally important to expressing goals is writing them down and sharing them. Knowing one another's goals, the team can devise a plan to achieve them. If I know you want to document and innovate the process, I can help. Having expressed, written, and exchanged goals and objectives, the team can understand and accept (or disagree with and resolve) them before proceeding. Later, when new teammates get added, they can join the mission as an informed team member.

Roles and Responsibilities

Teams that begin designing and building as a group without knowing who is on their team doom themselves to interpersonal failure. How can you interact if you don't who is there, what they can do, what their role or job is, and what motivates them? To fix this problem, find out. Get an organizational chart. Make time for it.

Have a party before the project starts, not after. Learn who "they" are as companies and individuals.[1] This process has many names: partnering, socializing, bonding among them. Without first setting up roles, responsibilities, and methods for working together, teams restrict workflow and confuse purpose. Even after you have listed all the players, things will change. On projects of any size, team members come and go regularly through attrition and other reasons. Keep your project directory current. Nurture and reestablish connections with existing and changing teammates. You will work together better. Connecting with your teammates is key to starting well.

Functions

The average owner or contractor teammate may not understand this: design firms have a host of job functions and people who perform them. The designer is the prime person and behavior type we discuss. Responsible for generating the building form, they are the spiritual leader that sets the project's plan and appearance. The project or production architect is responsible for producing a good set of documents. He or she is lasered-in on roof details, coordinated drawings, and communicating project scope and design intent to the contractor. Project managers are charged with leading their teams to

---

[1] When asked how to get to know your teammates better and to do what the team needs, Carol Wedge, CEO of Shepley Bullfinch, shared this tongue-in-cheek observation: "My first reaction is to get to know their boss." 6/23/18, AIA Conference on Architecture, New York.

on-time, on-budget completion, while keeping client, contractor and design teams happy and productive. All these individuals and a plethora of others like spec writers, staff designers, or technical experts can be leaders for their respective disciplines. Getting a schedule commitment from the design principal is less reliable than getting it from the project manager. Know which teammate you're dealing with to increase your chances of working together effectively. Finding teammates with skills in more than one of these functional areas is a welcome development.

✳ Tip: Responsibility Matrix

After listing the things to be done, a good way to assign responsibilities is to use a form like Table 18.1. It allows for primary and secondary responsibilities, support, approval/veto, and information for those who need only be kept abreast of progress. These letter codes allow for a nuanced way to assign responsibilities versus the simpler, binary "yours or mine." Create your own version for your project's activities. That way you will not leave anything out.

TABLE 18.1   Responsibility Matrix

Responsibility Matrix

Legend: R= Responsible, S= Support, I= Inform, A/V= Veto, ?=TBD

Responsibility	Owner	Architect	Contractor	Other
Geotechnical	R	S	S	
Survey	R	S		
Materials Testing	I	A/V	R	
Code Consultant		R		?

Communication Protocols

With the many players and kinds of expertise found in modern teams, many "languages" and "dialects" are spoken. Owner, architect, engineer, user, contractor, IT, and scheduler are just a few. To align and translate these ways of speaking, set preferences, hierarchies, alignment, and means in advance. Will tech-based collaboration tools be used to share data in a central location? How often are face-to face meetings? Who should call whom? Failing to plan for these processes is likely to overload unknowing teams in landslides of email, lost data, and miscommunication. Have a communication plan. Use the best tools you can.

How will you communicate? With whom? When? How often? What technology, software and tools will automate and connect the vast numbers of professionals that convene to design and build? Surely, you will not rely solely on old-school techniques. Is a project information portal in place? A central shared data repository? A conflict resolution hierarchy? Of course. Only a fool would embark upon a major project without the best technology and smartest tools, right? No one does projects by sending lots of indiscriminate emails without a plan, do they?

BIM/VDC/Digital Infrastructure

When we built projects with hand drawings and rotary dial phones, we did not need protocols. Now, with the multiplicity of software, platforms, digital data exchanges, and wealth of use cases, we need digital plans, protocols, and infrastructures. Why develop a digital model for sharing if some team members do not know how to use it, have that software, or the server capacity to access or accommodate it? The lead times to price, approve, buy, acquire, install, and learn these tools can be long. Start these enabling activities early.

✴ Tip: BIM Execution Plans

Teams that develop BIM Execution Plans (BxP) together ease workflows by setting goals and plans for digital data exchanges. They set themselves up to maximize BIM's potential over the greatest spectrum. As hard as the "people communication business" is, the business of computer interoperability can be harder. To maximize its potential, plan for it. Get help if you need it. Design is now done by machines talking to machines – and people talking to people. Leading BxP guides can be found in the National BIM Standard and via John Messner and his team's work at Penn State.

"A problem well stated is half solved."

– Charles Kettering

Programming and Research

Despite the AIA contract's longstanding assumption, few owners develop or provide adequate building programs. Some hire professionals to develop programs before design starts. Most owners rely on partners to help them figure out what they want and need during design and construction. Teams that recognize this need should jump at the chance to provide this valued service to owners – and get paid for it.

Before you start designing and building, and likely while doing so, define and articulate the owner's need by developing a program. No wonder designers are lagging – no one has determined what the owner needs! We've got to work together to create this often-lacking Project Design Control.

Beyond the program of requirements, teams should get smarter about research. In contrast to the self-informed design exploration and trial-and-error approach we have used for centuries, professionals should invest in capturing and sharing research. Those who do can add value and be more profitable. In any project endeavor, ask what has gone before. Use it as a benchmark. Site visits and comparable facility tours offer great value. They create a sense of shared history for the team, a common knowledge base that support their mission.

Project Analysis Kickoff Meeting

The project analysis kickoff meeting (see Figure 18.2) is a strategy to "launch" or "refresh" a project. It uncovers critical information, builds communication, and is proven to contribute to project success. This technique parallels "brainstorming," "partnering," "interdisciplinary work teams," and other management and problem-solving techniques. It gets projects off to their fastest, best start. Several key aspects make it work:

- *"All" team members are present*: a prerequisite under a team approach.
- *"All" information* available is present in a useful format.
- *Time is compressed,* capitalizing on and contributing to the synergy, communication and momentum that result in great value to the owner and to the team.
- *Lag time is eliminated* in favor of immediate understanding, consensus, and short decision cycles.
- *Rapid prototyping* of ideas is accomplished with immediate, multidisciplinary input.
- *A common team mission* is identified, and a bond is formed.
- *Synergy, expertise, interactivity, collaboration and greater value* are the result.
- *The project is defined*, balanced, and set up for success.
- *Project controls are set,* and possibly, design direction.

The agenda includes:

People. Companies, individuals, and relationships (the soft side) are introduced. These "people" issues set the stage for factual project issues.

Purpose. Purpose, vision, goals, expectations are shared and recorded.

Project. The project is defined, and measurable project controls are set: program, budget, scope, schedule, and documents. These tangible measurable limits form the baseline for design project management.

FIGURE 18.2   Project analysis.

---

[2] Heery's book blazed trails for accountable architecture. See p. 71 of his book for his original description of this process. With homage to Mr. Heery, I've renewed it with technologies and techniques that have emerged in the last five decades.

Possibilities. Possibilities are explored by analyzing design options and constraints in a multidisciplinary way including cost, schedule, constructability integrated with design potential. "Design" begins, with expert input.

Plan. The project plan is put into place by the team with work plans, action items, and next steps. This quickly builds a working team and well-defined project. Research shows that conducting a project analysis kickoff meeting during predesign, or as early as possible, is critical in successful, balanced projects. Projects that complete a predesign, feasibility (or project validation phase as it is called in IPD circles), have a grasp of key project issues. To start well, start with a project analysis.

1. Project Analysis Purpose, Organization, and Participants[2]

Much of the thinking and text in this section was borrowed from my days at Heery International. Its origins are the work of George Heery, FAIA (verbatim excerpts, updated to modern practice), from a process developed in his book *Time, Cost and Architecture* (McGraw-Hill, 1975), and related to earlier CRS "squatter sessions." (See Thomsen interview.) Heery called the process "predesign project analysis," because predesign is the most valuable time to do it – the right time. But too many miss the opportunity to do it during predesign. This is a shame – one of the biggest mistakes a team can make. In response, I have renamed it "project analysis," because, if necessary, it can be done anytime to reframe and reactivate a project. Regardless of where you are in the process, when new team members are added, or something changes significantly, schedule a project analysis to refresh your team. Since design happens throughout a project, a project analysis can happen almost any time, or repeat during any phase, but the earlier the better. Its purpose is to identify critical program, design, budget, schedule, scope, construction, procurement, and document considerations interactively and quickly. Since 80 percent of the direction, impact, and decisions affecting a project are made in the first 20 percent of the schedule, per the Pareto Principle, deal with pertinent data early, comprehensively, and efficiently to form a plan to balance and set up the project for success. Conducting the project analysis during predesign, planning, or programming *before* design begins separates problem definition and strategy from design process and solution. It also saves design and construction time due to its high-value planning and strategy early in the influence curve. In addition to being exposed to project analyses in Heery's book and conducting them as an employee in his firm, I have been lucky to do them in other contexts, including working with legendary multidisciplinary design think tank experts IDEO, and as a contractor and design manager. Borrowing and paraphrasing from all these groups, here is my recap:

The project analysis is a multidiscipline "think tank." Invite the owner, construction manager, architect, project manager, designer(s), engineers, landscape architect, planner, and interior designers, and all key experts as participants. Project size and nature governs who to invite. In organizing analysis sessions, don't let them become mere committee meetings, reporting sessions, or "typical" project kickoff meetings. Initiate sessions before design work starts in a way that creates a harmonious, productive, collaborative, interactive atmosphere. Challenge one another! If you are lucky enough to get your spot at the table, don't waste it. Speak up!

2. Preparing for the Project Analysis

Before the Session: Orientation/Preparation Package

Before convening a project analysis meeting, brief project team members to explain its purpose and nature, and their roles. Send this overview and the agenda to describe the process.

Preparation and attendance are always limited by time. Since this meeting is expensive and involves many people, it should be efficient and informative. To overcome that and stretch time, put pertinent information in participants' hands in advance by preparing a project orientation package. Compile existing pertinent project information (e.g. drawings, sketches, objectives, site and surrounding area data and photos, program of requirements, agenda, sketches, budget, schedule, zoning, agency approvals, and anything else helpful and necessary) for all attendees.

This package informs them in advance, so they come to the meeting ready for interaction and evaluation, not orientation. Your goal is to accelerate information flow to the point of overloading in advance of the meeting. Project analyses can have different focuses: visioning, planning, team building, orientation, or design opportunity analysis. Set your sights on your purposes and plan accordingly.

In addition to project information, distribute information management guidelines and exchange samples with participants in advance as test cases. Predetermine formats, BIM and VDC schemes and protocols, with software and hardware. Eliminate translation problems and maximize reuse of information generated in the project analysis. Hold a separate BIM Assessment, BIM Execution Planning (BxP) Session, or information or media exchange forum. Project managers leading a team through the digital design process for the first time should prepare and plan. In this case, project analysis sessions may be longer. Consider an introduction, a break of several days, and resume, with intervening work by all parties. Where projects are too large or agendas too long, break out less-interdisciplinary topics to be done offline with smaller expert groups.

2–3 weeks prior: Set the date and confirm the attendees. Who? Do I really have to? Principal-in-charge, consultants, PM, owner, PA, architects, interior designers, engineers, preconstruction team, scheduler. Verify facility location, adequacy, and size. Develop an agenda, with team buy in.

1–2 weeks prior: Compile Project Orientation Package (POP): What, why, background, site context, images, existing facilities, budget, program. Solicit input from others. Profile owner, organizational chart, annual report, and research data. Distribute POP to attendees with instruction and "homework assignment" to digest it before the meeting.

3. Conducting the Project Analysis

During the session: Managers leading their first project analysis will need extra preparation, and practice to develop their own styles, techniques, energy, and panache to make sessions productive and engaging. Project leadership should conduct the sessions jointly from a multiparty perspective.

Create an environment conducive to productivity. Coach participants to suggest partially developed ideas and accept constructive criticism with good humor. At this stage they should understand that comments and creative contributions outside their own disciplines are not only acceptable but encouraged. Meeting effectiveness is directly related to creating an intense, continuous, focused experience, including the people, resources, spontaneity, and atmosphere to support it. To be effective, facilitators should keep pace, elicit key opinions, and avoid getting stuck on any one detail. When necessary, set issues aside and come back to them. Defer judgment during the idea generation stage.

Prepare an agenda, so people can manage their time if they cannot attend the whole session. Prepare enlargements of the preparation package to display (or project them live) to illustrate objectives, facts, and visual relationships, and get buy in. Avoid solving tough, detailed issues. Concentrate on a wide range of parameters to get a feel for the entire project's key issues. Order in lunch to sustain focus (and the expensive group assembled). Keep together and productive.

(As a benefit, you'll get to know one another during the "work-through-lunch break.") Be visual. Use diagrams, images, and live screen capture of key points, data, and concepts.

After the session: Prepare meeting notes, distribute sketches, concepts, program items, and results in written and graphic form as a launching pad for design. Give project information (in the right form) to those who need it. This starts the Project Definition Package – the management baseline of a balanced project. Distribute meeting notes. Continue to develop a phase-end Project Definition Package. Conduct follow up meetings.

4. Duration, Agenda, and Results of a Project Analysis

Project analyses may run from hours to days. One or two days are typical, including follow-up documentation. "Only rare, simple projects can be covered in a half day."[3] "Typical" Project Analyses accomplish the following:

   a. *People*. Introduce team, firms, individuals, roles, responsibilities, interests and abilities. Use team orientation/workshop activities to form the team. Begin the project's "soft," "people," "relationship" side.

   b. *Purpose*. Establish and capture goals and objectives in terms of cost, function, time, quality, form. Articulate goals and objectives of owner, users, CM, architect, engineers, and so on.

   c. *Project*. Establish Project Design Controls: program, budget, scope, schedule, and documents, that will be used to design to, collaborate with, and manage project design.

   d. *Program*. Outline the following program of requirements types: space, site, systems, personnel program, operational, and other program requirements.

   e. *Budget*. Confirm, develop or analyze the total project budget analysis. Identify potential cost problems, cost reduction possibilities, and construction purchasing opportunities. "Gut check" the budget versus program (and versus design if design exists). Identify potential cost reductions, purchasing, and other factors that influence costs.

   f. *Scope*. Develop project scope and extent and systems to be included for building, site, parking, furnishings, fixtures and equipment (FFE), and so on as a guide to what must be designed, budgeted, and built. Use CSI checklists and other similar tools.

   g. *Schedule*. Develop overall project schedule, including design, project definition, construction, commissioning and other key milestones.

   h. *Documents*. Identify a project-specific approach to documents, and a list of key documents required, by phase.

   i. *Possibilities*. Explore/establish design concept(s), options, or at least the direction design work should proceed, plus areas of needed research.

   j. *Eliminate design blind alleys*. Identify constraints and opportunities. To the extent feasible, including construction, fabrication and delivery constraints, and labor, design, site, owner decision and financing, agency approval, and other administrative constraints. The ability to identify constraints comes with experience. Frequent constraints are site grading and foundations, long lead equipment, and permitting. Extensive earthwork, complex excavation, or foundation work may constrain or dictate design and construction sequence.

---

[3] G.T. Heery, *Time, Cost and Architecture* (McGraw-Hill, 1975), p. 73.

Other frequent constraints include structural frame fabrication/delivery/erection; labor contracts; site availability/zoning; major mechanical equipment; utility, street/site access work; electric switchgear or other electric equipment; elevator; lab equipment; furniture; special and unusual major equipment delivery/installation; agency approvals; materials availability (i.e. reinforcing steel, brick, etc.); owner's reviews and decisions, and environmental or sustainability issues can also drive timing.

    k. *Analyze potential building and engineering systems.* (Select them in some cases).

    l. *Verify owner requirements and preferences.* Discuss funding; financing; construction procurement policies, culture, objectives, interim deadlines, and key political issues.

    m. *Plan.* Set the management plan for procurement strategy, project duration and major activities in a critical date schedule. Develop a results-oriented plan for total project management (predesign through construction) to achieve the owner's goals – a project "road map" that defines the who, what, when, where, and how necessary to achieve the project's goals and objectives. Establish it after analyzing considerations from the programming and predesign phases. As the project proceeds, the management plan will feed the overall project schedule. Evolve the design, agency approval, purchasing, and occupancy schedules with the master schedule.

    n. *Energy/environmental/sustainable/LEED considerations.* Set team objectives, metrics, and strategies.

    o. *Design direction or options.* Each project has a point of diminishing return in the project analysis scope. For example, in a warehouse, or regular building form, a team might be able to evolve a design concept and go as far as selecting types of structural frame and wall construction systems. In a multi-building medical complex, probably only basic design options can be identified for later study.

    p. *Construction procurement.* What are the owner's, contractor's, design builder's, CMs, or IPD team's potential problems and opportunities in buying construction? Which subcontractors and other partners are likely to be responsive because of the prospect of repeat business?

    q. *Meeting notes/project analysis record documents.* The CM, PM, PA conducts the session with information distribution and reuse in mind. Prepare drawings and information in formats that can be quickly shared. Do meeting notes that capture program, objectives, schedule, budget, sketches, surveys, and action items as the point of design departure. Distribute them within two days to maintain momentum.

By the end of the project analysis, all major disciplines should be able to move forward earlier and more productively than in traditional design approaches. The multidiscipline team should be a visible cooperating force. Mutual respect and the ability to communicate should be established, and forces likely to produce results relevant to user needs set in motion. The project analysis provides an excellent opportunity to launch a project. It focuses on key aspects and identifies problem areas before time and manpower are expended. Projects that conduct project analyses have great success. Those that don't see the opposite result.

5. Design Charrettes.

    Project analysis turbocharges design efforts by *defining* the problem. Later, with "all" information in hand and all disciplines represented, conduct a *problem-solving* session or design workshop to synthesize a design effort. If done on the client's premises or away from the office, these have been called "squatter sessions," "design charrettes," and "deep dives."

✳ Tip: Project Analysis Techniques

Consider these innovative technologies and group problem-solving techniques:

"Wallpaper Job"	• Cover meeting room walls with drawings and data. Project them. • Use large "flip" charts to record decisions. • Use individual sticky notes, index cards, or digital versions. Allow individual data elements to be developed, moved, deleted, reconfigured, prioritized, and saved for translation, storage, and retrieval.
"BIM/Technology"	• Use Building Information Modeling: 3D, 4D, 5D. • Employ multiple screen projection for concurrent viewing.
"Time Warp"	• Get into "warp speed." • Start early, work through lunch, and/or stay late. • Do "mad-minute" reports – one minute per group. Cut to the chase. • Create an "acceleration ramp," "momentum," and "inertia." • Select an out-of-the-ordinary time and/or place to meet. • Break conventional thought barriers and discipline silos. • Do the people "soft" part in the park, outside, in a restaurant. • Do the paper/data part in a room with tables and computers. • Keep pace. Use bells or buzzers as timekeepers.
"Aphysical"	• Work live. Use e-mail, or interactive internet. • Stretch the work day. Be together while apart, before you're together.
"Asynchronous"	• Use after-hours messages, texts, recordings, videos, videoconference. • Send "one way" data in advance for prep and orientation. • Use the meeting for "two way"/interactive/collaborative work. • Watch comparable facility videos during lunch break.
"Machines"	• Use large-screen monitors, projectors, and displays. • Build budgets, schedules, goals, and other data "live" with the group. • Gather input, reach consensus, approve/revise/coordinate at once.
"Brainstorming"	• To have good ideas, have lots of ideas.
"Group wall voting"	• Use stickers or dots to rank or prioritize ideas.
"Attitude adjust"	• Dress comfortably. Bring unique food/beverages. Change the room.
"Partnering/Bonding"	• Use social sessions/group activities.
"Gaming/Activities	• Use active techniques, e.g. Marshmallow Throw," "3 Truths and a Lie."
"Alternate"	• Alternate warm, fuzzy, emotional data with hard data.
"Visioning/Role Play"	• Share aspirations, five-year future headline, best experience.
"Drawing/Diagram"	• Encourage nonvisual types to draw, collage. Elicit thoughts.

Teams that fail to make the time to do these things set a false economy under which they will labor inefficiently for months. Start well, please. You will not get a second chance. Level 1 Project Design Control activities form the foundation for the entire design effort.

Project Definition Package (PDP)

A PDP is a product, contributed to by the entire team, that captures or incorporates by reference all Project Design Controls. PDPs can be done before a project analysis, updated during, and be reissued after. If the notion of a package or document sounds dated, think of it as a central database that is constantly updated. With a PDP, a project can be managed, having had its controls set, balanced, and recorded at the outset. Traditionally, contractors document pricing and schedule, and note the architect's documents on which they are based. Expand team documentation to include program, scope, and clarifications. Add rules, roles, responsibilities and change management processes. Teams that do will have a written basis for all project aspects useable to track changes and manage design: Project Design Controls, written and accessible by all.

✳ Tip: "Team Metrics"

While program, budget, scope, schedule, and documents are the tangible project metrics, leading teams have also used the project analysis as a time to set objectives and frequency for measuring less-tangible factors, such as team performance. Using spider graphs, charts, flower diagrams, or scoresheets, teams have measured subjective factors such as trust, collaboration, communication, and overall performance. Agree on what's important and measure it. Teams that keep score know how they are doing and what to work on. Set rules of engagement. Define success. Project analyses and team metrics are project microcosms that set up project controls and processes. Please don't waste or rush through them. Come back to their output periodically to see how you are doing.

CHAPTER 19

Level 2: Structure (Measuring/Baseline)

Tangible, Measurable Project Design Controls: The "Structural" Baseline

As the management baseline for measurable design aspects, five "Structural" Project Design Controls offer potential. (See Figure 19.1.) They set objective limits, or target values. Once set, check them for balance and compatibility. Once balanced, use them as design performance indicators. Tangible, measurable Project Design Controls are:

1. Program
2. Budget
3. Scope
4. Schedule
5. Documents

Used with the softer, social, interpersonal processes in the other framework levels, they guide design management. The tangible project controls are listed in this logical order and diagrammed as a see-saw because:

- Program comes first and drives the need and the design. It's the starting point.
- Budget is a key control. It is (or should be) based on the program and validated as such. It governs how much can be spent.
- Scope is the physical manifestation of the design solution in response to program and budget. As the key cost driver, it occupies the central fulcrum point.
- Schedule is how long we've agreed it will take us to accomplish design.
- Documents are the traditional best way to "freeze" constantly evolving information to share, understand, and manage it. Documents are a written manifestation of design, the endpoint of design process, a snapshot in time. Digital models, data, and metadata are new document forms in addition to hard copies and digital drawings.

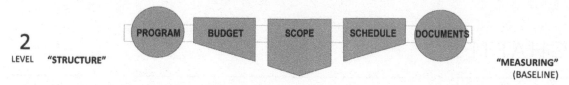

2
LEVEL   **"STRUCTURE"**

"MEASURING"
(BASELINE)

FIGURE 19.1   Level 2: "Structure" (Measuring/Baseline).

Program

Paradoxically, "programs" have gotten more complex and yet less developed. In simpler times, it may have been enough to give design teams a few criteria and let them use judgment to begin design. Times have changed. As specialization builds and time shrinks, we need to define problems before design and construction start or try to. But fewer owners and teams succeed in doing this. See for yourself in Table 19.1.

Meaningful program requirements should be qualitative, quantitative and include subjective goals and vision. Group program goals in categories: function, form, economy, time, and quality. Program types can include:

Area or Space Program
- How big is the building supposed to be to fulfill the owner's business need requirement? Are the areas stated as gross, net, usable, assignable, or rentable?

Building Systems Program
- What MEP systems are required to support the performance requirements of the owner's functional program? For example: "HVAC system shall provide a temperature range within x and y with a humidity level of z; lighting system shall provide a foot candle or lumens level of w. Reduce energy consumption by z%."

Site Program
- What functional purposes must the site accommodate in addition to the building? Future expansion? 4,000 parking spaces, a satellite dish field, employee amenities (such as ball fields, 12 acres of detention ponds, a structured underground utility corridor, or a 6-mile access road)?

Personnel Program
- How many people will be accommodated? How much space does each require based on what they do? Have all departments been considered? Are they job-sharing, benching, hoteling, or working remotely?

Operational Program
- During which hours will the building be open and operate? 9-5, 24x7? Is it a mission critical facility? A 100-year facility? Or a 10-year developer market-driven project with a short design life? Should the building provide an interactive, daylit, team space, or dark warehouse space? Is its business mission to sell products, yield a high security environment, or inspire great work?

Environmental Program
- LEED Certification? Carbon neutral? Net positive power generation? Building-wide recycling? Healthy, sustainable materials?

TABLE 19.1   Building Program Requirements Evolution

Past Practice ("Old")	Current Practice ("New")
Square Feet (Net or Gross)	Square Feet (Net or Gross)
Hours of Operation	Hours of Operation
# of Occupants	# of Occupants
Design Standards	Design Standards
	Performance Requirements
	MEP Requirements
	Redundancy
	Reliability
	Fault Tolerance
	Energy Use
	LEED/ Well Building
	Fire Protection
	BIM/VDC
	Audio Visual
	Security
	Energy Analysts
	Info Technology
	Low-Voltage Systems
	Wi-Fi
	Distributed Antenna Systems
	Interior Design
	Graphics
	Life Safety
	Fire Safety
	Ad infinitum

✳ Tip: Programming: A Shared Responsibility

While not the contractors' historical province, program controls design and drives budget. Design responds to program, and its resulting scope (and cost) drive projects. The documents produced, and the schedule that governs are also driven by program. While contractors have been "hands off" with programming, to manage design we need program awareness and documentation. Contractors can add expertise (such as constructability, schedule, cost, market conditions, options and trade expertise) to shape program and design and avoid "spikes" on other project controls. Programming should be a team responsibility. Owners should be willing to pay for their experts' skills in helping them make up their minds what they want and need to give them the flexibility today's business climate demands.

On Programs

How can teams begin to think about designing when they haven't defined which problems they're trying to solve? Granted, more owners and teams are now programming on the fly. The pace of change prevents them from slowing and committing to needs before projects start. New ways of working, and new technologies and materials defy even the experts to keep up.

"I don't know what I want or need, but I know it when I see it."

"You tell me what I need – you're the experts."

"What's everybody else doing?"

To cope, good teams develop and update programs as they go, using in-progress options analysis. They analyze alternatives and shape them into decision-ready form. Some common owner refrains are:

To manage design, first seek to understand. Listen and define the range of study to *limit* the range of investigation. In their quest for speed, out of desperation and the inexorable desire to get answers, teams rush in headlong with answers before they know the question. They derive hastily arrived-at schemes and pricing to quiet owner requests only to discover they've done the wrong thing. "That's not what we wanted." "It's too big." "It's in the wrong place." These responses deflate eager-to-please professionals who move too quickly. Ask and define first. Do later.

Volumes could be written on programming techniques, analytical tools, and big data to enable programming, but our focus here precludes it. (For a classic reference, see W. Pena, *Problem Seeking: An Architectural Programming Primer*.)

Budget

Budget is the amount of money allocated by the owner to spend for the project. Budget is distinct from estimate, cost, price, or funding. Manage variances between the budget and the current pricing. In the PDC framework, budget refers to all construction cost-related factors in the total project budget. Budget may be revised by a program change but only with matching adjustments in the other controls (e.g. documentation, schedule, and scope.)

✳ Tip: Common Construction Budget Misunderstandings

To use budget as a design control, design managers translate the language of construction cost into the language of designers. On past projects, when I haven't, I've fallen, or witnessed others fall, into the following budget pitfalls and misunderstandings:

- Neophyte owners, unfamiliar with terms, believed their budget to be the *total* project budget. Their contractors, after multiple attempts to confirm it, knew it to be the only construction budget. There was no meeting of the minds, despite repeated discussion on the subject. Disappointment followed.

- Designers, being eager to afford their proposed design solution believed the manufacturer's schematic phase budget concrete price. They didn't know any better. They neglected to include: formwork, both sides of the wall, labor, reinforcing, placement, finishing, scaffolding, waste, bond, insurance, markup, profit, contingency, and other factors. Little did they know the manufacturer was only quoting material costs and had no ability to provide a complete, in-place system price. Disappointment followed.

- An experienced owner, remembering his last project, insisted on reducing the roofing system budget to $10.00/sf. That's what he paid on his last job. He was unable to comprehend why this project – a small, 2,000 sf roof footprint, 20 stories up, in downtown Manhattan – needed to cost more. His obstinance in the face of facts won out. Union labor, ten years of escalation, night shifts, reverse economies of scale and a Cadillac specification weren't enough to convince him his roofing unit cost budget was too low due also to his stubborn insistence and refusal to listen to context and the expertise being offered. Disappointment followed.

Misunderstandings persist about construction budgeting. Most are perpetuated by those who haven't estimated or bought construction from trade contractors. These fallacies include:

Square Foot Pricing

Most of us are taught to do cost estimates by multiplying quantities (usually square feet) times unit prices. While it may be a good budgeting technique, this method is not how subcontractors price work. Trade contractors price materials, anticipated increases, labor rates, crew sizes, supervision, hoisting, equipment, market conditions, labor availability, numbers of mobilizations and demobilizations, unique design conditions and requirements, cleanup, safety requirements, closed vs. open specifications, competition, scale, general conditions requirements and their capacity and need for work. The total of these factors is the cost of the work. Divide that by a given area and the resultant is the proverbial square foot cost. Square foot costs are *resultants* or historical benchmarks, not bottom up pricing mechanisms.

Using a square-foot-based estimating approach can be problematic in a building information model. While it's useful to extract and share model quantities, models don't include factors like those listed above (e.g. waste, temporary conditions, or things not *in* the model.) Those who believe they can extract model quantities, apply unit costs from a generic database and be done will be wrong. They're missing context. Contractors, estimators, context and intangibles offer value to design partners. They help them see, understand, breakdown and use trade-contractor-based cost data to guide design efforts and stay in budget. Teams can also use this data to adjust systems budgets when necessary, shifting cost models to reflect team priorities.

Budget Validation

Budgets set by owners that haven't been updated for years since funding was allocated, validated by market input, or tested based on design or feasibility concepts, stand on the brink of disaster. Validate budgets when dormant projects start back up. Fix them early.

Budget Breakdowns, Target Value Design and Control

Design managers can enable budget success by breaking budgets into manageable pieces under design team control. Working backwards to extract general conditions, fees, temporary conditions, and contingencies can provide design target values. Transparency and detailed backup can too. Designers need to understand costs in tangible understandable terms to design to them. They don't need to worry about costs for things outside their control.

Scope

Scope is the extent, limit, amount and type of work to be included. While program dictates requirements, and how the building is to perform, scope dictates the amount of work to be included to accomplish that program, based on the design solution. If program is what the building is supposed to do, scope is how much of what kind of materials and equipment are included to fulfill the program. Often, scope growth is a direct result of program growth. Quantity takeoffs are nearly synonymous with scope, but various design solutions can increase scope for the same program based on their different approaches and solutions.

An example is a design scheme that explodes building form into many formal elements versus a neatly stacked box. Both may fulfill a 100,000 gross square foot program requirement, but the exploded-form scheme requires more scope to accomplish its solution – more site, structure, systems, ceilings, walls, soffits, waterproofing, glazing, and systems. Such an exploded approach may be the preferred solution depending on the balance of aesthetic, budget and other objectives.

Scope can be included by various project team members. Determines in whose budget each system falls. Cabling or site work pre-grading may be by owner, not in contract from a contractor's perspective. Use a differentiation matrix to indicate responsibility for design, budgeting, and installation of each discrete scope of work. Scope in this case refers to the entire *project* scope, not within one trade.

✳ Tip: "ScopeDoc"[1]

An approach I've used to capture and manage scope earlier is called ScopeDoc. ScopeDoc is a way to manage project costs. It repurposes design documents to manage scope by communicating design intent and extent. Its techniques help teams enable cost estimates.

The ScopeDoc process fully documents scope at each stage. Its ideas may challenge current practices, or "industry standards," but can define projects to great advantage. These quick, effective techniques save time and reduce unwanted value analysis. ScopeDoc proposes that the AECO Team:

a. Think Scope

Assign a "scope doctor" for each design system and discipline. Whether a project architect, manager, or designer, their mission is to lead and include all consultants in defining scope in early design phases.

b. Remember Less

Use industry standard checklists to augment memory, and select all systems to be included, rather than relying on what happens to be top of mind, recent or near.

c. Draw Less

Use small-scale diagrams to define scope. Draw what's needed and no more in early phases. Avoid needless detail, precision and waste (such as multiple, enlarged plan sheets with no information.) Work at small scale to migrate through entire plans – and more time for scope capture and design decisions, in lieu of sheet publishing.

---

[1] ScopeDoc is a proprietary technique, copyright M. LeFevre, 2005. Building on the success of the "ConDoc" method (Onkal Guzey et al., circa 1980s) in standardizing construction document methodology, I shift emphasis to optimizing *scope* documents via this term, versus ConDoc's emphasis on *construction* documents and offer the ScopeDoc mindset and technique for common use by all to improve their projects.

   d. Cut Words (Specify Less), Say "What"
      Use short form, lists, and spreadsheet style outline specs to specify in early phases. Avoid wordy, generalized specs and narratives. Be short, specific, and fast to communicate product intent.

   e. Draw More, Show "Where"

Document the extent of all systems at each phase, diagrammatically. You may need more diagrams than in traditional SD and DD packages, but they'll:

- Save time in communicating design intent and extent
- Avoid contractor and trade guessing, and scope and cost guesses
- Define scope to reduce Value Analysis later
- Use BIM and quick pdf markups or photocopy markups to note design extent.

Why Do We Need ScopeDoc?

Project budget overruns are more often the result of errors in scope than cost. When team members document project scope or can't read one another's minds, "gaps" are created. Guesses and assumptions follow. Eliminate them. Are our projects in budget the way we do it now? What have we got to lose? The ScopeDoc approach reduces value analysis and redesign. Since scope is the "middle ground" where design meets cost, let's manage it.

Some project teams are "separated" by different languages and communication modes. While owners are comfortable with words and numbers (dollars), the design team's primary medium is drawings, images and models. Cost and schedule have never been designers' preferred communication modes. Construction managers use all these forms and add schedule. Words, drawings, numbers, and schedules – and digital 3D models – are our primary team languages. Why don't we use them all where they're best suited? The more we speak common languages, the better we'll understand one another, and the more effective we'll be.

For layouts and extent, drawings and diagrams are most effective; they tell a thousand words. They show "where." Few projects describe their column grids in narratives. Rather, we draw them. Current drawing sets, owing to CAD and BIM practices that easily multiply views of a single database, are often vast paper wastes. Plan after plan of empty white space is created because it was easy. Many contain little information. Unknowing owners, too busy to look at drawings, see thick sets of plots and prints and assume the architect has done their job. Too many document sets omit information the contractor needs.

ScopeDoc proposes short cuts to help designers communicate – effective, creative ways customizable to firm practice. But without willingness to work in a different way, design teams are off and running – doing what they've always done, producing documents inadequate to describe project scope. The result? Contractors guess, projects creep over budget, and teams await the inevitable value analysis and redesign that erodes team morale, design quality, time and fees.

What Can ScopeDoc Do for Us?

ScopeDoc saves design time, controls scope and cost, and reduces the time and pain that comes later with value analysis scope reduction. ScopeDoc serves all parties: owners know what's in their job; designers know what's been (and needs to be) designed; and construction managers and trade

(Continued)

contractors know design intent and extent so they can price it. The result is more certainty, better balance, and "tighter" pricing.

Why Haven't We Tried ScopeDoc Before?

Architects and engineers are taught to design, NOT be concerned about quantities – "That's the contractor's job." Design's exploratory nature makes it impossible to know where you're going before you get there. Our industry hasn't rewarded managing scope; it's set up cultural barriers to defining it. As professionals, we should know what's in our projects. A doctor knows all his patient's systems (circulatory, respiratory, digestive, skeletal, and so on) We expect our mechanic to know all required in our car repair, and include its cost. Owners should expect their design build teams to know, manage, and solve their *entire* project.

Designers are schooled to analyze all forces that shape design: program, site, context, function, and human. What about scope and cost? Who among us disputes that these are major design determinants? Let's evolve the designer's (and their design manager partners') toolkits to address scope in early design documentation.

ScopeDoc Goals: What Are We Trying to Do?

ScopeDoc's goal is to help the team determine the scope, extent, and quality level of *all* building systems in early phases (program, project definition, SD, or DD) to manage scope and cost. If the team identifies all systems to be designed and makes placeholder decisions for each, scope is managed, and cost is controlled. Viewing design through the lens of the "Scope Doctor," the team knows what's included, how much of it there is, and the quality level for each system. Using these techniques (and tools such as the CSI MasterFormat or Uniformat checklists), teams ask: "Do we have one of these? If so, where, what, and how much of it is there?" In each phase, try these five things:

1. Describe and document *all* major, typical systems at appropriate detail and reliability. No matter the phase, document every system and scope, to some level. Leave no system behind.
2. Select and document their scope, extent, size, and material.
3. Develop "design" (i.e. "scope") completely, to a phase-appropriate level of certainty.
4. After DD, add only construction information such as dimensions, details, coordination, and atypical, unusual conditions for construction documents. Ideally, after DD stage, all systems intended are described, selected, and documented, no new systems are introduced, and scope remains constant, adjusted only to restore balance.
5. At early stages, rather than omit an unknown system, include it – at an "order of magnitude" level of description and depiction. Take your best guess. Think of scope as a "programming tool" to force ballpark decisions about what's in and what's out.

The goal: No needed or intended system should go missing. All should be shown, and their type and extent documented in some way. Are we asking for working drawings at concept stage? No. Are we asking for small scale diagrams, lists, and short product descriptions? Yes. The design team's refrain: "But we're only in SDs or DDs; we don't know that yet," is an excuse that permeates current practice. And it's a big reason projects go over budget! Teams need to "know" everything to be included, earlier, because scope is the major design cost driver. Recognize this and work differently. Rather than "faster horses" we need the "Model T," "iPhone," or quick, comprehensive, transformative solution of our era. ScopeDoc gives it to us.

How Can You Use ScopeDoc?

Concept, Meanings, and Approach

The term "ScopeDoc" is used as noun and a verb. As a noun, it refers to information produced in each phase as "scope documents," a record of design intent. Secondly, "scope doctoring" means active participation as a "scope doctor" to decide and define scope and guard against scope creep. "Preventative scope maintenance" is preferred to emergency triage, scope reduction, surgery, or amputation of design or scope. Have you done your ScopeDocs? Who is your ScopeDoc?

Use this approach as a "top-of-mind" tool to control scope. It's the "best-case scenario" to avoid the "worst-case scenario." Who is the "scope doctor" on your team – with ownership for producing ScopeDocs? Does someone take ownership for selecting every system? Do you accept "The Scope-Doc Challenge?"

Goal: Intent and Extent

Emphasize, record, and communicate scope in design to control scope and cost. Use line types and diagrammatic symbols to clearly delineate and quantify systems. Enable trade contractor understanding. Screen back or gray out background information to highlight and make visible the subject trade scope. Organize drawings with like systems in one location. Distinguish new work from existing via line weights or scope limit lines.

Single line drawings (even freehand) on Xerox backgrounds (or CADD/BIM images), or marked up PDFs communicate intended scope, extent, and systems routing, particularly for engineering systems. As a final check before drawing release, review documents for scope! Are all desired elements indicated? Is the extent clear? On drawings, in a list, or narrative? If an item isn't noted, intent or extent is unclear, and so is the cost basis.

"Design Range" Concepts

In early stages, if a material hasn't been selected yet, choose a "quality level," or "design range." Do this in a variety of ways:

1. Use a manufacturer's product name, e.g. "Alucobond" to establish a basis for design intent. (It can always be changed later.)
2. Target a unit price cost range allowance (e.g. "flat metal panel system or material to be determined in an installed cost range of $x to $y/SF).
3. If systems are not selected, use a "relative" approach to describe limits and extents:

Curtainwall System A: Highest quality at lobby areas; Curtainwall System B: Medium quality at office areas. Let the construction manager establish allowances or do it as a team.

(Continued)

ScopeDoc Deliverables

Teams should plan documents using documents guidelines to agree on which sheets will be done at each phase, based on design and construction needs. The following deliverables are suggested at all phases as checks of area, scope, design, and cost:

- Drawings, scope diagrams, and scope models
- Outline specs by CSI division (evel of detail commensurate with knowledge and phase)
- Code analysis summary (include life safety plans as applicable)
- Concept description (specific, sized systems, narratives to describe operational intent)
- Scope checklists, schedules, or notes
- Program area summary (required versus as drawn) to manage program scope creep
- Scope spec outline specifications format
- Use a tabular, spreadsheet, line item, outline format in lieu of traditional spec formats. It documents assumptions, intended products, and scope in a common format with minimum effort. Avoid long, generic, narrative-style specification formats at early stages to emphasize product and system selections.

Design Scheduling

Schedule is the time allocated or required to complete the project. Subsets of the overall project schedule include design schedule (including programming, reviews, presentations, coordination, documents, plotting, printing, approvals, and revisions), permitting, construction schedule, commissioning, move in, testing, and others. Track target versus actual schedules.

Schedule Forms

Design scheduling forms and practices vary. Forms include critical date lists, and task summaries of design tasks. Schedule-focused owners and constructors critical path views, including all consultants, key owner and contractor interfaces, with predecessors and links in ample detail to integrate into an overall project schedule. Design scheduling used to be optional. Now it drives project success – because we schedule projects aggressively. Design scheduling can't be ignored or left to the judgment of any single entity – it is too important. Table 19.2 illustrates the differences.

TABLE 19.2   Design Schedule Evolution

Aspect	Past Practice ("Old")	Current Practice ("New")
Participants	Architect w/ 5 consultant input	O, A, C, subs, et al.
Form	Paper/intuitive/unintelligent	Automated Intelligent Software
Method	Handcrafted, isolated	Lean pull scheduling
Update Frequency	None, or at phase ends	Daily/weekly
Team Reporting/Sharing	Phase starts	Daily/weekly
Phases	SD, DD, CD	Blurred, concurrent, ongoing

Developing Schedules

Managing the design and documents tardiness helps design partners think about, breakdown, plan, and schedule their processes. Planning design activities in sync with owner, contractor and team activities is a new idea to some designers. They may need help. Enlightened ones know it because they've felt the pain of not doing it. Now, they embrace it. Only by planning design activities in concert with the overall project can teams hope to manage collectively.

Well-managed teams break down, understand, and take ownership for schedules. Those who've never done design can't hope to schedule it. They don't know the steps, tools, and multi-path, cyclical, dependent nature of its myriad explorations. Using consensus lean pull scheduling techniques, we *can* establish interim steps and drawing due dates, while supporting processes like options analysis, design studies, presentations, pricing, and approvals, in advance of drawing, coordination and issuing for procurement. *Integrating* design into the overall project schedule supports completion. Using expert facilitation from experienced designers and builders, teams can develop inclusive, specific, measurable, achievable design schedules.

Calling It Complete

In her book *Architecture: The Story of Practice*, researcher Dana Cuff conducted her own search for design excellence, surveying leading firms. What she found supports the interviewees and my recommendations for early planning, goal setting, limit setting, consensus, and monitoring. Cuff shares:

After completing these two practices, teams faced challenges "finishing" due to the subjective nature of design product. Even after planning and managing well, the best teams had to "settle" for a work-around solution – a coping mechanism for managing design. Cuff explains:

> *"The approval then, not the conclusive development of the work itself, determines the end of a phase. The work is not intrinsically complete. The architect and client agree to call it complete."*
> – Dana Cuff[2]

> *"In excellent projects . . . [the team] established limits at the same time they set goals."*

> *"Once deadlines are created, someone must synchronize the work effort to adhere to them."*

Hope for The Best

Because design can always be made "better," as Cuff explains, it's a common practice to accept incomplete, imperfect documents to begin construction. Contracts and industry standards of care confirm this "imperfect" condition. After all, design documents are abstractions representing design intent, subject to the art and craft of the architect's firm standards, and industry norms and interpretation. Even the AIA offers few rules or standards for what to include at schematic, design development, or construction documents phases. It's varied and project-specific. Through document planning, management and review, a good architect/contractor team can tailor their design products to suit project context. They can and should.

---

[2] Dana Cuff, *Architecture: The Story of Practice* (MIT Press, 1993).

Limits

Design needs limits. As an inherently open-ended process, design can go on endlessly. Because design is a subjective pursuit, it's near impossible to tell when it's done. In his interview, Randy Deutsch points out, "There is no right or wrong answer." Design can always be said to be made better, or at least different. More study; more options; refine it further; enhance more details. In the 1970s, while a student at the University of Michigan's Taubman College of Architecture and Urban Planning, I heard the following analogy by legendary design methods experts Professors Colin Clipson and Joe Wehrer:

> *"Design is like railroad tracks."*
> *"You look out into the distance and see the tracks converging. The train is moving. You think you're nearing the end. But as you come over the hill and approach the horizon – it seems to have moved farther out, no closer than before. The design conductor yells: "Shovel more coal into the engine!" He adds more people. They work harder and longer. "More coal!" he shouts, and the horizon still pushes out."*
> *"Design is like railroad tracks."*

– Colin Clipson and Joe Wehrer[3]

Herein lies the challenge. Since design is never "done," or at best, is difficult to tell when it is, designers (and those who manage them) use several methods to bring it to a close, or "declare" it to be "done." They are:

a. Date Driven:

In this approach the team agrees on a certain date design will be "done." Or done enough for the interim purpose. Regardless of actual completion, the design team makes their best effort to "finish" design and documents by this date. They plot, post and release them, regardless of condition. In some cases, glaring omissions exist, or coordination hasn't been done. The documents represent a dis-integrated building, less than ideal for construction, bidding or other purpose. Architects' professional doctrines have stood behind this "We're human: we'll be less-than-complete, less-than-perfect" mantra for decades, and rightly so, but in a coping scenario, teams need to agree what "done enough" means for each interim date.

b. Fee Driven:

Usually dictated by the architect's project manager, the fee-driven limit takes the architect's fee, divides it by the number of hours or percentage of service allocated to this phase, and lets it dictate completion. Example: the architect has $20,000 to complete Construction Documents. With a five-person team, and an average $100/hour billing rate, the team has 40 hours to spend. At a full-time rate of 40 hours per week, the team has 1 week to finish, (contingencies, consultants, reimbursable costs, value and profit notwithstanding.) In new value-based fee models this method is obsolete or would be changed to produce great service, product, and value while earning a profit.

---

[3] Colin Clipson and Joseph Wehrer, Design methods class, University of Michigan, 1977.

c. Milestone, Work Product, or Deliverable Driven:
This approach sets an interim milestone as a stopping place. In the railroad analogy, the train clatters on until it reaches Peoria, and design is declared to be "done," or good enough for the stated purpose. "We're here. We've arrived. Get off the train." The design equivalent could be receiving an approval, finishing floor plans, submitting for permit, or issuing purchasing documents by a given date.

Destination or Journey?

The train analogy continues. Many great writers have said they have no idea where their stories will go when they begin them. They let the stories take them there. It's a process of unfolding. Design is no different. Whereas construction's destination is always known, design's never is. Having to lay the tracks as they go, architects locomote to unknown, unreachable horizons. In design, management takes on the aura of enjoying the journey and setting interim milestones instead of merely focusing on the destination.

Constraints

Many of the world's best designers agree that design *thrives* on adversity. Without the box itself, there's no "thinking outside the box." Without budget and time restrictions, design would go on interminably, with no connection to reality. Designers wouldn't be able to help themselves – they'd work without eating or sleeping, and keep on doing what they love. Without programmatic guidelines, and cost and time limits, designers – absent restrictions or controls – would scarcely be able to create a functional facility for normal use. Like the drug addict or kid in the candy store, their love for design and inability to control themselves would yield unsatisfactory results. That's why few projects are done that way. (Not many clients have unlimited time and money.) Architects are notorious not only for their inability to meet client's budgets, but for their own as well. There's an old, tired joke in the architecture community:

Q: "What would you do if you won a million dollars?"

A: "I'd open up an architecture firm and practice until I ran out of money."

Great artists, architects, engineers, and businesspeople have risen above limitations to create ingenious solutions. Whether limits of time, budget, or other type, unfettered design becomes problematic. As professionals hired for their abilities to push boundaries – much like teens exploring the limits of sex, drugs, music, and authority – designers rely on the give-and-take of clients and contractors to keep in line. While few admit it, some relish the conflict. It gives them connection – an enemy to rally against for their "cause célèbre": design excellence – particularly the unappreciated kind. They see design as their mission – against all odds. Sometimes, with unsympathetic owners and contractors, that's the case.

Don't Let Your Emotions Get the Best of You

When it comes to emotional attachment to their work, architects and contractors have radically different mindsets. "We're just the messengers," contractor colleague Mike Kenig said in our first encounter, in a way that sounded like he had said it many times. "We don't 'do' emotional. We provide information and let the owner decide." I've learned that while both sides are passionate, those who control their emotions are more effective.

✳ Tip: Setting Limits

When starting a project, how do designers, owners, or construction or preconstruction managers set design limits? From experience, only a few tangible project aspects can establish guidelines: program, budget, scope, schedule, and documents. As in any project management endeavor, the design manager, with the team, sets, tracks, manages to, and adjusts these limits – to manage design. Setting them with input and consensus from the designers who will live by them is advisable. Imposing arbitrary, unilateral limits is not.

✳ Tip: Don't Forget to Schedule . . .

In the days I managed myself as a designer and the years that followed, managing designers as a CM design manager partner, it was amazing to see how little thought I and my colleagues put into scheduling our work. Why was that the case? I think it was because we weren't trained to do it – because we loved the doing, not the management of the doing. Scheduling and managing of any kind were the tail of the design dog, a necessary evil. Because design process varied from firm to firm, project to project, and individual to individual, there was no set way to go about it.

Driven by passion and emotion, not business, we wanted to jump into the design sandbox as quickly as we could, rather than "waste time" trying to predict what we knew to be unpredictable. To this day, some design firms are bewildered when asked for their design schedules. "What do you mean?" they respond. Common justifications are: "No, we haven't done that yet," "We're busy trying to develop site concepts," or "We don't really do that."

Frequently, we planned for perfection. Take the design fee, divide it by an average hourly rate, allocate some to consultants and get the manhours available. That's how many of us were taught. Of course, nothing will ever go wrong or be late. We thought design schedules were about our design process. Far from being motivated to finish as quickly as possible, we wanted to know how *much* time we had – so we could be sure to spend all of it, and more. We were wrong.

In this internally focused scheduling mindset, or lack thereof, we typically forgot time for:

- Critical activities of others (owners, contractors, engineers, subcontractors)
- Data collection (program, surveys, geotechnical data, testing)
- Presentations
- Permit, funding, and approval milestones
- Regulatory steps and hurdles
- Need dates and reverse-driven deadlines
- Early trade procurement and construction in fast-track CM projects
- Contingencies
- Review, comment, incorporation of changes
- Engineer and consultant coordination
- Being over budget and having to develop value analysis solutions (since it happens on every project and is an apparently necessary phase, we should schedule it)

- Redesigning, reworking, or refining design to incorporate ideas to return to budget
- Changing undefined, evolving requirements, programs, and rework
- Pricing and proposal review
- Options analysis and constructability input
- Owner decision making and mind changing
- Bureaucracy and politics

Shockingly, almost all these things – and more – happen on *almost every* project. Eventually, after repeatedly forgetting or ignoring them, I began to list and suggest they be included in design schedules. If you want to manage design, you might too. Include them as activities, with starts, stops, and durations. Link them to predecessors and successors. Include time for study, exploration, multi-path, multi-variant options analysis, and circling back to revisit good ideas. Set limits and interim steps. The more detail the better. If you miss something, it's likely only a small step, not an entire phase. You'll do this knowing all the while that some giant curveball, hairball, or eight ball is still liable to throw the entire schedule out the window.

But if (when) that unpredictable event happens, your team will at least have set a plan, together – and written it for all to see – and can craft a recovery plan.

Some bristle at trying to schedule a mysterious, open-ended process like design in such a calculating, methodical way. But design without a schedule at your own risk. Working in a team, and knowing the muddling nature of design process, I'd rather start with a plan than not. Even if we miss some things or get their durations wrong, we're better off. In the words of John Maynard Keynes:

"I would rather be vaguely right, than precisely wrong."

Those who deny the need to schedule design set sail on an uncharted course are guaranteed to be precisely wrong – a course without destination, interim guideposts, route, or estimated arrival date. In these cases, others relying upon the designer's work product to price, buy or build, are helpless.

If you agree to schedule design, use tools and media of your choosing, and terms meaningful to you. Schedule both beloved and unavoidable activities. Seek facts and conflicts and address them. Design scheduling ostrichitis serves no one. Use software, spreadsheets, pencils, graph paper, sticky notes, and historical data as you like, but take a shot. Include all the tasks and players you can. Without all the puzzle's pieces you'll hardly solve it. In fact, you'll unsolve it with each piece that emerges. Without knowing where you're going, and when, you stand little chance of getting there in time. As professionals, we know our processes that repeat. Or, we should. We owe it to ourselves to plan for them and work to that plan. We owe it to our teammates too.

In recent collaborations I've found the best way to approach this subject is to give designers the benefit of the doubt. Ask if they have a design schedule. If they do, ask them to share it, then back-check it with the activities above. If they don't, work with them à la Lean pull scheduling, interactively. Question activities, needs, inputs, and exchanges. With group input and all activities included, your team has hope. That's better than having no design schedule at all or dictating an untenable one, only to watch it fail. Smart teams use multiple perspectives and experience frames to create consensus schedules. I hope you're one of them.

> ✳ **Tip: Scheduling What?**
>
> If you are stymied trying to schedule unpredictable design processes, remember this: you are better to have tried and failed than never to have tried at all. My approach to scheduling design, like the importance of managing scope in cost estimating, is to begin with a good activities list. Without all the puzzle pieces you won't be able to solve the puzzle. Start with what and why then migrate to putting them in order, assigning durations, and setting dependencies, predecessors and successors. Then you can test if the schedule works or adjust, iterate, reduce or parallel-process durations to meet deadlines. But if you've missed one or more activities you will have wasted the effort and will surely be wrong. Overrunning one or more durations comes at less cost than forgetting it altogether. Reuse your data. Use checklists. Get input from others. Then solve the schedule puzzle as best you can.

Design Process and Sequence

For newcomers to the field, it helps to understand a few design process fundamentals. The first is that it involves a never-ending cycle of collection, analysis, and synthesis. While these processes can happen currently, the classic diagram for this sequence looks like Figure 19.2.

The second principle is that design thrives in an alternating sequence of individual study followed by coming together to check, evaluate, or include more points of view. (e.g. meetings, feedback, reactions) as in this diagram. In contemporary practice, these cycles are becoming almost unrecognizable as teams iterate quickly and work live.

The third design process fundamental says that design is either diverging or converging. If more collection, analysis or synthesis is happening, it's diverging. More possibilities and options are being considered. Because work is ongoing, and more information is being processed, things aren't static. No decisions are being made, the team isn't turning the corner to reduce the study range or come closer to an endpoint. Know which direction you're headed. Act accordingly.

Knowing these three design process rudiments can help design schedulers include all the steps and know when to intervene, let their team go, or convene for decisions.

Monitoring and Reporting

Once an initial design schedule is set, share it. Monitor and report on it weekly. As things change, adjust it to keep balance. Can sequences be parallel-processed? Can steps be automated or cut? If design is king, extend the schedule as needed. Only by keeping design schedules visible and working faster or eliminating steps can schedules keep pace with change. If nobody does them or uses them, you're in for trouble.

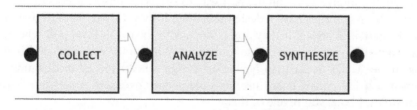

TIME ⟶

FIGURE 19.2   Collect | Analyze | Synthesize.

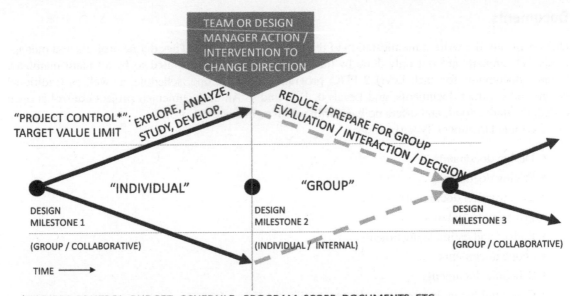

FIGURE 19.3   Individual | Group.

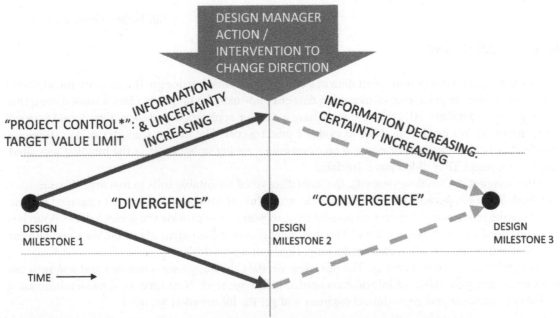

FIGURE 19.4   Diverge | Converge.

Documents

Documents are the written manifestation of the design *and all PDCs*. They define projects and manage design. Documents are not only done by the architect, they're contributed to by all team members. Prepare documents for each Level 2 PDC: program, budget, scope, schedule, as well as traditional design and contract documents, and Levels 0, 1, 3, and 4. An undocumented project control is open ended, misunderstood, and offers no basis with which to manage the project.

Example Document Types include:

- Design documents
- Pricing documents
- Construction documents
- Program documents
- Budget and estimate documents
- Scope documents
- Schedule documents
- Communication documents (roles, protocols, meeting notes)
- Change documents (documenting changes in all categories)

"Drawings are a set of instructions."

– Bill Halter, Cooper Carry

Data: The Lifeblood

You may not expect a discussion about data in a book about managing design. But data are the lifeblood of projects. While the principles of managing data may normally be taught in classes named computer programming or database 101, the core ideas have a simpler application in design. Ever found yourself in this situation? We need a site plan to evaluate grading cost. The design team draws one. Whoops. Upon further review the team discovered an easement or zoning setback where the building was sited. What went wrong? They didn't have the data.

After locating the building properly, the team discovered unsuitable soils in that area. Cost impact: $1 million. What happened? They didn't have the geotechnical data because it wasn't managed in time. What were the root causes? Unclear responsibility as to who was to provide the geotech data? Boundary survey not done in time? Missing data? The very inception of design ideas about building and site can be constrained or wasted when data are lacking. How can we fix this?

We can learn from our mistakes. The next time we start site design we remember and augment our memories by using checklists of information needed before we start. Next time we'll know whose job it is to hire the surveyor and geotechnical engineer and get the information we need.

Capturing Data?

In project situations, it's common to see groups of experienced professionals sitting in meetings unprepared and uninformed. Many have traveled from afar to attend. Others dial in on web-based telecommunications or videoconference platforms. What happens next? These same experienced professionals

✳ Tip: Normalizing Data

The next evolution in capturing data is to take "meeting minutes." This phrase seems like a holdover from church or PTA meetings (e.g. "At 8:32 Sarah Jones volunteered to bake cookies for the bake sale. 8:47 – motion to adjourn meeting."). To start with, let's call them what they are: Meeting Notes, or data. We don't need to write history or stories; we need actionable data.

The next issue with much data capture is that it's done longhand. "John will send the revised sketches to Mary by Tuesday." "Option C was approved but the contractor needs to verify the pricing quickly, or ASAP." What's wrong with those examples? They seem clear. But when are "quickly" or "ASAP"? By 8 a.m. tomorrow, or as soon as John can get around to it, maybe next month? Issues arise when there are many activities to collect, distribute, and act upon. In longhand form they're hard to manage. Who has time to cull through long emails to find their tasks? They're less actionable in this form. The time it takes to sort through the adverbs, adjectives, and imprecise words is waste. How can we do this better? By "normalizing" our data, as shown in Table 19.3.

Normalized data extracts common data types and organizes it in columns and rows at 90 degrees (a "normal" angle, in geometric terms) so they can be managed. It creates a database. In normalized data, recurring categories become headings ("fields" in data parlance.) They're the "fixed" information that repeats. Rows or line items are called "records" – unique information that changes by item. In normalized data, adjectives, adverbs, articles, and imprecise data are removed. The above action item simplified in a table or spreadsheet format looks like Table 19.3.

TABLE 19.3   Normalized Data

Responsibility	Action	To	Due
John	Submit Pricing	Mary	6/18/19

spend hours interacting and evaluating – yet nobody captures the valuable data they exchange. Despite their high hourly billing rates, they seem content to let their valuable words, ideas, and actions fall by the wayside. Next week, when the same unresolved issues arise in the team meeting again, the team must unearth them. "What was that issue? I thought Joe was following up on that? When do you need it? What are the options? What do they cost? Didn't we talk about this last week?" Unfortunately, these are common behaviors of experienced professionals. No wonder owners lose confidence.

Documents Planning, Management, Review, Tracking, and Incorporation

Good design managers manage documents over their information life cycle, from planning through management, review, and incorporation of comments. Some firms have techniques and templates for documents management. Many believe design managers should focus on documents review and checking. Here's a secret: even *greater* value lies in documents *planning*. Teams that wait until documents are done without discussing what they *should* contain miss their greatest impact. Advance conversations and consensus on drawing content and timing mean everything. Don't wait until the documents are done.

After planning, *management and monitoring* of drawing progress come next. No matter how much you trust your design team, don't wait until the deadline to see their result. Check in. Drop by their office, look online. Ensure they manage the scope, program and area, and incorporate input. These on-board reviews head off tangents and help teams stay on track. When I was a designer we resented this. Teams smarter than I was have learned to welcome it. Finally, to capitalize on all the attention given to creating them, good design managers track the incorporation of comments.

One problem remains: there's *never* enough time to do these processes. In the past, projects moved more slowly. Now, there's hardly time to stop design while drawings are reviewed. How to cope? Manage by priority and exception. If structure is being built *now*, focus on structural coordination items first. Divide and conquer with parallel drawing reviews. Split disciplines among team members. Finally, use live, on-screen web-based review and PDF markup sessions to cut lag time from printing, scanning and distributing review comments. Work "live" if you can. Tables 19.4, 19.5, and 19.6 highlight documents practice differences past and present.

TABLE 19.4 Documents Planning and Creation

Aspect	Past Practice ("Old")	Current Practice ("New")
AE/CM Team Documents Planning	No, by AE only	Yes, w/ CM/OAC support schedule, phase purchasing
Scope Focus	No	Yes, responsibility shift
By Phase	No	Yes
Documents Purpose(s)	Architect's discretion	Team derived
		Design/Presentation
		Permit/Budget
		Quantification/Purchasing
		Presentation/Coordination
		As-builts/facility operation
Form	2D, paper	2D, 3D, digital, models
w/ Data/Information Attributes	No	Yes
Interoperability	No	Yes

TABLE 19.5 Documents Management and Production

Aspect	Past Practice ("Old")	Current Practice ("New")
Periodic In-Progress CM Contact	No	Yes
Progress Prints	No/Occasional	Yes
Live, On-Screen Instant Review	No	Yes
Drawings 'Out' For Ongoing View	Yes, Stacked	No, Zoom, Access
Live Markup/Capture	No	Yes
Networked Web Review	No	Yes
Decision Tracking	No	Yes

TABLE 19.6   Documents Review

Aspect	Past Practice ("Old")	Current Practice ("New")
Duration	Weeks/Months	Hours/Days (not enough)
Project Design Work	Suspended?	Ongoing
Performed By #	1 to 3	Multiple/Many
Performed By (Experience)	Experienced	Less experienced
Performed By	Internal/Cross-Discipline	Independent
Review Scope	All	By trade package/urgency
How Transmitted	Printed/Ship Paper (Days)	Posted digital (instant)
Form	Paper Redline	PDF/Markup/Scan
Coordination	Light Table/Markers/2D	BIM/3D/Digital
Comment Incorporation/ Tracking	Yes	By exception

✳ Tip: Document Review Types

Contractors use multiple lenses to review documents. As architects, we were concerned with design, coordination, program, codes, and communicating design intent. We didn't worry about scope, purchasing, or quantities – that was "the contractor's job." In our quest to document efficiently, we approached cryptic, overly lean documentation levels. We were self-focused. We thought when the drawings and specs were done, that was the end of the convergence. We didn't know it was just the beginning.

Sitting at the contractor's table I've seen an army of supply network trade contractors, manufacturers, and vendors expend significant effort to win work. Bad documents wreak havoc in the bidding and construction that follow. One mis-coordinated note, missing system or spec section induces a nightmare of miscommunication. Whose flashing is it? Copper or bituminous? Is it in the curtainwall or masonry spec? RFI's, guessing and risk-taking follow. Completing documents at the end of a phase isn't the end of the convergence, it's the tip of an iceberg on which thousands of supply chain partners rely.

In the early days doing plan reviews under CM-at-Risk project delivery, I wondered why contractors and architects didn't review documents jointly. Now we do. How valuable it is to turn pages as a team and have design partners point out things that are changing, on hold, or didn't get done. These meetings point out inconsistencies and accelerate understanding. When the owner joins us, we gain a collective understanding of design, documents, risk, and work to be done. The foundations were preliminarily sized because the geotechnical work wasn't complete? Good to know. Owners and design partners rely on contractors' expertise. I did when I was the designer. As CMs, here are the documents review filters we use at Holder Construction:

1. **Design Status Review** Are documents done? Do they include what they need to? Project-specific information necessary to support purchasing and construction? Did they meet their target values and include what was planned? If not, what can we do to recover?

(Continued)

2. **Preconstruction Review** Is the scope of work clear? Can trades work be quantified and understood? Are there scope gaps, duplications, uncertainties, value analysis opportunities, or cost overages, options or alternates to balance the budget? What's the purpose of this set or digital model? Progress review and budget update? Permitting, purchasing, construction? (See Case Studies 3 and 4 for a look at my first subcontractor scope meetings as an architect.)

3. **Constructability Review** Do our preconstruction team and field construction experts understand sequencing, erection, and tolerances? How do safety, schedule, temporary construction, coordination and means and methods contribute? Can we suggest alternate detailing to enhance longevity, durability, cost or quality? Superintendents and trade contractors add wisdom in this phase. Design teammates need it. I did.

4. **Interdisciplinary Coordination Review** For work sufficiently developed, does architectural work align with structural, mechanical with electrical, and so on. Coordination reviews reveal much if done before construction. Now, we're forced to do it live, onscreen, by exception and priority field need, because time no longer permits weeks-long hard copy redline reviews.

Multiple perspectives help collaborative teams. I hope yours is one of them.

CHAPTER 20

Level 3: Systems (Relating/Collaboration)

Level 3 of the Project Design Controls framework houses the human/relational systems. (See Figure 20.1) We call them "systems" because they have complex moving parts and represent the framework's "communication," "circulatory," "personal," and "spiritual" systems. But unlike building and machine systems such as HVAC and electricity, these *human* systems factors rely on complex moving parts such as relationships, trust, and people to work.

Owner, Architect, Contractor: The Team

People come first, because without them, there is no judgment, expertise, or communication. Get to know the people on your team personally and professionally. Treat them as someone you value. Maybe they'll do the same for you. Make time for people and renew and refresh your efforts periodically. Invest in after-hours social time. It pays off. Trade organizational charts to ensure you know who they are and what they do. Invest in their preferences and accommodate them. Give.

Trust

No matter how well planned, organized, and documented your structural controls are, if the team doesn't have trust, the tangible factors are worthless. No one will believe or use them. Earn your teammates' trust. Don't be afraid to be vulnerable, admit when you've made a mistake, or need help.

Trust must be earned. Earn it by doing what you say you're going to. Under promise and overdeliver. Consider this: teams that trust one another preserve and maximize their productive work time. Teams that don't will rob themselves of time spent in backchecking, backstabbing, reworking teammates' efforts, and talking about them. You've done a cost estimate? Having earned my trust from past interactions, you grant me the freedom to do my work or solve some other team problem rather than question yours. Maybe I can return the favor? Sound like a fantasy? Ask those who've done it. It works.

Relationships

Relationships are the circulatory and respiratory systems of Project Design Controls: they drive the team understanding, communication, and trust through which all measurable controls flow. Develop

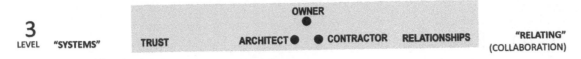

FIGURE 20.1 Level 3: "Systems" (Relating/Collaborating).

and nurture relationships with those you'll work with. When times are tough, your relationships will get you through the adversity. Relationships take intention, commitment and time. Do you have those? Give to relationships through actions. Create an atmosphere of giving and teamwork. The best teams find counterparts on other teams and align (e.g. designer to design manager, production detailer to facility manager, IT guru to BIM manager). See Figure 20.2.

"It's no wonder we sometimes have a failure to communicate. All too often, we're separated by a common language."

✳ Tip: Alignment

Ideally, you have already set up roles and responsibilities in the project planning phase. To promote better relationships, share organizational charts or project directories to find your counterparts in other organizations. Align people who do and care about the same things. Get them close, with direct communication, not buried in hierarchy. Whether it's your CEOs, administrative assistants, or entry-level technical staff, align them to speak a common language and get more done.

A second alignment is to consider a conflict resolution council of escalating layers of responsibility and oversight to resolve issues when needed. The goal: resolve issues on the front-line lowest levels without having to bump them upstairs.

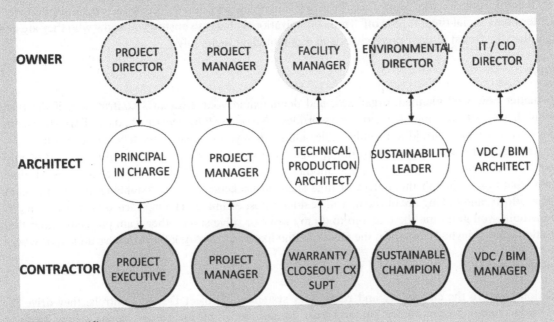

FIGURE 20.2 Alignment.

✳ Tip: Measuring Relationships

Compared to budgets and schedules, dynamic factors such as interpersonal relationships are harder to measure or control. How do you know if your relationship exists and is in good shape? What's the benchmark? How do you evaluate it? Some find it challenging. They entered design and construction because they liked to make things, not interact with people. How can we can measure the quality of our relationships? The easiest way is to ask.

Confronted with such a touchy-feely question, designers and builders can dismiss the question. "We're good!" they'll say. "Fine!" Others dodge the discomfort. "That's somebody else's job. I'm the [designer/detailer/insert role here]. I've got important work to do." The skeptics, cynics and jaded ones might even say: "What does it matter? These teammates are hopeless."

One technique is to reframe or breakdown the question. Rather than a single question, ask a series of more detailed ones. Instead of one unanswerable question, slice it into a series of binary yes/no queries, like the following.

Client and Partner Relationship Questions (Ongoing/Project Based)

1. Do you "like" them? More importantly, would they say they "like" you, or that you're doing a good job collaborating, caring and looking out for their interests?
2. Have you ever shared a meal or a beverage?
3. If you call your counterpart, will they readily, happily take your call (or quickly return it?)
4. If you ask them a favor, how will they react? Consider it? Reject it out of hand?
5. If you offer them a favor, will they accept (or consider) it?
6. If you send them imperfect information, will they review it, ask questions, offer corrections honestly?
7. Have they admitted a mistake to you? You to them?
8. Do you know who your counterpart is, and their role and responsibilities, professionally?
9. Do you know which project aspects motivate them? Schedule, budget, design, sustainability, process?) Are you working to support and realize those?
10. Do you know anything about them personally? (e.g. school, family, hobbies, where from, history, crucibles, how overcome, proud stories, foibles and failures?)
11. Do you know their preferred communication medium, frequency, and accessibility (e.g. email, meetings, calls, texts?)
12. Do you meet and/or communicate with them regularly?
13. Do the demonstrate by their actions that they are committed to the relationship?
14. Do you trust them? Do they trust you?
15. Do you feel like you exchange something of value between you? You to them, and vice versa?
16. Do you learn from one another? Grow, get better? Are you better together?
17. Do you feel they have the project's interest at heart? Or their own?
18. If you have faced conflict or disagreement, have you been able to work through it, reach understanding, use empathy to understand their reasoning, and move past it, stronger?

(Continued)

19. If you had the chance, would you seek to work with them again? Why? Why not? They you?

20. What can you do or what are you doing to you fix it? Addressing the issue? Changing something?

Questions like these asked of clients and partners can illuminate dark corners and reveal the state of our relationships. When reflecting on trust and familiarity, when we say the relationship is "good," what does that mean? More questions:

- In relationships, what does "good" look like? Do we know everything we can about their current workload, process, challenges and opportunities?
- What are the next steps with this client/partner/teammate on the project?
- Should we invite other relationship managers into the fold? Do others need to connect to make the relationship great?
- If leadership relationships are good, how is the working relationship? Project Manager? Day-to-day team members?
- Who is taking the lead with this relationship?

Common Languages

Since owners, designers, and contractors have unique languages, industry crusaders have developed common vocabularies. Owner organizations such as Construction Users Roundtable (CURT) and the Construction Owners of America (COAA) support committees for this subject. BIM leaders fight for COBie, LOD, and other data standard and interoperability protocols to push us toward common languages to exchange digital data between machines and people.

"To live a creative life, we must lose our fear of being wrong."

– Joseph Chilton Pearce

It's the same when humans talk. Without common understanding of terms, we resort to our own meanings. That's where the trouble starts. The architect has a "program." She means a "program of requirements" to design to. The owner has a "program." He means a repeatable process of building multiple projects. The HVAC subcontractor needs the "program." He means the digital control software to operate the HVAC system. It's no wonder we fail to communicate. Too often, we're separated by a common language.

✳ Tip: Nomenclature

Good project architects set advance standard project nomenclature and organize drawings using standard names and sheet organizations. On a recent project, the architects divided their plans into quadrants 1, 2, 3, and 4. The structural engineer split their plans diagonally at the expansion joints using N, E, W, S quadrants. The civil engineer used another designation. (And we wondered why coordination was hard.) Owners can lead in offering standard project names.

Hearing Differently

Even when we use the *same* word or phrase it can take on different interpretations and forms. Take "design schedule." We all know what that means, right? The words are clear. No multiple meanings. But let's look deeper. To a designer, design schedule may mean a three-activity macro look at schematic design, design development, and construction documents. Perhaps it doesn't include consultant coordination, contractor pricing, exploring options, or owner presentations. Faced with producing a design schedule, lacking these activities – and a constructor/owner dialogue – designers are likely to have a different view of their design workplan than their owner/contractor teammates.

"It's fundamentally unsettling to see the world as messy and chaotic. The mind is created to make sense of things. We are sense-making organisms."

– Daniel Kahneman

Here's the stereotype: within their studio enclave, the design team works their alchemy, making gold of common elements, speaking a language known only to those in the guild:[1] *mass models, elevation studies, maquettes, esquisses, parti, poche, materiality.*[2] Contractors and owners not privy to the high order of design rarely enter this sanctum. They don't know these veiled processes or how the magic happens.

"How do you get to know the people who design?" you ask. Spend time with them. Learn their processes, concerns, needs, and values. Ask to go into their studios. You'll need practice because they talk, hear, and see differently. Even if you don't *want* to know, do it to head off problems and help your teams. We need you. Outside our own organizations, designers and builders need each other symbiotically for survival despite our differences.

Ever had this experience? We attended the same meeting, yet left with wildly different recollections of what was said. Why? Because we heard and processed information through different filters and experience sets. For example:

"We're 10% over budget," is heard by the contractor as "Change the design, get in budget. Now."

"Not too bad," is heard by the architect as "Maybe the market will pick up and we'll come in under. The contractor probably has that much in contingency. Keep going with the current design."

Hopeful, optimistic, the architect is conditioned by a different belief set. Is it any wonder the design is still over budget at the next phase's pricing? It's the same building, more developed. Irreconcilable viewpoints continue in parallel while frustration and mistrust mount. Where's my budget-fixing superpower when I need it? Where's my language translator? Is there a Star Trek Vulcan mind meld in the

---

[1] For a peek at design process history and a famous owner/architect, relationship, read Franklin Toker's *Fallingwater Rising: Frank Lloyd Wright, E. J. Kaufmann, and America's Most Extraordinary House* (Knopf, 2007), the fascinating account of design and construction of Frank Lloyd Wright's Edgar Kaufman house.

[2] Vintage, vestige Beaux Arts terminology: maquette: small-scale concept model mockup; esquisse: sketch; parti: concept plan idea; poche: wall thickness.

house? That's Mr. Spock's alien empathy at the highest level – a knowledge transfer with no data loss. We all carry preconceptions and biases and hear in versions. We must ask and listen better to arrive at a common message.

On Management: Skeptics and Cynics

"When you look at a gamble you are evaluating states of wealth. One if you lose, another if you win. That's the foundation of prospect theory."[3]

Not many architects look at their work as gambles, prospects, or risk to be managed. They should. For all their powers of synthesis, too many architects are loath to integrate practicality, work planning, and economics – management – into their processes. That's when they (and their teammates) pay the price. With their broad training in arts, sciences, and humanities, architects excel at making sense of things. They study creative processes and apply them to bring order to chaos. But somewhere, the notions of meeting schedules, hitting budgets, and making money elude most of them.

Design drives project success, yet managing it remains sketchy, misunderstood, and under-performed, a seldom-mastered skillset. Improvement seems distant, even unattainable. Perhaps professionals are waiting for the mythical element spoken of by cynical engineers and scientists: "unobtainium."

Architects and engineers are a skeptical lot. They're trained that way. Their mission to effect change assumes that nothing exists that's good enough: they can always do better. Whether the subject is a design, calculation, drawing, or finished building, no one else's work is ever as good as their own. This innate drive to reinvent flies in the face of the business person's motivation: to develop a valuable product, service, or idea, and sell it widely for profit. For the average architect, efficiency and management are "unattainable" – *by design*. They reside in a distant, unfamiliar paradigm. For architects, already prone to think about their *inherent specialness*,[4] the pressure to be better at business can be bothersome. For decades the AIA's own canons advised against commerce as an objective – and we prided ourselves on obeying them. Most of us still do, even while lamenting our plight.

Collaboration

"Co-labor-ation" means "working together." Ideally, this produces synergy, not mere compromise or accommodation. Evaluate concepts such as co-location, face-to-face work, and sharing data. Not only can they be more fun, but you might learn something, make a friend, generate better solutions, and feel more fulfilled.

Team trust and relationships dictate performance capacity. The interpersonal behaviors and beliefs, through which the tangible project controls are interpreted is the team's collaboration lifeline. Even with measurable controls, without that lifeline, they'll be ineffective.

Even if you've been telling your architect she is over budget, if she doesn't want to listen, is too busy, mad at you, or speaking a different language, she may not hear or understand. The flow rate of communication and work within your team is a function of how well you collaborate. If you don't

---

[3] Daniel Khaneman, author of *Thinking, Fast and Slow*, on the 100th episode of NPR's "Hidden Brain," with Shankar Vedantam.
[4] After Rebecca W.E. Edmunds.

trust one another, you spend time micromanaging, documenting, or checking up on a teammate, cutting productivity. Successful collaborators identify peers with similar responsibilities and communicate.

Do they have authority to speak for their team on their issues? Learn their language. Develop a relationship and ensure all aspects are represented in the direction they give. Collaboration needs nurturing. Are you using colocation, weekly meetings, calls, social outings, checkpoints, and sharing imperfect information? Yes, I hope. Working together is decidedly better than the alternative.

✳ Tip: Culture

Good teams live their culture. At Holder Construction Company we work diligently at it. Culture and attitude are highly valued traits. We hire, train, and promote for it. When a recruit is being considered we ask: are they our kind of person? We believe we can develop their technical skills but face a greater climb to change their culture. In 2018, a task force produced the current version of the company's message on the subject, reduced to 5 core cultural tenets:

- Lead with integrity
- Perform together
- Care and connect
- Develop each other
- Improving relentlessly

Each contains a duality – two or more parties working together. Connection. While these tenets were developed to use *internally*, they were also intended to be shared *externally* with any partner in a kickoff meeting to make intentions explicit. Any partner not willing to stand behind these beliefs as we are is likely in need of discussion or trouble might loom. In using your PDC framework, establish and live by the same kind of cultural beliefs, so your human, relational systems flow.

CHAPTER 21

Level 4: Enclosure (Leading/Strategic)

Change, option and value analysis, consultant coordination, issue tracking, and decision support are the dynamic processes that manage project decisions. As Level 4 controls, they comprise the project's "brain, nervous and immune system" that keeps occupants safe, healthy, and out of the weather. As the framework's "enclosure," this level is a protective shield to guard against external forces, anticipate change, and bring closure. In a Maslow-ian sense it's the self-actualized management level. Once set, the Level 1, 2, and 3 controls inevitably change and evolve. It's the team's job to manage that change strategically, using Level 4 controls (Figure 21.1), as described in the following sections.

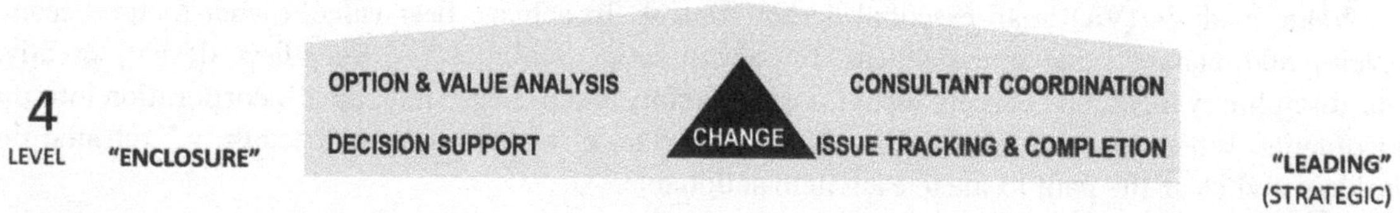

| **4**
LEVEL | **"ENCLOSURE"** | OPTION & VALUE ANALYSIS
DECISION SUPPORT | CHANGE | CONSULTANT COORDINATION
ISSUE TRACKING & COMPLETION | **"LEADING"**
(STRATEGIC) |

FIGURE 21.1 Level 4: "Enclosure" (Leading/Strategic).

Change

Change is the inevitable dynamic that *will* occur on your project and must be managed, including:

- Unplanned revisions, market conditions, errors, human nature and constant, well-intentioned miscommunications to project controls
- The resulting impacts and effects of one element's change on other project controls
- The management actions and reactions that must be made in response to these changes to keep or restore the project to balance.

Those surprised by project change reveal their limited experience. As a design manager, start by setting up change management systems (i.e. databases, cost, and list tracking). Manage them. Look ahead and prioritize issues. Use data to balance amid change. It's coming. Be ready.

Use owner decisions, design pending items, opportunities and exposures, and design criteria logs to track and manage change. You won't be able to do it in your head. You'll need databases and management systems. Use the core structural Project Design Controls in Level 2 as a baseline, with modifications. A change to any one Project Design Control will always have a related impact on one or more others. Another analogy: corks floating in water – push down on one and another must rise, or a cause a ripple effect.

Options and Value Analysis

Since projects are designed and built in dynamic conditions, owners want options open until the last minute. This can be problematic for designers looking for direction and contractors looking for documents to build by. Teams can help themselves with options analysis. Set up templates to analyze alternatives, complete with option names, short descriptions, supporting imagery, cost data, pro/con analysis, and schedule impacts. Teams that do stay ahead of the game. They create information in "decision-ready" form. They recognize and control it as part of their job. Those who lament they "can't get decisions" get left behind.

> ### ✳ Tip: Option Analysis Template
>
> Develop a standard one-page summary form for options analyses including decision factors. It reduces incomplete studies and yields faster decisions. One industry standard is the A3 template, an 11 × 17 sheet that outlines key aspects and plan of attack.

Value analysis (VA) is an essential design control. To achieve best value, owner-focused teams develop and manage value analysis ideas. To control design, embrace VA. Keep lists, develop creative interdisciplinary ideas, and classify them for acceptability and timing. Track their incorporation into the documents. When you hear phrases like "I'm waiting on x" or "I can't start because y," reframe the problem and clear the path to make each item actionable.

Decision Support: Issue Tracking and Completion

Decision support is a sister activity to options and value analysis. Beyond analyzing a single option, it's the ongoing management of pending decisions and priorities. When a teammate is overwhelmed or lacks information to make decisions, design managers help them get what they need. Prioritize, track, and manage these outstanding decisions; don't ignore them.

Consultant Coordination

In complex projects, the number of consultants grows by the minute. With not enough time to check their work, designers are challenged to coordinate document sets. Contractors and owners can help by connecting designers and their consultants with trade experts, users, and FM staff. Consultants design a major percentage of project cost. Help your team coordinate and integrate their work. Those who hide behind the "that's not my job" excuse defer the pain they'll face from uncoordinated work.

CHAPTER 22

Context:
Supply Network, Market Forces, Emerging Technology

A final "layer" completes the Project Design Controls framework – its context. (See Figure 22.1.) This "sphere of influence" shapes design. While much of it is "beyond" our ability to control, we reckon with context nonetheless. It's the environment in which the project ecosystem operates. Context includes the supply chain, supply network, and underlying support strata that comprise the design and construction industries-at-large. Context includes external forces, market conditions, geopolitical events, emerging technologies, and other unpredictable factors. These forces threaten to throw the Project Design Controls ecosystem into disequilibrium at a moment's notice. Don't ignore them. You can't control context; you can only react to or influence it. Beware and prepare.

Supply Network

Projects aren't possible without the complement of design consultants, manufacturers, suppliers, vendors, and trade contractors who support them. Manage in mind of them. Many a project has based its design on systems that aren't possible, available in the marketplace, or affordable. Ignore these ecosystem partners if you want trouble.

Market Forces

Be cognizant of market forces during design. Construction of another major project in your region, or overseas steel demand, could limit availability. Recessions or boom times can constrict labor or spike prices. Unaccounted for, these surprises can upset the best-managed designs.

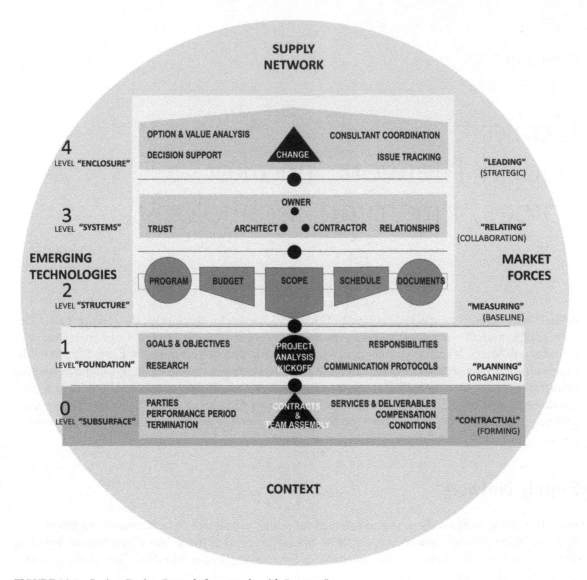

FIGURE 22.1 Project Design Controls framework, with "context"

Emerging Technologies

Good design managers invest in new tools as they emerge. Attempting to manage a large complex team without central shared communication and information systems in today's climate borders on lunacy. I've tried it. Commit to ongoing learning to stay abreast of new materials and technologies or pay a price in inefficiency.

Other Considerations

No Construction

Readers looking for how to manage construction won't find it in these pages. As Chuck Thomsen points out in the foreword, construction, as a more objective pursuit, is more easily managed. It can be quantified, scheduled, and resourced. As implementation and realization, it's a far cry from the exploration and investigation of design. Some of construction's biggest management challenges are the result of incomplete, late, or unmanaged design. That's our focus.

Design Management, Not Design

In digesting the Project Design Controls framework, note the conscious omission of the act of design. We're focused on managing it and helping designers set limits. The intent is not to convey the architect's or engineer's skill set, but rather, to set boundaries and support systems for them. Aesthetic studies, detailing, drawing, presenting, and the "pure" design activities in the AEs province are left to design professionals. Managing them is our mission.

Management Culture

Good designers have design management tenets as part of their culture. They know they're essential to good design and wouldn't think of excluding any of them. They know better because they've learned the hard way. We all have. Unmanaged, or unmanageable design firms think they're above these things and ignore or refute them or drag their feet. They're the kinds of firms that introduce design risk. Owners and contractors willing to embrace such risk can do so but be ready. Skipping any of these areas or letting them lapse at the expense of others is a recipe for unmanaged design, and a runaway train. Derailment is certain.

Within the PDC areas are countless tasks and tools to create and monitor for success. Deeper analysis could yield thousands of micro-duties, steps, and obligations. Managing the design of this book precludes us from covering them.

Love

Designers and builders love what they do more than most people. They're not in it for the money, they're in it to *make* things. They need to create. Bankers, lawyers, and businesspeople like money, transactions, and deals. Their stereotype is extrinsically motivated. The work is a necessary evil – a means to an end. Not so with those who conceive and create. For them, the work is everything. Think of it as a co-dependency problem. Many are willing to sacrifice salary and status for the opportunity to do better work, labor beside a master, or toil with peers who can improve their craft. I don't know a single person who enrolled in architecture school with the goal of becoming the next great capitalist. (Maybe they should have, but that's another discussion.) Great artist or architect, yes.

I'm not sure this is the case in the insurance business. Software may be a middle ground – it's a creative, indeterminate field, with potential to automate business and generate wealth. Its proponents are creative, geeky, yet business savvy in web-powered ways. Not enough architects get that yet.

I've seen architects ooh and ah about the elegance of a hardware set, reveal joint, scupper detail, or sumptuous material. They fawn over the expression of a design ideology – things others don't appreciate. Maybe the occasional broker gets excited about a beautiful contract, but I doubt it. Know this: design-focused architects march to the beat of a different drummer, the percussion and syncopation of design.

Designers are *intrinsically* motivated. They're driven by a higher cause. Their *design* investigations or social issues push them, not money or promotions. It should be no surprise, then, that getting done on time or earning a bonus won't matter if achieving those things mean compromising their work or ideals. In design, the work's the thing. While most architects know design's leverage points, they're less attuned to those of business.

Don't underestimate this difference. Those who understand it have taken the first step in knowing how to work with makers. Learning how to fuel those motivations and shore their weaknesses comes next. In his book *Flow*, Mihaly Csikszentmihalyi suggests that we find out what we're good at and like to do. And do it. Their convergence is the sweet spot, the intersection of motivation and capability, what he calls "flow."[1]

More than most workers, designers have found their sweet spots. What they may not have found is its integration with clients, schedules, budgets, and business. Or controlling the emotion that fuels it. That's where they may need help, and where supporting teammates can be valuable. These weaknesses have predictability in the design professions. Those who see them can call on the rest of us to help. Can we?

The Design Life Conundrum

A life in design and construction is difficult. It's a choice that confronts you with a myriad of lifestyle challenges. Making buildings is not a 9-to-5 job, so much so that maintaining relationships with family and friends outside the office can become problematic. Experts coach us on improving interactions with colleagues by taking time to get to know *them*, further challenging our home and personal lives. Yet the primacy of our design passion compels us forward. The conundrum persists.

Design's reach is vast. Its execution is demanding. Owners and contractors have no idea of the difficulties of reconciling code studies, consultant engagement, client listenership, digital contraptions and construction practicality, while continuing to serve at the pleasure of the almighty design master. Assimilating all that within a schedule and budget are almost unheard of. Doing it profitably is ever challenging. That's precisely why designers need help from their owner and contractor teammates.

Designers try valiantly, but by the time they're done trying, their fees and teams have long been exhausted. We've got to do the hard things. Together. It does no good to be a win-lose enterprise. All must win. Either/or was never the goal. In the good project teams of my career, the two co-existed: artistic talent and common sense. Enlightened architects found riches in team gettogethers. They listened. They grabbed the wisdom from their clients and contractors and turned it into gold. The palpable melodrama of exceeding the budget and deadlines was mercifully ended. We made it.

The Foolhardy

Who can expect to assemble experienced, opinionated experts who haven't worked together before and get them to function as a team? Each discipline thinks it's a little more important than the others. Many team leaders – often owners – are inexperienced in the very skills they need to lead such processes. How

---

[1] Mihaly Csikszentmihalyi, *Flow: The Psychology of Optimal Experience* (HarperCollins, 1991).

can they lead *well?* Most coaches had stellar careers playing their respective sport, learning the game's mechanics and strategies to prepare and direct their athletes. Mentored by coaches themselves, they internalize, mold and transfer that knowledge to motivate others. Many who design and build lack that essential background. Some practitioners even pride themselves on being "coach-averse." A Fountainhead-like individualist idealism overrides teamwork. In today's context, this approach is nothing less than foolhardy. That we get anything done at all comes largely thanks to grace, aplomb, and survivalist persistence, and the help of our teammates.

Embracing Limits and Leadership

What currency goes to the heart of why we fight for something we believe in? Surely, it's not management – staying between the lines, meeting the budget or getting work done on time? Deeper motivations, belief in greater causes: design, a vision, or a charismatic leader willing to fall on their sword – they're more like it. There's a dramatic difference between "energy vampires" who toil because they have to, are told to, or are micromanaged, and an enlightened, high-functioning team that works with vigor, driven by a greater cause toward a longer-term goal – because they follow a great leader or are in it together.

I believe all designers, and others working in related capacities to manage design and construction, would rather be *led* than managed. As creative self-starters and holistic thinkers, they prefer the freedom to decide their own approach over being micromanaged in task-specific ways. They want to embrace the restrictions imposed and employ them as an integral part of their craft. They know without constraints they're doomed to repeat the mistakes of the past. With them, they can rise to create a self-managed future.

A Bitter Pill

Do designers *want* to manage their efforts? Of course. In my days as a designer I tried. After being burned repeatedly, at the outset of each next design effort, I began with compliance and control top of mind. But deep down, other forces were always at play.

By its nature, design is exploration and rule-breaking. When an opportunity presents itself, it can't be turned away, can it? Even if it means going a little beyond the program, scope, or budget? Are we blinded by creative possibilities over cost? Does design trump all? Will luck or some unlikely windfall save the day? Not usually.

Design means working outside the lines, consequences be damned. It's how innovation occurs – and progress happens. For designers, this rule-breaking tendency can become an unsustainable norm despite the recurring lessons. On those occasions we must swallow the bitter pill – reworking, reducing, and diluting what we've created. Again, we must do the thing, in hindsight, we should have done in the first place. Failure can be constructive.

CHAPTER 23

Understanding and Using the Framework

Order and Logic: "Visual Onomatopoeia"

Think of the PDC management framework in stages, in the order and logic of construction. This thematic approach draws on familiar patterns, a "visual onomatopoeia" for designing and building, a metaphor with meaning, an allegory to construction.

Level 0 forms the "Subsurface," the formation of the team, their duties and contracts into which the other levels are built.

Level 1 lays the "Foundation," the planning and organizing activities that support the management structure.

Level 2 builds the "Structure," the columns and beams that frame the house: program, budget, scope, schedule, and documents.

Level 3 roughs in the "Systems," the human, relational aspects that power the project's management system. The things with complex moving human parts that make design management run: trust, relationships, and collaboration.

Level 4 adds the "Enclosure," that protects the project from stormy weather and keeps its occupants safe: issues tracking, decisions, critical thinking, coordination, and completion.

Within each of these levels, touch on their five checkpoints – a total of twenty-five things to consider, cycling through each quickly to ensure it's in place and aligned.

Too Simple? Change It

Design managers who think this management "playhouse" overly simplistic or a trope are welcome to view it under some other mental model, work intuitively, or use another method, but they should use its principles. My experience is most will develop a fluency in short order and understand its value. Means aren't important, results are. For every one of these tips, there's a way to break them, or do them in a different way.

✳ Tip: "Palace of Equals"

Rather than a hierarchy, think of the framework as a networked grid of nodes, each with potential to activate or imbalance another – a palace of equals. Like designers, design managers might leap directly from program to schedule to communication to documents without traversing a hierarchy. Would-be design managers need not learn a new system or form a new mental picture. Using this mnemonic device, they can rely on the one they already know, innate to their native language.

The Project Design Controls framework relies on newfound intrinsic motivation for use, particularly by victims of design over-indulgence or mismanagement, their own or others. Rather than a command and control mechanism, use it more like a Wikipedia, crowdsourced approach, with access, use, and contributions by many. Design managers should be like systems administrators: overseeing and letting the system self-level, as opposed to dictators.

Enter Slowly

For the timid, skeptical or reluctant, managing design doesn't have to be an instant, all-in proposition – on or off, all or nothing. More than a binary phenomenon, it allows a gradual entry into its new world. Try some of the metrics or soft skills. See if they help. Then more. You'll do best to use them all but start slowly if you must. Learn to trust it. Develop your relationship with this paradigm and migrate it across your team. The nexus of design and management can be a symbiotic coming together, not a clash. A both/and, not an either/or. I hope it is for you.

Processes: Repeatable, Shared, One Off?

Design and construction are "gig" economies. Their project-based, one-time endeavors contribute to our difficulties. Beyond language differences and first-time, unfamiliar teams, tools and processes aren't always aligned. Those without time to support shared standards but interested in improving methods use another team-limiting trick: they develop their *own* processes to cope.

What is a process? Nothing more than a series of steps repeated consistently and shared to ensure predictable results. Most of the ideas in this book are the result of a "process mindset" – an ability to see around corners or react after bumping into a corner. To a degree, it's doing something better next time.

Have a problem? Find a solution? Did you use a tool or process? Save it for reuse the next time the problem arises. If you can save it for others to repeat, then it's a process. One thing we all have in common: we need processes. We're better when we share them. If we can do that across company or industry lines, that's even better. But that takes time – and our project is happening now!

It doesn't take many projects for team members to find inefficiencies. "I'm only one year out of school, but I've seen that issue before. Where's my tool to fix it?" Even better: "My teammate was one step ahead of me. With better vision and planning skills, she gave me the tool before the problem happened." Process! Sharing! Teamwork! Does it matter whether your teammate got their paycheck from within your architectural firm, the owner, or the contractor? I think not. We are a *team*, aren't we?

Software developers have made careers by developing and sharing "freeware." Later, when everybody realizes how valuable it was, they get their rewards. Usually, it's by selling their solution to a bigger company expert at selling and marketing. Construction teams may not reap rewards like selling millions of software seats, but they can make their projects better by sharing.

What synergy we create when we share processes with mutual benefit! We should do more of it, but it takes desire. We've tried not sharing for centuries. It hasn't worked. That's why our industry is morphing to collaborative delivery models. As professionals, we should be skilled at sharing and efficiency. Knowing we need one another should inspire more sharing. It can't be any worse than what we're doing now, can it?

Twenty-Five Points

In total, twenty-five Project Design Control touchpoints track, manage, and adjust design. Forget one at your own peril. Combine, integrate and automate them if you can. Simplify and reduce them to their essence to focus on value-adding activities. Experienced, trusting teams don't waste time demanding or reminding one another to do them; they do them automatically.

Causes and Effects, Actions and Reactions

Project Design Controls are inextricably linked and affect one another. They are listed in levels and categories to provide a structure and language with which to manage design. When a cost estimate is done, and the building area has grown by 20,000 square feet since the last estimate, we ask why. It's certainly a Scope increase. It was reflected in the documents. If there was no commensurate Budget increase or owner-requested program change, we will face tough decisions. Why did this happen? What's the schedule impact? Where did our collaboration, communication, and change management systems break down to let this happen? Can we mitigate the effects? Like the McKinsey 7-S diagram, the connections are strong and powerful.

The Project Design Controls framework is a planning, management, and leadership matrix – a tool for oversight and leadership, strategy and intervention. Each of its intersections should be part of a continuous circuit; its nodes active parts of a dynamic grid. Without any one, the circuit is broken, failure the result.

Target versus Actual: Balance

Measure the targeted and actual values for each Project Design Control at each phase. Note variances between the two. Manage and act upon them as part of the change aspect of project controls to balance all factors. Use the three-question litmus test. (You can find it in Chapter 16, and again at the end of this chapter.)

Implicit in the idea of a matrix or circuit is continuous flow or balance. Project balance is a temporary condition, in which all project controls are present, compatible, and in equilibrium, for which anticipated changes are planned via contingencies.

Statics and Dynamics

Projects are never static. Balance is fleeting. Things are always in motion. Your design management mission (you have no choice but to accept it) is to return balance as quickly as possible to continue the flow. Forces seek to resolve themselves in the easiest way – or fail. Seconds after the pricing is presented and the team celebrates being in budget, the pendulum swings. Design begins anew. Risk rears its head. Soon after, a program change surfaces, a market shortage emerges, and prices spike. The roller coaster ride starts again. A prime consultant is fired, and a deadline is missed. This is where the conflict, and the fun, of managing design happens. Those who like it quiet and stable should change careers. They're not cut out for the tenuous balance and constant change of designing and building. That takes cross-trained, change-ready, smart, agile types.

Design Process: "The Integral"

Owners and contractors are sometimes baffled by the design team's slow or reluctant response to direction. "What's so hard?" they ask. "Just fix it and move on. We told you we cannot afford the balconies. Take them off."

What these questioners may not understand is that building design is holistic. Change a piece, change the whole. In the cost estimate, the balcony line-item costs can be deleted, and the estimate stays whole. The spreadsheet or cost database recalculates. In the design, maybe the balconies were a critical formal element, a needed articulation in the building composition. Perhaps they fulfilled a code requirement, area of refuge, or maintenance access. Deleting the balconies may dictate revisiting the floorplan, elevations, materials, details, or core idea of the building.

Situations like this can result in the design adding something in place of the balconies, or even a wholesale reconsideration of the design concept. Why? Because designers are trained, with good reason, that the building is an integral whole, not a sandwich that can be consumed regardless of whether or not the lettuce or tomatoes are present.

When these design revisitations or reinvestigations happen, teammates should understand the necessary ripple effects on time, cost, and design process. Beyond the emotional costs such direction may have, owners and contractors should allow for the implications such a seemingly simple change may have on design.

Design Process: "Muddling"

Those who have not studied or done design cannot possibly understand the uncharted and varied nature of its processes. Some prefer a calculated approach. Others intentionally begin without a plan. Almost all are taught that a certain amount of free play time is essential. Designers Gunnar Birkerts and Frank Lloyd Wright spoke of "marinating." That is, throw all the ingredients in the pot and let them soak for a while. These design masters would absorb the program data, then step away to let their minds work subconsciously. At some future time, having been precooked, ideas would emerge and be made explicit via drawings and models.

Others, such as my late mentor, renowned architect Terry Sargent, spoke of Charles Lindblom's famous phrase, "The Science of Muddling Through."[1] Originally used to describe the benefits of incrementalism in public policy, Sargent adapted it to describe his design process. Traveling unknown paths, circling back, and trying ideas left a circuitous trail, but yielded masterful work: more the result of exploration and persistence than planning, discipline or rigor. Whether subconscious or visible, ordered or playful, design can look like muddling, but there is a method to its madness.

---

[1] Charles Lindblom, *The Science of Muddling Through* (Public Administration Review, 1959).

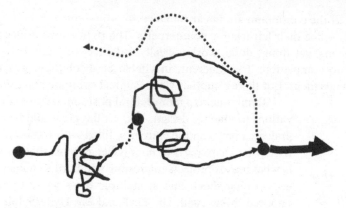

FIGURE 23.1   "The Science of Muddling Through," after T. Sargent and C. Lindblom.

Design Process: Leaps and Lights

Other phenomena underappreciated by people familiar only with "business" process, are leaps. Unlike proformas or schedules, design is characterized by alternating periods of refinement and exploring new ideas. This rapid prototyping can lead to creative discoveries or "leaps." These "aha" or "Eureka, I've found it!" moments are serendipitous. One could play and explore for weeks and not find one. Finding them under pressure, deadlines, or time constraints can be even harder. Teams that restrict or direct design partners run the risk of missing great solutions or settling for subpar work.

✳ Tip: Know When to Let Them Play, Supervise, or Join in

Don't be standoffish or dismissive of design process. Consider joining in. Add your perspective. Learn to frolic in designing Play-Doh and digital models. You might add something. You might keep the frogs in the wheelbarrow. Maybe you'll grow your understanding of this synergistic, cyclical process and become a better teammate:

- Designer has an idea.
- She draws it to represent, see, and share it.
- It tells her something.
- She knows more. She thinks.
- She draws something else.
- She draws, explores, studies, informs, draws.

Eventually a light turns on.

Conflicting Values

Value systems and paradigms of project players range across a broad continuum. On the left reside those who see it as their mission to create from scratch, all else be damned. They value the new, innovative, and designed, over time, money, and results. Their love of process – their greatest strength – is not surprisingly their greatest weakness.

At the right end of the continuum are linear thinkers, step-by-steppers who see the world as black or white, seldom gray – like their left-leaning counterparts. The right fringe prides themselves on their ability to make decisions, get things done, work logically and decisively, and drive out uncertainty in a militaristic march to completion. The inherent, inevitable clash of these generalized types is the essence of design management. But there's another view. Political columnist George Will put it thus:

"The world can be divided into two kinds of people – those that divide the world into two kinds of people, and those that don't."

To migrate to a more central position, let's explore the range of motivations to manage design. Yes, it's the great middle of the spectrum, the gray area between the extremes, filled with thousands of passionate centrists who share parts of *both* belief systems that's important. This range is what makes projects interesting and hard to manage: the fundamental, eye-opening shock that at multiple turns we're trying to do completely opposed things, with Dr. Jekyll and Mr. Hyde/Sybill-esque multiple personalities, styles, and methods in play. What fun!

On occasion, resolving these conflicts can be done simply by talking or spending time together. What a great surprise to find you were separated only by a small, correctable language barrier instead of a core belief. Maybe you both wanted desperately to resolve the issue or find a better way; you were just going about it differently. This range exists on all projects. It's the reason why no pat answer works for all situations

"Contractors have to build the whole building. Architects only have to draw part of it – the part they care about or have to draw. They leave the rest to us figure out."

For designed-focused architects, the threats of time and money limitations, angry clients, or putting one's career in jeopardy aren't powerful enough to resist architecture's allure. Its pull is too compelling. Design types are bred this way – educated, trained and encultured to be design-inclined. This is the perennial challenge of all who manage design. To compound things, even when they *want* to manage their process, it's very breadth, complexity, and integrated nature can render their attempts futile.

Contractors, charged with making design ideas real, are usually a more linear lot: "Tell me what to do and I'll do it." Theirs is not to set design direction, rather, they drive to it by any means necessary. But creativity can be found among contractors, engineers, and players typically cast as "set-in-their-ways" types. They've had to learn it to survive what they are given. A construction colleague, engineer and architect, David J. Miller, once said:

His words ring true. Projects need all the design and management creativity they can get, regardless of source. With this stereotypical framing of design management's extremes, and the great numbers of us in the middle trying to cope, Project Design Controls offer a survival system.

Design Risk

From contractors' and owners' perspectives, to manage design is to manage risk. Designers should adopt this outlook. A late, over budget, overly-innovative, or uncoordinated project drives costs skyward. Managing risk is the mission of everyone on the team, but designers haven't been schooled in risk. They're trained to induce it. Designers need to catch up in this regard. Each of the constructs offered as Project Design Controls are strategies to mitigate design risk – and yield dazzling, successful projects with happy clients and teams. Beyond the solutions offered by the Project Design Controls, the following issues are surfacing – the subjects of future books.

"The world can be divided into two kinds of people – those that divide the world into two kinds of people, and those that don't."

– George Will

Risk Trends

Shifts

"An ecosystem, as it shifts, demands that every part find its new fit. As our industry shifts, architects and members of building teams need to find theirs – to re-design or re-imagine the new fit."

"Designers can design all day long, and all night, yielding wild, innovative solutions with someone else's money, but they can't – or won't – do it for their own process, firm, life, or future."

– David Ferguson, Interim Dean, College of Architecture and Planning,
Ball State University, Muncie, Indiana

Why is that? Because they have no training, culture, skills, or motivation for it? No desire, responsibility or ability? By contrast, young graduates working in a construction company are immediately given responsibility to manage risk. They engage trade contractors, check scope and insurance, and procure work. Risk is trained for and controlling it is rewarded. To manage design, we *all* must embrace and manage risk.

To make sense of this yin-yang relationship, we need a dose of reality. I include myself in that broad condemnation. I saw it twenty years ago and did something about it. By associating myself with contractors who excel at managing risk every day, I have grown. I would like to think it's been a two-way street – that I have taught them a little about how design professionals think and work.

The industry trend is for more risk shedding for survival. But where does the risk go? If you're managing a team, beware of the following:

Delegated Design

This growing category stretches from simple delegated design (e.g. storefront glazing systems) to first-of-kind integrated solutions that tax the marketplace. (See the Wayne Wadsworth interview in Chapter 12.) Blurred responsibilities come with this tendency to delegate design responsibility.

Temporary Construction and Protection

Many structures are designed to be stable only in their final form, with the benefit of bracing and surrounding structure. Without engineer, contractor, and trade contractor agreement on a construction and erection sequence, managing design and construction of a structure can be risky – and dangerous. A sad example is the 2017 collapse of a pedestrian bridge at Florida International University, Miami, during construction. Clarify responsibility for interim structural conditions.

Warranties

The design profession has long steered clear of warranties such as the implied warranty of merchantability established by the Uniform Commercial Code that applies to products. As a service profession, design is governed instead by the common law "standard of care" of the design professional's performance. That is, what a reasonable professional practicing in a similar area would have reasonably done. As such, human error, gaps, errors, and omissions are expected in design services for customized projects. They're the norm, a far cry from product warranties that have had the benefit of years of refinement, mass production, feedback loops, and corrections. Be cognizant of the warranties attending to design and construction, and lack thereof.

Consultant Ownership

To cope with the growing number of consultants required, many architects are retracting their reach. In recent projects we've seen an attitude of architects separating from consultants rather than owning and integrating them. In these situations, engineering consultants are left on their own to coordinate or pass this duty to the contractor as means and methods. Where the architect has traditionally been contractually and spiritually responsible for design, coordination, and documentation of the entire building, a mentality of disowning consultants is replacing this responsibility.

Liability Shedding

As a defense mechanism and proof of the decline of consultant ownership, we have begun to see projects in which the architect states that he or she is responsible only for work on sheets bearing his or her professional registration and licensure seal. Buyer beware. In these cases, who then is responsible for the design, integration and coordination of the *whole* building including architecture, engineering, and other design disciplines? Address these areas of contract intent and responsibility in the contract and planning phase, not later.

When Does Design Management Happen?

Having learned Project Design Controls, our next challenge is when to apply them. Can we use them at the earliest possible moment, at project inception, predesign, design, and preconstruction, through construction, commissioning, and closeout?

Managing design happens throughout the project life cycle. It can start in the team assembly stage, between projects, in relationships with partners. Most often, it begins at the design or preconstruction stage. An often-missed opportunity – one we can create by scheduling it – is the planning or predesign phase, where we define and set projects up for success. We also control, and interact with design during construction and closeout, but in a different way, since design and programming should already be "complete." To be most effective, we should understand design during all project phases.

"Steps, Actions, Tasks" to Managing Design

Take the following steps to manage a design team. Repeat them throughout the project for each Project Design Control and task.

Step 1: Own It and Want It
Be aware of, and motivated to care about design. Having read the interviews in Part 1 and experienced these things yourself, your team should be primed to "own" design management – and want to fix it.

Step 2: Start
Conduct a Project analysis to launch your project well. Refresh it periodically.

Step 3: Define
Define the project. Collect and use Project Design Controls at all levels: forming, planning, structural, systems, and enclosure.

Step 4: Document
Record all project defining aspects as PDCs. Capture them in a project definition or project controls package. Periodically update and redocument them using the Project Design Controls framework.

Step 5: Monitor, Test, and Manage: Ask
Monitor and manage design as it advances. Do not assume that because you set something up early, that it's good for the duration. Do a litmus test on each control periodically. Ask your team at random. If any one of them doesn't know if a certain checkpoint is in place, what it's supposed to be, or if it is in line with the others, you are in trouble. Recheck things at a frequency that suits you. Continually test for balance, then act. Revisit baselines, understand and realign variances.

Step 6: Collaborate
Use soft skills such as design understanding, communication, and trust to interact with design partners. Employ relationship building and other soft skills. Nurture these skills. Reap the benefits and synergies that activate your hard metrics.

Step 7: Change and Adapt
Manage the change that *will* occur. Document all revisions and balancing actions. Be creative in your Darwinian approach to evolving your team.

Personalizing, Reporting, and Rewarding

Don't feel restricted to use the names or organization described in the PDC framework. Rename or restructure them as necessary. Automate and integrate them into your processes. Use lean techniques. Make them yours. Adapt them to your team's style.

Once you have created and used Project Design Controls, share and report them. They offer feedback and mutual benefit. Post design management metrics on the walls, on screens, and review them at weekly meetings. Make them visible, accessible, and an automatic part of your process. Use visual controls and dashboards; make them fun and easy. Help the team buy into and use them. After all, few of us really want to manage or be managed anyway, we'd rather "do."

Be accountable. Celebrate successes. Show progress against schedule. Highlight scope capture. Study metrics. Care about managing design. If you do it right as a team, it will reward you with a project and team you'll remember for a long time: the kind of thing you got into this business for.

Aspire to that.

Try, Team, and Talk

Don't become frustrated if a tool, process, or technique is not available, working, or suited to your need. Scale it or change it. We do not have all the answers. Apply judgment and experience in team-specific ways. Seek the advice, and lessons learned of peers and supervisors. Don't neglect an important resource – talk to your design partners! Learn their issues and obstacles. Help them overcome theirs and they may help with yours.

Reasons for Failure (and Success)

Some may think this model is too complex. Five layers. Twenty-five touchpoints. More issues to consider with context. Potentially hundreds of tools and tasks. It's not. The Project Design Controls framework is a time-tested assembly of interdependent aspects, each required and dependent upon the others. Remember, the Einsteinian goal: to make it "as simple as possible, and no more."

"For every complex problem there is a simple solution – and it's wrong."

Those who attempt to reduce design management's essence to one or five simple things are misguided. Einstein said it best:

Those who persist at design management oversimplification probably have not designed – they are linear thinkers challenged by complexity and ambiguity. They need simple answers – now – to resolve the uncertainty! And they are wrong. When it comes to design, there is no such thing.

Problems (and Solutions)

Experienced practitioners agree, on any project, every one of the Project Design Controls can be – and has been – a reason for dysfunction. Like these:

- Forgot to exchange a BIM/VDC execution plan? *Problem.*
- Set up measurable structural baselines (i.e. program, budget, scope, schedule, documents) but failed to develop a trusting relationship among the team within which to use them? *Problem.*
- Had all controls in place but failed to set up change mechanisms to track and manage owner decisions to completion? *Problem.*
- Forgot to carve out a planning phase? *Problem.*

Conversely,

- Set up and balanced all other PDCs? Mastered the people, relationship, trust and change process? *Solution!*
- Planned and conducted an exhilarating project analysis work session to kick off your project? Got to know those you will work with? Shared goals and objectives? Identified constraints and opportunities? *Congratulations – design managed! (For the time being.)*
- Tracked pending owner decisions to enable design starting and finishing? *Well done!*

How to Know

Now that you understand Project Design Controls and when to apply them, *how* do you know they are right? Some are objective and quantifiable, others are subjective and less discernable. Design controls will serve little if you can't put them into action and know they are working. Here's how you can know.

Level 1 Controls

Level 1 controls are task-based: classic management and planning activities. You'll know when you have completed them.

1. Schedule and conduct a full team, interactive, interdisciplinary project analysis work session. Share project information in advance. Get to know your team. Set Project Design Controls. Seek out a sample agenda for an interactive meeting like this. Prepare. Define all project aspects. Use the Project Design Controls to collect and classify project information. Get help from your team to gather the information. Get off to a great start. When you're done you can hold the work product in your hand, share it, and launch your team.

2. Develop a project communication plan. Who talks to whom? When will you meet and with whom? How will you resolve conflict? What communication tools will you use? Write it. Share it. "Why didn't you send your file? I didn't get the email."

3. Create a digital infrastructure and VDC/BIM execution plan. In today's practice, digital tools are the means of design production. Teams that don't plan for such things do themselves a disservice. Gather the information technology team. Discuss data sharing. Develop a BIM execution plan with hardware, software, infrastructure, training, and protocols. Don't doom teams to "spaghetti" file transfer, un-interoperable data, and non-standard nomenclature.

4. Develop a program of requirements if the owner doesn't have one. Analyze it together. Drive out generic goals and convert them into specific targets organized by category. Use research to inform your process rather than reinventing the wheel. Capture, share, and use your research.

Level 2 Controls

Set, record, and share tangible, measurable metrics for the five "Structural" design controls: program, budget, scope, schedule, and documents, as a baseline. Later, when they change, update them and reset the balance. You'll be able to because you have a baseline. Check and reconcile if the building is being drawn at the programmed size or has grown.

Level 3 Controls

Level 3 controls are more difficult to measure. How can you tell if you have trust? Do you really have a relationship? Who's to say whether you're collaborating or just working on the same project reluctantly? Here's a way to answer the litmus test for "softer" items. Ask questions. To test trust, ask yourself: When Joe the architect sends me a new concept, do I dread it, or look forward to it? Do I generally expect to believe and trust what he sends me? Why? Why not? Find out. Do we have a relationship? When you call her, does she take your call? Answer with a smile in her voice? Have you had coffee together? Do you know where she went to school? And vice versa. Those kinds of tests track the soft items. You'll know. Read Chuck Thomsen's interview in Chapter 2. The way they could tell if they were doing a good job was to ask their client. If the client said he liked a person, they had a good relationship. It was that simple, and still is.

Level 4 Controls

Managing the project's enclosure system is harder and riskier, but just as important. Like a human or building skin, this level keeps the weather out and uses sensory organs to perceive and regulate. Together with Level 3 controls, Level 4 controls make things happen and deploy cognitive "executive functions" to compartmentalize, decide and keep course. How can you use them? With three skills:

Collect: The first is collecting and tracking data, by looking, capturing, gathering, and recording problems and issues on lists.

Analyze: The second skill analyzes and prioritizes that data by importance and urgency. Which outstanding issue is most problematic or urgent? Address it first. Can we assign someone to deal with the other issues? Look at the current reality – the present (Where are we?) and project into the future. Look ahead for roadblocks and clear them. (Where are we going? Is the path clear?) Example: We still don't have owner direction on cafeteria program requirements – and we're building the building! We better do something. We have 97 outstanding issues and decisions required. The owner and design team say they're working on them but aren't keeping up. How can we help them focus and answer first things first? They can't work on them all at once.

Synthesize: The third skill involves soft skills to help teammates clear paths to get work done. Can't make that decision? Need more information? How can we help? Set up a meeting? Do an options analysis? Get pricing? Come up with more alternatives? Bring things together.

Listing, analyzing, prioritizing, and path clearing are prized Level 4 skills. You'll know if you're doing them right if you have such tools and use them to make decisions. Decision tools get design and construction completed. Managing others and their decisions is part of managing design.

How to Coach

Design managers find themselves in the roles of budget, schedule, and constructability mentors. How you handle that role is a personal choice. Like parents, coaches, or psychologists, some coax and cajole their charges positively: "Come on, you can do it! We gave you the targets. Let us help."

Others beat them into submission: "We've only got $150/SF to spend! The drawings were due last week. These documents are bad, uncoordinated, incomplete. We can't build from them!" Conflicted and unsupported, the design team, unable to reign in their tendencies to overdo, launch their teams into tumult once more. The miasma of "too much" returns. In choosing your management style, ask yourself which you'd prefer.

✳ Tip: Troubleshooting: Managing the Unmanageable

What can you do when you face an unmanageable project or team? You've put all the Project Design Controls in place, but design, direction, and progress are still floundering? Typical causes may look familiar:

Controls are not really in place. Things are moving, flexible, fluid. Not really set.

Project leadership (owner, CM, design team) is being too polite, nice, refusing to speak the truth or address conflict. Find solutions in these alternatives:

- Revisit the controls. Fix them, set them. Get buy-in. If everything is flexible what you have is Jell-O. You can't manage that.
- Change communication style. Raise the stakes. Work more privately or more publicly if you're a victim of groupthink.
- Do nothing and wait it out. (But it may be a long wait.)
- Wish for a miracle: more money, more time.
- Change team members or process(es).

Good luck if you find yourself in such a design management vortex.

✳ Tip: Overcoming Adversity: Sudden Change

Even if you've succeeded in having a great team, you're bound to run into roadblocks, detours and surprises along the way. What to do then? The first thing is to face each one squarely. As a good design manager, you should look for such conflicts, planning for what you'll do when they happen. A classic football coaching expression for this phenomenon is "sudden change." When your team fumbles or throws an interception, you don't get down. Having expected it, you work to turn it back around. It's an expected part of the game. It's no different in managing design. All projects have setbacks. When they hit, does your team get dejected? Or can you reframe these crucibles as learning experiences authentic to your project?

Motivating Designers

Designers are driven by the same forces as everyone else, except they value them differently. In his book *Drive: The Surprising Truth about What Motivates Us* (Riverhead Books, 2009), Dan Pink reveals the power of intrinsic motivation. More than being motivated by fear, money, or external pressures, the surprising truth is that most professionals are motivated by factors within themselves. Designers have this in spades. Stereotypically introverted and introspective, they thrive on synthesizing design solutions. It's who they are. Given the chance to exercise their most loved, highest aptitude skills, otherwise known as design, they anger when external forces get in the way of their self-realization, given the chance to shine.

When it comes to schedules and commitments, *ask* what they can do, don't *tell*. Having gotten their buy-in, you can give the support, structure and interim checkpoints they need. We're all better with a coach, someone to lead and push us. But what happens when we still fail? We agreed on a deadline and missed it? In those cases, reset and rethink. How can we adapt and overcome? What can we get by with? In these cases, the team fails on an interim milestone, not the entire project. Set targets, but don't be surprised when you miss a few. It's design, remember? Just think where you'd be if you hadn't set goals at all?

✳ Tip: Failed Attempts: Post Mortems

I've witnessed countless unknowing, well-intended "management types" fail in attempts to lead or manage designers because they didn't understand their intrinsic motivation or process. To understand, let's look at sample scenarios. Each is real but has had the names changed to protect the innocent.

Scenario 1: Owners to Architects

The owner's corporate facilities director, "Bubby," was an experienced architect himself. He kicked off our project like this:

"We need to get this building done for $100/square foot. I've done them all day long for that."

So much for setting goals, objectives, or aspiring to design synergy. So much for the people side of the business. Clearly, this project was to be a budget-driven death march. It was. With his citing of bygone pricing heroics from prior mediocre work, this owner couldn't have been clearer. Good owners reach higher and get more out of their design teams. They know how to engage them. They don't abandon their team.

Scenario 2: Contractors to Architects

In this case, the contractor's project manager, Biff, met his architectural teammates on the first day of the project. The enlightened owner had the foresight to introduce his team in a kickoff meeting, so they could get to know one another before design work began. Biff, a young, aggressive, rising star was experiencing his first opportunity as project lead. He never practiced design. When he was introduced to the design principal in charge, Brett, a 45-year-old firm owner dressed in black, Biff's first words were:

"Hey, Brett. I'm Biff. I'm the PM. Listen, we're going to need your early precast bid package in 6 weeks. Gotta buy that trade by 10/1. Can you handle that?"

What's was wrong with this approach? No personal niceties, no goals and objectives, no design respect, no consideration of context. Maybe architect Brett hadn't thought of the building being precast yet. Perhaps he had other innovative wall and structural systems he'd been exploring. Starting with having the solution dictated to him and having to execute it under pressure violated everything Brett stood for. What could have been a better approach for Biff? Try this:

"Hey Brett. I'm Biff. I'm the PM. I'm familiar with your work. I love your project at XYZ Tech. I'm looking forward to working with you and hearing some of your ideas about materials options. I'll have to admit the schedule is tight, but together, I think we can do it. [Time for Brett to respond . . .]

What would you think if we got together for a brainstorming session next week in your office over lunch – on us? If you like, or if you have any preliminary thoughts, we can invite a couple key trade contractors to provide cost and schedule input. Maybe we can start to develop a draft early phase design and purchasing schedule to set some short-term dates?"

One thing is certain: contractors would never procure or manage their steel subcontractor like they often do architects, with a hands-off, "good luck" approach. General contractors are intimately familiar with trade contractors' schedules, processes, interim deadlines, fabrication, shipping, and installation. They wouldn't think of not monitoring, owning, and caring about their subcontractor's progress. It should be so with design.

Wise contractors respect their design partners' motivations. They don't use schedule, budget or predetermined solutions like clubs. Try it.

Scenario 3: Program Manager/Developer to Architects

In this scene, the role of the design motivator is played by a person who hasn't practiced design or construction. Their real estate or management skills have positioned them to be in charge. Under contract to the owner, their job is pure delegated management, for a fee. Not responsible for design, nor accountable for construction or its cost, they create an added fourth node, converting the traditional OAC triangle into a square. The developer's lead PM, Beckie, introduces herself:

"Hi Brett, nice to meet you. Our owner has asked me to stay on you about budget and schedule. I haven't worked with your firm before and don't know much about the project, but I've drafted a preliminary design schedule we need you to meet. Can you do it?"

Having been in Brett's shoes, I can assure you that adding another party to "tell" me, as a designer, what to do and when to do it is an unwelcome development. Just one more management layer that needs more reporting, accounting, and meetings. I'd rather manage design with the team we have. That this new entity hasn't taken the time to get to know me, my firm, our work, or anything about the project does not delight me. Because of her adding little expertise that contributes to design or construction completion, this fourth-party owner's representative needs to show me more than emotionally detached information passing. If you find yourself in this position, bring something to the party, even if it's just a little courtesy, benefit of the doubt, emotional connection, homework, added value, or collaborative spirit.

Scenario 4: Architectural Principal/Division Manager to Architects

You would think within their own firm, among their own kind, that architects and engineers would know how to motivate designers. But that's not always the case. In this lookback, the supervisor, Bonzo, faced with the daunting task of herding multiple departmental design teams to perform on time and within their fee budgets, was clearly at the right side of the design-management continuum. Having not practiced as a designer, he was skilled at setting limits and using top down commands

(Continued)

to control his teams. One day, Bonzo called Brett into his office to lay down the fee and schedule expectations – to give him his "marching orders."

"As the design lead," he explained, "it's your job to do whatever it takes to get this done on time! We're not losing money on this one!" Using authoritative, militaristic metaphors, in no uncertain terms, he instructed him to "Climb the mountain, take the hill, go to war, all else be damned, to conquer the challenge. Work overtime. No whining. All in! Got it?"

"Yes sir," Brett said, and set off to do as directed. One month later, having given his all working late nights, he had succeeded in taking the hill. Bonzo called Brett into his office once more.

"What are you doing?" he screamed. "Look at all this overtime. You're eating up the man-hour budget and the fee! You've got to stop it now!" His vituperation having been concluded, Brett retreated to lick his wounds, dazed, confused, and directionless.

With no empathy for the situation, no interest in design, and no investigation of the process, nor its issues, obstacles, reasons, or team, this supervisor succeeded in only one thing: demotivating his designer completely. Had he shown even the slightest interest in helping clear the paths or caring about the team's design effort, they would have stayed motivated. His dispassion defused them all.

How to Succeed

If these examples show how *not* to motivate designers, can we offer any ways to do it successfully? Here are a few I've seen work well:

- Conduct an immersive, thought-provoking, aspirational kick off meeting (project analysis.)
- Share the project vision. Why is it important? What is it trying to do? Designers need to know.
- Look at comparable facilities to benchmark likes and dislikes and create shared experiences.
- Bring food and drinks. Keep the energy levels up. Have fun. Include social activities.
- Let the design team "go" in alternating sessions of convening and separating to explore design freely. Check in periodically to pull it back in.
- Review options with cost, schedule, function, aesthetic, and pro/cons. Track your progress.
- Join in. Be willing to get dirty hands. Participate in the mission rather than direct from afar.

Without sharing the "why" and making them "want to," your design team operates with handcuffs on, as technicians, not design professionals. Shun prescriptive mandates. Let your team grow from being laborers to the value-adding, wonder-creating professionals they want to be. Then, back them with the limits, structure, support and checkpoints they need. What if you fail? Reset and adjust. You'll only have missed an interim milestone. If they continue to fail to perform, they've earned the right to be supported differently or replaced. Use these scenarios to shed light on what to do and say to motivate designers.

Inclusion

When do we include more information, diverse perspectives, and people in our decisions and work? The following synopsis draws from a 2018 presentation by inclusion expert Dr. Steve Robbins to Holder Construction Company.

Inclusion, Teams, and Tribes: Ancient and Modern Brains

By now, we know designing and building are team sports. That's good, because, as humans, our DNA has hard-wired us to belong to a group, and to connect. Why? Because it makes us better and helps us survive. But we've evolved beyond basics. Now we have two modes of functioning. Our ancient brain is for survival: it's quick and instinctive. Our modern brain is refined. It's about quality. In doing their jobs, our brains build mental models to make assessments. The first one we learn for survival is: is danger present? Is a tiger waiting nearby to eat me? Does an enemy threaten me? Are you an insider or an outsider? We use superficial cues to decide such things because we're genetically programmed to, and quickly. How you look, how you dress, your stance, what you say, and how you say it are, or could be, clues as to who you might be. Our self-serving survival reaction to new information is to default to the negative and look for variances, to ask: What's in it for me and my tribe? In design and construction, we're conditioned to build and fix. Restated as ancient skills this was: Hunt. Kill. Solve. Get rewarded: Eat. Live. Sometimes we should pause and use our modern brains. That's what good designers, builders, and design managers do.

Seeing and Thinking

Scientists tell us that 30 percent of the brain's neurons are devoted to vision and only 2 percent to hearing. What does that tell us as managers? Look! Collect and process information. Write things down to see and share them better. We can use documented information to keep our distance from potential threats and to have time to reflect, react, and see. Used with the brain, seeing is the most important sense, Together, they have one job: to keep us alive. In the Project Design Control framework, Level 4 controls are our brains, and what they're trying to keep alive are our projects (and our happiness, careers, and personal lives, indirectly.) But in doing so, we face obstacles and distractions. Like noise. Noise can hide things, particularly the signal demanding focus in our heads. It's the manager's job to sort noise from clear, intended signals, so he or she can decide between blocking new information (possible noise) or being open and accepting it.

Open or Closed?

What does being open mean? Asking questions. Being curious. It's a hallmark of creative, empathic managers and leaders. But our brains operate on an efficiency principle. We like to operate in our comfort zone – the known, not the unknown. We have a bias toward the familiar. Bias is neutral – a cognitive shortcut the brain takes to conserve energy or achieve a goal. That sets up an inherent conflict: choosing a shortcut vs. a longer route out of our comfort zones. Because our bias is toward quick decisions, we tend to simplify questions. We have many biases, such as the availability bias: we choose from the options available to us. Familiarity bias says: I just saw x, so I'm thinking about x. (See Daniel Kahneman, *Thinking, Fast and Slow.*) While these biases exist for good reason – to help us process vast amounts of information quickly for survival – they can make it confusing, even dissonant to distinguish from our subconscious biases. Cognitive dissonance occurs when we're "caught between," in "information overload," when things "don't make sense." When that happens we don't like it, so we're motivated to get out of it fast. That leads to closed mindedness. We block the new. To change it, reframe it to explore context, new inputs, and changing conditions. In systems theory, closed systems eventually die. Those that don't adapt to changing environments go into entropy. Open systems take in new inputs and have outputs. They entertain new information and cast off some old.

What to Do?

What should we do to address these genetic traits? Listen more. Take in more information. It's the design manager's job. Look, see, then decide. What does the observational data tell you? What do you see? Collect more data. We need other perspectives to be smarter. Aristotle said, "The measure of a wise person is to entertain new ideas but not necessarily accept them." While it always takes less energy and ego to block new information, the question is: should we? Is *this* the time to stand pat and risk drawing the wrong conclusion? To be more inclusive, be less certain. Be more curious.

✳ Tip: Inclusion Questions

Innovation is change, finding new ways, and problem solving. The more ideas you have, the better your odds of having a good one. When faced with such a decision whether to entertain a new piece of information, resolve a situation, or maximize an interaction with someone outside your tribe, or the framework you're working within, ask three questions:

1. What journey did they walk that led them to their conclusion?
2. What journey did I walk that led me to my conclusion?
3. In shades of gray, why do I think I'm right when I can't prove it?

Finally, here are Dr. Robbins 3 Rs to practice for emotional intelligence and mindful engagement as an inclusive design manager:

1. Recognition: Self-awareness. Are you seeing the new input and assessing correctly?
2. Reflection: Pause to consider options and outcomes. Or your ancient brain will give you one.
3. Response: Does it come from a positive stance? Does it elevate the group and not just myself?

Goals

Throughout this book the goals have been constant:

- To help our industry get better at understanding designers and design process
- To better align our different tools, languages, processes, cultures, biases, and defaults
- To expose issues and actions to effect dramatic change in our disconnected profession
- To share a theoretical model and best practices as best-case Project Design Controls, for applied, tailored use to make your projects better

While the introduction framed issues and opportunities, Part 1, "Perspectives" (the interviews), and Part 2, "Project Design Controls," have been designed to serve these goals. I hope they've succeeded.

Self-Evaluation Quiz: Managing Design Litmus Test

As a final test before you return to your working life and apply – or forget – these principles, give *yourself* the "Managing Design Litmus Test," on your project, or for your firm:

1. Do you have Project Design Controls in place?
   In some form? Written? Can you hand them to a colleague to use? Enforce them?
2. Do they match what they're supposed to be?
   How are you doing? On time? In budget? Good drawings? Strong relationships? Can you point to the approved target and actual values that have been documented?
3. Are they aligned and balanced?
   Do some controls restrict or block others? Maybe good management metrics are in place, but no one trusts them or using them? Perhaps great camaraderie can be seen, but no limits are in place?

Scoring

How did you do? Readers who answered yes to all three questions on all Project Design Controls on all five levels – congratulations! You're managing design. (As well as anyone can.) The rest of you better get busy. You've got work to do.

CHAPTER 24

Case Studies

The following examples illustrate projects that have used Project Design Controls successfully.

CASE STUDY 1: GEORGIA TECH MANUFACTURING RESEARCH CENTER, ATLANTA

Integrating Design and Management

FIGURE CS1.1   Georgia Tech Manufacturing Research Center: *Progressive Architecture* cover; photo: Jonathan Hillyer.

In 1987, I left Heery International to join mentors Terry Sargent and Larry Lord at their new firm, Lord & Sargent. Having successfully tested my mettle as a project designer and project architect, I was ready to tackle the role of project architect/project manager. For my first assignment, I was entrusted to oversee the most significant project in the firm's young history, the Georgia Tech Manufacturing Research Center. This $12 million project was to house research labs for the "factory of the future." Little did I know, it would be the project of a lifetime.

It was my first opportunity to be responsible for "design management" in all aspects, including realizing the design vision, managing the consulting engineers, serving the client, and developing a thoroughly documented, technically sound building, within budget, on schedule, while making money. I took it seriously, having left a fine company to rejoin past mentors Terry Sargent and Larry Lord in their smaller, upstart firm. I staked my career on it.

This project was designed in a simpler time. The last gasps of the postmodern movement were audible. Serving our client involved only pleasing the Georgia Tech Campus architect, the Georgia Board of Regents and their guidelines, and the faculty advisory committee. Because it was a speculative building, there were no users, only a flexible research facility to house funded research. Just three program requirements were proffered, to design a facility that was:

- 100,000 gross square feet
- "Functionally flexible" to allow various configurations to accommodate future research
- Within its $12 million stated cost limitation

This last mandate was no idle request. In the state's design-bid-build contracts, architects were required to "own" the cost estimating responsibility. Based on a spate of overbudget public work, the state had developed prescriptive guidelines to guarantee stewardship of public funds. Architects were no longer allowed to design and hope their projects would bid within budget; they were required to hire cost estimators and submit estimates at every phase, certifying that design was in budget. In the folksy parlance of author, John Sims, "The state of an architect's mind is as much a fact as the Chattahoochee River." No more "guessing" designs to be in budget.

To address this charge, we did cost estimates at each phase. Though we didn't call it this at the time, we used the ScopeDoc technique. Because engineering systems were an integral part of the concept and a significant part of the cost, we cajoled the engineers, Newcomb & Boyd and Pond & Company, into drawing their systems in the early phases, so they could be estimated. We insisted they draw cross-sections to prove their systems would fit.

Led by Terry Sargent, the design expression was genius. Drawing its expression from our Detroit roots, historic references to Ford's Rouge River manufacturing plant melded with a building-as-machine modernist ethos to produce a wealth of visual metaphor. Cogs, wheels, gears, and a brick building skin suggesting computer dot matrix printer paper offered a rich palette. While I was hands-on in drawing, detailing, designing, and leading the team, on the management side I filled my hours with work plans, spreadsheets, schedules, and assigning roles. In hindsight, I was using Project Design Controls.

Amazingly, in the span of a few years, (including an eighteen-month hiatus while the owner cleared property easements) our design process used all available media. Initial concepts began as tracing paper sketches. We refined drawings in pencil on vellum, mylar and mylar lead, and pen and ink. We drew details of worm gears, and machine crafted elements by hand. Large scale physical modeling was enlisted to study exterior form and interior spaces. In a bold move, we showed our technology leadership by using the then new Autodesk AutoCAD software to produce all project plans and elevations – the perfect digital toolset to coordinate the project's mechanistic rhythms. We had some trouble with our technology experiment, given limited CAD drafting talent and dot matrix plotters. Near the project's end, our digital evangelist, Andy Smith, modeled the project in Bentley's Microstation software, producing an early 3D building information model, circa 1989 – 30 years ago.

As design neared completion, we entered it in Progressive Architecture's Design Awards Program for conceptual, unbuilt work. Our submittal included 10-foot-long pen-and-mylar drawings, hand drawn in the evenings, with 12-pack and Led Zeppelin accompaniment – a labor of love. Soon afterward, we were notified that we won!

At the end of one phase, in the throes of cost estimating, I was visibly worried. In my culturally conditioned individual behavior pattern I assumed a defensive position when queried by firm principal Larry Lord about cost. I objected. His reply: "Mike, I'm not trying to put you out on a limb, I'm trying to climb out there with you to help – and bring us both back." I breathed a sigh of relief. We reviewed the cost estimate together, collaboratively, benefitting from his budget experience and my project knowledge. It was one of my first management collaborations – we built the budget together. It made all the difference. It worked.

On bid day, we got lucky. Despite our exhaustive cost estimating, all bids except one exceeded the Stated Cost Limitation (SCL) by 5 to 15 percent. That's all we needed. A local contractor, Barge-Wagener, was inspired by the chance to build this intriguing design at the owner's alma mater. To win the work, they submitted a low bid of $11,995,000, $5,000 less than the SCL. Mercifully, we averted value analysis and redesign.

As construction began, Lord & Sargent merged with a venerable firm, decades-old Aeck Associates to add construction and technical capability. I managed the project during the design phase, but I handed the project off to experienced veterans for completion. Bringing it to reality was a challenge. The state's adversarial hard bid process set the stage for a prolonged battle throughout the ensuing two years of construction. Despite valiant efforts, the joys of design were tempered during the construction administration phase. Lacking the opportunity to work under a collaborative contract the team had only their professionalism and persistence to sustain them. The project remains a highly functional, award-winning accomplishment, one that left its protagonists with too much hair on their chests and too little profit. Since completion, the building has been widely recognized:

- Progressive Architecture 36th Annual Design Awards: Citation, January 1989
- Atlanta Urban Design Commission Award of Excellence, 1992
- AIA Atlanta Tour, 1992
- *Progressive Architecture,* cover story, "Technology Expressed," April 1992
- R&D Magazine Lab of the Year: High Honors, May 1993
- L'Industria Delle Costruzioni, "Centro di recirche ad Atlanta, Georgia," March, 1993
- AIA Georgia Design Honor Award, 1995

✳ Reflection

This project taught us to embrace aspects under our control. Despite its design-bid-build delivery and absence of owner, contractor, and architect synergy during design, we integrated design passion and management rigor to deliver a highly functional, award-winning project, in budget, and on schedule. In a confluence of stellar design, old-time collaboration, a common mission, simpatico team, and luck, we managed design. Although we didn't know them by that name then, we used every one of the Project Design Controls. It's not that easy anymore.

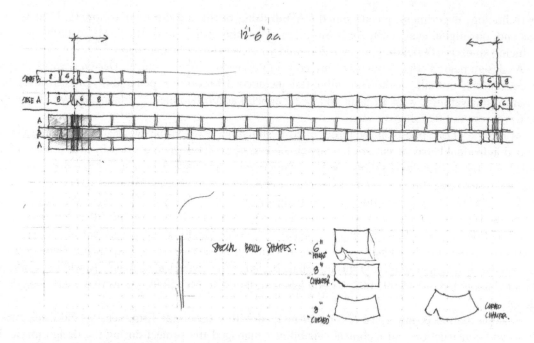

FIGURE CS1.2   Georgia Tech MARC: brick detail; author sketch.

FIGURE CS1.3   Georgia Tech MARC: physical model.

FIGURE CS1.4   Georgia Tech MARC: atrium; photo: Jonathan Hillyer.

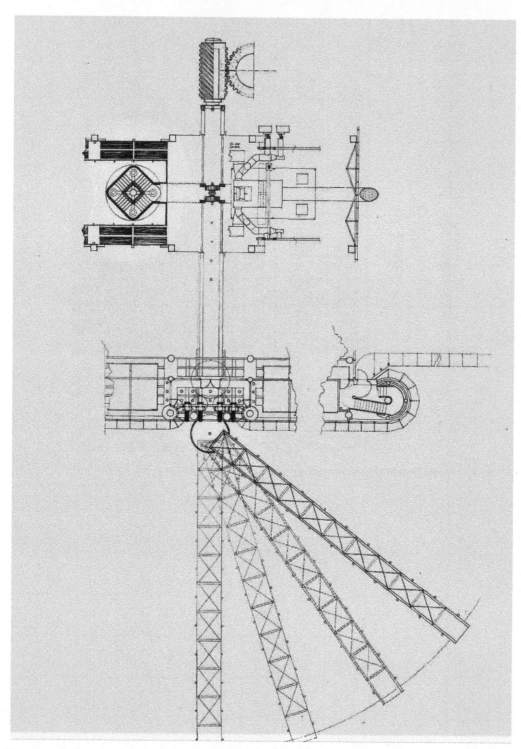

FIGURE CS1.5 Georgia Tech MARC: drawing, "Things That Move."

FIGURE CS1.6   Georgia Tech MARC: entry detail.

FIGURE CS1.7   Georgia Tech MARC: entry bridge.

FIGURE CS1.8   Georgia Tech MARC: entry bridge; photo: Jonathan Hillyer.

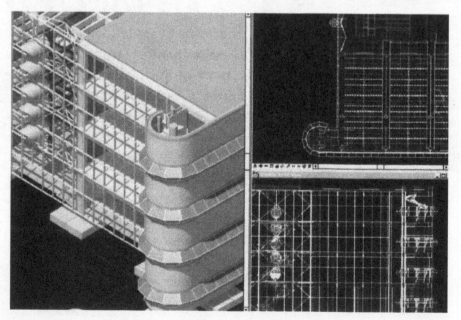

FIGURE CS1.9   Georgia Tech MARC: digital model; Courtesy Lord, Aeck & Sargent, a Katerra Company.

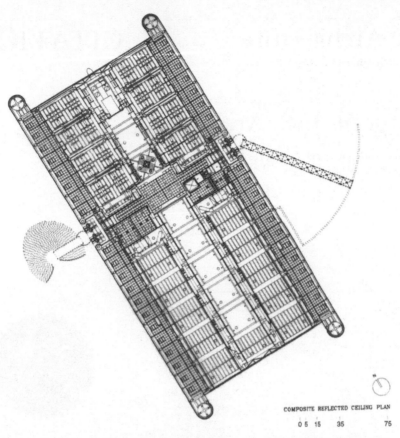

COMPOSITE REFLECTED CEILING PLAN

0 5 15   35   75

FIGURE CS1.10   Georgia Tech MARC: reflected ceiling; Courtesy Lord, Aeck & Sargent, A Katerra Company.

FIGURE CS1.11   Georgia Tech MARC: gear ceiling; photo: Jonathan Hillyer.

Progressive Architecture

The 36th Annual P/A Awards Program

CITATION

Lord & Sargent, Inc., Architects

Manufacturing Research Center, Georgia Institute of
Technology, Atlanta, Georgia

Larry Lord
Terrance E. Sargent
  Principals-in-charge
Michael LeFevre
  Project architect
F. Chip Bullock
David Butler
Allen Duncan
Michael Few
Jimmy Hawkins
David Hendershot
Klaudia Keilholz
Bert Lewars
Linda Seiz
Jack Owens
Valerie Von der Muhll
  Project team

Newcomb & Boyd
  *Mechanical and electrical
  engineers*
Bill Dean
  Engineering project manager

Armour, Cape and Pond, Inc.
  Structural, civil engineers
Charles P. Armour
  Structural engineer

Jack Owens
  Painter

The Board of Regents of the
University System of Georgia
  Client

Frederick D. Branch
  Vice chancellor for facilities
James R. van den Heuvel
  Architect

Georgia Institute of
Technology
  Client

Clyde D. Robbins
  Vice president for facilities
Jack P. Fenwick
  *Director of design and
  construction*
J. Bradley Satterfield
  *Manager of architectural
  services*
Don Alexander
  *Manager of services
  engineering*
Orlando Feorene
  *Chairman faculty advisory
  committee*
Michael E. Thomas
  *Acting director, manufacturing
  research center*

January 1989

FIGURE CS1.12 Georgia Tech MARC: Progressive Architecture Citation.

CASE STUDY 2: ZOO ATLANTA ACTION CONSERVATION RESEARCH CENTER, ATLANTA, GEORGIA

FIGURE CS2.1   Zoo Arc, Lord, Aeck & Sargent, Architects; photo: Jonathan Hillyer.

✳ Reflection

Our freeform environmental concepts made this project's design *almost* unmanageable.

We relied heavily on CM partner, Holder to collaborate in pricing unconventional never-before-built materials and constructing non-rectilinear forms.

Then still on the design side, it gave me an appreciation of just how valuable a good CM partner can be.

Understanding Contractor Processes

Before changing careers, I was Lord, Aeck & Sargent's (LAS) lead architect, project and program manager on a project for Zoo Atlanta in 1996. Years before LEED, this project, designed by Terry Sargent, had a green roof, locally sourced materials, and a coiled, copper-clad, metaphorical "snake" on the roof – and no right angles. It was a challenging design to construct, to say the least. We needed a good partner.

I was thrilled to recommend Holder Construction as construction manager. After concept drawings were complete, I left a late-Friday voice mail message for Holder's preconstruction manager, Doug Hunter:

"Hi Doug, this is Mike, the architect from LAS. We sent you our drawings yesterday. Can you have the cost estimate to us by Tuesday?" The next morning my phone rang. I answered.

"Mike, this is Doug Hunter at Holder. First, thanks for the heads up on this estimate for the zoo. But let me explain the process we go through to do an estimate. First, we have to get your drawings. Then we have to print and distribute them. That takes a few days. Then, we notify and prequalify subs. They need a few days for quantifying, pricing, and contacting material providers and suppliers. After they send their pricing, we analyze and level their proposals and have scope meetings. Then we check quantities, scope gaps, pull our proposal together, review the draft, and do value analysis to get to a plan to get back in budget. Then, we're finally ready to present it to you and the owner. All that typically takes 4 to 6 weeks. We can shorten it but here's what it entails, and what we sacrifice. Plus, we have unique things that have never been done before like copper 'snakes' on the roof, cedar 'curtainwalls,' and 'Coke bottle walls' that will take special attention since there is no precedent."

My response: *"Wow, Doug. I had no idea. Thanks! How much time do you need? 3 weeks. Okay, that will be great. Thanks for educating me. No problem."* His taking the time to politely educate me to his process made all the difference. Rather than call and curse me (which is probably what he wanted to do), he took the high road. He earned my trust. Holder became a great partner in affording and realizing our adventurous design. We couldn't have done it without them.

Scope Meetings, Value Engineering, and Trade Contractor Expertise

After we designed the sloped, curved, naturalistic cedar curtainwalls and sustainable concepts in this project, Holder, as contractor, was challenged to develop cost estimates for design solutions like these. No one had done anything like them before. Coiled copper snakes? Cedar curtainwalls? Stone Mountain granite rubble skin? Recycled Coke bottle partitions? Which database or experience set has the benchmark pricing for those?

To engage me in becoming part of the solution, Holder's preconstruction team invited me, as architect, to sit in their subcontractor scope meetings as we engrossed ourselves in value engineering. This cost reduction process was brought on after my firm's design, done in isolation before CM involvement, led us to be 40 percent over the owner's budget. I was invited to attend trade contractor scope meetings for the first time in my career.

As the meetings unfolded the trades' uncertainty became evident. But after we talked about our intent and I sketched and elaborated on our desired details, the dialogue opened. "You know, we would be perfectly happy with exposed fasteners here," I shared, sketching as I talked. The subcontractors' reactions were welcome: "Heck, I can just shoot a redhead there, that's way less expensive than what we had priced. I think we can cut our number in half. I can just counterbore a hole and power-fasten the base plate, then plug it." "That's perfect," I said, happy not only with direct, honest detailing solution, but also with its accompanying price reduction. Now all we had to do is repeat this process for a good number of the other trades and we'd be in budget. We did.

Summary

Here's the moral of the story: for unusual design work with little precedent, most design documents don't do justice to the information subcontractors need to price their work. The value that the CM brought, including access to their trusted, knowledgeable subs, and the synergy we had meeting face to face while drawing, talking, and sharing interactive feedback was extraordinary and project saving. When we were done, we had better understanding, common language, more information, and tighter pricing. We managed design by talking and working together. Since design and drawings are never "done" by definition, and budgets are always challenging to comply with, wouldn't this be a good way to work on all projects?

Our combined collaborative soft skills, and old-fashioned sketching, measurable against our hard systems of target budget values helped control otherwise separate, nonsynergistic efforts. Bringing them together was not only enlightening, but it rescued the project.

CASE STUDY 3: FLINT RIVERQUARIUM

"Turbo Value Analysis" and "After-the-Fact" Project Design Controls

FIGURE CS3.1   Flint RiverQuarium, Albany, Georgia, Antoine Predock Architects; photo: Tim Hursley.

In 2002, a potential client approached Holder with an opportunity. They had contracted with Antoine Predock Architects for the design of an aquarium and museum to be called the Flint RiverQuarium in Albany, Georgia. In this case, not all Project Design Controls had been set. Level 0 controls, contracts and team assembly" had been established by the owner without our involvement.

No collaborative agreements or incentives were in place. The architect worked for the owner and the CM was to be selected. Design was "complete" to design development level. A $14 million budget had been established, but significant issues had surfaced. Although onsite design visioning sessions had been conducted and a stunning contextual design had been approved, the project cost languished at $21 million dollars, 50 percent over budget. Even after working with two other construction managers, the team had been unable to resolve the budget overage. Two years in, the owner and architect came to us with a desperate challenge: find a way to get the project back in budget while preserving the design – within 2 weeks – and the project would be ours. This example offered a fine opportunity to manage design *after the fact,* in this case, one that had already been developed. Nevertheless, we set about implementing Project Design Controls. Revisiting Level 0 controls, we rebuilt the team to include ourselves. We added key trade contractor teammates. As a Level 1 Project Design Control, we began with a project analysis kickoff meeting to understand the objectives and key design features to be preserved. At Level 2 we reset measurable aspects such as program, scope, budget, schedule, and documents.

Due to the magnitude of the budget overage we used a highly collaborative process to address scope and documents. In a process we called "turbo value analysis" we employed a multidisciplinary team including preconstruction, MEP and field experts, our design partners, key trade contractors, an engaged owner, and me – as architectural interpreter and translator. We produced fast, freehand scope documents and value analysis sketches to convey the revised design intent and extent in quick sketch form – adding to shared understanding and re-pricing certainty. The design team willingly embraced and led the suggested construction options. Contractors and trade contractors respected design tenets.

✳ Reflection

Before this effort, we had never achieved such a dramatic value analysis effort in such a short time. Typically, it would have taken us several times as long to finish such a process. But the flexibility of being able to adapt our toolset, apply discipline, and work concurrently made the difference. In this case, we used Level 4 Project Design Control leadership strategies to focus on the "big picture," most important cost impact items and targeted them first. Being forced to work at order-of-magnitude scale in a hurry allowed us to succeed. It was a new form of balance, and a new way to accommodate change – and it took action.

In a series of intense, working meetings, the team proposed alternate construction means that preserved the design form and concept. The local limestone façade – the concept's essence – was preserved. In lieu of the cast-in-place integral concrete structure proposed we suggested more efficient, faster, more forgiving systems such as steel framing, precast concrete, and standard curtainwall systems in lieu of the custom, time-and-labor-intensive original, as-designed approach. Using live-sketching over design development drawings we elicited immediate feedback from trade contractor experts, and got immediate quantity and cost feedback, rebuilding the cost model.

With all parties present, we eliminated the time lag of preparing sketches, sending them out for pricing, waiting a few weeks, then hoping we had guessed right, only to need more iterations and time. Working concurrently and collaboratively we compressed the process to record levels.

At the end of this effort the owner remarked, "Your firm clearly has a passion for collaboration and preconstruction. We needed some of that." The architect said, "We got more done working with you guys in three days than we did with those other guys for three months."

The lessons in this story? Despite a late start and no input on team configuration or collaborative contracts, we were still able to employ Project Design Controls to rescue this project. Without such a combination of hard metrics and soft relational skills, the project would likely have been shelved. Today it stands as a testament to collaboration, as it has for a decade.

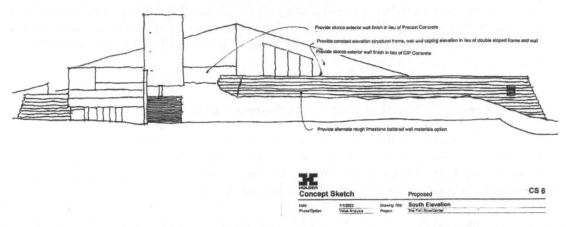

FIGURE CS3.2   Flint RiverQuarium: value analysis concept sketch.

CASE STUDY 4: HAYDEN LIBRARY REINVENTION, ARIZONA STATE UNIVERSITY, TEMPE, ARIZONA

FIGURE CS4.1   ASU Hayden Library; image courtesy Eric Zobrist, Ayers Saint Gross.

> **✳ Reflection**
>
> This project had an enlightened owner. They expected a higher-than-normal designer-contractor collaboration and got it.
>
> Because scope was a shared interest and strategy in early goal setting meetings, I had no reservations about pitching the Scope Doc technique to ASG – even though they didn't work for us. ASG was smart enough to embrace the idea. It resulted in a clear understanding of scope – and tighter pricing.

Design Partner Collaboration: Ayers-Saint-Gross (ASG)

As an example of a design firm that's easy to work with, try this one. This project, intended to reinvigorate an aging central library on the ASU campus and make it accessible to students, centered around project scope knowledge. Design partner Ayers Saint Gross (ASG) joined us and our client, Arizona State University, in using the following strategies to manage design:

"Partnering" Meeting

Before design started, ASG welcomed us to their office to get to know them. We discussed concerns, issues, roles and responsibilities. Not surprisingly, they had a clear strategy with which to approach design. We shared ours. They were aligned.

Scope Documents

To accelerate project knowledge and set up better budgeting, ASG embraced our suggestion to produce scope documents at schematic design. Holder's ScopeDoc whitepaper and sample formats were shared with the architect, their consultants, and the owner early on to educate, and share this state-of the art tool to enhance scope definition. As a result, SD and DD documents included excellent scope definition of design intent and extent to support balanced budgeting. Using our ScopeDoc format and approach, ASG and their consultants produced 204 sheets of schematic design drawings and a detailed outline specification to fully describe the project scope. It established a new high-water mark for SD package content and resulted in a thorough SD phase cost estimate to balance the project. Their effort stands as the benchmark for this practice.

Value and Options Analysis

Partially due to the thorough scope documents that informed the pricing, the schematic estimate exceeded the project budget, which had been set before design at the program stage. Because of the full scope definition, the team's confidence that all scope was included allowed informed, fast value analysis to return the project to budget at schematic stage. The SD budget was held as the cost model and the budget was maintained throughout the project. One additional cost spike arose at the signing of the guaranteed maximum price but was mitigated through contingency and final adjustments. The AE-CM-owner team worked integrally to identify, study, price and select early project options. An outcome was using a central external mechanical room and new stair towers to add efficiencies. Additive alternates and options were clearly documented in the DD documents to give the team packaging alternatives with which to balance the budget.

Data-Driven Building Type Experts

ASG's knowledge, databases, and collection of campus masterplan figure-ground drawings are widely recognized as evidence of their building type expertise for campus projects. The prints are coveted handouts at industry conferences.

Design Scheduling

With only suggestions and construction input from Holder Construction, ASG led the way in developing detailed design schedules. These were shared and celebrated in each monthly design workshop to guide OAC team process understanding and navigation through design. We always knew where we were, what was coming next, and where we were going, because the design team owned, shared, and managed their schedule.

Design Manager

I served in this role through design development to aid in communication between designers and contractors. By empathizing with, translating, and sticking up for the design team's challenges, I earned their trust and maybe helped a bit.

Documents Planning, Management and Review

To ensure quality content suitable for developing early budgets, Holder worked closely with architects Ayers Saint Gross (ASG) to develop sheet lists and deliverable plans to produce thorough document sets, including the work of the consultants. We conducted progress reviews, reviewed documents, and provided comments for design completion status, scope, trade packaging, construction, and interdisciplinary coordination.

Summary

This team's owner, architect, and contractor embraced design management and collaboration from the outset – an example of how attitude, shared goals, positive culture, and tracking tools led to a successful project.

CASE STUDY 5: EMORY UNIVERSITY CAMPUS LIFE CENTER, ATLANTA, GEORGIA

Value Analysis Incorporation and Collaboration – Duda|Paine Architects

FIGURE CS5.1   Emory CLC, image courtesy Duda|Paine Architects, rendering by top*side* front

✳ Reflection

It was a proud moment to see the design team and Holder Construction take ownership of the budget issues and assume responsibility to correct them. Holder's creation and sharing, with ideas from all parties, of the VA tracking tool from prior projects became the roadmap the team would use. This mature, informed approach addressed a team problem with a team solution.

Emory University's goal to replace their Campus Life Center was ambitious but essential. It would require demolition of the existing Dobbs University Center, construction of a temporary dining facility, and significant enabling civil and utilities work all in the center of an active campus. Based on their similar facilities experience, design excellence, and collaborative approach, Duda|Paine was selected as architect with Holder Construction as CM-at Risk for this $69M project.

As design progressed, the number of existing conditions and complexities grew, including extensive utility relocation, phasing of work to accommodate Emory's schedule, site expansion, structural systems analysis, review of finishes and exterior materials, constructability, and sequencing. Cost estimates, despite a close working relationship among the owner-architect-contractor (OAC) team, required value analysis (VA) to mitigate an 8 percent budget overage. The OAC team met in a series of intense meetings.

Difficult decisions were made, and ideas and target reductions were identified for a VA Roadmap of $5.5M. The next challenge was to guarantee this amount would be realized and incorporated. Holder called on their proven value analysis incorporation (VAI) practices, beginning with a process kickoff meeting with Duda|Paine's design principals to develop a VAI Plan. Items were categorized as "ready to incorporate" or "needing study, pricing or owner approval." The entire team presented an 8-week VAI plan and tracking log to Emory, and the design team took ownership of the process. Rather than requiring Holder's oversight, Duda|Paine self-led their process using the Holder-developed VAI tool to track and manage items across all consulting disciplines, producing documentation for Emory's sign-off. Alternates were identified to reduce initial costs including special lighting, finishes, and AV equipment. Reporting progress at weekly meetings, they incorporated all accepted items and alternates into the documents and were able to add most alternates back. The Emory project is a testament to design and construction team collaboration to identify savings and make decisions with one final goal in mind: success!

CHAPTER 25

Actions

What Works

For those consumed by projects and running their firms, without time to read, I've done it for you. This section summarizes the best practices shared by the interviewees. This book is about contemporizing and futureproofing two ancient professions: design and construction. The implication? We've got some work to do. If we want to be valued by our business partners, we need to change how we think and practice. Architecture and construction will always be valued services, not just indulgences for a privileged few. But only if we adapt. The following advice comes from the best experts I could find, and it's summarized to help you do just that.

"Go out on a limb. That's where the fruit is."

– Jimmy Carter

In Search of [Design] Excellence: [Designed and] Built to Last
Emerging Best Practices for Managing Design

Peters and Waterman's book *In Search of Excellence* offers wisdom based on extensive research. In *Built to Last,* Jim Collins and Scott Porras share practices of businesses that have prospered long term. This book's interviews, while a smaller sample size and not the absolute truths proposed by Peters, Waterman, Collins, and Porras, offer many insights.

"Do not fear mistakes. There are none."

– Miles Davis

1. **Strategic Services.** Designers and builders should develop and offer early strategic programming and advisory services to owners. The goal of these services is creating programs, design solutions, and budget parameters geared to shape smarter projects. Owners don't always know what they need; they don't do what we do for a living. Henry Ford said, "If I had asked my customers what they wanted, they'd have said: 'Faster horses.'" Just like Steve Jobs, Bill Gates, and Jeff Bezos, we need to figure out our value and give owners all things they often don't even know they need. (Serrato, Nall)

2. **Collaborate Appropriately.** All owners and projects are not the same. If you're executing a proven solution, it may be time to put your head down and get it done, not have more meetings. Tailor collaboration and communication models to the project. Match fees accordingly. (Moebes)

3. **The Right People.** Recruit and hire graduates who will foster a collaborative culture. Weed out the naysayers. Accentuate the positive. Find individuals with new skill sets. Your firm may not survive if you don't. (Styx, L'Italien, Paine)

4. **Mentor, Teach, and Train.** Ensuring a sustainable future for the industry requires that owners, designers and builders expose their staff to clients and teammates. Collaborative environments will breed life-long collaborators and thinkers. Educators must revise curricula to emphasize collaborative attitudes, behaviors and skills. Excising the loners, skeptics and malcontents will define our destiny. Start now. (Penney, Cheng, Marble, Deutsch, Bernstein, Simpson)

5. **Declare Victory.** Architects can reshape their future by "declaring victory" over old ways and thinking and leading with new mentalities and actions. No more grousing and griping. Out with the skeptics and "lone wolf" persona. The future needs positive, collaborative designers who aren't afraid of transparency and team work. Servant leadership takes courage. Begin by wanting it. Join the firms that have found balance. (Penney, Roberson, Moebes, L'Italien, Williams)

"I am always doing that which I cannot do, in order that I may learn how to do it."

– Pablo Picasso

6. **Be Bold. Be Profitable.** Discover ways to deliver more value with higher margins. Focus on profitability. Seed and fuel new value propositions, process retooling, research, recapitalization, and other steps to adapt, evolve, and survive. Try something different to get better outcomes. Discard archaic practices. Stay calm. (Fano)

7. **Owners Must Catch Up, Lead, and Engage.** Owners too busy to engage get subpar projects and business performance. Join skilled, passionate teams that want to give their all. Treat the process as a partnership. It is! Be willing to partner financially, contractually, and in every way. Renew and retool your skills (e.g. BIM/VDC, current processes) in owner responsibilities. Give direction, look at drawings and models to communicate with us, so we can deliver for you. (Moebes, Bryson)

8. **Support Diversity and Inclusion.** Put women, minorities, younger associates, and diverse views at the forefront. Enable and develop talent. Grow the perspective. (Grandstaff-Rice, Freelon, Williams)

9. **Embrace empowerment, delegation, and self-determination.** (Roberson, Freelon, Cousins)

10. **Redefine Design Excellence.** Demand holistic, operationally sound, cost-effective, high-performance, beautiful solutions. (Dumich, Willis, Williams, Freelon)

11. **Use Project Design Controls.** Build a structure of best practices. Use tools, checklists, and an integrated model to balance projects, including soft skills and tangible baseline metrics to track, adjust, and manage change. Adapt them to your practice. Make them part of your culture. (LeFevre)

Forty Questions

The following questions formed the core of the interviews. [Editorial comments added in brackets.] As question sets were individually tailored to interviewees, not all were asked all questions. Some were reverse-engineered – extracted from interviews – or asked or answered by those who spoke. In closing, I added a few and grouped them. Consider them in shaping your future as a design manager.

To plot *your* path, maybe you should ask these questions of yourself, your firm, and your clients and partners. Imagine what you might do together if you were to agree on the answers! Even if you don't, you'll each know where others are coming from. Maybe then you'll be able to work together better. That would be okay, wouldn't it? A little dialogue?

Management

1. **Oxymoron?** Is managing design an oxymoron? What has your approach to it been? What are you doing to manage design?

2. **Management Culture.** How is/was management addressed within the culture of your firm? Are design and managing design separable, conflicting, or an integrated yin/yang duality? Can design and management be done by the same person, or better served by focused experts? Embraced/shunned, integral/separate? Measured/addressed? Occasionally ignored? Do you offer management training in your firm?

3. **Management Structure.** How is your firm organized? Does it serve your culture and mission?

4. **Firm Size and Scale.** Do the resources of a large global firm help or hurt your process? With increasing globalization and specialization, what's the right firm size? Small/local, mid/regional, aggregated global giant?

5. **Specialization.** How can we cope with specialization? Is there an increasing need for generalists to manage it all?

6. **Risk.** How can we share, manage, and allocate risk and reward? Contracts, attitudes, processes, education? Are those good at these things willing to help partners who aren't? Are integrated practice or new delivery methods the answer?

Planning

7. **Planning.** Why don't we insist on project planning phases? [Who in their right mind thinks we can assemble large disparate groups with no common agenda or protocols, and work at high speeds to create the known from the unknown – then build it – given insufficient time, money

and resources?] Can we fix the too-fast project dilemma? How? Why don't we align goals, build teams, and set protocols before design starts?

8. **Visioning.** How can we get better at visioning, strategy, planning, and succession, to plot a course for change – and the future? Can we ask each other the right questions early enough?

9. **Future Vision.** Have you seen anything new, hopeful, or radical? If you started a project or firm tomorrow, what would you change? What would you do to make it a success? What are we facing next? What we will we see in 5, 10, or 15 years that will be radically different?

10. **Advice.** What advice would you give aspiring student entering the profession today? What are new generations showing us? How are they different from old guard designers and builders?

11. **Schedules.** Do you do design schedules and workplans to manage yourselves? How? Do they work?

Process

12. **Process.** How is design generated these days? Describe your process. Do you still use tools like hand sketches and physical models, or have they been automated by algorithms, artificial intelligence and digital modeling? [If we don't know, how can we manage?] Do you use a different design approach for different client and building types?

13. **Technology.** How is technology affecting your processes and people? What impact have machines had on your approach? BIM, VDC, prefabrication . . . Love 'em? Hate 'em? Unintended consequences? What are you doing about it? Scaling back? Investing more? Prioritizing? Is technology inseparable from process? Has it changed mindsets?

14. **Supply Chain Innovation and Prefabrication.** Are prefabrication and supply chain innovation changing thinking?

15. **Process Reinvention.** Are you seeing any step-skipping, game-changing processes resulting from automation or process reinvention?

16. **Lean.** Can Lean thinking and processes make their way into mainstream design and building?

Stewardship

17. **Stewardship.** How can architects overcome the perception that they're good at spending other people's money, and prone to going over budget? How do you estimate, manage, and control construction cost – the perennial elephant in the room?

18. **Management Strategies.** What strategies can teams employ to manage design upfront (i.e. "preventative design management and maintenance") to avoid the inevitable redesign, descoping, and demoralization that result from current processes?

Understanding

19. **Empathy.** How can we understand one another better? [clients, architects, contractors, one another] What do you need from me? What can I do to work with you better? What's frustrating you these days? Do you invest in building relationships, alliances, and trust with those outside your discipline? What can partners start or stop doing to work with you better?

20. **Alignment.** Can we get better at aligning objectives? [i.e. owners looking for "fast and cheap" should not work with "high design" firms aspiring to produce architecture, and vice versa.]

21. **Client Empathy.** How can service providers – designers, builders, and owner project managers alike, get better at understanding different user types, styles, needs, and cultures: client empathy?

22. **Understanding Design.** Can owners and builders learn to understand design process and that conventional management methods don't apply? Designers like to "do," not "manage." Can that change? Should it? Can partners understand design's nature: nonlinear, explorative, iterative, capable of radical re-redirection, throwing away and starting over at the last minute?

23. **Enabling Design.** How do owner and builder partners lead, enable, empathize with, and empower their designers, versus dictating and micromanaging them – to less effect and value? Can clients understand and collaborate with design partners to maximize collective value?

24. **Communication.** In complex teams spread out across the world, how can we regain the nuanced communication we had before email? [Building by email, pdfs, and RFIs is hardly working.]

25. **Perspective.** What drove you to design? Your career evolution? Share your perspective.

Information, Knowledge, Expertise

26. **Research.** Why don't designers, builders, and owners embrace research, reuse data and feedback? [Despite being the biggest industry, we're the least capitalized, most change resistant – and persist in doing things from scratch, ourselves.] Why do designers pursue one-off solutions? Why not add rigor, share research, get Lean? Can we? Should we?

27. **Contractor Expertise.** Why don't designers capitalize on contractor and trade contractor partners' cost and constructability expertise to inform design, and manage cost and schedule effectiveness?

Change

28. **Diversity and Inclusion.** How are you dealing with diversity, inclusion, and equity in your organization and partnerships, for race, gender, and all types?

29. **Advocacy and Change.** Where are the design profession's advocates and activists? Have we reached the pain point yet? How can those who aspire to be change advocates in this profession succeed? How and where do we begin? Who goes first? [The answer to "when" is *now*!] Courage!

30. **Challenges.** What are your biggest issues and challenges? How are you feeling about our collective profession? Do we need to change? What are the issues around collaboration? [Designing, building, or serving as owner managing these processes?] Share your perspective.

31. **Laggards.** How can the laggards – the change-resistant majority – get motivated and mobilized to change?

32. **Process Change.** When experienced, intelligent, passionate professionals get stuck in an inefficient process [i.e. designing and building most projects], why don't they change it? Or set their process to avoid it in the first place.

33. **Leading from Behind.** Forced to operate from a position of "leading from behind," how can engineers and second- and third-tier consultants cope and "manage" those leading design from above them: architects, owners, contractors?

34. **Owners.** How can building owners catch up, keep up, learn new tools and processes, lead, and get engaged in their projects? [They set overaggressive schedules, fail to plan, can't read drawings or use digital tools, are disengaged, and insist on buying professional services and buildings like first cost commodities rather than long-term investments.]

35. **Owner Leadership.** Are program owners and serial builders leading the way with R&D, process and policy leadership? [They could be, *should* be. Or don't they know how?]

Education

36. **Current Curricula.** How can educational systems increase their change rate and revamp their territorial curricula to keep up?

37. **Soft Skills and Collaboration Culture.** Beyond technical proficiencies, how do next generation practitioners acquire the skills they need (including soft skills, communication, collaboration, strategic planning, and for the business-challenged: business skills?) There isn't enough time in school. How can emerging leaders take these questions to heart and catalyze change? How can we fix the historically closed culture of designers that celebrates ego, individualism and talent over teamwork – and seldom teaches collaboration? Did you learn to collaborate, communicate, or be strategic in architecture school? [Ditto engineering, interiors, landscape architecture, and so on]

Entrepreneurship

38. **Entrepreneurship and Selling.** Can entrepreneurship become part of architectural thinking? Integration, vertical, horizontal, changing the game, and the market? [Traditionally, it's been scorned in the AIA canons, commercialism, etc.] How can architects learn to "sell," explain, defend their designs and speak their client's language?

39. **The Architect's Value.** How can architects change perception to regain their caché, respect, and increase their value? Can they move from contraction and risk avoidance modes? (I hope so. Some are.) Can architects care about making money and get better at it? Where can designers find "courage for the future," add value, and make money?

Design

40. **Architecture, Transdisciplinary, Performative Design.** How can architects reclaim their position as champions for art, beauty and architecture, aesthetic issues – via a new holistic architecture that adds value – energy, sustainability, happiness, productivity, everything? Is transdisciplinary, performative, machine-assisted, quantified design the new normal? Expected? Where does that leave the old guard? What does design mean to you?

My Take

Analysis of these questions and interview themes is revealing. Not surprisingly, in this book, the nature of design, and the challenge of managing it in conventional ways was the most frequent interview topic. It's the book's thesis. Reemphasizing customer focus almost matched it. Advocacy, activism, the need for change, and the impact of machines followed closely behind, as did leadership, knowledge sharing, and curricula change. Cultural alignment, communication and institutional change resistance, and mentoring and self-learning, collaboration, and entrepreneurial shifts were popular topics. Surprisingly, issues regarding finance, value propositions, and the perennial favorite, busting budgets, were less frequent. Issues of equity and diversity, while gaining momentum, were mentioned only a few times. Classic "architectural" concerns such as design excellence, holistic, sustainable, and transdisciplinary design, along with practice size and scale discussions, were also mentioned infrequently, and only by architects. In a mild validation of the book's fundamental issue, the concept of planning and scheduling – key ingredient of managing design – was mentioned by only a few experts, and even that was perhaps due to my prompting. Does this support the core premise that design scheduling is an afterthought?

I share the interviewees' sentiments and the themes they present. I've experienced them. In my time on both sides of the line, I admit to making plenty of mistakes. But instead of accepting the status quo, I've worked to overcome my shortcomings and get out of my comfort zone. There were good reasons for every mistake. Our culture, tradition, DNA, the nature of design, and changing market forces were, and are, reasons enough for the challenges we have faced. But we can change who we are. A hallmark of artists (and builders and owners as well) is that we are never satisfied. But making such change takes awareness, desire, and focusing on the right things. Here are a few key focus areas I have seen be effective.

Where to Focus: Drivers

Effective design management drivers are within reach. In the summary recommendations below, each assumes we're working as an owner-architect-contractor *team*. For reference, I've keyed them to interviewees that spoke to these issues in parentheses.

1. **Budgets.** Set smart cost models. Doing so without team input invites problems. Owners must value team input and use programming, options analysis, feasibility studies, and preplanning phases to set project budgets. Projects shelved and awaiting funding or approval need updated budgets for price escalation, market factors, program objectives, technology, and other issues that may have changed. Better yet, create a new, consensus budget with a team committed to stand behind their numbers. Forcing untested budget numbers on teams perpetuates problems. Use cost control tools to get – and keep – costs in line. All team members must get better at cost consciousness, as they are at home or in their business. (Penney, Paine, Styx, L'Italien)

2. **Schedules.** Plan the schedule. Don't set timelines without team input. Only the expert performing each work aspect knows their steps and how long they'll take. For fast-track schedules, devise a team plan to achieve them. Prefabrication? Modularization? A unitized systems approach? Concurrent, co-located design and construction teams? Plenty is possible with experienced, creative people, but not if they work in isolation. Consider the ideas and tools in this book to create integrated team schedules, with detailed steps and interim dates. Design may be difficult to do linearly, but interim guidelines and guardrails keep teams on track. Build in front-end planning time to set up projects and build teams. (Cousins, Carnegie, Paine, Thomsen)

3. **Team.** Get the right team first. Owners should select designers and builders that get what they're trying to do. If not of like mind you're in for trouble. While participant motivations and mindsets

can range widely, designers can display many mentalities. From service-oriented and repeat-building-type efficiency to cutting-edge design-daring-do, design firms are moved to do different things. Nothing sabotages a project faster than finding out the owner is going for fast and cheap, while their designer wants slow, new, expensive, and risky. How do you find out? Ask! (Thomsen, Williams, Gilmore)

4. **Programs.** Define problems. Help owners see what they need. Define team objectives and update them. Clear paths to allow divergent design explorations to converge into good documents and buildings. Reject consultants who throw up their hands in exasperation and declare, "I can't get decisions!" It's our job to make decisions happen. Get paid for it, since owners struggle to do it themselves. (Moebes, Thomsen)

5. **Design Approach and Scope.** After the right team is assembled, projects that build consensus budgets and schedules need designs and scopes that support them. A project needing fast-tracked packages and tight milestones is not suited to extended exploration of unique solutions. Designers must design their approach to suit the objectives, and it's their partner's jobs to help them. Look for incompatible ideas that stray from the plan and redirect them. Be courageous. Speak up or find a new design partner in synch with what the team is trying to do.

   Plenty of projects start with a program area requirement of x, only to see a design deconstruct it via terraces, balconies, soffits, plazas, atria, bridges, tunnels, forms, and surfaces that couldn't have been contemplated when the program and budget were set. It may be the right design move to destack or dematerialize a box, but when you do, you must do the same to the budget and schedule. Teams must design to "common" goals and limits. Do you really think a building budgeted by a university facility planner using normal comparable costs and assumptions: (i.e. stacked floors, moderately regular shape, historical SF costs) will stand up to being spread all over campus with three times as much roof, footings and skin? (Bryson, Cheng)

6. **Documents, Data, and Technology.** Get smarter. Be forward-thinking about how you'll exchange information. Work offline to develop standards and team sharing practices ready for teams when projects come. Develop the skillsets required for digital design and construction. On a project with a team? Work together to monitor and manage information efficiently. Let's not be the last profession on the planet to capitalize on the vast amounts of data and technology available. (Deutsch, Bernstein, Cheng, Kanner, Wolford)

It's Up to You

That you've read this book, or parts of it, says something: you're aware of your limitations. You want to arm yourself to improve your working relationships with partners. I've provided a lot to absorb, so take what works. Skeptics might say, "We've heard it all before." They may be right, but that doesn't alter the fact that working with designers presents challenges aplenty:

- Each may have a different culture, language, focus, and mindset. (e.g. form, function, materials, sustainability, engineering systems, digital data, and a host of others.)
- Designers are not always under contract to the teammates that bear the daily responsibility to manage them. If they don't work for us, why would they want to listen to us?

- "Nondesigners" (i.e. owners and contractors) stay busy doing their primary jobs. Where in the world are they going to find the time to become practiced at design management skills? If our supervisors don't expect or demand we use them, can't we just leave these design management tools and lectures on the shelf?

Great points. Welcome to the realities of collaboration. Here are more:

- Owners aren't likely to get better at managing designers anytime soon. The responsibility seems to be falling to designers and contractors. (Owners: please prove me wrong!)
- Design partners face growing complexity in practicing their trade.
- Projects aren't likely to get simpler, nor schedules longer. Budgets are always taut.

You do the math. Read the tea leaves. Smell the collaboration coffee. Get with the program or get left behind. That's the real question in managing design: What are *you* going to do about it?

Design Leadership

In a 2017 article in a Fast Company e-newsletter, Michael DuTillo[1] said that design leaders are characterized by four traits:

- Expansive thinking
- Dissatisfaction with the status quo
- A teacher's disposition
- Managing up as well as down (They demand a seat at the table and are willingly assume the responsibilities that come with it.)

I challenge you to enlist each of these characteristics in working collaboratively to manage design.

Good Enough. Fail.

You'll never solve all problems incumbent in the designer-contractor-owner relationship. Using the ideas offered here can help. Fail, early and often. It's how you learn. It's what defines you more than your successes. Try. Take chances. Calculated risks. You'll regret not taking them. Tailor communication strategies to reach those on other teams. You'll get a few things wrong, but that's okay. Ultimately, it's the only way to get it right.

The Ideal Project

This exposition of industry problems and attitudes begs a few last questions: What would you do if you had it your way? Do you have better ideas? If I started a project tomorrow and had the influence, here's what I'd do to set up the ideal project:

---

[1] Fast Company Newsletter: Design Thinking, August 2017.

1. Team Assembly

My first act would be to assemble the team. If I didn't have one I'd worked with before, I'd scour the country for the best, interview them, and conduct a simulated work session to get a feel for their style.

I would *not* select my team based on cost, fee, or qualifications alone. My team is going to pour their guts out working together for a long time. Why would I want to buy them like a commodity based on lowest price or only some of the criteria?

I would *not* select them in a design competition or imply they bring a free design to the interview. That's an abuse of their professional skills and services, and a ridiculous way to start. Not only does it mean they developed a design in isolation, without my input as a client, without getting to know me, but they most likely did it with no construction or cost input – a recipe for certain disaster.

If we truly believe it takes a team, let's not cast the die before we come together. I've done it and it's no fun. The owner falls in love with an ill-conceived design with uninformed budgets and schedules. It's risky. A design competition might work for a sculpture, a landmark or memorial, but a building is more than an object. Get your team together first and set objectives – together.

2. Program, Planning, Predesign and Project Definition Phase

Before design starts, I'd convene my team to select a site, develop a program, set goals, and conduct a Planning Phase for predesign, programming, and project definition. This would allow us to explore options, establish protocols, and define a path to success. We would include issues or sustainability, resilience, and building and human performance as key design determinants. Pro sport leagues have training camps and preseasons, project teams should too. We'd do this as a group. Then, and only then, would we start design. We'll never get a second chance to plan.

3. Project Analysis

After the planning phase, I'd restart the official design phase by conducting a week-long squatter session with all key players in the room. We'd have read a project orientation package in advance, so we could come ready to interact and add value. I'd set an agenda with these steps:

- People (Roles, responsibilities, styles. Who's in the room? What do we do and care about?)
- Project (Facts, givens, requirements, common goals and objectives, budget, schedule, etc.; what we know)
- Possibilities (Exploring options)
- Plan (Workplan and schedule for short and midterm design efforts, and design and construction of the entire project. How will we get there together?)

Why a squatter session or project analysis? Reconvening the team, armed with information from the planning phase, and holding them captive in an energized environment capitalizes on the contributors' synergy and expertise. With the immediate, interactive feedback from being together, great things result. After it's over, I'd send the output to the next wave as the team grows so they can join us in earnest.

4. Team Learning and Growth

I'd be transparent. To make the process fun, I'd advocate on the job learning and growth for participants through intentional project sharing, research and development. Personal and team growth would be explicit goals: sharing useful things gets rewarded, good deeds get done.

5. Place: Co-location: "The Team Room"

The next thing I'd do is set up a team workspace. "Facilities" facilitate. We should have a great collaboration space to work in, big enough for everyone who needs to participate – one big central room with easy access, based on Christopher Alexander's pattern "Central Room with Alcoves" in his book *A Pattern Language: Towns, Buildings, Construction*. To do away with the warring factions, conflicting cultures, and objectives that plague us, I'd give this room a name: "Team-ville," "Expert-ania," or simply "The Team Room." It would be a conflict-resolution zone where collaboration is the rule, and where grace, mercy, and acts of kindness thrive. I've been in these settings. They work. The team prospers. Occasional dustups would be unavoidable, but we'd work them out quickly, in the room. After our day was over, I'd encourage us all to go home to our families and friends to find balance.

This space would create the atmosphere professionals thrive in – as excited to come to work as a child on Christmas morning. I'd do away with the demoralizing, adversarial ways that infect many projects. In that room, I'd set the schedule and budget with input from my expert consultants: enough time and money to do the kind of project we all want, one that will serve as a great place to work for years.

In projects set up that way the team's energy is obvious. They have a smile on their face. Like in a football game, the team that "believes" has a quickness in their step as they break the huddle. They've got "momentum." They charge to the line of scrimmage visibly eager. They feel the team dynamic. Setting that atmosphere is the owner's responsibility. They set the rules of the game. Owners who set adversarial rules waste money on their so-called teams.

The team room would be outfitted with technology, infrastructure, and screens (what we used to call the BIM room or BIG room) to connect remote teammates so they feel like they're "in the room."

On a recent major project, we co-located onsite in huge modular trailers. Each traditional team member was housed in their own: "the architect trailer," "the owner trailer," "the contractor trailer," and "the subcontractor's trailers." It saved travel time, but the continued territorial divides did little to bring us together. The cultures within each building remained separate.

6. Design

As the owner, I'd state my architectural objectives, something like the work of Norman Foster: thoughtful, sophisticated, orderly, integrated, and of the machine age, with regular geometries, and richly detailed proportions and materials to last for the ages. I'd promote a design philosophy tailored to the project's, mine, and our objectives rather than anyone's singular agenda.

7. Budget and Schedule

If I was under pressure by my board of directors to design and build super-fast or cheap, or both, I'd still do the same thing – bring the team in, and with these additional team objectives, derive a budget, schedule, and concept that achieves them. We wouldn't work separately in opposition.

8. Program

I'd spend time on programming and pay for its development. We'd establish performance criteria for all building systems. I'd leave some things open ended to allow exploring options and new ideas. I'd pay my team to explore, because it's my job to tell them what I want. I'd be smarter with their input and would update the program in process, because things change.

9. Process, Communication, and Accountability

As an owner demonstrating leadership, I'd bring "program-developed" protocols, guidelines, and processes to bear, derived from project feedback. I'd have decision-making discipline using logs and templates. We'd record everything that happened automatically, so we could talk, interact, and work without having to stop, politic, write letters, posture, compile RFIs, cause friction, and perpetuate bad processes. I was part of a recent project that agreed meeting face to face was the equivalent of giving legal notice. It worked because it shifted the team's focus to getting things done, not writing letters and covering fannies.

10. Leader Styles

If I was to cast characters in this project play, I'd choose Tom Hanks to be the lead architect. He's smart, affable, and leader enough to create the kind of team it takes to pull great things from all of us. He's got a great sense of humor too. That helps. I'd cast Robin Wright as a team member. She's an example of a smart, strong leader. (Hopefully, she wouldn't play the manipulator role as she does in HBO's *House of Cards*.) She'd probably demand to play the owner. In the old-school adversarial days, Ernest Borgnine would've been cast as the contractor. In our new vision, Denzel Washington would be better. His even-tempered, empathic listening style would keep us all on track. If I couldn't afford this cast, I'd still make sure my team had these leadership qualities. That's my idealistic process for setting up right to manage design of a great project. Idealism aside, this list presents strategies for kicking off a project the right way.

Take Action

If you've read this book, your challenge is to take action. Do something! Use the wisdom offered by these experts. They've made the mistakes. Their life lessons provide strategies to a better future. Use their advice to make long-term change. Use the Project Design Controls framework – the

"We want the Big Ten championship and we're gonna win it as a team. They can throw out all those great backs, and great quarterbacks, and great defensive players, throughout the country and in this conference, but there's gonna be one team that's gonna play solely as a team. No man is more important than the team. No coach is more important than the team. The team, the team, the team, and if we think that way, all of us, everything that you do, you take into consideration what effect does it have on my team? Because you can go into professional football, you can go anywhere you want to play after you leave here. You will never play for a team again.

parts that make sense for *you*. This book has exposed you to a range of outlooks. In the end, there's only one way to manage design: work as a team. Find an ally who shores up your deficiencies. Or let the lone genius of one individual rise to the fore and let the team carry their breakthrough idea over the finish line. Maybe both at the right times. That's for you to decide. You're the leader. The book passes the "baton" to you.

> **Owners:** Lead. Engage with your teams and direct us. We'll serve you better and create better projects. Learn the new tools and processes architects and contractors use to serve you.
>
> **Architects:** Value yourselves. Change your attitudes to be truly team focused. Reinvent yourselves to be profitable to continue practicing in the future.
>
> **Contractors:** Be kindler gentler partners. With your newfound understanding of design, help your teammates help themselves!

If we can each can do these small things, we just might make it. I believe we can!

The Team

When you're working in teams, individualism is for losers and loners. It's not fun. And it's not as powerful or productive as being a part of a team. If you've ever had a good team experience, you know what I mean. Every team needs a leader – a coach. One who had great influence on me was the legendary coach of the University of Michigan football team, the late Glenn "Bo" Schembechler.[2] He led Michigan's team from 1969 through 1989. In a timeless talk before the 1983 Ohio State game, he delivered a memorable oratory. Even if you're not a sports fan, or you're not a fan of Michigan, his comments apply:

A Final Request

If you've made it this far, I ask one thing. In your part as a team member and manager of design, no matter your role, do everything in your power to make it be a team. A team. A team. A team. You'll remember the experience for a long time.

So will "they."

You'll play for a contract. You'll play for this. You'll play for that. You'll play for everything except the team and think what a great thing it is to be a part of something that is, the team. We're gonna win it. We're gonna win the championship again because we're gonna play as team, better than anybody else in this conference, we're gonna play together as a team. We're gonna believe in each other, we're not gonna criticize each other, we're not gonna talk about each other, we're gonna encourage each other. And when we play as a team, when the old season is over, you and I know, it's gonna be Michigan again. Michigan."

---

[2] With apologies to colleagues from competing universities, my passion for the Maize and Blue runs deep. Translate this message to your team, your legendary coach, mentor, project leader, or yourself. It will mean more. Whatever colors the speaker may wear, the words ring true.

Epilogue

Future Vision

COLLABORATIVE CONSTRUCTION

Team Members Seek Ways Out of the Building Modeling Haze

Designers and constructors struggle to craft kinder, gentler business models (Reprinted courtesy of *Engineering News-Record*, copyright © BNP Media, June 5, 2006. All rights reserved.)

By Nadine M. Post, Engineering News Record

> *"We're at the kick-off meeting for a great project. All the players are present. The surfaces of the room are filled with digital screens, caves, images (think Tom Cruise . . . Minority Report). The group can see and access all project data compiled to date to get a jump-start. We begin to use BIM on day one, by capturing goals and expectations, developing and validating program and scope. We gut-check the budget and even begin to generate project and systems options in 3D, 4D and 5D. Everyone on the team sees this integrated data live as we generate it. With immediate consensus and no time lag waiting for meeting notes, we get done the equivalent of an entire phase in a day or two.*
>
> *"After a BIM-enabled social team-building session, we resume the next steps and evolve the BIM through what used to be known as 'phases.' We use it throughout the entire concurrent design, construction and commissioning process.*
>
> *With the time saved from submittals, review and rework, we finish early and under budget. The owner moves in, we deliver the digital asset for facilities management and celebrate with a weekend on the owner's yacht."*
>
> *This particular building information modeling (BIM) utopia is in the mind of Michael LeFevre, Director of Planning & Design Support Services for Holder Construction Co., Atlanta. But LeFevre, an architect, thinks his perfect project scenario may actually happen by 2020.*
>
> *That's because a revolution is beginning in the buildings sector and the catalyst is computer-enabled BIM. The coming kinder, gentler BIM business models are expected to produce better buildings, faster, at lower cost, with fewer claims and less agitation."*

Armed with the radical shifts in attitude that began with the publishing of *Managing Design,* and other industry developments, we joined the growing movement to collaborative teams. Having seen IPD, hyper-tracked, high-performing design-build teams, and the sudden change on projects all over the country, we gelled as an industry to create new synergies. Using the best technology had to offer, with old-fashioned social skills, we built newfound optimism to serve our clients and deliver greatness. With fewer RFIs and enlightened owners, we met our common objectives, and we made money. You see, it's 2020 or 2025. We're seeing clearly now – and a lot has changed.

Life is good.

Is this time travel? Sci-fi? No, it's real. We haven't yet found Shangri-La, but we're managing – to manage design. We're calling our own plays now. Why not you? Managing design is a wonderful thing.

There's a protagonist in this story. It's you. Your fate matters most. Beyond the voices and conversations, the real question is: *what will **you** do differently?* How will you use these tools? My goal has been to help you find your voice and leverage in managing design. Can you make a difference? I think you can.

Prognostications and Advice

I hold these issues to be self-evident after a half-century in design and construction. Others in this book seem to agree.

1. Design

Design is never "done," but can and must be managed. It takes courage, involves risk, and will be pursued by designers long after clients and contractors are satisfied, even if it means losing money. Owners and contractors wishing to engage designers should do so knowingly, and not be surprised by the immutable principles that reside in the minds and culture of those who design. To handle it, set limits and use interim milestones to keep course. It's worth it. You want an incredible result? Step up.

2. Fees and Profitability

Fees and profitability are not enough for design firms under present models. New value propositions are required with owners and contractors willing to support them. If you had a rough time on your last project, consider new incentives to help your design professionals deliver value. Past and current models and compressed schedules have combined to create the current zeitgeist and substandard service and document levels. Faced with these pressures, designers have few alternatives but to do less. Until we figure out a way to pay them more, we'll get what we've always gotten. To owners and contractors who resist, saying, "That's their problem," I argue this: It's *our* problem. For architects to clamber back to a position of value, a collective effort is necessary. The current situation isn't sustainable. When it comes to managing fees, services, profitability, and value we need new approaches. To aspire to more value, teams must not only *want* to be businesslike, but must have deals and contexts that allow it. Alignment of all three has been rare.

3. Time

Owners have reduced project timelines to ridiculously low levels. They're asked to execute facilities in untenably fast cycles. In turn, they ask it of their teams. Projects that used to take a year or two, or more, are being compressed to months. They demand irrevocable fast-track decisions in what used to be slow, thoughtful processes. Machines can only do so much. We would all do well to create timelines with planning, project definition, and team building up front; ultimately, they get the shortest overall schedules. To schedule design, break processes into small, tangible bites. Shorter durations can be corrected more easily. Integrated into overall project objectives, dates can be met. Design can be managed.

4. Budgets

Going over budget is the elephant in the design and construction room. Those determined to address it have a fighting chance. To meet budgets, the shocking method is only one: hope! At best, budgets can be 'managed' via educated dice rolls by hard working teams. Risk, market conditions, and surprises are ever present, and shouldn't surprise experienced teams. We're always at the market's mercy. Every project is first-of-kind. In a project's time and situation, regardless of rigor, we're left to rely on chance. We do our best, then cross our fingers hoping for a favorable result. When it doesn't come, we share the pain of rework. We can reduce that pain by streamlining processes (or pursuing more prosaic designs, but few of us want that).

We can use contingencies and strategies to mitigate damage. Rather than ask why we're "going over budget" again, smart teams expect and plan for it. Managed risk and cost models help too.

Tools and armor can gird us for budget battle but fight we must. In all my years on thousands of projects I've never seen it any other way. Anyone in the business knows it to be true. We have limited budget control at best. We'd better be ready with new attitudes, strong processes, and committed teams.

5. People

People will continue to be cut from different cloths and cultures with different value sets. Owners will scurry about in corporate/institutional business and bureaucratic frenzies. They will have neither time nor knowledge to empathize with designers and contractors using unfamiliar tools and skills. They'll always be too busy dealing with their own crises. Smart contractors and designers will embrace this and develop service models to help them cope. They'll teach them about the lesser-known, seldom encountered, messy processes of design and construction. Empathetic trusted allies will prosper.

Extol your teammates' virtues. Listen to and get to know them. Build a trust culture. In the end, no matter how automated our processes, or how many machines are engaged, design and construction are human acts. Only we have the judgment, creativity, and capacity to oversee creating a first-time facility. Lean in, listen, hear, and do.

6. Prejudice, Insiders, and Outsiders: How We're Wired

Try as we might, we will never rid ourselves of the biases, prejudices, and reactions we have toward those not like us. Why? Because we're *designed* to have them. Each of us relies on mental models and patterns to make it through the day. We're wired that way for survival. We don't have time to process subtleties. If something exhibits similarities to previous patterns, we declare it to be the same and put it into that set. Call it "stereotyping for survival." We like to work in our comfort zone with familiar situations and people. They're safer, easier, and require less energy. Why wouldn't we?

Bias is unstoppable. Its triggers are automatic. While we can't control bias, we *can* control our reactions to it. In projects demanding faster-than-humanly-possible results, how do we react when faced with instant decisions? We go for results. What's lost? The nice-to-have gentler responses and the time it takes to listen, learn, get out of our safe place and engage outsiders.

Brain studies show when we enter a room, we instinctively scan it to assess if danger is present. Is it filled with friends or foes? We're genetically predisposed to do this, lest we be eaten by a tiger or beaten by a rival. There's not enough time between stimulus and response. And it's only getting worse. To wit: the judgment time while texting and driving and listening to Bruno Mars blasting on your car stereo to react to that "idiot" who pulled out in front of you. (You make that rash judgment because you saw his license plate holder proudly displaying the name and colors of your hated, rival university.) The "fool" left you insufficient time, so you resort to shortcuts to cope: kneejerk stereotyping and character judgment, horn honking, finger flipping, bad words, then a stomp on the brakes.

Under pressure on the jobsite, we react to a man wearing a hard hat, spitting tobacco, and swearing. Must be a contractor. Turns out he's the owner. A youngish person wearing all black with circular black glasses? No question – an architect. Wrong again. She's the manufacturer's rep for the porcelain flooring for the lobby.

"I sent you that text three minutes ago with an RFI about the edge-of-slab dimension! Did you get it? Why haven't you responded?" These communications, sent by a trusted colleague, might garner this response: "Slow down. I was in the bathroom, [or on the phone with the client]." Sent by someone who may be an "outsider," they might provoke anger, profanity, or argument as response – because they're not one of "us." We have inherent trust for "insiders." They're "us," members of our tribe. Outsiders carry a greater burden of proof. For our own survival, mind you.

Context matters. Scientists talk about conditional diversity, cognitive interference, cascading associative memory, aging, and increased brain viscosity[1] contributing to clouded judgment. We can't unlearn our experience or biases. What we can do is pause, recognize, and react differently. With practice we can change. To manage the design of our interpersonal relationships we need to know and manage ourselves. Coping with emotional intelligence factors such as vulnerability, empathy, joy, fear, and anger go a long way. When faced with someone or something you don't understand, agree with, or can't categorize into a familiar pattern, consider this observation from F. Scott Fitzgerald:

> *"The true test of intelligence is the ability to hold contradictory ideas in the mind at the same time and still function."*

7. Reactions

Several interviewees waxed rhapsodic about the good old days or the joys of design. Those days are gone. When design euphoria is met by the harsh reality of getting back to budget or needing answers in the field, things can break down. Each act has the potential to set conflict in motion. When teamwork devolves, infuriation results.

Most of us are quick to "pull the trigger" when faced with disappointment. We're predictable. Conditioned by inefficiencies and scarred by past encounters, we lapse quickly into judgment. In doing so, aren't we just as guilty as those we accuse? The owner's rep who throws a fit when the architect designs over budget, the contractor who becomes enraged by late drawings, the architect who bad-mouths the builder for proposing a cost cut? Being part of a team, like other life relationships, means accepting our mates, warts and all. That's not easy.

Experience should help us see these issues coming and head them off. If we try and still fail, can we summon the maturity to suppress knee-jerk reactions? Is lashing out the answer? Do we think emotional

---

[1] Steven Robbins, PhD, Leadership presentation to Holder Construction Company, 2/6/2018.

pique or threats help? The best owners select fanatical collaborators instead of raging lunatics as team-mates. Teams need good "placators," reasoners and connectors who can set aside put differences and biases. Did we learn to do that in school? Like it or not, creating buildings is human. Political. Bipartisan providers should be the "new normal" to supplant past divisiveness. Doing away with self-serving behavior can resolve the conundrum of managing design. Inexplicably, we're still looking for those who have mastered it.

8. Freedom, Voices, Actions

On projects, thrown into the maelstrom of information overload and rapid judgment, we look for acceptance – words, actions, and beliefs in consonance with our own. When we don't find them, in our battle to manage the unmanageable we're faced with a choice: remain enslaved to enmity or choose freedom.

Organizational Systems Thinking: The 7-S+1 Model

For those who seek change in their project, firm, or process, I return to a favorite paradigm. Lest I lapse into a business lecture to an unwilling audience, I'll offer this: search for "McKinsey's 7 S Organizational Effectiveness Model," or its source article, "Structure Is Not Organization," by Robert Waterman, Thomas Peters, and Julien Phillips.[2] In a still-timely-after-38-years stroke of genius, the model describes 7 factors that must align for an organization to function well. They are:

- Structure
- Strategy
- Systems
- Staff
- Skills
- Style
- Shared values

The model has a single premise: for an organization to succeed (or if you want to change it), all 7 factors must align. A change in one affects the others. Use this model to guide your change efforts. You'll be glad you did. The idea is so simple, even architects, engineers, owners, and builders can understand it. Some factors are "hard," objective and measurable (i.e. structure, strategy, systems), and some are "soft," subjective and less measurable (i.e. skills, style, staff). The model's primary author, Tom Peters, says, "Hard is soft. Soft is hard,"[3] pointing out the easy-to-measure, objective nature of the "hard" factors, and the more difficult metrics of the human, "soft" factors. Even if you don't consider yourself a business strategy wonk, if you create buildings, the elegance of this model should inspire you.

---

[2] Robert H. Waterman, Jr., Thomas J. Peters, and Julien R. Phillips, "Structure Is Not Organization," *Business Horizons*, June 1980.

[3] Tom Peters, "A Brief History of the 7-S Model," January 9, 2011, https://tompeters.com/2011/03/a-brief-history-of-the-7-s-mckinsey-7-s-model/.

When faced with a broken process or team, odds are good that one or more of these seven factors are misaligned. Fix them and see if things get better. That's management – and leadership. To apply this model in real world project contexts, I took liberty with the diagram's internal focus and added an eighth "S," the supply network, or context in which organizations operate.

Reach and Closure: Design Futures Council Summit on the Future of Architecture, 2018

To extend this book's breadth beyond U.S. soil and complete the exploration, in October 2018, I attended the Design Futures Council's "Summit on the Future of Architecture." This event, held in Venice October 9–11, 2018, concurrently with the Venice Biennale di Architettura 2018, was attended by 50 industry leaders and guests. Highlights from its notable global cohort of presenters and guests include the following.

Opening remarks from host and DesignIntelligence/Design Futures Council CEO Dave Gilmore, in which he challenged attendees to consider:

- *The accelerating pace of change*
- *The ever-growing specter of irrelevance for architecture firms, and outsider encroachment*
- *Reactive responses versus responsive relationships*
- *The challenge and opportunity to create our futures*

Jim Anderson, OAA, AAA, AIBC, MRAIC, AIA, is Partnership Chair of the multidisciplinary Canadian design practice Dialog. His honest, introspective sharing of notable stories from his firm included:

- *In Canada, the six urban centers are known as VECTOM (Vancouver, Edmonton, Calgary, Toronto, Ottawa, and Montreal, coined by Centre for the Study of Commercial Activity (CSCA) at Ryerson University re: Canada's six largest urban markets with populations over 1 million.*
- *Canadian architectural graduates come from twelve schools of architecture, not all located in these cities. To attract them, firms must offer compelling purpose and culture.*
- *Start with why.*
- *To get better, start keeping score.*
- *A firm intern shared this: "The most important thing I learned this summer was that I need to understand different people's perspectives before I start solving the problem." (During his short tenure he developed an app to provide metrics.)*
- *The firm has developed a community well-being framework. (You can't figure out what a community needs without asking it.)*
- *The firm's new IT leader, Roberta Kowalishan, has a masterful grasp of technology and also of people.*

- *Instead of listing our many design disciplines (e.g. architecture, planning, engineering, interiors, graphics), we simply call ourselves a design practice.*
- *We don't think of one hire, we think, "If this, then that," considering the systems effects this hire could have on the firm.*
- *We enjoy an inherent Canadian optimism (We want to punch above our weight class.)*
- *We take the 100-year view.*
- *A reluctant CEO, thrust into the role by the untimely passing of the firm's managing partner, Jim and colleagues practice collective leadership: "We think in terms of a mosaic: each piece retains its identity, but collectively the group produces more."*

Francis Gallagher, managing principal of HKS London, presented: "View from the U.K." His outlook discussed Brexit, and the growing interest in Socialist candidates by the country's millennials and its resulting potential impact on the U.K.'s global economic strategies and development.

Laura Lee, has taught professional practice and ethics at Carnegie Mellon University for several years and is a leading researcher. She served as Thinker-in-Residence for the Australian Government and is working as a strategy and systems designer for the public and private sector in Europe. In her recent work, her mission is to demonstrate the value of design and the impact of built environment on the quality of life. Her stunning presentation offered Buckminster Fuller-esque systems thinking to point the way for practitioners "In Search of Lost Humanism." She shared these points:

- *Are we relevant, resilient, and responsible as a profession?*
- *Architecture serves only 1 percent of developed work worldwide and is largely seen as a "product."*
- *Current and projected distribution of the world's wealth (the 1 percent versus 99 percent) is also not sustainable.*
- *What must happen for architecture to disruptively evolve from 1 percent to meeting the needs of 10 billion in 2050?*
- *Integrated service, strategic, and systems design must connect to realize meaningful and convergent effects.*
- *We need incremental but clear evolutionary steps, to move from Now to New, and toward a full revolution.*
- *Architecture must develop from mainly private transactional to transition to public transformational value.*
- *Architects need to understand the upstream potential and downstream consequences of their work.*

The profession needs to develop an open, shared **knowledge** platform based on cases which demonstrate the value of collaboration in tackling complex problems for diverse contexts and a range of scales.

Mission-oriented practice-based design science **research** leading to industry **innovation** is becoming a differentiator for value creation in the profession and provides intelligence for evidence-based decision-making.

Radical reform of architecture **education** is urgent to reset the culture of the profession toward the public interest which is transdisciplinary and fully integrates research and practice from day one.

- *Redefining building as commodity to architecture as outcomes and performance is the way forward.*
- *Design's greatest future potential is our ability to operate as a meta-discipline to construct conversations that lead to greater public productivity and prosperity.*

Paul Doherty is CEO of the Digit Group. Over decades, his firm and work have transcended traditional design practice, to master digital data, development, and facility management to position him as a global leader in the Smart cities movement. Now he deploys skills in politics and diplomacy with his digital designer and entrepreneur's toolset to realize change on a global scale. His presentation "The Perspective from China" outlined his work in the Smart Cities initiative and guiding principles including:

- *Energy (e.g. use of piezo electric power generation to produce electricity from people walking on surfaces)*
- *Information Communications Technology*
- *Migration of 8 million people in China within the next 12 months (New cities are required.)*
- *Public Safety and Security*
- *Green/Sustainable Buildings*
- *Water, waste, education, healthcare*
- *Transportation (including transportable buildings and carless cities)*
- *Architecture must break free and emancipate itself from traditional shackles.*

Adrian Parr, Dean of the University of Texas at Arlington's College of Architecture, and UNESCO Chair of Water and Human Settlements, an Australian with global perspective, shared these points:

- *Global wealth inequity is worsening, to irreparable levels. As designers, responsible for physical and social change we bear a transboundary, transgenerational responsibility.*
- *Manfredo Tafuri's book "Architecture and Utopia: Design and Capital Development" (MIT Press, 1976) spoke to the relationship architecture and capitalism.*
- *In continuing to take work on reactive terms, designers are being "fatally complacent." Venture capital firms are using us as tools to achieve their goals; designers overcatering to markets is problematic on a world scale.*
- *In her work to prepare students for their professional futures, she sees typical graduates facing six-figure debt, yet possessed by social responsibility.*
- *The notion of architects returning to roles as leaders or master builders is nostalgic, and no longer possible. But we have historically offered and still offer unique skills: we expand problem bases, redefine problems, and transcend simple solutions. The result can be that we improve peoples' quality of life on a large scale.*
- *How we ask, answer, or reframe the question can reframe and begin to solve problems on a more socially responsible scale.*

Peter MacKeith, Assoc. AIA, Dean of the Fay Jones School of Architecture at the University of Arkansas, shared these thoughts:

- *His mission is to build a culture of collaboration at Arkansas.*
- *Coming into a practice now means giving up ownership of a name, ego, and historic design notions to pursue a higher purpose and selling that our projects and work will make a difference.*

From the University of Cincinnati College of Architecture, former and incoming Deans Robert Probst, FAIA, and Tim Drachna shared, with a rapidly emerging digital future and artificial intelligence on the horizon, the 13 Deans of the University of Cincinnati, discussed and envisioned the future of education and speculated on subjects of greatest importance to teach:

- *Consensus areas included communication, collaboration, empathy, compassion, ethics, trust, and truth.*
- *For impact, designers must learn to act through and in concert with others.*

A few romantic notions persisted. One example was client and founder of the White City Project, Elena Olshanskaya, who is developing new cities in Africa. Her recounting of her comprehensive education as a shoe designer in Russia that included means of production, factory lines, materials, aesthetics, and business rang true in its recalling Bauhaus-ian learn-by-doing pedagogy. But she and we all had to agree that while we share frustration about the growing number of parties and experts now needed on our teams, the fact remains that we have many more kinds of "shoes" to deal with these days in making buildings. And no one of us can know all that is necessary about them all.

In response to an audience question about who will be in charge in these future teams, I shared my perspective as architect-turned-contractor, turned design-construction connector: "As CMs, we don't care who's in charge." If the team is functioning well and succeeding, we don't care whether the owner, designer, contractor, or someone else leads. Of course, every team leads a leader and so does every discipline. Eventually even in alternative, shared-incentive contracts, the collective team needs a leader to lead, report and be accountable. But, in day-to-day work, any of us can rotate into the leader's chair at rapid intervals when we have something of value to offer.

President and CEO of DC Strategies, Barbara Heller, FAIA, reflected on the adverse impact of the insurance industry on design practice, in effect co-opting the profession to risk aversion. I added that, as contractors, we embrace risk and manage it every day – a radically different mindset than most design firms have today. Jeffrey Stouffer, executive vice president of HKS Dallas, reinforced that, too often, architects are seated, or have allowed themselves to be seated, at the "kids table," and must return to the "adult table" through their actions. Darryl Condon, managing principal of HCMA, cautioned architects who believe they are immune to commoditization, urging them to keep the "50-year" view.

Scott Simpson, FAIA, DFC Senior Fellow, moderated a panel discussion with Gary Wheeler, FASID, FIIDA, DFC Senior Fellow, HDR's Interior Design and Global Workplace Strategy Leader, and Dr. Ted Landsmark, Kitty and Michael Dukakis Center for Urban and Regional Policy, and former head of NAAB and ACSA. In this discussion, practitioners' needs were compared against the speculated future abilities of the academy to meet them. Findings included practitioners' needs were compared against the speculated future abilities of the academy to meet them. Findings included:

- *Architecture is how we tell our stories. How we produce it must change. The 1 trillion construction economy is 30 percent inefficient. Forces such as consolidation, scale, integration, speed, AI, robotics, drones, wellness, sustainability, materials science, and so on will have major impacts on the profession.*
- *We need new business models (e.g. royalties, or value-based compensation).*
- *Design education program length should be reevaluated. Its needs will come from practice.*
- *Schools should institute tenure term limits and periodic contract renewals to avoid turf protection and stale curricula and introduce students and interns to design management practices.*
- *Learn to demonstrate outcomes and use metrics instead of guessing and talking to ourselves.*
- *Use more case studies to learn from failure versus approaching design as a "start from scratch" approach.*
- *In collaborative teams, the response time is becoming dramatically shorter. Teams working together in real time are achieving short, fast, informed cycles. This contrasts with their former non-value-added time-lagged, wasteful, call-and-response modes. Speed demands it. Value results from it.*

Dave Gilmore and Jamie Frankel's conference summation and consulting work with more than 50 firms this year included reminders to understand, own and act on the following advice:

- *Business reinvention will be required.*
- *Demonstrable value creation is a prerequisite.*
- *Value exchange starts with want or need.*
- *Financial modeling is a given.*
- *Anecdotal decision making is flawed. (Three data points is not a trend.)*
- *Energy modeling is a new minimum expectation.*
- *Intellectual property and new products should be created, managed, and sold.*
- *Data: Do firms know they have it? Do they harvest it? Do they understand the ascending scale of epistemological knowledge: from data, to information, to knowledge, to understanding, to wisdom, all shared.*
- *Plot a course and reconnect the dots daily to remain flexible, strategic and responsive.*
- *Finds ways to break free of the "project/fee cycle" for revenue. Devise new revenue models that create value while we sleep.*

Continuing

Based on this event and the experiences and opinions shared in this book, closure is not the point. How to continue and how to create new way ways to manage design are the objectives. Managing design has been and will be our purpose in two ways:

First, in the sense of the future of the profession, and how to increase the levels of self-awareness, change-agency, entrepreneurism, and social responsibility within the design professions.

Second, by discovering how to recondition the minds of those who do design and those who manage it, to *want* to work within schedule, budget, while balancing other objectives such as clients, moral and ethical issues, business, competitive cache, and be able to do so.

In that light, a few predictions can be drawn about the future of managing design and the profession:

- Those who are unable to learn to work collaboratively, across disciplines, will be left behind.
- Faster, rapid prototyping and strategic live work and decisions will replace the Diverge|Converge and Group|Individual models (See Figures 19.3 and 19.4.) Cycles will become fast and frequent. "I'll get back to you," or "Let us study that for 3 weeks then price it," will become unacceptable responses. Team are increasingly working live, real time with near-continuous cost estimating.
- A growing multiplicity of roles for designers will proliferate, including:
  - Specialists
  - Generalists
  - Enablers to connect and manage them
  - Most likely, a few vestigial small practitioners will still do small residential and commercial work, possibly by hand and head, without machine assistance.
- At the other extreme, major corporate giants of the digital age are likely to continue to expand their global dominance by expanding into the technologically lagging design and construction industries to acquire and transform firms. Some have already entered the development sector. In doing so, they threaten to relegate non-value-adding providers to being mere tools of production, deployed for the benefit of these more capitalized, larger companies. If the digital giants don't drive such acquisition, large private equity firms likely will. Perhaps both.
- Firms and teams that reject this prospect should act now to create value or be prepared to accept this possible new role as "producer" – someone who acquiesces to needing babysitting or management.
- The essential, invariant skills of architects are likely to continue, including the abilities to:
  - See and think differently.
  - Reframe problems at multiple scales.
  - Think in system terms to connect factors (art, science, technology, and now, business, environmental and political issues). If this growing list of design determinants is real, by definition, the relative weight of purely formal aesthetic design issues must be reduced in favor of balancing the larger list. They must be considered together.)
  - Those who can do these things in new ways wearing multiple hats have the potential to be valuable. We must "keep the magic" that design professionals can bring.

While I can't foretell the future of the profession, I can speculate it will be different. Aggregation, specialization, automation, and atomization all seem inevitable. As warned by some, extinction of some of the design profession's current, non-value-adding, non-technology-enabled forms is possible. Small firms will continue in varying forms, having adapted for survival. Like others of my kind, my pathos is connection. Like other servant leaders, my purpose is to mend the splits, connect, be tolerant, be my best, and help others be theirs. To achieve that, we need not only to be smarter, we need each other.

Constants and Encouragement

In talking to industry leaders I've confirmed a suspicion: designing and building are emotional pursuits. Labors of love. We have issues, but things change. People and attitudes do too. Tolerance, gratitude, and compassion can be learned. While my digging unearthed chronic issues, it also uncovered a community. Few other businesses are made up of such passionate people. Dedication, love of craft, outcome orientations, and a higher calling are common characteristics of designers and builders.

Another constant is connection. While many of us do this work because we love the doing, even brash project managers and lone wolf designers do it for another reason: connection. We're motivated by acceptance and recognition. Even egalitarian or selfless acts, on the surface, seek the positive experience of belonging and self-esteem. In other words, we want to be liked and to feel good. The result? Another continuum: self versus group.

One revelation was unanticipated: the common desire for change. Despite our lingering differences, each interviewee shared a desire to transform the industry. This commonality energized the dialogue. My hope, in sharing this change collective, is that it restores faith and catalyzes action.

I hope I've persuaded you to join me. Now it's your turn. In his foreword, Randy Deutsch describes this book as "like eavesdropping on a stimulating conversation among industry stalwarts at a dinner party." But the party's over. Time to get to work. Let's do what we love: design and build, only better. While it didn't start that way, the book's purpose became to give voice to voices. I started with one – mine. What resulted is a chorus.

What is your story? What can you add to the conversation? What will you do next? Let's bring hubris back to designing and building. Let's come together and manage design. Maybe more expert encouragement will get us going:

"Only those who will risk going too far can possibly find out how far it is possible to go."
> – T.S. Eliot

"What you have to do and the way you have to do it is incredibly simple. Whether you are willing to do it is another matter."
> – Peter Drucker

"A man would do nothing, if he waited until he could do it so well that no one would find fault with what he has done."
> – Cardinal Newman

"I can accept failure. Everybody fails at something. But I can't accept not trying. Fear is an illusion."
> – Michael Jordan

Answers

What are the answers to the book's questions: Can design be managed? Are you managing design? Can we create a better future? My answer to all is yes. For most of us, it involves changing behavior, processes, and our minds. Some will use structured tactics like the Project Design Controls framework. Structure and limits are helpful. Others will use empathy. Migrating to the center can help us understand others and cope. Both seem like good starts compared to uncompromising, extremist positions, and outmoded, inefficient processes, with no controls in place.

Can we change the design and construction continuum? I think so. Engaging the here and now could be the answer, but not by keeping busy and doing things the way we always have. It takes will and intention. Throughout this book I've challenged you to change and given you a framework to apply. You might ask: "What are you – the author – doing about all this?"

"I like to live the life that I sing about – in my song."

– B.B. King

Yes, I'm preaching it, writing it, singing it, and living it. I will be for a while.

Are you with me?

Yes, managing design is an oxymoron, but can be done by those who want to. And that's what makes this business fun. I hope you use these ideas to make it so for those you work with, and that we can continue the conversation.

Good luck.

Acknowledgments

This book is a tribute to many people who have worked to achieve better collaboration between design partners, owners, and contractors, and tried to manage design.

Literati: This book would not have been possible without the seed planted by Margaret Cummins, Editorial Director at Wiley Publishing. Her nurturing of the idea shaped its form and made it better. The expertise and support of Associate Editor Kalli Schultea and Project Editor Blesy Regulas were enabling forces. Rebecca W. E. Edmunds, your coaching, passion for collaboration and keen editorial skills were a serendipity. Your suggestion to broaden the group was game changing. Bob Fisher at DesignIntelligence offered timely guidance. Longtime collaborator Thomas Powers's first read, frank advice, and IPA-fueled edits were invaluable. Talented ubermensch friends Susan Liebeskind, Jeffrey Donnell, and Marcy Louza championed valuable final edits.

Holder Associates: To Tommy Holder, who, in 1997, was willing to take a leap of faith on my idea to join the company to create a new position, thanks for bringing me aboard, and putting up with me for twenty-plus years. To Dave Miller, who supported my right-brained creative approach and balanced it with stellar left-brained operational leadership. Thanks to Mike Kenig, who helped blaze the trail, kicking open doors and clearing obstacles so I could walk in and do my thing. His passion was exhausting. His energy and level of care were contagious. To Wayne Wadsworth, who provided counsel, guidance, and support for the crazy ideas and tools I had over the years, you were an extraordinary mentor, helping to guide me the through the hazards of the constructor's culture when I was ready to leap off the edge. You taught me there are no limits to optimism, positive attitude, and forward thinking. To Beth Lowry, who supported the contrarian creative view I brought, and supported and me and my writing. To Doug Hunter and the preconstruction leadership, thanks for learning to become design managers in your own right and spreading the word. To Jon Lewis, for his field-based, forward-focused superintendent's perspective. To Mike Burnett, David J. Miller, Christina Byrnes, Janie Mills, and other architectural brethren and members of the Holder Planning & Design Support Services department, and Jonathan Platz and the Creative Services team, thanks for helping to share these ideas and being a part of the collaboration cadre. Carry the effort forward nobly!

To Holder Operations leadership, for helping create tools and processes to better our projects. To hundreds of Holder associates who have used these ideas to collaborate with design partners better, thanks! To more who will follow in the footsteps of these teachings and wonder: who was that crazy old fool of an architect, working within a CM firm, who left cryptic scriptures behind on the ancient art and science of collaboration? Use the force! You're fortunate to work for a great company. Make it better!

Mentors and Mentees: To my architectural mentors, Larry Lord, Tony Aeck, Terry Sargent, and Tivadar Balogh, who got out on the limb *with* me, for showing me the way to leadership potential and supporting my research as a student of the game. To Al Balta, Rick Herrmann, and Woody Holman, thanks for giving me plenty of rope.

To the next generation leaders who followed in my footsteps to lead Lord, Aeck & Sargent, John Starr, Joe Greco, and Scott O'Brien, for making me look good.

To longtime Heery colleagues, architect pals who shaped ideas through years of projects and beach-walk industry analysis: Mike Holleman, Scott Dreas, Mike Miller.

Thanks to all interviewees. You gave your time, experience, and wisdom for the same reason: you saw issues and wanted to do something about them. I learned much talking with you. I hope I gave something back and that our work will endure. Thanks for sharing your ideas in our collective quest to improve our industry. Chuck Thomsen, you are an inspiration, still going at 85. Your 1989 books on *Managing Brainpower* were precedents. Randy Deutsch, you influence many. Beverly Willis and Jim Cramer, you are exemplars. Thank you Thom Penney and John Busby, past AIA presidents, and Bob Carnegie, Peter Styx, Marc L'Italien, Jeffrey Paine, Allison Williams, Phil Freelon, Matthew Dumich, Chad Roberson, Renee Cheng, Scott Marble, Phil Bernstein, Dan Nall, Kurt Swensson, John Rapaport, John Lord, David Scognamiglio, Jeremy Moskovitz, Jeff Giglio and Don Davidson, Arol Wolford, David Fano, Josh Kanner, Dave Gilmore, Margaret Serrato, Bruce Cousins, Agatha Kessler, Barbara White Bryson, John Moebes, Art Frazier, Simon Joaquin Clopton, and Emily Grandstaff-Rice.

No end of thanks goes to the many customers who have embraced Holder's approach to design and construction collaboration while entrusting us to build their buildings over the years. Without trusted partners like you there would be no stories or case studies to share. Design partners welcomed our involvement and shaped these practices. Thank you all for your friendship and business.

Thanks to new friends made during these interviews and architectural colleagues from great firms across the country who tolerated attempts to "enable" their design processes and bridge the architect/contractor gap – your willingness to listen and try new things made our projects better. Go forth and collaborate!

To my wife, Anita, who has put up with my idiosyncrasies for five decades of being an architect. This despite her engineer father telling her, "Oh God, honey, all those guys are fruitcakes!" To Danielle, Char, Mom, and Dad, thanks for the support only a family can give.

Finally, to practice what I preach, a heartfelt thank you to all readers who have given time and attention in joining me on this journey – one that has been far more sociological, inclusive, and rewarding than ever I imagined.

About the Author

Michael Alan LeFevre, FAIA, NCARB, LEED AP BD+C, is Vice President, Planning & Design Support at Holder Construction Company (Atlanta, Charlotte, Dallas, Denver, Phoenix, San Jose, Washington, D.C.) Holder provides clients with construction management services through a team approach and has annual revenues of $3.5 billion. He is responsible for company-wide design collaboration, developing new systems, education, and processes for working with design partners, trade contractors, and owners.

Le Fevre holds a Master's in Architecture degree (with High Distinction) from the University of Michigan (1977) and is a winner of awards for his work in design, collaboration, and technology, including the AGC "Best Information Technology Solutions" (2007), and an AIA BIM Award (2006). He is a two-time winner of *Progressive Architecture* Design Award Citations (1972, 1989), and winner of High Honors in *R&D* magazine's "Lab of the Year" in 1993. He was formerly a principal at Lord Aeck Sargent Architects, Project Designer/Project Architect with Heery International, adjunct instructor in Architecture at the University of Michigan's Taubman College of Architecture & Urban Planning, associate design director at Herrmann & Holman, and intern architect with Tivadar Balogh.

In 2012, he was elevated to the American Institute of Architects' College of Fellows for his contributions in "advancing the practice of Architecture."

A member of the 2009 AIA Technology in Architectural Practice national BIM Awards Jury, he is a top-rated keynote speaker and has presented on Design Collaboration and Building Information Modeling to over 150 groups, conferences, universities, and companies, nationally and internationally, including the AIA, AGC, McGraw-Hill/ENR Megatrends that Matter, AIA Future of Professional Practice Conference, Ontario Association of Architects, Construction Specifications Institute, Society for College and University Planning, BIM4Builders Conference, McGraw-Hill/ACE Arizona BIM Conference, KA Connect, University of Southern California School of Architecture, SCAD, and others. He is a past Industry Advisory Board member at Georgia Institute of Technology, Kennesaw State, and the University of Michigan's Taubman College of Architecture and Urban Planning's Alumni Board of Governors and a member of the Design Futures Council. Recent design management efforts include a new corporate campus on the West Coast and the Mercedes-Benz Stadium in Atlanta.

A frequently cited resource, he has written cover story features and articles in AIA publications, *DesignIntelligence*, *Design-Build Dateline*, and *National Journal of Building Information Modeling*. He was a contributing author to *Lessons from the Future*, by James Cramer and Scott Simpson, Wren Press, 2018. In this debut book with Wiley Publishing, he shares industry perspectives, experience, and principles to catalyze industry change and improve collaboration.

To continue the discussion, contact the author at lefevremichael1@gmail.com, on LinkedIn, or visit the book's website at www:Wiley.com/managingdesign.

Bibliography

The following references may help you understand how to work with and manage designers.

Alexander, Christopher, Sara Ishikawa, and Murray Silverstein. *A Pattern Language: Towns, Buildings, Construction*. Oxford University Press, 1977.

Allison, M., H. Ashcraft, et al. *Integrated Project Delivery: An Action Guide for Leaders*. Charles Pankow Foundation, 2018.

Ashcraft, Howard. *Integrating Project Delivery*. Wiley, 2017.

Austin, S. et al. *Design Chains: A Handbook for Integrated Collaborative Design*. Loughborough University, 2001.

Bernstein, Phil. *Re-integrating Architecture*. Birkhauser, 2018.

Best, Kathryn. *Design Management: Managing Design Strategy, Process and Implementation*. Bloomsbury, 2015.

Broshar, Strong, Friedman, et al. *AIA Report on Integrated Practice*. AIA, 2006.

Brown, Tim. *Change by Design: How Design Thinking Transforms Organizations and Inspires Innovation*. Harper Business, 2009.

Bryson, Barbara White, and Canan Yetmen. *The Owner's Dilemma*. Ostberg Press, 2010.

Boyd, Danah. *It's Complicated: The Social Lives of Networked Teens*. Yale University Press, 2014.

Camus, Albert. *The Myth of Sisyphus and Other Essays*. Vintage/Random House, 1955.

Charles Pankow Foundation. *Professional's Guide to Managing the Design Phase of a Design-Build Project*. Charles Pankow Foundation, 2014.

Clayton M. *The Innovator's Dilemma*. Harvard Business School Press, 1997.

Construction Specifications Institute. *CSI MasterFormat*. 2004 edition.

Collins, Jim, and Jerry I. Porras. *Built to Last*. Collins, 1994.

Cramer, James. *Design+Enterprise*. Greenway Consulting/Ostberg 2002.

Cramer, James, and Scott Simpson. *The Next Architect: A New Twist on the Future of Design*. Ostberg, 2004.

Cramer, James, and Scott Simpson. *How Firms Succeed: Field Guide to Design Management*. Ostberg, 2007.

Cramer, James, and Scott Simpson. *Lessons from the Future*. Wren Press, 2018.

Cuff, Dana. *Architecture: The Story of Practice*. MIT Press, 1992.

Construction Users Roundtable (CURT) Whitepaper 1202. "Collaboration Integrated Information and the Project Life Cycle in Building Design, Construction and Operation," August 2004.

deBono, Edward. *Six Thinking Hats*. Little, Brown & Co., 1984.

Delong, Thomas J., John J. Gabarro, Robert J. Lees. *When Professionals Have to Lead: A New Model for High Performance*. Harvard Business School Press, 2007.

Depree, Max. *Leadership is an Art*. Dell, 1990.

Deutsch, Randy. BIM & Integrated Design: Strategies for Architectural Practice; John Wiley & Sons, 2015.

Deutsch, Randy. Data-Driven Design and Construction; John Wiley & Sons, 2015.

Deutsch, Randy. Convergence: The Redesign of Design; John Wiley & Sons, 2017.

Deutsch, Randy. *Superusers: An Architect's Guide to the Coming Tech Transformation*. Routledge, 2019.

Duffy, Francis. *Architectural Knowledge: The Idea of a Profession*. E&FNSpon/Routledge, 1998.

Emmitt, Stephen. *Design Management for Architects*. Wiley/Blackwell 2007.

Florman, Samuel C. *The Existential Pleasures of Engineering*. Thomas Dunne, 1996.

Foer, Joshua. *Moonwalking with Einstein*. Penguin Press, 2011.

Frederick, Matthew. *101 Things I Learned in Architecture School*. MIT Press, 2009.

Friedman, Daniel, Phil Bernstein, Joe Burns, et al. *Goat Rodeo: Practicing Built Environments*. Fried Fish Publishing, 2016.

Gates, William. *Business at the Speed of Thought*. Warner Books, 1999.

Gensler, M. Arthur. *Art's Principles*. Wilson Lafferty, 2015.

George, Bill. *Discover Your True North: Becoming an Authentic Leader*. Wiley, 2015.

Goleman, Daniel. *Emotional Intelligence*. Bantam Books, 1995.

Gutman, Robert. *Architectural Practice: A Critical View*. Princeton Architectural Press, 1988.

Heery, George. *Time, Cost and Architecture*. McGraw-Hill, 1975.

Holston, David. *The Strategic Designer: Tools and Techniques for Managing the Design Process*. How Books, 2011.

Jenkins, Henry, Mizoko Ito. and Danah Boyd. *Participatory Culture in a Networked Era*. Polity, 2016.

Jones, J. Chris. *Design Methods*. Wiley, 1970.

Kelley, Tom, and Jonathan Littman. *The Art of Innovation*. Doubleday, 2001.

Kieran, Stephen, and James Timberlake. *Refabricating Architecture: How Manufacturing Technologies Are Poised to Transform Building Construction*. McGraw Hill, 2004.

Kahneman, Daniel. *Thinking, Fast and Slow*. Farrar, Straus and Giroux, 2011.

Kenig, Michael E.; *Project Delivery Systems*. AGC of America, 2011.

King, Ross; Brunellschi's Dome: *How a Renaissance Genius Reinvented Architecture*. Penguin Books, 2000.

Kostof, Spiro. *The Architect*. Oxford University Press, 1977.

Lama, Dalai, Desmond Tutu, and Douglas Abrams. *The Book of Joy*. Avery Press, 2016.

Lamott, Ann. *Hallelujah Anyway: Rediscovering Mercy*. Riverside Books, 2017.

Lawson, Bryan, and Kees Dorst. *Design Expertise*. Architectural Press Elsevier, 2009.

LeFevre, Michael. "No BIM for You: The Case for Not Doing BIM? Leverage Points, Reframing and Key Decision Factors in BIM." *National Building Sciences Journal of Building Information Modeling*, 2011.

LeGuin, Ursula. *No Time to Spare: Thinking About What Matters*. Houghton Mifflin Harcourt, 2017.

Lepatner, Barry. *Broken Buildings, Busted Budgets*. University of Chicago Press, 2007.

Lesser, Wendy. *You Say to Brick: The Life of Louis Kahn*. Farrar, Strauss and Giroux, 2017.

Lewis, Roger K. *Architect? A Candid Guide to the Profession*. MIT Press, 2013.

Maligrave, Harry Francis. *The Architect's Brain*. Wiley-Blackwell, 2011.

Martin, Roger. *The Design of Business: Why Design Thinking is the Next Competitive Advantage*. Harvard Business School Press, 2009.

May, Rollo. *The Courage to Create*. Bantam Books, 1975.

McDonough, William, and Michael Braungart. *Cradle to Cradle: Remaking the Way We Make Things*. North Point Press, 2002.

Meadows, Donella; *Leverage Points: Places to Intervene in a System*. Sustainability Institute, 1999.

Peters, Thomas J., Robert H. Waterman. *In Search of Excellence*. Warner Books, 1982.

Peters, Thomas, J. "A Brief History of the 7-S Model." January 9, 2011, https://tompeters.com/2011/03/a-brief-history-of-the-7-s-mckinsey-7-s-model/.

Pfammeter, Ulrich. *The Making of the Modern Architect and Engineer*. Birkhauser, 2000.

Pink, Daniel. *A Whole New Mind*. Riverhead Books, 2005.

Post, Nadine M. "Paradigm Shifting: Digital Modeling Mania Upends Entire Building Team." *Eng. News Record*/McGraw-Hill, June 2006.

Rand, Ayn. *The Fountainhead*. Signet, 1943.

Rowe, Peter G. *Design Thinking*. MIT Press, 1987.

Schrage, Michael. Serious Play: How The World's Best Companies Simulate to Innovate, Harvard Business School Press 2000.

Sperber, Esther. "To End Abuse in Architecture, Start with the Lone Wolf Myth." *Architect Newswire,* August 3, 2018.

Susskind, Richard, and Daniel Susskind. *The Future of the Professions: How Technology Will Transform the Work of Human Experts*. Oxford University Press, 2015.

Thomsen, Chuck. *Managing Brainpower*. AIA Press, 1989.

Thomsen, Chuck. *Program Management*. CMAA, 2008.

Toker, Franklin. *Fallingwater Rising*. Alfred A. Knopf, 2004.

Venturi, Robert. *Complexity and Contradiction in Architecture*. Museum of Modern Art, 1966.

Vedantum, Shankar. *The Hidden Brain*. Spiegel & Grau, 2010.

Waterman, Robert H., Thomas J. Peters, Julien R. Philips. "Structure Is Not Organization" *Business Horizons,* June 1980.

Photo Credits

Figure CS1.1: Georgia Tech Manufacturing Research Center, Entry p. 299, Jonathan Hillyer

Figure CS1.2: Georgia Tech Manufacturing Research Center, Brick Detail, p. 302

Figure CS1.3: Georgia Tech Manufacturing Research Center, Physical Model, p. 302

Figure CS1.4: Georgia Tech Manufacturing Research Center, Atrium, p. 303, Jonathan Hillyer

Figure CS1.5: Georgia Tech Manufacturing Research Center, Things That Move, p. 304

Figure CS1.6: Georgia Tech Manufacturing Research Center, Entry Detail, p. 305

Figure CS1.7: Georgia Tech Manufacturing Research Center, Entry Bridge, p. 305

Figure CS1.8: Georgia Tech Manufacturing Research Center, Entry Bridge, p. 306, Jonathan Hillyer

Figure CS1.9: Georgia Tech Manufacturing Research Center, Digital Model, p. 306, Lord, Aeck & Sargent, a Katerra Company

Figure CS1.10: Georgia Tech Manufacturing Research Center, Ceiling, p. 307, Lord, Aeck & Sargent, a Katerra Company

Figure CS1.11: Georgia Tech MARC, Gear Ceiling, p. 307, Jonathan Hillyer

Figure CS1.12: Georgia Tech MARC Progressive Architecture Citation, p. 308

Figure CS2.1: Zoo Atlanta ARC, p. 309, Jonathan Hillyer

Figure CS3.1: Flint RiverQuarium, p. 311, Tim Hursley

Figure CS3.2: Flint RiverQuarium, Value Analysis Concept Sketch, p. 313

Figure CS4.1: ASU Hayden Library, p. 313, Ayers Saint Gross

Figure CS5.1: Emory University CLC, p. 316, Duda|Paine, top*side* front

Interview Photos

Agatha Kessler photo p. 66 © Chris Humphreys
Allison Grace Williams photo p. 134 courtesy Allison Williams
Arol Wolford photo p. 175 courtesy Arol Wolford
Arthur Frazier III photo p. 29 by Stan Kaady, Photographer
Barbara White Bryson photo p. 19 courtesy Barbara Bryson
Beverly Willis photo p. 15 courtesy Beverly Willis
Bob Carnegie photo p. 105 courtesy HOK
Bruce Cousins photo p. 91 courtesy Bruce Cousins
Casey Robb photo p. 180 courtesy Casey Robb
Chad Roberson photo p. 96 © Warner Photography Inc.
Chuck Thomsen photo p. 9 courtesy Chuck Thomsen
Dan Nall photo p. 139 courtesy Dan Nall
David Gilmore photo p. 81 courtesy David Gilmore
David Fano photo p. 197 courtesy WeWork
Don Davidson photo p. 161 courtesy Don Davidson
Emily Grandstaff-Rice photo p. 205 courtesy Arrowstreet
Jeff Giglio photo p. 161 courtesy Jeff Giglio
Jeffrey Paine photo p. 117 by Lindsay Vigue Photography
Jeffrey Paine staff photo p. 117 by Patrick Davison

Illustrations

Figures

Tables

Index

HISTOIRE DE L'ÉGLISE

****
******
****
**
*

HISTOIRE DE L'ÉGLISE
DEPUIS LES ORIGINES JUSQU'A NOS JOURS

FONDÉE PAR AUGUSTIN FLICHE ET VICTOR MARTIN

21

Le pontificat de Pie IX

(1846-1878)

par

R. AUBERT

Professeur au grand Séminaire de Malines

BLOUD & GAY

1952

A NOS LECTEURS

Le 19 novembre 1951, Augustin Fliche s'est éteint à Montpellier où il était né 67 ans auparavant, jour pour jour.

A aucun moment de nos longues années de collaboration, la crainte ne nous avait effleurés qu'il pourrait ne pas voir l'achèvement de sa monumentale Histoire de l'Église. *Comment aurait-on associé l'idée de défaillance physique à l'image d'un homme chez qui le goût de la vie et une santé joyeuse paraissaient être une sorte d'apanage familial ! Pourtant, en moins de cinq mois, la maladie a eu raison d'une si exceptionnelle vitalité.*

Augustin Fliche a vu venir la mort en chrétien. Ceux qui l'ont approché au cours de ces derniers mois ont été frappés par son courage, sa sérénité. Il se tenait prêt, avec des lueurs d'un tenace espoir.

*L'*Histoire de l'Église *dont il a conçu le plan, recruté les collaborateurs, dirigé l'exécution, lui paraissait la grande œuvre de sa vie : la synthèse de sa foi et de sa science, un témoignage d'une probité entière et l'affirmation d'une surnaturelle espérance.*

Sa disparition nous frappe tous, éditeurs, collaborateurs, lecteurs, élèves, amis. L'équipe qu'il a réunie achèvera l'œuvre commencée, dans l'esprit de son fondateur : la vie et l'histoire de l'Église, c'est la vie et l'histoire mêmes de toute la société chrétienne. Son successeur, secondé par son co-directeur et son ami, l'abbé Jarry, veillera avec nous sur la dernière étape des travaux et les mènera à leur terme.

LA LIBRAIRIE BLOUD ET GAY

BIBLIOGRAPHIE GÉNÉRALE [1]

I. — Sources.

1) Actes du Saint-Siège :

Acta Pii IX, Pontificis Maximi, t. I (1848-54), t. II (1855-57), t. III (1858-64), t. IV (1865-68), t. V (1869-71), t. VI (1871-74), t. VII (1875-78), Rome, s. d., à compléter par *Recueil des allocutions consistoriales, encycliques et autres lettres apostoliques des souverains pontifes citées dans l'encyclique et le « Syllabus » du 8 décembre 1864,* Paris, 1865 ; Marconi, *La parola di Pio IX ovvero discorsi e detti di S. Santità,* Gênes, 1864 ; P. de Franciscis, *Discorsi del S. P. Pio IX, pronunziati in Vaticano dal principio della sua prigionia,* Rome, 1872-1876, 4 vol.

Pour les documents émanant des Congrégations romaines, A. de Roskovany, *Romanus Pontifex tamquam primas Ecclesiae et princeps civilis e monumentis omnium saeculorum demonstratus,* t. VI à XI, Nitre, 1867-1876 ; Nussi, *Conventiones de rebus ecclesiasticis inter S. Sedem et civilem potestatem,* Mayence, 1870 ; *Jus Pontificum S. Congregationis de Propaganda Fide,* t. VI et VII, Rome, 1894-1909 ; et les *Acta Sanctae Sedis,* annuels, à partir de 1865.

2) Concordats :

E. Muench, *Vollständige Sammlung alter und neuer Konkordate,* Leipzig, 1859 ; A. Mercati, *Raccolta di Concordati su materie ecclesiastiche tra la S. Sede e le autorità civili,* Rome, 1919.

3) Conciles provinciaux :

Collectio lacensis. Acta et decreta S. Conciliorum recentiorum, t. II (patriarcats orientaux), t. III (Amérique du Nord et Empire britannique), t. IV (France), t. V (Allemagne, Autriche-Hongrie et Pays-Bas), t. VI (Italie et Amérique latine), Fribourg-en-Br., 1876-1890. Voir aussi Mansi, *Sacrorum conciliorum nova et amplissima collectio,* t. XL-XLVIII (édit. Petit et Martin), Leipzig, 1909-1922.

4) Autres sources concernant les Églises particulières :

On trouvera un certain nombre de documents dans l'ordre chronologique dans Chantrel-Chamard, *Annales ecclésiastiques,* Paris, 1887-1893, mais la

---

(1) Nous donnons ici la liste des signes et abréviations utilisés dans les notes :

Ann. É. Sc. P.	*Annales de l'École libre des Sciences politiques* (Paris).
Civ. catt.	*Civiltà cattolica* (Rome).
Coll. Lac.	*Collectio Lacensis. Acta et decreta conciliorum recentium.*
D. A. C. L.	*Dictionnaire d'archéologie chrétienne et de liturgie.*
Denz.	Denzinger, *Enchiridium Symbolorum...,* 18e éd., Fribourg-en-Br., 1932.
D. H. G. E.	*Dictionnaire d'Histoire et de Géographie ecclésiastiques.*
D. T. C.	*Dictionnaire de Théologie catholique.*
H.-P. Bl.	*Historisch-Politische Blätter* (Munich).
Rass. St. R.	*Rassegna storica del Risorgimento* (Rome).
R. H. E.	*Revue d'histoire ecclésiastique* (Louvain).
R. H. É. Fr.	*Revue d'Histoire de l'Église de France* (Paris).
Riv. St. C. It.	*Rivista di Storia della Chiesa in Italia* (Rome).
T. Q. S.	*Theologische Quartalschrift* (Tubingue).

Les principaux dépôts d'archives utilisés sont désignés comme suit :

Arch. Mal.	Archives de l'archevêché de Malines.
Arch. Min. A. É. Brux.	Archives du Ministère des Affaires Étrangères de Bruxelles.
Arch. Min. A. É. Paris	Archives du Ministère des Affaires Étrangères de Paris.
Arch. Nat.	Archives Nationales, à Paris.
Arch. S. Sulp.	Archives du Séminaire Saint-Sulpice, à Paris.

plupart des sources sont encore inédites ou du moins n'ont pas été réunies en recueils commodes (par exemple, les mandements épiscopaux). On tiendra compte du fait que certains travaux, presque contemporains des événements, doivent être à certains égards considérés comme des sources, et que, d'autre part, les travaux récents, notamment les biographies, contiennent souvent de nombreux extraits de sources inédites.

Les principales sources imprimées seront indiquées en tête des différents chapitres. Nous donnerons ici quelques indications plus détaillées relatives aux :

5) Sources concernant l'histoire de l'Église de France :

Outre les archives du ministère des Cultes, bien exploitées dans les ouvrages indiqués ci-dessous de Ch. Pouthas et J. Maurain, il faut surtout consulter :

a) les mandements des évêques, surtout de Dupanloup, Darboy, Pie, Plantier et Régnier (recueil commode des « plus remarquables mandements ou discours de la plupart de nos seigneurs archevêques ou évêques de France, de Savoie et de Belgique » dans la IIe Série des *Orateurs sacrés* de Migne, Paris, 1856-1866), et leur correspondance, généralement inédite (voir toutefois *Lettres choisies de Mgr Dupanloup*, édit. F. Lagrange, Paris, 1888, 2 vol. et *Correspondance du cardinal Pie et de Mgr Cousseau*, édit. Hiou, Paris, 1894), mais dont on trouve de nombreuses citations dans les biographies ;

b) les correspondances de L. Veuillot (Paris, 1931-1932, 12 vol.), d'A. Cochin (Paris, 1926, 2 vol.), de Mgr Gay (Paris, 1893, 3 vol.), de Mgr de Ségur (Paris, 1882, 2 vol.), de l'abbé Perreyve (Paris, 1875), de Mme Swetchine (t. II et III, Paris, 1873), de Lacordaire (avec Mme Swetchine, Paris, 1861 ; avec Mme de La Tour du Pin, Paris, 1906 ; *Lettres inédites*, Paris, 1881) ; de Montalembert (avec L. Cornudet, Paris, 1905 ; avec l'abbé Texier, Paris, 1899 ; voir aussi *Revue de Paris*, 1er décembre 1906 et *Revue historique*, t. CXCII, 1941, p. 253-289) ; du côté des adversaires de l'Église, celle de Renan et Berthelot (Paris, 1898), ainsi que de Mérimée et Panizzi (Paris, 1881, 2 vol.) ;

c) les souvenirs de Mgr Mabile (édit. P. Mabile, *Mgr Mabile, évêque de Versailles*, Paris, 1926, 2 vol.), de Falloux (*Mémoires d'un royaliste*, Paris, 1888, 2 vol.), de Melun (édit. Le Camus, Paris, 1891, 2 vol.), de Kolb-Bernard (*Souvenirs intimes*, édit. Le Liepvre, Paris, 1922), de Mgr de Ségur (*Souvenirs et récits d'un frère*, Paris, 1882, 2 vol.), de l'abbé Subileau (*Cinquante ans de ministère paroissial et d'autorité épiscopale en Anjou, 1844-1885*, Paris, 1894) ; et ceux, tendancieux mais très instructifs, d'Hyacinthe Loyson (A. Houtin, *Le P. Hyacinthe dans l'Église romaine*, Paris, 1920 ; *Le P. Hyacinthe, réformateur catholique*, Paris, 1922), de W. Guettée (*Souvenirs d'un prêtre romain devenu prêtre orthodoxe*, Paris, 1899), de V. Duruy (*Notes et Souvenirs*, Paris, 1901, 2 vol.), de Taine (*Carnets de voyage, notes sur la province, 1863-1865*, Paris, 1897), de F. Sarcey (*Journal de jeunesse*, Paris, 1903) ; voir également É. Ollivier, *L'Empire libéral*, Paris, 1895-1910, 16 vol.) ;

d) les documents officiels : *Lois, décrets et règlements relatifs à l'administration des Cultes (1851-1853)*, Paris, 1854 ; *Circulaires, instructions et autres actes relatifs aux affaires ecclésiastiques*, t. II et III, Paris, 1888 ; *Statistique comparée de l'enseignement primaire*, t. II, Paris, 1880 ; *Statistique de l'enseignement secondaire en 1865*, Paris, 1868 ; *L'administration de l'instruction publique de 1860 à 1869*, Paris, 1870 (discours et rapports de Duruy) ; O. Gréard, *La législation de l'instruction primaire en France* (recueil de lois, décrets et ordonnances), t. III (1848-1863) et IV (1863-1879), Paris, 1890 et 1896.

e) les périodiques catholiques, notamment *L'Univers*, *Le Monde* et *L'Ami de la Religion* (quotidiens), les *Annales de la Charité*, la *Revue de l'enseignement chrétien*, *Le Correspondant*, les *Études*, les *Annales de philosophie chrétienne*, l'*Almanach du Clergé* et *La France ecclésiastique*.

6) Sur la vie romaine sous Pie IX :

Outre les nombreux mémoires et correspondances utilisés et cités par S. Negro, *Seconda Roma*, Milan, 1942, voir G. Moroni, *Dizionario di erudizione storico-ecclesiastica*, Venise, 1840-1878, 109 vol. (mine précieuse de renseignements, mais sans esprit critique), ainsi que les journaux du temps : *Diario di Roma* (jusqu'en 1848), *Gazzetta di Roma* (1848), *Monitore Romano* (1849), *Giornale di Roma* (1849-1870), *Osservatore Romano* (depuis 1860).

L'annuaire officiel de la Curie a plusieurs fois changé d'appellation : *Notizie dell'anno* jusqu'en 1862, puis *Annuario pontificio* jusqu'en 1872 et enfin *Gerarchia cattolica*.

II. — Travaux.

1) Histoires générales de l'Église :

J. Schmidlin, *Papstgeschichte der neuesten Zeit*, t. II, *Pius IX und Leo XIII*, Munich, 1934 (surtout pour les relations Église-État) ; Ch. Pouthas, *Le pontificat de Pie IX* (Cours de Sorbonne), Paris, 1945 (point de vue non-catholique, mais très objectif et perspicace) ; L.-A. Veit, *Die Kirche im Zeitalter des Individualismus*, 2. Hälfte : *1800 b. z. Gegenwart* (t. IV, 2, de la *Kirchengeschichte* sous la direction de J.-P. Kirsch), Fribourg-en-Brisgau, 1933 (excellents aperçus synthétiques sur le mouvement catholique) ; *Histoire illustrée de l'Église*, sous la direction de G. de Plinval et R. Pittet, t. II, Genève-Paris, 1948 (*idem*) ; F. Mourret, *Histoire générale de l'Église*, t. VIII, Paris, 1928 (manuel commode, mais nombreuses inexactitudes de détails) ; A. Boulenger, *Histoire générale de l'Église*, vol. IX, Lyon-Paris, 1947-1950 (plus complet que le précédent, surtout pour les pays autres que la France) ; H. Stephan et H. Leube, *Die Neuzeit* (t. IV du *Handbuch der Kirchengeschichte für Studierende*, sous la direction de G. Krueger), Tubingue, 1931, p. 188-307 (protestant) ; Mac Caffrey, *History of the Catholic Church in the XIXth century*, Dublin, 1909, 2 vol. ; E. Jarry, *L'Église contemporaine*. Paris, 1935, 2 vol. ; Ch. Poulet, *Initiation à l'histoire ecclésiastique*, t. II, Paris, 1946 (exposé clair des principaux faits de la période contemporaine). L'ouvrage de J.-B. Bury, *History of the papacy in the XIXth century (1864-1878)*, Londres, 1930, est fort tendancieux, mais présente un certain intérêt pour l'histoire doctrinale.

2) Biographies de Pie IX :

Il n'existe pas encore de biographie critique de Pie IX, car les sources restent encore en grande partie inaccessibles (voir cependant p. 290, n. 2). La plupart des biographies parues du vivant de Pie IX ou peu après sa mort (on trouvera les principales citées dans J. Schmidlin, *Papstgeschichte...*, t. II, p. XIV-XVIII) ne sont guère que des panégyriques. On retiendra cependant, à cause des nombreux détails qu'elles contiennent et d'un minimum de critique, Ch. Sylvain, *Histoire de Pie IX le Grand et de son pontificat*, Paris, 1885, 3 vol. ; A. Pougeois, *Histoire de Pie IX, son pontificat et son siècle*, Paris, 1877-1886, 6 vol. ; Wappmannsperger, *Leben und Wirken des Papstes Pius IX*, Ratisbonne, 1879 ; Stepischnegg, *Papst Pius IX und seine Zeit*, Vienne, 1879, 2 vol.

Parmi les ouvrages plus récents, on retiendra surtout : A. Monti, *Pio IX nel risorgimento italiano*, Bari, 1928 (utilise certains documents inédits) ; E. Vercesi, *Pio IX*, Milan, 1930 ; F. Hayward, *Pie IX et son temps*, Paris, 1948 (superficiel pour l'histoire des Églises particulières, mais très bon pour la Question romaine et la personnalité de Pie IX et de ses collaborateurs).

3) Études relatives aux Églises particulières :

a) *Pour la France* (1) :

Très bonne synthèse par Ch. Pouthas, *L'Église et les questions religieuses*

(1) On se reportera avec intérêt à la bibliographie critique dressée par G. Weill, *Le catholicisme français au XIXe siècle*, dans *Revue de synthèse historique*, déc. 1907.

en France de 1848 à 1877 (Cours de Sorbonne), Paris, 1945. Les ouvrages suivants reposant comme le précédent sur une documentation de première main (surtout le premier) sont également fondamentaux : J. Maurain, *La politique ecclésiastique du Second Empire de 1852 à 1869*, Paris, 1930 (excellent ; nuance l'ouvrage de tendance anticléricale de A. Debidour, *Histoire des rapports de l'Église et de l'État en France de 1789 à 1870*, Paris, 1898) ; J. Brugerette, *Le Prêtre français et la société contemporaine*, t. I (*1815-1871*) et II (*1871-1908*), Paris, 1933-1935 ; E. Lecanuet, *L'Église de France sous la IIIᵉ République*, t. I (*1870-1878*), Paris, 1907 (à comparer, pour les relations entre l'Église et l'État, avec A. Debidour, *L'Église catholique et l'État sous la IIIᵉ République*, t. I, Paris, 1906). Voir en outre *L'épiscopat français depuis le Concordat jusqu'à la séparation*, Paris, 1907.

Parmi les aperçus d'ensemble, de préférence à H. Guillemin, *Histoire des catholiques français au XIXᵉ siècle*, Genève, 1947 (très unilatéral et incomplet), on se reportera à G. Goyau, *Histoire religieuse de la nation française* (t. VI de l'*Histoire de la Nation française*, sous la direction de G. Hanotaux), Paris, 1922, livre V ; Mgr Baunard, *Un siècle de l'Église de France : 1800-1900*, Paris, 1900 (intéressant pour le mouvement catholique ; mais trop panégyrique) ; A. Dansette, *Histoire religieuse de la France contemporaine*, t. I, Paris, 1948 ; Ch. Poulet, *Histoire de l'Église de France*, t. III, *Époque contemporaine*, Paris, 1949 (point de vue plus conservateur que le précédent) ; W. Gurian, *Die politischen und sozialen Ideen des französichen Katholizismus 1789-1914*, München-Gladbach, 1929.

Sur les congrégations : P. Nourrisson, *Histoire légale des congrégations religieuses en France depuis 1789*, Paris, 1928, 2 vol. ; J. Maillaguet, *Le miroir des ordres et instituts religieux de France*, Avignon, 1865 ; E. Keller, *Les congrégations religieuses en France. Leurs œuvres et leurs services*, Paris, 1880.

Sur l'enseignement chrétien : A. des Cilleuls, *Histoire de l'enseignement libre dans l'ordre primaire en France*, Paris, 1898 ; G. Weill, *Histoire de l'enseignement secondaire en France (1822-1920)*, Paris, 1921.

Sur le mouvement charitable : J.-B. Duroselle, *Les débuts du catholicisme social en France (1822-1870)*, Paris, 1951.

Parmi les très nombreuses biographies, on retiendra surtout les suivantes, rangées par ordre alphabétique (1) : Mgr Foulon, *Histoire de la vie et des œuvres de Mgr Darboy*, Paris, 1889 ; **F. Lagrange, *Vie de Mgr Dupanloup*, Paris, 1883-1884, 3 vol. ; C. de Ladoue, *Mgr Gerbet, sa vie, ses œuvres*, Paris, 1870, 3 vol. ; **Dom Delatte, *Dom Guéranger, abbé de Solesmes*, Paris, 1909, 2 vol. ; J. Paguelle de Follenay, *Vie du cardinal Guibert*, Paris, 1896, 2 vol. ; *J.-T. Foisset, *Vie du R. P. Lacordaire*, Paris, 1870, 2 vol. ; **G. Bazin, *Vie de Mgr Maret*, Paris, 1891-1892, 3 vol. ; F. Besson, *Vie de S. E. le cardinal Mathieu*, Paris, 1882, 2 vol. ; *H. Boissonnot, *Le cardinal Meignan*, Paris, 1899 ; *Mgr Baunard, *Le vicomte Armand de Melun*, Paris, 1880 ; **E. Lecanuet, *Montalembert, d'après ses papiers et sa correspondance*, Paris, 1896-1902, 3 vol. ; *Ch. Guillemant, *P.-L. Parisis*, Paris, 1916-1925, 3 vol. ; *Mgr Baunard, *Histoire du cardinal Pie*, Paris, 1893, 2 vol. ; J. Clastron, *Vie de Mgr Plantier, évêque de Nîmes*, Paris, 1882, 2 vol. ; C.-J. Destombes, *Vie de S. E. le cardinal Régnier, archevêque de Cambrai*, Paris, 1885, 2 vol. ; *H. Chaumont, *Mgr de Ségur, directeur des âmes*, Paris, 1884, 2 vol. ; F. Poujoulat, *Vie de Mgr Sibour*, Paris, 1857 ; *E. Veuillot, *Louis Veuillot*, Paris, 1902-1913, 4 vol.

b) *Pour l'Allemagne* :

H. Brueck-J.-B. Kissling, *Geschichte der katholischen Kirche in Deutschland im XIX. Jh.*, t. III et IV, Munster, 1902-1908 (point de vue trop étroitement confessionnel) ; G. Goyau, *L'Allemagne religieuse, Le Catholicisme,*

---

(1) Les astérisques indiquent celles qui présentent un intérêt particulier (*) ou exceptionnel (**).

Paris, 1905-1909, 4 vol. (très belle synthèse qui garde sa valeur malgré la publication de nombreuses monographies apportant des nuances sur certains points) ; K. Bachem, *Vorgeschichte, Geschichte und Politik der deutschen Zentrumspartei zugleich ein Beitrag zur Geschichte der katholischen Bewegung*, t. I à IV, Cologne, 1926-1929 ; J.-B. Kissling, *Geschichte der deutschen Katholikentage*, Munster, 1920-1921, 2 vol.

Parmi les biographies, on retiendra surtout O. Pfuelf, *Kard. Joh. von Geissel*, Fribourg-en-Br., 1895-1896, 2 vol. ; Id., *Bischof Ketteler*, Fribourg-en-Br., 1899 ; F. Vigener, *Ketteler. Ein deutsches Bischofsleben des XIX. Jh.*, Munich, 1924 ; Stamm, *Conrad Martin*, Paderborn, 1892, 3 vol. ; J. Friedrich, *Ignaz von Doellinger*, Munich, 1899-1901, 3 vol. ; L. v. Pastor, *August Reichensperger*, Fribourg-en-Br., 1899, 2 vol.

c) *Pour l'Angleterre* :

P. Thureau-Dangin, *La Renaissance catholique en Angleterre au XIXe siècle*, t. II et III, Paris, 1903-1906 (excellente synthèse gardant toute sa valeur) ; É. Halévy, *Histoire du peuple anglais au XIXe siècle*, t. IV, Paris, 1946, l. III ; *The English Catholics 1850-1950. Essays to commemorate the Centenary of the restauration of the Hierarchy*, sous la direction de Mgr G. A. Beck, Londres, 1950.

Pour une étude plus détaillée, on se reportera aux biographies, tout particulièrement à C. Butler, *The Life and Times of bishop Ullathorne*, Londres, 1926, 2 vol. (excellents aperçus synthétiques sur les différents problèmes soulevés) ; en outre : W. Ward, *Life and Times of cardinal Wiseman*, Londres, 1897, 2 vol. (trad. franç., Paris, 1900 ; à compléter entre autres par les lettres publiées dans *Dublin Review*, janvier 1919 ; voir aussi D. Gwynn, *Cardinal Wiseman*, Dublin, 1950) ; E. Purcell, *Life of cardinal Manning*, Londres, 1895, 2 vol. ; W. Ward, *The Life of J. H. cardinal Newman*, Londres, 1912 ; Id., *William Georges Ward and the catholic Revival*, Londres, 1893 ; J.-G. Snead-Cox, *The Life of cardinal Vaughan*, Londres, 1910.

d) *Pour l'Italie* :

S. Jacini, *La politica ecclesiastica italiana da Villafranca a Porta Pia*, Bari, 1938 (objectif) ; A.-C. Jemolo, *Chiesa e Stato in Italia negli ultimi cento anni*, Rome, 1949, chap. I à III (point de vue catholique-libéral ; voir les observations faites d'un point de vue traditionnel par S. Lener, dans *Civ. catt.*, 1949, vol. I, p. 295-309) ; A. Della Torre, *Il cristianesimo in Italia dai filosofisti ai modernisti*, Palerme, 1912 (anticlérical modéré ; très détaillé sur les courants d'idées religieux).

e) *Pour la Belgique* :

E. de Moreau, art. *Belgique*, dans *D. H. G. E.*, t. VII, col. 727-747 ; Id., *Adolphe Dechamps*, Bruxelles, 1911 ; A. Simon, *Le cardinal Sterckx et son temps*, Wetteren, 1950, 2 vol. (capital ; très documenté) ; E. Rempry, *Les remaniements de la hiérarchie épiscopale et les sacres épiscopaux en Belgique au XIXe siècle*, Bruxelles, 1904 ; *Un siècle de l'Église catholique en Belgique (1830-1940)*, sous la direction de C. Joset, Courtrai, 1934 (trop élogieux, mais nombreux renseignements utiles).

f) *Pour les autres pays* :

La bibliographie sera donnée dans le cours du volume.

CHAPITRE PREMIER

LES DÉBUTS DU PONTIFICAT DE PIE IX [1]

L'ÉTAT DES ESPRITS A ROME
A LA MORT DE GRÉGOIRE XVI
Envisagé avec le recul du temps, le pontificat de Grégoire XVI apparaît aujourd'hui fécond en initiatives heureuses du point de vue religieux, et plus d'une mesure de ce pape, que les historiens du *Risorgimento* présentèrent comme le type du pontife réactionnaire, révèlent davantage un homme de l'avenir qu'un homme du passé. Mais les habitants de l'État romain, plus soucieux de leurs intérêts immédiats que des progrès de l'Église universelle, reprochaient à leur souverain et à son secrétaire d'État, le réactionnaire cardinal Lambruschini, leur répugnance à réorganiser l'État pontifical sur des bases plus modernes ; la bourgeoisie et la jeunesse intellectuelle surtout s'insurgeaient de plus en plus contre un régime qui écartait systématiquement l'élément laïque des charges importantes au profit d'ecclésiastiques souvent mal préparés. Ce mécontentement était encore accru par l'attitude toute négative du pape et de son ministre vis-à-vis des aspirations des patriotes à libérer la péninsule de l'ingérence autrichienne et à réaliser, sous une forme ou sous une autre, l'unité italienne. Le caractère révolutionnaire et anticlérical du programme unitaire de Mazzini, l'inspirateur de la « Jeune Italie », avait d'abord écarté de nombreux catholiques, mais ils témoignaient de plus en plus de sympathies pour le mouvement national depuis que celui-ci avait pris une allure plus modérée, sous l'influence de gentilshommes piémontais et d'intellectuels toscans, d'un christianisme un peu vaporeux mais sincèrement religieux, et qui, se rendant

---

(1) BIBLIOGRAPHIE. — I. SOURCES. — P. BALLERINI, *Les premières pages du pontificat du pape Pie IX*, Rome, 1909 (écrit en 1867, réunissant de nombreux documents et témoignages contemporains) ; *Pie IX e l'Italia, ossia storia della sua vita*, Milan, 1848 (livre anonyme précieux à cause des documents qui y sont rassemblés) ; A. MANNO, *L'opinione religiosa e conservatrice in Italia dal 1830 al 1850*, Turin, 1910 (correspondance de Mgr Corboli-Bussi) ; A. DE BROGLIE, *Mémoires*, t. I, Paris, 1938, chap. V. On trouvera en outre de nombreux documents inédits dans A.-M. GHISALBERTI, *Nuove ricerche sugli inizi del pontificato di Pio IX e sulla Consulta di Stato*, Rome, 1940 ; P. PIRRI, *La politica unitaria di Pio IX dalla Lega doganale alla Lega italica*, dans *Riv. St. C. It.*, t. II, 1948, p. 183-214 ; et H. D'IDEVILLE, *L'ambassade du comte Rossi et les débuts du pontificat de Pie IX*, Lyon, 1885.

II. TRAVAUX. — Sur Pie IX avant son élection : Aux biographies signalées p. 7, ajouter G. PONTRANDOLFI, *Pio IX e Volterra*, Volterra, 1928 (sur sa jeunesse) ; P. LETURIA, *El viaje a America del futuro Pontifice Pio IX (1823-1825)*, dans *Xenia Piana Pio Papae XII dicata* (*Miscellanea Historiae Pontificiae*, vol. VII), Rome, 1943, p. 367-444 ; G. MARGOTTI, *Pie IX ed il suo episcopato nelle diocesi di Spoleto ed Imola*, Turin, 1877 ; G. MAIOLI, *Pio IX. Da Vescovo a Pontefice. Lettere al cardinale Amat (1832-1848)*, Modène, 1949.

Sur le conclave de 1846 : E. CIPOLLETTA, *Memorie politiche sui conclavi da Pio VII a Pio IX*, Milan, 1863 ; T. BUTTINI, *La morte di Gregorio XVI e l'elezione di Pio IX*, dans *Rass. St. R.*, t. XXVII, 1940, p. 41-68.

Sur les deux premières années du pontificat : outre les biographies de Pie IX, notamment A. MONTI, *Pio IX nel Risorgimento italiano*, Bari, 1928 et l'ouvrage cité de A.-M. GHISALBERTI, *Nuove ricerche...*, voir R. PALMAROCCHI, *Alcuni aspetti della politica di Pio IX nei primi anni di governo*, dans *Rass. St. R.*, t. XXIII, 1936, p. 695-718.

bien compte que la fidélité au catholicisme constituait en Italie la première
des traditions nationales, cherchaient en conséquence à réaliser l'unité
de la patrie non pas contre l'Église, mais avec son appui. Un livre notamment avait profondément remué les esprits, *Il Primato civile degli Italiani* (1843) [1], dans lequel l'abbé Gioberti avait exposé avec enthousiasme le programme « néoguelfe » d'une fédération des souverains italiens, affranchis de l'influence autrichienne, groupés sous la présidence
du pape et appuyés sur la force militaire du Piémont.

Le désir de réforme et les aspirations nationales s'étaient développés
à un point tel, pendant les derniers temps du pontificat de Grégoire XVI,
que, au dire d'observateurs impartiaux, les populations étaient « partout
frémissantes sous leur frein actuel et impatientes de le briser » [2] ; la
joie exubérante avec laquelle les prisonniers politiques et leurs familles
accueillirent la nouvelle de la mort du pape, le 1er juin 1846 [3], donne
d'ailleurs une idée de la haine qui s'était accumulée dans certains milieux
contre le pape et le régime qu'il incarnait. Nombreux étaient ceux qui,
même dans le clergé, faisaient écho à la brochure de Massimo d'Azeglio,
publiée quelques mois plus tôt, sur les *Récents événements des Romagnes*,
dans laquelle ce patriote, après avoir stigmatisé la brutalité du gouvernement pontifical dans cette province, appelait de ses vœux un pape
libéral et réformateur et désignait pour ce rôle le cardinal Gizzi.

L'ÉLECTION DU CARDINAL MASTAI   On conçoit que, dans cette atmosphère tendue, les facteurs politiques
aient joué un rôle prépondérant au moment de l'élection du nouveau
pape [4]. « Pas un mot ne se disait plus autour de moi des questions religieuses qui faisaient naguère le fond de toutes les conversations, notait
avec étonnement un diplomate français. On n'entendait plus parler que
des questions les plus exclusivement temporelles du gouvernement du
Saint-Siège [5]. »

Tandis que les sympathies de la population romaine se portaient
vers le cardinal Gizzi, « le pape d'Azeglio », ou vers le vieux cardinal
capucin Micara, très populaire mais d'une santé trop précaire pour avoir
des chances sérieuses, le Sacré Collège se partageait entre deux tendances.
Les *zelanti* ou intransigeants se groupaient autour de Lambruschini
et faisaient valoir qu'avec lui l'appui de l'Autriche demeurerait acquis
pour la répression des tendances révolutionnaires. Mais les autres, conscients de l'impopularité que laissait derrière lui le précédent gouver-

(1) Cf. t. XX, p. 438.
(2) Rapport du ministre de Hollande du 18 mars 1844, dans A.-M. Ghisalberti, *Cospirazioni
del Risorgimento*, Palerme, 1938, p. 8-11.
(3) On peut en voir un exemple dans les souvenirs d'un détenu du château Saint-Ange,
G. Galletti, *La mia prigionia*, Bologne, 1870, p. 246-248.
(4) Sur les positions à la veille du conclave, voir, outre J. Schmidlin, *Papstgeschichte...*, t. II,
p. 12-15, les dépêches du ministre de Hollande dans A.-M. Ghisalberti, *Nuove ricerche...*, p. 16-
17 ; de l'ambassadeur de Naples, dans E. Cipolletta, *Memorie politiche...*, p. 224 et 229 ; de l'ambassadeur de Piémont dans l'art. cit. de T. Buttini, dans *Rass. St. R.*, t. XXVII, 1940, p. 45 et
suiv. ; de Campbell-Scarlett, dans *Correspondence respecting the affairs of Italy, 1846-1847*,
Londres, 1849, t. I, p. 18.
(5) A. de Broglie, *Mémoires*, t. I, p. 130.

nement, estimaient qu'il fallait cette fois se soustraire à toute influence étrangère et faire quelques concessions à l'esprit du temps. Dirigés par le cardinal Bernetti, que l'hostilité de Metternich avait obligé Grégoire XVI en 1836 à écarter de la secrétairerie d'État, ils préconisaient un pape originaire des États pontificaux et ouvert aux idées modernes, ajoutant qu'il convenait qu'il fût habitué au maniement des affaires et des hommes plutôt qu'issu d'un ordre religieux. Plusieurs d'entre eux avaient lu et apprécié la brochure d'Azeglio, mais craignaient que Gizzi ne représentât des idées trop avancées. Parmi les candidats plus modérés sur lesquels se portaient leurs préférences, l'un des plus souvent cités était le cardinal Mastai, archevêque d'Imola [1], qui avait su se rendre sympathique dans des régions où le gouvernement pontifical était le moins bien vu, et qui était chaleureusement recommandé par un religieux célèbre de tendances libérales, le P. Ventura.

Vu la gravité de la situation, on décida, après un moment d'hésitation, d'ouvrir le conclave sans attendre l'arrivée des cardinaux étrangers. Dès le début, les votes se concentrèrent sur les noms de Lambruschini et de Mastai : quinze au premier et treize au second [2]. La crainte de voir Lambruschini triompher, peut-être aussi celle de voir un cardinal autrichien arriver porteur d'un *veto* impérial, incita les autres votants à grouper sans plus tarder leurs voix sur le concurrent le mieux placé de l'ancien secrétaire d'État. Aussi Mastai eut-il dix-sept voix au deuxième scrutin et vingt-sept au troisième, tandis que Lambruschini n'en avait plus que onze ; dès le 16 juin au soir, il atteignait la majorité des deux tiers après un conclave qui n'avait, contre toute attente, duré que quarante-huit heures. Très maître de lui [3], il déclara accepter la lourde charge qui lui incombait et choisir, en souvenir de son bienfaiteur Pie VII, le nom de Pie IX.

Un peu déconcertée au premier moment à la nouvelle de l'élection d'un prélat qu'elle ne connaissait guère, puisqu'il vivait en province depuis plus de quinze ans, la population romaine fit cependant bon accueil au nouveau pape : certains rappelaient le temps où, jeune prêtre, il se dévouait dans les quartiers populaires de la ville ; d'autres avaient été conquis par ses manières affables et sa voix chaude et musicale lors de sa première bénédiction ; d'autres enfin colportaient le bruit qu'il était de tendance libérale et cela suffisait à éveiller tous les espoirs : le secrétaire de l'ambassade de France nous a conservé le souvenir de

---

(1) On a souvent présenté l'élection du cardinal Mastai comme une surprise. Les documents contemporains le montrent au contraire comme l'un des principaux *papabili* : voir les dépêches de l'ambassadeur de France dans L. LEDERMANN, *Pelegrino Rossi*, Paris, 1929, p. 333 ; de l'ambassadeur de Naples, dans E. CIPOLLETTA, *Memorie politiche...*, p. 225, et surtout de l'ambassadeur de Piémont, dans T. BUTTINI, *art. cit.*, p. 45 et du ministre de Hollande dans A.-M. GHISALBERTI, *op. cit.*, p. 9 ; déjà en 1842, l'ambassadeur d'Autriche désignait Mastai comme l'un des *papabili* (*ibid.*, p. 15).

(2) Fac-simile des listes des quatre scrutins dans A. MONTI, *op. cit.*, p. 60.

(3) La plupart des historiens reprennent une tradition d'après laquelle Mastai, qui était par hasard l'un des trois scrutateurs et devait donc lui-même proclamer les résultats, manqua s'évanouir d'émotion. Mais le récit qui paraît le plus exact, celui fait par le cardinal Fieschi au chroniqueur romain Roncalli (reproduit pour la première fois d'après le manuscrit intégral par A.-M. GHISALBERTI, *op. cit.*, p. 12), le montre au contraire parfaitement calme et nous apprend que c'est Lambruschini qui perdit connaissance en sortant de la salle du conclave.

l'explosion de joie qui accueillit à Civita-Vecchia le voyageur de la diligence de Rome annonçant en jurant de plaisir : « *Il papa è fatto e liberale, coglione !* [1] » Quant aux chancelleries étrangères, l'accueil fut en général favorable, car ce pape, que les diplomates décrivaient comme étant de tendances modérées, satisfaisait à la fois la France, qui avait craint un pape inféodé au parti autrichien, et Metternich, qui avait appréhendé l'élection de Gizzi [2].

LA PERSONNALITÉ DU NOUVEAU PAPE — Jean-Marie Mastai était né à Sinigaglia, le 13 mai 1792, dans une famille de petite noblesse, et il avait été élevé dans une atmosphère d'ardente piété. Comme il avait souffert dans sa jeunesse d'une maladie de caractère épileptique [3], on avait hésité un moment à l'admettre aux ordres sacrés, mais la faveur de Pie VII lui avait permis de surmonter cet obstacle. Après avoir achevé ses études théologiques au Collège romain, il fut ordonné prêtre le 10 avril 1819 et s'occupa pendant quelques années de l'orphelinat de *Tata Giovanni*, où il fit l'admiration de tous par son zèle et sa générosité. Initié à la vie apostolique et aux méthodes de spiritualité ignatiennes par le saint cardinal Odescalchi, avec qui il demeura toujours très lié, et dont il appréciait hautement les qualités spirituelles [4], l'abbé Mastai songea sérieusement à entrer dans la Compagnie de Jésus, mais les avis répétés de son confesseur, le chanoine Starace, et du maître des novices de Saint-André du Quirinal l'en dissuadèrent [5]. Désigné peu après, en 1823, pour accompagner à titre d'auditeur le délégué pontifical au Chili, il revint de ce lointain voyage, qui lui avait permis de saisir le problème missionnaire, avec un zèle apostolique encore avivé, mais sans la moindre ambition de continuer sa carrière dans la diplomatie pontificale. Nommé à la tête de l'hospice Saint-Michel, où, comme à *Tata Giovanni*, il eut vite fait de s'attirer la sympathie de tous, il y fit preuve de réels talents d'administrateur et, deux ans plus tard, le 24 avril 1827, Léon XII le nommait archevêque de Spolète.

Là, il sut faire face avec une douce fermeté à une situation délicate, au moment des troubles révolutionnaires de 1831, et, lorsqu'il eut été transféré en 1832 dans le diocèse plus important d'Imola, où les éléments radicaux étaient particulièrement nombreux, il réussit à conquérir tous les cœurs par sa bonté inépuisable et sut se faire apprécier des milieux

---

(1) A. DE BROGLIE, *Mémoires*, t. I, p. 127.

(2) H.-R. VON SBIRK, *Metternich. Der Staatsmann und der Mensch*, Munich, 1925, t. II, p. 128. Le 26 juin, Metternich écrivait à son ambassadeur à Rome que cette élection lui avait causé « une satisfaction aussi vive que légitime ». (*Mémoires, documents et écrits divers*, t. VII, Paris, 1883, p. 248.) Le soi-disant *veto* contre le cardinal Mastai, dont le cardinal Gaysruck aurait été porteur et que seule son arrivée tardive aurait empêché d'exercer, est une fable : Gaysruck avait seulement pour mission de s'opposer à l'élection d'un pape appuyant ouvertement les aspirations italiennes à l'indépendance. (A. MONTI, *op. cit.*, p. 58-59 ; A.-M. GHISALBERTI, *op. cit.*, p. 20 et surtout S. BORTOLOTTI, *Metternich e l'Italia nel 1846*, Turin, 1945, p. 114-119.)

(3) Sur la nature de cette maladie et la réalité des phénomènes épileptiques, voir A. MONTI, *op. cit.*, p. 18-24 et surtout ID., *Pio IX*, Milan, 1943, p. 6-7, où l'on trouvera un nouveau document fort explicite.

(4) P. PIRRI, *Vita del Servo di Dio Carlo Odescalchi*, Isola del Liri, 1935, p. 77.

(5) CLERICI, *Pio IX*, Milan, 1928, p. 22 ; Mgr CANI, *Procès romain pour la cause de béatification de Pie IX*, Paris, 1910, p. 14, n. 31.

libéraux par ses qualités administratives, sa bonne volonté et son absence totale d'esprit de parti. Le caractère conciliant du prélat, joint à la tendance modérément réformatrice de la famille Mastai [1], devait naturellement le porter à désapprouver la politique dénuée de souplesse de Grégoire XVI et de Lambruschini, et nous savons par les *Mémoires* du P. Curci qu'il ne s'en cachait guère. En outre, doué d'une certaine perspicacité naturelle, il se rendait compte du caractère suranné de beaucoup d'institutions pontificales et souffrait de la « muraille de bronze qui se dressait entre les libéraux et la papauté », selon l'expression d'un témoin peu suspect, l'historien Spada. Aussi, tout en jugeant sévèrement l'action du parti révolutionnaire, n'hésitait-il pas à prendre la défense des « modérés », qui souhaitaient voir le pape entrer dans la voie des réformes, et il avait lui-même esquissé, au cours de l'année 1845, un programme de réformes administratives dont il devait s'inspirer au début de son pontificat [2].

Faut-il aller plus loin et dire, comme on le fait parfois, qu'il s'était rallié au programme libéral et national, à la suite de ses entretiens avec son diocésain le comte Pasolini, qui aurait réussi à lui faire partager son enthousiasme pour les idées de Gioberti ? La chose paraît peu probable. D'après le fils de Pasolini, le cardinal Mastai aurait lu la brochure d'Azeglio, les *Speranze d'Italia* de César Balbo, les actes des congrès de savants italiens qui, depuis quelques années, se réunissaient annuellement pour préparer la renaissance nationale, ainsi que le *Primato* de Gioberti, et il aurait marqué une sympathie de plus en plus marquée pour plusieurs des thèses essentielles défendues dans ces ouvrages, au point qu'en partant pour le conclave il les aurait fait mettre dans ses malles « afin de les offrir au nouveau pape » [3]. Mais les historiens récents mettent généralement en doute l'exactitude de ces détails, en particulier du dernier, et les seuls faits certains sont les suivants : le cardinal avait lu la brochure d'Azeglio, dont il disait que « parmi beaucoup de mensonges et de calomnies effrontées, elle contient quelques vérités » [4] ; il possédait dans sa bibliothèque un exemplaire du *Primato* [5] ; enfin la question de savoir dans quelle mesure il avait effectivement lu des ouvrages de ce genre importe assez peu, car il est bien évident que, dans une région comme la Romagne, où le sentiment national et libéral était très vif, l'évêque d'Imola a dû avoir souvent l'occasion d'entendre exposer et défendre avec ferveur les idées de Gioberti, de Balbo et d'Azeglio.

Quoi qu'il en soit, Mastai semble ne s'être jamais rallié personnellement au programme néoguelfe et, par ailleurs, son prétendu libéralisme

(1) Voir le témoignage de Doubet dans J. Gay, *Les deux Rome et l'opinion française*, Paris, 1931, p. 64, et le rapport de Rossi du 17 juin 1846, dans F. Guizot, *Mémoires*, t. VIII, Paris, 1861, p. 341-342. La boutade, d'ailleurs exagérée, de Grégoire XVI : *In casa dei Mastai, tutti sono liberali, fino al gatto*, paraît authentique. (F. Mourret, *Histoire générale de l'Église*, t. VIII, p. 337, n. 1.)

(2) Le P. Pirri en a retrouvé le texte autographe dans les archives du Vatican. (Cf. *Riv. St. C. It.*, t. I, 1947, p. 38.)

(3) G. Pasolini, *Memorie raccolte da suo figlio*, Turin, 1887, p. 52 et suiv.

(4) Mastai à L. Taparelli, 19 avril 1846, dans P. Pirri, *Carteggio del P. Luigi Taparelli d'Azeglio*, Turin, 1922, p. 182.

(5) L. Sandri, *La biblioteca privata di Pio IX*, dans *Rass. St. R.*, t. XXV, 1938, p. 1426-1432.

se réduisait en fait, d'une part à une libéralité d'âme qui le portait à estimer qu'il vaut mieux désarmer l'esprit révolutionnaire par la douceur qu'essayer de le dompter par la force, surtout lorsque le souverain est en même temps prêtre, d'autre part à un désir très sincère de s'attaquer aux abus de l'administration pontificale et d'introduire certaines réformes, pourvu qu'elles n'aboutissent pas à accorder à la population une part effective dans le gouvernement, ce qui lui paraissait incompatible avec le caractère religieux de celui-ci.

LES PREMIÈRES MESURES   Le nouveau pape était donc désireux de remédier à la triste situation de l'État pontifical et conscient de la nécessité d'une pacification rapide des esprits ; mais il était en même temps très soucieux de ne pactiser en aucune manière avec les idées libérales avancées et, dès les premiers mois de son pontificat, son attitude fut dominée par cette double préoccupation. Au début, toutefois, beaucoup ne remarquèrent que la première et lui donnèrent vite une portée qu'elle n'avait pas, habitués qu'ils étaient à considérer comme « libéral » tout prélat disposé à faire preuve d'esprit de conciliation.

L'accueil fait au décret d'amnistie du 17 juillet [1] est caractéristique à cet égard. Il était d'usage que l'avénement du pape fût accompagné de mesures d'indulgence en faveur des condamnés politiques, mais la majorité des cardinaux de même que l'ambassadeur d'Autriche conseillaient de se borner à quelques grâces individuelles, tandis que d'autres, tel Gizzi, conseillaient une amnistie générale. Après quelques hésitations, Pie IX se décida pour la seconde solution, mais en tenant compte des observations de la majorité, et les termes du décret réduisirent en fait la portée réelle de la mesure à un point tel que Metternich lui-même s'en déclara satisfait. Telle quelle cependant, elle souleva un enthousiasme indescriptible dans la population des États pontificaux, qui voulut y voir le désaveu des méthodes en usage sous le gouvernement précédent, et elle augmenta encore la popularité du pape auprès du petit peuple de Rome, déjà gagné par ces dons qui plaisent aux foules et que Pie IX possédait en abondance : son regard empreint de bonté, sa parole tour à tour familière et vibrante, sa bonhomie et sa simplicité d'allures, qui tranchaient sur la froideur distante de son prédécesseur.

Bientôt d'autres décisions semblèrent confirmer la première impression et annoncer un esprit nouveau : la nomination du cardinal Gizzi comme secrétaire d'État ; le choix, comme conseiller intime, de Mgr Corboli-Bussi, jeune prélat de grand mérite, ouvert aux idées nouvelles ; la constitution, dès la fin de juillet, d'une commission chargée d'examiner un programme de réformes administratives. Très significative aussi apparut la nouvelle que Pie IX était décidé à introduire dans ses États l'éclairage au gaz et les chemins de fer, auxquels Grégoire XVI s'était toujours opposé.

Mais si ces premières mesures remplissaient d'espoir les libéraux modé-

(1) L. SALVATORELLI, *L'amnistia di Pio IX*, dans *Prima e dopo il Quarantotto*, Turin, 1948.

rés, nombreux dans la bourgeoisie des États pontificaux, qui croyaient avoir affaire au pape prédit par Gioberti, elles inquiétaient fort les partisans du régime précédent, les *gregoriani*, encore nombreux dans la Curie et le clergé, et moins de six semaines après l'avénement du nouveau pape, certains aumôniers de religieuses invitaient déjà leurs dirigées à prier pour détourner de l'Église les maux dont elle était menacée « sous le gouvernement d'un pontife libéral »[1]. La situation du jeune pape, pris entre les espoirs excessifs de la majorité de ses sujets et la résistance systématique des milieux réactionnaires, était délicate, et l'on comprend que, sous l'effet de ces pressions contradictoires, sa politique ait pris, après les débuts prometteurs des premières semaines, « une allure un peu hésitante, assez conforme en vérité à la nature intime du bon pape »[2]. Celui-ci avait pu, grâce à ses dons personnels, agir comme pacificateur sur un terrain restreint, mais il était dénué des qualités essentielles de l'homme d'État et se laissait trop facilement influencer par son tempérament émotif.

Dès que la population romaine s'aperçut de ces hésitations, elle s'empressa de réagir de manière assez originale. Pendant les premiers mois du nouveau règne, la joie populaire s'était exprimée à plus d'une reprise par de bruyantes manifestations de sympathie envers le pontife et les plus avisés s'étaient rendu compte à quel point celui-ci était sensible aux vivats et aux applaudissements. Des agitateurs eurent vite fait dès lors d'utiliser ces manifestations pour faire pression sur le pape, provoquant artificiellement sur son passage, tantôt une froideur qui l'affligeait profondément, tantôt au contraire un enthousiasme exubérant, suivant qu'il avait accordé ou non telle ou telle mesure réclamée par l'opinion publique.

Ce manège avait à première vue quelque chose d'amusant : « Il y a dans cette foule, écrivait Albert de Broglie à son père au début de janvier 1847, un mélange d'enfants jouant à la procession et de politiques spéculant sur la faiblesse de leur souverain et voulant le séduire par la vanité, de badinage puéril et de calcul profond, qui est merveilleux. » Mais quelques mois plus tard il décelait mieux le danger de ces manœuvres équivoques : « Le pape ne voit pas que le public est plus fin que lui et lui tire les concessions les unes après les autres par des compliments, exactement comme dans la fable du renard et du corbeau[3]. » Si un diplomate d'orientation libérale s'exprimait de la sorte, on juge de ce que devait être la réaction des milieux conservateurs : ils blâmaient de plus en plus ouvertement la faiblesse du pape, et gémissaient de voir à la tête de l'Église un homme « chaud de cœur, mais faible de conception et sans esprit de gouvernement », comme disait à présent Metternich[4]. Celui-ci s'inquiétait surtout en constatant les répercussions, dans les autres États de la péninsule, des réformes arrachées peu à peu au pape par ses sujets.

---

(1) Dépêche du 30 juillet 1846, citée dans A.-M. GHISALBERTI, *op. cit.*, p. 27. Voir sur ces résistances : L. SALVATORELLI, *L'amnistia di Pio IX...*, p. 25 et 30.
(2) F. HAYWARD, *Pie IX et son temps*, p. 53.
(3) Lettres des 8 janvier et 28 avril 1847, dans A. DE BROGLIE, *Mémoires*, t. I, p. 143 et 145.
(4) METTERNICH, *Mémoires, documents et écrits divers*, t. VII, p. 476.

Tous les efforts faits jusqu'alors par les conservateurs pour entourer de prestige l'autorité absolue du pontife-roi se retournaient à présent contre eux : « Nous avions tout prévu, sauf un pape libéral ! », avouait le vieux chancelier.

PIE IX ET LA QUESTION ITALIENNE   Le ministre autrichien avait d'autant plus de motifs de s'alarmer qu'il se rendait bien compte que les *Viva Pio Nono !* qui retentissaient à travers toute l'Italie s'adressaient, plus encore qu'au prince réformateur, à celui qu'on commençait à considérer comme le champion de l'indépendance nationale.

Effectivement, s'il ne songeait nullement à faire sien le programme néoguelfe, qu'il estimait utopique et surtout incompatible avec sa mission sacerdotale, Pie IX était à la fois conscient de ce qu'il y avait de légitime dans l'aspiration nationale et soucieux de prévenir le danger de la voir devenir le monopole des partis avancés, et c'est pourquoi il se décida assez vite, sur les conseils notamment de Mgr Corboli-Bussi, à la favoriser dans la mesure conciliable avec son rôle spirituel. C'est ainsi qu'il protesta énergiquement lorsque, en juillet 1847, à la suite des troubles qui avaient éclaté à Rome à l'occasion d'un soi-disant complot du parti grégorien, les Autrichiens occupèrent Ferrare à l'improviste, mesure qui porta à son comble l'exaspération des patriotes italiens.

Or, étant donné l'excitation des esprits, il suffisait de quelques paroles et d'un geste tranchant sur la réserve hostile de Grégoire XVI pour que beaucoup d'Italiens se persuadent que le nouveau pape reprenait à son compte la totalité du programme national et s'apprêtait, conformément aux vœux de Gioberti, à prendre la tête de la croisade qui allait expulser l'Autriche de la Péninsule. De nombreux prêtres, entre autres, qui avaient jusque-là hésité à se rallier au mouvement pour l'unité italienne à cause de l'anticléricalisme de certains patriotes radicaux, se sentirent rassurés par la consécration religieuse que semblait lui conférer la faveur pontificale, et ce ralliement du clergé, fort important dans un pays catholique comme l'Italie, contribua beaucoup à rendre la cause nationale populaire dans les campagnes. Pie IX devenait ainsi, malgré lui, le symbole du réveil italien : situation qui, dans l'immédiat, renforçait son prestige, mais qui s'avèrerait lourde de danger le jour où les circonstances l'obligeraient à dissiper l'équivoque sur laquelle reposait en partie la popularité dont il bénéficiait.

L'APPROBATION DE LA POLITIQUE DE MONTALEMBERT   Tandis que les Romains applaudissaient le « pape réformateur » et que les Italiens tressaillaient à l'idée que le *statu quo* de 1815, maintenu avec tant de peine par les partisans du légitimisme, était sur le point de s'écrouler, « abattu, dirait-on, par cette main même qui en avait constitué jusque-là le plus solide soutien »[1], les catholiques

---

(1) S. JACINI, dans *Actes du Congrès pour le centenaire de la Révolution de 1848*, Paris, 1949, p. 162.

qui, un peu partout en Europe, faisaient confiance à la liberté pouvaient constater que, sur le plan ecclésiastique également, il y avait quelque chose de changé.

En France notamment, Grégoire XVI et surtout son secrétaire d'État avaient, dans les derniers temps, témoigné à plusieurs reprises de la froideur aux dirigeants du parti catholique, auxquels ils reprochaient de compromettre, par leur bruyante campagne pour la liberté d'enseignement, les efforts de Rome en vue de rétablir l'ancienne alliance entre l'Église et l'État, à la faveur de l'évolution conservatrice du gouvernement Guizot. *Parla bene, ma devrebbe parlare meno*, disait Grégoire XVI de leur chef Montalembert [1]. Très inquiet, celui-ci décida, dès l'avénement du nouveau pape, de prendre les devants afin d'éviter qu'il ne se laissât circonvenir par l'ambassadeur de Louis-Philippe et, au début de septembre, il lui faisait remettre, par son ami l'abbé Dupanloup, un mémoire sur la situation du catholicisme en France [2] dans lequel il montrait le danger d'une politique qui chercherait, par des concessions onéreuses, à obtenir pour l'Église un appui gouvernemental qui ne pourrait être que décevant. La jeunesse française, écrivait-il, « élevée presque exclusivement dans les idées libérales, au premier rang desquelles se place l'indépendance réciproque de l'Église et de l'État, professe la plus vive répugnance pour un système qui tendrait, même indirectement, à faire regarder le prêtre comme un fonctionnaire ». Il concluait, par un mot de Mgr Sibour : « L'État ne peut plus être aujourd'hui que le protecteur de la liberté de l'Église. »

Or Pie IX se trouvait préparé à entendre parler de la sorte, car il subissait fortement à cette époque l'influence de son ancien compagnon d'études, le P. Ventura [3], l'éloquent disciple de Lamennais, qui devait, quelques mois plus tard, célébrer dans l'oraison funèbre d'O'Connell l'alliance du catholicisme et de la liberté. Aussi Dupanloup fut-il reçu de la façon la plus encourageante. Le pape fit à plusieurs reprises l'éloge de Montalembert : « Depuis deux ans, je lis ses discours avec un très grand plaisir... C'est le champion de la bonne cause. » Et il affirma sans hésiter qu'il fallait continuer à réclamer énergiquement la liberté d'enseignement ; il conseilla seulement, faisant allusion à la tendance représentée par Veuillot, d'éviter d'être trop intransigeant, et il approuva la position plus modérée adoptée par Dupanloup dans sa brochure sur *La Pacification religieuse* : « Je désire, lui dit-il, que tous les prêtres entrent dans vos sentiments et que tous ceux qui défendent la liberté le fassent comme vous l'avez fait, suivent la même voie que vous, la voie de la fermeté et de la conciliation [4]. »

L'ENCYCLIQUE « QUI PLURIBUS »   Pie IX se montrait donc moins méfiant que Grégoire XVI à l'égard des libertés modernes. Rien ne serait plus faux cependant que de parler

---

(1) F. VEUILLOT, *L. Veuillot*, t. II, p. 90-99 ; E. LECANUET, *Montalembert*, t. II, p. 286-292.
(2) Analysé dans E. LECANUET, *op. cit.*, t. II, p. 312-315.
(3) H. de Riancey à Combalot, 18 décembre 1846, dans A. RICARD, *L'abbé Combalot*, p. 352. Sur Ventura (1792-1861), de l'ordre des théatins, cf. A. RASTOUL, *Le P. Ventura*, Paris, 1906.
(4) E. LECANUET, *op. cit.*, t. II, p. 316-318, d'après une lettre de Dupanloup à Montalembert.

d'une rupture complète avec la politique religieuse de son prédécesseur. Il est symptomatique de constater que le premier document doctrinal du nouveau pape, l'encyclique *Qui pluribus*, du 9 novembre 1846, fut rédigée par le cardinal Lambruschini [1], l'ancien secrétaire d'État de Grégoire XVI, pour lequel il se montrait plein de ménagements et qu'il consultait volontiers [2], dans l'espoir chimérique de désarmer les préventions du parti conservateur. Cette première encyclique, qui constitue une excellente synthèse des positions doctrinales défendues tout au long du pontificat précédent, porte avant tout sur les rapports entre la raison et la foi et condamne les deux excès opposés du rationalisme et du fidéisme, mais elle dénonce également en passant les principes fondamentaux du libéralisme religieux, « cet épouvantable système d'indifférence, qui ôte toute distinction entre la vertu et le vice, la vérité et l'erreur », et qui prétend appliquer au catholicisme lui-même la théorie du progrès absolu dans l'humanité, « comme si cette religion était l'œuvre des hommes et non de Dieu ».

L'ENTHOUSIASME DANS LE MONDE   Pie IX n'entendait donc nullement se rallier aux principes du libéralisme, et même quant à leurs applications politiques, il était décidé à procéder avec la plus grande prudence. Mais dans l'enthousiasme du premier moment, on ne remarqua que ses allures bienveillantes et ses paroles de compréhension et d'apaisement. Il semblait personnifier ce courant d'aspirations imprécises où s'unissaient, à la veille de 1848, un souffle de christianisme et un souffle de démocratie. Aussi les débuts de son pontificat furent-ils, pour tous ceux qui espéraient voir l'Église se désolidariser définitivement de l'ancien régime, une période éphémère de rêves et d'illusions qui finirent, pendant quelques mois, par gagner le pape lui-même. Ozanam résumait l'impression de beaucoup quand il écrivait à dom Guéranger, au début de 1847 :

> Ce pontife qu'on rencontre à pied dans les rues, qui cette semaine s'en allait un soir visiter une pauvre veuve et la secourir sans se faire connaître, qui prêchait, il y a quinze jours, au peuple assemblé à Saint-André della Valle, ce courageux réformateur des abus du gouvernement temporel, semble vraiment envoyé de Dieu pour conclure la grande affaire du XIXe siècle, l'alliance de la religion et de la liberté [3].

L'opinion non catholique elle-même regardait vers Pie IX avec admiration. Un grand journal anglais saluait en lui « le prince le plus éclairé du siècle » et les participants d'un meeting réuni à New-York lui adressaient « le témoignage d'une sympathie sans borne, non point comme catholiques, mais comme fils d'une république et comme amis de la liberté » [4]. En septembre 1847, le gouvernement britannique trouvait

---

(1) H. Schroers, *J. W. J. Braun*, Bonn, 1925, p. 402-403. Texte dans *Acta Pii IX*, t. I, Rome, 1864, p. 6 et suiv.
(2) A. Dyroff, *Rosmini*, Munich, 1906, p. 78.
(3) Lettre du 29 janvier 1847, dans dom Delatte, *Dom Guéranger*, t. I, p. 410.
(4) Message cité dans G. Desdevises du Dézert, *L'Église et l'État en France*, t. II, Paris, 1908, p. 116.

un moyen élégant de tourner la loi interdisant toute relation diplomatique avec le Saint-Siège et chargeait lord Minto de se rendre à Rome
pour aider Pie IX à se libérer de l'influence autrichienne et l'encourager
dans ses tendances réformistes. Un peu plus tard, favorablement impressionné par l'allure libérale que semblait prendre la politique pontificale,
le gouvernement des États-Unis décidait d'établir à Rome une légation [1].

C'est dans cette atmosphère que Massimo d'Azeglio écrivait à l'un
de ses amis français :

> Voilà Pie IX le promoteur de tout le mouvement libéral et la papauté à
> la tête du siècle. Qui l'eût dit, il y a dix-huit mois !... Si Pie IX continue (et
> pourquoi non ?), il devient le chef moral de l'Europe et il fera ce que n'ont pu
> faire ni Bossuet ni Leibnitz, il rétablira l'unité du christianisme [2].

Ces considérations, où se mêlaient l'exagération et l'utopie, semblaient alors d'autant plus justifiées que Pie IX venait de remporter
en Orient deux importants succès qui accroissaient son prestige : la reprise
des relations avec la Turquie et la signature d'un concordat avec la
Russie.

LE RAPPROCHEMENT   Les autorités turques supportaient de plus en
AVEC LA TURQUIE   plus impatiemment la politique des ambassa-
                      deurs d'Angleterre, de Russie et surtout de France,
qui, sous prétexte de protéger leurs coreligionnaires, travaillaient souvent
à favoriser les intérêts temporels de leurs gouvernements. S'étant rendu
compte de ce mécontentement, un prêtre napolitain en résidence à Constantinople avait suggéré au grand-vizir que son maître pourrait prendre
l'initiative de confier au Saint-Siège, puissance avant tout spirituelle,
la protection des chrétiens, ce qui enlèverait aux diplomates étrangers
tout prétexte d'intervenir. L'idée plut au sultan qui, au début de 1847,
chargea l'ambassadeur turc à Vienne d'aller assurer au pape que le
sultan « désirait vivre en amitié » avec Sa Sainteté et qu'il « saurait protéger les chrétiens qui habitaient ses vastes États ». C'était la première
fois, depuis la fin du XVe siècle, qu'une mission de ce genre se présentait
au Quirinal et le sultan fut enchanté de l'accueil à la fois cordial et brillant
qu'elle reçut de Pie IX, grâce à l'habileté du P. Ventura et en dépit
des efforts des diplomates français, qui avaient intérêt à la voir échouer.
Les premiers résultats ne se firent pas attendre : le 4 octobre Pie IX
pouvait annoncer que le gouvernement ottoman autorisait le rétablissement d'un patriarcat latin à Jérusalem, dont le titulaire, Mgr Valerga,
aurait des pouvoirs étendus sur tous les catholiques de l'empire.

Espérant que la perspective de partager la situation privilégiée qu'on
escomptait pour les catholiques pourrait influencer les chrétiens séparés
du Proche-Orient, le pape estima même l'occasion favorable pour adresser aux autorités orthodoxes, par l'intermédiaire de l'archevêque de

---

(1) L.-F.-S. STOCK, *United States Ministers to the Papal States, 1848-1868*, Washington, 1933.
Sans doute le délégué était-il accrédité auprès du souverain temporel, non auprès du chef de l'Église,
mais l'innovation, à cette date, n'en est pas moins caractéristique.

(2) Lettre à E. Rendu du 20 septembre 1847, dans H. D'IDEVILLE, *L'ambassade du comte Rossi*,
p. 38.

Saïda, envoyé à Constantinople au début de 1848 pour remercier le sultan, un appel à l'union, qui n'eut d'ailleurs pas de suite [1].

LE CONCORDAT AVEC LA RUSSIE — En même temps que la situation de l'Église se renforçait dans l'empire turc, elle s'améliorait dans l'empire russe, grâce à l'heureux aboutissement des négociations entamées à la suite de la visite du tsar Nicolas à Grégoire XVI, en 1845. La question était d'importance car, outre un certain nombre de catholiques dans les provinces occidentales et méridionales de l'empire, elle intéressait plusieurs millions de Polonais. Dirigés du côté russe par le comte Bloudov et du côté du Saint-Siège par le cardinal Lambruschini et Mgr Corboli-Bussi, les pourparlers se poursuivirent au Quirinal du 19 novembre 1846 au 1er mars de l'année suivante, puis, après un moment de crise, ils reprirent le 15 juin [2]. Les délégués de l'empereur avaient été intraitables sur plusieurs points, la question des mariages mixtes entre autres, et avaient refusé de revenir sur la suppression brutale de l'Église uniate d'Ukraine en 1839. Dans ces conditions, Lambruschini, qui s'était montré sceptique dès le début, inclinait pour la rupture, mais Pie IX et plusieurs de ses conseillers furent de l'avis contraire, à cause de l'extrême gravité de la situation : à l'exception d'un seul évêque, fort âgé, il ne restait plus en Russie que des administrateurs diocésains sans caractère épiscopal et les catholiques à l'abandon étaient guettés par la menace du schisme. A présent que l'empereur faisait des avances, acceptant de pourvoir les sièges vacants et promettant certaines concessions, notamment concernant les séminaires, il paraissait déraisonnable de laisser passer l'occasion. C'est ainsi que put être mise au point, le 3 août 1847, une convention [3] dont on indiquait explicitement le caractère incomplet, mais qui apparut cependant, dans l'état actuel des choses, comme un succès pour le Saint-Siège.

LA GUERRE DU SONDERBUND — L'Église voyait donc s'ouvrir à l'Est des perspectives réconfortantes, quand éclata en Suisse [4] une crise qui éveilla de profonds échos dans la conscience européenne et qui apparaît à distance comme l'un des signes avant-coureurs du grand conflit, appelé à se développer tout au long du pontificat de Pie IX, entre l'Église et le radicalisme libéral.

Tandis que l'hostilité séculaire entre catholiques et protestants, déjà fort atténuée au cours du XVIIIe siècle, continuait à s'apaiser, l'opposition entre les catholiques et les radicaux était allée au contraire en

---

(1) Cf. chap. XIV, p. 480.

(2) A. Boudou, *Le Saint-Siège et la Russie*, t. I, Paris, 1922, p. 508-556.

(3) Texte dans A. Boudou, *op. cit.*, p. 559-564.

(4) Sur la crise du *Sonderbund*, on consultera J. Dierauer, *Geschichte der schweizerische Eidgenossenschaft bis 1848*, t. V, Gotha, 1917 (trad. franç., Lausanne, 1917-1918) ; E. Staehelin, *Der Jesuitenorde und die Schweiz*, Bâle, 1923, à compléter par O. Pfuelf, *Die Anfänge der deutschen Provinz der neuerstandenen Gesellschaft Jesu und ihr Wirken in der Schweiz*, Fribourg-en-Br., 1922 ; le numéro spécial de la *Schweizer Rundschau*, t. XLVII, 1947, p. 241-396, en particulier O. Vasella, *Zur historischen Würdigung des Sonderbunds*. On ne peut négliger les ouvrages de C. Siegwart-Mueller, *Ratsherr J. Leu von Ebersol*, Altdorf, 1863 ; *Der Kampf zwischen Recht und Gewalt*, ibid., 1864 ; *Der Sieg der Gewalt über das Recht*, 1866 (plaidoyers, mais contenant de nombreux documents), ni l'apologie des jésuites par J. Crétineau-Joly, *Histoire du Sonderbund*, Paris, 1850. (Voir G. Castella, *Comment fut composée l'histoire du « Sonderbund » de Crétineau-Joly*, dans *Mélanges M. Godet*, Neuchâtel, 1937.)

s'aggravant au cours des dernières années du pontificat de Grégoire XVI. Il serait inexact de n'y voir qu'un conflit politique, mettant aux prises, d'une part, le parti conservateur et fédéraliste, s'appuyant surtout sur les masses paysannes, dont les catholiques formaient un important contingent, et d'autre part, le parti progressiste, représentant la bourgeoisie libérale, qui aspirait à un État centralisé, dans l'intérêt du commerce et de l'industrie. Les motifs proprement religieux prenaient en effet une part croissante au fur et à mesure que l'aile extrémiste du libéralisme, sous l'influence d'émigrés proscrits d'Allemagne, d'Italie ou de Pologne, et inféodés à la franc-maçonnerie, affichait des prétentions totalitaires à ne souffrir à côté d'elle aucune autre conception du monde et multipliait avec un véritable fanatisme les brimades à l'égard des catholiques et surtout des ordres religieux : c'est pour sauvegarder leurs libertés religieuses tout autant que pour défendre leurs libertés locales que les catholiques suisses soutenaient activement le parti fédéraliste, convaincus que le jour où la Confédération helvétique formerait un État centralisé dirigé par les radicaux, ceux-ci s'empresseraient d'imposer leur idéologie antireligieuse même aux cantons catholiques.

La question prenait toutefois un aspect politique par suite du fait que les principaux *leaders* catholiques ne concevaient la sauvegarde de leurs croyances que dans la perspective de l'ancien régime : au lieu de faire appel, comme les catholiques belges ou comme les catholiques français groupés autour de Montalembert, au respect de la liberté de conscience, avec toutes les conséquences qui en découlent, et de s'attirer ainsi la sympathie des libéraux modérés, ils entendaient continuer à jouir d'institutions officiellement catholiques, ce qui impliquait la prolongation du système de l'État confessionnel et donc, dans un pays de religion mixte, la juxtaposition de deux blocs territoriaux dont les constitutions s'inspireraient, dans l'un, des principes réformés, dans l'autre, des principes catholiques. A la solution du *parti* catholique, on préférait celle de *l'État* catholique autonome, sans se rendre compte que l'on allait ainsi à l'encontre du sens de l'histoire et que l'on devait inévitablement provoquer la réaction unanime de la bourgeoisie urbaine, gagnée aux idées modernes. C'est ce qui arriva lorsque, à l'instigation de Lucerne, où se trouvait alors le véritable centre du catholicisme suisse, et sous la direction du juriste Siegwart-Müller [1], les sept cantons catholiques décidèrent, en 1845, de former une « alliance séparée » (*Sonderbund*), prélude à la constitution, au sein de la Confédération, d'un solide État catholique et paysan de tendance à la fois démocratique et antilibéral.

A cette erreur de conception politique était venue s'ajouter une grave faute de tactique. Dans l'intention de fortifier la résistance catholique en vue de la lutte idéologique qui s'annonçait, une partie des dirigeants du canton de Lucerne désirait depuis plusieurs années faire appel aux jésuites, qui avaient déjà plusieurs établissements florissants à Fribourg

---

[1] Sur Constantin Siegwart, qui incarna avec Joseph Leu la résistance des catholiques suisses au radicalisme à l'époque du *Sonderbund*, dont il fut le véritable inspirateur, la meilleure étude reste celle de SEGESSER, dans *Sammlung Kleiner Schriften*, t. II, 1879, p. 447 et suiv.

et dans le Valais, mais les supérieurs de la Compagnie, soucieux de ne pas accentuer les méfiances protestantes à leur égard, s'étaient toujours dérobés. Or, brusquement, en 1844, les radicaux profitèrent de certaines imprudences des jésuites valaisans pour déclencher contre eux une campagne qui les présentait comme les perturbateurs de la paix publique et proposait de leur interdire le séjour en territoire suisse [1] ; ils espéraient, une fois les masses protestantes ameutées, conquérir aisément la majorité en se présentant comme les défenseurs de la démocratie menacée. Indignés par tant de mauvaise foi et sans se rendre compte qu'ils allaient faire le jeu de leurs adversaires, les catholiques lucernois, même ceux qui s'étaient jusqu'alors montrés réticents, voire hostiles à faire appel aux jésuites, voulurent montrer qu'ils ne se laisseraient pas intimider et, forçant la main aux autorités de la Compagnie [2], ils décidèrent, le 24 octobre 1844, de leur confier le Séminaire et la Faculté de théologie.

C'était une grave imprudence et les radicaux surent si bien l'exploiter à leur avantage que, dès le printemps de 1847, ils avaient la majorité à la Diète fédérale. Le 20 juillet, celle-ci déclarait le *Sonderbund* illégal et en exigeait la dissolution ; le 3 septembre, elle ordonnait en outre l'expulsion des jésuites de tout le territoire suisse. Les cantons catholiques refusèrent de s'incliner et ce fut bientôt la guerre civile. Les catholiques, forts de leur bon droit, espérèrent jusqu'à la dernière minute un miracle, qui aurait pu se produire sous la forme d'une intervention des puissances conservatrices, France et Autriche, désireuses d'éviter un triomphe radical [3]. Mais le général Dufour réussit à écraser en trois semaines les armées du *Sonderbund* et les radicaux vainqueurs purent imposer leurs conditions : expulsion de tous les religieux, installation de gouvernements libéraux dans tous les cantons catholiques, puis, un peu plus tard, une nouvelle constitution substituant un État fédéral à la fédération d'États souverains, et la liberté des cultes au principe de l'État confessionnel.

LA POLÉMIQUE ANTIJÉSUITIQUE EN ITALIE

L'hostilité contre les jésuites, qui, sans en être la véritable cause, avait joué un rôle capital dans la crise suisse, n'était pas un phénomène isolé. Un peu partout en Europe, la Compagnie de Jésus se trouvait, à la veille de 1848, en butte à d'âpres critiques à cause de son trop grand attachement aux conceptions politiques de l'ancien régime. Montalembert n'avait pas tort quand il regrettait, dans une lettre à son ami le P. de Ravignan, l'obstination de beaucoup de jésuites « à confondre avec les égarements révolutionnaires

---

(1) Voir F. STROBEL, *Die Jesuitenfrage zur Sonderbundszeit*, dans *Schweizer Rundschau*, t. XLVII, 1947-1948, p. 269-282 (à nuancer toutefois par P. J. KAEMPFEN, *Das Wallis und das Sonderbund*, *ibid.*, p. 299-304).

(2) « *Consensus importunis precibus extorsus fuit* », écrivait l'un des dirigeants de la Compagnie le 22 décembre 1844, soit près de trois ans avant que les événements aient pris une allure catastrophique. (F. STROBEL, *art. cit.*, p. 279.)

(3) Sur les efforts de Guizot qui, tout en combattant les jésuites en France, soutenait au fond leur parti en Suisse, mais fut empêché par l'Angleterre d'intervenir plus activement, voir L. BURGENER, *La politique suisse de la France en 1847*, dans *Zeitschrift für schweizerische Geschichte*, t. XXVII, 1947.

cette tendance invincible qui porte le monde moderne à substituer le
principe et la pratique de la souveraineté nationale à la monarchie absolue » [1], et c'est en vain qu'au début de 1847 le P. Taparelli d'Azeglio,
bien modéré pourtant, s'était efforcé de convaincre son général, le
P. Roothaan, de la nécessité d'une évolution [2]. En présence de la réconciliation idyllique qui s'esquissait un peu partout entre le catholicisme et
la liberté, les jésuites, dans leur ensemble, prenaient une attitude négative et apparaissaient à beaucoup « comme le seul nuage sombre dans
un ciel serein » [3]. Ceci explique du reste en partie l'attitude très réservée
prise par Pie IX à l'égard du *Sonderbund* [4], au grand scandale des catholiques suisses, incapables de faire sur le moment la distinction entre
l'aspect religieux et l'aspect politique du conflit dans lequel ils étaient
engagés et où les jésuites suisses et autrichiens furent loin d'avoir toujours un rôle pacificateur.

En Italie, cette tendance conservatrice de la Compagnie devait inévitablement la faire apparaître comme ayant partie liée avec l'Autriche,
malgré les sympathies manifestées par quelques-uns de ses membres
au mouvement néoguelfe [5] et ces préventions se trouvèrent portées à
leur comble par les violentes attaques du philosophe et écrivain politique Gioberti [6]. Dans le complément donné en 1845 à son célèbre *Primato*, celui-ci avait dénoncé la Compagnie de Jésus comme le principal
obstacle au relèvement civil et religieux de l'Italie et à la fusion har-
monieuse de la religion et de la civilisation moderne. Après un vain
essai de mettre fin à cette campagne par des interventions privées [7],
les jésuites décidèrent de relever le défi et c'est pour leur répliquer que
Gioberti rassembla, avec l'aide de ses amis, l'arsenal le plus complet
qu'on puisse rêver de critiques et de calomnies contre les jésuites : *Il
Gesuita moderno*, qui, par delà la Compagnie de Jésus, atteint au fond
tout le catholicisme postridentin et révèle le caractère peu chrétien de
son auteur, qui devait par la suite évoluer de plus en plus vers vers le
panthéisme. L'ouvrage, écrit avec un remarquable talent de pamphlétaire,
parut à Lausanne en mai 1847 et se répandit aussitôt à travers toute
l'Italie, et pas seulement dans les milieux anticléricaux. Le P. Ventura,
par exemple, malgré quelques réserves, le trouvait écrit « *non sine aliquo
afflatu divino* » [8].

(1) Lettre citée dans E. LECANUET, *Montalembert*, t. II, p. 272-273.
(2) Lettre du 2 mars 1847, dans *Civ. Catt.*, 1948, vol. III, p. 498-502.
(3) L. SALVATORELLI, *La Rivoluzione europea*, Milan, 1949, p. 92.
(4) Il faut ajouter qu'« il semble bien que dans les affaires suisses, le Saint-Père ait été tenu
dans l'ignorance de la véritable situation ». (G. CASTELLA, dans *Mélanges Godet*, p. 204, en note.)
(5) Voir sur ces exceptions A. MESSINEO, *Il P. Luigi Taparelli d'Azeglio e il Risorgimento italiano*, dans *Civ. Catt.*, 1948, vol. III, p. 380-381.
(6) Sur Gioberti (1801-1852), l'ouvrage essentiel est celui de U. PADOVANI, *Vincenzo Gioberti
ed il cattolicismo*, Milan, 1927. Gioberti avait d'abord espéré gagner les jésuites à son projet d'une
fédération italienne présidée par le pape, mais il se rendit bientôt compte du caractère chimérique
de cet espoir en même temps que de la nécessité de donner des gages à la fraction la plus avancée
du parti libéral qui lui reprochait de ménager la Compagnie détestée. (Cf. sur ce dernier point *Civ.
catt.* ,1928, vol. I, p. 423.) Plusieurs libéraux modérés, par contre, se désolidarisèrent des attaques
de Gioberti contre les jésuites, Tommaseo, Cantù et surtout Silvio Pellico. (Cf. U. PADOVANI,
op. cit., p. 310-318.)
(7) Cf. U. PADOVANI, *op. cit.*, p. 322-323, 327-328, 356-357.
(8) Dans une lettre à Gioberti reproduite dans *Ricordi biografici e carteggio di V. Gioberti*,
édit. G. MASSARI, t. III, Naples, 1868, p. 230.

Le roi Charles-Albert de Piémont, qui protégeait la Compagnie, après avoir dû renoncer à interdire la vente de l'ouvrage de Gioberti dans ses États, essaya d'en obtenir la condamnation par le pape. Mais celui-ci, depuis plusieurs mois déjà, témoignait de beaucoup de froideur à l'égard des jésuites. Aussi, tout en faisant savoir à Gioberti qu'il aimerait lui voir modifier certaines affirmations, il se rallia à l'avis du préfet de la Congrégation de l'*Index* qui estimait que, le dogme n'étant pas en cause, il n'y avait pas à intervenir [1]. Bien plus, non content d'affirmer que, lui vivant, l'ouvrage ne serait jamais mis à l'*Index* [2], il pria même le P. Roothaan de renoncer à publier les quelque soixante lettres obtenues de divers évêques italiens en faveur de sa Compagnie [3].

Cette attitude a de quoi surprendre à première vue, quand on se souvient des anciennes sympathies de l'abbé Mastai pour la Compagnie de Jésus. Elle est moins étonnante si l'on songe que, conscient de la sourde opposition que sa politique rencontrait chez beaucoup de ses membres, et peu fait, avec son tempérament jovial et spontané, pour apprécier la réserve compassée du P. Roothaan, il devait être assez disposé à ajouter foi aux nombreuses critiques colportées contre les jésuites par une partie de son entourage [4]. Les choses allèrent si loin que, dans le courant de 1847, le bruit courut à Rome que Pie IX songeait à supprimer une nouvelle fois la Compagnie. Cette nouvelle, d'ailleurs non fondée, prit suffisamment de consistance pour que les provinciaux, réunis à Rome à la fin de l'année pour l'assemblée triennale, jugeassent utile de présenter au pape une adresse de totale soumission, très habilement rédigée par le P. Beckx, délégué de la province d'Autriche, qui avait lancé l'idée [5].

Les attaques de Gioberti ne tardèrent pas à produire leurs effets. Diffusées partout par les adversaires du « parti austro-jésuite », elles soulevèrent en quelques mois l'opinion publique dans l'Italie entière, et, bien qu'ils conservassent partout de fervents défenseurs, les jésuites se virent successivement expulsés du royaume de Naples, puis du Piémont. A Rome même, où on les accueillait ironiquement au cri de : *Viva Gioberti, il filosofo cristiano !*, les menaces de violence se multiplièrent et, le 28 mars 1848, le pape faisait savoir au P. Roothaan que n'étant plus en mesure de garantir leur sécurité, il les priait de quitter les États pontificaux, tout en rendant hommage aux mérites de la Compagnie.

Bien qu'il fût dès lors très inquiet sur l'évolution de la situation, Pie IX ne se rendait sans doute pas compte de la gravité de la crise, mais on peut dire que, dès ce moment, on vivait à Rome dans une atmosphère prérévolutionnaire et le pape n'allait pas tarder à se trouver complètement débordé.

(1) Cf. *Civ. Catt.*, 1916, vol. IV, p. 433-441 et U. Padovani, *op. cit.*, p. 388-391.
(2) D'après V. Gioberti, *Ricordi biografici e carteggio*, t. IV, p. 66-67.
(3) Lettre du P. Roothaan à Mgr Pecci, citée dans G. de Vaux et H. Riondel, *Le P. J. Roothaan*, Paris, 1936, p. 146-147.
(4) P. Albers, *De H.-E.-P. Roothaan*, t. II, Nimègue, 1912, p. 398 ; G. de Vaux et H. Riondel, *op. cit.*, p. 147.
(5) A.-M. Verstraeten, *Leven van den H. E. P. Petrus Beckx*, Anvers, 1899, p. 299-301 ; le texte de l'adresse est reproduit p. 552-553. Sans doute cette même appréhension fut-elle pour une part dans la publication par un laïc français dévoué à la Compagnie, J. Crétineau-Joly, d'un ouvrage sur *Clément XIV et les jésuites*, qui se terminait par le vœu de ne plus jamais avoir un pape dont le cœur serait plus grand que la tête.

CHAPITRE II

LA CRISE DE 1848

§ 1. — La révolution romaine [1].

LES PREMIÈRES DÉSILLUSIONS Il faut remonter quelques mois en arrière pour comprendre l'agitation qui régnait dans les États pontificaux au printemps de 1848 et dont les émeutes contre les jésuites n'étaient qu'une manifestation parmi d'autres. Un bon observateur écrivait, en avril 1847 :

Il ne faut pas se dissimuler que l'Italie subit en ce moment une véritable crise morale, dont Rome, chose étrange, est devenue le centre d'action, et que, d'une extrémité de la Péninsule à l'autre, deux idées dominent les populations, l'une d'obtenir des garanties constitutionnelles et l'autre de voir l'étranger repasser les Alpes [2].

L'immense popularité de Pie IX lui était venue de ce que l'on s'était persuadé qu'il était prêt à prendre la tête de ce double mouvement. Quelques gestes du nouveau pape, plus timides peut-être que ceux accomplis par son prédécesseur, avaient, dans l'atmosphère embrasée de l'époque, été interprétés dans le sens d'une réconciliation de la papauté avec les aspirations libérales et nationales et, partout en Italie, c'est au cri de : *Viva Pio Nono !* que se déroulaient désormais toutes les manifestations contre la domination des Habsbourg ou contre les régimes absolutistes issus de la Restauration. Le mythe du « pape libéral » agissait comme un « catalyseur » [3] sur les éléments disparates qui constituaient l'opinion progressiste italienne à la veille de 1848 et tous se trouvèrent un instant

(1) BIBLIOGRAPHIE. — I. SOURCES. — A celles indiquées dans G. MOLLAT, *La Question romaine de Pie VI à Pie XI*, Paris, 1932, p. 225, 242, 259 et 268, ajouter surtout A. DE LIEDEKERKE, *Rapporti delle cose di Roma (1848-1849)*, édit. A.-M. GHISALBERTI, Rome, 1949 (dépêches du ministre de Hollande) ; A. CAPOGRASSI, *La conferenza di Gaeta del 1849 e Antonio Rosmini*, Rome, 1941 (procès-verbaux des séances) ; D'ALOE, *Diario del soggiorno in Napoli di S. S. Pio IX*, Rome, 1850 ; *L'Orbe cattolico a Pio IX esulante da Roma*, 1850, 2 vol. (texte des adresses des évêques et des fidèles).

II. TRAVAUX. — C. SPELLANZON, *Storia del Risorgimento e dell'unità d'Italia*, t. IV, Milan, 1938, qui signale l'abondante bibliographie antérieure. Compléter par D. MASSÈ, *Pio IX e il gran tradimento del' 48*, Alba, 1948 (bonne synthèse faite d'un point de vue catholique) ; L. SALVATORELLI, *Prima e dopo il Quarantotto*, Turin, 1948 ; P. PIRRI, *La politica unitaria di Pio IX dalla Lega Doganale alla Lega Italica*, dans *Riv. St. C. It.*, t. II, 1948, p. 183-214, et *La Missione di Mons. Corboli-Bussi in Lombardia e la crisi della politica italiana di Pio IX*, ibid., t. I, 1947, p. 38-84 (utilisant de nombreux documents inédits provenant des archives du Vatican) ; G. QUAZZA, *La Questione Romana nel 1848-49 da fonti inedite*, Modène, 1947 (d'après les archives de Turin) ; D. DEMARCO, *Pio IX e la rivoluzione romana del 1848*, Modène, 1947 ; R. MOSCATI, *Austria, Napoli e gli stati conservatori italiani, 1849-1852*, Naples, 1942 ; P. DELLA TORRE, *Pio IX e la restaurazione del 1849-1850*, dans *Aevum*, t. XXIII, 1949, p. 267-298 (plaidoyer) ; A. GHISALBERTI, *Una restaurazione « reazionaria e imperita »*, dans *Archivio della Soc. rom. di storia patria*, t. LXXII, 1949, p. 139-178 (sévère mais juste).

(2) Rapport du ministre de Hollande dans A. GHISALBERTI, *Nuove ricerche...*, p. 35.

(3) S. JACINI, dans *Actes du Congrès historique pour le centenaire de la Révolution de 1848*, Paris, 1949, p. 161.

unis dans une commune espérance. La déception n'en allait être que plus
forte quand on dut se rendre à l'évidence : les actes du pape ne répondaient pas à l'attente.

C'est dans le domaine des réformes intérieures que les désillusions
commencèrent. Le pape n'avait pas seulement à tenir compte de l'opposition, qui ne cessait d'augmenter, de la plupart des prélats de la Curie,
encouragés en sous-main dans leur résistance par les ambassadeurs
d'Autriche et de Bavière, mais lui-même, s'il était sincèrement désireux
d'améliorer la situation de ses sujets, n'entendait pas aller au delà de ce
que nous pourrions nommer un paternalisme ecclésiastique : convaincu
de la nécessité de réformes administratives, il répugnait par contre à
l'idée de réformes d'ordre constitutionnel, craignant, en cédant à des
laïcs quelque chose de sa royauté sacerdotale, de limiter l'indépendance
dont le Saint-Siège avait besoin pour l'accomplissement de sa mission
spirituelle. Aussi, en dépit des sages avis de l'ambassadeur de France,
Rossi, qui lui rappelait à chaque occasion que l'introduction de l'élément
laïque dans la direction des affaires était le problème essentiel, préfé-
rait-il encore accorder à ses sujets des libertés dangereuses comme la
liberté de presse ou la liberté de réunion, plutôt que de leur confier une
part des responsabilités gouvernementales.

Sous la pression des circonstances, il avait bien dû pourtant s'engager
dans cette direction. Après six mois de tergiversations, il avait institué
sous la présidence du cardinal Antonelli, par un *motu proprio* du 14 octobre 1847, une *Consulta* ou conseil de vingt-quatre notables admis à délibérer en matière législative, administrative et militaire, mais en insistant
sur le rôle purement consultatif du nouvel organisme [1] ; puis, au début
de janvier 1848, il acceptait enfin de faire une place effective à l'élément
laïque dans le gouvernement ; enfin, au lendemain de la chute de Louis-
Philippe, il se décida précipitamment à octroyer, le 14 mars, la constitution réclamée depuis des mois, mais celle-ci n'était de nouveau qu'une
demi-mesure [2].

L'incohérence de cette politique de concessions au compte-gouttes et
de réformes réalisées par à-coups, le plus souvent sous l'effet des manifestations de rue, commençait à exaspérer les libéraux modérés ; elle inquiétait en même temps tous les observateurs sérieux, qui comprenaient le
danger de « ces alternatives de recherche de la popularité et de molles
tentatives de réaction, qui sont le véritable moyen de hâter les révolutions et jouent comme un soufflet de forge pour enflammer un brasier » [3].
Il devenait de plus en plus évident que, « ferme et doué d'une grande
énergie dans tout ce qui tenait au domaine de l'Église et à son pouvoir
spirituel », Pie IX était par contre, comme prince temporel, « incliné
à la faiblesse et à l'irrésolution » [4].

---

(1) Sur cette affaire de la *Consulta*, voir A. Ghisalberti, *Nuove ricerche...*, chap. III.
(2) *Statut fondamental pour le gouvernement temporel des États de la Sainte Église*, texte dans
Atti del S. Pont. Pio IX, Rome, 1857, Pars II[a], t. I, p. 222-238. Cf. L. Wollenberg, *Lo Statuto
Pontificio nel quadro costituzionale del 1848*, dans *Rass. St. R.*, t. XXII, 1935, p. 527-594.
(3) A. de Broglie, *Mémoires*, t. I, p. 165.
(4) E. de Ligne, *Le pape Pie IX à Gaëte. Souvenirs d'un diplomate belge*, dans *Le Correspondant*,
t. CCCXV, 1929, p. 194 et 173.

L'ÉCHEC DE LA POLITIQUE
ITALIENNE DU PAPE

Ce caractère hésitant du pape, incapable de choisir entre les conseils qui lui venaient de droite et de gauche, dominé par le désir utopique d'agir de manière à ne déplaire à personne, devait se manifester tout autant, et avec des conséquences plus graves encore, dans son attitude à l'égard du mouvement italien.

Il est inexact de prétendre, avec beaucoup d'historiens du *Risorgimento*, que Pie IX ait commencé par soutenir d'abord le programme libéral d'unification italienne, pour l'abandonner par la suite ; mais il l'est tout autant d'affirmer, comme l'ont fait souvent les historiens catholiques, qu'il n'eut jamais de politique italienne bien arrêtée et se borna à quelques gestes vagues auxquels on donna à l'époque une portée qu'ils n'avaient pas. La vérité est entre ces deux positions extrêmes. Pie IX ne pouvait évidemment pas accepter l'idée mazzinienne d'une république italienne unitaire, puisque celle-ci impliquait la suppression de la souveraineté pontificale, et l'on a vu qu'il trouvait le programme néoguelfe trop ambitieux et incompatible avec sa mission spirituelle. Toutefois, après quelques tâtonnements, il s'était rallié aux conceptions de certains modérés qui, envisageant la possibilité de diminuer l'influence autrichienne en Italie à l'occasion d'un remaniement pacifique de la carte d'Europe, cherchaient à préparer cette éventualité en resserrant les liens entre les différents États italiens. C'est dans cet esprit qu'il avait immédiatement cherché à donner une orientation politique aux négociations qui s'engagèrent au mois d'août 1847 en vue de la conclusion d'une ligue douanière avec la Toscane et le Piémont, et qu'il accueillit favorablement l'idée, venue de Florence et fort appuyée par Mgr Corboli-Bussi, d'unir les princes italiens dans une ligue défensive.

La politique pontificale, qui tendait à promouvoir une solution pacifique de la question italienne, allait malheureusement se trouver rapidement dépassée par les événements. La fièvre montait, en effet, dans toute l'Italie et, tandis que la Sicile se soulevait contre l'absolutisme des Bourbons et que, dans le Nord, la révolte grondait contre l'Autriche, certains agitateurs cherchaient à entraîner dans le mouvement les sujets du pape en brandissant le spectre d'un coup de force autrichien et en les persuadant que la meilleure sauvegarde pour défendre leur liberté menacée consistait à prendre les armes préventivement et à participer, en étroite union avec les autres États italiens, à la guerre imminente. C'est dans ces circonstances que Pie IX prononça, le 10 février 1848, une allocution où, après avoir rappelé qu'à l'époque des grandes invasions ce n'était pas la force militaire mais le prestige spirituel de Rome, centre de la catholicité, qui avait permis d'éviter la ruine totale, il concluait dans une éloquente envolée : « Bénissez donc, ô grand Dieu, l'Italie, et conservez-lui ce don précieux entre tous, la foi ! », cette foi chrétienne qui avait toujours été et serait encore à l'avenir la meilleure sauvegarde des populations de l'État pontifical.

De ces paroles, une seule chose frappa les auditeurs, non seulement la foule électrisée par des tribuns du genre du fameux Ciceruacchio, mais

même des esprits réfléchis : Pie IX appelait les bénédictions du ciel sur
l'*Italie*, cette Italie dont Metternich prétendait qu'elle n'était qu'une
expression géographique et qui aspirait au contraire à renaître unifiée
après des siècles de décadence et de divisions. Dès lors, « dans le pape
qui priait Dieu de conserver la foi à l'Italie, on voulut voir un Jules II
bénissant la guerre » [1] et, bien malgré lui, Pie IX contribua ainsi à accentuer encore l'équivoque et à accroître, dans des proportions jamais
encore atteintes, l'exaltation populaire du nord au sud de la péninsule.
« Cette bénédiction donnée à l'Italie équivaut à une malédiction pour
l'Autriche, à une croisade », écrivait plein d'enthousiasme à l'un de ses
amis le patriote Farini, et ils étaient des milliers à penser comme lui.

On comprend dans ces conditions que, lorsqu'à la fin de mars, à la
suite de la révolution de Vienne et du soulèvement de la Lombardie
contre l'Autriche, le gouvernement piémontais prit les armes pour se
porter au secours des insurgés, les Italiens, et les Romains plus que personne, étaient persuadés que le pape allait s'engager activement dans la
« guerre sainte » dont l'avenir de la patrie était l'enjeu. Tandis que l'archevêque de Florence invitait à exposer le Saint-Sacrement et à prier Dieu
pour le triomphe de la cause italienne, et que celui de Milan, après avoir
autorisé ses séminaristes à s'inscrire dans les bataillons de volontaires
estudiantins, promettait la coopération du clergé à l'œuvre de « la libération complète de l'Italie » [2], le gouvernement provisoire adressait
au pape, dès le 25 mars, un message annonçant que les insurgés s'étaient
battus au cri de « Vive Pie IX ! ».

La situation de ce dernier était dramatique : le père commun des
fidèles pouvait-il déclarer la guerre à une nation catholique, au risque de
pousser vers le schisme l'ensemble des pays germaniques, que les prélats
austrophiles lui présentaient habilement comme plus travaillés encore
qu'ils ne l'étaient par les tendances joséphistes ? Mais, par ailleurs, le
prince italien pouvait-il se désintéresser de la cause nationale, à un
moment surtout où il paraissait difficile de contenir l'ardeur belliqueuse
de beaucoup de ses sujets ? Le problème aurait été plus facile à résoudre
si le Piémont n'avait pas fait échouer, par crainte de limiter sa liberté
d'action, le projet de ligue défensive grâce auquel l'intervention des
troupes pontificales ne serait plus apparue comme le résultat d'une
décision délibérée du pontife, mais comme la conséquence automatique
d'un traité qui le liait. On décida donc à Rome de faire une dernière
tentative en ce sens et, le 10 avril, Mgr Corboli-Bussi partait pour le camp
de Charles-Albert, officiellement pour lui exposer les raisons morales et
les difficultés matérielles qui rendaient impossible une déclaration de
guerre immédiate, mais en fait dans le but de le gagner à l'idée de la
ligue. Le Piémont toutefois, présumant de ses forces et espérant exploiter

---

(1) A. GHISALBERTI, art. *Pio IX*, dans *Enciclopedia Italiana*, t. XXVII, p. 321. Voir le texte
de la proclamation du 10 février dans L. FARINI, *Lo Stato Romano dall'anno 1815 al 1850*, Florence, 1850, t. I, p. 340-342 ; sur la portée exacte du passage : *O benedite, gran Dio, l'Italia*, cf. D.
MASSÈ, *op. cit.*, p. 36-39.

(2) L'appui du clergé lombard fut effectivement enthousiaste et efficace : cf. A. MARAZZA,
Il clero lombardo nella rivoluzione del' 48, Milan, 1948.

à son profit exclusif le réveil italien, ne voulait pas renoncer au rôle prépondérant qu'il avait tenu jusqu'alors : il souhaitait un appui militaire des autres États italiens, mais il désirait être seul à mener le jeu politique. Aussi se déroba-t-il à nouveau, d'autant plus que le délégué pontifical, personnellement favorable à l'intervention, outrepassa ses instructions et donna l'impression que le pape était prêt à laisser l'armée pontificale passer spontanément à l'action [1].

L'ALLOCUTION DU 29 AVRIL ET SES SUITES   Or en fait, malgré sa sympathie profonde pour la cause italienne, Pie IX était bien décidé à se désolidariser ouvertement d'une intervention militaire qui lui paraissait manifestement incompatible avec sa mission religieuse, mais que, débordé par ses ministres laïques, il se voyait de plus en plus impuissant à empêcher, comme il le confiait, le 25 avril, à un diplomate :

> Mon autorité s'affaiblit chaque jour, le pouvoir que j'exerce, en ce qui concerne le temporel, n'est pour ainsi dire plus que nominal. Ne veulent-ils pas, ces hommes dont le patriotisme exalté ne connaît plus aucun frein, me faire déclarer la guerre, à moi chef d'une religion qui ne veut que la paix et la concorde ? Eh bien ! je protesterai ; l'Europe saura la violence qu'on m'a faite, et si l'on veut continuer à exiger de moi des choses que ma conscience repousse, je me retirerai dans un couvent pour y pleurer sur les malheurs de Rome [2].

Quelques jours plus tard, le 29 avril, il convoquait à l'improviste un consistoire et déclarait solennellement que, pasteur suprême, il ne pouvait déclarer la guerre à une nation dont les ressortissants étaient ses fils spirituels :

> Fidèle aux obligations de Notre suprême apostolat, nous embrassons tous les pays, tous les peuples, toutes les nations, dans un égal sentiment de paternel amour [3].

Cette allocution, qui dissipait brutalement une équivoque trop longtemps entretenue, produisit chez les patriotes une déception d'autant plus vive que les milieux conservateurs s'empressèrent de l'exploiter en prétendant y voir une désapprobation de la lutte nationale dans son ensemble, ce qui n'était certainement pas l'intention du pape. Très étonné de l'interprétation qu'on donnait à ses paroles, celui-ci voulut faire une mise au point ; malheureusement, le texte de celle-ci, rédigé par Mgr Pentini de manière à rassurer les patriotes, fut retouché subrepticement par le cardinal Antonelli [4] et ne produisit pas l'effet attendu. En même temps, soucieux d'apporter au mouvement national tout l'appui qu'il jugeait compatible avec ses fonctions religieuses et son rôle supranational, le pape tenta une démarche directe auprès de l'empereur d'Au-

---

(1) P. Pirri, *La missione di Mons. Corboli-Bussi...*, p. 57, 66 et 69.
(2) A. de Liedekerke, *Rapporti delle cose di Roma*, p. 39.
(3) Texte dans *Acta Pii IX*, t. I, p. 92-98. Cf. J. Muller, *Die Allokution Pius IX vom 29 April 1848*, Bâle, 1928 et L. Salvatorelli, *L'allocuzione di Pio IX del 29 Aprile 1848*, dans *Prima e dopo il Quarantotto*, p. 127-131. Sur les origines de l'allocution, cf. D. Massè, *op. cit.*, p. 115-127 et P. Pirri, *La missione di Mons. Corboli-Bussi...*, p. 63-64, 68 et 80.
(4) Cf. G. Pasolini, *Memorie raccolte da suo figlio*, p. 103 ; C. Spelanzon, *Storia del Risorgimento*, t. IV, p. 381. Texte définitif de la proclamation dans L. Farini, *Lo Stato romano...*, t. II, p. 106-108.

triche. Sa lettre, datée du 3 mai, mais dont la première idée avait été suggérée par Mgr Corboli-Bussi antérieurement à la fameuse allocution du 29 avril [1], montre combien Pie IX se trouvait encore sous l'influence de l'idéologie du moment, car — détail trop rarement remarqué — elle reconnaissait implicitement la supériorité des droits des nationalités sur le droit divin des rois et le caractère intangible des traités [2]. Mais cet essai de médiation, qui échoua complètement et fut même exploité par certains contre le pape, ne lui fit pas regagner les sympathies que l'allocution lui avait fait perdre. L'idéal néoguelfe d'une fédération italienne présidée par le pape ne disparut sans doute pas aussi immédiatement qu'on le dit d'ordinaire, mais, dès ce moment, beaucoup commencèrent à déclarer ouvertement que, s'il y avait une incompatibilité entre les fonctions religieuses du pape et ses obligations de prince italien, il n'y avait qu'à en tirer les conséquences et à condamner une souveraineté temporelle qui s'avérait nuisible à la patrie. La question romaine, telle qu'elle allait se poser pendant les vingt années suivantes, était virtuellement ouverte.

A Rome même, la situation politique, déjà très tendue, s'aggrava brusquement. Les éléments avancés, dont beaucoup partageaient l'idéal républicain de Mazzini, s'empressèrent de profiter de la profonde déception causée par ce que l'on considérait comme la volte-face — certains disaient plus : la trahison — de Pie IX, pour substituer auprès des patriotes leur influence à celle des libéraux modérés, qui se trouvaient discrédités pour avoir jusque-là fait confiance au pape, et, sous l'effet de l'agitation qui s'ensuivit, celui-ci se vit contraint d'accepter un ministère soumis en fait à la pression constante des « clubs » radicaux, bien que son président, le comte Mamiami, fût un homme relativement modéré. Il n'est guère étonnant, dans ces conditions, que l'anarchie n'ait fait que se développer au cours de l'été : à Rome, à certains jours, le pape semblait « presque prisonnier dans son propre palais », tandis que, en province, les attentats politiques se multipliaient et tout se passait de plus en p'us « comme si le souverain pontife, absolument dépouillé de son autorité, n'était plus là que pour la forme » [3].

L'AGGRAVATION DE LA CRISE　　Les difficultés provoquées par la guerre contre l'Autriche et la quasi-impossibilité de trouver une formule viable de papauté constitutionnelle auraient suffi à rendre la situation redoutable. Elle était rendue plus inextricable encore par le mécontentement né de la crise économique qui sévissait dans l'État pontifical comme dans toute l'Europe. Le petit peuple, victime du chômage dû à l'arrêt du tourisme et à la mévente des objets d'art, après avoir souffert l'année précédente de la hausse des denrées alimentaires, était une proie facile pour les agitateurs, qui s'en ser-

---

(1) Cf. P. PIRRI, *La missione di Mons. Corboli-Bussi*, p. 67-69, 76-79. Le texte de la lettre est publié par F. GENTILI, *La lettera di Pio IX all'Imperatore d'Austria*, dans *Nuova Antologia*, 1er août 1914, p. 458-459.
(2) L. SALVATORELLI, *La rivoluzione europea*, Milan, 1949, p. 211-212.
(3) A. DE LIEDEKERKE, *Rapporti delle cose di Roma*, p. 65 et 72.

vaient habilement pour faire pression sur le pape et lui arracher, sous la crainte de l'émeute, de nouvelles concessions. Inversement, les propriétaires fonciers, le clergé en particulier, sévèrement touchés par les mesures financières nécessitées par la crise, en rendaient responsable la politique belliqueuse du gouvernement et s'indignaient de ce que le pape n'intervînt pas avec plus d'énergie pour s'opposer à celle-ci [1].

Pour reprendre en main une situation à ce point compromise, il eût fallu un homme d'État de génie. Le pauvre Pie IX en était bien incapable, surtout en ce moment où, profondément abattu par sa soudaine impopularité, « ayant presque entièrement perdu cette gaîté douce et bienveillante qui prêtait à ses entretiens un charme tout particulier » [2], il était devenu plus hésitant que jamais. La fraction réactionnaire de son entourage l'entretenait savamment dans la crainte qui le rongeait, de compromettre, par des concessions politiques, le caractère sacré de son autorité pontificale et l'indépendance nécessaire à sa position religieuse. Cependant, le comte de Liedekerke voyait juste quand il écrivait à la fin de mai :

Son cœur est si bon, il est au fond si Italien, il a si besoin d'être entouré de l'amour et de la confiance de ses sujets, il attache tant de prix à être salué par leurs acclamations, que je ne serais pas surpris de le voir, un de ces jours, poser des actes détruisant peu à peu la portée de l'allocution du 29 avril [3].

La faveur témoignée à Antonio Rosmini, venu à Rome comme délégué du gouvernement de Turin, est une preuve parmi d'autres de l'exactitude de ce diagnostic et du désir, encore profondément ancré dans l'âme de Pie IX, de ne pas renier l'idéal qu'il avait incarné pendant les deux premières années de son pontificat. Il insista en effet pour que Rosmini restât à Rome, une fois sa mission terminée, lui faisant part de son désir de faire appel à ses services et lui laissant même entrevoir la possibilité de le nommer cardinal et secrétaire d'État ; or l'abbé piémontais ne s'était pas contenté de prôner des vues réformatrices hardies dans son opuscule sur *La Constitution selon la Justice sociale*, mais il avait, dans le courant de mai, par l'intermédiaire du cardinal Castracane, conseillé à plusieurs reprises au pape de se prononcer en faveur de la guerre contre l'Autriche [4]. En même temps, toutefois — nouvelle preuve de l'incapacité de Pie IX à se décider pour une ligne de conduite bien définie, — celui-ci confiait la direction du ministère à l'ancien ambassadeur de Louis-Philippe, le comte Pelegrino Rossi, libéral modéré lui aussi, mais hostile à la politique italienne du Piémont et connu pour être partisan d'un régime constitutionnel à la française, au sujet duquel Rosmini faisait les plus expresses réserves [5].

---

(1) Cet aspect économique et social de la révolution romaine, négligé jusqu'à présent, a été bien mis en lumière par D. DEMARCO, *op. cit.* (surtout p. 133-140).
(2) A. DE LIEDEKERKE, *op. cit.*, p. 61.
(3) *Ibid.*, p. 51.
(4) On trouvera les lettres où il exposait son point de vue dans *La vita di A. Rosmini scritta da un sacerdote dell'Istituto della Carità*, Turin, 1897, t. II, p. 169 et suiv. (Cf. D. MASSÈ, *op. cit.*, p. 154-155.)
(5) Sur l'opposition de points de vue entre Rosmini et Rossi, cf. W. LOCKHART, *Ant. Rosmini-Serbati*, trad. SEGOND, Paris, 1889, p. 255-256 et A. CAPOGRASSI, *La Conferenza di Gaeta e Antonio Rosmini*, Rome, 1941, p. 5-6.

En faisant appel à Rossi, le pape espérait donner satisfaction à la fois aux gens d'ordre et à certaines revendications de la bourgeoisie libérale, mais le choix n'était pas heureux, car Rossi n'était populaire ni à droite ni à gauche. Son libéralisme doctrinaire et son intention d'abolir certains privilèges ecclésiastiques le rendaient suspect au clergé, tandis que ses allures cassantes et les dispositions sévères qu'il prit pour mettre fin à l'agitation endémique le firent traiter de réactionnaire et même de « dictateur » par les éléments avancés. Son hostilité déclarée à la reprise de la guerre lui aliénait en outre les sympathies de beaucoup de patriotes. Aussi Rossi ne fut-il guère regretté lorsque, le 15 novembre, il fut assassiné à l'entrée du Parlement, au moment où il s'apprêtait à exposer son programme de réformes [1].

LA FUITE DU PAPE — Ce meurtre, perpétré en plein jour et avec la complicité de nombreux assistants, précipita les événements. Le lendemain, les émeutiers assiégeaient le pape dans son palais du Quirinal en exigeant la convocation d'une Constituante et la déclaration de guerre à l'Autriche, tandis qu'en ville, cardinaux et prélats étaient l'objet de menaces de tout genre. Beaucoup d'entre eux s'empressèrent de quitter la ville sous des déguisements divers. Quant au pape, profondément impressionné par ces événements et persuadé désormais qu'il était vain d'espérer transiger avec la révolution, il confiait, dès le premier soir, à l'ambassadeur d'Espagne : « Il vaudra mieux abandonner la place ! » Une partie de son entourage le poussait d'ailleurs dans ce sens, escomptant ainsi compromettre les démocrates en les faisant passer pour des persécuteurs et spéculant sur les désordres qui suivraient le départ des autorités régulières pour discréditer complètement les libéraux et frayer les voies à la réaction. Personnellement étranger à ces intrigues politiques, Pie IX, ballotté entre des conseils contradictoires, hésita pendant quatre jours, puis se décida pour le départ. Après avoir un moment songé à se rendre aux Baléares, il accepta la proposition de l'ambassadeur de France qui lui offrait l'hospitalité de son pays. Le comte de Spaur, ministre de Bavière, se chargea d'organiser l'évasion et de conduire le pape jusqu'au port napolitain de Gaëte, où un navire français viendrait le chercher. Le 24 novembre, à la fin de l'après-midi, le pape, vêtu comme un simple ecclésiastique et portant une paire de lunettes foncées, s'enfuyait de son palais et réussissait à quitter la ville sans encombre [2].

A peine le roi de Naples eut-il connaissance de la présence de Pie IX dans ses États qu'il s'empressa d'accourir et insista vivement pour qu'il n'allât pas plus loin. Cette proposition fut vivement appuyée par les

---

(1) Sur les antécédents de l'assassinat de Rossi et sur les responsabilités directes et indirectes, cf. R. GIOVAGNOLI, *Pellegrino Rossi e la rivoluzione romana*, Rome, 1898-1911, 3 vol., ainsi que G. BRIGANTE COLONNA, *L'uccisione di Pellegrino Rossi*, Milan, 1938.

(2) Le meilleur exposé des faits est celui de G. MOLLAT, *La fuite de Pie IX à Gaëte*, dans *R. H. E.*, t. XXXV, 1939, p. 266-282 ; l'auteur donne, en commençant, un exposé critique des sources, dont la plus importante est la relation du comte Spaur lui-même (publiée par L. SIMEONI dans *Rivista storica del Risorgimento italiano*, t. XIX, 1932, p. 252-263). Il faut ajouter à présent P. PIRRI, *Relazione inedita di Seb. Liebl sulla fuga di Pio IX a Gaeta*, dans *Miscellanea Pio Paschini*, Rome, 1949, t. II, p. 421-451.

ministres de Bavière, de Portugal et de Belgique, ainsi que par le cardinal Antonelli. Ils craignaient qu'un séjour en France n'encourageât les velléités libérales du pape et n'apparût comme une consécration du régime républicain. Ils lui firent remarquer le double intérêt qu'il avait à ne pas « indisposer les cours monarchiques » et à ne pas s'éloigner de ses États. Rosmini, qui avait rejoint le pape, objecta bien que rester à Gaëte serait se compromettre avec un prince détesté par les patriotes italiens à cause de sa politique réactionnaire et de ses accointances avec l'Autriche, mais Antonelli finit par l'emporter, et le pape fut bientôt décidé à ne pas aller plus loin, tout en laissant entendre à l'ambassadeur de France qu'il ne renonçait pas complètement à son premier projet [1].

A GAETE Le duel d'influence entre Rosmini et Antonelli se poursuivit pendant les premières semaines du séjour à Gaëte. Antonelli n'avait cependant pas fait partie jusqu'alors de la fraction réactionnaire ; il s'était montré favorable en général au mouvement de réformes inauguré par Pie IX et rien ne prouve qu'il n'était pas sincère ; il avait joué un rôle important dans l'élaboration du *Statuto* et appuyé la politique de Rossi. Mais, depuis l'assassinat de ce dernier, il avait changé de tactique et n'attendait plus le salut que d'un rétablissement de l'ancien régime avec l'aide de l'Autriche conservatrice [2]. Aussi, tandis que Rosmini conseillait au pape de ne pas rompre les ponts avec le Parlement de Rome, Antonelli refusa-t-il brutalement de recevoir une délégation envoyée par celui-ci pour prier le pape de rentrer dans sa capitale, et, le 4 décembre, il invitait les puissances européennes à intervenir par les armes afin de rétablir le pouvoir temporel du souverain pontife. Puis il détourna le pape d'adresser à ses sujets la proclamation conciliante qu'avait rédigée Rosmini et le poussa au contraire à désavouer solennellement, le 17 décembre, le gouvernement provisoire, qui répondit en convoquant une Constituante : la rupture était consommée.

Par tempérament, Pie IX aurait pourtant incliné plutôt vers la conciliation, mais il subissait une pression constante de la part du roi de Naples et de tout son entourage, du très antilibéral ministre de Bavière, et de nombreux prélats pétris d'ancien régime, qui tous approuvaient hautement la manière de faire d'Antonelli. « Pour autant que j'aie pu jusqu'à présent en juger, notait avec regret, le 13 décembre, un observateur modéré, l'air que l'on respire autour du pape m'a semblé bien *réactionnaire* ; l'on ne voit de salut, de garantie, de durée que dans l'emploi de la force [3]. » C'est en vain que Rosmini, se rendant compte du danger qu'il y avait à identifier la cause pontificale à celle de l'Autriche et des puissances conservatrices, conseilla au pape de recourir à une médiation piémontaise plutôt qu'à des troupes étrangères ; Antonelli, qui désirait absolument l'intervention autrichienne, n'eut pas de peine à exploiter les

---

(1) Y. DE CHAMBRUN, *Un projet de séjour en France du pape Pie IX*, dans *Revue d'histoire diplomatique*, t. L, 1936, p. 322 et suiv. et 481 et suiv., et E. MICHEL, *Documenti relativi al mancato viaggio in Francia di Pio IX*, dans *Rass. St. R.*, t. XXIII, 1936, p. 945-956.
(2) E. SODERINI, art. *Antonelli*, dans *Enciclopedia Italiana*, t. III, p. 547-548.
(3) E. DE LIEDEKERKE, *Rapporti delle cose di Roma*, p. 130.

préventions de Pie IX contre le gouvernement de Turin [1]. Aussi, sentant décliner son crédit, Rosmini se décida à quitter Gaëte et, bientôt, les machinations d'Antonelli l'obligèrent même à fuir le royaume de Naples, après avoir perdu le peu qui lui restait de la faveur pontificale [2].

A Rome, la Constituante élue le 21 janvier s'était empressée de confier le pouvoir exécutif à un triumvirat dirigé par Mazzini et, quinze jours plus tard, de proclamer la République dans les termes suivants :

Art. I. — Le pape est déchu de fait et de droit du gouvernement temporel de l'État romain.

Art. II. — Le pontife romain aura toutes les garanties d'indépendance nécessaires pour l'exercice de sa puissance spirituelle.

Art. III. — La forme du gouvernement de l'État romain sera la démocratie pure et prendra le nom glorieux de République romaine.

Cette évolution des choses était inévitable, car l'intransigeance de l'attitude pontificale avait achevé de discréditer les modérés, disposés à collaborer, et laissé le champ libre aux républicains avancés. Il faut noter toutefois que cette république, qui entendait rester catholique, ne prit à aucun moment un caractère extrémiste et que, les premiers moments d'excitation passés, prêtres et prélats n'eurent plus à craindre de se voir molestés.

L'EXPÉDITION DE ROME — A peine la République romaine avait-elle été proclamée que, le 14 février, le pape protesta solennellement en présence du Sacré Collège et du corps diplomatique et, quatre jours plus tard, le cardinal Antonelli adressait un nouvel appel aux puissances, conviant l'Autriche, la France, l'Espagne et Naples à intervenir par les armes. Dans l'esprit de Pie IX, il s'agissait d'une sorte de croisade à but purement religieux ; Antonelli y voyait surtout la seule solution permettant à la fois la restauration du pape et l'abolition, qu'il jugeait indispensable, du *Statuto*.

Le gouvernement français, dans le but de gagner du temps, commença par proposer la réunion d'une conférence des quatre puissances à Gaëte [3]. La situation de la France était en effet délicate : les catholiques n'avaient porté Louis-Napoléon à la présidence qu'avec la promesse d'un appui au pape et ils avaient dans le gouvernement un organe influent en la personne de Falloux, mais par ailleurs l'opinion démocrate et républicaine était nettement hostile à une intervention militaire dirigée contre un gouvernement constitutionnel et national, qui se réclamait des principes de la Grande Révolution, et on ne voulait pas la heurter de front à la

---

(1) A cause de sa politique de laïcisation et à cause des rapports officieux qu'il entretenait avec le gouvernement révolutionnaire de Rome. (Cf. P. Pirri, *Pio IX e Vittorio Emanuele II dal loro carteggio privato*, t. I, *La Laicisatione dello stato sardo*, Rome, 1944, p. 4*-24*.)

(2) Rosmini a raconté lui-même les faits dans un mémoire anonyme : *Della missione a Roma di Ant. Rosmini-Serbati negli anni 1848-1849*, Turin, 1881. Cette narration est confirmée et complétée par les documents provenant des archives napolitaines utilisés par A. Capograssi, *op. cit.*, en particulier p. 81-92 et 219-233. D'autres qu'Antonelli avaient du reste travaillé à le discréditer auprès du pape. (Cf. P. Rocfer, *Souvenirs d'un prélat romain...*, Paris, 1895, p. 38-40 et A. Dyroff, *Rosmini*, Munich, 1906, p. 83-84.)

(3) Il y eut 13 séances, du 30 mars au 22 septembre. La première idée était venue du roi de Naples qui, en janvier, avait suggéré de réunir les représentants de toutes les puissances, même non catholiques, afin d'examiner la question italienne dans son ensemble, mais le refus de la Russie avait fait échouer ce premier projet. (Cf. A. Capograssi, *op. cit.*, p. 22-23.)

veille des élections de mai, dont on attendait un revirement vers la droite. Ces préoccupations de politique intérieure expliquent les hésitations de la politique française dans la conduite de l'affaire italienne. Dans l'espoir que la seule présence des troupes françaises suffirait à décourager Mazzini et les extrémistes, on se décida bien, vers le milieu d'avril, à l'envoi d'un corps expéditionnaire, en présentant la mesure comme destinée à empêcher l'Autriche de prendre trop d'influence en Italie ; mais dès qu'il apparut que les adversaires du « gouvernement des prêtres » s'appuyaient sur une importante fraction de la population, plutôt que d'enjoindre au général Oudinot d'activer les opérations militaires, on préféra dépêcher aux Romains un diplomate, le jeune Ferdinand de Lesseps, chargé de négocier une transaction raisonnable [1]. C'est seulement après que les élections eurent amené à l'Assemblée législative une majorité conservatrice que les troupes françaises reçurent l'ordre d'attaquer et de rétablir par la force le gouvernement du pape : le 3 juillet, elles s'emparaient de Rome, tandis que les trois autres puissances occupaient le reste de l'État pontifical.

Le gouvernement français espérait toutefois que la restauration du pape s'accomplirait dans une atmosphère libérale et, à Gaëte, ses diplomates avaient insisté depuis le début pour que les libertés constitutionnelles précédemment accordées ne fussent pas retirées. Mais, dès le 4 juin, le nonce à Paris avait fait savoir, en s'appuyant sur une déclaration de Falloux, que l'on n'insisterait pas outre mesure [2] et, un peu plus tard, Louis-Napoléon fut obligé, devant l'indignation des catholiques français, de désavouer sa lettre au colonel Ney, dans laquelle il disait entre autres : « La République française n'a pas envoyé une armée à Rome pour étouffer la liberté italienne... Je résume ainsi le rétablissement du pouvoir temporel du pape : amnistie générale, sécularisation de l'administration, code Napoléon et gouvernement libéral [3]. »

Antonelli, soutenu par l'Autriche et les autres puissances, estima dès lors qu'il pouvait se montrer irréductible et Pie IX oublia d'autant plus aisément ses promesses formelles de ne pas toucher au *Statuto* qu'il restait sous l'impression de l'échec de Rossi, celui précisément qui l'avait pendant deux ans poussé dans la voie des réformes libérales, et qu'il était par ailleurs frappé de constater que plusieurs de ses anciens conseillers, Corboli-Bussi par exemple, devenaient eux-mêmes très réservés [4]. La période libérale de son règne était définitivement close.

LA RÉACTION CONSERVATRICE   L'évolution des idées se traduisit rapidement dans les faits. Déjà au consistoire du 20 avril 1849, le pape s'était exprimé en termes sévères sur les écrits

---

(1) Sur les circonstances de cette mission, voir G. EDGAR-BONNET, *Ferdinand de Lesseps. Le diplomate. Le créateur de Suez*, Paris, 1951, p. 80-114.
(2) P. PIRRI, *Pio IX e Vittorio Emanuele...*, t. I, p. 28*-29*.
(3) Texte complet de cette lettre du 18 août dans A. DE FALLOUX, *Mémoires d'un royaliste*, t. I, p. 528.
(4) Sur les motifs qui influencèrent l'évolution de Pie IX, voir entre autres A. MONTI, *Pio IX nel Risorgimento*, p. 112 : A. CAPOGRASSI, *op. cit.*, p. 69 et P. PIRRI, *Pio IX e Vittorio Emanuele...*, t. I, p. 26*-30*.

de Gioberti et, le 30 mai, la Congrégation de l'*Index*, réunie à sa demande, condamnait *Il Gesuita moderno* en même temps que les deux ouvrages où Rosmini avait exposé ses projets de réforme et qu'un opuscule du P. Ventura [1]. « Vous voyez, gémissait ce dernier, nous étions en Italie trois prêtres renommés, ils nous ont condamnés tous les trois ! [2] »

Mais ce n'était là qu'un prélude. Dans le courant du mois d'août, les premières mesures prises par les trois cardinaux chargés de restaurer l'autorité pontificale — le « triumvirat rouge » — indiquèrent clairement l'orientation de la nouvelle politique, et cette impression fut plus que confirmée par la publication d'une amnistie qui n'épargnait en somme que ceux qu'il était impossible de frapper à cause de leur grand nombre, et surtout par le *motu proprio* du 18 septembre, qui ne soufflait mot des libertés politiques et présentait un programme établi « sur des bases en retard de dix-huit années sur les nécessités du moment » [3], puisque ce n'était au fond « qu'une édition revue et corrigée du *Memorandum* de 1831 » [4]. Les seules réformes annoncées étaient d'ordre administratif et même des observateurs peu suspects de sympathies libérales, comme le consul de Wurtemberg, devaient reconnaître que « l'on en revenait ouvertement et sans réserve à l'ancien système de l'absolutisme pur et simple » [5].

Ceux qui, sous la conduite d'Antonelli, présidaient à cette « réaction aveugle et passionnée » [6], avaient préféré retarder le retour du pape. Dès le mois de février précédent, le comte de Liedekerke avait décelé les motifs de cette politique :

Sachant combien (le Saint-Père) est peu fait, par la bonté de son cœur, pour ordonner des mesures de rigueur, et combien ce même cœur est accessible à toutes les infortunes, quelle qu'en soit l'origine, on voudrait, une fois Rome occupée par les forces étrangères, l'en tenir longtemps éloigné pour n'être pas contrarié par sa présence dans le développement du plan réactionnaire que la camarilla s'est proposé de poursuivre et dont on ne lui a probablement laissé entrevoir que les dispositions les moins acerbes, afin qu'il ne se refusât pas à le sanctionner [7].

Ce n'est qu'en avril 1850 que Pie IX rentra enfin dans sa capitale, où l'accueil fut plutôt morne, et lorsque le nouveau régime, qui « portait le cachet du caractère dominateur du cardinal Antonelli » [8] en favorisant

---

(1) F. REUSCH, *Der Index der verbotenen Bücher*, t. II, Bonn, 1885, p. 1132-1134, 1137-1138, 1141. Le cardinal Mai avait poussé à la condamnation de Gioberti après l'avoir déconseillée dix-huit mois plus tôt (U. PADOVANI, *V. Gioberti*, p. 394-395), mais il préféra donner sa démission comme préfet de l'*Index* plutôt que de signer celles de Rosmini et de Ventura. Il s'agissait d'un discours de ce dernier en l'honneur des morts de la révolution de Vienne, du projet de constitution romaine de Rosmini, et des *Cinque piaghe della Santa Chiesa* (les cinq plaies en question étaient l'abandon de la langue vulgaire dans la liturgie, l'éducation fermée du clergé, l'isolement des évêques, l'absence de participation des laïcs et du bas-clergé dans l'élection de ceux-ci, et l'administration défectueuse des biens d'Église).

(2) J. GAY, *Les deux Rome et l'opinion française*, Paris, 1931, p. 84.

(3) A. GHISALBERTI, *Una restaurazione « reazionaria e imperita »*, dans *Archivio della società romana di Storia patria*, t. LXXII, 1949, p. 145.

(4) Rapport d'A. de Liedekerke du 22 septembre 1849, cité *ibid.*, p. 150.

(5) Rapport du 20 septembre 1849, cité par R. MOSCATI, *La diplomazia europea e il problema italiano nel 1848*, Florence, 1946, p. 180-184.

(6) Rapport d'A. de Liedekerke du 2 janvier 1850, dans A. GHISALBERTI, *art. cit.*, p. 155.

(7) A. DE LIEDEKERKE, *Rapporti delle cose di Roma*, p. 158.

(8) G. MOLLAT, *La question romaine*, p. 276.

la concentration de l'autorité entre les mains du secrétaire d'État, eut été définitivement mis au point, nombreux furent les Romains qui, oubliant la profonde différence de tempérament entre les deux pontifes, s'écrièrent avec l'ancien chef libéral Mamiami : « Grégoire XVI est ressuscité ; il règne à nouveau au Quirinal ! [1] »

LA FONDATION DE
LA « CIVILTA CATTOLICA »

Si Pie IX avait, jusqu'en 1848, manifesté une certaine indulgence à l'égard des *institutions* libérales, il n'avait jamais éprouvé la moindre sympathie pour les *principes* libéraux, et la manière dont les événements avaient évolué ne pouvait que renforcer ses convictions. On en eut un indice parmi d'autres dans l'invitation qu'il adressa en 1850 au clergé romain de remplacer la « tenue d'abbé » — culottes serrées au genou, redingote et tricorne — par la soutane longue (*toga talaris*, ou *abito piano*), afin de mieux marquer la distance entre les hommes d'Église et « les hommes du siècle, infestés de principes révolutionnaires » [2].

D'une portée autrement considérable fut la part qu'il prit, avant même son retour à Rome, dans la fondation de la revue des jésuites italiens, *La Civiltà cattolica*, dont le rôle devait être si important pendant tout son pontificat. Depuis 1847, l'idée d'un organe périodique dirigé par la Compagnie était dans l'air, mais le P. Roothaan inclinait plutôt pour une revue savante rédigée en latin. Au lendemain des orages de 1848, un jeune père napolitain, Carlo Curci, estima qu'il serait plus utile de lancer une revue bimensuelle de culture générale, s'adressant avant tout aux laïcs et destinée à servir d'antidote à la propagation des idées révolutionnaires en rappelant et en appliquant à tous les domaines les principes chrétiens. Le P. Général, qui craignait d'attirer sur la Compagnie les foudres des pouvoirs publics, surtout si la revue abordait les questions politiques, commença par faire des difficultés, mais, impatient d'agir, Curci sollicita une audience de Pie IX et lui exposa son plan. Il mit tant d'éloquence à le faire valoir que le pape, qui venait par un *motu proprio* d'inviter les évêques du monde entier à défendre la vérité par la presse, approuva aussitôt le projet et offrit même de prendre à sa charge les frais du premier numéro. Comme le P. Roothaan renouvelait ses objections et faisait entre autres observer que les Constitutions de saint Ignace interdisaient aux jésuites de traiter d'affaires politiques, Pie IX répondit que les inconvénients n'étaient pas à comparer avec les avantages d'une revue si utile à l'Église et qu'il avait le droit de dispenser d'un point de la règle. Le général s'inclina et adressa alors une circulaire aux provinciaux d'Italie, pour les encourager à soutenir la nouvelle œuvre.

(1) Lettre à E. Rendu, 23 octobre 1849, dans J. GAY, *Les deux Rome et l'opinion française*, p. 50. Il faut cependant, pour juger avec impartialité, reconnaître avec G. MOLLAT, *op. cit.*, p. 277, que « l'éducation politique du peuple était toute à faire et que le parti libéral, privé de chefs capables, avait montré son impuissance à triompher de la démagogie révolutionnaire. On comprend que dans ces conditions, au lendemain de troubles violents, Pie IX ait craint les pires aventures ». C. Balbo, qui s'était vainement efforcé, de mai à juillet 1849, d'amener le pape à aligner sa politique sur celle du Piémont, s'abstint cependant de le juger sévèrement, contrairement à ce qu'on dit souvent. (Cf. P. PIRRI, *op. cit.*, p. 32* et 22.)

(2) Cf. S. NEGRO, *Seconda Roma*, Milan, 1943, p. 150. La mesure, d'abord peu suivie, se généralisa lorsque, en 1851, le pape fut revenu à la charge.

Au début d'avril 1850 paraissait à Naples le premier numéro, tiré à
4.200 exemplaires. Six mois plus tard, le siège de la revue fut transféré
à Rome à cause de la censure tracassière du gouvernement napolitain
et bientôt la revue compta 12.000 abonnés, dont beaucoup en dehors
de l'Italie ; car on savait que la revue était suivie de près par le pape,
et le talent journalistique du P. Curci, qui écrivait avec beaucoup de
verve en une langue ample et sonore, ajoutait encore au succès du nouvel
organe officieux du Saint-Siège [1].

§ 2. — L'Église de France pendant la Seconde République [2].

*L'ÉGLISE DE FRANCE A
LA VEILLE DE LA RÉVOLUTION*

On était loin en 1848 de « l'alliance du
trône et de l'autel » qui avait été si dom-
mageable à l'Église au moment de la ré-
volution de 1830. Le gouvernement Guizot, inquiet devant l'agitation
démocratique et désireux de pouvoir compter sur la grande force d'ordre
qu'est l'Église, avait bien tenté un rapprochement au cours des dernières
années de la Monarchie de Juillet, mais les rapports entre les deux pouvoirs demeuraient froids pour divers motifs. D'abord, les sympathies
légitimistes, qui restaient vives dans le vieux clergé et chez de nombreux
curés de campagne, influencés par la noblesse retirée dans ses terres.
Ensuite, l'attitude de l'archevêque de Paris, Mgr Affre, qui réagissait
contre la « mentalité concordataire », assez répandue en France depuis
Napoléon, tendant à faire des évêques des « préfets violets ». « Homme
de son temps », selon l'expression d'une revue allemande au lendemain
de sa mort, Affre avait pris conscience de l'évolution en cours depuis
le début du siècle et du danger de lier la cause de l'Église à celle d'une
dynastie ou d'un régime politique, et surtout il se refusait à toute entente
qui se ferait aux dépens de la liberté de l'Église. Aussi ne manquait-il
pas une occasion de revendiquer, au grand déplaisir du roi, l'indépendance du pouvoir spirituel et de protester contre la prétention de l'État
à maintenir l'Église en tutelle ; il avait même fait parvenir à Pie IX,

---

(1) Sur les débuts de *La Civiltà cattolica*, voir *Il nostro centenario*, dans *Civ. catt.*, 1949, vol. II,
p. 5-40, et P. PIRRI, *Il P. Roothaan*, p. 463 et suiv.
(2) BIBLIOGRAPHIE. — I. SOURCES. — *Notes et lettres de Montalembert* [en 1848-1849], édit.
A. TRANNOY, dans *Revue historique*, t. CXCII, 1941, p. 253-289 ; *Correspondance de Montalembert
et de l'abbé Texier*, Paris, 1899 ; A. DE FALLOUX, *Mémoires d'un royaliste*, 2 vol., Paris, 1888, et
Le parti catholique, Paris, 1856 ; A. DE MELUN, *Mémoires*, t. II, Paris, 1891 ; T. W. ALLIES, *Journal
in France in 1845 and 1848*, Londres, 1849. En outre les collections de *L'Univers* et de *L'Ère nouvelle*, ainsi que H. WALLON, *Revue critique des journaux publiés à Paris depuis la Révolution de
février jusqu'à la fin de décembre*, Paris, 1849.
II. TRAVAUX. — Outre les ouvrages généraux et les biographies de Montalembert, Veuillot,
Dupanloup, Lacordaire, Maret, Melun, Sibour, Combalot, voir J. LEFLON, *L'Église de France
et la révolution de 1848*, Paris, 1948 ; F. MOURRET, *Le mouvement catholique en France de 1830 à 1850*,
Paris, 1917 ; J.-B. DUROSELLE, *Les débuts du catholicisme social en France*, Paris, 1951, IIe partie ;
A. TRANNOY, *Responsabilités de Montalembert en 1848*, dans *R. H. E. F.*, t. XXXV, 1949, p. 177-
206. A consulter aussi la précieuse collection de la revue *La Révolution de 1848*, ainsi que G. LEDOS,
Morceaux choisis et bibliographie de Lacordaire, Paris, 1923 et E. GALOPIN, *Essai de bibliographie
chronologique sur Ozanam*, Paris, 1933.
Pour la situation en province : H. CABANE, *Histoire du clergé de France pendant la Révolution
de 1848*, Paris, 1908 ; P. GENEVRAY, *Le clergé et les catholiques de Toulouse et de la Haute-Garonne
sous la République de 1848*, dans *La révolution de 1848 à Toulouse*, sous la dir. de J. GODECHOT,
Toulouse, 1948 ; A. DROULERS, *Le cardinal d'Astros et la République de 1848*, dans *Bulletin de littérature ecclésiastique*, 1950, p. 88-112 ; A. CHARLES, *La révolution de 1848 et la seconde République
à Bordeaux*, Bordeaux, 1946.

au lendemain de son avénement, un long mémoire élaboré de concert avec ses suffragants et contresigné par un certain nombre d'autres évêques, dans lequel il dénonçait les multiples tracasseries administratives dont l'Église avait à pâtir non seulement dans le domaine scolaire, mais lors des nominations de vicaires généraux, de curés, d'aumôniers de collège, et même à l'occasion de certains actes purement spirituels de son ministère [1]. Il faut enfin tenir compte de l'allure combative des laïcs groupés autour de Montalembert et de *L'Univers*, qui, encouragés par Pie IX [2], s'en prenaient avec une âpreté croissante au monopole universitaire devant l'obstination des partis au pouvoir, représentant la haute bourgeoisie voltairienne, à refuser la liberté de l'enseignement secondaire aux catholiques [3].

Cet engagement au service de l'Église de laïcs qui sont inévitablement amenés à prendre des initiatives et à influencer le cours des relations entre l'Église et l'État est évidemment un signe de vitalité, mais c'est aussi une nouveauté qui inquiète plus d'un évêque. Beaucoup en effet estiment que les intérêts de l'Église dans la vie publique devraient continuer à être défendus exclusivement par des tractations directes entre la hiérarchie et les membres du gouvernement. En présence des tentatives pour régler ces questions en faisant appel à l'opinion publique soit par la voie parlementaire soit par la presse, certains objectent, comme l'archevêque de Rouen, que « les laïcs n'ont pas mission pour s'occuper des affaires de l'Église » et commencent même à dénoncer ce qu'ils nomment le « laïcisme » comme la nouvelle hérésie, sans se rendre compte que l'éloquence d'un Montalembert ou la verve d'un Veuillot font plus pour l'Église que bien des mandements épiscopaux. Il faut d'ailleurs reconnaître que leurs craintes trouvent une apparence de raison dans l'esprit un peu trop indépendant de l'un ou l'autre des nouveaux catholiques d'action, et aussi dans les exagérations de quelques esprits avancés, groupés autour d'un certain Bordas-Dumoulin, partisan d'une réforme qui rendrait aux laïcs des pouvoirs d'enseignement et de gouvernement [4].

LA RÉVOLUTION DE FÉVRIER ET L'ÉGLISE   Malgré le manque de cordialité entre l'Église et le gouvernement de Louis-Philippe, les catholiques furent atterrés à la nouvelle du renversement de la monarchie. Par delà le souvenir encore proche des manifestations anticléricales qui avaient accompagné la révolution de 1830, la proclamation de la république évoquait les pires excès qui avaient marqué, dans l'ordre social et religieux, la Première République. « Nous ne pouvons plus nous représenter tout ce que ce mot de République disait alors à l'imagination, devait expliquer plus tard

(1) Sur ce mémoire, voir L. ALAZARD, *D.-A. Affre, archevêque de Paris*, Paris, 1905, p. 571 et suiv.
(2) Cf. *supra*, chap. I, p. 19.
(3) Sur ces controverses pendant les deux dernières années de la Monarchie de Juillet, voir, outre les biographies de Montalembert, de Veuillot et de Dupanloup, L. FOLLIOLEY, *Montalembert et Mgr Parisis d'après des documents inédits* (1843-1848), Paris, 1902, chap. VI à X.
(4) Voir sur ce mouvement, dont l'influence demeure très restreinte, L. SÉCHÉ, *Les derniers jansénistes*, Paris, t. III, p. 4 et suiv.

Albert de Broglie. Les souvenirs de Quatre-vingt-treize étaient les seuls vivants [1]. »

Pourtant, contrairement à ce qu'on appréhendait, les journées de février ne présentèrent aucun caractère hostile à la religion, au contraire. A Paris, tandis que le gouvernement provisoire faisait appel aux prières de l'Église, les émeutiers saluaient les prêtres avec sympathie, les faisaient venir pour administrer les mourants sur les barricades, et improvisaient même une procession pour accompagner le crucifix et les vases sacrés évacués de la chapelle des Tuileries, en criant : « Vive le Christ ! Vive la liberté ! Vive Pie IX ! » Et si, en province, il y eut, par ci par là, quelques incidents, dans la plupart des villages les curés furent associés aux fêtes et invités à bénir les arbres de la liberté [2].

Cette attitude respectueuse des révolutionnaires à l'égard de l'Église ne s'explique pas uniquement par la froideur des rapports qui avait subsisté jusqu'à la fin entre elle et le gouvernement qui venait de s'écrouler, mais aussi par le fait que, à la différence des anciens jacobins ou des bourgeois libéraux de 1830, les « quarante-huitards », héritiers directs du Romantisme, sont pénétrés de sentimentalité chrétienne, admirent l'idéal de fraternité et d'égalité des premiers chrétiens, et saluent en Jésus-Christ « le premier des socialistes ». L'Avenir, avec sa devise : « Dieu et liberté ! », leur a révélé un catholicisme pour qui les « droits de l'homme » ne s'opposent pas nécessairement aux droits de Dieu, et l'euphorie des deux premières années du pontificat de Pie IX semble confirmer tous les espoirs. On n'a guère d'écho encore des premières désillusions italiennes et l'on va répétant le mot du P. Ventura, annonçant que le pape s'apprête à « baptiser la démocratie, cette héroïne sauvage », et celui d'Ozanam, qui, dans Le Correspondant, après avoir évoqué l'attitude de la papauté du VIIIᵉ siècle dégageant son destin du « trône vermoulu des Césars de Byzance » pour se tourner vers les jeunes nations germaniques, vient d'engager les catholiques français, placés devant un dilemme analogue, à agir de même : « Passons aux barbares et suivons Pie IX ! »

Les milieux prolétariens eux-mêmes, déjà profondément déchristianisés, ne sont pas encore coupés de l'Église. Il est caractéristique de constater que le seul des journaux ouvriers qui réussisse à vivre de 1840 à 1848, L'Atelier, soit de tendance nettement catholique [3], et si, dans la région lyonnaise, on sent une sourde hostilité contre le clergé [4], il n'en va pas de même ailleurs, à Lille, à Marseille, et spécialement à Paris ; là, les ouvriers, tout en restant méfiants à l'égard de l'Église officielle, ont été favorablement impressionnés par le fait que le clergé, moins

---

(1) A. DE BROGLIE, Mémoires, t. I, Paris, 1938, p. 185.

(2) Voici la formule de bénédiction prescrite par l'évêque de Quimper à cette occasion : « Sub illius tegmine germinet libertas, vigeant aequalia civium jura, regnet charitas fraternitatis, ut sit memoriale vitalis illius ligni in quo pendens Filius tuus veram nobis sanguine suo comparavit libertatem et fraternitatem. » (Cité dans H. GUILLEMIN, Histoire des catholiques français au XIXᵉ siècle, p. 131.)

(3) Sur L'Atelier et son animateur Buchez, voir A. CUVILLIER, Un journal d'ouvriers : L'Atelier, Paris, 1914, et Buchez et les origines du socialisme chrétien, Paris, 1948.

(4) Ch. DUTACQ, Histoire politique de Lyon pendant la Révolution de 1848, p. 100 et suiv.

préoccupé de politique que sous la Restauration, s'est rapproché du
peuple, et surtout par les initiatives charitables qui se sont développées
autour d'Armand de Melun, assisté de Mme Swetchine et de la sœur
Rosalie, d'une part, autour de Frédéric Ozanam et de la Société de Saint-
Vincent-de-Paul, de l'autre : ces initiatives contrastent en effet avec
l'indifférence du gouvernement et la dureté de la grande bourgeoisie
libérale envers le sort misérable des travailleurs [1].

LE RALLIEMENT
A LA RÉPUBLIQUE

Traitée avec respect par la Révolution, l'Église,
qui garde rancune à la Monarchie de Juillet, accueille
la jeune République avec sympathie, après le pre-
mier moment d'inquiétude. Tandis que les quelques catholiques démocrates et républicains de la veille expriment ouvertement leur joie et
que le plus marquant d'entre eux, le P. Lacordaire, qui prêche à Notre-
Dame le 27 février, proclame ses espérances dans une magnifique envolée
oratoire, la grande masse se rassure. Falloux rappelle aux légitimistes
de l'Ouest que « la religion fleurit dans les républiques américaines »
et conseille en conséquence au clergé « d'éclairer les habitants de nos
campagnes en rassurant leur piété au lieu de l'alarmer » [2]. Montalembert
lui-même, soucieux avant tout des intérêts de l'Église, fait taire ses rancunes personnelles : il détermine au ralliement le Comité catholique
puis le nonce et il y exhorte publiquement les catholiques, bien que,
très méfiant à l'égard de la démocratie, il soit atterré par la disparition
de la monarchie constitutionnelle de type aristocratique qui restera
toujours son idéal, et stigmatise dans ses lettres privées « les indignes
flagorneries de l'éloquent et démagogique P. Lacordaire » [3]. Quant à
Veuillot, excessif comme toujours, il s'empresse de piétiner le régime
tombé :

Qui songe aujourd'hui en France à défendre la monarchie ? Qui peut y songer ?
La France croyait encore être monarchique, elle était déjà républicaine [4].

Quelques jours plus tard, il résume l'opinion de beaucoup en ces premiers moments d'euphorie quand il écrit :

Nous avons longuement appris à préférer les travaux, les périls mêmes de
la liberté aux embûches de la protection. (...) Dieu dans le ciel, la liberté sur la
terre, voilà toute notre charte en deux mots [5].

La hiérarchie réagit de la même façon. Dès le 24 février, l'archevêque
de Paris assure le gouvernement provisoire de son « loyal concours »,
rappelle qu'en Amérique les gouvernements démocratiques n'ont jamais
rencontré la moindre opposition de la part du clergé, rédige une lettre
pastorale prescrivant de modifier la prière *Domine, salvum fac Francorum
regem* en *Salvam fac Francorum gentem* et, les jours suivants, il multiplie

---

(1) Voir, sur l'état d'esprit des ouvriers, les observations de J.-B. DUROSELLE dans : *1848*,
Révolution créatrice, Paris, 1948, p. 207.
(2) A. DE FALLOUX, *Mémoires d'un royaliste*, t. I, p. 287.
(3) Lettre du 13 mars 1848, dans *Revue historique*, t. CXCII, 1941, p. 260.
(4) *L'Univers*, 26 février 1848.
(5) *Ibid.*, 14 mars 1848.

les marques de bienveillance à l'égard du nouveau régime. En province, si Mgr Sibour [1], ardent républicain, est le seul à oser faire ouvertement l'éloge de la démocratie, tous les évêques sont cependant unanimes à prôner le ralliement [2], certains, comme Mgr de Bonnechose, par résignation à l'inévitable, beaucoup avec l'espoir de voir s'ouvrir une ère de liberté civile et religieuse qui ne pourra qu'être favorable à l'Église.

Les élections à la Constituante, qui, pour la première fois en France, ont lieu au suffrage universel, confirment du reste ces espoirs : l'Église, naguère dépourvue d'action sur le pays légal quand il ne comptait que 240.000 électeurs, bourgeois irréligieux pour la plupart, dispose dorénavant d'une influence politique énorme grâce aux masses paysannes, dont la majorité est encore prête à suivre ses consignes : en maints endroits, les électeurs sont même conduits au scrutin par leur curé, tambour en tête.

Quoi qu'aient prétendu, à la suite de Karl Marx, quelques historiens récents, il n'y a aucune raison sérieuse de suspecter la sincérité du ralliement de l'ensemble des catholiques au début de la seconde République. Ils se sont laissé emporter par cet « esprit de 1848 » qui s'était lentement élaboré au cours des années précédentes dans une atmosphère baignée d'idéalisme sentimental et de préoccupations religieuses, et qui se traduit par une aspiration à voir toutes les classes fraternellement unies travailler en commun au bonheur de tous [3].

LES CATHOLIQUES ET LE PROBLÈME SOCIAL La révolution de 1848 posait toutefois à la conscience des catholiques français un problème plus complexe que la simple acceptation de la République : « Derrière la révolution politique, il y a une révolution sociale, expliquait Ozanam à un correspondant de province. Derrière la question de la République, qui intéresse les lettrés, il y a les questions qui intéressent le peuple, pour lesquelles il s'est armé, les questions de l'organisation du travail, du repos, du salaire. Il ne faut pas croire que l'on puisse échapper à ces problèmes [4]. »

Or, il est indéniable que le ralliement aux idées de réforme sociale fut loin d'être aussi net que le ralliement au nouveau régime politique. Si Veuillot, qui avait diagnostiqué dès le 24 février au soir le caractère « socialiste » de la révolution, s'intéresse à plusieurs reprises au problème dans *L'Univers*, la masse des catholiques y reste indifférente. Il y a du reste beaucoup plus d'ignorance que de mauvaise volonté dans cette attitude. Dans une France encore essentiellement rurale et où l'industrie naissante est encore surtout artisanale, les centres prolétariens sont très limités, et comme la haute bourgeoisie industrielle et financière vit depuis plus d'une génération en dehors de l'Église, il n'est pas étonnant

(1) Alors évêque de Digne. Il sera nommé à Paris le 10 juillet 1848.
(2) Il existe une collection abondante sinon complète des mandements épiscopaux provoqués par l'avénement de la République aux *Arch. Nat.*, F[19], 5488 et 5489. On en trouve de nombreux extraits caractéristiques dans G. BAZIN, *Vie de Mgr Maret*, t. I, p. 190-205.
(3) Cf. J.-B. DUROSELLE, *L'esprit de 1848*, dans *1848, Révolution créatrice*, p. 187-189, 202-207, 211-215.
(4) Lettre du 6 mars 1848 dans *Ozanam. Livre du Centenaire*, Paris, 1913, p. 350.

que le tragique de la misère ouvrière reste incompris des catholiques, qu'on trouve surtout dans les milieux paysans, dans la petite bourgeoisie provinciale, parmi les propriétaires terriens et, dans une proportion minime, dans les professions libérales. Le clergé lui-même partage cette ignorance et pour les mêmes raisons : le bas-clergé se recrute presque exclusivement à la campagne et, même lorsqu'il exerce son ministère dans un centre industriel, il juge facilement en rural l'ouvrier turbulent, buveur et peu économe ; quant au haut-clergé, il conserve en bien des cas les réactions de la classe à laquelle il se rattache par son origine ou sa formation : « petits bourgeois parvenus à de hautes dignités, grisés par elles, n'ayant pas par ailleurs ce goût des choses surnaturelles qui les eût poussés au renoncement et rendus attentifs aux besoins de leur peuple, braves gens au demeurant, avec d'excellentes dispositions, mais *si bourgeois* » [1].

Il y a toutefois des catholiques, des prêtres, des évêques aussi, — Mgr Affre en est — qui ont pris conscience du problème au cours de cet âge d'or du capitalisme libéral que constituent les dix-huit années de la Monarchie de Juillet. Les uns se groupent autour de la Société d'Économie charitable, qui se recrute essentiellement dans les milieux légitimistes ; les mesures proposées à l'Assemblée législative par son principal animateur, le vicomte de Melun, paraissent aujourd'hui bien fragmentaires et surtout « profondément imprégnées de paternalisme » [2], mais on ne peut nier le désintéressement de ses membres, qui accueillent les tendances sociales de la jeune République avec un préjugé favorable. De ces hommes d'œuvre conservateurs mais sincèrement préoccupés du sort malheureux du prolétariat et soucieux d'y remédier, se distinguent un certain nombre de prêtres et de catholiques de tendance nettement démocrate, convaincus de la nécessité de réformes sociales profondes, qui n'atténueront pas seulement la dureté du régime du travail, mais modifieront radicalement les rapports entre le capital et le travail. On trouve parmi eux d'anciens saint-simoniens et d'anciens fouriéristes, d'anciens lecteurs de *L'Avenir*, qui ont été frappés par les vues sociales hardies de Lamennais, des membres de la Société de Saint-Vincent-de-Paul, qui se sont rendu compte que la charité ne constituait pas l'unique solution, des membres de cette association originale, la Société de Saint-François-Xavier patronnée par Mgr Affre et dirigée par l'abbé Ledreuille, « le prêtre des ouvriers », qui, en 1846, groupait déjà près de 15.000 travailleurs parisiens. Ces milieux sont très actifs au lendemain de la révolution : les uns collaborent au journal ouvrier, *L'Atelier* ; d'autres, comme le journaliste nantais Chevé, s'inspirent nettement des idées de Proudhon et collaborent même parfois au journal de celui-ci ; l'abbé Chantôme fonde la *Revue des réformes et du progrès*, Victor Calland la *Revue du Socialisme chrétien*. On pouvait se demander si ce catholicisme social et démocratique n'allait pas se rapprocher si étroitement du socialisme qu'une synthèse finirait par s'opérer entre les deux courants.

---

(1) Y. R., dans *R. H. E.*, t. XLII, 1947, p. 264.
(2) J.-B. Duroselle, *Les catholiques et le problème ouvrier en 1848*, dans *Actes du Congrès historique du Centenaire de la révolution de 1848*, p. 270.

LE GROUPE DE
« L'ÈRE NOUVELLE »

Un jeune professeur à la Sorbonne, l'abbé Maret, sympathisait avec ces tendances et souhaitait voir l'ensemble des catholiques s'y rallier. Il regrettait le caractère limité des objectifs poursuivis jusqu'alors par le parti catholique qui, trop exclusivement préoccupé de la liberté d'enseignement et de la question des congrégations religieuses, « s'était montré peu jaloux du sort des classes déshéritées et avait méconnu, du moins en pratique, la tendance inévitable de la société moderne »[1]. Au début d'avril, avec la collaboration d'Ozanam, qui partageait entièrement ses sympathies républicaines et ses préoccupations sociales, il fonda un nouveau quotidien dont Lacordaire, le plus grand nom de l'Église de France avec Montalembert, accepta de prendre la direction et dont le titre était à lui seul un programme. L'Ère nouvelle n'entendait pas seulement défendre la part de vérité des principes de 1789 et les avantages du régime républicain, mais également diverses réformes sociales jugées révolutionnaires à l'époque : la participation des ouvriers aux bénéfices et à la conduite de l'entreprise, l'organisation de l'arbitrage, l'assistance aux chômeurs.

Ce programme, approuvé par l'archevêque de Paris, rencontra des adhésions enthousiastes, surtout dans le jeune clergé ; dès le premier trimestre, le journal avait 6.000 abonnés et, en juin, il tirait à 20.000 exemplaires, ce qui montre qu'il y avait plus qu'une poignée de partisans derrière les animateurs de cette première ébauche de démocratie chrétienne. Malheureusement, cette équipe, aux idées souvent justes et aux intentions généreuses, manquait de cohésion : Lacordaire, démocrate par sentiment plutôt que par raison, était au fond un modéré qui allait vite s'effrayer des allures anarchiques que prenait la République et chercher à donner au journal une allure antisocialiste ; Maret, au contraire, emporté par un enthousiasme quelque peu candide, considérait la République comme le régime idéal et en célébrait le caractère évangélique, ce qui discréditait ses principes auprès des gens pondérés ; plus rassis, Ozanam voyait dans la République sociale le seul régime en rapport avec l'évolution de l'histoire et les aspirations du temps, mais son optimisme était à longue échéance et il s'attendait dans l'immédiat à des abus, inévitables de la part d'un peuple encore inhabile à gérer lui-même ses intérêts[2]. C'est précisément ce que ne devait pas comprendre la masse des catholiques : épouvantés par ces abus, ils allaient se laisser entraîner par la réaction et se détourner de ceux qui s'efforçaient de promouvoir une république chrétienne dont les réformes sociales seraient basées sur les principes de l'Évangile.

LE REVIREMENT
VERS LA DROITE

L'intérêt suscité par L'Ère nouvelle et par tel ou tel journal catholique de province défendant des idées analogues[3] ne peut faire oublier que la plupart des catholiques, comme la grande majorité de la nation, du

---

(1) Note du 24 février 1848, dans G. BAZIN, Vie de Mgr Maret, t. I, p. 225.
(2) Ces divergences de vues sont finement analysées par J. LEFLON, L'Église de France et la révolution de 1848, p. 71-77.
(3) Tel Le Réveil du Midi à Toulouse, organe de ceux que les républicains anticléricaux de l'en-

reste, s'ils s'étaient résignés sans peine à la disparition momentanée de la monarchie, craignaient par dessus tout de voir la république bourgeoise se transformer en une république démocratique et sociale. Sans partager les vues simplistes et partiales exposées récemment par Henri Guillemin [1], il faut toutefois reconnaître que, très soucieux du maintien de l'ordre et du caractère intangible du « dogme sacré de la propriété », ne songeant pas à faire la distinction entre les exigences extrémistes des partisans du communisme et les réformes sociales pénibles mais raisonnables, ils n'étaient que trop portés à considérer les réformes de structures prônées par les démocrates comme une inadmissible prétention à s'emparer par la violence du bien d'autrui. Leur méfiance, accrue par les mesures financières que le gouvernement dut prendre pour faire face aux premières mesures sociales et à la crise économique aggravée par l'agitation politique, se renforça évidemment après l'envahissement de la Chambre, le 15 mai, par la foule ameutée, et tourna à la panique après les émeutes de juin [2]. Du coup, la masse du pays, conservatrice mais non réactionnaire et qui avait été un moment sensible aux appels à la fraternité des classes, perdit de vue la solution de la question sociale sous la terreur de l'épouvantail rouge, les insurgés lui apparaissant comme des « destructeurs, non seulement des commodités bourgeoises, mais de l'ordre, mais de la civilisation, satisfaisants pour 90 pour cent des Français » [3].

Or, au même moment, la mort de Mgr Affre, tué par un énergumène au moment où il essaie d'intervenir en médiateur auprès des insurgés [4], prive de son principal appui *L'Ère nouvelle*, où Maret et Ozanam essaieront vainement d'expliquer que la violence même du soulèvement populaire montre la profondeur de la misère ouvrière et l'urgence de mesures tendant à y remédier. Le courageux journal, qui s'était heurté dès le premier moment à la sourde opposition de ceux qu'inquiétait son « enthousiasme démocratique », doit faire face à présent à une hostilité déclarée : mal vu par la majorité des évêques et ouvertement désavoué par quelques-uns qui en interdisent la lecture à leurs prêtres, il est attaqué vivement dans *L'Univers*, *L'Ami de la Religion* et *Le Correspondant*. Veuillot inaugure ses violents articles contre les socialistes, qui contribueront à faire son succès auprès du bas-clergé, conservateur dans l'ensemble. Montalembert,

droit nommaient « les néocatholiques révolutionnaires ». (Cf. P. DROULERS, dans *Bulletin de littérature ecclésiastique*, t. LI, 1950, p. 96, 99 et 103, n. 75). Voir un aperçu d'ensemble sur la presse démocrate chrétienne de province dans J.-B. DUROSELLE, *Les débuts du catholicisme social*, p. 335-358.

(1) Spécialement dans son *Histoire des catholiques français au XIXe siècle*.

(2) On sait que l'occasion de ces émeutes fut la liquidation brusquée des Ateliers nationaux. On a voulu imputer à Falloux, l'un des chefs catholiques les plus influents, la responsabilité de la brutalité avec laquelle cette mesure fut appliquée. (H. GUILLEMIN, *M. de Falloux et les journées de juin*, dans *La Vie intellectuelle*, janvier 1948, p. 46-58.) Toute la clarté n'est pas encore faite sur ces événements (cf. Ch. SCHMIDT, *Des Ateliers nationaux aux barricades de juin*, Paris, 1948), mais l'accusation semble bien exagérée. (Cf. J. LECLER, dans *Études*, t. CCLVI, 1948, p. 154-155, n. 3.)

(3) A. TRANNOY, *La « République sociale » de 1848 en France*, dans *Esprit et Vie* (Maredsous), t. I, 1948, p. 505.

(4) Ici aussi les passions ont troublé le jugement historique. Un examen objectif des meilleurs témoignages conduit à conclure que l'archevêque fut très vraisemblablement victime d'une balle tirée par un énergumène du côté des insurgés, mais que ceux-ci réprouvèrent ce crime avec indignation et manifestèrent clairement leur sympathie pour la victime. (Cf. R. LIMOUZIN-LAMOTHE, *Un archevêque aux barricades : Mgr Affre*, Paris, 1948, spécialement p. 19-35.)

qui d'ailleurs, on l'oublie trop aujourd'hui, continue à s'intéresser au soulagement de la misère ouvrière, mais qui, chevalier du moyen âge égaré dans le monde moderne, ne comprend rien au problème social posé par cette misère, prend maintenant la tête du mouvement de réaction contre les partis démocratiques dont les tendances politiques lui semblent mettre en danger la vraie notion de la liberté. La crainte d'un retour aux plus mauvais jours de la Révolution, le spectre de la Terreur et des massacres de septembre, lui donnent l'impression qu'il défend l'Église, en agissant ainsi. Il dénoncera bientôt le « catholicisme démocratique et social » comme « le plus grand de tous les dangers »[1] et, en attendant, il mène contre L'Ère nouvelle, accusée de vouloir pactiser candidement avec les pires ennemis de l'Église[2], une campagne acharnée, parfois peu honnête : Lacordaire, qui désavoue pourtant en secret l'allure de plus en plus avancée imprimée au journal par Maret, devait reprocher à Montalembert et à ses amis d'avoir eu recours « à une tactique plus odieuse encore que celle qui fut employée contre L'Avenir »[3]. Mais, répétons-le, il serait injuste d'oublier les motifs religieux qui inspirent cette politique conservatrice pour n'y voir qu'une défense de privilèges sociaux : « Les sentiments violemment antisocialistes d'un Veuillot ou d'un Montalembert sont très différents de ceux d'un Cousin ou d'un Thiers. Sincèrement, ils estiment la religion menacée avec l'ordre social, cet ordre social dont ils pensent par ailleurs que la destruction signifierait anarchie. Il n'y a pas l'ombre d'égoïsme de classe chez un Veuillot, et, s'il n'est pas contestable que Montalembert soit influencé par des préjugés, ce serait mal le connaître que de mettre en cause sa bonne foi : dans ses pires emportements réactionnaires, il croit toujours combattre pour le Christ[4]. »

Le malheur est qu'en croyant défendre les valeurs religieuses, Montalembert et Veuillot en réalité les compromettent. Lacordaire, pourtant bien revenu de ses premières illusions depuis les journées de mai et de juin, s'en rend compte : « L'Ami de la Religion et L'Univers, écrit-il, seront cause qu'à la prochaine émeute on tombera sur les églises et les prêtres. » Et lorsque Montalembert fait à la tribune sa célèbre déclaration : « Je ne connais qu'une recette pour faire croire à la propriété ceux qui ne sont pas propriétaires : c'est de leur faire croire en Dieu... qui a dicté le décalogue et punit éternellement les voleurs », il s'attire à bon droit cette réplique de la gauche : « C'est ravaler la religion[5] ! »

Cette attitude réactionnaire, qui ne voit dans les revendications populaires qu'un « désir immodéré de la jouissance » ou un « esprit de révolte contre l'autorité », n'est d'ailleurs pas seulement le fait de quelques riches propriétaires. La petite bourgeoisie et la paysannerie dans son

---

(1) Montalembert au P. d'Alzon, 23 mars 1849, dans *Actes du Congrès historique du Centenaire de la révolution de 1848*, p. 275.
(2) *L'Ère nouvelle*, écrit-il entre autres à dom Guéranger, ne fait que délayer « le mot indigne et si peu catholique d'Ozanam : *Passons aux barbares* », alors que « l'Église a toujours résisté aux barbares au lieu de les courtiser ». (Lettre du 5 juillet 1848, dans dom DELATTE, *Dom Guéranger*, t. I, p. 419.)
(3) Lettre du 1er mai 1849, dans *Lettres inédites*, Paris, 1874, p. 187.
(4) A. DANSETTE, *Histoire religieuse de la France contemporaine*, t. I, p. 365.
(5) *Moniteur* du 21 septembre 1848.

ensemble, et même un certain nombre d'ouvriers, impressionnés par les diatribes des agitateurs contre « la famille et la propriété », pensent de la même manière et sont prêts à répéter avec Veuillot : « La misère est la loi d'une partie de la société ; c'est la loi de Dieu à laquelle il faut se soumettre. » L'épiscopat ne réagit malheureusement pas assez contre cette mentalité et, négligeant les causes matérielles de la révolte populaire, se contente trop souvent de préconiser le retour à la religion. Ainsi Mgr d'Astros :

Nous adjurons ici tous les hommes impartiaux de nous dire si cette seconde expérience, car nous en fîmes une première en 93, ne les convainc pas de cette vérité que la religion est la base indispensable de la société. Il n'est pas question ici de son origine céleste...[1].

A des désordres dont « le vice » leur paraît la principale cause, ils ne voient de remède que dans un retour du peuple à la foi, qui lui fera accepter les nécessaires disciplines sociales. Ces évêques, assurément, ne songent nullement à encourager l'égoïsme bourgeois et beaucoup attirent l'attention des riches sur leurs devoirs de charité, précisant même que lorsque la misère est trop grande, il y a un devoir strict de donner plus que son superflu, de donner « à proportion de la misère du pauvre »[2]. Mais ils se montrent plus que méfiants à l'égard de toute tentative de réforme institutionnelle : hostilité à l'égard des « nouveautés » chez les uns, incapables d'imaginer que la société puisse légitimement évoluer au cours de l'histoire ; « pessimisme temporel » hérité de Bossuet chez les autres, persuadés que l'injustice introduite dans la vie sociale par le péché originel y régnera toujours. La plupart sont en outre péniblement impressionnés par l'irréligion agressive ou la religiosité frelatée ou la morale antifamiliale de certains doctrinaires socialistes. Mgr Sibour, appelé à succéder à Mgr Affre grâce à l'intervention de l'abbé Maret, qui a fait valoir auprès du gouvernement ses tendances démocratiques, commence bien son épiscopat à Paris par des visites dans les quartiers populaires et les fabriques, sous la direction de sœur Rosalie, et fait preuve de compréhension[3]. Mais l'un des prélats les plus influents, Mgr Gousset, archevêque de Reims, exprime une opinion qui se généralise chaque jour quand il affirme : « La démocratie est l'hérésie de notre temps. Elle sera aussi dangereuse et aussi difficile à extirper que le jansénisme[4]. » Et l'abbé Combalot, le célèbre prédicateur et missionnaire diocésain, n'hésite pas à traiter les ouvriers révoltés de « hordes anthropophages » et leurs chefs d'« apôtres du terrorisme, de la spoliation et du brigandage »[5].

Affolés par la « peur rouge », certains catholiques vont jusqu'à accuser les hommes d'œuvres d'exciter l'ambition du peuple en lui suggérant, par leurs mesures charitables, l'idée de droits imaginaires[6] et, pour

(1) Mandement du 9 juillet 1848, cité par P. DROULERS dans *Bulletin de littérature ecclésiastique*, t. LI, 1950, p. 104.
(2) Mandement de Mgr d'Astros, du 2 janvier 1850, cité *ibid.*, p. 107.
(3) Voir J. DANIELO, *Visites pastorales de Mgr Sibour*, Paris, 1854.
(4) Cité d'après le journal de Montalembert, par A. TRANNOY, dans *Revue historique*, t. CXCII, 1941, p. 276.
(5) A. RICARD, *L'abbé Combalot*, p. 392 et 395.
(6) J. SCHALL, *Adolphe Baudon*, Paris, 1897. p. 132.

le même motif, l'Assemblée législative repousse le plan d'assistance publique, pourtant bien timide, élaboré par Armand de Melun. Seuls quelques hommes perspicaces, résistant à l'entraînement général et convaincus que « si les remèdes (proposés par les démagogues) sont faux, les souffrances ne le sont pas »[1], pensent avec le P. Gratry que « l'ignorance du devoir social est la source du sang dont Paris fume encore »[2]. Comme Lacordaire découragé et Ozanam vaincu par la maladie s'effacent, c'est autour d'un homme nouveau qu'ils se regroupent, Frédéric Arnaud de l'Ariège, à qui Maret a fait appel pour accentuer l'orientation sociale de *L'Ère nouvelle* et qui fondera bientôt, avec les encouragements de Mgr Sibour, le *Cercle de la Démocratie catholique*, centre de ralliement des chrétiens sociaux de Paris et de province qui, comme le dit le prospectus initial, aspirent à ce que la religion d'amour « ne soit plus considérée par le peuple comme la complice de l'intérêt égoïste »[3].

LE GRAND PARTI DE L'ORDRE   En présence de la collusion croissante de l'Église avec la réaction sociale, et bientôt politique, il importe de ne pas oublier que les responsabilités sont partagées et que, pour une bonne part, ce sont les réactionnaires qui cherchent à s'annexer l'Église. Tocqueville devait noter que « la crainte du socialisme a produit momentanément sur les classes moyennes un effet analogue à celui que la Révolution avait produit jadis sur les hautes »[4] : parmi ceux qui louent à présent « le sens austère et consolateur du christianisme » et montrent en lui la dernière sauvegarde de l'ordre bourgeois, beaucoup sont d'anciens anticléricaux qui n'ont guère de titres à parler au nom de l'Église. On connaît la boutade d'Ozanam observant que depuis que les possédants se sentent menacés, « il n'y a voltairien affligé de quelque mille livres de rente qui ne veuille envoyer tout le monde à la messe à condition de n'y pas mettre les pieds ». Le cas le plus typique de ce « cléricalisme sans Dieu »[5] est celui de Thiers ; cet adversaire acharné de l'enseignement libre déclare à présent :

> Je demande formellement autre chose que ces détestables petits instituteurs laïques ; je veux des frères, bien qu'autrefois j'aie pu être en défiance contre eux ; je veux encore là rendre toute puissante l'influence du clergé ; je demande que l'action du curé soit forte, beaucoup plus forte qu'elle ne l'est, parce que je compte beaucoup sur lui pour propager cette bonne philosophie qui apprend à l'homme qu'il est ici-bas pour souffrir[6].

L'Église n'est pas responsable de ces déclarations cyniques, mais elle n'est toutefois pas sans reproches. D'abord, une partie importante du clergé et de la hiérarchie a le tort d'encourager sans faire les réserves nécessaires cette « régénération religieuse » de caractère si intéressé. Mgr Pie, par exemple, après avoir noté avec joie qu'« on a vu, dans quelques

(1) *L'Ère nouvelle* du 25 juin 1848.
(2) A. GRATRY, *Demandes et réponses sur les devoirs sociaux*, Paris, 1848.
(3) Sur cette activité sociale d'Arnaud de l'Ariège, voir J.-B. DUROSELLE, *op. cit.*, p. 374-383.
(4) Lettre de septembre 1851 citée par G. WEILL, *Histoire du catholicisme libéral*, p. 100.
(5) L'expression est de H. GUILLEMIN, *Histoire des catholiques français*, p. 181.
(6) *La Commission extraparlementaire de 1849*, Paris, 1937, p. 31.

paroisses, le maire, le percepteur, le notaire, le médecin, les divers propriétaires s'avancer vers la table eucharistique à la tête de leurs concitoyens », rappelle aux hésitants que s'ils veulent que l'Église défende efficacement la propriété, il faut qu'ils donnent aux prolétaires l'exemple de la pratique religieuse [1].

Ensuite, les chefs du parti catholique, estimant sans hésiter que, selon le mot du *Correspondant*, « aujourd'hui, la cause de la religion et celle de la propriété sont connexes », acceptent sans difficulté de répondre aux avances de la bourgeoisie orléaniste, convaincus qu'il faut « se la concilier à tout prix » parce que, comme l'écrivait Montalembert à son ami l'abbé Texier, « si elle a besoin de nous, nous ne pouvons pas non plus sauver le pays sans elle » [2]. Leur calcul est habile sur le plan parlementaire, mais, absorbés par l'aspect politique des problèmes, ils ne soupçonnent guère le danger de « cette sorte de pacte entre la religion et les intérêts capitalistes » [3]. Sans doute, lorsqu'ils s'unissent au groupe des Thiers et des Molé, connu sous le nom de *Comité de la rue de Poitiers*, pour constituer avec eux le « Grand Parti de l'Ordre », Montalembert et Falloux ont le souci de promouvoir les intérêts de l'Église et pas seulement de défendre les intérêts conservateurs, et ils obtiendront effectivement pour elle de substantiels avantages ; mais en même temps, ils la rendent inévitablement solidaire des mesures de plus en plus antidémocratiques par lesquelles l'Assemblée, au cours des années 1849 et 1850, s'efforce de réduire l'influence des éléments de gauche [4] et bientôt les catholiques français n'apparaîtront plus seulement comme antisocialistes, mais aussi comme antirépublicains.

Leur attitude à l'égard de Louis-Napoléon Bonaparte ne peut que confirmer cette impression. Une fois de plus, d'ailleurs, cette attitude est complexe. Certes, beaucoup de catholiques, suivant le courant général de désillusion, ont contribué à porter Bonaparte à la présidence, le 10 décembre 1848, parce qu'ils ne voulaient pas de Cavaignac, trop républicain à leurs yeux ; et Montalembert, qui se déclare « ravi » de cette « défaite du rationalisme démocratique par un nom », ne se fait aucune illusion sur le caractère dictatorial et antirépublicain de cette élection [5]. Mais on ne peut oublier que jusqu'au dernier moment, le même Montalembert a négocié avec les deux candidats en offrant l'appui des catholiques à celui qui promettait la liberté d'enseignement et s'engagerait à restaurer l'autorité temporelle du pape [6], car la révolution romaine et l'exil de Pie IX, tout en renforçant la conviction de ceux qui ne croient plus à la possibilité d'un compromis entre l'Église et la démocratie, avaient

---

(1) Lettre pastorale pour le Carême 1851.
(2) Lettre de juillet 1849 dans *La Correspondance de Montalembert et de l'abbé Texier*, p. 305.
(3) A. TRANNOY, *Responsabilités de Montalembert en 1848*, dans *R. H. E. F.*, t. XXXV, 1949, p. 186.
(4) Lois des 19 juin, 27 juillet et 9 août 1849, 31 mai, 6 juin et 16 juillet 1850.
(5) *Journal*, 11 décembre 1848 et lettre du 18 décembre 1848, cités par A. TRANNOY, dans *Revue historique*, t. CXCII, 1941, p. 277-278.
(6) Sur les négociations menées par Montalembert avec les deux candidats au nom des catholiques, voir les passages de son *Journal* reproduits par A. TRANNOY, *art. cit.*, p. 273 et suiv. Les lettres-manifestes au *Constitutionnel* et au nonce sont reproduites dans H. THIRRIA, *Napoléon III avant l'Empire*, t. I, p. 459.

fait passer la Question romaine au premier plan des préoccupations des catholiques français [1]. Des raisons religieuses ont donc incontestablement joué un rôle dans l'élection de Louis-Napoléon et pas seulement les préférences pour un régime politique autoritaire. Il reste indéniable, cependant, que, au fur et à mesure que les mois passent, et que le prince-président évolue dans un sens dictatorial, on sent croître la sympathie des prêtres et des évêques pour un type de gouvernement qui avait une longue tradition derrière lui, et l'attitude de Mgr Sibour, archevêque de Paris, qui voudrait au contraire voir les catholiques rester davantage à l'écart de la politique, est désapprouvée publiquement par une grande partie du clergé [2].

LES CATHOLIQUES ET LE DEUX DÉCEMBRE

L'attitude des catholiques à l'égard du coup d'État est dès lors facile à prévoir. A quelques exceptions près — Lacordaire, qui refusera en signe de protestation de prêcher encore à Paris, Ozanam, et le petit groupe des démocrates chrétiens, mais aussi Dupanloup — ils se rallient après un moment d'hésitation à l'avis de Veuillot : « Il n'y a de choix possible qu'entre Bonaparte empereur et la République socialiste. » Montalembert, qui deviendra bientôt l'un des grands adversaires de Napoléon III et dénoncera avec sa passion coutumière « la grande palinodie des catholiques français », commence par conseiller de voter *Oui* au plébiscite « pour défendre nos églises, nos foyers, nos femmes contre ceux dont les convoitises ne respectent rien » [3]. Tandis qu'à Rome, Pie IX se réjouit de cette nouvelle victoire de la « cause de l'ordre » et du « principe d'autorité » [4] et estime que « le ciel vient de payer à la France la dette de l'Église » [5], la plupart des évêques, parmi lesquels se distinguent Mgr de Salinis, candidat républicain quatre ans auparavant, et Mgr Parisis, le panégyriste des libertés publiques sous la Monarchie de Juillet, « célèbrent en 1852 l'alliance de l'Église et de l'Empire comme ils ont célébré en 1848 l'alliance de l'Église et de la République » [6]. Le bas-clergé, qui appréhendait une reprise de l'agitation socialiste en province, suit le rythme sans se faire prier.

Pour juger la situation sans passion, il faut d'ailleurs ajouter que presque toute la bourgeoisie française abandonne sans regret la cause

---

(1) Deux tendances s'opposaient à l'Assemblée : les républicains voulaient se borner à assurer la liberté personnelle du pape et à protéger Rome contre une intervention autrichienne, tandis que les monarchistes désiraient rétablir le pouvoir temporel du pape. Cavaignac partageait la première opinion ; en se déclarant, sur l'invitation expresse de Montalembert, « décidé à garantir efficacement la liberté *et l'autorité* du gouvernement pontifical », Louis-Napoléon laissait entendre qu'il appuierait la seconde.

(2) Voir, à titre d'exemple, la violente polémique déclenchée par Combalot dans A. RICARD, *L'abbé Combalot*, p. 402-438.

(3) *L'Univers*, 14 décembre 1851. Sur les conseils du nonce et d'évêques influents et malgré les adjurations de Dupanloup, il accepte même, espérant servir les intérêts catholiques, d'entrer dans la commission consultative, mais il démissionnera dès le mois de janvier.

(4) Voir les dépêches de l'ambassadeur de France éditées dans J. MAURAIN, *Le Saint-Siège et la France de décembre 1851 à avril 1853*, Paris, 1930, p. 17-29. Même son de cloche dans une lettre de dom Guéranger à Mgr Pie du 3 janvier 1852, dans dom DELATTE, *Dom Guéranger*, t. II, p. 51-52.

(5) Cité par E. LECANUET, *Montalembert*, t. III, p. 37, n. 2.

(6) A. DANSETTE, *Histoire religieuse de la France contemporaine*, t. I, p. 374.

parlementaire discréditée et que le peuple lui-même reste indifférent [1]. Mais les *Te Deum* qui associent devant l'opinion le clergé au coup d'État rendent furieux les chefs socialistes et républicains. Ceux-ci, qui avaient déjà protesté avec fureur contre « la croisade contre la République romaine », vont désormais dénoncer sans se lasser « l'alliance du sabre et du goupillon » : les sympathies religieuses, assez générales chez eux en 1848, vont désormais faire place à un violent anticléricalisme, qu'ils réussiront peu à peu à inculquer à leurs troupes.

LES BÉNÉFICES DE L'ÉGLISE   Toutefois, pour le moment, ce danger est encore lointain et la Seconde République se solde apparemment, pour l'Église de France, par un bilan nettement favorable. Outre de nombreux retours à la pratique religieuse dans la bourgeoisie, les avantages d'ordre institutionnel sont sensibles. Sans doute, l'Église n'a pas obtenu la modification des articles organiques malgré les efforts de Mgr Parisis, qui présidait le Comité des Cultes [2], mais la Constitution a réglé au mieux de ses intérêts plusieurs points importants : l'article 7 proclame la liberté complète en matière de religion, tout en reconnaissant en même temps le droit des ministres des cultes à être rétribués par l'État ; l'article 8 reconnaît sans restrictions la liberté d'association (intéressante pour les religieux) et la liberté de réunion ; l'article 9 enfin affirme en principe que « l'enseignement est libre ». Favorisées par une législation moins tatillonne, les congrégations religieuses font de rapides progrès : on compte 207 nouvelles autorisations en trois ans contre 384 pour les dix-huit années de la Monarchie de Juillet, et le nombre total des religieux passe en conséquence de 28.000 en 1848 à 37.357 en 1851. Pendant ce temps, la fortune de l'Église continue à se reconstituer : les dons et legs se montent à 14.666.911 fr. contre 36 millions et demi pour les dix-huit années précédentes, et les nouvelles acquisitions à 11.119.752 fr. contre 25 millions environ [3].

Les rapports avec les pouvoirs publics sont aisés et souvent cordiaux. Invité à « donner une consécration religieuse aux institutions nouvelles », l'archevêque de Paris a célébré la messe au milieu de la place de la Concorde, le 12 novembre 1848, lors de la proclamation de la Constitution. La plupart des hommes politiques sont, sinon catholiques, du moins sympathiques au christianisme et « jamais on n'a tant parlé de Dieu et de l'Évangile à propos des affaires de l'État » [4]. Falloux, conseillé par Dupanloup, a fait d'excellentes nominations épiscopales et les évêques, surveillés de près par le nonce, s'habituent peu à peu à régler les affaires ecclésiastiques avec Rome sans intervention du gouvernement, car

(1) J. MAURAIN, *Un bourgeois français au XIXe siècle : Baroche*, Paris, 1938, p. 502 ; A. DUVEAU, *La vie ouvrière en France sous le Second Empire*, Paris, 1946, p. 71.

(2) Sur les travaux du Comité des Cultes, voir l'ouvrage de son secrétaire P. PRADIÉ, *La question religieuse en 1682, 1790, 1802 et 1848 et historique complet des travaux du Comité des Cultes de l'Assemblée constituante de 1848*, Paris, 1849. Sur les efforts de Mgr Parisis en vue de mettre sur pied un nouveau régime contractuel entre l'Église et l'État, cf. *Arch. Nat.*, Nouv. versement de la Chambre des députés, C 270.

(3) Ch. POUTHAS, *L'Église et les questions religieuses en France de 1848 à 1877*, p. 73-74.

(4) P. BASTID, *Doctrines et institutions de la Seconde République*, t. I, p. 102.

celui-ci leur laisse beaucoup plus de liberté que sous la monarchie dans l'administration de leurs diocèses. Après un moment d'hésitation, il se laisse même forcer la main par Mgr Sibour et autorise la tenue de conciles provinciaux. Il n'y en avait plus eu depuis 1727, mais l'idée était dans l'air depuis la fin du règne de Louis-Philippe. Après le refus opposé par Rome à un concile national, l'archevêque de Paris invita ses suffragants à se réunir en concile du 17 au 28 septembre 1849 et, arguant de la liberté de réunion, il refusa, malgré les instances du gouvernement, de demander une autorisation qu'on lui promettait d'ailleurs. Le gouvernement, pour sauver la face, autorisa alors spontanément les conciles provinciaux pour un an et, de septembre 1849 à octobre 1850, douze conciles se réunirent, dans le but entre autres d'obvier à la grande diversité de liturgie et de discipline qui régnait encore en France à cette époque[1].

C'est toutefois dans le domaine de l'enseignement que l'Église retira le principal avantage de sa situation sous la Seconde République.

LA LOI FALLOUX[2]   Depuis des années, les catholiques réclamaient la liberté d'enseignement. La loi Guizot leur avait donné satisfaction pour le degré primaire, mais, jusqu'à la chute de la Monarchie de Juillet, le monopole universitaire avait subsisté pour le secondaire. Maintenant que la Constitution de 1849 avait reconnu la liberté, il s'agissait de l'organiser. Montalembert, qui avait été depuis quinze ans à la pointe du combat, se rendit compte qu'il était préférable pour la cause qu'il entendait servir de s'effacer devant un homme moins compromis et, avec l'aide de Dupanloup, il réussit à pousser au ministère de l'Instruction Publique, en décembre 1848, le comte de Falloux[3], un catholique très dévoué à la cause de l'Église et de l'enseignement chrétien, négociateur habile à qui ses ennemis reprochaient « une fourberie peu commune et très efficace » et dont les amis admettaient qu'il préférait, parmi les vertus chrétiennes, la prudence du serpent à la simplicité de la colombe.

Une double tâche attendait le nouveau ministre : remanier le projet sur l'enseignement primaire officiel déposé récemment par Carnot, qui était loin de plaire aux catholiques ; et mettre en application pour le secondaire l'article de la constitution accordant la liberté « sous la garantie des lois et la surveillance de l'État ». Plutôt que de se borner, pour ce dernier point, à autoriser par décrets ministériels l'ouverture de collèges libres, il préféra organiser par une loi générale à la fois l'exercice de l'enseignement public et les conditions de l'enseignement privé, estimant que cette solution, qui promettait d'être plus durable, avait en outre l'avantage de favoriser d'utiles contacts entre les deux enseignements.

---

(1) Les décrets en sont publiés dans *Coll. lac.*, t. IV ; voir aussi G. DARBOY, *Les conciles récemment tenus en France*, dans *Le Correspondant*, 25 novembre 1850.

(2) Voir H. MICHEL, *La loi Falloux*, Paris, 1906 et J. LECLER, *La loi Falloux*, dans *Études*, t. CCLXVI, 1950, p. 3-25. On peut désormais suivre la préparation de la loi grâce à *La Commission extraparlementaire de 1849. Texte intégral inédit des procès-verbaux*, publié par G. CHENESSEAU, Paris, 1937 (qui rend inutile H. DE LACOMBE, *Les débats de la Commission de 1849*, Paris, 1879).

(3) On peut suivre les tractations difficiles qui ont abouti à ce résultat très favorable aux catholiques dans les *Mémoires d'un royaliste* de Falloux et dans le *Journal* de Montalembert, à compléter par F. LAGRANGE, *Vie de Mgr Dupanloup*, t. I, p. 469-476.

Falloux confia la préparation de la loi à deux commissions, formées en majeure partie d'extraparlementaires [1], dont les membres avaient été fort habilement choisis : on n'y trouvait aucun partisan du maintien du monopole universitaire, mais on remarquait aussi l'absence de Veuillot, jugé trop compromettant [2], et celle de Mgr Parisis, l'un des plus brillants champions de la liberté d'enseignement cependant, que Falloux avait écarté pour que Dupanloup soit seul à parler au nom de l'Église et puisse déployer à l'aise ses talents de conciliateur. Car il faudrait négocier, Falloux s'en rendait compte, étant donné les méfiances de nombreux membres de l'Assemblée qui « auraient volontiers accordé la liberté à tout le monde excepté au clergé » [3] : il serait impossible de trouver une majorité si l'on ne parvenait pas à rallier Thiers en se montrant modéré. D'ailleurs, contrairement aux intransigeants comme Veuillot et une bonne partie du clergé, le ministre ne regrettait guère de devoir transiger avec les droits de l'État et de l'Université : il jugeait l'Église incapable de remplacer entièrement l'Université en matière d'enseignement et il appréhendait, comme il devait le reconnaître plus tard, « que l'éducation donnée par les ecclésiastiques ne répondît pas aux exigences de l'esprit moderne ».

Le travail de la commission extraparlementaire (les deux commissions ayant très vite fusionné) se poursuivit pendant quatre mois. On eut la surprise de voir Thiers, qui restait sous l'impression des journées de juin et redoutait la diffusion des idées révolutionnaires par les instituteurs, se déclarer « prêt à donner au clergé tout l'enseignement primaire » [4], mais les catholiques refusèrent eux-mêmes ce cadeau : « Ce que je demande pour le clergé, c'est la liberté de l'influence, mais non la domination, exclusive de la liberté », répliqua Montalembert [5], et plusieurs personnalités ecclésiastiques, qui vinrent témoigner devant la commission, estimèrent que les reproches adressés aux instituteurs étaient exagérés.

Par contre, en ce qui concerne le secondaire, Thiers, qui trouvait que « l'État a un peu le droit de frapper la jeunesse à son effigie » [6], posait une série de conditions qui limitaient sérieusement la liberté et qui, entre autres, revenaient pratiquement à empêcher les jésuites d'ouvrir des collèges. C'est alors que Dupanloup eut à intervenir : il finit par l'emporter après une discussion serrée et il réussit de même à obtenir de Cousin plus de concessions de la part de l'Université que celui-ci, sincèrement désireux d'une entente pourtant, n'était disposé à en faire au début.

Dès ce moment, l'essentiel était fait. La refonte du projet par une commission parlementaire où les catholiques n'avaient la majorité que de

---

(1) Neuf appartenaient à la Constituante, entre autres Thiers et Montalembert, 4 venaient de l'Université, notamment Cousin ; 2 de l'enseignement privé ; Dupanloup et un pasteur siégeaient à titre d'ecclésiastiques ; 5 membres enfin étaient des personnalités s'intéressant aux questions d'éducation.

(2) Montalembert aurait préféré qu'on fît appel à Veuillot afin de ne pas le pousser dans l'opposition, mais, explique Falloux, « après mûre réflexion, j'aimai mieux l'exposer à la tentation de critiquer les choses faites sans lui que l'armer du droit d'empêcher de les faire ». (*Mémoires d'un royaliste*, t. I, p. 423.)

(3) A. DE MELUN, *Mémoires*, t. II, p. 68.

(4) *La Commission extraparlementaire*, p. 30.

(5) *Ibid.*, p. 33.

(6) *Ibid.*, p. 193.

justesse, sa discussion à la Chambre, où les attaques des universitaires et de la gauche manquèrent d'adresse et de brillant, sa modification momentanée par le Conseil d'État, où prédominaient les partisans d'une centralisation étatique de l'enseignement, ne modifièrent guère l'économie générale de la loi, qui fut votée le 15 mars 1850 par 399 voix contre 237 [1].

Cette loi aboutissait à une réforme d'ensemble de l'instruction publique et privée à partir d'un double principe : liberté de l'enseignement privé d'une part, moyennant quelques conditions assez bénignes ; influence de l'Église sur l'enseignement public d'autre part. Ce fut surtout ce second aspect qui frappa les contemporains et provoqua de violentes réactions : l'instituteur nommé par le conseil municipal et non plus par les autorités universitaires, soumis de surplus à la surveillance du curé, semblait « mis en tutelle » [2] ; et la refonte des Conseils académiques aboutissait à une nette diminution de l'influence de l'Université en faisant perdre beaucoup de son prestige au recteur départemental, écrasé entre le préfet et l'évêque. Si l'Église obtenait d'une Chambre qui n'était pourtant guère cléricale de pareils avantages, c'était pour une bonne part grâce à l'atmosphère de « défense sociale » qui se développait de plus en plus et que surent habilement exploiter Thiers et Montalembert :

Qui donc défend l'ordre et la propriété dans nos campagnes, s'écriait ce dernier. Est-ce l'instituteur ? Non, c'est le curé (...). Aujourd'hui le curé, le clergé en général, et celui des campagnes en particulier, représentent à la fois l'ordre moral, l'ordre politique et l'ordre matériel (...). Il y a en France deux armées en présence. Elles sont chacune de 30 à 40.000 hommes ; c'est l'armée des instituteurs et l'armée des curés (...). Celle-ci fonctionne admirablement dans sa mission sociale [3].

On conviendra avec le P. Lecler « qu'en 1850, le chef du parti catholique semble oublier son bel idéalisme, celui de 1830, et rabaisse dangereusement le débat » [4], et que cette loi de 1850 avait indéniablement un relent de réaction et de cléricalisme. Mais ceci ne peut faire oublier l'essentiel, à savoir qu'elle accordait enfin une liberté scolaire conforme à la fois à l'idéal libéral et aux intérêts de l'Église, et qu'elle marquait entre les aspirations des catholiques et les prétentions de l'Université un sérieux effort de conciliation, particulièrement souhaitable en présence de la profonde division morale de la France après quarante ans de monopole universitaire.

LA DISCORDE DANS LE CAMP CATHOLIQUE Les catholiques remportaient une grande victoire et pourtant beaucoup d'entre eux furent mécontents. Tous ceux qui avaient souhaité « la liberté de l'enseignement comme en Belgique », c'est-à-dire sans la moindre surveillance de l'État, étaient déçus par cette loi de transaction qui mêlait partiellement l'Université, bastion de l'antichristia-

(1) A ce moment Falloux n'était plus ministre ; il avait été remplacé à la fin de l'année précédente, lors du changement de cabinet, par M. de Parieu, qui sut conduire habilement la défense de la loi dans l'esprit de son prédécesseur.
(2) Ch. Pouthas, *L'Église et les questions religieuses en France de 1848 à 1877*, p. 82.
(3) Cité par H. Michel, *La loi Falloux*, p. 373.
(4) J. Lecler, dans *Études*, t. CCLXVI, 1950, p. 21.

nisme d'après eux, à la vie des futures maisons libres. Et surtout, une partie importante du clergé ne pouvait se résoudre à admettre que l'Église qui, un demi-siècle plus tôt, possédait encore le monopole de l'enseignement, fût seulement autorisée à partager ce monopole avec l'Université : affirmant les droits de l'Église enseignante, ils en déduisaient les applications « avec toute la rigueur de l'esprit géométrique »[1], sans tenir aucun compte des possibilités. Plus réalistes, les défenseurs de l'œuvre de Falloux et de Dupanloup — car la loi était autant l'œuvre de celui-ci que de celui-là — répondaient qu'il fallait proportionner les demandes aux chances que l'on avait de les faire écouter et se montrer conciliant sur l'application des principes ; et estimant qu'il était impossible d'obtenir plus, ils se réjouissaient avec Lacordaire de cet « édit de Nantes du XIXe siècle ».

Ces oppositions de points de vue étaient encore aggravées par l'incompatibilité de caractère qui éclatait chaque jour davantage entre le champion le plus éloquent de l'intransigeance : Louis Veuillot, plébéien au tempérament de polémiste, et les protagonistes de la loi : Falloux, machiavélique et distingué, Dupanloup, le tacticien de la conciliation, Montalembert, l'aristocrate modéré refusant de se lancer dans une opposition qui eût pu affaiblir le parti de l'ordre. L'Univers, qui avait tout fait pour empêcher la loi de passer, la maudit une fois votée et les discussions devinrent bientôt si violentes que le pape, alerté par Montalembert et Dupanloup, dut intervenir pour engager l'épiscopat à accepter la loi, bien qu'elle ne fût pas en pleine conformité avec les principes.

La scission du « bloc catholique », dont les signes avant-coureurs étaient apparus vers la fin de la Monarchie de Juillet, était dorénavant consommée et le moindre incident allait concourir à l'aggraver, par exemple la discussion sur l'emploi des classiques païens, soulevée par l'abbé Gaume [2], qui était peu de chose à l'origine mais devint rapidement, par suite de l'intervention de L'Univers, une querelle entre le journalisme laïque et les évêques, et contribua à envenimer encore l'opposition qui se réveillait entre gallicans et ultramontains [3].

§ 3. — La Révolution de 1848 et l'Église dans les pays germaniques [4].

LES QUESTIONS RELIGIEUSES AU PARLEMENT DE FRANCFORT

Les questions religieuses n'avaient joué aucun rôle dans les révolutions qui se succédèrent en Allemagne au cours du mois de mars 1848 ni dans les campagnes électorales qui suivirent.

---

(1) E. AMANN, art. *Veuillot*, dans *D. T. C.*, t. XV, col. 2803.
(2) Dans son livre : *Le ver rongeur des sociétés modernes* (1851), une thèse véhémente contre l'emploi des auteurs païens dans les collèges catholiques ; voir A. RICARD, *Étude sur Mgr Gaume, ses œuvres, son influence, ses polémiques*, Paris, 1872.
(3) Cf. *infra*, chap. IX, p. 270-276.
(4) BIBLIOGRAPHIE. — On trouvera une étude critique des sources et une bibliographie complète dans V. VALENTIN, *Geschichte der Deutschen Revolution (1848-1849)*, Berlin, 1930-1931, 2 vol. A noter spécialement H. BRUECK-J.-B. KISSLING, *Geschichte der katholischen Kirche in Deutschland im XIX. Jhdt*, t. III, Mayence, 1905, chap. I à IV ; G. GOYAU, *L'Allemagne religieuse*, t. II, Paris, 1905, chap. V ; R. LEMP, *Die Frage der Trennung von Kirche und Staat im Francfurter Parlament*, Tubingue, 1913 ; F. SCHNABEL, *Der Zusammenschluss des politischen Katholizismus in*

Très vite cependant il apparut que les transformations politiques obligeaient à reconsidérer le problème des rapports de l'Église et de l'État.

Au « Parlement » qui réunit à Francfort, à partir de mai, les délégués des États allemands chargés d'élaborer une constitution nationale, deux conceptions extrêmes étaient en présence : les radicaux demandaient, conformément aux idées du jour, la séparation de l'Église et de l'État ; les partisans du régime antérieur souhaitaient au contraire voir l'Église demeurer sous la surveillance étroite de l'État. Aucun des deux systèmes ne pouvait satisfaire les catholiques, qui s'insurgeaient depuis longtemps contre la tutelle d'une bureaucratie tatillonne d'inspiration joséphiste, mais qui craignaient d'autre part que la séparation n'aboutît à priver l'Église de toute influence dans la vie publique et à accélérer le processus de laïcisation de la société.

Cette dernière considération influençait spécialement le clergé bavarois, qui faisait observer que la renaissance catholique des dernières années s'était développée à la faveur du Concordat et de la protection royale [1]. A quoi d'autres répondaient que les autres souverains d'Allemagne n'étaient pas catholiques et que d'ailleurs le brusque revirement du roi de Bavière, l'année précédente, à la suite de l'affaire Lola Montès, montrait le danger de s'en remettre au bon vouloir d'un prince ; ne valait-il pas mieux, dès lors, profiter de l'évolution politique qui s'amorçait et revendiquer pour l'Église, non certes la séparation pure et simple, mais du moins la liberté par rapport à l'État, à condition d'apprendre à se servir efficacement de celle-ci pour faire pénétrer les principes catholiques dans l'ensemble de la vie ? Telle était la thèse que défendaient les agents les plus actifs du mouvement catholique : le publiciste badois Buss ; le chanoine Lennig, de Mayence, et son groupe où les anciens mennaisiens n'étaient pas rares ; Staudenmaier, le plus estimé des disciples de Moehler, qui, dans une brochure : *La situation actuelle de l'Église*, fort appréciée par l'archevêque de Cologne, expliquait que la vraie liberté n'est pas irréligieuse. L'un des plus décidés parmi les partisans de l'indépendance de l'Église était même un bavarois, le chanoine Doellinger, membre très en vue du cercle de Goerres, où ces idées étaient depuis longtemps classiques, et qui, à Francfort, au cours d'une discussion remarquée, défendit son point de vue en invoquant l'exemple des États-Unis et surtout de la Belgique [2].

C'est à la thèse de la liberté de l'Église que se rallièrent la plupart des délégués catholiques au Parlement national [3]. Ils n'obtinrent pas

Deutschland im Jahre 1848, Heidelberg, 1910 ; en outre, les ouvrages de K. BACHEM et de J.-B. KISSLING, ainsi que les biographies de Geissel et de Doellinger, signalées dans la bibliographie générale, p. 9.

Pour l'Autriche : *Aktenstücke die bischöfliche Versammlung zu Wien betreffend*, Vienne, 1850, et I. KANKOFFER, *Handbuch der Patente, Gesetze und Verordnungen welche für Kultus und Unterricht vom 2 Dez. 1848 bis Ende Dez. 1854 erschienen sind*, Vienne, 1855, ainsi que H. FRIEDJUNG, *Œsterreich von 1848-1860*, t. I : *Die Jahre der Revolution und der Reform, 1848-1851*, Vienne, 1908.

(1) Certains prêtres, à Cologne et en Bavière, craignaient en outre que les évêques, affranchis de la tutelle gouvernementale, n'abusassent de leur autorité à l'égard du bas-clergé (J. FRIEDRICH, *Ign. von Doellinger*, t. II, p. 386-387).

(2) Voir son discours dans J. FRIEDRICH, *op. cit.*, t. II, p. 396-407.

(3) Bien que le Parlement représentât non seulement l'Allemagne mais aussi l'Autriche, les députés catholiques n'atteignaient pas 20 % par suite de la place restreinte que les intérêts reli-

la pleine indépendance qu'ils demandaient, ni la liberté de collation des charges ecclésiastiques, ni la suppression du *placet*, mais du moins, après huit séances de discussions très animées [1], ils purent obtenir, le 11 septembre, une déclaration qui, malgré certaines réserves, semblait bien mettre un terme au système joséphiste :

Toute société religieuse ordonne et gouverne ses affaires avec autonomie, mais reste, comme toute autre société dans l'État, soumise aux lois de l'État.

Ce n'était encore là qu'un demi-succès, d'autant plus que, quinze jours plus tard, l'Assemblée, où les adversaires de l'Église étaient nombreux [2], retirait au clergé, par 316 voix contre 76, le droit de surveiller l'école [3].

L'ORGANISATION DES CATHOLIQUES ET LE CONGRÈS DE MAYENCE   Les dirigeants catholiques, qui attendaient des délibérations de Francfort la liberté pour l'Église, savaient que celle-ci n'était qu'un moyen qui vaudrait suivant le parti qu'on en saurait tirer. Il fallait au plus vite former une opinion catholique sur laquelle on pût compter. Le chanoine Lennig [4], que ces problèmes préoccupaient depuis longtemps, s'empressa, pour éclairer cette opinion, de fonder un quotidien, le *Mainzer Journal* [5], imité peu après par les catholiques de Cologne, et de réunir quelques laïcs, afin de discuter en commun sur la situation nouvelle que les événements allaient faire à l'Église. Ils furent bientôt quatre cents, et pour caractériser la tendance de l'association, on la mit sous le patronage de celui que l'Europe applaudissait encore comme « le pape libéral ». L'idée du *Piusverein für religiöse Freiheit* convenait au tempérament allemand et aux aspirations du moment et trouvait par ailleurs, en certains endroits, le terrain préparé par des initiatives antérieures [6]. Aussi des groupements analogues apparurent-ils vite à travers toute l'Allemagne et, dès le mois d'août, on décida d'unifier ces initiatives spontanées.

Du 3 au 5 octobre, les délégués de toute l'Allemagne se rencontrèrent à Mayence [7]. La présence de 23 députés du Parlement de Francfort contri-

gieux avaient tenue dans les préoccupations des électeurs. Ces députés catholiques, qui se répartissaient de l'extrême-droite à l'extrême-gauche, ne formaient pas un parti, mais ils avaient des contacts lorsque les questions religieuses étaient en jeu ; ils s'inspirèrent plus d'une fois des méthodes employées par Montalembert.

(1) On en trouve un bon résumé et des citations bien choisies dans WICHMANN, *Denkwürdigkeiten aus dem ersten deutschen Parlament*, p. 179-201.

(2) Radowitz se plaignait vivement de l'atmosphère hostile et du « réel chemin de croix » que constituaient pour les catholiques certaines séances. (F. MEINECKE, *Radowitz und die deutsche Revolution*, Berlin, 1913, p. 166.)

(3) Un amendement, accepté le 30 novembre, précisa que cette exclusion s'entendait « sous réserve de l'enseignement religieux ». Le même jour, on supprima également de la constitution l'article voté antérieurement qui excluait d'Allemagne les jésuites et les rédemptoristes.

(4) Sur Lennig (1803-1866), l'animateur, après Liebermann et avant Ketteler, du groupe de Mayence, voir H. BRUECK, *A. Fr. Lennig, Generalvikar und Domdekan von Mainz*, Mayence, 1870.

(5) A. DIEHL, *Das « Mainzer Journal » im Jahre 1848*, Mayence, 1911.

(6) Sur ces initiatives en Bavière, à Cologne et dans le grand-duché de Bade, voir les quatre premiers chapitres de J.-B. KISSLING, *Geschichte der deutschen Katholikentage*, t. I, Munster, 1920.

(7) Cf. *Verhandlungen der 1. Versammlung des katholischen Vereines Deutschlands*, Mayence, 1848 (trad. franç. par M. BESSIÈRES, *Congrès catholique de Mayence, 1848*, Paris, 1907) et T. PALATINUS, *Entstehung der Generalversammlungen der Katholiken Deutschlands und die erste grundelegende zu Mainz*, Wurzbourg, 1894.

bua à souligner l'importance de la réunion et Doellinger, dans un toast qui inquiéta certains ultramontains, y vit un symbole de la résurrection d'une Église allemande unifiée, ayant sa personnalité propre au sein de l'Église universelle et capable de tenir tête victorieusement aux tendances régaliennes des gouvernements. On décida de fusionner tous les groupements divers en une vaste *Association catholique d'Allemagne*, qui tiendrait périodiquement ses assises. Elle resterait soigneusement à l'écart des controverses politiques et ferait porter son effort sur un double but : continuer la lutte contre les tendances joséphistes en vue d'obtenir la pleine indépendance de l'Église et la liberté d'enseignement ; et, plus positivement, « viser à introduire, dans toutes les orientations de la vie, des principes catholiques et travailler à résoudre le grand problème du présent, qui est la question sociale » [1].

Cette insistance sur le problème social fut l'une des caractéristiques essentielles du congrès. La peur du socialisme avait provoqué en Allemagne dans les milieux bourgeois les mêmes effets qu'en France : *Gegen Demokraten helfen nur Soldaten*, y affirmait-on volontiers au lendemain des émeutes de septembre 1848, faisant écho aux applaudissements qui avaient salué à Paris Cavaignac et Lamoricière après les journées de juin. Mais tandis que les catholiques français dans leur ensemble se solidarisaient avec cette réaction bourgeoise en face des revendications populaires qui commençaient à se faire entendre, à Mayence, au contraire, les délégués catholiques des régions ouvrières attirèrent l'attention sur les responsabilités des chrétiens et ils rallièrent sans trop de peine l'assemblée à leur point de vue. « Nous devons, conclut le président, annoncer le socialisme du christianisme, non pas avec des paroles, mais avec des actes vivants, avec du dévouement, avec des sacrifices [2]. » Et Lennig, dans la préface qu'il donna aux *Actes* du Congrès, constatant que jusqu'à présent les laïcs qui s'étaient dévoués à la cause catholiques étaient tenus trop éloignés des masses populaires, affirma que désormais « dans l'*Association*, le peuple catholique apprendrait à connaître ses hommes » et découvrirait « le vrai démocratisme chrétien » [3].

L'ASSEMBLÉE DE WURZBOURG Trois semaines après le congrès de Mayence, qui marque une date capitale dans l'histoire de l'Action catholique au XIXe siècle, l'épiscopat allemand se réunissait à son tour, pour la première fois depuis la réunion d'Ems de 1786, afin d'examiner les problèmes soulevés par les bouleversements récents. Il s'agissait de s'entendre sur une attitude commune en matière de politique ecclésiastique, mais aussi de prendre position à l'égard de l'organisation de l'action catholique au sein des *Vereine*, et des demandes de rétablissement des synodes diocésains [4], où certains craignaient de

(1) *Verhandlungen...*, p. VIII.
(2) *Ibid.*, p. 76-77.
(3) *Ibid.*, p. XIV.
(4) Le mouvement synodal, qui se manifesta un peu partout de 1848 à 1850, était soutenu par des hyper-orthodoxes comme le curé Binterim, mais il avait pour chef de file le vieux professeur Hirscher, dont l'ouvrage *Die kirchlichen Zustände der Gegenwart* fut mis à l'*Index* en octobre 1849. Contrairement à ce qu'on a parfois prétendu, c'était un homme bien intentionné et d'une

discerner une tendance à modifier, sous l'influence des idées démocratiques, la constitution même de l'Église en diminuant l'autorité épiscopale au profit des prêtres, voire des laïcs.

L'idée de voir les évêques allemands se réunir, à l'imitation de ce que faisaient régulièrement les évêques belges, était dans l'air depuis quelque temps. Lancée par Lennig au lendemain de la révolution, elle avait eu tout de suite la faveur de l'archevêque de Cologne, Geissel, et le vieil archevêque de Breslau, Diepenbrock, s'y était rallié, de même que les plus représentatifs des délégués catholiques à Francfort, consultés sur le conseil de ce dernier. Aussi, passant outre aux timidités de l'épiscopat bavarois, qui craignait d'exciter les radicaux, de heurter le gouvernement royal et de déplaire à Rome, Geissel décida de brusquer les choses et, sans délégation papale, mais sûr de l'appui du nonce, il convoqua ses collègues à Wurzbourg pour le 22 octobre [1]. Seuls les évêques des régions périphériques — Silésie polonaise, Bohême, Autriche — firent défaut et, pendant trois semaines, l'épiscopat allemand, assisté par quelques théologiens, délibéra sur un programme très vaste, que Geissel avait exposé dans un remarquable mémoire composé avec l'aide de Doellinger [2] : l'Église et l'État, la question scolaire, les rapports avec les non-catholiques, les synodes diocésains, l'organisation des catholiques et l'action sociale.

Trois hommes dominèrent les débats : l'évêque d'Augsbourg, Richarz, qui représentait le point de vue des évêques bavarois, hostiles aux innovations et redoutant par dessus tout la dénonciation du concordat ; Doellinger, qui exposa à nouveau, sans grand succès, son idée d'une Église allemande, avec primat et concile national, mais qui apparut surtout comme l'ardent champion de l'indépendance de l'Église, prêt, avec un optimisme que certains trouvèrent un peu chimérique, à accepter la séparation si l'État voulait abuser de l'union pour opprimer l'Église ; enfin Geissel, partisan lui aussi de la liberté mais avec plus de modération, et qui, à travers des escarmouches constamment renouvelées, réussit, grâce à son ascendant croissant, à rallier l'assemblée à ses vues en lui faisant entre autres désavouer le droit de *placet* et de patronat et affirmer la nécessité, en conséquence des décisions de Francfort, d'organiser comme en Belgique un enseignement libre catholique. Sa maîtrise incontestable sut transformer l'Assemblée de Wurzbourg, où les divergences de vue des participants étaient flagrantes, en une imposante manifestation d'unité.

LA CONSTITUTION PRUSSIENNE   A Wurzbourg comme à Mayence, on avait dû se borner à des vœux, et l'évolution politique allait rendre bientôt caduques les décisions de principe de Francfort. Mais ces vœux et ces principes devinrent des

orthodoxie foncière, mais qui manquait parfois de prudence et ne comprenait pas que le retour aux institutions de l'Église primitive et médiévale était une chimère au XIXᵉ siècle, alors que l'évolution de l'Église allait dans le sens opposé.

(1) On en trouvera les actes officiels dans la *Coll. Lac.*, t. V, col. 942-1144 et un résumé détaillé dans *Archiv für Kirchenrecht*, t. XIV et XV, 1869. Cf. J.-B. SAUZE, *L'assemblée épiscopale de Wurzbourg*, Paris, 1907 et H. STORTZ, *Staat und Kirche im Lichte der Wurzburger Bischofsdenkschrift von 1848*, Bonn, 1934.

(2) Texte dans la *Coll. lac.*, t. V, col. 946-958.

réalisations de fait dans le plus important des États allemands. En effet, la constitution octroyée, le 5 décembre 1848, par le roi de Prusse assurait aux catholiques, conformément aux conclusions de la Commission parlementaire, non seulement le libre exercice du culte, mais l'indépendance à l'égard de l'État, avec, pour conséquence, la liberté de communication avec les chefs ecclésiastiques, la libre collation des charges, la suppression du *placet*, la liberté d'enseignement et le droit d'association. L'archevêque de Cologne avait raison de voir « un événement d'une incalculable portée »[1] dans cette constitution qui allait permettre, à la faveur de la liberté, un rapide épanouissement de la vie religieuse dans les provinces occidentales de la Prusse, et que les catholiques des autres États allemands citeraient bientôt en modèle à leurs gouvernements. « Moins de dix ans auparavant, l'on plaignait les catholiques de Prusse ; l'heure était venue de les envier. Le centre de gravité du catholicisme allemand était déplacé ; ce n'est plus en Autriche qu'il fallait le chercher, mais en Prusse rhénane[2]. »

1848 EN AUTRICHE — En Autriche aussi, cependant, l'année 1848 avait marqué pour l'Église le début d'un réveil. A la veille de la révolution, trois tendances se partageaient les esprits : le joséphisme, appuyé par la bureaucratie et par une partie du clergé, habitué à une tutelle devenue pour lui commode ; le mouvement de restauration catholique, issu des efforts du bienheureux Hofbauer et de ses rédemptoristes, représenté notamment par le précepteur du prince héritier, Rauscher, partisan d'une intime union entre l'Église et l'État, mais dans laquelle l'État serait, comme au moyen âge, au service de l'Église ; enfin la tendance plus libérale des amis du philosophe Gunther, qui souhaitait l'indépendance de l'Église et attendait son redressement non d'un appui officiel, mais d'un renouveau de la vitalité catholique. La Révolution de mars balaya la bureaucratie, principal soutien du joséphisme, mais elle se montra en même temps fort hostile au groupe de Rauscher, compromis par ses relations trop étroites avec le régime de Metternich, et prit très vite une attitude inquiétante au point de vue religieux. Les évêques, très gouvernementaux, étaient déconcertés et plus que tout autre le pusillanime et conservateur archevêque de Vienne, Mgr Milde, type parfait du prélat d'ancien régime.

En présence du danger, le groupe de Gunther passa immédiatement à l'action, à l'imitation de ce qui se passait en Allemagne. Le célèbre converti et prédicateur Veith prit la direction du mouvement, organisa, avec l'aide de Sébastien Brunner, une presse catholique et provoqua la constitution d'une « association catholique » réunissant les laïcs cultivés pour la défense des intérêts catholiques, en union étroite avec l'association allemande correspondante. En même temps, un de ses amis, W. Gaertner, essayait, au grand mécontentement de l'archevêque, d'amener celui-ci à sortir de sa réserve[3].

(1) Geissel au nonce, cité dans O. PFUELF, *Kard. J. v. Geissel*, t. I, p. 661-662.
(2) G. GOYAU, *L'Allemagne religieuse*, t. II, p. 404.
(3) Voir, sur ces différents efforts, E. WINTER, *Die geistige Entwicklung A. Gunther und seiner Schule*, Paderborn, 1931, *passim*. Au même moment, en Hongrie, l'aspiration des laïcs à parti-

Ce n'est qu'à la fin de l'année, lorsque la réaction absolutiste s'amorça, que Milde, qui avait boudé la conférence de Wurzbourg, trop « allemande » à son gré, fit proposer à l'empereur de convoquer une réunion analogue de l'épiscopat autrichien, qui se tint à Vienne du 27 avril au 17 juin 1849, sous la présidence du cardinal Schwarzenberg. Beaucoup des trente-cinq évêques conservaient des sympathies joséphistes, mais Schwarzenberg, ami de Gunther, et Rauscher, le véritable animateur de la conférence, soutenus par la forte personnalité de Diepenbrock, dont une partie du diocèse était autrichien, défendirent les droits de l'Église dans l'esprit de Wurzbourg. L'assemblée, tout en entendant bien conserver les privilèges dont l'Église avait joui jusqu'alors, prit acte des promesses de liberté ecclésiastique contenues dans la constitution du 4 mars. Elle émit une série de vœux précis dans ce sens et désigna un comité épiscopal chargé d'en poursuivre la réalisation par des négociations avec le gouvernement. Le jeune empereur François-Joseph, l'élève de Rauscher, marqua tout de suite son désir d'arriver le plus vite possible à un accord complet et il inaugura la nouvelle politique par les ordonnances des 18 et 23 avril 1850, qui accordaient à l'Église de substantielles satisfactions, en matière scolaire notamment.

§ 4. — Le rétablissement de la hiérarchie aux Pays-Bas [1]

LA SITUATION APRÈS L'ACCORD DE 1841 La Prusse ne fut pas le seul État protestant où les catholiques tirèrent avantage de l'évolution des institutions dans un sens plus libéral à la suite des événements de 1848. Ce fut aussi le cas au Danemark, où la constitution de 1849 proclama la liberté de culte, après trois siècles de proscription du catholicisme, et surtout aux Pays-Bas.

Les catholiques hollandais étaient un peu plus d'un million, le tiers de la population. Surtout concentrés dans les deux provinces méridionales du Limbourg et du Brabant, il y en avait toutefois près de 500.000 répartis parmi les protestants dans les provinces centrales [2] et, rien qu'à Amsterdam, on en comptait environ 50.000. Mais, en dépit de leur importance numérique, ils restaient, légalement et en fait, dans une condition inférieure et la Hollande demeurait toujours un pays de mission, dépourvu de diocèses canoniquement organisés. A l'exception des religieux, dont beaucoup préféraient ce régime qui leur laissait plus d'autonomie, la plupart des catholiques souhaitaient la disparition de cette

ciper plus activement à la vie de l'Église se manifestait par le souhait de voir instituer des conseillers ecclésiastiques laïques. (Cf. STEPISCHNEGG, *Papst Pius IX und seine Zeit*, t. I, p. 66 et suiv.)

(1) BIBLIOGRAPHIE. — I. SOURCES. — *Interpellatie in de zitting der Staten Generaal van den 13 april 1853 omtrent de instelling van bisdommen in de Nederlanden*, La Haye, 1853, ainsi que les documents relatifs à la dénonciation du concordat de 1827 publiés dans le *Nederlandsche Staatscourant* du 27 avril 1853.

II. TRAVAUX. — P. ALBERS, *Geschiedenis van het herstel der hierarchie in de Nederlanden*, Nimègue, 1903-1904, 2 vol. (nombreux documents), à compléter par J. WITLOX, *Studien over het herstel der hierarchie in 1853*, Tilbourg, 1928, et A. COMMISSARIS, *Van toen wij vrij werden*, t. I (*1795-1853*), La Haye, 1928.

(2) Sur l'organisation, assez précaire, de l'Église dans les provinces où les catholiques étaient en minorité, voir le *Handboekje voor de zaken der Roomsche Katholieke Eeredienst*, t. I, La Haye, 1847, p. 3 et suiv.

situation anachronique, mais on hésitait sur la meilleure solution. Les uns désiraient voir mettre enfin en application le concordat, conclu en 1827 mais resté lettre morte, qui prévoyait entre autres l'érection de deux évêchés, l'un à Amsterdam, l'autre à Bois-le-Duc. D'autres estimaient qu'un seul diocèse pour la partie sud était beaucoup trop peu et qu'il fallait absolument les multiplier, même si l'on devait pour cela renoncer aux avantages du concordat : certains du reste, influencés par Lamennais et son école, — c'était le cas entre autres du célèbre publiciste Le Sage ten Broek — n'étaient guère favorables à un régime concordataire et faisaient valoir les avantages de la liberté de l'Église.

A Rome, les préférences allaient à la solution concordataire, à condition que certaines stipulations du concordat de 1827 fussent modifiées, notamment celle concernant la nomination des évêques, jugée à présent trop favorable au gouvernement. Ce dernier, de son côté, voulait bien concéder un second évêché dans le Limbourg, mais à la condition que l'on renonçât à créer un diocèse dans le Nord, dont les calvinistes ne voulaient à aucun prix. Les négociations, entamées sur ces bases en 1841 par Mgr Capaccini, n'avaient abouti qu'à un demi-résultat ; tout en maintenant le principe du concordat, on s'était décidé, à cause de l'opposition enragée des milieux protestants, à retarder encore sa mise en application ; en attendant, le nombre des vicaires apostoliques dans le Sud avait été augmenté. Le Saint-Siège, qui avait obtenu que le roi renonçât à toute intervention dans les nominations épiscopales, était très satisfait, mais la hiérarchie ordinaire n'était toujours pas rétablie, au grand regret de ceux qui, comme le turbulent avocat van der Horst, l'ancien collaborateur de l'école de Malines, attaché à présent à la direction du culte catholique à La Haye, estimaient que le principal but à atteindre était « l'introduction de l'épiscopat, avec ou sans le concordat »[1].

LA CONSTITUTION DE 1848   C'est sur ces entrefaites que, en mars 1848, sous l'influence des troubles qui agitaient de proche en proche l'Europe entière, le roi Guillaume II, esprit large et tolérant, nomma une commission de cinq libéraux, dont l'âme fut le professeur Thorbecke, afin de préparer la revision de la constitution. Les catholiques firent immédiatement parvenir à celle-ci une liste de leurs desiderata, qui se résumaient en trois points : suppression du placet royal, liberté d'enseignement, liberté d'association (c'est-à-dire en fait suppression des entraves à l'établissement des ordres religieux). Les protestants essayèrent d'abord de s'opposer à des mesures qui revenaient en fait à concéder la liberté complète aux catholiques et à leur permettre de communiquer sans entraves avec Rome. Mais les catholiques ne se laissèrent pas intimider : soutenus par quelques publicistes méritants, en particulier l'abbé Broer, l'animateur de la revue De Katholiek, et surtout l'abbé Smits, le principal rédacteur du quotidien De Tijd[2], ils réa-

---

(1) Van der Horst au cardinal Sterckx, 9 avril 1841, dans A. SIMON, Le cardinal Sterckx et son temps, t. II, p. 390.
(2) Voir à leur sujet W.-G. VERSLUIS, Geschiedenis der emancipatie der Katholieken van 1795 tot heden, Utrecht, 1948, p. 59-65 et 87-92.

girent vigoureusement au cri de « Maintenant ou jamais ! » et Thorbecke, qui avait besoin de leur appui pour que son projet passe en dépit de l'opposition des conservateurs, s'efforça de leur donner satisfaction. Finalement, les catholiques durent se contenter d'un compromis en matière d'enseignement, mais pour le reste, la nouvelle constitution fut pour eux un grand succès car, en dépit de tous ceux qui auraient souhaité voir la Hollande rester « la nation protestante », elle proclamait que « tous les groupes religieux sont égaux devant la loi » et que « toute société religieuse doit régler elle-même ses affaires intérieures ». Les catholiques se voyaient donc enfin « relevés de la demi-incapacité qui, depuis le début du siècle, avait remplacé pour eux l'incapacité complète »[1].

LA RESTAURATION DE LA HIÉRARCHIE — Le rétablissement d'une organisation ecclésiastique régulière devait apparaître comme le couronnement normal de cette complète émancipation légale, et la question semblait à première vue facile à résoudre, à présent que les différents cultes étaient libres de s'organiser comme ils l'entendaient. Toutefois, certains milieux ecclésiastiques se demandaient s'il valait vraiment la peine d'exciter les passions antiromaines pour ce qui leur paraissait plutôt une question de formes, surtout après les améliorations consécutives à l'accord de 1841 ; et à Rome, où l'on évoquait encore parfois le fantôme du jansénisme hollandais, on craignait un peu que des évêques n'aient tendance à faire preuve de trop d'indépendance. Ce furent les laïcs qui insistèrent, soutenus d'ailleurs par certains vicaires apostoliques, Mgr van Wyckerslooth notamment, et par l'archevêque de Malines, le cardinal Sterckx, qui estimait avoir une responsabilité spéciale à l'égard de cette Église de Hollande si proche de la sienne[2]. Ces laïcs se rendaient compte de la nécessité d'avoir sur place une autorité capable de les guider dans les questions politiques qui touchaient aux intérêts de l'Église. Déjà en 1847, ils avaient en vain adressé une pétition à Pie IX, et ils n'eurent pas plus de succès quand ils revinrent à la charge en 1849, au lendemain du vote de la constitution. Ils réussirent pourtant deux ans plus tard à convaincre les vicaires apostoliques de faire une nouvelle démarche et celle-ci trouva cette fois meilleur accueil à Rome où l'on avait fini par être impressionné par les avis convergents de l'internonce, du cardinal Sterckx, de l'évêque de Liége, un Hollandais d'origine, et de Mgr Zwijsen, l'une des personnalités ecclésiastiques les plus en vue des Pays-Bas.

Mais le Saint-Siège une fois convaincu, c'est du côté du gouvernement que vinrent les difficultés. Lorsque, en décembre 1851, l'internonce annonça l'intention du souverain pontife d'user de la liberté d'organisation des cultes pour établir quatre ou cinq diocèses, le gouvernement hollandais rappela que le concordat de 1827, toujours en vigueur bien que n'ayant jamais été mis en application, ne prévoyait que deux diocèses.

(1) P. Verschave, *La Hollande politique*, Paris, 1910, p. 221.
(2) Sur les interventions répétées de Sterckx en faveur du rétablissement de la hiérarchie, voir A. Simon, *Le cardinal Sterckx*, t. II, p. 382-383.

On ne fit pas difficulté à Rome pour reconnaître la primauté d'une convention diplomatique sur le droit constitutionnel, et des conversations s'engagèrent pour essayer d'amener le gouvernement à modifier le texte du concordat conformément aux désirs des catholiques. Mais le gouvernement, qui savait combien les protestants étaient hostiles à ce concordat, se montra intransigeant, dans le but d'amener Rome à y renoncer spontanément afin de supprimer l'obstacle à la multiplication des diocèses. Le Saint-Siège finit par s'y résigner, bien qu'il eût préféré rétablir la hiérarchie tout en conservant le concordat, et, à l'automne 1852, le concordat fut dénoncé par consentement mutuel.

Fidèle à son idée maîtresse qu'il fallait l'entente entre l'Église et l'État, le cardinal Sterckx avait insisté pour que, à défaut de concordat, la hiérarchie fût du moins rétablie dans l'entente avec le gouvernement. Mais devant la mauvaise volonté que semblait marquer celui-ci, obligé de tenir compte de l'opinion protestante, le Saint-Siège, estimant qu'on avait déjà traîné suffisamment, décida de brusquer les choses, au risque de le regretter ensuite [1].

Divers projets avaient été présentés à Rome, dont l'un, établi par Zwijsen et l'internonce, prévoyait quatre diocèses : Bréda, Roermond, Haarlem et Bois-le-Duc ; mais l'historien Alberdingk Thijm, qui collaborait activement au mouvement catholique néerlandais, insista pour qu'on soulignât qu'il s'agissait d'une *restauration* et, afin que la continuité avec le passé apparût mieux, il proposa de fixer le siège de l'archevêché à Utrecht, l'ancien siège métropolitain fondé par saint Willibrord. C'est à cette dernière solution que se rallia la Propagande et Pie IX ratifia cette décision le 20 décembre 1852. Le 4 mars 1853, il signait la bulle d'érection *Ex qua die* [2] ; trois jours plus tard, il annonçait la nouvelle aux cardinaux réunis en consistoire et, le 4 avril, les titulaires des nouveaux diocèses étaient désignés ; comme il fallait s'y attendre, Mgr Zwijsen était nommé archevêque d'Utrecht et prenait ainsi la tête de la hiérarchie.

Ce fut une explosion de fureur dans le monde protestant, connue sous le nom d'« agitation d'avril » ; elle fut aggravée par les adversaires du gouvernement qui saisirent l'occasion pour provoquer la chute du cabinet Thorbecke. En chaire, dans les réunions publiques, dans la presse, les attaques contre le catholicisme se multiplièrent, tandis que les adresses affluaient à la Cour et aux Chambres en vue d'empêcher l'érection des évêchés (l'une d'elles, rédigée à Utrecht, citadelle de l'aristocratie antilibérale et de l'antipapisme, portait 200.000 signatures). Les catholiques conservèrent leur calme et, après s'être réunis en conférence à Tilbourg, les 7 et 8 avril, les nouveaux évêques prirent en main l'administration de leurs diocèses. Le gouvernement essaya en vain d'obtenir qu'au moins les sièges d'Utrecht et de Haarlem fussent transférés dans d'autres villes. Puis un projet de loi, élaboré par un ancien partisan du *Placet*, pour régler

---

(1) Le 4 mai 1853, le ministre de Belgique à Rome écrivait : « L'affaire de la hiérarchie hollandaise continue à faire du bruit. On avoue ici qu'on a été trop brusque. On regarde l'avenir des catholiques néerlandais comme assez compromis. » (Dans A. SIMON, *op. cit.*, t. II, p. 394, n. 6.)

(2) *Acta Pii IX*, t. I, p. 416 et suiv.

la police des cultes, sembla remettre en question toutes les dispositions libérales de la constitution de 1848 ; mais le roi Guillaume III, bien que moins sympathique que son prédécesseur à l'Église romaine, jugea qu'il serait dangereux pour l'unité du pays de pousser les catholiques à bout. Aussi la loi fut-elle modifiée, non sans maintenir plusieurs dispositions vexatoires, comme l'interdiction de sortir hors des églises en ornements liturgiques, ou comme l'élévation au rang de culte reconnu de la petite Église schismatique janséniste [1].

Le gouvernement finit par se décider à reconnaître officiellement les nouveaux évêques, le 24 septembre, après avoir obtenu quelques minimes concessions : le pape imposait aux évêques un serment de fidélité au roi, sur le modèle de celui approuvé pour les évêques d'Irlande ; et il était entendu que les évêques d'Utrecht et de Haarlem ne résideraient pas en permanence dans leur ville épiscopale. Une étape capitale dans l'histoire de la Hollande catholique était ainsi franchie sans encombres.

§ 5. — Le rétablissement de la hiérarchie en Angleterre [2].

CROISSANCE DE L'ÉGLISE EN ANGLETERRE

Au moment où Pie IX se décidait à réorganiser l'Église des Pays-Bas en lui accordant à nouveau le régime d'une Église adulte, il venait de prendre une décision analogue pour une Église voisine, qui avait, elle aussi, été la victime de la Réforme protestante, celle d'Angleterre. Toutefois, dans ce dernier cas, l'occasion ne fut pas fournie comme aux Pays-Bas par une évolution plus favorable de la législation à la suite des événements de 1848, mais par le rapide développement de la petite communauté catholique anglaise qui, au cours des premières années du pontificat de Pie IX, s'était profondément transformée sous l'action de trois facteurs.

L'ACTION DES CONVERTIS ET DES RELIGIEUX

D'abord la vague de conversions, provoquée par le mouvement d'Oxford [3] et dont les points culminants se situent en 1845-1846 et en 1850-1851, amène dans l'Église un nombre respectable d'universitaires influents et de clergymen de grande valeur intellectuelle et reli-

---

(1) Le *Oud Bischoppelijke Klerezij* (vieux clergé épiscopal) s'était séparé de Rome en 1702, lorsque le vicaire apostolique Codde, ayant été déposé par Rome comme janséniste, un certain nombre de prêtres et de fidèles avaient refusé de se soumettre. Le schisme s'était perpétué avec des évêques validement consacrés et il comptait encore en 1853 une trentaine de prêtres et environ 10.000 fidèles, dirigés par deux évêques ayant leurs sièges à Utrecht et à Haarlem ; ceux-ci avaient protesté officiellement auprès du gouvernement contre la nomination par Rome de prélats à des sièges déjà occupés par des successeurs « réguliers » des anciens évêques. (Cf. C. B. Moss, *The Old Catholic Movement*, Londres, 1948, chap. VII, VIII et XII, à mettre au point par P. Albers, *op. cit.*, t. I, p. 253-257 et t. II, p. 433-434.)

(2) BIBLIOGRAPHIE. — Voir surtout W. Ward, *Life and Times of cardinal Wiseman*, Londres, 1897, t. I, chap. XV à XVIII (cité dans la traduction française, Paris, 1900), où l'on trouvera de nombreux documents, et *The English Catholics 1850-1950. Essays to commemorate the Centenary of the restauration of the Hierarchy*, sous la direction de Mgr G. A. Beck, Londres, 1950. Également l'autobiographie de W. Ullathorne, *From Cabin-boy to Archbishop*, Londres, 1943, chap. XXXII à XXXVII ; du même, *History of the Restauration of the Cath. Hierarchy in England*, Londres, 1871. Exposé d'ensemble dans P. Thureau-Dangin, *La Renaissance catholique en Angleterre au XIXᵉ s.*, t. II, Paris, 1903, chap. I et IV.

(3) Cf. t. XX, p. 484-490.

gieuse, à propos desquels Wiseman n'avait pas hésité jadis à écrire :
« Je suis prêt à reconnaître qu'en toute chose, sauf le bonheur de posséder
la vérité..., nous leur sommes inférieurs. J'ai dit depuis longtemps à
ceux qui m'entourent que si les théologiens d'Oxford entraient dans
l'Église, nous devrions être prêts à retomber dans l'ombre et à passer
au second plan [1]. »

En même temps qu'il accueille à bras ouverts ces nouveaux converti-
tis, Wiseman se préoccupe de réveiller le zèle un peu endormi d'un clergé,
pieux et consciencieux certes, mais aux vues étroites et routinières et
qui ne songe guère qu'à préserver son petit troupeau des influences
mauvaises du dehors. En vue de réagir contre cette passivité, il appelle
le plus possible de religieux étrangers dans le district de Londres, dont il
a pratiquement la direction depuis 1847 [2] et, sous l'impulsion de ceux-ci,
le catholicisme anglais sort peu à peu de sa torpeur et de son effacement.
Plusieurs des nouveaux convertis entrent du reste dans ces congrégations, tandis que le plus célèbre d'entre eux, Newman, après un séjour
de quelques mois à Rome, où tout en se préparant au sacerdoce, il a
pris contact avec les disciples de saint Philippe Neri, fonde à Birmingham,
au début de 1848, la première maison de l'Oratoire [3]. Cette nouvelle
congrégation prit vite de l'extension grâce à l'appoint d'un groupe réuni
antérieurement autour d'un autre converti, William Faber, qui, en 1850,
ouvrira un second Oratoire en plein Londres et en fera rapidement le
foyer le plus ardent de la vie catholique dans la capitale [4].

L'ÉMIGRATION IRLANDAISE   On exagérerait difficilement les conséquences de cet afflux de sang nouveau
qui relève sensiblement le *standing* intellectuel et la vitalité de l'Église
catholique en Angleterre. Mais un troisième élément vient encore accroître
la transformation. Depuis le début du siècle, les effectifs catholiques
augmentaient régulièrement, par suite de l'installation d'émigrés irlandais, attirés par la révolution industrielle, et la communauté catholique,
jadis axée autour des manoirs de quelques gentilshommes, dont le catholicisme s'expliquait parfois tout autant par tradition politique et familiale que par conviction religieuse, se transformait peu à peu en une
Église composée pour une bonne part de prolétaires résidant dans les
grandes villes ou dans les ports, cédant facilement à l'ivrognerie, mais
dont l'exubérance celtique et la dévotion enthousiaste s'accommodaient
fort mal des allures de catacombes de l'ancien catholicisme anglais.

Or cette évolution se précipite avec le brusque accroissement de cette
émigration lors de la grande famine de 1845-1847. Sur le gros million
d'Irlandais forcés de s'expatrier, nombreux sont ceux qui se fixent en

---

(1) Lettre de 1841, citée dans E. PURCELL, *Life and Letters of A. Phillips de Lisle*, t. I, Londres, 1900, p. 290.
(2) Il avait été nommé en août 1847 provicaire apostolique et devint vicaire en titre en février 1849, à la mort de Mgr Walsh.
(3) Cf. W. WARD, *The Life of J.-H. cardinal Newman*, t. I, Londres, 1912, chap. v à vii.
(4) Cf. *Life and Letters of F. W. Faber*, Londres, 1888. En dépit de sa mauvaise santé, Faber devait être jusqu'à sa mort (1863) d'une activité extrême, prêchant, confessant et publiant de nombreux livres de spiritualité fort appréciés.

Angleterre, où la population catholique dépasse bientôt les 700.000 âmes (sur une population totale de 18 millions d'habitants), dont 250.000 seulement sont d'origine anglaise. Les autorités religieuses se voient obligées de multiplier les paroisses, d'ouvrir de nouvelles églises (en se contentant souvent de locaux de fortune) et de faire appel à des prêtres venant d'Irlande, pour assurer un ministère auquel le clergé local est absolument incapable de suffire [1].

PREMIÈRES TRACTATIONS EN VUE DU RÉTABLISSEMENT DE LA HIÉRARCHIE   Cette rapide évolution devait inévitablement poser la question d'une réorganisation de l'administration ecclésiastique du pays. Tandis que l'Irlande avait conservé une hiérarchie épiscopale, les catholiques anglais, au contraire, étaient, depuis la Réforme, au régime des pays de mission. Nombreux étaient ceux qui, à Rome comme en Angleterre, estimaient que ce régime ne correspondait plus à la situation nouvelle [2] et qu'il fallait rétablir la hiérarchie régulière. C'était en particulier l'avis de Wiseman, qui attachait à la pompe extérieure et aux honneurs qui environneraient une hiérarchie régulière une importance que certains trouvaient un peu exagérée. Malgré les objections de ceux des catholiques qui souffraient encore de la timidité des siècles de persécution et qui se savaient appuyés à la Curie romaine par un prélat anglais, le cardinal Acton, l'idée d'un retour au régime normal faisait son chemin et, à Pâques 1847, lors de la séance annuelle des vicaires apostoliques, Wiseman fut désigné avec un autre de ses collègues pour aller discuter la chose à Rome.

Leurs arguments parurent convaincants et le principe fut admis du rétablissement de la hiérarchie. Mais comme rien ne venait, les vicaires apostoliques revinrent à la charge vers le milieu de 1848. Leur délégué, Mgr Ullathorne, grâce à son tact et à son sens des affaires, réussit, au bout de neuf semaines de négociations, à écarter les dernières difficultés et il quitta Rome en novembre avec l'assurance que les documents officiels étaient en préparation. La fuite du pape à Gaëte vint toutefois arrêter l'exécution du projet et celui-ci dormit encore pendant deux ans dans les bureaux de la chancellerie pontificale.

DISSENSIONS PARMI LES CATHOLIQUES   La révolution romaine n'était d'ailleurs pas la seule cause de ce retard. L'un des arguments mis en avant quelques années auparavant par le cardinal Acton pour déconseiller de rétablir la hiérarchie et de donner ainsi plus d'indépendance aux catholiques de son pays était la propension de ceux-ci, disait-il, à se diviser en factions rivales [3]. Or préci-

---

(1) Cf. D. GWYNN, *The famine and the Church in England*, dans *Irish Ecclesiastical Record*, t. LXIX, 1947, p. 896-909. D'après Mgr Ward, l'afflux des Irlandais chassés par la famine « eut plus d'influence sur l'avenir du catholicisme en Angleterre que le mouvement d'Oxford lui-même ».

(2) Outre les inconvénients très réels du système (qu'on trouvera exposés dans l'autobiographie d'Ullathorne, p. 246, et surtout dans le *memorandum* de Wiseman au pape de juillet 1847, reproduit dans W. WARD, *op. cit.*, t. I, p. 486-488), on faisait remarquer que des Églises plus jeunes, comme celles d'Amérique du Nord ou d'Australie, possédaient déjà une hiérarchie.

(3) W. ULLATHORNE, *From Cabin-boy to Archbishop*, p. 247-248.

sément, à partir de 1848, les symptômes se multiplient d'une indéniable tension.

D'une part, les gentilshommes campagnards, jaloux de leurs anciens privilèges, regardent avec méfiance la part croissante que prend dans la vie catholique l'élément irlandais et reprochent notamment aux prêtres venus d'Irlande d'intervenir en politique dans un sens opposé aux intérêts conservateurs [1].

Mais c'est surtout l'attitude des nouveaux convertis qui provoque des difficultés. La plupart de ceux-ci appuient, parfois avec un peu d'intempérance, les efforts de Wiseman pour rétablir les manifestations oubliées de la piété catholique. Ils s'associent avec une ardeur de néophytes aux tentatives des religieux pour promouvoir des formes de dévotion en usage dans les pays méridionaux. Ainsi Faber, qui « partait de cette idée que le modèle de la vie catholique devait être cherché, non dans un pays où la persécution avait forcé le catholicisme en quelque sorte à se déguiser, mais dans ceux où il avait pu s'épanouir, et particulièrement dans sa vraie patrie, à Rome » [2]. Or les convertis de la génération précédente, groupés autour de lord Shrewsbury et de l'architecte Pugin, fervents du style gothique [3], tout comme les catholiques de vieille souche, à la religion austère et formaliste, sont craintifs devant tout ce qui pourrait exciter les protestants, hostiles à toute innovation, portés à s'opposer à ce qu'ils considèrent comme l'introduction en Angleterre d'une « religion italienne », qui répugne au tempérament national [4]. Même un esprit ouvert comme Mgr Ullathorne s'inquiète du manque de discrétion de gens dont « l'inexpérience est aussi grande que le zèle », et il reproche à Newman de sembler approuver le projet conçu par Faber de traduire de l'italien une série de vies de saints qui lui paraissent manquer de la sobriété désirable [5].

Ces divergences de vues étaient presque inévitables dans une Église formée d'éléments hétérogènes qui n'avaient pas encore eu le temps de fusionner, mais le coup d'œil génial de Wiseman lui avait montré dans quelle direction se trouvait l'avenir et il encourageait ouvertement les entreprises des convertis et des religieux. Cette attitude n'avait pas manqué de lui aliéner les sympathies d'une partie de l'ancien clergé, et l'on comprend dès lors que celui-ci craignait de le voir, comme vicaire apostolique de Londres, devenir le chef de la nouvelle hiérarchie et étendre à toute l'Angleterre la nouvelle politique qu'il favorisait dans la capitale.

---

(1) D. GWYNN, art. cit., p. 904. Cela n'empêcha du reste pas ces mêmes gentilshommes de contribuer généreusement à la construction des églises nouvelles nécessitées par l'afflux irlandais.
(2) P. THUREAU-DANGIN, op. cit., t. II, p. 303.
(3) Sur le rôle, généralement méconnu, de ce groupe qui, avant les convertis d'Oxford, contribua efficacement au renouveau catholique, cf. D. GWYNN, Lord Shrewsbury, Pugin and the catholic Revival, Londres, 1946.
(4) Sur les « anciens catholiques » (old catholics), cf. D. MATHEW, Acton. The formative Years, Londres, 1946, p. 38-46 ; sur leurs divergences de vues avec les nouveaux convertis et les religieux italiens, cf. W. WARD, op. cit., chap. XXIV ; Mgr WARD, Sequel to cath. Emancipation, chap. XXIX et J.-G. SNEAD-COX, Life of cardinal Vaughan, Londres, 1910, chap. IV.
(5) Voir sur cet incident caractéristique C. BUTLER, The Life and Times of Bishop Ullathorne, t. I, Londres, 1926, p. 152-162 ; Life and Letters of F. W. Faber, p. 342-358 ; Mgr WARD, Sequel..., t. II, p. 243-252 ; W. WARD, Vie de Wiseman, t. II, p. 241-247 et Life of Newman, t. I, p. 206-214.

WISEMAN CARDINAL Soucieuses de ne pas aggraver ces dissensions, les autorités romaines crurent prudent de ne pas brusquer les choses et l'on envisagea un moment, pour atténuer les oppositions, d'écarter Wiseman en le nommant cardinal de Curie. Ce fut un coup terrible pour celui-ci, qui voyait déjà brisée l'œuvre de sa vie. Mais lorsque la décision devint publique, en juillet 1850, elle provoqua de telles protestations de la part des catholiques clairvoyants que le futur cardinal, en arrivant à Rome, n'eut guère de peine à convaincre le pape de le laisser en Angleterre. Du coup, le rétablissement de la hiérarchie en fut précipité, car il était difficile de revenir sur la décision de l'élever au cardinalat et impossible de conférer cette dignité à un simple vicaire apostolique : il ne restait d'autre solution que de le voir le plus vite possible archevêque de Westminster.

LE RÉTABLISSEMENT DE LA HIÉRARCHIE ET LA RÉACTION EN ANGLETERRE Le 29 septembre 1850, le pape publiait le bref remplaçant les huit vicariats apostoliques par un archevêché et douze évêchés suffragants. Le lendemain, Wiseman recevait le chapeau et le 7 octobre, dans une lettre pastorale enthousiaste, il laissait éclater sa joie de voir « l'Angleterre catholique retrouver son orbite dans le firmament religieux d'où sa lumière avait longtemps disparu ».

La décision pontificale et surtout le ton triomphant de la lettre du nouveau cardinal provoquèrent en Angleterre une explosion de colère antipapiste assez inattendue[1]. On prétendit y voir une bravade et la presse entama une violente campagne contre cette « usurpation étrangère ». La populace brûla en effigie le pape et le cardinal, tandis que le premier ministre, lord Russel, faisait voter un *bill* dénonçant « l'agression papale » et interdisant aux nouveaux évêques de porter leur titre.

Ceux des catholiques qui s'étaient opposés au rétablissement voyaient dans ces événements la confirmation de leurs craintes et même les amis de Wiseman étaient déconcertés au point de lui conseiller de temporiser et de ne pas paraître en Angleterre pour le moment. Mais celui-ci décida de faire front. Il rentra à Londres et, après avoir envoyé au gouvernement les explications nécessaires sur la portée de la décision pontificale, il rédigea en quelques jours, en s'aidant de notes préparées par Mgr Ullathorne, un *Appel au peuple anglais* qui retourna complètement l'opinion. Il acheva son succès par une série de conférences publiques au cours desquelles il réussit très habilement à mettre en lumière le ridicule de toute cette agitation, de sorte que cette crise n'eut d'autre résultat que de renforcer singulièrement son prestige.

(1) Peu d'années auparavant, le premier ministre, consulté sur l'éventualité de l'instauration de la hiérarchie au Canada et en Australie, avait répondu : « Qu'est-ce que cela peut bien nous faire que vous vous appeliez vicaires apostoliques ou évêques ? » (W. WARD, *Vie de Wiseman*, t. I, p. 485-486.)

L'ÉGLISE ET L'ITALIE JUSQU'EN 1870 [1]

§ 1. — Les débuts de la laïcisation en Italie [2].

*LE SENTIMENT RELIGIEUX EN
ITALIE AU MILIEU DU
XIXᵉ SIÈCLE*

Le mouvement de déchristianisation qui avait sévi au XVIIIᵉ siècle en France et dans les pays germaniques n'avait pas atteint l'Italie dans les mêmes proportions. Qu'il habitât dans les villes populeuses du Piémont ou de la Lombardie-Vénétie, dans les duchés cultivés du centre, dans les campagnes arriérées du Midi napolitain, ou dans les États de l'Église [3] où toute la vie civile restait encore réglée comme au moyen âge en fonction de la vie religieuse, l'Italien demeurait sincèrement attaché au catholicisme, en dépit d'infiltrations voltairiennes dans une partie de la classe cultivée.

La religion du peuple, malgré son ignorance et sa moralité défectueuse, était sérieuse, tout éloignée qu'elle fût de la « respectabilité » protestante ou du conformisme de la Contre-Réforme : sans parler des pratiques de dévotion, qui conservaient tout leur attrait sur les masses, les très nombreuses confréries, auxquelles on tenait surtout par esprit de corps, maintenaient tout compte fait un lien étroit entre l'Église et

(1) BIBLIOGRAPHIE. — 1° Sur la question romaine. On trouvera une liste très détaillée des sources et des travaux dans G. MOLLAT, *La question romaine de Pie VI à Pie IX*, 2ᵉ éd., Paris, 1932. On notera spécialement l'utile recueil de documents de H. BASTGEN, *Die Römische Frage. Dokumente und Stimmen*, Fribourg-en-Br., 1917-1919, 3 vol. ; les ouvrages de R. DE CESARE, *Roma e lo Stato del Papa dal ritorno di Pio IX al 20 settembre*, Rome, 1907 et de A. MONTI, *Pio IX nel Risorgimento italiano*, Bari, 1928 ; S. JACINI, *Il tramonto del potere temporale nelle relazioni degli ambasciatori austriaci a Roma*, Bari, 1931 ; J. MAURAIN, *La politique ecclésiastique du Second Empire*, chap. XIII-XV, XVII, XIX, XXII et XXV. Ajouter A.-C. JEMOLO, *La questione romana*, Milan (1938) et V. DEL JUDICE, *La questione romana e i rapporti fra Stato e Chiesa fino alla Conciliazione*, Rome (1948).
Nombreux matériaux dans *La Civiltà cattolica* (depuis 1850 ; point de vue romain) et dans la *Nuova antologia* (depuis 1866 ; point de vue italien).
2° Sur la politique ecclésiastique et la vie catholique en Italie. Aux ouvrages généraux de Jacini, Jemolo et Della Torre signalés p. 9, ajouter : *L'episcopato e la rivoluzione in Italia*, Mondovi, 1867, 2 vol. ; ainsi que : A.-C. JEMOLO, *La questione della proprieta ecclesiastica nel Regno di Sardegna e nel Regno d'Italia, 1848-1888*, Turin, 1911.
(2) BIBLIOGRAPHIE. — I. SOURCES. — P. PIRRI, *Pio IX e Vittorio Emanuele II, dal loro carteggio privato*. I. *Laicizzazione dello Stato Sardo (1848-1856)*, Rome, 1944 ; *Lettres de Mgr Billiet*, archevêque de Chambéry, à Mgr Rendu, évêque d'Annecy, 1844-1899, publiées par R. AVEZOU, dans *Mém. et Doc. publiés par la Soc. savoisienne d'histoire*, t. LXXXVIII, Chambéry, 1936, p. 41-111.
II. TRAVAUX. — Aux ouvrages cités à la note précédente, ajouter : T. CHIUSO, *La Chiesa in Piemonte*, t. III, Turin, 1889 ; A.-C. JEMOLO, *Il « partito cattolico » piemontese nel 1855*, dans *Il Risorgimento italiano*, t. XI, 1918-19, p. 1-52 ; C. COLONIATI, *Mgr Luigi Fransoni, arcivescovo di Torino 1832-1862*, Turin, 1902 et A. OMODEO, *L'opera politica del conte di Cavour*, t. I (1848-1857), Florence (1940).
(3) Il ne faut pas oublier en effet que, jusqu'en 1860, l'Italie reste une « expression géographique », groupant 7 États : le royaume de Piémont ou de Sardaigne (4.650.000 habitants), le Nord Lombardo-Vénitien, sous la domination autrichienne (4.800.000 habitants) ; le grand-duché de Toscane (1.700.000 habitants) ; les deux duchés de Modène et de Parme et Plaisance (500.000 habitants chacun), l'État pontifical (3 millions d'habitants) et le royaume de Naples ou Deux-Siciles (8 millions 300.000 habitants).

le peuple. De leur côté, les intellectuels et la bourgeoisie souhaitaient bien, du moins dans le Nord et en Toscane, une réforme de l'Église et surtout tendaient à ne plus tenir compte de ses exigences concernant l'organisation de la vie publique, mais leur sensibilité catholique devait les rendre à peu près imperméables à la propagande protestante, qui tentera vainement sa chance à partir de 1850 [1].

Pourtant, depuis 1848, la situation n'est plus intacte. De différents côtés on signale que l'indifférence religieuse progresse dans le peuple : par exemple à Bologne, en 1853, « il y a un grand nombre de familles d'ouvriers où personne ne prend part aux fêtes et réunions dans les églises » [2] et don Bosco constate le même phénomène à Turin. Dans les milieux cultivés, le rationalisme incrédule commence à pénétrer lentement tandis que Bianchi-Giovini s'inspire de l'exégèse radicale allemande dans sa *Critica degli Evangeli* (1853), un jeune prêtre qui avait fait défection lors de la crise de 1848, Bonavino, plus connu sous le nom d'Ausonio Franchi, fonde en 1854 un hebdomadaire, *La Ragione*, destiné à propager l'idée que le rationalisme doit devenir la religion du XIXe siècle et se substituer non seulement au « catholicisme jésuitique », mais même au catholicisme libéral et au protestantisme.

Le véritable danger toutefois est ailleurs. Un grand nombre d'Italiens rendent Pie IX et l'Église responsables de l'échec du mouvement national de 1848 et l'hostilité croissante témoignée par les milieux catholiques dirigeants envers les conceptions libérales ne fait que renforcer le malaise. Aussi constate-t-on un refroidissement marqué des relations de la bourgeoisie avec le clergé et, bien que la franc-maçonnerie, contrairement à ce qu'on dit souvent, ait été peu active avant 1859 [3], l'anticléricalisme progresse, favorisé par des journaux comme *L'Opinione* dont l'esprit, naguère sympathique à l'Église, se modifie radicalement en quelques années. La plupart entendent bien, au début, concilier leur opposition au catholicisme officiel et aux tendances politico-religieuses dominantes à Rome avec la foi catholique et la pratique des sacrements, mais il est inévitable que les rancœurs envers le pape et la Curie ne conduisent peu à peu à l'indifférence à l'égard de la doctrine elle-même. L'Église, au moment où elle achève, en France, de perdre la classe ouvrière, est en passe, en Italie, de perdre la bourgeoisie.

LE CLERGÉ En présence de cette inquiétante évolution des esprits, le clergé n'est guère à même de réagir. Il n'y a que peu de choses à attendre du côté des religieux, malgré leur nombre. Trop de *frati* mènent dans leurs couvents, dont les biens-fonds sont considérables, une vie nonchalante, sans préoccupations apostoliques [4]. Les jésuites

(1) Les « Vaudois », à qui la liberté fut octroyée par le Piémont en 1848, devinrent aussitôt missionnaires ; leurs efforts, soutenus par la Suisse et surtout par l'Angleterre, inquiétèrent vivement les autorités ecclésiastiques, mais dès 1860 l'échec du mouvement était patent. (Voir A. DELLA TORRE, *Il cristianesimo in Italia*, p. 146-155, 190-199.)

(2) Doubet à Rendu, janvier 1854, dans J. GAY, *Les deux Rome et l'opinion française*, p. 67.

(3) Cf. A. LUZIO, *La Massoneria e il Risorgimento italiano*, Bologne, 1925, 2 vol.

(4) En 1870 encore, M. Icard s'étonnait des doléances de l'abbé cistercien de Sainte-Croix-en-Jérusalem, à Rome, qui craignait de voir le concile interdire le pécule des religieux : « Selon le

font preuve d'un beau zèle, mais ils sont détestés à cause de leurs relations avec le parti austrophile et du manque de nuances de leur antilibéralisme. Le travail le plus fécond est accompli par les congrégations venues de France, frères des Écoles chrétiennes [1], dames du Sacré-Cœur, sœurs de Saint-Vincent-de-Paul. Le cardinal Racanati, capucin, le reconnaît lui-même : « En fait d'œuvres pratiques, il n'y a que les œuvres françaises et dirigées par les Français qui aient un véritable succès [2]. »

Quant au clergé séculier, dont le recrutement est aisé et qui compte vers 1850 plus de 60.000 prêtres pour une population qui n'atteint pas 25 millions d'habitants, il est loin de donner ce qu'on pourrait en attendre. D'abord, beaucoup de prêtres ne font pas de ministère actif et se contentent, comme sous l'ancien régime, de gérer leur patrimoine familial ou de servir de précepteur ou de chapelain dans quelque famille noble ; nombreux sont d'ailleurs, surtout dans le royaume de Naples, ceux qui gardent à l'égard de leur évêque une liberté d'allure presque complète [3]. Ensuite, sauf en quelques régions comme la Toscane, où le niveau culturel du clergé est nettement supérieur, la formation intellectuelle très rudimentaire reçue dans les séminaires et le manque d'adaptation à l'esprit moderne ne permettent pas à la plupart des prêtres de conseiller avec fruit la bourgeoisie dans la crise de conscience avec laquelle elle est aux prises. Quelques-uns encouragent en sous-main les tendances nationales et libérales avec un esprit frondeur qui fait plus de mal que de bien ; les autres se confient aveuglément à la presse conservatrice et paraissent aisément fanatiques. Beaucoup ne comprennent rien au désir du monde laïque d'obtenir une certaine autonomie, et leur prétention à continuer à régenter comme autrefois toute la vie individuelle et sociale contribue à les rendre impopulaires dans les régions touchées par le libéralisme. Par contre, les mœurs trop faciles d'une partie du clergé de l'Italie méridionale et centrale sont acceptées assez aisément par les fidèles, comme le relèvent souvent les observateurs étrangers : « On ne se fait aucune idée, note l'un d'eux, de la tolérance qui existe sous ce rapport en Italie, où l'on se dit, et fort sagement peut-être, que l'habit ne défend pas toujours l'homme de certaines faiblesses et qu'il faut donc traiter celles-ci avec indulgence aussi longtemps qu'elles ne dégénèrent pas en scandales [4]. »

Ces ombres réelles ne peuvent toutefois faire perdre de vue tout ce qui subsiste de vertu sacerdotale et d'ingéniosité apostolique dans le

---

P. Abbé, le pécule est un *stimolo* dont les hommes ont besoin dans leurs travaux ; il ne faut pas leur imposer l'héroïsme de la vertu et voir les choses dans l'abstrait mais dans le concret, comme il nous disait. Cela ne donne pas une grande idée de la perfection religieuse dans les couvents. » (*Journal de mon voyage à Rome*, p. 270, *Arch. S. Sulp.*, Fonds Icard.)

(1) Qui se sont vus obligés, en 1850, de remplacer le vicaire italien de la province romaine par un Français, pour réagir contre la décadence menaçante. (Cf. G. RIGAULT, *Histoire générale de l'Institut des Frères des Écoles Chrétiennes*, t. VI, Paris, 1947, p. 17-26.)

(2) Dans J. GAY, *op. cit.*, p. 85.

(3) Les aspirants au sacerdoce n'étaient même pas obligés de fréquenter régulièrement le séminaire. A Naples, par exemple, les élèves de la ville pouvaient rester dans leur famille et se contenter de passer les examens et de faire une retraite de dix jours chez les lazaristes avant chaque ordination ; les élèves de la campagne devaient passer deux ans comme internes au séminaire, puis ils étaient autorisés à loger en ville et à suivre les cours comme externes. (Renseignements recueillis par M. Icard, *Journal de mon voyage à Rome*, p. 313-314 et 320, *Arch. S. Sulp.*, Fonds Icard.)

(4) A. DE LIEDEKERKE, *Rapporti delle cose di Roma*, p. 56.

clergé italien. Don Bosco, qui a reçu l'ordination sacerdotale en 1841, n'est pas un isolé à Turin, et nous savons l'aide précieuse qu'il a reçue dans ses débuts de saints prêtres comme don Cafasso, l'abbé Borel et bien d'autres encore. De même à Rome, à côté des prêtres oisifs qui mendient des intentions de messe ou des *monsignori* mondains, qui retiennent surtout l'attention des touristes, la chronique locale nous a transmis les noms de bons prêtres qui ont voué leur vie à confesser ou à prêcher des missions dans les quartiers populaires, et la mort, en 1850, de don Vincent Palotti, l'un des précurseurs de l'Action catholique, ne met pas un terme à son influence. De même encore si, parmi les 220 évêques d'Italie, il y a trop de prélats, pieux et respectables certes, mais qui ne sont que de médiocres pasteurs d'âmes, on relève des noms qui font honneur à l'épiscopat, tel Mgr Moreno, évêque d'Ivrée de 1838 à 1876, dont les lettres pastorales sont lues avec intérêt dans toute l'Italie, et qui, tout en se préoccupant de relever le niveau intellectuel et moral de son clergé, devient un initiateur dans le domaine des œuvres de presse ; tel encore le cardinal Riario Sforza, archevêque de Naples de 1845 à 1877, qui rappelle en de nombreux points saint Charles Borromée et sera peut-être, lui aussi, élevé un jour sur les autels [1].

LA POLITIQUE ECCLÉSIASTIQUE DANS LES DIFFÉRENTS ÉTATS ITALIENS

Si la situation religieuse était dans l'ensemble assez semblable du nord au sud de la péninsule, il n'en allait pas de même quant à l'attitude des gouvernements italiens envers l'Église.

Dans le royaume de Naples, la position privilégiée de l'Église, héritage de l'ancien régime, fut maintenue sans difficulté jusqu'à l'annexion par le Piémont et les autorités s'appliquèrent même à n'exercer leurs droits d'intervention dans la vie ecclésiastique qu'en parfait accord avec le pouvoir religieux. Dans la Lombardie et la Vénétie, le concordat conclu avec l'Autriche en 1855 mit fin au régime d'inspiration joséphiste et, là aussi, jusqu'aux événements de 1859, l'Église se trouva dans une situation légale très favorable, bien que l'opinion cultivée, même croyante, se montrât rétive envers un régime qui lui paraissait trop clérical.

En Toscane, à part le pieux grand-duc, très respectueux des désirs du Saint-Siège, les milieux dirigeants souhaitaient maintenir le contrôle du pouvoir civil sur l'Église, mais la nécessité pour le gouvernement d'obtenir l'appui du clergé l'obligea, en 1848, à entrer en négociation avec Rome et, après des tractations difficiles, à signer, en 1851, sans grand enthousiasme, une convention où il renonçait en fait, pour l'essentiel, à la législation d'inspiration joséphiste — léopoldine, disait-on là-bas — qui avait prévalu jusqu'alors dans le grand-duché [2].

Le Saint-Siège avait donc lieu d'être satisfait de la position de l'Église dans l'Italie méridionale et centrale ainsi que dans les provinces autri-

---

(1) Voir E. Federici, *Sisto Riario Sforza*, Rome, 1945.
(2) Cf. A.-M. Bettanini, *Il concordato di Toscana, 25 aprile 1851*, Milan, 1933, et R. Mori, *Il concordato del 1851 tra la Toscana e la Santa Sede*, dans *Archivio storico italiano*, t. XCVIII, 1940, p. 41-82 et t. XCIX, 1941, p. 133-146.

chiennes ; l'évolution de la situation dans le royaume de Piémont lui causait par contre de graves soucis.

LES PREMIÈRES MESURES DE LAICISATION EN PIÉMONT Les concordats de 1828 et de 1841 avaient confirmé à l'Église la plupart des privilèges dont elle avait joui sous l'ancien régime : possession de biens-fonds considérables, qui continuaient à s'accroître par des dons et legs (autorisés à partir de 16 ans) ; maintien du système de la dîme en Sardaigne ; droit, pour les évêques dont les ressources étaient insuffisantes, d'imposer des charges supplémentaires à leurs diocésains ; sanctions légales contre ceux qui ne respectaient pas le repos dominical ou offensaient la religion de quelque façon que ce fût ; inspection des écoles par le clergé ; reconnaissance légale du seul mariage religieux, les curés étant chargés des registres de l'état-civil ; atténuation du code pénal en faveur des prêtres et religieux ; enfin privilège du for ecclésiastique, les curies épiscopales étant seules compétentes pour les causes relatives aux outrages à la religion, aux dîmes, aux mariages et pour toutes celles où un clerc se trouvait impliqué.

L'opposition générale de l'opinion progressiste italienne au maintien de pareil état de choses se trouvait renforcée dans le royaume piémontais par plusieurs éléments : l'appui souvent passionné apporté par la majorité du clergé à la politique réactionnaire menée à l'époque de la Restauration par des hommes comme Solaro della Margarita ou l'archevêque de Turin, Fransoni, avait laissé de profondes rancœurs dans les milieux libéraux ; celles-ci avaient encore été attisées par des influences françaises, celles de Michelet, et d'Eugène Sue surtout, qui avaient créé un courant nettement jacobin et anticlérical, dont on ne trouvait pas l'équivalent dans les autres États italiens ; enfin, alors que dans ces autres États, où, après 1848, les partisans de l'absolutisme étaient revenus au pouvoir, les gouvernements cherchaient l'appui de l'Église, garante du « principe d'autorité », en Piémont, où le système constitutionnel s'était maintenu, les milieux politiques se méfiaient au contraire de l'influence du clergé, considéré comme hostile à tout régime parlementaire et comme ayant partie liée avec l'Autriche.

Dès 1848, le gouvernement sarde, qui se sentait obligé de tenir grand compte de l'aile gauche de la Chambre, était décidé à appliquer un programme de laïcisation qui mettrait fin à la situation prépondérante du clergé dans l'État et aux inconvénients économiques de la main-morte, auxquels la bourgeoisie se montrait très sensible. Quelques mesures furent prises immédiatement : « émancipation » légale des protestants et des juifs et suppression de la Compagnie de Jésus ; puis contrôle gouvernemental sur les établissements d'enseignement libre et retrait aux autorités ecclésiastiques du droit de surveiller l'école. Il était plus délicat d'abolir unilatéralement le for ecclésiastique et les immunités, qui avaient fait l'objet explicite du concordat de 1841. Le gouvernement se montra disposé à négocier avec le Saint-Siège un nouveau statut qui sauvegarderait les droits essentiels de celui-ci, en mettant toutefois comme condition que les évêques les plus intransigeants, tel Mgr Fransoni, soient écartés au préalable, mais Pie IX et Antonelli, prêts à faire certaines concessions à propos du privilège du for, jugèrent pareille prétention inadmissible ; du reste, l'orientation libérale du gouvernement de Turin et son attitude plus qu'équivoque au moment de la Révolution romaine ne les incitaient guère à lui faire confiance.

Après dix-huit mois de tractations sans résultat, où les rôles principaux furent tenus, du côté piémontais, par l'abbé Rosmini, puis par Siccardi, un jurisconsulte partisan d'une modernisation des institutions mais qui n'avait rien d'un sectaire, et, du côté pontifical, par Mgr Charvaz, l'ancien précepteur du roi Victor-Emmanuel, le gouvernement dirigé par Massimo d'Azeglio prit le parti d'aller de l'avant, en dépit des protestations officielles du souverain pontife et de ses interventions secrètes auprès du roi, lequel se trouvait tiraillé entre les exigences gouvernementales et le caractère apparemment raisonnable de la nouvelle politique ecclésiastique, d'une part et, d'autre part, le désir d'éviter la rupture avec Rome [1]. Sur l'initiative de Siccardi, devenu ministre de la Justice, deux lois furent votées, les 8 avril et 5 juin 1850, supprimant le for et les immunités ecclésiastiques et rendant beaucoup plus difficile l'acquisition de biens-fonds par l'Église ; l'année suivante, les dîmes furent abolies en Sardaigne ; puis, en juin 1852, fut déposé un projet, qui n'aboutit d'ailleurs pas, introduisant le mariage civil.

AGGRAVATION DE LA TENSION   Dans l'entretemps cependant les négociations avec Rome se poursuivaient, car le gouvernement piémontais, après avoir pris des mesures de rigueur contre les prélats les plus intransigeants, hésitait à heurter de front l'ensemble du clergé qui gardait une grande influence à la campagne et jouissait de nombreuses sympathies dans l'aristocratie et l'entourage du roi : ce dernier, auquel Mgr Charvaz reprochait d'être « totalement sous la dépendance de ses ministres » [2], intervint d'ailleurs personnellement à plus d'une reprise, notamment en 1852, dans un sens modérateur. Certains estimaient qu'il devait être possible, moyennant quelques concessions de forme, de s'entendre avec Rome sur des lois qui, comme l'écrivait Cavour, « tendaient à introduire, dans une mesure modérée, ce qui existait depuis un demi-siècle dans tous les autres États catholiques » [3]. En réalité, cependant, les possibilités d'entente n'existaient guère. La plupart des ministres sardes étaient convaincus de l'impossibilité de mener une politique libérale et italienne en accord sincère avec le Saint-Siège, dominé, estimait-on, par le parti absolutiste et pro-autrichien, et ils croyaient

---

(1) Sur la position du roi Victor-Emmanuel II, voir A.-C. JEMOLO, *Chiesa e Stato in Italia...*, p. 132 et suiv. et P. PIRRI, *Pio IX e Vitt.Em. II, passim*, notamment p. 57*. Le roi fut à plusieurs reprises, alors et dans la suite, ennuyé par l'évolution de la situation, mais il n'eut pas, semble-t-il, de véritable crise de conscience : ses convictions religieuses étaient assez superficielles et il avait en outre l'approbation secrète de quelques évêques moins intransigeants. (Cf. P. PIRRI, *op. cit.*, p. 73-74, note.)

(2) Mgr Charvaz à Antonelli, 19 février 1850, dans P. PIRRI, *op. cit.*, p. 65.

(3) Lettre du 3 janvier 1853, dans P. MATTER, *Cavour et l'unité italienne*, t. II, Paris, 1927, p. 262.

qu'avec un peu de patience et d'habileté, on surmonterait la crise de
conscience des catholiques piémontais. Le Saint-Siège, de son côté, se
rendant compte de l'influence exercée en Piémont par la minorité jacobine, jugeait inutile de faire des concessions pour obtenir un accord dont
l'esprit ne serait quand même pas respecté, et qui mécontenterait les
autres Italiens auxquels on refusait de pareilles concessions, bien qu'ils
fussent plus respectueux des droits de l'Église. On n'admettait d'ailleurs
pas la thèse de Turin invoquant la situation de la France ou de la Belgique,
car, comme le reconnaissait l'ambassadeur de France lui-même, « en
prétendant faire reconnaître par Rome le mariage civil, le gouvernement
piémontais demande beaucoup plus qu'il n'a été fait en France » où Pie VII
s'était borné à ne pas exiger l'abolition de ce qui avait été introduit sans
son assentiment par la Révolution [1].

La situation s'aggrava de façon définitive lorsque la direction de
la politique échappa aux modérés par suite de l'alliance de Cavour avec
Ratazzi et la gauche anticléricale. Celle-ci n'était qu'une minorité dans
le pays, mais une minorité dynamique qui constituait en fait l'aile marchante du mouvement constitutionnel et unitaire et il n'est pas étonnant
qu'elle ait réussi à imposer ses vues radicales en dépit des efforts catholiques pour organiser la résistance sur le plan parlementaire et de la
répugnance de nombreux libéraux, aux croyances assez floues, à l'égard
d'une politique d'hostilité déclarée envers l'Église [2]. C'est sous la pression
de cette gauche radicale que fut votée, le 22 mai 1855, à une faible majorité de onze voix, et malgré une contre-proposition raisonnable faite
au nom de l'épiscopat par Mgr Nazari di Callabiana, la fameuse « loi
des couvents », qui supprimait une partie des chapitres collégiaux et
tous les ordres religieux autres que ceux s'adonnant au soin des malades
ou à l'enseignement, en attribuant le produit de leurs biens sécularisés
à l'entretien du clergé paroissial. L'application de cette loi, qui touchait
604 établissements jouissant d'un revenu de plus de deux millions de
lires, ne donna lieu à aucune réaction violente ni de la part des intéressés, ni de la part du peuple des campagnes, contrairement à ce que certains avaient laissé entrevoir [3], mais les ponts paraissaient désormais
rompus et Pie IX se décida à frapper de nullité toute la législation élaborée au mépris des droits de l'Église et à prononcer l'excommunication
majeure contre tous ceux qui avaient concouru à son élaboration.

Cette législation, il importe de le remarquer, s'inspirait beaucoup
moins du principe libéral de la séparation de l'Église et de l'État que
de la tradition jacobine et régaliste, soucieuse d'affirmer les droits sou-

---

(1) Dépêche de Rayneval, 2 septembre 1852, dans J. MAURAIN, *Le Saint-Siège et la France de
décembre 1851 à avril 1853*, p. 117.
(2) Tel Massimo d'Azeglio, sur l'attitude duquel on consultera S. JACINI, *La politica ecclesiastica
italiana*, p. 39-42 et A.-M. GHISALBERTI, *L'intervento di M. d'Azeglio nella crisi politico-religiosa
del 1855*, dans *Ricerche Religiose*, t. XVIII, 1947, p. 40-45.
(3) S'il n'y eut nulle part d'opposition ouverte, il faut pourtant noter que les tracasseries
anticléricales provoquèrent parmi les catholiques de Savoie, très dociles au clergé, un profond
mécontentement qui contribua à détacher ces régions du gouvernement de Turin et à les tourner
vers la France. (Cf. R. AVEZOU, *La Savoie depuis les réformes de Charles-Albert jusqu'à l'annexion
à la France*, dans *Mémoires de la Société savoisienne d'histoire*, t. LXIX, 1932, p. 1-176 ; t. LXX,
1933, p. 73-247.)

verains de l'État sur la réglementation des questions politico-ecclésiastiques et sur l'utilisation du patrimoine ecclésiastique. L'esprit de la loi des couvents, par exemple, se trouve éclairé par les paroles de Ratazzi à don Bosco lorsqu'il incita ce dernier, en 1857, à fonder lui-même une congrégation religieuse :

— Excellence, c'est vous qui me parlez de congrégation, alors que la loi...
— Oh ! la loi, la loi, je la connais parfaitement et j'en sais la portée. C'est aux biens de main-morte qu'elle en veut et aux anciens Ordres vivant en marge de la législation. Mais faites-moi une société où chacun des membres conserve ses droits civils, se soumette aux lois de l'État, paie l'impôt personnellement, une société qui ne soit au fond qu'une association de citoyens libres vivant ensemble dans un but de bienfaisance, et je vous garantis qu'il n'y a pas un gouvernement régulier et sérieux qui puisse vous gêner. Au contraire, s'il est juste, il vous devra protection comme aux autres sociétés, qu'elles soient commerciales, industrielles ou de secours mutuels [1].

LA POLITIQUE ECCLÉSIASTIQUE DE CAVOUR La condamnation publique du gouvernement piémontais ne mit pas fin aux tractations officieuses et notamment à cette correspondance personnelle entre Pie IX et Victor-Emmanuel, dont le ton paternel d'une part, la bonne volonté respectueuse de l'autre, tranchent avec l'intransigeance des déclarations officielles. C'est que le Saint-Siège désirait limiter dans la mesure du possible la portée pratique des mesures dommageables à l'Église, tandis que le roi aspirait à voir cesser un conflit qui l'ennuyait beaucoup. Cavour, de son côté, ne désirait nullement prolonger la lutte contre l'Église. Il avait bien patronné la politique religieuse des partis de gauche parce qu'il avait besoin de l'appui de ces derniers, mais il différait profondément des jacobins anticléricaux par ses sentiments intimes et par ses conceptions politico-ecclésiastiques [2]. C'était un homme de juste milieu qui, sans avoir probablement retrouvé la foi catholique pleine et entière après la crise de rationalisme qui avait marqué sa jeunesse, croyait toutefois aux valeurs permanentes du christianisme et admirait l'Église comme une force civilisatrice utile à la société ; au lieu de vouloir comme les anticléricaux la juguler et entraver systématiquement son influence, il désirait au contraire la voir progresser à condition qu'elle se cantonne sur le terrain proprement religieux.

Ses relations de famille avec des protestants de « l'Église libre » fondée par Vinet en Suisse romande, puis ses contacts avec des catholiques libéraux français avaient fait de Cavour un adepte de la formule : l'Église libre dans l'État libre, et un adversaire des concordats, qui lui paraissaient une entrave à la pleine liberté de l'Église et supposent par ailleurs que l'Église est une « société parfaite » comme l'État, négociant avec celui-ci un partage d'influence sur un terrain commun, alors que, selon lui, le spirituel et le temporel appartenaient à des domaines tout à fait

---

(1) Rapporté par A. AUFFRAY, *S. Jean Bosco*, 6e édit., Paris, 1947, p. 195.
(2) Sur les sentiments religieux de Cavour et son attitude à l'égard de l'Église, cf. M. MAZZIOTTI, *Il conte di Cavour e il suo Confessore. Studio storico con documenti e carteggi inediti*, Bologne, 1915 ; S. JACINI, *op. cit.*, p. 22-23 et 81 ; A.-C. JEMOLO, *Chiesa e Stato...*, p. 134-143, ainsi que F. RUFFINI, *La giovinezza di Cavour*, Turin, 1937-1938, 2 vol.

distincts, il importait de séparer nettement les compétences civiles et
politiques d'avec les compétences religieuses. A Rome, les idées n'étaient
pas mûres, surtout au lendemain du concordat autrichien de 1855, si
favorable à l'Église, pour qu'on s'accommodât sans plus du point de
vue libéral, mais sans doute un *modus vivendi* aurait-il pu intervenir, une
fois passée l'amertume laissée par le caractère unilatéral de la nouvelle
législation ecclésiastique piémontaise, si le développement de la question
romaine à partir de 1859 n'était pas venu opposer irrémédiablement le
Piémont au Saint-Siège.

§ 2. — La guerre d'Italie et la question romaine [1].

LA QUESTION ROMAINE   A partir de 1859, la question romaine devait
                          passer pour une vingtaine d'années au premier
plan de l'actualité. Comme elle relève plus de l'histoire diplomatique
et de la chronique militaire que de l'histoire ecclésiastique, on se bornera
à un rapide rappel des faits, mais il importe de ne jamais perdre de vue que
cette question exerça sur l'évolution de l'histoire de l'Église pendant
la plus grande partie du pontificat de Pie IX une influence psychologique
qu'on pourrait difficilement exagérer.

Ses répercussions furent considérables, en France et en Italie, sur
la politique ecclésiastique suivie par les gouvernements, et même sur la
vie religieuse, car en bloquant les énergies des catholiques sur la solution d'un problème avant tout politique, elle les détourna pour de longues
années des problèmes proprement religieux [2]. Mais surtout, elle accentua le
raidissement du pape et de nombreux catholiques à l'égard du libéralisme
et contribua de la sorte à faire apparaître l'Église comme foncièrement
hostile aux idées modernes.

En effet, il était évident, pour quiconque réfléchissait, que le maintien du petit État pontifical tel qu'il avait survécu à la Révolution était
devenu un anachronisme et que sa disparition n'était « que le dernier
terme d'une évolution sociale plusieurs fois séculaire, qu'un cas particulier
d'une loi générale inflexiblement appliquée à toute l'Europe, à Cologne
et à Liége comme à Avignon » [3], la sécularisation des États de l'Église
devant inévitablement suivre celle des évêchés, réalisée dès le début du
siècle. Mais les milieux catholiques responsables refusèrent de l'admettre
parce que, préoccupés à juste titre de maintenir intacte la pleine indépendance du Saint-Siège, ils ne comprirent pas que le problème devait

(1) BIBLIOGRAPHIE. — Cf. p. 72, n. 1. On tiendra spécialement compte, parmi les sources,
de M. MINGHETTI, *Miei ricordi*, t. III ; de J. GAY, *Les deux Rome et l'opinion française*, Paris, 1931,
p. 26-100 (lettres de Doubet à Rendu en 1853) et de *La Questione Romana negli anni 1860-1861.
Carteggio del conte di Cavour con D. Pantaleoni, C. Passaglia e O. Vimercati*, Bologne, 1929, 2 vol.

(2) Augustin Cochin faisait preuve de perspicacité quand il écrivait : « Cette déplorable question
de la souveraineté temporelle a été le grand malheur des catholiques de France. Cette question
secondaire (...) partage tellement nos forces que toutes les autres questions, intérêts religieux,
œuvres de charité, etc., sont négligées et parfaitement abandonnées. Elle a en outre ce grand inconvénient de nous placer pieds et poings liés entre les mains du gouvernement : tout le monde doit
être à ses pieds de peur qu'il ne retire ses troupes de Rome. » (Cité dans A. BATTANDIER, *Le cardinal Pitra*, p. 490.)

(3) A. LEROY-BEAULIEU, *Un empereur, un roi, un pape*, Paris, 1879, p. 210-213.

être repensé sur de nouvelles bases, laissant ainsi aux adversaires de l'Église le mérite d'être à peu près les seuls à proclamer la nécessité d'un changement inévitable. Bien plus, se rendant bien compte que c'était au nom de la conception libérale de l'État et du droit des peuples à disposer d'eux-mêmes que la forme d'existence traditionnelle de l'État pontifical était attaquée, ces mêmes responsables multiplièrent leurs protestations contre les principes de 1789, sans se soucier de distinguer la part de vérité qu'ils contenaient, donnant ainsi l'impression que la conception catholique de l'État devait nécessairement s'identifier avec un régime théocratique. Par hantise de la « révolution », menace pour l'existence de l'État romain, le Saint-Siège se solidarisa de plus en plus avec les gouvernements conservateurs dont l'appui paraissait, dans l'immédiat, la garantie la plus efficace pour l'Église.

L'ÉTAT PONTIFICAL   Le rétablissement de l'autorité du pape
APRÈS LA CRISE DE 1848   dans ses États en 1849 n'avait été, selon
le mot de Mgr Corboli-Bussi repris par
M. Ghisalberti, qu'une « restauration réactionnaire et maladroite », doublement maladroite car elle entendait, malgré la promesse formelle de ne pas toucher au *Statuto*, rétablir l'absolutisme pur et simple sous sa forme la plus odieuse, celle que l'opinion libérale dénonçait comme « un despotisme clérical », et cela grâce à la protection des troupes étrangères : garnison française de Rome et régiments autrichiens cantonnés dans les Légations. La présence de ces derniers surtout, détestés par les patriotes et dont les procédés tracassiers exaspéraient les populations, contribua beaucoup à déconsidérer le gouvernement pontifical.

Les premières années semblèrent pourtant donner tort aux pessimistes qui jugeaient le pouvoir temporel irrémédiablement compromis. Les assassinats politiques, fréquents au lendemain de la restauration, diminuèrent ; un complot organisé en 1853 par les partisans de Mazzini échoua lamentablement ; la situation économique s'améliora grâce à diverses mesures administratives intelligentes ; Pie IX retrouva même une partie de sa popularité et connut de 1851 à 1857 la période la plus sereine de son règne. Mais si son système de gouvernement paternaliste, très soucieux du bien-être de ses sujets, se plaisant à encourager l'introduction des inventions modernes, était de nature à satisfaire le petit peuple qui voyait son niveau de vie s'élever et appréciait la simplicité débonnaire de son souverain, la bourgeoisie par contre ne pouvait se déclarer satisfaite d'un régime qui non seulement ne laissait aucune responsabilité politique aux citoyens, mais dont la législation continuait à s'inspirer pour l'essentiel du droit canon médiéval et limitait strictement les libertés civiles.

Le malaise était encore accru par l'infériorité où l'élément laïque continuait à se trouver par rapport aux ecclésiastiques dans l'administration temporelle de l'État, en dépit des améliorations apportées en 1850. Les défenseurs du régime objectaient bien que les ecclésiastiques ne représentaient pas 5 pour cent du personnel, mais ils ne tenaient pas compte de deux choses : d'abord, pour le ministère de l'Intérieur, c'est-à-dire pour la direction générale de l'État, la proportion dépassait 10 pour cent ; et surtout, les ecclésiastiques, même dans les services où ils étaient en petit nombre, occupaient les postes de commande[1], y compris des fonctions aussi peu religieuses que celles de ministre du commerce ou de directeur général de la police. Ces *abbati* et *monsignori*, dont plusieurs n'étaient d'ailleurs pas prêtres, avaient en général une sérieuse formation juridique, et certains ne manquaient pas de compétence administrative. Toutefois, au dire d'un défenseur du pouvoir temporel, le comte de Rayneval, le gouvernement pontifical, dans son ensemble, apparaissait comme un gouvernement « composé de Romains et fonctionnant à la romaine », c'est-à-dire, précisait-il, « manquant d'énergie, d'activité, d'initiative, de fermeté, comme la nation elle-même »[2]. Les ecclésiastiques étant à la tête, il était inévitable qu'on leur fît endosser la pleine responsabilité des atermoiements en matière de réformes et celle des abus, nombreux comme presque partout en Italie à cette époque. Si le général de Lamoricière, le vaillant champion du pape, se laissait aller à dire dans un moment de colère : « On ne fera rien à Rome tant qu'on n'aura pas pendu quatre *monsignori* aux quatre coins de la ville », on conçoit quelle devait être la mentalité des libéraux.

LE CONGRÈS DE PARIS   Or tandis que le gouvernement pontifical s'obstinait à demeurer un gouvernement d'ancien régime, dont les méthodes apparaissaient aux contemporains plus arriérées encore que les conceptions, le Piémont, le seul État italien où le régime constitutionnel ait continué à fonctionner depuis 1848, était devenu, sous l'intelligente impulsion de Massimo d'Azeglio puis de Cavour, un pays fort et prospère, en qui mettaient plus que jamais leurs espoirs tous ceux qui, du Nord au Sud, aspiraient à voir l'Italie enfin unifiée et administrée conformément aux exigences modernes. L'existence des États de l'Église, dont les dirigeants déclaraient le statut intangible, était un obstacle majeur à la réalisation de ces espoirs, mais l'habile Cavour, qui savait qu'en dépit du calme apparent, le mécontentement continuait à couver dans l'État pontifical, entreprit d'exploiter cette situation au profit de ses projets[3].

On devait lui reprocher d'avoir, chez ceux dont il convoitait le bien, « créé le désordre pour avoir le droit de rétablir l'ordre ». Il serait plus

---

(1) Le *Dictionnaire encyclopédique de la théologie catholique* (trad. franç. de WETZER et WELTER, art. *Italie*, t. XII, 1870, p. 24) donne les chiffres suivants : il y avait 243 fonctionnaires ecclésiastiques contre 5.059 laïques ; mais alors qu'ils n'étaient que 3 contre 2.017 aux Finances, ils étaient 156 contre 1.411 à l'Intérieur. Les chiffres des traitements sont significatifs également : à l'Intérieur, 52.000 écus pour les ecclésiastiques contre 254.000 pour les laïques ; aux Finances, les 3 ecclésiastiques touchaient 5.680 écus contre 514.772 aux 2.017 laïques ; au Commerce, l'unique ecclésiastique 2.000 écus contre 13.136 pour les 61 laïques ; à la Police, 4.119 écus pour 2 ecclésiastiques contre 75.052 écus pour 404 laïques.

(2) Rapport du 14 mai 1856, cité par G. MOLLAT, *La question romaine*, p. 301, note.

(3) Il est bien évident qu'il s'agissait d'exploiter une situation qui facilitait l'entreprise de Cavour, mais sans laquelle le problème se serait malgré tout posé. Cochin remettait bien les choses au point dans une lettre à Rendu du 24 décembre 1862 : « Les misères de ce gouvernement ont été une des causes de sa chute, nullement la principale. M. de Cavour ni Garibaldi, M. Ricasoli ni M. d'Azeglio même ne se seraient arrêtés devant un pape réformateur. » (*Augustin Cochin. Sa vie, ses lettres*, t. I, p. 314.)

exact de dire qu'il sut avec adresse exploiter le désordre latent, en encourageant l'agitation dans les États de la péninsule et en posant le problème du malaise italien devant l'opinion européenne. A partir de 1856, il soutint en secret l'action de la *Société nationale italienne*, fondée par l'exilé La Farina dans l'intention de promouvoir l'unité italienne autour de la maison de Savoie ; par des tracts et des réunions clandestines, cette nouvelle organisation acheva de miner l'autorité des gouvernements de l'Italie centrale et méridionale ; dans les Marches et la Romagne en particulier, elle supplanta rapidement les doctrines républicaines de Mazzini en ralliant à son programme l'ensemble des libéraux, radicaux et modérés.

Pendant que se poursuivait ce travail de sape, Cavour avait alerté les diplomates réunis en congrès à Paris au printemps de 1856 en leur communiquant un rapport de l'ancien ministre de Pie IX, Minghetti, sur la triste situation des États de l'Église et l'urgence qu'il y avait à y introduire des réformes ; si le gouvernement français, qui tenait à ménager les catholiques, se montra réservé, le ministre anglais, lord Clarendon, appuya l'attaque et dénonça en termes acerbes le gouvernement romain, « le pire qui fut jamais ». Ces critiques eurent un grand retentissement et, peut-être à la demande du secrétaire d'État, l'ambassadeur de France à Rome entreprit de les réfuter dans un long rapport [1], où il établissait que, depuis le retour de Gaëte, le gouvernement pontifical « avait marché résolument dans la voie des réformes et des améliorations, qu'il avait réalisé de considérables progrès » ; il mettait en lumière la répression sévère du brigandage, dont on avait exagéré l'importance, l'exécution de nombreux travaux d'utilité publique, les encouragements à l'agriculture, à l'industrie et au commerce, les réformes financières et douanières, l'équilibre du budget, la réorganisation des administrations municipales.

Tout cela était en grande partie exact et c'est le mérite du récent ouvrage de P. Dalla Torre [2] d'avoir pour la première fois établi de manière objective et détaillée le bilan des réalisations administratives du gouvernement de Pie IX depuis 1850. Seulement le véritable problème n'était pas là. L'exécution, d'ailleurs un peu lente, d'un programme qui rappelait le despotisme éclairé du xviiie siècle et répondait en gros aux vœux du *Memorandum* de 1831, ne correspondait plus à la situation réelle : le problème, surtout depuis 1848, était devenu politique, dominé qu'il était par la double aspiration que rappellerait un jour l'inscription du monument Victor-Emmanuel à la place de Venise : *Patriae Unitati. Civium Libertati.* Ce fut le drame de l'Italie et du Saint-Siège au xixe siècle que cette double aspiration, qui nous semble si légitime aujourd'hui, apparût alors aux yeux de Pie IX et de ses principaux conseillers, par suite des circonstances, comme incompatible avec les exigences d'indépendance spirituelle du gouvernement suprême de l'Église.

---

(1) Imprimé dans le *Recueil des traités, conventions et actes diplomatiques concernant l'Autriche et l'Italie*, Paris, 1859, p. 697-726.

(2) P. DALLA TORRE, *L'opera riformatrice ed amministrativa di Pio IX fra il 1850 e il 1870*, Rome, 1945.

Rien de plus caractéristique, pour toucher du doigt les causes de cette incompréhension, que les entretiens du pape, au cours du voyage qu'il fit à travers ses États en 1857 [1], avec des modérés qui n'auraient pas demandé mieux que d'unir leur patriotisme et leur libéralisme avec leur foi catholique et leur dévouement au pontife romain [2]. A son vieil ami d'Imola, le comte Pasolini, qui avait amené la conversation sur la question des réformes politiques, Pie IX répliqua que si un gouvernement libéral devait ressembler à celui du Piémont, il ne pourrait être qu'anti-chrétien. Et à Minghetti, dont il connaissait les sympathies, il déclara avec humeur :

Le Piémont ? On s'y trouve fort mal. On y persécute la religion. On ne laisse passer aucune occasion d'outrager l'Église. Le roi... Pauvre homme ! il ferait mieux d'aller battre le blé ! Il a un ministre incroyant, Ratazzi ; Cavour est intelligent, mais je crains fort que lui aussi ait peu de religion [3].

Le scandale que causait la politique ecclésiastique piémontaise faisait apparaître au pape comme inadmissibles les conceptions unitaires de Cavour. Quant au régime constitutionnel, le moment ne lui paraissait pas opportun pour introduire des réformes — « le monde est trop agité », expliquait-il, — et surtout la tragédie de 1848 continuait à le hanter : « Les peuples sont incontentables. J'ai fait une expérience par trop douloureuse [4]. »

Ces déclarations expriment pour une bonne part le fond de l'âme de Pie IX, mais ses jugements étaient encore renforcés par la manière unilatérale dont il était conseillé et renseigné. Son entourage réactionnaire ne manquait pas une occasion de raviver dans son âme impressionnable les souvenirs sanglants de la révolution romaine [5] et de lui montrer comme la meilleure sauvegarde de l'ordre établi la persistance en Italie de l'influence autrichienne que la politique d'indépendance du Piémont cherchait au contraire à éliminer. L'un des plus habiles à raffermir sans cesse Pie IX dans ces convictions fut à coup sûr son secrétaire d'État, le cardinal Antonelli, que les représentants du gouvernement de Vienne considéraient comme « un ami plein de bienveillance pour l'Autriche » [6] et qui,

---

(1) Pie IX ayant fait le vœu d'aller en pèlerinage à Lorette, on en profita pour donner à son voyage, à l'instar de la tournée que François-Joseph venait de faire en Lombardie-Vénétie, le caractère d'une visite officielle de ses États, destinée à la fois à affirmer ses droits souverains et à donner aux populations des provinces l'occasion d'affirmer leur loyalisme. Antonelli régla tout minutieusement à l'avance, interdisant aux municipalités de présenter au pape des revendications d'ordre administratif ou politique et filtrant soigneusement ceux qui seraient admis en sa présence. Il réussit dès lors à ancrer davantage encore Pie IX dans l'idée que les idées réformatrices n'étaient le fait que de quelques individualités isolées. (Voir notamment la lettre de Pie IX à son frère, du 29 juillet 1857, dans A. Monti, *Pio IX nel Risorgimento*, p. 256-257.) Sur ce voyage, qui dura du 4 mai au 5 septembre 1857, voir le recueil *Pio IX ed i suoi popoli nel 1857*, Rome, 1860, ainsi que G. Maioli, dans *Atti e Memorie della Deputazione di Storia patria per le Marche*, sér. IV, vol. IV, 1927, p. 117 et suiv.

(2) Voir G. Pasolini, *Memorie raccolte da suo figlio*, t. I, p. 255 et suiv. et V. M. Minghetti, *Miei ricordi*, t. III, Turin, 1890, p. 177-194.

(3) V. M. Minghetti, *op. cit.*, t. III, p. 193.

(4) *Ibid.*, t. III, p. 184.

(5) Antonelli invoquait fréquemment l'exemple de Rossi quand il voyait Pie IX sur le point de céder quelque chose à l'opinion réformiste : « Que Votre Sainteté se souvienne : le comte lui aussi voulait accorder des concessions et prendre de semblables mesures. Le pape sait quelle fut sa récompense. » (Cité dans H. d'Ideville, *Pie IX*, p. 39-40.)

(6) Fessler à Rauscher, 17 mai 1863, dans C. Wolfsgruber, *J. O. Rauscher*, p. 186. Voir aussi S. Jacini, *Il tramonto del potere temporale*, *passim*.

en 1870 encore, expliquait, comme le dernier mot de la sagesse politique, que « toutes les forces vives du conservatisme devaient se coaliser afin d'en imposer aux passions subversives » [1].

LE CARDINAL ANTONELLI Jacques Antonelli [2], secrétaire d'État de Pie IX de 1849 à 1876, porte plus qu'aucun autre la responsabilité de la politique suivie par le Saint-Siège en matière temporelle. Pie IX en effet avait apprécié à Gaëte le savoir-faire du jeune cardinal et son incontestable dévouement aux intérêts du Saint-Siège. Très vite, désireux de se consacrer à la direction religieuse de l'Église et conscient de son propre manque de capacités politiques, il s'était déchargé sur lui de la conduite des affaires temporelles. Prétendre que l'emprise du secrétaire d'État sur le pape fût totale serait inexact, surtout lorsque des intérêts religieux étaient en cause ; mais Antonelli joua incontestablement pendant des années « un rôle de véritable dictateur » [3], ne laissant guère d'initiative aux ministres chargés de le seconder, reléguant le Sacré Collège dans un rôle effacé, et réussissant souvent à suggérer au pape ses propres vues grâce à une coterie fort bien organisée.

Ce prince de l'Église qui resta diacre toute sa vie fut l'un des derniers représentants d'une lignée largement représentée à la cour pontificale durant l'époque moderne, un de ces prélats tout laïques de sentiments, pour qui les intérêts de ce monde comptaient plus que ceux de l'autre, mais dont les mœurs faciles allaient pourtant de pair avec une foi sincère [4]. Fils d'un marchand de biens, Antonelli resta toute sa vie un parvenu avide d'argent, dominé par l'idée très italienne de *fare una famiglia*, que l'opinion publique accusait « d'avoir remplacé le népotisme des papes par le népotisme des secrétaires d'État » [5] et qui devait effectivement laisser une belle fortune à ses enfants naturels [6]. Son esprit positif et pratique et le sens des affaires qu'il tenait de son père le servirent dans sa carrière et son ascension fut rapide, mais il ne semble pas avoir été un grand administrateur, aux vues larges et aux impulsions fécondes, et le mérite des essais de modernisation de l'État pontifical à partir de 1850 ne lui revient que partiellement [7].

(1) Dépêche du ministre de Belgique à Rome, 4 mai 1870, *Arch. du Min. des Aff. Étrangères de Bruxelles*, St-Siège, t. XIII, 2.

(2) La personnalité d'Antonelli (1806-1876) reste pour une bonne part énigmatique ; ses papiers personnels sont toujours inaccessibles et l'on ne possède aucune biographie. Voir, en attendant, P. RICHARD, art. *Antonelli*, dans *D. H. G. E.*, t. III, col. 832-837 (trop élogieux) ; E. SODERINI, art. *Antonelli*, dans *Enciclopedia Italiana*, t. III, p. 547-548 ; MORONI, *Dizionario di erudizione storico-ecclesiastica*, passim (voir le t. I de l'*Index*).

(3) F. HAYWARD, *Pie IX et son temps*, p. 175.

(4) Sur les sentiments religieux d'Antonelli, voir E. SODERINI, *art. cit.*, p. 547. Il ne faut pas donner une portée trop pessimiste au fait que Pie IX ait poussé un soupir de soulagement quand il apprit que son secrétaire d'État, quelques jours avant sa mort, avait demandé les derniers sacrements.

(5) Dépêche de l'ambassadeur d'Autriche, décembre 1861, dans S. JACINI, *Il tramonto...*, p. 82.

(6) Il ne légua rien en mourant, ni au pape, ni aux pauvres, ni aux ordres dont il était le protecteur : « Il n'y a pas, dans les annales du Sacré Collège, testament où les choses ecclésiastiques et pieuses aient été plus oubliées. » (L. TESTE, *Préface au conclave*, p. 67.) Sa fortune, à vrai dire, n'atteignait pas les chiffres fabuleux que citèrent les journaux anticléricaux : d'après le ministre de Belgique, sa succession s'élevait « au plus à 7 millions de francs » (dépêche du 15 novembre 1876, *Arch. Min. A. É. Brux.*, St-Siège, t. XV) et, d'après E. Ollivier, elle n'aurait pas dépassé 423.000 fr. (ce qui, en francs-or, représente déjà une somme respectable).

(7) Tout en l'appréciant beaucoup, le cardinal de Reisach reconnaissait qu'il n'avait rien d'un

Fut-il du moins un homme d'État ? Certains contemporains l'ont nié :
« La voilà, la grande incapacité méconnue », disait le ministre de Prusse
lors de l'entrée des Italiens à Rome en 1870 en désignant les fenêtres du
cardinal [1]. Mais d'autres, au contraire, appréciaient au plus haut point
ses talents. Insensible à l'éloge et au blâme, imperturbable et insinuant
tout à la fois, il était toujours avec les diplomates « de la plus fabuleuse
amabilité » [2]. Cet homme rusé, qui savait être « d'une câlinerie féline » [3],
excellait à parler des heures sans rien dire en se donnant des airs d'abandon et de confidence qui trompaient les plus perspicaces. « Ne confiez
pas ce projet à votre ambassadeur à Rome, écrivait Cavour ; gardât-il
le secret, Antonelli est si fin qu'il le devinerait. » Et Emile Ollivier, qui
l'avait connu de près, faisait le plus grand cas de ses capacités :

Il savait peu ce que les livres contiennent, mais il était très érudit dans la
science que les choses apprennent, et il devait à l'acuité d'un esprit alerte et
libre de ne rien ignorer de ce qui se devine [4].

En réalité cependant, ses talents étaient limités. De bons observateurs
se rendaient compte qu'il ne dirigeait guère les événements et que son
art consistait surtout à « continuellement transiger », à aplanir des difficultés de détails, à « recoller » [5] les cassures après les saillies de langage
d'un pontife aisément oublieux du terne voulu des formules de chancellerie. Le ministre de Belgique pouvait écrire en 1868 :

Tout en reconnaissant avec unanimité à l'Éminence en question une grande
intelligence, on s'accorde aussi à lui attribuer plus de finesse native que de
savoir, plus de dextérité dans la conduite des affaires intérieures que de largeur
et d'ensemble dans ses vues en matière de politique ou d'administration générale [6].

Ces appréciations de diplomates contemporains légitiment pleinement
le jugement du P. Boudou qui, le comparant au génial secrétaire d'État
de Pie VII, conclut : « Consalvi fut un grand homme ; Antonelli fut un
homme habile [7]. » Habileté qui supposait certes une rare souplesse d'esprit, mais ne suffisait pas à vaincre les difficultés de fond, lesquelles
n'étaient que reculées. Dans la question romaine en particulier, on doit
bien reconnaître qu'Antonelli se contenta, somme toute, de retarder
le plus longtemps possible une catastrophe qu'il prévoyait inévitable [8],
en s'efforçant de continuer avec adresse mais sans beaucoup d'originalité
la politique de Lambruschini.

organisateur (*Janssens Briefe*, édit. PASTOR, t. I, p. 254) et H. d'Ideville opposait son « immo-
bilisme » à la constante préoccupation de Mgr de Mérode d'introduire à Rome des réformes utiles
et des améliorations matérielles (*Pie IX*, p. 115-117).
(1) Rapporté par L. TESTE, *Préface au conclave*, p. 47.
(2) K. VON SCHLOEZER, *Römische Briefe*, p. 126.
(3) S. NIGRA, *Seconda Roma*, p. 165.
(4) E. OLLIVIER, *L'Église et l'État au concile du Vatican*, t. I, p. 505.
(5) K. VON SCHLOEZER, *Römische Briefe*, p. 127.
(6) Dépêche du 7 juillet 1868, *Arch. Min. A. É. Brux.*, St-Siège, t. XIII, 2,
(7) A. BOUDOU, *Le Saint-Siège et la Russie*, t. II, p. 456.
(8) Soderini estime toutefois que ce n'est qu'à partir de 1868 qu'il se rendit compte que ce dénoue-
ment inévitable était tout proche.

LA GUERRE D'ITALIE ET
SES CONSÉQUENCES

Au Congrès de Paris, Cavour avait posé les premiers jalons de son œuvre italienne, mais il était impuissant tant qu'il restait seul en face de l'Autriche. Or, il réussit à circonvenir Napoléon III, en qui subsistaient toujours d'anciennes sympathies italiennes, et à l'intéresser à la cause de l'unification. L'empereur, pour des raisons variées dont il est difficile de préciser l'importance relative, désirait le maintien du pouvoir temporel, mais il estimait sans intérêt le problème de l'étendue de l'État pontifical et répugnait particulièrement au caractère antimoderne de son gouvernement. Ses dernières hésitations tombèrent lorsque, en 1858, l'affaire Mortara lui permit de constater que son hostilité contre « le gouvernement des prêtres » était partagée par la masse des Français : la servante chrétienne d'un riche commerçant juif de Bologne ayant secrètement baptisé le fils de celui-ci au cours d'une maladie qui paraissait devoir l'emporter, le Saint-Office fit enlever à ses parents cet enfant de trois ans afin de l'élever chrétiennement [1] ; cette manière d'agir, théoriquement conforme au droit canon, souleva l'indignation des libéraux, qui organisèrent autour d'elle « un de ces charivaris comparable à l'orchestration, quarante ans plus tard, de l'affaire Dreyfus » [2] et beaucoup de catholiques se sentirent fort mal à l'aise à l'idée que ce n'était là, comme l'écrivait l'ambassadeur de France, « qu'un incident entre mille qui peuvent surgir tous les jours par l'application maladroite ou rigoureuse d'une législation incompatible avec la forme actuelle de la société » [3].

Le 21 juillet 1858, au cours d'une entrevue secrète à Plombières, Napoléon et Cavour arrêtaient les grandes lignes d'un plan destiné à éliminer les Autrichiens de la péninsule, à soustraire à l'administration cléricale les trois quarts de l'État pontifical et à constituer une confédération italienne analogue à la confédération germanique et dont on donnerait la présidence au pape. Au début de 1859, une brochure intitulée : *L'empereur Napoléon III et l'Italie* [4] préparait l'opinion française à une intervention au delà des Alpes et, en avril, éclatait la guerre résolue à Plombières. L'effondrement autrichien à la suite des rapides victoires de Magenta et de Solferino n'eut pas seulement pour conséquence d'unir définitivement la Lombardie au Piémont ; il provoqua aussi le soulèvement des duchés de l'Italie centrale et de plusieurs provinces pontificales. Les autorités romaines étouffèrent rapidement l'insurrection dans les Marches, mais elles ne purent empêcher les Romagnols de voter leur

---

(1) Voir les détails de l'affaire dans R. DE CESARE, *Roma e lo Stato del Papa*, t. I, p. 278-294 ; sur l'appréciation théologique, voir *La Civiltà cattolica*, sér. III, t. XII, 1858, p. 385 et suiv., 529 et suiv., et en sens contraire, la brochure de l'abbé DELACOUTURE, *Le droit canon et le droit naturel dans l'affaire Mortara*, Paris, 1858. L'attitude d'ensemble du gouvernement de Pie IX à l'égard des juifs fut pourtant toujours bienveillante. (Cf. E. LŒVINSON, *Gli Israeliti nello Stato pontificio*, dans *Rass. St. R.*, t. XVI, 1929, p. 768 et suiv.)

(2) F. HAYWARD, *Pie IX et son temps*, p. 195.

(3) Dépêche du 5 octobre 1858 dans J. MAURAIN, *La politique ecclésiastique du Second Empire*, p. 230.

(4) Inspirée par l'empereur, elle fut rédigée par un publiciste à sa dévotion, le vicomte de la Guéronnière, avec la collaboration d'un des rares catholiques au courant de la situation réelle de l'Italie, Eugène Rendu. (Voir la lettre de ce dernier à Chiala, du 25 août 1883, dans *Lettere edite ed inedite di C. Cavour*, t. III, Turin, 1887, p. 385-396.)

rattachement à la monarchie piémontaise. Or, bien que Napoléon III refusât officiellement, lors de l'armistice de Villafranca puis du traité de Zurich, de reconnaître l'extension du Piémont en Italie centrale, il ne fit cependant rien pour l'empêcher, mais au contraire, comme il était question de réunir une conférence européenne pour régler définitivement le problème italien, il inspira une nouvelle brochure : *Le Pape et le Congrès*, où l'on pouvait lire entre autres que, s'il importait de réaffirmer le principe de la souveraineté pontificale, « la ville de Rome en résume surtout l'importance, le reste n'est que secondaire (...). Plus le territoire sera petit, plus le souverain sera grand » [1].

Se sentant tacitement encouragé, Victor-Emmanuel invita le pape, quelques semaines plus tard, non seulement à accepter le fait accompli en Romagne, mais à lui concéder le gouvernement effectif des Marches et de l'Ombrie, dont il conserverait la souveraineté nominale, et à accorder aux habitants de Rome et du territoire qui demeurerait sous son autorité directe les mêmes droits civils et politiques que dans le royaume piémontais. Mais Pie IX, qui était pourtant loin de se désintéresser de la cause italienne, estimait qu'en l'occurrence il lui était impossible de céder : « Pour s'en convaincre, expliquait-il à Napoléon III, il suffit de réfléchir à ma situation, à mon caractère sacré et aux droits du Saint-Siège, droits qui ne sont pas ceux d'une dynastie, mais de tous les catholiques. Les difficultés sont insurmontables parce que je ne peux céder ce qui ne m'appartient pas [2]. » On a vu du reste qu'à cette raison de principe — qui, chez cet homme de foi, l'emportait sur tous les calculs de l'opportunité politique — s'en ajoutait une autre, de nature religieuse également, la crainte de voir étendue aux États de l'Église la politique de laïcisation menée depuis dix ans par le gouvernement de Turin. Aussi se refusa-t-il à toute transaction et, après avoir, dans l'encyclique *Nullus certi*, du 19 janvier 1860, flétri à la face du monde catholique « les attentats sacrilèges commis contre la souveraineté de l'Église romaine », exigé la « restitution pure et simple » de la Romagne et fait répéter par le cardinal Antonelli qu'il n'y avait qu'une solution : « rendre ce qui a été pris » [3], il fulmina solennellement l'excommunication majeure contre les usurpateurs des droits du Saint-Siège [4].

LA RÉACTION DANS LE MONDE CATHOLIQUE

Les vicissitudes de l'État pontifical posaient aux catholiques un épineux problème. En Italie, beaucoup de patriotes sincèrement croyants se demandaient pourquoi le pape n'accepterait pas une solution analogue à celle qui devait être ratifiée soixante ans plus tard au traité du Latran. En Angleterre, dans les milieux peu sympathiques à l'ultramontanisme, en Allemagne dans le monde universitaire, un bon nombre pensaient de même [5]. En France, Lacordaire écrivait :

(1) La brochure, parue le 22 décembre 1859, était écrite comme la précédente par La Guéronnière.
(2) Lettre du 8 janvier 1860, citée dans F. HAYWARD, *Pie IX et son temps*, p. 230.
(3) *La Questione romana negli anni 1860-1861*, t. II, p. 14-17.
(4) Bulle du 26 mars 1860, dans *Acta Pii IX*, t. III, p. 137-147.
(5) Pour l'Angleterre, voir notamment *La Questione romana...*, t. I, p. 154 et E. PURCELL,

Je suis pour le Saint-Siège contre ses oppresseurs, je crois à la nécessité morale de son domaine temporel (...), mais en même temps je désire l'affranchissement de l'Italie, des modifications sérieuses dans le gouvernement des États romains [1].

Mgr Maret, après avoir écrit à Napoléon III que sa politique italienne était excellente, notait pour son propre compte que si « toute attaque *injuste* contre les propriétés et les droits temporels de l'Église et du Saint-Siège est coupable », on pouvait se demander si l'on n'avait pas affaire en l'occurrence à un soulèvement légitime de la part des populations « régies par le droit d'un autre âge » [2]. L'un ou l'autre allait encore plus loin, des laïcs comme Doubet, Rendu ou même Cochin, des prêtres comme l'abbé Deguerry, curé de la Madeleine, ou le futur cardinal Meignan :

Je crois, écrivait ce dernier, que les conditions actuelles ont de grands inconvénients pour la religion. On dit qu'elles assurent l'indépendance spirituelle du pape. Je ne veux pas le nier absolument. Néanmoins je trouve encore le pape trop dépendant et je m'imagine que Dieu, bientôt peut-être, lui assurera une meilleure indépendance que celle d'un prince toujours flanqué de baïonnettes étrangères, toujours obligé à se défendre contre des populations qui le subissent à regret (...). Lorsque Dieu déchaîna la révolution française et enleva au clergé ses privilèges temporels, on crut tout perdu. Dieu ne se laissa point émouvoir par des supplications aveugles, et il fit bonne justice. J'espère qu'il en sera de même et qu'il délivrera le pape *malgré lui* et *malgré nous* [3].

Pourtant, surtout en France, ces voix étaient des exceptions, et qui osaient à peine se faire entendre, car les masses catholiques réagissaient au contraire vigoureusement en faveur du pouvoir temporel. Tandis que les fidèles d'Autriche puis d'Allemagne adhéraient en foule à l'Œuvre Saint-Michel, destinée à procurer à la cause pontificale une aide spirituelle et matérielle, et qu'à Gand un groupe de catholiques belges fondait le Denier de Saint-Pierre, en s'inspirant d'une idée lancée dix ans auparavant par Montalembert [4], le cardinal Rauscher, archevêque de Vienne, faisait signer par cent cinquante évêques d'Autriche, d'Allemagne, de Suisse, de Belgique, de Hollande, de Grande-Bretagne, une protestation destinée au Congrès projeté par Napoléon [5], et de toute part les adresses de fidélité couvertes de milliers de signatures affluèrent à Rome.

La réaction fut particulièrement vive en France [6]. Veuillot et Dupanloup, pour une fois d'accord, tentèrent d'ameuter l'opinion publique ; Mgr Parisis, Mgr Pie, d'autres encore suivirent, et les sermons plus ou moins ouvertement hostiles à l'empereur coupable de trahir le pape se multiplièrent. Une partie de la population ne s'émut guère, vu le peu

Life of Manning, t. II, p. 386-399 (pour réagir contre cette mentalité, Manning et Ward firent de la défense du pouvoir temporel l'un des principaux objectifs de la *Dublin Review*). En Allemagne, la position que Doellinger devait développer dans ses fameuses conférences de l'Odéon (cf. *infra*, p. 204-205) ne lui était pas personnelle (voir par ex. *Janssens Briefe*, édit. Pastor, t. I, p. 137).

(1) Lacordaire à Montalembert, février 1860, dans E. Lecanuet, *Montalembert*, t. III, p. 218-219.

(2) Maret à Napoléon III, 12 avril 1859, dans G. Bazin, *Vie de Mgr Maret*, t. II, p. 64-67 ; notes personnelles de janvier 1860, *ibid.*, p. 67-70.

(3) Meignan à Montalembert, 28 mars 1859, dans E. Lecanuet, *Montalembert*, t. II, p. 218-219 en note.

(4) Voir A. M. von Steinle, *Der Peterspfennig*, Fribourg-en-Br., 1892 et C. Daux, *Le Denier de Saint-Pierre*, Paris, 1907.

(5) C. Wolfsgruber, *J. O. Rauscher*, p. 413.

(6) Voir J. Maurain, *La politique ecclésiastique du Second Empire*, chap. XIV et XV.

de popularité du système de gouvernement pontifical, mais il en alla tout autrement des fidèles fervents, dont certains désignaient avec émotion les provinces pontificales comme « les États du roi mon père »[1], et des légitimistes, très influents dans le monde catholique, qui virent dans la campagne d'agitation déclenchée par le clergé une occasion de manifester leurs sentiments antibonapartistes. C'est d'ailleurs de ces milieux légitimistes, plutôt que de l'épiscopat, que partit l'initiative de quêtes au profit du trésor pontifical[2] et ce sont eux également qui fournirent le principal appoint français à l'armée de volontaires étrangers que le Saint-Siège s'efforça de constituer.

LA RÉORGANISATION DE L'ARMÉE PONTIFICALE — Pie IX, que l'attitude passive du gouvernement impérial avait déçu et irrité et qui subissait de plus en plus des influences hostiles à Napoléon III, désirait substituer aux régiments français des troupes à lui, sur lesquelles il pourrait compter pour appuyer la défense de ce qui restait des États de l'Église. Ce plan était surtout préconisé par Mgr de Mérode, un ancien officier belge à l'esprit entreprenant, qui, depuis dix ans qu'il vivait dans l'intimité du pape, avait réussi à gagner la confiance de celui-ci par son désintéressement, sa droiture et sa franchise de langage, lesquelles tranchaient sur la cupidité et les allures feutrées de nombreux prélats romains, Antonelli en tête. Il était assez téméraire de prétendre recruter, équiper et former une armée en quelques mois et le secrétaire d'État, qui craignait de pousser à bout le gouvernement français, dont l'appui lui paraissait autrement sérieux, s'opposa au projet, mais le pape passa outre et, en février 1860, désigna Mérode comme prominitre des armes. Celui-ci, avec un zèle jamais découragé, qui devait persévérer des années durant, se mit aussitôt à la tâche et, pour commencer, fit appel à un officier français de valeur mais connu pour son hostilité au régime bonapartiste, le général de Lamoricière : « C'est un petit soufflet que nous donnons à l'empereur », reconnaissait le nouveau ministre des armes, qui était tout sauf un diplomate[3].

Lamoricière trouvait l'armée pontificale dans une situation lamentable. Elle comportait en tout onze bataillons de 600 hommes, mal armés et mal équipés ; l'artillerie, le train et les ambulances étaient quasi inexistants ; le général décida de porter aussitôt les effectifs à 25.000 hommes et adressa un appel aux catholiques d'Europe. En quelques semaines, il put disposer de 5.000 Autrichiens, entraînés et encadrés, envoyés secrètement par leur gouvernement, de près de 4.000 Suisses, de 3.000 Irlandais et de plusieurs centaines de Belges et de Français, ces derniers appartenant surtout à l'aristocratie légitimiste de l'Ouest[4]. Mais la

---

(1) E. LAFOND, *Rome. Lettres d'un pèlerin*, Paris, 1857, titre du chap. II.
(2) Les évêques (dont certains d'ailleurs, trouvant que Pie IX se montrait trop intransigeant, auraient préféré rester sur la réserve) ne prirent en main l'organisation du Denier de Saint-Pierre qu'un peu plus tard, la plupart après Castelfidardo. Les sommes recueillies de la sorte de 1860 à 1870 paraissent avoir atteint annuellement environ 300.000 fr.-or à Paris, 200.000 fr. dans les diocèses riches et pieux, 3 à 4 millions pour l'ensemble de la France. (J. MAURAIN, *op. cit.*, p. 431.)
(3) Dépêche de l'ambassadeur de France, 10 avril 1860, dans J. MAURAIN, *op. cit.*, p. 408.
(4) Sur l'armement et l'organisation de l'armée pontificale, voir le *Rapport du général Lamori-*

bravoure et l'enthousiasme ne pouvaient malheureusement pas compenser le manque d'homogénéité et le caractère improvisé de cette armée ; les événements n'allaient pas tarder à le prouver.

CASTELFIDARDO   L'émotion causée par la perte de la Romagne n'était pas encore calmée qu'un nouveau coup de théâtre se produisit. Le hardi condottiere Garibaldi avait débarqué en mai en Sicile avec un millier de partisans ; trois mois plus tard, il passait le détroit de Messine avec ses « chemises rouges », mettait en déroute l'armée napolitaine, provoquait la chute du roi François II et annonçait son intention de marcher sur Rome après avoir substitué le régime républicain à la monarchie bourbonnienne. Cavour, soucieux de le gagner de vitesse et de profiter de la révolution pour adjoindre la partie méridionale de la péninsule au royaume d'Italie en train de se constituer, réussit à arracher l'accord tacite de Napoléon à une occupation des Marches et de l'Ombrie afin de permettre à l'armée piémontaise de descendre par là vers le Sud. Le 7 septembre, sous prétexte de réprimer des troubles qu'il avait lui-même fomentés, il adressait un double ultimatum à la cour de Rome et au général de Lamoricière, leur enjoignant de dissoudre immédiatement les corps « mercenaires » et « étrangers » qui menaçaient « d'étouffer dans le sang italien toute manifestation du sentiment national »[1]. Croyant pouvoir compter sur l'armée française, Antonelli fit pr uve d'intransigeance. Induit en erreur par les illusions du secrétaire d'État et de Mgr de Mérode, Lamoricière perdit du temps et les Piémontais n'eurent pas de peine à tailler en pièces, en quelques jours, ses forces dispersées, notamment le 18 septembre, à Castelfidardo, où s'illustra la brigade du général de Pimodan. Le 29 septembre, Lamoricière, bloqué à Ancône par la flotte de l'amiral Persano, capitulait.

A Rome, la consternation était à son comble. Le pape, très monté contre les Français, qu'il accusait de jouer un double jeu, envisageait de s'embarquer pour l'Espagne ou pour Trieste, et déjà une corvette autrichienne l'attendait à Civita-Vecchia ; il était poussé dans ce sens par certains prélats pris de panique à l'idée de tomber aux mains des Piémontais, mais aussi par Mgr de Mérode et son groupe, qui escomptaient la réaction de l'opinion catholique en faveur du pape « chassé de Rome ». L'ambassadeur de France réussit à faire ajourner cette décision qu'il redoutait, car il se rendait compte que les Piémontais n'attendaient que le départ du Saint-Père pour occuper Rome et il entrevoyait l'effet funeste qui en résulterait pour le gouvernement impérial. Napoléon ordonna du reste à ses troupes de défendre efficacement ce qui subsistait de l'État pontifical et ce n'est que deux mois plus tard qu'il en vint à souhaiter à son tour un départ du souverain pontife qui aurait permis de liquider la question romaine [2] ; mais à ce moment les idées à Rome avaient

cière sur les opérations de l'armée pontificale dans les Marches et l'Ombrie, Paris, 1861 ; cf. F. HAYWARD, *Pie IX et son temps*, p. 233-237.
(1) Le texte s'en trouve dans A. DE SAINT-ALBIN, *Histoire de Pie IX*, t. II, p. 394-395.
(2) C'est ce que Thouvenel, qui était ministre des Affaires Étrangères, a appelé « le secret de l'empereur ».

évolué et le pape était maintenant décidé à ne s'en aller que si on l'expulsait vraiment.

À LA RECHERCHE D'UN COMPROMIS — En quelques mois, les données de la question romaine avaient changé du tout au tout. Réduit aux environs de Rome, c'est-à-dire à un territoire exigu et assez pauvre, l'État pontifical n'était plus viable, et il n'y avait guère de chances de récupérer jamais les provinces perdues. En effet les plébiscites organisés dans les Marches et l'Ombrie en novembre, où l'annexion avait été approuvée par 138.000 voix contre 1.200 et 97.000 contre 300, montraient que le nouvel état de choses répondait aux vœux de la population ; le roi de Naples avait, de son côté, renoncé à la résistance organisée et, quelques semaines plus tard, le royaume d'Italie allait être officiellement proclamé. Le moment semblait donc venu de s'incliner devant les réalités et de rechercher une transaction raisonnable. C'était l'avis du gouvernement français, mais aussi d'un bon nombre de catholiques et même d'ecclésiastiques italiens, qui auraient volontiers fait leur l'opinion exposée par Cavour à Passaglia :

> Cette noble fermeté qui, tant que la lutte était encore indécise, avait provoqué l'admiration de tout le monde et même des adversaires, devenait une opiniâtreté aveugle et par conséquent coupable, du moment que la victoire était restée à l'Italie une et indépendante. Vouloir continuer seul et contre le gré de l'Europe une résistance stérile et faite seulement pour aigrir les esprits et déchaîner le fanatisme des classes ignorantes, essayer, sans pouvoir l'empêcher, de retarder l'œuvre de la réconciliation générale, ce serait commettre une faute grave, qui tôt ou tard, mais infailliblement, tournerait au détriment de l'Église elle-même [1].

A Rome aussi, certains réalistes commençaient à penser de même. Non seulement des hommes un peu suspects à cause de leurs idées avancées, comme le cardinal d'Andrea, grand seigneur jaloux de l'influence du « paysan » Antonelli et en froid avec les jésuites, ou l'oratorien allemand Theiner, le savant préfet des archives vaticanes, connu lui aussi pour son hostilité envers la Compagnie de Jésus. Mais aussi des conseillers écoutés de Pie IX, occupant des postes influents : l'ex-jésuite Passaglia, qui conservait une partie du crédit théologique qu'il s'était acquis à l'occasion de la définition du dogme de l'Immaculée Conception ; le prédicateur apostolique, le capucin Louis de Trente, hostile au parti d'Antonelli et très estimé par Pie IX ; le cardinal Amat, un vieil ami du pape, l'un des rares membres de la Curie vraiment intelligent et ouvert aux idées modernes ; peut-être les cardinaux de Silvestri et Bofondi ; en tout cas Mgr Franchi, secrétaire aux Affaires ecclésiastiques extraordinaires, et le cardinal Santucci, personnage de faible envergure mais esprit conciliant et qui n'avait pas peur de parler nettement au pape pour le mettre en garde contre les avis du parti opposé.

Les adversaires de toute conciliation ne manquaient d'ailleurs pas non plus : les jésuites de *La Civiltà cattolica*, qui représentaient l'intransi-

---

(1) D'après une dépêche de l'ambassadeur d'Autriche du 26 février 1861, dans S. JACINI, *Il tramonto del potere temporale...*, p. 49.

geance des principes et la mentalité conservatrice ; certains cardinaux
réactionnaires, incapables de comprendre l'évolution qui s'accomplissait sous leurs yeux ; l'entourage immédiat du Saint-Père, Mgr Ricci,
Mgr Stella, Mgr Talbot, tous ceux qui allaient pendant vingt ans espérer
contre toute espérance qu'un effondrement du nouveau royaume ou un
sursaut d'indignation des nations catholiques permettrait, comme en
1849, une restauration intégrale des droits du souverain pontife ; Mgr de
Mérode enfin. Ce soldat fougueux partagea lui aussi au début ces illusions, mais surtout, il considérait comme une honte de capituler devant
l'adversité et, méprisant les Italiens autant qu'il détestait Napoléon III,
il estimait qu'il fallait repousser avec éclat toute proposition quelle qu'elle
fût, émanant de Turin ou de Paris. Plus modéré dans la forme, plus diplomate, Antonelli, sans refuser a priori toute conversation, était d'avis lui
aussi qu'un rapprochement avec l'Italie n'offrait guère d'avantage et
que la moins mauvaise solution était encore de spéculer sur l'embarras
que causerait à Napoléon III l'opposition catholique au cas où il cesserait
de soutenir la cause pontificale, et sur l'appui diplomatique, affaibli
mais non négligeable, qu'il était possible d'attendre de l'Autriche [1].

Ballotté entre ces tendances divergentes, Pie IX hésitait sur l'attitude
à prendre. Excité par les témoignages de fidélité indignée — purement
platoniques, d'ailleurs, pour la plupart — qui lui parvenaient du monde
entier, subissant de façon croissante l'ascendant de Mgr de Mérode
dont le dévouement chevaleresque et le courage en ces moments pénibles
plaisaient à sa nature généreuse, il était plutôt porté à prêter une oreille
complaisante aux conseils d'intransigeance et, sous le coup de l'excitation, « sa nature impressionnable le poussait parfois à des sorties véhémentes » [2] qui pouvaient donner l'impression que c'était de lui que partait
la résistance la plus opiniâtre à toute concession. Mais par ailleurs, il
avait trop de finesse pour ne pas entrevoir l'irrémédiable décadence
du pouvoir temporel, et trop de sens patriotique pour ne pas souffrir
profondément à certains jours de devoir se mettre en travers de l'œuvre
d'unification nationale en train de s'accomplir [3].

Cavour, qui avait à Rome ses agents d'information, savait pouvoir
compter dans la Curie sur les « sympathies existantes et peu déguisées

---

(1) L'importance du facteur français dans la question romaine de 1860 à 1870 a été maintes
fois mise en lumière. (On peut suivre les péripéties presque au jour le jour dans l'ouvrage cité de
J. MAURAIN.) On oublie trop souvent que, du moins jusqu'en 1867, date où les libéraux prirent
la direction des affaires en Autriche, le Saint-Siège considéra la monarchie des Habsbourg comme
un appui plus sûr bien que moins puissant, et que le ton confiant des relations du pape et de son
secrétaire d'État avec l'ambassadeur d'Autriche contraste singulièrement avec le caractère souvent
tendu de celles avec les diplomates français. (Voir en particulier S. JACINI, Il tramonto..., p. x,
169, 172 et A. MONTI, Pio IX nel Risorgimento, p. 139-145 et 182.)

(2) F. HAYWARD, Pie IX et son temps, p. 267.

(3) On a relevé une série d'indices permettant de conclure avec assurance que les sympathies
italiennes de Pie IX avaient survécu à la crise de 1848. En voici un entre plusieurs rapporté par
H. d'Ideville : en septembre 1864, au cours d'une conversation où il était question de Cavour,
« tout à coup, d'une voix sourde et basse, comme s'il se fût parlé à lui-même, sans se préoccuper
de ma présence, il murmura ces mots : Ah, comme il a aimé son pays, questo Cavour, questo Cavour.
Cet homme était vraiment italien. Dieu lui aura certainement pardonné comme nous lui pardon-
nons... Je me souviens qu'ayant raconté le fait à Mgr de Mérode, le prélat, qui était loin de partager cette indulgence italienne, me parut contrarié de l'aveu échappé au pape, sans en être étonné. »
(H. d'IDEVILLE, Pie IX, p. 55-56.) Cf. aussi G. MAIOLI, Pio IX. Da Vescovo a Pontefice, p. 55 et
n. 114 et surtout A. MONTI, Pio IX nel Risorgimento italiano, p. 7-10.

pour la cause italienne » que l'ambassadeur d'Autriche notait avec dépit [1], et il avait bon espoir d'arriver, grâce à elles, au double but qu'il poursuivait : la réalisation du dernier point du programme national, « Rome capitale », satisfaction à laquelle la majorité du « pays réel » aurait à la rigueur renoncé mais que l'opinion libérale, prépondérante au Parlement, n'entendait sacrifier à aucun prix ; puis la réconciliation avec le souverain pontife et l'apaisement religieux dans le Royaume, ce qui lui permettrait de rassurer l'opinion catholique européenne et de désarmer l'opposition cléricale italienne, dangereuse tant que la situation du nouvel État ne serait pas stabilisée.

L'habile ministre prépara le terrain en multipliant les menus dons d'argent aux agents subalternes de la cour pontificale et en laissant entrevoir « aux gros poissons » [2], y compris le cardinal Antonelli [3], des compensations « propres à satisfaire leur ambition et leur intérêt personnel » [4]. Mais il avait fort bien compris, comme il l'écrivait en décembre à Napoléon III, que Pie IX, quels que fussent ses sentiments personnels et les avis de son entourage, était dominé dans toute cette affaire par « une conviction religieuse profonde » et qu'il n'accepterait dès lors d'envisager une transaction « qu'après avoir acquis la conviction que non seulement il peut le faire sans manquer à sa conscience et aux devoirs de son ministère sacré, mais que sa renonciation au pouvoir temporel serait utile à l'Église et servirait aux intérêts véritables de la religion » [5]. Il envisagea dès lors de lui proposer un accord sur la base suivante : le pape renoncerait franchement au pouvoir temporel destiné de toute façon à disparaître assez prochainement ; en revanche, l'Italie renoncerait à tout ce qu'il y avait encore d'inspiration régaliste ou joséphiste dans sa législation ecclésiastique et accorderait à l'Église la liberté pleine et entière : « Ces concessions, estimait-il, sont énormes. Leur application produirait une véritable révolution religieuse en Italie [6]. »

ÉCHEC DES NÉGOCIATIONS   Les négociations, amorcées en décembre 1860 grâce au cardinal Santucci et à Passaglia, se heurtèrent presque aussitôt au mauvais vouloir d'Antonelli. Elles se terminèrent au mois de mars par un échec complet, sans avoir même jamais pris une tournure vraiment officielle [7].

Cet échec n'a rien d'étonnant. En effet, bien que l'ambassadeur d'Autriche, hostile à un accord qui aurait affermi la position de l'Italie, ait exprimé la crainte que « le cœur de Pie IX soit plus accessible que ne l'a été celui de son premier ministre aux idées mi-libérales, mi-religieuses,

---

(1) Dépêche du 26 février 1861, dans S. Jacini, *Il tramonto...*, p. 54.
(2) *La Questione romana...*, t. I, p. 279.
(3) *Ibid.*, t. II, p. 3. Il est par contre fort douteux qu'Antonelli, par l'intermédiaire de son secrétaire Aguglia, ait fixé lui-même, comme prix du marché, des avantages pour sa famille et 3 millions d'écus pour lui. (Cf. *ibid.*, t. I, p. 222.) Plusieurs points demeurent obscurs dans les multiples négociations menées parallèlement au cours de ces semaines.
(4) *La Questione romana...*, t. II, p. 144.
(5) *Ibid.*, t. II, p. 142.
(6) *Ibid.*, t. II, p. 144.
(7) La plupart des documents relatifs à ces négociations sont publiés dans *La Questione romana...*; on peut y ajouter F. Salata, *Per la storia diplomatica della questione romana*, t. I, Milan, 1929 et D. Pantaleoni, *L'idea italiana nella soppressione del potere temporale dei papi*, Turin, 1884.

si éloquemment représentées par l'abbé Passaglia » [1], l'opinion moyenne à la Curie n'était pas prête à envisager une solution aussi radicale que celle proposée par Cavour, moins que jamais depuis que l'arrivée du roi François II et des exilés napolitains avait renforcé l'influence de l'élément antilibéral. Cavour avait en outre eu le tort de mêler trop de monde à son entreprise, qui était devenue le secret de Polichinelle, alors que tant de gens avaient intérêt à la faire échouer ; il n'avait d'ailleurs pas été très heureux dans le choix de ses intermédiaires : non seulement le bavard avocat Bozino, le craintif P. Molinari, pour ne pas parler de personnages aussi équivoques qu'Isaia ou Aguglia ; mais même les principaux négociateurs : le docteur Pantaleoni, entreprenant mais trop optimiste et trop imaginatif, et Passaglia, « riche d'érudition, hardi de pensées, indépendant d'opinion » [2] mais fort prétentieux et « prenant facilement des vessies pour des lanternes » [3], étaient trop maladroits pour se mesurer avec un homme de la force d'Antonelli. Il faut enfin tenir compte du fait que si Cavour était sincère quand il promettait la liberté religieuse et s'il souhaitait, comme on l'a vu, la prospérité de l'Église, il avait dû tenir compte du fanatisme jacobin des partis de gauche et s'était vu obligé sous leur pression d'étendre immédiatement l'essentiel de la législation ecclésiastique piémontaise, y compris la loi des couvents, aux provinces annexées, sans tenir compte des concordats existants et de prendre des mesures de rigueur contre les évêques, les prêtres et les religieux qui essayaient de résister : cette manière d'agir n'était évidemment pas de nature à créer l'atmosphère de confiance indispensable à la réussite des pourparlers entamés.

On comprend dans ces conditions le ton tranchant de l'allocution *Jamdudum cernimus* que Pie IX prononça au consistoire du 18 mars [4] : tout en se déclarant prêt à pardonner à ceux qui le haïssaient, « aussitôt qu'ils viendraient à résipiscence », il se déclarait décidé à ne souscrire à aucune « concession injuste » et mettait en garde ceux qui, « séduits par l'erreur ou entraînés par la crainte, voudraient donner des conseils favorables aux désirs des injustes perturbateurs de la société civile ».

Cette dernière phrase visait à coup sûr les nouveaux efforts tentés par Napoléon III qui, après avoir inspiré une troisième brochure : *La France, Rome et l'Italie*, s'efforçait de promouvoir entre Turin et Rome un accord qu'il désirait ardemment, car il souhaitait évacuer Rome, mais se rendait compte que « la majorité du pays verrait avec regret et même avec un certain mécontentement l'armée (française) s'éloigner de Rome tant que l'existence indépendante de la papauté ne serait pas complètement assurée » [5].

Les tractations actives qui se poursuivaient entre la cour impériale et celle de Turin furent interrompues par la mort inopinée de Cavour, survenue le 6 juin, et les exigences accrues de son successeur en rendirent

(1) Dépêche du 26 février 1861, dans S. JACINI, *Il tramonto...*, p. 54.
(2) Note de Mgr Plantier en 1858 dans J. CLASTRON, *Vie de Mgr Plantier*, t. I, p. 350.
(3) S. JACINI, *La politica ecclesiastica italiana*, p. 67-69.
(4) *Recueil des allocutions consistoriales, encycliques...*, p. 434-445.
(5) Thouvenel à Gramont, 24 février 1861, dans J. MAURAIN, *op. cit.*, p. 506.

impossible la continuation. Napoléon se résigna à ajourner la liquidation
de la question romaine jusqu'à la mort du pape, qu'il estimait prochaine.
Pie IX était en effet âgé de 69 ans et les événements des derniers mois
semblaient avoir sérieusement ébranlé sa robuste constitution. L'éventualité d'un changement de pontificat et l'espoir de voir succéder à
Pie IX un pape plus libéral, disposé à traiter avec l'Italie, parurent à
Napoléon un motif suffisant pour ne pas brusquer les choses et pour
maintenir les troupes françaises à Rome jusqu'à nouvel ordre.

L'ASSEMBLÉE DES   Pie IX, après les hésitations des premières se-
ÉVÊQUES DE 1862   maines, avait opté pour l'intransigeance [1] ; on se
                          persuadait en effet dans son entourage que le nou-
veau royaume d'Italie ne tarderait pas à se disloquer, faute de pouvoir
assimiler les Deux-Siciles, où le roi François II organisait des guérillas
avec l'appui du cardinal Antonelli. En attendant le moment où le
pouvoir temporel pourrait de la sorte être intégralement rétabli, le
pape estimait nécessaire de voir réaffirmés ses droits avec le plus de
solennité possible, en marquant nettement qu'il ne s'agissait pas de
revendiquer un quelconque droit historique, mais de défendre un droit
religieux, résultant de la fonction suprême qu'il devait pouvoir exercer
en toute indépendance, et en soulignant d'autre part comment, derrière
la campagne de dénigrement contre l'État pontifical, il y avait, d'après
lui, toute l'opposition qui dressait la conception laïque de l'État contre
la conception chrétienne de la société. Il avait déjà fait rassembler et
éditer par le P. Curci tous les écrits publiés depuis trois ans par les évêques
du monde entier concernant la souveraineté temporelle du pape, afin de
mettre en lumière le témoignage de l'Église universelle [2], mais il voulut
faire un pas de plus.

La canonisation de trente chrétiens martyrisés au Japon au XVIᵉ siècle
devait avoir lieu à la Pentecôte de 1862. Pie IX invita tous les évêques
qui le pourraient à venir à Rome à cette occasion pour solenniser
davantage la cérémonie. Bien qu'Antonelli, interrogé à ce sujet par
l'ambassadeur de France, l'eût assuré qu'aucune question étrangère
à la canonisation ne serait traitée à la réunion, sans d'ailleurs accepter
d'en donner confirmation par écrit [3], l'intention du pape était claire.
Aussi le gouvernement italien interdit-il à ses évêques de participer à la
réunion et le gouvernement français, qui craignait que la manifestation
préparée par le pape ne fût pas dirigée seulement contre la politique
italienne de l'empereur, mais aussi contre les principes de l'État moderne,
essaya de faire pression sur les évêques pour qu'ils s'abstiennent. Il
n'y en eut pas moins à Rome, au mois de mai, 323 cardinaux, archevêques
et évêques, dont une cinquantaine venus de France, accompagnés de
plus de 4.000 prêtres et de 100.000 fidèles.

---

(1) L'une des conséquences en fut la fondation, en juin 1861, de l'*Osservatore Romano*, dont la
direction fut confiée à deux exilés politiques avec mission de donner au journal une allure combative, que la presse romaine avait au contraire systématiquement évitée jusqu'alors. (Voir l'article d'ALFA reproduit dans la *Documentation catholique* du 30 juillet 1950, col. 1007-1010.)
(2) C. CURCI, *Il moderno dissidio tra la Chiesa e l'Italia*, Florence, 1877, p. 59-60.
(3) J. MAURAIN, *op. cit.*, p. 603.

Le 29 mai, jour de l'Ascension, Pie IX donna la bénédiction solennelle *urbi et orbi* à cette foule en délire [1]. Le 6 juin, il adressa aux prêtres étrangers rassemblés dans la chapelle sixtine un vibrant discours et lorsque, à la fin, l'un des assistants entonna spontanément la formule liturgique : *Oremus pro Pontifice nostro Pio*, tous répondirent en chœur : *Dominus conservet eum et vivificet eum et beatum faciat eum in terra et non tradat eum in manibus inimicorum ejus*. Le 8 juin eut lieu la canonisation, mais comme l'écrivait Cochin, « on ne pensait guère au Japon » [2]. C'est le lendemain qui fut la journée essentielle. Le pape prononça devant les évêques réunis en consistoire l'allocution *Maxima quidem* dans laquelle il dénonça les erreurs du rationalisme et du matérialisme contemporain, puis flétrit le gouvernement italien en déclarant qu'il continuerait à défendre avec fermeté son pouvoir temporel. Le cardinal Matei répondit au nom des évêques par une adresse dont la rédaction n'avait pas été sans peine [3]. Le pape en effet aurait désiré qu'un désaveu des libertés modernes fût joint aux considérations sur le pouvoir temporel et le cardinal Wiseman avait rédigé un projet dans ce sens, mais Dupanloup, dont le prestige était grand depuis ses éloquentes interventions en faveur des droits du pape, s'y opposa et proposa un autre texte, qui rendait entre autres hommage à l'aide française. Après des discussions assez vives, on décida de faire « un composé de ces deux pièces » pour aboutir à un texte assez diffus qu'Antonelli aurait souhaité plus énergique [4] : on déclarait le pouvoir temporel une institution providentielle, indispensable au bien de l'Église, sans en faire toutefois une vérité de foi ; on félicitait Pie IX de l'avoir défendu avec intransigeance et les évêques se déclaraient prêts à le suivre dans cette voie « jusqu'à la prison et jusqu'à la mort ».

L'effet moral de cet ensemble de cérémonies fut considérable et accru encore par l'adhésion que quatre cents autres évêques, ceux d'Italie entre autres, donnèrent au texte de l'adresse. Cochin, de retour à Paris, traduisait l'impression générale quand il écrivait : « Cette démonstration de la puissance, de l'union, de l'étendue de l'Église, a beaucoup frappé. On sent que le jour des funérailles d'un vivant si vivant et si colossal n'est pas venu [5]. »

§ 3. — Le royaume d'Italie et le Saint-Siège [6].

LE CLERGÉ FACE A LA SITUATION NOUVELLE

Les événements de 1859-1860 plaçaient le clergé italien dans une situation difficile. Un certain nombre de prêtres voyaient avec joie se réaliser l'unité de la patrie, mais avaient-ils le droit d'approuver

---

(1) L. Veuillot a décrit la scène dans *Les Parfums de Rome*. Sur l'ensemble des cérémonies de 1862, voir la brochure du card. Wiseman, *Rome et l'épiscopat catholique à la Pentecôte de 1862*.

(2) Cochin à Montalembert, 7 juin 1862, dans *Augustin Cochin. Ses lettres*, t. I, p. 284.

(3) Voir entre autres une lettre de Mgr Lavigerie à Rouland, du 10 juin 1862, dans J. Maurain, *op. cit.*, p. 612-615.

(4) D'après une dépêche de l'ambassadeur d'Autriche du 14 juin 1862, dans S. Jacini, *Il tramonto...*, p. 93-94. Dans la rédaction définitive, un rôle important revint au sous-secrétaire d'État pour les Affaires ecclésiastiques, Mgr Franchi, homme très modéré.

(5) Cochin à Dupanloup, 25 juin 1862, dans *Augustin Cochin. Ses lettres*, t. I, p. 286.

(6) Bibliographie. — Aux ouvrages mentionnés en tête du chapitre, p. 72, n. 1, ajouter pour les débuts de l'action catholique en Italie, F. Olgiati, *La storia dell'azione cattolica in Italia, 1865-1914*, 2e éd., Milan, 1922, chap. I et II.

des annexions qui se faisaient au détriment des anciens souverains légitimes, surtout quand l'un de ces souverains était le pape ? et quelle attitude prendre à l'égard de la législation anticléricale que les Piémontais introduisaient dans leurs nouvelles provinces ?

Dans les anciennes provinces pontificales et dans les duchés du centre, la grande majorité du clergé, et surtout du haut clergé, se fit remarquer « par ses principes conservateurs et son attachement au Saint-Siège »[1]. L'archevêque de Florence adressa bien à Victor-Emmanuel un discours de bienvenue qui fit scandale à Rome, mais plusieurs évêques refusèrent de participer à la fête nationale piémontaise et beaucoup s'opposèrent avec éclat à l'introduction des mesures de laïcisation, notamment à l'application de la loi des couvents qui, dans les territoires nouvellement annexés, devait entraîner la suppression de 721 établissements et la disparition de 12.000 religieux. Le gouvernement réagit par des peines d'exil ou de prison et même des cardinaux furent victimes de ces mesures de rigueur, ce qui amena, au cours de l'hiver 1860, plusieurs centaines de prêtres des Marches et de l'Ombrie à signer une note de protestation rédigée sous l'inspiration romaine et destinée à Napoléon III[2].

L'attitude du clergé napolitain fut tout autre. L'ambassadeur d'Autriche notait avec dépit que, « dans presque toutes les localités », les prêtres avaient pris fait et cause pour les Piémontais et qu'on avait même vu des moines « combattre dans les rangs des insurgés contre les troupes royales »[3]. Le mécontentement qu'on en témoigna à Rome ne ralentit guère ce ralliement que plusieurs évêques, comme celui d'Ariano, encouragèrent.

LE MOUVEMENT « CONCILIATEUR » Cette tendance au « ralliement » était encouragée par le changement de mentalité de nombreux patriotes qui commençaient à regretter l'allure anticléricale et antiromaine, presque inévitable d'ailleurs[4], prise depuis 1849 par le mouvement national et qui risquait de l'acculer à une impasse par suite de la réprobation du monde catholique. On ne pouvait d'ailleurs manquer d'observer que, dans la situation politique européenne du XIXe siècle, le pouvoir temporel, loin d'être une garantie de liberté pour le souverain pontife, le rendait au contraire étroitement dépendant de la France et de l'Autriche et que, dès lors, il n'y avait guère d'inconvénient à y renoncer pourvu que le pape reçût les garanties suffisantes. La voie semblait dès lors ouverte à un rapprochement et divers projets de solution transactionnelle commencèrent à circuler, les uns défendus par de purs théoriciens, dont le plus en vue était Mamiani[5], les autres tenant davantage compte des contingences pratiques, comme ceux de Jacini et surtout de Gennarelli[6]. Ces projets furent accueillis

(1) Dépêche du 13 juillet 1860, dans S. JACINI, *Il tramonto...*, p. 27.
(2) *La Questione romana negli anni 1860-1861*, t. I, p. 154-155, 164.
(3) Dépêche du 13 juillet 1860, dans S. JACINI, *Il tramonto...*, p. 27.
(4) Voir sur ce caractère inévitable quelques pages très éclairantes de A. C. JEMOLO, *Chiesa e Stato in Italia...*, p. 182-187.
(5) T. MAMIANI, *Della rinascenza cattolica* (1862).
(6) Voir notamment A. GENNARELLI, *Le dottrine religiose della corte di Roma in ordine al dominio temporale*, Florence, 1862.

avec faveur par de nombreux ecclésiastiques piémontais et lombards, non seulement par ceux qui avaient conservé de la formation fébronienne reçue au séminaire ou à l'université de Turin une défiance marquée à l'égard de Rome, mais par tous ceux qui, prêts à reconnaître l'autorité du pape dans sa plénitude, souhaitaient toutefois que celui-ci, en acceptant de sanctionner le fait accompli, rendît possible un accord qui permettrait de tirer le plus de profit possible d'une évolution inévitable.

C'est dans le but de propager ces idées et de réagir contre la collusion, qu'on prétendait favorisée par les jésuites, entre la souveraineté du pape et l'influence autrichienne en Italie, qu'un groupe de prêtres appartenant à la *Société ecclésiastique* de Milan fonda en 1860 le journal *Il Conciliatore*. Dans le même esprit furent créées à Florence et à Naples deux autres associations sacerdotales destinées à chercher, en même temps qu'une solution à la question romaine, un terrain d'entente avec les aspirations libérales. Le mouvement trouva jusque dans le clergé romain des appuis non négligeables. Le plus en vue fut Passaglia [1], que l'échec des tractations Cavour-Pantaleoni n'avait pas découragé. Il commença par publier des brochures anonymes engageant le clergé à ne pas se montrer hostile à la patrie reconstituée, puis, ayant été obligé de quitter Rome, il se réfugia à Turin, où il devait diriger de 1862 à 1866 le journal *Il Mediatore*, et il organisa dans le clergé un mouvement d'adresses au pape en faveur de l'entente avec le nouveau royaume d'Italie [2]. Un prélat romain, savant et pieux mais un peu exalté, Mgr Liverani, un ancien protégé de Pie IX, donna ouvertement son appui à l'action de Passaglia et, de Florence où il s'était replié, il publia un pamphlet très vif contre le pouvoir temporel, le déclarant irréformable depuis qu'il était devenu la proie de la clientèle d'Antonelli, de cardinaux incapables et de jésuites réactionnaires aux ordres de l'Autriche [3].

Les autorités ecclésiastiques ne tardèrent pas à intervenir. L'ouvrage de Liverani ne fut pas mis à l'*Index* grâce à l'intervention du cardinal d'Andrea, mais Pie IX ne cacha pas le mécontentement que lui causait le journal *Il Conciliatore*, qui cessa de paraître en juillet 1861 ; la même année, l'archevêque de Florence interdit la société de prêtres qui s'était formée dans son diocèse et, quelques mois plus tard, devant l'insistance du vicaire capitulaire, la *Société ecclésiastique* de Milan cessa également ses activités. Ces interventions, l'action de la presse ultramontaine, dont les dirigeants se dépensèrent en ces années avec un zèle jamais découragé, l'influence enfin des séminaires sur la jeunesse cléricale, rallièrent peu à peu la majorité du clergé au point de vue du Vatican. Pourtant les sympathies nationales furent loin de disparaître et Passaglia essaya, en 1863, de remettre sur pied une nouvelle association ecclé-

---

(1) Sur l'activité politique de Passaglia, cf. PASQUALE D'ERCOLE, *Carlo Passaglia*, dans *Annuario della R. Università degli Studi di Torino*, 1887-1888, p. 127-175.

(2) Il recueillit en quelques semaines les signatures de 8.176 prêtres diocésains et de 767 religieux.

(3) *Il papato, l'impero e il regno d'Italia*, Florence, 1861. Le cas Liverani, estime Jacini, est « important et caractéristique de l'attitude d'une minorité dans le haut clergé italien ». (*Il tramonto...*, p. 60.)

siastique, groupant toutes les bonnes volontés du Nord au Sud [1]. Le gouvernement, de son côté, favorisa systématiquement, en les nommant dans des chaires universitaires ou à des postes lucratifs, les prêtres frappés par les autorités ecclésiastiques pour s'être affichés trop ouvertement comme partisans du nouveau régime.

A côté d'ambitieux, qui semblaient prêts à aller jusqu'au schisme s'il le fallait, il y avait parmi ces prêtres « conciliateurs » beaucoup d'hommes tout à fait respectables, qui se faisaient peut-être quelque illusion sur la complexité des problèmes à résoudre, mais qui ne se départirent jamais d'un grand respect pour l'autorité hiérarchique. Ils cherchaient sincèrement à sauvegarder les droits de l'Église tout en satisfaisant les aspirations nationales et libérales, convaincus qu'ils étaient de voir l'intransigeance romaine tourner au détriment de la religion catholique en Italie. Tel était notamment le cas du fameux P. Tosti [2], un moine du Mont-Cassin, qui, avec un optimisme indomptable, devait multiplier des années durant les suggestions d'accommodement, allant jusqu'à préparer une rencontre personnelle entre Pie IX et Victor-Emmanuel, au grand mécontentement des extrémistes de droite et de gauche.

LES FLUCTUATIONS DE LA POLITIQUE ECCLÉSIASTIQUE ITALIENNE — L'hésitation d'une partie du clergé italien à s'enfermer dans l'attitude d'intransigeance préconisée par le Vatican s'explique d'autant mieux que la politique du gouvernement italien à l'égard de l'Église fut loin d'être aussi uniformément hostile à celle-ci qu'on le croit souvent, évoluant au contraire constamment entre l'idéal libéral d'une séparation entre l'Église et l'État qui laisserait la première libre et même prospère dans son domaine propre, d'une part, et, d'autre part, la conception, restée vivace dans l'Italie du Nord, d'un régalisme joséphiste, qui prenait facilement les allures d'un anticléricalisme jacobin.

Ricasoli était un libéral, partisan d'un régime de séparation laissant à l'Église sa pleine liberté d'action, mais, au lieu d'avoir subi, comme Cavour, l'influence de Gioberti et de Montalembert, il appartenait au groupe toscan de Gino Capponi et de Raphaël Lambruschini, modernistes avant la lettre, qui prônaient une réforme de l'Église dans un sens plus spirituel et moins juridique. Cette insistance sur la nécessité d'une réforme interne du catholicisme, fréquemment soulignée par Ricasoli, notamment dans sa lettre au pape du 10 septembre 1861, indisposa grandement contre lui Rome et le clergé qui le considérèrent souvent, bien à tort, comme passé au protestantisme.

C'est en contradiction avec ses principes libéraux et sous la pression

---

(1) Lorsque, après 1866, Passaglia découragé eut renoncé à son œuvre conciliatrice, un certain nombre de prêtres libéraux la poursuivirent, groupés à Florence autour du périodique l'*Esaminatore* dont l'âme était Bianciardi, et à Naples autour de la *Società nazionale emancipatrice*, qui devait s'agiter au moment du concile du Vatican et dont une partie des membres finirent par se rallier au vieux-catholicisme.

(2) Cf. F. QUINTAVALLE, *La conciliazione fra l'Italia ed il Papato nelle lettere del P. L. Tosti e del Sen. G. Casalti*, Milan, 1907 ; J. GAY, *Les deux Rome...*, chap. III ; A. QUACQUARELLI, *Il P. Tosti nella politica del Risorgimento*, Rome, 1945.

des partis de gauche, que Ricasoli se vit obligé de couvrir des manifestations violemment anticléricales et des mesures arbitraires telles que des perquisitions chez les dirigeants de la Société de Saint-Vincent-de-Paul, à l'imitation de ce qui se faisait en France. C'est au contraire en pleine conformité avec ses principes régalistes que Rattazzi, « un type caractéristique de la mentalité bourgeoise en Piémont à cette époque »[1], multiplia les mesures anticléricales et, sans être probablement lui-même affilié à la franc-maçonnerie[2], apparut comme l'exécuteur par excellence de la politique de celle-ci.

Quant à Minghetti et à Menabrea, c'étaient des libéraux dans la ligne de Cavour et ils s'efforcèrent de réagir contre les radicaux de plus en plus nombreux qui, tout en continuant à répéter l'ancienne formule : l'Église libre dans l'État libre, visaient en réalité non plus la séparation, mais l'effacement pur et simple de l'Église ; ils durent toutefois se résigner, eux aussi, à prendre certaines mesures anticléricales, parce que la confiscation des biens d'Église apparaissait comme le seul moyen de remédier à une situation financière lamentable, et parce que la gauche, excitée par les manifestations d'intransigeance du Saint-Siège, réclamait à grands cris des représailles.

Ces alternances de libéralisme, de réformisme et d'anticléricalisme systématique dans la politique religieuse intérieure du royaume d'Italie devaient se renouveler jusqu'en 1870, au hasard des changements de ministères. Elles eurent leurs répercussions sur l'évolution de la question romaine.

LA CONVENTION DE SEPTEMBRE   Plus les mois passaient et plus Napoléon III souhaitait liquider la question romaine et mettre fin à l'intervention française, mais pour des motifs complexes, — crainte de mécontenter l'opinion catholique en France, influence de l'impératrice et d'une partie de son entourage, sentiment personnel à l'égard de l'Église ou de Pie IX, — il entendait assurer au pape la possession du petit territoire auquel, depuis Castelfidardo, se réduisait sa puissance temporelle et, malgré des sautes d'humeur périodiques, il restait fidèle à la ligne de conduite qu'il avait exposée à Victor-Emmanuel en 1861 :

Je laisserai mes troupes à Rome tant que Votre Majesté ne se sera pas réconciliée avec le pape ou que le Saint-Père sera menacé de voir les États qui lui restent envahis par une force régulière ou irrégulière[3].

Mais les tentatives de promouvoir une réconciliation sur la base de ce que Cochin nommait « une sorte d'Avignon dans Rome »[4] se heurtèrent à l'intransigeance du Saint-Siège qui se retranchait derrière la constitution de Pie V interdisant d'aliéner la moindre parcelle des biens de l'Église. D'autre part, après plusieurs alertes, la santé de Pie IX s'améliora et la perspective de voir un pape plus souple lui succéder sous

(1) S. JACINI, *La politica ecclesiastica italiana*, p. 211-212.
(2) Voir A. LUZIO, *Aspromonte e Mentana*, Florence, 1935, p. 74.
(3) Lettre du 12 juillet 1861, citée dans *La Questione romana...*, t. II, p. 251.
(4) Cochin à Dupanloup, 10 novembre 1862, dans *Augustin Cochin. Ses lettres*, t. I, p. 306.

peu apparut moins probable [1]. L'empereur chercha alors à obtenir des Italiens l'assurance que, pour un certain temps du moins, ils mettraient un terme à leurs convoitises. C'était beaucoup leur demander, car la fraction la plus dynamique de l'opinion nationale n'admettait pas que l'Italie ressuscitée pût avoir une autre capitale que Rome ; toutefois, la manière dont le gouvernement de Turin avait barré la route de Rome à Garibaldi lors de l'échauffourée d'Aspromonte, en août 1862, permettait d'espérer qu'il observerait d'éventuels engagements à l'égard du Saint-Siège. Les négociations secrètes menées personnellement par Napoléon III avec les représentants italiens finirent par aboutir, à la fin de l'été 1864, à un accord connu dans l'histoire sous le nom de *Convention de septembre* : l'Italie s'engageait à prendre à sa charge une partie de la dette pontificale, à respecter ce qui restait des États de l'Église et, pour preuve de sa bonne volonté, à transférer sa capitale à Florence ; moyennant quoi les troupes françaises quitteraient Rome dans un délai de deux ans. Cet accord, qui avait été conclu tout à fait à l'insu du pape et qui apparaissait à beaucoup comme un abandon à peine déguisé, plongea dans la consternation tous ceux qui désiraient le maintien du pouvoir temporel.

Le délai de deux ans fixé pour l'évacuation de Rome laissait toutefois un certain répit que la Cour de Rome essaya de mettre à profit. « On tablait sur l'action de la Providence et sur celle du temps » [2] : on espérait toujours la dissolution prochaine du nouveau royaume d'Italie ; on espérait un appui plus efficace de la part de l'Autriche, bien que celle-ci se bornât toujours à de bonnes paroles ; on espérait enfin que la campagne menée par le clergé français et les discussions engagées au Corps législatif, où les partisans du pape gagnaient en influence au détriment des anticléricaux, obligeraient Napoléon III à poursuivre sous une forme ou sous une autre la défense de l'État pontifical en dépit de la Convention.

LA CHUTE DE *MONSEIGNEUR DE MÉRODE*	Dans ces conditions, l'habile Antonelli, dont l'influence depuis 1860 avait été de plus en plus supplantée par celle de

Mérode, au point qu'on avait plusieurs fois cru sa démission inévitable [3], redevint l'homme indispensable. Mérode [4], depuis Castelfidardo,

(1) Comme Pie IX n'écoutait guère les avis de ses médecins, sa santé subissait de fréquents avatars (Cf. S. NIGRA, *Seconda Roma*, p. 204-205), mais il était au fond de robuste constitution, comme le relevait H. d'Ideville en mai 1864 : « Ce retour à la vie étonne et, avouons-le, irrite bien du monde. Les cardinaux ne savent plus que penser de ces alertes continuelles. Quant à moi, je crois que le pape Pie IX les enterrera tous. N'a-t-il pas encore son frère, qui a près de 84 ans ? Son père est mort à 92 ans. » (*Journal d'un diplomate en Italie : Rome*, p. 170.)

(2) G. MOLLAT, *La question romaine*, p. 351. A Pie IX qui s'exprimait dans ce sens, lord Clarendon répliquait sarcastiquement : « En effet, la Providence peut faire des miracles et depuis dix ans elle en a fait beaucoup ; mais tous en faveur de l'Italie. » (Cité dans S. JACINI, *Il tramonto...*, p. 237.)

(3) Il en est constamment question dans les mémoires et les correspondances entre 1860 et 1865. Les attaques contre Antonelli ne provenaient d'ailleurs pas seulement de Mgr de Mérode, mais d'une partie du Sacré Collège groupée autour du cardinal Altieri (et plus tard, après la mort de ce dernier, autour du cardinal Sacconi). Mais von Schlozer voyait juste quand il écrivait, en juin 1864 : « Une rupture totale entre le pape et son secrétaire d'État, comme Mérode le souhaiterait, n'est pas possible. Leur mariage a déjà duré depuis trop longtemps, et la position d'Antonelli à l'égard du corps diplomatique et des cours est si bonne qu'il serait difficile au pape de le remplacer à ce point de vue. » (*Römische Briefe*, p. 75.)

(4) Sur Xavier de Mérode (1820-1874) et son action à Rome, voir surtout Mgr BESSON, *Frédéric-François-Xavier de Mérode*, Paris, 1886.

ne se faisait plus guère d'illusion sur les chances de sauver le pouvoir temporel, mais il était d'avis qu'il fallait « tomber avec honneur », en se défendant en chevalier plutôt que de mendier l'appui de Napoléon III dont il sentait bien qu'il finirait par faire défaut, et en modernisant autant que possible ce qui restait de l'État pontifical de manière à rendre inexcusables ceux qui invoquaient son caractère retardataire pour en justifier la suppression. Mais il s'était heurté dans ses efforts de réforme à l'indolence romaine ou à la résistance opiniâtre de ceux qui profitaient des nombreux abus existants [1], et, dans sa politique de fière indépendance à l'égard du gouvernement français, aux prélats qui pensaient avec Antonelli qu'il ne fallait pas se montrer trop difficile et saisir toutes les occasions de prolonger la situation existante, semblables, disait Hübner, à « un condamné à mort qui n'ose plus espérer sa grâce, mais qui se flatte encore d'obtenir un sursis » [2]. C'est ce dernier parti qui finit par l'emporter et, en octobre 1865, le pape pria Mgr de Mérode d'offrir sa démission [3]. Il faut avouer du reste que, malgré ses vertus, son désintéressement et son ardeur infatigable au service de la cause pontificale, il avait fini par devenir par trop envahissant et qu'il méritait le jugement qu'avait porté sur lui dès 1860 l'ambassadeur d'Autriche :

Ce prélat est certainement très dévoué au Saint-Père, mais il manque complètement de ce tact et de cette modération qui ont formé si longtemps l'apanage des hommes d'État romains et qui ne contribuaient pas peu à leur assurer de l'influence auprès des cabinets de l'Europe [4].

TENTATIVES D'ACCORD AVEC L'ITALIE SUR LE PLAN RELIGIEUX — Pie IX avait dit et répété qu'il ne pouvait être question de transiger avec le gouvernement italien en renonçant à ses droits sur les provinces usurpées : « Ni moi ni mes successeurs ne voudrions ni ne pourrions le faire », affirmait-il à l'ambassadeur de France, qui tentait d'obtenir qu'il prît une attitude plus souple [5]. Par contre, il était disposé à se montrer conciliant si l'on pouvait, en laissant de côté les questions politiques, arriver à un *modus vivendi* concernant les problèmes proprement religieux. Un point surtout le préoccupait, la vacance de nombreux sièges épiscopaux — 108 en avril 1865, dont Turin, Milan, Bologne — par suite de l'exil ou de la mort de leurs pasteurs et de l'impossibilité de procéder à leur remplacement, vu les exigences gouvernementales, sans reconnaître indirectement la légitimité du nouvel État italien.

(1) « Parler réforme à Rome, disait-il dans un moment d'humeur, est aussi ridicule que de vouloir nettoyer une pyramide avec une brosse à dents. » (Rapporté par K. von Schloezer, *Römische Briefe*, p. 168.) L'existence de ces abus était admise même par des hommes aussi peu suspects de sympathie libérale que le cardinal de Reisach ou l'ambassadeur d'Autriche. (Cf. par ex. *Janssens Briefe*, édit. Pastor, t. I, p. 254 ou S. Jacini, *Il tramonto...*, p. 100.)

(2) Dépêche du 2 décembre 1866, dans S. Jacini, *Il tramonto...*, p. 197.

(3) Retiré de la politique, Mgr de Mérode ne resta pas inactif. Il poursuivit la tâche qu'il avait entreprise « d'hausmanniser Rome » (H. d'Ideville, *Mgr de Mérode*, Paris, 1874, p. 25) et créa notamment un nouveau quartier autour de la *via nazionale* qui est son œuvre ; après 1870, la majeure partie des bénéfices de ces spéculations immobilières fut employée par le généreux prélat au profit d'institutions religieuses victimes des spoliations gouvernementales.

(4) Dépêche du 17 novembre 1860, dans S. Jacini, *Il tramonto...*, p. 39. Mgr Talbot écrivait au lendemain de sa chute : « Mérode était devenu le roi de Rome, son influence lui avait tourné la tête au point qu'il n'obéissait même plus au pape *and had reduced the Council of Ministers to a beargarden.* » (Talbot à Manning, 26 décembre 1865, dans E. Purcell, *Life of Manning*, t. II, p. 267.)

(5) Cité dans H. Bastgen, *Die Römische Frage*, t. II, p. 369.

En mars 1865, sur l'initiative de don Bosco et d'un évêque franciscain, Mgr Ghilardi, une négociation fut amorcée par une lettre autographe de Pie IX à Victor-Emmanuel, dans laquelle il demandait au roi, afin « d'essuyer quelques larmes aux yeux de l'Église d'Italie », d'envoyer à Rome un laïc de confiance avec lequel on pourrait envisager les moyens de rétablir la paix religieuse. Cette suggestion fut accueillie favorablement non seulement par le roi, mais par le gouvernement, où les hommes religieux prédominaient, et le président du conseil, La Marmora, désigna comme chargé d'affaires un ancien collaborateur de Cavour, bon catholique et expert en droit ecclésiastique, le député Vegezzi [1]. Bien que les instructions eussent été conçues dans un sens assez joséphiste, on aboutit presque, mais l'accord, que Pie IX souhaitait vivement, fut torpillé par la double opposition de la gauche anticléricale à Florence et des adversaires acharnés de l'Italie à Rome : « Les jésuites et les mazziniens jubilent », notait Gregorovius à l'annonce de l'échec de la négociation [2]. Le gouvernement, qui avait été près de céder sur les questions très épineuses de l'*exequatur* et du serment des évêques, se déroba et cette tentative n'aboutit qu'au retour des évêques exilés, ce qui était quelque chose, mais assez peu.

Ceux qui, à Rome comme dans le royaume, souhaitaient un accord ne perdirent pas courage, en dépit de nouvelles mesures de laïcisation : mise en application, en janvier 1866, du nouveau code qui introduisait le mariage civil, et loi du 19 juin 1866, qui confisquait la plus grande partie de la fortune ecclésiastique. A la fin de 1866, de nouveaux pourparlers furent engagés entre Florence et Rome à l'initiative de Ricasoli, qui avait repris la direction des affaires et entendait régler dans un sens libéral, au mieux de leurs intérêts réciproques, les relations de l'Église et de l'État. Il députa le commandeur Tonello à Rome, où la défaite de l'Autriche à Sadowa avait fait perdre aux intransigeants une partie de leur assurance. Les négociations se déroulèrent dans une atmosphère beaucoup plus détendue que l'année précédente et, sans qu'il fût question d'aboutir à un concordat, on se mit d'accord verbalement sur un certain nombre de points importants [3]. L'intervention discrète du saint fondateur des Salésiens, don Bosco, qui, très apprécié par le pape, jouissait en même temps de l'estime du roi Victor-Emmanuel et de la plupart des chefs politiques piémontais [4], permit de trouver une solution immédiate à la difficile question des nominations épiscopales et, dès le printemps de 1867, Pie IX put pourvoir un tiers des diocèses vacants. Puis Ricasoli

---

(1) Sur cette mission, voir entre autres C. Boncompagni, *La Chiesa e lo Stato in Italia*, Florence, 1866 et S. Jacini, *Due anni di politica italiana*, Milan, 1868, ainsi que les douze dépêches de l'ambassadeur d'Autriche (hostile à la négociation) dans S. Jacini, *Il tramonto...*, p. 139-166.

(2) *Römische Tagebücher*, Stuttgart, 1892, p. 301. K. von Schlözer rapporte que les « noirs » (ou « papalins », c'est-à-dire le parti intransigeant, par opposition aux « bleus » ou Romains amis de l'Italie et libéralisants) avaient été atterrés de voir le pape recevoir un envoyé de Victor-Emmanuel et prêt à certaines concessions : « Ça recommence comme en 1847 », gémissait un prélat. (*Römische Briefe*, p. 212-213.)

(3) Sur la mission Tonello, voir le recueil *Il Ministro Ricasoli e le relazioni della Chiesa collo Stato. Discussione alla Camera dei Deputati intorno alla missione Tonello a Roma con documenti e note*, Florence, 1867.

(4) Sur les interventions de don Bosco au cours de ces années difficiles et sur ses excellentes relations avec le monde politique italien, voir A. Auffray, *Saint Jean Bosco*, chap. xiii.

entreprit de régler par une loi l'ensemble des problèmes religieux pendants, liquidation des biens d'Église, question du *placet* et de l'*exequatur*, relations entre le droit canon et la nouvelle législation. Mais l'opposition des radicaux fit échouer cette tentative et le second cabinet Ratazzi inaugura une nouvelle période d'anticléricalisme féroce, qui aboutit à la loi du 15 août 1867, d'inspiration nettement jacobine, dont les conséquences devaient peser des années durant sur la politique ecclésiastique du royaume. Des efforts furent bien tentés, en 1868, sous le gouvernement de Menabrea, pour reprendre le contact avec le Saint-Siège (continuation de la correspondance secrète entre Pie IX et le roi, mission Fè d'Ostiani à Rome), mais les nouveaux développements de la question romaine à la fin de 1867 rendirent les rapports de plus en plus difficiles.

MENTANA ET SES SUITES   Le départ des derniers bataillons français à la fin de 1866 avait achevé de jeter la consternation dans les milieux romains, déjà très déprimés par l'effondrement autrichien et par la cession de la Vénétie à l'Italie, que le pape considérait comme « l'arrêt de mort de Rome »[1] ; le nouvel auditeur de Rote français, arrivé à Rome en novembre, ne voyait pas d'autre attitude à prendre que celle d'« une famille qui assiste à une extrémisation »[2].

Les premiers mois de 1867 furent cependant très calmes et la petite armée pontificale, réorganisée par le général Kanzler avec l'aide de volontaires étrangers[3], suffit à maintenir l'ordre. A la Pentecôte, les grandes fêtes du centenaire des Apôtres purent se dérouler avec un éclat extraordinaire. Mais bientôt Garibaldi rentra en scène et, avec la complicité du gouvernement Ratazzi, se décida à marcher sur Rome, où ses émissaires avaient préparé une insurrection. La situation n'était toutefois plus la même qu'en 1860 : Napoléon III se sentait obligé de ménager l'opinion catholique, et il était d'ailleurs excité par l'impératrice « dont la dévotion tout espagnole et quelque peu exaltée faisait une ennemie implacable de la cause italienne et surtout des chemises rouges »[4]. Aussi, après avoir vainement essayé de calmer les Italiens, il se décida à intervenir, bien qu'il eût préféré l'éviter et, le 27 octobre, un corps expéditionnaire s'embarquait à Toulon. Grâce aux hésitations de Garibaldi, la brigade du général de Failly eut le temps de faire sa jonction avec les troupes pontificales et de mettre l'envahisseur en déroute, les 3 et 4 novembre, à Mentana.

Ce succès des chassepots français souleva l'enthousiasme dans le monde catholique où l'on célébra à l'envi les *Gesta Dei per Francos*[5], et la seconde expédition de Rome, « qui sacrifiait au Saint-Siège l'Italie démocratique et libérale »[6], fut généralement bien accueillie en France

(1) Dépêche de Hübner du 31 juillet 1866, dans S. Jacini, *Il tramonto...*, p. 191.
(2) Mgr Isoard à H. Loyson, 24 novembre 1866, dans A. Houtin, *Le P. Hyacinthe Loyson dans l'Église romaine*, p. 172-173.
(3) Les contemporains furent surtout frappés par la présence dans ce corps de nombreux représentants de l'armorial français, belge ou irlandais, mais le peuple chrétien s'y trouvait aussi représenté ; c'est ainsi qu'un petit pays comme la Hollande donna de 1864 à 1870 environ 5.000 zouaves. (Cf. C. de Vieux-Bois, *Van een Romeinsch dorpje en pauselijke zouaven*, 1925.)
(4) F. Hayward, *Pie IX et son temps*, p. 336.
(5) La bibliothèque de Pie IX, actuellement au Séminaire du Latran, comporte 1.428 brochures écrites à cette occasion.
(6) J. Maurain, *La politique ecclésiastique du Second Empire*, p. 821.

par tous les conservateurs. Le 5 décembre, le ministre Rouher prononçait
au Parlement son mot célèbre :

> Nous le déclarons au nom du gouvernement français : l'Italie ne s'emparera
> pas de Rome ! Jamais la France ne supportera cette violence faite à son honneur
> et à la catholicité.

Cet engagement solennel, que la France s'était toujours jusque-là
refusée à prendre publiquement, marquait l'abandon de la politique italienne de Napoléon III, obligé désormais de tenir compte de l'opinion
catholique sur le plan parlementaire. Aussi, jusqu'à la guerre de 1870,
la question romaine allait demeurer stationnaire.

LA VIE CATHOLIQUE
DANS LE ROYAUME D'ITALIE
Si l'attitude intransigeante de Rome dans
la question de l'unité italienne éloigne de
l'Église une grosse partie de la bourgeoisie,
celle-ci garde cependant dans son ensemble des sentiments chrétiens
diffus et l'incrédulité proprement dite demeure rare. Elle existe pourtant
et, tandis que la franc-maçonnerie, dont l'influence croît à partir de 1860,
contribue à répandre un voltairianisme vieillot, la traduction de la *Vie
de Jésus* de Renan ou l'ouvrage *Raison et Dogme* de Philippe de Boni
rencontrent un certain succès ; un peu partout apparaissent des sociétés
de libres penseurs qui, depuis 1866, ont leur journal, *Il libero Pensiero*.

En face de l'indifférence religieuse qui se développe dans les milieux
cultivés et commence à déborder sur les masses, en face surtout de l'incrédulité militante qui se groupe dans le « parti de l'action » autour de Garibaldi et des politiciens anticléricaux, clergé et fidèles se sentent déconcertés. Un catholique belge notait avec étonnement, en 1863, à propos de la
délégation italienne au congrès de Malines : « Les catholiques italiens
ressemblent réellement à des hommes qui ont été surpris pendant leur
sommeil par le flot révolutionnaire ; ils sont complètement débordés [1]. »
L'exemple des catholiques allemands avait pourtant retenu leur attention
et, en 1849 déjà, les évêques piémontais avaient eu l'idée d'une « Association catholique italienne » sur le modèle du *Katholisch Verein* ; Pie IX
consulté avait encouragé l'initiative, tout en conseillant d'en confier la
direction à l'épiscopat plutôt qu'aux laïcs [2]. Ce n'est toutefois qu'une
quinzaine d'années plus tard, au retour de Malines, où ils avaient pris
plus directement contact avec les réalisations belges et allemandes [3],
que les catholiques italiens commencèrent vraiment à s'organiser.

La première impulsion vint de l'avocat Casoni, de Bologne, qui fonda
en 1865 une « Société catholique italienne pour la défense de la liberté
de l'Église », laquelle eut bientôt des sièges locaux à Rome, Milan, Florence, Livourne, Naples, Reggio de Calabre ; reconnue par le pape, elle
disparut malheureusement à cause des troubles consécutifs à la guerre
contre l'Autriche. Mais en 1867, les dirigeants de deux groupements de

---

(1) Souvenirs d'A. Delmer dans *La Revue générale* (Bruxelles), t. XC, 1909, p. 346.
(2) Voir les trois documents publiés dans P. Pirri, *Pio IX et Vitt.-Em. II dal loro carteggio
privato*, t. I, p. 244-248 (cf. aussi p. 57*-58*).
(3) Sur le congrès de Malines de 1863, voir *infra*, p. 170-171.

jeunesse, le comte Acquaderni de Bologne et Mario Fani de Viterbe, se rencontraient à Rome à l'occasion des fêtes du centenaire des apôtres et, avec l'aide du jésuite Pincelli, jetaient les bases de la *Società della Gioventù cattolica italiana* ; il s'agissait de grouper aussi bien des étudiants que des jeunes ouvriers afin de soutenir leurs sentiments catholiques et d'exalter leur dévouement à la cause pontificale. Dès janvier 1868, un jeune professeur de Brescia, Jérôme Lorenzi, qui avait envisagé la fondation d'un groupement analogue au *Studentenverein* allemand, lançait le premier hebdomadaire destiné à éclairer les jeunes catholiques sur les questions politico-religieuses, *Il Giovane cattolico*, auquel collaborèrent plusieurs notabilités du catholicisme italien d'alors, tel Cesare Cantù.

L'ŒUVRE DE DON BOSCO   Tandis que les laïcs sortaient peu à peu de leur engourdissement, le clergé ne restait pas inactif. Sans doute, il est inévitable que dans un pays où l'on compte environ 1 prêtre pour 400 habitants — et dans certains diocèses, notamment dans le Sud, la proportion est encore plus forte [1] — un certain nombre d'ecclésiastiques n'aient guère d'esprit apostolique, mais beaucoup, surtout dans le Nord, font plus que leur devoir et la belle résistance des catholiques piémontais aux menées anticléricales est là pour le prouver.

Parmi ces prêtres, il en est un dont le zèle entreprenant et surtout la sainteté de vie font la gloire du clergé italien. Jean Bosco [2] était né en 1815 dans une modeste famille de paysans des environs de Turin. Son patronage improvisé s'est peu à peu transformé en une œuvre dont la renommée dépasse de loin les frontières du royaume et il s'est révélé un pédagogue hors ligne, en qui s'unissent le bon sens, le goût du risque et le sens de l'apostolat, dominé par l'idée que l'éducateur est un ami qui guide l'enfant pour prévenir les manquements au devoir plutôt qu'un maître qui surveille pour punir les transgressions. C'est le même souci de permettre aux enfants de se développer dans un climat familial, dans une atmosphère de joie et de gaieté, qui l'amène à entreprendre, en 1855, ce qu'on a pu appeler la première colonie de vacances du continent. Bientôt, l'extension de son œuvre l'oblige à faire appel à des collaborateurs, qui deviendront une congrégation florissante, dont les règles, soumises une première fois au pape en 1858, sont approuvées en 1869 et définitivement confirmées en 1874 ; en 1872, une congrégation féminine, les *Filles de Marie Auxiliatrice*, vient s'y ajouter et, en 1876, le saint fondateur complète son œuvre par un tiers-ordre laïque bien moderne, l'*Union des coopérateurs Salésiens*. L'Église d'Italie, éprouvée par une douloureuse crise nationale et par une politique anticléricale, n'est pas près de mourir.

---

(1) D'après les statistiques de 1881, il y avait par exemple dans le diocèse de Naples 2.803 prêtres pour 700.000 habitants répartis en 85 paroisses ; dans celui de Salerne, 538 prêtres pour 145.000 habitants ; dans celui de Sorrente, 403 prêtres pour 51.000 habitants. (WERNER, *Katholischer Kirchenatlas*, Fribourg-en-Br., 1888, p. 3-24.)

(2) Sur don Bosco (1815-1888), voir A. AUFFRAY, *Un grand éducateur. Saint Jean Bosco*, 6e édit., Lyon-Paris, 1947.

L'ÉGLISE EN FRANCE SOUS LE SECOND EMPIRE [1]

§ 1. — La situation privilégiée de l'Église.

LA PROTECTION OFFICIELLE AU DÉBUT DE L'EMPIRE

L'Église de France, que le premier consul avait trouvée presque moribonde à l'aube du xixᵉ siècle, avait vu sa situation se redresser en cinquante ans d'une manière inespérée et elle se trouvait en pleine expansion au moment de l'avénement de Napoléon III. Elle était sortie renforcée de la crise de 1848, ayant gagné sur les deux tableaux. En effet, la révolution, en proclamant la liberté d'association, avait indirectement favorisé le développement des congrégations, et surtout, en introduisant le suffrage universel dans un pays dont les masses paysannes étaient encore généralement sous l'influence du clergé, elle assurait à l'Église une influence politique autrement efficace que sous le régime censitaire. D'autre part, la réaction suscitée par les excès de la révolution avait attiré à l'Église la sympathie de tous ceux — et on ne les trouvait pas seulement dans la classe riche — qui avaient soudain apprécié en elle la seule garantie vraiment efficace pour le maintien de l'ordre social, et son influence sur l'enseignement s'en était trouvée sensiblement renforcée.

Au cours du Second Empire, l'Église allait profiter largement de ces divers avantages, grâce à l'appui, ou du moins à la neutralité plus ou moins bienveillante, dont elle bénéficia de la part des pouvoirs publics pendant la majeure partie de ces dix-huit années. Non pas que l'empereur ou ses conseillers fussent particulièrement soucieux de favoriser l'Église. Si l'on met à part l'impératrice, qui était une catholique convaincue, et le prince Napoléon, qui était un anticlérical décidé, c'étaient des opportunistes, plutôt sceptiques en matière religieuse, mais préoccupés de l'intérêt du gouvernement et appréciant les avantages qui devaient résulter de l'influence morale et sociale de la religion au détriment de la propagande révolutionnaire ; en outre, le concours de l'Église leur paraissait utile pour rallier à l'Empire les milieux légitimistes, avec lesquels le clergé restait en rapports étroits [2].

(1) BIBLIOGRAPHIE. — Les principales sources ont été indiquées en tête du volume de même que les ouvrages généraux, p. 6-8 (tenir surtout compte du cours de Ch. Pouthas et du volume de J. Maurain) et les biographies (tenir surtout compte de celles de Maret, de Dupanloup, de Régnier, de Veuillot et de Melun). On ajoutera, sur le mouvement charitable, J.-B. DUROSELLE, *Les débuts du catholicisme social en France (1822-1870)*, Paris, 1951, IIIᵉ Partie.

(2) Les motifs des sympathies légitimistes du clergé français étaient complexes. Beaucoup d'adversaires des tendances libérales voyaient dans le comte de Chambord le modèle du prince chrétien, qui rendrait à l'Église la place qu'elle avait tenue dans la société sous l'ancien régime. Mais d'autres motifs s'ajoutaient, sur lesquels une lettre de Mgr Pie, de janvier 1852, jette quelque

Sans doute, l'Église, malgré ses instances, n'obtint ni la suppression de l'antériorité obligatoire du mariage civil par rapport au mariage religieux, ni l'abrogation des articles organiques. Mais le gouvernement impérial au début n'insista pas plus sur l'application de ces derniers que ne l'avait fait la Seconde République et, comme elle, il laissa l'Église resserrer ses liens avec le centre de la Chrétienté conformément au programme ultramontain, et ferma les yeux sur l'extension des congrégations. Tout en autorisant libéralement les dons et legs aux établissements religieux, il augmenta sensiblement le budget des cultes [1]. Il interpréta la loi militaire d'une manière très favorable aux ecclésiastiques. Il réprima les tendances irréligieuses qui s'étaient manifestées dans l'enseignement public, fit poursuivre les mauvais livres par les tribunaux et établit un régime de presse libéral à l'égard de l'Église, mais sévère pour ses adversaires ; sans aller jusqu'à imposer, comme on le lui demandait, le repos dominical, il autorisa les préfets à ordonner la fermeture des cabarets pendant la grand'messe, encouragea la participation de l'armée et des fonctionnaires aux processions et invita les autorités religieuses à assister aux cérémonies civiles, poursuivant ainsi « une politique que l'entrée des cardinaux au Sénat avait inaugurée et qui consistait à associer aux yeux des populations le prestige de l'Église à celui de l'État » [2].

L'Église obtenait de la sorte, « dans une société opprimée, les libertés les plus étendues dont elle eût joui depuis le Concordat » [3], conservant le bénéfice de celui-ci tout en voyant se restreindre peu à peu, en fait, les prérogatives qu'il conférait au pouvoir civil. Très sensibles aux avantages apostoliques que semblait comporter la situation, les catholiques accentuèrent leur ralliement à l'Empire, tandis que, dans les mandements ou discours épiscopaux, se multipliaient les apologies de Napoléon III et les éloges du nouveau régime. La guerre de Crimée, qui prit aux yeux de beaucoup l'allure d'une croisade contre la Russie orthodoxe, resserra encore l'entente du gouvernement avec le clergé ainsi qu'avec la Cour de Rome.

Cette entente un peu trop affichée avec un régime qui avait fait bon marché des libertés publiques, et dont les relations avec la bourgeoisie affairiste étaient connues, n'allait pourtant pas sans danger et le mot de Changarnier qui stigmatisait le Second Empire comme « un tripot béni par les évêques » était révélateur de l'amertume que causait aux anciens républicains l'attitude d'une partie notable de la hiérarchie. En outre,

lumière : « Vous dites des légitimistes des choses vraies, mais exagérées et dangereuses. Comment parler avec cette sévérité des seuls hommes riches qui accomplissent leur devoir religieux dans les 9/10 de la France, des seuls hommes dont la fortune contribue au soutien des œuvres religieuses ?... Voulez-vous que je mette sur le même rang les bourgeois enrichis dont tout le mérite est de me donner un bon dîner quand je vais dans leur paroisse et de venir à l'église ce jour-là ? Voulez-vous que je les mette sur le rang des vrais chrétiens que l'on rencontre toujours devant les autels, qui soutiennent nos séminaires, qui répondent à tous nos appels ? » (Cité dans dom DELATTE, *Dom Guéranger*, t. II, p. 52.)

(1) Celui-ci passe de 39.518.544 fr. en 1852 à 45.967.896 fr. en 1859, tandis que — comparaison significative — le budget de l'Instruction publique est réduit pendant la même période de 22 millions 958.893 fr. à 20.996.163 fr. (Ch. NICOLAS, *Les budgets de la France depuis le commencement du XIXe siècle*, Paris, 1882, p. 276 et 280.)

(2) J. MAURAIN, *La politique ecclésiastique du Second Empire*, p. 70.

(3) *Ibid.*, p. 80.

à y regarder de plus près, les avantages dont profitait l'Église étaient précaires, car le gouvernement restait libre de remettre en vigueur une législation qu'il n'appliquait plus guère, mais qu'il s'était refusé à abroger. Or précisément, plus la menace révolutionnaire s'atténuait et plus il prenait ombrage de l'influence grandissante du clergé dans la vie publique. Cette influence lui paraissait d'autant plus inquiétante que, par suite des efforts systématiques du parti ultramontain [1], il voyait le Saint-Siège accroître d'année en année son autorité sur l'Église de France aux dépens du pouvoir civil. Excitée par la politique maladroite du nonce Sacconi, la tradition gallicane du ministère des Cultes, contenue pendant plusieurs années, recommença dès lors à se manifester, surtout lorsque Rouland eut succédé à Fortoul comme ministre.

LA POLITIQUE ECCLÉSIASTIQUE DES DIX DERNIÈRES ANNÉES   Dès 1856, on pouvait pressentir qu'une crise se préparait, mais seuls quelques esprits avisés s'en rendaient compte. Clergé et fidèles, dans leur ensemble, n'en avaient cure et le voyage des souverains en Bretagne, durant l'été de 1858, apparut comme une « retentissante manifestation d'alliance entre l'Empire et l'Église » [2]. C'est l'évolution de la question romaine à la suite de la guerre d'Italie qui fit éclater brusquement le conflit.

Quand on sut que Napoléon III acceptait le démembrement de l'État pontifical, beaucoup d'évêques et la majorité du clergé, excités par *L'Univers*, se retournèrent avec indignation contre l'empereur, « brusquement précipité du rang de Charlemagne et de saint Louis dans les séjours infernaux » [3]. Oubliant pour un temps leurs divisions, les ultramontains de *L'Univers*, qui avaient été ses plus chauds partisans, et les amis de Mgr Dupanloup essayèrent de développer à travers le pays une agitation susceptible d'embarrasser le gouvernement. Mais, en plusieurs régions, les masses catholiques ne réagirent guère parce que, le culte n'étant pas entravé, elles ne sentaient pas l'Église menacée. Le gouvernement en conclut qu'« un pouvoir favorable à la religion mais contraire à la domination du clergé serait soutenu par la population » [4] et il décida dès lors de réagir contre « les envahissements de l'Église », en appliquant systématiquement le plan de campagne qu'avait dressé Rouland dans un *Mémoire* remis à l'empereur en avril 1860 [5] : nomination d'évêques opposés aux interventions croissantes de la Curie romaine dans la vie de l'Église de France ; arrêt du développement des congrégations ou associations religieuses ; appui plus marqué à l'école publique. L'empereur donna aisément son accord, car il avait été blessé par les attaques du

(1) Sur les progrès du parti ultramontain pendant le Second Empire et sur le réveil de l'opposition gallicane qui en fut la conséquence, cf. *infra*, p. 298-300 et p. 305-309.
(2) J. MAURAIN, *op. cit.*, p. 228.
(3) A. DANSETTE, *Histoire religieuse de la France contemporaine*, t. I, p. 393. Voir pour plus de détails, *supra*, p. 89-90.
(4) Rapport du préfet d'Ille-et-Vilaine, 10 avril 1859, *Arch. Nat.*, F¹ᶜ III.
(5) *Mémoire remis à l'Empereur par un de ses ministres des Cultes sur la politique à suivre vis-à-vis de l'Église*, publié par L. PAGÈS, Bourges, 1873. On en trouvera des extraits dans J. MAURAIN, *op. cit.*, p. 451-460. (Sur son authenticité, cf. *ibid.*, p. 460-463 ; sur sa préparation, p. 446-451.)

clergé et par ce qu'il considérait comme l'ingratitude du pape à son égard [1] ;
il était surtout profondément mécontent de l'allure de manifestation
légitimiste que prenait souvent en France la défense du Saint-Siège.

Grâce à la liberté de parler et d'écrire, de se réunir et de s'associer,
que les catholiques conservaient seuls en France, ils avaient pu organiser ce qu'aucun autre parti n'avait tenté depuis le coup d'État : une
campagne d'opposition. La nouvelle politique allait s'efforcer, par une
série de vexations et de tracasseries [2], de ruiner cette redoutable influence
du clergé, tout en continuant à protéger ostensiblement la religion dans
le pays [3] et à l'étranger [4] afin de garder la confiance des masses catholiques.

Mais cette façon de faire n'eut d'autre résultat que d'accroître encore
l'hostilité du clergé [5] et l'attitude d'opposition de celui-ci entraîna la
défection des « catholiques avant tout » et de nombreux légitimistes
ralliés jusqu'alors à l'Empire. Or l'influence électorale de ces notables
et du clergé était considérable [6] et le gouvernement impérial fut d'autant
moins en mesure de la négliger que les élections de 1863 révélèrent les
progrès rapides du parti républicain : l'unité du « grand parti de l'ordre »
était plus que jamais nécessaire [7]. Pour se concilier à nouveau l'appoint
du « parti clérical », il fallut donc se résigner à en revenir à une attitude
moins hostile : on commença par des adoucissements de forme, mais
bientôt le changement se précisa. On n'en revint pas à la protection
officielle des premières années et le gouvernement continua à laisser
la presse anticléricale attaquer l'Église, espérant grâce à cette « soupape

(1) Effectivement, Pie IX, très déçu par la politique ecclésiastique et italienne de Napoléon III,
subissait de plus en plus des influences hostiles à l'empereur, entre autres celles de Mgr de Mérode,
beau-frère de Montalembert, et celle du nonce Sacconi, qui fréquentait beaucoup les salons de
l'opposition. Le gouvernement impérial finit par inspirer une telle méfiance au Saint-Siège que
celui-ci le soupçonna de préparer un schisme.

(2) Limitation de la diffusion des mandements épiscopaux en les soumettant à la loi du timbre ;
suppression de leur traitement aux membres du clergé qui s'opposaient publiquement à la politique
impériale ; rupture totale des relations avec les deux évêques qui s'étaient le plus illustrés dans
la défense de la cause pontificale, Mgr Dupanloup et Mgr Pie ; interdiction de la société de Saint-
Vincent-de-Paul (cf. infra, p. 122) ; modification de la jurisprudence en matière de dons et legs,
très favorable jusqu'alors ; fréquentes interdictions aux ordres religieux d'ouvrir de nouvelles
maisons ; abandon des ménagements dont la justice avait jusqu'alors usé à l'égard des ecclésiastiques compromis dans des affaires de mœurs ; liberté de plus en plus grande laissée à la presse anticléricale ; appui donné à la pièce d'Émile Augier, Le Fils de Giboyer, qui ridiculisait Veuillot
et les « cléricaux » ; heurts fréquents avec Rome à propos des nominations d'évêques, amenant
parfois des vacances de plusieurs années.

(3) Par exemple en suspendant le cours de Renan au Collège de France, en février 1862 ; ou en
affectant d'importants crédits à la construction et à la réparation d'églises.

(4) En Syrie, où les troupes françaises protègent les Maronites ; en Chine et en Cochinchine,
où elles défendent les missionnaires persécutés ; au Mexique, où elles essaient de substituer l'empire
catholique de Maximilien à la république anticléricale de Juarès, le gouvernement de l'empereur
se pose en défenseur de l'Église, dont les intérêts se confondent d'ailleurs avec ceux de la France.

(5) Ses griefs sont résumés dans le Mémoire remis par Dupanloup au ministre Drouyn de Lhuys
à la fin de 1862 (reproduit dans F. Lagrange, Vie de Mgr Dupanloup, t. II, p. 494-500). La campagne électorale de 1863 montra à quel point le clergé était devenu hostile au gouvernement dans
l'ensemble de la France.

(6) Ce fait, souvent affirmé par Maurain, est confirmé par les monographies. Voir par exemple,
J. Lacouture, Histoire religieuse et politique des Landes de 1800 à 1870, Belhade, 1940 ; R. Cuzacq,
Les élections législatives à Bayonne et au pays basque de 1848 à 1870, Bayonne, 1948.

(7) Voir l'analyse du résultat des élections dans J. Maurain, op. cit., p. 643-667. Au dire de
celui-ci, le clergé apparut dès lors comme « l'élément essentiel du vieux parti de l'ordre, dont les
élections de 1863 marquaient le réveil et qui, jouant un rôle de plus en plus important à mesure
que le gouvernement impérial s'affaiblissait, allait, après sa chute, gouverner la France. Tous les
succès remportés par le clergé français de 1863 à 1876 ont leur origine dans la part qu'il prit aux
élections de 1863 ». (P. 667.)

de sûreté » donner l'illusion que la presse était redevenue libre. Mais, par contre, les ministres soutinrent de plus en plus mollement la politique de Duruy à l'Instruction publique [1], cherchant, en donnant de diverses manières des gages de leur dévouement aux intérêts de la religion, à se concilier à nouveau la confiance du clergé.

Si la politique gallicane de Rouland fut poursuivie par son successeur Baroche, c'est qu'elle visait moins à combattre l'influence de l'Église dans la société que celle du Saint-Siège dans l'Église de France. On espérait qu'un clergé plus indépendant de Rome serait pour le gouvernement un collaborateur plus docile et que, dans la mesure où il cesserait de s'aligner sur les positions toujours plus intransigeantes du Vatican à l'égard des tendances politiques modernes, il y aurait moins d'inconvénients à lui rendre une partie de son influence, jugée par ailleurs salutaire du point de vue social. Il y avait dans ce raisonnement une bonne part d'illusion, car il majorait indûment l'influence réelle que les adversaires de l'ultramontanisme conservaient encore dans l'Église de France. Comme l'a judicieusement noté Maurain, dans un état basé sur le suffrage universel et qui évoluait de plus en plus vers le parlementarisme libéral, « le gallicanisme de Baroche devait nécessairement décliner au profit de deux autres tendances : le cléricalisme ultramontain, soutenu par la majorité des classes dirigeantes et par une grande partie des paysans ; et l'anticléricalisme radical qui prévalait auprès des intellectuels, des ouvriers, des artisans et des paysans plus évolués » [2]. Il faut ajouter, toutefois, que si les seize évêques nommés de 1865 à 1869 dans un esprit d'opposition à Rome ne réussirent pas à arrêter l'évolution de l'Église de France dans le sens ultramontain, ils devaient être parmi les principaux soutiens de la politique plus conciliante à l'égard de la société moderne suivie par Léon XIII après la mort de Pie IX, et l'on est amené, de ce point de vue, à juger moins sévèrement qu'on ne l'a fait souvent la politique religieuse préconisée par les principaux conseillers ecclésiastiques du gouvernement impérial à cette époque, le doyen de la faculté de théologie de la Sorbonne, Mgr Maret, et l'archevêque de Paris, Mgr Darboy, auxquels on peut adjoindre Mgr Lavigerie, évêque de Nancy depuis 1863.

La Convention de septembre et le mécontentement qu'elle suscita empêcha la politique d'apaisement de porter immédiatement ses fruits. Mais le « Jamais ! » de Rouher, en 1867 [3], rendit enfin possible, malgré certains heurts qui continuèrent jusqu'à la fin de l'Empire, une réconciliation que souhaitaient depuis plusieurs années les évêques gouvernementaux, hostiles à la politique de Pie IX et qui estimaient fort utile l'appui extérieur apporté par le pouvoir civil à l'Église. La peur commune de l'opposition républicaine rapprochait une nouvelle fois les deux puissances : elle rejetait l'Empire déclinant vers la grande force conservatrice constituée par le clergé, lequel s'effrayait de son côté de l'anticléricalisme

---

(1) Cf. *infra*, p. 120.

(2) J. MAURAIN, *Un bourgeois français au XIXᵉ siècle. Baroche*, Paris, 1936, p. 451-452. De formation juridique comme Rouland, Baroche était au moins aussi gallican, bien qu'il eût désapprouvé comme excessives et maladroites certaines mesures de son prédécesseur.

(3) Cf. *supra*, p. 105-106.

affiché par les républicains et de leurs préférences pour une séparation de l'Église et de l'État. « Ainsi l'empereur finit-il par où le prince-président a commencé : après quelques années de brouille, il contracte une nouvelle alliance avec l'Église dont il a d'abord été le protecteur [1]. »

L'INFLUENCE SOCIALE DE L'ÉGLISE

L'appui gouvernemental n'est pas le seul avantage dont jouisse l'Église de France sous le Second Empire. Son influence sur les classes dirigeantes en est un autre [2]. L'élection de Mgr Dupanloup, en 1854, à l'Académie française, où plus aucun évêque n'avait siégé depuis la mort de Mgr de Quélen, souligne le regain de prestige de l'Église. Dix ans plus tard, lors de ses tournées d'inspection dans les principales villes de France, Taine sera stupéfait du rôle tenu par le clergé dans la vie provinciale. Non seulement la noblesse lui reste acquise, mais la bourgeoisie riche, naguère hostile, subit de plus en plus son ascendant. Grâce à la générosité de ces milieux, le patrimoine de l'Église continue à se reconstituer rapidement : de 1852 à 1860, on compte 27.331.851 fr. de dons et legs et 27.214.620 fr. d'acquisitions, soit 62 pour cent de ce que l'Église avait acquis depuis 1801 [3]. A la campagne, où les municipalités se recrutent souvent parmi les notables conservateurs, le clergé en profite pour faire confier aux congrégations de nombreuses écoles publiques et pour obtenir des communes de gros subsides. Dans les régions industrielles, il fait subventionner par les patrons catholiques les œuvres destinées à maintenir l'influence de l'Église sur les ouvriers.

L'Église peut en outre compter sur l'appui de beaucoup de hauts fonctionnaires, dont la puissance est considérable dans un régime autoritaire : « La plupart des diplomates, des officiers, des magistrats inamovibles, une partie du personnel des administrations centrales, un certain nombre de préfets, d'inspecteurs d'académie, et beaucoup de maires » sont acquis aux intérêts de l'Église, estime M. Maurain qui ajoute : « Ils faussent la politique ecclésiastique du gouvernement impérial en exagérant sa bienveillance pour le clergé pendant la période d'alliance, en faisant plus ou moins défection pendant les périodes de conflit [4]. »

Sans doute faut-il faire la part des choses lorsque les préfets dénoncent l'ingérence abusive des curés dans la vie des communes ou présentent l'évêque comme le véritable maître du département, affirmant par exemple qu'à Rennes, « il était présenté, même pour les affaires administratives et politiques, au moins autant de pétitions à l'évêché qu'à la préfecture » [5].

(1) A. DANSETTE, *Histoire religieuse de la France contemporaine*, t. I, p. 432.
(2) On trouvera de nombreux exemples de cette influence dans l'ouvrage de Maurain, en particulier au chap. XII (Les questions religieuses et la vie publique locale).
(3) Ch. POUTHAS, *L'Église et les questions religieuses...*, p. 103-104. A titre d'exemple, le procureur général de Toulouse signalait, en 1856 : « Il y a ici plus qu'ailleurs des traditions invétérées de patronage entre les familles dites patriciennes et le clergé. C'est chez elles qu'il trouve argent, appui, accueil. Il ne se fait rien non plus dans ces familles sans que le directeur de conscience soit consulté ou employé. Les mariages, les testaments, les affaires, tout est soumis à un examen à contrôle religieux, et ensuite soutenu par l'influence religieuse. Il y a eu à Toulouse (c'est un fait judiciairement constaté) un conseil d'ecclésiastiques casuistes où l'on consultait sur la quotité des legs à faire, etc. » (*Arch. Nat.*, BB30 388.)
(4) J. MAURAIN, *op. cit.*, p. 956-957.
(5) Rapport du 8 octobre 1858, *Arch. Nat.*, F1c III, Ille-et-Vilaine.

Il semble toutefois incontestable que, dans certaines régions, le clergé ne garda pas toujours la discrétion souhaitable et justifia par là-même les accusations de « cléricalisme », auxquelles l'opinion française s'est toujours montrée si sensible.

LA CONSOLIDATION DES CADRES ECCLÉSIASTIQUES — Le regain de prestige de la religion et la meilleure appréciation de la situation du clergé, conséquence de son influence sociale et de la faveur gouvernementale, ainsi que les effets de la loi Falloux, ne tardèrent pas à avoir leur influence sur le recrutement sacerdotal. A première vue, on aurait plutôt l'impression contraire en constatant que le chiffre des ordinations a tendance à diminuer jusqu'en 1861 [1]. Mais il faut se souvenir que la quasi-totalité des candidats au sacerdoce viennent d'un petit séminaire, et qu'il faut dès lors une dizaine d'années pour que l'amélioration des entrées se traduise dans le chiffre des ordinations. Effectivement, à partir de 1862, celui-ci se relève sensiblement, passant de 1.196 en 1861 à 1.503 en 1864 et à 1.753 en 1868. Cette évolution favorable n'est toutefois pas générale et, dans certaines régions comme le Centre, la Bourgogne, et surtout la Champagne et les environs de Paris, la situation commence même à devenir inquiétante, bien que les régions riches en vocations permettent des compensations là où un fléchissement apparaît.

Comme le clergé français s'était surtout reconstitué de 1820 à 1840, la mortalité n'augmente encore que faiblement [2] et le nombre des prêtres progresse en 18 ans de près de 20 pour cent, passant de 46.969 en 1853 à 56.295 en 1869 [3], ce qui permet aux évêques de pourvoir de nombreux postes de vicaires restés jusque-là sans titulaires, de faire passer de 1.541 à 2.467 le nombre d'aumôniers dans les services publics, et de créer 1.600 paroisses nouvelles. Le Second Empire apparaît de la sorte « comme une apogée entre une période de médiocre développement, qui était celle de la Monarchie de Juillet, et une période au contraire de décroissance progressive qui sera celle de la Troisième République » [4]. On peut se demander si la hiérarchie utilisa de la façon la plus judicieuse les effectifs accrus dont elle put disposer : en effet, tandis qu'elle multiplie imprudemment à la campagne les lieux de culte, dont le service deviendra pour les générations ultérieures une charge terriblement lourde [5] et qu'elle engourdit les jeunes prêtres dans des postes de vicaires de villages, l'augmentation des paroisses urbaines est minime malgré le développement rapide des quartiers périphériques des villes. Dans un pays où l'industrialisation fait en vingt ans des progrès considérables, l'Église conserve

(1) Après avoir plafonné un peu au-dessus de 1.300 jusqu'en 1857, il descend de 1.267 en 1858 à 1.196 en 1861. (Voir le tableau des ordinations annuelles dans F. BOULARD, *Essor ou déclin du clergé français ?* Paris, 1950, p. 465.)

(2) En 1850, le nombre de sexagénaires dans le clergé n'atteignait que 7 %. Il montera à 22 % à la fin du Second Empire et atteint aujourd'hui 35 %. (F. BOULARD, *op. cit.*, p. 81.)

(3) En 1870, il y avait de la sorte en France 1 prêtre pour 730 habitants. De 1830 à 1870, l'effectif du clergé s'est accru de 39 %, pendant que la population croissait seulement de 17 %.

(4) Ch. POUTHAS, *L'Église et les questions religieuses...*, p. 109.

(5) Voir sur ce point les observations de F. BOULARD, *op. cit.*, p. 367 et suiv.

une organisation adaptée à la structure exclusivement rurale qui avait prédominé jusqu'alors.

Si la population des séminaires augmente, les méthodes de formation ne changent guère. Le cardinal de Reisach reprochait aux séminaires français de faire la part trop grande au « dressage »[1]. Le mot est un peu dur et les appréciations de Renan sur le séminaire Saint-Sulpice sont autrement nuancées. Il faut pourtant reconnaître que dans la plupart des séminaires la formation est avant tout centrée sur la multiplication des exercices de dévotion, la mémorisation des traités vus en classe et une discipline minutieuse laissant le moins possible à l'initiative individuelle. Ceci se comprend d'ailleurs par la conception que l'on se fait du prêtre idéal au XIXe siècle : un pasteur, habitué à une vie retirée, qui attend les fidèles dans son église ou son presbytère plutôt que l'animateur d'une paroisse missionnaire, entraînant les laïcs à l'apostolat et partant lui-même à la recherche des brebis perdues.

Que cette formation traditionnelle offre de sérieux avantages, c'est ce que prouvent non seulement la haute moralité et ce qu'on pourrait appeler la conscience professionnelle de l'ensemble du clergé français, mais aussi le nombre de prêtres éminents dans les diverses fonctions du ministère que l'on rencontre au cours de ces années : pasteurs de campagne comme le saint curé d'Ars ou le curé du Mesnil-Saint-Loup ; pasteurs de grandes villes comme M. Hamon, le curé de Saint-Sulpice ; apôtres du monde ouvrier comme Antoine Chevrier, le fondateur du Prado de Lyon ; animateurs de bonnes œuvres comme l'abbé Lelièvre, le protecteur des Petites Sœurs des Pauvres ; écrivains spirituels ou directeurs de conscience comme Mgr de Ségur, Mgr Gay ou le délicat abbé Perreyve[2].

Mais ces belles figures de prêtres ne peuvent voiler les inconvénients du système. On doit, après avoir fait l'éloge du zèle et de la respectabilité du clergé du Second Empire, regretter que son genre de vie n'apparaisse pas toujours suffisamment évangélique aux observateurs du dehors. Les fonctionnaires locaux, qui sont loin d'être toujours malintentionnés, lui reprochent souvent son manque de tact, une certaine arrogance et une trop grande préoccupation de sa situation matérielle[3]. En plusieurs régions, les paysans ne se gênent pas pour dire que « le sacerdoce est un métier de paresseux et que leur pasteur, né comme eux sous le chaume, porterait le poids du jour et de la chaleur s'il ne s'était fait homme d'Église ». L'abbé Combalot, qui rapporte le mot, doit avouer « qu'un grand nombre de curés de villages éparpillent leur existence dans des visites, des courses, des délassements, des repas où l'esprit sacerdotal

---

(1) Cité dans *Janssens Briefe*, édit. Pastor, t. I, p. 205.
(2) F. Trochu, *Le curé d'Ars. Saint J.-M. Vianney*, Paris, 1925 ; B. Maréchaux, *Le Père Emmanuel*, Mesnil-Saint-Loup, 1935 (il s'agit de l'abbé André, curé du Mesnil, qui devint sur le tard olivétain) ; L. Branchereau, *Vie de M. Hamon*, Paris, 1877 ; A. Lestra, *Le P. Chevrier*, Lyon, 1945 ; Mgr Baunard, *Ernest Lelièvre*, Paris, 1906 ; H. Chaumont, *Mgr de Ségur, directeur des âmes*, Paris, 1884 ; B. du Boisrouvray, *Mgr Gay, évêque d'Anthédon*, Tours, 1921, 2 vol. ; Cl. Peyroux, *Henri Perreyve*, Paris, 1934. Voir aussi *Biographie des prêtres du diocèse de Cambrai (1847-1887)*, Paris, 1890, 3 vol.
(3) Cf. J. Maurain, *op. cit.*, p. 320-321.

s'éteint » [1]. Par ailleurs, dans les villes, les prêtres les plus zélés donnent trop souvent l'impression de manquer de contact avec le peuple et de restreindre leur action à quelques groupes plus fervents [2]. Tandis qu'ils se dépensent à protéger les « bien pensants » en des cercles toujours plus fermés, la foi baisse lentement dans la masse, influencée par le milieu ambiant.

Enfin, observateurs français et étrangers sont d'accord pour déplorer l'infériorité très nette du clergé diocésain sous le triple point de vue littéraire, scientifique et théologique, et l'amoindrissement d'influence qui en résulte auprès du public cultivé. Cette situation s'explique un peu par le recrutement de ce clergé où les enfants de la noblesse et de la bourgeoisie sont l'exception [3], mais surtout par le caractère arriéré de l'enseignement de trop de grands séminaires, qui « dépasse à peine le niveau d'un grand catéchisme de persévérance fait en latin » [4]. Quelques évêques commencent à apercevoir le problème : Mgr Dupanloup publie en 1855 son *Règlement relatif aux études ecclésiastiques dans le diocèse d'Orléans*, Mgr Plantier encourage dans son séminaire les études d'Écriture sainte et favorise la bibliothèque, un peu partout on augmente le nombre des professeurs [5], mais presque personne parmi les responsables ne remarque encore que rien ne sera fait tant que ces professeurs n'auront pas reçu une formation scientifique adéquate et il faudra attendre la fin du siècle pour constater un réel changement sur ce point [6].

On ne s'étonnera guère si ce clergé peu cultivé, dans un pays où il n'y a encore ni congrès religieux ni revues ecclésiastiques sérieuses, n'ait pratiquement aucun moyen d'information ou de contrôle sur la situation générale de la France et s'inspire le plus souvent de préjugés locaux. Par contre, sous l'influence de *L'Univers*, dont il suit les consignes avec passion, il reconnaît avec une ardeur toujours croissante l'autorité du Siège de Rome sur l'Église de France.

Cet enthousiasme ultramontain du clergé inférieur, qui remonte aux campagnes de *L'Avenir*, est renforcé par l'espoir de trouver dans la Curie

---

(1) Cité par A. RICARD, *L'abbé Combalot*, p. 393-394.

(2) *Janssens Briefe*, t. I, p. 188-189. Évidemment, il y a des exceptions, tels ces directeurs d'œuvres pour jeunes ouvriers que nous fait connaître J.-B. DUROSELLE, *Les débuts du catholicisme social en France*, Paris, 1951, III° partie, chap. II. L'un de ces prêtres est même suffisamment perspicace pour se rendre compte qu'en présence de la déchristianisation des milieux ouvriers les ressources du ministère paroissial ordinaire sont insuffisantes : « Il s'agit ici de conquérir. Envoyez d'abord des chasseurs d'âmes, le pasteur s'avancera quand il y aura un troupeau à garder. » (Cité p. 576.)

(3) Cette abstention presque totale des classes dirigeantes s'explique par diverses causes. Mgr Bougaud, dans son ouvrage bien documenté sur la crise des vocations : *Le grand péril de l'Église de France au XIX° siècle*, Paris, 1878, accuse la diminution des naissances et l'indifférence religieuse encore grande dans ces milieux, et aussi le fait que presque tous les prêtres devant être dirigés vers le ministère paroissial, les jeunes gens attirés par un apostolat plus intellectuel s'orienteraient plutôt vers les ordres religieux (Il faut noter pourtant qu'à partir de 1850, on voit augmenter peu à peu le nombre des « prêtres habitués », fixés dans les villes, où ils s'occupent d'étude, de prédication ou de bonnes œuvres). Sans doute faut-il tenir compte également du fait que le sacerdoce diocésain n'apparaissant pas, dans l'Église paisible du Second Empire, comme un apostolat conquérant, les plus généreux préfèrent s'orienter vers les missions. Enfin, il faut noter que dans les collèges libres, où sont élevés les enfants de la bourgeoisie, on ne se préoccupe guère d'encourager les vocations ecclésiastiques, les autorités estimant que les petits séminaires y suffisent.

(4) *Le Correspondant*, t. CVI, 1877, p. 162.

(5) Celui-ci, pour les petits et grands séminaires réunis, passe de 1.699 en 1848 à 3.086 en 1870.

(6) Cf. F. GARILHE, *Le clergé séculier français au XIX° siècle*, Paris, 1898, en particulier p. 39-48.

romaine un contrepoids à la tendance plus autoritaire de beaucoup d'évêques. En effet, si tous les prélats ne se comparent pas aussi crûment que le cardinal de Bonnechose à un général qui sait faire « marcher » son régiment, la plupart des évêques, qui sont d'anciens vicaires généraux, cherchent à centraliser l'administration diocésaine, à renforcer les contrôles, ils multiplient les ordonnances et les règlements concernant la discipline ecclésiastique ou le ministère paroissial, laissant de moins en moins à l'initiative de leurs prêtres [1]. Ils utilisent largement le droit, que leur a donné le concordat, de déplacer les prêtres dont ils sont mécontents. Ils suivent de très près leur clergé non seulement au cours de tournées pastorales plus fréquentes, mais aussi grâce à la généralisation des conférences diocésaines [2].

Plus centralisée à l'intérieur du diocèse, l'administration épiscopale devient aussi plus uniforme à l'intérieur de la France, grâce entre autres aux conciles provinciaux, dont le nombre ne diminue guère pendant les premières années de l'Empire. Il s'établit ainsi « une sorte de pratique commune de l'épiscopat », caractérisée par la reconstruction de nombreuses églises, par la prescription de méthodes plus modernes, telle l'obligation pour les curés de tenir un *status animarum*, par le souci de remédier à l'ignorance religieuse en augmentant le nombre des missionnaires diocésains et en faisant prêcher le catéchisme, qu'on cherche par ailleurs à uniformiser, par l'encouragement aux formes nouvelles de dévotion et aux nombreuses œuvres qui vont en se multipliant.

Cette activité systématique et consciencieuse suffit à prouver que le temps des prélats grands seigneurs est définitivement révolu et que, à quelques exceptions près, les évêques nommés par Louis-Philippe ou Napoléon III sont avant tout des hommes d'Église qui, même lorsqu'ils interviennent dans les affaires publiques d'une manière que certains trouvent un peu abusive, n'ont en vue que les intérêts spirituels de la religion et qui consacrent sans compter à cette tâche leur temps et leurs peines. Malheureusement, on ne retrouve que bien rarement parmi eux, tant à Paris qu'en province, la lucidité dont Mgr Affre avait fait preuve en présence des transformations sociales et intellectuelles qui caractérisent le milieu du xixe siècle. On peut regretter que la hiérarchie ne fasse pas preuve de plus d'imagination pastorale et que, au lieu de repenser profondément les méthodes d'apostolat, elle cherche plutôt à institutionaliser celles qui avaient fait leurs preuves pendant la première moitié du siècle.

L'ESSOR DES CONGRÉGATIONS   L'action des évêques et du clergé diocésain est grandement facilitée par l'appui que leur apportent les congrégations religieuses. Le Second Empire est

---

(1) A titre d'exemple, le cardinal Régnier adressa 96 mandements à ses prêtres au cours des 31 ans qu'il dirigea le diocèse de Cambrai ; en 1855, il publia de nouveaux statuts diocésains extrêmement précis.
(2) Avant 1848, ces conférences n'existaient que dans 15 diocèses. Depuis 1864 une revue mensuelle, *Les Conférences ecclésiastiques*, publiée à Paris, indique les sujets à traiter et publie des réponses modèles. C'est à Paris aussi que paraît la première *Semaine religieuse* hebdomadaire, en 1853.

pour celles-ci une période de grand développement et de prospérité. Elles bénéficient largement de la générosité des classes riches [1] et multiplient les constructions au point que certains raillent cette « maladie de la pierre », tandis que d'autres commencent à reparler de la « main-morte ». Quant à la politique gouvernementale, elle est au début « complaisante sinon complice » [2] et évolue par la suite entre une bienveillance mesurée et une tolérance boudeuse, en dépit de quelques menaces de 1860 à 1865 [3]. Aussi la proportion des religieux, qui était de 1 pour 957 habitants en 1851, passe-t-elle en 1861 à 1 pour 348 et sera en 1877 de 1 pour 250 [4].

Le développement des congrégations de femmes, en particulier, fut favorisé, d'un côté par le décret-loi du 31 janvier 1852, qui, en pratique, n'exigeait plus, pour l'autorisation, qu'un simple décret au lieu d'une loi ; d'un autre côté par le zèle apostolique de nombreux curés ou directeurs de conscience qui orientent systématiquement vers la vocation religieuse les jeunes filles pieuses dont ils ont la charge ; sans doute faut-il faire aussi sa part à l'attrait qu'exerçait sur beaucoup de jeunes paysannes la dignité de la cornette. A côté de quelques congrégations qui prennent de l'ampleur et débordent même les frontières de la France, comme les petites sœurs des Pauvres [5], on constate un pullulement de menues congrégations diocésaines, destinées surtout à l'enseignement primaire féminin et au soin des malades. En vingt-cinq ans, l'effectif total des religieuses aura quadruplé, passant de 34.208 en 1851 à 89.243 en 1861 pour atteindre 127.753 en 1877, malgré la perte de l'Alsace-Lorraine.

De leur côté, les religieux, qui dépassaient à peine les 3.000 en 1851, sont 17.656 en 1861 et 30.203 en 1877. Ici aussi, on assiste à de nouvelles fondations, comme celle des assomptionnistes, par le P. d'Alzon [6], qui tente de reprendre à son compte l'ancien projet de Lamennais d'un ordre voué à l'apostolat intellectuel ; à la résurrection d'ordres anciens, comme celui des carmes, qui s'étend rapidement grâce à l'illustre converti, Herman Cohen, et dont les belles chapelles attirent un public distingué ; enfin au développement de congrégations déjà existantes, tels les frères des Écoles chrétiennes, qui sous la conduite d'un homme de première valeur, le frère Philippe, triplent leurs effectifs en vingt ans [7], ou les jésuites [8], qui demeurent congrégation non autorisée, mais qui, « toujours habiles à s'approcher des véritables détenteurs du pouvoir, cherchent

---

(1) Sur les dons et legs autorisés de 1852 à 1860, 49 % vont à des fabriques d'église, 37 % à des congrégations et 14 % à l'enseignement (c'est-à-dire en fait, également à des congrégations). En 1860, la fortune des congrégations représente le tiers de la fortune totale de l'Église de France.

(2) Ch. POUTHAS, *L'Église et les questions religieuses...*, p. 156.

(3) Même pendant la période où Rouland manifesta à l'égard des congrégations une hostilité systématique, il fut parfois contrecarré par d'autres influences ; comme il était de plus très difficile de les surveiller efficacement, le gouvernement ne put que ralentir leurs progrès, sans parvenir à les arrêter.

(4) Ch. POUTHAS, *op. cit.*, p. 158. On ne possède pas de statistique précise entre 1861 et 1877.

(5) Les débuts remontent à la Monarchie de Juillet, mais c'est sous l'Empire que la congrégation reçut l'existence légale et l'autorisation canonique. Voir A. LEROY, *Histoire des Petites Sœurs des Pauvres*, Paris, 1902 ; et, sur leur grand bienfaiteur et protecteur, qui fit passer le nombre de leurs maisons de 30 à 260, Mgr BAUNARD, *Ernest Lelièvre*, Paris, 1907.

(6) Cf. S. VAILHÉ, *Vie du P. Emmanuel d'Alzon*, Paris, 1927-1934, 2 vol.

(7) Voir G. RIGAULT, *Histoire générale de l'Institut des Frères des Écoles chrétiennes*, t. V, Paris, 1945.

(8) Voir les t. III et IV de J. BURNICHON, *La Compagnie de Jésus en France, Histoire d'un siècle*, Paris, 1919 et 1922.

à acquérir sur la bourgeoisie l'influence qu'ils possédaient autrefois sur les rois et leur entourage » [1]. Il faut enfin ne pas oublier le magnifique développement que commencent à prendre les congrégations missionnaires, dont il sera question dans un autre volume.

LES PROGRÈS DE L'ENSEIGNEMENT CATHOLIQUE   Le développement rapide des congrégations fournit à l'Église le personnel nécessaire pour mener à bien la double entreprise de pénétration de l'enseignement de l'État et de développement de l'enseignement libre en profitant de l'attitude « à la fois protectrice et libérale » du gouvernement impérial.

L'influence chrétienne s'était sensiblement accrue dans l'enseignement public au début de l'Empire, par suite de l'épuration qui frappa les maîtres suspects de tendances révolutionnaires ou irréligieuses, du renforcement du rôle des aumôniers dans les écoles normales, de la présence de nombreux ecclésiastiques dans l'Université, et de la soumission pratique des instituteurs au curé en beaucoup de localités. L'enseignement religieux avait en outre été rendu obligatoire dans les lycées et l'Église obtint même à plusieurs reprises que le gouvernement s'opposât à une orientation trop antichrétienne dans l'enseignement supérieur [2]. Le clergé estima cependant ces avantages insuffisants et chercha, surtout dans les villes et les gros bourgs, à confier à des religieux les écoles publiques. Les populations, qui acceptaient volontiers la chose pour les filles, semblent l'avoir moins souhaitée pour les garçons, mais le clergé pouvait compter sur l'appui des municipalités, dirigées souvent par de grands propriétaires ou des industriels favorables à l'Église, et qui trouvaient un avantage financier dans l'établissement d'écoles congréganistes. De 1850 à 1854, le clergé se fit ainsi céder une cinquantaine de collèges secondaires communaux, tandis que de 1850 à 1863 le nombre d'écoles publiques de garçons confiées aux frères passait de 1.227 à 3.038, représentant 20 pour cent de la population scolaire, et celles de sœurs de 5.237 à 8.061, avec 697.000 élèves contre 636.000 dans les écoles tenues par des institutrices laïques.

En même temps, mettant à profit la loi Falloux et une jurisprudence très favorable en matière de dons et legs, l'Église multipliait les établissements libres. Ici, c'est surtout sur le terrain de l'enseignement secondaire que se porta le principal effort, du moins au début [3]. Dès 1854, on comptait 249 maisons, dont 150 environ ne comprenant que les petites classes, 67 collèges complets tenus par le clergé séculier et 33 autres dirigés par des religieux. Ces collèges devaient « grandement contribuer à changer l'esprit de la bourgeoisie française et à préparer la naissance de nos *bien-pensants* modernes » [4], car les manières et les relations

(1) A. DANSETTE, *Histoire religieuse de la France contemporaine*, t. I, p. 376.
(2) En 1862 et en 1864, à l'occasion du cours de Renan au Collège de France ; en 1868, avec moins de succès, à l'occasion de l'enseignement matérialiste de la Faculté de Médecine.
(3) Voir le *Rapport au Comité de l'Enseignement libre sur l'exécution et les effets de la loi organique du 15 mars 1850*, Paris, 1853, rédigé par le comte Beugnot. Le P. d'Alzon avait fondé en 1851 la *Revue de l'Enseignement chrétien*, pour exposer les moyens de mettre à profit la loi Falloux.
(4) E. BEAU DE LOMÉNIE, *Les responsabilités des dynasties bourgeoises*, t. I, Paris, 1947, p. 138.

que l'on se créait dans ces établissements, dont la clientèle se recrutait pour une bonne part dans l'aristocratie et la haute bourgeoisie, exerçaient un gros attrait sur beaucoup de parents. Pourtant, tandis que l'enseignement des frères des Écoles chrétiennes était d'une très haute tenue, au point que l'Université s'inspira parfois de ses méthodes [1], la qualité de cet enseignement secondaire, qu'il avait fallu rapidement improviser avec des moyens insuffisants, laissait parfois à désirer [2] ; c'est ainsi qu'un rapport de 1867 signalait que les maisons ecclésiastiques « ne font arriver qu'un très petit nombre de leurs élèves aux écoles d'accès vraiment difficile et d'où sortent les hommes qui exercent de l'influence sur les lettres, les sciences et l'industrie : pas un à l'École normale, 15 sur 145 à l'École polytechnique, 22 sur 235 à l'École centrale » [3].

L'importance et la rapidité des progrès de l'Église en matière scolaire inquiétèrent vite le gouvernement. Rouland déjà essaya de limiter ces progrès, mais ce fut surtout Victor Duruy, ministre de l'Instruction publique de 1863 à 1869, « libre penseur jusqu'à la moelle » au dire de son ami Jules Simon, qui, en dépit des protestations réitérées de l'épiscopat, travailla à fortifier l'enseignement public laïque, à le défendre contre la concurrence de l'enseignement libre et à arrêter la conquête des écoles municipales par les congrégations. En multipliant les écoles laïques et en étendant la gratuité, il réussit à diminuer l'importance relative de l'enseignement catholique, mais les progrès acquis par celui-ci furent maintenus et, dans les régions les plus chrétiennes, ils se poursuivirent, bien qu'à une allure ralentie.

Sur un point particulier, l'action laïque de Duruy aboutit à un échec à peu près complet. En 1867, le ministre essaya, avec beaucoup de prudence, d'organiser des cours destinés à compléter la formation intellectuelle des jeunes filles de la bourgeoisie, qui échappait totalement au contrôle de l'État. Mais l'Église, qui avait renoncé à contre-cœur à reconquérir le monopole de l'enseignement des garçons, n'entendait pas se laisser déposséder de celui qu'elle possédait encore de fait. Mgr Dupanloup, approuvé bientôt par quatre-vingts prélats et par Pie IX, entama contre Duruy une polémique extrêmement vive, et le clergé français, à peu près unanime pour une fois, organisa une opposition qui, même dans les régions anticléricales, compromit sérieusement le succès des nouveaux cours : « Autour du petit noyau formé par les filles d'universitaires et de protestants, ils ne groupèrent qu'un nombre d'élèves insuffisant [4]. »

Enfin, l'Église devait, *in extremis*, remporter un dernier succès. En 1867, Pie IX, puis, à sa suite, le cardinal Contarini, avaient suggéré aux évêques français de fonder une Université catholique comme à Louvain. La pénétration des idées rationalistes dans l'enseignement supérieur

---

(1) Voir notamment G. RIGAULT, *Histoire générale de l'Institut des Frères des Écoles chrétiennes*, t. V, p. 348-352.
(2) J. BRUGERETTE, *Le prêtre français et la société contemporaine*, t. I, p. 181. Dans les collèges de jésuites, cependant, on tâchait de rester fidèle aux anciennes traditions classiques. (Cf. J. BURNICHON, *op. cit.*, t. III, p. 463-481.)
(3) Rapport à l'empereur du 21 novembre 1867, dans V. DURUY, *Notes et Souvenirs*.
(4) J. MAURAIN, *op. cit.*, p. 855.

officiel fit rebondir l'idée et Veuillot en fit l'un des articles du programme catholique lors des élections de 1869 : le 25 février 1870, à la
suite d'une campagne de pétitions, le ministre de l'Instruction publique
annonçait l'accord de principe du gouvernement.

LA VITALITÉ DES ŒUVRES Le gouvernement refusa de se laisser
déposséder par l'Église de l'Assistance
publique, comme l'aurait voulu Armand de Melun [1], mais il fit appel à la
collaboration des catholiques dans son œuvre sociale. Aussi vit-on se
développer, à côté des hospices et des hôpitaux, des institutions catholiques de tout genre : associations de bienfaisance et œuvres des malades
pauvres ; sociétés de secours mutuels, spécialement florissantes dans le
Midi ; patronages et autres œuvres, s'adressant aux jeunes travailleurs [2].
Les animateurs les plus en vue de ces dernières œuvres furent les abbés
Timon-David, auteur d'une *Méthode de direction des œuvres de jeunesse* ;
Le Boucher, à qui l'on doit le premier journal des patronages, *Jeune
Ouvrier*, et les premiers congrès de directeurs d'œuvres (Angers, 1858 ;
Paris, 1859) ; Maignen enfin, le fondateur du *Cercle des Jeunes Ouvriers*,
qui rêvait de voir se réaliser en France un « compagnonnage chrétien »
analogue à celui de Kolping en Allemagne.

Certaines congrégations se vouaient spécialement à cette forme d'apostolat, ainsi les Frères de Saint-Vincent-de-Paul de M. Le Prévost [3].
Mais la part de l'élément laïque était prépondérante dans la plupart de
ces œuvres, dont beaucoup étaient des émanations soit de la Société d'Économie charitable, soit de la Société de Saint-Vincent-de-Paul. La Société
d'Économie charitable avait été restaurée en 1855, à l'occasion de l'Exposition universelle. Elle a toujours pour cheville ouvrière le vicomte
Armand de Melun, dont la vie continuera (il meurt en 1877) à se
confondre avec le mouvement charitable en France [4]. C'est sous ses
auspices qu'est créé, en 1859, un « Comité consultatif des Œuvres et
Instituts charitables » destiné à centraliser les multiples initiatives catholiques dans ce domaine en même temps qu'à leur procurer des renseignements, des conseils et l'appui de son influence ; la présidence d'honneur
en est confiée à Mgr de Ségur, dont le rôle est si important dans le mouvement catholique à cette époque. Malheureusement les résultats de
l'immense effort de la Société d'Économie charitable sont plutôt décevants,
faute d'une doctrine sociale sérieusement élaborée. Quant à la Société
de Saint-Vincent-de-Paul, dirigée depuis 1847 par Adolphe Baudon [5],
alors qu'elle ne comptait pas cinq cents conférences au moment du coup

---

(1) A. de Melun avait publié un article programme : *De la liberté de la charité religieuse* dans *Les
Annales de la Charité* du 29 février 1852 et soumis au gouvernement un projet de loi. Les critiques
adressées par l'administration se trouvent aux *Arch. Nat.*, AB XIX, 523.
(2) Voir à leur sujet J.-B. Duroselle, *op. cit.*, IIIe partie, chap. i et ii.
(3) La première idée de J.-L. Le Prévost avait été de grouper des frères en habit laïque pour
s'occuper des malheureux et des apprentis des faubourgs de Grenelle et Montparnasse. C'est en
1852 qu'il en fit une congrégation religieuse. (Voir *Vie de M. Le Prévost*, Paris, 1890.)
(4) Cf. Mgr Baunard, *Vie du vicomte Armand de Melun*, Paris, 1881 et J.-B. Duroselle,
op. cit., IIIe partie, chap. III.
(5) J. Schall, *Un disciple de saint Vincent de Paul au XIXe siècle : Ad. Baudon*, Paris, 1897.

d'État, elle en groupe 1.300 en 1860, ce qui représente environ 30.000 membres [1]. Mais on constate bientôt en beaucoup d'endroits une désaffection de la jeunesse catholique à l'égard de la société, par suite du « caractère rébarbatif » que prennent certaines réunions dirigées par des vieillards [2].

Le gouvernement finit par prendre ombrage de cette « administration charitable à côté des institutions publiques » [3] et, à la fin de 1861, le ministre de l'Intérieur prit prétexte des tendances légitimistes de certains membres influents de la Société de Saint-Vincent-de-Paul pour intervenir dans la direction de celle-ci. Ces mesures, qui désorganisèrent pour plusieurs années la société, mécontentèrent les catholiques et provoquèrent de la part de l'évêque de Nîmes, Mgr Plantier, une protestation extrêmement violente.

Il est d'ailleurs exact que les associations charitables, où se rencontraient les catholiques les plus riches et les plus influents et où les nobles étaient nombreux, avaient parfois tendance à devenir une puissance sociale où certains n'entraient pas seulement par piété, mais par snobisme ou par intérêt. Pourtant leur développement provenait aussi d'un réel réveil religieux et la preuve en est qu'il va de pair avec celui de nombreuses autres œuvres, telle l'œuvre du Calvaire, qui rayonne en dehors de Lyon à partir de 1853, ou l'œuvre de Saint-François-de-Sales, fondée par Mgr de Ségur en 1857 pour lutter contre l'incrédulité et la mauvaise presse. On admirera spécialement l'épanouissement des œuvres missionnaires : Propagation de la Foi, Sainte-Enfance, Œuvre d'Orient, et celui des œuvres de dévotion : adoration perpétuelle, adoration nocturne, unions de prières, congrégations de la Sainte Vierge, etc. Si la bourgeoisie catholique française n'entrevoit pas encore le véritable terrain où doit se poser et se résoudre le problème social, du moins redécouvre-t-elle des valeurs chrétiennes authentiques qu'elle se soucie de plus en plus de faire pénétrer dans sa vie quotidienne [4].

On regrettera toutefois deux choses. D'abord qu'en dépit de la création du « Comité consultatif des Œuvres », il n'y ait guère de direction d'ensemble dans le mouvement catholique et que très peu d'œuvres réussissent à s'étendre à l'échelle nationale ; alors que les catholiques allemands ont leurs assemblées annuelles depuis 1848 et que les catholiques belges cherchent à les imiter à partir de 1863, on ne constate rien de semblable en France. On regrettera surtout que l'efflorescence rapide d'œuvres paroissiales toujours plus nombreuses aboutisse le plus souvent à la constitution d'un « milieu catholique » isolé, en marge de la civilisation profane qui retient de plus en plus l'attention tant des masses que de l'élite intellectuelle.

---

(1) Le chiffre de 100.000 membres, avancé par le ministre Persigny et repris par J. Maurain, est manifestement très exagéré. (Cf. J.-B. Duroselle, op. cit., p. 550.)
(2) *Bulletin de la Société de Saint-Vincent-de-Paul*, 1869, p. 280 et suiv.
(3) Rapport de Hamille à Rouland du 10 décembre 1859, *Arch. Nat.*, F[19] 1.932.
(4) Le souci de promouvoir une vie chrétienne fervente parmi les laïcs anime un nombre croissant d'ecclésiastiques. Parmi les directeurs les plus influents, il faut signaler Mgr de Ségur et Mgr Dupanloup, qui publie entre autres dans ce but, en 1854, sa *Vie de Madame Acarie*.

LA SAINTETÉ SOUS
LE SECOND EMPIRE
L'allure bourgeoise — au plus mauvais sens du mot — du catholicisme français sous le Second Empire, qui a été abondamment relevée ces dernières années, ne peut faire négliger les trésors de vie chrétienne et même de sainteté qui se dissimulent souvent sous cette apparence. Les laïcs éminents que furent Léon Dupont, « le saint homme de Tours », Philibert Vrau ou Camille Féron, « les saints en redingote », dont le procès de béatification est en cours [1], et beaucoup d'autres, prouvent qu'il y avait autre chose que des « chrétiens sociologiques » à cette époque ; Mgr de Ségur ou le P. Chevrier, l'apôtre de la pauvreté sacerdotale, eurent des émules anonymes plus nombreux qu'on ne le croit souvent ; et que dire des nombreux fondateurs et fondatrices de congrégations religieuses, qui furent sans doute aidés par des circonstances extérieures favorables, mais dont l'œuvre ne fut cependant réalisée qu'au milieu d'épreuves de tout genre, qui supposent une abnégation peu commune [2].

Mais le Second Empire n'a pas connu seulement de saintes âmes ; il a eu ses saints au sens propre du mot. Le plus grand d'entre eux, Jean-Marie Vianney, l'humble curé d'Ars (†1859), voit son rayonnement encore accru grâce au développement des chemins de fer — plus de 100.000 pèlerins par an vers la fin ! — et il continue plus que jamais à « rouvrir dans les cœurs voltairiens les sources de la ferveur » [3]. A côté de lui, mentionnons parmi ceux qui ont mérité les honneurs des autels : Michel Garicoïts (†1863), fondateur des Prêtres du Sacré-Cœur-de-Betharam ; Pierre Julien Eymard (†1868), fondateur des Pères du Saint-Sacrement ; Madeleine-Sophie Barat (†1865), la fondatrice des Dames du Sacré-Cœur ; Marie Pelletier (†1868), la fondatrice des Sœurs du Bon Pasteur d'Angers ; Catherine Labouré (†1876), l'humble Fille de la Charité qui est à l'origine de la médaille miraculeuse ; Pauline Jaricot (†1862), l'animatrice de l'Œuvre de la Propagation de la Foi et du Rosaire Vivant.

§ 2. — Les signes avant-coureurs de la crise religieuse [4].

LA RELIGION VÉCUE
DU PEUPLE FRANÇAIS
Tout ce qui précède a montré suffisamment combien la France du Second Empire demeure encore un pays catholique, un pays où la situation de l'Église doit sans conteste être qualifiée de brillante. Mais

(1) P. JANVIER, *Léon Dupont, le saint homme de Tours*, Tours, 1882, 2 vol. ; Mgr BAUNARD, *Les deux frères. Cinquante ans de l'Action catholique dans le Nord*, Paris, 1910.
(2) Voir notamment Ch. POULET, *La sainteté française contemporaine*, Paris, 1946 ; Mgr D'HULST, *La vie surnaturelle en France au XIXe siècle*, dans *La vie chrétienne dans l'histoire*, Paris, 1896, l. X, chap. v.
(3) M.-H. VICAIRE, dans *Histoire illustrée de l'Église*, sous la direction de G. DE PLINVAL, t. II, p. 270.
(4) BIBLIOGRAPHIE. — Sur la pratique religieuse du Second Empire, consulter, d'une part, les rapports des préfets, recteurs et procureurs généraux conservés aux Archives Nationales et utilisés par J. MAURAIN, *op. cit.*, p. 238-324, d'autre part, les observations des prédicateurs de missions (notamment celles publiées à partir de 1862 dans la revue *Les Missions de la Congrégation des Oblats de Marie Immaculée*) utilisées par G. LE BRAS, dans les *Notes de statistique religieuse* de la *R. H. E. F.*, depuis 1933. On trouvera un tableau utile quoique tendancieux dans E. VACHEROT, *La Religion*, Paris, 1868.
Sur les progrès de l'anticléricalisme, l'ouvrage fondamental est celui de G. WEILL, *Histoire de*

les institutions ne sont qu'un cadre. Que valent la foi et la religion vécue du peuple français ? La situation varie suivant les régions et aussi suivant les classes sociales. Dans l'ensemble cependant, elle commence à devenir inquiétante.

LE RECUL DE LA PRATIQUE DANS LES CAMPAGNES D'importantes régions de la France conservent une immense majorité de pratiquants et même une ferme dévotion. C'est le cas de l'Ouest (Basse-Normandie, Bretagne, Maine, Anjou, Vendée), où depuis longtemps le clergé uni à la noblesse dirige la vie publique comme la vie privée des fidèles ; du Nord, qui devient à cette époque l'un des grands centres du catholicisme français [1] ; de l'Alsace et de la Lorraine, où le clergé s'efforce de protéger ses ouailles contre les mauvaises influences en s'opposant à l'expansion du français [2] ; enfin des régions montagneuses, isolées et pauvres.

Parfois cependant, on peut se demander ce que recouvre exactement la pratique extérieure du culte et, après avoir fait la part de l'esprit partisan, on peut admettre qu'il y a du vrai dans des remarques comme celle que fait en 1859 le préfet des Landes : « Le Landais pratique très exactement ses devoirs. Il fait le dimanche 20 km. et plus à travers la lande, en tous les temps, pour aller à la messe, mais est-ce l'heure qu'il passe à l'église ou la journée qu'il perd au cabaret qui l'attirent ? [3] »

D'ailleurs, dans une bonne partie des régions rurales, la déchristianisation progresse à partir de 1850 : « C'est alors que commence vraiment l'époque moderne... Le peuple se détache de ses coutumes et de ses costumes, de ses superstitions et de ses routines, et de ses mœurs chrétiennes déjà ébranlées [4]. » Bien des évêques pourraient répéter, après celui de Chartres, que si « la foi n'est pas encore éteinte dans ces âmes droites et simples », par contre, « ses observances et ses pratiques sont abandonnées par le plus grand nombre » [5]. Si l'action patiente d'un curé d'Ars ou d'un curé du Mesnil-Saint-Loup réussit à transformer quelques paroisses, ce qui frappe surtout c'est le résultat très limité des nombreuses missions, qui ont repris depuis 1848 sous l'impulsion de l'abbé Combalot et dont se chargent spécialement les Eudistes, les Lazaristes, les Oblats de Marie et, en Vendée, les Montfortains [6]. Le cas n'est pas

l'idée laïque en France au XIXe siècle, Paris, 1929. Voir aussi J. MAURAIN, *op. cit.*, chap. XXIII et XXVI et E. LECANUET, *L'Église de France sous la IIIe République*, t. I, Paris, 1907, chap. I.

(1) Deux prélats énergiques, Mgr Parisis à Arras, Mgr Régnier à Cambrai, ont beaucoup contribué à ce résultat. (Voir Ch. GUILLEMANT, *P.-L. Parisis*, t. III, Paris, 1925 et C.-J.-B. DESTOMBES, *Vie de S. Ém. le cardinal Régnier*, Paris, 1885.) De même, sur un autre plan, un frère Adrien, un frère Messien, un frère Eleuthérius.

(2) On trouvera quelques détails dans J. MAURAIN, *op. cit.*, p. 271-276, 768-770. Voir aussi G. MAY, *La lutte pour le français en Lorraine avant 1870*, Paris, 1912. Cette attitude a parfois fait juger sévèrement l'évêque de Strasbourg, par ex. F. HAUVILLER, *Un prélat germanisateur dans l'Alsace française : Mgr Rass*, dans *Revue historique*, t. CLXXIX, 1937, p. 98-121 ; il était en réalité plus francophile qu'on ne le dit souvent. (Cf. F. L'HUILLIER dans *Études alsaciennes*, Strasbourg, 1947, p. 245-262.)

(3) Lettre du 21 juin 1859 aux *Arch. Nat.*, F19 2480.

(4) E. SEVRIN, *La pratique des sacrements et observances au diocèse de Chartres sous l'épiscopat de Mgr Clausel de Montals (1824-1852)*, dans *R. H. E. F.*, t. XXV, 1939, p. 319.

(5) Cité *ibid.*, p. 342.

(6) Ces derniers, toutefois, qui se recrutent sur place et sont très populaires, continuent avec succès l'œuvre de restauration de la vie religieuse qu'ils poursuivent depuis le XVIIIe siècle.

exceptionnel de ces huit communes du diocèse de Soissons où, sur 6.357 habitants, 1.110 femmes et seulement 126 hommes se confessent au cours de la mission [1].

Beaucoup, du moins parmi les hommes, se contentent désormais d'un conformisme saisonnier : baptême, administré généralement dans la semaine qui suit la naissance ; première communion, entre dix et treize ans ; mariage, et enterrement religieux [2]. Si les exceptions sur ces points sont très rares et font scandale, l'observance dominicale, et plus encore le précepte pascal, sont très négligés. Déjà alors, la première communion est en beaucoup d'endroits la dernière, « une fin d'école, une fin de catéchisme, une fin de religion » [3]. Dans le diocèse de Soissons, à peine 5 pour cent des hommes font leurs Pâques, et les femmes elles-mêmes sont une minorité [4]. Mgr Dupanloup considère comme un beau résultat d'avoir fait remonter les communions pascales de 30.000 à 40.000 dans son diocèse qui compte 350.000 âmes [5]. Il y a même, dans les Charentes, de vrais pays de mission où l'on trouve des paroisses ne comptant pas un seul pascalisant, dont la majorité des hommes, y compris parfois le sacristain, n'ont pas fait leur première communion et dont la moitié des habitants ne sont pas mariés religieusement [6].

Cette baisse religieuse, qui se poursuit en dépit de la multiplication des écoles catholiques, s'explique entre autres par l'action néfaste de la presse : journal local anticlérical, romans d'Eugène Sue, brochures impies ou ordurières, qu'on trouve au cabaret. Mais il faut également tenir compte d'autres facteurs : l'insistance du clergé à obtenir sans cesse des communes de nouveaux subsides pour la paroisse [7] ; le caractère négatif de la religion prêchée par beaucoup de curés qui condamnent sans nuance bals, cabarets et amusements de toute espèce considérés, dans les régions dites « évoluées », non plus seulement comme des choses plaisantes, mais comme étant dans la ligne du Progrès et de la libération de l'individu ; enfin la collusion du clergé de campagne et des châtelains légitimistes, qui entretient l'inquiétude de ceux qui ont profité de la Révolution et qui provoque un raidissement croissant chez les bénéficiaires de l'évolution économique et sociale : petits propriétaires, dans les régions vinicoles notamment, ou petite bourgeoisie rurale, marchands de biens, vétérinaires, boutiquiers.

L'HOSTILITÉ CROISSANTE DES MILIEUX OUVRIERS Les milieux ouvriers avaient commencé bien plus tôt à s'éloigner de l'Église et déjà sous la Monarchie de Juillet la plupart des travailleurs avaient cessé de fréquenter les sacrements, du moins dans les

---

(1) *R. H. E. F.*, t. XXIV, 1938, p. 321. A Chézy-sur-Marne, en 1869, 3 hommes se confessent sur 1.264 habitants.
(2) G. Le Bras, *Introduction à l'histoire de la pratique religieuse en France*, t. I, Paris, 1941, p. 104-106.
(3) E. Sevrin, *art. cit.*, p. 327.
(4) *R. H. E. F.*, t. XXIV, 1938, p. 321 ; t. XXVI, 1940, p. 70.
(5) *Janssens Briefe*, édit. Pastor, t. I, p. 230.
(6) *R. H. E. F.*, t. XXVI, 1940, p. 70-71 ; J. Maurain, *op. cit.*, p. 311-312 ; *Correspondance du cardinal Pie et de Mgr Cousseau*, édit. Hiou, Paris, 1894, *passim*.
(7) Cette dernière cause est bien mise en lumière par E. Dupont, *La part des communes dans les frais du culte paroissial pendant l'application du Concordat*, Paris, 1906.

villes. « L'immense majorité des Parisiens ne pratique pas », relevait l'ambassadeur d'Autriche à Pâques 1851 [1]. La même constatation était faite cinq ans plus tard par un prêtre en contact avec les milieux populaires de la capitale : « Près d'un million de travailleurs et de petits commerçants vivent pour la plupart éloignés des instructions et des pratiques de la religion [2]. »

Pourtant, si l'on met à part Lyon et Saint-Étienne, où les ouvriers se montraient dès 1840 formellement hostiles [3], on put constater en 1848 que les masses ouvrières restaient animées de sentiments religieux et même de certaines sympathies pour les prêtres. Cet atavisme chrétien se maintient encore chez beaucoup pendant le Second Empire, surtout en dehors des grands centres, et la foi semble généralement plutôt endormie qu'inexistante. Une évolution s'amorce toutefois au lendemain de la répression violente des émeutes de juin 1848 : dès septembre 1850 le journal de Proudhon rompt avec le socialiste catholique Chevé et la correspondance d'Armand de Melun à l'époque de l'Assemblée législative contient de nombreuses allusions au développement rapide d'un anticléricalisme ouvrier [4]. Cette situation nouvelle se précise au cours des années 1860-1870, qui voient se confirmer l'apostasie de la classe ouvrière. L'incrédulité voltairienne gagne maintenant les ateliers, notamment là où les ouvriers ont le temps de réfléchir : tailleurs, cordonniers, etc. Mais le facteur principal de cette évolution semble avoir été l'attitude de la bourgeoisie catholique, dont l'indifférence en matière sociale ou les allures paternalistes déçoivent profondément l'élite du monde ouvrier ; celle-ci commence alors à remplacer la foi religieuse par la foi au progrès social [5] et s'habitue à ranger les prêtres parmi ses principaux adversaires dans sa lutte contre les forces du conservatisme réactionnaire [6].

Des prêtres et des laïcs à l'âme d'apôtre essaient vainement d'enrayer cette apostasie des masses. Des œuvres sont fondées pour protéger les jeunes travailleurs ; une littérature populaire très abondante de propagande religieuse se développe « à l'ombre de Mgr de Ségur » [7], dont les *Réponses aux objections contre la Religion* sont répandues à 700.000 exemplaires de son vivant. Mais cette action, qui n'est pas sans résultat — témoin le succès rencontré par l'hebdomadaire *L'Ouvrier* — reste pourtant

---

(1) Von Huebner, *Neuf ans de souvenirs d'un ambassadeur d'Autriche à Paris*, t. I, Paris, 1904, p. 15-16. Il ajoutait avec beaucoup de perspicacité : « Pendant la semaine sainte, les églises regorgeaient de monde ; seulement leur nombre est comparativement petit. » La situation est la même à Lyon. (Cf. Proudhon, *Correspondance*, t. II, p. 134.)

(2) Mémoire adressé à Mgr Sibour et cité dans *R. H. E. F.*, t. XXXIV, 1948, p. 55-56.

(3) Notamment parce qu'ils reprochaient aux ateliers dirigés par des religieux, et destinés à pourvoir à l'instruction professionnelle des enfants pauvres, de contribuer à avilir les salaires en vendant leur production à bas prix. (Cf. Ch. Dutacq, *Histoire politique de Lyon pendant la révolution de 1848*, Paris, 1910, p. 64-75 et 107 et suiv.)

(4) J.-B. Duroselle, *Les débuts du catholicisme social en France*, Paris, 1951, p. 494.

(5) C'est ce qui apparaît bien dans l'ouvrage d'un ancien collaborateur de *L'Atelier*, Anthime Corbon, *Le Secret du peuple de Paris*, Paris, 1863.

(6) Voir E. Dolléans, *Histoire du travail*, Paris, 1943, p. 171 et G. Duveau, *La Vie ouvrière en France sous le second empire*, Paris, 1946, ch. iv (les mœurs), p. 419-537, *passim*.

(7) J.-B. Duroselle, *Les débuts du catholicisme social en France*, p. 686-689, qui suggère l'intérêt qu'il y aurait, pour faire une étude exhaustive de la propagande catholique auprès des masses, à examiner les listes de livres pour les bibliothèques paroissiales publiées dans le *Bulletin de la Société de Saint-Vincent-de-Paul*.

superficielle ; elle est incapable d'atteindre réellement le prolétariat, qui se constitue progressivement en une classe à part et se sent de plus en plus étranger, sociologiquement et psychologiquement, au milieu catholique traditionnel. Dès 1856, un prêtre clairvoyant le reconnaissait : « La tâche imposée à notre siècle, c'est le retour à la religion des classes populaires... Si cruel qu'en soit l'aveu, il faut pourtant bien en convenir, les travaux sont grands et le succès d'une nullité désolante... A part quelques rares individualités que l'on parvient à enlever de temps à autre, la masse est inébranlable et ne se laisse point entamer [1]. »

LA MENTALITÉ RELIGIEUSE DANS LA BOURGEOISIE   Si la situation religieuse est inquiétante à la campagne et franchement mauvaise dans le prolétariat ouvrier, elle paraît meilleure dans les hautes classes. Au cours du Second Empire, on assiste dans la bourgeoisie aisée à une évolution analogue à celle qui s'était produite un demi-siècle plus tôt dans l'aristocratie : les nobles étaient revenus à la pratique religieuse par esprit de tradition ; les bourgeois riches, qui, sous la Monarchie de Juillet, éprouvaient pour l'Église un mépris supérieur, saluent en elle, depuis la grande peur de 1848, une puissante force de conservation sociale et leurs enfants subiront fortement l'influence des collèges libres et des congrégations religieuses qui exercent leur ministère dans la bonne société : l'échec de Duruy dans sa tentative d'organiser un enseignement laïc pour jeunes filles est assez significatif.

Il faut cependant se garder des illusions. Cette société, où les chrétiens d'élite ne manquent du reste pas, est souvent « plus cléricale que croyante » [2]. S'ils apprécient pour les autres l'utilité sociale de la religion, beaucoup de ses membres ne s'embarrassent guère de la pratique intégrale, ni même de la morale chrétienne ; ils souhaitent voir le catholicisme évoluer dans un sens plus libéral et leur vie familiale reste le plus souvent viciée par l'institution du mariage d'argent dont l'importance va croissant.

Dans la moyenne bourgeoisie, la situation est encore moins satisfaisante. Le goût de la vie facile, des jouissances matérielles et du luxe se développe et ne tarde pas à se marquer dans les chiffres de la natalité. La baisse de la moralité est sensible de l'Exposition universelle de 1855 à celle de 1867 : « En dansant sur des flonflons d'Offenbach le quadrille de la *Vie Parisienne*... la France de l'Exposition étale avec ostentation sa décadence et sa légèreté [3]. » Assurément, le voltairianisme se démode d'une génération à l'autre et la religion, en province surtout, commence à apparaître comme une marque extérieure de la respectabilité. Mais si les fonctionnaires fréquentent davantage l'église sous Napoléon III, « c'est pour plaire à leurs chefs plutôt qu'à Dieu » [4] et les professions

---

(1) L'abbé PICHERIT, dans *Le Jeune Ouvrier*, t. I, 1856, p. 148-180.
(2) J. MAURAIN, *op. cit.*, p. 860.
(3) M. BOURCET, *Psychologie de quatre-vingts ans d'expositions françaises*, dans *Études*, t. CCVIII, 1931, p. 535 et 537.
(4) G. LE BRAS, *Introduction à l'histoire de la pratique religieuse en France*, t. I, Paris, 1941, p. 105.

libérales continuent de se tenir à distance. La médecine en particulier apparaît comme le fief du matérialisme et l'athéisme de la Faculté de Médecine à la fin de l'Empire cause de sérieuses inquiétudes à l'épiscopat, de même du reste que l'hostilité contre toute forme de religion qui s'affirme parmi les étudiants en général à partir de 1865 [1]. L'Église perd du terrain dans le monde intellectuel.

L'ÉVOLUTION DE LA LITTÉRATURE ET DE LA PHILOSOPHIE
C'est que, en France comme en Allemagne, l'atmosphère relativement favorable à la religion engendrée par le romantisme dans les milieux cultivés se transforme rapidement ; à partir de 1850, les progrès de la science et de la technique orientent les esprits vers une conception plus matérialiste de la vie.

En littérature, le réalisme s'affirme avec Flaubert, le positivisme inspire les *Poèmes antiques* et les *Poèmes barbares* de Leconte de Lisle. Victor Hugo fait des *Misérables* une « grande calomnie par réticence » contre l'Église [2] qui n'aurait rien fait pour combattre l'ignorance et le paupérisme, et, dans ses poésies, il oppose aux dogmes périmés et à la morale étouffante du catholicisme un spiritualisme (voire même un spiritisme) humanitaire qui jouira d'une popularité durable dans les milieux petits-bourgeois et parmi les instituteurs de la Troisième République. La Congrégation de l'*Index* est alertée et, après avoir condamné en décembre 1863 les œuvres de George Sand, elle englobe, le 27 juin 1864, dans un décret collectif les œuvres les plus marquantes de la littérature française du temps : *Les Misérables*, *Madame Bovary*, *Salammbô*, *Le Rouge et le Noir* et le reste de l'œuvre de Stendhal, les romans de Balzac, ceux de Mürger, et quelques œuvres légères moins connues [3].

Ce ne sont du reste pas seulement les littérateurs, mais les savants comme Berthelot ou Paul Bert, les philologues comme Renan ou Havet, les philosophes comme Littré ou Vacherot, les critiques comme Taine ou Sainte-Beuve, qui s'attaquent ouvertement aux dogmes chrétiens. Ils prennent pied à l'Institut, malgré l'opposition de Mgr Dupanloup, et dans des organes influents comme la *Revue des Deux Mondes* [4] ou la *Revue de Paris* et y écrivent en faveur de la libre pensée contre l'oppression des consciences symbolisée par le *Syllabus*, tandis que Sainte-Beuve défend la même cause au Sénat, où il parle un jour du « grand diocèse qui s'étend par toute la France…, qui gagne et s'augmente sans cesse…, qui compte par milliers des déistes, des spiritualistes et adeptes de la religion dite naturelle, des panthéistes, des positivistes… et des sectateurs de la science pure » [5].

---

(1) J. MAURAIN, *op. cit.*, p. 754-756, 838-839, 856-857, 861.
(2) L'expression est de L. GAUTIER, *Études littéraires pour la défense de l'Église*, Paris, 1865, p. 95.
(3) Texte du décret dans *La Civiltà cattolica*, sér. V, vol. XI, 1864, p. 227-228.
(4) Après 1867 cependant, celle-ci se rapproche des catholiques libéraux. (Voir l'ouvrage commémoratif : *Un siècle de vie française à la Revue des Deux Mondes*, Paris, 1929.)
(5) *Journal officiel*, 19 mai 1868.

LES PROGRÈS DE L'IDÉE LAIQUE ET DE L'ANTICLÉRICALISME — A la révolution qui s'accomplit dans les esprits au détriment des idées religieuses vient s'ajouter, dans les milieux de gauche, une renaissance de l'anticléricalisme qu'expliquent suffisamment les complaisances de l'Église à l'égard du régime autoritaire issu du 2 décembre. On s'offusque à la fois d'une alliance qui semble, à en juger par les commentaires de Veuillot, « ouvertement conclue contre la liberté »[1], et de la tendance à vouloir imposer d'autorité les principes de l'Église à la société civile. Dès lors, un nombre croissant de Français, confirmés encore dans leurs opinions par des incidents comme l'affaire Mortara, croient déceler chez le clergé une volonté de s'opposer au progrès ou à la civilisation moderne et ils viennent grossir les rangs de ceux qui avaient représenté l'esprit laïque sous la Seconde République et ressenti la loi Falloux comme un camouflet : anciens collaborateurs de la revue *La liberté de penser* ; anciens étudiants du Quartier Latin qui, entraînés par des Normaliens comme Taine, About ou Sarcey, se donnaient jadis rendez-vous au cours de Michelet en entonnant la chanson de Béranger : *Hommes noirs, d'où sortez-vous ?* ; instituteurs socialistes ou lecteurs de *L'Enseignement du peuple* d'Edgar Quinet, où l'on trouve déjà tout le programme de l'école laïque[2]. Tout ce monde, exaspéré par la « réaction cléricale » des premières années de l'Empire, accueille avec faveur les ouvrages où Proudhon et Vacherot, le premier avec une fougue de pamphlétaire, le second avec l'austérité d'un philosophe, dénoncent l'incompatibilité entre la démocratie, fondée sur la liberté et la justice, et le catholicisme, fondé sur l'autorité et l'arbitraire[3].

Les conflits violents provoqués par la question romaine, où s'opposent au fond la conception laïque et la conception catholique traditionnelle de l'État, ainsi que les compromissions fréquentes du clergé avec la cause légitimiste, qui apparurent au grand jour à cette occasion, contribuèrent à fortifier l'impression d'une antipathie foncière des milieux catholiques pour « les principes qui forment la base de la société moderne ». On le vit bien à l'occasion des élections de 1863, où les catholiques étaient dans l'opposition, lorsque certains d'entre eux envisagèrent une alliance avec les Républicains, sous le nom d'Union libérale. Aux démocrates tentés par cette proposition, Alphonse Peyrat répondit :

Réfléchissez-y bien et vous verrez qu'il n'existe aucun terrain où puisse s'opérer une coalition entre le parti réactionnaire et le parti démocratique. Il y a longtemps que vous discutez contre les légitimistes et les cléricaux : les trouvez-vous aujourd'hui plus libéraux qu'il y a dix ans ?... Au fond, croyez-le bien, ils sont toujours les mêmes, aimant ce que nous détestons, détestant ce que nous aimons. Pour résumer ma pensée en un seul mot, *c'est là l'ennemi*[4].

Reprise par Gambetta en 1876, la formule devait devenir le cri de guerre du parti républicain. Détail significatif : l'adjectif « anticlérical »,

(1) G. WEILL, *Histoire de l'idée laïque en France au XIXe siècle*, p. 124.
(2) *Ibid.*, chap. v.
(3) P.-J. PROUDHON, *De la justice dans la Révolution et dans l'Église*, Paris, 1858 ; E. VACHEROT, *La Démocratie*, Paris, 1859. Les deux ouvrages furent condamnés par les tribunaux.
(4) Lettre ouverte à Nefftzer, publiée dans *Le Temps* du 9 avril 1863.

9

qui apparaît pour la première fois en 1852, se répand à partir de 1859, et c'est entre 1860 et 1863 que le mot « cléricaux » commence à être employé substantivement [1].

Les adversaires de l'Église peuvent compter sur une presse dont les chiffres du tirage, considérables par rapport à ceux des autres journaux et en augmentation constante, montrent les progrès de l'opinion anticléricale dans le pays. A côté du *Journal des Débats*, organe des salons orléanistes, qui recommence à lancer l'idée d'une séparation de l'Église et de l'État, oubliée depuis un quart de siècle, deux journaux parisiens sont à la tête du combat : *La Presse*, d'Émile de Girardin, dont les collaborateurs pour les questions religieuses sont nettement hostiles au catholicisme, et surtout *Le Siècle*, « le journal quasi officiel de l'anticléricalisme », que l'on trouvait dans les cafés et les cabarets, ce qui augmentait encore le nombre de ses lecteurs [2]. Plus tard viendront s'ajouter encore trois autres feuilles, influencées par le protestantisme libéral : *L'Opinion nationale* de Guéroult, *Le Temps* de Nefftzer et *L'Avenir national* de Peyrat.

A la fin de l'Empire, les anticléricaux commencent en outre à avoir leurs associations : la franc-maçonnerie passe peu à peu sous leur contrôle et, en 1866, un franc-maçon républicain, Jean Macé, fonde la Ligue de l'Enseignement [3], destinée à défendre et à propager l'enseignement laïque. Malgré les condamnations retentissantes de Mgr Dupont des Loges et de Mgr Dupanloup, elle se répand rapidement en province, où il y a bientôt toute une organisation de conférences, de bibliothèques populaires, de colportage de tracts, et où se développe parallèlement une propagande en faveur des enterrements civils, qui effraie le clergé.

LES DIVISIONS DES CATHOLIQUES   En dépit d'une situation apparemment brillante, l'Église de France a donc à faire face, pendant le Second Empire, à un début de crise dont on ne peut sous-estimer la gravité. Or elle se trouve, pour l'aborder, dans des conditions défavorables, par suite de l'intensification des dissentiments entre catholiques, qui se font à présent jour dans tous les domaines. L'allure agressive des champions de l'ultramontanisme mécontente les partisans des coutumes traditionnelles de l'ancien clergé de France et les pousse à appuyer la politique gallicane du gouvernement, dont les conséquences sont cependant parfois dommageables pour la religion [4]. Devant la baisse de la pratique religieuse et les progrès de l'anticléricalisme, les uns ne voient de salut que dans une protection plus marquée de la part du gouvernement ; pour les autres, au contraire, seule une indépendance

---

(1) G. WEILL, *op. cit.*, p. 179 et n. 1 ; J. MAURAIN, *op. cit.*, p. 960.

(2) Son tirage était de 22.521 en juillet 1853, 35.677 en janvier 1855, 44.000 en juillet 1866. En 1855, *L'Univers* tirait à 3.500 exemplaires et l'ensemble des journaux de tendance catholique de Paris et de province atteignait 30.000 exemplaires contre 90.000 pour la presse anticléricale. (Voir G. WEILL, *op. cit.*, p. 129-134 ; statistiques dans J. MAURAIN, *op. cit.*, p. 159-161, 171,749.)

(3) Voir J. MACÉ, *Les origines de la ligue de l'enseignement (1867-1870)*, Paris, 1891 ; A. DESSOYE, *Jean Macé et la fondation de la Ligue de l'Enseignement*, Paris, 1883.

(4) C'est ainsi que plusieurs des évêques nommés par Baroche étaient loin de présenter toutes les garanties désirables. Même des évêques hostiles au progrès de l'ultramontanisme s'en inquiétaient. (Cf. J. PAGUELLE DE FOLLENAY, *Vie du cardinal Guibert*, t. II, Paris, 1896, p. 420-421.)

plus grande de l'Église à l'égard du pouvoir civil lui permettrait de retrouver l'estime des nouvelles élites intellectuelles. Intransigeants et libéraux s'accusent mutuellement de l'aggravation de la situation. Les premiers estiment que les concessions au libéralisme sont la cause de tout le mal : « Ils sont effrayants, écrit l'un d'eux, avec leur culte violent des libertés et leur grotesque vénération pour la raison, le progrès, que sais-je ? Des évêques y perdent les mâles accents et l'intelligence surnaturelle [1]. » Les seconds gémissent au contraire de ce que « en peu de temps les intempérants, les exagérés ont dissipé ce qui semblait définitivement acquis à la cause de Dieu. Ce que les Maret, les Lacordaire, les Ozanam avaient réussi à capitaliser, les fous furieux l'ont dilapidé » [2].

D'accord pour dénoncer l'intégrisme ultramontain, dont les figures marquantes sont Mgr Gerbet, Mgr Parisis, Mgr Mabile, Mgr Berteaud, et de plus en plus Mgr Pie, sans oublier Louis Veuillot, les libéraux ne s'entendent même pas entre eux : les amis de Mgr Dupanloup et de Montalembert s'affichent en toute occasion comme les adversaires irréconciliables de Napoléon III et les champions du pouvoir temporel du pape ; mais beaucoup d'autres, et notamment les prélats sortis du groupe de Mgr Maret, avec à leur tête Mgr Darboy, archevêque de Paris depuis 1863, répondent que les droits de la conscience moderne, ignorés par les institutions archaïques de l'État pontifical, sont suffisamment sauvegardés dans la constitution impériale.

Absorbés par ces querelles intestines, la plupart ne se rendent pas compte que l'avenir, non seulement du catholicisme mais de la religion, est en jeu et qu'une adaptation urgente s'imposerait sur le double terrain intellectuel et social, où le retard reste considérable [3].

(1) Mgr Berteaud au cardinal Pitra, été 1863, dans R. FAGE, *Lettres inédites de Mgr Berteaud au cardinal Pitra*, Brives, 1919, p. 5.
(2) Mgr Meignan à Mgr Maret, 2 janvier 1873, dans H. BOISSONOT, *Le cardinal Meignan*, Paris, 1899, p. 260-261.
(3) Il sera plus longuement question de ces différents points dans les chapitres ultérieurs : chap. VII, p. 211-217, pour l'infériorité intellectuelle du catholicisme français ; chap. VIII, p. 229-236 et p. 249 et suiv., *passim*, pour la question du catholicisme libéral ; chap. IX, p. 273-276, p. 298-300 et p. 305-309, pour celle de l'ultramontanisme, et chap. XIV, p. 492-493 pour l'attitude sociale des catholiques.

CHAPITRE V

L'ÉGLISE DANS LES PAYS GERMANIQUES
DE 1850 A 1870

§ 1. — L'Église dans l'Empire des Habsbourg [1].

LE CONCORDAT DE 1855 Les ordonnances d'avril 1850 étaient loin d'avoir satisfait les promoteurs de la restauration catholique en Autriche, qui réclamaient avant tout une profonde modification de la législation matrimoniale, imprégnée de l'esprit du XVIIIe siècle. Rauscher, qui était revenu à la charge dès les premiers mois de 1851, réussit à convaincre le gouvernement impérial de conclure un accord avec Rome à ce sujet, en faisant du reste espérer d'assez larges concessions, et les négociations officielles s'engagèrent au début de 1853.

Très vite, il apparut qu'on serait entraîné beaucoup plus loin qu'on ne l'avait prévu d'abord. D'une part, en effet, Rauscher et tous ceux qui avaient subi l'influence du romantisme catholique allemand désiraient voir l'Autriche revenir à l'idéal médiéval de l'État chrétien organisant selon les vues de l'Église toute la vie sociale et familiale, et souhaitaient dès lors un accord qui, au delà des questions particulières comme celle du mariage, porterait sur l'ensemble des rapports entre l'Église et l'État ; Rauscher estimait en outre qu'un accord général, valable pour toutes les parties de l'Empire, y compris la Hongrie, contribuerait à renforcer l'unité de la monarchie et assurerait l'appui de l'Église à la politique de restauration menée depuis 1848. D'autre part, la Curie romaine était engagée au même moment dans des négociations analogues avec d'autres pays ; comme elle se rendait compte que l'empereur, poussé par sa mère, l'archiduchesse Sophie, tenait à tout prix à aboutir, elle voulut profiter de cet avantage pour faire du concordat autrichien un concordat modèle, entièrement conforme aux exigences de la doctrine catholique et du droit canon. Aussi non seulement repoussa-t-elle jusqu'à la fin tous les amendements suggérés par Rauscher, mais elle insista pour que les diverses stipulations apparussent explicitement non comme des concessions de l'État, mais comme l'application des principes ultramontains clairement formulés et officiellement reconnus par l'Autriche.

(1) BIBLIOGRAPHIE. — G. WOLFSGRUBER, *Kirchengeschichte Œsterreichs-Ungarn*, Vienne, 1909, p. 71-130 (avec une bibliographie détaillée, p. 148-184) ; J.-F. v. SCHULTE, *Lebenserinnerungen*, Giessen, 1908-1909, 3 vol. (tendancieux mais bien documenté) ; *Coll. lac.*, t. V. En outre, les biographies de Rauscher et de Schwarzenberg, citées p. 133, n. 5 et 134, n. 2.
Sur le Concordat : M. HUSSAREK, *Die Verhandlung des Konkordats von 18 Aug. 1855*, dans *Archiv für oesterreichische Geschichte*, t. CIX, 1922, p. 447-811 (très nombreux documents) à nuancer par H. SINGER, *Das Konkordat von 1855*, dans *Mitteilungen des Vereines für Geschichte der Deutschen in Böhmen*, t. LXII, 1924, p. 95-116 et 165-262.
Sur l'Église en Hongrie : J. KARACSONYI, *Magyarország egyháztörténete főbb vonàsaiban 970 -től 1900 -ig*, Nagy-Vàrad, 1915, IVe P. ; voir aussi J. SIMOR, *Epistolae pastorales et instructiones selectae ab a. 1857-1882*, Estergom, 1882-1883, 5 vol.

L'humiliation paraissait grande à tous ceux — et ils étaient nombreux, même dans l'épiscopat — qui conservaient des sympathies pour le système joséphiste, mais après deux ans et demi de tractations difficiles, Rauscher, le délégué officiel de l'empereur, qui avait cédé à toutes les exigences romaines, réussit à forcer la main aux autorités de Vienne, peut-être un peu par souci personnel de réussite, mais surtout dans la conviction de servir à la fois son Église et sa patrie. Le Concordat, signé le 18 août 1855, reconnaissait les grands principes défendus par l'école ultramontaine : la primauté de juridiction du pape et la valeur autonome du droit ecclésiastique admis comme ayant force de loi ; il soulignait la position nettement privilégiée de l'Église dans l'État, confirmant entre autres la surveillance par les évêques de l'enseignement à tous les degrés, déclarant les tribunaux ecclésiastiques seuls compétents pour toutes les affaires matrimoniales, et limitant assez sensiblement les droits des non-catholiques.

On devine le mécontentement de l'opinion libérale, qui commençait alors à s'éveiller en Autriche : « J'eus l'impression, déclarait par la suite Auersperg, d'un Canossa où l'Autriche du XIXe siècle venait sous le sac et la cendre faire pénitence pour le joséphisme du XVIIIe siècle [1]. » Mais c'est à l'étranger surtout, où il fut considéré comme « un coup de massue sur la tête de la France » [2], championne des principes de 1789, que ce « concordat du moyen âge » [3] fit sensation. C'était une grande victoire apparente pour le Saint-Siège, mais il tenait trop peu compte des principes politiques modernes et, surtout dans sa formulation, donnait trop l'impression de s'inspirer d'une conception théocratique de la société pour ne pas inquiéter les esprits prudents qui estimaient qu'il eût mieux valu se contenter d'un succès moins spectaculaire et procéder progressivement à l'élimination de ce qui restait de joséphisme dans la législation autrichienne. L'avenir devait justifier ces craintes, mais le concordat eut en tout cas le grand avantage de rendre à l'Église autrichienne une réelle autonomie par rapport à la bureaucratie gouvernementale, ce qui permit un certain épanouissement de la vie catholique, une reprise des conciles provinciaux, interrompus depuis des siècles, et surtout une orientation de plus en plus décidée vers le centre de la Chrétienté.

L'ÉGLISE D'AUTRICHE SOUS LA « DICTATURE » DE RAUSCHER — La mise en application du concordat fut immédiatement entreprise avec le concours dévoué du ministre Léo Thun [4] sous la direction de Rauscher devenu archevêque de Vienne depuis 1853 et élevé au cardinalat en lendemain de la signature [5].

---

(1) Cité dans C. WOLFSGRUBER, *Kirchengeschichte Œsterreichs*, p. 90.
(2) E. de Ségur à Thouvenel, 28 juillet 1855, dans J. MAURAIN, *La politique ecclésiastique du Second Empire*, p. 100, n. 2.
(3) Le mot est de l'impératrice Eugénie à l'ambassadeur d'Autriche. (J.-A. von HUEBNER, *Neuf années de souvenirs*, t. I, Paris, 1904, p. 372.)
(4) Celui-ci s'appliqua particulièrement à la réorganisation du système scolaire. Cf. S. FRANKFURTER, *Graf Leo Thun, Fr. Exner und H. Bonitz. Beiträge zur Geschichte der oesterreichischen Unterrichtsreform*, Vienne, 1893.
(5) Sur Rauscher (1797-1875), voir C. WOLFSGRUBER, *Joseph Othmar Rauscher*, Vienne, 1888,

Ni génial ni brillant, privé des qualités qui attirent la sympathie, Rauscher s'imposait à l'admiration de ses collègues par son austérité et sa vertu, mais surtout par son intelligence claire et son érudition déconcertante. « On n'exagère pas, écrivait un de ses adversaires, en disant qu'il était supérieur non seulement aux différents membres de l'épiscopat autrichien pris un à un, mais à tout l'épiscopat, et on peut se demander s'il existait dans l'Église un autre évêque aussi savant que lui [1]. » Servi par un tempérament énergique et autoritaire, il fut pendant vingt ans le chef incontesté de l'épiscopat autrichien, qu'il réunissait régulièrement à Vienne en Assemblées générales où il était pratiquement le seul à avoir un rôle actif. Les évêques supportaient souvent avec peine ses allures despotiques, mais n'osèrent jamais s'insurger contre sa direction : même le cardinal Schwarzenberg [2], qui savait être indépendant à l'occasion, s'inclinait devant sa supériorité.

L'idée maîtresse de sa politique ecclésiastique était la nécessité d'une collaboration étroite entre l'Église et l'État, dans l'intérêt de l'une et de l'autre. Dévoué corps et âme à l'Autriche et à son empereur, absolutiste et centraliste en politique, il considérait qu'à côté de l'armée et de l'administration, une Église puissante et unifiée était la meilleure garantie contre l'extension des idées démocratiques et révolutionnaires et contre les aspirations autonomistes des diverses nationalités de la Monarchie. Mais, patriote et réactionnaire, Rauscher était avant tout prêtre, et soucieux des intérêts de la religion. Il travailla inlassablement à ce que l'État, à son tour, apportât tout son appui à l'Église, laissée par ailleurs libre de déterminer elle-même ses buts et ses besoins.

Plus soucieux de donner aux institutions un caractère chrétien que de favoriser parmi les catholiques les initiatives et les personnalités, il n'encouragea que modérément le mouvement d'associations catholiques, mal vu d'ailleurs par le gouvernement à cause de son caractère plus démocratique et de la possibilité de déviations politiques ; il chercha en tout cas à soustraire le mouvement à la direction des laïcs et des esprits ouverts aux idées modernes, au grand désespoir de Veith [3], qui avait espéré, grâce aux associations, réveiller les catholiques autrichiens de leur torpeur et réagir contre les progrès sensibles de l'indifférence religieuse dans les classes cultivées. Par contre, il s'appliqua à la réorganisation de l'ensei-

à nuancer par l'esquisse très suggestive du vieux-catholique J.-F. v. SCHULTE, art. *Rauscher*, dans *Allgemeine Deutsche Biographie*, t. XXVII, p. 449-457.

(1) J.-F. v. SCHULTE, *art. cit.*, p. 450.

(2) Après Rauscher, Frédéric Schwarzenberg (1809-1885), archevêque de Prague depuis 1850, fut le prélat le plus en vue de l'Empire des Habsbourg pendant cette période. Grand seigneur, il appartenait encore à l'ancien régime par bien des aspects et, comme beaucoup de prélats autrichiens, il considérait avant tout l'Église comme une force sociale qui devait s'imposer par son prestige extérieur. Bien moins doué que Rauscher, il était cependant plus sensible que lui à certaines tendances modernes en matière scientifique. Cf. C. WOLFSGRUBER, *Kardinal Schwarzenberg*, Vienne, 1905-1917, 3 vol. E. Ollivier en a tracé, d'après les souvenirs de la princesse Wingenstein, un portrait très exact. (*L'Église et l'État au concile du Vatican*, t. II, p. 10-11.)

(3) On ne saurait exagérer l'influence que Veith exerça, jusqu'à sa mort en 1876, dans les milieux cultivés, tant par sa prédication à Vienne que par ses innombrables livres de dévotion, les plus lus de toute l'Autriche. Rauscher lui reprochait toutefois ses sympathies gunthériennes et surtout la persistance de ses tendances constitutionnelles en politique, alors que lui-même s'orientait plus que jamais vers l'absolutisme et une union étroite entre l'Église et l'État. (Cf. J. VIDMAR, *Joseph-Emmanuel Veith*, Vienne, 1887.)

gnement officiel sous la direction de l'Église, à la construction de nouvelles
églises et aussi au développement des œuvres de charité et à la multiplication des missions paroissiales. Malheureusement, préoccupé d'action
gouvernementale plutôt que d'action pastorale directe, il ne fit rien pour
réagir contre l'une des grandes lacunes de l'épiscopat autrichien : le
manque de contact avec les masses populaires.

OMBRES INQUIÉTANTES   Officiellement, d'après le concordat, l'Empire
des Habsbourg était un État catholique.
Il comportait toutefois d'assez notables minorités dissidentes : plusieurs
millions d'orthodoxes en Galicie et dans les Balkans ; près de 200.000
luthériens, à Vienne et en Bohême surtout ; 15 pour cent de calvinistes
en Hongrie, occupant souvent des positions sociales en vue. Mais ce n'était
pas là le plus grave : si les campagnes demeuraient croyantes et prati-
quantes, l'indifférence et même l'incrédulité progressaient rapidement
parmi les intellectuels, dans la bourgeoisie aisée et dans l'aristocratie,
théoriquement catholiques, au point que, vers 1850 déjà, Rauscher déplorait que les gens vraiment pieux et dévoués à l'Église s'y sentissent réellement isolés, et qu'en 1869, il estimait que la situation était aussi grave
que celle de la France à la fin du XVIIIe siècle [1].

Il y avait bien quelques timides tentatives de réaction, à l'imitation
de l'action catholique allemande, autour de quelques évêques à l'âme
apostolique et sous l'influence des jésuites ou des rédemptoristes, qui
avaient été admis à nouveau dans le pays dès 1852. Dans l'ensemble
pourtant, malgré son extérieur brillant, le catholicisme autrichien reste
superficiel. Les évêques sont préoccupés de la baisse du recrutement sacerdotal et, sauf dans des régions ferventes comme le Tyrol, on atteint
rarement la proportion d'un prêtre pour 1.000 habitants, parfois à peine
1 pour 1.500. En bien des endroits, on ressent les inconvénients inhérents
au sytème bénéficial de l'ancien régime où le curé, étant en même temps
à la tête d'une exploitation agricole, s'intéresse parfois plus à ses foins
ou à ses blés qu'à ses ouailles.

Si la moralité des prêtres est généralement honorable, leur genre de
vie est le plus souvent très bourgeois, parfois même fastueux. C'est
surtout le cas dans le haut clergé où subsistent des institutions comme
le chapitre d'Olmütz, richement prébendé, dont le doyen vint à Rome
en 1867 plaider le droit à rester jalousement réservé à des chanoines
ayant seize quartiers de noblesse ; ce doyen était lui-même caractéristique,
type de prélat mondain, collectionneur, chasseur et gastronome, fort pieux
par ailleurs [2]. Les évêques, même lorsqu'ils sont dévoués à leur tâche,
continuent à mener la vie opulente d'autrefois : bâtiments imposants,
vastes salles meublées luxueusement, troupes de laquais en livrée
galonnée, gardes de hussards. L'inconfort des voyages dans des régions
arriérées n'encourageait évidemment guère ces évêques grands seigneurs
à multiplier les visites pastorales : Schwarzenberg regrettait de n'avoir

(1) C. Wolfsgruber, *J.-O. Rauscher*, p. 308 et 359.
(2) S. Negro, *Seconda Roma*, Milan, 1943, p. 169-170 ; *Janssens Briefe*, édit. Pastor, t. I, p. 345.

plus visité certaines parties de son diocèse depuis douze ans et à Carlsbad, en 1869, il y avait onze ans que la confirmation n'avait plus été administrée [1]. Dans certaines régions, comme en Bohême, les évêques ne connaissant que l'allemand étaient d'ailleurs incapables d'adresser la parole en public à leurs fidèles tchèques.

Le rapide développement des congrégations féminines enseignantes et hospitalières, et l'action des jésuites dans la haute société, sont des éléments plus réconfortants, mais les anciens ordres sont en pleine décadence : la visite canonique décidée par les évêques en 1849 et qui se prolonge de 1852 à 1859 n'apporta guère de modifications et releva surtout les manquements fréquents des religieux au vœu de pauvreté.

L'ÉGLISE EN HONGRIE   Plus encore qu'en Autriche, l'Église en Hongrie a gardé une allure d'ancien régime. Si le clergé autrichien a conservé le faste et beaucoup des avantages sociaux des siècles antérieurs, il a subi comme celui de France et d'Allemagne, depuis la période napoléonienne, une notable diminution de son influence politique ; dans le royaume magyar, au contraire, la représentation ecclésiastique à la « table des magnats » garde une place prépondérante et les évêques, aux revenus considérables, apparaissent encore comme des princes territoriaux ou comme des mécènes. Et si Montalembert, lors de son passage en Hongrie en 1861, loue le libéralisme de certains prêtres hongrois qui le consolait du conservatisme du clergé autrichien, il s'inquiétait de leur « mondanité » et de leur peu de souci de protéger leurs fidèles contre les exactions des juifs [2], tout autant que de leur trop grande indifférence à l'égard de Rome [3].

Effectivement, en dépit de certaines initiatives heureuses, comme la fondation, en 1848, par le futur évêque de Fogarassy, de la Société Saint-Étienne pour la diffusion des bons livres, les mœurs et la piété laissaient souvent à désirer dans le royaume. Pourtant, après la crise de 1848, où le gouvernement avait pris à l'égard de l'Église une attitude assez inquiétante, la Hongrie connut une longue période de paix religieuse. Deux conciles provinciaux (Gran, 1858 et Kolocsa, 1863) s'occupèrent de l'application du concordat de 1855 ; celui-ci n'avait d'ailleurs été accueilli qu'avec réserve par l'épiscopat hongrois, qui répugnait aux tendances unificatrices de Rauscher et du gouvernement de Vienne.

L'enseignement fut réorganisé comme en Autriche par le ministre Thun dans un esprit très catholique, et cette orientation subsista lorsqu'en 1867 la Hongrie retrouva un statut indépendant et s'orienta dans un sens libéral. Ce fut en effet un catholique admirateur de Montalembert qui devint ministre de l'Instruction publique. En dépit de la forte influence des protestants et de la franc-maçonnerie juive, la Hongrie ne connut que dans une mesure très atténuée la réaction anticoncordataire qui

---

(1) C. Wolfsgruber, *Kard. Schwarzenberg*, t. III, p. 166. Voir aussi un rapport adressé à Mgr Dupanloup, aux *Arch. Nat.*, AB XIX, n° 522.
(2) Souvent du reste, ils confient eux-mêmes à des juifs la gestion des domaines ecclésiastiques.
(3 E. Lecanuet, *Montalembert*, t. III, p. 276.

devait se développer en Autriche vers 1860 ; elle le dut pour une bonne part à la modération des deux primats, Mgr Scitovsky (1849-1866) et Mgr Simor (1866-1891), qui évitèrent de donner l'impression d'une trop grande intervention de la hiérarchie dans la vie de l'État. Les évêques acceptèrent même volontiers, en octobre 1869, une nouvelle réglementation faisant une place importante aux laïcs dans l'examen de toutes les matières ecclésiastiques qui n'étaient pas purement spirituelles.

§ 2. — L'épanouissement du catholicisme allemand au lendemain de 1848 [1].

LA SITUATION ENVIABLE DES CATHOLIQUES PRUSSIENS

Le morcellement de l'Allemagne en un grand nombre de petites souverainetés rendait la situation de l'Église très variable suivant les régions. C'est ainsi que beaucoup de petits princes protestants, rétifs à toute idée de tolérance, continuèrent jusqu'à la veille de 1870 leurs mesquines tracasseries contre le catholicisme [2], tandis que la Prusse, championne du protestantisme allemand contre l'Autriche catholique, était saluée à juste titre par Montalembert comme étant, « après la Belgique, le pays où les intérêts catholiques sont le mieux compris et garantis » [3].

Effectivement, l'épiscopat prussien, qui sans perdre le contact avec l'âme populaire, s'était tenu à l'écart du radicalisme révolutionnaire, sut faire valoir son loyalisme et sa modération pour encourager les dis-

---

(1) BIBLIOGRAPHIE. — I. SOURCES. — Les rapports (*Verhandlungen...*) des réunions annuelles du *Katholischer Verein* et MARX, *Generalstatistik der Katholischen Vereine Deutschlands*, Trèves, 1871 ; B. WEBER, *Charakterbilder : Kartons aus dem deutschen Kirchenleben*, Mayence, 1858 ; *Schriften und Reden von Joh. card. von Geissel*, édit. DUMONT, Cologne, 1869, 3 vol. ; *Briefe von und an Ketteler*, édit. RAICH, Mayence, 1879 ; C. STAMM, *Urkundensammlung zur Biographie des Dr Conrad Martin*, Paderborn, 1892 et *Aus der Briefmappe des Bischofs C. Martin, ibid.* ; L. BERG-STRAESSER, *Der politische Katholizismus. Dokumente zu seiner Entwicklung*, t. I (*1815-1870*), Munich, 1921 ; et les collections du *Katholik*, des *Historisch-politische Blätter* et de l'*Archiv für katholisches Kirchenrecht*.

En outre, pour la Prusse : les décrets des conciles de Cologne dans *Col. lac.*, t. V, col. 251-382 ; DUMONT, *Sammlung kirchlicher Erlasse in der Erzdiözese Köln*, Cologne, 1891 ; les *Beiträge zum preussischen Kirchenrechte des Katholischen Kirchen- und Schulwesens*, Paderborn, 1854, 3 vol. ; E. RENDU, *De l'éducation populaire dans l'Allemagne du Nord*, Paris, 1855.

Pour la Bavière : *Systematische Zusammenstellung der Verhandlungen des bayerischen Episkopates mit der königlich-bayerischen Staatsregierung von 1850 bis 1859*, Fribourg-en-Br., 1905.

Pour la province du Haut-Rhin : *Officielle Aktenstücke über die Kirchen- und Schulfrage in Baden*, Fribourg-en-Br., 1864 et suiv., 6 vol. ; M. LIEBER, *In Sachen der oberrheinischen Kirchenprovinz*, Fribourg-en-Br., 1853.

II. TRAVAUX. — Outre les ouvrages généraux de H. BRUECK et de G. GOYAU, ainsi que les biographies de Geissel, de Martin, de Ketteler, de Doellinger et de Reichensperger, signalés dans la bibliographie générale, p. 9, voir J.-B. KISSLING, *Geschichte der deutsche Katholikentage*, t. I, Munster, 1920 (bonne synthèse de l'ensemble du mouvement catholique) ; V. CRAMER, *Bücherkunde zur Geschichte der katholischen Bewegung in Deutschland im XIX. Jdht.*, München-Gladbach, 1914 ; A. DIEHL, *Zur Geschichte der katholischen Bewegung im XIX. Jhdt.*, Mayence, 1911 ; J.-H. SCHUETZ, *Das segensreiche Wirken der Orden und Kongregationen der katholischen Kirche in Deutschland*, Fribourg-en-Br., 1926 ; W. LIESE, *Wohlfahrtspflege und Caritas im Deutschen Reich...*, München-Gladbach, 1914.

Sur les luttes religieuses dans la province du Haut-Rhin : A. HAGEN, *Staat und katholische, Kirche in Wurtenberg in den Jahren 1848-1862*, Stuttgart, 1928, 2 vol. (fondamental) ; STUTZ, *Die Einführung des allgemeinen Pfarrkonkurses in Grosshertogtum Baden*, dans *Festgabe P. Krüger*, Berlin, 1911, p. 97 et suiv., à compléter par A. ROESCH dans *Archiv für kath. Kirchenrecht*, t. XCVI, 1916, p. 203 et suiv. ; H. BRUECK, *Die oberrheinische Kirchenprovinz*, Mayence, 1868, IIIe P. (unilatéral et polémique).

(2) On en trouvera quelques exemples dans G. GOYAU, *op. cit.*, t. IV, p. 33-36.

(3) Cité dans L. v. PASTOR, *Aug. Reichensperger*, t. I, p. 354.

positions tolérantes du roi Frédéric-Guillaume IV, qui ne demandait pas mieux que de renforcer l'unité morale du royaume en gagnant les sympathies des catholiques rhénans. Aussi la constitution de 1850, qui remplaça celle de 1848, confirma-t-elle l'affranchissement de l'Église, et les règlements scolaires qui la complétèrent, en accentuant de plus en plus le caractère confessionnel de l'enseignement officiel, donnèrent au clergé assez d'influence sur l'école publique pour qu'il jugeât inutile d'organiser un enseignement catholique privé. L'Église avait ainsi les avantages de la liberté sans subir les inconvénients de la séparation, et l'entente avec l'État fut d'autant plus aisée que celui-ci confia à des fonctionnaires dévoués à l'Église la section du ministère des Cultes chargée de traiter les affaires catholiques (*Katholische Abteilung*).

A la faveur de la liberté religieuse, la vie catholique prospéra grâce aux efforts conjugués du clergé paroissial, des congrégations religieuses et des diverses associations laïques qui se multipliaient au point que *La Civiltà cattolica* pouvait écrire, en 1851 : « Chaque mois, on voit là-bas la naissance d'une institution catholique nouvelle [1]. » Le recrutement sacerdotal était aisé : en 1865, on comptera un prêtre pour 775 catholiques dans le diocèse de Cologne, un pour 650 dans celui de Paderborn et, bien que les populations catholiques, plus pauvres, profitassent moins de l'enseignement secondaire que les protestants, il y avait pour l'ensemble de la Prusse 96 prêtres pour 100.000 catholiques contre 60 pasteurs pour 100.000 protestants [2]. L'essor des congrégations religieuses fut particulièrement remarquable : dès 1848, tandis que les sœurs de Saint-Charles, de Nancy, s'installaient à Trèves et que se développaient à Aix-la-Chapelle les Pauvres Sœurs de saint François de Françoise Schervier, une amie de celle-ci, Clara Fey, dont le livre d'*Exercices spirituels* devait avoir une influence si profonde et si durable en Allemagne, fondait les Sœurs du pauvre Enfant Jésus. L'année suivante, la sœur du *leader* catholique Mallinckrodt fondait à Paderborn les Sœurs de la Charité. Puis, tandis que les anciens ordres voués à la prédication, franciscains, lazaristes, rédemptoristes, jésuites surtout, multipliaient leurs couvents, les fondations de nouvelles congrégations continuèrent, celle des Frères de la Miséricorde, de Trèves, par Pierre Friedhofen, celle des Pauvres Frères de Saint-François, par Jean Hoever, bien d'autres encore [3]. Dans le seul diocèse de Cologne, en vingt ans, le nombre des religieuses passa de 240 à 2.726 et, pour l'ensemble de la Prusse, on compterait en 1872 104 couvents d'hommes et 851 de femmes, groupant plus de 11.000 religieux et religieuses [4]. Les jésuites, dont la catholique Bavière n'avait pas voulu, s'établirent à Cologne dès le mois d'août 1848 : deux ans plus tard, ils ouvraient en Westphalie leur premier noviciat et ils allaient

---

(1) *Civ. catt.*, 1851, vol. v, p. 380.
(2) J.-F. v. Schulte, *Lebenserinnerungen*, t. II, Giessen, 1908, p. 25-26, 135.
(3) Voir notamment B. Gossens, *Die gottselige Mutter Franziska Schervier*, Munich, 1932 ; Schw. Adalberta Maria, *Mutter Klara und ihre Werk für die Kinder*, Munster, 1926 ; J. Kroell, *Pater Friedhofen*, Fribourg-en-Br., 1926 ; H. Schiffers, *Joh. Höver*, Fribourg-en-Br., 1930. Tableau d'ensemble dans H.-C. Wendlandt, *Die weibliche Orden und Kongregationen der katholische Kirche und ihre Wirksamkeit in Preussen von 1818 bis 1918*, Paderborn, 1924.
(4) Bongarz, *Die Klöster in Preussen und ihre Zerstörung*, Berlin, 1880, p. 46-49.

jouer un rôle de premier plan dans les missions populaires qui, de la Silésie à la Rhénanie, réveillèrent la ferveur des masses catholiques, trop souvent assoupie ou desséchée par l'*Aufklärung*.

Dès 1849 se fonda l'*Association Saint-Boniface*[1], chargée de soutenir matériellement et moralement les catholiques disséminés dans les régions protestantes d'Allemagne, et dont le centre était à Paderborn. En 1858, fut restauré l'évêché d'Osnabrück, qui devint une seconde capitale pour les missions du Nord de l'Allemagne. Ce renouveau conquérant du catholicisme prussien attira bientôt les convertis du protestantisme au point d'inquiéter les autorités[2].

Tout ce mouvement catholique plein d'espérance était présidé, surtout depuis la mort de Diepenbrock en 1853, par une personnalité de premier plan, le cardinal archevêque de Cologne, Jean Geissel[3], qui s'acquittait de sa tâche avec un tact parfait, « aimant mieux continuer à vaincre que se flatter d'avoir vaincu »[4] et réussissant à conserver à la fois l'estime de la cour de Berlin et la pleine confiance de Rome. Élevé en Rhénanie sous l'occupation française, puis au séminaire de Mayence, auprès de Liebermann, il avait acquis une haute idée de la mission du prêtre et de l'Église, mais aussi une vision claire des conditions dans lesquelles cette mission devait s'exercer dans le monde moderne, et il compensait une formation théologique assez rudimentaire par une raison admirablement équilibrée et « le coup d'œil sommaire et sûr d'un homme de gouvernement »[5]. Au delà de son diocèse, admirablement réorganisé dans un sens très centralisé (car il avait vite compris que l'ancien régime était définitivement révolu), au delà de sa province ecclésiastique, dont le concile de Cologne de 1860 révéla la cohésion, son influence s'étendait à toute l'Allemagne du Nord et de l'Est.

La liberté que la constitution reconnaissait aux catholiques prussiens, et dont ils tiraient un si heureux profit, restait pourtant menacée par les retours offensifs d'un gouvernement et d'une administration où l'élément protestant prédominait. C'est pour appuyer au Parlement les protestations des évêques contre l'un de ces retours, les édits Raumer de 1852[6], que les députés catholiques, sur les conseils de Montalembert,

(1) Sur le *Bonifatiusverein*, fondé à l'initiative de Doellinger lors des *Katholikentage* de Ratisbonne (1849) et de Linz (1850), voir la biographie de son premier président par O. Pfuelf, *Graf zu Stolberg-Westheim*, Fribourg-en-Br., 1913, ainsi que A.-J. Kleffner et F.-W. Wolker, *Der Bonifatiusverein. 1849-1899*, Paderborn, 1899. L'actif évêque de Paderborn, Mgr Conrad Martin, présida, avec un zèle tout spécial, le *Bonifatiusverein*, de 1859 à 1875.

(2) Cette inquiétude était d'ailleurs injustifiée, car si les catholiques progressent légèrement dans les régions à majorité protestante, ils reculent par contre un peu dans celles à majorité catholique. C'est ainsi qu'à Berlin, ils passent de 4,16 % de la population en 1858 à 7,18 % en 1880, mais en Westphalie, ils reculent de 55,95 % à 52,37 % de 1846 à 1880, et en Rhénanie de 75,07 % à 72,27 %. (H.-A. Krose, *Konfessionsstatistik Deutschlands*, Fribourg-en-Br., 1904, tableaux 30 et suivants.)

(3) Sur Geissel (1796-1864), voir O. Pfuelf, *Kard. J. v. Geissel*, Fribourg-en-Br., 1895-1896, 2 vol. ; A. Beck, *Die Kirchenpolitik des Erz. J. v. Geissel*, Giessen, 1905, ainsi que les documents publiés par le chanoine Dumont et cités p. 137, n. 1.

(4) G. Goyau, *L'Allemagne religieuse*, t. III, p. 231.

(5) *Ibid.*, p. 237.

(6) Une circulaire du 22 mai interdisait les missions dans les paroisses situées dans des provinces protestantes ; une autre, du 16 juillet, refusait aux jésuites étrangers l'accès en Prusse et interdisait aux clercs d'étudier dans un séminaire dirigé par les jésuites (notamment au Collège germanique de Rome). On en trouvera le texte, ainsi que le compte rendu des débats parlementaires qu'ils provoquèrent, dans *Die Ministerial-Erlasse von 22 Mei und 16 Juli 1852*, Paderborn, 1853.

décidèrent, tout en gardant leur pleine liberté sur le terrain purement politique, de se grouper en une « fraction catholique » qui défendrait au Parlement les intérêts de l'Église. La vitalité du catholicisme prussien se manifestait ainsi sous une nouvelle forme, et celle-ci ne disparut pas lorsque, en 1858, le prince-régent Guillaume affirma son désir de substituer la parité des confessions à la conception qui considérait l'État prussien comme évangélique en son essence ; pressentant qu'il y aurait bientôt d'autres combats à livrer, on se borna à atténuer le caractère confessionnel du groupe et à lui donner le nom plus neutre de *Centre* [1].

HEURS ET MALHEURS DE LA RÉSISTANCE CATHOLIQUE EN ALLEMAGNE DU SUD   Tandis que le principal État protestant d'Allemagne reconnaissait à l'Église l'essentiel de la liberté qu'elle demandait, la catholique Bavière et les petits États protestants du Sud essayaient au contraire de ne pas tenir compte du nouvel esprit issu des événements de 1848 et de maintenir un régime de surveillance et d'intervention administrative.

En Bavière, le caractère officiellement catholique du gouvernement rendait moins pénible la persistance des conceptions joséphistes, concrétisées dans l'Édit de Religion qui s'était superposé au Concordat. Pourtant les plaintes unanimes contre la situation morale et religieuse des gymnases officiels, l'estime des membres du cercle de Goerres pour la liberté religieuse et le souci du droit canon chez l'archevêque de Munich, Charles-Auguste de Reisach [2], très romain de formation, déclenchèrent, au lendemain de la conférence de Wurzbourg, un mouvement de réaction. Pendant plusieurs années, le roi et son gouvernement se dérobèrent aux demandes réitérées des évêques, qui avaient mis au point leurs revendications au cours d'une réunion pleinière à Freysing, en octobre 1850 [3]. Ce n'est qu'à l'automne de 1854 que le roi accepta de céder sur quelques points importants comme l'éducation des clercs. Encore obtint-il de Rome, en compensation, l'éloignement de Reisach, qui devint cardinal de Curie, et l'indifférence de l'opinion catholique, accrue par la complicité des prêtres qui craignaient le despotisme d'une hiérarchie affranchie du gouvernement, permit à l'arbitraire administratif de continuer à sévir en bien des domaines.

Les évêques de la troisième province ecclésiastique de l'Allemagne méridionale, celle du Haut-Rhin, dépendaient d'États protestants : Wurtemberg, Bade, Hesse, Nassau, dont les souverains prétendaient intervenir dans les affaires de l'Église catholique avec la même autorité que dans les communautés protestantes. Là aussi, l'influence de la Conférence de Wurzbourg et l'exemple de la Prusse ne tardèrent pas à se faire

---

(1) Voir L. BERGSTRAESSER, *Studien zur Vorgeschichte der Zentrumspartei*, Tubingue, 1910, et H. DONNER, *Die Katholische Fraktion in Preussen, 1852-1858*, Leipzig, 1909.

(2) Sur Reisach et ses difficultés avec le gouvernement bavarois, voir les articles de K. HOLL et de A. DOEBERL, dans *H.-P. Bl.*, t. CLXII, 1918, p. 269 et suiv., 341 et suiv., 417 et suiv., 469 et suiv., 558 et suiv., 669 et suiv.

(3) Le *Mémoire* épiscopal, rédigé par Windischmann et Doellinger (lequel joua à Freysing comme à Wurzbourg un rôle de premier plan), est reproduit dans *Coll. lac.*, t. V, col. 1162-1189.

sentir, d'autant plus que les aspirations à la liberté de l'Église allaient trouver un porte-parole de valeur en la personne du nouvel évêque de Mayence, Ketteler, que Lennig et ses amis réussirent à faire nommer en 1849 [1]. En mars 1851, les évêques, réunis à Fribourg autour du vieil archevêque Vicari, adressèrent un *Mémoire* à leurs gouvernements respectifs, mais sans résultat appréciable. Encouragés par Rome, aidés par le juriste catholique Lieber, excités par Ketteler, qui revivait l'atmosphère du conflit de Cologne de 1837, ils revinrent à la charge en 1853 et, n'obtenant toujours rien, décidèrent de passer outre et de ne plus accepter d'ingérences gouvernementales pour les nominations de curés ou la formation des clercs, ni dans le règlement des mariages mixtes.

Dans le grand-duché de Bade, le conflit prit vite un caractère aigu et les interventions pacifiantes de Ketteler, qui faisait de plus en plus figure de chef effectif de l'épiscopat du Haut-Rhin, furent contrecarrées par l'action de Bismarck, délégué prussien à la diète de Francfort, qui craignait qu'une victoire catholique ne servît les intérêts autrichiens. Mais Vicari sentant derrière lui ses fidèles, ses collègues, et bientôt l'opinion catholique mondiale, ne céda pas [2], même lorsque, en mai 1854, il eut été menacé de poursuites et mis aux arrêts.

La signature, au mois d'août suivant, d'un accord assez favorable entre Ketteler et le grand-duc de Hesse [3] montra que les gouvernements commençaient à se rendre compte de la nécessité de concessions. Peu à peu, ils entrèrent en négociations avec Rome, surtout lorsque le concordat avec l'Autriche, qui semblait sonner le glas du joséphisme, fut venu encourager encore la résistance catholique, qui s'affirma au grand jour en juin 1855, à l'occasion des fêtes de Fulda pour le onzième centenaire du martyre de saint Boniface. Le Saint-Siège désirait une solution concordataire. Il obtint, après d'interminables pourparlers, la signature de conventions très satisfaisantes avec le Wurtemberg en 1857, avec Bade en 1859 et avec Nassau en 1861, mais ce succès fut vite suivi d'une amère déception, car l'opposition protestante, soutenue par les juristes anticléricaux, profita de l'affaiblissement du prestige de l'Autriche après la guerre d'Italie de 1859 pour obtenir la dénonciation de ces conventions. Toutefois, les lois gouvernementales ou les accords verbaux qu'ils leur substituèrent pour régler indépendamment de Rome les rapports de l'Église et de l'État maintinrent l'essentiel des satisfactions accordées et, sauf à Bade, assurèrent pour de longues années la paix religieuse. En

---

(1) Le gouvernement, conseillé par le chapitre, avait d'abord désigné Léopold Schmid, « un joséphiste conscient et un rationaliste qui peut-être s'ignorait encore ». (G. GOYAU, *op. cit.*, t. IV, p. 9.) L'opposition du groupe du *Piusverein*, soutenue par Doellinger (cf. J. FRIEDRICH, *Ign. v. Döllinger*, t. II, p. 499-507), lui fit refuser l'investiture par le pape et aboutit à la nomination de Ketteler (voir F. VIGENER, *Die Mainzer Bischofswahl von 1849-1850*, dans *Zeitschrift der Savigny-stiftung. Kan. Abt.*, t. XI, 1921, p. 351-427.)

(2) Il alla même dans son intransigeance jusqu'à excommunier les fonctionnaires catholiques qui appliquaient les mesures étatistes, décision regrettable, car elle amena le gouvernement, au grand détriment de l'Église, à ne plus choisir que des non-catholiques comme hauts fonctionnaires, afin d'être sûr d'eux. (Cf. H. BAIER, dans *Historisches Jahrbuch*, t. LV, 1935, p. 410-422.)

(3) Accord d'ailleurs assez mal accueilli par ses collègues, qui lui reprochaient de briser l'unité d'action, et surtout par Rome, qui estimait que c'était au Saint-Siège et non pas aux évêques à mener des négociations de ce genre.

définitive, au terme de dix années de lutte, les catholiques du Haut-Rhin n'avaient « rien perdu, beaucoup gagné » [1].

UN ANIMATEUR : KETTELER   Les libertés conquises par l'Église auraient été peu de choses si elles ne s'étaient accompagnées d'un approfondissement de la vie religieuse. Car la brillante renaissance catholique qui s'était développée autour de Sailer, Hirscher, Moehler, Goerres ou Liebermann, n'avait atteint qu'une élite et les traces de la décadence du xviiie siècle étaient loin d'être effacées. Un prêtre français de passage à Munich en 1842 nuançait à juste titre les avis trop optimistes qui avaient cours sur l'Allemagne du Sud : « Les catholiques ont gagné en ferveur et en zèle ; mais, comme en France, ils ont perdu en nombre. On parle beaucoup de certains retours et rarement on compte les pertes. Les préjugés exploités en France par Eugène Sue trouvent ici de l'écho et son œuvre est aussi populaire à Munich qu'à Paris [2]. » Dans les campagnes, si certaines coutumes religieuses locales avaient survécu à un siècle de rationalisme, la foi du peuple était superficielle et la pratique religieuse était souvent médiocre. Beaucoup de prêtres, touchés par l'esprit de l'*Aufklärung*, étaient dépourvus de piété et de zèle ; n'ayant guère le sens de l'Église et se dispensant aisément de la messe quotidienne ou même du bréviaire, ils se comportaient davantage en fonctionnaires qu'en pasteurs d'âmes. La situation était d'autant plus dangereuse que l'ambiance culturelle ne jouait plus en faveur de l'Église comme pendant la première moitié du siècle, depuis qu'aux sympathies romantiques pour la Chrétienté médiévale se substituait une atmosphère matérialiste et laïque, contre laquelle il était urgent de prémunir les masses.

Parmi ceux qui comprirent le plus vite qu'une nouvelle époque commençait et que des méthodes nouvelles s'imposaient, l'évêque du plus petit diocèse d'Allemagne occupe une place à part et son influence devait de proche en proche marquer profondément la vie ecclésiastique de l'Allemagne tout entière. Guillaume Emmanuel von Ketteler [3], haut fonctionnaire prussien devenu prêtre à trente ans et évêque de Mayence à trente-sept, avait conservé de ses origines de noblesse westphalienne et de ses contacts avec le cercle de Goerres une admiration sans borne pour la vieille Allemagne catholique et féodale. L'idéal de sa vie fut la restauration intégrale de l'esprit catholique et de la pénétration chrétienne des divers secteurs de l'activité humaine, qui avaient caractérisé la civilisation médiévale et que le laïcisme moderne battait en brèche. Mais, en même temps, il sentait bien plus vivement que beaucoup de ses contemporains tout ce qui séparait irrémédiablement son époque du moyen âge

(1) *Der Katholik*, 1863, vol. I, p. 4.
(2) Cité dans H. Boissonot, *Le cardinal Meignan*, Paris, 1899, p. 72.
(3) Sur Ketteler (1811-1877), voir surtout la biographie fondamentale par le protestant F. Vigener, *Ketteler. Ein deutsches Bischofsleben des XIX. Jhdts*, Munich, 1924, et celle, fort utile à cause des nombreux documents reproduits, par le jésuite O. Pfuelf, *Bischof Ketteler*, Fribourg-en-Br., 1899, 3 vol. (Cf. à son sujet S. Merkle dans *Deutsche Literaturzeitung*, N. F. II, 1925, col. 1449 et suiv.) ; en outre G. Goyau, *Ketteler*, Paris, 1908 et L. Lenhart, *Bischof W. E. Freiherr von Ketteler*, Revelaer, 1937. Voir aussi chap. viii, p. 242 et 259-260, et chap. xiv, p. 489-491.

et la nécessité de servir son idéal médiéval par des moyens nouveaux, dont il croyait trouver la meilleure formule dans le catholicisme post-tridentin. Aussi fit-il largement appel, pour l'aider dans son œuvre, aux services de la Compagnie de Jésus, dont les principes correspondaient à son idéal de piété austère et de donation totale à l'Église, comme à son tempérament énergique et à son sens parfois un peu despotique de la hiérarchie. Ces derniers traits de caractère expliquent aussi son souci de substituer à la mentalité irénique et à la pédagogie religieuse toute en nuances d'un Sailer, un esprit plus combatif accusant volontiers les angles, et une direction plus autoritaire et plus systématique, qui devaient caractériser le catholicisme allemand de la seconde moitié du siècle, au grand déplaisir des tenants de l'ancienne école [1].

Pour l'aider dans sa tâche, Ketteler trouvait à Mayence des auxiliaires dont les vues pouvaient parfois différer des siennes, mais qui étaient acquis d'avance au même idéal que lui, le doyen du chapitre Lennig, son neveu Moufang, président du séminaire et animateur de multiples œuvres, et l'infatigable professeur de dogmatique, Heinrich, tous trois prêtres exemplaires, d'une activité débordante et d'un dévouement presque trop intransigeant à la cause catholique [2].

L'un des premiers soucis de Ketteler fut de remédier au manque de formation sacerdotale des séminaristes, obligés, selon le système alors en vigueur, de faire leurs études à l'Université de Giessen, dans une atmosphère qui se ressentait encore de l'*Aufklärung*. Sans chercher à améliorer le régime universitaire, il se rallia à l'avis un peu unilatéral de Lennig et d'Heinrich et, en mai 1851, après avoir retiré ses séminaristes de Giessen, il rouvrit à Mayence, sans autorisation du gouvernement, un grand séminaire [3]. Le retentissement de cette décision fut grand dans toute l'Allemagne catholique, où, pendant des années, on discuta les avantages et les inconvénients de cette éducation en vase clos, très critiquée par les intellectuels ; en tout cas, malgré les sympathies de plusieurs évêques pour la nouvelle formule [4], il fallut attendre jusqu'en 1865 pour voir une seconde réalisation, à Spire [5]. Moins spectaculaires, mais plus efficaces encore furent les efforts de Ketteler pour accroître la ferveur et le zèle de

---

(1) L'un de ceux qui, par ses conseils et ses lettres de direction, aida le plus efficacement le clergé de l'Allemagne du Sud à s'adapter à cette transformation fut le théologien Denzinger. (Voir les souvenirs de son frère dans *Der Katholik*, 1883, vol. II, p. 638 et suiv.)

(2) Sur Lennig, voir chap. II, p. 59, n. 4 ; sur Christophe Moufang (1817-1890), voir *Der Katholik*, 1890, vol. I, p. 481 et suiv., vol. II, p. 1 et suiv. ; sur Jean-Baptiste Heinrich (1816-1891), voir *Der Katholik*, 1891, vol. I, p. 289 et suiv., 403 et suiv.

(3) F. VIGENER, *Die katholische theologische Fakultät in Giessen und ihr Ende,* dans *Mitteilungen der oberhessischen Geschichtsvereins*, t. XXIV, 1922 ; sur les antécédents de la décision, cf. H. BRUECK, *Die oberrheinische Kirchenprovinz*, p. 285-290 et F. SCHNABEL, *Deutsche Geschichte im XIX. Jhdt.*, t. IV, p. 192.

(4) La question avait déjà été agitée, sur le désir du pape, à la conférence de Freysing, où Doellinger s'était vivement opposé à l'avis favorable de Windischmann. (Cf. O. PFUELF, dans *Stimmen aus Maria Laach*, t. XLIII, 1892, p. 44-65.) Outre l'inconvénient de confier leurs séminaristes à des professeurs dont ils pouvaient difficilement contrôler l'enseignement, les évêques reprochaient à la formule universitaire, très onéreuse, d'obliger les futurs prêtres à s'endetter, et ils croyaient, à tort d'ailleurs, qu'elle ne répondait pas aux prescriptions du concile de Trente. (Cf. S. MERKLE, *Das Konzil von Trient und die Universitäten*, Berlin, 1905.)

(5) Sur l'ouverture du séminaire de Spire et les controverses auxquelles elle donna lieu, voir F.-X. REMLING, *Nik. v. Weis, Bischof zu Speyer*, t. I, Spire, 1871, p. 302-323, 428-436 et J. FRIEDRICH, *Ign. v. Döllinger*, t. III, p. 392-401.

ses prêtres qui se concrétisèrent entre autres dans l'instauration, dès 1850, de retraites sacerdotales annuelles [1] et, un peu plus tard, de conférences diocésaines.

Tous les moyens furent mis en œuvre par l'évêque de Mayence pour affermir la piété des fidèles : organisation de conférences religieuses pour la masse, de retraites pour les hommes et jeunes gens, en particulier pour les instituteurs ; multiplication des missions populaires ; encouragements répétés aux confréries du Cœur de Marie, aux pèlerinages à la Vierge, à la récitation du chapelet, alors peu en usage, au culte du Sacré-Cœur et du Saint-Sacrement ; contact fréquent avec son peuple par lettres pastorales ou brochures, mais aussi personnellement à l'occasion des tournées de confirmation. En même temps, soucieux de réagir dans tous les domaines contre la séparation entre la religion et la vie, il cherchait par tous les moyens à accroître l'influence de l'Église sur l'école, encourageait autour de lui toutes les œuvres qui tendaient à pénétrer d'esprit chrétien les diverses activités temporelles, et mettait un soin particulier à éveiller les catholiques à leurs responsabilités sociales. Après avoir montré la voie dans la lutte victorieuse contre les prétentions étatistes, Ketteler devenait ainsi, pour beaucoup de prêtres et même d'évêques allemands, un guide et un modèle d'esprit sacerdotal et de zèle pastoral.

LES MULTIPLES ASPECTS DE L'ACTION CATHOLIQUE — Sous l'influence d'hommes comme Geissel, Ketteler, Lennig, Moufang, les méthodes apostoliques, basées sur l'action individuelle et les contacts personnels, qu'avaient préconisées les Sailer, les Hirscher, les Diepenbrock, firent place de plus en plus à un catholicisme de masse, dont les principales manifestations furent les missions populaires, la presse et les associations (*Vereine*), en attendant qu'il s'exerçât également, à la grande fureur des adversaires du « catholicisme politique », sur le terrain parlementaire.

L'Allemagne n'avait guère connu les missions avant 1848, mais dès que l'ostracisme à l'égard des ordres religieux se fût atténué, elles prirent un développement considérable [2]. A côté des franciscains et des rédemptoristes, les jésuites, malgré l'hostilité d'une partie du clergé, y tinrent le rôle principal. « Des orateurs comme le P. Hasslacher, le P. Roder, le P. Roh, le P. de Waldburg-Zeil, sont demeurés célèbres ; et l'on garda longtemps le souvenir des grandes missions prêchées à Cologne en 1850, à Heidelberg en 1851, à Francfort en 1852, à Augsbourg en 1853. Martin, l'évêque de Paderborn, disait que la renaissance de l'Allemagne catholique était la gloire du P. Roh et du P. Hasslacher [3]. » On vit se renouveler des scènes des anciens âges, « des villages se vider une ou deux fois par

---

(1) L'habitude des retraites sacerdotales se répandait à la même époque en Rhénanie grâce à l'impulsion de Geissel, mais celui-ci ne fit pas appel aux jésuites dès le début comme Ketteler.
(2) Voir surtout A. GROETEKEN, *Die Volksmissionen der norddeutschen Franziskaner*, Munster, 1910 et B. DUHR, *Aktenstücke zur Geschichte der Jezuitenmissionen in Deutschland, 1848-1872*, Fribourg-en-Br., 1903.
(3) G. GOYAU, art. *Allemagne*, dans *D. H. G. E.*, t. II, col. 582. Cf. J. KNABENBAUER, *Petrus Roh*, Fribourg-en-Br., 1872 ; J. HERTKENS, *Erinnerungen an P. Hasslacher*, Munster, 1879 et surtout J. MUNDWILER, *P. Georg von Waldburg Zeil*, Fribourg-en-Br., 1906.

jour et déverser sur la paroisse voisine, où prêchaient les missionnaires, le flot de leurs habitants ; des missionnaires parlant en plein air pour évangéliser la foule que l'église ne pouvait contenir ; d'autres cernés au confessionnal par des rassemblements de pénitents qu'aucune attente ne lassait ; et d'interminables rangs de communiants, à jeun parfois depuis la veille au soir, s'échelonner à des heures tardives de la matinée depuis la place du village ou depuis le fond du cimetière » [1].

Les jésuites prolongèrent l'action de leurs missions en mettant au point des livres d'instruction religieuse solides et complets, répondant aux préoccupations pédagogiques de leur temps, en particulier le catéchisme du P. Deharbe [2] et le manuel de religion du P. Wilmers [3], qui connurent vite une diffusion considérable. On doit toutefois regretter que le souci de la précision et de la correction théologique ait fait abandonner la méthode kérygmatique de Hirscher, basée sur l'Écriture et l'histoire du salut, et ait amené pour de longues années « l'irruption de la scolastique dans la catéchèse » et, de là, dans la prédication [4].

Les animateurs les plus perspicaces du mouvement catholique, Lennig et Doellinger, avaient compris très tôt l'importance de la presse, mais les difficultés à vaincre étaient grandes : manque d'hommes et d'argent, tracasseries administratives, hostilité d'une partie du clergé qui craignait l'immixtion des journalistes dans les affaires de l'Église. Aussi, en dépit de nombreux dévouements, si l'on disposait de quelques revues bien faites : le *Katholik* de Mayence, et surtout les *Historich-politische Blätter* de Munich, plus accessibles au grand public [5], la situation de la presse quotidienne laissa longtemps à désirer et, en 1857, sur 1.441 journaux, on n'en comptait que 87 catholiques, presque tous strictement locaux. La fondation d'un *Pressverein* fut suivie de quelques initiatives heureuses : multiplication de journaux populaires à bon marché et de brochures religieuses de vulgarisation, et surtout lancement, en 1860, par le libraire Bachem, ami de Montalembert, d'un journal de grand style, les *Kölnische Blätter* [6].

Le *Pressverein* n'était qu'un cas particulier d'un mouvement spécifiquement allemand qui aboutit à la création, en quelques années, d'un nombre surprenant d'associations très vivantes, répondant chacune à une nécesssité précise [7] : associations purement religieuses, comme les

---

(1) G. Goyau, *Bismarck et l'Église*, t. II, Paris, 1911, p. 18.

(2) Rédigé en 1847, remanié en 1853, il fut adopté avec de légères modifications, d'abord par les évêques bavarois, puis dans la plupart des diocèses allemands et imité ensuite à l'étranger. (Cf. F.-X. Thalhofer, *Entwicklung des katholischen Katechismus in Deutschland von Canisius bis Deharbe*, Fribourg-en-Br., 1899.)

(3) *Handbuch zu Deharbes Katechismus und ein Lehrbuch zum Selbstunterricht*, publié à partir de 1851 et qui devint par la suite le célèbre *Lehrbuch der Religion*.

(4) Voir F. Arnold, dans *Lumen vitae*, t. III, 1948, p. 496-499.

(5) Malheureusement un peu trop *süddeutsch* au gré des catholiques de Prusse.

(6) Devenu depuis 1869 la *Kölnische Volkzeitung*. (Cf. L. Bachem, *J. Bachem, seine Familie und die Firma J. P. Bachem in Koln*, Cologne, 1912, 2 vol.). Un précédent essai par Beda Weber, en 1855, avait été sans lendemain. (Cf. H. Schnee, *Die Zeitung « Deutschland »*, dans *Historisches Jahrbuch*, t. LII, 1932, p. 477-494.) Sur l'ensemble de ces efforts, cf. K. Loeffler, *Geschichte der katholischen Presse Deutschlands*, München-Gladbach, 1924.

(7) On trouvera un bon aperçu de l'ensemble du mouvement dans J.-B. Kissling, *Geschichte der deutschen Katholikentage*, t. I, Munster, 1920. Voir, à titre d'exemple, L. Werthmann, *Fünfzig Jahre Raphaelsverein*, 1919 ; A. Schnuetgen, *Der Verein vom hl. Karl Borromäus geschichtlich gewürdigt*, 1924.

congrégations mariales ; œuvres de jeunesse, où se distingue notamment
le P. von Doss ; associations d'étudiants, associations d'artistes chrétiens,
Cäcilienverein pour les amateurs de musique religieuse, association
Saint-Charles-Borromée pour la diffusion des bons livres ; associations
s'occupant des missions, de la Terre-Sainte, des émigrés ; conférences
de Saint-Vincent-de-Paul, très actives[1], et tout le réseau d'œuvres charitables qui se développe à l'entour ; enfin et surtout, associations à objet
directement social, groupant les apprentis (*Gesellenverein* de Kolping),
les paysans, les ouvriers, pour la défense de leurs intérêts professionnels
en même temps que pour leur soutien moral et religieux. Ce caractère
social du catholicisme allemand, qui refuse de se cantonner dans les
œuvres de pure bienfaisance, comme on le fait trop volontiers en France,
lui permettra de conserver de profondes attaches dans les masses populaires et d'y trouver un appui lorsqu'il devra, à l'époque du *Kulturkampf*,
tenir tête à la bourgeoisie radicale.

Cet ensemble d'œuvres, qui tient du prodige et dont on mesurerait
difficilement la somme de dévouement et de sacrifices pécuniaires qu'il
exigea, avait son organe unificateur dans l'*Association catholique d'Allemagne*, qui se réunissait chaque année en un congrès, *Katholikentag*, où les
problèmes les plus brûlants et les plus actuels étaient examinés et des
appels adressés aux initiatives qui s'avéraient nécessaires. Ces congrès,
comme les associations elles-mêmes, étaient en bonne partie aux mains
des laïcs, bien que témoignant toujours d'une grande déférence à l'égard de
la hiérarchie, et ils s'abstenaient soigneusement de s'immiscer dans les
questions politiques, malgré le désir contraire manifesté au début par les
catholiques rhénans.

Ce mouvement si vivant avait cependant un défaut : exclusivement
tourné vers l'action, il négligea trop le terrain culturel qui fut ainsi complètement dominé par des auteurs protestants ou même incroyants[2]. On
trouvait bien, à l'origine, parmi les animateurs de l'Association catholique, à côté d'hommes d'action comme Lennig ou Buss, des intellectuels
et même des savants, tels Doellinger ou Knoodt. Mais très vite, dès le
congrès de Linz en 1850, ce fut la tendance des amis de Lennig qui l'emporta et l'on doit regretter leur exclusivisme. Car la lutte entre la foi
et l'incroyance s'annonçait dès ce moment de plus en plus redoutable. En
quelques années, le matérialisme théorique et pratique, répandu dans le
peuple par les chefs socialistes et dans la bourgeoisie par les hommes
de science, allait faire perdre à l'Église beaucoup plus de terrain qu'elle
n'en avait gagné à la faveur du romantisme, et bientôt, sauf dans quelques
régions comme la Rhénanie ou le Palatinat, les philosophes, historiens,
savants, médecins, juristes ou hommes de lettres catholiques devinrent
l'exception. Conscient du danger que constituait son infériorité dans le

---

(1) Reichensperger avait signalé au Congrès de Mayence de 1848 la fécondité de la formule
française. Le développement en Allemagne fut très rapide, mais à plusieurs reprises on discuta
l'opportunité de rester uni au centre de Paris. (Voir *Vinzenzgeist und Vinzenzarbeit*, sous la direction de H. Bolzau, 1933.)

(2) Voir H. Rost, *Die wirtschaftliche und kulturelle Lage der deutschen Katholiken*, Cologne,
1911.

domaine culturel, le catholicisme, pour essayer d'échapper à une atmosphère dangereuse, se crut obligé de se replier sur lui-même ; mais tout en constituant une protection provisoire, pareille attitude engendrait une mentalité de *ghetto*, qui devait limiter singulièrement ses possibilités de rayonnement, en même temps qu'elle risquait de le faire apparaître aux yeux de certains comme un corps étranger dans la nation.

§ 3. — Les préludes du Kulturkampf [1].

GERMANISME ET ROMANISME Le danger d'isolement qui menaçait les catholiques était ressenti avec une particulière acuité dans les milieux universitaires, où l'on avait vu avec dépit la place que prenaient les hommes d'œuvres de Mayence, au détriment des intellectuels, dans la direction du mouvement catholique. A leur avis, le danger était encore accru par la multiplication des interventions romaines dans la vie de l'Église d'Allemagne parce que, ne tenant pas compte des besoins et des caractères propres du catholicisme germanique, elles risquaient de mettre celui-ci en état d'infériorité.

Cette opposition croissante à l'ultramontanisme trouva, à partir de 1860, un centre de ralliement en la personne de Doellinger, le savant catholique le plus en vue d'Allemagne, le champion de toutes les causes catholiques durant les vingt dernières années [2]. Elle fut pour le renouveau ecclésiastique une cause d'affaiblissement, non seulement parce qu'elle amena certains catholiques influents, surtout en Allemagne du Sud, à soutenir à plus d'une reprise les tentatives gouvernementales en vue d'enrayer des initiatives catholiques jugées inopportunes, mais surtout parce qu'elle contribua notablement à renforcer les préjugés antiromains des milieux non catholiques.

RECRUDESCENCE DE L'OPPOSITION CONFESSIONNELLE L'indifférentisme religieux de l'*Aufklärung*, les sympathies romantiques pour le moyen âge allemand, l'orientation biblique et mystique de l'école de Sailer, avaient sensiblement diminué l'hostilité protestante à l'égard du catholicisme. Divers facteurs contribuèrent au contraire à la réveiller à partir du milieu du siècle : l'orientation plus polémique prise par l'école de Munich, où Doellinger — qui sera l'un des premiers à regretter le durcissement confessionnel —

---

(1) BIBLIOGRAPHIE. — Aux sources et travaux mentionnés en tête de l'art. II, ajouter : L. BERGSTRAESSER, *Politischer Katholizismus. Dokumente zu seiner Entwicklung*, t. I (1820-1870), Munich, 1921.

Pour la Prusse : K. BACHEM, *Vorgeschichte, Geschichte und Politik der deutschen Zentrumspartei*, t. I, Cologne, 1926, et H. WENDORF, *Die Fraktion des Zentrums im preussischen Abgeordnetenhause, 1859-1867*, Leipzig, 1916 ; E. SCHMIDT, *Bismarks Kampf mit dem politischen Katholizismus*, t. I (*1848-1870*), Hambourg, 1942 (polémique).

Pour la Bavière : *Denkwürdigkeiten des Fürsten Chl. zu Hohenlohe-Schillingsfürst*, édit. F. CURTIUS, t. I, Stuttgart, 1905.

Pour l'Autriche : outre les biographies de Rauscher et surtout de Schwarzenberg, signalées p. 133, n. 5 et p. 134, n. 2, M. HUSSAREK, *Die Krise und die Lösung des Konkordats*, dans *Archiv für oesterreichische Geschichte*, t. CXII, 1932, p. 211-480 (nombreux documents) et K. MEINDL, *Bischof F. J. Rudigier*, Linz, 1891-1892, 2 vol.

(2) Cf. *supra*, p. 58-61 et *infra*, p. 196-197.

publie ses ouvrages acerbes et sarcastiques contre Luther et la Réforme [1] ; le redressement de la conscience catholique à la suite des
missions prêchées par les jésuites, qui donne aux protestants l'impression
d'une offensive conquérante, alors qu'en réalité la proportion entre les
deux confessions reste remarquablement stable [2] ; le groupement des
fidèles en associations nettement confessionnelles ; le renforcement
de l'influence romaine en Allemagne et les campagnes de revendications
pour la liberté de l'Église qui indisposent les gouvernements protestants.
Deux autres éléments accentuèrent encore l'évolution : la rivalité croissante entre la Prusse protestante et l'Autriche catholique, et la prétention
des radicaux, favorablement accueillie par les protestants libéraux du
Protestantenverein, de mettre sous le patronage de Luther leur lutte
contre l'Église au nom de la liberté de pensée [3].

NATIONALISME ET LIBÉRALISME   Les événements de 1848 avaient fait
apparaître au grand jour la tendance
à unifier l'Allemagne autour de la Prusse, sans l'Autriche et contre l'Autriche, mais cette école favorable à la « Petite Allemagne » se heurtait à
la résistance de l'idée « Grande Allemagne », que l'historiographie romantique, tant protestante que catholique, avait remise en honneur en exaltant
le Saint Empire, et qui gardait en outre la faveur des partisans du principe de légitimité. Or, si quelques catholiques prussiens, dont le général
Radowitz était le plus marquant, appuyaient la première solution [4],
la très grande majorité des catholiques allemands, *leaders* en tête [5],
restèrent longtemps des partisans résolus de la solution autrichienne.
Ils y étaient souvent poussés par des raisons politiques et historiques,
que leurs adversaires oublient trop facilement, mais la part des motifs
religieux fut certainement fort grande, surtout depuis que la signature
du concordat de 1855 par François-Joseph avait encore augmenté les
sympathies catholiques : « Avec l'Autriche au sommet, le corps germanique faisait figure catholique ; amputé de l'Autriche et cherchant à
Berlin son point d'appui, il prendrait l'aspect d'une puissance protestante [6]. » C'est du reste ce que confirmait l'école historique prussienne,
en cherchant à démontrer que la mission unificatrice de la Prusse résultait
de l'équivalence entre germanisme et protestantisme.

(1) Cf. *infra*, p. 196.
(2) En 1822, il y avait en Allemagne 9.091.500 catholiques contre 16.193.000 protestants, soit
35,42 % contre 63,79 % de la population ; en 1858, ils étaient 12.212.607 contre 22.640.070, soit
34,52 % contre 63,99 % ; en 1871, après l'annexion de l'Alsace-Lorraine, ils seront 14.869.292
contre 25.581.685, soit 36,21 % contre 62,31 %. (H.-A. KROSE, *Konfessionsstatistik Deutschlands*,
Fribourg-en-Br., 1904, tableau 9.)
(3) L'un des premiers indices de cette nouvelle tendance fut le pamphlet de Bunsen (*Die Zeichen
der Zeit*, Leipzig, 1855.) Au début, certains protestants croyants, comme Stahl ou Leo, dénoncèrent
le caractère équivoque de pareille alliance.
(4) J. GRISAR, *Das preussische Unionsprojekt und die Katholiken Preussens (1849-1850)*, dans
Stimmen der Zeit, t. CXIII, 1927, p. 380-392. Sur Radowitz, l'un des catholiques allemands les
plus en vue du milieu du siècle, voir F. MEINECKE, *Radowitz und die deutsche Revolution*, Berlin,
1913.
(5) Non seulement les Bavarois du cercle de Goerres, Doellinger en tête, et le groupe de Mayence,
qui se trouvaient d'accord sur ce point malgré leur antagonisme sur tant d'autres, mais aussi
beaucoup de sujets du roi de Prusse : le chef de la fraction *Grossdeutsch* à Francfort était même
le rhénan Reichensperger.
(6) G. GOYAU, *L'Allemagne religieuse*, t. III, p. 14.

L'inquiétude des catholiques allemands devant l'affaiblissement autrichien consécutif à la guerre d'Italie de 1859 [1] fut d'autant plus vive que les ambitions prussiennes n'étaient pas seulement encouragées par la réaction protestante, mais bruyamment soutenues par les libéraux de l'Association Nationale Allemande (*Nationalverein*), violemment hostiles à la « prêtraille » et désireux de voir disparaître, avec l'Autriche, « la forteresse de l'ultramontanisme ». Car le libéralisme avait évolué depuis 1848 : une forme nouvelle apparaissait, plus philosophique que politique, que Ketteler devait dénoncer dans un ouvrage retentissant [2] comme le plus grave danger qui menaçât l'âme allemande. Ce libéralisme extrême à base naturaliste trouvait une large audience dans une bourgeoisie profondément atteinte par le matérialisme théorique et pratique des années cinquante, soucieuse d'éliminer toute influence religieuse de la vie publique et privée, et décidée, en présence des efforts catholiques pour enrayer ce processus de laïcisation, à courber l'Église sous le joug de l'État « aussi despotiquement que l'avaient fait les bureaucraties vaincues » [3].

L'ALLEMAGNE APRÈS 1866 La guerre austro-allemande de 1866 devait nécessairement, dans ces conditions, prendre une portée idéologique. Tandis que les protestants la présentaient comme une « guerre de religion » dirigée contre « le catholicisme papal abrupt qui empêche la liberté de penser » [4], Bluntschli, l'un des coryphées du libéralisme nationaliste, montrait dans l'Autriche vaincue par la Prusse l'État médiéval, clérical et féodal, battu par l'État moderne [5]. Aussi l'accablement des catholiques fut-il profond à la nouvelle de Sadowa, considéré par l'opinion publique comme une victoire protestante et libérale.

Mais à quoi bon des récriminations stériles contre l'inévitable, qui ne servaient qu'à rendre suspect le loyalisme des catholiques allemands ? Ketteler, qui avait tenté, en 1862, de faire le partage entre le vrai et le faux libéralisme, entreprit cette fois, malgré la peine profonde qu'il ressentait, de dissocier le fait *kleindeutsch*, auquel il fallait se résigner sans arrière-pensée, et tout le courant d'idées anticatholique qui l'avait précédé et escorté. Il écrivit en quelques semaines une brochure intitulée : *L'Allemagne après la guerre de 1866* où, après avoir fait sommairement l'historique de la crise, il invitait les catholiques à ne se laisser « surpasser par personne en amour de la patrie allemande, de son unité et de sa grandeur », tout en souhaitant voir la Prusse se désolidariser de ceux qui voulaient donner à sa « vocation allemande » une signification confes-

---

(1) Ces revers déclenchèrent chez les catholiques allemands une violente hostilité contre la France, principal instrument de la défaite autrichienne.

(2) Avec la publication, en février 1862, de *Freiheit, Autorität und Kirche*, Ketteler sortait délibérément du cercle de son diocèse ou de la province ecclésiastique du Haut-Rhin, pour s'adresser à toute l'Allemagne. Cette orientation nouvelle allait de plus en plus se préciser au cours des années suivantes.

(3) G. Goyau, *Ketteler*, p. XIX.

(4) Paroles d'un dignitaire luthérien citées par E. Ringseis, *Erinnerungen an J.-N. Ringseis*, t. IV, 1892, p. 381.

(5) J.-K. Bluntschli, *Denkwürdiges*, t. III, p. 145.

sionnelle ou philosophique. Ces vues réalistes heurtaient trop le sentiment pour être acceptées immédiatement. Aussi les catholiques mirent-ils du temps à se rendre à l'évidence, fournissant de la sorte à leurs adversaires un prétexte facile à critiques et une apparente justification à l'accentuation de leur politique anticléricale.

Car cette politique se faisait plus menaçante à mesure que les années passaient, spécialement dans le domaine de l'enseignement. Depuis 1860 les critiques contre le caractère confessionnel de l'école publique étaient devenues plus fréquentes en Prusse, en Hesse, à Bade, en Bavière, comme d'ailleurs en Autriche, et les tendances à la laïcisation scolaire commençaient à produire leurs effets dans la législation [1]. Si le ministre prussien Muehler proclamait encore en 1869 l'impossibilité de rompre « une alliance intime et plus que millénaire entre la culture et la religion, entre l'École et l'Église » [2], il se heurtait à la Chambre à l'opposition croissante des libéraux, tandis qu'en Bavière un projet scolaire très favorable aux idées de laïcisation était déposé, à la fin de 1867, avec l'approbation d'un premier ministre catholique mais anticlérical, le prince de Hohenlohe. Sur d'autres terrains encore, les catholiques se sentaient menacés : en Prusse, la modification de l'organisation de la « division catholique », en 1861, et la mort de Geissel, en 1864, les laissaient affaiblis contre l'offensive antireligieuse qui se préparait et, un peu partout, l'hostilité renaissait contre les ordres religieux, les jésuites en particulier.

Devant le danger, les tendances caractéristiques de la décade précédente s'accusèrent davantage : entente plus étroite entre les évêques des différentes régions d'Allemagne, qui décidèrent même en 1867, à l'instigation de Ketteler, de se réunir désormais régulièrement à Fulda, autour du tombeau de saint Boniface ; regroupement des forces catholiques dans un sens plus combatif lors des *Katholikentage*, en particulier à Innsbruck en 1867 et à Bamberg en 1868, où la lutte scolaire fut à l'avant-plan des préoccupations ; développement de l'action sociale ; organisation politique enfin. Sur ce dernier point comme sur tant d'autres, les catholiques de Prusse étaient en avance et si les conflits de Bismarck avec la Chambre condamnèrent le parti du Centre à une éclipse temporaire, ses dirigeants, Malinckrodt, Reichensperger et les autres, purent, dans la retraite, tirer parti des premières expériences et préparer, notamment au cours des réunions tenues à Soest à l'initiative de Hueffer, l'évolution qui devait faire d'un parti religieux un parti politique en contact étroit avec les préoccupations de l'opinion publique. Pendant ce temps, en Bavière, une partie du clergé, soutenue par les *Historisch-politische Blätter*, s'appliquait à faire « apparaître les adversaires de l'Église comme étant aussi les ennemis des libertés bavaroises » [3] et réussissait, en spéculant sur les préoccupations autonomistes, à soulever contre les projets scolaires du ministère Hohenlohe une opinion catholique jusqu'alors indifférente.

---

(1) Voir les faits dans Brueck-Kissling, *Geschichte der Katholischen Kirche im XIX. Jhdt.*, t. III, 4e partie, La lutte pour l'école. Cf. aussi G. Goyau, *op. cit.*, t. III, p. 282 et suiv. ; t. IV, p. 92, 101 et suiv., 167 et suiv.

(2) Cité dans Brueck-Kissling, *op. cit.*, t. III, p. 462.

(3) G. Goyau, *L'Allemagne religieuse*, t. IV, p. 126.

LE « KULTURKAMPF » BADOIS En Bavière comme en Prusse, les difficultés auxquelles avaient à faire face les catholiques n'étaient encore que les premières escarmouches d'une lutte qui ne deviendrait aiguë que le jour où Bismarck, ayant mené à bonne fin sa guerre contre la France, n'aurait plus besoin de leur appui. Dans le grand-duché de Bade, au contraire, l'offensive contre l'Église se développa ouvertement dès les années soixante, surtout depuis qu'au ministre anticlérical Lamey avait succédé, en 1866, un ministre plus anticlérical encore, Jolly : laïcisation de l'école primaire par la loi de 1864, aggravée par celle de 1868 et par diverses mesures administratives ; laïcisation des fondations pieuses ; institution du mariage civil obligatoire ; mesures contre les congrégations ; prétention de contrôler l'éducation des clercs afin de « germaniser les prêtres » qui voulaient « romaniser les masses » [1] ; pression sur le chapitre de Fribourg, à la mort de Mgr Vicari, pour influencer le choix du nouvel archevêque.

Jolly, encouragé en sous-main par Bismarck, mettait à la base de sa politique « le grand principe que l'Église est dans l'État et soumise à l'État » et, opposant la « logique laïque » à la logique romaine, il estimait qu'il fallait « remettre en honneur la pensée fondamentale du joséphisme et trouver de nouvelles formes » [2]. La seule différence pour l'Église, c'est qu'au lieu d'être opprimée comme jadis par une bureaucratie aux ordres d'un prince, elle l'était à présent par une majorité parlementaire : le grand-duché de Bade devenait de la sorte « une sorte de terrain d'expérience sur lequel s'essayaient à l'avance les maximes et les méthodes du *Kulturkampf* » [3]. Mais les catholiques à leur tour, sous la direction du député Lindau, l'organisateur du *parti catholique populaire*, apprenaient à résister aux bourgeois libéraux en s'appuyant sur les paysans et les artisans.

LA POLITIQUE ANTICONCORDATAIRE EN AUTRICHE Le concordat de 1855 s'était imposé aux milieux dirigeants d'Autriche à la faveur de la réaction consécutive à 1848, mais il n'avait jamais été populaire [4]. Même de bons catholiques le considéraient comme une capitulation excessive du gouvernement en faveur d'une conception théocratique périmée ; *à fortiori*, tous ceux qui restaient imprégnés des tendances joséphistes souhaitaient-ils voir restaurer les droits de l'État sur l'Église, et l'opinion libérale s'insurgeait en outre contre le caractère officiellement catholique maintenu à de nombreux aspects de la structure sociale de l'État autrichien ; enfin, la manière trop raide dont Rauscher fit appliquer certains articles mit en relief aux yeux des protestants les restrictions apportées en matière de liberté des cultes.

C'est sur le terrain de la parité de droit entre les deux confessions

---

(1) Baumgarten-Jolly, *Staatsminister Jolly*, Tubingue, 1897, p. 110.
(2) *Ibid.*, p. 44.
(3) G. Goyau, art. *Allemagne*, dans *D. H. G. E.*, t. II, col. 583.
(4) Voir notamment *Das Tagebuch des Polizeiministers Kempen von 1848 bis 1859*, édit. J.-K. Mayer, p. 48 et suiv. et surtout p. 368-397 ; cf. H. Singer (*art. cit.* à la p. 132, n. 1), p. 205-209.

que l'opposition se plaça d'abord. Les évêques, encouragés par Pie IX, refusèrent de céder en invoquant le fait que l'Autriche était un État catholique, où l'Église romaine avait droit à des égards spéciaux. Personnellement, l'empereur François-Joseph était prêt à les soutenir, très conscient qu'il était de ses responsabilités à l'égard de Dieu et de son devoir de protéger l'Église [1], mais il fut obligé, pour des raisons de politique intérieure, de céder aux libéraux et au premier ministre von Beust. L'attitude de ce dernier n'était d'ailleurs pas inspirée, comme on l'a dit parfois, par ses convictions protestantes, ni par une hostilité systématique à la religion, mais par des motifs purement politiques : il devait ménager les forces anticatholiques dont la presse et la représentation parlementaire l'emportaient de loin sur leur importance réelle dans le pays, et il recherchait l'alliance des libéraux d'Allemagne du Sud en même temps qu'un rapprochement avec l'Italie nouvelle [2].

On essaya d'abord d'obtenir du Saint-Siège qu'il renonçât à l'amiable à certains points du concordat, en matière de mariages mixtes notamment, mais les tractations menées dans ce but par Fessler en 1863 se heurtèrent à l'intransigeance de Rome, où l'on invoquait les stipulations du concile de Trente et l'impossibilité de reconnaître aux hérétiques les mêmes droits qu'aux fidèles. Rauscher s'efforça de gagner du temps, mais les défaites militaires de 1866 vinrent fournir un aliment nouveau à l'opposition, qui entreprit d'en rejeter la responsabilité sur l'éducation que recevaient les jeunes Autrichiens dans les « écoles concordataires ». Il y avait beaucoup de mauvaise foi dans cette campagne, car l'influence de l'Église sur l'école datait de bien avant le concordat et si, effectivement, les autorités ecclésiastiques se préoccupaient souvent plus de la formation religieuse que du progrès scientifique, le caractère arriéré de l'enseignement tenait surtout à l'insuffisance des moyens financiers que lui consacrait le gouvernement : par rapport à la première moitié du XIXe siècle, l'amélioration depuis le concordat était incontestable, encore qu'insuffisante [3]. Mais en mettant en relief dans un texte officiel le caractère « clérical » du système scolaire autrichien, le concordat offrait une prise facile aux critiques libérales et il allait finalement entraîner pour un temps la suppression du caractère confessionnel de l'école, qui put au contraire être maintenu dans les autres pays germaniques.

En effet, à la fin de 1867, le gouvernement déposa trois projets de loi, en contradiction manifeste avec le concordat, restituant la juridiction matrimoniale aux tribunaux civils, accordant aux communautés protestantes l'égalité de droits, et rendant à l'État la direction et la surveillance exclusive de l'enseignement public, tout en laissant à l'Église

(1) Sur l'attitude religieuse de l'empereur, voir M. HUSSAREK, Die Krise und die Lösung des Konkordats (cité p. 147, n. 1), p. 388-391.
(2) Voir F.-W. EBELING, Fr. F. Graf von Beust, t. II, Leipzig, 1871, p. 436 et suiv., 448 et suiv.) 457, 467 et suiv., 484. Beust lui-même expose son point de vue dans ses Mémoires (Paris, 1888. et reproduit, t. II, p. 133-135, une lettre significative écrite le 10 octobre 1867 au cardinal Rauscher.
(3) Cf. H. SINGER (art. cit. à la p. 132, n. 1), p. 229-234. Sur l'ensemble du problème scolaire, cf. STRAKOSCH-GRASSMANN, Bibliographie zur Geschichte des österreichischen Unterrichtswesens, Kornenburg, 1901 et suiv. et A. GRUNDL, A. Krombholz (1790-1869), ein deutscher Priester und Schulorganisator aus Böhmen, Prague, 1937.

le droit d'ouvrir des écoles libres. On essaya encore d'obtenir le consentement du Saint-Siège à ces modifications, mais celui-ci avait de la peine à se rendre compte de l'évolution des idées et il estima impossible de renoncer spontanément à un concordat qui faisait de l'Autriche, comme le disait Rauscher, « le seul pays où les droits de l'Église étaient encore reconnus dans leur intégralité »[1]. A l'ambassadeur qui faisait observer que Rome avait accepté le mariage civil dans le concordat français de 1801, Antonelli répondit qu'il s'était alors agi de reconstruire en sauvegardant ce qui pouvait encore être sauvé, tandis que, dans le cas de l'Autriche, on ne pouvait renoncer à une situation encore intacte[2]. Le gouvernement décida donc de mettre Rome devant le fait accompli et, le 25 mai 1868, les trois lois furent votées.

Certains évêques ne se contentèrent pas de protestations platoniques, mais refusèrent de transférer aux magistrats civils les actes des procès matrimoniaux ; ce fut le cas dans le Tyrol, où l'on était très ultramontain, et l'évêque de Linz, Mgr Rudigier, fut même condamné par les tribunaux. Mais Rauscher, qui tenait à tout prix au système concordataire, manœuvra de façon à ne pas pousser le gouvernement à bout. A Rome aussi, on désirait éviter la rupture avec l'un des rares États dont on espérait malgré tout quelque appui dans la question romaine[3] : on ne rappela pas le nonce, contrairement à ce qui avait été annoncé d'abord, et la modération du ton de la protestation pontificale du 22 juin trancha sur la sévérité qui avait accueilli l'introduction de mesures analogues en Italie et au Mexique.

(1) Rauscher à Schwarzenberg, 6 mars 1867, dans C. WOLFSGRUBER, *Kardinal Schwarzenberg*, t. III, p. 3.
(2) Lettre de Rome du 20 février 1868, dans C. WOLFSGRUBER, *op. cit.*, t. III, p. 26-27.
(3) Cf. S. JACINI, *Il tramonto del potere temporale*, p. 239 ; voir aussi p. 242.

CHAPITRE VI

L'ÉGLISE DANS LE RESTE DE L'EUROPE OCCIDENTALE JUSQU'AU CONCILE DU VATICAN

§ 1. — Les Iles Britanniques [1]

LE « SECOND PRINTEMPS » DE L'ÉGLISE D'ANGLETERRE

La décision de créer d'emblée treize sièges épiscopaux en Angleterre était un acte de confiance en l'avenir, que certains trouvaient un peu prématuré : « Nous ne sommes pas mûrs pour la hiérarchie, écrivait Newman à un ami en février 1851 ; à présent qu'ils l'ont obtenue, ils sont tout à fait incapables de pourvoir aux sièges [2]. » Effectivement, la chose n'alla pas sans peine, ce qui se comprend si l'on songe que le clergé ne comptait guère plus de 800 prêtres, assez divisés au surplus sur les méthodes à suivre. Tout finit cependant par s'arranger et, en juillet 1852, put se réunir à Oscott le premier concile de la province de Westminster [3], dont le but était de rétablir dans tous les domaines de la vie ecclésiastique le régime normal prévu par le droit canon et que le cardinal Wiseman, au zénith de sa carrière et débordant d'optimisme, dirigea avec une maîtrise incomparable.

Au cours du sermon d'ouverture, Newman avait célébré « le second printemps de l'Église d'Angleterre » [4]. Les années qui suivirent ne réalisèrent pas toutes les espérances conçues en ces jours d'euphorie, mais le catholicisme anglais connut cependant un développement sensible, dont Wiseman put tracer le tableau avec fierté lors du congrès de Malines de 1863 [5]. Tandis que les conversions continuaient, bien qu'entourées de moins de publicité, et commençaient à s'étendre au monde des affaires et du barreau qui n'avait guère été touché au début par le mouvement d'Oxford, l'afflux des Irlandais s'était poursuivi. Des villes comme Manchester ou Liverpool devenaient de plus en plus catholiques et, grâce à la générosité combinée des riches convertis et des pauvres immigrants, les églises commençaient à se multiplier. Leur nombre était passé de 597 en

(1) BIBLIOGRAPHIE. — Sur l'Église en Angleterre : voir les ouvrages généraux ainsi que les biographies de Wiseman, Manning, Newman, Ward, et surtout Ullathorne, signalés p. 9 ; ajouter E. PURCELL, *Life and Letters of Ambrose Phillips de Lisle*, Londres, 1900.

Sur l'Église en Irlande : A. BELLESHEIM, *Geschichte der katholischen Kirche in Irland*, t. III, Mayence, 1891, p. 487-600 ; en outre P. CULLEN, *Pastoral Letters and other Writings*, Dublin, 1882, 3 vol.

Sur l'Église en Écosse : A. BELLESHEIM, *Geschichte der katholischen Kirche in Schotland*, t. II, Mayence, 1883, chap. XII, à compléter par J.-H. HANDLEY, *The famine and the developpement of the Church in Schotland*, dans *Irish ecclesiastical Record*, t. LXIX, 1947.

(2) Newman à Capes, 18 février 1851, dans W. WARD, *Life of Newman*, t. I, p. 260.

(3) Cf. *Coll. lac.*, t. III, col. 895-970.

(4) *Sermons preached on various occasions*, n° 10.

(5) *Situation des catholiques en Angleterre*, dans *Assemblée générale des catholiques en Belgique. Première Session à Malines*, t. I, Bruxelles, 1864, p. 241-264.

1850 à 872 en 1863, tandis que celui des prêtres passait de 826 à 1.242, celui des couvents d'hommes de 11 à 55 et celui des maisons de religieuses de 59 à 108.

« Notre côté faible, ajoutait Wiseman [1], c'est l'éducation des enfants que notre pauvreté nous empêche de cultiver comme nous le voudrions. » Du moins le *Comité des Écoles pauvres*, présidé par un homme d'œuvres au dévouement inlassable, Charles Langdale, avait-il réussi à obtenir des pouvoirs publics quelques avantages : un certain crédit pour les écoles catholiques destinées aux enfants du peuple, et la mise à part des catholiques dans les maisons de rééducation pour jeunes délinquants et enfants abandonnés. En outre, ce comité, qui était devenu l'organe officiel des catholiques auprès du gouvernement, avait su obtenir de ce dernier plusieurs concessions appréciables : en 1858, des aumôniers catholiques pour l'armée (dont le quart de l'effectif était formé d'Irlandais) ; en 1862, des chapelains catholiques dans certaines prisons et dépôts de mendicité (*Workhouses*).

LA « *POLITIQUE DE PRÉSENCE* » DE WISEMAN

Persuadé que l'une des grandes faiblesses du catholicisme anglais était son isolement à l'égard du reste de la nation, Wiseman s'appliqua inlassablement à promouvoir un rapprochement. Il multiplia ses conférences sur l'art, la science ou les réformes philanthropiques dans le but de montrer aux anglicans « que nous pouvons aussi bien qu'eux donner au public un régal intellectuel et que nous ne nous intéressons pas moins qu'ils ne le font au progrès du peuple » [2]. L'opinion publique découvrait avec étonnement dans cet ecclésiastique catholique un *gentleman* anglais et sa popularité rejaillissait sur son Église. « Les conférences de Votre Éminence, lui écrivait Ambrose Phillips, font plus — mille fois plus — que toutes les controverses du monde pour gagner le cœur de la vieille Angleterre [3]. »

En même temps, Wiseman continuait à encourager autant qu'il le pouvait l'action analogue menée par les convertis, notamment par Newman, qui dans ses célèbres *Lectures* sur les *Difficultés soulevées par les Anglicans* ou sur *La situation présente des catholiques en Angleterre* s'en prit aux préjugés de John Bull à l'égard de ses nouveaux coreligionnaires et réussit, en maniant avec une maîtrise supérieure toutes les nuances de l'ironie, à mettre les rieurs de son côté. Wiseman appréciait la culture supérieure, les vues progressistes et l'enthousiasme conquérant des convertis, qui tranchaient sur la timidité des « anciens catholiques ». En dépit d'oppositions dont certaines demeurèrent irrésistibles, il s'efforça de hâter la fusion entre ces deux éléments si différents par l'origine et la formation, soutenu d'ailleurs sur ce point par l'aide intelligente de son collègue de Birmingham, Mgr Ullathorne.

(1) *Situation des catholiques en Angleterre*, dans *Assemblée générale des catholiques en Belgique, Première session à Malines*, t. I, Bruxelles, 1864, p. 245.
(2) Wiseman au secrétaire de l'Institut de Leeds, 10 janvier 1853, dans W. WARD, *Vie du cardinal Wiseman*, t. II, Paris, 1900, p. 63.
(3) Cité *ibid.*, t. II, p. 173.

L'UNIVERSITÉ DE DUBLIN Wiseman se préoccupait particulièrement de relever le niveau intellectuel du catholicisme anglais. C'est dans ce but qu'il avait nommé, au grand déplaisir de plusieurs, Ward, converti et père de famille, comme professeur de philosophie puis de théologie dogmatique dans son séminaire, dont il déplorait l'enseignement routinier.

C'est dans ce but aussi qu'en 1851 il poussa Newman à répondre à l'appel de l'épiscopat irlandais qui lui offrait le rectorat de l'Université catholique de Dublin, dont la fondation avait été décidée l'année précédente, conformément aux vœux de la Propagande [1]. Le succès de l'Université libre de Louvain, dans un pays moins peuplé que l'Irlande, semblait indiquer que le projet n'était pas chimérique et on pouvait espérer que la nouvelle université pourrait devenir un foyer de formation pour les jeunes catholiques anglais tout aussi bien que pour ceux d'Irlande : pour Wiseman comme pour Newman, il s'agissait au fond d'établir en Irlande un nouvel Oxford à l'usage des catholiques de langue anglaise.

Quelques mois plus tard, Newman exposa son programme dans une série de conférences remarquables [2] : « Une université n'est ni un couvent, ni un séminaire ; c'est un lieu où l'on prépare des hommes du monde, pour le monde. » Malheureusement ni le clergé irlandais, plus zélé que cultivé, ni les évêques, dont aucun n'avait reçu de formation universitaire, n'étaient préparés à entrer dans ces vues [3] et Newman ne tarda pas à se rendre compte qu'il travaillait au milieu de l'indifférence générale, bien plus, que l'épiscopat semblait moins préoccupé de soutenir ses efforts que de l'empêcher de prendre des initiatives trop personnelles. La méfiance séculaire entre Anglais et Irlandais acheva de rendre l'entreprise inviable. Tandis que les Irlandais s'offusquaient de la préférence donnée aux convertis anglais dans la composition du corps professoral, Newman dut bientôt se rendre à l'évidence : des *gentlemen* anglais n'enverraient jamais leurs fils dans cette Irlande méprisée. Ce n'est toutefois qu'en 1858 qu'il abandonna une œuvre à laquelle il s'était donné sans compter pendant sept ans [4] et qu'il chercha à réaliser sur d'autres terrains ce qui lui appa-

---

(1) Au moment de l'élection de Pie IX, l'épiscopat irlandais se trouvait divisé sur l'attitude à prendre à l'égard de la nouvelle loi sur l'enseignement supérieur déposée par Peel le 9 mai 1845. La plupart des évêques, soutenus par le clergé et l'opinion publique, estimaient dangereux pour la foi ce projet qui prévoyait l'établissement, dans les Universités officielles, de « colleges » non confessionnels, tandis que les archevêques de Dublin et d'Armagh et quatre autres évêques, trouvant cette opposition déraisonnable, essayaient de s'entendre avec le gouvernement. Le 9 octobre 1847, un rescrit de la Propagande désavouait ces derniers, tout en reconnaissant leur bonne foi (*Coll. lac.*, t. III, col. 802), mais ils n'en continuèrent pas moins leurs tractations avec le vice-roi. Aussi, au printemps de 1848, une délégation conduite par l'archevêque de Tuam remit au pape un mémoire, signé par 19 évêques, reprenant toutes les objections contre la loi et déplorant l'attitude de leurs collègues. Ceux-ci essayèrent d'agir de leur côté à Rome, où ils avaient des intelligences, mais le 11 octobre 1848, un nouveau rescrit confirmait le précédent et condamnait à nouveau les collèges royaux (*Coll. lac.*, t. III, col. 803-804). L'archevêque de Dublin, Murray, essaya bien deux ans plus tard d'agir une nouvelle fois à Rome dans le sens de la conciliation, mais l'insuccès complet de sa démarche le décida enfin à se rallier à l'avis de la majorité. (Voir sur cette question A. BELLESHEIM, *Gesch. der kath. Kirche in Irland*, t. III, p. 472-480 et 515.)

(2) Réunies plus tard en volume sous le titre : *On the Idea of a University*.

(3) « *Their idea was a glorified seminary for the laity* », déclare Butler, en se basant sur les souvenirs de son père, qui fut professeur et vice-recteur à l'Université de Dublin et très lié avec Newman. (*Lifes and Times of B. Ullathorne*, t. II, p. 312-313.)

(4) Sur tout cet épisode, consulter W. WARD, *Life of Newman*, t. I, chap. XI à XIII.

raissait comme l'œuvre essentielle de sa vie : se faire l'apôtre de la pensée, de l'éducation et de la culture catholiques en Angleterre.

LA CONVERSION DE MANNING — Ce n'étaient pas seulement des convertis éprouvés comme Newman ou Ward que Wiseman poussait en avant. Il agissait de même avec ceux de la veille. Le 15 juin 1851, il ordonnait prêtre l'ancien archidiacre de Chichester (l'équivalent d'un vicaire général catholique), Henry Édouard Manning [1], qui avait abjuré l'anglicanisme deux mois auparavant, et il n'allait pas tarder à en faire l'un de ses principaux collaborateurs.

Rien de plus dissemblable pourtant que ces deux hommes : « D'une part Wiseman, avec son imposante corpulence, son mélange de bonhomie et de somptuosité, son exubérance de dons naturels, son esprit ouvert au point d'être un peu dispersé, son imagination toujours en mouvement, sa sensibilité généreuse mais facile à blesser, son inaptitude administrative, sa volonté mobile et prompte à se dérober devant les obstacles ; d'autre part Manning, avec ses qualités d'homme d'État et de diplomate, sa puissance et sa fixité de volonté, son autorité facilement impérieuse, sa dignité austère, sa distinction un peu froide, et ce je ne sais quoi, gardé de son ancien état, qui faisait dire de lui au cardinal : du sommet de la tête à la pointe des pieds, il est un *parson* [2]. » Toutefois, en dépit de ces différences marquées de tempérament, Wiseman et Manning s'étaient dès le premier moment sentis en parfaite communion d'idées quant à la politique ecclésiastique à suivre. En outre, après trois années partagées entre l'apostolat en Angleterre et l'étude à Rome, où il fut vite remarqué par Pie IX, Manning s'était décidé à réaliser l'un des projets les plus chers du cardinal, déçu par le peu d'intérêt des ordres religieux pour le ministère populaire [3] : la fondation d'une communauté de prêtres rappelant les missionnaires diocésains français, et s'inspirant des règles de saint Charles Borromée, d'où le nom d'*Oblats de Saint-Charles*.

L'AFFAIRE ERRINGTON — L'influence croissante de Manning n'allait pas sans susciter le mécontentement du vieux clergé, groupé autour du coadjuteur de Westminster, Mgr Errington, administrateur consciencieux mais esprit étroit et obstiné. Celui-ci croyait déceler chez l'ancien archidiacre anglican un esprit d'intrigue et d'ambition, mais il y avait surtout, à la base de son opposition, une hostilité de principe contre l'enthousiasme romain des convertis, qui les poussait à introduire de nouvelles formes de piété et à souhaiter la multiplication des interventions du pape dans la vie catholique anglaise.

Wiseman ne tarda pas à se rendre compte de l'erreur de jugement

---

(1) Sur Manning (1808-1892), l'ouvrage de base reste encore, malgré son caractère tendancieux, celui de E. Purcell, *Life of Manning*, Londres, 1896, 2 vol., précieux par les très nombreux documents inédits qu'il reproduit ; voir aussi H. Hemmer, *Vie du cardinal Manning*, 2e éd., Paris, 1897. Le portrait donné par Lytton Strachey (*Victoriens éminents*, Paris, 1933, p. 19-142) contient certains traits justes, mais est trop sévère.

(2) P. Thureau-Dangin, *La renaissance catholique en Angleterre*, t. II, p. 286-287.

(3) Voir à ce sujet une longue lettre de Wiseman au P. Faber du 27 octobre 1852 dans W. Ward, *Le cardinal Wiseman*, t. II, p. 136-149.

qu'il avait commise en appelant un tel homme à lui succéder et il aurait
volontiers donné raison à Manning lorsque celui-ci écrivait, à propos
d'Errington : « Son accession au siège de Westminster déferait tout
l'ouvrage accompli par Wiseman depuis le rétablissement de la hiérarchie
et ferait reculer le progrès du catholicisme pour toute une génération [1]. »
Aussi, dès le début de 1859, le cardinal entreprit-il des démarches à
Rome pour obtenir l'éloignement de son coadjuteur, mais la chose n'alla
pas sans peine, car Errington, profondément froissé de ce qu'on semblait
suspecter ses sentiments à l'égard du Saint-Siège, entendait faire valoir
son bon droit. Il ne céda que sur un ordre formel du pape, en juillet
1860 [2].

LES DERNIÈRES ANNÉES   Wiseman souffrit d'autant plus de ce conflit
   DE WISEMAN   qu'il se rendait compte que son coadjuteur
                        était appuyé à Rome par une partie des
évêques anglais. Ceux-ci, qui estimaient que l'affaire avait été menée
sans beaucoup de *fair-play*, supportaient de plus en plus difficilement
les manières d'agir du cardinal que son tempérament autocratique,
excité par Manning, dont la nature était encore plus autoritaire que la
sienne, poussait à imposer ses vues à ses suffragants sans trop tenir
compte des règles canoniques. A partir de 1859, la résistance des
évêques, groupés autour de Mgr Ullathorne et de Mgr Clifford, devint
aiguë. Finalement, grâce à l'appui du cardinal Barnabo, préfet de la
Propagande, et malgré la sympathie personnelle de Pie IX pour Wiseman
et Manning, ils obtinrent gain de cause à Rome sur plusieurs points
importants [3].

Cet échec assombrit les dernières années du cardinal, dont les forces
déclinaient rapidement. Il eut pourtant encore, avant de mourir, la
grande joie d'assister, en 1864, au succès extraordinaire de l'*Apologia
pro vila sua* de Newman. Relevant le défi de Kingsley, qui avait mis en
doute sa loyauté, Newman entreprit d'exposer l'histoire de sa vie et les
motifs de sa conversion afin de faire l'opinion publique juge d'un différend
dont la portée profonde dépassait de loin son cas personnel. La sincérité de
son accent toucha le cœur de la nation anglaise et, en quelques semaines,
il regagna auprès de ses compatriotes anglicans une faveur qui ne le
quittera plus et dont le catholicisme bénéficia largement. L'un des journalistes anglicans les plus en vue, Hutton, estimait que l'*Apologia* avait

---

(1) Lettre à Mgr Talbot, dans E. PURCELL, *Life of Manning*, t. II, p. 171.
(2) Le meilleur exposé de cette affaire se trouve dans W. WARD, *Le cardinal Wiseman*, t. II,
chap. XXIV, XXV et XXVII. On y trouvera de nombreux documents (de même que dans les appen-
dices D, E et F de l'édition anglaise) auxquels il faut ajouter un mémoire de F. Rymer publié
dans C. BUTLER, *Life and Times of B. Ullathorne*, t. I, p. 278-306 et les lettres publiées dans LESLIE,
Life of Manning (en appendice), dans *Dublin Review*, janvier 1923, ainsi que dans E. PURCELL,
Life of Manning, t. II, chap. V, mais l'exposé de l'affaire dans ce dernier ouvrage contient plusieurs
grosses inexactitudes (cf. W. WARD, *Life and Times of cardinal Wiseman*, t. II, p. 579-585), concernant le rôle de Manning, notamment. (Cf. aussi C. BUTLER, *op. cit.*, t. I, p. 271-273.)
(3) Consulter surtout C. BUTLER, *op. cit.*, t. I, p. 217-256 et p. 276. L'auteur, qui s'appuie sur
un dossier de lettres inédites, estime que si la plupart des évêques manquaient un peu d'envergure et n'avaient ni le coup d'œil ni l'allant de Wiseman, ils étaient cependant des pasteurs zélés
et consciencieux, et que leurs griefs contre Wiseman étaient fondés. Les points faibles de Wiseman comme administrateur sont également bien mis en lumière par D. GWYNN, *Cardinal Wiseman*, Dublin, 1950, chap. XI et XII.

plus fait que toute la littérature religieuse contemporaine pour venir à bout de la méfiance que l'on éprouvait en Angleterre à l'égard de l'Église romaine. Le rêve de Wiseman commençait à se réaliser, les catholiques recommençaient à faire entendre leur voix dans la nation.

LA SUCCESSION DE WESTMINSTER A la mort de Wiseman, en février 1865, le chapitre de Westminster, appuyé par plusieurs évêques, marqua son opposition à la politique du défunt cardinal en proposant comme successeur Mgr Errington. Ce geste indisposa fortement Pie IX et l'un de ses conseillers les plus écoutés, Mgr Talbot, ami personnel de Manning, en profita pour lui suggérer habilement la candidature de ce dernier, qui était en même temps appuyée auprès du cardinal de Reisach par le P. Coffin, un converti de 1845 devenu provincial des rédemptoristes d'Angleterre. Mais le cardinal Barnabo, redoutant d'accroître encore la tension au sein du catholicisme anglais, obtint de la Propagande qu'elle recommandât au pape de porter plutôt son choix sur Mgr Ullathorne, évêque de Birmingham, que Manning avait, deux ans auparavant, proposé comme coadjuteur à Wiseman et qui, à mesure que les forces de ce dernier déclinaient, avait pris dans l'épiscopat anglais un rôle prépondérant. Le pape, après plusieurs semaines d'hésitation, décida de passer outre à l'avis de la Congrégation et, le 30 avril, par une sorte de petit coup d'État qui fit sensation, il nomma Manning archevêque de Westminster. On eut d'ailleurs l'agréable surprise de voir cette décision pontificale accueillie par tous, évêques, prêtres et laïcs, avec une parfaite bonne grâce, qui montrait qu'on s'était exagéré leurs sentiments d'indépendance à l'égard de Rome.

ACTIVITÉ PASTORALE DE MANNING Sans manifester la moindre rancune, Manning s'empressa de faire appel à toutes les bonnes volontés pour poursuivre la réorganisation de la vie catholique à laquelle il collaborait depuis déjà dix ans. L'augmentation constante de la population catholique ne lui faisait pas illusion : il se rendait compte que le nombre des émigrants irlandais abandonnés à eux-mêmes qui perdaient la foi dans les grands centres d'Angleterre l'emportait nettement sur celui des conversions. Aussi, soucieux de toucher le plus efficacement possible les catholiques disséminés dans les quartiers populaires, multiplie-t-il les chapelles et les écoles, dont Wiseman avait déjà doublé le nombre, divisant les grandes paroisses en « missions » de peu d'étendue, dont les prêtres peuvent connaître personnellement leurs ouailles. Dès 1866, il met sur pied le *Westminster Diocesan Education Fund*, la plus réussie peut-être de ses entreprises, qui lui permit de faire passer de 11.245 à 22.580 le nombre d'enfants recevant une éducation catholique.

N'oubliant pas qu'il ne suffit pas de multiplier les centres religieux, il réussit à porter à 350 le nombre des prêtres de son diocèse, qui avait déjà passé de 120 à 210 sous son prédécesseur, et comme le nombre n'est pas tout, il s'inquiète aussi de la formation de ce clergé, n'hésitant pas à lui proposer un haut idéal de sainteté qu'il développera dans un de ses livres les plus réputés : *Du sacerdoce éternel.*

Le souci de l'apostolat populaire ne détourne cependant pas Manning des autres aspects du problème catholique en Angleterre qui avaient tant préoccupé son prédécesseur. S'il s'était lancé dans l'affaire Errington avec un acharnement qui avait de quoi surprendre, c'est qu'il y voyait un élément décisif pour la solution de la question fondamentale, celle de « savoir si, oui ou non, l'Église d'Angleterre se contenterait de se confiner dans une dispensation un peu meilleure des sacrements à la petite communauté catholique établie dans le pays, ou bien si elle se mêlerait à la vie de la nation, agissant sur son esprit par une sérieuse culture catholique et sur sa volonté par un plus large et plus vigoureux emploi des énergies mises en jeu par la restauration de la hiérarchie » [1]. Or, ici aussi s'imposait une amélioration dans la formation du clergé, non plus sur le plan surnaturel cette fois. Le souvenir que Manning avait gardé de ses anciens collègues anglicans lui faisait sentir ce qui manquait au clergé catholique, et particulièrement aux prêtres irlandais, nombreux en Angleterre, « non pas certes comme zèle et dévouement, mais comme distinction sociale et culture intellectuelle » [2]. Il devait rester préoccupé jusqu'à la fin de sa vie de cette cause de faiblesse qui empêchait le clergé d'avoir prise sur la société anglaise.

LA QUESTION DES UNIVERSITÉS   Toutefois, si Manning décèle aussi clairement que Wiseman les lacunes du catholicisme anglais et s'il cherche avec une activité inlassable à en relever le niveau, il est par contre bien plus intransigeant que lui à l'égard de tout rapprochement avec l'anglicanisme. Il l'avait déjà montré sous Wiseman à propos de l'*Association en vue de promouvoir l'union des chrétiens,* qu'il avait beaucoup contribué à faire désavouer par Rome [3]. Il fit preuve de la même intransigeance dans la question de la fréquentation des universités par les catholiques [4].

Certaines familles, qui attachaient un grand prix aux avantages intellectuels et surtout sociaux assurés par un passage à Oxford ou Cambridge, commençaient à y envoyer leurs fils. Ceux qui souhaitaient voir les catholiques sortir de leur isolement et considéraient l'échec de l'Université de Dublin comme une expérience décisive, regardaient cette évolution comme un moindre mal, à condition que des mesures efficaces fussent prises pour protéger la foi des jeunes gens. Telle était la position de Wiseman [5], celle aussi d'Ullathorne, homme de tradition, hostile aux idées avancées, mais à qui l'intransigeance répugnait tout autant. Newman, qui se rendait compte mieux que personne des dangers qu'entraînerait pour les

(1) Mémoire justificatif composé par Manning à l'occasion de l'affaire Errington, dans W. WARD, *Le cardinal Wiseman,* t. II, p. 393.
(2) P. THUREAU-DANGIN, *La Renaissance catholique...,* t. III, p. 13.
(3) Cf. *infra,* p. 484.
(4) On consultera sur cette question W. WARD, *Life of cardinal Newman,* t. II, chap. XXI et XXIV à XXVI, ainsi que E. PURCELL, *Life of Manning,* t. II, chap. XIII, mais le meilleur exposé d'ensemble se trouve dans C. BUTLER, *Life and Times of B. Ullathorne,* t. II, chap. XIII.
(5) C. BUTLER, *op. cit.,* t. II, p. 1-2, signale de lui des lettres inédites qui sont absolument claires.

jeunes catholiques la fréquentation des universités anglicanes, mais qui savait aussi qu'il était vain de vouloir élever en serre chaude ceux qui étaient appelés à agir demain au milieu du monde, voyait la solution dans un apostolat spécialisé auprès des étudiants et proposa, au début de 1864, de venir s'établir à Oxford pour s'occuper d'eux.

Manning était, lui, radicalement hostile à l'idée de voir des étudiants catholiques à Oxford, inquiet à l'idée qu'au contact des milieux protestants, l'élite catholique anglaise risquait de « devenir semblable à celle de France, catholique de nom mais indifférente, laxiste et libéralisante »[1]. Tandis qu'il usait de toute son influence sur le cardinal et qu'il encourageait Ward à entreprendre une campagne dans la *Dublin Review*, il agissait à Rome grâce à son ami Talbot, afin d'obtenir une condamnation nette, analogue à celle qui avait interdit la fréquentation des *Queen's Colleges* en Irlande, quelques années auparavant[2]. A Rome cependant, on estima prudent de ne pas aller aussi loin et on se borna à ratifier la décision de la réunion des évêques du 13 décembre 1864 qui avait, à la quasi-unanimité, *déconseillé* la présence d'étudiants catholiques à Oxford.

Newman renonça immédiatement à son projet, mais son évêque, Mgr Ullathorne, revint à la charge et finit par obtenir de la Propagande, en décembre 1866, l'autorisation d'établir un Oratoire à Oxford, avec la réserve toutefois que, si Newman désirait s'y fixer, il faudrait l'en dissuader *blande suaviterque*[3]. Newman, déçu et froissé, eut d'ailleurs l'occasion de constater que Pie IX aussi bien que le cardinal Barnabo, informés unilatéralement par Manning et Talbot, étaient plus hostiles que jamais à l'idée d'une formation universitaire mixte.

ACCENTUATION DES DIVERGENCES ENTRE CATHOLIQUES
Les préventions de Manning contre Newman avaient certainement joué un rôle important dans toute cette affaire. Manning, l'homme d'action, facilement impérieux, absolu dans ses idées, parfois même fanatique quand il se croit guidé par une inspiration du Saint-Esprit, incapable de sympathiser avec les difficultés intellectuelles d'autrui, pouvait difficilement comprendre Newman, le penseur subtil, tout en nuances, excellant à saisir les points de vue différents du sien et compréhensif devant les perplexités de l'esprit humain[4]. Si Manning n'allait pas jusqu'à penser comme Talbot que « le Dr Newman était l'homme le plus dangereux d'Angleterre »[5], il croyait cependant de son devoir de ruiner une influence qu'il estimait pernicieuse :

Je crains un catholicisme anglais dont Newman est le plus haut type, écrivait-il. C'est le vieux ton anglican, patristique, littéraire, d'Oxford, transplanté dans

---

(1) Note autobiographique dans E. PURCELL, *Life of Manning*, t. II, p. 349.
(2) Cf. *supra*, p. 156, n. 1.
(3) La méfiance des autorités romaines datait de l'époque où Newman s'était occupé de la revue *The Rambler* et avait été compromis par l'orientation trop libérale du groupe d'Acton (cf. *infra*, p. 242-244).
(4) Voir l'étude pénétrante de H. BREMOND, *Manning et Newman*, dans *L'inquiétude religieuse*, 1re série, Paris, 1930, p. 230-256.
(5) Talbot à Manning, 25 avril 1867, dans E. PURCELL, *Life of Manning*, t. II, p. 318.

l'Église… La chose qui nous sauvera des vues minimalistes sur la Mère de Dieu et sur le Vicaire de Notre-Seigneur, c'est le million d'Irlandais qui sont en Angleterre… Je suis heureux d'apprendre qu'ils n'ont aucun goût pour le catholicisme coupé d'eau, littéraire, mondain, de certains Anglais [1].

Par malheur, ces divergences de vues furent bientôt de notoriété publique et provoquèrent contre Manning et son entourage de vives indignations. Alors que les anciennes oppositions entre catholiques de naissance et convertis s'effaçaient peu à peu, de nouveaux partis allaient de la sorte se former, à l'occasion des controverses autour du libéralisme d'abord, puis à propos de l'infaillibilité pontificale, les esprits intransigeants se groupant autour de l'archevêque et de ses partisans les plus excessifs, tels Ward ou Vaughan, tandis que les autres, plus ouverts, dont le jeune Acton était l'un des types les plus représentatifs, espéraient trouver en Newman un chef de file, malgré le désir très net de ce dernier de rester le plus possible à l'écart des luttes intestines.

L'ÉGLISE CATHOLIQUE
EN IRLANDE
Tandis que le catholicisme se développait en Grande-Bretagne, en partie grâce à l'afflux des émigrés irlandais, l'Église d'Irlande connaissait une nouvelle vitalité sous la direction énergique du cardinal Cullen [2]. Jouissant de la pleine confiance des autorités romaines, il avait, quelques mois à peine après son élévation au siège archiépiscopal d'Armagh, été désigné comme délégué apostolique, avec mission de convoquer un concile plénier chargé d'adapter autant que possible les usages irlandais au droit canon traditionnel et de régler en particulier le problème scolaire conformément aux décisions romaines [3]. Le concile, auquel prirent part quatre archevêques et vingt-quatre évêques, se réunit à Thurles, le 22 août 1850, et ses décrets, approuvés par Pie IX le 4 mai 1851, furent mis en application grâce à une série de conciles provinciaux [4].

Cullen, devenu en 1852 archevêque de Dublin, non content de condamner le mouvement Fenian, insista beaucoup, malgré l'avis contraire de Mgr Mac Hale, pour que le clergé irlandais n'intervînt pas dans les questions politiques où les intérêts religieux n'étaient pas directement engagés, ce qui le fit accuser par les milieux patriotes d'être à la solde du gouvernement anglais. Pourtant, il n'hésita pas à lutter pied à pied contre ce même gouvernement pour obtenir une amélioration du régime scolaire et, s'il échoua dans ses efforts pour créer une Université catholique, il finit par obtenir sur les autres points des résultats substantiels.

Les progrès du catholicisme se marquèrent bientôt dans tous les do-

(1) Manning à Talbot, 25 février 1866, *ibid.*, t. II, p. 322-324.
(2) Né en 1803, il avait fait à Rome de très brillantes études. Nommé par Léon XII professeur au collège de la Propagande, puis, en 1832, recteur du collège irlandais de Rome, il était souvent consulté par la Curie sur les problèmes de son pays. Désigné le 19 décembre 1849 comme archevêque d'Armagh et transféré le 1ᵉʳ mai 1852 au siège de Dublin, il fut le premier prélat irlandais depuis la Réforme à être élevé au cardinalat, en 1867. Il mourut en 1878. (Notice dans *The Catholic Encyclopaedia*, t. IV, p. 564-566.)
(3) Sur ces décisions et les discussions qui les occasionnèrent, cf. *supra*, p. 156, n. 1.
(4) Cf. *Coll. lac.*, t. III, col. 771 et suiv. Voir la remarquable pastorale publiée par les évêques à l'issue du concile dans *Dublin Review*, t. XXX, 1865, p. 176.

maines : construction d'églises, d'écoles et d'institutions de bienfaisance ; multiplication des couvents ; développement des missions populaires, où se distinguèrent particulièrement les lazaristes. Les évêques s'appliquèrent aussi à relever le niveau d'un clergé qui n'était pas toujours à la hauteur de sa mission [1], notamment en améliorant le séminaire central de Maynooth [2] et en développant l'usage des retraites sacerdotales.

ET EN ÉCOSSE Le rétablissement de la hiérarchie catholique en Angleterre ne s'était pas étendu à l'Écosse, qui resta jusqu'en 1878 administrée par des vicaires apostoliques, dont le plus remarquable fut Mgr James Gillis, mort en 1864 vicaire apostolique d'Édimbourg, où il introduisit plusieurs congrégations religieuses.

En Écosse comme en Angleterre, l'immigration irlandaise joua un rôle important, au point qu'en 1878, sur 380.000 catholiques, il n'y en avait pas 25.000 d'origine écossaise [3]. Il fallut multiplier les églises et les écoles paroissiales, construites au début avec les *pennies* des pauvres immigrants, mais pour lesquels on obtint par la suite quelques subsides officiels. Il fallut surtout faire appel à des prêtres venant d'Irlande, et ce fut l'origine de dissensions, attisées par un journal local, *The Free Press* : les prêtres irlandais reprochaient aux vicaires apostoliques écossais de leur réserver la tâche difficile de fonder de nouvelles missions parmi les immigrants, puis, une fois la paroisse établie et l'église construite, de les envoyer recommencer ailleurs leur œuvre de pionniers, tandis qu'un prêtre d'origine écossaise recueillait le fruit de leurs travaux. Vers 1865, dans le district de Glasgow, le conflit prit de telles proportions qu'il fut porté devant la Propagande. Celle-ci, en 1867, chargea Manning d'enquêter. Sur son conseil l'évêque et son coadjuteur furent écartés et remplacés par Mgr Eyre, ancien vicaire général de Newcastle, homme habile et énergique, qui réussit assez vite à rétablir la situation.

§ 2. — L'Église en Belgique [4].

LE CARDINAL STERCKX La réorganisation de l'Église belge, à laquelle les évêques s'étaient attachés dès le lendemain de la révolution de 1830, se poursuivit activement pendant les premières années du pontificat de Pie IX sous la sage direction du cardinal Sterckx,

(1) Voir les plaintes recueillies à Rome en 1864 dans *Janssens Briefe*, édit. PASTOR, t. I, p. 260.
(2) Cf. O'DEA, *Maynooth and the University Question*, Dublin, 1903.
(3) *The Catholic Encyclopaedia*, art. *Scotland*, t. XIII, p. 622. Il y avait eu près de 100.000 émigrants au moment de la grande famine et il y en eut plus de 200.000 entre 1850 et 1880. (Cf J.-H. HANDLEY, dans *Irish ecclesiastical Record*, t. LXIX, 1947, p. 913-914.)
(4) BIBLIOGRAPHIE. — I. SOURCES. — Outre les mandements et lettres pastorales des différents évêques, on trouvera de très nombreux renseignements dans les Actes des Congrès de Malines, publiés sous le titre : *Assemblée générale des catholiques en Belgique* (Malines, 1864, 1865 et 1868). Voir également les deux recueils périodiques : *Revue catholique* et *Précis historique*.
Pour les relations entre l'Église et l'État : *La Belgique et le Vatican. Documents et travaux législatifs...*, Bruxelles, 1880-1881, 3 vol. ; Ch. WOESTE, *Mémoires pour servir à l'histoire contemporaine de la Belgique*, t. I (*1859-1894*), Bruxelles, 1927.
II. TRAVAUX. — A ceux indiqués p. 9, ajouter : M. DEFOURNY, *Les congrès catholiques en Belgique*, Louvain, 1908 ; et *Prêtres de Belgique*, numéro spécial de la *Nouvelle Revue théologique*, t. LVII, 1930, p. 683 et suiv.

archevêque de Malines de 1831 à 1867. Si, selon le mot très exact de
M. Pouthas, l'Église de Belgique, vers le milieu du XIXᵉ siècle, « est devenue,
par sa vigueur et son indépendance, un modèle, une sorte d'idéal pour les
autres Églises européennes, notamment pour l'Église de France » [1], elle
le doit sans doute à toute une élite de prêtres et de laïcs dévoués qui
surent comprendre qu'une loyauté inconditionnée envers la foi chrétienne,
l'Église et le pape, était parfaitement compatible avec une ouverture d'esprit sympathique à l'égard des aspirations de leur temps et notamment
à l'égard de ce qu'il y avait de sain et même de chrétien dans les intui-
tions fondamentales du libéralisme moderne. Mais l'œuvre de ces hommes
méritants eut vraisemblablement été bien moins féconde s'ils n'avaient
pu s'appuyer sur la direction prudemment audacieuse de l'homme providentiel que fut Engelbert Sterckx. C'est lui, pour une bonne part, qui
réussit, malgré le régime de séparation légale, à donner à l'Église de Belgique la place qu'elle occupe encore aujourd'hui dans l'État, « non pas
celle d'un corps constitué, mais celle d'une présence et d'une autorité
spirituelles, reconnues et respectées » [2]. Avec son respect de l'autonomie
de l'État dans son domaine propre et son attitude compréhensive à l'égard
de la pensée moderne d'une part, avec son attachement dévoué au Saint-
Siège et ses réalisations apostoliques variées, désintéressées et fermes,
d'autre part, ce prélat pieux et laborieux apparaît comme « un précurseur
remarquable des évêques de l'époque contemporaine, celle de l'Église
libre dans l'État libre » [3].

L'influence de Sterckx déborde largement son diocèse, le plus important
du pays d'ailleurs, car grâce à la réunion annuelle des évêques à Malines,
à laquelle le nonce participe fréquemment, grâce aussi à son prestige
personnel et à « une pression plus que persuasive sur la volonté de ses
suffragants » [4], il réussit à assurer une action commune de l'épiscopat
belge, en dépit du fait que plusieurs de ses collègues eussent conservé plus
que lui la nostalgie de l'ancien régime.

LA VITALITÉ CHRÉTIENNE   Si les efforts des évêques en vue d'assurer
l'influence de l'Église s'inspirent peut-être
trop, chez l'un ou l'autre d'entre eux, du vieil esprit théocratique de
domination sur la société civile, leur but dernier est en tout cas clair :
permettre à l'Église d'exercer dans les meilleures conditions sa mission
toute spirituelle de sanctification des âmes. On ne trouve pas, parmi eux,
de ces prélats qui se laissent si bien absorber par le souci de promouvoir
des conditions favorables à l'apostolat qu'ils n'ont plus le temps de s'intéresser à cet apostolat lui-même ou à la direction proprement religieuse de
leurs diocèses. Malgré l'emmêlement du politique et du religieux, qui
restera longtemps caractéristique de la situation religieuse en Belgique, ces
évêques sont avant tout des pasteurs d'âmes.

(1) Ch. Pouthas, *L'Église catholique de l'avènement de Pie VII à l'avènement de Pie IX* (Cours
de Sorbonne), Paris, 1945, p. 294.
(2) A. Simon, *Le cardinal Sterckx et son temps*, t. II, p. 444.
(3) *Ibid.*, t. II, p. 446.
(4) *Ibid.*, t. II, p. 280.

Ils peuvent compter, pour l'accomplissement de leur mission pastorale, sur la collaboration d'un clergé suffisant en nombre [1] et généralement très soumis. Au lendemain de 1830, par suite de la grande liberté d'action dont il avait joui durant les longues vacances des sièges épiscopaux, puis sous la pression des idées mennaisiennes et démocratiques, le clergé belge s'était montré turbulent et assez indépendant, notamment dans les Flandres. Mais les évêques, qui avaient une conscience très nette de leur autorité, avaient vite repris les choses en main et, dès le milieu du siècle, les prêtres belges acceptent avec beaucoup plus de docilité que leurs confrères français la centralisation croissante de la vie diocésaine, ainsi que le souci des évêques de systématiser et de surveiller de plus en plus minutieusement le ministère sacerdotal.

Formé dans des séminaires dont les programmes ont été remaniés et unifiés de 1842 à 1848, au cours de réunions interdiocésaines suscitées par Sterckx, soutenu par des retraites annuelles et par des « conférences » ou récollections studieuses de prêtres d'un même doyenné, réorganisées dès la fin des années trente, ce clergé belge, dont la culture moyenne restera longtemps assez rudimentaire, se distingue par sa grande sévérité de mœurs, qui n'exclut pas un usage modéré des joies de l'existence, et par sa piété peu mystique mais très méthodiquement réglée. Il se distingue encore par son orientation vers l'action, par le réalisme de ses méthodes et aussi par sa bonhomie et sa simplicité d'allures. Avec cela, un standing de vie très bourgeois, que les évêques souhaitaient, comme un moyen d'accroître le prestige de l'Église dans la société du temps.

L'épiscopat trouve également une aide efficace auprès des congrégations religieuses : de 12.000 environ en 1846, c'est-à-dire déjà autant que sous Marie-Thérèse, aux plus beaux jours du patronage officiel, l'effectif des religieux passe à 18.000 en 1866 pour atteindre 25.326 en 1880 [2]. Mais la tendance à soumettre étroitement l'activité des religieux aux directives épiscopales afin d'unifier l'action apostolique n'alla pas toujours sans heurts : c'est ainsi que les tentatives des jésuites pour ouvrir dans différentes villes des cours de philosophie qui auraient fait concurrence à l'Université de Louvain donnèrent lieu à un conflit sérieux et prolongé entre le cardinal Sterckx et la Compagnie de Jésus, dont la forte influence dans les hautes classes de la société suscitait d'ailleurs pas mal de jalousies [3]. Dans l'ensemble toutefois, la Belgique apparaît comme un des

(1) On comptait, en 1840, un prêtre rétribué par l'État (c'est-à-dire pratiquement, dans le ministère paroissial) pour 938 habitants et, en 1860, un pour 971 ; il faut ajouter les prêtres utilisés dans l'enseignement libre. Le chiffre des vocations sacerdotales a une tendance à augmenter, moins rapidement toutefois que l'augmentation de la population, mais par ailleurs, le chiffre annuel des décès est de loin inférieur à celui des ordinations. Voir notamment, pour le diocèse de Malines, L. LECLERCQ, *Un siècle de recrutement sacerdotal*, dans *Collectanea Mechliniensia*, t. IV, 1930, p. 558-563. Si l'on tient compte des religieux prêtres, on arrive à des chiffres très satisfaisants ; c'est ainsi qu'en 1862, il y avait, dans le diocèse de Malines, 2.200 prêtres séculiers et réguliers pour 1.100.000 habitants.

(2) E. DE MOREAU, dans *D. H. G. E.*, t. VII, col. 749. A titre d'exemple, plus de 160 établissements nouveaux furent ouverts dans le diocèse de Malines de 1832 à 1867 ; ce sont surtout les communautés enseignantes qui se multiplient. (A. SIMON, *Le cardinal Sterckx*, t. II, p. 96-97, en note.)

(3) Voir notamment les lettres du recteur de la maison de Louvain et du provincial de Belgique au P. Roothaan, des 15 janvier et 9 février 1846, *Archives des Jésuites à Rome*, Belgique, 2-XI-18*a* et 18*b*.

pays où les deux clergés, séculier et régulier, marchent le plus la main dans la main : ils se rencontrent continuellement dans le ministère auprès des laïques et c'est aux religieux que les évêques confient les retraites et les récollections sacerdotales.

Sous la direction attentive des évêques, prêtres et religieux fondent et animent des œuvres de tout genre. La période de 1830 à 1846 les avait surtout vus s'intéresser à l'organisation de missions paroissiales et à la fondation de bibliothèques catholiques destinées à lutter contre la diffusion des mauvaises lectures dans le peuple. A partir de 1847, outre le lent développement des journaux catholiques, qui restent malheureusement trop cantonnés au terrain politique et s'entre-déchirent souvent, il faut surtout signaler la multiplication des œuvres de bienfaisance [1], appelées par la grande crise agricole de 1845 à 1848 et par la misère croissante de la classe ouvrière ; les laïcs y tiennent la place principale et rachètent un peu par leur grande générosité leur manque de perspicacité devant le caractère nocif du libéralisme économique intégral.

Enfin, longtemps avant de devenir une nation coloniale, la Belgique catholique porte un intérêt croissant à l'expansion missionnaire : jusque vers 1880, celui-ci se porte surtout vers les missions d'Amérique — le légendaire P. De Smet eut de très nombreux confrères wallons et surtout flamands — mais, peu à peu, d'autres champs d'apostolat retiennent l'attention, en particulier la Chine, depuis qu'en 1862 un vicaire de Bruxelles, l'abbé Verbist, a fondé la congrégation des Missionnaires de Scheut [2].

LES PROGRÈS DE L'INDIFFÉRENCE RELIGIEUSE

La vitalité remarquable dont fait preuve l'Église de Belgique au cours des trente années du pontificat de Pie IX ne doit pas faire illusion : moins touché que la France, le pays n'échappe pas au mouvement général de déchristianisation qui caractérise la seconde moitié du XIXe siècle. Assurément le chiffre des non-baptisés restera extrêmement faible jusqu'à la guerre de 1914 et le sentiment religieux demeure encore trop vivace pour qu'on rompe déjà totalement avec le culte traditionnel, tant dans le peuple que dans la classe bourgeoise, mais l'émancipation à l'égard de l'Église se développe.

C'est surtout après 1880 que s'accomplira le divorce entre l'ouvrier belge et la religion chrétienne, spécialement sous l'inspiration du socialisme, dont les publications avaient commencé à s'en prendre violemment à l'Église au cours des années soixante. Pourtant des documents sûrs prouvent que, dès 1830, même dans les régions les plus catholiques, une fraction notable de la classe ouvrière s'était éloignée de l'Église et Kersten se plaignait dès 1834 de ce qu'une bonne partie de la population en Wallonie ne fréquentât plus régulièrement les sacrements. Dans les villes, l'absentéisme à la messe dominicale ou même à la communion pascale

---

(1) Voir DE HAERNE, *Tableau de la charité chrétienne en Belgique*, Bruxelles, 1857, ainsi que les rapports des Congrès de Malines, surtout celui de 1867.
(2) Cf. J. RUTTEN, *Les missionnaires de Scheut et leur fondateur*, Louvain, 1930.

commence à sévir dans une proportion inquiétante [1] et l'enseignement rationaliste des Universités d'État et de celle de Bruxelles exerce une influence de plus en plus néfaste dans le monde des médecins et des avocats [2]. Quant aux campagnes, où la pratique extérieure se maintient, on constate un relâchement des mœurs et souvent un certain indifférentisme, favorisé par l'action des « libres penseurs », encadrés par les francs-maçons, dont l'influence était sensible malgré leur petit nombre et qui réussissaient aisément par leurs critiques à saper l'autorité du curé auprès des simples.

LE RAIDISSEMENT LIBÉRAL Dans le peuple, ces progrès de l'indifférence religieuse ne se traduisent encore que par une passivité plus grande à l'égard des prescriptions ecclésiastiques. Mais dans la bourgeoisie libérale, on voit réapparaître l'anticléricalisme militant qui s'était fort atténué pendant une quinzaine d'années. La plupart des libéraux au lendemain de 1830 étaient encore croyants, parfois même pratiquants et cherchaient à « allier la fidélité catholique à la lutte contre les prétendues prétentions civiles du clergé » [3], c'est-à-dire contre les efforts de l'Église pour utiliser l'État au profit de sa mission spirituelle. Or, il est incontestable que beaucoup de catholiques et spécialement la hiérarchie, tout en se ralliant par « tactique » aux libertés constitutionnelles, qui leur apparaissaient comme un excellent moyen d'éviter l'emprise de l'État sur l'Église, n'avaient pas oublié l'ancien régime et conservaient le souci de maintenir dans la pratique l'influence prépondérante de l'Église dans l'État ; ils avaient bien abandonné l'idée d'une religion d'État, mais estimaient cependant que les institutions devaient non seulement éviter d'entraver, mais favoriser positivement la religion catholique, « dominante de fait ». C'était précisément ce dont les libéraux ne voulaient à aucun prix. Aussi, de plus en plus mécontents de ce qu'ils considéraient comme « les empiétements du clergé », en particulier des revendications répétées des évêques en matière d'enseignement, exaspérés en outre par la circulaire épiscopale de 1838 contre les francs-maçons, qui rendait beaucoup plus délicate la situation du « libéral qui va à la messe », ils décidèrent de passer à l'action, et, au cours d'un congrès réuni en juin 1846, formulèrent nettement leur programme : « indépendance réelle » du pouvoir civil à l'égard de l'Église et tout spécialement « organisation d'un enseignement public à tous les degrés sous la direction exclusive de l'autorité civile... en repoussant l'intervention d'un ministre

---

(1) En ce qui concerne par exemple Anvers, où la situation est cependant meilleure qu'à Bruxelles, voir F. Prims, *Geschiedenis van Antwerpen*, t. XXVIII, Anvers, 1949, p. 46-47.

(2) Certains évêques essayèrent d'alerter l'opinion publique et, en 1857, la question de la liberté scientifique des professeurs fit à la Chambre l'objet d'un violent débat. (Cf. G. Jacquemyns, *La condamnation de l'Université de Gand par les évêques belges en 1856. L'affaire Brasseur*, dans *Revue de l'Université de Bruxelles*, t. XXXVIII, 1932, p. 45 et suiv. ; E. de Moreau, *Ad. Dechamps*, p. 239 et suiv. ; A. de Trannoy, *Jules Malou*, Bruxelles, 1905, p. 300 et suiv.) Le Saint-Siège, qui s'inquiétait de la situation, aurait souhaité voir créer dans chaque diocèse une école de philosophie catholique, mais les évêques, qui jugeaient la mesure inefficace, craignaient en outre qu'elle n'affaiblît la position de l'Université de Louvain et ne renforçât l'influence des jésuites. (Cf. A. Simon, *Le cardinal Sterckx*, t. II, p. 197-212.)

(3) A. Simon, *L'Église catholique et les débuts de la Belgique indépendante*, p. 72.

des cultes *à titre d'autorité* dans l'enseignement organisé par le pouvoir civil » [1].

PRÉLUDE A LA GUERRE SCOLAIRE — La loi scolaire de 1850 fut la première manifestation de la nouvelle politique, anticléricale et laïcisante, qui devait aller en s'accentuant pendant une trentaine d'années. Depuis 1830, l'influence du clergé était devenue prépondérante dans l'enseignement secondaire, la formule la plus en faveur étant celle de collèges subventionnés par la commune, mais où la nomination des professeurs dépendait pour une bonne part des évêques. Sans vouloir exclure la religion de l'école, les libéraux entendaient organiser un enseignement officiel qui fût indépendant des interventions ecclésiastiques et qui, s'inspirant de la lettre de la constitution, mettrait, théoriquement du moins, tous les cultes sur le même pied. Passant outre au mécontentement des évêques, qui, invoquant l'esprit de la constitution, se croyaient en droit d'exiger que l'atmosphère de l'école publique restât catholique, ils réussirent à faire voter une loi qui « au point de vue religieux, sans qu'il y parût trop, était toute une révolution » [2].

Plusieurs évêques, surtout ceux des Flandres et celui de Tournai, théologiens un peu raides et combatifs, estimaient que les ponts étaient rompus et qu'il n'y avait d'autre solution admissible que de boycotter l'enseignement officiel en développant en face de lui des collèges libres. Mais le cardinal Sterckx était d'un autre avis. C'était un esprit modéré, comprenant beaucoup mieux que ses collègues les susceptibilités libérales et les conditions nouvelles auxquelles l'Église devait s'adapter ; il était convaincu de l'intérêt qu'avait l'Église à s'entendre avec l'État afin de pouvoir exercer sur lui le maximum d'influence. Prêt dans ce but à faire de larges concessions, il ne désespérait pas d'arriver, malgré tout, à un *modus vivendi* qui permettrait d'assurer une éducation et une formation religieuse aux nombreux enfants qui fréquenteraient de toute façon les écoles de l'État. Son sens de la conciliation, sa bonhomie qui n'excluait ni l'autorité ni la ténacité, finirent par triompher de la double résistance du gouvernement et de ses collègues de l'épiscopat : en 1854, un accord, sanctionné par arrêté royal, connu sous le nom de *Convention d'Anvers*, précisait les dispositions de la loi de 1850 en assurant au clergé de substantielles garanties.

L'OFFENSIVE ANTICLÉRICALE — Cette belle victoire du sage cardinal n'était pourtant qu'une accalmie avant l'orage. Lorsqu'il avait pris le pouvoir en 1847, avec l'intention d'entreprendre une œuvre de sécularisation, le ministère Rogier espérait encore — et beaucoup de libéraux avec lui — qu'il serait possible de la réaliser d'accord avec l'Église, celle-ci renonçant à toute autorité dans le domaine civil, l'État s'engageant de son côté à rester respectueux à l'égard de la religion. Mais l'attitude intransigeante de la plupart des évêques lors

(1) Art. 2 et 3 du programme dans *Le Congrès libéral de Belgique*, Bruxelles, 1846.
(2) E. DE MOREAU, *L'Église en Belgique*, Bruxelles, 1944, p. 240.

de la loi sur l'enseignement moyen avait dissipé ces illusions et l'attitude de Rome à cette occasion devait encore accroître la déception.

Le gouvernement avait en effet un moment espéré qu'il serait plus aisé de s'entendre avec le « pape libéral » qu'avec les évêques belges, jugés trop étroits d'esprit. Mais Pie IX, qui suivit de près les débats sur la loi de 1850, estima, peut-être sous l'influence de Mgr Malou, de Bruges, en tout cas malgré les avis rassurants de Sterckx, qu'il lui fallait marquer publiquement sa désapprobation : il était convaincu, comme il le déclarait au représentant belge à Rome, que « la question de l'enseignement est la question religieuse dans une de ses applications les plus importantes »[1], il venait de rappeler aux évêques d'Irlande les principes dont l'Église estimait ne pas pouvoir se départir en cette matière, et, en outre, il interprétait la loi comme le premier élément d'un ensemble destiné à diminuer la liberté de l'Église en Belgique. Le 20 mai 1850, il condamna pratiquement la loi et cette intervention, que rien ne laissait prévoir, devait avoir de graves conséquences, car « de même que la circulaire contre les francs-maçons avait, en 1838, creusé le fossé entre les libéraux et les évêques belges, l'allocution de 1850 modifie les rapports entre le gouvernement libéral et le Saint-Siège »[2] en enlevant désormais aux libéraux tout espoir d'une entente raisonnable avec l'Église.

Décidés à poursuivre leur œuvre de sécularisation sans elle et éventuellement contre elle, les libéraux entrèrent dès lors résolument dans la voie du sectarisme. Tandis qu'ils se mettaient à dénoncer le prêtre comme un danger public et à proclamer la nécessité « d'arracher les intelligences aux ténèbres de l'obscurantisme »[3], au Parlement la lutte se porta peu à peu sur tous les terrains. Tour à tour surgirent la question des fondations charitables, celle des cimetières, celle du temporel du culte, celle des bourses d'études à transférer de l'enseignement catholique à l'enseignement officiel. Le ministère Frère-Orban, qui se maintint au pouvoir de 1857 à 1870, entreprit de les résoudre dans un sens nettement laïque, soutenu par un corps électoral dont il exploitait habilement la méfiance à l'égard de la main-morte et des couvents omnipotents et surtout l'obsession des fameux « empiétements du clergé ». Il fut excité plus d'une fois, d'ailleurs, par les exigences et l'agressivité de certains évêques qui partageaient les sympathies d'une partie de l'opinion publique catholique pour les thèses extrémistes de Louis Veuillot et qui ne semblaient pas se rendre compte que beaucoup de ces questions présentaient un aspect bien plus politique que religieux[4].

---

(1) Cité par A. SIMON, *Le cardinal Sterckx*, t. I, p. 479.
(2) A. SIMON, *L'Église catholique et les débuts...*, p. 119.
(3) C'est alors que se constituent des groupements de libres-penseurs en vue d'assurer à leurs membres un enterrement civil malgré l'opposition de leur famille. Cette propagande n'obtiendra toutefois des résultats sensibles que dans les dernières années du siècle. A Anvers, en 1867, pour 2.996 enterrements à l'église, on ne compte encore qu'un seul enterrement civil. (F. PRIMS, *Geschiedenis van Antwerpen*, t. XXVIII, p. 46.)
(4) L'ancien ministre catholique De Decker s'en plaignait amèrement à Doellinger lors du passage de celui-ci à Bruxelles en 1858, notamment à propos des évêques de Bruges et de Gand : « Ces évêques ne comprennent pas que des questions de ce genre, en Belgique, sont aux trois quarts politiques et non religieuses. » (Lettre de Doellinger, citée dans J. FRIEDRICH, *Ign. v. Doellinger*, t. III, p. 208.)

LES CONGRÈS DE MALINES   Après toute une vie de conciliation, le cardinal Sterckx lui-même se trouvait « acculé à la lutte » [1]. Longtemps, malgré les avertissements de ses collègues, de l'évêque de Liége, Van Bommel, notamment, il s'était refusé à organiser systématiquement les forces catholiques pour la défense des droits de l'Église. Certes, il avait toujours fait appel à la collaboration des laïcs, tant pour la lutte parlementaire que pour l'action dans les œuvres ; c'est même précisément pour s'assurer d'une élite catholique au sein des carrières libérales, sur lesquelles était axée la société du xixᵉ siècle, qu'il consacra tant de temps et d'efforts à la défense et à la prospérité de l'Université catholique de Louvain, qui fut sans doute, de toutes ses œuvres, celle qui lui tint le plus à cœur. Mais cette action des laïcs au service de la foi, il avait longtemps préféré qu'elle s'exerçât en ordre dispersé, craignant d'exciter les libéraux en leur donnant l'impression d'une offensive dirigée contre eux et d'aboutir à la constitution de deux fronts laïques, l'un catholique, l'autre libéral, au grand détriment du travail apostolique.

A partir de 1857, il lui fallut bien se rendre à l'évidence. Le front libéral qu'il avait voulu éviter s'était malgré tout constitué et, grâce à la timidité et à la dispersion des catholiques, il devenait tout-puissant au Parlement. Et au moment même où il fallait ainsi renoncer à l'espoir de voir les cadres politiques faciliter l'action de l'Église sur la société, celle-ci, travaillée par le matérialisme et impressionnée par les progrès des sciences, se détachait de plus en plus de Dieu. De nouvelles formules s'imposaient.

Or, un professeur de Louvain, Jean Moeller, Allemand de naissance, qui suivait de près tout ce qui se passait dans sa patrie d'origine et constatait les heureux résultats des grandes assemblées annuelles des catholiques allemands, souhaitait voir se constituer en Belgique une organisation analogue. Il réussit à gagner à ses vues quelques laïcs dévoués, Barthélemy Dumortier, Adolphe Dechamps, Prosper de Haulleville et surtout Édouard Ducpétiaux, organisateur de génie qui s'était déjà distingué par la direction de plusieurs congrès internationaux [2]. Au début de 1863, ils décidèrent d'organiser au cours de l'été un grand congrès catholique en Belgique dans le but de coordonner l'action des différentes œuvres et de déclencher un puissant mouvement d'opinion capable d'influencer l'action parlementaire.

Avec un sens de l'adaptation peu commun, surtout chez un vieillard, Sterckx donna immédiatement son accord. Il comprenait que, le prestige de l'épiscopat étant dorénavant insuffisant auprès du gouvernement, il fallait appeler à la rescousse le laïcat organisé, le peuple chrétien, qui devait se préparer à user des libertés constitutionnelles pour défendre sa oi face à d'autres citoyens hostiles à celle-ci.

Le premier congrès, qui se tint à Malines du 18 au 23 août 1863, fut

---

(1) A. SIMON, *Le cardinal Sterckx*, t. I, p. 545.
(2) Sur Ducpétiaux (1804-1861), voir E. RUBBENS, *Edouard Ducpétiaux*, Louvain, 1922-1934, 2 vol.

un remarquable succès. Des décisions utiles y furent prises pour la coordination des œuvres et surtout le fameux discours prononcé par Montalembert [1], tout en envenimant de vieilles querelles, contribua à galvaniser les énergies du public catholique, « lui permettant ainsi de faire la liaison entre le libéralisme de 1830, timide, prudent, et le libéralisme catholique de 1880 et des années suivantes, lorsque la formule de l'Église libre dans l'État libre fut pratiquement admise par le Saint-Siège » [2].

Sterckx s'était contenté d'approuver le congrès de 1863. Conscient du bien qui en résultait, il favorisa nettement la nouvelle réunion prévue pour 1864 et la défendit victorieusement contre les réticences de Rome, où le discours de Montalembert avait fait mauvaise impression. Il veilla à lui imprimer une allure plus pratique et à la centrer sur le terrain des bonnes œuvres, de manière à éviter toute polémique doctrinale autour du catholicisme libéral.

C'est la même orientation qui prévalut encore lors du congrès de 1867, où l'intérêt se porta surtout sur les œuvres sociales, dont l'industrialisation rapide du pays soulignait l'urgence, et sur la question de l'enseignement libre, qui prenait une importance croissante avec la laïcisation de l'enseignement public [3]. Toutefois, la défense des intérêts et des libertés religieuses, à laquelle le congrès conviait les catholiques, devait inévitablement, par suite du complexe politico-religieux belge, se développer sur le terrain parlementaire et l'on peut à bon droit dire que le principal résultat des trois congrès de Malines, bien que ce résultat n'eût pas été positivement recherché par la plupart des participants, fut la constitution d'un parti catholique organisé.

TENSION CROISSANTE En attendant, les libéraux intensifiaient leur politique anticléricale. Longtemps, l'épiscopat avait pu compter, dans sa résistance, sur l'appui du roi Léopold Ier, qui, tout protestant qu'il était, marqua toujours une « particulière bienveillance » [4] envers l'Église catholique, considérée par lui comme la grande force d'ordre dans le jeune royaume. Mais lorsque, en novembre 1865, le roi moribond eut consenti à nommer au ministère de la Justice le plus ardent champion du radicalisme, Bara, aucun obstacle ne sembla plus arrêter l'œuvre de laïcisation et la résistance opiniâtre du cardinal ne réussit qu'à retarder de quelques années les projets sur les bourses d'études et sur le temporel du culte. Dès avant 1870, les objectifs de la politique nouvelle étaient en bonne partie atteints.

(1) Sur ce discours et ses répercussions, cf. *infra*, p. 250-252.
(2) A. SIMON, *Le cardinal Sterckx*, t. II, p. 115.
(3) C'est à partir du congrès de 1867 que, contrairement à l'intention des constituants et de la première génération catholique, se développa le slogan qui devait devenir pour des dizaines d'années la plate-forme du parti catholique : l'enseignement officiel ne doit avoir qu'un rôle *supplétif* par rapport à l'enseignement libre.
(4) A. SIMON, *L'Église catholique et les débuts...*, p. 76. Sur l'attitude sympathique de Léopold II à l'égard de l'Église et des évêques, cf. *ibid.*, p. 76-81.

§ 3. — L'Église aux Pays-Bas et en Suisse [1]

LE CATHOLICISME NÉERLANDAIS   Sans se laisser déconcerter par les manifestations protestantes, les nouveaux évêques nommés en 1853 s'étaient empressés d'organiser leurs diocèses conformément au droit canon. Un point surtout présentait des difficultés, la détermination des circonscriptions paroissiales, car, jusqu'alors, surtout dans les grandes villes du centre, il n'existait pas de paroisses fixes, chacun s'attachant au curé qui lui convenait le mieux ; de plus, à Amsterdam notamment, beaucoup de religieux avaient joui jusqu'alors des droits paroissiaux et durent y renoncer. Ce fut un bonheur pour l'Église néerlandaise d'avoir à sa tête, pour présider à cette réorganisation délicate, un homme que ses remarquables qualités devaient faire désigner un jour comme « le fondateur de la Hollande catholique contemporaine », l'archevêque d'Utrecht, Mgr Zwijsen [2]. Jadis ami du prince Guillaume, à l'époque où il était curé de Tilburg, il était devenu en 1842 coadjuteur du vicaire apostolique de Bois-le-Duc, mais ses soucis pastoraux ne l'avaient pas empêché de continuer à s'intéresser de près aux intérêts religieux de l'ensemble des Pays-Bas, notamment aux problèmes scolaires. Cet homme d'action, d'une grande fermeté de principes mais qui savait aussi être diplomate à l'occasion, était tout désigné pour prendre la tête de la nouvelle hiérarchie.

Sous sa sage direction, le catholicisme continua à enregistrer des progrès modérés, mais continus. Le nombre total des fidèles ne s'accrut que de 15 pour cent en 25 ans, passant de 1.215.000 en 1853 à 1.395.000 en 1878 (sur une population totale de 4 millions) ; mais la proportion s'améliora sensiblement dans les grandes villes du centre : en 1865, Amsterdam dépassait largement les 60.000 catholiques ; La Haye, la capitale, en comptait 23.000 sur 85.000 habitants ; Leyde, la ville universitaire, un tiers de la population ; Arnhem, 12.000 catholiques au lieu de 500 en 1810. Tout en continuant à s'appuyer avant tout sur les masses paysannes des provinces méridionales et sur une partie de la classe moyenne, le catholicisme voit son influence augmenter dans la bourgeoisie industrielle et commerçante, grâce à l'immigration d'un certain nombre d'Allemands entreprenants venant de la région de Munster. Par contre, le grand commerce maritime et colonial reste presque exclusivement aux mains des protestants et, parmi les fonctionnaires et les officiers de l'armée, les catholiques sont à peine un dixième. Quant à la noblesse, si la majeure partie, qui descend des Gueux, est protestante, il y a aussi une aristocratie catholique, originaire de Gueldre ou de West-

---

(1) Bibliographie. — Sur l'Église aux Pays-Bas : *Het Katholiek Nederland, 1813-1913*, Nimègue, 1913, 2 vol. ; C.-J. Commissaris, *Van toen wij vrij werden*, t. II, La Haye, 1929 ; D. Langedijk, *De schoolstrijd in de eerste jaren na de wet van 1857*, Kampen, 1937 ; G. Brom, *Romantiek en Katholicisme in Nederland*, Groningue, 1926, 2 vol. ; P. Verschave, *La Hollande politique*, Paris, 1910.

Sur l'Église en Suisse : K. Mueller, *Die katholische Kirche in der Schweiz seit dem Ausgang des XVIII. Jh.*, Einsiedeln, 1928 ; T. Schwegler, *Geschichte der katholischen Kirche in der Schweiz*, Stans, 1945.

(2) Sur Mgr Zwijsen (1794-1877), voir J. Witlox, *Mgr Joannes Zwijsen, een levens- en karakterbeeld*, Bois-le-Duc, 1927.

phalie, très riche et qui apporte un appui non négligeable aux œuvres catholiques. La générosité de l'ensemble des fidèles, y compris les plus modestes, est d'ailleurs remarquable ; c'est sur elle que repose l'entretien de l'église, du presbytère, du cimetière, éventuellement des hôpitaux et des écoles, et personne ne s'étonne de voir dans ce but les quêtes se multiplier ou les places à l'église mises en location à un prix plus ou moins élevé suivant l'éloignement de l'autel (coutume qui scandalisait tellement Lacordaire qu'il refusa, dit-on, de prêcher à Rotterdam, à moins que, pour cette occasion, toutes les places ne fussent gratuites [1]). Aussi, bien que le gouvernement n'intervienne que pour une petite part dans la rétribution du clergé, la situation matérielle de celui-ci est-elle largement assurée.

Un indice de la vitalité du catholicisme néerlandais, c'est l'augmentation rapide du nombre de prêtres et de religieux. On comptait moins de 1.500 prêtres à l'avénement de Pie IX ; ils seront environ 2.200 trente ans plus tard. Cette augmentation facilite le ministère paroissial, qui est basé sur la visite du prêtre à domicile, très importante chez un peuple qui a le goût de la vie de famille. Quant aux religieux, leur développement est plus remarquable encore et plusieurs congrégations florissantes de sœurs de charité et de religieuses enseignantes sont nées dans le pays.

Toute l'activité de ce peuple, où le « catholique non pratiquant » reste une rarissime exception, et de ce clergé pieux et actif, est contrôlée et dirigée de près par un épiscopat dont le pouvoir ne trouve guère de contrepoids, dégagé qu'il est de toute tradition susceptible de lui créer des obligations déterminées et jouissant d'une liberté complète à l'égard du gouvernement. Ceci ne va pas sans limiter quelque peu les initiatives : « Une surabondance de *charismata* n'est point à redouter tant que tous les projets sont immédiatement canalisés dans la voie tracée par la hiérarchie. Si, dans une société, tout était en règle à force d'avoir été réglementé, la Hollande serait un paradis [2]. » Mais du moins l'Église des Pays-Bas au XIXᵉ siècle présente-t-elle le grand avantage d'avoir échappé presque totalement aux déchirements intérieurs que l'on constate à la même époque en France, en Allemagne ou en Angleterre ; elle ne connaît pas les luttes entre personnalités qui épuisent le plus clair de leurs talents à se combattre les unes les autres et, d'ailleurs, dans ce petit pays où les chefs se connaissent tous individuellement, les égards et les ménagements s'imposent fatalement : cela a pu retarder plus d'une fois le progrès, mais ç'a été au profit de la sûreté d'allure et l'on a pu comparer à juste titre la Hollande catholique à « une maison où l'on se trouve peut-être un peu à l'étroit, mais où règne en tous cas l'esprit de famille ».

LES CATHOLIQUES ET LA VIE CULTURELLE — La plus grande faiblesse du catholicisme hollandais était son infériorité dans le domaine culturel. Habitués pendant deux siècles à passer pour des parents pauvres et élevés dans un classicisme borné, les catholiques n'exerceront pendant longtemps aucune influence dans

(1) G. Brom, art. *Pays-Bas*, dans *D. T. C.*, t. XII, col. 81.
(2) *Ibid.*, col. 94.

la vie scientifique, artistique ou littéraire du pays. Jusqu'en plein
xxᵉ siècle, on constatera l'absence d'hommes de valeur catholiques dans la
plupart des professions intellectuelles et l'on a fini par se rendre compte
que la vraie cause n'en était pas leur exclusion systématique par les
protestants, mais bien l'infériorité réelle des catholiques eux-mêmes.

On doit apprécier d'autant plus les quelques hommes qui tranchent
un peu sur cette médiocrité générale. On a déjà signalé à propos de la
campagne pour la constitution de 1848 les vaillants publicistes toujours
sur la brèche que furent Mgr Broer et Mgr Smits. Ce dernier fut, jusqu'à
sa mort en 1872, l'animateur du grand quotidien catholique, *De Tijd*,
qu'il sut diriger à la fois avec ardeur et modération et sans prétendre
imposer une direction au catholicisme hollandais. Il faut leur adjoindre
deux laïcs extrêmement méritants : Nuyens, polémiste de talent et fondateur de l'*Œuvre catholique des brochures*, destinée précisément à relever
le niveau culturel du peuple catholique ¹, et surtout Joseph Alberdingk
Thijm ², l'apôtre ardent de la renaissance catholique, qui devait rester
pendant un demi-siècle le pivot de l'action catholique hollandaise. Cet
ancien négociant autodidacte, qui prépara par son cercle littéraire et son
amitié avec le poète flamand Guido Gezelle l'avénement d'une nouvelle
époque en littérature, parvint à s'imposer par sa vaste érudition en dehors
des milieux catholiques et, grâce à ses qualités de cœur et d'intelligence,
à remplir pendant des années le rôle d'agent de liaison entre les catholiques
et les protestants. Auteur de travaux estimés sur l'histoire de l'art et de la
littérature néerlandaise, sa vie et son œuvre constituent la meilleure
réfutation de l'affirmation classique qui voulait voir dans la Hollande
« la nation protestante » par excellence et il fut fidèle jusqu'à sa mort à la
définition qu'il donnait de lui-même dès 1857 : « Plus j'avance et plus je
me sens devenir catholique jusqu'au tréfonds de l'âme en même temps que
néerlandais jusqu'à la moelle des os ³. »

LES CATHOLIQUES DANS
LA VIE PUBLIQUE
La revision constitutionnelle de 1848, si
favorable à l'Église, avait été l'œuvre des
libéraux, tandis que le parti conservateur,
héritier des traditions séculaires antipapistes, après avoir pendant longtemps maintenu les catholiques dans une injurieuse tutelle, continuait
à s'opposer de toutes ses forces à la réalisation de leurs revendications.
Il était normal, dans ces conditions, que, sur le plan parlementaire, les
catholiques appuyassent le parti libéral, d'autant plus que celui-ci, vers
1850, n'avait pas encore l'allure doctrinaire et sectaire qu'il devait prendre
par la suite. L'alliance allait se prolonger pendant dix-huit ans et assurer
aux catholiques, jadis quantité négligeable dans la nation, le prestige
que confère toujours la participation à l'action gouvernementale.

(1) Sur Guillaume Nuyens (1823-1894), qui, modeste médecin de village, réussit à composer
une histoire estimable de la révolution du xviᵉ siècle où il réfutait nombre d'assertions fausses
colportées par les ennemis de l'Église, voir G. Gorris, *Dr W.-J.-F. Nuyens, beschouwd in het licht
van zijn tijd*, Nimègue, 1908.
(2) Cf. C. Alberdingk Thijm, *J.-A. Alberdingk Thijm*, Amsterdam, 1896.
(3) Prospectus du *Volksalmanak voor Nederlandsche Katholieken*, pour 1857.

Peu à peu toutefois, des divergences de vue irréductibles apparurent entre les deux partenaires. La sympathie déclarée de la presse libérale pour les agissements du Piémont contre le Saint-Siège blessa profondément les catholiques hollandais, très attachés à la papauté ; puis le *Syllabus* de 1864 et la campagne déclenchée contre lui dans les milieux anticléricaux firent réfléchir plus d'un des nombreux partisans des idées catholiques libérales et renforcèrent la position de ceux qui soulignaient l'incompatibilité entre la conception catholique de la vie et les principes libéraux issus de la Révolution française. D'ailleurs l'évolution du libéralisme vers un radicalisme sectaire s'accusait de plus en plus. Lorsqu'à ces diverses causes de friction vint s'ajouter un profond désaccord de principe sur la question scolaire, la rupture devint inévitable. Par dépit contre les libéraux qui avaient trahi leurs espérances, par suite aussi d'une méfiance croissante à l'égard des idées « modernes » dont ils n'avaient d'abord remarqué que les aspects libérateurs, les catholiques se décidèrent alors, en 1866, à tendre la main aux conservateurs, qui, « modérés, calmés et matés » [1], semblaient avoir enfin renoncé à leurs vieux préjugés antiromains. Mais cette alliance, qui ne dura guère, était au fond une faute : les catholiques n'en retirèrent que « peu de profit et beaucoup d'inimitié » [2] et, en matière scolaire, la principale cause de la rupture avec les libéraux, elle ne leur procura aucun avantage.

LA QUESTION SCOLAIRE Au lendemain de la constitution de 1848, qui avait reconnu le principe de la liberté de l'enseignement, mais sans admettre que les écoles libres pussent bénéficier de subventions des pouvoirs publics, les catholiques s'étaient divisés sur la politique à suivre en matière scolaire. Les uns estimaient qu'il fallait tout mettre en œuvre pour obtenir des subsides à l'enseignement privé, afin de pouvoir multiplier les écoles catholiques, les seules admissibles à leurs yeux. D'autres, considérant que cet idéal serait malgré tout trop coûteux et difficile à réaliser, estimaient plus réaliste de chercher à améliorer le statut de l'école publique. Or, étant dans l'impossibilité d'imposer leurs vues propres — l'organisation d'écoles confessionnelles catholiques pour les catholiques — et placés devant l'alternative de choisir entre la politique scolaire préconisée par les protestants, qui voulaient conserver à l'école publique son caractère nettement calviniste, et celle des libéraux, qui réclamaient la neutralité de l'enseignement officiel, beaucoup considéraient que c'était cette dernière qui leur serait le plus favorable. Elle aurait l'avantage de supprimer le caractère confessionnel protestant de l'école publique et, en outre, ils espéraient que la neutralité serait en fait d'inspiration chrétienne, et même catholique dans les provinces de Limbourg et de Brabant où la presque totalité de la population et des autorités locales étaient catholiques. C'est dans cet espoir que plusieurs parlementaires catholiques, désapprouvés d'ailleurs par une partie de leurs coreligionnaires, aidèrent les libéraux à faire passer la loi du 13 août

---

(1) G. Douwes, *Staatkundige Geschiedenis van Nederland*, Amsterdam, 1906, p. 39.
(2) P. Verschave, *La Hollande politique*, Paris, 1910, p. 231.

1857, qui apportait de notables améliorations matérielles à l'organisation de l'enseignement public, mais établissait, au grand scandale des protestants, la neutralité scolaire.

Après quelques années, les catholiques qui avaient fait confiance à la loi de 1857 durent se rendre à l'évidence : l'enseignement neutre devenait, en fait, de plus en plus un enseignement foncièrement laïc, voire même irréligieux. Ils commencèrent alors à se rapprocher des protestants croyants et à réclamer de concert avec eux une revision de la loi dans un sens plus chrétien. Et lorsque, en 1868, le ministre de l'intérieur Fock, parlant au nom du gouvernement, eut déclaré en termes formels qu'il ne pouvait en être question, les évêques, qui avaient déjà précisé leur point de vue lors du concile provincial de 1865, estimèrent le moment venu de se prononcer : le 23 juillet, dans un mandement collectif [1], ils affirmèrent solennellement le droit de l'enfant catholique à être élevé dans une école catholique, et le devoir corrélatif qui incombait aux parents catholiques. Conformément à cette nouvelle orientation, on commença, à partir de 1865, à multiplier les écoles libres, puisque la loi les autorisait, et l'on put faire face aux lourds sacrifices que cette politique entraînait grâce à l'institution dans les paroisses de comités scolaires très efficients.

LE CATHOLICISME SUISSE AUX PRISES AVEC LE LIBÉRALISME ANTICLÉRICAL
La guerre du *Sonderbund* n'avait pas eu pour les catholiques suisses les conséquences catastrophiques qu'on aurait pu craindre. Ils avaient certes dû renoncer à leur vieil idéal de l'État confessionnel, mais la nouvelle constitution fédérale de 1848 leur reconnaissait partout la pleine liberté des cultes. Dans ces conditions, le catholicisme allait pouvoir se développer, même dans les cantons protestants. La construction, aidée par la générosité des catholiques étrangers, des belles églises Notre-Dame à Genève et Saint-Pierre à Berne, au cœur de cantons jadis farouchement calvinistes, fut une preuve tangible de l'évolution de la situation.

Toutefois, comme aux Pays-Bas, le véritable ennemi n'était plus le protestantisme, qui, sous l'influence d'Alexandre Vinet, évoluait de plus en plus vers la tolérance, mais bien le libéralisme antireligieux, qui allait continuer, pendant plus d'un quart de siècle, dans les cantons dominés par les radicaux libres-penseurs, à entraver le libre épanouissement de l'Église. Les couvents et monastères qui subsistaient encore furent presque tous supprimés et leurs biens confisqués par l'État ; même dans le catholique Tessin, le gymnase dirigé par les bénédictins dut disparaître dès 1852 ; en 1856, les écoles catholiques cantonales de Saint-Gall disparurent à leur tour et d'autres cantons supprimèrent tout enseignement confessionnel. En plusieurs endroits, les biens ecclésiastiques furent soumis à l'administration de l'État. A Bâle, on obligea l'évêque à renoncer dans son séminaire au manuel de théologie morale de Gury ; ailleurs, on alla parfois jusqu'à interdire aux prêtres de prendre part à des retraites ecclésiastiques !

(1) On en trouvera le texte dans G. Douwes, *op. cit.*, p. 45

Le conflit le plus spectaculaire fut celui qui opposa l'évêque de Lausanne-Genève aux gouvernements des cantons de Genève, Fribourg, Berne, Vaud et Neuchâtel, sur lesquels s'étendait son diocèse. Ces cinq cantons avaient conclu entre eux, le 15 août 1848, un accord dénommé *Concordat* par lequel ils prétendaient réglementer unilatéralement la vie ecclésiastique dans le diocèse, se réservant de nommer l'évêque, exigeant le *placet* cantonal pour une série d'actes épiscopaux et pour la publication des documents pontificaux, obligeant à adapter les statuts synodaux aux lois civiles et soumettant les futurs prêtres à un examen devant une commission de laïcs délégués par le gouvernement.

Des mesures de ce genre n'étaient pas pour déplaire à quelques ecclésiastiques imprégnés d'idées joséphistes, disciples de Wessemberg, dont l'influence resta très forte en Suisse jusqu'à sa mort en 1860 [1] ; elles plaisaient aussi à certains catholiques qui ne se rendaient pas compte que leur enthousiasme pour l'idéal démocratique de l'antique Helvétie ne pouvait aller jusqu'à vouloir modifier la constitution divine de l'Église. Mais Mgr Marilley, appuyé par la masse de son clergé et de ses fidèles, protesta et interdit à ses prêtres de prêter serment au nouveau règlement. Quelques semaines plus tard, le Saint-Siège protestait à son tour contre le droit que les autorités cantonales voulaient s'arroger de nommer les évêques. Le gouvernement fit aussitôt arrêter Mgr Marilley et l'expulsa de Suisse, après un internement de deux mois dans la forteresse de Chillon. Bien que les pourparlers avec Rome eussent repris dès 1852, sous la pression du mécontentement catholique, ce n'est qu'en 1856 qu'on arriva à s'entendre sur un *modus vivendi* et que le prélat put enfin rentrer dans son diocèse [2].

LA VITALITÉ CATHOLIQUE   Les Suisses, avant tout pratiques et positifs, n'ont jamais eu un penchant très marqué pour les spéculations philosophiques ou théologiques ; d'ailleurs, à part Fribourg, que Veuillot caractérisait comme « une petite Rome silencieuse et cachée », les régions traditionnellement catholiques — le Valais et les cantons boisés entourant le lac de Lucerne — sont pauvres et essentiellement ruraux, ce qui ne contribue guère à favoriser l'épanouissement intellectuel. Certes, la Suisse catholique peut être fière d'hommes tels que le P. Gall Morel, considéré par ses contemporains comme un « Goethe catholique » [3], le comte de Scherer-Boccard, journaliste et historien, fondateur de l'*Académie Saint-Charles-Borromée* [4], ou Mgr Greith, évêque

---

(1) Sur cette persistance des idées joséphistes en Suisse, qui allait apparaître au grand jour après la proclamation, en 1870, de la primauté pontificale, cf. E. CAMPANA, *Il Concilio Vaticano*, t. II, Lugano, 1926, chap. x.

(2) L'accord signé deux ans plus tard avec le canton de Fribourg concernant l'administration des biens ecclésiastiques constitua une nouvelle étape sur la voie de l'apaisement. (Cf. H. MARMIER, *La convention du 23 avril 1858 entre l'évêque de Lausanne et Genève et l'État de Fribourg*, Fribourg, 1938.)

(3) Sur le P. Gall Morel, O. S. B. (1803-1872), cf. *infra*, p. 178, n. 2.

(4) Sur Theodor von Scherer (1816-1885), qui eut également une grande activité politique jusqu'à l'échec du *Sonderbund* et qui devint par la suite président du *Piusverein*, cf. J.-G. MAYER, *Graf Th. v. Scherer-Boccard*, Fribourg, 1900.

de Saint-Gall de 1862 à 1882, élève préféré de Goerres à Munich [1]. Ce sont malgré tout des exceptions. Ce n'est pas dans le domaine de la vie de l'esprit que se distinguent les catholiques suisses.

Les mesures vexatoires de tout genre auxquelles ils étaient soumis de la part des radicaux les obligèrent en effet à s'organiser en fonction des nécessités modernes. L'exil ou la confiscation de biens qui avaient frappé une partie de leurs dirigeants à la suite du *Sonderbund* constituaient un handicap supplémentaire, mais ils surent en triompher en s'inspirant des exemples étrangers. Sur le plan politique, les catholiques se mirent à l'école de Montalembert, qu'ils vénéraient pour le dévouement chevaleresque à leur cause dont il avait fait preuve en 1847 ; ils comprirent assez vite la fécondité des idées libérales du grand leader catholique français dans la situation nouvelle de la Suisse et ce sont elles qui devinrent le programme du parti « libéral-conservateur ». L'influence de Montalembert domina également dans le *Studentenverein*, réplique catholique à la société estudiantine radicale *Zofingia*, tandis que le clergé suivait avec admiration les publications de Dupanloup, dont l'action était renforcée par les fréquents passages de l'évêque d'Orléans à la célèbre abbaye d'Einsiedeln, l'un des centres de direction intellectuelle et spirituelle du catholicisme suisse [2].

C'est par contre à l'Allemagne que fut empruntée l'idée d'une organisation centralisée des œuvres destinée, en groupant les forces catholiques, à soutenir et à multiplier l'action individuelle : le *Piusverein*, organisé au cours des années soixante, devint vite florissant dans les vieux cantons catholiques où s'épanouissaient depuis le milieu du siècle les œuvres de dévotion et de bienfaisance. Parmi tous ceux, laïcs et religieux, qui se dévouèrent à ces dernières, il faut tout spécialement mentionner le capucin Théodose Florentini [3], apôtre de la charité et réformateur social, à l'esprit d'entreprise toujours en éveil et au zèle inlassable, qui peut sans doute être considéré comme le plus grand philanthrope suisse du XIXe siècle. Il fut aidé, dans ses multiples fondations d'écoles et d'hôpitaux, par la Mère Marie-Thérèse Scherer [4], supérieure des sœurs de charité de la Sainte-Croix d'Ingenbohl, qu'il avait fondées et qui, sous la conduite intelligente de cette religieuse d'élite, étaient appelées à prendre un développement considérable dans toute l'Europe centrale. Cette efflorescence d'œuvres sociales, qui fut spécialement marquée dans les cantons industriels, où les catholiques se trouvaient dispersés parmi les protestants, devait contribuer à faire de la Suisse catholique, quelques années plus tard, un des foyers du catholicisme social à ses débuts.

---

(1) Sur Karl-Johann Greith (1807-1882), voir J. Osch, *K. J. Greith*, St-Gall, 1909.
(2) L'école abbatiale d'Einsiedeln connut sous Pie IX un rayonnement tout spécial grâce aux efforts combinés du P. Henri Schmid, abbé de 1846 à 1874, du converti Karl Brandes et surtout du P. Gall Morel, pédagogue expérimenté, poète religieux estimé par ses contemporains, en même temps qu'historien et érudit de valeur, qui réussit à faire de la bibliothèque du monastère un centre de recherches scientifiques. (Cf. B. Kuehme, *P. Gall Morel. Ein Monchsleben aus dem XIX. Jh.*, Einsiedeln, 1875.)
(3) Sur le P. Théodose (1808-1865), voir P. Veit Godient, *Der Caritasapostel, Theodosius Florentini*, Lucerne, 1944.
(4) Sur M. Th. Scherer (1825-1888), voir L. Collin, *Une fleur des Alpes*, Ingenbohl, 1931.

MONSEIGNEUR MERMILLOD L'un des agents les plus actifs de l'adaptation du catholicisme suisse aux circonstances nouvelles fut incontestablement Gaspard Mermillod [1]. Brillant orateur, c'était en même temps un réalisateur, à l'esprit entreprenant et d'une activité débordante, qui se fit rapidement remarquer comme vicaire puis comme curé de Genève, où il multiplia les œuvres et les écoles grâce à l'argent récolté au cours de ses tournées de prédication à travers l'Europe. Polémiste toujours sur la brèche, il fonda en 1862 une revue mensuelle, puis, cinq ans plus tard, un journal qui devint bientôt quotidien, *Le Courrier de Genève*, afin de soutenir les catholiques dans la controverse qu'ils devaient mener continuellement contre leurs adversaires. C'était enfin un ardent ultramontain, et l'on comprend que Pie IX ait songé à lui, à tous ces titres, lorsque les catholiques genevois, dont le nombre allait croissant par suite de l'immigration savoyarde, demandèrent à être constitués en diocèse indépendant. Comme la constitution helvétique interdisait l'érection de nouveaux diocèses, le pape se borna, pour commencer, à le nommer évêque auxiliaire de Mgr Marilley pour le canton de Genève et, plein d'un optimisme qui devait s'avérer quelque peu prématuré, lui donna comme consigne, au cours de la consécration épiscopale qui eut lieu à Rome même, le 25 septembre 1864 : « Allez, montez sur le siège de saint François de Sales, allez vers cette Genève qui n'a pas craint de s'appeler la Rome protestante. Portez-lui le trésor de mon amour et convertissez-la ! »

§ 4. — L'Église en Espagne et au Portugal [2].

L'ÉVOLUTION DE LA POLITIQUE RELIGIEUSE ESPAGNOLE La défaite d'Espartero et la victoire des modérés aux élections d'octobre 1843 avaient mis fin, en Espagne, à la politique anticléricale poursuivie depuis dix ans par le parti libéral, et des négociations avaient été aussitôt engagées avec le Saint-Siège en vue de la conclusion d'un concordat qui, tout en reconnaissant la place privilégiée de l'Église, entérinerait certaines réformes dont elle n'avait pas su prendre elle-même l'initiative. La méfiance persistante de Grégoire XVI à l'égard du nouveau gouvernement, encore trop libéral à ses yeux, empêcha un premier projet d'aboutir, mais Pie IX prit dès son avénement une attitude plus conciliante en rétablissant les relations diplomatiques et en ratifiant les nominations d'évêques faites par le gouvernement d'Isabelle. La chose était d'ailleurs urgente car, en 1843,

(1) Sur Mgr Mermillod (1824-1892), voir L. Jeantet, *Le cardinal Mermillod*, Paris, 1906 et Ch. Comte, *Mermillod d'après sa correspondance*, Paris, 1924.

(2) Bibliographie. — *a*) Pour l'Espagne : V. La Fuente, *Historia eclesiastica de España*, t. VI, Madrid, 1875 ; P. Gams, *Kirchengeschichte von Spanien*, t. III, Innsbruck, 1879 ; P. de Luz, *Isabelle II, reine d'Espagne*, Paris, 1935 ; J. Becker, *Relaciones diplomaticas entre España y la Santa Sede durante el siglo XIX*, Madrid, 1909 ; M. Gonzalez-Ruiz, *Vicisitudes de la propriedad eclesiastica en España durante el siglo XIX*, dans *Revista española de derecho canonico*, t. I, 1946, p. 383-424 ; Lopez Pelaez, *El derecho español en sus relaciones con la Iglesia*, Madrid, 1902. En outre, sur les conflits idéologiques, Menendez y Pelayo, *Historia de los Heterodoxos españoles*, t. VII, 2e édit., Madrid, 1932.

b) Pour le Portugal : F. de Almeida, *Historia da Igreja em Portugal*, t. IV, vol. 1 à 3, Coimbre, 1917-1923 ; M. de Oliveira, *Historia eclesiastica de Portugal*, Lisbonne, 1940.

il y avait eu jusqu'à 47 sièges vacants sur 62, et, malgré le retour des évêques exilés, il restait de nombreux vides à combler.

Peu après, les négociations en vue du concordat reprenaient et, grâce aux efforts du nonce Brunelli et du savant archevêque de Tolède, le cardinal Costa y Borras [1], elles aboutirent, le 16 mars 1851, à une convention en 46 articles [2]. Le catholicisme était déclaré religion d'État, l'exercice public du culte était interdit à toute autre confession religieuse, et les évêques recevaient le droit de veiller à ce que l'enseignement, même dans les universités, fût conforme à la doctrine catholique. Les souverains conservaient le droit de nommer les évêques, et le nombre et les limites des diocèses et des paroisses étaient modifiés en tenant compte des situations nouvelles. L'Église perdait le privilège du for ecclésiastique, mais par contre les évêques voyaient leur autorité renforcée par la suppression de la plupart des juridictions privilégiées et des exemptions, si nombreuses auparavant, dont jouissaient les chapitres. L'interdiction des ordres religieux était partiellement levée. Enfin, l'Église recevait à nouveau le droit de posséder ; elle renonçait toutefois aux biens confisqués antérieurement et déjà vendus, mais, en compensation, l'État prenait à sa charge les traitements du clergé, à vrai dire assez modestes.

Une certaine restauration religieuse s'en suivit. Les séminaires notamment se repeuplèrent, mais, malgré tout, en 1868, le total des prêtres diocésains n'atteignait plus que 44.735 [3], contre 57.892 quarante ans plus tôt ; ce chiffre était du reste encore relativement considérable : 1 prêtre pour 380 habitants. Le relèvement des ordres religieux ne s'effectua par contre qu'au compte-goutte : au lieu des 92.627 religieux répartis en plus de 3.000 maisons, que comptait l'Espagne en 1826, au moment des lois d'abolition, il n'y avait encore en 1862 que 1.746 religieux et 13.347 religieuses. Pour ces dernières, une difficulté spéciale venait de ce que les couvents devant vivre de l'argent apporté par les postulantes, les jeunes filles sans fortune étaient souvent écartées ; des fondations pieuses en vue de leur procurer des dots essayèrent de remédier à cet inconvénient.

Si le concordat apparut dans l'ensemble comme favorable à l'Église, il rendait pourtant celle-ci beaucoup plus dépendante de l'État qu'auparavant et, de plus, certains articles ne furent guère observés, entre autres celui relatif à la surveillance de l'enseignement public par l'Église. Mais le principal inconvénient fut que, n'ayant pas été reconnu par les libéraux antigouvernementaux, il amena l'Église à prendre plus que jamais parti pour les conservateurs dans les luttes politiques, puisque le retour au pouvoir des radicaux devait signifier la rupture de l'accord.

C'est ce qui arriva effectivement lorsque, en 1854, Espartero reprit le pouvoir : il s'empressa de dénoncer le concordat et les mesures vexatoires se multiplièrent [4] en dépit de l'opposition de la reine, impressionnée

---

(1) Cf. F. Costadellas, *El arzobispo Costa y Borras*, Barcelone, 1947. Il fut un zélé défenseur de la papauté et un adversaire ardent de la politique libérale. Il prépara une édition, publiée après sa mort, des conciles de la province de Tarragone.

(2) Texte dans V. La Fuente, *op. cit.*, t. VI, p. 387-400.

(3) *Guia del estado eclesiastico de España para el año de 1868*, Madrid, 1868.

(4) Il faut noter toutefois que la décision de mettre en vente tous les biens ecclésiastiques ne fut pas provoquée uniquement par l'anticléricalisme, car les propriétés des universités, des écoles

par l'énergique protestation du pape dans son allocution consistoriale du 26 juillet 1855. Mais dès l'automne 1856, le gouvernement repassait aux mains des conservateurs, dirigés par Narvaez, et l'Église connut de nouveau jusqu'à la révolution de 1868 une paix relative, qui fut sanctionnée par un nouvel accord avec le Saint-Siège, en avril 1860, complétant les dispositions du concordat. L'Église pouvait compter sur les sympathies de la haute aristocratie et du peuple, encore profondément chrétien, ainsi que sur la protection de la reine Isabelle, qui unissait curieusement à des mœurs qui défrayaient la chronique scandaleuse une piété tournant parfois à la bigoterie et une docilité totale aux conseils de son confesseur en matière de politique religieuse. L'adhésion de la reine aux principes du *Syllabus* et ses encouragements aux efforts de Narvaez pour maintenir, au besoin par la force, l'ordre moral, de même que son attitude dans la question romaine, poussèrent même Pie IX en 1867 à conférer la rose d'or à cette souveraine qui était l'une des dernières en Europe à pratiquer encore une politique officiellement catholique. Mais cette politique ne put empêcher l'Espagne de s'entr'ouvrir toujours davantage aux idées libérales et, au contraire, l'appui affiché que l'Église apportait à un gouvernement de plus en plus réactionnaire acheva d'exaspérer contre elle les rancunes de la bourgeoisie des villes et de la petite noblesse libérale, encadrées par la franc-maçonnerie.

LA VIE CATHOLIQUE EN ESPAGNE Tandis que l'Église d'Espagne se compromet sur le terrain politique, le clergé encourage, sur le plan culturel, un parti-pris antimoderne, allant parfois jusqu'au fanatisme, qui va devenir caractéristique du catholicisme espagnol et qui aura pour conséquence de limiter dangereusement son influence sur le monde intellectuel, où les pratiques religieuses subsistent à cause de l'ambiance officielle, mais où la foi commence à diminuer de façon sensible. La science catholique apparaît par ailleurs purement ecclésiastique : œuvres théologiques, inspirées de saint Thomas, du dominicain Pascal et de ses élèves, les futurs cardinaux Cuesta et Gonzales ; éditions remarquables des grands mystiques du XVIᵉ siècle et des autres écrivains spirituels espagnols, travaux de mystique spéculative. Mais on ne rencontre plus de penseurs catholiques comme Balmès ou Donoso Cortes, disparus vers 1850, qui aient l'audience du grand public. Ce fait est d'autant plus regrettable que les cercles cultivés commencent à subir l'influence des erreurs philosophiques allemandes, kantisme et hégélianisme, sous la forme du système de Krause notamment, qui, introduit par Sanz del Rio, connaît une vogue d'où le snobisme n'est pas absent.

Les erreurs de tactique, les fautes de jugement et l'insuffisance intellectuelle de l'Église d'Espagne ne peuvent faire oublier tout ce qui subsistait encore de vie chrétienne authentique et de dévouement à l'Église dans ce pays où le catholicisme plonge des racines si profondes. S'il faut regretter l'indolence d'une trop grande partie du clergé inférieur,

ou des hôpitaux furent touchées par la même mesure : le gouvernement s'inspirait des conceptions des économistes libéraux qui considéraient les biens de mainmorte comme une institution anachronique et nuisible à un fonctionnement normal du régime capitaliste.

on peut admirer par contre, surtout en pays basque et en Catalogne,
des modèles de dévouement et d'austérité, telles la sainte comtesse de
Jorbalan, plus connue sous le nom de *Madre Sacramento*, fondatrice des
Adoratrices du Saint-Sacrement, ou l'énergique animatrice de la congrégation hospitalière des Servantes de Marie, la madrilène Maria Desolata
Torres Acosta ; tels l'abbé Lopez de Novoa, qui, en prenant modèle sur
la France, fonde les Petites Sœurs des Vieux abandonnés, ou saint Antoine
Marie Claret y Clara, qui après s'être dévoué avec un zèle infatigable au
service des classes populaires de la région de Barcelone, devint le fondateur d'un ordre missionnaire prospère, les Fils du Cœur Immaculé de
Marie, et trouva le moyen, au milieu d'une activité multiforme, de prêcher
plus de 10.000 fois et d'écrire près de deux cents ouvrages de spiritualité
populaire ou sacerdotale [1].

AU PORTUGAL La situation politique et religieuse du Portugal pendant la première moitié du XIXe siècle avait présenté
beaucoup d'analogie avec celle de l'Espagne et, là aussi, les catholiques
s'étant trouvés dans le même camp que les absolutistes et en lutte avec les
libéraux appuyés sur les sociétés secrètes, ils avaient eu fort à souffrir
de la victoire de ces derniers. Pendant les dernières années de Grégoire XVI
cependant, la situation s'était améliorée et, le 21 octobre 1848, le comte
de Tomar, ministre plénipotentiaire, signait avec l'internonce une convention réglant entre autres les questions des séminaires, des bénéfices et
du for ecclésiastique [2] ; les confiscations de biens ecclésiastiques y étaient
entérinées de même que la suppression de la dîme, et il était décidé que
dorénavant les curés seraient payés par une contribution communale.
La convention de 1848 fut complétée en 1857 par un accord concernant le
droit de patronage royal aux Indes et en Chine.

La même année 1848, un jeune jésuite portugais, exilé en Italie, ren-
trait dans son pays, suivi bientôt par quelques autres ; cinq ans plus tard,
il ouvrait un collège, et peu à peu la Compagnie reprit pied dans le
royaume, tandis que, à partir de 1858, les autres ordres réapparaissaient
eux aussi. Mais bientôt l'agitation anticléricale, soutenue par les francs-
maçons, reprit et aboutit à de nouvelles sécularisations [3], tandis qu'une
violente campagne était entamée à partir de 1865 en faveur du mariage
civil, car, comme en Espagne, si le petit peuple restait fermement attaché
aux traditions catholiques, la bourgeoisie cultivée se laissait de plus en plus
gagner par le rationalisme antichrétien, mais sous des influences françaises plutôt qu'allemandes, cette fois.

La politique anticléricale, poursuivie avec plus d'esprit de suite au
Portugal qu'en Espagne, eut entre autres pour conséquence une diminution beaucoup plus sensible du nombre des prêtres ; ceux-ci, qui étaient
environ 18.000 en 1822, ne seront plus que 6.841 en 1888 [4]. Chose plus

(1) Cf. C. Fernandez, *El beato Padre Antonio M. Claret. Historia documentada de su vida y
empresas*, Barcelone, 1948, 2 vol. Il fut le confesseur et conseiller de la reine Isabelle de 1857 à 1868.
(2) Texte dans Borges de Castro, *Colecçao dos tratados*, t. VII, p. 221-223.
(3) Lois des 4 avril 1861, 22 juin 1866 et 28 août 1869.
(4) O. Werner, *Orbis terrarum catholicus*, Fribourg-en-Br., 1890, p. 52. Le chiffre de 1888 donne

grave, la qualité de ces prêtres est également déficiente : le haut clergé, à quelques exceptions près, se comporte à l'égard de Rome avec une très grande froideur, que l'affaire du schisme de Goa contribue à entretenir ; et le bas clergé manque d'instruction et laisse souvent à désirer du côté des mœurs. Sur ce dernier point, toutefois, la situation s'améliore un peu à partir de 1870, grâce à la réorganisation des séminaires, qui s'était poursuivie lentement depuis le concordat de 1848, en dépit des entraves maçonniques et des interventions constantes du gouvernement. Celles-ci allaient si loin que, non content de vouloir réglementer les programmes et la vie des établissements, il décréta même, en 1861, que, dans les diocèses où il n'y avait pas encore de séminaires, les candidats pourraient être nommés aux postes ecclésiastiques sans avoir fait d'études complètes.

§ 5. — En Scandinavie [1].

LE CATHOLICISME EN SUÈDE   L'édit de tolérance de 1783, qui n'accordait qu'aux seuls étrangers le droit de pratiquer publiquement une autre religion que le luthéranisme, resta en vigueur jusqu'en 1860 et en 1859 encore, six femmes furent condamnées au bannissement pour s'être converties au catholicisme. Mais les répercussions de ce jugement en Europe provoquèrent un nouvel examen de la question, qui aboutit à l'arrêté royal du 13 octobre 1860, complété ensuite par les lois de 1870 et de 1872. Par ces différentes mesures, les lois d'exception étaient abrogées et les Suédois âgés de plus de 18 ans autorisés à changer de religion. Dès lors Mgr Studach, vicaire apostolique de 1833 à 1874, put exercer son apostolat dans des conditions moins défavorables et construire à Göteborg en 1865 et à Malmö en 1872 deux nouvelles églises, mais les conversions demeurèrent insignifiantes.

EN NORVÈGE   En Norvège, où la messe fut célébrée publiquement pour la première fois depuis la réforme en 1843, le roi octroya officiellement la tolérance religieuse dès 1845. Rome en conçut de grandes espérances et s'empressa d'envoyer des missionnaires, mais, pas plus qu'en Suède, le catholicisme ne réussit à s'imposer à l'attention.

AU DANEMARK   Les progrès furent un peu plus marquants au Danemark où la constitution de 1849 proclama la pleine liberté religieuse, mais l'augmentation du nombre des fidèles fut surtout due aux immigrants étrangers.

En fait, dans les trois pays scandinaves réunis, le catholicisme, à la mort de Pie IX, ne comptera pas encore 3.000 fidèles et il continuera à apparaître comme une forme étrangère du christianisme, peu adaptée aux habitudes du pays.

une moyenne de un prêtre pour 666 habitants, avec d'assez fortes variations suivant les régions : 4.928 prêtres pour 2.607.000 habitants dans la province ecclésiastique de Braga ; 434 prêtres pour 540.952 habitants dans la province d'Evora.

(1) BIBLIOGRAPHIE. — L. CROUZIL, *Le catholicisme dans les pays scandinaves*, Paris, 1902, 2 vol. ; J. METZLER, *Die apostolischen Vikariate des Nordens*, Paderborn, 1911 ; K. KJELSTRUP, *Norvegia catholica 1843-1943*, Rome, 1947 ; L. F. DELTOMBE, *L'Église catholique en Suède*, dans *Nouvelle Revue théologique*, t. LXXII, 1950, p. 715 et suiv.

LES SCIENCES ECCLÉSIASTIQUES JUSQU'AU CONCILE DU VATICAN [1]

§ 1. — Les études à Rome et la restauration de la scolastique [2].

L'INFÉRIORITÉ INTELLECTUELLE A ROME SOUS PIE IX « La théologie romaine est trop insouciante de ce qui se passe autour d'elle. Le rationalisme y est en général mal connu et futilement combattu. L'histoire n'a pas de représentant célèbre. La linguistique est négligée. La médecine est arriérée. Le droit est resté ce qu'il était avant le mouvement qui lui a été imprimé par les découvertes dont Savigny s'est fait l'habile propagateur [3]. » A ces plaintes du futur cardinal Meignan, de passage à Rome en 1846, font écho tout au long du pontificat de Pie IX bien d'autres témoignages provenant non seulement de partisans de l'unité italienne ou d'adversaires des méthodes romaines, mais également d'hommes entièrement dévoués au Saint-Siège. En 1854, Flir, le recteur autrichien de l'*Anima*, tout en appréciant une grande habileté en matière de casuistique et la vaste érudition de quelques orientalistes, constatait avec étonnement l'absence totale « de ce qu'on nomme science en Allemagne » [4], et l'ancien nonce à Vienne, le cardinal Viale Prela, lui avouait qu'il serait effectivement fort utile pour la science romaine de se revigorer au contact de la littérature scientifique germanique [5]. La situation n'avait guère changé, dix ans plus tard, lors du

(1) BIBLIOGRAPHIE. — J. BELLAMY, *La théologie catholique au XIXᵉ siècle*, Paris, 1904 ; M. GRABMANN, *Geschichte der katholischen Theologie*, Fribourg-en-Br., 1933, IIIᵉ Part. ; E. BRÉHIER, *Histoire de la Philosophie*, t. II, fasc. 4, Paris, 1938 ; E. CL. SCHERER, *Geschichte und Kirchengeschichte an den deutschen Universitäten*, Fribourg-en-Br., 1927. Notices utiles dans H. HURTER, *Nomenclator litterarius theologiae catholicae*, Innsbruck, 1903-1913, 5 vol. et C. SOMMERVOGEL, *Bibliothèque de la Compagnie de Jésus*, Bruxelles, 1890-1932, 11 vol.

Sur le mouvement rationaliste à cette époque : J.-M. ROBERTSON, *A History of Free Thought in the XIXth century*, Londres, 1929 et A.-D. WHITE, *A History of the Warfare of Science and Theology*, Londres, 1896, 2 vol. (adapt. franç., Paris, 1899) ; bon aperçu synthétique dans le chapitre *Le conflit de la science et de la croyance* par J. MAURAIN dans le t. XVII de la collection « Peuples et Civilisations » : *Du Libéralisme à l'Impérialisme*, Paris, 1939.

Pour l'Allemagne, en particulier, A. SCHWEITZER, *Geschichte der Lebensjesuforschung*, Tubingue, 1913 ; pour la France, les chap. VII et VIII de G. WEILL, *Histoire de l'idée laïque en France au XIXᵉ siècle*, Paris, 1929 ; pour l'Italie, A. DELLA TORRE, *Il Cristianesimo in Italia dai filosofisti ai modernisti*, Palerme, 1912.

(2) BIBLIOGRAPHIE. — Sur les études à Rome : *L'Università Gregoriana del Collegio Romano*, Rome, 1925 ; A. KERKVOORDE, *La formation théologique de M. J. Scheeben à Rome (1852-1859)*, dans *Ephemerides theologicae lovanienses*, t. XXII, 1946, p. 174-193 ; T. ORTOLAN, art. *Italie*, dans *D. T. C.*, t. VIII, col. 235-242.

Sur la renaissance thomiste : A. PELZER, *Les initiateurs italiens du néothomisme*, dans *Revue des Sciences phil. et théol.*, t. XVIII, 1911, p. 230-254 ; P. DEZZA, *Alle origine del neotomismo*, Milan, 1940 ; P.-A WALZ, *Il tomismo dal 1800 al 1879*, dans *Angelicum*, t. XX 1943, p. 300-326.

(3) Lettre du 4 avril 1846 dans H. BOISSONOT, *Le cardinal Meignan*, p. 116-118.

(4) A. FLIR, *Briefe aus Rom*, p. 14-15 (lettre du 15 septembre 1854).

(5) *Ibid.*, p. 64 (lettre du 16 octobre 1856).

séjour à Rome du pieux historien Janssen : il était scandalisé par la
négligence des Romains à exploiter leurs riches archives, par l'organisation défectueuse des bibliothèques, par le manque d'intérêt des autorités
pour l'enseignement supérieur ou le travail scientifique, et il rapportait
à ses amis des réflexions comme celle d'un jeune savant italien : « Ici,
les études sont mortes, il n'y a que la pratique qui compte », ou celle du
très ultramontain cardinal de Reisach : « A part J.-B. de Rossi, il n'y a
personne dans la Commission archéologique pontificale qui y comprenne
quelque chose ! [1] »

Ces plaintes sont trop nombreuses et trop uniformes pour ne pas
correspondre en bonne partie à la réalité, mais elles doivent pourtant
être nuancées. Sans parler des sciences profanes, où l'on pourrait signaler
des exceptions comme le P. Angelo Secchi, le célèbre astronome et phy-
sicien du Collège romain, de réputation mondiale, la tradition érudite du
XVIIIe siècle n'a pas disparu tout à fait, l'archéologie chrétienne est
brillamment représentée et, dans le domaine théologique et philosophique,
on assiste à un réveil incontestable.

TRAVAUX D'ÉRUDITION   Le cardinal Mai [2], le génial déchiffreur de
palimpsestes, l'infatigable éditeur de textes
classiques et patristiques, appartient à peine au pontificat de Pie IX,
puisqu'il meurt en 1854. Mais son collaborateur, le barnabite Vercellone,
publie de 1857 à 1869 quelques travaux estimables de critique textuelle
biblique et édite magnifiquement le *Codex Vaticanus*, tandis que l'ouvrage
de Cavedoni sur les monnaies dans la Bible a les honneurs d'une traduction allemande. D'autre part, un savant de grande classe, le bénédictin français Pitra [3], qui avait été appelé à Rome en 1858, mène à bon
terme, au prix d'un labeur déconcertant, des travaux de grande valeur,
bien que trop peu remarqués, sur le droit canonique oriental et sur l'hym-
nographie byzantine, pour ensuite continuer patiemment ses publications
d'inédits de Pères anténicéens.

Dans le domaine de l'histoire ecclésiastique, les Romains en restent
encore généralement à la vieille conception de l'*historia magistra vitae*,
où les soucis de présentation l'emportent sur ceux de la critique ; mais
quelques érudits allemands, en résidence à Rome, commencent à travailler
selon les exigences scientifiques modernes. Ainsi l'abbé Thiel, qui édita
en 1867 un premier volume de lettres des souverains pontifes jusqu'au
pontificat d'Hormisdas, et surtout l'oratorien Theiner, préfet des Archives
vaticanes depuis 1855, esprit brouillon et parfois négligent, mais travailleur actif, qui publie, outre de nombreux ouvrages d'histoire moderne,
neuf volumes d'inédits : *Hungaria sacra, Polonia sacra, Hibernia sacra*, etc. [4].

---

(1) *Janssens Briefe*, édit. PASTOR, t. I, p. 229, 275-277, 285.
(2) Sur Angelo Mai (1782-1854), voir les articles de E. AMANN, dans *D. T. C.*, t. IX, col. 1650-1653
et de H. LECLERCQ, dans *D. A. C. L.*, t. X, col. 1196-1202, qui relève aussi les défauts de ce grand
travailleur.
(3) Sur J.-B. Pitra (1812-1889), voir les deux biographies, qui se complètent, de F. CABROL,
Histoire du cardinal Pitra, Paris, 1891 et d'A. BATTANDIER, *Le cardinal Pitra*, Paris, 1896.
(4) Parus de 1859 à 1864. C'est lui aussi qui, après être tombé en disgrâce, publia les actes du
concile de Trente (2 vol., Agram, 1874).

L'ARCHÉOLOGIE CHRÉTIENNE   Si la mise en valeur des manuscrits du Vatican sous Pie IX est surtout le fait d'érudits étrangers, l'archéologie chrétienne, par contre, est une science purement romaine [1]. Elle s'était réveillée pendant le pontificat de Grégoire XVI, grâce aux encouragements de ce pape éclairé et aux efforts trop oubliés du chanoine Settele, puis du P. Marchi. Ce dernier, après avoir réorganisé le musée Kircher, avait inauguré, avec une certaine maladresse encore, l'étude scientifique des catacombes, tombée en profonde décadence depuis le xvie siècle, et Pie IX sanctionna ce réveil en créant en 1852 une *Commission d'archéologie sacrée*.

L'œuvre du P. Marchi fut poursuivie par le P. François Tongiorgi, qui professa jusqu'en 1886 au Collège romain, où séjournait également un autre archéologue, au talent plus inégal, le P. Garrucci, mais l'élève de prédilection et le véritable continuateur de Marchi fut Jean-Baptiste de Rossi [2]. Si l'on n'a plus le droit, après les travaux du P. Fausti, de dire que de Rossi a créé l'archéologie chrétienne, il l'a en tout cas établie sur des bases inébranlables et dotée d'une méthode rigoureuse, en même temps qu'il précisait les règles de l'épigraphie chrétienne au point d'en paraître le fondateur. Les Allemands, si méprisants pour la science italienne, n'ont pu que s'incliner : « Avant de Rossi, disait Mommsen, l'archéologie chrétienne n'était qu'un passe-temps d'amateur ; avec lui, elle est devenue une science [3]. » Et, en 1854, l'Académie de Berlin s'assurait sa collaboration pour le *Corpus inscriptionum latinarum*. De Rossi réussit à vaincre toutes les difficultés, y compris les objections de ceux qui trouvaient périlleux de voir un laïc s'occuper d'histoire ecclésiastique [4] et, servi par une érudition et une mémoire prodigieuses et par une attention à laquelle aucun détail n'échappait, il élabora une vaste synthèse, la *Roma sotterranea cristiana* [5], où revivait toute l'histoire de l'Église romaine primitive, de sa doctrine, de sa hiérarchie, de sa liturgie et de son art, et qui constituait, selon le mot du cardinal Pie, un « lieu théologique » nouveau, bien que, avec une probité scientifique exemplaire, il se refusât absolument à majorer la portée d'un fait dans un but apologétique. De Rossi eut en outre le mérite non seulement d'intéresser à ses découvertes, par ses conférences et ses visites guidées, un très vaste public

---

(1) Voir G. FERRETTO, *Note storico-bibliografiche di archeologia cristiana*, Cité du Vatican, 1942 et R. FAUSTI, *Il P. Giuseppe Marchi e il rinnovamento degli studi di Archeologia cristiana*, dans *Xenia Piana Pio XII dicata*, Rome, 1943, p. 445-514. De Rome, l'archéologie chrétienne se répandit à l'étranger. C'est ainsi qu'en France, Edmond Le Blant, qui avait fait un séjour à Rome en 1847, publia à partir de 1856 les *Inscriptions chrétiennes de la Gaule*. (Cf. H. LECLERCQ, dans *D. A. C. L.*, t. VIII, col. 2143-2218.)

(2) Sur J.-B. de Rossi (1822-1894), voir H. LECLERCQ, dans *D. A. C. L.*, t. XV, col. 18-100.

(3) Cité par L. DUCHESNE, dans *Bulletin critique*, t. XV, 1894, p. 372.

(4) Mgr Capalti l'avait un jour averti en ces termes des obstacles qui l'attendaient : « L'usage maintient une foule de vieux récits auxquels personne ne croit. Vos études vous amèneront à les examiner de près. Si vous les présentez comme vrais, vous passerez, non pour un sot, car cela n'est pas possible, mais pour un homme dépourvu de probité scientifique. Si vous les écartez, il se trouvera des hypocrites pour crier au scandale et des imbéciles pour les croire ; là, pour vous, beaucoup d'ennuis. » (L. DUCHESNE, dans *Revue de Paris*, t. V, 1894, p. 720-721.) Ces mots sont révélateurs de l'état d'esprit qui régnait à Rome au milieu du xixe siècle.

(5) En 3 volumes qui parurent respectivement en 1863, 1867 et 1877. Il avait déjà publié, en 1857, les *Inscriptiones christianae Urbis Romae*, et fondé, en 1863, le *Bolletino di archeologia cristiana*.

mais surtout de former des élèves qui devaient compléter et perfectionner
encore son œuvre.

LA THÉOLOGIE AU
COLLÈGE ROMAIN

L'enseignement du Collège romain, qui avait été
réorganisé au lendemain des troubles de 1848, s'il
présentait une allure plus scolaire et moins scientifique que celui des universités allemandes [1], a cependant été souvent exagérément déprécié. Certes, les professeurs superficiels n'y manquaient pas, tel celui de droit canon, le futur cardinal Tarquini, plus
soucieux de virtuosité dialectique que d'une étude historique des sources
du droit. Mais le P. Patrizi, qui enseignait l'Écriture Sainte depuis 1835,
était un travailleur consciencieux, dont l'ouvrage sur le sens typique
fut la première monographie catholique approfondie sur la question et
dont le *De Evangeliis*, publié en 1853, constitua longtemps l'étude la plus
documentée sur la chronologie de la vie du Christ. Le P. Ballerini, qui
occupa la chaire de morale pendant un quart de siècle, à partir de 1856,
se distinguait à la fois par l'étendue de ses connaissances et par la perspicacité de son jugement. En dogmatique, le P. Perrone, qui cessa d'enseigner en 1853 mais resta préfet des études jusqu'en 1876, n'avait pas
cherché à faire œuvre scientifique, mais plutôt de vulgarisation et de
controverse contre les erreurs du jour ; toutefois, ouvert aux idées du
temps, il avait pressenti l'importance de la théologie positive, qui allait
occuper pendant trente ans une place de choix au Collège romain, grâce
aux Pères Passaglia, Schrader et Franzelin.

Passaglia, le plus brillant élève de Perrone, était soucieux d'appuyer
sa théologie sur une enquête scripturaire et patristique approfondie,
et sa connaissance des Pères grecs était remarquable ; il la complétait
cependant par une spéculation inspirée de la théologie post-tridentine,
de Petau et de Thomassin en particulier. Bien que ne connaissant pas
l'allemand, il se faisait tenir au courant de tout ce qui paraissait dans
cette langue par les élèves du Collège germanique, dont il était le préfet,
et surtout par son jeune collègue, le P. Schrader, qui, avec plus d'étroitesse d'esprit et dans une forme plus scolastique, s'inspira pourtant
de très près, par la suite, de la méthode et de la doctrine de son aîné [2].
Lorsque, en 1857, Schrader fut nommé professeur à Vienne et que Passaglia eut abandonné sa chaire, Franzelin poursuivit leur œuvre, « édifiant
une théologie, moins brillante peut-être que celle de Passaglia, mais plus
solide et plus rigoureuse, appuyée sur la critique des textes, des monuments et des faits » [3], en mettant à profit les récentes découvertes archéologiques, ses connaissances approfondies des langues orientales et les travaux
de l'école historique allemande.

---

(1) Les souvenirs d'Hettinger (*Aus Welt und Kirche*, t. I, Fribourg-en-Br., 1911, chap. i), qui
fut élève au Collège romain de 1841 à 1845, fournissent des indications intéressantes sur la manière
dont l'enseignement était organisé.
(2) Sur Carlo Passaglia (1812-1887) et Clément Schrader (1820-1855), voir H. SCHAUF, *C. Passaglia und Cl. Schrader. Beitrag zur Theologie des XIX. Jhdts*, Rome, 1938.
(3) A. KERKVOORDE dans *Ephemerides theologicae lovanienses*, t. XXII, 1946, p. 181-182. Sur
Jean-Baptiste Franzelin (1816-1886), voir G. BONAVENIA, *Raccolta di memorie intorno alla vita dell'
Em. cardinale Franzelin*, Rome, 1877.

Désireux de faire une large place à la théologie positive, concevant le travail spéculatif comme un effort de synthèse organique des données de la foi à partir des images bibliques, plutôt que comme une pénétration philosophique des vérités révélées, préoccupés en outre de garder un certain contact avec la pensée moderne, tout en s'exprimant dans le style scolastique, les théologiens du Collège romain ne s'intéressèrent guère à la renaissance thomiste. D'ailleurs, même à la faculté de philosophie, où les P. Tongiorgi et Palmieri professaient une philosophie éclectique, le caractère caduc de la physique aristotélicienne, vivement ressenti par quelques jésuites au courant des sciences modernes, jetait le discrédit sur l'ensemble de la philosophie de saint Thomas.

LES DÉBUTS DE LA RENAISSANCE THOMISTE

Si l'Université grégorienne ne fut pour rien dans le mouvement qui devait aboutir sous Léon XIII à l'encyclique *Aeterni Patris*, c'est cependant aux jésuites, plus qu'aux dominicains, que revient le mérite d'avoir, par delà la scolastique suarézienne ou le vague spiritualisme issu de Descartes, remis en honneur le thomisme authentique.

Celui-ci avait été réintroduit en Italie, vers 1810, par l'abbé Buzzetti, professeur au séminaire de Plaisance, sous la double influence de l'ex-jésuite espagnol Masdeu et de ses lectures et réflexions personnelles. Il avait su gagner à ses idées deux jeunes jésuites, Dominique et Séraphin Sordi, dont le premier avait été chargé, en 1831, de l'enseignement de la philosophie au scolasticat de Naples par le P. Taparelli d'Azeglio. Celui-ci en effet, non content de s'inspirer de saint Thomas dans ses célèbres travaux de droit naturel [1], déploya toute sa vie une inlassable activité pratique en faveur du renouveau thomiste et il y avait déjà, pendant son rectorat au Collège romain, intéressé confidentiellement quelques élèves, dont le P. Curci, le futur fondateur de *La Civiltà cattolica*. Cette revue devint d'autant plus aisément une tribune du néo-thomisme que Curci avait parmi ses principaux collaborateurs Séraphin Sordi et Taparelli lui-même, qui souligna l'avantage d'opposer à la philosophie hégélienne, génératrice de révolutions, le merveilleux équilibre de la philosophie sociale de saint Thomas ; ils convertirent en outre à leurs vues un autre collaborateur, le P. Liberatore, qui, à partir de 1853, devint l'un des plus ardents protagonistes de la renaissance thomiste [2].

Les efforts des jésuites de la *Civiltà* n'étaient pas isolés. Il y avait en Italie d'autres groupes de fervents thomistes : à Plaisance, où le souvenir de Buzzetti ne s'était pas perdu ; à Naples, où le chanoine Sanseverino, vieil ami de Liberatore, avait suivi la même évolution que ce dernier ; à Pérouse, autour de l'archevêque Pecci ; un peu plus tard, au scolasticat des jésuites de la Province de Lombardo-Vénétie, l'*Aloisianum*, où le P. Cornoldi avait été nommé professeur en 1859. Dès 1850, un éditeur de

(1) Sur Louis Taparelli d'Azeglio (1793-1862) et son œuvre considérable dans le domaine de la morale sociale et internationale, voir R. Jacquin, *Taparelli*, Paris, 1943 (sur sa contribution au renouveau thomiste, voir surtout, p. 51-61 et 113-117).

(2) On fait parfois remonter le thomisme du P. Liberatore à une date beaucoup plus ancienne, mais à tort comme l'a montré G. Van Riet, *L'épistémologie thomiste*, Louvain, 1946, p. 37-39.

Parme avait entrepris la réédition des œuvres complètes de saint Thomas
et, à partir de 1862, les manuels s'inspirant du docteur angélique se multiplièrent. C'est vers la même époque que les dominicains, dont la ferveur
thomiste n'avait guère jusqu'alors rayonné en dehors de leur ordre,
commencèrent à contribuer efficacement au mouvement, grâce à quelques
auteurs de renom, tels les futurs cardinaux Zigliara, en Italie, et Gonzalez,
en Espagne.

Mais celui qui contribua le plus à la renaissance du thomisme en dehors
de l'Italie fut encore un jésuite, le P. Kleutgen, d'origine allemande,
mais résidant à Rome [1]. Il avait pris contact à l'Université de Munster
avec le kantisme et l'hégélianisme ainsi qu'avec les systèmes théologiques d'Hermes et de Gunther. Mais dès son scolasticat, à Fribourg,
il avait été gagné par la philosophie traditionnelle et il essaya de communiquer ses convictions à ses compatriotes en leur offrant dans leur
langue une confrontation de la *Théologie d'autrefois* et de la *Philosophie
d'autrefois* [2] avec les systèmes catholiques qui prétendaient chercher
leur inspiration dans la philosophie moderne. Si ses ouvrages ne furent
guère pris en considération par le monde universitaire, ils eurent par
contre dans les milieux ecclésiastiques un grand retentissement, encore
accru par la traduction du second en italien et en français. Kleutgen, qui
avait repensé avec originalité et profondeur la doctrine du Docteur Angélique, méritait vraiment le titre de *Thomas redivivus*, bien qu'il n'eût pas
hésité à aborder des problèmes insoupçonnés de saint Thomas ou même, à
l'occasion, à s'inspirer plutôt de Suarez.

Le thomisme progressait donc lentement, mais il ne triomphera vraiment que sous Léon XIII. Dès les années cinquante, au contraire, la
scolastique, conçue dans un sens plus général, emporte nettement les
faveurs des milieux romains où la réaction va s'accentuer contre les
systèmes qui, au cours de la première moitié du siècle, avaient tenté de
la discréditer ou de la remplacer : le traditionalisme et l'ontologisme, en
Italie, en France et en Belgique ; le gunthéranisme dans les pays germaniques.

*NOUVEAU DÉSAVEU
DU TRADITIONALISME*
La condamnation de Lamennais et les difficultés de Bautain avaient fait perdre pas mal
de prestige au traditionalisme, qui, soucieux
de réagir contre l'orgueil de la raison, affirmait que seul le recours à la
révélation primitive justifiait la certitude des vérités métaphysiques fondamentales. Il survivait pourtant sous une forme mitigée. Le P. Ventura [3],
brillant publiciste et orateur italien que les événements de 1848 avaient

---

(1) Sur Joseph Kleutgen (1811-1883) qui, de 1843 à 1870, s'occupa des jésuites allemands fréquentant la Grégorienne, voir F. LAKNER, *Kleutgen und die kirchliche Wissenschaft Deutschland
im XIX. Jhdt*, dans *Zeitschrift für katholischen Theologie*, t. LVII, 1933, p. 161-214. Son grand mérite
consista à montrer comment la doctrine traditionnelle était capable de répondre aux problèmes
nouveaux et de la présenter dans la perspective synthétique des grands systèmes modernes, dégagée de la langue latine et des procédés vieillis qui la rendaient inassimilable aux esprits cultivés.
(2) *Die Theologie der Vorzeit*, Munster, 1853-1870, 5 vol. ; *Die Philosophie der Vorzeit*, Munster,
1860-1863, 2 vol.
(3) Sur Joachim Ventura (1792-1861), voir A. RASTOUL, *Le P. Ventura*, Paris, 1906 et P. Sé-
JOURNÉ, art. *Ventura*, dans *D. T. C.*, t. XV, col. 2635-2639.

contraint de s'exiler en France, tout en reconnaissant que l'existence de Dieu, l'immortalité de l'âme et les bases de la morale pouvaient, une fois connues, être démontrées par la raison, maintenait contre « les semi-pélagiens de la philosophie » la nécessité de la révélation pour leur première acquisition et l'importance du langage, don de Dieu aux hommes, pour leur élaboration. Ces idées, bien accueillies dans les milieux ultramontains, où les anciens amis du maître de la Chênaie n'étaient pas rares, rejoignaient celles développées par Augustin Bonnetty [1], un laïc méritant, d'une puissance de travail étonnante, qui, dans sa revue les *Annales de Philosophie chrétienne*, s'attachait avec beaucoup d'érudition à relever dans les mythes antiques et les superstitions des peuples sauvages des survivances de la révélation primitive faite par Dieu au genre humain, tout en critiquant âprement l'influence néfaste du rationalisme d'Aristote sur l'enseignement philosophique des collèges catholiques et jusqu'au sein de la théologie scolastique, à laquelle il reprochait son « langage si peu chrétien ».

A partir de 1850, Bonnetty, qui ne manquait pas de partisans, même au sein de l'épiscopat, fut vivement pris à partie par un jésuite, le P. Chastel, qui réussit à obtenir pour son livre *De la valeur de la raison humaine* l'approbation des P. Liberatore et Passaglia, ainsi que du Maître du Sacré Palais. Rome se décida à intervenir en 1855, mais d'une manière assez bénigne, car la controverse avait interféré avec des discussions entre partisans et adversaires d'une extension de l'influence romaine en France, et le parti dirigé par Mgr Sibour espérait, en faisant condamner les idées de Bonnetty, atteindre les protecteurs de ce dernier, *L'Univers* et les évêques ultramontains, dont les sympathies traditionalistes étaient apparues à l'occasion du concile d'Amiens de 1853. On évita donc d'exprimer un désaveu formel, mais Bonnetty fut invité à souscrire quatre propositions destinées à écarter toute équivoque [2]. Les trois premières s'inspiraient de celles signées par Bautain vingt ans plus tôt, touchant les rapports de la raison, de la révélation et de la foi ; la dernière concernait la scolastique :

La méthode dont se sont servis saint Thomas, saint Bonaventure et d'autres scolastiques après eux ne conduit pas au rationalisme et n'est point cause que dans les écoles contemporaines la philosophie est tombée dans le naturalisme et le panthéisme. Par conséquent il n'est point permis de faire un crime à ces docteurs et à ces maîtres d'avoir employé cette méthode, surtout avec l'approbation ou du moins la tolérance de l'Église.

LA RÉACTION CONTRE
L'ONTOLOGISME

Des critiques analogues contre la scolastique, accusée d'être trop influencée par le « psychologisme » d'Aristote et le rationalisme de Descartes, étaient formulées par les tenants de l'ontologisme. Ce plato-

---

(1) Sur Bonnetty (1798-1879), voir DADOUE, *Augustin Bonnetty, sa vie, ses travaux, ses vertus*, dans *Annales de Philosophie chrétienne*, t. XCVI, 1879, p. 348-441. Sur les controverses soulevées en France par ses idées, voir J. BURNICHON, *La Compagnie de Jésus en France*, t. IV, p. 37-46 et J. MAURAIN, *Le Saint-Siège et la France, passim*.

(2) Texte dans *Acta Sanctae Sedis*, t. III, Rome, 1867, p. 224 et dans H. DENZIGER, *Enchiridion symbolorum*, n°s 1649-1652. Il semble que ce soit Darboy, alors vicaire général de Paris, qui ait eu l'idée de ce formulaire et l'ait proposé à Rome. (Cf. SOMMERVOGEL, dans *Les Études*, t. XXI, 1872, p. 164, n. 1.)

nisme chrétien, qui se situait dans la tradition des Pères alexandrins, de saint Augustin et de saint Bonaventure, affirmait la possibilité d'une perception immédiate de Dieu par l'intelligence humaine, qui verrait en lui les essences métaphysiques ou idées universelles, sans aucun intermédiaire créé (*species intelligibilis*). Combiné avec le traditionalisme, le système fut professé à Louvain jusqu'à la veille du concile du Vatican. Il était fort répandu en France : dans la Compagnie de Jésus, où il fut ouvertement enseigné au scolasticat de Vals jusqu'en 1850 et où il conserva longtemps des sympathies plus ou moins tacites ; dans beaucoup de séminaires sulpiciens, grâce à l'influence de Baudry puis de Branchereau, l'auteur de la *Philosophie de Clermont* ; à la Sorbonne, avec Maret et Hugonin, le directeur des Carmes ; et même dans le public cultivé, grâce à l'appui de *L'Ami de la Religion* et du *Correspondant*. Enfin, en Italie, l'ontologisme, conçu avec un profond sens religieux par Rosmini, puis combiné avec certains thèmes hégéliens par Gioberti et Mamiani, était apparu à la fois comme la philosophie chrétienne par excellence et comme la philosophie nationale, prolongeant la lignée des platoniciens des xve et xvie siècles et susceptible de concourir à l'unification spirituelle de la péninsule [1].

Les doctrines ontologiques se heurtèrent très vite à l'hostilité de la Compagnie de Jésus. Au souci de réagir contre les tendances panthéistes du giobertisme et de défendre la philosophie d'inspiration péripatéticienne devenue traditionnelle dans l'Église depuis des siècles, s'ajoutaient des motifs de nature moins doctrinale : le mouvement national italien n'avait guère les sympathies des jésuites de la péninsule et, de plus, certains d'entre eux n'étaient pas exempts de jalousie à l'égard de Rosmini, dont l'apostolat s'exerçait avec succès sur un terrain analogue au leur [2]. Le discrédit qui, après 1848, frappa tout ce qui était de tendance libérale et « moderne » fut mis à profit par les adversaires de l'ontologisme, qui réussirent sans trop de peine à faire mettre à l'*Index* les œuvres philosophiques de Mamiani et de Gioberti [3], tandis que le général des jésuites prenait des mesures pour mettre un terme au succès des doctrines ontologistes au scolasticat de Vals [4].

On essaya aussi d'atteindre Rosmini [5], mais après un examen de ses œuvres qui dura plus de trois ans, la Congrégation de l'*Index* conclut, le 3 juillet 1854, par une sentence de *Dimittantur*, c'est-à-dire d'acquittement, à la grande joie du pape qui interdit de renouveler encore des accusations contre un homme dont il appréciait fort la générosité d'âme et la foi profonde. Au cours des années qui suivirent, le succès de la philosophie rosminienne ne fit que s'affirmer en Italie, où son influence fut considé-

(1) Voir A. Fonck, art. *Ontologisme*, dans *D. T. C.*, t. XII, col. 1014-1046 (*L'ontologisme au XIXe s.*), et, pour plus de détails, K. Werner, *Die italienische Philosophie des XIX. Jhdts*, Vienne, 1884-1886, 5 vol. (surtout le t. I : *Rosmini und seine Schule*), ainsi que U. Padovani, *V. Gioberti ed il cattolicismo*, Milan, 1927.
(2) Voir à ce propos A. Dyroff, *Rosmini*, Munich, 1906, p. 60-85.
(3) Décrets des 12 janvier 1850 (Mamiani) et 14 janvier 1852 (Gioberti).
(4) Sur cet incident et sur l'ordonnance du 6 janvier 1850, qui en fut la conclusion, voir J. Burnichon, *La Compagnie de Jésus en France*, t. III, p. 140-161.
(5) H. Reusch, *Der Index der verbotenen Bücher*, t. II, Bonn, 1885, p. 1142-1144.

rable dans les séminaires, dans plusieurs congrégations religieuses et sur quantité de prêtres et de laïcs, philosophes, savants ou hommes de lettres. Mais les défenseurs de la scolastique ne se tinrent pas pour battus et, surtout à partir de 1858, les attaques reprirent sous la direction du P. Liberatore et de *La Civiltà cattolica*. Toutefois, du vivant de Pie IX, elles n'aboutirent à aucune condamnation et, en 1876, un avis officiel du Maître du Sacré Palais vint même préciser qu' « il n'était pas licite d'infliger une censure théologique aux doctrines soutenues par Rosmini dans celles de ses œuvres examinées par la S. Congrégation de l'*Index* ».

Si les néoscolastiques durent attendre le pontificat de Léon XIII pour voir condamner le système de Rosmini, ils furent plus heureux dans leurs attaques contre les ontologistes français et belges. En 1861, ils obtinrent le désaveu par le Saint-Office de sept propositions assez caractéristiques [1]. Les ontologistes essayèrent bien, en s'appuyant sur une parole de Pie IX à l'archevêque de Tours, d'expliquer qu'on avait visé les panthéistes allemands plutôt que des philosophes catholiques, mais l'année suivante, Branchereau, qui avait soumis au Saint-Siège un résumé de sa doctrine en quinze thèses, fut officiellement avisé que le jugement de 1861 s'appliquait également à lui, et en 1866, Hugonin, proposé pour un évêché, dut signer une rétractation reconnaissant qu'il était, lui aussi, atteint par le décret.

L'AFFAIRE UBAGHS   C'est également l'ontologisme, mais combiné avec le traditionalisme en un « singulier accouplement », qui était devenu la doctrine dominante à Louvain, tant à la faculté de théologie, avec Tits, puis Laforêt, qu'à celle de philosophie avec Ubaghs [2]. Dès 1843, la congrégation de l'*Index* avait porté un jugement défavorable sur les cours de ce dernier, mais le Secrétaire d'État avait préféré ne pas l'en avertir directement et se borner à lui suggérer certaines modifications par l'intermédiaire de l'archevêque de Malines. Ubaghs introduisit quelques légères corrections dans sa *Théodicée* et crut de bonne foi, après un échange de correspondance avec Rome, que tout était désormais en ordre, en dépit de l'hostilité persistante des jésuites, mais la question rebondit à partir de 1858 à l'instigation des évêques de Bruges et de Liège, anciens élèves de la Grégorienne, qui s'assurèrent l'appui du P. Perrone et de *La Civiltà cattolica*. On réussit à obtenir du pape qu'il transférât l'examen de la cause de l'*Index* au Saint-Office, parce que le cardinal d'Andrea, préfet de l'*Index*, adversaire des jésuites, se montrait trop favorable aux louvanistes. Les interventions du cardinal Sterckx, soucieux de sauver la réputation de l'Université, retardèrent la décision romaine pendant plusieurs années, mais en 1864 le Saint Office déclara que les corrections

---

(1) Décret du 18 septembre 1861. (Texte dans *Acta Sanctae Sedis*, t. III, p. 204-205 et dans H. Denziger, *Enchiridion*, nos 1659-1665.) Sur les discussions qui suivirent, voir A. Fonck, art. *Ontologisme*, dans *D. T. C.*, t. XII, col. 1047-1055, ainsi que l'ouvrage, bien renseigné sur les dessous de l'affaire, de J. Kleutgen, *L'ontologisme jugé par le Saint-Siège*, trad. Sierp, Paris, 1867.
(2) Voir J. Henry, *Le traditionalisme et l'ontologisme à l'Université de Louvain*, dans *Annales de l'Institut Supérieur de Philosophie*, t. V, 1922, p. 44-149, à compléter, pour ce qui regarde la condamnation, par J. Kleutgen, *op. cit.*, et par A. Simon, *Le cardinal Sterckx et son temps*, t. II, p. 162-197.

apportées jadis par Ubaghs étaient insuffisantes et, en 1866, désavoua ses ouvrages de manière définitive, en se référant notamment à la condamnation des sept thèses ontologistes de 1861. L'année suivante, le professeur Ledoux, qui, avec le chanoine Lupus, avait été parmi ses adversaires les plus acharnés, établissait en ces termes le bilan de la controverse : « On se ferait illusion si l'on croyait que les doctrines incriminées ont disparu entièrement. On ne croit pas à l'infaillibilité du pape dans ces sortes de décrets ; on attribue beaucoup à des influences qui dominent à Rome. Il est évident néanmoins que le résultat obtenu est très important : le traditionalisme et l'ontologisme sont définitivement exclus de l'enseignement de l'Université [1]. »

§ 2. — Le conflit entre la scolastique et la théologie universitaire en Allemagne [2].

LA THÉOLOGIE DANS LES UNIVERSITÉS ALLEMANDES

Le brillant développement pris dans les universités allemandes depuis le début du XIXe siècle par la spéculation philosophique et les études historiques avait incité les intellectuels catholiques à repenser leur foi en fonction des problèmes et des difficultés modernes. Un travail fécond avait été accompli, qui donnait à l'Allemagne catholique, par rapport aux autres nations, une prépondérance marquée dans le domaine scientifique.

Tout n'était cependant pas d'égale valeur dans ces tentatives, dont plusieurs, dès le milieu du siècle, appartenaient déjà au passé, telle celle de Baader, dont l'influence ne se faisait plus guère sentir que chez quelques philosophes et théologiens munichois, Oischinger, Michelis, Frohschammer, Deutinger. Quant à Hermes, l'énorme succès qu'il avait rencontré n'avait pas empêché son système d'être désavoué par l'Église et les dis-

---

(1) Ledoux à Perrone, 9 juin 1867, Archives des jésuites à Rome, *Belg.*, 3.XIV.9.

(2) BIBLIOGRAPHIE. — I. SOURCES. — J.-E. JOERG, *Erinnerungen*, dans *H.-P. Bl.*, 1890 ; A.-M. WEISS, *Lebensweg und Lebenswerk*, Fribourg-en-Br., 1925, chap. I et II ; G. VON HERTLING, *Erinnerungen aus meinem Leben*, t. I, Fribourg-en-Br., 1919 ; J. F. VON SCHULTE, *Lebenserinnerungen*, Giessen, 1908-1909, 3 vol. ; *Briefe Döllingers an eine junge Freundin*, édit. H. SCHROERS, Kempten, 1914 ; *Briefwechsel zwischen Michelis und Döllinger*, édit. MENN, dans *Internationale Kirchliche Zeitschrift*, N. F., t. II, 1912, p. 319 et suiv., t. III, 1913, p. 62 et suiv. ; *Janssens Briefe*, édit. PASTOR, t. I, Fribourg-en-Br., 1920 ; *Briefe von und an Ketteler*, édit. RAICH, Mayence, 1879 ; A. FLIR, *Briefe aus Rom*, Innsbruck, 1864 ; *Verhandlungen der Versammlung Katholischer Gelehrten in München*, Ratisbonne, 1863. En outre, les collections du *Katholik*, de *La Civiltà cattolica*, des *Stimmen aus Maria Laach* et du *Theologisches Literaturblatt*.

II. TRAVAUX. — Deux livres contemporains des controverses gardent leur intérêt : A. VON SCHMID, *Wissenschaftliche Richtungen auf dem Gebiet des Katholizismus in neuesten und gegenwärtigen Zeit*, Munich, 1862, et K. WERNER, *Geschichte der katholischen Theologie (Deutschlands) seit dem Trienter Konzil*, Munich, 1866. L'ouvrage du vieux-catholique J. FRIEDRICH, *Ignaz von Doellinger. Sein Leben auf Grund seines schriftlichen Nachlasses*, Munich, 1899-1901, 3 vol., est fondamental bien que partial ; celui du catholique A. KANNENGIESER, *Les origines du vieux-catholicisme et les Universités allemandes*, Paris, 1901, est tendancieux et superficiel.

Parmi les travaux plus récents, retenir surtout M. SCHMAUS, *Die Stellung M. J. Scheebens in der Theologie des XIX. Jhdts*, dans *M. J. Scheeben. Der Erneuerer katholischen Glaubenswissenschaft*, Mayence, 1935 ; A. KERKVOORDE, *Scheeben et son époque*, Introduction du *Mystère de l'Église* (Unam Sanctam, XV), Paris 1946 ; F. VON BEZOLD, *Geschichte der rheinischen Friedrich-Wilhelm-Universitäts zu Bonn (1818-1918)*, Bonn, 1920 ; S. MERKLE, *Die Vertreter der Kirchengeschichte in Wurzburg b. z. Jahre 1879*, dans *Aus der Vergangenheit des Universitäts Wurzburg*, Berlin, 1932 ; P. FUNK, dans *T. Q. S.*, t. CVIII, 1927, p. 209-220 (sur l'école de Tubingue) ; en outre, sur la condamnation de Günther, l'ouvrage définitif de E. WINTER, *Die geistige Entwicklung A. Günthers und seiner Schule*, Paderborn, 1931.

ciples du « Kant de la théologie » avaient vainement cherché en 1846 à profiter de l'encyclique *Qui pluribus* pour relever la tête [1]. En fait, vers 1850, seules deux grandes tendances restaient en présence, celle de Günther et celle de l'école de Tubingue.

Günther avait eu le mérite d'écarter quantité de jeunes catholiques de la séduction du panthéisme hégélien et de rendre à beaucoup d'entre eux l'estime de la spéculation théologique. Un réel génie métaphysique et le charme de ses relations personnelles lui avaient conquis de nombreux adhérents, et la propagation de ses idées avait encore été favorisée, au cours des années quarante, par la protection de plusieurs évêques et du cardinal de Schwarzenberg, son ancien élève, ainsi que par la nomination de nombreux disciples comme professeurs de théologie ou de philosophie, spécialement en Prusse, où le gouvernement évitait de nommer des hermésiens, pour ne pas heurter de front les autorités ecclésiastiques, mais ne voulait pas non plus, au lendemain de l'affaire Droste zu Vischering, choisir des hommes provenant de l'entourage de l'archevêque Geissel. Au lendemain de la crise de 1848, où ils avaient joué un rôle éminent dans la direction du mouvement catholique, la fondation d'un annuaire philosophique, *Lydia*, avait marqué l'intention des gunthériens de s'affirmer toujours plus énergiquement.

Leur rayonnement n'éclipsait cependant pas le prestige de l'école de Tubingue, d'orientation moins exclusivement spéculative et centrée davantage sur l'Écriture et la Tradition. Elle aussi rayonnait au dehors, à Fribourg notamment, avec des hommes comme Hirscher [2], « prêtre d'une piété ardente, mais d'une intelligence parfois aventurée » [3], dont la réputation souffrit longtemps des critiques injustes de Kleutgen, et Staudenmaier [4], un esprit spéculatif qui eut le grand mérite d'opposer à la philosophie hégélienne de l'histoire, purement dialectique, une théologie chrétienne de l'histoire, dégageant de la Révélation les étapes de l'action vivante et libre de Dieu parmi les hommes. Le chef incontesté de cette école, pendant le deuxième tiers du siècle, c'est Kuhn [5], professeur de dogmatique à Tubingue de 1839 à 1882. Surtout spéculatif, il enthousiasmait élèves et lecteurs par la clarté et l'éclat de la présentation, par la vigueur de sa dialectique et par la profondeur de ses vues. Bien que quelques-unes de ses idées, sur la grâce en particulier, aient été parfois un peu hasardeuses, le fond de sa pensée était profondément catholique et il se montra toujours nettement opposé aux tendances semi-rationalistes. Toutefois, pour penser sa foi, il entendait faire appel avant tout

(1) Ils avaient pris prétexte du passage où Pie IX insistait sur la démonstration rationnelle des *Praeambula fidei*, mais l'archevêque de Cologne, Geissel, était aussitôt intervenu auprès du nonce à Munich et du cardinal Lambruschini et il avait réussi à obtenir, le 25 juillet 1847, un bref désavouant une nouvelle fois les prétentions hermésiennes. (Les faits sont racontés en détail au chap. xi de l'ouvrage de H. SCHROERS, *Ein vergessener Führer aus der rheinischen Geistesgeschichte des XIX. Jhdts*, J. W. Braun, Bonn, 1925.)

(2) Sur J.-B. Hirscher (1788-1865), voir F. LAUCHERT, dans *Revue internationale de théologie*, t. II, 1894, p. 627 et suiv., t. III, 1895, p. 260 et suiv. et 723 et suiv., t. IV, 1896, p. 151 et suiv.

(3) E. MANGENOT, art. *Hirscher*, dans *D. T. C.*, t. VI, c. 2512.

(4) Sur Antoine Staudenmaier (1800-1856), voir P. WEINDEL, *Das Verhältnis von Glauben und Wissen nach der Theologie von Fr. A. Staudenmaier*, Dusseldorf, 1940.

(5) Sur Jean Kuhn (1806-1887), voir P. SCHANZ, dans *T. Q. S.*, t. LXIX, 1887, p. 531-598, et P. GODET, dans *Annales de Philosophie chrétienne*, t. LXXVIII, 1907, p. 26-47 et 163-182.

à la philosophie moderne, sans d'ailleurs s'associer au mépris systématique des gunthériens pour la scolastique ; celle-ci lui apparaissait, dans sa perspective dominée par l'idée de développement, comme une étape utile, mais dépassée, de l'histoire de la pensée chrétienne.

LE PRESTIGE DE LA
THÉOLOGIE HISTORIQUE
A côté du souci de réinstaller résolument la théologie dans la foi et de présenter celle-ci, de même que les réalités connexes de Révélation, de Tradition et d'Église dans une lumière intrinsèquement surnaturelle, le grand mérite des fondateurs de l'école de Tubingue, Drey et Moehler, avait été de repenser le dogme catholique avec un sens aigu de la « dimension historique » et du développement [1]. A leur suite, divers théologiens, s'intéressant davantage à la vie du dogme qu'à sa métaphysique, s'appliquèrent à développer les conséquences de l'insertion du fait chrétien et de la révélation dans l'histoire.

Cette orientation de la théologie dans un sens moins spéculatif fut d'ailleurs accélérée par des préoccupations apologétiques. En effet, les progrès de la critique historique, la meilleure connaissance de l'histoire et des littératures de l'Ancien Orient, les incessantes découvertes d'œuvres liturgiques ou patristiques, d'inscriptions et de textes renouvelant les connaissances sur l'Église ancienne, posaient quantité de problèmes que la polémique protestante ou rationaliste s'empressait d'exploiter : tandis que les historiens protestants mettaient au point une méthode critique qui leur permettait d'ébranler certaines positions traditionnelles concernant les origines chrétiennes ou le moyen âge, se développait dans certaines facultés, surtout à Tubingue, une exégèse radicale, mettant en danger les bases de la foi, et une interprétation hégélienne de l'histoire des dogmes amorcée par Christian Baur. En étroit contact avec leurs adversaires dans les universités, les théologiens catholiques comprirent vite la nécessité de les suivre sur le terrain nouveau où s'engageait la lutte, de les combattre avec les armes qu'ils employaient, c'est-à-dire les faits, les textes et les documents, et éventuellement de reviser certaines positions qui se révélaient inconciliables avec les faits.

Dans le domaine de l'exégèse, le travail fourni, quoique non négligeable [2], resta pourtant assez en arrière sur ce qui se faisait du côté protestant. Par contre, l'histoire des dogmes et l'histoire de l'Église connurent une efflorescence remarquable : à Tubingue, d'abord, où Hefele [3] commença à publier en 1855 sa monumentale *Histoire des conciles*, tandis qu'une pléiade de jeunes travailleurs, gravitant autour de la *Theologische Quartalschrift*, mettaient à profit les nouvelles méthodes critiques pour retracer les étapes du progrès de la pensée chrétienne ; à Munich,

---

(1) Voir J.-R. GEISELMANN, *Lebendiger Glaube aus geheiligter Ueberlieferung. Der Grundgedanke der Theologie J. A. Moehlers und der katholischen Tübinger Schule*, Mayence, 1942.

(2) On trouvera mentionnés les principaux travaux dans l'article *Allemagne* par E. MULLER, dans *D. T. C.*, t. I, col. 863-865.

(3) Sur Karl-Joseph Hefele (1809-1893), voir P. GODET, dans *Revue du Clergé français*, t. L, 1907, p. 449-474.

surtout, grâce au rayonnement croissant de Doellinger, le chef incontesté de l'école historique catholique allemande.

DOELLINGER
ET L'ÉCOLE DE MUNICH

Vers 1850, Joseph Ignace Doellinger [1] avait déjà tout un passé de gloire derrière lui. Rêvant depuis sa jeunesse d'un apostolat intellectuel, excité davantage encore dans son ardeur combattive par les coryphées du groupe de Mayence, Klee, Weiss et Raess, avec lesquels il s'était lié peu après son ordination, il avait trouvé définitivement sa voie lorsque, en 1826, nommé professeur à Munich, il était devenu un intime du cercle de Goerres. Nourrissant auprès de ce dernier et de ses amis son amour de l'Église et l'intelligence des conditions nécessaires à son développement et à sa vitalité dans les sociétés modernes, il avait, comme la plupart d'entre eux, mené de front, pendant vingt ans, l'enseignement universitaire et la polémique journalistique en faveur de la liberté de l'Église. Ses exceptionnels mérites dans les deux domaines n'avaient pas tardé à attirer l'attention sur lui, et les traductions anglaise, française et italienne de son *Manuel d'Histoire ecclésiastique*, qui tranchait par l'originalité de la conception et la clarté de l'exposition, avaient étendu sa réputation loin au delà des pays de langue allemande. A vrai dire, admirablement doué pour l'analyse des doctrines, sa vraie vocation, au dire de son ami Acton, eût dû être la théologie dogmatique, mais, conscient de ce que la principale faiblesse des catholiques à l'égard des protestants était leur infériorité dans le domaine historique, son zèle pour l'Église l'avait orienté vers l'histoire ecclésiastique. Il manqua cependant toujours, en dépit de sa froideur apparente, de l'objectivité désintéressée du pur savant et fit toute sa vie de l'histoire à thèse, au profit de l'Église pendant les trente premières années, contre les idées dominantes à Rome par la suite et contre l'Église pour finir.

En 1846, gagné par l'atmosphère antiprotestante qui se développait dans le groupe de Goerres sous l'influence des convertis Jarcke et Philips, il avait entrepris de réfuter l'œuvre de Ranke et, bien que ses trois volumes sur *la Réforme* n'eussent pas la même envergure et fussent moins une fresque historique qu'un choix tendancieux de citations mises en œuvre dans un but polémique, cette peinture du luthéranisme par ses premiers adhérents, où n'étaient retenues que les ombres, avait rencontré dans les milieux catholiques un succès proportionnel à la fureur des protestants et achevé de faire reconnaître son auteur comme le principal champion de l'Église. Aussi ne faut-il pas s'étonner qu'au cours de la crise de 1848, il fût apparu comme un des principaux chefs catholiques, tant sur le terrain politique que religieux : inspirateur, avec Radowitz, du groupe catholique au Parlement de Francfort ; défenseur de la liberté religieuse au Landtag bavarois, où il siégea de 1849 à 1851 ; conseiller,

---

(1) Sur Doellinger (1799-1890), outre l'ouvrage capital de J. Friedrich (cité p. 193, n. 2), voir surtout F. Vigener, *Drei Gestalten aus dem modernen Katholizismus*, Munich, 1926 ; également E. Michael, *Ignaz v. Doellinger, eine Charakteristik*, Innsbruck, 1893 (polémique) ; J. Acton, *Doellingers historical Work*, dans *English Historical Review*, t. V, 1890, p. 700-744 et P. Godet, *Doellinger*, dans *Revue du Clergé français*, t. XXXVI, 1903, p. 17 et suiv., 125 et suiv., 366 et suiv.

plus actif qu'aucun vicaire général et que beaucoup d'évêques, à l'Assemblée épiscopale de Wurzbourg [1] ; animateur enthousiaste des premiers congrès catholiques, salué à celui de Ratisbonne comme l'homme « dont la parole fait autorité dans l'ensemble du monde catholique » [2].

Bientôt toutefois, déçu par l'orientation que prenait le mouvement catholique en Allemagne sous la conduite des hommes de Mayence, mécontent de l'influence croissante des jésuites dans la conduite générale de l'Église et des progrès de la centralisation romaine, qui allait à l'encontre de son idéal de large autonomie nationale, il abandonna l'action religieuse pour se consacrer de plus en plus exclusivement au travail scientifique [3], toujours avec une préoccupation apologétique d'ailleurs : *Hippolyle et Calliste* (1853), un chef-d'œuvre de critique historique applaudi par toute la science allemande, est destiné à rassurer les consciences catholiques inquiètes des objections contre la papauté que des protestants comme Bunsen avaient cru trouver dans les *Philosophoumena*, récemment découverts ; *Paganisme et Judaïsme* (1857), qui met le sceau à sa réputation, a pour but de montrer à la lumière de l'histoire qu'aucun développement naturel, grec ou juif, ne saurait expliquer l'apparition du christianisme. Plus encore que par ses publications, qui ne recueillent qu'une minime partie de son immense érudition sans cesse alimentée par de nouvelles lectures et par de nouvelles recherches dans les dépôts de manuscrits [4], son influence, qui va grandissant et dont on a peine aujourd'hui à se faire une idée, s'exerce par ses cours, dont l'action se prolonge grâce à ses élèves dans de nombreuses chaires d'Allemagne, d'Autriche et de Suisse ; par des contacts personnels, où la richesse d'idées, la vivacité du regard et la simplicité du ton séduisent l'interlocuteur ; par une correspondance enfin, qui s'étend à tout le monde savant germanique, anglais et français.

C'est pourtant à ce moment, où son prestige est au plus haut, que Doellinger va commencer à être en butte aux suspicions des milieux ultramontains, dont le centre le plus actif se trouve à Mayence.

L'ÉCOLE DE MAYENCE
ET LA RESTAURATION SCOLASTIQUE
En 1848, les hommes de Mayence, Lennig, Heinrich, Moufang, avaient travaillé la main dans la main avec Doellinger et suivi avec sympathie les efforts parallèles de Günther, en Autriche, pour organiser les catholiques. Leur but paraissait identique : affranchir l'Église du joug des bureaucraties joséphites et resserrer ses contacts avec le monde laïc menacé par une ambiance de moins en moins favorable au catholicisme. Mais très vite, les divergences étaient apparues. Pour Doellinger, Günther et leurs amis, le monde laïc, c'étaient avant tout

---

(1) Voir chap. ii, p. 60-61.
(2) *Verhandlungen der dritten Generalversammlung*, Ratisbonne, 1849, p. 117.
(3) Il avait catégoriquement refusé l'archevêché de Salzbourg en 1850 en arguant de sa vocation de savant : « *Ich bin nicht da um pompam facere* », expliquait-il à ses amis. (J.-E. JOERG, *Erinnerungen*, dans *H.-P. Bl.*, 1890, p. 242.)
(4) Dans ce dernier domaine cependant, Doellinger resta longtemps de la vieille école et ce n'est qu'assez tard qu'il comprit toute l'importance des inédits. (Cf. J. ACTON, dans *English Hist. Review*, t. V, 1890, p. 733-734.)

les classes dirigeantes, les intellectuels catholiques, qu'il fallait délivrer
du complexe d'infériorité éveillé par l'efflorescence de la science protestante et rationaliste, en leur apprenant, à son contact même, à rivaliser
à armes égales avec elle et en leur donnant le sentiment d'une complète
liberté scientifique en dehors des questions, relativement peu nombreuses,
où le dogme était clairement en jeu. Ils espéraient de la sorte conquérir
pour l'Église, dans le monde de la pensée, une influence analogue à celle
qu'elle était en train de conquérir dans la vie publique par son action
politique. Au contraire, le monde laïc auquel s'intéressaient les hommes
d'œuvre de Mayence, c'étaient les masses catholiques — paysans, artisans,
classes moyennes — dont les convictions chrétiennes, raffermies avec
l'aide d'un clergé plus pieux et plus zélé, s'extérioriserait dans un puissant
mouvement d'associations bien disciplinées, filialement soumises au
Saint-Siège et capables de faire passer dans les différents secteurs de la
vie courante les mots d'ordre de la hiérarchie.

On comprend dès lors l'importance qu'on attachait à Mayence à ce
que la formation des futurs prêtres et des futurs dirigeants du mouvement
catholique se fasse dans une atmosphère franchement ecclésiastique,
aussi à l'abri que possible des influences dissolvantes du rationalisme
moderne. Plus soucieux d'avoir de bons prêtres que des prêtres savants,
les partisans de cette tendance étaient résolument hostiles au système
germanique qui obligeait le jeune clergé à faire ses études dans des facultés
de théologie annexées aux universités d'État ; ils souhaitaient y substituer le régime des séminaires diocésains, en vigueur en France et en
Italie, que Ketteler s'était empressé de rétablir à Mayence dès son arrivée [1]. Beaucoup d'entre eux souhaitaient en outre que même les jeunes
laïcs fussent soustraits à l'ambiance dangereuse des universités, dont
le corps professoral était protestant dans son immense majorité [2] et
qui paraissaient peu propres à favoriser la formation d'une opinion
publique catholique ; aussi préconisaient-ils la création d'une université
libre, qui remplirait auprès des catholiques allemands le rôle bienfaisant
joué en Belgique par l'Université de Louvain. L'idée, lancée par Buss
à la conférence de Wurzbourg, était appuyée par les congrès catholiques
ainsi que par les revues et les groupements qui cherchaient leur inspiration
dans l'entourage de Ketteler. Elle faisait peu à peu son chemin [3], à la
grande indignation de professeurs comme Doellinger, qui se sentaient
personnellement offensés par les plaintes formulées contre l'enseignement
universitaire, mais qui avaient surtout un sens très aigu de la nécessité
d'avoir un clergé à la hauteur des exigences intellectuelles modernes

---

(1) Voir chap. v, p. 143.

(2) Au *Katholikentag* de 1864, Moufang citait des chiffres impressionnants : en Prusse, alors
que la proportion des habitants était de 7 millions de catholiques contre 10 millions de protestants, on ne comptait que 55 professeurs catholiques sur 556, et encore plusieurs n'étaient catholiques que de nom.

(3) Ce n'est toutefois qu'à partir de 1861 que des mesures concrètes furent envisagées et des
difficultés diverses retardèrent la décision jusqu'en 1869, puis la guerre et le *Kulturkampf* obligèrent à renoncer au projet. (Cf. G. RICHTER, *Der Plan zur Errichtung einer katholischen Universität zu Fulda im XIX. Jhdt*, Fulda, 1922.) En attendant, à partir de 1863, on s'appliqua, sous l'impulsion de G. von Hertling, à développer les *Katholischen studentischen Korporationen*. (Cf. J.-B.
KIESSLING, *Geschichte der deutschen Katholikentage*, t. I, chap. xvi.)

et du danger d'élever en vase clos les jeunes catholiques, en les privant des ressources scientifiques que seules, à leur avis, les vieilles universités étaient à même de leur fournir.

Les partisans de la formation universitaire craignaient d'autant plus de voir les catholiques coupés de la vie scientifique de leur temps si les vues de leurs adversaires prévalaient, que ceux-ci, généralement peu avertis de l'urgence des problèmes critiques et surtout convaincus que toute la pensée spéculative allemande depuis Kant était engagée dans une voie sans issue, préconisaient ouvertement le retour à la scolastique. C'était devenu le mot d'ordre de la revue mayençaise, *Der Katholik*, que dirigeaient les professeurs nommés par Ketteler en 1849, lors de la réouverture du séminaire : ceux-ci s'étaient rapidement dégagés des dernières influences de Tubingue et avaient repris la tradition introduite au séminaire, trente ans auparavant, par le supérieur Liebermann. Le mouvement, qui avait son centre à Mayence, ne tarda d'ailleurs pas à s'étendre : il était vigoureusement appuyé, de Bonn, par un jeune laïc batailleur, Clemens [1], qui avait découvert Suarez au cours d'un séjour à Rome et que des déconvenues personnelles avaient achevé d'indisposer contre les tenants des systèmes modernes ; il était soutenu, à Munich, par les *Historisch-politische Blätter*, par l'archevêque Reisach, ancien élève du séminaire germanique de Rome, par le vicaire général Windischmann, qui attirait depuis longtemps l'attention sur les dangers du gunthérianisme ; il était encouragé en Autriche par le cardinal Rauscher et il devait trouver, à partir de 1856, un foyer actif à la faculté de théologie de l'Université d'Innsbruck, confiée aux jésuites ; ces derniers comptaient partout parmi les plus fervents protagonistes de la restauration scolastique, au succès de laquelle les œuvres de Kleutgen, publiées à partir de 1853, contribuèrent pour leur part.

Bientôt, un certain nombre d'ecclésiastiques et même d'intellectuels furent convaincus de l'impossibilité de concilier les systèmes modernes avec la doctrine de l'Église et du danger que présentait l'orientation des facultés universitaires. A la vérité, l'enseignement théologique des universités souffrait de réelles déficiences, par suite du caractère hétérodoxe des philosophies dominantes, de la tendance à réduire la théologie à la philosophie ou à l'histoire, et de l'oubli du rôle directif qui doit revenir au magistère ecclésiastique guidé par l'Esprit Saint. Le malheur voulut que la réaction, légitime et même indispensable, fût en partie menée par des hommes qui n'étaient pas seulement inquiets des excès de la science allemande, mais encore indifférents ou même hostiles à la science tout court.

Ils ne l'étaient certes pas tous. Ce n'était le cas ni de Reisach ni surtout de Rauscher, qui avaient pour la science, à condition qu'elle fût orthodoxe, une admiration toute germanique. Ce n'était pas non plus le cas de Schrader ni de Guidi, appelés de Rome par Rauscher à l'Université de Vienne en 1857, ni même de Heinrich, le professeur de dogmatique

---

(1) Sur Franz Jakob Clemens (1815-1862), catholique très zélé et membre actif du *Katholisch Verein*, voir le panégyrique publié dans *Der Katholik*, 1862, vol. I, p. 257-280.

de Mayence, qui entendaient bien compléter leurs exposés scolastiques par une étude sérieuse de l'Écriture et des Pères. Mais déjà Ketteler, qui n'avait rien d'un intellectuel, voyait trop la vie universitaire à travers les beuveries d'étudiants auxquelles il avait participé dans sa jeunesse. Et surtout, beaucoup d'anciens élèves du Collège romain n'avaient guère profité des perspectives qu'ouvrait l'enseignement d'un Passaglia ou d'un Franzelin, et, soit étroitesse, soit paresse d'esprit, semblaient ignorer le renouvellement de point de vue qu'imposait en bien des domaines le progrès des méthodes historiques ; portés à confondre toute position classique avec la Tradition, leur oracle était *La Civiltà cattolica*, dans laquelle ils puisaient un conformisme s'accommodant volontiers des opinions toutes faites. Il y avait enfin les survivants du courant fidéiste ou traditionaliste, si influent dans l'entourage de l'archevêque Geissel à l'époque de la controverse hermésienne ; convaincus de l'inutilité et même du danger d'une étude scientifique des vérités religieuses, ils étaient encore moins disposés que les anciens « Romains » à apprécier le souci des « théologiens allemands » de confronter la pensée chrétienne avec les grands systèmes postkantiens et surtout d'entreprendre une vérification sérieuse, conforme aux exigences de la critique moderne, des textes invoqués communément comme arguments par les catholiques en faveur de leurs positions traditionnelles. Comme, par surcroît, la plupart attachaient une grande importance aux dévotions extérieures et aux manifestations du merveilleux, on comprend à quel point ils devaient exaspérer des savants dont l'éducation ecclésiastique avait été faite dans l'esprit de Sailer.

Tout ce monde, poursuivant la tradition inaugurée jadis par des gens comme Binterim, dénonçait à Rome avec un sectarisme intransigeant tous ceux qui ne pensaient pas comme lui, et cette pratique, courante de 1840 à 1865, contribua, beaucoup plus qu'on ne le dit souvent, à envenimer les rapports entre Rome et la science allemande : on doit le regretter d'autant plus qu'elle ne procédait pas seulement d'un zèle, souvent mal éclairé, pour la vérité, mais parfois aussi d'intrigues et de rivalités personnelles. Dès 1854, des attaques injustifiées parties de Mayence obligèrent de la sorte un des meilleurs historiens de l'Église, Schwarz, à abandonner sa chaire à l'Université de Wurzbourg [1], mais le premier grand succès de la réaction scolastique fut la condamnation du günthéranisme en 1857.

LA CONDAMNATION DE GUENTHER   On savait gré à Günther, dans les milieux ecclésiastiques, du zèle apostolique avec lequel il s'était dévoué à délivrer les intellectuels catholiques de la fascination panthéiste. Mais plus le péril hégélien s'éloignait, plus on s'inquiétait de la hardiesse de ses idées, surtout dans les milieux hostiles à tout compromis avec la philosophie moderne. L'arrogance de ses disciples avait d'ailleurs contribué à accroître le mécontentement et,

(1) Cf. S. MERKLE, dans *Aus der Vergangenheit der Universität Würzburg*, p. 154-181.

dès 1848, on avait vu se reformer contre Günther l'ancien front anti-
hermésien, particulièrement en Rhénanie, où la maladresse de Knoodt,
qui avait fait de Bonn un centre günthérien très actif, avait dressé contre
lui l'archevêque Geissel et son entourage.

Les événements de 1848 mirent une sourdine à la controverse, mais
celle-ci reprit bientôt de plus belle. La réaction absolutiste en Autriche
était défavorable aux disciples de Günther, qui s'étaient montrés sympathiques aux tendances libérales [1] et de plus, Rauscher, hostile depuis
vingt ans, devenait archevêque de Vienne en 1853. C'est toutefois de
Rhénanie que partit l'attaque décisive. Geissel ayant été consulté par
Rome en décembre 1853 à propos d'un professeur günthérien du séminaire
de Trèves, il confia la rédaction du rapport à Clemens, antigünthérien
acharné, et de plus, ennemi personnel de Knoodt ; il profita de l'occasion
pour dénoncer Günther, dont les œuvres furent en conséquence déférées
à la Congrégation de l'*Index*. Dès qu'on apprit la chose, les attaques
redoublèrent ; tout le monde théologique et philosophique d'Allemagne
et d'Autriche fut bientôt en effervescence et la polémique dépassa vite
en âpreté ce qu'on avait vu quinze ans plus tôt lors de la querelle hermésienne.

Günther pouvait toutefois compter sur de puissantes protections qui
surent convaincre Pie IX de l'excellence de ses intentions : le cardinal
Schwarzenberg, qui restait profondément dévoué à son ancien maître
et qui sentait le danger de heurter de front les milieux universitaires,
déjà trop portés à bouder l'Église ; le vieux cardinal Diepenbrock, de
Breslau, qui n'approuvait pas l'exclusivisme passionné de Geissel ; à
Rome même, l'abbé bénédictin de Saint-Paul et deux hommes de confiance
du pape, Flir, le recteur de l'*Anima*, et Mgr de Hohenlohe. Les choses
auraient sans doute pu s'arranger si les délégués de Günther à Rome, les
professeurs Baltzer et Knoodt, n'avaient tout gâté par leur attitude
désagréable à l'égard des jésuites et par leur mépris affiché pour l'état
de la philosophie à Rome. Malgré tout, on hésitait à condamner, et,
pendant deux ans, l'affaire resta pendante. Mais les adversaires de Günther
revenaient sans cesse à la charge : Rauscher, dont l'influence depuis le
concordat allait croissant au détriment de celle de Schwarzenberg ;
Viale Prela, l'ancien nonce à Vienne ; Reisach, devenu cardinal de Curie ;
et surtout Geissel, qui faisait de la condamnation une question de prestige.
Le pape se vit enfin forcé de trancher, mais, en dépit des pressions qui
s'exerçaient, il le fit de la façon la plus conciliante possible [2] : il se contenta
d'une mise à l'*Index* des œuvres de Günther, qui fut averti au début de
1857 par une lettre aimable du cardinal d'Andrea. Après un moment
d'hésitation, il trouva dans son passé de dévouement à l'Église la force
de se soumettre malgré son amertume.

Le décret ne condamnant aucune doctrine particulière, ses amis prétendirent d'abord que l'on pouvait continuer à défendre l'essentiel du
système. Geissel, appuyé par Reisach, obtient alors un bref précisant la

---

(1) Voir *supra*, chap. ii, p. 62.
(2) Flir relève expressément la chose. (*Briefe aus Rom*, p. 72.)

condamnation [1]. Les plus modérés s'inclinèrent, tandis que les radicaux, comme Baltzer, Knoodt ou son élève Reinkens, se raidirent, estimant leur honneur scientifique en jeu, et finirent par quitter l'Église. C'est parmi eux que le vieux-catholicisme devait recruter par la suite ses adhérents les plus actifs.

NOUVELLES INTERVENTIONS ROMAINES

La condamnation de Günther encouragea les néoscolastiques à s'en prendre aux philosophes et aux théologiens qui prétendaient poursuivre leurs travaux sans tenir compte des directives ecclésiastiques. L'ouvrage publié par Clemens l'année précédente, *De scolasticorum sententia philosophiam esse theologiae ancillam*, devint le manifeste du parti et lorsque Kuhn voulut réagir et défendre les droits de la philosophie à l'autonomie, *Der Katholik*, qui suivait de plus près les courants doctrinaux depuis sa réorganisation en 1858 avec l'aide de Clemens, soutint à fond celui-ci [2]. Les mayençais avaient d'ailleurs la satisfaction de voir s'ouvrir une brèche dans l'enseignement universitaire lui-même : en 1854, un ancien élève de la Grégorienne, Denzinger [3], avait obtenu, non sans peine, la chaire de dogmatique à Wurzbourg et il avait été rejoint peu après par deux autres « Romains », l'historien de l'Église Hergenroether [4] et l'apologiste Hettinger [5]. Tous trois entendaient travailler dans un esprit de pleine soumission à l'autorité ecclésiastique et, l'année même de sa nomination, Denzinger, en publiant son fameux *Enchiridion Symbolorum, definitionum et declarationum*, avait rappelé aux théologiens allemands, si portés à l'oublier, l'importance qu'il fallait attribuer aux décisions, même non infaillibles, du magistère ordinaire.

Dans leurs attaques, toujours plus âpres, contre leurs adversaires, les scolastiques étaient aidés par le nonce à Munich, qui se faisait volontiers l'écho de leurs dénonciations, et, dans les milieux de la Curie, par Reisach et par Kleutgen, dont l'influence était d'autant plus redoutable qu'il était consulteur à l'*Index*, et qui contribua beaucoup, par ses critiques, pas toujours impartiales, à discréditer aux yeux de Rome les philosophes et théologiens allemands. Les mises à l'*Index* qui se succédèrent à partir de 1857 [6] étaient relativement peu nombreuses en comparaison

(1) Texte dans les *Pii IX Acta*, t. II, p. 587 et suiv. ; extraits dans H. DENZINGER, *Enchiridion*, nᵒˢ 1655-1658.

(2) Sur la vive controverse qui opposa Kuhn à Clemens jusqu'à la mort prématurée de celui-ci en 1862, voir G.-B. GUZZETTI, *La perdita della fede nei cattolici*, Venegono (Varese), 1940, p. 79-86.

(3) Sur Henri Denzinger (1819-1883), voir les souvenirs de son frère dans *Der Katholik*, 1883, vol. II, p. 428 et suiv., 523 et suiv., 638 et suiv. C'était un prêtre zélé en même temps qu'un professeur né et un travailleur consciencieux.

(4) Sur Joseph Hergenroether (1824-1890), voir S. MERKLE, dans *Aus der Vergangenheit der Universitäts Würzburg*, p. 186-214. Professeur peu brillant, c'était un savant de réelle valeur qui avait complété sa formation romaine par un doctorat conquis à Munich sous la direction de Doellinger et dont les 3 volumes sur Photius, commencés dès 1855 et publiés de 1867 à 1869, assirent la réputation.

(5) Sur François Hettinger (1819-1890), voir l'introduction par E. MUELLER en tête de la 10ᵉ édit. de son *Apologie des Christentums* (1914). Celle-ci, publiée en 1863, était une œuvre « où la science allemande, l'école romaine et l'esprit plus populaire et plus littéraire à la fois de l'apologétique française ont laissé de profondes traces ».

(6) En 1857, un livre de Frohschammer sur l'origine de l'âme ; en 1858, un livre du günthérien Trebisch ; en 1859, les apologies de Günther par Knoodt et Baltzer et une critique de la théologie thomiste par Oischinger ; en 1860, Huber, un élève de Doellinger ; en 1861, Lassaulx, un autre munichois (cf. F.-H. REUSCH, *Der Index der verbotener Bücher*, Bonn, 1885, p. 1125-1132).

des dénonciations [1], mais, jointes aux critiques souvent simplistes qui les accompagnaient, elles mécontentèrent profondément les milieux universitaires catholiques qui, au nom de la liberté de la science, commencèrent à s'insurger, souvent avec arrogance, contre « un esclavage de la pensée » qui obligerait le théologien ou le philosophe à s'incliner devant une autre autorité que celle de ses pairs, alors qu'un dogme défini n'était pas en jeu.

Pareilles prétentions ne pouvaient que déplaire profondément à Rome où les préoccupations intellectuelles étaient faibles et où la réaction contre le libéralisme battait son plein. Les milieux dirigeants, et Pie IX en tout premier lieu, estimaient qu'en présence de la lutte entamée contre l'Église au nom de la science moderne, la stricte discipline de l'état de siège convenait mieux que les encouragements à la libre recherche, et on se méfiait d'autant plus des revendications des savants allemands que, si beaucoup d'entre eux étaient fort bien intentionnés et souvent victimes des excès de zèle de ceux qui les dénonçaient, il y en avait d'autres dont l'attitude montrait bien qu'on ne s'inquiétait pas en vain.

C'est ainsi que Frohschammer [2], de Munich, dont les tendances rationalistes s'étaient accentuées depuis sa mise à l'*Index*, non content de publier en 1861 un ouvrage tendant à justifier la légitimité d'une apostasie basée sur des recherches scientifiques, défendait dans d'autres publications des opinions fort proches du protestantisme libéral. A la fin de 1862, le pape condamnait sévèrement les doctrines de l'aventureux professeur, mais on remarqua surtout que, dans la lettre adressée à ce propos à l'archevêque de Munich [3], il regrettait qu'il fût loin d'être le seul à revendiquer « une liberté d'enseigner et d'écrire inconnue jusqu'à présent dans l'Église ». La distance aidant, les défiances romaines à l'égard de la science allemande s'aggravaient rapidement et commençaient à englober même les plus illustres, à commencer par Doellinger.

DOELLINGER SUSPECT   Il y avait quelques années déjà que le célèbre historien était en butte à certaines suspicions. Ses paroles anticoncordataires de 1848 et ses discours en faveur d'une « Église nationale allemande » avaient été mal accueillis par certains, et dès 1851, Windischmann, dont il avait contrecarré les efforts pour multiplier les séminaires diocésains, avait fait quelques réserves à son sujet [4]. Lorsque, un peu plus tard, Doellinger manifesta son déplaisir de la définition de l'Immaculée-Conception, au nom d'une théorie fixiste de la Tradition qui prétendait s'opposer à tout enrichissement du dogme,

---

(1) Le sérieux et la prudence avec lesquels procéda l'*Index* furent reconnus par plusieurs savants peu suspects de sympathies pour les scolastiques, tels Flir (*Briefe aus Rom*, p. 47) ou Kuhn (cf. *T. Q. S.*, t. CVIII, 1927, p. 215).

(2) Sur Jakob Frohschammer (1821-1893), outre sa biographie par B. MUENZ (1894), voir l'ouvrage de G.-B. GUZZETTI (cité à la p. 202, n. 2), p. 87-105.

(3) Lettre *Gravissimas inter*, du 11 décembre 1862, dans *Pii IX Acta*, t. III, p. 548 et suiv. ; extraits dans H. DENZINGER, *Enchiridion*, nos 1666-1676. Frohschammer ayant refusé de se soumettre, ses pouvoirs sacerdotaux lui furent retirés et, peu après, il abandonna définitivement l'Église.

(4) « Il s'infiltre jusque dans les meilleures têtes — je n'excepte pas même Doellinger — un esprit qui peut nous conduire aux plus déplorables excès. » (Windischmann à Ketteler, 5 août 1851, dans *Briefe von und an Ketteler*, édit. RAICH, p. 225.)

il scandalisa plusieurs de ses amis. D'autres furent heurtés par son hostilité contre la double tendance, favorisée par les jésuites, à accroître les prérogatives pontificales et à revendiquer la suprématie de l'Église sur le pouvoir civil, et le canoniste Philips, très curialiste de tendance, avec qui il avait de fréquentes discussions sur ce sujet, lui reprocha d'être touché par l'esprit protestant.

Ces accusations, répétées, trouvèrent crédit parmi les ultramontains de la nouvelle génération et il y eut probablement, conformément au procédé fréquent dans ce groupe, l'une ou l'autre dénonciation. En tout cas, lors de son voyage à Rome, en 1857, Doellinger s'y sentit l'objet d'une certaine méfiance et il en souffrit d'autant plus, lui dont toute la vie était vouée à la défense de la cause catholique, qu'il fut en outre blessé dans sa vanité de savant en constatant le peu de cas qu'on semblait y faire de sa renommée. Aussi prêta-t-il une oreille complaisante aux lamentations de Theiner sur l'absence de vie scientifique à Rome et sur l'ignorance de ceux qui prétendaient condamner la science allemande et, dès ce moment, alors qu'il devait une bonne partie de sa formation aux anciens historiens italiens de l'école de Baronius, de Muratori et de Benoît XIV [1], et qu'en 1854 encore, il considérait Kleutgen comme « un théologien capable et circonspect » [2], il n'eut plus de paroles assez méprisantes pour le niveau des études en Italie et pour la scolastique en particulier. Il rapporta d'ailleurs de Rome une impression très pessimiste sur la situation politique des États pontificaux et sur l'évolution probable de la question romaine, à laquelle il avait été initié par son élève de prédilection, le jeune John Acton, très libéral de tendance et ardent partisan de l'unité italienne. Tout ceci n'était pas fait pour le rapprocher des anciens élèves du Collège romain à propos desquels courait dans son entourage la fameuse boutade : « *Doctor romanus, asinus germanus*, un docteur de Rome n'est plus qu'un âne en Allemagne », et qui se sentaient froissés de cette prétention à ne voir de science ecclésiastique sérieuse que dans les universités germaniques et à considérer l'influence croissante prise par eux dans la vie ecclésiastique allemande comme une victoire non seulement du jésuitisme, mais de l'obscurantisme.

Malgré tout, si les plus zélés des ultramontains regardaient avec une certaine méfiance ce théologien aux méthodes trop peu romaines, les autorités ecclésiastiques conservaient dans leur ensemble leur estime au savant professeur, lorsque se produisit, en 1861, l'incident des Conférences de l'Odéon. Les événements consécutifs à Castelfidardo ayant profondément ému l'opinion catholique, Doellinger décida de consacrer à la question romaine deux conférences où il montrerait la légitimité des droits historiques du pape sur ses États et l'utilité du pouvoir temporel, mais aussi la possibilité, moyennant certaines conditions, d'en envisager la suppression. Doellinger, convaincu de la disparition imminente des États romains, avait simplement pour but de « fortifier les esprits

---

(1) Acton a bien mis en lumière le fait que la première formation de Doellinger avait été très peu allemande. (*English Hist. Review*, t. V, 1890, p. 708 et suiv.)

(2) Lettre du 31 janvier 1854, dans J. Friedrich, *op. cit.*, t. III, p. 134.

contre les espérances insultantes des protestants » [1], en faisant voir que le pouvoir temporel n'était pas essentiellement lié à la notion catholique de la papauté. Mais c'était déjà trop pour l'opinion ultramontaine, qui fut surtout choquée de ce que, voulant montrer que les catholiques n'étaient pas aveugles, il eût expliqué l'hostilité contre la souveraineté pontificale en invoquant les défauts du gouvernement des États de l'Église, archaïque et clérical. Le nonce quitta ostensiblement la séance, Mgr Laurent, évêque de Luxembourg, compara le maître de Munich à Cham dévoilant les faiblesses de Noé et à Judas trahissant son Seigneur, et plusieurs sautèrent sur l'occasion pour essayer, en organisant une campagne de protestations, de discréditer Doellinger aux yeux des masses catholiques, très dévouées au pape dans l'ensemble.

Déconcerté, Doellinger essaya de mettre les choses au point, d'abord en écrivant à divers amis des lettres qu'il leur demanda de propager, un peu plus tard en faisant une déclaration apaisante au *Katholikentag* de Munich, enfin et surtout en publiant à la fin de l'année un gros ouvrage, *L'Église et les Églises. La papauté et le pouvoir temporel*, dans lequel il reprenait les idées exposées à l'Odéon, mais en adoucissant les angles et en ayant l'habileté d'y joindre un savant parallèle entre l'Église catholique et les Églises séparées, qui constituait une admirable apologie du Saint-Siège. En dépit de considérations sur « l'administration cléricale » et les « prêtres policiers », que Montalembert lui-même trouvait déplacées, et malgré une invitation assez maladroite au pape, pour le cas où il devrait fuir devant les Italiens, à venir chercher en Allemagne non seulement un asile mais un complément d'éducation pour la Curie, l'ouvrage fut accueilli assez favorablement par Pie IX [2]. Mais ce ne fut pas le cas dans les milieux intransigeants et dès lors, les critiques contre Doellinger et la tendance théologique qu'il représentait se firent plus ouvertes, assurées qu'elles étaient de rencontrer désormais un accueil favorable auprès de ceux qui mettaient la défense des droits du pape au premier rang de leurs préoccupations.

LA CONFÉRENCE DES SAVANTS CATHOLIQUES — Dépité de voir son prestige entamé auprès des catholiques et exaspéré d'entendre mettre en doute sa valeur scientifique par des gens parfaitement incompétents, Doellinger, toujours dévoué à l'Église malgré son opposition croissante à la politique de la Curie, s'inquiétait en outre de voir les forces intellectuelles du catholicisme allemand divisées contre elles-mêmes, alors qu'il eût fallu faire front plus que jamais contre les attaques, toujours plus radicales, de la science incroyante. Il fallait absolument tenter une réconciliation.

Or, depuis plusieurs années, l'idée de réunir les savants catholiques allemands en congrès était dans l'air. Doellinger, après quelques hésitations, finit par se laisser convaincre par l'éditeur Herder et surtout

_____

(1) Doellinger à Meignan, avril 1861, dans H. Boissonot, *Le cardinal Meignan*, p. 208, n. 1.
(2) Il confiait à Ullathorne que, sans être d'accord en tout point, il le considérait comme un bon livre qui ne pourrait faire que du bien. (Acton à Doellinger, 26 août 1862, dans J. Friedrich, *op. cit.*, t. III, p. 269.)

par le vieil Hirscher qu'il y aurait là un moyen de rapprocher les points de vue et, malgré le scepticisme des professeurs de Tubingue, qui estimaient qu'une pareille réunion n'aboutirait qu'à de nouvelles disputes, il fit sonder les Mayençais par un de ses élèves, Janssen, un homme de juste milieu, et par son ancien secrétaire Joerg, qui devait passer peu après dans le camp ultramontain. En octobre 1862, au cours d'une rencontre à Fribourg avec Hirscher et Alzog, un modéré, un programme fut élaboré, mais Heinrich, l'animateur de la revue de Mayence, tout en acceptant le principe de la réunion, souleva plusieurs objections concernant les objectifs [1]. Hirscher et Kuhn estimèrent que, dans ces conditions, il valait mieux ne pas insister et Doellinger n'était pas loin de penser de même quand l'intervention d'un de ses amis, Michelis, ressuscita le projet en y intéressant le nonce à Vienne. C'est dans ces conditions que, au début d'août 1863, Doellinger, Alzog et l'orientaliste Haneberg invitèrent les savants catholiques d'Allemagne, d'Autriche et de Suisse à se réunir à Munich en septembre.

Malheureusement, dans l'entre-temps, les défiances contre Doellinger n'avaient fait que s'aggraver. On lui reprochait de nouvelles déclarations, moins respectueuses encore que les précédentes, à l'égard de la congrégation de l'*Index* [2], et surtout son dernier livre, *Papstfabeln*, où il dénonçait le caractère légendaire de plusieurs traditions relatives aux papes du moyen âge et exploitait contre la thèse de l'infaillibilité les défaillances des papes Libère et Honorius. Dans ces conditions, l'annonce d'un congrès, où il était à prévoir qu'il jouerait un rôle directeur, n'était pas pour plaire à tout le monde. Le nonce à Munich, déjà très monté contre Doellinger, fut fort mécontent de ce qu'il considérait comme « une sorte de synode » [3] convoqué sans l'autorisation de l'autorité ecclésiastique, et l'évêque de Paderborn, Mgr Martin, interdit même à ses prêtres d'y assister [4]. Les encouragements d'autres évêques contrebalancèrent toutefois ces oppositions et lorsque l'assemblée s'ouvrit, le 28 septembre, le nombre des participants, 84, dépassait les prévisions. A part les jésuites, irréductiblement hostiles, et les Tubinguiens, qui, sentant venir l'orage, ne voulaient pas se compromettre, presque tout ce que l'Allemagne catholique comptait comme théologiens, philosophes ou historiens de valeur était présent, y compris les coryphées de l'ultramontanisme et de la scolastique, Moufang, Heinrich, Phillips, ainsi que Scheeben, qui venait d'inaugurer son enseignement au grand séminaire de Cologne et de se faire connaître par la publication dans Der Katholik d'une première esquisse des *Mystères du christianisme* [5].

(1) Voir sa lettre dans J. Friedrich, *op. cit.*, t. III, p. 288-294.

(2) Ces reproches reposaient partiellement sur une méprise : la diffusion par Frohschammer d'un texte fautif et remanié de la réponse donnée par Doellinger à un de ses élèves à propos de l'*Index* avait donné l'impression qu'il prenait parti pour le professeur condamné, alors qu'il n'en était rien. (Cf. J. Friedrich, *op. cit.*, t. III, p. 309-311 et 693, n. 13, et *Janssens Briefe*, t. I, p. 245-246.)

(3) Haneberg à Doellinger, 12 septembre 1863, dans J. Friedrich, *op. cit.*, t. III, p. 306.

(4) Sur la réaction hostile de certains milieux catholiques, voir K. Mueller, *Leben und Briefe von Joh. Theod. Laurent*, t. III, Trêves, 1889, p. XIII-XXVIII.

(5) Il sera plus longuement question, dans le volume suivant, de l'œuvre théologique de M. J. Scheeben (1835-1888).

En guise d'introduction, Doellinger, élu président par ses anciens élèves et admirateurs, qui formaient la majorité de l'assemblée, prononça un remarquable discours sur le *Passé et l'Avenir de la Théologie* [1], que Goyau nomme à juste titre « une déclaration des droits » de la théologie. Il y développait un certain nombre d'idées justes sur l'union des Églises, sur l'utilité pour la théologie de ne pas se borner à la méthode analytique médiévale, mais de chercher à présenter le donné révélé dans sa plénitude organique et dans la perspective du développement historique, sur la nécessité de distinguer les données révélées de ce qui n'est que présentation accidentelle ou hypothèse provisoire. Mais d'autres passages étaient plus équivoques. Non content de revendiquer pour le théologien, là où la foi n'était pas directement en cause, une totale « liberté de mouvement », « aussi indispensable à la science que l'air au corps humain », et d'exiger que l'on combattît les erreurs théologiques avec des « armes scientifiques » plutôt qu'avec des censures ecclésiastiques, il affirmait que « tout comme chez les Hébreux le prophétisme coexistait avec la hiérarchie sacerdotale, ainsi devait-il y avoir dans l'Église, à côté des pouvoirs ordinaires, une puissance extraordinaire, l'opinion publique », et que c'était à la théologie à former celle-ci ; comme il dénonçait par ailleurs la complète décadence théologique des pays latins et considérait que seule l'Allemagne possédait vraiment « les deux yeux de la théologie, à savoir la philosophie et l'histoire », et pouvait prétendre à devenir en ce domaine « l'institutrice des nations », il donnait l'impression de revendiquer pour les théologiens allemands la véritable direction de l'Église catholique.

Les théologiens de Mayence et de Wurzbourg ne pouvaient laisser passer sans protester pareilles affirmations, mais on finit par se mettre d'accord sur une formule un peu ambiguë, concernant les droits respectifs de l'autorité et de la liberté. Lorsque, le dernier jour, Doellinger revint sur la profonde différence qui séparait au point de vue scientifique les deux écoles, « l'allemande » et « la romaine », dont l'une, disait-il, défendait le catholicisme « avec des canons », tandis que l'autre se contentait « d'arcs et de flèches », il concéda la légitimité des deux méthodes, mais insista pour que cessent à l'avenir les suspicions et les dénonciations qui décourageaient les jeunes travailleurs [2].

Malgré la tension continuelle entre les deux écoles, on n'alla cependant pas jusqu'à la rupture. Au contraire, au cours du banquet final, Moufang félicita Doellinger d'avoir suscité et dirigé « cette entreprise vraiment bienfaisante et hautement méritoire » [3] et beaucoup purent croire que le but principal de la réunion, l'entente avec le « parti romain », était atteint.

LA RÉACTION DE ROME   Le lendemain, un télégramme, rédigé de concert par Doellinger et Moufang, priait Mgr de Hohenlohe d'avertir le pape de ce que les débats s'étaient déroulés « dans

---

(1) On en trouvera le texte, légèrement atténué, dans les *Verhandlungen der Versammlung...*, p. 25-59.
(2) *Verhandlungen...*, p. 130-133.
(3) *Ibid.*, p. 139-140.

l'esprit de l'Église » et de ce que l'épineuse question des rapports entre la philosophie et le magistère ecclésiastique avait été résolue « dans le sens d'une complète soumission à l'autorité ». Tout content et n'écoutant, comme toujours, que sa première impression, le pape, à qui on avait fait craindre que la réunion ne tournât en une levée de boucliers contre les Congrégations romaines, chargea le prélat de répondre qu'il bénissait l'Assemblée et l'engageait « à continuer son œuvre vraiment catholique » [1]. Doellinger, qui s'empressa de communiquer cette réponse à la presse, pouvait à juste titre écrire à une confidente : « Étant données les circonstances, c'est là une importante victoire [2]. »

On devine la fureur du nonce. Ses doléances, appuyées par des rapports beaucoup moins bienveillants que ceux de Hohenlohe sur l'atmosphère du congrès et les discours prononcés par Doellinger, eurent vite fait de modifier les sentiments du pape, qui pria Hohenlohe de déclarer que son télégramme avait exagéré la portée des paroles pontificales ; mais celui-ci, estimant qu'il n'en était rien, refusa de se rétracter, en dépit d'une scène extrêmement violente [3].

Cet incident, caractéristique des procédés qui, à plus d'une reprise, sous le pontificat de Pie IX, provoquèrent l'indignation d'hommes moins susceptibles que Doellinger, n'était qu'un premier indice du mécontentement romain. Les appréciations méprisantes du maître de Munich pour la science italienne avaient évidemment froissé, et le mauvais effet produit par la hardiesse de certaines de ses affirmations était encore accru par le fait qu'elles se produisaient un mois après le discours de Montalembert à Malines [4] : après l'apologie des libertés modernes en politique, la revendication de la liberté pour les sciences sacrées !

Certains ne parlaient de rien de moins que de mettre à l'*Index* les Actes du Congrès. Reisach, très monté contre Doellinger, mais soucieux d'éviter un pareil désaveu à l'ensemble de la science allemande, s'interposa [5] et finalement l'idée prévalut de se borner à une lettre du pape à l'archevêque de Munich qui, datée officiellement du 21 décembre 1863, parut effectivement vers la fin de janvier et à la rédaction de laquelle Reisach prit une part importante [6]. Tout en exprimant l'espoir de voir les bonnes intentions des congressistes produire de bons fruits, le pape blâmait les attaques contre la scolastique et déplorait qu'une assemblée de théologiens se fût réunie sans mandat de la hiérarchie, à qui cependant « il appartient de diriger et de surveiller la théologie ». Il précisait en outre que le savant catholique n'était pas seulement lié par les définitions

---

(1) Texte des deux télégrammes dans J. Friedrich, *op. cit.*, t. III, p. 335-336.

(2) *Briefe Doellingers an eine junge Freudin*, édit. H. Schroers, p. 160.

(3) Tant que les archives du Vatican restent inaccessibles, nous ne sommes renseignés sur cet incident que par Friedrich (*op. cit.*, t. III, p. 335-340), mais d'après divers recoupements les faits rapportés semblent substantiellement exacts.

(4) *La Civiltà cattolica* fit immédiatement le rapprochement dans une série d'articles sur *Il Congresso dei dotti cattolici in Monaco di Baviera e le Scienze Sacre*, t. LVI et LVII, 1864. (Voir surtout t. LVI, p. 386-387.)

(5) Voir les lettres écrites de Rome par Janssen les 12 et 13 décembre 1863, dans *Janssens Briefe*, édit. Pastor, t. I, p. 201-202.

(6) Lettre de Janssen du 20 janvier 1864, *ibid.*, t. I, p. 236. Voir le texte de la lettre *Tuas libenter* dans *Pii IX Acta*, t. III, p. 638 et suiv.

solennelles, mais qu'il devait aussi tenir compte du magistère ordinaire, des décisions des Congrégations romaines et de l'enseignement commun des théologiens. Bien que l'historien Janssen, qui passait l'hiver chez Reisach, eût réussi à arracher à celui-ci plusieurs atténuations, ce n'en était pas moins un désaveu, que Scheeben lui-même « aurait souhaité un peu plus clément » [1]. Ce désaveu fut encore complété, quelques mois plus tard, par une réglementation très stricte, qui rendait pratiquement impossibles de nouvelles réunions de ce genre.

La tentative de Doellinger de « clore entre les théologiens catholiques d'Allemagne une véritable guerre civile et de concentrer leurs forces vives contre l'ennemi commun » [2] avait échoué : ses adversaires étaient plus aigris que jamais par ses sarcasmes contre la scolastique, et encouragés dans leurs attaques contre la « science allemande » maintenant qu'ils savaient que Pie IX s'exprimait sur celle-ci « peu, très peu favorablement » [3] ; et lui-même n'avait plus de mots assez méprisants pour ceux qui faisaient si peu de cas de la théologie historique, le plus beau titre de gloire de l'Allemagne catholique, à son avis, et qu'il accusait de partager avec le nonce et les jésuites la responsabilité du blâme qui le frappait, incapable de comprendre que la réaction romaine était inévitable contre sa prétention à vouloir ressusciter, en plein XIXe siècle, au profit de la science allemande « le rôle imposant et légèrement impérieux de la vieille Sorbonne » [4].

LES DEUX CAMPS Désormais les jeux était faits et l'âpreté de la polémique allait redoubler. D'un côté, ceux qui estimaient qu'il fallait avant tout reconquérir à l'Église l'estime des milieux cultivés par une utilisation loyale des méthodes historiques et par une présentation du dogme catholique adaptée à la mentalité philosophique moderne. Comme les Tubinguiens, bien que partisans de ce double idéal, n'approuvaient pas le radicalisme de l'école de Munich et se tenaient de plus en plus à l'écart, l'aile marchante de ce groupe était constituée par les amis de Doellinger, qui eurent bientôt leur revue à eux, le *Theologisches Literaturblatt*, fondé en 1865 par un jeune professeur de Bonn, Reusch.

En face d'eux, l'école « romaine » ou scolastique, pour qui la tâche essentielle consistait à présenter aux catholiques un système doctrinal complètement élaboré et d'une orthodoxie indiscutable. Au séminaire de Mayence, où l'équipe du *Katholik* était plus ardente que jamais ; au séminaire de Cologne, où Scheeben renouvelait l'enseignement de la dogmatique ; au séminaire d'Eichstatt, où Stoeckl inaugurait un enseignement philosophique d'inspiration non seulement scolastique mais strictement thomiste ; au scolasticat des jésuites, où le P. Riess avait fondé en 1864 les *Slimmen aus Maria Laach* pour commenter la doctrine du *Syllabus*, on proclamait qu'aucune des deux conditions désirées n'était réalisée

(1) A. Kerkvoorde, dans l'Introduction à Scheeben, *Le Mystère de l'Église* (*Unam Sanctam*, XV), p. 31, n. 1.
(2) P. Godet, dans *Revue du Clergé français*, t. XXXVI, 1903, p. 371.
(3) Lettre de Janssen du 27 janvier 1864, dans *Janssens Briefe*, t. I, p. 243.
(4) G. Goyau, *L'Allemagne religieuse*, t. IV, p. 261.

dans l'enseignement des théologiens universitaires. On leur reprochait de ne jamais traiter que des points de détail, de se borner à des travaux d'exégèse ou de théologie positive considérés, à la suite de *La Civiltà cattolica*, comme des à-côté, peu utiles et méritant à peine le nom de théologie. On leur reprochait surtout leur esprit d'indépendance à l'égard du Saint-Siège, et certains ne manquaient pas, « dans leurs commentaires sur les condamnations de la veille, de prévoir et de préparer les condamnations du lendemain » [1]. Ils dénonçaient dans les congrès catholiques le danger qu'il y avait à confier à des maîtres aussi suspects la jeunesse catholique et surtout la jeunesse cléricale, et ces accusations, qui, par dessus la tête de quelques professeurs trop audacieux, atteignaient l'ensemble des théologiens universitaires allemands, portaient à leur comble la fureur de ceux-ci, déjà outrés de s'entendre dire qu'ils enseignaient à leurs élèves « beaucoup de superflu sans se préoccuper de l'essentiel » [2] et qui répliquaient en dénonçant à leur tour avec indignation les Mayençais, les *Germaniker* et les jésuites comme les fourriers d'une nouvelle inquisition.

Rares furent ceux qui, comme l'excellent Janssen, le futur historien de l'Allemagne à la veille de la Réforme, surent rester en bons termes avec les deux partis. « L'heure était aux mêlées théologiques, avec toutes les injustices de jugement, toutes les violences de plume qu'entraînait la passion même de la lutte [3]. » C'est en vain qu'un ultime effort pour rapprocher les points de vue fut tenté par l'un ou l'autre : Werner, par exemple, un ancien günthérien, l'un des rares théologiens autrichiens de marque à cette époque, qui avait découvert la valeur permanente de la scolastique au cours de travaux historiques qui en font un précurseur de Denifle et de Baeumker ; Aloïs Schmidt, dont l'ouvrage : *Les tendances scientifiques dans le domaine du catholicisme contemporain*, ne manquait pas de clairvoyance, malgré des vues discutables sur la légitimité des doutes contre la foi ; Dieringer de Bonn, qui reconnaissait l'intérêt de l'œuvre de Kleutgen, mais demandait aux théologiens de se rendre compte de la nécessité de compléter pour le XIXe siècle les solutions du moyen âge. Ils se heurtèrent soit à l'indifférence, soit à des critiques sans nuances. Les positions étaient désormais trop arrêtées pour qu'aucun des deux partis consentît à admettre que ses préventions étaient exagérées et qu'il pouvait avoir à apprendre quelque chose dans le camp d'en face pour compléter ce que son point de vue avait d'unilatéral.

La tension s'aggravait d'année en année et le moindre incident fournissait l'occasion d'âpres polémiques où la vraie science ne trouvait pas plus son compte que la charité. L'un des grands responsables de cette situation était certainement Doellinger. De plus en plus aigri au fur et à mesure qu'il sentait fondre sa popularité, on eût dit qu'il cherchait à plaisir à légitimer les attaques de ses adversaires par ses compromissions

---

(1) G. Goyau, *L'Allemagne religieuse*, t. IV, p. 230.
(2) Mgr Meglia, nonce à Munich, au cardinal Antonelli, 15 janvier 1869, dans T. Granderath, *Histoire du concile du Vatican*, trad. franç., t. I, p. 180.
(3) G. Goyau, *op. cit.*, t. IV, p. 269.

toujours plus affichées avec des savants protestants ou des ministres anticléricaux, et par ses jugements toujours plus malveillants sur tout ce qui venait de Rome. Et que dire de l'arrogance de beaucoup de ses amis et du manque d'esprit catholique de certains de ses disciples, Huber, Friedrich, Pichler, « la jeune école munichoise », comme on les appelait ?

Les responsabilités étaient toutefois partagées et l'intransigeance ou l'étroitesse d'esprit de plusieurs défenseurs des tendances romaines contribuèrent notablement à envenimer une situation déjà délicate. Il y avait certes parmi eux des esprits modérés, Hergenrœther, par exemple, ou Scheeben, mais ce n'était pas le cas pour tous. Les *zelanti* romains avaient leur pendant dans ceux que Ritter surnommait les *Zornigen*, les « furieux »[1]. Reisach lui-même, témoin peu suspect, se plaignait du tort causé au mouvement néoscolastique par les interventions « à tort et à travers » de Plassmann, spécialiste des dénonciations à l'*Index*[2]. Ne faudrait-il pas en dire autant, malgré l'excellence de leurs intentions, des polémiques déclenchées par Clemens et plus tard par Constantin von Schaezler[3] ? Ce sectarisme de beaucoup de champions de la néoscolastique eut des conséquences fort dommageables. D'abord, parce que l'œuvre, autrement estimable, de Scheeben, s'en trouva compromise et ne réussit pas à conquérir l'intérêt du public cultivé. Ensuite, parce qu'en s'en prenant avec violence à l'œuvre, sans doute plus discutable, de Kuhn, il incita les « bien-pensants » à englober l'école de Tubingue dans la même méfiance que le reste de la théologie allemande, rendant ainsi impossible l'influence équilibrée qu'aurait pu exercer cette école qui ne souffrait ni de l'historicisme ni du rationalisme qui viciaient l'école de Munich, et qui aurait pu, d'autre part, compléter ce qui manquait de biblique et de concret ainsi que de sens du mystère à la scolastique du xixe siècle.

§ 3. — Le retard des sciences ecclésiastiques en France[4].

LE RENOUVELLEMENT DES SOURCES DE L'INCROYANCE

Le public catholique cultivé fut plus touché en France qu'en Allemagne par le changement d'atmosphère qui suivit le passage de l'époque romantique à l'époque positiviste, et un nombre croissant d'intellectuels se rallièrent à une conception rationaliste de l'univers et de l'histoire religieuse de l'humanité, d'où tout surnaturel était exclu.

Deux hommes surtout contribuèrent au succès d'une philosophie

(1) G. von Hertling, *Erinnerungen*, t. I, p. 179.
(2) *Janssens Briefe*, édit. Pastor, t. I, p. 258.
(3) G. Haefele, *Const. von Schäzler*, dans *Divus Thomas* (Frib.), t. V, 1927, p. 411-448.
(4) Bibliographie. — Consulter sur l'infériorité de la pensée catholique en France, Ch. Pouthas, *L'Église et les questions religieuses en France de 1848 à 1877*, p. 223-260, à illustrer par E. Vacherot, *La Théologie catholique en France*, dans *Revue des Deux Mondes*, t. LXX, 1868, p. 294-318 ; At, *Les apologistes français au XIXe siècle*, Paris, 1909 ; A. Houtin, *La question biblique chez les catholiques de France au XIXe siècle*, Paris, 1902. Voir par contre une note plus optimiste dans A. Baudrillart, *Le renouveau intellectuel du clergé français au XIXe siècle*, Paris, 1903.
En ce qui concerne la Belgique : le numéro spécial des *Ephemerides theologicae lovanienses*, consacré à la Faculté de théologie de Louvain au xixe siècle, t. IX, 1932, p. 608-704, ainsi que E. van Roey, *Les sciences théologiques*, dans *Le mouvement scientifique en Belgique, 1830-1905*, t. II, Bruxelles, 1908, p. 483-523.

dégagée de toute préoccupation religieuse : Taine, qui dès 1857 avait exécuté l'éclectisme cousinien dans son spirituel ouvrage : *Les Philosophes classiques du XIXᵉ siècle en France*, et Littré, qui révéla Auguste Comte au grand public. Leur influence sur les jeunes fut encore augmentée par le prestige de leur opposition au régime impérial, et l'ardente campagne menée par Mgr Dupanloup en 1862 pour empêcher l'élection de Littré à l'Académie française lui fit une sorte de popularité. Une partie de la jeunesse ne tarda pas à prendre une position plus radicale encore, en accueillant avec faveur les théories matérialistes venues d'Allemagne ou en se ralliant à l'athéisme militant de Proudhon et de Blanqui. Pour répondre au reproche de ruiner les bases de la morale, les adversaires de l'Église s'efforcèrent en même temps de constituer une « morale indépendante »[1] de toute base religieuse, à laquelle Renouvier, le « Kant républicain », donnera en 1869 son expression classique.

Parallèlement à cette évolution de la philosophie, le progrès des sciences naturelles, historiques et philologiques mettait en danger l'autorité traditionnelle de la Bible. Les hommes du Second Empire suivent avec un intérêt passionné les découvertes de Boucher de Perthes, le créateur de la science préhistorique, et les travaux de Darwin sur l'*Origine des Espèces*, qui leur semblent ruiner l'autorité des premiers chapitres de la Genèse. Les données de l'histoire biblique primitive paraissent de moins en moins conciliables avec les progrès de l'égyptologie et de l'assyrologie. Les débuts de l'étude comparée des religions, sur laquelle la célèbre *Introduction à l'histoire du bouddhisme indien* de Burnouf attire l'attention, mettent en question la transcendance du christianisme. Enfin et surtout, la critique biblique rationaliste de Strauss et de l'école radicale de Tubingue, longtemps ignorée en France, commence à y pénétrer par la Faculté de théologie protestante de Strasbourg, puis, à partir de 1858, grâce à la *Revue germanique*. Brusquement, en 1863, ses conclusions sont révélées au public cultivé par la *Vie de Jésus* d'Ernest Renan[2], le meilleur spécialiste français des questions bibliques. Cette œuvre de vulgarisation fut critiquée à l'étranger par des savants non catholiques, mais ses qualités littéraires ainsi que la spectaculaire levée de boucliers de l'épiscopat français lui assurèrent un succès sans précédent. Aussi atteignit-elle pleinement son but qui était d'éveiller l'intérêt du grand public pour les problèmes bibliques et de lui montrer dans l'origine du christianisme un phénomène purement humain.

UNE « THÉOLOGIE ORATOIRE »   Comment le clergé français s'adapte-t-il à l'évolution scientifique du temps et riposte-t-il en présence du développement inquiétant du rationalisme et du positivisme ? Il semble inconscient du danger. Lorsque l'abbé Meignan, éclairé par son contact avec l'Allemagne, essaya en 1859 d'attirer

---

(1) *La morale indépendante* est aussi le titre de la revue hebdomadaire fondée en 1865 par Massol et son disciple Henri Brisson.
(2) Sur la crise religieuse d'Ernest Renan, qui quitte le séminaire Saint-Sulpice en 1845, voir, outre les *Souvenirs d'enfance et de jeunesse*, P. LASSERRE, *La jeunesse d'Ernest Renan*, Paris, 1925-1932, 3 vol., et J. POMMIER, *La jeunesse cléricale d'Ernest Renan*, Paris, 1933.

l'attention sur ce point, ses articles, publiés dans *Le Correspondant* sous
le titre : *D'un mouvement antireligieux en France,* furent si mal accueillis
qu'il fut obligé de changer son dessein primitif et de conclure par un
parallèle optimiste entre la crise où le rationalisme contemporain acculait
l'Église anglicane et la force de résistance du catholicisme en France.
Et lorsque, à la suite de la publication de la *Vie de Jésus,* des progrès de
la libre-pensée et de la campagne en faveur de la morale indépendante,
les yeux commencèrent à s'ouvrir et que Mgr Dupanloup et Mgr Darboy
signalèrent enfin la gravité de la situation dans des mandements remarqués [1], beaucoup crurent encore qu'il suffisait de protester avec indignation
contre « l'orgueil de l'esprit ».

Dans un article publié en 1868 sur *La théologie catholique en France,*
Etienne Vacherot, après avoir énuméré les noms du P. de Ravignan, du
P. Lacordaire, de Mgr Dupanloup, de Mgr Pie, de Mgr Darboy, de Mgr
Maret, du P. Hyacinthe Loyson, du P. Félix, des abbés Bautain, Freppel
et Perreyve, du P. Perraud, du P. Gratry, écrivait : « Si l'on ne voyait
que le talent et le succès, on pourrait se croire revenu aux beaux jours
de la théologie chrétienne, au siècle des Arnauld, des Bossuet, des Fénelon. » Mais il ajoutait que ces œuvres « d'éloquence passionnée », capables
de faire « bondir d'enthousiasme et d'indignation » le public des cathédrales et des salons, n'apportaient pas même un début de réponse aux
travaux où les affirmations de la foi traditionnelle se trouvaient contre-
dites au nom de « l'esprit historique et critique qui est le véritable esprit
du siècle » [2]. Pour l'un ou l'autre des noms cités, pour Mgr Maret notamment, le jugement est trop sévère, mais, dans l'ensemble, il est juste et
met le doigt sur le drame de la pensée catholique française au milieu
du XIXe siècle : elle en reste aux méthodes oratoires du romantisme,
alors que les esprits qui pensent sont de plus en plus impressionnés par
les résultats des sciences positives ou par les minutieuses analyses de
la critique historique.

Les célèbres *Instructions synodales sur les erreurs du temps* de Mgr Pie,
composées, sous la conduite de dom Guéranger, par celui que ses admirateurs appelaient « le saint Hilaire du XIXe siècle », sont de beaux monuments d'éloquence, mais assez pauvres en fait d'esprit scientifique et,
au surplus, dirigés contre des auteurs qui, comme Cousin, n'exerçaient
plus d'influence sur les jeunes générations. Quant au P. Gratry [3], il fut
un grand animateur, mais s'il triomphe là où suffit la pénétration psychologique ou l'attention ardente aux aspirations du cœur humain vers
l'idéal, et s'il entrevoit la voie où s'engageront plus tard Ollé-Laprune
et Blondel, « il expose trop en littérateur et pense trop en poète pour
être mis au nombre des grands philosophes » [4].

Dans le domaine biblique, si quelques rares prêtres, tels l'abbé Meignan

---

(1) Mgr Dupanloup, *Avertissement à la jeunesse et aux pères de famille* (1863) ; *Les malheurs et
les signes des temps* et *L'Athéisme et le péril social* (1866) ; Mgr Darboy, *Instructions pastorales* pour
l'avent 1865 et le carême 1867.

(2) *Art. cit.,* p. 297-303.

(3) Sur Alphonse Gratry (1805-1872), professeur de religion à l'École normale supérieure de
1846 à 1851, professeur à la Sorbonne depuis 1863, voir A. Chauvin, *Vie du P. Gratry,* Paris, 1911.

(4) A.-D. Sertillanges, *Le christianisme et les philosophies,* t. II, Paris, 1941, p. 347.

ou l'abbé Vollot, prennent contact avec la science allemande, leur influence reste très limitée et le seul foyer d'études un peu actif est le séminaire Saint-Sulpice, où se conserve la tradition de M. Garnier et où enseigne, jusqu'à sa mort, en 1868, M. Le Hir, un hébraïsant, le maître de Renan. Mais leurs méthodes intellectuelles, héritées du siècle précédent, étaient alors suspectes de rationalisme aux yeux de beaucoup et pourtant elles étaient encore bien insuffisantes pour répondre aux problèmes posés par l'exégèse radicale allemande. Faut-il s'étonner dès lors si l'on ne trouve à opposer à l'ouvrage de Renan que « de vieilles dissertations de séminaristes, peu propres à entraîner la conviction »[1] ? La réfutation ne paraissait pourtant guère difficile aux exégètes d'Outre-Rhin qui trouvaient bien superficielle la fameuse *Vie de Jésus* : « Je viens de faire un voyage en Allemagne, écrivait l'abbé Meignan à un ami : on se moque de Renan, mais aussi de nous[2]. »

INFÉRIORITÉ MARQUÉE DANS LE DOMAINE HISTORIQUE

Ce qui frappe surtout, par opposition avec l'Allemagne, c'est le petit nombre de travaux sérieux de théologie positive. Les œuvres de Freppel ne méritent guère ce titre[3]. Celles même de dom Guéranger, quelle qu'ait été leur utilité, trahissaient une érudition un peu courte et des préoccupations polémiques qui nuisent à l'objectivité. Si Ginouilhac publie en 1852 une *Histoire du dogme catholique pendant les trois premiers siècles*, qui n'est pas dépourvue de valeur, si l'ouvrage de Maret sur *Le Pape et le Concile*, publié à l'occasion du concile du Vatican, révèle une bonne connaissance de l'antiquité chrétienne et certaines préoccupations critiques, dom Pitra est le seul à consacrer sa vie à ce genre de recherches[4]. Sa renommée ne tarde d'ailleurs pas à le faire appeler à Rome et il ne crée pas d'école en France, où Migne continue pourtant inlassablement à mettre à la disposition de travailleurs éventuels le précieux instrument que constitue sa patrologie : il achève les 217 volumes de la série latine en 1855, puis publie les 162 volumes de la série grecque de 1857 à 1866[5], et du moins le grand nombre de ses souscripteurs indique-t-il dans le clergé de France un goût renouvelé des choses de l'esprit et un désir d'aller aux sources.

Mais le travail vraiment scientifique est à peu près nul. L'histoire de la philosophie scolastique devait certes profiter des études de Barthélemy Hauréau comme celle de la liturgie des travaux érudits de Léopold Delisle, mais ces précurseurs, qui n'étaient pas précisément des hommes d'Église, ne furent guère suivis et souvent à peine appréciés dans les milieux ecclésiastiques. Quant à l'histoire de l'Église proprement dite,

---

(1) A. DANSETTE, *Histoire religieuse de la France contemporaine*, t. I, p. 426.
(2) Meignan à Lagrange, 23 septembre 1863, dans H. BOISSONNOT, *Le cardinal Meignan*, p. 192.
(3) S'il n'a pas laissé d'œuvre scientifique et s'il n'a pas eu d'idées neuves ni en théologie ni en critique, Freppel n'en fut pas moins un admirable vulgarisateur, doué d'une intelligence vive et d'une mémoire prodigieuse, connaissant à fond les œuvres des Pères. Sur la période studieuse de sa vie, voir le t. I d'E. TERRIEN, *Mgr Freppel*, Angers, 1931.
(4) Encore se plaignait-il de ce que ses travaux fussent moins lus par ses compatriotes catholiques que par les protestants allemands et anglais. (Lettre du 24 avril 1873, dans F. CABROL, *Le cardinal Pitra*, p. 171.)
(5) Voir les articles *Migne* du *D. T. C.*, t. X, col. 1722-1740 et du *D. A. C. L.*, t. XI, col. 941-957.

la situation est tout simplement lamentable. Au moment même où se réorganise en Belgique la Société des Bollandistes, destinée à fournir les bases d'une hagiographie critique, on voit renaître en France les légendes concernant l'origine apostolique de certains diocèses, dont les grands érudits des xviie et xviiie siècles semblaient cependant avoir fait définitivement justice. Ressuscitée en 1835 par un sulpicien, M. Faillon, la thèse de l'apostolicité des Églises de Gaule est à nouveau enseignée dans les séminaires, pénètre dans les manuels d'histoire, ne rencontre bientôt plus guère d'opposition et semble triompher définitivement avec l'apparition, en 1877, de l'ouvrage de dom Chamard, *Les Églises du monde romain* [1]. Par ailleurs, le projet, conçu vers 1865, par M. Palmé, de rééditer les *Acles des Conciles* dut être bientôt abandonné, et les moines de Solesmes ne réussirent pas à continuer la publication de la *Gallia christiana* comme ils en avaient eu l'intention : dom Pitra mis à part, les collaborateurs firent totalement défaut. Il y a bien, de ci de là, un certain nombre de prêtres qui utilisent leurs loisirs à des travaux d'érudition, mais ils se cantonnent presque toujours dans les recherches d'historiographie et surtout d'archéologie locales et, faute de formation technique, leurs publications sortent rarement de la médiocrité [2].

Très en retard du côté de l'érudition, le clergé du Second Empire ne se rachète même pas du côté de la synthèse et le succès qui accueillit l'*Histoire générale de l'Église depuis la création jusqu'à nos jours* de l'abbé Darras « fut considéré à l'étranger comme la preuve la plus significative de la décadence des études historiques dans le clergé français » [3].

UN ESPOIR MANQUÉ :   Bref, absence quasi totale dans le domaine positif
L'ORATOIRE   et, en matière spéculative, pour quelques œuvres
                         plus brillantes que profondes ou pour un ouvrage
qui s'impose par ses qualités bien françaises de clarté et de modération, le *Compendium Theologiae moralis* du P. Gury [4], destiné du reste aux seuls professionnels, que de non-valeurs !

Conscient des nécessités de la controverse philosophique et religieuse, Gratry avait bien songé à donner corps à une idée que Lamennais puis Bautain avaient déjà cherché à réaliser : grouper quelques prêtres d'élite en une « sorte d'atelier apologétique ». L'un de ses amis, le chanoine de Valroger, lui suggéra la formule de l'Oratoire de France, disparu depuis la Révolution, et, s'en remettant pour les questions d'organisation à un curé de Paris, l'abbé Pététot, ils débutèrent en 1852. Ils purent bientôt compter sur quelques recrues de valeur, mais malheureusement le supérieur avait des vues plus pratiques et préféra orienter ses sujets

---

(1) Cf. A. Houtin, *La controverse de l'apostolicité des Églises de France au XIXe siècle*, Laval, 1900.

(2) On trouvera la liste de ces travaux dans le recueil de R. de Lasteyrie, *Bibliographie générale des travaux historiques et archéologiques publiés par les sociétés savantes de la France*, t. I à IV, Paris, 1888-1904, en particulier t. I, p. 223-319 (Bulletins de la Société française d'archéologie).

(3) H. Leclercq, art. *Historiens du christianisme*, dans *D.A.C.L.*, t. VI, col. 2643.

(4) Professé d'abord au scolasticat jésuite de Vals, il fut réédité 17 fois de 1850 à 1866, et se répandit à Rome et dans de nombreux séminaires étrangers.

vers l'enseignement secondaire, en dépit des protestations du P. Gratry et du P. de Valroger, qui virent ainsi s'évanouir l'espoir de reprendre la tradition des grands travaux scientifiques des oratoriens du passé [1].

LA CAUSE DU MAL : L'ABSENCE
D'UN ENSEIGNEMENT SUPÉRIEUR

La réorganisation d'un enseignement secondaire catholique présentait certes des avantages en orientant vers l'étude une partie du clergé français, trop exclusivement absorbé jusqu'alors par le ministère paroissial ; elle allait lui permettre, en faisant la preuve d'une culture littéraire remarquable, de reprendre de l'influence sur les esprits de la classe bourgeoise. Mais en mobilisant pour l'enseignement de la jeunesse tout ce que le clergé et les ordres religieux comptaient d'intellectuels, elle rendait par le fait même plus difficile l'éclosion d'une véritable activité scientifique au sein du clergé français. Là n'est pourtant pas la cause profonde du mal : plus que le temps, ce qui lui manque c'est une initiation à l'esprit critique et aux méthodes universitaires nouvelles.

C'est ce qu'avait compris l'homme perspicace qu'était Mgr Affre lorsqu'il avait, en 1845, réorganisé l'École des Carmes, afin de faciliter aux jeunes prêtres la préparation d'une licence ou d'un doctorat à la Sorbonne [2]. Mais la plupart des évêques ne voyaient pas l'importance qu'il y avait à donner à l'élite de leur clergé une formation universitaire dont ils se défiaient du reste. *A fortiori* ne comprenaient-ils pas qu'il aurait été opportun d'en envoyer une partie suivre les cours des facultés allemandes, étant donné que, même à la Sorbonne, les méthodes scientifiques modernes commençaient à peine à pénétrer. C'est que ces méthodes et l'esprit critique qui est à leur base apparaissaient alors aux yeux d'un trop grand nombre de catholiques éminents comme irrémédiablement suspects de rationalisme et d'hétérodoxie. Un jésuite dont la sûreté de doctrine ne peut être mise en doute, le P. Matignon, l'observait discrètement : « A côté de l'autorité qui ordonne, il faut la science qui démontre... Je ne crains point d'insister sur cette nécessité parce que nous avons en France certains catholiques qui n'en paraissent pas convaincus [3]. » Il faut d'ailleurs reconnaître que l'usage que faisaient des méthodes nouvelles les ennemis de la foi n'était pas pour inspirer confiance dans la Science aux esprits timorés ; et que de plus certains jeunes prêtres semblaient ne considérer les études supérieures que comme un moyen d'arriver, au témoignage même d'un des rares hommes ouverts aux nécessités scientifiques : « Il faut des hommes de travail mais désintéressés. L'ambition et l'intrigue perdent et gâtent tout. Le peu que j'ai appris là-dessus me dégoûte. On veut à trente ans être connu, avoir des relations brillantes, et l'épiscopat à l'horizon. Dès qu'on y touche, on ne fait plus rien ; de là cette pénurie d'hommes sérieusement et solidement instruits [4]. »

---

(1) On consultera la troisième partie de l'ouvrage de A. PERRAUD, *L'Oratoire de France au XVIIe et au XIXe siècle*, Paris, 1866, les chap. ix et x de A. CHAUVIN, *Le P. Gratry*, Paris, 1901 et G. DE VALROGER, *Le P. de Valroger*, Paris, 1911.

(2) Voir le volume précédent, p. 482-483.

(3) Dans *Les Études*, t. IV, 1864, p. 277.

(4) L'abbé Vollot à l'abbé Laroche, 20 juin 1862, dans A. CROSNIER, *Souvenirs de l'abbé Vollot*, Angers, 1896, p. 280.

Malgré le nombre limité de ses élèves, l'École des Carmes rendit de grands services. Mais elle ne put devenir « l'École Normale de professeurs destinés aux grands séminaires » que l'abbé Darboy appelait de ses vœux en 1847. En effet, la Faculté de théologie de la Sorbonne n'était pas à la hauteur de sa tâche. Mgr Maret, qui avait étudié en Allemagne et continuait à suivre de près le mouvement intellectuel d'Outre-Rhin, avait bien essayé d'enrayer la décadence où l'avait laissée tomber son prédécesseur, l'abbé Glaire, en groupant autour de lui, dès sa nomination comme doyen en 1853, quelques professeurs de valeur [1]. Mais l'enseignement gardait une allure avant tout littéraire, et surtout l'établissement attirait peu de candidats, car il ne pouvait conférer les grades canoniques. Les négociations entreprises avec Rome à ce sujet et poursuivies par intermittence durant tout le Second Empire n'aboutirent pas à cause de la crainte du Saint-Siège, partagée par plusieurs évêques et très justifiée d'ailleurs, de voir les facultés d'État devenir des foyers de gallicanisme [2]. Ce n'est qu'avec la fondation d'universités catholiques, à la faveur de la loi de 1875 sur la liberté de l'enseignement supérieur, que la France devait enfin être dotée de facultés canoniques de théologie qui allaient permettre de remédier, avec un demi-siècle de retard, aux lacunes de l'Église, sur le plan intellectuel.

UNE CONTRE-ÉPREUVE :
L'UNIVERSITÉ DE LOUVAIN
Alors que le niveau moyen de la vie intellectuelle en Belgique, vers le milieu du XIXe siècle, était nettement inférieur à celui de la France, la formation scientifique du clergé y apparaît au contraire meilleure. On en trouve une preuve parmi d'autres dans la qualité et le succès de la revue fondée en 1847 par les abbés Loiseaux et Falise, les *Mélanges théologiques*, qui deviendront, en 1869, la *Nouvelle Revue théologique* [3].

Cette différence de situation s'explique sans doute par des contacts plus étroits avec l'Allemagne, mais pour une bonne part aussi grâce à Louvain. L'éloge du vieux-catholique Schulte, si exigeant, mérite d'être retenu :

Le clergé belge possède en l'Université catholique de Louvain un établissement d'instruction comme n'en possède aucun pays hors de l'Allemagne et de l'Autriche. Ici, on a conservé le contact de jadis entre les différentes facultés et rendu possible de cette manière une formation complète des canonistes dans les sciences auxiliaires du droit canonique. Bien que direction et fin de l'École de Louvain soient absolument cléricales, on ne saurait contester que les publications canoniques de cette école dépassent les travaux des universités non allemandes [4].

Mais le droit canon, brillamment professé de 1850 à 1865 par Henri Feye, sous la direction duquel paraissent plusieurs thèses qui attirent

(1) Cf. G. BAZIN, *Vie de Mgr Maret*, t. I, chap. xxvi et t. II, chap. i.
(2) Sur ces difficiles tractations, voir J. MAURAIN, *La politique ecclésiastique du Second Empire*, p. 104-110, 202-206 et 688-692.
(3) Cf. J. LEVIE, *La Nouvelle Revue théologique. A travers 60 années*, dans *Nouvelle Revue théologique*, t. LVI, 1929, p. 785-799.
(4) J.-F. VON SCHULTE, *Die Geschichte der Quellen und Literatur des canonischen Rechts*, t. III, Stuttgart, 1880, p. 295.

l'attention du monde savant, n'est pas le seul à bénéficier de ce contact avec d'autres facultés. La théologie en profite aussi et, grâce au développement des sciences auxiliaires, archéologie chrétienne et paléographie avec Reusens, orientalisme surtout avec Beelen et Lamy, elle peut s'établir sur des bases solides, et sans produire des hommes de la valeur de Kuhn, Hefele ou Franzelin, se tenir à la hauteur des exigences scientifiques en matière d'exégèse ou de patrologie, conformément à la tradition inaugurée par Malou [1], dont l'érudition patristique et le sens de la tradition apparurent en pleine lumière dans son célèbre ouvrage sur l'Immaculée Conception, publié en 1857.

LA RÉSURRECTION DES BOLLANDISTES   C'est grâce à l'intervention du recteur de l'Université de Louvain que les jésuites belges avaient repris, en 1837, l'œuvre des anciens bollandistes, interrompue depuis un demi-siècle [2]. Les premiers temps furent difficiles : la bibliothèque n'existait plus, la tradition était rompue, plusieurs collaborateurs de valeur disparurent prématurément, la piété peu éclairée d'une partie du clergé français et italien s'inquiétait de voir certaines traditions plus ou moins vénérables bousculées par les nouveaux « dénicheurs de saints ». L'œuvre réussit pourtant peu à peu à se réorganiser grâce au P. Victor de Buck, « le plus puissant travailleur que le bollandisme ait possédé depuis Papebroch ». Sous sa direction parurent successivement les cinq derniers tomes des *Acta Sanctorum* d'octobre, tandis que l'horizon s'élargissait au delà du monde gréco-romain au profit des sources slaves ou de l'hagiographie celtique. Par la trempe exceptionnelle de son talent, son activité infatigable, la génialité de son coup d'œil, sa faculté d'assimilation indéfinie toujours prête aux tâches les plus imprévues, le P. de Buck fut « dans toute la force du terme l'homme providentiel, taillé à souhait pour atténuer les inconvénients de la production hâtive et surmenée à laquelle lui et ses compagnons se trouvaient condamnés » [3].

Mais, à sa mort, en 1876, on dut bien reconnaître que le bollandisme n'avait pas encore retrouvé sa tradition d'école et que, jusque-là, son activité avait trop reposé « sur les inspirations soudaines et la prodigieuse rapidité de travail d'un génial improvisateur » [4]. Le terrain était toutefois préparé pour lui donner son assiette définitive : ce devait être l'œuvre du P. de Smedt [5], dont les *Principes de critique historique*, publiés une première fois en 1869 dans *Les Études* de Paris, avaient amorcé pour les pays de langue française la révolution entamée depuis une vingtaine d'années en Allemagne et en Angleterre par des hommes comme Doellinger ou John Acton.

---

(1) Sur l'œuvre théologique de Jean-Baptiste Malou (1809-1864), professeur de dogmatique à Louvain de 1837 à 1849, puis évêque de Bruges, voir *Annuaire de l'Université de Louvain*, 1865, p. (257)-(272). C'est Malou qui appela comme professeur à son séminaire de Bruges l'allemand Bernard Jungmann, un élève de Franzelin, qui devait, à partir de 1871, renouveler l'enseignement de l'histoire ecclésiastique et de la patrologie à Louvain.
(2) P. Peeters, *L'œuvre des Bollandistes*, Bruxelles, 1942, chap. vii à ix.
(3) Id., *Figures bollandiennes contemporaines*, Bruxelles-Paris, 1948, p. 19.
(4) Id., *L'œuvre des Bollandistes*, p. 111.
(5) Sur Charles de Smedt (1831-1911), voir P. Peeters, *Figures bollandiennes contemporaines*, p. 11-26.

§ 4. — Les efforts d'adaptation de l'apologétique : Dechamps et Newman.

LE XIXe SIÈCLE, SIÈCLE DE L'APOLOGÉTIQUE

Le XIXe siècle avait débuté sous le signe d'une révolte du monde cultivé contre le christianisme, héritage de l'action conjuguée des déistes anglais et des encyclopédistes français. Après le bref réveil de religiosité provoqué par le romantisme, ce mouvement, qui procédait moins d'une difficulté morale à se plier devant les exigences de la religion que de difficultés proprement intellectuelles, reprend de plus belle et, tandis que redoublent les objections contre les notions de surnaturel et de révélation, les titres historiques du christianisme sont à nouveau mis en discussion au nom des exigences de la critique, en attendant que les découvertes de l'orientalisme et les débuts de l'histoire comparée des religions posent le problème de sa transcendance. En même temps, favorisée par les aspirations sociales qui se font jour, s'élabore peu à peu une « religion de l'humanité » dont l'idéal, tout terrestre, s'oppose nettement à l'idéal chrétien. De tout cela résulte dans beaucoup d'esprits l'impression que le christianisme n'est qu'une manifestation de la pensée humaine, intéressante peut-être, mais dépassée et ne répondant plus aux exigences de l'homme de l'âge scientifique. Bientôt, ce seront les bases mêmes du théisme et l'idée de religion qui seront mises en question sous l'influence du positivisme.

Le centre principal de ce courant antichrétien se trouve à présent dans les universités allemandes : « L'Allemagne est la patrie de toutes les espèces d'incroyances, écrivait en 1861 l'un des observateurs les plus perspicaces du mouvement des idées en Europe. De tous les pays, c'est celui qui est le plus avancé sur la voie de l'incroyance ; elle nourrit l'irréligion des autres pays [1]. » Tandis que l'hégélianisme continue à recueillir de nombreux suffrages dans le monde philosophique et que la jeune gauche hégélienne, avec Feuerbach, prépare les voies à l'athéisme de Marx, les explications matérialistes du monde et de la vie sont répandues dans le grand public par des hommes de science comme le médecin Büchner, dont le livre célèbre *Force et Matière* est de 1855, le physiologiste Vogt et bientôt le biologiste Haeckel.

Parallèlement, Stuart Mill en Angleterre commence en 1843 à publier son *System of Logik*, Spencer publie de 1852 à 1857 ses *Principes de Psychologie*, et ces représentants d'un rationalisme scientifique, qui prolonge la tradition de l'empirisme anglais, ont un immense succès qui dépasse largement les frontières de leur pays. Ils contribuent à répandre une interprétation transformiste de l'univers inspirée de Darwin et, en réduisant pratiquement la philosophie à la psychologie expérimentale, ils popularisent l'idée que les réalités spirituelles et Dieu en particulier constituent pour l'homme le domaine de « l'inconnaissable ». En France aussi, nous l'avons déjà vu, s'élabore, sous le Second Empire, une philosophie

---

(1) John Acton à Simpson, 3 septembre 1861, dans dom GASQUET, *Lord Acton and his Circle*, Londres, 1906, p. 193.

positiviste où se rejoignent le courant issu de Comte, l'agnosticisme anglais et le matérialisme allemand. Même en Italie, où beaucoup de penseurs restent pourtant fidèles à la tradition de Gioberti ou de Rosmini et s'opposent à « l'invasion des barbares », l'hégélianisme allemand pénètre cependant, entre autres à Naples avec de Sanctis, tandis que le positivisme se développe sous l'influence combinée de Comte, de Haeckel et de Spencer.

Il ne faut pas s'étonner dans ces conditions de la prépondérance prise par l'apologétique au XIXe siècle. L'approfondissement du dogme contemplé pour lui-même, qui avait jusqu'alors paru la tâche essentielle de la théologie et dont les promoteurs de la théologie scolastique s'efforcent de souligner l'importance, semble à la plupart des penseurs catholiques une tâche moins urgente que celle qui consiste à défendre les bases du christianisme, et bientôt de la religion, contre leurs adversaires. Même lorsqu'ils font de la théologie proprement dite, ils le font encore dans une perspective apologétique. Les spéculations hardies d'Hermes ou de Günther, qui se rattachaient formellement à la dogmatique, avaient été en réalité commandées par le souci de rendre le dogme acceptable aux esprits conquis par la philosophie moderne ; et les savants travaux d'exégèse ou de théologie positive entrepris en Allemagne sous la direction des maîtres de Tubingue et de Munich visent moins à un « resourcement » au contact de l'Écriture et des Pères qu'à une défense des grandes thèses chrétiennes ou catholiques contre la critique rationaliste ou protestante.

Quant aux ouvrages d'apologétique proprement dite, leur nombre sans cesse grandissant témoigne d'un zèle apostolique des plus louables chez les écrivains catholiques, mais la plupart souffrent d'un manque presque total d'adaptation. Plus solides en Allemagne, plus superficiels en France, ils ressassent désespérément les mêmes arguments classiques, qui ne manquent souvent pas de valeur objective, mais ne mordent plus guère sur les esprits de moins en moins préparés à les recevoir par suite de l'atmosphère intellectuelle du temps.

Quelques auteurs toutefois font exception, et tentent de se placer au point de vue de ceux qu'ils cherchent à convaincre, annonçant ainsi le renouveau des méthodes apologétiques qui sera l'œuvre de la génération suivante. Il faut surtout signaler, parmi les praticiens, quelques prédicateurs français et, parmi ceux qui sont en même temps des théoriciens, le rédemptoriste belge Victor Dechamps et, les dominant tous de loin par la puissance de son génie et la fécondité de ses vues d'avenir, Newman.

L'APOLOGÉTIQUE MORALE
EN FRANCE
Personne en France, au cours du Second Empire, n'aborde avec compétence le problème capital de la conciliation entre la foi chrétienne et la mentalité scientifique nouvelle. Plusieurs par contre cherchent à montrer par l'appel à l'expérience des faits que le christianisme est loin d'être aussi démodé que le prétendent ses adversaires. Ainsi le P. Félix, qui de 1853 à 1869 esquisse dans la chaire de Notre-Dame une « apologétique réaliste » [1]. Constamment centré sur les réalités de

(1) P. FERNESSOLE, *Les Conférenciers de Notre-Dame*, t. II, Paris, 1936, p. 122.

son temps, dont il sait faire de pénétrantes analyses, il s'efforce, en confrontant le christianisme avec l'idée de Progrès, qui soulève l'enthousiasme de ses contemporains, de montrer que, loin de contrarier aucune aspiration légitime de l'humanité, il est au contraire la seule voie par laquelle celle-ci pourra atteindre ce vers quoi elle tend obscurément. Souligner l'harmonie entre le christianisme et les grandes aspirations de l'homme moderne, tel est également l'un des sujets préférés du P. Hyacinthe Loyson dans les sermons très goûtés qu'il prêcha jusqu'à la veille de son apostasie.

LA « MÉTHODE DE LA PROVIDENCE » DU P. DECHAMPS

Dechamps[1] est aussi un prédicateur, mais qui cherche à utiliser les leçons de son expérience pastorale, reflet des voies de la Providence, pour renouveler l'apologétique courante. S'adressant à des déistes qui rejettent *a priori* la possibilité d'une révélation surnaturelle et refusent en conséquence de reconnaître l'origine divine de l'Église, il élabore progressivement à partir de 1837 une méthode qui s'inspire de saint Augustin, de Pascal, de Bossuet, et qui n'est pas sans analogie avec celle que Lacordaire développe à la même époque d'une manière indépendante. Cette méthode qu'il exposera, puis défendra contre des critiques parfois assez vives dans une série d'ouvrages publiés de 1857 à 1874, se ramène essentiellement à la constatation de deux phénomènes corrélatifs : « Il n'y a que deux faits à vérifier, un en vous et un hors de vous. Ils se recherchent pour s'embrasser et, de tous les deux, le témoin, c'est vous-même[2]. »

Dechamps a compris que rien n'intéresse plus l'homme moderne que lui-même et qu'une apologétique adaptée doit partir des aspirations de l'homme. Aussi commence-t-il par lui faire prendre conscience de son impuissance à accomplir le Bien et à résoudre les questions fondamentales de la destinée, pour l'aider ensuite à découvrir dans son âme un désir de « relations positives, directes, vivantes avec Dieu » ; tel est le « fait intérieur » qui constitue la « pierre d'attente » de la révélation surnaturelle. La question religieuse posée de la sorte, de manière non plus seulement théorique mais vivante, pourra être résolue grâce au « fait extérieur », qui est constitué par l'Église catholique dans sa réalité actuelle : celle-ci apparaît manifestement comme la réponse du ciel à l'appel de l'homme, non seulement parce que son message répond à nos questions et apaise notre inquiétude, mais surtout parce qu'elle « porte ses lettres de créance, écrites de la main du Dieu vivant, dans les caractères mêmes dont il l'a revêtue »[3]. Ces caractères, qui font preuve, non en tant qu'ils sont signalés par l'Écriture, mais parce qu'ils constituent un miracle moral, impliquant l'intervention divine, sont la catholicité ou l'unité de l'Église dans le temps et dans l'espace, et surtout la sainteté de l'Église.

---

(1) Sur Victor Dechamps (1810-1883), l'un des plus célèbres prédicateurs belges, devenu successivement provincial des Rédemptoristes, évêque de Namur et enfin, en 1867, archevêque de Malines, voir, en attendant la biographie que prépare le P. Becqué, H. Saintrain, *Vie du cardinal Dechamps*, Tournai, 1884 ; M. Becqué, *L'apologétique du cardinal Dechamps*, Paris-Louvain, 1949 ; A. Deboutte, *De apologetische methode van Kardinaal Dechamps*, Bruges-Louvain, 1945.

(2) Épigraphe des *Entretiens sur la démonstration catholique de la révélation chrétienne*.

(3) *Entretiens sur la démonstration catholique...*, *Œuvres complètes*, t. I, Malines, 1874, p. XIII.

Malgré le souci d'introduire dans l'apologétique catholique l'appel
à la psychologie et le point de vue du sujet, ce qui intéresse pourtant
avant tout Dechamps, c'est le fait extérieur, le fait de l'Église comme
principal motif de crédibilité rendant inutile le recours à l'enquête érudite
que suppose nécessairement la méthode suivie par les manuels classiques [1].
Les réflexions sur le fait intérieur ne sont qu'une introduction, une « pièce
secondaire » utile mais non indispensable. Le grand mérite de Newman
sera précisément de centrer résolument son étude de l'acte de foi sur les
conditions subjectives de celui-ci et d'en présenter une analyse plus
nuancée que celle de Dechamps, qui paraît souvent « rapide et superficielle, moins l'œuvre d'un spécialiste que d'un prédicateur » [2].

*NEWMAN ET LA
GRAMMAIRE DE L'ASSENTIMENT »*
L'analyse psychologique de l'adhésion inconditionnée du croyant et la
justification rationnelle de celle-ci
ont fait l'objet des préoccupations de Newman à travers toute son existence [3]. Déjà à Oxford, il avait abordé la question devant les héritiers du
rationalisme du xviiie siècle dans des sermons très remarqués, réunis en
volume en 1843. Il y revint avec une attention croissante au fur et à mesure
que ses contacts apostoliques et sa volumineuse correspondance lui révélaient les progrès d'un esprit de scepticisme à base de science et de critique,
prétendant rejeter toute certitude qui ne fût pas du type mathématique
et considérant que seuls des facteurs d'ordre sentimental pouvaient expliquer la fermeté de la foi des catholiques à un système doctrinal considéré
comme intangible. En 1864, après qu'il eut, dans l'*Apologia*, décrit minutieusement son itinéraire spirituel et ses raisons de croire, on l'invita
à écrire une œuvre où il montrerait « comment il établit le pont par lequel
il s'avance si délibérément de l'état de doute qui semble s'attacher inévitablement à des conclusions fondées sur de simples probabilités, jusqu'à
l'état d'absolue certitude qu'il semble y substituer » [4]. Newman releva
le défi et rédigea une première ébauche pendant l'été de 1866, mais le
travail ne le satisfit pas ; il le reprit jusqu'à dix fois et ce n'est qu'au début
de 1870 qu'il le publia enfin, sous le titre un peu ésotérique de *Grammaire
de l'Assentiment*.

Newman n'est pas un constructeur de système, il est trop attiré par
l'observation du concret. Aussi cet ouvrage, d'une composition assez
déconcertante, est-il « l'un des plus obscurs qui aient jamais été écrits » [5].

(1) Si Dieu s'est vraiment manifesté à l'humanité, écrit Dechamps, « il ne se peut qu'il faille
s'enfermer dans une bibliothèque pour s'en apercevoir ». (*Lettres philosophiques et théologiques*, O. C.,
t. XVI, p. 192.)
(2) M. Becqué, *Le cardinal Dechamps et le cardinal Newman*, dans *Nouvelle Revue théologique*,
t. LVII, 1945, p. 1146.
(3) Sur le thème newmanien de la connaissance religieuse, voir surtout, outre l'article *Newman*,
dans le *D. T. C.*, M. Nédoncelle, *La philosophie religieuse de J.-H. Newman*, Strasbourg, 1946,
et H. Walgrave, *Kardinaal Newman's theorie over de ontwikkeling van het dogma in het licht van
zijn kennisleer en zijn apologetiek*, Anvers, 1944.
(4) W. Froude à Newman, octobre 1864, dans G.-H. Harper, *Cardinal Newman and William
Froude. A Correspondence*, Baltimore, 1933, p. 180-181. W. Froude était le type du savant agnos-
tique et la correspondance que Newman entretint avec lui pendant de longues années éclaire admirablement le contexte intellectuel dans lequel fut écrite la *Grammar of Assent*.
(5) J. Bacchus, *How to read the Grammar of Assent*, dans *The Month*, t. CXLIII, 1924, p. 106.

Une première partie cherche à répondre à la question, souvent posée sous l'influence du protestantisme libéral : si le but de la religion est d'exciter des sentiments de dévotion envers Dieu, à quoi bon les formules dogmatiques abstraites auxquelles l'Église semble attacher tant d'importance ? Le principe de la réponse consiste à montrer, à l'aide de la célèbre distinction entre assentiments « réels » et « notionnels », que dans la foi dogmatique l'adhésion ne porte pas uniquement ni même principalement sur des abstractions.

Puis vient l'examen de la double difficulté soulevée par les rationalistes : de quel droit l'Église exige-t-elle du croyant d'adhérer à ses dogmes avec une certitude absolue, alors que jamais nous ne pouvons démontrer de manière strictement scientifique toutes les prémisses sur lesquelles se base l'apologétique ? Et *a fortiori*, comment peut-on considérer comme légitime aux yeux de la raison la foi des innombrables croyants trop peu instruits pour procéder à un examen scientifique des preuves du christianisme ? La solution de Newman ne consiste nullement, comme on l'a dit souvent, à sacrifier la nécessité d'une justification rationnelle de la foi pour lui substituer une quelconque illumination ou un coup de pouce donné par la volonté sous l'influence de l'amour. Ses soi-disant sarcasmes contre la raison ne sont que des critiques du rationalisme et ce qu'il sacrifie, c'est uniquement l'expression technique et savante du raisonnement. A son avis, la pensée réelle est autrement complexe que le raisonnement verbal, exprimé, qui ne la traduit jamais que partiellement et superficiellement : elle prend pour prémisses l'expérience totale de la vie et saisit la convergence de multiples indices grâce à une faculté de synthèse spontanée, dénommée *Illative sense*.

De même qu'il ne sacrifie pas la vraie raison, Newman ne rejette pas non plus les arguments de l'apologétique classique, les preuves historiques. Mais il a compris qu'ils ne pouvaient convaincre concrètement que dans la mesure où ils viennent à la rencontre de la « dialectique de la conscience », des appels et des pressentiments qu'éveille en l'homme la considération du tragique de sa situation dans le monde. Newman élabore de la sorte une véritable apologétique « existentielle » [1], parce que, plus clairement que Dechamps, il a compris que le vrai problème ne consiste pas seulement à éveiller l'intérêt, mais à rendre possible l'assimilation par le sujet des preuves objectives.

Malheureusement, vers 1870, personne n'était préparé à s'inspirer de ces vues dont Newman, avec une rare divination, avait pressenti l'importance pour l'époque qui venait, et c'est avec une nostalgie mêlée d'amertume qu'on se dit avec Jean Guitton : « Newman n'a pas eu d'écho. S'il avait créé une école, on aurait vu se développer une psychologie de la pensée implicite et de la vie profonde, une logique de la conviction, une sociologie de l'idée et de l'influence, une métaphysique de l'ordre moral, une spiritualité de l'individuel, une histoire du sentiment religieux, une théologie à la fois dogmatique, psychologique et positive [2]. »

---

(1) J.-H. WALGRAVE, *Newman's verantwoording van het geloof in de Kerk*, Anvers, 1946, p. 30.
(2) J. GUITTON, *La Philosophie de Newman*, Paris, 1933, p. XXXIX-XL.

CHAPITRE VIII

CATHOLICISME ET LIBÉRALISME AU MILIEU DU XIXe SIÈCLE [1]

§ 1. — Controverses entre catholiques autour du libéralisme.

CATHOLICISME ET LIBÉRALISME AU LENDEMAIN DE 1848

Commencée en idylle, avec le mouvement réformiste et national dans la péninsule italique et avec la victoire de la démocratie romantique sur la bourgeoisie anticléricale à Paris,

(1) BIBLIOGRAPHIE. — I. SOURCES. — On consultera surtout :

a) les journaux et revues, notamment *L'Ami de la Religion*, le *Français*, le *Correspondant*, *Les Études*, le *Mediator*, le *Rambler*, le *Journal de Bruxelles* du côté libéral, et dans l'autre camp *La Civiltà cattolica*, l'*Unità*, la *Dublin Review*, *Le Bien Public* (Gand), *Le Monde* et *L'Univers* (les principaux articles de Veuillot ont été reproduits, parfois avec quelques modifications, dans les *Mélanges*, nouv. édit. en 14 vol., Paris, 1933-1940) ;

b) les œuvres des principaux protagonistes, notamment Mgr Pie (*Œuvres*, Paris, 1879, 9 vol.), Mgr Dupanloup (*Œuvres choisies*, Paris, 1861, 4 vol. ; *Nouvelles œuvres choisies*, Paris, 1873-1875, 4 vol.), Montalembert (*Œuvres polémiques et diverses*, t. II et III, Paris, 1868), Falloux (*Discours et Mélanges*, Paris, 1882), J. Donoso Cortes (*Obras*, Madrid, 1903-1904, 5 vol.), Ketteler (*Freiheit, Autorität und Kirche*, Mayence, 1862), Gioberti (*Opere inedite*, édit. MASSARI, Turin, 1856-1861, 6 vol.) ; ajouter J. MOREL, *Somme contre le catholicisme libéral*, Paris, 1877, 2 vol. ; J. COGNAT, *Polémique religieuse. Quelques pièces pour servir à l'histoire des controverses de ce temps*, Paris, 1861, et les différents *Mémoires* rédigés par Mgr Maret (reproduits dans G. BAZIN, *Vie de Mgr Maret*, Paris, 1891, 3 vol.) ;

c) les correspondances de Veuillot (Paris, 1931-1932, 12 vol.), de Mgr Pie et Mgr Cousseau (édit. HIOU, Paris, 1894), de Mgr Berteaud (R. FAGE, *Lettres inédites de Mgr Berteaud au cardinal Pitra*, Brives, 1919), de Mgr Dupanloup (édit. LAGRANGE, Paris, 1888, 2 vol.), d'Aug. Cochin (édit. H. COCHIN, Paris, 1926, 2 vol.), de Lacordaire et Mme Swetchine (édit. FALLOUX, Paris, 1864), de Montalembert (spécialement à l'abbé Texier, Paris, 1899, à Léon Cornudet, Paris, 1905, et à l'abbé Delor, dans *Revue de Paris*, juin 1902), de John Acton (cf. p. 244, n. 1), de Pantaleoni (*La Questione Romana negli anni 1860-1861. Carteggio Cavour-Pantaleoni*, Bologne, 1929, 2 vol.) ;

d) les Mémoires d'A. de Broglie (t. I, Paris, 1938, chap. IX : Le groupe et l'action du Correspondant), d'A. de Falloux (*Mémoires d'un royaliste*, Paris, 1888, 2 vol., à confronter avec E. VEUILLOT, *Le comte de Falloux et ses Mémoires*, Paris, 1888), ainsi que les souvenirs de Mgr Mabile (édit. P. MABILE, Paris, 1926, 2 vol.) ; en outre pour l'Italie, ceux de Massimo d'Azeglio (*I miei ricordi e lettere*, édit. BARBERA, Florence, 1920) et de MINGHETTI (*Miei ricordi*, Turin, 1889-1890, 3 vol.).

II. TRAVAUX. — Bon exposé d'ensemble par C. CONSTANTIN, art. *Libéralisme catholique*, dans *D. T. C.*, t. IX (1926), col. 506-629.

Pour la France : outre les biographies de Veuillot, Pie, Guéranger, Gerbet, Dupanloup, Lacordaire, Ravignan, Darboy, Maret (indiquées, p. 8) et celle, capitale, de Montalembert par E. LECANUET (t. III : *L'Église et le Second Empire*, Paris, 1902), voir G. WEILL, *Histoire du catholicisme libéral en France*, Paris, 1909 ; J. FÈVRE, *Histoire critique du catholicisme libéral en France jusqu'au pontificat de Léon XIII* (réquisitoire partial mais bien informé) ; J. MAURAIN, *La politique ecclésiastique du Second Empire*, Paris, 1930, ainsi que L. OLLÉ-LAPRUNE, dans *La Vitalité chrétienne*, Paris, 1901, p. 41-59 (jugement nuancé).

Pour la Belgique : les biographies d'Ad. Dechamps et du cardinal Sterckx (Cf. p. 9).

Pour l'Italie : outre les ouvrages de A. DELLA TORRE et de S. JACINI, signalés p. 9, voir surtout A.-C. JEMOLO, *Chiesa e Stato in Italia negli ultimi cento anni*, Turin, 1949, chap. II et III, et A. PIETRA, *Storia del Movimento cattolico liberale*, Milan, 1948.

Pour l'Allemagne : les ouvrages généraux de G. GOYAU et de BRUECK-KIESSLING, ainsi que les biographies de Ketteler et de Doellinger, indiqués p. 9.

Pour l'Angleterre : W. WARD, *W. G. Ward and the catholic Revival*, Londres, 1893, chap. VI à VIII ; M. WARD, *The Wilfrid Wards and the Transition*, Londres, 1934, chap. I à III ; A. GASQUET, *Lord Acton and his Circle*, Londres, 1906, ainsi que les biographies de Manning, Newman et Ullathorne (cf. p. 9).

la révolution européenne de 1848 s'était achevée en tragédie avec les journées de juin en France, les convulsions sanglantes de Vienne, de Francfort et de Hongrie, la révolution romaine et l'exil de Pie IX. Et comme les mouvements révolutionnaires, qu'ils fussent de nature politique, sociale ou nationale, s'étaient partout réclamés ouvertement des principes de 1789, la plupart de ceux pour qui l'ordre social était la préoccupation fondamentale se tournèrent à nouveau vers les « prophètes du passé », qui n'avaient cessé de vanter la prépondérance reconnue par l'Ancien Régime au « principe d'autorité » et de revendiquer pour la vérité, représentée par l'Église et sa doctrine, un régime privilégié par rapport à l'erreur, source de catastrophes.

Cette crise ne pouvait que rendre plus aigu le grand problème avec lequel la pensée catholique se trouvait confrontée depuis un demi-siècle : l'attitude à prendre à l'égard du monde issu de la Révolution. Pouvait-on s'en accommoder ou devait-on le rejeter comme intrinsèquement mauvais ? Plus spécialement, comment fallait-il se comporter à l'égard du régime des libertés modernes, libertés politiques, liberté de presse, de conscience, des cultes : était-ce un progrès à encourager, une situation de fait inévitable à utiliser au mieux des intérêts de l'Église, ou un mal à combattre sans compromission ?

Le problème, toujours épineux de nos jours, l'était bien plus encore il y a un siècle, car les théologiens commençaient à peine à repenser les principes qui permettraient, moyennant les discernements et les purifications nécessaires, d'assimiler au christianisme les idées de démocratie, d'égalité, ou de liberté, et ce travail d'adaptation était d'autant plus délicat qu'il ne pouvait être question, comme quelques-uns se le figuraient candidement, de reprendre telles quelles ces idées, nées en dehors de l'Église et qui s'étaient souvent développées dans un esprit hostile à son égard [1]. Il ne faut pas oublier, en effet, que l'ancien régime était pétri d'influences chrétiennes et qu'il faisait à l'Église sa place au cœur même de la vie nationale. Les promoteurs d'un régime politique et social nouveau « ne pouvaient (donc) guère aboutir qu'en combattant les influences d'Église, et parfois même le christianisme, qui adhéraient à ce passé ». Par contre, beaucoup de catholiques, hypnotisés par le souvenir de la réussite exceptionnelle de la chrétienté médiévale, étaient portés à croire que la restauration chrétienne, dont l'urgence apparaissait plus grande encore au sortir de la crise de 1848, supposait nécessairement un retour intégral à ce passé, dont les institutions de l'ancien régime leur apparaissaient comme le prolongement. Ce retour en arrière paraissait d'autant moins invraisemblable, au milieu du XIXe siècle, que, disparu depuis quelques années à peine en Europe occidentale, le régime ancien subsistait encore partiellement dans l'Espagne d'Isabelle II, dans l'Empire des Habsbourg, en Bavière, dans l'Italie centrale et méridionale, et surtout dans l'État pontifical.

---

(1) Cf. Y. Congar, *Vraie et fausse Réforme dans l'Église*, Paris, 1950, p. 345-346, 562-569, 604-622. J'emprunte plusieurs expressions à cet ouvrage perspicace.

LA RÉACTION ANTILIBÉRALE   C'est en partie sous l'effet de la peur, ou du moins de la prudence, que de nombreux catholiques et la hiérarchie dans son ensemble s'associèrent à la vague de réaction qui submergea l'Europe au lendemain de 1848. Mais des penseurs entreprirent de justifier et de confirmer cette attitude instinctive par des considérations doctrinales et théologiques. Les plus influents furent l'Espagnol Donoso Cortès, l'équipe des jésuites italiens de *La Civiltà cattolica* et le célèbre journaliste français Louis Veuillot.

Donoso Cortès [1] s'était laissé tenté dans sa jeunesse par les idées révolutionnaires, mais pour les abandonner bientôt, quand il eut constaté que, loin de libérer l'homme, elles n'aboutissaient qu'à substituer à l'absolutisme du gouvernement légitime une dictature plus pesante encore. « Puisqu'il faut opter entre la dictature du poignard et la dictature du sabre, proclama-t-il dès lors, mon choix est fait : je choisis celle du sabre. » Mais, élargissant le problème au delà de l'aspect politique, il entreprit de dénoncer, avec une tournure d'esprit rappelant les théologiens de la contre-réforme espagnole du xvɪᵉ siècle, « l'abîme insondable, l'antagonisme absolu » existant entre la civilisation moderne et le christianisme [2]. Son *Essai sur le catholicisme, le libéralisme et le socialisme* (1851), rapidement traduit en français, puis en allemand, chaudement recommandé par son ami Veuillot, produisit partout une profonde impression.

Ce catholicisme intransigeant et autoritaire devait trouver son organe officiel dans la revue romaine *La Civiltà cattolica*. L'idéal qui avait présidé à la fondation de celle-ci était la restauration intégrale des principes chrétiens dans la vie individuelle, familiale, sociale et politique, qui apparaissait à Pie IX comme la tâche essentielle de son pontificat, et la revue a certainement fourni une importante contribution à cette œuvre grâce à des collaborateurs comme le P. Taparelli d'Azeglio, l'un des plus remarquables philosophes politiques catholiques du siècle dernier. Mais on doit déplorer la fâcheuse tendance de plusieurs rédacteurs à confondre trop souvent la restauration chrétienne ou la fidélité catholique avec le conservatisme politique [3].

LOUIS VEUILLOT   Le même idéal de restauration chrétienne sous le signe de l'absolutisme et de l'ancien régime inspire Louis Veuillot [4], le rédacteur en chef de *L'Univers*, dont l'influence

---

(1) Sur Juan Donoso Cortès, marquis de Valdegamas (né en 1809, mort comme ambassadeur de France à Paris, le 3 mai 1853), voir É. Schramm, *Donoso Cortès. Su vida y su pensamiento*, Madrid, 1936.

(2) Donoso Cortès, *Œuvres*, Paris, 1862, t. I, p. 337 et 340.

(3) Voir les articles, à tendance apologétique, sur *La Civiltà cattolica e l'assolutismo politico*. *Ricordi*, dans *La Civiltà cattolica*, 1924, vol. II, p. 219 et suiv., 397 et suiv., 505 et suiv., et l'ouvrage polémique de B. Spaventa, *La politica dei Gesuiti nel sec. 16 e nel sec. 19*, édit. G. Gentile, Rome, 1911.

(4) On commence aujourd'hui à juger avec plus de nuances ce journaliste de génie (né en 1813, mort en 1883), mais une biographie impartiale manque encore. On trouvera une excellente esquisse par E. Amann, art. *Veuillot*, dans *D. T. C.*, t. XV, col. 2799-2835, et de nombreux matériaux dans l'apologie écrite par son frère E. Veuillot, *Louis Veuillot*, Paris, 1899-1913, 4 vol. Ses œuvres complètes ont été récemment republiées à la librairie Lethielleux : *Œuvres diverses*, 1924-1930, 14 vol. ; *Correspondance*, 1931-1932, 12 vol. ; *Mélanges*, 1933-1940, 14 vol.

déborde largement les frontières françaises et qui peut se vanter sans
mentir de la sympathie, de plus en plus déclarée, de Pie IX et de son
entourage [1]. Autoritaire par tempérament, en dépit de ses origines plé-
béiennes, Veuillot prône comme idéal le pouvoir absolu s'appuyant sur
l'Église et l'Église s'abritant sous le pouvoir absolu. Toutes les occasions
lui sont bonnes pour développer cette thèse. Il loue Charlemagne pour
avoir édicté la peine de mort contre les Saxons qui refusaient le baptême,
et Louis XIV pour avoir révoqué l'édit de Nantes [2]. Impressionné par
le recul de la pratique religieuse et par les progrès de l'immoralité dans
le peuple au cours des dernières décades, il réclame le retour à l'antique
union des pouvoirs spirituel et temporel, jusqu'à la religion d'État inclu-
sivement, et il anathématise les « libertés modernes », dont il dénonce les
points faibles et les conséquences dommageables avec un incomparable
talent de pamphlétaire. Hostile à toute compromission, il confesse sa
foi « non seulement sans respect humain, mais avec bravade » [3], hérissant
la vérité pour la mieux défendre, la déformant même parfois à force
d'outrances, au risque de faire passer l'Église « pour l'ennemie de tout
ce qui passionne le siècle » [4].

Il importe, pour être juste, de noter que cette hostilité sans nuances
contre les principes de 1789 et les institutions modernes, si elle témoigne
parfois d'un esprit partisan et d'une certaine étroitesse intellectuelle,
provient avant tout, chez Veuillot, comme chez les jésuites de la *Civiltà*
et beaucoup de leurs partisans, d'un zèle alarmé pour la défense des valeurs
essentielles du catholicisme, que le mouvement d'idées issu de la Révolution paraissait mettre en danger. Certes, d'assez nombreux catholiques
s'exagéraient le danger par suite d'une mentalité « de droite » qui les
poussait à préférer l'immuable et le « tout fait » imposé d'en haut par
voie d'autorité, à la nouveauté, aux initiatives qui ne sont encore qu'aspiration qui se cherche, ou aux requêtes présentées au nom des « droits
de l'homme ». Mais il faut reconnaître que tout n'était pas chimérique
dans leurs appréhensions : *Quatre-vingt-neuf*, pour l'opinion libérale
d'alors, représentait généralement bien plus qu'un simple régime politique,
c'était un véritable « mythe », comportant entre autres le rejet de toute
soumission à une autorité dépassant la conscience individuelle, ce qui
impliquait le rejet de la souveraineté de Dieu sur la vie sociale et celui
de l'autorité de la Révélation et de l'Église sur la vie intellectuelle. C'est
ce que ne cessaient de rappeler les « chiens de garde de l'orthodoxie »,
avec une indignation d'où le sectarisme n'était malheureusement pas
absent, mais qui avait au moins le mérite d'attirer l'attention sur certaines
exigences imprescriptibles de la vérité que d'autres catholiques étaient
parfois tentés de perdre de vue.

(1) Pie IX n'ignorait pas le caractère excessif de la polémique de Veuillot et lui adressa à plusieurs reprises des conseils de modération, mais son dévouement au Saint-Siège touchait au plus
intime de lui-même l'homme essentiellement émotif qu'était le pape, au point qu'on a pu considérer
le grand polémiste français comme « l'enfant chéri de Pie IX ». (F. HAYWARD, *Pie IX*, p. 154.)
(2) Voir en particulier *L'Univers* des 6 et 12 novembre 1852 : *De la liberté sous l'absolutisme*.
(3) P. DE LA GORCE, *Histoire du Second Empire*, t. II, Paris, 1896, p. 154.
(4) L. OLLÉ-LAPRUNE, *La vitalité chrétienne*, p. 57.

LA PROTESTATION DES CATHOLIQUES LIBÉRAUX

Jusqu'à la fin du pontificat de Pie IX, *La Civiltà* et *L'Univers* allaient répéter, avec une intransigeance frisant parfois le fanatisme, leurs condamnations en bloc du monde moderne, radicalement vicié par l'idéologie libérale, et présenter comme les seules compatibles avec l'orthodoxie leurs conceptions politico-religieuses visant à obtenir pour l'Église un régime de privilèges au sein d'un État officiellement catholique et soustrait à la pression de l'opinion publique. Ces idées, qu'on savait partagées par les cercles influents de Rome, rencontrèrent une faveur marquée en de nombreux milieux, sans rallier pourtant l'unanimité des catholiques.

Les plus clairvoyants, en effet, faisaient observer qu'une partie importante de la classe dirigeante ayant en fait cessé d'être croyante, c'était une utopie que d'espérer encore de l'État aide et protection désintéressée pour l'Église et qu'une neutralité bienveillante était ce qu'on pouvait attendre de mieux. Or, sentant que la réaction consécutive à 1848 n'empêchait pas le libéralisme d'avoir pour lui l'avenir, ils s'inquiétaient de voir l'Église se solidariser avec un stade de civilisation qui serait bientôt dépassé et, en semblant poser en thèse l'incompatibilité de la civilisation moderne et du catholicisme, renouveler, sur un plan plus vaste, les vieilles difficultés soulevées jadis à propos de l'accord entre la raison et la foi. Quelques-uns entrevoyaient en outre, avec plus ou moins de lucidité, les valeurs authentiques et le réel progrès humain que représentait le courant libéral, en dépit de ses excès ; ils prenaient conscience du fait que la société civile parvenait lentement à sa majorité, et qu'un nouvel « humanisme » s'élaborait au milieu de tâtonnements inévitables. Ils étaient prêts à accueillir avec sympathie une conception plus moderne de l'homme, plus respectueuse des droits de la personnalité et de sa libre détermination, plus ouverte à ce que nous appellerions aujourd'hui « l'autonomie du temporel », sans d'ailleurs remarquer toujours le danger qu'il y avait d'exagérer cette autonomie et de revendiquer pour l'homme une trop grande indépendance, incompatible avec les droits de Dieu.

Deux groupes de catholiques allaient de la sorte s'opposer progressivement, aussi soucieux l'un que l'autre de servir l'Église, mais concevant de manière différente la façon la plus opportune de la servir. Les uns, entendant par « monde moderne » leur siècle en tant qu'époque historique ayant ses institutions, ses valeurs, ses espérances, estimaient qu'il fallait aller au devant de lui et lui montrer l'Église prête à s'incarner en lui comme elle l'avait fait aux époques antérieures avec la civilisation gréco-romaine, avec le mouvement communal et l'enthousiasme aristotélicien du XIIIᵉ siècle, avec les aspirations humanistes de la Renaissance. Les autres, entendant par « monde moderne » la part anticatholique de la Révolution, croyaient préférable, pour être sûrs de ne point pactiser avec l'erreur, de rompre les ponts et de renforcer les défenses afin d'empêcher ce « monde moderne » d'entrer dans l'Église, indifférents au risque qu'il y avait, en agissant ainsi, de se couper de ses contemporains et de renforcer encore l'équivoque en vertu de laquelle de nombreux libéraux se figuraient sincèrement qu'on ne pourrait bâtir une société conforme aux aspirations modernes qu'après avoir privé l'Église catholique de toute influence.

Entre ces deux groupes, les conflits étaient inévitables. Ils furent rendus plus aigus encore par l'interférence d'autres discussions qui, au même moment, divisaient également les élites catholiques des principaux pays : en France, où les problèmes politiques prennent facilement une portée philosophique et religieuse, par la question du ralliement à l'Empire après le coup d'État ; en Italie, par la Question romaine, étant donné que le gouvernement piémontais cherchait à légitimer son œuvre d'unification de la patrie et de laïcisation de ses institutions au nom des principes modernes ; en Allemagne, par l'opposition croissante entre les théologiens universitaires et les défenseurs de la scolastique, où l'on retrouvait la même opposition de mentalité entre ceux qui faisaient confiance au progrès et avaient le sens de l'histoire, et ceux qui préféraient se fier à des méthodes ayant déjà fait leurs preuves et estampillées par l'autorité hiérarchique.

On finit par en arriver à la constitution de véritables partis, ce qui acheva d'exacerber les passions, car, pour reprendre une observation du P. Congar, « de personne à personne, l'opposition qui tient à la manière de sentir est à peu près insurmontable ; elle le devient tout à fait quand les personnes forment un groupe : un groupe en face d'un autre groupe apparaît toujours comme redoutable ; il suscite la psychose du complot et de la méfiance » [1]. Tout au long du pontificat de Pie IX, on put constater ce qui devait se reproduire encore à plusieurs reprises par la suite : « D'un côté, les catholiques intégristes redoutent sans cesse que par les ponts-levis baissés, l'ennemi ne pénètre dans la place ; ils craignent que les autres ne pactisent avec l'erreur et flairent partout des relents d'hérésie. De l'autre côté, les catholiques ouverts, avertis par des expériences qui n'ont rien d'imaginaire, soupçonnent sans cesse les intégristes de les dénoncer à Rome : ce qui, il faut bien le comprendre, suscite en eux des sentiments de méfiance nuancée d'un certain mépris ; mépris qui se nourrit volontiers du sentiment qu'il y a, chez les dits intégristes, une énorme ignorance de l'histoire et des *a priori* que permet seule cette ignorance. Ce dont, à leur tour, les intégristes ont sourdement conscience, souffrent, s'irritent, et cherchent une surcompensation par un redoublement de rigueur dogmatique, de méfiance et d'avertissements. »

LA DIVISION DES CATHOLIQUES EN FRANCE

Cette constitution de partis rivaux se produisit en France dès la fin de la Seconde République, à la suite de la loi Falloux. Celle-ci ayant été, comme on l'a vu, désavouée par Veuillot et ceux qui ne pouvaient se résigner à un compromis où l'Église n'obtenait pas pleine satisfaction, les promoteurs de la loi entreprirent de la défendre et, pour la justifier, Dupanloup et Montalembert furent bientôt amenés à porter la question sur le terrain des principes. Les vieilles

---

(1) Y. CONGAR, *Vraie et fausse Réforme dans l'Église*, p. 612.

divergences de vue sur la société chrétienne idéale, qui étaient passées à l'arrière-plan depuis la condamnation de Lamennais, réapparaissaient de la sorte en pleine lumière. Elles s'accentuèrent encore au lendemain du coup d'État, lorsque la satisfaction avec laquelle avait été accueilli l'établissement du régime autoritaire se transforma chez certains catholiques et dans le clergé en une véritable adulation provoquée par l'espoir d'obtenir de la faveur impériale un régime privilégié qui mettrait l'Église en mesure de faire éventuellement appel à la contrainte légale pour remplir plus aisément sa mission.

Convaincu de ce que l'évolution des esprits rendait illusoire pareille tactique, Montalembert la dénonça, dans une brochure publiée en septembre 1852, *Les Intérêts catholiques au XIXᵉ siècle*, où il essayait de montrer, en simplifiant un peu les données historiques, que la renaissance religieuse en Europe au cours des cinquante dernières années avait été le résultat exclusif de la liberté dont l'Église avait joui sous les régimes parlementaires. Il serait d'autant plus impardonnable de l'oublier, ajoutait-il, que ceux-ci représentent l'avenir : « La cause de l'absolutisme est une cause perdue. Malheur à ceux qui voudraient enchaîner à cette idole décrépite les intérêts immortels de la religion ! »

Veuillot s'empressa de répliquer à cette « Marseillaise parlementaire » et désormais toutes les occasions lui seront bonnes pour dénoncer les « catholiques libéraux » au nom de ceux qu'il appelle les « catholiques sans épithète », représentant ses adversaires comme des gens qui, après avoir mesuré ce que la société moderne peut encore supporter de christianisme, voudraient inviter l'Église à s'y réduire, répétant sans se lasser qu'il faut exiger le respect intégral des droits de l'Église et le retour au régime privilégié qui fut jadis le sien, au lieu de se borner à souhaiter une extension de la liberté que la vérité partagerait avec l'erreur.

L'ÉCOLE DU « CORRESPONDANT »   Montalembert, fidèle à l'idéal exprimé par la formule célèbre : *L'Église libre dans l'État libre*[1], était doublement exaspéré par une conception qui non seulement voulait à tout prix obtenir de l'État un appui illusoire et funeste à ses yeux, mais qui recherchait cet appui auprès d'un État qu'il détestait entre tous, le régime dictatorial de Napoléon III. Sa seule consolation était de constater que ses invectives contre « cette école fanatique et servile, qui cherche à identifier partout l'Église avec le despotisme »[2], trouvaient un écho favorable chez ceux des catholiques qu'inquiétait la renaissance de l'anticléricalisme provoquée dans les milieux libéraux par les mesures gouvernementales en faveur de l'Église, mais, tant que dura la « lune de miel de l'Église et de l'Empire »[3], ceux qui pensaient ainsi ne représentèrent guère qu'un état-major sans troupes.

---

(1) Cette formule, que Montalembert employa pour la première fois dans une lettre ouverte à Cavour, datée du 18 octobre 1860, lui avait été suggérée par son ami Cochin (*Aug. Cochin. Ses lettres et sa vie*, t. I, p. 233) ; voir aussi E. GIACOMETTI, *Die Genesis von Cavour's Formel : Libera Chiesa in libero Stato*, Zurich, 1918.
(2) Montalembert à Cantù, 14 décembre 1854, dans E. LECANUET, *Montalembert*, t. III, p. 95.
(3) A. DANSETTE, *Histoire religieuse de la France contemporaine*, t. I, p. 373.

Ardemment désireux de contrebalancer l'influence de *L'Univers* et d'atteindre un public plus vaste que la petite élite des salons parisiens et de l'Institut, Montalembert décida, en 1855, de reprendre en mains *Le Correspondant*, une revue mensuelle languissante, dont le programme avait toujours été, depuis vingt-cinq ans, de prêcher l'alliance de l'Église et de la liberté, et il réussit à porter en peu de temps le nombre des abonnés de 700 à 3.000 [1], grâce à l'appui fidèle d'anciens amis et de forces jeunes : Falloux, toujours maladif et surtout toujours circonspect, mais que des instances répétées ont finalement décidé à sortir de sa retraite silencieuse ; Foisset, pieux magistrat dijonnais, modéré et clairvoyant, ami et confident de Lacordaire, qui, lui, depuis ses déceptions de 1848 et de 1851, se tient à l'écart ; Albert de Broglie, jeune historien de talent, que ses traditions de famille orientaient déjà vers le libéralisme et qui s'applique à montrer, dans ses études sur *L'Église et l'Empire romain au IVe siècle*, que la victoire du christianisme sur le paganisme, loin d'être le résultat d'une protection officielle ou d'une opposition violente à la civilisation romaine, fut due uniquement à des influences morales et au prestige de la vérité sur les âmes ; Augustin Cochin, enfin, chrétien exemplaire, homme d'œuvres d'une charité inépuisable et éclairée, caractère d'élite, esprit remarquable, « sachant tout exposer comme il savait tout comprendre » [2], modeste collaborateur de ses grands amis auxquels il fournira plus d'une fois la matière de leurs discours ou de leurs brochures [3], et entre lesquels il réussit à force de tact et d'abnégation à maintenir l'unité de vues et d'action. Tous ces hommes trouvent un conseiller, souvent aussi un chef, dans un prélat dont l'influence en France et bientôt en Europe s'accroît de jour en jour, Monseigneur Dupanloup.

MONSEIGNEUR DUPANLOUP Éducateur sans égal au dire de Renan, extraordinaire éveilleur d'âmes, confesseur très goûté du faubourg Saint-Germain, prédicateur apprécié quoique son éloquence nous paraisse aujourd'hui bien romantique, écrivain fécond exposant en larges développements oratoires des lieux communs entremêlés d'idées neuves et originales, Dupanloup était déjà célèbre quand Falloux le fit nommer, en 1849, à l'évêché d'Orléans [4]. Depuis lors, tout en administrant son diocèse d'une façon modèle — avec une autorité un peu despotique qui lui aliène certaines sympathies dans son clergé — il suit de près la vie de l'Église de France et de l'Église universelle, au point de se faire traiter par un de ses adversaires d'« évêque surtout des affaires étrangères à son diocèse » [5].

---

(1) La nouvelle série s'ouvrit, en janvier 1856, par un article programme, véritable manifeste, dû à Albert de Broglie (reproduit dans *Questions de religion et d'histoire*, t. II, Paris, 1860, p. 145 et suiv.). On trouvera une histoire de la revue dans *Le Correspondant* du 11 octobre 1929.

(2) L. Ollé-Laprune, *La vitalité chrétienne*, p. 50.

(3) Voir notamment sa lettre à Montalembert du 7 février 1863, dans *Augustin Cochin. Ses lettres et sa vie*, t. I, p. 326.

(4) Une biographie impartiale et exhaustive de Félix Dupanloup (1802-1878) fait encore défaut. En attendant, l'ouvrage fondamental reste celui de Mgr Lagrange, *Vie de Mgr Dupanloup*, Paris, 1883-1884 (à confronter avec le livre passionnément hostile d'U. Maynard, *Mgr Dupanloup et son historien*, Paris, 1884) ; voir aussi E. Faguet, *Mgr Dupanloup*, Paris, 1914.

(5) U. Maynard, *Mgr Dupanloup et son historien*, p. 66.

Homme d'action au tempérament impétueux et tout en contrastes, « prêchant la conciliation avec une vivacité sans pareille » [1], d'une vitalité
prodigieuse et d'une énergie jamais lasse, « sous la robe épiscopale, c'est
un soldat » [2] dont les brochures et les mandements « partent comme des
obus » [3] et qui, de l'aveu même d'un de ses intimes, « avec des intentions
pures, a parfois employé des moyens qui ont froissé » [4]. Passionnément
dévoué à l'Église, il est toujours parmi les premiers à dépister les nouveaux
dangers qui la menacent : il le montrera à propos de la Question romaine,
à propos des projets laïques de Duruy et en bien d'autres occasions :
à l'heure où Mgr Pie s'acharne encore contre le vieux Cousin, il s'insurge
déjà contre Taine et Littré.

Quoi qu'on en ait dit, il était au fond, d'esprit et de formation, plutôt
conservateur, détestant la démocratie, n'ayant guère de sympathie
pour la Révolution française et guère de compréhension pour les nouveautés intellectuelles. Mais, comme l'a fort bien diagnostiqué Faguet, si
« son esprit était autoritaire, son âme était tendre, d'abord, et de plus avait
une délicatesse, une pudeur, qui lui défendait, à l'égard d'une autre âme,
toute violence, toute précipitation indiscrète » [5]. Il se souvenait du mot
de saint Athanase : « Ce n'est pas avec l'aide des soldats et des javelots
qu'on prêche la vérité, mais par la persuasion et le conseil. » Et, comme il
le déclarait dans son discours de réception à l'Académie, dont l'esprit
de tolérance scandalisa dom Guéranger, ce qu'il cherchait avant tout,
« ce n'est pas ce qui sépare, mais ce qui rapproche, les points de départ
communs » [6]. Il lui répugnait par ailleurs de donner l'impression de n'avoir
jadis demandé la liberté que pour conquérir le pouvoir, et il comprenait
en outre, comme ses amis, à quel point une domination effective de l'Église
sur les pouvoirs publics était dorénavant impossible. Tout ceci explique
son inquiétude et son indignation en face des diatribes intransigeantes
de *L'Univers*, où l'on ne songeait qu'à exalter l'Inquisition, la théocratie
médiévale et la religion d'État. Lui, au contraire, s'appliquait inlassablement à rapprocher l'Église et le monde moderne et à montrer que le libéralisme, auquel celui-ci tient tant, peut être compris « en un sens parfaitement chrétien », se gardant bien, d'ailleurs, de faire rentrer dans les
points acceptables du programme libéral la sécularisation de l'État et
en particulier la laïcisation de l'enseignement, qu'il devait combattre
sans répit sa vie durant.

Malheureusement, avec ces idées somme toute fort modérées, Dupanloup donna plus d'une fois l'impression, dans l'entraînement de la polémique, de ne pas garder l'exacte mesure et d'avoir tendance à sacrifier
les exigences de la vérité au souci de la liberté. Sa formation théologique
assez rudimentaire l'exposait du reste à négliger parfois certaines distinctions ou à employer des expressions de nature à inquiéter les hommes

(1) F. Mourret, *Le mouvement catholique en France de 1830 à 1850*, Paris, 1917, p. 224.
(2) E. Ollivier, *L'Église et l'État au concile du Vatican*, t. I, p. 443.
(3) E. de Pressensé, dans *La Revue bleue*, t. XXXIV, 1884, p. 583.
(4) Gaduel à Lagrange, 26 novembre 1882, *Arch. Nat.*, AB XIX, 527.
(5) E. Faguet, *Mgr Dupanloup*, p. 149.
(6) Cité dans F. Lagrange, *Vie de Mgr Dupanloup*, t. II, p. 155.

de doctrine. De là les méfiances persistantes contre lui et ceux dont il apparaissait comme l'inspirateur, méfiances bien exprimées dans la lettre suivante de Mgr Parisis, son ancien compagnon de lutte dans la campagne pour la liberté d'enseignement :

Il y a longtemps que je déplore les illusions et les ravages de cette école des accommodements composée d'hommes généralement honorables, croyants et pratiquants, mais peureux et que Notre-Seigneur eût appelés *modicae fidei*, qui voyant la puissance et l'extension du rationalisme, tremblent pour l'Église de Dieu et se persuadent qu'elle doit par prudence faire à cet ennemi, prétendu nouveau, des concessions qu'elle n'a faites dans aucun temps [1].

L'ÉCOLE DE « L'UNIVERS »   Le groupe du *Correspondant* voyait croître son influence : il était bien accueilli par l'opinion éclairée et cinq de ses chefs entrèrent successivement à l'Académie, jadis foyer de voltairianisme ; il pouvait compter sur des sympathies dans l'aristocratie orléaniste et, ce qui était plus important, sur l'appui de plusieurs évêques ; il avait des adhérents dans les grands ordres, chez les dominicains, grâce à Lacordaire, à l'Oratoire autour du P. Gratry, même chez les jésuites, avec le P. de Ravignan et les premiers rédacteurs des *Études*. Malgré tout, les catholiques libéraux restèrent une minorité. D'assez nombreux catholiques de province et surtout la grande masse du clergé, dénués du sens des nuances nécessaire pour déceler la part de vérité qui se cherchait dans le libéralisme, ne pouvaient admettre que l'Église dût renoncer à des droits et privilèges qu'elle avait possédés sans conteste pendant des siècles ; leur simplisme tout autant que leur souci d'une rigoureuse orthodoxie, dont ils majoraient volontiers les exigences, s'accommodaient mieux des positions intransigeantes de doctrinaires comme Mgr Pie ou dom Guéranger, qui, plutôt que de chercher les possibilités d'adaptation à des situations concrètes éminemment complexes, préféraient « se placer toujours au centre des principes pour en déduire les conséquences logiques » [2].

Mgr Pie [3] est un homme de tradition. On s'en douterait rien qu'à voir son allure majestueuse, qui contraste avec celle de Dupanloup, « l'homme au pas rapide ». Tandis que celui-ci plaide la cause du Christ et de l'Église dans des articles de journaux, des brochures ou des livres qui ne sont que des improvisations prolongées, Mgr Pie enseigne doctoralement en évêque et en juge de la foi. Sa formation théologique, sans avoir aucune originalité, est incontestablement plus solide que celle de l'évêque d'Orléans, et il a un sens très juste des principes. Mais s'il sait faire preuve dans l'action d'une modération dont manquent beaucoup de ses amis, que Montalembert n'avait pas tort de baptiser « nos modernes inquisiteurs », il est au fond pénétré, lui aussi, « d'un redoutable intégrisme » [4]. Ardent légitimiste, il espère bien voir réalisé un jour par le comte de Chambord l'État chrétien tel que le moyen âge l'avait conçu et, en attendant, il

---

(1) Parisis à Guéranger, 26 octobre 1856, dans dom Delatte, *Dom Guéranger*, t. II, p. 146-147.
(2) C. Constantin, art. *Libéralisme catholique*, dans *D. T. C.*, t. IX, col. 581.
(3) Sur Louis Pie (1815-1880), voir Mgr Baunard, *Histoire du cardinal Pie*, Paris, 1885, 2 vol.
(4) E. Amann, dans *D. T. C.*, t. XV, col. 2826.

dénonce inlassablement le « naturalisme », celui qui nie l'ordre surnaturel, mais aussi celui qui veut écarter Dieu et l'Église des affaires de ce monde et dont il croit voir dans le libéralisme catholique une manifestation d'autant plus dangereuse qu'elle est moins apparente. Quelques expressions imprécises ou équivoques de Dupanloup, regrettables de la part de tout catholique et bien davantage de la part d'un évêque, le confirment dans ses soupçons, et il y est d'ailleurs encouragé par son ami dom Guéranger, l'un des principaux inspirateurs de ses fameuses *Instructions synodales sur les erreurs du temps présent*.

Dom Guéranger a réalisé en France une grande œuvre [1] : son plus récent historien, qui est loin d'être un panégyriste, estime qu'en dehors de Lamennais, il ne voit personne « qui ait, plus fortement que lui, marqué de son empreinte la vie catholique de son temps » [2]. Mais l'abbé de Solesmes avait de gros défauts : outre certaines faiblesses de caractère, il avait l'esprit étroit, « acceptant trop dans leur rigueur des doctrines susceptibles de plus de modération » [3] et se laissant trop souvent aller, en toute bonne foi d'ailleurs, « à manipuler les textes historiques dans le sens de ses idées préconçues » [4]. Jugeant tout d'un point de vue absolu, « sans se préoccuper du mode et de l'opportunité » [5], convaincu en outre qu'il n'y avait de salut pour la France « que sous le régime despotique » [6], il ne pouvait supporter les catholiques libéraux, chez qui il découvrait sans cesse de nouveaux indices de leur naturalisme, jusque dans l'effort d'Albert de Broglie pour mettre en lumière dans l'histoire de l'Église le jeu complexe des causes secondes [7].

Dom Guéranger — dom Guerroyer, disaient ceux qui ne l'aimaient pas — aurait suffi pour transformer en polémique batailleuse la discussion d'idées entre les catholiques libéraux et les intransigeants. Mais celui qui envenima la querelle d'une manière définitive fut Louis Veuillot. Certes, nul n'a pourfendu avec plus de vigueur la libre pensée sous toutes ses formes, ni travaillé avec plus de succès, par ses sarcasmes impitoyables à l'adresse des ennemis de l'Église, à délivrer le catholique moyen du complexe d'infériorité qui avait longtemps pesé sur lui. Mais le mot d'Ozanam n'est pas tout à fait faux, qui disait de lui et de ses partisans enthousiastes : « Ils ne se proposent pas de ramener les incroyants, mais seulement d'ameuter les passions des croyants » [8], et les conseils de

---

(1) Sur son œuvre de restauration monastique et liturgique, voir le tome précédent, p. 508-509 et dans ce volume, p. 459-460 et 472 ; sur son rôle dans la défaite du gallicanisme, voir p. 271 et 300. La biographie classique de dom Guéranger (1805-1875) par dom DELATTE, *Dom Guéranger, abbé de Solesmes*, Paris, 1909-1910, 2 vol., est à nuancer par l'ouvrage d'E. SEVRIN, *Dom Guéranger et Lamennais*, Paris, 1933 et celui d'A. LEDRU, *Dom Guéranger et Mgr Bouvier*, Paris, 1911.
(2) E. SEVRIN, *op. cit.*, p. 345.
(3) Lacordaire à Guéranger, 26 avril 1838, dans dom DELATTE, *op. cit.*, t. I, p. 232.
(4) A. LEDRU, *op. cit.*, p. 120.
(5) Falloux à Guéranger, octobre 1856, dans dom DELATTE, *op. cit.*, t. II, p. 135.
(6) Guéranger à Montalembert, 22 novembre 1852, *ibid.*, t. II, p. 68-69.
(7) « Dans une nouveauté féconde, il lui sembla voir une innovation dangereuse » (P. ALLARD *Le duc de Broglie, historien de l'Église*, dans *Le Correspondant*, 1901, t. CCII, p. 473), à savoir la négation du rôle de la Providence, comme il l'expliqua dans *L'Univers* en 26 articles qu'il réunit ensuite en volume : *Du Naturalisme dans l'histoire*, Paris, 1858. Plus compréhensif, Pie IX demanda à Mgr Pie de conseiller à dom Guéranger de cesser ses attaques et témoigna sa bienveillance à Albert de Broglie. (Dom DELATTE, *op. cit.*, t. II, p. 155-156, 200 ; A. DE BROGLIE, *Mémoires*, t. I, p. 302.)
(8) Cité dans A. DANSETTE, *Histoire religieuse de la France contemporaine*, t. I, p. 335.

Mgr Pie ou de dom Guéranger ne pouvaient qu'encourager ses sympathies naturelles pour un catholicisme autoritaire ou ses dispositions à renchérir encore sur les rigueurs de l'orthodoxie, en maudissant à toute occasion la science, le progrès et les libertés modernes.

Pour défendre ses idées excessives et monnayer, avec une terrible injustice parfois, les jugements de ses amis théologiens sur les catholiques libéraux, Veuillot dispose, dans *L'Univers*, d'une tribune quotidienne. Sans doute, le haut clergé, même quand il partage ses idées, se montre souvent réticent à l'égard de ce journaliste qui prétend en remontrer aux évêques en fait d'orthodoxie, mais il devient par contre en province l'oracle de nombreux prêtres, qui apprécient son langage populaire et sa verve incisive [1]. Aussi peut-on, à juste titre, le considérer comme « le principal responsable de l'attitude politique du clergé et des catholiques pendant le Second Empire » [2], du moins de ceux qui se plient encore docilement aux directives ecclésiastiques : en leur communiquant le caractère absolu et l'allure agressive de son propre tempérament, il a contribué plus qu'aucun autre à la naissance en France d'un « esprit clérical » et à la constitution d'une « école arrogante et inexpérimentée, intolérante de langage bien plus que de cœur, maudissant en bloc le siècle et les contemporains » [3], provoquant ainsi les adversaires à la riposte et à la violence.

Les controverses interminables et toujours renaissantes entre les tenants de *L'Univers* et les amis du *Correspondant* ont été racontées bien des fois [4]. Elles ne cessèrent pas même lorsque l'évolution de la Question romaine, à la suite de la guerre d'Italie, eut amené les premiers à se ranger avec fracas parmi les adversaires de Napoléon III, et les seconds, à quelques exceptions près, comme Lacordaire [5], à se lancer à fond dans la défense de la cause pontificale, montrant par là la sincérité de leurs sentiments catholiques.

UN TROISIÈME GROUPE :
MGR MARET ET SES AMIS

Le talent et la distinction d'esprit des collaborateurs du *Correspondant*, ainsi que l'importance des services rendus à l'Église par plusieurs d'entre eux, ont souvent fait perdre de vue qu'ils n'étaient pas les seuls catholiques à l'époque à préconiser une attitude plus libérale. Un groupe d'ecclésiastiques de valeur et de prélats en étroite unité de vues avec le doyen de la faculté de théologie de la Sorbonne, Mgr Maret, condamnait encore plus sévèrement l'hostilité du « parti ultracatholique » à l'égard des principes de 1789, « qui, *bien entendus*, découlent du chris-

---

(1) Sur l'influence de *L'Univers*, voir E. Amann, art. *Veuillot*, dans *D. T. C.*, t. XV, col. 2831, G. de Liboux, dans *Revue augustinienne*, t. IV, 1903, p. 260-261 et J. Maurain, *La politique ecclésiastique du Second Empire*, p. 158-159, p. 161, p. 806-807.

(2) Ch. Pouthas, *L'Église et les questions religieuses de 1848 à 1878*, p. 240.

(3) P. de la Gorce, *Histoire du Second Empire*, t. II, p. 159. Cf. E. Amann, dans *D. T. C.*, t. XV, col. 2831-2832.

(4) L'une des plus pénibles fut celle occasionnée par le pamphlet *L'Univers jugé par lui-même*, composé par l'abbé Cognat sur les instigations de Dupanloup et diffusé avec l'appui de Mgr Sibour. (Voir, outre les biographies de Veuillot et de Dupanloup, A. Moser, *M. Cognat*, Paris, 1889.)

(5) Cf. *supra*, p. 88-89.

tianisme comme de la raison philosophique » [1]. Mais comme, d'après eux, l'essentiel de ces principes, et en particulier la liberté de conscience, se trouvait sauvegardé dans la constitution impériale, ils n'approuvaient pas l'hostilité des amis de Montalembert à l'égard du régime napoléonien. Après 1859, ils ne s'associèrent pas non plus à leur attitude intransigeante dans la question romaine, refusant d'être « libéraux en France et absolutistes à Rome » [2] ; relevant l'inconséquence de ceux qui s'étaient faits les champions de l'Irlande et de la Pologne et que tout semblait rapprocher de Massimo d'Azeglio ou de Minghetti, ils ne cachaient pas leurs sympathies pour les aspirations constitutionnelles et nationales des Romains et insistaient d'autre part sur le poids que représentait la souveraineté temporelle du pape au détriment de sa mission spirituelle.

Or, une partie importante de la bourgeoisie catholique partageait au fond cette opinion, car il ne faut pas oublier que le tapage fait par les catholiques intransigeants, ainsi que leur influence dans l'Église, s'explique davantage par leur agressivité que par leur nombre. On peut dès lors parler, surtout à partir de 1860, d'un troisième parti qui, pour être moins nettement délimité que les deux autres, n'en était pas le moins nombreux, surtout si on y rattache les très nombreux catholiques tièdes, prêts à en prendre et à en laisser dans l'enseignement officiel de l'Église au cas où celui-ci heurterait par trop leurs conceptions intellectuelles ou politiques. « Le très petit nombre des catholiques libéraux conscients, le nombre immense des catholiques modérés, c'est-à-dire inconsciemment libéraux, sont en effet deux caractéristiques essentielles du catholicisme français au XIXe siècle [3]. »

LE LIBÉRALISME CATHOLIQUE HORS DE FRANCE

C'est en France que les controverses entre catholiques à propos du libéralisme furent les plus bruyantes, par suite de la forte personnalité des protagonistes dans les deux camps, par suite aussi du caractère passionné qu'y prenaient les oppositions politiques. Mais le problème sous-jacent, l'accord de la civilisation moderne et de la foi chrétienne, était trop fondamental pour que des discussions analogues ne surgissent pas un peu partout en Europe occidentale, nuancées différemment d'ailleurs suivant les préoccupations propres à chaque pays.

LES « CATHOLIQUES CONSTITUTIONNELS » EN BELGIQUE

En Belgique comme en France, on discutait surtout la question de savoir dans quelle mesure le catholicisme pouvait s'accommoder du régime des libertés modernes. Le cardinal Sterckx, dont l'influence était prépondérante, restait fidèle aux principes de

---

(1) *Mémoire* de Mgr Maret au ministre des Cultes, du 7 janvier 1857. (Cité dans G. Bazin, *Vie de Mgr Maret*, t. I, p. 425.)
(2) *Mémoire* de Mgr Maret à Napoléon III, fin 1863. (Cité *ibid.*, t. II, p. 292.)
(3) J. Maurain, *La politique ecclésiastique du Second Empire*, p. 713, en note, où se trouve citée également une appréciation du chargé d'affaires, à Rome, en 1865, estimant qu'il y avait en France « un ou deux dixièmes de catholiques fervents, deux dixièmes de schismatiques, de dissidents ou d'incrédules et six dixièmes de catholiques tièdes ».

« l'École de Malines », dont il avait été l'un des inspirateurs sous le régime hollandais et lors de l'élaboration de la constitution très libérale de 1830 [1]. Moins doctrinaires que les mennaisiens français, Sterckx et ses amis, de même qu'un certain nombre d'hommes politiques catholiques groupés autour d'Henri de Mérode, s'étaient placés exclusivement sur le terrain de la pratique, acceptant les libertés modernes et même la séparation légale entre l'Église et l'État parce qu'ils y avaient vu le meilleur moyen de favoriser, à l'abri des entraves de l'État, la vie et le développement de l'Église. Il serait faux de dire que leur désir de la liberté n'était pas sincère, « mais manifestement, du moins le plus grand nombre, ils ne la désiraient pas pour elle-même » [2] ; ils étaient en outre convaincus que, étant donnée la puissance du catholicisme dans le pays, l'État devrait en tenir compte en interprétant et en appliquant les lois dans un sens catholique, en dépit de la séparation constitutionnelle, et que, dès lors, celle-ci n'était pas incompatible avec un accord de fait entre l'Église et l'État.

Il était toutefois inévitable que, sous l'effet des âpres polémiques qui se développaient en France, la question passât, en Belgique aussi, du plan de la tactique à celui des idées. Certains catholiques, s'enthousiasmant pour les libertés comme telles, finirent par exalter la constitution comme un idéal, faisant remarquer par exemple, comme Barthélemy Dumortier, que, grâce à la séparation, la Belgique était le seul pays catholique où « c'est le pape qui gouverne l'Église : et c'est là un grand bien » [3]. D'autres, au contraire, réfléchissant aux condamnations prononcées par Grégoire XVI dans l'encyclique *Mirari vos*, avaient tendance à considérer la constitution de 1830 comme un moindre mal dont il fallait souhaiter l'amendement dès que les circonstances le permettraient. Jusque vers 1860, sous l'influence de Veuillot, ce point de vue prévalut dans la presse catholique belge. Mais à partir de 1856, Adolphe Dechamps, l'un des chefs du parti catholique et ami de Montalembert, entreprit de réagir contre ces « ultras » qui faisaient le jeu du parti libéral, hostile à l'Église. Il transforma le *Journal de Bruxelles* en organe des « catholiques constitutionnels » et il gagna assez vite aux idées catholiques libérales modérées la majorité des esprits, surtout dans la jeunesse, tandis que les intransigeants, surnommés « veuillotistes », se groupaient autour du journal gantois *Le Bien Public*, dont plusieurs collaborateurs étaient des amis personnels du célèbre journaliste français [4].

LIBÉRALISME ET UNITÉ NATIONALE EN ITALIE La question du libéralisme catholique allait se poser en Italie en des termes assez différents. Dès la première moitié du siècle, une partie notable de la classe intellectuelle et de la bourgeoisie urbaine s'était sentie attirée vers l'idéologie libérale. Mais il s'agissait généralement

---

(1) Voir A. Simon, *Le cardinal Sterckx et son temps*, t. I, p. 65-155 et spécialement p. 138-141.
(2) A. Simon, *L'Église catholique et les débuts de la Belgique indépendante*, p. 13-14.
(3) Lettre de février 1857, dans A. Simon, *Le cardinal Sterckx*, t. I, p. 234.
(4) Voir E. de Moreau, *Adolphe Dechamps*, p. 313-320 et 423-438.

d'un libéralisme très modéré et, au surplus, dans ce pays où l'incroyance n'avait guère pénétré et qui savait tout ce que représentaient l'Église et la papauté dans l'héritage national, les sympathies libérales s'accompagnaient rarement d'anticléricalisme avant la crise de 1848. Rosmini, par exemple, recueillait l'approbation de nombreux lecteurs lorsque, dans le grand journal libéral *Il Risorgimento*, il se désolidarisait des principes voltairiens à la base de la Révolution française tout en faisant l'éloge de la vraie liberté et des aspirations constitutionnelles.

Les conditions paraissaient donc particulièrement favorables à une synthèse du vrai libéralisme et du catholicisme, et beaucoup de prêtres, en Italie du Nord, s'y montraient favorables [1], mais deux éléments particuliers vinrent compliquer la situation.

En premier lieu, certains libéraux italiens, qui subissaient la lointaine influence du jansénisme de la fin du XVIIIe siècle et celle, plus proche, du protestantisme suisse, n'hésitaient pas à envisager la question d'une réforme du catholicisme dans ses institutions, voire même dans son dogme, afin de s'adapter à l'évolution moderne. En Toscane surtout, autour de Gino Capponi, de Raphaël Lambruschini, de Tommaseo, s'étaient constitués, vers 1840, des groupes de véritables modernistes avant la lettre [2] ; la même tendance apparaît dans les œuvres posthumes de Gioberti, qui révèlent le caractère au fond très peu chrétien de sa pensée religieuse. Rosmini lui-même, mais cette fois dans un sens foncièrement catholique, avait exposé dans *Les cinq Plaies* tout un programme de réformes de l'Église inspiré de l'esprit évangélique ; avec moins de nuances, certains prêtres ou religieux réclamaient une « démocratisation » de la discipline ecclésiastique, une plus grande indépendance à l'égard de la hiérarchie ou des supérieurs. Bien que ce courant réformiste eût perdu beaucoup de son importance après 1850, ses tendances inquiétantes devaient fatalement compromettre le mouvement libéral italien aux yeux des autorités ecclésiastiques et de tous les défenseurs de la tradition, au premier rang desquels il faut placer la Compagnie de Jésus, dont l'influence doctrinale allait en grandissant dans les sphères officielles. Les diatribes de Gioberti contre les jésuites ne contribuèrent pas peu, on s'en doute, à indisposer ces derniers contre le courant catholique libéral qui avait profondément reçu la marque de l'abbé piémontais.

Mais la raison profonde qui devait rendre inévitable la rupture avec le libéralisme en Italie est ailleurs, dans l'union intime qui existait dans ce pays entre le mouvement libéral et les aspirations à l'unité nationale. Or les déceptions de 1848 et la soi-disant « trahison » de Pie IX discréditèrent le programme néoguelfe d'une fédération italienne présidée par le pape, auquel s'étaient ralliés la plupart des libéraux modérés,

---

(1) Ainsi lorsqu'en 1848 on organisa en Piémont des pétitions pour donner la parité de droits civils et politiques aux non-catholiques, de nombreux prêtres donnèrent leur signature, notamment le futur archevêque de Turin, Gastaldi, et deux autres futurs évêques, en dépit de la réticence de l'épiscopat. (T. Chiuso, *La Chiesa in Piemonte*, t. III, p. 219 et suiv. ; cf. A.-C. Jemolo, *Chiesa e Stato in Italia*, p. 73-74.)

(2) Sur ce mouvement, voir F. Landogna, *Saggio sul cattolicesimo liberale in Italia nel sec. XIX*, Livourne, 1925, et P. Fossi, *Italiani dell'Ottocento : Rosmini, Capponi, Lambruschini, Tommaseo, Manzoni*, Florence, 1941.

et l'on en revint à la solution unitaire de Mazzini, reprise par Cavour
en faveur de la maison de Savoie. Dans cette solution, l'État pontifical
devait disparaître. Comme une partie de ses sujets, excédés depuis des
années par le « gouvernement des prêtres », ne demandaient pas mieux,
les patriotes invoquèrent contre le principe du légitimisme le droit des
citoyens à disposer d'eux-mêmes. En même temps ils prenaient prétexte
du caractère très peu libéral des institutions romaines pour dénoncer
à l'Europe le caractère anachronique de la souveraineté temporelle du
pape. Certains libéraux, influencés par le protestantisme, ajoutaient
que rien dans l'Évangile n'indiquait que le pape devait posséder la puissance politique pour remplir dignement sa mission [1].

C'était plus qu'il n'en fallait pour exciter contre le mouvement libéral
l'indignation de Pie IX et de tous ceux qui considéraient le maintien
des États de l'Église comme la condition indispensable de l'indépendance
du souverain pontife, et cette indignation était d'autant plus vive que
les radicaux piémontais, dont l'anticléricalisme s'accentuait avec l'intransigeance de Rome, imposaient une législation qui, au nom de la formule
des catholiques libéraux français : « L'Église libre dans l'État libre »,
lésait gravement les droits acquis de l'Église et limitait en fait sa liberté
d'action, même dans le domaine spirituel [2].

Cette évolution de la situation rendait fort délicate la situation des
catholiques italiens. Les plus soumis à la hiérarchie n'entendaient pas
transiger avec les droits du pape et de l'Église et ils s'inquiétaient en
outre de voir que la liberté accordée à la propagande religieuse favorisait
le développement du protestantisme dans le Nord. Mais beaucoup d'autres
restaient des disciples enthousiastes de Gioberti dont le *Rinovamento
civile d'Italia* (1851) eut des milliers de lecteurs et des dizaines d'imita-
teurs. Surnommés *Neocattolici* par leurs adversaires intransigeants, ces
hommes, en qui subsistait l'esprit de 1848, le désir ardent d'unir le catholicisme avec le libéralisme modéré et le patriotisme italien, prenaient
leur inspiration auprès des survivants du mouvement libéral catholique
de la génération précédente, Massimo d'Azeglio notamment et surtout
Tommaseo, qui, dans son livre *Rome et le Monde* (1851), avait affirmé que
« le vrai prêtre doit non pas dominer mais prêcher », tout en suggérant
d'ailleurs de laisser au pape une petite ville, « un nouveau Saint-Marin »,
pour garantir son indépendance. Ils avaient leur revue, *Il Cemento*, qui
s'attacha d'abord surtout à la partie négative du programme, la lutte
contre ceux qui voulaient faire du catholicisme l'allié du despotisme,
puis fusionna en 1856 avec un organe giobertiste.

Mais la plupart n'osaient pas exprimer ouvertement leurs sympathies
à cause de l'opposition catégorique marquée par le Saint-Siège, dont on

---

(1) Sur l'influence du protestantisme sur le libéralisme italien, déjà relevée par A. Della Torre,
voir M. Petrocchi, *I reflessi europei sull'48 italiano*, Florence, 1946.

(2) Montalembert releva vivement la nette opposition qu'il y avait sous l'identité de formule
entre le programme des radicaux italiens et celui de ses amis, qui voulaient la liberté au service
de l'Église (voir E. Lecanuet, *Montalembert*, t. III, p. 223-228). Cavour lui-même, bien qu'il ne
partageât pas l'anticléricalisme de la gauche radicale, « considérait la liberté comme une valeur
absolue à laquelle l'Église devait s'adapter en modifiant son esprit ». (L. Salvatorelli, dans *Ricerche religiose*, t. XIX, 1948, p. 139.)

savait qu'il encourageait les attaques acharnées de la presse catholique
contre la politique de Cavour et de ses successeurs. Lorsque le directeur
de l'*Armonia* de Turin, le chanoine Audisio, que son attitude proromaine
avait fait chasser de sa patrie, conseilla en 1862 à l'abbé Margotti de
ne pas prendre trop à la lettre les excitations du Vatican, il se fit traiter
par Pie IX de *Piemontesaccio* au cours d'une scène violente, et le pape
invita Margotti à fonder un nouveau journal, plus agressif encore, l'*Unità
cattolica* [1]. Le grand nombre d'abonnés qu'obtint rapidement ce dernier
quotidien prouva que la tendance qu'il représentait allait en augmentant
depuis les événements de 1859-1860, qui heurtèrent profondément les
bons catholiques et détachèrent du gouvernement de Turin la majorité
du clergé.

Malgré tout cependant les partisans d'une réconciliation de l'Église
avec le programme libéral et unitaire du gouvernement italien continuèrent à faire entendre leur voix, proposant comme solution de toutes les
difficultés pendantes « *la libertà come nel Belgio* » et le retour du pape à sa
mission purement spirituelle. Ces thèses, qui conservèrent quelque sympathie dans le clergé, parmi les adversaires des méthodes de la Curie et
les héritiers des « réformistes » du début du siècle [2], furent même parfois
appuyées, avec beaucoup de prudence, par l'un ou l'autre évêque, comme
Mgr Losanna, grand admirateur de Dupanloup, ou Mgr Nazari, futur
archevêque de Milan. La plupart de ces ecclésiastiques se bornaient d'ailleurs à souhaiter l'établissement d'un régime parlementaire ou l'introduction de réformes politiques et administratives dans l'État pontifical ;
ceux d'entre eux qui étaient prêts à transiger avec l'*idéologie* libérale
proprement dite n'étaient vraiment qu'une petite minorité, dépourvue
au surplus d'influence, car les hommes politiques les trouvaient encore
trop cléricaux, tandis que les désaveux répétés du pape les rendaient
suspects aux yeux des catholiques.

LES CATHOLIQUES ALLEMANDS
ET LE LIBÉRALISME

Le vieux leitmotiv des disciples de
Lamennais : « La liberté comme en
Belgique », avait aussi résonné en
Allemagne de 1840 à 1860, et en dépit des sympathies profondes d'une
partie de l'Allemagne méridionale pour une étroite alliance de l'Église et
de l'État, les chefs du mouvement catholique y avaient plus d'une fois
suivi une ligne de conduite analogue à celle préconisée par Montalembert
et ses amis.

Il ne faudrait pas cependant se laisser prendre aux apparences et
identifier sans plus leur attitude avec celle des catholiques libéraux
français [3]. Non seulement, la plupart étaient loin de partager l'enthousiasme parlementaire de ceux-ci, mais, en outre, ils ne vibrèrent jamais

(1) Sur l'*Armonia* et l'*Unità*, voir A. DRESLER, *Geschichte der italianischen Presse*, t. II, Munich,
1934, p. 115-116.
(2) Voir plus haut, p. 98-100. On trouvera des détails intéressants dans l'introduction de
F. QUINTAVALLE, *La conciliazione fra l'Italia e il Papato*, Milan, 1907, et dans S. JACINI, *Il tramonto
del potere temporale...*, p. 72-77.
(3) Voir quelques pages très éclairantes de F. SCHNABEL, *Deutsche Geschichte im XIX. Jahrhun-
dert*, t. IV, *Die religiöse Kräfte*, Fribourg-en-Br., 1937, p. 189-202.

pour les principes de 1789, dans lesquels ils ne voyaient qu'une importation étrangère, et quand ils parlaient de liberté, ils pensaient moins à la liberté individuelle qu'à la liberté corporative du moyen âge, où l'individu restait soumis aux exigences supérieures d'un ordre organique. Ensuite, et c'était là une différence fondamentale avec la France, pendant longtemps presque personne en Allemagne ne songea à la possibilité d'un État non chrétien, mais seulement d'un État non confessionnel ; dès lors, quand les catholiques affirmaient que l'Église n'avait besoin d'aucun privilège, mais seulement de la liberté, la liberté qu'ils réclamaient s'inscrivait à l'intérieur d'une conception générale de la vie qui restait chrétienne. Par ailleurs, les catholiques allemands n'eurent jamais l'idée de présenter la séparation de l'Église et de l'État comme un idéal, mais seulement comme un moindre mal en présence des prétentions joséphistes, surtout lorsque celles-ci étaient exercées par un État protestant. Enfin, étant en minorité, lorsqu'ils prônaient la liberté des cultes ou de conscience, ils ne semblaient pas, comme en France, faire une concession à l'erreur, mais au contraire plaider en faveur du catholicisme.

On comprend, dans ces conditions, qu'à l'encontre de ce qui se passait en France à la même époque, les militants catholiques, s'ils font bon marché des privilèges de l'Église et exaltent le régime de la tolérance légale, dénoncent en même temps avec virulence le caractère antichrétien du libéralisme idéologique, tandis que les hommes de doctrine qui condamnent ce dernier comme l'une des manifestations les plus inquiétantes du naturalisme, approuvent cependant ceux qui renoncent à revendiquer pour l'Église les droits lui revenant en tant que société parfaite et demandent la liberté de toutes les confessions. Le cas le plus caractéristique est celui de Ketteler. Il considérait l'évolution du parti libéral allemand depuis 1848 comme un immense danger, non seulement pour l'Église, mais pour toutes les traditions nationales germaniques, et il trouvait tragique de le voir bénéficier indûment des sympathies de l'époque pour la liberté et de la réaction antiabsolutiste. C'est pour essayer de détourner ces sympathies dans le sens d'un sain régime de liberté ordonnée, tel qu'on le trouvait au moyen âge, qu'il écrivit son ouvrage *Liberté, Autorité et Église*, où il opposait une conception chrétienne et germanique de la liberté à la conception non chrétienne et étrangère issue de la Révolution française. Mais dans le même ouvrage, il ne faisait aucune difficulté pour reconnaître que si l'Église avait jadis joui de privilèges, « effet naturel de l'unité de foi », elle n'en avait pas besoin pour exister et que c'était une grave erreur « d'attendre le salut des événements extérieurs et surtout de l'avénement de quelque prince illustre et saint » ; et il ajoutait que « nul principe religieux ne défend à un catholique de croire qu'il est des circonstances où l'État ne peut rien faire de mieux que d'accorder une entière liberté de religion » (ajoutant d'ailleurs que cette tolérance ne pouvait s'étendre à des sectes athées ou immorales) [1].

---

(1) L'ouvrage, publié en février 1862, fut aussitôt traduit en français. *Les Études* (1862, t. I, p. 230-240) et *Le Correspondant* (t. LV, 1862, p. 604-642) s'empressèrent de souligner la confirmation que les catholiques libéraux français croyaient trouver de leurs idées dans le dernier chapitre.

*LA LIBERTÉ SCIENTIFIQUE
ET LA THÉOLOGIE*

Beaucoup moins hypnotisés que les catholiques français par la question du libéralisme politique et facilement d'accord dans la pratique sur la question de la liberté religieuse, les catholiques allemands commençaient par contre à aborder de front la question de la liberté de l'intellectuel par rapport au magistère de l'Église en face des problèmes posés par la critique, question que les catholiques libéraux français ne soulevèrent guère avant le dernier quart du XIXᵉ siècle. Or, sur ce point, l'esprit du libéralisme avait pénétré profondément dans les milieux intellectuels allemands, parfois à leur insu, comme le montre le ton des adversaires de l'école romaine dans les controverses qui opposaient günthériens et scolastiques, ou Munichois et Mayençais. Il a été longuement question au chapitre précédent de cet esprit d'indépendance qui se développait dans les universités, excité parfois par l'étroitesse de vue de certains disciples des jésuites : l'attitude de Doellinger à l'assemblée des savants catholiques de 1863 en est une manifestation typique.

*LE GROUPE DU « RAMBLER »
EN ANGLETERRE*

En Angleterre aussi, c'est sur la question de la liberté intellectuelle du catholique que les discussions s'engagèrent. En 1848, un petit groupe de convertis avaient fondé une revue, *The Rambler*, qui entendait réagir contre l'infériorité intellectuelle manifeste du catholicisme anglais et aborder, conformément aux exigences scientifiques, les problèmes posés par la critique moderne. Partant du principe que chaque science, l'histoire aussi bien que la physique, ne doit relever que de ses propres méthodes, ils entendaient s'adonner à ces études avec une liberté absolue, sans préoccupation des résultats, même si ceux-ci étaient opposés à l'une ou l'autre opinion traditionnelle parmi les catholiques. C'était là, pensaient-ils, la condition *sine qua non* pour que le catholicisme s'imposât à l'attention du monde savant et de l'opinion cultivée.

Mais le *Rambler* ne se borna pas à aborder dans cet esprit d'indépendance totale des sujets d'histoire. Il touchait aussi à l'occasion des questions d'ordre théologique [1] et il prétendit même prendre position sur les problèmes brûlants de l'heure, comme par exemple le rôle des journalistes catholiques, ou l'infériorité des méthodes d'enseignement et d'éducation dans les collèges catholiques. Or, c'étaient là des questions que les évêques entendaient se réserver, Wiseman aussi bien que les prélats les plus traditionalistes. Le cardinal avait bien approuvé l'intention du *Rambler* de faire de l'histoire avec impartialité et non pas en homme de parti, ce parti fût-il l'Église catholique, mais il n'admettait pas que des laïcs prissent des initiatives dans les questions intéressant la défense de l'Église : leur vraie place, estimait-il, est dans la politique, le commerce, l'armée ou la marine, la science, l'art [2].

---

(1) Catholiques fervents, parfaitement soumis en matière de dogme, les rédacteurs du *Rambler* croyaient, comme beaucoup de théologiens allemands, qu'ils étaient tout à fait libres dans les questions que l'Église n'avait pas encore tranchées solennellement.

(2) Voir sa brochure *Words of Peace and Justice* (1848).

Les collaborateurs du *Rambler*, qui entrevoyaient ce qu'il y avait de restrictif dans cette conception du laïcat chrétien, eurent le tort de vouloir discuter avec leurs chefs spirituels d'égal à égal et parfois sans beaucoup d'égards, ne voyant qu'étroitesse d'esprit chez les « anciens catholiques » d'Angleterre et ne dissimulant pas qu'à leurs yeux tout ce qui venait de Rome était vieilli et démodé. Dans ces conditions, « plus encore par le ton irritant avec lequel il traitait les sujets délicats que par les thèses soutenues », le *Rambler* devait inévitablement être regardé par les autorités ecclésiastiques « comme un enfant terrible » [1].

L'INTERVENTION DE NEWMAN   Avant de sévir, les évêques décidèrent toutefois, au début de 1859, de faire une dernière tentative en sollicitant l'intervention de Newman. On savait en effet combien le ton et certains procédés du *Rambler* l'indignaient, et l'on espérait en même temps qu'il pourrait se faire écouter, parce qu'il n'avait pas caché sa sympathie pour la tâche de rénovation entreprise et pour les principes qui l'inspirait, tout comme il avait montré l'intérêt qu'il portait à l'œuvre féconde accomplie dans le champ de l'histoire de l'Église par l'un des principaux inspirateurs du groupe, le savant Doellinger [2]. Newman parvint à convaincre Simpson, le directeur de la revue, de se retirer et il prit lui-même en main la direction. Mais l'excitation des esprits de part et d'autre rendait bien difficile de mener à bien l'œuvre de pacification et, après deux numéros, l'évêque de Newman, Mgr Ullathorne, lui demanda d'y renoncer.

Malheureusement, le nom de Newman était désormais lié à celui du *Rambler* dans l'esprit de beaucoup, d'autant plus qu'il continua par la suite à aider de ses conseils certains des rédacteurs qui étaient ses amis et dont il comprenait les préoccupations scientifiques. Ward, avec son dévouement fougueux et parfois bien intempestif à l'Église, se mit à dénoncer son ancien maître comme un libéral de la pire espèce, celle qui ne s'affiche pas, tandis que le cardinal Wiseman, vieilli, laissait faire, et que son successeur, Manning, qui trouvait trop tiède l'ultramontanisme de Newman, contribuait encore à le desservir auprès des autorités romaines [3].

L'illustre converti était pourtant éloigné autant qu'on peut l'être du libéralisme religieux, lui qui, depuis les débuts du mouvement d'Oxford, n'avait cessé de réagir, au nom de la vérité révélée et du magistère infaillible de l'Église, contre la prétention orgueilleuse de son siècle à faire de l'esprit humain la norme de toute vérité. Mais tout soumis qu'il fût à l'autorité, il répugnait à ce qui l'exagérait ; respectueux de la tradition, il souhaitait cependant voir l'Église faire bon accueil à la science moderne dans ce que ses efforts avaient d'honnête, et il estimait que les vues étroites

---

(1) Dom GASQUET, *Lord Acton and his Circle*, p. XL.

(2) Voir divers témoignages dans J. FRIEDRICH, *Ign. von Doellinger*, t. II, p. 106-107 et dans W. WARD, *Life of Newman*, t. I, p. 444-445 et 504.

(3) La méfiance romaine débuta en 1860, lorsque, par suite d'une négligence de Wiseman, on eut l'impression que Newman refusait de fournir au Saint-Office les explications qu'on attendait (voir sur ce regrettable incident W. WARD, *Life of Newman*, t. II, p. 128, 157, 170-173, 179, 187, 549 et C. BUTLER, *Life and Times of Bishop Ullathorne*, t. I, p. 315-321).

et intransigeantes prônées par certains catholiques comme une preuve de fidélité à l'Église faisaient au contraire le plus grand tort à celle-ci. La tournure d'esprit de cet intellectuel de race, à qui rien ne déplaisait davantage que les positions préconçues et les jugements « tout faits », devait inévitablement exciter la méfiance d'hommes à l'esprit rigide, ralliés de confiance à une scolastique étroite et soucieux de voir l'autorité restreindre sans cesse le champ de la libre recherche de l'esprit humain.

JOHN ACTON   En septembre 1859, le *Rambler* passait aux mains de celui en qui devait bientôt s'incarner le libéralisme catholique anglais, sir John Acton [1]. C'était un catholique de naissance, ancien élève de Wiseman, envers qui il garda toujours, même au milieu des plus vives polémiques, un affectueux respect. Mais il n'avait rien d'un « ancien catholique ». Fils d'un baronnet anglais et d'une noble allemande dont le père avait représenté Louis XVIII au Congrès de Vienne, petit-fils d'un ministre du roi de Naples et neveu d'un cardinal de Curie, il devait à ses origines une indépendance des préjugés insulaires de ses concitoyens, qui fut encore complétée par sa formation cosmopolite. Des voyages aux États-Unis et en Russie et l'influence du second mari de sa mère, lord Granville, avaient renforcé son goût très vif pour la liberté politique, tandis que les années passées sous la direction de Doellinger, dont il fut le disciple de prédilection, l'avaient rempli d'admiration pour le travail accompli à Munich et à Tubingue par les savants catholiques allemands en vue d'adapter l'enseignement ecclésiastique aux exigences de la critique moderne. Il était rentré en Angleterre en 1856 avec deux grandes ambitions : écrire une *Histoire de la liberté* en vue de laquelle il finit par accumuler dans sa bibliothèque d'Aldenham près de 60.000 volumes ; et prendre la direction d'une revue qui, par son impartialité et sa valeur scientifique, forcerait l'audience du public non catholique.

Acton mit au service du *Rambler* son érudition prodigieuse servie par une intelligence perspicace, ainsi qu'une conscience professionnelle exceptionnelle, et la revue connut vite un succès qu'aucune publication catholique n'avait jamais atteint en Angleterre. Toutes les œuvres importantes paraissant en France, en Italie et surtout en Allemagne y étaient l'objet de longues recensions critiques fort appréciées même dans les milieux anglicans ; on comptait parmi les collaborateurs les catholiques étrangers les plus en vue, Montalembert, Dupanloup, Gratry, Doellinger, de Rossi. Mais, sous la direction de celui qui devait se définir un jour « un homme qui a renoncé à tout ce qui dans le catholicisme est incompatible avec la liberté et à tout ce qui dans la politique est incompatible avec la foi

---

(1) On ne dispose pas encore d'une biographie complète de John Acton (1834-1902), qui serait passionnante, et la publication de sa correspondance a été interrompue ; trois volumes en ont paru : *Letters to Mary Gladstone*, édit. H. PAUL, Londres, 1904 (voir à ce propos *The Tablet*, octobre et novembre 1906) ; *Lord Acton and his Circle*, édit. GASQUET, Londres, 1906 ; *Selections from the Correspondance of lord Acton*, édit. FIGGIS et LAURENCE, Londres, 1917. Outre l'*Introductory Memoir* de H. PAUL en tête des lettres à M. Gladstone, on consultera D. MATHEW, *Acton. The formative Years*, Londres, 1946 (jusqu'en 1864) et U. NOACK, *Katholizität und Geistesfreiheit nach den Schriften von J. Dalberg-Acton*, Francfort, 1936.

catholique » [1], les tendances libérales et indépendantes de la revue ne firent que s'accentuer. Or, les esprits étaient moins préparés que jamais à discerner sous les intempérances de langage la valeur de l'œuvre entreprise, à ce moment où la réaction contre toutes les formes de libéralisme se développait rapidement dans les sphères ecclésiastiques officielles. En avril 1862, devant la réprobation marquée publiquement par l'épiscopat envers certains articles sur le pouvoir temporel du pape, le *Rambler* dut disparaître.

Trois mois plus tard, toutefois, Acton lançait un nouveau périodique, *The Home and Foreign Review*, dont les rédacteurs et le programme étaient pratiquement les mêmes. Manning venait précisément de réorganiser la *Dublin Review*, l'ancienne revue de Wiseman, où Ward menait depuis des années contre le *Rambler* et son esprit une campagne où son tempérament combatif et sans nuances s'en donnait à cœur joie. Les polémiques reprirent de plus belle, attisées par les exagérations mêmes de Ward, jusqu'au moment où parut le bref pontifical à l'archevêque de Munich à la suite de l'Assemblée des savants catholiques de 1863. Le pape y énonçait, sur le respect de la méthode scolastique et sur la soumission due par les écrivains catholiques à l'autorité ecclésiastique, des principes où Acton et ses amis virent à juste titre la réprobation de certaines de leurs idées. Dans un article intitulé *Conflits avec Rome*, Acton expliqua que les principes défendus jusqu'alors lui paraissaient toujours vrais, mais qu'il se refusait à « opposer une résistance active et persévérante aux volontés connues du Saint-Père » et que, dès lors, la revue cessait de paraître.

§ 2. — La réaction de Rome : le Syllabus [2].

INQUIÉTUDES ROMAINES Les sympathies réformistes de Pie IX pendant les premiers mois de son pontificat n'impliquaient, on l'a vu, aucune concession à l'idéologie libérale : l'impression pénible produite à Rome par la constitution toscane où il était question de « cultes autorisés » au lieu de « cultes tolérés » [3] est symptomatique à cet égard ; bien plus, le « pape libéral », tout en accordant divers avantages d'ordre administratif aux juifs de ses États, n'envisagea pas un instant de leur accorder l'émancipation, c'est-à-dire l'égalité de droits civils et politiques.

(1) Acton à Lady Blennerhasset, 1879, dans *Selections from the Correspondence*, édit. FIGGIS et LAURENCE, p. 54.

(2) BIBLIOGRAPHIE. — Aux sources et travaux indiqués en tête du chapitre, et aux principales biographies de Pie IX, ajouter, en ce qui concerne l'élaboration du *Syllabus* et les réactions provoquées par celui-ci :

I. SOURCES. — *Recueil des allocutions consistoriales, encycliques et autres lettres apostoliques des souverains pontifes citées dans l'Encyclique et le Syllabus du 8 décembre 1864*, Paris, 1865 ; A. DE SAINT-ALBIN, *L'Encyclique et les évêques de France*, Paris, 1865 ; C. PASSAGLIA, *Sopra l'enciclica publicata il giorno 8 dec. 1864*, Turin, 1865 ; liste des brochures et lettres pastorales provoquées par l'Encyclique dans les pays germaniques dans le *Literarischer Handweiser* de 1865 et 1866.

II. TRAVAUX. — Le meilleur exposé est celui de L. BRIGUÉ, art. *Syllabus*, dans *D. T. C.*, t. XIV, col. 2877-2923 ; voir aussi P. HOURAT, *Le Syllabus*, Paris, 1904 ; A. QUACQUARELLI, *La crisi della religiosità contemporanea dal Sillabo al Concilio Vaticano*, Bari, 1946, chap. I à III ; R. AUBERT, *Les réactions suscitées par la publication du Syllabus*, dans *Collectanea Mechliniensia*, t. XIX, 1949, p. 309-317 (documents inédits).

(3) A.-M. BETTINI, *Il concordato di Toscana*, Milan, 1933, p. 6-12.

La crise de 1848, dont les effets avaient été particulièrement pénibles pour le souverain pontife, ne pouvait que renforcer encore la méfiance romaine à l'égard de principes dont les conséquences dangereuses venaient d'apparaître au grand jour. Sans se solidariser pleinement avec les écrivains réactionnaires qui tentaient de présenter la monarchie héréditaire absolue comme le seul régime politique vraiment conforme à l'idéal catholique, et tout en continuant, pendant les premières années, à témoigner d'une réelle bienveillance envers des catholiques libéraux comme Falloux ou Montalembert, Pie IX était désormais convaincu qu'il existait une connexion intime entre les principes de 1789 et la destruction des valeurs traditionnelles dans l'ordre social, moral et religieux.

Par ailleurs, les autorités ecclésiastiques, « plus sensibles peut-être au scandale des faibles qu'à celui des forts »[1], n'avaient pas été sans remarquer que les masses, surtout les masses paysannes, qui formaient encore l'immense majorité de la population européenne, abandonnaient moins facilement une Église honorée et reconnue par le pouvoir. Certes, dès l'époque de Grégoire XVI, on avait admis, dans la pratique, l'acceptation par les catholiques de formes gouvernementales basées sur la reconnaissance des libertés modernes, allant même jusqu'à la séparation légale de l'Église et de l'État, et l'expérience belge, notamment, avait été acceptée, après quelques hésitations, grâce aux explications apaisantes du cardinal Sterckx ; mais le Saint-Siège espérait bien que cette expérience ne se généraliserait pas et entendait conserver, partout où la chose serait possible, les avantages apostoliques qu'il voyait au catholicisme d'État. Les heureux résultats que semblait procurer à première vue la politique de concordats poursuivie avec ardeur à partir de 1850[2], et dont l'apogée devait être atteinte avec le concordat autrichien de 1855, paraissaient du reste donner un démenti aux apologistes de « l'Église libre dans l'État libre », d'autant plus qu'en Belgique même, dont on invoquait si complaisamment l'exemple en faveur des mérites de la formule, l'évolution anticléricale du gouvernement libéral, à partir de 1855, montrait que ni le régime parlementaire ni la proclamation des libertés modernes ne mettaient l'Église à l'abri des attaques de ses adversaires[3].

Bien plus, c'était au nom de ces mêmes libertés modernes que la souveraineté du pape sur ses États était mise en question et que le gouvernement piémontais poursuivait une politique de laïcisation qui, non seulement évinçait l'Église de domaines qu'elle considérait depuis des siècles comme lui revenant de droit, mais qui aboutissait à expulser les religieux, à emprisonner les prêtres et à laisser le champ libre à la propagande protestante. Ce dernier point impressionnait d'autant plus, qu'au

---

(1) Y. Congar, dans *Histoire de l'Église,* sous la direction de G. de Plinval, t. II, p. 383, n. 1.
(2) En 1851, avec la Toscane (et un accord provisoire avec l'Espagne) ; en 1855 avec l'Autriche ; en 1857 avec le Wurtemberg et Naples (et un accord provisoire avec le Portugal) ; en 1859 avec Bade et l'Espagne. On les trouvera rassemblés dans Nussi, *Conventiones de rebus ecclesiasticis,* Mayence, 1870.
(3) Voir notamment la lettre de Pie IX à son frère du 29 juillet 1857, dans A. Monti, *Pio IX nel Risorgimento,* p. 257 : « Voyez ce qui arrive à présent en Belgique... », ainsi que sa riposte à Minghetti la même année : « Voyez, en Belgique aussi, les coquins (*i birbanti*) commencent à triompher ! » (M. Minghetti, *Miei ricordi,* t. III, p. 193).

premier moment, on pouvait craindre que le protestantisme ne rencontrât un certain succès dans la bourgeoisie italienne cultivée, déçue par la « trahison » du pape, mais désireuse de demeurer dans une atmosphère chrétienne [1].

On comprend mieux dans ces conditions l'inquiétude grandissante de Pie IX et de son entourage en constatant que l'émoi suscité par les excès de la démocratie révolutionnaire en 1848 n'avait pas réussi à enrayer la pénétration des idées libérales dans les milieux catholiques. Les indices se multipliaient : la sympathie témoignée par une partie importante de la classe dirigeante française à l'égard de la nouvelle politique ecclésiastique de Napoléon III, beaucoup moins favorable à l'Église ; l'indulgence, pour ne pas dire plus, de tant de catholiques italiens pour la politique de sécularisation de Cavour et pour son œuvre d'unification italienne considérée pourtant comme un terrible danger pour l'indépendance du souverain pontife ; les déclarations équivoques de Doellinger relatives au pouvoir temporel ; l'indépendance d'esprit à l'égard du magistère ecclésiastique manifestée par les *Deutsche Theologen* et par les rédacteurs du *Rambler*.

Ces allures indépendantes paraissaient particulièrement inquiétantes, car elles indiquaient que, progressivement, le virus libéral passait du domaine politique ou social au domaine proprement religieux, et l'on pouvait se demander si certains catholiques, non contents de critiquer la façon traditionnelle de concevoir les rapports entre l'Église et l'État, n'envisageaient pas une réforme de la constitution même de l'Église dans un sens plus démocratique, mettant par là en danger la nature hiérarchique et dogmatique du catholicisme. En réalité, s'il y avait quelques velléités en ce sens chez l'un ou l'autre, spécialement en Italie, dans le groupe de « modernistes » avant la lettre qui évoluaient autour de Lambruschini ou de Gino Capponi, le soupçon était parfaitement injuste pour l'ensemble des catholiques libéraux et surtout pour les plus marquants d'entre eux. Toutefois, les intégristes, dont l'influence à Rome allait en augmentant à cause de leur ferveur ultramontaine, avaient beau jeu d'exciter ces méfiances en exploitant les déclarations imprudentes ou trop peu nuancées de leurs adversaires, dont certains, la chose est indéniable, en arrivaient à aimer la liberté pour elle-même et non plus seulement comme un moyen de servir l'Église, même quand ils n'allaient pas jusqu'à écrire, comme Léopold de Gaillard dans *Le Correspondant*, qu'il y avait trois lieux saints de l'humanité, Rome, Jérusalem et... Washington. « Peut-être, devait écrire un peu plus tard Adolphe Dechamps à Montalembert, avons-nous été ou paru être trop penchés à ériger la séparation en progrès moral et social et en principe absolu. Nous avions peut-être l'air de parler la langue de ce libéralisme qui ne voit le progrès social que dans la sécularisation universelle, dans l'état rationaliste [2]. »

_____

(1) « C'est à présent l'âge d'or pour le protestantisme », écrivait en 1853 le rationaliste Franchi. (A. Della Torre, *Il Cristianesimo in Italia*, p. 192.)

(2) Lettre du 24 janvier 1865, citée dans E. de Moreau, *Adolphe Dechamps*, p. 428.

PREMIÈRE ÉBAUCHE
DU « SYLLABUS »

C'est dans cette atmosphère que parvint à Rome une *Instruction sur les erreurs du temps présent* due à Mgr Gerbet, évêque de Perpignan, un ancien mennaisien devenu l'un des protagonistes de la réaction antilibérale en France ; l'évêque avait joint à son mandement une liste de 85 propositions erronées [1]. Or, il y avait des années qu'il était question à Rome d'une condamnation des principales erreurs modernes en matière politique et sociale [2]. L'orientation prise par la Question romaine depuis 1859 avait rendu la chose très actuelle : dès la fin de 1860, le parti hostile à toute conciliation avec l'Italie cherchait à provoquer une déclaration confirmant les droits de l'Église sur la société civile [3]. Le 10 mars 1861, le pape lui-même avait, dans l'allocution *Jamdudum*, fait une amère critique de la société moderne, car, à mesure qu'il vieillissait et que les déceptions s'accumulaient, il était porté à ne plus voir du monde moderne que les aspects inquiétants pour l'avenir de l'Église et considérait qu'il était de son devoir de les dénoncer. Estimant que la liste dressée par Mgr Gerbet constituait une excellente base de départ pour le document solennel qu'il projetait, il chargea une commission de faire un choix parmi les 85 propositions, en leur appliquant la note théologique convenable.

Les théologiens consulteurs, après avoir mis rapidement au point un catalogue (*Syllabus*) qui s'inspirait de très près de celui de Mgr Gerbet, s'appliquèrent, en une trentaine de séances, à censurer les propositions, réduites d'abord à 70 puis à 61. Lorsqu'en février 1862, ils remirent au pape le résultat de leurs travaux, l'idée d'une condamnation du libéralisme avait fait son chemin. En dépit des marques de bienveillance témoignées par Pie IX aux rédacteurs du *Correspondant* pour leur dévouement dans la Question romaine, Cochin de passage à Rome avait raison d'écrire à ses amis : « Il est trop clair que les influences et la majorité ne sont pas de notre côté [4]. » C'est ce qu'indiquait d'ailleurs la mise à l'*Index* du livre de l'abbé Godard sur *Les principes de 1789 et la doctrine catholique.*

LA CANONISATION
DES MARTYRS JAPONAIS

Le bruit courut bientôt que le pape profiterait, pour faire une déclaration solennelle, de la présence à Rome de nombreux évêques convoqués à l'occasion de la canonisation de martyrs du Japon, afin de manifester en faveur du pouvoir temporel [5]. Effectivement, le projet primitif d'adresse au souverain pontife, rédigé par le cardinal Wiseman, parlait, au témoignage de Lavigerie, « non seulement du pouvoir temporel », mais

(1) Texte dans P. HOURAT, *Le Syllabus*, t. I, p. 42-56.
(2) A la suite d'une suggestion faite par le concile de Spolète de 1849 et reprise par *La Civiltà cattolica*, Pie IX chargea en 1852 le cardinal Fornari de consulter secrètement à ce sujet un certain nombre d'évêques et de laïcs. Il chargea ensuite la commission dogmatique qui avait élaboré la bulle relative à l'Immaculée-Conception de préparer le nouvel acte, mais jusqu'en 1860 le travail ne progressa guère. (P. HOURAT, *op. cit.*, t. I, p. 6-23.)
(3) D'après une lettre de Pantaleoni à Cavour du 6 décembre 1860, dans *La Questione romana negli anni 1860-1861*, t. I, p. 117-118.
(4) Cochin à Montalembert, 3 mai 1862, dans *Augustin Cochin...*, t. I, p. 278.
(5) Sur cette assemblée d'évêques, voir plus haut, p. 96-97.

encore des principes des sociétés modernes, qu'il qualifiait ainsi : *Libertates illas ridiculas quibus modernae nationes gloriantur* [1]. Mais le parti de la modération l'emporta grâce à l'intervention de Dupanloup, grâce aussi, peut-être, à Antonelli, soucieux de ne fournir au gouvernement français aucun prétexte d'abandonner la protection militaire de Rome. « L'avis général, écrivait encore Cochin, est que l'air n'est pas assez calme pour qu'on dogmatise [2]. » Toutefois, le pape critiqua sévèrement, dans son allocution *Maxima quidem*, les principes modernes visant à affranchir la philosophie, la morale et la politique du contrôle de la religion [3], et il fit remettre à chaque évêque le projet de *Syllabus* en leur demandant de transmettre leur avis à Rome dans un délai de deux mois. Nous savons que Dupanloup fit prévoir « un orage » au cas où l'on publierait pareil document [4], mais l'ensemble des avis fut favorable ; certains évêques suggérèrent même de compléter la condamnation, Ketteler notamment qui insista pour qu'on censurât aussi le rationalisme larvé de certains théologiens, tentés d'exagérer l'indépendance de la science à l'égard de l'autorité ecclésiastique [5].

AJOURNEMENT DE LA CONDAMNATION

La communication avait été faite sous e sceau du secret, mais il y eut des fuites. Dès le 19 juillet, l'ambassadeur de France communiquait la liste des propositions à son gouvernement [6], puis, au mois d'octobre, un hebdomadaire de Turin publia le texte intégral en en faisant une âpre critique, et toute la presse anticléricale eut vite fait d'emboîter le pas. Ce fut peut-être cette divulgation intempestive qui poussa le pape à donner une autre forme à son intervention. En tout cas, il chargea une nouvelle commission d'extraire de ses allocutions, lettres et encycliques antérieures les passages où il avait déjà condamné les erreurs relevées dans le premier projet, et la mise au point de ce nouveau texte se prolongea jusqu'au début de 1864.

RÉACTION DU GROUPE DU « CORRESPONDANT »

Cet ajournement de la condamnation projetée coïncidait avec une certaine détente, en France, entre les dirigeants du *Correspondant* et leurs adversaires intégristes. En effet, l'évolution de la politique ecclésiastique de Napoléon III qui, après s'être rapproché en France des milieux anticléricaux, venait de laisser Cavour s'emparer d'une bonne partie des États pontificaux, avait profondément déçu les admirateurs du « nouveau saint Louis ». Soucieux de favoriser ce revirement d'opinion,

---

(1) Lavigerie à Rouland, 10 juin 1862, dans J. MAURAIN, *La politique ecclésiastique du Second Empire*, p. 613 (confirmé par une lettre de Cochin, *op. cit.*, t. I, p. 346).
(2) Cochin à Montalembert, 7 juin 1862, *op. cit.*, t. I, p. 284.
(3) Texte dans *Acta Pii IX*, t. III, Rome, 1865, p. 451 et suiv. Lavigerie jugeait comme suit cette allocution : « Elle ressemble à tous les actes de la même espèce. Elle emploie, pour frapper les erreurs opposées à la doctrine de l'Église et malheureusement aussi aux simples préjugés de quelques-uns, les adjectifs les plus formidables. » (Lettre à Rouland, 10 juin 1862, dans J. MAURAIN, *op. cit.*, p. 614, n. 3.)
(4) F. LAGRANGE, *Vie de Mgr Dupanloup*, t. II, p. 455.
(5) G. KRUEGER, dans *Theologische Literaturzeitung*, 1924, col. 355.
(6) J. MAURAIN, *La politique ecclésiastique du Second Empire*, p. 614, n. 1.

Albert de Broglie avait insisté pour que la revue prît un ton modéré, évitât de s'engager sur le terrain dangereux des affirmations de principe et se bornât à souligner en toute occasion comment, dans la pratique, leur point de vue apparaissait comme le plus profitable à l'Église et combien l'évolution des événements justifiait en fait leurs prévisions. Il espérait également par là désarmer les préventions romaines à leur égard, déjà atténuées grâce à leur attitude très nette dans la défense de la souveraineté temporelle du pape.

Mais en réalité, dans le groupe du *Correspondant*, en dépit de l'amitié très réelle qui liait ces hommes entre eux et qui devait encore se manifester de façon touchante au cours d'une réunion intime à La Roche-en-Brény, en octobre 1862 [1], il y avait, sous une apparente unité, dualité de tendance. Les uns, comme Dupanloup, Falloux ou de Broglie, étaient des conservateurs modérés, qui ne voyaient, dans l'attitude conciliante à l'égard des libertés modernes, qu'un moyen de servir l'Église ; les autres, tout aussi conservateurs au point de vue social, prônaient par contre, avec un enthousiasme romantique, la liberté pour elle-même comme un progrès en soi et posaient en thèse que toute action coercitive ou impérative de l'Église était partout et toujours répréhensible. Ainsi pensait notamment Montalembert, qui ne pouvait supporter la « timidité » de ses amis et enrageait de ne pouvoir clamer ses convictions et fustiger ses adversaires avec « cette éloquente surabondance d'adjectifs infamants » que lui reprochait amicalement Cochin [2]. Or, soudain, une occasion inespérée s'offrit.

LE CONGRÈS DE MALINES   Au printemps de 1863, le « noble comte » fut invité à venir prendre la parole au premier congrès des catholiques belges [3], et l'un des organisateurs, Adolphe Dechamps, l'encouragea à parler franchement :

> Une tribune retentissante s'ouvre pour vous, un auditoire composé de catholiques de toutes nations, de cardinaux, d'évêques, de prêtres, de religieux... Cette tribune, il faut vous en emparer, cet auditoire, il faut vous en servir *au profit de notre commune cause*. Il importe au plus haut point que le résultat soit libéral et que le programme qui en sortira soit le vôtre : le catholicisme et la liberté [4].

C'était bien ainsi que Montalembert envisageait les choses : « Je ne parlerai pas seulement pour la Belgique, répondait-il à Dechamps, mais pour les Italiens, les Anglais, pour mes compatriotes surtout [5]. » Et il précisait quelques jours plus tard : « Il s'agit moins pour moi de persuader

---

(1) Sur cette entrevue, qui fut plus tard l'objet d'une polémique assez vive avec Veuillot, cf. E. LECANUET, *Montalembert*, t. III, p. 330-332 ; Mgr LAGRANGE, *Vie de Mgr Dupanloup*, t. II, p. 392-396 ; E. VEUILLOT, *Louis Veuillot*, t. III, p. 487-492.

(2) Cochin à Montalembert, 22 février 1862, dans *Augustin Cochin...*, t. I, p. 271.

(3) Voir sur cet épisode R. AUBERT, *L'intervention de Montalembert au congrès de Malines en 1863*, dans *Collectanea Mechliniensia*, t. XX, 1950, p. 525-551.

(4) Ad. Dechamps à Montalembert, 20 juillet 1863, dans E. DE MOREAU, *Adolphe Dechamps*, p. 440-441. A vrai dire, d'autres organisateurs redoutaient au contraire que Montalembert « ne vienne casser les vitres dans l'Assemblée de Malines en dirigeant une charge à fond contre les veuillotins français et belges ». (*Souvenirs d'A. Delmer*, dans *Revue générale*, t. XC, 1909, p. 332.)

(5) Montalembert à Dechamps, 27 juillet 1863, dans E. DE MOREAU, *op. cit.*, p. 439.

l'Assemblée de Malines et d'y être applaudi que de faire ainsi en public mon testament politique [1]. »

Montalembert parla les 20 et 21 août [2] devant un auditoire « d'abord ahuri, puis attaché, puis entraîné, subjugué, passionné » [3]. On comprend aisément l'ahurissement de l'auditoire au premier moment : l'éloquent champion du libéralisme catholique semblait prendre plaisir à jouer avec le feu en s'avançant sur le terrain des principes, alors que l'on savait que Rome songeait sérieusement à une condamnation des fameuses libertés modernes. Après un vibrant éloge de la Belgique « catholique et libérale », il reprocha aux catholiques des autres pays d'être encore souvent « par le cœur, par l'esprit, et sans trop s'en rendre compte, de l'ancien régime, c'est-à-dire du régime qui n'admettait ni l'égalité, ni la liberté politique, ni la liberté de conscience ». Or, les faits sont là : « Dans une moitié de l'Europe, la démocratie est déjà souveraine ; elle le sera demain dans l'autre. » Le temps de la théocratie est fini pour de bon, « la simple apparence d'une alliance trop intime de l'Église avec le trône suffit pour la compromettre et l'affaiblir ». L'Église n'a plus à se placer dès lors aujourd'hui que sur le terrain du droit commun, où sa liberté aura comme garantie « la liberté générale » des citoyens. Et il n'y a pas lieu de s'en affliger : « Pour moi, j'avoue franchement que, dans cette solidarité de la liberté et du catholicisme, je vois un progrès réel. » Bien plus, poursuivit-il le second jour, la plus délicate des libertés modernes, la liberté de culte, constitue elle aussi un progrès. Tout en repoussant explicitement « la ridicule et coupable doctrine que toutes les religions sont également bonnes », il n'hésita pas à affirmer, en dépit des conseils réitérés de son ami de Broglie, que le régime moderne de la tolérance civile de l'erreur était préférable à l'ancien, blâmant la révocation de l'édit de Nantes et jugeant sévèrement l'Inquisition :

L'Italie, l'Espagne et le Portugal sont là pour prouver l'impuissance radicale du système oppressif de l'ancienne alliance de l'autel et du trône pour la défense du catholicisme...

L'inquisiteur espagnol disant à l'hérétique : *la vérité ou la mort !* m'est aussi odieux que le terroriste français disant à mon grand-père : *la liberté, la fraternité ou la mort !* La conscience humaine a le droit d'exiger qu'on ne lui pose plus jamais ces hideuses alternatives.

L'éloquence de Montalembert fut saluée par un tonnerre d'applaudissements et, bien que l'orateur eût déclaré vouloir faire « non de la théologie, mais de la politique et surtout de l'histoire », le cardinal-archevêque de Malines l'assura qu'il avait parlé « en parfait théologien ». Mais ce n'était pas l'avis de tous les assistants, et le nonce envoya à Rome un rapport défavorable, dont l'effet fut sans doute accru par les commentaires de la délégation italienne à Malines, qui constituait « la quintessence du cléricalisme intégriste » [4]. Pourtant, la première réaction de

(1) Montalembert à Foisset, début août 1863, dans E. LECANUET, *op. cit.*, t. III, p. 348.
(2) Texte dans *Assemblée générale des catholiques en Belgique. Première session à Malines. 1863*, Bruxelles, 1864, p. 168-190, 303-327.
(3) Cochin à Falloux, 28 août 1863, dans *Augustin Cochin...*, t. I, p. 346-347 (cf. p. 343).
(4) S. JACINI, *Il tramonto del potere temporale...*, p. 112.

La Civiltà cattolica ne parut pas trop défavorable et, tout en rappelant avec insistance que les libertés de conscience et de presse étaient condamnables quand on les considérait comme des principes universels, *en thèse*, elle ajoutait toutefois que « *à titre d'hypothèse*, c'est-à-dire considérées comme des dispositions appropriées aux conditions spéciales de tel ou tel peuple, elles peuvent être légitimes et les catholiques peuvent les aimer et les défendre »[1].

MONTALEMBERT BLAMÉ   Mais les adversaires veillaient, dépités qu'ils étaient de voir les partisans du libéralisme catholique se prévaloir du succès du discours de Malines, publié sous le titre provoquant : *L'Église libre dans l'État libre*, et relever la tête, en se disant, comme Cochin, qu'il était « difficile de blâmer non pas un homme, mais 4.000 hommes qui l'ont acclamé »[2].

Ces adversaires trouvaient du reste un terrain favorable pour répandre leurs critiques. Non seulement le pape et son entourage restaient sous la pénible impression causée par l'introduction de l'indifférence religieuse dans la législation des régions d'Italie récemment annexées par le Piémont, là où, la veille encore, l'union des pouvoirs spirituel et temporel était la plus étroite. Mais de plus, Mgr Pie venait précisément, en condamnant la *Vie de Jésus* de Renan, de dénoncer comme l'un des plus grands scandales imputables au régime moderne le fait qu'un tel livre pût librement se répandre, « sans réclamation d'aucune autorité, avec l'applaudissement de la foule des journalistes », et nombreux étaient les catholiques que ce cas concret mettait en défiance contre une conception qu'un ami belge de Veuillot, le comte du Val de Beaulieu, allait dénoncer peu après dans une brochure qui fut fort goûtée à Rome : *L'erreur libre dans l'État libre*.

Le Monde commença par multiplier les allusions insidieuses au discours de Malines, sans toutefois en publier le texte, de sorte que ses lecteurs pouvaient imaginer le pire. Puis Mgr Pie, qui était en France le champion de l'école intégriste, et dont le crédit était grand auprès du pape, suggéra à celui-ci d'intervenir officiellement.

L'embarras de Pie IX était grand. Il n'avait pas lu le discours[3], mais il en connaissait la tendance et trouvait les conceptions prônées par Montalembert inadmissibles dans leur forme absolue. « L'Église, affirmait-il à un visiteur, n'admettra jamais comme un bien et un principe que l'on puisse prêcher l'erreur et l'hérésie à des peuples catholiques. Le pape veut bien la liberté de conscience en Suède et en Russie ; mais il ne la veut pas en principe[4]. » Il était d'autant plus mécontent que, comme

---

(1) *Civ. catt.*, 17 octobre 1863, p. 118 et suiv., en particulier p. 147. Cet article était le premier à formuler de manière aussi nette la distinction destinée à devenir classique. On sait que, à Paris, beaucoup commencèrent par sourire de cette distinction qu'on jugeait bien subtile. Faisant allusion à la vie mondaine de Mgr Chigi, ils s'en allaient répétant : « La thèse, c'est quand le nonce dit qu'il faut brûler les juifs ; l'hypothèse, c'est quand il dîne chez M. de Rothschild ! »

(2) Cochin à Falloux, 28 août 1863, dans *Augustin Cochin...*, t. I, p. 346.

(3) Témoignage cité par W. WARD, *W. G. Ward and the Catholic Revival*, p. 243.

(4) L'abbé de Briey à Mgr Pie, 30 octobre 1863, dans Mgr BAUNARD, *Histoire du cardinal Pie*, t. II, p. 215.

beaucoup d'autres à Rome, il rapprochait la manifestation malinoise
des allures indépendantes que venaient d'afficher à l'égard du Saint-Siège
les théologiens allemands réunis à Munich autour de Doellinger. Mais par
ailleurs, un homme aussi sensible que Pie IX pouvait difficilement oublier
le dévouement chevaleresque témoigné par Montalembert à l'Église
depuis tant d'années et il subissait en outre les assauts répétés de Dupanloup, qui passa tout l'hiver à Rome afin d'essayer de conjurer le double
danger qu'il redoutait : un blâme personnel pour son ami ; une condamnation explicite des libertés modernes, que le nouvel incident risquait
de précipiter. Le crédit de l'évêque d'Orléans auprès du pape était grand
depuis qu'il avait mis toute son ardeur fougueuse à ameuter l'opinion
catholique en faveur du pouvoir temporel [1]. Mais il avait affaire à forte
partie : « Les passions, écrivait-il, ne peuvent pas être plus ardentes
qu'elles ne le sont. Je parle des passions belges, françaises et anglaises [2]. »
Et à Rome même, les intransigeants comme dom Pitra appuyaient les
dénonciations qui venaient de l'étranger.

Finalement, après des alternatives d'espoir et d'inquiétude, dues à
l'esprit changeant de Pie IX, Dupanloup ne put empêcher un blâme
discret à l'endroit de Montalembert. Au début de mars 1864, le cardinal
Antonelli était chargé d'envoyer deux lettres, l'une à l'archevêque de
Malines, pour lui faire part en termes assez vagues du déplaisir de Rome,
l'autre, plus explicite, à l'orateur lui-même [3]. Ces lettres ne furent pas
rendues publiques, mais le coup, pour être atténué, n'en fut pas moins
amèrement ressenti, d'autant plus qu'Antonelli laissait entrevoir que
ce désaveu personnel serait bientôt suivi d'une condamnation moins
discrète des doctrines libérales.

VERS LA PUBLICATION D'UNE ENCYCLIQUE La publication de ce document solennel fut
cependant retardée jusqu'à la fin de l'année
sous l'effet de diverses influences.

En Belgique, le cardinal Sterckx et les hommes politiques catholiques
craignaient qu'une condamnation des libertés modernes ne fît le jeu
du parti libéral, qui affirmait que les catholiques n'acceptaient pas sincèrement la constitution. A la fois pour éclairer Rome et pour apaiser les
controverses qui s'étaient rallumées entre le *Journal de Bruxelles* et
Le Bien Public après le discours de Montalembert, Sterckx publia en
mars une justification de la constitution qui fit une grosse impression
en France comme en Belgique, et cette prise de position embarrassa le
Saint-Siège [4]. D'autre part, Adolphe Dechamps, qui, au lendemain du

---

(1) On a pu composer un ouvrage de près de 300 pages avec les brefs de reconnaissance et de
félicitation adressés à Dupanloup par Pie IX. Celui-ci, tout en l'appréciant fort, regrettait toutefois
l'insuffisance de ses positions doctrinales et son enthousiasme pour les idées modernes : « *È eccelente,
ma troppo francese* », confiait-il au cardinal de Reisach en décembre 1863 (*Janssens Briefe*, édit.
PASTOR, t. I, p. 242 ; cf. p. 233).
(2) Cité dans F. LAGRANGE, *Vie de Mgr Dupanloup*, t. III, p. 436-437.
(3) Voir la lettre à Sterckx dans *Collectanea Mechliniensia*, t. XX, 1950, p. 550, n. 1, et celle à
Montalembert dans E. LECANUET, *Montalembert*, t. III, p. 373-374.
(4) Voir sur l'incident A. SIMON, *Le cardinal Sterckx et son temps*, t. I, p. 237-251. La brochure
de Sterckx était intitulée *La Constitution belge et l'Encyclique de Grégoire XVI. Deux lettres sur
nos libertés constitutionnelles*. Fidèle à sa ligne de conduite, le cardinal se cantonnait sur le terrain

fameux discours de Malines, avait déjà averti Antonelli qu'une condamnation de celui-ci équivaudrait à un blâme de la constitution, dont il
n'était « que le commentaire » [1], revint à la charge au mois de mars, pour
répondre aux instances de ses amis du *Correspondant* ; il adressa au pape
un mémoire où il énumérait les dangers qu'entraînerait une encyclique
condamnant les principes à la base des constitutions modernes. Puis,
comme à Paris on trouvait qu'il ne suffisait pas d'écrire, il chargea le
comte de Liedekerke d'aller exposer personnellement à Rome les appréhensions des milieux catholiques belges à la veille d'élections importantes [2].
Une lettre de Léopold Ier à Pie IX appuyait cette démarche [3]. Enfin,
au même moment, le gouvernement français, alerté par le groupe de
Mgr Maret, ordonnait à son ambassadeur d'intervenir auprès du cardinal
Antonelli, qui donna l'impression de comprendre le caractère inoppor-
tun de la condamnation projetée [4].

L'encyclique fut ajournée. Mais les pressions en sens contraire se
faisaient de plus en plus vives. Les évêques ultramontains de France
réclamaient une condamnation, le Saint-Office y poussait, de même que
certains cardinaux [5]. Lorsque la convention de septembre rendit manifeste qu'on ne gagnait pas grand'chose en ménageant le gouvernement
de Napoléon III, l'un des principaux arguments d'Antonelli pour engager à la prudence disparut, et Pie IX, au début de décembre, se rallia à
l'avis de ceux qui, comme Mgr de Mérode [6], le poussaient à se prononcer
enfin sans équivoque sur le plan des principes.

L'ENCYCLIQUE « QUANTA CURA »
ET LE « SYLLABUS »   Datée de la fête de l'Immaculée-
Conception, l'encyclique tant atten-
due parut en réalité vers le milieu
de décembre [7]. Le pape y condamnait, en des termes où l'on sentait par
moment vibrer l'indignation, les principales erreurs modernes : le rationalisme, qui va jusqu'à nier la divinité du Christ ; le gallicanisme, qui
exige une sanction du pouvoir civil pour l'exercice de l'autorité ecclésiastique ; l'étatisme, qui vise au monopole de l'enseignement et supprime les
ordres religieux ; le socialisme, qui prétend soumettre totalement la
famille à l'État ; la doctrine des économistes, qui considèrent l'organisation de la société comme n'ayant d'autre but que l'acquisition des

Pratique. Il ne reprenait ni les idées de Montalembert, ni même la distinction entre la thèse et l'hypothèse, mais montrait que, en fait, « les libertés religieuses consignées dans la constitution belge
ont eu pour but principal de favoriser la religion catholique » (p. 30).

(1) *Mémoire* du 15 septembre 1863, « pour être mis sous les yeux de Sa Sainteté », dans E. DE
MOREAU, *Adolphe Dechamps*, p. 445-448.

(2) Voir sur cette seconde intervention de Dechamps, E. DE MOREAU, *op. cit.*, p. 455-462. Falloux lui écrivait, le 29 mars 1864 : « Si Dieu permet que nous échappions au péril, c'est par vous,
Monsieur, qu'il aura opéré ce miracle. » (*Ibid.*, p. 460.)

(3) E. DE MOREAU, *op. cit.*, p. 459.

(4) J. MAURAIN, *op. cit.*, p. 702 ; G. BAZIN, *Vie de Mgr Maret*, t. II, p. 71-74. Le 16 avril, le ministre de Belgique à Rome écrivait : « L'encyclique *paraît* ajournée pour quelques temps.... L'ambassadeur de France m'a dit avoir conseillé de la faire aussi modérée que possible. » (*Arch. Min. A. É.*
Bruxelles, Saint-Siège, t. XII.)

(5) Voir à titre d'exemple le *Mémoire* du cardinal Pitra, cité dans A. BATTANDIER, *Le cardinal*
Pitra, p. 490, où l'école catholique libérale est comparée à Port-Royal.

(6) Sur l'opposition entre Mgr de Mérode et Antonelli sur ce point, voir R. AUBERT dans *Collec-*
tanea Mechliniensia, t. XIX, 1949, p. 310-311, en note.

(7) Texte dans *Acta Pii IX*, t. III, Rome, 1865, p. 701 et suiv.

richesses ; enfin et surtout le naturalisme, qui considère comme un progrès que la société humaine soit constituée et gouvernée sans tenir compte de la religion et qui revendique dès lors comme un idéal la laïcisation des institutions, la séparation de l'Église et de l'État, la liberté de la presse, l'égalité des cultes devant la loi, la liberté de conscience totale, regardant comme le meilleur régime « celui où l'on ne reconnaît pas au pouvoir le devoir de réprimer par la sanction des peines les violateurs de la religion catholique ».

A cette encyclique était annexé un catalogue de 80 propositions jugées inacceptables [1], désigné sous le nom de *Syllabus*. Elles concernaient le panthéisme et le naturalisme ; le rationalisme, qui revendique notamment pour la philosophie et la théologie une indépendance absolue par rapport au magistère ecclésiastique ; l'indifférentisme, qui considère que toutes les religions se valent ; le socialisme, le communisme et la franc-maçon-nerie ; le gallicanisme ; les fausses doctrines sur les relations de l'Église et de l'État ; les conceptions morales erronées sur le mariage chrétien ; la négation du pouvoir temporel des papes ; enfin le libéralisme moderne. Ces propositions avaient été extraites par la commission dont il a été question plus haut de divers documents pontificaux qui les avaient déjà condamnées, mais, retirées ainsi de leur contexte, elles présentaient parfois un aspect déconcertant. Telle la dernière qui désavouait l'affirmation suivante :

80. Le pontife romain peut et doit se réconcilier et transiger avec le pro-grès, avec le libéralisme et la civilisation moderne [2].

PREMIÈRES RÉACTIONS   L'encyclique *Quanta cura*, protégée par son style solennel, n'attira guère l'attention en dehors des milieux ecclésiastiques, mais il en alla tout autrement du *Syllabus*, où les mêmes doctrines étaient formulées d'une manière tranchante et brève, aisément accessible au grand public. Jamais peut-être un document pontifical ne souleva pareille émotion.

S'il fut accueilli avec enthousiasme par les ultramontains ardents, qui se croyaient désormais en mesure d'acculer l'école adverse à sortir de l'Église ou à rallier leur point de vue, il heurta dès sa publication l'opinion moyenne. La majorité des catholiques étaient stupéfaits en croyant voir le pape se prononcer contre le progrès et la civilisation moderne, et les milieux anticléricaux, aidés involontairement par les commentaires sans nuances de certains journaux intégristes [3], s'empres-sèrent d'exploiter contre l'Église certaines propositions qui, isolées de leur contexte, devaient inévitablement faire scandale : un journal pié-

---

(1) Sur la valeur dogmatique exacte de ce document de forme insolite, cf. L. Choupin, *Valeur des décisions doctrinales et disciplinaires du Saint-Siège*, 3e édit., Paris, 1929, p. 119-157. On trou-vera dans le même ouvrage, aux p. 187 à 415, un commentaire détaillé des 80 propositions.

(2) Le contexte de l'allocution *Jamdudum*, d'où cette proposition est extraite (Cf. H. Denzin-ger, *Enchiridion*, nos 1777-1780), montre clairement que ce qui est rejeté, c'est le progrès et la civilisation moderne *tels que les entendent les ennemis de l'Église*, c'est-à-dire consistant essentielle-ment à rejeter l'influence de l'Église dans la société.

(3) Le 13 janvier, *Le Monde* reproduisait avec éloges le passage suivant du *Pensamiento español* : « Notre foi unique est de stigmatiser comme anticatholique le libéralisme, le progrès et la civili-sation moderne. Nous condamnons comme anticatholiques ces avortons de l'enfer. »

montais n'alla-t-il pas jusqu'à affirmer que le pape, qui venait de condamner les découvertes de la science moderne, allait supprimer dans ses États les trains, le télégraphe, les machines à vapeur et l'éclairage au gaz [1] ! La presse irréligieuse ne cachait pas sa joie d'une pareille aubaine. Quant au gouvernement de Napoléon III, comme il semblait clair que les nouveaux documents visaient avant tout la France [2], il prétendit interdire aux évêques « la publication de ces actes qui contiennent des propositions contraires aux principes sur lesquels repose la constitution de l'Empire » [3].

A Rome, on était un peu décontenancé par tout ce bruit. Si le pape ne s'émouvait guère et répétait que, si c'était à refaire, il n'agirait pas autrement, plusieurs membres de la Curie ne dissimulaient pas leurs réserves concernant la forme rigide et absolue de l'intervention pontificale [4]. Antonelli prodiguait aux diplomates des déclarations lénifiantes, déclarant que « c'était contre l'esprit du socialisme et contre les mauvaises passions du siècle qu'était dirigé l'anathème du Vatican et qu'il s'étonnait que l'empereur, qui représente les idées conservatrices, ne vît pas tout de suite ce qu'il y avait de conservateur dans les doctrines émises par le Saint-Père », ajoutant encore que celles-ci formulaient la thèse, l'idéal de l'Église, nullement l'hypothèse, la règle de conduite tenant compte des circonstances [5]. Les diplomates ne firent d'ailleurs aucune difficulté pour admettre cette interprétation et ce n'est certainement pas eux qui jetèrent de l'huile sur le feu [6].

C'est aussi de cette façon que l'encyclique et le *Syllabus* furent compris par les catholiques libéraux de Belgique, où Dechamps écrivait : « L'encyclique du 8 décembre nous laisse dans la même position que l'encyclique de 1832 nous avait faite [7]. » Mais en France, tous ceux qui souhaitaient un rapprochement entre l'Église et le monde moderne étaient consternés. Mgr Darboy et Mgr Maret s'empressèrent, en vain d'ailleurs, de conseiller au ministre Baroche d'engager des négociations avec Rome pour obtenir des modifications ou au moins des éclaircissements, tandis que d'autres évêques, moins gouvernementaux, s'adressaient directement au pape pour lui demander une interprétation rassurante de l'encyclique [8]. Dans

---

(1) Sur les réactions de la presse anticléricale, cf. A. QUACQUARELLI, *La crisi della religiosità contemporanea*, p. 18-20, 45-46.

(2) À vrai dire, le pape en voulait au moins autant, sinon plus, à l'Italie. Le 31 décembre 1864, au sortir d'une audience, le curé Debeauvais écrivait à Dupanloup : « Il ne veut, dit-il, s'en prendre dans les détails qu'au royaume d'Italie et aux associations clérico-libérales du Piémont ; il a répété au duc Salviati qu'il désirait que cela fût bien su en France. » (*Arch. S. Sulp.*, fonds Dupanloup.)

(3) Circulaire de Baroche du 1ᵉʳ janvier 1865. Les évêques n'en tinrent pratiquement pas compte.

(4) Voir les dépêches du ministre de Belgique à Rome des 7 et 14 janvier 1865 et celles de l'ambassadeur de France des 27 et 31 décembre 1864, citées par R. AUBERT, dans *Collectanea Mechliniensia*, t. XIX, 1949, p. 312-313.

(5) Dépêches de l'ambassadeur de France des 3 et 7 janvier 1865, dans J. MAURAIN, *op. cit.*, p. 714.

(6) Voir, pour l'ambassadeur de France, R. AUBERT, *art. cit.*, p. 313-314 ; pour celui d'Autriche, S. JACINI, *Il tramonto del potere temporale*, p. 134.

(7) Adolphe Dechamps à Cochin, 4 janvier 1865, dans E. DE MOREAU, *Adolphe Dechamps*, p. 464. Dechamps devait toutefois reconnaître par la suite que, s'il avait raison sur le plan pratique et politique, il devait par contre, sur le plan des idées, nuancer et corriger un peu ses théories. (E. DE MOREAU, *op. cit.*, p. 428-429.)

(8) J. MAURAIN, *op. cit.*, p. 708-709, 716.

le groupe du *Correspondant*, le désarroi était extrême : Montalembert et Albert de Broglie estimaient que tout était fini, malgré l'avis contraire de Falloux et de Foisset. Le découragement n'était pas moindre parmi les disciples : « En province, écrivait-on à Montalembert, la situation des catholiques libéraux est fort triste ; les gens du *Monde* les traitent comme des hérétiques, comme des *pestiférés*, et les incrédules avec une commisération outrageante [1]. »

C'est à ce moment, où les exagérations des partisans de Veuillot risquaient de compromettre gravement le catholicisme, que se produisit l'intervention de Mgr Dupanloup qui renversa la situation.

LA BROCHURE DE DUPANLOUP   Au moment de la publication du *Syllabus*, l'évêque d'Orléans venait d'achever un écrit protestant contre la convention de septembre. Dès qu'il se rendit compte de l'émotion soulevée, il se remit à l'œuvre et, travaillant jour et nuit avec la collaboration de Cochin, il improvisa rapidement une seconde partie où il replaçait dans leur contexte les propositions les plus discutées en les éclairant à l'aide de la distinction entre la thèse et l'hypothèse. Il les rendait de la sorte parfaitement acceptables, même la fameuse proposition 80, à propos de laquelle il expliquait que le pape n'a à « se réconcilier » ni avec ce qui est bon dans la civilisation moderne, puisqu'il n'avait jamais cessé de l'encourager, ni avec ce qui est mauvais et qu'il devait évidemment continuer à condamner. Faguet exagère quand il dit que Dupanloup « fait du *Syllabus* et de l'encyclique un commentaire qui est une condamnation respectueuse de l'encyclique et du *Syllabus* » [2], mais il est exact que « l'encyclique ainsi commentée devenait le document le plus inoffensif et les pointes les plus aiguës étaient toutes émoussées » [3].

Albert de Broglie, qui le constate avec étonnement, continue : « L'écrit parut, et je n'ai de ma vie vu pareil effet. La société, qui ne respirait plus, eut un soulagement comme un homme sur le point d'étouffer à qui on aurait coupé la corde qui lui serrait la gorge. » Sans doute, d'autres mandements épiscopaux étaient déjà entrés dans cette voie, mais tous les contemporains s'accordent à souligner l'effet extraordinaire de la brochure de Dupanloup, publiée sous le titre : *La convention du 15 septembre et l'encyclique du 8 décembre*. Annoncée le 23 et mise en vente le 26 janvier, la première édition fut enlevée en deux heures, et trois semaines plus tard 100.000 exemplaires étaient en circulation, sans parler des multiples traductions. Les nonces de Lisbonne, de Munich et de Vienne écrivaient à celui de Paris le bien que l'écrit faisait autour d'eux et ce dernier s'était empressé dès le premier jour de féliciter l'auteur. De tous les pays, même d'Amérique et d'Océanie, 630 évêques lui écrivirent pour le remercier d'avoir si bien traduit leurs propres sentiments et le pape lui-même confiait à un ami de l'évêque d'Orléans : « Il a expliqué et fait comprendre l'encyclique comme il faut qu'on la comprenne. »

---

(1) Cité dans E. Lecanuet, *Montalembert*, t. III, p. 384.
(2) E. Faguet, *Mgr Dupanloup*, p. 162.
(3) A. de Broglie, *Mémoires*, t. I, p. 313-314.

Dépités, les journaux anticléricaux reprochaient à Dupanloup d'avoir « transfiguré » l'encyclique [1], tandis que les intransigeants, très mécontents, l'accusaient de l'avoir défigurée. Veuillot, qui se trouvait précisément à Rome, fit immédiatement remettre au pape des notes à ce sujet, mais Pie IX savait gré à Dupanloup d'avoir apaisé la tempête et il avait surtout apprécié la critique de la convention de septembre et la protestation indignée contre l'interdiction de publier l'encyclique, dont l'évêque avait eu l'habileté d'encadrer son commentaire. Il lui adressa, le 4 février, un bref de remerciement conçu en termes prudents et se terminant par le souhait « qu'il saurait d'autant mieux par la suite exposer la véritable pensée contenue dans l'encyclique qu'il avait réfuté avec plus d'énergie les interprétations erronées ».

L'école intransigeante crut déceler une réserve dans ces derniers mots et s'opposa autant qu'elle le put à l'interprétation des documents pontificaux proposée par Dupanloup. Mgr Pie notamment critiqua ceux qui s'appliquaient « à établir qu'après l'encyclique, il n'y a pas plus de lumière qu'auparavant, et que toutes les mêmes opinions peuvent être aussi librement soutenues... L'acte du 8 décembre est dirigé contre les adversaires, contre ceux du dehors, c'est vrai, mais il s'adresse encore plus, s'il est possible, à ceux de la maison » [2]. Quant à Veuillot, il publia, l'année suivante, une brochure extérieurement semblable à celle de l'évêque d'Orléans, qu'il intitula *L'illusion libérale*, et dont le but, au témoignage de son frère, « était moins de réfuter l'ennemi que de prendre à partie les catholiques libéraux », accusés de n'être ni catholiques ni libéraux. On y retrouvait sa disposition foncière à voir tout déterminé d'en haut, ainsi que sa méfiance à l'égard de tout ce que l'homme fait par lui-même, y compris les inventions et le mouvement du siècle, telle qu'il l'avait déjà exprimée, quelques années auparavant, dans *Les Parfums de Rome*. Le pape en regretta la forme un peu cassante, mais aurait ajouté : « Ce sont absolument mes idées [3]. »

Après le *Syllabus* comme avant, les deux partis se retrouvaient donc sur leurs positions. La condamnation pontificale avait certes fait réfléchir, mais son caractère trop rigide et surtout l'interprétation que les intégristes avaient voulu en donner avaient fourni par contre-coup à Dupanloup l'un de ses plus beaux triomphes.

LES REMOUS EN ITALIE   L'agitation autour du *Syllabus* fut moins profonde en Italie qu'en France, mais les réactions y furent assez semblables. A Turin, tandis que 150.000 catholiques manifestaient au pape leur gratitude pour son intervention, Passaglia, dont l'influence restait grande sur le clergé à tendance libérale, entreprit dans son hebdomadaire *Il Messagero* une vive critique du

(1) Tel était aussi l'avis du ministre des Cultes. (Voir sa lettre à son fils du 28 janvier 1865, dans J. Maurain, *Un bourgeois français au XIXᵉ siècle : Baroche*, Paris, 1936, p. 325.) C'était même le sentiment de Montalembert, qui écrivait, le 5 février 1865 : « Je suis de ceux qui trouvent que vous avez changé l'enfant en nourrice et, comme dit le *Journal des débats*, que vous faites bénir par Pie IX Jacob au lieu d'Esaü. » (*Arch. S. Sulp.*, fonds Dupanloup.)
(2) *Entretiens avec le clergé* (10 juillet 1865), dans *Œuvres*, t. V, p. 436.
(3) E. Veuillot, *Louis Veuillot*, t. III, p. 500-503.

document, à la préparation duquel il avait jadis collaboré. A Naples et
à Palerme, les francs-maçons brûlèrent publiquement les deux documents
pontificaux et, d'une manière générale, la presse anticléricale les exploita
comme en France pour discréditer l'Église.

ET EN ALLEMAGNE   En Autriche, le gouvernement, où l'influence libérale allait croissant, songea lui aussi, un instant,
à en interdire la publication, mais il recula devant une violation aussi
nette du concordat. L'émotion était d'ailleurs beaucoup moins vive
dans les pays germaniques. L'école de Mayence, qui, en dehors des
milieux intellectuels, influençait la grande masse des catholiques,
accueillit le *Syllabus* avec joie, mais en l'interprétant généralement dans
le sens le plus modéré. Au *Katholikentag* de Trèves, en septembre 1865,
Heinrich parla avec éloges de ce document, « l'acte le plus important du
XIXe siècle, et peut-être de nombreux siècles »[1], mais il le présenta comme
un remède, amer peut-être mais efficace, aux profondes erreurs du libéralisme antichrétien tel qu'il s'incarnait dans les partis radicaux et dans
la philosophie matérialiste et panthéiste du temps ; il ne chercha donc
pas à en faire un instrument de lutte contre les catholiques qui considéraient les libertés civiles comme favorables à l'Église dans la pratique.

Certains jésuites, pourtant, essayèrent d'aller plus loin. De 1865 à
1869, les *Stimmen aus Maria Laach* publièrent autour du *Syllabus* de
nombreux articles assez peu nuancés. Tel fut aussi le cas de la série de
brochures : *Der Papst und die modernen Ideen*, éditées à Vienne de 1864
à 1867 sous la direction du P. Schrader : elles reflétaient l'opinion de
la majorité du clergé autrichien, généralement conservateur et porté
à interpréter le *Syllabus* dans un sens maximaliste. Schrader, non content
de présenter celui-ci comme un document infaillible, en traduisit les
idées condamnées en propositions affirmatives d'une façon qui n'était
pas toujours exacte et qui fut blâmée par Scheeben, peu suspect cependant[2]. Il écrivit en outre, à propos des propositions 77 à 79 : « En face
de ces propositions, on ne peut plus dire aujourd'hui, comme on l'a
répété tant de fois, qu'il n'existe aucun principe ecclésiastique empêchant
un catholique de penser qu'il est des circonstances où l'État ne peut
rien faire de mieux que d'accorder une pleine liberté religieuse[3]. » C'était
rejeter la doctrine de l'hypothèse et critiquer la position prise par Ketteler
dont ce passage reproduisait textuellement une phrase.

Ces exagérations poussèrent l'évêque de Mayence à s'expliquer une
nouvelle fois, et il crut pouvoir maintenir la même position qu'avant le
Syllabus. Dans le chapitre intitulé *Le libéralisme et l'encyclique* de son
livre sur *L'Allemagne après la guerre de 1866*, il défendit les libertés de
conscience et de culte, non pas, expliquait-il, par indifférentisme religieux
ou comme un idéal, mais parce que ces libertés « entendues dans le sens
de la constitution prussienne étaient comme la meilleure réglementation

----

(1) *Verhandlungen...*, Trèves, 1865, p. 63-74.
(2) Cf. S. MERKLE, dans *Deutsche Literaturzeitung*, 1925, N. F., t. II, col. 1457-1458.
(3) *Die Encyclica vom 8 dezember 1864*, Vienne, 1865, p. 33.

des affaires ecclésiastiques de ce pays et même comme une nécessité ».

Les seuls en Allemagne à réagir violemment contre les documents pontificaux furent les membres de l'école de Munich, mais cette opposition ne se manifesta guère au grand jour et Doellinger, dont l'évolution antiromaine en fut accélérée, n'osa pas publier le violent pamphlet [1] où, au lieu de chercher, comme Dupanloup, à atténuer la portée des condamnations en en expliquant le sens, il accusait au contraire les innovations théologiques qu'il croyait découvrir dans cette « charte du parti ultramontain partant à la conquête de l'Allemagne ».

LE LIBÉRALISME CATHOLIQUE EN FRANCE AU LENDEMAIN DU « SYLLABUS »

A la suite du *Syllabus*, nombre de catholiques estimèrent que leur enthousiasme pour les idées modernes avait eu quelque chose d'excessif et se rapprochèrent de l'école de *L'Univers* [2]. D'autres jugèrent plus nécessaire que jamais de continuer à résister aux thèses intransigeantes qui, d'après eux, acculaient la religion à une impasse.

Le groupe du *Correspondant*, auquel étaient venus se joindre, avec Gratry, quelques jeunes ecclésiastiques de talent comme le P. Didon ou l'abbé Perreyve, continua à se tenir en contact étroit avec les catholiques constitutionnels de Belgique, et il s'empressa de faire connaître en France le nouvel ouvrage de Ketteler. Pourtant, ses membres les plus modérés, surtout Foisset et de Broglie, conscients du fait que leur attitude, « sans être précisément censurée, n'avait pas l'approbation du chef de l'Église » [3], estimèrent prudent d'éviter d'aborder dans leur revue, dont le succès allait croissant [4], les questions polémiques ou trop controversées, au grand mécontentement de Montalembert, qui en était arrivé à considérer Veuillot comme « l'ennemi le plus redoutable de la religion que le XIXᵉ siècle ait produit » [5], et qui aurait voulu pourfendre publiquement les jésuites de *La Civiltà cattolica*, « qui prennent chaque jour à tâche, en défendant l'Église et le Saint-Siège, d'outrager la raison, la justice et l'honneur » [6]. Sous l'influence conjuguée de la maladie et des déceptions, l'entraîneur enthousiaste de jadis devenait chaque jour plus amer, allant jusqu'à traiter de « prévaricateurs » [7] ses anciens compagnons de lutte.

Peu porté aux violences d'expression d'un Montalembert, le cercle qui gravitait autour de Mgr Maret [8] n'en tenait pas moins sur le fond

---

(1) Publié après sa mort par A. REUSCH, *Kleinere Schriften gedrückte und ungedrückte von Doellinger*, Stuttgart, 1892, p. 197-227. Cf. J. FRIEDRICH, *Ign. von Doellinger*, t. III, p. 395-401.

(2) Notamment les jésuites qui dirigeaient la revue *Les Études*. Ils s'étaient jusqu'alors montrés sympathiques aux tendances catholiques libérales, mais les articles du P. Matignon sur la notion catholique de liberté marquèrent un net changement d'attitude.

(3) A. DE BROGLIE, *Mémoires*, t. I, p. 314.

(4) Dans les mois qui suivirent le *Syllabus*, le nombre des abonnés, qui avait déjà passé en dix ans de 800 à 4.500, s'accrut encore de plusieurs centaines. (E. LECANUET, *Montalembert*, t. III, p. 398-399.) L'historien allemand Janssen, si méprisant pour la production française du temps, en parle, en 1868 et 1869, en termes très élogieux. (*Janssens Briefe*, édit. PASTOR, t. I, p. 356 et 369.)

(5) Montalembert à Foisset, 26 décembre 1866, dans E. LECANUET, *op. cit.*, t. III, p. 417.

(6) Extrait d'un article (que *Le Correspondant* n'osa pas publier) sur *L'Espagne et la liberté*.

(7) Lettre du 16 janvier 1870, dans A. HOUTIN, *Autour d'un prêtre marié*, Paris, 1909, p. 238.

(8) S'y rattachaient entre autres, à des titres divers, Mgr David, évêque de Saint-Brieuc, Mgr Gi-

des choses des positions plus avancées que les amis de Dupanloup. Dès
la fin de janvier 1865, Maret avait composé, à l'intention du pape et de
l'épiscopat, un long mémoire où, plutôt que de recourir à la distinction
entre la thèse et l'hypothèse, il préférait montrer le caractère contingent
de la thèse elle-même, et présentait plusieurs des principes libéraux qui
sont à la base des constitutions modernes comme le résultat heureux
des progrès de la conscience humaine [1]. En parfaite communion avec
ces idées où l'on voit déjà s'esquisser un nouvel « humanisme » qui ne
deviendra pleinement conscient de lui-même qu'au cours du XXe siècle,
l'archevêque de Paris, Mgr Darboy, avait terminé sa réfutation des
critiques adressées par la presse anticléricale au *Syllabus* par un appel,
que certains jugèrent impudent, à la modération du souverain pontife :

> Vous venez de signaler et de condamner les principales erreurs de notre
> époque. Tournez maintenant vos yeux vers ce qu'elle peut avoir d'honorable et
> de bon et soutenez-la dans ses généreux efforts (...). C'est à vous qu'il appar-
> tient de réconcilier la raison avec la foi, la liberté avec l'autorité [2].

Quelques-uns n'hésitaient pas à aller plus loin encore dans la voie
de la conciliation avec le monde moderne et le progrès des sciences.
Sans parler des nombreux laïcs, qui étaient au fond plus libéraux que
catholiques, il s'était formé à Paris, grâce à la tolérance des derniers
archevêques, un milieu ecclésiastique très hostile au Saint-Siège, d'une
grande hardiesse d'idées et parfois même de conduite. Bien que n'appartenant pas strictement à ce milieu, on peut considérer comme représentatif
de cette dernière catégorie le carme Hyacinthe Loyson, prédicateur très
influent, soutenu par Mgr Darboy, que son libéralisme de plus en plus
accentué devait mettre bientôt en conflit avec les autorités de son ordre,
et qui, plutôt que de s'incliner, préféra quitter l'Église au cours de l'été
1869 [3]. Il devait par la suite caractériser très exactement la situation :
« Au fond, il y avait un abîme entre les catholiques libéraux, y compris
Montalembert, et moi. Ils étaient résolument et avant tout catholiques
romains ; moi, je ne l'étais plus, alors même que j'en prenais encore et
sincèrement le nom [4]. »

A Rome, où ces distinctions n'apparaissaient guère, les excès manifestes de ce dernier groupe contribuaient à discréditer les précédents,
qui devinrent d'ailleurs d'autant plus suspects qu'au cours des dernières
années du Second Empire ils laissèrent apparaître des sympathies gallicanes de plus en plus prononcées.

nouilhac, évêque de Grenoble, les abbés Meignan et Hugonin, bientôt évêques eux-mêmes, Mgr
Isoard, auditeur de Rote, qui envoyait de Rome, sur les milieux intransigeants de là-bas, d'acerbes
réflexions (reproduites dans A. HOUTIN, *Le P. Hyacinthe dans l'Église romaine*, Paris, 1920).
    (1) Ce *Mémoire*, fort intéressant et très modéré de ton, est reproduit *in extenso* dans G. BAZIN,
Vie de Mgr Maret, t. II, p. 321-344 (voir spécialement p. 330 et 337-339).
    (2) Cité dans Mgr FOULON, *Histoire de Mgr Darboy*, Paris, 1889, p. 329-330.
    (3) Sur Loyson (1827-1912), voir A. HOUTIN, *Le P. Hyacinthe*, 3 vol., Paris, 1920-1924 (nombreux
documents) et G. RIOU, *Le P. Hyacinthe et le libéralisme d'avant le concile*, Paris, 1910.
    (4) Loyson à Houtin, septembre 1909, dans A. HOUTIN, *Autour d'un prêtre marié*, p. 223, n. 1.

CHAPITRE IX

LES PROGRÈS DE L'ULTRAMONTANISME [1]

§ 1. — Le courant ultramontain au début du pontificat de Pie IX [2].

LA RENAISSANCE DE L'ULTRAMONTANISME PENDANT LA PREMIÈRE MOITIÉ DU XIXᵉ SIÈCLE

Par une conséquence imprévue, la crise révolutionnaire, comme on a pu le voir dans le volume précédent, avait singulièrement contribué à relever le prestige de la papauté, tant en France que dans les pays germaniques. Tandis que le pape apparaissait soudain aux yeux de toute l'Europe comme le symbole du principe d'ordre et d'autorité en face de la Révolution menaçante, l'Église gallicane avait vu s'effondrer, avec la monarchie, son principal appui. Le clergé français, dépouillé de ses privilèges, fut vite amené à constater que la meilleure politique consistait à présent à se serrer étroitement autour du chef de l'Église et les préjugés contre Rome se dissipèrent d'autant plus aisément que la persécution des prêtres réfractaires, punis pour avoir obéi au Saint-Siège, fortifia leur dévouement, tandis que les violences du Directoire et de Napoléon à l'égard de Pie VI et de Pie VII contribuaient à les rendre populaires. Par ailleurs, en obligeant Pie VI, lors du Concordat, à remanier les diocèses et à exiger la démission de tous les évêques, le premier consul avait porté un coup mortel à la théologie gallicane en amenant le pape à affirmer par ces actes l'étendue réelle des pouvoirs

(1) BIBLIOGRAPHIE. — W. WARD, *W. G. Ward and the Catholic Revival*, Londres, 1893, chap. v ; C. BUTLER, *The Vatican Council*, t. I, Londres, 1930, chap. II à IV ; E. CAMPANA, *Il Concilio Vaticano*, vol. I : *Il Clima del Concilio*, Lugano, 1926 ; J. FRIEDRICH, *Geschichte des vatikanischen Konzils*, t. I : *Vorgeschichte b. z. Eröffnung des Konzils*, Bonn, 1877 (très tendancieux) ; J. VON SCHULTE, *Geschichte der Quellen und Literatur des canonischen Rechtes*, t. III, Stuttgart, 1880 (tendancieux).

En ce qui concerne spécialement la France : A. KELLER, *La fin du gallicanisme et Mgr Maret*, Alençon, 1900 ; P. DE QUIRIELLE, *Pie IX et l'Église de France*, dans *Ann. E. Sc. P.*, t. V, 1890, p. 490-514 ; t. VI, 1891, p. 111-148 ; E. MICHAUD, *De la falsification des catéchismes français et des manuels de théologie par le parti romaniste de 1670 à 1868*, Paris, 1872 (très tendancieux) ; J. MAURAIN, *La politique ecclésiastique du Second Empire*, Paris, 1930, chap. III, V, IX, XVI, XVIII, XXI, XXII, XXIV, XXVII.

En outre, les biographies de Guéranger, de Veuillot, de Mgr Maret, de Mgr Mathieu et de Mgr Dupanloup, signalées p. 8, ainsi que W. GUETTÉE, *Souvenirs d'un prêtre romain devenu prêtre orthodoxe*, Paris, 1889 (tendancieux).

En ce qui concerne l'Allemagne : F. VIGENER, *Gallikanismus und episkopalistische Strömungen im deutschen Katholizismus*, dans *Historische Zeitschrift*, t. CXI, 1913, p. 495-581, ainsi que les biographies (partiales dans un sens ou dans l'autre) de Geissel, de Ketteler et surtout de Doellinger, signalées, p. 9.

En ce qui concerne l'Angleterre : W. WARD, *The Life and Times of cardinal Wiseman*, Londres, 1897, chap. XVII et XXIV, ainsi que les biographies de Manning, de Newman et d'Ullathorne, signalées p. 9.

(2) Outre les ouvrages signalés à la note précédente, consulter, en ce qui concerne la France : J. MAURAIN, *Le Saint-Siège et la France de décembre 1851 à avril 1853. Documents inédits*, Paris, 1930 ; J. MARTIN, *La nonciature de Paris et les affaires ecclésiastiques de France sous le règne de Louis-Philippe*, Paris, 1949 ; A. HOUTIN, *Un dernier gallican : Henri Bernier*, Paris, 1904 ; F. GOUSSET, *Le cardinal Gousset*, Besançon, 1893, ainsi que les biographies de Combalot, de Mgr Gerbet, de Mgr Guibert, de Mgr Sibour et de Mgr Parisis, signalées p. 8.

qu'il avait bien conscience de posséder, mais qu'il n'avait jamais osé jusque-là exercer ; et les articles organiques, en investissant les évêques d'un pouvoir discrétionnaire sur les desservants, allaient amener le bas clergé à chercher à Rome une protection contre l'arbitraire épiscopal et à affirmer par là, non plus en théorie mais en pratique, les droits du Saint-Siège sur les Églises particulières.

Pendant ce temps, en Allemagne, les nombreuses vacances de sièges prolongées pendant des années, en obligeant à demander à Rome des dispenses que les évêques s'étaient jusque-là réservées, furent l'occasion pour beaucoup de redécouvrir la vieille thèse théologique de « l'épiscopat universel » du souverain pontife. D'autre part, tandis que les sécularisations détruisaient la puissance politique des archevêques de la vallée du Rhin, jadis si hautains dans leurs rapports avec le Saint-Siège, les remaniements territoriaux réduisaient partout, sauf en Bavière, les catholiques au rang de minorité au sein d'États protestants et les amenaient dès lors tout naturellement à chercher auprès du pape un appui contre leurs gouvernements. Ces gouvernements de leur côté, désirant éviter d'avoir en face d'eux une Église forte et unie, crurent habile de ne pas favoriser la réorganisation d'une Église nationale allemande et de régler plutôt la situation de leurs sujets catholiques par des accords avec Rome, qui semblait encore faible et lointaine ; ce faisant, ils soulignaient à nouveau l'étendue des droits du pape dans l'Église.

La restauration des Bourbons en France et la nomination de nombreux évêques choisis dans l'ancien clergé gallican semblaient devoir enrayer pour un temps le réveil de l'ultramontanisme, mais celui-ci put compter sur l'appui des jésuites et surtout sur l'influence de deux écrivains prestigieux, qui se firent les apologistes de la monarchie pontificale, Joseph de Maistre et Lamennais. Le premier publia, en 1819, son livre *Du Pape*, qui allait devenir pendant un demi-siècle la bible de l'ultramontanisme ; il y prônait l'infaillibilité du pape et son pouvoir absolu dans l'Église, moins par des arguments théologiques que par des considérations *a priori* sur la nécessité de l'ordre et de l'absolutisme, qui rencontrèrent un vif succès dans les milieux de la Restauration. Quant à Lamennais, on sait la passion qu'il mit à combattre l'épiscopat gallican, auquel il reprochait de sacrifier la liberté véritable de l'Église aux fallacieux avantages de la protection gouvernementale. Tout en habituant son entourage à traiter les évêques avec une rare impertinence, il réussit, en quelques années, à enthousiasmer pour ce qu'on appelait en France « les doctrines romaines » la majorité du jeune clergé et les laïcs les plus ardents, qui souffraient de la situation pénible du catholicisme sous la Restauration. On ignore trop que les principaux protagonistes de l'ultramontanisme en France jusqu'au concile du Vatican furent d'anciens disciples de Lamennais, dont beaucoup, comme dom Guéranger, Gerbet, Salinis ou l'abbé Combalot, devenus par la suite de farouches adversaires des idées libérales, s'efforcèrent de jeter un voile sur leurs relations de jeunesse avec le prophète de la Chênaie. L'indifférence hostile de la Monarchie de Juillet à l'égard de l'Église acheva d'orienter vers Rome tous ceux qui s'inquiétaient de voir les évêques, trop gouvernementaux à leur gré, résister insuffisamment aux progrès de la déchristianisation. « Le glaive spirituel s'est pour ainsi dire rouillé dans les mains (de l'épiscopat), s'écriait l'un d'eux. Un pape seul peut lui rendre cette énergie de la puissance qu'il n'a plus. Des torrents d'erreurs ruissellent et débordent de la presse, des chaires de philosophie, des classes universitaires, et les évêques n'osent plus les saisir corps à corps, les étreindre, les étouffer [1]. » La lutte pour la liberté de l'enseignement fut pour les chefs du parti catholique dirigés par Montalembert l'occasion de rallier à ces convictions le grand public catholique et désormais l'ultramontanisme ne devait plus cesser de trouver dans l'élite du laïcat français un soutien des plus ardents.

Il n'avait pas fallu vingt ans pour rendre populaire dans l'ancienne citadelle du gallicanisme l'ultramontanisme théorique et pratique, la croyance en l'infaillibilité personnelle du souverain pontife et l'acceptation d'interventions toujours plus directes de la Curie romaine dans les affaires intérieures de l'Église de France. L'évolution fut plus lente dans les pays germaniques, et surtout plus complexe. Là aussi, dans les différents « cercles » de catholiques militants, à Munster autour de la princesse Gallitzin, à Munich autour de Goerres, à Vienne autour du saint rédemptoriste Clément-Marie Hofbauer et du converti Frédéric Schlegel, un peu plus tard à Cologne autour du combatif archevêque Droste zu Vischering, il y avait, sous l'influence des causes rappelées ci-dessus, une nette tendance à s'appuyer de plus en plus sur Rome pour résister aux gouvernements d'inspiration joséphiste, qui prétendaient limiter la liberté de l'Église. Là aussi les catholiques proclamaient plus volontiers que jadis leur croyance en la primauté du pape et plusieurs canonistes, dont le rôle est aujourd'hui négligé, mais dont l'influence fut profonde et durable, tels Frey et son continuateur Scheill, osaient affirmer à nouveau avec énergie les droits de l'Église à l'égard de l'État ou combattre les plans trop indépendants d'un Dalberg ou d'un Wessenberg [2]. Montalembert, au retour d'un voyage en Allemagne en 1833, ne pouvait que confirmer ce qu'avait déjà noté le protestant Perthes : « L'Église catholique est devenue plus romaine et met davantage l'accent sur le principe hiérarchique [3]. »

Mais, tout en confessant hautement que le pape est le chef de l'Église et en se montrant résolument hostiles au fébronianisme, la plupart des « ultramontains » allemands — le nom n'avait guère alors, dans les pays germaniques, d'autre signification que celle de : catholique convaincu [4] — restaient généralement partisans d'un gallicanisme modéré, où il n'était plus question de supériorité du concile sur le pape, mais bien de la nécessité d'une étroite union du pape avec l'épiscopat pour que son enseignement soit infaillible. Concevant l'Église comme une « monarchie tempérée »

(1) Combalot à Salinis, 15 juin 1846, dans A. RICARD, L'abbé Combalot, p. 350.
(2) Cf. J. v. SCHULTE, Geschichte der Quellen des can. Rechtes, t. III, p. 307 et 337-339.
(3) Cité par L. A. VEIT, Die Kirche im Zeitalter des Individualismus, p. 105.
(4) La chose est bien mise en lumière par L. A. VEIT, op. cit., p. 106-107 et surtout par S. MERKLE, dans T. Q. S., t. CXIX, 1938, p. 60-108.

et ne voulant rien admettre en elle qui évoquât l'absolutisme, ils s'opposaient à la thèse, classique depuis Bellarmin, de « l'infaillibilité séparée », que les théologiens du Collège romain s'efforçaient de remettre en lumière, de même que celle de la juridiction universelle et souveraine du pontife romain. *Le Pape* de Joseph de Maistre, où des hommes ayant une certaine formation scientifique ne voyaient que « niaises déclamations de dilettante » — ainsi s'exprimait la *Theologische Quartalschrift* de Tubingue — ne pouvait qu'accentuer leur méfiance pour des idées qui n'avaient plus eu de représentants en Allemagne depuis la disparition des jésuites, un demi-siècle plus tôt. L'ouvrage, traduit en allemand dès 1822, n'exerça d'abord quelque action que parmi les laïcs, qui, moins au fait des subtilités théologiques, considéraient comme allant de soi que la primauté impliquât l'infaillibilité séparée, et qui sympathisaient d'autant plus volontiers avec le courant venant de France [1] que leur enthousiasme romantique les portait à idéaliser le régime théocratique médiéval.

Peu influencé par le courant français, le clergé allemand allait pourtant à son tour se rallier en partie aux pures doctrines romaines. Le mouvement partit du séminaire de Mayence, où le supérieur, Liebermann, ancien élève des jésuites de Strasbourg, enseignait la théorie de Bellarmin, et dont la revue *Der Katholik*, fondée en 1821, devint vite l'un des principaux organes de la propagande ultramontaine [2]. L'action exercée, assez prudemment au début, par l'école de Mayence, fut peu à peu renforcée par celle de prêtres qui avaient fait leurs études à Rome, où le Collège germanique, dirigé par les jésuites, avait rouvert ses portes en 1819. Diverses circonstances favorables contribuèrent à gagner à leurs vues la sympathie de beaucoup de ceux qui se préoccupaient moins de science théologique rigoureuse que des progrès du mouvement catholique. On ne tarda pas à remarquer que, dans les conflits qui opposaient les catholiques aux gouvernements pour la défense des libertés ecclésiastiques, les gens de Mayence et leurs amis étaient souvent parmi les plus acharnés à combattre pour les droits de l'Église, et la confiance qu'ils gagnèrent par là devait assez normalement rejaillir sur les doctrines et les conceptions ecclésiastiques dont ils se faisaient les protagonistes. Comme, parallèlement, dans la controverse passionnée autour des théories d'Hermes, ce furent les Mayençais et les « Romains » qui les dénoncèrent avec le plus d'intransigeance, la condamnation de ces doctrines en 1835 augmenta encore leur influence au détriment des théologiens hostiles à la théologie romaine, dont l'attitude avait été beaucoup moins tranchée et parfois nettement favorable au professeur allemand. Un ultramontanisme populaire, alimenté par des ouvrages de vulgarisation comme celui du jésuite Weninger [3], se développa de la sorte à partir des années

---

(1) La part exacte des influences françaises, qui eurent un rôle incontestable sur la restauration catholique allemande, n'est pas encore tout à fait éclaircie : cf. E. FLEIG, *Zur Geschichte des Einströmens französischen Restaurationsdenkens nach Deutschland*, dans *Historisches Jahrbuch*, t. LV, 1935, p. 500-520.

(2) Voir G. KRUEGER, *Der Mainzer Kreis und die Katholische Bewegung*, dans *Preussisches Jahrbuch*, t. CXLVIII, 1912, p. 395-414.

(3) *Die apostolische Vollmacht des Papstes in Glaubensentscheidungen*. L'ouvrage, paru en 1841, eut une 2e édition dès l'année suivante.

trente, dont les effets allaient se faire nettement sentir dès le début des congrès catholiques, en 1848.

LA PERSISTANCE DES TENDANCES GALLICANES

Les historiens ont noté maintes fois la surprenante rapidité avec laquelle l'ultramontanisme réussit à triompher, au cours des premières décades du XIXe siècle, des puissantes traditions gallicanes et joséphistes si longtemps prépondérantes en France et en Allemagne. Mais, dans un excès de simplification, ils oublient souvent d'ajouter que ces anciennes traditions ne disparurent pas sans résistance et que l'on peut en relever des indices, plus nombreux qu'on ne veut bien le reconnaître, jusqu'à l'époque du concile du Vatican. A fortiori, celles-ci avaient elles encore une réelle vivacité au moment de l'avénement de Pie IX.

Pour comprendre la situation, il importe de ne jamais perdre de vue le double aspect du gallicanisme : à côté de la tendance à restreindre les prérogatives du pape au profit de l'épiscopat, celui-ci comportait un aspect politico-ecclésiastique, déniant à l'autorité spirituelle toute autorité sur le temporel tout en accordant au pouvoir séculier des droits importants dans l'organisation des Églises nationales. La réaction ultramontaine, en conséquence, se développa sur un double plan : tendance à reconnaître au pape dans l'Église une autorité spirituelle totale ; revendication pour l'Église de l'indépendance par rapport au pouvoir civil et même d'un certain pouvoir, au moins « indirect », sur l'État. Or, il est indéniable que c'est ce dernier aspect surtout qui expliquait l'enthousiasme suscité par le courant ultramontain : on y voyait d'abord les grands avantages apostoliques qu'avait l'Église à s'affranchir de la tutelle des gouvernements. Par contre, si beaucoup appréciaient également, toujours dans le même but apostolique, une concentration plus grande des forces catholiques autour du souverain pontife, plus d'un se demandait s'il était opportun que cette concentration se traduisît par une centralisation si poussée qu'elle aurait inévitablement pour conséquence de réduire l'autorité réelle des évêques, et par une uniformisation de la discipline, de la liturgie, voire de la piété, qui obligerait à renoncer à des coutumes locales vénérables et à adopter dans l'Église entière un « style de vie » religieuse analogue à celui de l'Italie.

Dès lors, tandis que sous son aspect politico-religieux le mouvement ultramontain rencontrait une faveur générale, il était normal que, sous son aspect proprement ecclésiastique, il se heurtât au contraire à diverses résistances, non seulement à celle de théologiens qui, n'ayant guère le sens du progrès dogmatique, ne concevaient pas que le pape puisse prendre une place plus importante que dans l'Église des premiers siècles, mais aussi à celle d'hommes attachés aux usages du passé, où le particularisme avait dominé, plutôt que tournés vers l'avenir, qui s'annonçait toujours plus international, et à celle de membres du haut clergé inquiets de perdre leurs prérogatives traditionnelles et se demandant même parfois si la notion de l'épiscopat de droit divin n'était pas mise en danger. Ces résistances étaient d'autant plus explicables que, contrairement à l'opinion commune, la Déclaration des Quatre Articles de 1682 n'avait jamais été condamnée, mais seulement déclarée nulle et non avenue [1].

C'est ainsi qu'en France, où l'approbation par soixante évêques de la condamnation du *Manuel de droit ecclésiastique* de Dupin montrait le complet discrédit où était tombé le gallicanisme politique, le gallicanisme théologique modéré conservait au contraire, vers 1845, des sympathies non négligeables, qui apparaissent dans le ton de plusieurs adhésions épiscopales au mandement de Mgr Guibert contre les frères Allignol, dans l'attitude des évêques en conflit avec dom Guéranger à propos de l'introduction de la liturgie romaine, dans les déclarations de Mgr Affre sur l'exemption des religieux, dans les lettres pastorales où les évêques de Montpellier et de Chartres font l'éloge de Bossuet défendant les principes de l'ancienne Église gallicane, dans les approbations de plusieurs évêques à l'*Humble Remontrance au R. P. dom Prosper Guéranger* où le vicaire général d'Angers affirmait la parfaite orthodoxie des articles de 1682. Les prélats qui pensaient de la sorte étaient tenus en suspicion par le jeune clergé, mais ils se sentaient appuyés par beaucoup de prêtres plus âgés, qui demeuraient attachés aux traditions de l'ancien clergé de France formé par Saint-Sulpice, à ce « catholicisme austère et archaïque » avec « sa conception particulière de la hiérarchie, de la liturgie et de la dévotion », qui devait constituer l'obstacle le plus sérieux à la victoire totale des idées romaines en France [2].

En Grande-Bretagne également, les partisans d'une théologie gallicane modérée étaient loin d'avoir disparu, malgré un recul incontestable : la *Théologie* de Bailly resta en usage au séminaire irlandais de Maynooth jusqu'à sa mise à l'*Index* en 1853 [3] et le séminaire Saint-Edmond eut à sa tête jusqu'en 1851 Edouard Cox, l'un des membres les plus en vue du « clergé antiromain » d'Angleterre. Le désir de s'émanciper le plus possible du contrôle du Saint-Siège et de réduire les interventions du pape au minimum s'était affirmé pendant deux siècles et Wiseman eut fort à faire pour venir à bout, avec l'aide de certains convertis, de l'esprit d'indépendance à l'égard de Rome des catholiques de vieille roche. Comme en France d'ailleurs, cet « antiromanisme », que l'on trouvait jusqu'au Collège anglais de Rome [4], provenait moins d'une doctrine arrêtée que d'un attachement aux anciens usages ecclésiastiques et d'une méfiance à l'égard des pratiques et dévotions d'origine italienne [5].

Les influences proprement doctrinales étaient plus sensibles en Europe centrale. En Lombardie, les séminaires et les universités, soumis à l'influence autrichienne, avaient enseigné pendant de longues années les théories de Febronius, reprises par Tamburini, et il en restait des traces

(1) Voir les remarques pertinentes de S. MERKLE, dans *T. Q. S.*, t. CXIX, 1938, p. 61-62 et 69.
(2) P. DE QUIRIELLE, dans *Ann. E. Sc. P.*, t. V, 1890, p. 513-514.
(3) Dr NEVILLE, *Theology, Past and Present, at Maynooth*, dans *Dublin Review*, octobre 1879, p. 449 et suiv. ; W. J. WALSH, *The alleged Gallicanism of Maynooth and the Irish Clergy*, *ibid.*, avril 1880, p. 210 et suiv.
(4) E. PURCELL, *Life of Manning*, t. II, p. 307-308.
(5) Talbot le relevait encore en 1859 dans une lettre à Manning citée dans E. PURCELL, *op. cit.*, t. II, p. 100.

dans l'esprit du clergé, qui souhaitait notamment prendre part à nouveau à l'élection des évêques. En Autriche surtout, même parmi les antijosé-phistes tout dévoués au Saint-Siège, nombreux étaient ceux qui restaient à leur insu sous l'influence des anciennes conceptions. Rauscher lui-même, l'un des agents les plus actifs de l'ultramontanisme à Vienne, avait laissé dans les premiers projets de concordat des traces non équivoques d'épis-copalisme ou de nationalisme ecclésiastique. « *Josephinismum sapit !* » murmuraient les canonistes romains [1]. En Allemagne enfin, sans parler de Wessemberg, qui ne devait mourir qu'en 1860, entouré jusqu'à la fin de nombreux disciples et amis, la plupart des théologiens continuaient, au nom de ce qu'ils considéraient comme la tradition ancienne de l'Église, à présenter la thèse de l'infaillibilité personnelle du pape comme une doctrine libre à laquelle beaucoup d'entre eux refusaient d'adhérer. Parallèlement, certains milieux ecclésiastiques, au fur et à mesure qu'ils prenaient conscience des inconvénients qu'entraînaient inévitablement les progrès de l'ultramontanisme, commençaient à s'inquiéter.

LE PARTI ULTRAMONTAIN EN ALLEMAGNE A LA VEILLE DE 1848 — Ces réticences s'expliquent si l'on songe au caractère complexe que prenait en Allemagne, tout comme en France et en Angleterre, le mouvement ultramontain. Il comportait à la fois des doctrines précises concernant les prérogatives du souverain pontife et la supériorité de l'Église sur le pouvoir civil, un programme de modification de l'organisation ecclésiastique dans un sens plus autoritaire et plus centralisé, une tendance à restreindre la liberté de la recherche scientifique en matière philosophique ou théo-logique, enfin une conception renouvelée de la piété, considérée à tort ou à raison comme d'inspiration jésuite, où l'on insistait moins sur la dévotion intérieure que sur la réception fréquente des sacrements et la multiplication des exercices extérieurs.

A mesure que se précisaient ces diverses tendances, qui n'étaient apparues d'abord que de manière diffuse, s'accentuait l'opposition des partisans de l'ancien ordre de choses. Certains évêques constataient que les interventions romaines tenaient parfois trop peu compte de la situation particulière des régions protestantes ; aussi tentaient-ils de se réserver comme par le passé la décision de cas que le Saint-Siège esti-mait lui revenir et ils y avaient d'autant plus tendance que la propagande en faveur de recours plus fréquents à Rome était souvent le fait d'hommes qui ne se distinguaient ni par l'intelligence ni par la modération et qui, « bons catholiques mais mauvais chrétiens », se faisaient une spécialité de dénoncer aux Congrégations romaines, sans souci de la justice ni de la vérité, tous ceux avec qui ils étaient en désaccord [2]. Par ailleurs, beau-

---

(1) M. HUSSAREK, *Die Verhandlung des Konkordats*, p. 577-580 ; H. SINGER, dans *Mitteilungen des Vereines für Gesch. der Deutschen in Böhmen*, t. LXII, 1924, p. 107, n. 11 ; A. FLIR, *Briefe aus Rom*, p. 23, 25 et suiv., 46.

(2) Le curé Binterim, de Cologne, dont on a voulu faire au siècle passé l'un des coryphées de la renaissance catholique en Allemagne, est un modèle du genre ; voir sa biographie, qui remet les choses au point, par C. SCHOENIG, *A.-J. Binterim (1779-1855) als Kirchenpolitiker und Gelehrter*, Dusseldorf, 1933.

coup de prêtres ne réalisaient pas encore à quel point l'évolution du monde exigeait un renouvellement des méthodes et dénonçaient ce qu'ils croyaient être une opposition nationale entre le génie germanique et les conceptions italiennes et romaines, alors qu'il s'agissait avant tout, selon l'observation très juste de Schnabel, d'un conflit de générations : c'était l'Allemagne elle-même qui, pressée par la complexité croissante des problèmes, évoluait de l'ancien système collégial et corporatif du moyen âge vers la centralisation à la française [1]. Si l'un ou l'autre ultramontain avait parfois tendance à faire écho à Reisach écrivant à Lambruschini qu'il pensait à Rome « comme un pauvre exilé loin de sa vraie patrie » [2], la plupart d'entre eux, et Reisach tout le premier, étaient d'excellents Allemands, convaincus seulement que dans les circonstances présentes il n'y avait de vitalité possible pour le catholicisme allemand que dans un contact plus étroit avec Rome.

Parmi ceux qui discernaient moins clairement cette nécessité et voyaient avec mauvaise humeur les efforts entrepris pour uniformiser davantage la vie catholique sur le modèle romain, beaucoup ne manquaient ni de piété ni de dévouement à l'Église. Mais la résistance qu'ils opposaient aux ultramontains semblait à ceux-ci faire le jeu de certains dignitaires ecclésiastiques médiocrement zélés, ou des gouvernements peu soucieux de voir l'Église s'affranchir enfin de leur tutelle. L'incompréhension réciproque aboutit bientôt, comme c'est souvent le cas, à un raidissement des attitudes et c'est ce qui explique que, vers 1848, les ultramontains commencèrent à apparaître comme un véritable « parti » dans l'Église, avec tout ce que cela suppose d'intransigeance d'une part, d'organisation d'autre part.

Ils pouvaient envisager l'avenir avec d'autant plus de confiance qu'ils pouvaient désormais compter sur de puissants appuis. Depuis 1841, Geissel était archevêque de Cologne et son action épiscopale, qui s'étendait à toute l'Allemagne occidentale, s'inspirait des principes qu'il avait reçus au séminaire de Mayence. La même année, l'archevêque de Munich avait reçu comme coadjuteur Charles-Auguste de Reisach, qui devait lui succéder en 1846. Ancien élève du Collège germanique à Rome, puis régent des études au Collège de la Propagande, ce qui l'avait mis en contacts étroits avec le cardinal Cappellari, le futur Grégoire XVI, Reisach, qui était à la fois un homme de caractère et un homme de principes, fut pendant plus de vingt ans l'un des agents les plus dévoués de l'influence romaine en Allemagne du Sud [3], habilement secondé d'ailleurs par le nonce à Vienne, Mgr Viale Prela [4], dont l'action en faveur de l'ultramontanisme, plus souple et plus discrète que celle de l'archevêque de Munich, n'en fut pas moins efficace, en Bavière et dans la province du Haut-Rhin aussi bien qu'à Vienne.

(1) J. Schnabel, *Deutsche Geschichte im XIX. Jh.*, t. IV : *Religiöse Kräfte*, p. 267-268.
(2) Lettre du 8 janvier 1838, dans *Römische Quartalschrift*, t. XXXIV, 1926, p. 204.
(3) Une bonne biographie de Reisach (1800-1869) fait encore défaut. On peut consulter en attendant C. Moufang, *Carl August cardinal von Reisach*, dans *Der Katholik*, 1870, vol. I, p. 129-150, et, pour la période des débuts, J.-B. Goetz, *Kardinal K. A. Graf v. Reisach als Bischof von Eichstätt*, Eichstätt, 1901.
(4) Pour Viale Prela (1798-1860) également, il faut se contenter de l'ouvrage provisoire de Fantoni, *Della vita del cardinale Michele Viale Prela*, Bologne, 1861.

D'autre part, si les travaux théologiques en faveur des thèses ultramontaines continuèrent à faire défaut, ce qui n'était pas pour dissiper le discrédit où les tenaient les milieux universitaires, un appoint non négligeable leur fut par contre apporté par un canoniste laïc, Georges Phillips [1], dont l'enseignement à Munich et à Vienne, et les œuvres, publiées à partir de 1845, contribuèrent beaucoup à populariser dans les pays germaniques la doctrine du magistère infaillible du pape et de son épiscopat universel, ainsi que celle du pouvoir indirect de l'Église sur l'État. Malgré des différences réelles, Vigener n'a pas tort de le considérer « comme un de Maistre allemand » [2] : mêmes outrances et même influence. On retrouve chez lui une tendance analogue à exagérer l'absolutisme pontifical et à identifier le pape avec l'Église contrairement à la position plus nuancée que défendait Perrone à Rome ; on sent, à côté d'une façon toute juridique et extérieure d'envisager l'Église et son unité, l'intransigeance du converti — il avait abjuré le protestantisme en 1828 — et l'enthousiasme du romantique pour la théocratie médiévale. Or, son œuvre, sans être de première valeur au point de vue scientifique, supposait beaucoup d'érudition, et comme elle était écrite en allemand, en un style clair et élégant, qui trahissait par moment son profond dévouement à l'Église, elle conquit de nombreux lecteurs et joua dans la victoire de l'ultramontanisme un rôle qu'on ne soupçonne plus guère aujourd'hui.

L'OFFENSIVE ULTRAMONTAINE EN FRANCE SOUS LA NONCIATURE DE FORNARI

Au moment où les forces ultramontaines achevaient de se grouper en Allemagne, le mouvement s'affirmait en France avec une résolution qui n'allait plus cesser de croître. Pendant des années, c'étaient des causes spécifiquement françaises qui avaient amené la partie la plus ardente du clergé français et les dirigeants du parti catholique à découvrir le danger d'une Église nationale isolée en face d'un gouvernement hostile, et à prôner la nécessité de s'appuyer sur le Saint-Siège. Celui-ci, inquiet de l'origine mennaisienne de cette évolution des esprits et soucieux de ne pas envenimer les rapports avec le gouvernement, s'était d'abord tenu sur la réserve. Son représentant à Paris, l'internonce Garibaldi, se caractérisait d'ailleurs par « un esprit de conciliation attentif à ne pas réveiller les oppositions doctrinales toujours fâcheuses » [3] et jugeait préférable de laisser mourir lentement le gallicanisme plutôt que d'aborder de front ce qui en subsistait encore. Son successeur, Mgr Fornari, qui occupa la nonciature de Paris de 1843 à 1850, adopta une politique opposée et, facilement excessif et « désireux de jouer un rôle » [4].

---

(1) Sur Phillips (1804-1872), professeur à Munich de 1834 à 1847, puis à Vienne de 1851 à sa mort, voir surtout J. v. SCHULTE, Geschichte der Quellen des canonischen Rechtes, t. III, p. 375-387.
(2) F. VIGENER, Gallikanische und episkopalistische Strömungen..., p. 577-578.
(3) J. MARTIN, La nonciature de Paris..., p. 333 ; cf. p. 274-275.
(4) A. SIMON, Le cardinal Sterckx et son temps, t. II, p. 341. L'ambassadeur de France à Rome devait apprécier son œuvre en ces termes : « Il s'est conduit dans les affaires bien plus en homme de théorie qu'en homme d'État. En poursuivant l'application absolue des principes romains sans s'inquiéter des traditions locales, sans compter avec les personnes et ne songeant jamais qu'aux principes abstraits, il a puissamment contribué à réveiller d'anciennes passions qui, à plusieurs

il soutint résolument les ultramontains, même dans ce que leurs procédés avaient parfois de déplaisant. Non content de s'opposer aux prétentions de plusieurs évêques à l'indépendance en matière liturgique et canonique, il appuya volontiers les réactions violentes que ces prétentions provoquaient chez certains prêtres ou religieux qui, anciens disciples de Lamennais pour la plupart, avaient hérité de leur maître une arrogance insolente à l'égard des évêques. Fornari ne s'en offusquait guère, car il se sentait plein de mépris pour les lacunes de la formation canonique et théologique de beaucoup d'évêques français et avait du reste fait preuve lui-même, lors de sa nonciature en Belgique, d'une certaine désinvolture à l'égard des prérogatives épiscopales [1].

C'est dans tous les domaines que se développa, au cours de la nonciature de Fornari, ce que l'on serait tenté d'appeler l'offensive ultramontaine. Un missionnaire diocésain au cerveau incandescent et au zèle parfois intempestif, tenté de considérer la courtoisie envers les évêques comme « un cinquième article des libertés gallicanes » [2], l'abbé Combalot, se fait le propagateur des idées romaines dans les chaires et les presbytères, tandis que dom Guéranger, dont le monastère de Solesmes devient l'un des principaux centres du mouvement, considère sa campagne en faveur de la liturgie romaine comme dirigée avant tout contre les sympathies gallicanes du haut clergé. L'abbé Rohrbacher, qui a conçu depuis 1826 l'idée de récrire à l'usage de la nouvelle génération cléricale l'*Histoire ecclésiastique* de Fleury dans un sens ultramontain, réalise son projet de 1842 à 1849 [3] avec l'appui du P. Gaultier, savant religieux de la congrégation du Saint-Esprit. La cellule de ce dernier était « le salon romain de Paris » [4], sa bibliothèque, la mieux fournie qu'on pût rêver sur la question gallicane, et c'est chez lui que descendaient, lors de leurs passages à Paris, les deux plus ardents défenseurs du mouvement ultramontain dans l'épiscopat français, Mgr Gousset et Mgr Parisis. Le premier, tout en exerçant comme archevêque de Reims, de 1840 à 1866, une influence prépondérante, devait beaucoup contribuer, par ses manuels de théologie en français, à vulgariser une dogmatique antigallicane et une morale moins rigoriste que celle prônée par Saint-Sulpice et le clergé des générations précédentes. Quant au second, connaissant le désir du Saint-Siège de voir les diocèses français renoncer à leur droit coutumier, il inaugura l'usage d'interroger régulièrement les Congrégations romaines sur des questions relatives au culte ou à la discipline ecclésiastique, au grand mécontentement de ceux qui jugeaient que la Curie n'avait pas à intervenir dans les affaires internes de l'Église de France.

reprises, ont menacé de faire explosion au détriment de la cause même qu'il voulait servir. » (Dans J. MAURAIN, *Le Saint-Siège et la France*, p. 78, n. 2.)

(1) A. SIMON, *L'Église catholique et les débuts de la Belgique indépendante*, p. 64-67.

(2) A. HOUTIN, *Un dernier gallican*, p. 151. « Ce pauvre énergumène, qui veut être jusqu'à son dernier souffle l'*insulteur* public des évêques », écrivait l'évêque de Chartres à Dupanloup, le 18 avril 1851. (*Arch. S. Sulp.*, fonds Dupanloup.)

(3) Travailleur acharné mais médiocrement original, Rohrbacher (1789-1856) s'attacha d'abord à Lamennais, puis devint professeur au grand séminaire de Nancy et passa les dernières années de sa vie à Paris. (Notice biographique par Mgr RICARD, Paris, 1885.)

(4) *Notice sur le R. P. Gaultier*, Paris, s. d., p. 42. Sur le cercle du P. Gaultier, voir aussi F. CABROL, *Le cardinal Pitra*, p. 205-206.

C'est aussi dans le cercle du P. Gaultier que le nonce trouva son principal auxiliaire lorsque, en 1849, après avoir réussi, avec l'aide de Mgr Gousset, à empêcher la réunion d'un concile national proposée à Pie IX par deux archevêques et onze évêques [1], il lui fallut s'opposer aux prétentions de Mgr Sibour de convoquer un concile provincial sans solliciter l'autorisation pontificale : à sa demande, l'abbé Bouix exposa dans *L'Univers* les principes canoniques en la matière et développa ensuite ces articles en un gros traité qui mécontenta vivement l'archevêque [2].

Les conciles provinciaux furent l'occasion d'une nouvelle et importante avance des idées romaines, d'abord parce que, dans plusieurs provinces, les évêques ultramontains, que les nominations faites par Falloux sur les indications des dirigeants du parti catholique avaient renforcés, purent faire prévaloir leurs idées, mais aussi parce que le Saint-Siège imposa, en dépit d'une vive répugnance et parfois d'une longue opposition [3], la modification de certains décrets, obtenant par là l'affirmation explicite de certaines de ses prérogatives et — ce qui importait plus encore — la reconnaissance implicite de son droit de contrôle. Comme *L'Univers*, inspiré par Parisis et Gousset, se chargeait de souligner avec complaisance tout ce qui allait dans ce sens, l'impulsion qui, depuis des années, orientait les catholiques vers Rome en fut accrue d'autant.

Mgr Fornari quitta la nonciature de Paris au début de 1851, mais, de Rome, il continua jusqu'à sa mort, en 1854, à appuyer de toutes ses forces les efforts des ultramontains français, protégeant *L'Univers* contre les attaques répétées de Mgr Sibour ; soutenant devant les Congrégations les prêtres français qui s'insurgeaient au nom du droit canon contre une décision de leur évêque ; encourageant la *Correspondance de Rome*, une revue où un jeune ecclésiastique remuant, l'abbé Chaillot, se chargeait de fournir un aliment à ces revendications en commentant les décisions des Congrégations et les textes des canonistes [4]; appuyant enfin les dénonciations d'ouvrages à tendance gallicane et contribuant à obtenir leur mise à l'*Index*. L'ambassadeur de France se plaignit à diverses reprises du rôle de Fornari qui, disait-il à Antonelli,

s'était fait le centre, depuis son retour à Rome, d'une petite Église composée d'un certain nombre de prêtres français, honorables sans doute, mais très exagérés dans leurs opinions et, s'aidant de leur concours, continuait de loin la croisade dont, sur les lieux mêmes, il s'était fait l'apôtre [5].

Les mises à l'*Index*, qui marquèrent le début d'une nouvelle étape dans la lutte contre les dernières traces du gallicanisme, avaient com-

---

(1) Sur cet essai manqué, où l'archevêque de Tours Mgr Morlot appuyait Mgr Sibour, cf. F. Poujoulat, *Vie de Mgr Sibour*, p. 280 ; A. Houtin, *Un dernier gallican*, p. 242-244 et F. Gousset, *Le cardinal Gousset*, p. 220.

(2) Voir sur l'incident E. Naz, art. *Bouix*, dans *Dictionnaire de Droit canonique*, t. II, col. 970, et J.-F. v. Schulte, *Geschichte der Quellen des can. Rechtes*, t. III, p. 669, n. 1.

(3) Ce fut le cas notamment dans la province de Paris. Voir quelques détails dans E. Michaud, *De la falsification des catéchismes...*, p. 254-259.

(4) Sur Ludovic Chaillot, voir J. Maurain, *Le Saint-Siège et la France*, p. 227-233. L'action excitatrice de la *Correspondance de Rome* fut souvent dénoncée par l'épiscopat français, par le gouvernement et par l'ambassadeur à Rome. (Cf. J. Maurain, *Le Saint-Siège et la France*, passim, spécialement p. 54-66.)

(5) Cité dans J. Maurain, *op. cit.*, p. 84. Cf. aussi p. 66, 67, 78, 87.

mencé en 1850 avec deux opuscules du chanoine Bernier [1]. Puis ce fut un mandement de Mgr Clausel de Montals, affirmant que, lorsque le pape se trompe, c'est aux évêques à le redresser et accusant les adversaires des principes gallicans de n'être que des mennaisiens déguisés et insoumis, adversaires du pouvoir épiscopal ; le décret ne fut pas rendu public, mais Mgr Fornari en avertit l'évêque au nom du pape [2]. Quelques mois plus tard, l'*Index* frappait l'*Histoire de l'Église de France* de l'abbé Guettée, qui avait reçu l'approbation de 42 évêques [3], et le traité de droit canon de l'abbé Lequeux, trop favorable à l'autonomie épiscopale et au droit coutumier [4]. Comme ce dernier ouvrage était très répandu dans les séminaires, la condamnation fit sensation, d'autant plus qu'elle apparut comme un désaveu d'un des chefs de la résistance gallicane, l'archevêque de Paris, dont l'auteur était vicaire général. Les adversaires de Mgr Sibour réussirent même à accentuer le blâme indirect en faisant condamner, peu après, une encyclopédie qui avait été publiée avec son approbation [5]. Encouragés par dom Guéranger, ils répandaient en même temps le bruit que le même sort attendait le manuel de théologie de Mgr Bouvier, l'un des évêques hostiles à la campagne menée par l'abbé de Solesmes en faveur de la liturgie romaine ; la nouvelle fut bientôt démentie, mais l'évêque s'empressa de publier une nouvelle édition amendée [6], ce qui était un nouveau succès pour les ultramontains. D'ailleurs, si cette menace-là n'était pas fondée, l'*Index* frappait, à la fin de 1852, un autre manuel très en vogue, la *Théologie* de Bailly, remaniée pourtant par Receveur en 1842 en tenant compte du progrès des idées romaines [7].

LA RÉACTION GALLICANE ET L'ENCYCLIQUE « INTER MULTIPLICES »

L'enthousiasme avec lequel *L'Univers* saluait ces mesures, destinées à éliminer toute trace de gallicanisme de l'enseignement des séminaires, acheva d'exaspérer les adversaires de l'ultramontanisme, qui tentèrent une contre-offensive. Celle-ci paraissait avoir des chances de succès, car, sans partager nécessairement les thèses gallicanes sur les limites de l'infaillibilité et de la juridiction pontificale, une partie du clergé regrettait la « standardisation » [8] de la liturgie et la disparition de coutumes vénérables, tandis que

---

(1) *Humble remontrance au R. P. dom Prosper Guéranger et L'État et les cultes.* Cf. A. HOUTIN, *Un dernier gallican*, p. 261-273.

(2) Lettre publiée dans H. ANDRÉ, *Cours alphabétique et méthodique de droit canon*, t. IV, Paris, 1859, p. 340.

(3) F.-H. REUSCH, *Der Index der verbotenen Bücher*, t. II, 2, p. 1106-1107.

(4) Bien que Mgr Sibour ait engagé l'auteur à résister, celui-ci se soumit, mais il fit de vains essais à Rome pour obtenir que la condamnation soit levée après correction de son Manuel. (Voir *L'Observateur catholique*, t. XVI, p. 121 ; *La Réforme catholique*, 1877, p. 88 et surtout J. MAURAIN, *op. cit.*, p. 38, 53, 74, 77, 94, 106, 221 et 240-247.)

(5) Il s'agit du *Dictionnaire Universel* de Bouillet. (Voir sur cette affaire J. MAURAIN, *op. cit.*, *passim*, en particulier, p. 93-99.)

(6) Dom DELATTE, *Dom Guéranger*, t. II, p. 56, à nuancer par A. LEDRU, *Dom Guéranger et Mgr Bouvier*, p. 212-225 et 259. Voir aussi J. MAURAIN, *op. cit.*, p. 140-141 et *passim*.

(7) Contrairement au cas précédent, la dénonciation, cette fois, ne venait pas de France ni de Fornari, mais du pape, qui avait appris que l'ouvrage servait de base à l'enseignement des séminaires irlandais. (Cf. J. MAURAIN, *op. cit.*, p. 138, 145, 173-174 et *Maynooth Commission. Report of Her Majesty' Commissionners*, Dublin, 1855, II, p. 88.) Mais c'était un nouveau coup pour les gallicans de France.

(8) É. AMANN, art. *Veuillot*, dans *D. T. C.*, t. XV, col. 2832.

dans l'épiscopat beaucoup s'inquiétaient de la place importante que prenaient de simples fidèles dans le mouvement ultramontain. Habitués depuis des siècles à être les seuls responsables des intérêts religieux, de nombreux prélats étaient déconcertés devant la naissance d'une action catholique laïque qui cherchait à défendre l'Église par des armes inédites, la tribune parlementaire et la presse ; et l'arrogance avec laquelle certains journalistes, Veuillot en tête, entendaient dicter leur devoir aux évêques et discuter avec eux d'égal à égal n'était pas faite pour les amadouer, d'autant plus que, sans parler du mouvement presbytéral qui s'était esquissé en certaines régions [1], on voyait un peu partout des prêtres, soutenus par des canonistes comme l'abbé André [2] ou même par un journal spécialisé, Le Rappel, prétendre en appeler à Rome des moindres décisions épiscopales qui ne leur plaisaient pas. Plus d'un évêque croyait pouvoir dénoncer chez certains coryphées des droits du Saint-Siège un esprit d'insubordination qui leur faisait préférer une autorité lointaine à une surveillance plus immédiate, ou même — et sans doute n'avaient-il pas tout à fait tort — des conceptions inexactes sur la place accordée par le Christ à l'épiscopat dans l'Église.

Une coalition se forma de la sorte entre tous ceux qu'indisposait l'allure intransigeante et tapageuse que prenait parfois l'ultramontanisme, et ils essayèrent de résister d'une double manière : en présentant une justification théorique de leurs positions, et en réduisant au silence les ultramontains les plus bruyants.

A l'instigation d'une quinzaine d'évêques, parmi lesquels les plus actifs furent Mathieu, Sibour et peut-être Dupanloup, un théologien parisien rédigea un long mémoire [3] destiné à défendre les prérogatives épiscopales et les anciens usages de l'Église de France, à propos des principales questions alors en discussion : droit coutumier, conciles provinciaux, réforme liturgique, intervention des Congrégations romaines, journalisme catholique. Le premier point était en réalité le nœud du problème. Beaucoup de diocèses ayant leurs règles particulières, non conformes au droit canon ou aux décisions des Congrégations, il s'agissait de savoir si cette discipline nationale, relativement autonome et diversifiée, devait disparaître définitivement devant le droit de l'Église universelle, ce qui, pour certains, revenait à se demander si l'Église ne pouvait pas être conçue comme une fédération d'Églises unies en fait de doctrine, mais particularisées en fait de discipline. Comme plusieurs évêques craignaient de se compromettre, le mémoire fut imprimé anonymement et distribué par les soins de l'archevêque de Paris, durant l'automne 1852. Il fit sensation car, venant après les menaces de Sibour de censurer les ouvrages

---

(1) Il avait débuté dans le diocèse de Viviers avec les frères Allignol, qui revendiquaient l'inamovibilité pour les curés du fait que ceux-ci seraient d'institution divine comme les évêques. Ces idées avaient eu un grand retentissement dans toute la France et plusieurs journaux ecclésiastiques de Paris leur avaient fait écho : La Voix de la Vérité, avec l'abbé Migne, La Gazette de France, avec l'abbé de Genoude, Le Bien social surtout, que Mgr Affre condamna, le 26 mai 1845. Mgr Sibour, alors évêque de Digne, avait suivi de près toute cette polémique.

(2) Voir à son sujet le Dictionnaire de Droit canonique, t. I, col. 519-520.

(3) Mémoire adressé à l'épiscopat sur la situation présente de l'Église gallicane relativement au droit coutumier, Paris, 1852. Voir à son sujet A. MATER, L'Église catholique, sa constitution, Paris, 1906, p. 56-59 et P. MABILE, Vie de Mgr Mabille, t. II, Paris, 1926, p. 173.

de Bouix et de Rohrbacher ainsi que la traduction française du droit
canon de Phillips, il marquait la volonté de faire front à l'avance ultramontaine, mais il allait trop à contre-courant pour ne pas se heurter à
une vive opposition : Mgr Pie le dénonça à l'*Index*, tandis que le cardinal
Gousset, non content d'en publier une longue réfutation, convoquait
à Amiens un concile dont les participants, tous ardents ultramontains,
eurent vite fait de le condamner.

La publication du *Mémoire sur le droit coutumier* ne fut pas la seule
manifestation de résistance. Depuis des mois, en s'adressant directement
à Rome ou — ce qui froissa vivement le pape — en sollicitant l'intervention du gouvernement auprès du Saint-Siège, plusieurs évêques s'efforçaient de contrecarrer l'action des ecclésiastiques français de l'entourage
du cardinal Fornari, tels que l'abbé Bouix, l'abbé Chaillot, ou Mgr de
Ségur, et ils finirent par obtenir, en juin 1852, la suppression de la *Correspondance de Rome*, qu'ils avaient dénoncée publiquement deux mois
auparavant dans un mémoire sur *Le journalisme dans l'Église*. Restait
L'Univers. Mgr Sibour avait déjà essayé de l'atteindre en 1850, mais il
avait dû reculer devant une intervention de Mgr Gousset, soutenu par
Fornari. En 1852, l'impertinence de Veuillot envers les évêques qui ne
partageaient pas les vues de l'abbé Gaume dans la querelle des classiques
païens fournit l'occasion de frapper un nouveau coup. Ce fut Dupanloup
qui s'en chargea cette fois : sans être à proprement parler gallican, il n'a-
vait pas pardonné au journaliste son opposition lors de la loi Falloux et ne
supportait pas le peu de cas qu'il semblait faire de l'autorité épiscopale.
Il écrivit à tous les évêques individuellement pour les engager à souscrire
une déclaration habilement rédigée qui désavouait pratiquement Veuillot
en rappelant que les actes épiscopaux étaient justiciables du Saint-
Siège et non des journalistes. Mais il ne parvint à réunir qu'une quarantaine de signatures, car, une nouvelle fois, Gousset s'empressa d'intervenir, d'abord auprès des évêques, pour les dissuader de signer, puis à
Rome, pour éclairer les dessous de l'affaire. Pie IX, qui déclarait à Mgr de
Ségur : « Toute cette affaire ne tend qu'à arrêter le mouvement régéné-
rateur de l'unité romaine »[1], remercia Gousset de son intervention et
le cardinal secrétaire d'État fit savoir qu'on ne prisait guère à Rome cet
essai de « concile national par voie postale ou télégraphique »[2].

Ce nouvel échec ne découragea pas l'opposition. Au début de 1853,
elle essaya d'obtenir de Rome le désaveu du concile d'Amiens, en exploitant certaines déclarations favorables au traditionalisme, pour lequel les
ultramontains depuis Bonald avaient toujours manifesté des sympathies.
En même temps, elle s'en prit une fois de plus à Veuillot, qui venait de
s'engager dans une nouvelle polémique, à propos de Donoso Cortès, cette
fois. Le moment semblait bien choisi, car beaucoup d'évêques, même non
gallicans, commençaient à être excédés de ce qu'ils nommaient les prétentions de l'Église enseignée à se substituer à l'Église enseignante ; mais
tandis que la plupart adressaient leurs plaintes secrètement au pape

(1) Rapporte par Mgr BAUNARD, *Histoire du cardinal Pie*, t. I, p. 487.
(2) Le mot est de Dom Delatte. (*Dom Guéranger*, t. II, p. 64.)

et que Mgr Guibert se bornait à un avertissement modéré, Sibour, fougueux et maladroit comme toujours [1], décida d'interdire *L'Univers*. Cette mesure fit une déplorable impression à Rome où le crédit du journal ne faisait qu'augmenter. Aussi le pape, qui avait jusque-là montré qu'il ne désirait pas brusquer les choses [2], se vit-il obligé d'intervenir. Au début d'avril, il publiait sur les problèmes agités en France l'encyclique *Inter multiplices* [3], où, tout en conseillant aux journalistes plus de modération dans la forme, il reconnaissait leur utilité dans l'Église, en même temps qu'il louait les évêques qui avaient réintroduit la liturgie romaine et blâmait le *Mémoire sur le droit coutumier*. C'était la confirmation du concile d'Amiens, en même temps qu'un désaveu très net non seulement du gallicanisme, même modéré, mais aussi de tous ceux qui s'opposaient, pour quelque raison que ce fût, au courant ultramontain, et ceux-ci ne s'y trompèrent pas [4].

L'INTERVENTION SYSTÉMATIQUE DE ROME

Dès ce moment, en effet, on était bien décidé, à Rome, à prendre en main la direction du mouvement ultramontain et à le favoriser positivement. Pendant la première moitié du XIXe siècle, c'était avant tout sous des influences purement locales que la France, puis l'Allemagne, avaient recommencé à se tourner vers Rome. Le Saint-Siège avait suivi attentivement les progrès du mouvement, mais sans en prendre au début l'initiative, et l'on comprend le mot de Tocqueville disant que « le pape fut plus excité par les fidèles à devenir le maître absolu de l'Église qu'ils ne le furent par lui à se soumettre à cette domination. L'attitude de Rome fut plus un effet qu'une cause » [5]. Comme la plupart des grands mouvements de l'histoire de l'Église, le courant ultramontain du XIXe siècle est parti de la « base », de la conscience du peuple chrétien. Mais, comme toujours aussi, la hiérarchie ne tarda pas à s'y intéresser et le pontificat de Grégoire XVI, lequel avait jadis composé un ouvrage sur l'infaillibilité réédité peu après son élection [6], marqua une nette tendance à réagir contre l'effacement auquel la papauté avait dû se résigner au cours des siècles précédents, et à refaire de l'Église catholique une Église fortement centralisée comme au XIIIe siècle, mais s'étendant cette fois au monde entier grâce au mouvement missionnaire, et en mesure, par suite du développement des moyens

(1) Sur cette maladresse de Sibour, voir les dépêches de l'ambassadeur de France dans J. MAURAIN, *Le Saint-Siège et la France*, p. 106-107, 175-179 et une lettre de Maret à Ozanam dans G. BAZIN, *Vie de Mgr Maret*, t. II, p. 452-453.

(2) Cette modération du pape et de son secrétaire d'État est signalée à diverses reprises par l'ambassadeur de France (J. MAURAIN, *op. cit.*, passim) ; voir aussi une note de Contencin aux *Arch. Nat.*, F[19] 1947.

(3) Texte dans *Acta Pii IX*, t. I, p. 439 et suiv. L'encyclique est datée du 23 mars, mais fut en réalité publiée un peu plus tard (J. MAURAIN, *op. cit.*, p. 191) ; elle avait été rédigée par le cardinal Recanati, un capucin très au courant des choses françaises. (J. GAY, *Les deux Rome et l'opinion française*, p. 53 et 85-86.)

(4) Le nonce avait du reste été chargé de le leur faire savoir explicitement comme le montre par exemple sa lettre à Dupanloup, du 7 avril 1853. (*Arch. St-Sulp.*, Fonds Dupanloup.)

(5) Tocqueville à Reeve, 7 novembre 1856, dans É. OLLIVIER, *L'Église et l'État au concile du Vatican*, t. I, p. 314.

(6) *Il triomfo della S. Sede e delle Chiesa contro gli assalti dei Novatori*, Rome, 1799 ; 3e édit., Venise, 1832 ; traduction allemande, Augsbourg, 1838 et 1848.

de communication, de rendre cette centralisation plus étroite et plus effective.

La nouvelle politique s'accentua rapidement sous Pie IX, après la crise de 1848. On n'échappe pas à l'impression qu'à partir de ce moment, dans le but de renforcer le prestige et la force de résistance de l'Église en face de la Révolution montante, un plan s'élabora, auquel les jésuites, devenus très influents à la Curie, ne furent sans doute pas étrangers. Il s'agissait de systématiser la méthode de Fornari : profiter de toutes les circonstances pour soutenir ou stimuler, de Rome, les efforts déployés dans les divers pays par les ultramontains en vue de resserrer les liens avec le centre de la chrétienté et de renforcer l'autorité pontificale. C'était parfait, mais on n'évita pas certaines outrances, en visant à supprimer toutes les nuances régionales de la vie ecclésiastique.

Déjà l'approbation donnée, après un moment d'hésitation, à la campagne de dom Guéranger en faveur du retour à la liturgie romaine avait été inspirée par ce souci. Lorsque, en 1849, il fut question, en France et en Allemagne, de réunir un concile national, on s'empressa de décourager ces velléités, tandis qu'après avoir refusé à l'archevêque de Baltimore les privilèges de la primatie, on l'invitait discrètement à réduire au minimum les exceptions au droit commun, afin que l'Église américaine n'apparaisse pas comme une Église nationale avec ses coutumes propres [1]. Par contre, on encouragea partout la tenue de conciles provinciaux, afin d'accélérer le retour à l'observation intégrale du droit canon universel. Dans le même but, ainsi que pour contrecarrer l'esprit d'indépendance de certains évêques et habituer les esprits au droit souverain d'intervention du pape dans toute l'Église, on encouragea, on provoqua même au besoin fidèles, prêtres et évêques à recourir à la Curie pour toute espèce de questions. Les contemporains relevèrent également comme caractéristique de la nouvelle politique la nomination de Ketteler à Mayence, en 1849, où Pie IX, sans qu'il y ait eu un procès informatif en forme et sans fournir de base canonique, refusa l'investiture à Schmid, l'élu du chapitre.

Sur le plan doctrinal également, les positions se firent soudain plus nettes. Sans doute, il y avait des années que les professeurs du Collège romain exposaient, sans la moindre réticence, les doctrines de saint Thomas et de Bellarmin sur les prérogatives pontificales et notamment sur l'infaillibilité ; et c'était en 1847 déjà que l'on avait réédité les *Vindiciae* de Ballerini contre Febronius, l'un des ouvrages sur lesquels la propagande ultramontaine allait s'appuyer volontiers. Mais ce n'est qu'après le milieu du siècle que ce travail positif se doubla d'une action répressive qui frappa beaucoup les contemporains. Les condamnations d'ouvrages français, dont il a été question plus haut, en furent la manifestation la plus caractéristique, mais il y en eut d'autres. Par exemple, la mise à l'*Index*, en 1851, des traités de droit canon du professeur Nuyts, de Turin [2], qui défendait du reste des thèses bien plus radicales.

---

(1) GUILDAY, *A History of the councils of Baltimore*, p. 157 et 181.
(2) Sur Népomucène Nuyts, dont les doctrines s'inspiraient à la fois du libéralisme et du joséphisme, et qui mourut en 1876 sans s'être réconcilié avec l'Église, cf. E. CAMPANA, *Il Concilio Vaticano*, t. I, p. 840-842, et A. QUACQUARELLI, *La crisi della religiosità contemporanea*, p. 12-13.

On peut également envisager dans la perspective de l'affirmation de plus en plus énergique des thèses ultramontaines la proclamation, en 1854, du dogme de l'Immaculée-Conception, car cet événement, qui impressionna vivement le monde catholique, contribua singulièrement à mettre en évidence les prérogatives du pape et l'importance croissante de son rôle dans l'Église.

LA PROCLAMATION DE L'IMMACULÉE-CONCEPTION [1]   L'apparition de la Vierge à Catherine Labouré, en 1830, et la diffusion de la médaille miraculeuse avaient été l'occasion pour beaucoup d'évêques de demander d'introduire le terme « immaculée » dans la préface de la fête de la Conception de la Vierge et d'ajouter aux litanies une invocation mentionnant ce privilège ; puis à partir de 1840 avaient commencé les requêtes en vue d'obtenir que la doctrine elle-même fût définie comme dogme de foi, la première provenant d'un groupe de 51 prélats français. Grégoire XVI toutefois avait jugé préférable de ne pas y donner suite à cause de l'opposition des milieux jansénisants et des réticences des évêques d'Angleterre, d'Irlande et surtout d'Allemagne, où les Facultés de théologie devaient se montrer jusqu'à la définition défavorables à une doctrine qu'elles jugeaient trop peu scientifiquement établie.

Les demandes reprirent avec l'avénement de Pie IX et, en 1847, le P. Perrone publia un mémoire très remarqué où il répondait affirmativement à la question : « L'Immaculée-Conception de la bienheureuse Vierge Marie peut-elle faire l'objet d'une définition dogmatique ? » Pie IX désigna en conséquence, le 1er juin 1848, une commission de vingt théologiens chargée d'étudier « s'il y avait lieu d'accéder aux vives instances d'un très grand nombre d'évêques ». La fuite à Gaëte et les préoccupations du souverain pontife n'arrêtèrent pas les travaux, au contraire : le 2 février 1849 le pape adressait à tous les évêques l'encyclique *Ubi primum* où il leur demandait de prier pour la définition et de donner leur avis sur l'opportunité de celle-ci. Sur plus de 600 réponses, les neuf dixièmes furent favorables et, parmi les opposants, seuls quelques-uns, dont l'archevêque de Paris Mgr Sibour [2], estimaient que la doctrine n'était pas définissable, tandis que les autres considéraient simplement la définition comme inopportune : il s'agissait surtout d'évêques de pays protestants, mais on en trouvait d'autres, comme le primat de Belgique, qui craignaient d'exciter inutilement les milieux libéraux.

Le pape fit alors préparer une bulle exposant les fondements de la doctrine, mais les deux premiers projets, dus l'un au P. Perrone, l'autre au P. Passaglia, ne furent pas agréés. Une nouvelle commission de théologiens, désignée en mai 1852, mit au point un troisième projet, après

---

(1) BIBLIOGRAPHIE. — I. SOURCES. — *Pareri dell' Episcopato cattolico, di capituli, di congregazioni, di università... etc., sulla definizione dogmatica dell'immacolato concepimento della B. V. Maria*, Rome, 1851-1854, 10 vol. ; V. SARDI, *La solenne definizione del dogma dell'immacolato concepimento di Maria santissima, Atti e documenti*, Rome, 1905.

II. TRAVAUX. — X. LE BACHELET, art. *Immaculée-Conception*, IV. La définition, dans *D. T. C.*, t. VII, col. 1189-1218.

(2) *Pareri dell'Episcopato...*, t. III, p. 310.

avoir étudié à fond la question de principe — la possibilité de définir une vérité qu'on ne trouve pas affirmée dans l'Écriture — et la question de fait — les arguments de tradition en faveur de l'Immaculée-Conception. Dans cette commission, un rôle de premier plan revint à Passaglia, grand connaisseur de la tradition grecque et orientale.

Une fois la proclamation solennelle fixée au 8 décembre 1854, le pape invita l'épiscopat de chaque nation à se faire représenter à cette cérémonie. Le 20 novembre, les évêques déjà arrivés à Rome furent réunis et le cardinal Brunelli leur déclara que le Saint-Père, sans autoriser une discussion ni sur le fond ni sur l'opportunité, désirait cependant entendre leur avis sur le texte de la bulle. Les observations furent plus nombreuses qu'on ne l'avait prévu et la discussion se prolongea jusqu'au 24, alimentée notamment par l'évêque de Bruges, Mgr Malou, spécialiste de la question, et par les archevêques de Baltimore, de Munich, de Prague et de Vienne ; elle reprit encore lors du consistoire des cardinaux, le 1er décembre, et plusieurs modifications furent de la sorte introduites *in extremis*, surtout sous l'influence de Rauscher. Le pape fut, paraît-il, un peu mortifié de voir des étrangers prétendre faire la leçon aux théologiens romains, mais il estima qu'« il fallait accepter cette humiliation afin qu'on ne dise pas que tout dépendait des jésuites » [1].

Le 8 décembre, en la basilique Saint-Pierre, au cours d'une cérémonie prestigieuse, en présence de près de deux cents cardinaux, archevêques et évêques — on n'en avait plus vu autant rassemblés depuis le concile de Trente — Pie IX lut le décret « d'une voix claire et sonore », mais son « émotion était telle qu'il dut s'arrêter jusqu'à trois reprises » [2]. Dans les jours qui suivirent, tout le monde catholique fut en fête : tandis que la reine d'Espagne envoyait au Saint-Père une tiare valant deux millions de réaux et que le roi de Naples faisait annoncer la définition dans un ordre du jour à l'armée au son de salves d'artillerie, des manifestations mariales se succédaient jusque dans les moindres villages, contribuant singulièrement à accentuer la dévotion populaire envers la Vierge [3].

Mais là n'était pas le seul intérêt de la définition. « Voyez-vous, aurait confié Mgr Talbot quelques jours auparavant, le plus important n'est pas le nouveau dogme en lui-même, mais la manière dont il est proclamé [4]. » Effectivement, lorsque, le 20 novembre, deux évêques avaient demandé au cardinal président s'il ne convenait pas de faire mention dans la bulle de l'approbation de l'épiscopat, il leur fut répondu que « si le souverain pontife prononçait seul la définition, à laquelle tous les fidèles adhéreraient spontanément, son jugement fournirait une démonstration pratique de l'autorité souveraine de l'Église en matière de doctrine *et de*

(1) Rapporté par A. FLIR, *Briefe aus Rom*, p. 25.
(2) Lettre du neveu de l'archevêque de Malines, du 9 décembre 1854. (Archives privées de la famille Devoghel.)
(3) Les manifestations d'opposition furent extrêmement réduites. Seuls quelques prêtres jansénisants firent entendre leurs protestations (cf. F.-H. REUSCH, *Der Index der verbotener Bücher*, t. II, 2, p. 1152-1156 ; L. SÉCHÉ, *Les derniers jansénistes*, t. III, p. 24-26) ; les universitaires allemands mécontents s'inclinèrent en silence (leur attitude est éclairée par la lettre de Doellinger à Michelis, du 31 janvier 1854, reproduite dans J. FRIEDRICH, *Ign. v. Doellinger*, t. III, p. 132-134).
(4) Knoodt à Doellinger, 10 juillet 1866, dans J. FRIEDRICH, *Ign. v. Doellinger*, t. III, p. 146.

l'infaillibilité dont Jésus-Christ a investi son vicaire sur la terre »[1]. Certes, Pie IX n'avait pas manqué de consulter l'épiscopat et les gallicans pouvaient toujours affirmer que c'était l'approbation effective par celui-ci qui conférait sa pleine valeur à la promulgation pontificale ; mais il est clair que l'acte final, auquel les évêques assistaient en simples spectateurs, mettait singulièrement en relief la personne du pape ; c'était un nouveau coup pour les tendances épiscopaliennes et les ultramontains allaient s'empresser d'en mettre en lumière toutes les conséquences.

§ 2. — Pie IX et la Curie romaine vers 1860 [2].

LA CURIE SOUS PIE IX — En abandonnant au cardinal Antonelli tout ce qui regardait le gouvernement temporel de ses États pour se consacrer exclusivement à la direction religieuse de l'Église universelle, Pie IX, si peu « moderne » par ailleurs, s'était orienté délibérément dès 1850 dans la voie qui sera celle de la papauté contemporaine. La Curie subit au cours de son pontificat une évolution analogue, moins rapide cependant. La force des habitudes et l'existence jusqu'en 1870 d'une administration civile dirigée par des prélats et même des cardinaux dont plusieurs ne sont pas prêtres, expliquent la persistance de certains usages qui rappellent l'ancien régime. C'est ainsi qu'en 1862, Mgr Berardi [3] ayant été désigné comme nonce alors qu'il n'était que tonsuré, se vit en quelques jours conférer tous les ordres, y compris l'épiscopat. De même, on rencontre encore des ecclésiastiques sans grandes préoccupations religieuses et aux allures plutôt mondaines, mais les touristes protestants, qui s'en scandalisaient volontiers, ont exagéré l'importance du phénomène ; l'attaché français notait au contraire, en 1865 :

> Ce que j'ai observé depuis trois ans que j'habite à Rome, c'est la régularité parfaite des mœurs du clergé romain. On prétend que le Saint-Père, si indulgent d'ordinaire, est d'une sévérité inflexible pour tout ce qui concerne ce chapitre. Les cardinaux et les *monsignori* vont rarement dans le monde et on ne rencontre généralement dans les ambassades et chez les princes romains que quelques-uns d'entre eux [4].

(1) Cité par I. GOSCHLER, art. *Vierge* (Immaculée-Conception), dans *Dictionnaire encyclopédique de la théologie catholique*, t. XXV, Paris, 1871, p. 280-281.

(2) BIBLIOGRAPHIE. — I. SOURCES. — A. FLIR, *Briefe aus Rom*, Innsbruck, 1864 ; H. D'IDEVILLE, *Journal d'un diplomate en Italie*. II. *Rome (1862-1864)*, Paris, 1873 ; E. LAFOND, *Rome. Lettres d'un pèlerin*, 2e édit., Paris, 1864, 2 vol. (enthousiaste) ; K. VON SCHLOEZER, *Römische Briefe (1864-1869)*, Stuttgart, 1913 (protestant) ; GREGOROVIUS, *Diari Romani*, Milan, 1895 (*idem*) ; C. CURCI, *Memorie*, Florence, 1891 (à utiliser avec prudence) ; *La Curia Romana e i Gesuiti*, Florence, 1861 (polémique) ; *La crise de l'Église*, Paris, 1878 (*idem*) ; les extraits du journal d'Acton à Rome (1857) publiés dans *Cambridge Historical Journal*, t. VIII, 1946, p. 186-204 ; les lettres de Doubet à Rendu (1853-1854) dans J. GAY, *Les deux Rome et l'opinion française*, Paris, 1931 ; les dépêches de l'ambassadeur d'Autriche à Rome dans S. JACINI, *Il tramonto del potere temporale (1860-1870)*, Bari, 1931. Les *Souvenirs d'un prélat romain* (Mgr Chaillot) édités par P. ROCFER, Paris, 1895, fourmillent d'inexactitudes.

II. TRAVAUX. — Outre les biographies de Pie IX, signalées p. 7, S. NEGRO, *Seconda Roma (1850-1870)*, Milan, 1941 ; L. TESTE, *Préface au conclave*, Paris, 1876 ; J. BASTGEN, *Die römische Kurie, ihre gegenwärtige Zusammensetzung und ihr Geschäftsgang*, Munster, 1854.

(3) Juriste compétent et « l'un des meilleurs hommes de gouvernement du dernier siècle de la Rome pontificale » (S. NEGRO, *op. cit.*, p. 435), Mgr Berardi, avant de devenir, en 1868, cardinal et ministre des Travaux publics et du Commerce, avait été, depuis 1856, le principal collaborateur du cardinal Antonelli à la secrétairerie d'État.

(4) H. D'IDEVILLE, *Journal d'un diplomate... Rome*, p. 351.

Le sérieux croissant de la Curie va d'ailleurs de pair avec l'accentuation, plus nette chaque année, de son rôle de centre nerveux de la chrétienté ; les fonctionnaires ecclésiastiques chargés de l'administration civile voient leur importance diminuer au profit des prélats pieux et des canonistes ou théologiens attachés aux Congrégations, qui secondent le pape à des titres divers dans l'exercice quotidien et de plus en plus effectif de sa primauté de juridiction sur les catholiques du monde entier.

Malheureusement, ces progrès de l'esprit religieux à la Curie s'accompagnent souvent d'un recul sur le plan intellectuel [1] et surtout d'une inquiétante incompréhension à l'égard du monde moderne et des évolutions nécessaires : « C'est la cour de Louis XVI avec plus de vertu », écrivait Cochin [2], qui ajoutait quelques jours plus tard :

Beaucoup de largeur, de gravité, de science, avec une ignorance complète du temps présent, la superstition de ce qui est vieux en tout, depuis le costume jusqu'aux opinions, depuis l'étiquette jusqu'à la théologie [3].

Les témoins les moins suspects sont d'accord pour relever et déplorer la méfiance et l'hostilité qui régneront à la Curie jusqu'à la mort de Pie IX, non seulement, ce qui se comprendrait encore, à l'égard des idées libérales [4], mais également à l'égard des méthodes critiques et des résultats de la science contemporaine. On ose à peine imaginer la réaction qui eût accueilli, dans l'atmosphère romaine des années soixante, l'*Histoire des papes* de Pastor, et le cardinal Gasselini n'exagérait guère lorsqu'il gémissait : « L'influence de l'Église dans le passé tient en grande partie à ce qu'elle était à la tête de la science. Aujourd'hui, elle est à la queue [5]. »

Chose plus grave encore, l'étroitesse d'esprit des bien-pensants, des *zelanti*, leur faisait accueillir sans précaution et même encourager les dénonciations qui, de France et d'Allemagne surtout, tendaient à compromettre tous ceux qui, soit en théologie soit en politique ecclésiastique, n'adoptaient pas le point de vue strictement conformiste qui avait les préférences des milieux romains. Le pontificat de Pie IX n'a certes pas été, comme Dupanloup l'écrivait dans un moment de mauvaise humeur, « le règne et le triomphe de la calomnie » [6], mais la calomnie y a joué un trop grand rôle et la mentalité qui régnait à la Curie y a largement contribué.

---

(1) Voir *supra*, p. 184-185. Déjà en 1853, le catholique libéral Doubet déplorait « le manque de capacité dans la prélature (...). Le clergé est plus pieux, plus moral, c'est vrai. Mais les hommes habiles n'y trouvent plus de carrière et n'y entrent pas. D'abord plus de collège ou académie de prélats, puis les cardinaux n'ont plus le moyen de payer des consulteurs, qui formaient une pépinière ». (Lettre du 29 juillet 1853, dans J. GAY, *Les deux Rome...*, p. 57.)

(2) Cochin à Montalembert, 3 mai 1862, dans *Augustin Cochin. Ses lettres*, t. I, p. 278.

(3) Cochin à Corcelle, 27 mai 1862, *ibid.*, t. I, p. 281.

(4) Quelques-uns cependant, les cardinaux Altieri, Santucci, Di Pietro ou Mgr Franchi, par exemple, étaient considérés à Rome comme ayant des sympathies pour les institutions libérales ; encore faut-il se souvenir que sous le régime Antonelli, on était vite considéré comme tel. Le cas du cardinal d'Andrea est différent : ce grand seigneur était en réaction ouverte contre l'esprit qui régnait à Rome. (Voir l'art. *Andrea*, dans *D. H. G. E.*, t. II, col. 1736-1737.)

(5) Déclaration notée par Hyacinthe Loyson dans son journal, le 18 août 1868. (Dans A. HOUTIN, *Le P. Hyacinthe dans l'Église romaine*, p. 177, note.)

(6) *Arch. Nat.*, AB XIX, n° 524.

*LA DIMINUTION D'INFLUENCE
DU SACRÉ COLLÈGE*
Le pontificat de Pie IX voit se produire à Rome une évolution analogue à celle qui, dans les diocèses, réduisait à presque rien le rôle jadis si grand des chapitres métropolitains au profit de l'autorité centralisatrice de l'évêque, assisté de ses vicaires généraux et de ses secrétaires : l'influence du Sacré Collège diminue sensiblement, tant dans la direction temporelle de l'État pontifical que dans la direction religieuse de l'Église.

Les cardinaux ne sont consultés que « très rarement », écrivait en 1860 un correspondant romain de Cavour et, quand ils le sont, « les choses sur lesquelles le pape les interroge sont pour la plupart des choses abstraites, des questions générales » ; les décisions effectives ne leur appartiennent plus [1]. Dix ans plus tard, l'ambassadeur de France constatait : « Jamais le rôle des cardinaux n'a été aussi effacé qu'il l'est aujourd'hui et leur influence aussi insignifiante [2]. » Les contemporains s'accordent pour attribuer la principale responsabilité de cette situation à Antonelli, qui ne fit du reste qu'accentuer la tradition de « despotisme bureaucratique » [3] inaugurée sous Grégoire XVI par Lambruschini :

A tort ou à raison, observait l'ambassadeur d'Autriche, les cardinaux lui reprochent de les avoir privés de toute influence réelle dans la direction des affaires de l'État, d'avoir placé presque exclusivement des nullités dévouées à lui dans les postes éminents et lucratifs, afin de tenir éloignés du Saint-Père tous ceux dont l'indépendance de vues et de conseils pourrait contrarier l'ascendant de son influence.

Dans ces conditions, poursuivait-il,

les cardinaux actuels sont des prélats très respectables, très consciencieux dans l'accomplissement de leurs fonctions et devoirs particuliers, mais dont la moitié à peine et peut-être moins encore ne s'élève au-dessus de la médiocrité [4].

Cet avis était partagé même par des amis sincères du Saint-Siège :

Les cardinaux, sauf trois ou quatre exceptions, estimait l'un d'eux, ne comptent parmi eux que des médiocrités insignifiantes, d'excellents moines, de bons prêtres érudits, d'habiles théologiens, sans doute, tous gens auxquels l'expérience des grandes affaires, le tact politique et la connaissance des hommes manquent totalement [5].

Pie IX ne voyait pas d'un mauvais œil cette évolution qui transformait peu à peu une assemblée composée surtout jusqu'alors de représentants de l'aristocratie romaine et de hauts fonctionnaires remarquables par leurs mérites politiques ou administratifs, en un groupement d'hommes d'Église qui s'étaient distingués par leur zèle pastoral, leur

(1) Anonyme à Cavour, 10 novembre 1860, dans *La Questione romana negli anni 1860-1861*, t. I, p. 74-75.
(2) Dépêche du 2 juin 1869, *Arch. Min. A. É. Paris*, Rome, 1043, f⁰ 406.
(3) Pantaleoni à Cavour, 2 octobre 1860, dans *La Questione romana...*, t. I, p. 46.
(4) Rapport de décembre 1861 dans S. JACINI, *Il tramonto del potere temporale*, p. 82-84. Même son de cloche dans la lettre de Pantaleoni citée à la note précédente et dans les souvenirs d'H. d'Ideville : « Le cardinal Antonelli, depuis près de vingt ans qu'il dirige les conseils du pape, a vu se renouveler presque entièrement le Sacré Collège et son plus grand soin, dans une pensée un peu personnelle, a été d'écarter tous les hommes dont le mérite pouvait lui porter ombrage ou dont l'influence aurait pu contre-balancer son crédit. » (*Journal d'un diplomate... Rome*, p. 210.)
(5) H. D'IDEVILLE, *op. cit.*, p. 210.

science ecclésiastique ou leurs tendances ultramontaines. C'est même à son initiative personnelle que furent introduits dans le Sacré Collège des théologiens ou des canonistes comme Bilio, Pitra, Tarquini, Guidi, Franzelin, et dès 1850, en internationalisant davantage le Sacré Collège par la création de dix cardinaux étrangers contre seulement quatre italiens, il avait clairement montré que, pour lui, leur tâche essentielle ne résidait plus dans l'administration des États romains.

Contrairement à ce qu'on aurait toutefois pu attendre, l'influence du Sacré Collège comme tel resta assez limitée, même dans la direction religieuse de l'Église. Cela s'explique sans doute pour une part par le manque de personnalités d'envergure, mais surtout par suite de l'intervention très personnelle de Pie IX dans la conduite de l'Église : il aimait à se rendre compte des choses par lui-même et à prendre des décisions sans craindre de passer par-dessus les instances régulières. Aussi les consistoires se firent-ils plus rares, de même que les réunions de cardinaux chargés de délibérer sur tel ou tel problème intéressant la vie de l'Église. Il faut toutefois tenir compte de l'influence personnelle exercée par certains cardinaux soit sur l'esprit du pape, soit directement sur la conduite des affaires. Le cardinal Bedini, qui avait beaucoup voyagé et possédait un bon jugement, le cardinal dominicain Gaude, en matière théologique, le cardinal de Reisach, pour les questions intéressant les pays germaniques, jouèrent à certains moments un rôle non négligeable. Le plus marquant paraît avoir été le cardinal Barnabo [1], préfet de la Propagande de 1856 à 1874, qui, à ce titre, tout en ayant la responsabilité d'un certain nombre d'Églises organisées comme celle d'Angleterre, présida avec compétence, prudence et énergie à la merveilleuse expansion missionnaire qui caractérisa le pontificat de Pie IX. Ses allures un peu despotiques ne peuvent faire oublier l'honnêteté foncière de cet homme, qui garda toujours son franc-parler même avec le pape, lequel d'ailleurs trancha plus d'une fois dans un sens opposé à ses conseils, lors de la nomination de Manning par exemple.

L'ENTOURAGE DU PAPE   L'influence de son entourage sur un souverain absolu est toujours importante. Elle l'est tout particulièrement quand il s'agit, comme c'était le cas pour Pie IX, d'un homme facilement impressionnable, enclin à n'envisager les problèmes que sous un angle restreint. Quels étaient les hommes qui jouissaient de la confiance du souverain pontife ?

A son retour à Rome en 1850, Pie IX, désireux de réagir contre la politique d'italianisation suivie par son prédécesseur, avait fait choix de trois camériers participants étrangers, un prince bavarois, Gustave-Adolphe de Hohenlohe, un anglais récemment converti, Georges Talbot, et un noble belge très lié à la France, Xavier de Mérode. Le premier, excellent cœur mais timide et peu intelligent — « *un ragazzo matto ma*

---

(1) Le portrait qu'en a tracé É. Ollivier (*L'Église et l'État au concile du Vatican*, t. II, p. 185) doit être corrigé par les appréciations de Mgr Talbot (dans E. PURCELL, *Life of Manning*, t. II, p. 210), du P. Hecker (dans W. ELLIOTT, *Le P. Hecker*, trad. franç., p. 233), de Flir (*Briefe aus Rom*, p. 119), de Banneville (*Arch. Min. A. É. Paris*, Rome, 1043, fº 431 vº) et de C. KOROLEVSKY (art. *Audo* dans *D. H. G. E.*, t. V, col. 341).

un angelo », comme l'appelait le pape [1] — n'eut jamais qu'un rôle effacé et, bien qu'élevé au cardinalat en 1866, il finit par perdre tout crédit à cause de ses relations avec l'opposition anti-ultramontaine. Mgr Talbot, au contraire, fut pendant près de vingt ans le confident et le favori le plus intime de Pie IX, auquel il était lui-même attaché corps et âme. Bien qu'il eût tendance à s'exagérer son influence, celle-ci était grande et l'on a pu comparer son rôle « à la position des confesseurs du roi sous l'ancien régime : influence sans responsabilité » [2]. Or, même s'il faut admettre que l'image que nous en a laissée Purcell a été poussée au noir [3], on doit regretter que trop souvent les jugements du pape, en particulier sur les hommes et les choses d'Angleterre, sur Newman par exemple, n'aient été que le reflet des vues unilatérales et sans nuances de ce prélat un peu exalté [4].

Quant à Mgr de Mérode, nous le connaissons déjà [5]. Son crédit, à lui aussi, était grand auprès de Pie IX qui sympathisait avec son tempérament primesautier et caustique, mais qui appréciait surtout sa loyauté et son désintéressement. Soldat plutôt que diplomate, il encouragea souvent le pape dans la voie de l'intransigeance et ses préventions contre le régime de Napoléon contribuèrent à aggraver la méfiance à l'égard du gouvernement de l'empereur et des évêques qui, comme Darboy, cherchaient à rester en bons termes avec lui. Il est certes dommage que l'un des principaux conseillers de Pie IX ait été un homme que son caractère emporté, son esprit « plus pétillant que juste » [6] et son ardeur un peu brouillonne avaient fait surnommer « l'écervelé » (*il matto*). Mais il ne faut pas oublier que ce sobriquet lui avait été décoché par ses ennemis, et Dieu sait s'il en avait, tant parmi les prélats qui ne pouvaient supporter ses efforts pour réformer les abus administratifs, que dans le parti avancé qui voyait en lui, à cause de son attitude dans la Question romaine, l'incarnation de l'esprit antirévolutionnaire. Ce dernier jugement manquait pour le moins de nuances et, sans parler de l'intérêt du prélat pour les inventions modernes, qu'il cherchait à introduire à Rome, on retiendra le jugement d'un diplomate qui n'avait rien d'un réactionnaire : « Je dois dire que dans cette affaire comme dans beaucoup d'autres, c'est Mgr de Mérode qui se montre le plus libéral des conseillers de Pie IX [7]. »

(1) Notes d'Acton en 1857 dans *Cambridge Historical Journal*, t. VIII, 1946, p. 190. Mgr de Hohenlohe (1823-1896) avait une phobie presque maladive des jésuites, ce qui explique pour une part sa réserve croissante à l'égard des idées dominantes à Rome.
(2) C. Butler, *Life and Times of B. Ullathorne*, t. I, p. 228.
(3) C'est l'avis du cardinal Gasquet dans *Dublin Review*, t. CLXIV, 1919, p. 1-2. D'après lui, les nombreuses lettres inédites des évêques anglais et irlandais à Talbot obligeraient à nuancer fort l'impression qui se dégage de la correspondance entre ce dernier et Manning publiée par Purcell. En 1857, lors de son séjour à Rome avec Doellinger, Acton le considérait comme « le plus judicieux des camériers ». (Dans *Cambridge Historical Journal*, t. VIII, 1946, p. 190.)
(4) Atteint de maladie mentale, Mgr Talbot dut être interné à partir de 1868.
(5) Cf. *supra*, p. 102-103.
(6) Dépêche du 20 novembre 1865 de l'ambassadeur d'Autriche qui, après avoir loué « son activité et son énergie prodigieuse », concluait : « Mais son mérite et ses vertus sont contre-balancés par de grands défauts et, somme toute, on peut dire, sans être trop sévère, que c'était un collègue détestable, un ami peu commode et bien souvent un enfant terrible. » (Dans S. Jacini, *Il tramonto del potere temporale*, p. 174.)
(7) Dépêche du ministre de Belgique à Rome, 15 novembre 1862, *Arch. Min. A. É. Bruxelles*,

Parmi les camériers italiens qui avaient la faveur de Pie IX, les contemporains mentionnent surtout, outre Mgr Borromeo et Mgr Ricci, dont les avis étaient souvent d'inspiration réactionnaire, Mgr Stella, dont Doubet écrivait, lors de son passage à Rome :

> Je l'estime pour sa droiture, sa conscience, son dévouement, mais pas pour ses lumières ni pour sa manière d'entendre les choses religieuses. C'est un confesseur de sœurs cloîtrées, qui veut quelquefois avoir ses pensées sur les choses de ce monde [1].

Il faut enfin ajouter à ces jeunes confidents un ami très écouté du pape pendant plus de trente ans, le cardinal Patrizzi [2], dont tous les esprits non prévenus admiraient la charité, l'austérité et la piété. Les partisans des idées libérales dénonçaient en lui « la quintessence de l'ignorance de l'âne combinée avec l'entêtement de la mule, le tout uni à une forte dose de dévotion qui dépasse la bigoterie pour atteindre au fétichisme » [3], mais avec plus de mesure, un bon observateur des choses romaines notait, au lendemain de sa mort : « Le Sacré Collège et l'Église n'ont pas perdu en lui une grande lumière, mais un grand exemple de piété et de vertu sacerdotale [4]. »

Ce fut le malheur de Pie IX que, parmi les prélats pieux et dévoués aux intérêts du Saint-Siège dont il aimait à s'entourer, il n'y en eût aucun qu'on pût qualifier de « grande lumière », ni même d'esprit vraiment perspicace. La prudence oblige évidemment à n'accueillir qu'avec les plus expresses réserves les critiques d'adversaires comme le docteur Pantaleoni, d'après lequel « il serait difficile d'imaginer un personnel plus stupide, plus déraisonnable, plus excessif dans les idées » [5], ou comme le comte de Gramont, qui après avoir dénoncé la « déloyauté » et la « décrépitude » de la « coterie du Vatican », concluait : « On éprouve un mélange de pitié et de dégoût en sondant cet abîme : c'est ici que l'Orient commence [6]. » On ne peut toutefois s'empêcher de constater la fréquence de plaintes de ce genre, quoique moins violentes, de la part d'hommes profondément attachés au Saint-Siège. Le zèle très réel des familiers de Pie IX ne pouvait remplacer la compétence et la pondération de jugement qui leur manquaient trop souvent, de même que l'envergure des conceptions et le sens de la complexité des situations concrètes dont est faite la vie quotidienne de l'Église.

Saint-Siège, t. XI. Cette observation est corroborée par H. D'IDEVILLE, *Mgr de Mérode*, p. 41 : « En politique, Mgr de Mérode appartenait à l'école parlementaire la plus avancée ; aussi n'était-il jamais d'accord avec le cardinal Antonelli qui appartient à l'école autoritaire ». (Voir également, du même, *Journal d'un diplomate... Rome*, p. 234.)

(1) Doubet à Rendu, juillet 1853, dans J. GAY, *Les deux Rome...*, p. 53.

(2) Constantin Patrizzi (1798-1876) était le frère du savant exégète jésuite Francesco Patrizzi. Il occupait le poste de cardinal-vicaire et comme l'a observé L. Teste, « ce n'est pas un des spectacles les moins singuliers que celui de Pie IX traversant trente années entre le cardinal Antonelli et le cardinal Patrizzi, entre la Ruse et la Piété ». (*Préface au conclave*, p. 43.)

(3) Pantaleoni à Cavour, 6 novembre 1860, dans *La Questione romana...*, t. I, p. 70.

(4) L. TESTE, *Préface au conclave*, p. 80-81.

(5) Pantaleoni à Cavour, 6 novembre 1860, dans *La Questione romana...*, t. I, p. 70.

(6) Dépêche du 30 octobre 1860, dans J. MAURAIN, *La politique ecclésiastique du Second Empire*, p. 424.

L'INFLUENCE DES JÉSUITES A en croire plus d'un observateur, la place prépondérante prise par les jésuites dans les conseils du pape et dans les Congrégations à partir de 1850 s'expliquerait précisément par l'infériorité flagrante qui régnait à la Curie, tant dans le clergé séculier que parmi les représentants des anciens ordres.

Par tempérament, Pie IX n'aimait guère les jésuites. Nous ne sommes pas obligés de croire Theiner, qui prétend lui avoir entendu dire un jour vers 1855 : « Je ne pensais pas que les procédés peu charitables dont use cette Compagnie pouvaient atteindre de telles proportions » [1] ; mais il est certain qu'il n'appréciait guère leurs allures compassées et qu'il les trouvait trop intrigants. Son attitude à leur égard, au début de son pontificat, avait été plus que réservée et, même après son retour de Gaëte, ses rapports avec les autorités de la Compagnie demeurèrent froids tout un temps. Pourtant, il ne pouvait s'empêcher de rendre justice à leurs vertus et à leurs mérites ; il se rendait compte de l'œuvre spirituelle très féconde qu'ils accomplissaient dans toute l'Église, notamment en Allemagne ; il leur était reconnaissant de l'appui qu'ils apportaient à la réaction contre les idées libérales, à la défense du pouvoir temporel et au progrès du courant ultramontain. Par ailleurs, plus les questions à examiner augmentaient avec les progrès de la centralisation, plus il appréciait le sérieux de la formation des jésuites et l'avantage que présentaient leurs relations internationales ; c'était entre autres l'avis que Doellinger recueillait lors de son passage à Rome en 1857 : « Les jésuites sont cosmopolites, souples, expérimentés ; aussi apparaissent-ils comme des conseillers indispensables [2]. »

Quoi qu'il en soit des motifs — et sans doute faut-il aussi compter parmi eux tout simplement l'augmentation rapide des effectifs de la Compagnie [3] qui permettait à celle-ci de se mettre largement à la disposition du Saint-Siège — le fait est incontestable : l'influence des jésuites à la Curie ne cessa d'augmenter. Les avis de leur général et des membres influents de l'ordre furent de plus en plus écoutés par le pape, leurs théologiens apparurent de plus en plus comme des oracles dans les Congrégations, ils jouèrent un rôle très actif dans la préparation du *Syllabus* puis du concile du Vatican, enfin leur revue *La Civiltà cattolica*, suivie de très près par Pie IX, fut considérée partout comme l'organe officieux du Saint-Siège. Il serait certes fort exagéré de parler avec un mémoire anonyme publié à la mort de Pie IX d'« une sorte de gouvernement occulte, composé de jésuites de *La Civiltà* », ou de dire comme lord Acton que « Pie IX a fait des jésuites le canal de son influence et est devenu l'instrument de la leur » [4]. Mais il est bien certain que, dans les différents domaines de la politique ecclésiastique, de l'organisation de l'action catholique, de la théologie et de la spiritualité, ce sont les points de vue prônés par les jésuites, surtout par les jésuites romains, qui s'imposèrent toujours davantage au cours du pontificat de Pie IX.

---

(1) Rapporté dans *Janssens Briefe*, édit. L. PASTOR, t. I, p. 237.
(2) Cité dans J. FRIEDRICH, *Ign. von Doellinger*, t. III, p. 183.
(3) Cf. *infra*, p. 458.
(4) J. ACTON, *The Vatican Council*, dans *North British Review*, t. LIII, 1870, p. 187.

PROGRÈS CROISSANTS
DE LA CENTRALISATION

On a déjà signalé dans l'article précédent la tendance à concentrer toujours davantage à Rome la direction de l'Église entière, à renforcer l'action directe du Saint-Siège dans les diocèses et à uniformiser le plus possible les usages ecclésiastiques jusque dans le détail. Les moyens mis en œuvre dans ce but à partir de 1850 furent systématiquement développés et de nouveaux facteurs accélérèrent l'évolution.

Le pape procéda de plus en plus aux nominations épiscopales sans tenir compte des propositions faites par le haut clergé local et en s'inspirant surtout de la formation romaine et de la docilité des candidats, n'hésitant pas, à plus d'une reprise, à préférer à des prêtres méritants, jugés trop indépendants, des hommes de second plan dont on était sûr. Or, grâce à la durée de son règne, Pie IX eut l'occasion de renouveler à peu près complètement le corps épiscopal : en 1869 déjà, sur les 739 évêques de la Chrétienté, 81 seulement dataient de Grégoire XVI. Même là où par suite des concordats, sa liberté de choix était réduite, il usa plus d'une fois de la ressource qu'il avait de refuser un candidat suspect en invoquant quelque motif canonique ; en outre, il restait libre d'élever au cardinalat qui il voulait, et c'était un moyen efficace de renforcer l'autorité des évêques les plus dévoués à Rome. Il en résulta rapidement une évolution de mentalité qui se manifesta entre autres à l'occasion des conciles provinciaux ; les évêques firent de moins en moins de difficultés pour reconnaître au Saint-Siège le droit de contrôler et éventuellement de modifier leurs décisions, témoignant ainsi en pratique, comme les théologiens ne tarderont pas à le souligner, du caractère limité de leurs pouvoirs ; certains allèrent même plus loin, en soumettant à l'avance à l'agrément de Rome les matières sur lesquelles ils comptaient délibérer. Nombreux furent d'ailleurs les conciles qui reconnurent spontanément, d'une manière ou d'une autre, l'infaillibilité du souverain pontife ou sa primauté de juridiction [1].

Lorsque certaines velléités d'indépendance s'affichaient trop ouvertement, Pie IX n'hésitait d'ailleurs pas, pour y mettre fin, à convoquer l'intéressé, fût-il archevêque de Paris ou revêtu de la dignité patriarcale. Quelques-unes de ces entrevues furent plutôt orageuses, mais le pape préférait d'ordinaire laisser agir son charme personnel, et les victoires obtenues de cette manière ne se comptent pas ; il y eut même des évêques, arrivés à Rome pleins de réticence, qui, conquis par la bonté et l'absence de rancune de Pie IX, s'en retournèrent chez eux ardents champions de l'ultramontanisme. Ces contacts personnels avec les évêques, dont il avait pu constater l'efficacité, Pie IX chercha du reste à les multiplier en faisant revivre l'obligation des visites périodiques *ad limina*, tombée en désuétude, et en provoquant les grandes assemblées de 1854, 1862 et 1867, dont il sut faire d'imposantes manifestations de l'unité catholique en même temps que des apothéoses du pouvoir pontifical.

---

(1) Outre huit conciles tenus en France de 1849 à 1851, mentionnons ceux de Québec (1851 et 1868), Halifax (1857), Thurles (1850), Armagh (1854) et Tuam (1854) en Irlande, Westminster (1850 et 1855), Cologne (1860), Colocsa (1863), Utrecht (1865), Baltimore (1866), Quito (1869).

Cet épiscopat, dont la déférence à l'égard des ordres et même des désirs du Saint-Siège va croissant d'année en année, est à la fois conseillé, dirigé et surveillé par les nonces, dont la fonction primitive d'agents diplomatiques accrédités auprès des gouvernements passe au second plan, tandis que leur rôle dans la vie interne des Églises ne fait qu'augmenter. Ils interviennent constamment comme intermédiaires entre le pape et les évêques, voire entre le pape et le clergé local. Car c'est une tactique fréquente, pour faire échec aux résistances de certains évêques moins dociles aux nouvelles tendances centralisatrices, de favoriser les rapports directs entre le clergé inférieur et le Siège de Rome, « reproduction, dans l'ordre religieux, notaient certains observateurs, du système politique qui a prévalu à une autre époque, à savoir l'alliance de la Monarchie avec le Tiers-État contre l'Aristocratie »[1].

Cette politique, qui fut surtout appliquée en France, réussit d'autant mieux que Pie IX n'hésitait pas à donner tort à un supérieur, fût-il évêque, lorsqu'il était saisi de la plainte d'un subordonné qui lui semblait justifiée ; et il savait en tout cas accueillir de simples prêtres de si bonne grâce qu'il avait vite fait de gagner leur dévouement total. Ces prêtres, conquis par le sentiment, devenaient les agents les plus zélés de l'ultramontanisme, accueillant avec joie tout nouveau progrès de l'influence romaine. C'est grâce à eux notamment que put être mené à bien ce vaste « travail d'assimilation des Églises de la catholicité tout entière au rite, à la liturgie, aux coutumes, à l'habillement sacerdotal de Rome »[2], qui constitue l'un des traits marquants du pontificat de Pie IX. C'est en partie grâce à eux également que les nonces réussirent finalement à imposer le respect des décisions des Congrégations romaines, dont l'activité dans la vie de l'Église devenait d'année en année plus considérable, en dépit de l'opinion, très répandue jusqu'au milieu du siècle, qui prétendait établir une nette distinction entre l'autorité purement consultative de ces Congrégations et l'exercice de la primauté pontificale.

Dans le but de multiplier les prêtres influents sur le concours desquels on pourrait compter, Pie IX encouragea autant qu'il le put prêtres et séminaristes à fréquenter les Universités romaines. Il se rendait compte de l'utilité de pareils séjours, non seulement pour les affermir dans les pures doctrines ultramontaines, mais surtout pour les familiariser avec les manières de penser et le « style de vie » romains, et réagir ainsi contre les tendances particularistes encore vivaces dans les différents clergés nationaux[3]. Tout en exhortant les évêques à envoyer des sujets toujours plus nombreux dans les collèges existant déjà, il insista pour que de nouveaux séminaires nationaux s'y ajoutent, revenant à la charge sans se lasser, comme dans le cas du Séminaire américain, ouvert finalement

---

(1) Dépêche du ministre de Belgique à Rome du 25 mai 1870. (*Arch. Min. A. É. Bruxelles*, Saint-Siège, t. XIII, 2.)

(2) Dépêche de l'ambassadeur de France du 23 janvier 1864 dans J. MAURAIN, *La politique ecclésiastique du Second Empire*, p. 695.

(3) Un groupe d'élèves du Séminaire français ayant été reçu en audience, l'un d'eux écrivait, le 29 novembre 1859 : « Le Saint-Père nous a félicité de notre séjour à Rome parce qu'on y puise : sainteté, science et unité. Il a insisté surtout sur *unité*. » (*Arch. Nat.*, AB XIX, nᵒ 517.)

en 1859 [1] ; il participa personnellement à la fondation du *Collegium Pium*, pour l'Amérique latine, en 1852, et à celle du Séminaire polonais en 1866, de même qu'à la reconstitution du Collège irlandais en 1862 ; en 1867, au Collège anglais, dont l'esprit ne lui plaisait pas, il nomma de sa propre autorité, sans consulter l'épiscopat, un nouveau recteur, recommandé par Talbot et Manning [2].

Un moyen souvent employé par Pie IX pour renforcer le prestige des prêtres formés à Rome ou dévoués aux conceptions romaines fut de les élever à la prélature, ce qui était alors assez inusité : à lui seul, il nomma plus de *monsignori* étrangers que ses prédécesseurs en l'espace de deux siècles et plus d'un contemporain dénonça dans cette « inondation de prélats romains » une nouvelle façon de faire échec à l'autorité des évêques, qui souvent n'étaient même pas consultés.

D'autres forces encore furent utilisées pour stimuler le courant centralisateur, en particulier les ordres religieux et la presse. On a pu comparer à l'action exercée au XIIIe siècle par les ordres mendiants le rôle que les principaux journaux catholiques, soutenus plus ou moins ouvertement par les nonces, ont joué sous Pie IX pour rendre populaire au plan local la cause pontificale [3]. Quant aux ordres religieux, internationaux par essence et possédant le plus souvent leur centre à Rome [4], ils devaient tout naturellement concourir à la même tâche. Les plus actifs furent les jésuites, dont l'action a d'ailleurs été parfois exagérée et surtout dénaturée [5], et leur principale revue, *La Civiltà cattolica*, alla jusqu'à conseiller de faire le vœu de soutenir « *usque ad effusionem sanguinis* » la doctrine de l'infaillibilité du pape [6].

Lorsqu'on a énuméré les différents moyens, parfois très humains, dont se servirent le pape et la Curie pour assurer le succès de l'influence romaine, le principal n'a pourtant pas encore été dit. L'élément déterminant dans le triomphe rapide et définitif de l'ultramontanisme fut la personnalité même de Pie IX et le prestige immense, dépassant de loin celui de n'importe lequel de ses prédécesseurs depuis des siècles, dont le pontife jouit auprès des masses catholiques et du clergé pendant plus d'un quart de siècle. C'est ce qu'il nous faut à présent essayer de comprendre.

OMBRES ET LUMIÈRES DE PIE IX   Sous beaucoup de rapports, pourtant, Pie IX n'était pas une personnalité de premier plan. Un faible et un médiocre, disaient de lui ses adversaires ; ou encore : un autocrate vaniteux. Le jugement était profondément injuste, mais beaucoup d'esprits non prévenus pensaient à peu près comme Meignan, qui écrivait, au lendemain de son élection :

(1) Voir D. SHEARER, *Pontificia americana*, p. 296, 303-304, 310, 318-319.
(2) E. PURCELL, *Life of Manning*, t. II, p. 365-377.
(3) A. QUACQUARELLI, *La crisi della religiosità contemporanea*, p. 54.
(4) Pie IX s'appliqua du reste avec persévérance à convaincre les congrégations dont le centre était encore à l'étranger d'établir à Rome leur maison généralice.
(5) Friedrich notamment a voulu y voir la cause essentielle du triomphe de l'ultramontanisme. Même pour l'Allemagne, où cette action semble avoir été plus marquée qu'ailleurs, c'est là une présentation des faits manifestement tendancieuse : bien d'autres éléments ont joué dans le même sens.
(6) *Un nuovo tributo a San Pietro*, dans *Civ. catt.*, sér. VI, vol. X, 1867, p. 641-651.

Je l'ai visité jadis à Rome ; c'est un homme pieux et plein de générosité ; mais il n'avait été distingué jusqu'ici ni par les qualités de l'esprit, ni par ses connaissances. On disait beaucoup de bien de son caractère conciliant et de sa bonne humeur (...). Le nouveau pape sera-t-il à la hauteur des difficultés qui surgissent de toute part ? [1]

Que faut-il en penser, pour autant que l'état actuel de notre documentation nous permet de porter un jugement ? Beaucoup d'appréciations contemporaines sont déformées par la passion, dévotion enthousiaste chez les uns, hostilité farouche chez les autres, et la plupart de celles qui sont publiées proviennent de gens qui n'ont pas vécu dans l'intimité immédiate de Pie IX. Les témoignages les plus sincères, de même que la majeure partie de la correspondance et des papiers personnels du pape sont encore inédits et pratiquement inaccessibles [2]. Certaines données peuvent toutefois dès à présent être dégagées avec des garanties suffisantes d'objectivité.

Par suite de sa santé précaire et des troubles de la période napoléonienne, l'abbé Mastai avait dû se contenter d'études assez sommaires, et il n'eut jamais que des notions superficielles en histoire, en droit canon et même en théologie. Pourtant, sans être à proprement parler un intellectuel, il s'intéressait aux choses de l'esprit : du temps où il s'occupait de l'hospice de *Tata Giovanni*, il s'était inscrit à l'*Académie des Savants*, dont il suivait régulièrement les conférences ; pendant son épiscopat à Imola, il lisait volontiers et prenait plaisir aux discussions d'idées ; devenu pape, il aimait à se renseigner sur les progrès des inventions modernes et suivait de près les fouilles ou les restaurations de monuments, sans avoir du reste sur ce dernier point la compétence de son prédécesseur. Il avait surtout cette finesse italienne qui, même sans beaucoup d'érudition, comprend beaucoup de choses et, si rien dans ses encycliques, qui sont souvent d'une grande banalité de pensée, ne révèle l'ampleur et la fermeté de vue de Léon XIII, on ne peut lui dénier une intelligence vive et pratique qui le mettait à même d'apprécier avec bon sens des situations concrètes, du moins lorsque celles-ci — ce ne fut malheureusement pas toujours le cas — lui étaient présentées avec exactitude.

Non dénué d'intelligence, Pie IX était cependant avant tout un émotif. Tous ceux qui l'ont connu s'accordent sur ce point. C'est un homme, écrivait par exemple le P. Hecker après un an de séjour à Rome, « d'un esprit large, d'un cœur plus large encore, et qui se laisse diriger par ses impulsions plutôt que par son jugement ; mais ses impulsions sont grandes, nobles, universelles » [3]. Elles l'étaient du moins fréquemment, mais plus

---

(1) Meignan à Maret, 1846, dans H. Boissonot, *Le cardinal Meignan*, p. 118-119.
(2) L'ouverture des archives du Vatican serait à ce point de vue d'un grand profit et permettrait une appréciation plus objective de la figure de Pie IX. Plusieurs publications de sources faites ces dernières années ont toutefois déjà permis de redresser des jugements trop sévères, notamment ses notes de voyage au Chili (publiées par le P. Leturia, dans *Xenia Piana Pio XII dicata*, Rome, 1943, p. 367-444) ; sa correspondance avec le cardinal Amat, publiée par G. Maioli, *Pio IX. Da Vescovo a Pontefice*, Modène, 1949 ; sa correspondance avec Victor-Emmanuel, publiée par le P. Pirri, *Pio IX e Vittorio Emanuele II dal loro carteggio privato*, t. I, Rome, 1944 (t. II et III à paraître) ; quelques lettres, adressées principalement à son frère et publiées par A. Monti, *Pio IX nel Risorgimento italiano*, Bari, 1928, p. 227-272. Voir également Mgr Cani, *Procès romain pour la cause de béatification de Pie IX*, Paris, 1910.
(3) Cité dans W. Elliot, *Le P. Hecker*, trad. franç., p. 250.

d'une fois elles trahirent aussi la nervosité excessive et la mobilité d'humeur que le pontife avait conservées de sa maladie d'enfance. Son extrême impressionnabilité explique ces accès de colère aussi violents que soudains, dont ceux qui tentaient de lui résister furent plus d'une fois victimes [1] ; elle explique ces paroles regrettables dont le pape ne mesurait pas toujours au moment même la portée ; elle explique surtout ces variations dans les décisions ou les jugements, d'après le dernier avis entendu, qui donnait du dehors l'impression d'un caractère peu loyal [2] ou irrésolu. Il ne faudrait d'ailleurs pas exagérer ce dernier trait : en matière politique certes, Pie IX fit souvent preuve d'une indécision qui le poussait à se contenter de demi-mesures ; mais il savait être ferme lorsqu'il estimait que son devoir de pasteur était en jeu et rien ne permet de le considérer comme un instable. On relevait au contraire comme une caractéristique de Pie IX « une tendance qui le porte à affronter les situations difficiles quand il croit servir les intérêts de l'Église » [3].

Servir l'Église, telle fut la seule ambition de Pie IX. Elle s'enracinait dans une foi profonde et dans une ardente piété. Ses adversaires ont raillé ce qu'ils nommaient son « mysticisme », sa tendance à tout ramener à l'action de la Providence, l'importance qu'il attachait aux prophéties et autres manifestations du merveilleux [4], et peut-être n'était-il pas exempt, de fait, d'une certaine crédulité. Mais on ne peut qu'admirer sa confiance totale en Dieu, son souci de ne prendre aucune décision importante sans avoir invoqué la Vierge, la large part qu'il accorda toujours à la prière dans sa vie : scrupuleusement fidèle à sa méditation quotidienne, il aimait à réciter son bréviaire à genoux, à prier longuement devant le Saint-Sacrement, à entrer dans les églises au cours de ses promenades. Sans doute est-il bon de nuancer l'affirmation de Manning : « C'est la personne la plus surnaturelle que j'ai jamais vue » [5], par la remarque de Talbot, qui se vantait de le connaître mieux que personne au monde : « Le Saint-Père est un excellent homme, mais ce n'est pas un saint ; il a ses faiblesses [6]. » Il est clair en tout cas que non seulement sa ferveur, qui faisait l'admiration de ceux qui le voyaient officier, répondait à l'état de ses sentiments intimes, mais surtout que son souci dominant fut toujours d'agir avant tout en prêtre, en homme d'Église, responsable devant Dieu de la défense des valeurs chrétiennes menacées par le développement de l'impiété, du laïcisme et du rationalisme. Ce n'est pas par ambition personnelle qu'il encouragea toujours plus nettement les progrès

---

(1) Le clergé d'Imola avait surnommé *Mastaina* ces brusques explosions, qui ne duraient d'ailleurs jamais longtemps et n'avaient guère de conséquence pour l'intéressé (sauf lorsque, comme ce fut parfois le cas pour des prélats âgés, ils en sortaient avec une crise cardiaque).

(2) Forçant un peu la réalité, le docteur Pantaleoni écrivait en 1861 : « Le pape, comme tous les gens faibles, finit (passez-moi le mot un peu trop cru) par mentir, sans le vouloir, je l'admets, mais ce n'est pas moins un mensonge. » (Lettre du 22 janvier 1861 à Cavour dans *La Questione romana...*, t. I, p. 212.)

(3) Dépêche du ministre de Belgique à Rome du 26 janvier 1870, *Arch. Min. A. É. Bruxelles*, Saint-Siège, XIII, 2.

(4) On en trouvera quelques exemples dans les ouvrages, très tendancieux du reste, de J. FRIEDRICH, *Geschichte des vatikanischen Konzils*, t. I, IIIᵉ P., chap. xv et de J.-B. BURY, *History of the Papacy*, p. 51-53.

(5) Manning à lady Herbert, 1ᵉʳ juin 1865, dans E. PURCELL, *Life of Manning*, t. II, p. 230.

(6) Talbot à Manning, 26 décembre 1865, *ibid.*, p. 267.

du courant ultramontain [1], mais parce qu'il y voyait à la fois la condition de la restauration et de l'épanouissement de la vie catholique là où les interventions gouvernementales dans la vie de l'Église menaçaient d'étouffer l'ardeur apostolique, et le meilleur moyen de grouper toutes les forces vives du catholicisme pour réagir contre la vague montante du libéralisme antichrétien.

Un excellent homme, comme disait Talbot, un homme qui, sans être un génie, ne manquait ni de talents ni de caractère, un pasteur d'âmes surtout, zélé et consciencieux, tel nous apparaît Pie IX. Mais ceci suffirait-il à rendre compte de l'influence hors pair dont il jouit dans le monde entier durant tout son pontificat ? Une page d'Émile Ollivier vaut d'être citée à ce propos, qui situe bien le pape avec ses qualités et ses limites :

Sa vie est d'une sainteté édifiante, son âme d'une pureté angélique, mais bien des papes ont été saints et purs. Il est désintéressé, mais il y a longtemps que le népotisme ne s'est pas assis dans la chaire de Pierre. Son esprit est d'une spontanéité charmante, mais ce n'est pas sans exemple et on cite encore au Vatican les bons mots de Benoît XIV. Il saisit rapidement les idées, mais d'autres les ont creusées. Il est fin, mais d'autres ont été profonds. Il voit avec justesse le côté des choses qu'on lui montre, mais d'autres ont deviné celui qu'on leur cachait. Influencé toujours, il n'est jamais dominé, parce que, grâce à la mobilité de ses impressions, il ne subit pas longtemps la même influence, mais d'autres n'ont dû leur indépendance qu'à la fermeté d'une volonté toujours maîtresse d'elle-même. Il est doux dans les grandes choses, mais d'autres n'ont pas été impatients dans les petites. Il corrige par sa bonne grâce et la bonté de son accueil les duretés qu'un entourage passionné arrache parfois à sa distraction, mais d'autres n'ont rien accordé aux passions de leur cour. Il ne connaît pas la haine, mais d'autres ont ignoré même les rancunes passagères [2].

A quoi tient donc cette popularité exceptionnelle que Pie IX sut mettre avec tant d'à-propos au service de la cause ultramontaine ?

UN PAPE POPULAIRE   Tous ceux qui ont approché Pie IX ont été unanimes à confesser le charme fascinateur qui émanait de sa personne et que de nombreux éléments contribuaient à créer : une belle prestance ; des allures d'aristocrate [3] s'alliant à une très grande simplicité — « noble et digne avec désinvolture, désinvolte avec dignité », disait Spada [4] — ; une voix musicale, enchanteresse, qui secondait merveilleusement des dons innés d'orateur, capable de passer sans effort de la familiarité au sublime ; une âme tendre, ouverte aux sensations, vibrant spontanément avec ceux qui l'écoutaient ; un naturel enjoué, un sourire plein de finesse, une douceur, une sérénité, qui faisaient dire à un vieux militaire : « Ça fait du bien rien que de voir cette figure-là.

---

(1) Tout au plus peut-on dire que dans les dernières années de sa vie s'affirma davantage une tendance à se laisser griser par les flatteries et à ne plus supporter aucune contradiction ; mais il serait peu judicieux de prétendre apprécier l'action de Pie IX en ne tenant compte que des faiblesses d'un vieillard de 80 ans.

(2) Le portrait de Pie IX par É. OLLIVIER (*L'Église et l'État au concile du Vatican*, t. I, p. 307-311), encore que tendancieux, présente un intérêt particulier, car il a été composé à l'aide de notes fournies par la princesse Sayn de Wittgenstein, qui vécut pendant des années à Rome en contact étroit avec la cour pontificale (je dois à l'amabilité de Mme Troisier-Ollivier d'avoir pu prendre connaissance des lettres de la princesse).

(3) On notera à ce propos qu'avec un train de vie très simple, Pie IX était très soigneux de sa personne et de ses vêtements, prenait grand souci de ses mains et avait même un faible pour l'eau de Cologne. (F. HAYWARD, *Pie IX et son temps*, p. 169.)

(4) Cité dans G. MAIOLI, *Pie IX. Da Vescovo a Pontefice*, p. 56.

Quand on sort de voir le pape, on emporte du calme et de la joie pour le reste de la journée » [1] ; une bonne grâce accueillante, une affabilité qui avait des trouvailles exquises et sur laquelle on ne tarissait pas. En lui, selon le mot de Mgr Gerbet, Dieu avait couronné de la tiare le génie de la bonté, et même ceux qui étaient en désaccord d'idées avec lui ne pouvaient s'empêcher de le reconnaître : « Autant de cerveau qu'une sauterelle mais un cœur d'ange », disait déjà du cardinal Mastai l'avocat libéral Mancurti [2], et l'évêque de Nîmes, parti pour Rome en 1858 plein de préventions gallicanes, notait dans son journal de voyage :

> Sa tête, régulière dans les traits, noble dans son ensemble, se distingue surtout par un visage plein de douceur et respirant la plus suave bonté. Rien n'est limpide et bienveillant comme son regard, rien n'est paternel et gracieux comme son sourire. Dans sa voix qui ne manque pas de puissance, il entre un accent de tendresse admirable ; on sent à sa parole, comme à l'aspect de sa figure, qu'il porte dans le cœur un immense trésor d'amour [3].

On comprend qu'un flamand réaliste comme le cardinal Sterckx ait pu, au sortir d'une audience, écrire à ses vicaires généraux : « Cette audience a été vraiment *délicieuse*. Je commencerais à douter si le pape n'est pas un ange du Ciel [4] ! »

Or, ce pape dont émanait une si étonnante séduction a multiplié, d'une manière inouïe jusqu'alors, les contacts humains au cours desquels ce charme pouvait agir. Émile Ollivier a fort bien compris la portée du phénomène :

> Une bonne partie du long pontificat de Pie IX aura été employée à haranguer et à causer. La plupart des papes dans leurs rapports privés se sont montrés bons et simples ; mais ils n'étaient pas aisément accessibles. La foule ne les voyait guère que claquemurés dans une étiquette froide et compassée, au milieu d'un entourage solennel, et ne recueillait de leur bouche que des paroles brèves à la façon des oracles. Pie IX a brisé cette clôture, secoué cette contrainte et changé cette rhétorique. Se sentant plein de vie, toujours prêt à la riposte, à l'exhortation persuasive et aux mouvements qui emportent, il a aimé, sans un trop grand souci de l'étiquette, à montrer la papauté aussi aimable sur la place publique qu'elle l'avait été jusqu'alors dans l'intimité du Vatican [5].

Pie IX, à l'inverse de Grégoire XVI, aimait la vie sociale. Il recevait volontiers à sa table, contrairement à l'usage, l'un ou l'autre invité ; il organisa même, à l'occasion, de grands banquets, pour les élèves des séminaires étrangers, pour les évêques lors de la réunion de 1862, pour les artistes et archéologues l'année suivante. Surtout, il accordait sans compter des audiences : audiences collectives, où il laissait déborder son cœur en d'émouvantes improvisations ; audiences privées, dont bénéficiaient tour à tour des prêtres, des séminaristes, des journalistes, des hommes d'œuvres — il finit par connaître personnellement tous les dirigeants du catholicisme français — des pèlerins et leur famille, et jusqu'à de vieilles anglaises protestantes qui n'étaient pas les moins ravies. Ces audiences commençaient à 10 heures ; interrompues à 2 heures

(1) Rapporté par E. LAFOND, *Rome. Lettres d'un pèlerin*, t. I, p. 490.
(2) Cité par G. MAIOLI, *op. cit.*, p. 22.
(3) Cité dans J. CLASTRON, *Vie de Mgr Plantier*, t. I, p. 343.
(4) Lettre du 20 novembre 1854 aux *Arch. Mal.*, Fonds Sterckx, Voyages à Rome.
(5) É. OLLIVIER, *L'Église et l'État au concile du Vatican*, t. I, p. 310.

pour le déjeuner, elles reprenaient vers 4 ou 5 heures après la promenade
pour se prolonger souvent jusque vers 9 ou 10 heures du soir. Les visiteurs étaient tout de suite à l'aise grâce à la bonhomie du pape et à sa
conversation pleine d'abandon, émaillée de mots d'esprit dont il raffolait.
Il n'attendait d'ailleurs pas que les visiteurs vinssent à lui. Il aimait à
se promener familièrement dans les rues de Rome et à se laisser aborder
par les quémandeurs, tout heureux de faire plaisir, au point qu'on a
pu dire qu'il était « le souverain le plus occupé du monde, mais aussi le
plus accessible à ses sujets »[1]. Il avait des gestes charmants, entrant
dans une boutique pour acheter un flacon de vin et le tendre à un gosse
en larmes qui venait de laisser choir celui qu'il portait à sa mère ; descendant de voiture pour suivre jusqu'au cimetière le corbillard d'un pauvre
hère qu'aucune famille n'accompagnait ; acceptant de dire la messe
pour un soldat français qui l'avait accosté, et de la remettre d'un jour
parce que le brave garçon avait une petite fête avec des amis.

Tous ces traits étaient immédiatement colportés par la ville et les
pèlerins, plus nombreux chaque année grâce au développement des
chemins de fer, en multipliaient la répercussion autour d'eux une fois
revenus dans leur patrie. Il faut ajouter, pour comprendre le rayonnement
de Pie IX sur ceux qui ne le connurent jamais que de loin, l'universelle
sympathie que lui valurent dans le monde catholique ses malheurs,
depuis l'exil à Gaëte jusqu'à l'agression brutale et toujours à nouveau
menaçante du gouvernement piémontais, ainsi que « sa noble attitude,
patiente et ferme, de souverain désarmé devant l'insolente Révolution »[2].
Aux yeux de nombreux fidèles, le pieux pontife, dont les vertus étaient
encore idéalisées par la distance, faisait littéralement figure de martyr
et de saint.

Bref, si diverses causes rendaient inévitable le triomphe de l'ultramontanisme et si plusieurs circonstances l'ont grandement facilité, il n'en
reste pas moins que ce triomphe « fut autant le triomphe d'un homme
que celui d'une doctrine ; ç'a été l'exaltation du pape en même temps
que celle de la papauté »[3] et certains ultramontains en avaient bien
conscience :

Voir Jésus dans son vicaire, observait l'un d'eux, c'est un acte de foi méritoire, mais qui a moins de mérite quand on a vu le pape dans Pie IX. Le Christ
est dans chacun de ses vicaires en plénitude de doctrine et d'autorité, mais la
personne de Pie IX prête en quelque sorte trop de force à la cause de Rome[4].

Cette « dévotion au pape », qui se développa d'une manière particulièrement spectaculaire en France, est un phénomène nouveau dont
l'importance ne peut être sous-estimée. Elle explique la facilité avec
laquelle l'ensemble du clergé et des fidèles se rallia avec enthousiasme
à la doctrine, obscurcie depuis des siècles, de l'infaillibilité personnelle
du souverain pontife. Elle explique aussi le préjugé favorable dont, sauf

(1) G.-F. MAGUIRE, Roma, il suo Sovrano e le sue istituzioni, Florence, 1858, p. 157.
(2) F. MOURRET, Le concile du Vatican, p. 183.
(3) P. DE QUIRIELLE, dans An. E. Sc. P., t. V, 1890, p. 502.
(4) E. LAFOND, Rome. Lettres d'un pèlerin, t. I, p. 496.

en Orient, où les tendances nouvelles heurtaient trop de vénérables traditions pour ne pas rencontrer les plus vives résistances [1], devait nécessairement bénéficier, du moment qu'il était inspiré et souhaité par le Saint-Père lui-même, le mouvement de centralisation et d'uniformisation de l'Église, qui allait donner sa physionomie caractéristique au catholicisme contemporain.

§ 3. — Néo-ultramontanisme et résistances gallicanes à la veille du concile du Vatican [2].

PROGRÈS CROISSANTS DU COURANT ULTRAMONTAIN

L'enthousiasme des masses catholiques pour Pie IX facilita singulièrement l'élimination des derniers centres de résistance à l'ultramontanisme. En divers pays du reste, ceux-ci étaient déjà pratiquement inexistants. Ainsi aux Pays-Bas, où, contrairement à ce qui s'était passé en Angleterre, et sans doute parce que l'Église avait dû s'y affirmer en face du schisme janséniste, évêques et fidèles avaient toujours fait preuve d'un sentiment papal très prononcé. De même en Belgique où, en dépit de van Espen, les thèses gallicanes n'avaient jamais ébranlé sérieusement le dévouement envers « le pape de Rome » ; tout au plus les évêques, très conscients d'être les « Ordinaires du lieu », joignaient-ils à une déférence filiale à l'égard du Saint-Siège le souci de « passer les directives romaines par leur jugement et leur interprétation avant de les appliquer aux fidèles » [3]. En Espagne aussi, malgré des sympathies épiscopaliennes qui s'étaient fait jour à la fin du XVIIIe siècle dans une petite partie du haut clergé et chez quelques théologiens, les sentiments romains, restés vivaces dans l'ensemble de l'Église, avaient vite repris le dessus et, dès les années trente, la persécution libérale avait achevé de rapprocher les évêques du Saint-Siège et celui-ci pouvait en outre compter sur l'appui de la reine Isabelle. Mais la situation ne se présentait pas partout de manière aussi favorable.

PROGRÈS DU MOUVEMENT EN AUTRICHE

Les préjugés antiromains s'étaient maintenus beaucoup plus longtemps en Autriche qu'en Allemagne, même dans le clergé, en dépit de l'influence exercée par les disciples de Clément-Marie Hofbauer. Les choses ne commencèrent vraiment à changer qu'après les événements de 1848 qui favorisèrent le parti réformateur. Celui-ci remporta bientôt une grande victoire avec le concordat de 1855, qui reconnaissait explicitement comme de droit divin le primat de juridiction du souverain

---

(1) Cf. chap. XIII, art. 3.

(2) BIBLIOGRAPHIE. — Aux sources et travaux indiqués en tête du chapitre, ajouter J.-F. VON SCHULTE, *Erinnerungen*, Giessen, 1908-1909, 3 vol., et A. HOUTIN, *Le P. Hyacinthe dans l'Église romaine*, Paris, 1920 (à utiliser tous deux avec prudence).

En ce qui concerne spécialement la France : *Mgr Darboy et le Saint-Siège*, dans *Revue d'histoire et de littérature religieuse*, t. XII, 1907, p. 240-281 ; R. DURAND, *Mgr Darboy et le Saint-Siège*, dans *Bulletin de la Société d'histoire moderne*, t. VII, 1907-1908, p. 6-10 ; P. MABILE, *Vie de Mgr Mabile*, Paris, 1926.

(3) A. SIMON, *L'Église catholique et les débuts de la Belgique indépendante*, p. 58.

pontife (art. 2) et dont plusieurs articles allaient permettre d'introduire les stipulations du droit canon dans la législation civile. Le concordat permit surtout de réorganiser l'enseignement ecclésiastique conformément aux désirs du Saint-Siège : envoi plus fréquent d'étudiants ecclésiastiques à Rome, orientation ultramontaine des études dans les séminaires diocésains ; à l'Université de Vienne, Phillips secondé par Fessler put former une nouvelle génération de canonistes, tandis qu'à l'Université d'Innsbruck, où la Faculté de théologie fut confiée aux jésuites en 1857, l'*Archiv für katholischen Kirchenrecht* du canoniste Moy allait devenir, surtout lorsque Vering s'adjoignit à la rédaction en 1861, un organe influent de l'ultramontanisme. Dans l'ancienne citadelle du joséphisme et du fébronianisme, un ancien professeur au Collège romain, le jésuite Schrader, réussissait, au cours de son professorat à Vienne de 1857 à 1863, à faire partager aux milieux influents de l'aristocratie son ardeur pour les thèses ultramontaines les plus radicales, et quelques années plus tard, en 1867, Mgr Roskovany allait inaugurer son monumental recueil des actes du Saint-Siège, *Romanus Pontifex tamquam Primas Ecclesiae et princeps civilis e monumentis omnium saeculorum demonstratus*, qui devait rencontrer un vif succès dans le jeune clergé.

Les évêques à leur tour étaient gagnés par le courant, certains même avec intransigeance, comme Rudigier, Gasser et Fessler, qui avaient mûri leur zèle ultramontain du temps où ils étaient collègues au séminaire de Brixen [1]. Mais celui qui contribua le plus efficacement au triomphe des idées romaines dans la monarchie des Habsbourg fut un homme qui n'admettait cependant pas les positions théologiques ultramontaines dans leur totalité, le cardinal Rauscher, archevêque de Vienne depuis 1853. C'est lui qui, avec le comte Thun, ministre des Cultes, accusa, dans l'application, le caractère « clérical » du concordat ; c'est lui qui orienta dans un sens ultramontain les associations catholiques et la presse religieuse ; c'est avec son appui que se multiplièrent les missions paroissiales prêchées par les jésuites ou les rédemptoristes, si importantes pour la diffusion du nouveau type de piété ; et c'est à la suite des visites de monastères dont il prit l'initiative que les vieux ordres religieux, jusque-là très indépendants, commencèrent à se laisser pénétrer d'esprit romain.

ET EN GRANDE-BRETAGNE   En Angleterre, Manning fut, sans conteste, le chef du mouvement ultramontain. Wiseman avait certes beaucoup fait pour dissiper les préventions contre Rome, mais il conserva toujours de ses contacts avec le groupe de Goerres une certaine modération en ce domaine. Manning, au contraire, s'y appliqua avec un zèle de néophyte, qui n'était pas exempt d'une certaine outrance. Ami personnel de Pie IX, converti au catholicisme par le désir de trouver une règle infaillible de foi, convaincu de la

---

(1) On peut suivre leur action dans les biographies, malheureusement trop peu nuancées, de K. MEINGL, *Bischof Rudigier*, Linz, 1891-1892, 2 vol. ; J. ZOBL, *Vinzenz Gasser, Fürstbischof von Brixen*, Brixen, 1883 ; A. ERDINGER, *Dr. J. Fessler*, Brixen, 1874.

nécessité de renforcer le principe d'autorité en face d'une société
ébranlée par l'esprit révolutionnaire, il ne pouvait comprendre l'attitude des évêques qui lui déclaraient comme Mgr Clifford : « Rome
tend de plus en plus à réduire les anciens privilèges des évêques et nous
sommes soucieux de n'être limités que le moins possible [1]. » Aidé par
quelques collaborateurs dont l'intransigeance égalait malheureusement
le zèle, Faber, Coffin, Vaughan, Ward surtout, sans parler de Mgr Talbot,
son confident à Rome, il s'appliqua avec une énergie indomptable à
extirper ce qu'il nommait « l'esprit antiromain et antipapal du catholicisme anglais » [2]. Son acharnement dans l'affaire Errington provenait
en partie de là : il était convaincu que le coadjuteur et la fraction de
l'ancien clergé qui le soutenait étaient « teintés de gallicanisme » — ce
qui était exact — et « déloyaux envers le Saint-Siège » [3] — ce qui était
profondément injuste.

En dépit de l'opposition du groupe libéral d'Acton et du *Rambler*,
malgré l'hésitation de Newman, toujours inquiet de voir présenter la
vérité sans nuances, l'action persévérante de l'archevêque de Westminster
et de son groupe réussit en quelques années à gagner à leurs vues la
majorité des fidèles et du clergé d'Angleterre. Comme sur le continent,
un rôle non négligeable revint dans ce succès à la presse religieuse :
Manning disposait en effet depuis 1862 de la *Dublin Review*, et il pouvait
se reposer en toute confiance sur son rédacteur en chef, Ward, qui,
joignant à l'intransigeance du converti la terrible logique du mathématicien, excellait à pousser à leurs conséquences extrêmes les principes
ultramontains.

Un mouvement parallèle se produisait d'ailleurs au même moment
en Irlande sous la direction du cardinal Cullen, totalement acquis aux
conceptions centralisatrices du Saint-Siège et qui avait adopté pour
son diocèse le catéchisme de Mayence. Le clergé irlandais fit d'autant
moins de difficulté pour se rallier que, même avant la réouverture du
collège irlandais de Rome, l'esprit du séminaire central de Maynooth
avait déjà achevé de se modifier : en 1853, les ouvrages de Perrone et
de saint Alphonse de Liguori y avaient remplacé le manuel condamné
de Bailly, et l'influence du professeur O'Hanlon, partisan de longue
date de l'infaillibilité pontificale, l'avait dès lors emporté [4].

NOUVEAUX SUCCÈS
EN ALLEMAGNE

En Allemagne où, dès 1848, l'ultramontanisme
occupait déjà d'importantes positions, les progrès
furent d'autant plus faciles. A côté de Cologne,
où Geissel est à la pointe du combat jusqu'à sa mort en 1864 et réussit à
faire du concile provincial de 1860 une synthèse des idées ultramontaines, Mayence demeure toujours l'un des centres du mouvement et

---

(1) Cité dans C. BUTLER, *The Life and Times of B. Ullathorne*, t. I, p. 237. Mgr Clifford devait
rester jusqu'au concile du Vatican l'un des adversaires les plus décidés des tendances représentées
par Manning.
(2) E. PURCELL, *Life of Manning*, t. II, p. 81.
(3) *Ibid.*, p. 87.
(4) Voir HEALY, *Maynooth College. Its Centenary History*, Dublin, 1895.

certains considèrent même Ketteler comme le véritable chef de file du « parti » pendant cette période. A la vérité cependant, les vrais animateurs du mouvement sont plutôt ses collaborateurs, Moufang et Heinrich, dont l'action prolonge celle de Lennig, car si Ketteler est très dévoué au pape et approuve les catholiques qui vont chercher à Rome les mots d'ordre pour la renaissance catholique, s'il favorise les activités de la Compagnie de Jésus et s'il appuie de toute son autorité la réaction contre l'indépendance d'esprit des *Deutsche Theologen*, il commence par contre à s'inquiéter de ce qu'il considère comme des empiétements de la Curie sur les attributions épiscopales.

D'année en année les hommes de Mayence voient se multiplier ceux qui travaillent dans le même sens qu'eux : les jésuites, qui augmentent en nombre et en influence et réussissent à faire passer peu à peu dans l'enseignement catéchistique les thèses romaines sur les prérogatives pontificales ; les *Germaniker*, anciens élèves du Collège germanique de Rome [1], dont trois représentants pénètrent même, en 1854, à la Faculté de théologie de Wurzbourg ; les prêtres de plus en plus nombreux qui vont faire leurs études à Innsbruck, sous la direction des jésuites.

Tout ce monde se retrouve régulièrement à l'occasion des congrès catholiques annuels, qui deviennent de vraies manifestations ultramontaines, et leurs efforts sont efficacement appuyés par les deux hommes qui assurent pendant cette période la liaison directe entre Rome et l'Allemagne : le nonce à Munich et Reisach, qui, cardinal de Curie depuis 1856, rallie tous ceux, évêques, hommes d'œuvres ou intellectuels, qui désirent voir l'Allemagne catholique sortir de l'isolement où elle s'était tenue si longtemps à l'égard du centre de la Chrétienté.

LA VICTOIRE DÉFINITIVE DE L'ULTRAMONTANISME EN FRANCE

En France, l'encyclique *Inter multiplices*, de 1853, avait marqué un tournant décisif. Quelques années suffirent pour que capitulent les derniers centres de résistance à l'unification liturgique : en 1863, onze diocèses seulement conservaient encore leur ancienne liturgie et neuf d'entre eux avaient déjà promis en principe de l'abandonner prochainement ; l'année suivante, le cardinal de Bonald, de Lyon, qui était pourtant soutenu par son clergé unanime, contrairement à ce qui se passait ailleurs, dut s'incliner à son tour à la suite d'un blâme où Pie IX lui signifiait qu'en matière liturgique le Saint-Siège ne se borne pas à conseiller les évêques, mais qu'il peut décider souverainement.

D'autre part, en semblant donner raison à Veuillot contre l'épiscopat, le pape avait en fait encouragé le journaliste à faire de *L'Univers* un organe de combat qui allait jouer, sous l'Empire, en faveur de l'ultramontanisme, le même rôle que *L'Avenir* au début de la Monarchie de Juillet. Dévoué corps et âme à l'Église et à son chef, mais justifiant souvent le reproche que lui faisaient ses adversaires de poser au « pape laïc des Gaules »,

(1) Cf. A. STEINHUBER, *Geschichte des Collegium Germanicum-Hungaricum in Rom*, Fribourg-en-Br., 1895, t. II, chap. VI (Das Collegium Germanicum im XIX. Jh.)

Veuillot, encouragé par ses amis, par dom Guéranger en particulier, allait pendant vingt ans clouer au pilori tous ceux qui regrettaient l'uniformisation radicale de la liturgie ou l'évolution de la piété moderne, et vociférer, à la grande joie du bas clergé, contre « l'arbitraire » des évêques, dont il oubliait aisément qu'ils ont été établis par le Christ comme pasteurs légitimes de leurs diocèses. La justice oblige du reste à reconnaître que cette œuvre négative, dont on voit mieux aujourd'hui ce qu'elle avait de discutable, s'accompagnait d'une œuvre positive : conquis personnellement par « les parfums de Rome », touché par les épreuves de Pie IX, enthousiasmé par l'intrépidité avec laquelle celui-ci se faisait le défenseur des principes, Veuillot sut communiquer son culte de la papauté aux Français et, selon le mot très juste de Mgr Amann, « si en France on aime le pape plus encore qu'on ne le vénère, c'est un résultat précieux de la grande action de *L'Univers* »[1].

On exagérerait difficilement l'influence de Veuillot, qui n'eut son équivalent en aucun autre pays. Mais on ne peut négliger les autres éléments dont l'action s'ajouta à la sienne en faveur de la cause ultramontaine. En premier lieu, la neutralité bienveillante du gouvernement au début du Second Empire, qui permit à des prélats comme Mgr Gousset ou Mgr Parisis d'exercer une influence sensible sur les nominations épiscopales.

Ensuite, l'appui de plus en plus décidé de la part des autorités romaines à une campagne dont les buts étaient louables, mais dont les moyens ne l'étaient pas toujours. Jusque vers 1860, si les surenchères ultramontaines étaient appuyées par le groupe de prêtres français qui s'agitaient à Rome autour de Mgr de Ségur, puis de Mgr Chaillot[2], par contre Antonelli et Pie IX lui-même, préoccupés de ne pas accroître les dissensions intérieures en France, avaient à plusieurs reprises conseillé la modération, tout en faisant observer ce qu'il y avait de paradoxal à ce que ce soit « le Saint-Siège lui-même qui doive contenir un mouvement entièrement dirigé en sa faveur »[3]. Mais l'attitude plus que réservée que prirent dans la question romaine la plupart des gallicans, généralement très gouvernementaux, causa à Rome une profonde indignation, qui s'accrut encore lorsque, un peu plus tard, on s'imagina que le gouvernement impérial préparait en sous-main un schisme. L'insistance d'un certain nombre d'évêques pour maintenir à l'égard de la Curie un peu de leur ancienne autonomie devait nécessairement apparaître dans cette perspective comme extrêmement redoutable et, dès ce moment, la faveur de Pie IX et de son entourage fut définivement acquise sans la moindre réserve à leurs adversaires. On accueillit désormais les yeux fermés les rapports unilatéraux de certains milieux ultramontains et du nonce Chigi sur les membres du clergé français accusés de tiédeur à l'égard du courant romain. L'un des principaux informateurs du nonce était le supérieur des Pères du Saint-Esprit,

(1) Art. *Veuillot*, dans *D. T. C.*, t. XV, col. 2833.

(2) Mgr Chaillot, qui devait reprendre en 1860 la publication de la *Correspondance de Rome*, publiait déjà depuis 1855 un organe analogue, les *Analecta juris pontificii*.

(3) Antonelli à l'ambassadeur de France (copie d'une dépêche du 27 juillet 1857 aux *Arch. Nat.*, f[19], 1947).

dont la congrégation, avec l'aide des anciens élèves du Séminaire français de Rome dont elle avait la direction, jouait à ce sujet un rôle « discret mais important »[1] qui lui valut d'être dénoncée comme « le bureau d'information ou l'inquisition du parti en France »[2].

Le triomphe ultramontain fut encore favorisé du fait que, en l'absence de Facultés canoniques de théologie en France, Rome était le seul centre possible de hautes études ecclésiastiques. Un nombre croissant de jeunes prêtres revenaient du Séminaire français, définitivement organisé en 1853, « avec les idées, les coutumes et même l'habillement des prêtres italiens »[3] et les séminaires diocésains de France suivirent rapidement. Tandis qu'en certains endroits, comme à Reims, on adoptait franchement les *Praelectiones* de Perrone, ailleurs on corrigeait, « conformément aux observations proposées par quelques théologiens romains », les anciens manuels épargnés par les condamnations de 1852, la *Théologie de Toulouse* et celle de Mgr Bouvier, ou les leçons de droit canon de M. Icard. Bien plus, la Compagnie de Saint-Sulpice, qui, très attachée aux traditions de l'ancien clergé de France, avait longtemps boudé le courant nouveau[4], s'y ralliait à son tour. En mars 1860, son supérieur général, le vieux M. Carrière, s'était décidé à faire le voyage de Rome, mandé par Pie IX à qui on avait présenté la Société comme « le boulevard du gallicanisme »[5], et dès son retour il renforça les instructions déjà données en 1856 afin que l'enseignement de la Compagnie fût strictement conforme aux doctrines romaines[6] ; à sa mort, en 1864, le pape obtint en outre que son successeur fût un ultramontain décidé[7].

Parallèlement se poursuivait la substitution de toute une littérature ecclésiastique d'assez faible valeur scientifique, mais franchement ultramontaine, aux ouvrages de Bossuet, des Mauristes et des Messieurs de Port-Royal dont s'était nourri le clergé des générations antérieures et que *L'Univers* s'était chargé de discréditer avec plus d'éloquence que de compétence. A l'exception de ceux, très rares, qui se tenaient au courant de la production allemande, les prêtres français n'eurent bientôt plus à leur disposition, pour s'initier à la connaissance du passé de l'Église, que les diatribes de l'abbé Davin contre Bossuet, le médiocre plagiat de l'abbé Jager intitulé *Histoire de l'Église catholique en France*, ou la tendancieuse *Histoire universelle de l'Église* de Rohrbacher, dont Veuillot disait qu'aucun ouvrage n'avait autant fait pour la cause de l'infaillibilité[8]. Seul dom Guéranger a quelque valeur et encore son érudition est-elle loin de valoir celle de Mabillon ; il l'utilise d'ailleurs trop souvent pour la propagande et le combat au détriment de l'objectivité.

(1) J. MAURAIN, *La politique ecclésiastique du Second Empire*, p. 283 ; voir aussi p. 481 et 723.
(2) Dans le Mémoire anonyme *La crise de l'Église* rédigé dans l'entourage de Dupanloup.
(3) A. DANSETTE, *Histoire religieuse de la France contemporaine*, t. I, p. 411.
(4) Il ne faudrait d'ailleurs pas exagérer le gallicanisme de Saint-Sulpice, surtout au xixe siècle ; voir le plaidoyer de M. ICARD, *Observations sur quelques pages de la continuation de l'Histoire de l'Église de M. l'abbé Davin*, Paris, 1886.
(5) Lettre du 27 novembre 1859 à M. Boiteux, aux *Arch. Nat.*, AB XIX, n° 517.
(6) Circulaire du 17 juin 1860 aux *Arch. S. Sulp.*, fonds Carrière.
(7) Voir J. MAURAIN, *La politique ecclésiastique du Second Empire*, p. 438, n. 2 et p. 688-689.
(8) L. VEUILLOT, *Rome pendant le concile*, t. II, p. 443.

EXAGÉRATIONS ET DÉVIATIONS   Les ultramontains avaient le sens et la sensibilité catholiques. Ils percevaient plus clairement que leurs adversaires ce qu'étaient les prérogatives conférées par le Christ à son Vicaire sur terre ; ils voyaient le bien-fondé des interventions romaines dans la vie des Églises particulières qui, laissées à elles-mêmes, auraient pu difficilement se dégager des emprises gouvernementales et de la tentation d'une vie séculière ; ils désiraient une centralisation accentuée, en pressentant qu'elle était inévitable pour résoudre les problèmes religieux sur leur vrai plan : le plan supra-national. Leur tort fut de présenter des idées justes avec des arguments souvent trop peu étayés scientifiquement et de provoquer ainsi l'opposition de milieux qui, pour n'être pas systématiquement ultramontains, ne manquaient pourtant pas de sens catholique.

Parfois, aussi, ils avaient l'art de rapetisser les questions : le rabat était-il un péché gallican ? Les souliers à boucles et les culottes courtes du clergé romain étaient-ils, comme le laissaient entendre des prêtres anglais[1], un *test* d'esprit catholique ? Cela n'était qu'enfantin. Il y avait l'odieux. Les diatribes de Veuillot, quelque injustes qu'elles aient été parfois, avaient au moins le mérite de la franchise. Mais que penser des dénonciations en sous-main, si fréquentes au cours de cette période, dont certains ultramontains, en Allemagne, en Angleterre ou en France, se faisaient une vraie spécialité ? Ces hommes n'hésitaient pas à jeter le soupçon, avec une inconcevable légèreté, non seulement sur l'orthodoxie, mais même sur l'honneur sacerdotal de ceux dont ils croyaient utile de briser la carrière. On comprend la méfiance que devaient soulever les idées propagées parfois par de tels défenseurs.

LE NÉO-ULTRAMONTANISME   Les idées elles-mêmes n'étaient pas toujours au point et l'on voyait se préciser un ensemble de tendances que Wilfrid Ward, et dom Butler à sa suite, ont proposé de désigner sous le nom de « néo-ultramontanisme » pour les distinguer du sain ultramontanisme qui devait finalement s'imposer à la suite du concile du Vatican ; ces tendances allaient effectivement beaucoup plus loin que les thèses de Bellarmin reprises par les ultramontains de la première moitié du siècle.

C'est ainsi qu'un journal religieux de Nîmes, évoquant la définition de l'Immaculée-Conception, écrivait en 1865 :

Louis XIV avait prononcé ce mot célèbre : *L'État, c'est moi !* Pie IX a fait plus : il a dit en action, avec plus de raison que lui : *L'Église, c'est moi !* [2]

Certains théologiens faisaient écho à ces exagérations et minimisaient le rôle des évêques dans l'Église au point de prétendre définir celle-ci comme « la société des fidèles gouvernés par le pape »[3]. Un modéré comme M. Icard le relevait avec inquiétude :

(1) W. WARD, *Le cardinal Wiseman*, trad. franç., t. II, p. 462.
(2) Cité dans G. BAZIN, *Vie de Mgr Maret*, t. II, p. 356.
(3) Cf. F. MOURRET, *Le concile du Vatican*, p. 217-218.

On entend dire à des hommes d'une certaine école que le pape est tout à la fois la source de toute juridiction et de toute infaillibilité doctrinale. Lui seul enseigne l'Église universelle et les évêques n'ont d'autre mission que de transmettre aux fidèles la doctrine qu'ils reçoivent du Saint-Siège. La majorité des évêques n'a pas d'autre infaillibilité que celle que Dieu a donnée à la majorité des fidèles, l'infaillibilité passive, qui consiste à se laisser enseigner et à adhérer à la doctrine du Souverain Pasteur... Ils croient glorifier et exalter le chef de l'Église en le séparant ainsi du reste des pasteurs, à peu près, disais-je un jour à un nonce, Mgr Sacconi, comme si on croyait honorer la tête d'un grand personnage en la détachant de son cou pour l'élever bien haut, tandis que son corps serait couché sur le sol [1].

D'autres théologiens ultramontains présentaient comme de droit divin les thèses les plus extrêmes de la théocratie médiévale telles qu'elles avaient été formulées par Grégoire VII ou Boniface VIII, affirmant par exemple que le Christ devait nécessairement rendre saint Pierre et ses successeurs participants de ses droits sur le temporel des rois et même sur les biens des particuliers. D'autres encore, plus préoccupés de dilater l'infaillibilité que d'en préciser soigneusement les conditions et les limites, voulaient étendre ce privilège bien au delà des seules définitions *ex cathedra*, à toute encyclique, à toute affirmation du pape en matière religieuse, voire même en matière politico-religieuse.

Ces conceptions extrémistes étaient vulgarisées à grand bruit par la presse ultramontaine. Ward par exemple, dont l'idéal eût été, suivant sa propre formule, de « toucher chaque matin à son déjeuner, avec le numéro du *Times*, une nouvelle bulle pontificale », s'en faisait le défenseur dans la *Dublin Review*, refusant aux théologiens le droit de discuter ou d'interpréter le sens et la portée des actes pontificaux, revendiquant pour certains d'entre eux un degré d'autorité supérieur à celui qu'à Rome même les théologiens leur attribuaient, et dénonçant comme un *minimiser* quiconque ne pensait pas comme lui. *L'Univers*, évidemment, renchérissait et parlait de gallicanisme « dès là qu'on ne considérait pas comme faisant loi les désirs, même secrets, du pape et les intentions de ceux qui l'approchaient de plus ou moins près » [2]. Et que dire de ceux qui confondaient l'infaillibilité avec l'inspiration, voire avec l'impeccabilité !

La dévotion au pape, elle aussi, prenait parfois des formes bien discutables que l'archevêque de Reims dénonçait comme « une idolâtrie de la papauté » [3]. Les uns, afin de mieux confesser que, pour eux, le pape était le vicaire de Dieu sur la terre — le « vice-Dieu de l'humanité » disait l'un d'eux [4] — croyaient bon de lui appliquer des hymnes qui, dans le bréviaire, s'adressent à Dieu lui-même :

> *Rerum*, Pius, *tenax vigor,*
> *Immotus in te permanens*
> *Da verba vitae quae regant*
> *Agnos, oves et saeculum.*

ou encore :

---

(1) Notes préparatoires au concile du Vatican (hiver 1868-1869), *Arch. St-Sulp.*, fonds Icard.
(2) É. AMANN, art. *Veuillot*, dans *D. T. C.*, t. XV, col. 2807.
(3) Mgr Landriot à Montalembert, dans E. LECANUET, *Montalembert*, t. III, p. 468, en note.
(4) E. LAFOND, *Rome. Lettres d'un pèlerin*, t. I, Paris, 1864, p. 487.

A Pie IX, Pontife-Roi :

Pater Pauperum
Dator munerum
Lumen cordium
Emitte coelitus
Lucis tuae radium.

D'autres le saluaient des titres attribués au Christ par la Sainte Écriture : « *Pontifex sanctus, innocens, impollutus, segregatus a peccatoribus et excelsior coelis factus.* » Or, ces inconvenances n'étaient pas seulement boutades d'irresponsables. Un article de *La Civiltà cattolica*, la revue des jésuites romains, expliquait que « lorsque le pape médite, c'est Dieu qui pense en lui »; l'un des évêques ultramontains les plus en vue de France, Mgr Bertaud, de Tulle, présentait le pape comme « le Verbe incarné qui se continue »; et l'évêque de Genève, Mgr Mermillod, n'hésitait pas à prêcher sur « les trois incarnations du Fils de Dieu » : dans le sein d'une vierge, dans l'eucharistie et dans le vieillard du Vatican.

Toutes ces extravagances et ces adulations, contre lesquelles Pie IX ne réagissait pas suffisamment, se trouvaient reproduites en bonne place dans *L'Univers*, à la grande indignation de ceux qui ne savaient pas reconnaître sous ces formules malheureuses l'expression maladroite de la foi populaire. Elles contribuaient à entretenir, spécialement autour de Munich et de Paris, quelques centres de résistance qui n'étaient plus « que des notes discordantes isolées dans un concert d'acclamations », mais dont l'hostilité s'alimentait aux rancœurs qu'avaient suscitées contre les autorités romaines leur attitude peu compréhensive à l'égard de la science ecclésiastique allemande et leur intransigeance à l'égard de ceux qui cherchaient à réconcilier l'Église avec la société moderne.

LA RÉSISTANCE DES UNIVERSITÉS ALLEMANDES L'attitude boudeuse des Facultés de théologie d'Allemagne était presque inévitable. Leurs perspectives ecclésiologiques s'inspiraient de celles de Moehler qui, pour orthodoxes qu'elles fussent, étaient assez différentes de celles de l'école de Bellarmin, centrée sur l'organisation externe et l'unité juridique de l'Église. Ensuite, la supériorité réelle — qu'ils s'exagéraient encore avec un naïf orgueil — de la formation historique des théologiens allemands les rendait très sensibles aux objections tirées de l'antiquité chrétienne contre l'infaillibilité du pape : cas du pape Libère, question d'Honorius, etc., ainsi qu'au caractère précaire de beaucoup d'arguments invoqués par les promoteurs de l'ultramontanisme, dont la science n'était pas toujours à la hauteur de leur sens catholique. Enfin et surtout, l'opposition entre les partisans de la scolastique et les théologiens universitaires, férus de théologie positive et de philosophie moderne, n'avait pas tardé à prendre l'allure d'un conflit entre Rome et l'Allemagne. C'est de Rome que les champions de la « théologie d'autrefois » avaient ramené leur enthousiasme ; c'est de Rome que partaient des condamnations, souvent justifiées, mais qui exaspéraient d'autant plus des travailleurs dévoués à l'Église que la procédure archaïque des Congrégations romaines permettait aux protestants de discréditer à bon compte toute la science catholique. Et quelques scolastiques ayant eu le tort de faire écho à la boutade d'un prélat romain à la science plutôt superficielle : « *I Tedeschi sono tutti un poco eretici* », les représentants de la pensée germanique se sentaient offensés à la fois dans leur prestige scientifique et dans leur fierté nationale, à un moment où l'unité allemande se faisait et où les catholiques étaient accusés par leurs adversaires de manquer de patriotisme. Aux ultramontains de la nouvelle génération, qui oubliaient un peu vite l'appui que les universitaires avaient apporté au mouvement catholique quinze ans auparavant, lorsque « les hommes de science, les hommes d'œuvre et les hommes de lutte fraternisaient dans les meetings » [1], et qui se présentaient maintenant comme les seuls catholiques dignes de ce nom, leurs adversaires répliquaient qu'ils n'étaient qu'un « parti » [2] de fanatiques, forts sans doute de l'appui d'un « nombre respectable de pieux laïcs » [3], mais occupés à discréditer l'Église aux yeux des classes dirigeantes.

L'opposition des savants catholiques aux progrès de l'ultramontanisme eût toutefois pu se borner à une résistance passive, comme ce fut le cas à Tubingue. Elle prit l'allure d'une lutte ouverte qui devait aboutir au schisme vieux-catholique, sous l'influence du vieux maître de Munich, l'illustre Doellinger, qui avait pris la tête du mouvement de réaction contre l'invasion des « Romains » dans la science catholique et dans la direction de l'Église d'Allemagne. Le dernier mot n'a pas encore été dit sur son évolution religieuse au cours de ces années [4] et l'on a trop tendance, du côté catholique, à simplifier les choses. A l'époque des conférences de l'Odéon [5], les critiques de Doellinger contre Rome visaient avant tout les déficiences des États pontificaux, les méthodes archaïques de la Curie et le peu d'esprit scientifique qui y régnait ; et dans ses *Légendes pontificales du moyen âge*, l'étude critique sur la donation de Constantin, invoquée à tort par les défenseurs du pouvoir temporel, avait autant d'importance que l'examen du cas d'Honorius. Mais après l'assemblée de Munich de 1863, quand il devint clair que le courant scolastique voulait écraser l'école historique en faisant intervenir l'autorité du Saint-Siège, Doellinger, convaincu que l'avenir de la théologie était en jeu, commença à considérer comme un immense danger la tendance ultramontaine à exalter l'infaillibilité du pape et à étendre sans cesse le champ de ses interventions. Lorsque, quelques mois plus tard, le *Syllabus* lui parut vouloir imposer comme de foi le système théocratique médiéval, son inquiétude s'accrut encore et il lui sembla que l'avenir

---

(1) G. Goyau, *L'Allemagne religieuse*, t. IV, p. 200.
(2) Un ami de Doellinger publia en 1865 une brochure assez impertinente intitulée *Kirche oder Partei ?* Ce fut l'occasion d'une polémique ; Hergenröther notamment répondit par *Kirche und nicht Partei*, Wurzbourg, 1865.
(3) Doellinger à A. Gramisch, dans F. Vigener, *Drei Gestalten...*, p. 158.
(4) Sur l'évolution des sentiments religieux de Doellinger voir notamment une lettre de lord Acton, qui fut l'ami le plus intime de son âge mûr, dans *Selections from the Correspondence from the first lord Acton*, édit. Figgis et Laurence, p. 57-58. Vigener montre également très bien (*Drei Gestalten...*, passim, en particulier p. 135-137) comment Doellinger passa insensiblement d'une opposition *théologique* à une opposition *dogmatique* ; pendant longtemps, il crut de très bonne foi ne s'attaquer qu'à une « opinion d'école » qui voulait s'imposer par la force.
(5) Cf. *supra*, p. 204-205.

même du catholicisme serait compromis si un changement dans la conduite de l'Église n'intervenait pas. C'est dans ces dispositions d'esprit
qu'il se remit à l'étude de l'évolution des prérogatives pontificales, en
utilisant, beaucoup plus qu'il ne l'avait fait jusqu'alors, divers inédits
dont ses correspondants lui avaient signalé l'existence, et il crut pouvoir
conclure de ce nouvel examen que la doctrine de l'infaillibilité du pape,
dont on voulait faire un dogme, n'était qu'une création du moyen âge.
L'indignation de Doellinger fut à son comble lorsque, en 1867, il apprit
qu'on se proposait à Rome de canoniser l'inquisiteur Arbues : partageant avec les catholiques libéraux l'horreur des méthodes de coerci-
tion utilisées jadis par l'Église, il commença dès lors à se demander
si, au delà des institutions ecclésiastiques, ce n'était pas l'idéal religieux
lui-même que le parti dominant dans l'Église était en train de fausser.
De là le ton toujours plus violent de ce qu'il écrit à partir de ce moment [1].

La violence même du ton de la polémique menée par Doellinger et
son groupe et leur prétention à vouloir porter celle-ci devant l'opinion
publique indisposa vivement les autorités ecclésiastiques, comme on
le vit notamment à l'occasion de la conférence de Fulda de 1867. Aussi,
bien que plusieurs évêques, Rauscher et Ketteler y compris, ne fussent
guère favorables aux tendances centralisatrices qui s'accentuaient à
Rome depuis quelques années, l'opposition demeura en Allemagne
exclusivement universitaire. Il n'en alla pas de même en France, où se
reconstitua, au sein de l'épiscopat, pendant les dernières années du
Second Empire, un groupe d'adversaires décidés de l'ultramontanisme.

LE GROUPE DE MGR MARET — Une opposition universitaire ne se concevait guère
en France, où les Facultés de théologie étaient prati-
quement inexistantes. Il subsistait pourtant un petit
foyer actif de gallicanisme théologique à la Sorbonne, autour d'un des
très rares théologiens de valeur que la France ait connus au XIXe siècle,
Mgr Maret [2]. Ce prélat pieux, qui n'avait rien de l'arrogance des professeurs allemands, professait pour les vues ecclésiologiques de Bossuet
une admiration sans mélange qu'il s'appliquait à justifier avec érudition. Le néo-ultramontanisme, avec ses tendances à l'absolutisme et
à la théocratie, heurtait cet esprit mesuré qui sympathisait fort avec
les aspirations démocratiques de son temps et chercha toute sa vie à
rapprocher l'Église et la société moderne. Adversaire des polémiques
bruyantes, son moyen d'action préféré était les *Mémoires* confidentiels,
adressés tantôt à l'empereur, tantôt à l'épiscopat, tantôt au souverain
pontife, dans lesquels il exposait ses vues et signalait les dangers qu'il
redoutait avec une remarquable clarté.

A la Faculté de théologie, dont il était doyen depuis 1853, Maret s'était
entouré d'un petit groupe de prêtres intelligents, de sentiments libéraux

---

(1) Une partie parut sous forme d'articles anonymes dans la libérale *Allgemeine Zeitung* d'Augsbourg. D'autres articles, demeurés inédits, ont été publiés par Reusch (*Kleinere Schriften gedrückte
und ungedrückte von J. J. I. v. Doellinger*, Stuttgart, 1892).
(2) Sur Henri Maret (1805-1884), voir G. BAZIN, *Vie de Mgr Maret*, Paris, 1891, 3 vol., à nuancer
par E. AMANN, art. *Maret*, dans *D. T. C.*, t. IX, col. 2033-2037.

et de tendance modérément gallicane, qu'il s'efforçait de faire parvenir à l'épiscopat grâce à ses relations étroites avec le ministre des Cultes et avec Napoléon III, car, fidèle à la tradition des gallicans des siècles antérieurs, il cherchait dans l'appui du gouvernement un refuge contre les interventions romaines. Cet appui gouvernemental fut faible jusque vers 1860, par crainte de voir le clergé ultramontain appuyer l'opposition légitimiste ; il ne devint effectif qu'avec la modification de la politique ecclésiastique impériale provoquée par l'agitation autour de la Question romaine, qui allait permettre, pendant une dizaine d'années, à Rouland puis à Baroche, franchement gallicans comme presque tout le personnel du ministère des Cultes, de contrecarrer systématiquement l'action du parti romain. Le premier, tout en multipliant à partir de 1860 les mesures tracassières, réussit à désorganiser les blocs ultramontains que formaient les provinces de Reims et de Bordeaux. Quant au second, ministre des Cultes de 1863 à 1869, il renforça puissamment, par ses dix-huit nominations épiscopales, le groupe des évêques d'orientation libérale et gallicane[1]. Il se faisait d'ailleurs moins d'illusions que Maret sur l'efficacité immédiate de cette politique, sachant combien les velléités de résistance de ces évêques étaient freinées par l'enthousiasme romain de leur clergé. Toutefois, ne réalisant pas à quel point celui-ci répondait à une réaction spontanée du sentiment catholique et à une évolution inéluctable, s'imaginant comme beaucoup qu'il tenait avant tout à la personnalité de Pie IX, il estimait que l'essentiel consistait à préparer la réaction libérale qui suivrait sa mort en se bornant, dans l'immédiat, à freiner autant que possible les « entreprises d'un pape intransigeant et passionné »[2].

Cette résistance limitée elle-même, il ne put la pousser aussi loin qu'il l'eût désiré, par suite de la nécessité pour le gouvernement impérial de tenir compte de l'influence électorale des masses ultramontaines. « Le gallicanisme comptait de nombreux adhérents parmi les magistrats et les hauts fonctionnaires ; c'est pourquoi il avait pu prévaloir sous l'empire autoritaire. Mais il n'avait ni dans les classes riches ni dans le peuple assez de partisans pour constituer une force électorale. Au contraire, la politique gallicane mécontenterait les éléments ultramontains du clergé, c'est-à-dire ceux dont l'influence électorale était la plus grande : les congrégations et le bas clergé[3]. » Malgré tout, les nominations épiscopales de Baroche inquiétèrent beaucoup les ultramontains, car les amis de Maret, en relations épistolaires constantes entre eux, formaient un réseau bien uni d'adversaires de la politique romaine. Il était à craindre qu'en exploitant les susceptibilités de beaucoup d'évêques à l'égard des allures dictatoriales de Veuillot, ils ne ralliassent à leur cause un bon nombre de modérés qui n'avaient pas pardonné à Pie IX d'avoir, en

---

(1) Il importe de noter que si ces évêques gallicans n'hésitaient pas à recourir à l'appui du gouvernement contre ce qu'ils considéraient comme des interventions abusives du Saint-Siège dans l'Église de France, ils étaient par contre résolument hostiles au « gallicanisme parlementaire » dont les tendances étaient encore vivantes au ministère des Cultes et qui prétendait soumettre l'exercice de l'autorité épiscopale au contrôle de l'État.
(2) J. MAURAIN, *Un bourgeois français au XIXᵉ siècle : Baroche*, p. 450.
(3) *Ibid.*, p. 451.

1853, « humilié l'épiscopat français devant un pamphlétaire » et qui rappelaient mélancoliquement le mot de Grégoire le Grand : « L'honneur de mes frères dans l'épiscopat, c'est mon propre honneur », en regrettant que « les temps fussent changés » [1]. Il était aussi à craindre de les voir renforcer l'influence de centres comme Lyon, où les traditions d'autonomie demeuraient vivaces. Mais ce qui inquiétait le plus les ultramontains, c'était l'influence grandissante de Mgr Darboy [2], que Maret avait fait nommer en 1863 archevêque de Paris.

MONSEIGNEUR DARBOY « Si Mgr Darboy n'est pas le chef avoué de cette coalition, il en est l'âme et le centre », écrivait Mgr Mabile dans une note sur « la réaction gallicane » [3]. Esprit clair et perspicace, conscient de la crise religieuse qui mûrissait sous ses yeux et du danger qu'il y avait à heurter de front les sympathies libérales de l'opinion dirigeante française, convaincu que le bien de l'Église exigeait une entente étroite avec le gouvernement impérial, plein d'admiration du reste pour Napoléon III dont il approuvait la politique italienne, gagné au gallicanisme, après avoir été ultramontain dans sa jeunesse, sous l'influence de Mgr Sibour et de son vicaire général Lequeux, excédé d'ailleurs par les procédés de Veuillot et de ses amis, ces « étranges catholiques dont la piété consiste principalement à saluer le pape de loin pour insulter les évêques de près » [4], Darboy ne pouvait éprouver que de l'aversion pour les tendances qui prévalaient à Rome sous Pie IX et il le fit sentir avec une désinvolture à laquelle on n'était plus habitué. Non seulement il prétendit agir dans l'administration de son diocèse avec une indépendance qui ne tenait aucun compte des progrès réalisés par l'influence romaine au cours des dernières années, mais lorsque le pape lui adressa des reproches à ce sujet, il répondit par une justification dont le ton, tantôt raide et tantôt persifleur, dut froisser profondément Pie IX. Il y protestait contre « l'ingérence directe et immédiate de la papauté » dans l'administration épiscopale et contre la tendance à introduire subrepticement en France « le régime des pays de mission » ; tant que le Saint-Siège n'aura pas modifié en termes explicites et solennels la constitution actuelle de l'Église, concluait-il,

je suis prêt à m'opposer autant qu'il m'est possible à ce déplorable système d'annexion, à cette confiscation inméritée de l'autorité épiscopale et je crois rendre ainsi un véritable service à l'Église et au Saint-Siège. Je ne doute pas, au reste, que mes vénérables collègues les évêques de France, dès qu'ils seront mis au courant de cette affaire, ne m'accordent, avec leurs sympathies, l'appui de leurs conseils et de leur autorité [5].

---

(1) Landriot au cardinal Pitra, 27 février 1865, dans J. MAURAIN, *La politique ecclésiastique du Second Empire*, p. 789-790, en note.
(2) On ne possède pas encore de biographie définitive de Georges Darboy (1813-1870). La meilleure, due à Mgr J. FOULON (Paris, 1889), est à mettre au point par les brochures ****La vérité sur Mgr Darboy*, Gien, 1889 ; E. AMBROISE-DARBOY, *Notes sur Mgr Darboy*, Orléans, 1923-1924, et surtout par la publication de documents inédits : *Mgr Darboy et le Saint-Siège*, citée p. 295, n. 2.
(3) Écrite en 1867 et reproduite dans P. MABILE, *Vie de Mgr Mabile*, t. II, p. 171-175.
(4) Darboy à Maret, 20 janvier 1863, dans G. BAZIN, *Vie de Mgr Maret*, t. II, p. 238.
(5) Lettre du 1ᵉʳ septembre 1864, dans *Revue d'histoire et de littérature religieuse*, t. XII, 1907, p. 240-255.

Le commentaire libéral que Darboy donna au *Syllabus*, le discours prononcé au Sénat, où il faisait au fond l'apologie des Articles organiques, une nouvelle lettre passablement irrespectueuse où il disait au pape : « Ceux qui savent n'ont pas votre confiance et ceux qui l'ont ne savent pas » [1], d'autres incidents encore, achevèrent d'irriter le Saint-Père qui adressa à l'archevêque, le 26 octobre 1865, une lettre lui reprochant entre autres d'avoir émis « des opinions tout à fait contraires à la divine primauté du pontife romain sur l'Église universelle tout entière » et de « favoriser, par ses façons d'agir et de sentir, les doctrines fausses et erronées de Febronius » [2].

Darboy se contenta d'une réponse « déférente mais sans rétractation aucune » [3] ; mais le pape n'insista pas, espérant qu'une entrevue personnelle permettrait d'arranger les choses. Ce n'est toutefois qu'en 1867 que Darboy se décida enfin à faire sa visite *ad limina* et il réussit assez habilement à éviter toute explication. On tenta alors d'obtenir de lui qu'il désavouât son attitude, en mettant à ce prix, au grand déplaisir du gouvernement français, l'obtention du cardinalat, mais, sollicité à plusieurs reprises, il refusa avec hauteur. Pie IX, indigné, allait jusqu'à dire, en 1868, à un religieux français : « Si vous voyez votre empereur et l'archevêque de Paris, dites-leur que le pape se porte bien et qu'il prie tous les jours *ut inimicos S. Ecclesiae humiliare digneris* [4]. » Pendant ce temps, par une manœuvre où le nonce Chigi joua sans doute un rôle, les adversaires de l'archevêque essayaient d'ébranler son autorité en France en faisant publier le blâme pontifical de 1865, sans souci de son caractère confidentiel. Ils réussirent partiellement, mais accrurent en même temps les rancœurs des milieux parisiens et de beaucoup de libéraux contre ce qu'ils nommaient les « méthodes romaines ».

LES CATHOLIQUES LIBÉRAUX   Car, depuis quelques années, les catholiques libéraux se montraient, eux aussi, fort réservés à l'égard du mouvement ultramontain et commençaient à se rapprocher du groupe de Maret et Darboy, en dépit de leur hostilité persistante à l'égard de l'empire et des évêques gouvernementaux. Jadis pourtant, au temps de *L'Avenir* et jusqu'aux débuts du Second Empire, ils avaient compté parmi les adversaires décidés du gallicanisme : « Ils en voulaient à Bossuet, explique Albert de Broglie, de n'avoir résisté au pape que pour obéir plus docilement au roi et, confondant l'infaillibilité du pape avec l'indépendance du pouvoir spirituel, ils s'y étaient rattachés comme à une garantie de la liberté de leur propre conscience [5]. » Mais à la revendication de la liberté au nom de l'ultramontanisme avait succédé la méfiance à l'égard de l'ultramontanisme au fur et à mesure que celui-ci leur était apparu « représenté et personnifié par *L'Univers*

---

(1) Lettre du 19 février 1865, dans *Rev. d'hist. et de littér. relig.*, t. XII, 1907, p. 256-258.
(2) Le texte latin en est publié dans *L'Univers* du 18 mars 1869 et la traduction française dans le numéro du 22 mars.
(3) J. Maurain, *La politique ecclésiastique du Second Empire*, p. 783.
(4) A. Houtin, *Le P. Hyacinthe dans l'Église romaine*, p. 203.
(5) A. de Broglie, *Mémoires*, t. I, p. 357.

et *La Civiltà* »[1]. Ils se souvenaient d'un mot de Mgr Sibour reprochant à la nouvelle école ultramontaine « une double idolâtrie . idolâtrie du pouvoir temporel et du pouvoir spirituel »[2]. Ils dénonçaient une centralisation qui leur paraissait « au moins aussi menaçante et aussi funeste dans l'Église que dans l'État »[3], et qu'ils appréhendaient d'autant plus qu'elle leur semblait dirigée par des hommes sans connaissance du véritable état de l'Europe et des besoins de la société contemporaine. Des hommes aussi dévoués au pape que Dupanloup, qui avait à plus d'une reprise affirmé sa foi en l'infaillibilité, craignaient à présent qu'on ne définît celle-ci, parce qu'ils redoutaient de voir par là renforcée l'autorité du *Syllabus* ou même de certains documents anciens comme la bulle *Unam Sanctam*, où étaient exposées des vues politico-religieuses incompatibles avec la mentalité moderne.

Cette orientation nouvelle du libéralisme catholique allait se manifester clairement lors de la réunion solennelle de plusieurs centaines d'évêques à Rome à la fin du mois de juin 1867, qui marqua une étape importante dans l'histoire de l'ultramontanisme.

L'ASSEMBLÉE DES ÉVÊQUES DE 1867 — L'idée d'affirmer solennellement l'infaillibilité du pape était dans l'air depuis quelque temps. Les uns y voyaient surtout un moyen de dédommager Pie IX des déboires qu'il avait éprouvés dans sa souveraineté temporelle, mais d'autres voyaient plus loin : les menaces qui pesaient sur ce qui restait de l'État pontifical et l'attitude de plus en plus hostile à l'Église manifestée par les gouvernements contemporains faisaient craindre que les papes ne fussent bientôt dans l'impossibilité de se concerter librement avec l'épiscopat avant de prendre position dans des questions vitales pour l'Église ; de l'avis de beaucoup, le meilleur moyen de remédier à ce péril consistait à affirmer clairement que « les papes, même exilés et captifs, possédaient toute l'autorité dans l'Église dispersée et muette »[4]. Lorsque Pie IX annonça son intention de profiter du dix-huitième centenaire du martyre des apôtres Pierre et Paul pour procéder, comme en 1862, à « une revue générale de l'armée catholique »[5], beaucoup estimèrent qu'il fallait saisir cette occasion pour proclamer la croyance de l'Église à l'infaillibilité du pape.

Pareille suggestion se heurta à l'opposition non seulement de ceux qui persistaient à nier que le pape jouît du privilège de l'infaillibilité indépendamment d'une ratification par le corps épiscopal, mais aussi de nombreux catholiques libéraux qui admettaient pourtant l'infaillibilité personnelle. Beaucoup de modérés trouvaient d'ailleurs peu indiqué de vouloir trancher une question aussi importante presque par surprise, sans discussion sérieuse de la part du corps épiscopal, et d'une manière

(1) Montalembert à M. Lallemand, février 1870, dans E. LECANUET, *Montalembert*, t. III, p. 467.
(2) Mgr Sibour à Montalembert, 10 septembre 1853, *ibid.*, t. III, p. 104.
(3) Montalembert au P. Félix, 27 septembre 1864, *ibid.*, t. III, p. 379.
(4) J. BRUGERETTE et É. AMANN, art. *Vatican*, dans *D. T. C.*, t. XV, col. 2540.
(5) F. HAYWARD, *Pie IX et son temps*, p. 329. Sur les fêtes de 1867, voir P. KARLBRANDES, *Der hl. Petrus in Rom oder Rom ohne Petrus*, Einsiedeln, 1867.

qui ne permettrait pas de stipuler les conditions précises dans lesquelles le charisme d'infaillibilité intervient effectivement. Dupanloup qui, comme en 1862, joua un rôle important dans la commission chargée d'élaborer l'adresse au pape, réussit à obtenir de Mgr Franchi que le mot « infaillible », qui se trouvait à plusieurs reprises dans le projet primitif, fût supprimé ; malgré les protestations de Manning, il fit prévaloir son point de vue après une âpre discussion et il obtint en outre que fût ajouté au passage sur le magistère pontifical ces mots qui en limitaient la portée : *ad custodiendum depositum* [1].

C'était un succès pour ceux qui voulaient éviter de brusquer les choses. Il n'en reste pas moins vrai de dire avec Manning que tout au long des solennités grandioses qui se succédèrent depuis le 20 juin, jour de la Fête-Dieu, jusqu'au 1er juillet, et auxquelles prirent part près de cinq cents évêques, tous les patriarches orientaux, plus de 14.000 prêtres et 130.000 pèlerins étrangers, « on ne fêta pas seulement le martyre de Pierre, mais sa primauté sur le monde » [2]. Pèlerins et évêques apportèrent au pape de tous les coins de la terre des cadeaux somptueux et surtout d'importantes sommes d'argent destinées à renflouer le budget de l'État pontifical. Parmi les saints canonisés le 29 juin, on comptait les martyrs de Gorcum, victimes des protestants, saint Josaphat, apôtre de l'Union des Ruthènes au siège romain, Léonard de Port-Maurice, l'un des grands protagonistes de la doctrine de l'infaillibilité pontificale au siècle précédent. Pour la première fois depuis le xviie siècle, la chaire de saint Pierre fut retirée du reliquaire du Bernin et exposée à la vénération des fidèles. Enfin et surtout, l'adresse qui fut présentée au pape le 1er juillet [3], si elle ne contenait pas, contrairement à l'attente de certains, l'affirmation explicite de l'infaillibilité, l'insinuait assez clairement, en exaltant le magistère pontifical et l'autorité de la chaire de Pierre.

---

(1) C. BUTLER, *The Vatican Council*, t. I, p. 86-87. Comparer la version de E. MANNING, *The true Story of the Vatican Council*, p. 52-55, avec le document reproduit par É. OLLIVIER, *L'Église et l'État au concile du Vatican*, 4e édit., t. II, p. 601, la lettre d'Ullathorne du 26 octobre 1869 dans C. BUTLER, *op. cit.*, t. I, p. 147 et les détails donnés par F. LAGRANGE, *Vie de Mgr Dupanloup*, t. III, p. 58-59.

(2) E. MANNING, *The true Story of the Vatican Council*, p. 45.

(3) Texte dans *Coll. Lac.*, t. VII, col. 1033-1042.

CHAPITRE X

LE CONCILE DU VATICAN [1]

§ 1. — La préparation du concile et les premières polémiques [2].

LA GENÈSE DU CONCILE Pie IX profita des fêtes grandioses de 1867 pour annoncer officiellement son intention de convoquer à Rome un concile œcuménique. C'est à la fin de 1864 qu'il avait fait part pour la première fois au Sacré Collège de ce projet qu'il mûrissait depuis plusieurs années, encouragé par les suggestions de plusieurs évêques. Il s'agissait, dans le cadre de la restauration générale de la société chrétienne entreprise depuis les débuts de son pontificat, de compléter et de confirmer l'œuvre d'exposition doctrinale esquissée dans le *Syllabus*, en faisant contre le rationalisme théorique et pratique du XIXe siècle ce que le concile de Trente avait fait au XVIe siècle contre le protestantisme.

Une commission cardinalice instituée en mars 1865 conseilla de deman-

(1) BIBLIOGRAPHIE. — I. SOURCES. — MANSI, *Amplissima Collectio Conciliorum*, t. 49-53, Paris, 1923-1927 (cité *A. C. C.*), à compléter, pour les documents non officiels, par *Collectio Lacensis*, t. VII : *Acta et Decreta S. Conc. Vaticani cum permultis aliis documentis ad concilium ejusque historiam spectantibus*, Fribourg-en-Br., 1892. En outre R. AUBERT, *Documents concernant le Tiers-Parti au concile du Vatican*, dans *Abhandlungen über Theologie und Kirche* (Mélanges Karl Adam), Dusseldorf, 1951, p. 241-259 ; F. GUÉDON, *Autour du concile du Vatican*, dans *Les Lettres*, 1928, vol. II, p. 19 et suiv., 190 et suiv., 314 et suiv. (lettres de Mgr Foulon) ; J. FRIEDRICH, *Tagebuch während des vatikanischen Concils*, 2e édit., Nordlingen, 1873 (vieux-catholique) ; T. MOZLEY, *Letters from Rome*, Londres, 1891, 2 vol. (anglican) ; L. VEUILLOT, *Rome pendant le concile*, Paris, 1872, 2 vol. (enthousiaste).

J'utilise également le *Journal de mon voyage à Rome* de M. Icard (*Arch. S. Sulp.*, fonds Icard) ; les *Souvenirs du concile du Vatican*, de Mgr Collet (copie communiquée par M. le chanoine Guédon) ; les procès-verbaux des réunions du groupe français de la minorité (copie communiquée par le même) ; les papiers de Darboy à l'archevêché de Paris ; les dépêches de l'ambassadeur de France à Rome (*Arch. Min. A. E. Paris*, Rome, 1043 à 1046) et celles du ministre de Belgique (*Arch. Min. A. É. Bruxelles*, St-Siège, t. XIII, 2).

II. TRAVAUX. — Le meilleur exposé est celui de C. BUTLER, *The Vatican Council. The Story told from inside in Bishop Ullathorne's letters*, Londres, 1930, 2 vol. Le plus complet, du point de vue catholique, est celui de T. GRANDERATH, *Geschichte des vatikanischen Konzils*, Fribourg-en-Br., 3 vol., 1903-1906, cité dans la trad. franç., Bruxelles, 1908-1919 (trop apologétique) ; du point de vue vieux-catholique, celui de J. FRIEDRICH, *Geschichte des Vatikanischen Concils*, Nordlingen, 1877-1887, 3 vol. (très partial). Voir également F. MOURRET, *Le concile du Vatican*, Paris, 1919 ; É. OLLIVIER, *L'Église et l'État au concile du Vatican*, Paris, 1877, 2 vol. ; F. BRUGERETTE-É. AMANN, art. *Vatican*, dans *D. T. C.*, t. XV, col. 2536-2585 ; J. ACTON, *The Vatican Council*, dans *North British Review*, t. XIV, 1870, p. 183-229.

Les biographies, sur ce chapitre, sont généralement incolores et décevantes ; voir cependant, dans la majorité, celles de Manning (par E. PURCELL, Londres, 1895), de Pie (par Mgr BAUNARD, Paris, 1901) et de Spalding (par J.-L. SPALDING, New York, 1873) ; dans la minorité, celles de Dupanloup (par F. LAGRANGE, Paris, 1881), de Ketteler (par O. PFUELF, Mayence, 1899, et par F. VIGENER, Munich, 1924) et surtout de Schwarzenberg (par C. WOLFSGRUBER, Vienne, 1916) et de Maret (par G. BAZIN, Paris, 1891) ; celle de Doellinger (par J. FRIEDRICH, Munich, 1901), ainsi que A. HAGEN, *Hefele und das vatikanischen Konzil*, dans *T. Q. S.*, t. CXXIII, 1942, p. 223 et suiv.

(2) BIBLIOGRAPHIE. — Aux indications données à la note précédente, ajouter : comme recueil de sources, E. CECCONI, *Storia del concilio vaticano scritta su i documenti originali*, Florence, 1873-1879, 4 vol., trad. franç., Paris, 1887, 4 vol. ; comme exposé d'ensemble de la situation à la veille du concile, E. CAMPANA, *Il Concilio Vaticano*, t. I : *Il Clima del Concilio*, Lugano, 1926, 2 vol. ; en outre E. TERRIEN, *Monseigneur Freppel*, t. I, Angers, 1931, chap. XX et XXI.

der confidentiellement l'avis d'une quarantaine d'évêques et la plupart
des prélats consultés se déclarèrent favorables au projet [1]. Quelques-uns
y voyaient une occasion de « mettre en évidence cette vérité que Rome
n'appartient pas aux Italiens, mais au monde catholique » [2] ; d'autres
suggéraient de s'appliquer à préciser les rapports entre l'Église et l'État
en tenant compte de l'état de choses engendré par la Révolution française ;
la majorité s'inquiétait surtout d'adapter le droit canon aux transformations de la société depuis le concile de Trente et de réagir contre les
erreurs relatives aux principes fondamentaux du christianisme : panthéisme, naturalisme, rationalisme, indifférentisme ; beaucoup proposaient de prendre comme base du programme l'encyclique *Quanta cura*.

Le pape avait d'abord envisagé la convocation du concile pour l'anniversaire du martyre des apôtres, mais il apparut vite que les préparatifs
ne pourraient pas être finis à cette date ; puis l'aggravation de la situation
politique à la suite de Sadowa et de l'évacuation de Rome par les troupes
françaises amena même à interrompre les travaux préliminaires. Ceux
qui, à Rome, étaient opposés à l'idée du concile en profitèrent pour faire
valoir leurs arguments. Antonelli craignait qu'il ne suscitât de nouvelles
difficultés politiques au Saint-Siège. D'autres jugeaient un concile inutile :
« N'avons-nous pas le pape ? Est-ce que le pape n'a pas le pouvoir de
décider toutes les questions ? [3] » Et le cardinal Pitra se faisait l'écho de
nombreux prélats et religieux qui craignaient de voir restreindre le
rôle du clergé romain dans l'Église quand il avouait candidement : « Quoi !
Convoquer un concile ! Mais les théologiens français et allemands viendraient bouleverser nos Congrégations ! [4] » Aussi Pie IX resta-t-il hésitant jusqu'à la veille des fêtes de 1867. Il semble que ce fut Dupanloup,
arrivé à Rome un bon mois à l'avance, qui réussit à vaincre ses derniers
doutes. En tout cas, au début de juin, le pape était décidé [5] et le 26,
dans une allocution aux évêques réunis à Rome, il annonça officiellement
sa décision.

LES COMMISSIONS
PRÉPARATOIRES

On s'appliqua dès lors à mettre sur pied les commissions chargées de déterminer les matières à discuter
et de préparer des projets de décrets [6]. Soixante
consulteurs devaient être choisis sur place et trente-six autres appelés
de l'étranger. Malheureusement, le recrutement de ces derniers, dont la
désignation avait été laissée aux nonces, « ne fut pas toujours des plus
judicieux » [7]. On commença en effet par ne désigner que des hommes
ouvertement ralliés à l'ultramontanisme, aux compétences parfois assez
relatives ; l'exclusion notamment de tous les théologiens des universités

(1) On trouvera les réponses dans MANSI, *A. C. C.*, t. 49, col. 9-207.
(2) Réponse de Mgr Guibert, *ibid.*, col. 116.
(3) F. LAGRANGE, *Vie de Mgr Dupanloup*, t. III, p. 55.
(4) Mgr COLLET, *Souvenirs du concile du Vatican* (inédits).
(5) Dupanloup l'annonçait au cardinal Sterckx dès le 2 juin. (*Arch. Mal.*, fonds Sterckx, Voyages
à Rome.)
(6) Elles étaient au nombre de cinq : une pour les questions dogmatiques, une pour la discipline
ecclésiastique et le droit canon, une pour les ordres religieux, une pour les missions et les Églises
orientales, une pour les relations entre l'Église et l'État.
(7) F. BRUGERETTE-É. AMANN, art. *Vatican*, dans *D. T. C.*, t. XV, col. 2537.

allemandes, à l'exception de Hettinger et Hergenroether, anciens élèves du Collège germanique, provoqua des protestations de divers évêques, y compris Ketteler, dont les préventions contre les *Deutsche Theologen* étaient pourtant connues. On élargit alors quelque peu le choix en faisant entre autres appel à Newman (qui refusa pour raison de santé), à Freppel, ami de Mgr Maret et des catholiques libéraux, et, pour l'Allemagne, à Dieringer, de Bonn, à Alzog, de Fribourg, à Haneberg, de Munich, et à Hefele, de Tubingue, le grand spécialiste de l'histoire des conciles, qu'il était inconcevable de ne pas avoir désigné plus tôt. Par contre, en dépit de l'insistance des archevêques de Prague et de Breslau, Reisach s'opposa formellement à ce qu'on invitât Doellinger [1], dont Mgr Dechamps, tout ultramontain qu'il était, affirmait : « On ne peut pas ne pas l'écouter » [2] ; on contribuait de la sorte à aigrir encore davantage contre Rome le chef de l'école historique allemande.

Le « parti-pris » [3] qui avait présidé au choix des consulteurs fit d'autant plus mauvaise impression que, dans certains milieux, on était mécontent du secret qui entourait la préparation du concile et son programme. On y voyait une marque de défiance à l'égard de l'épiscopat et une tentative de la Curie à vouloir « tenir le concile avant le concile ».

LA CONVOCATION — Un an après la première annonce, le 29 juin 1868, Pie IX publiait la bulle *Aeterni Patris* [4] convoquant les évêques et autres ayants-droit à participer au concile qui s'ouvrirait dans la basilique vaticane le 8 décembre 1869, quinzième anniversaire de la proclamation de l'Immaculée-Conception.

La détermination de ceux qui pouvaient prendre part au concile en dehors des évêques résidentiels donna lieu à quelques discussions. Pour les évêques titulaires, on trancha par l'affirmative malgré une intervention du pape qui aurait aimé écarter des prélats comme Mgr Maret ; pour les abbés, on décida d'admettre, outre les généraux d'ordre et les abbés *nullius*, les supérieurs de congrégations, ce qui permettait à dom Guéranger, l'un des chefs de l'opposition antilibérale en France, d'y assister. La question la plus délicate était celle des chefs d'État catholiques : les exclure, c'était rompre avec une vieille tradition suivie par tous les conciles antérieurs ; mais était-il possible de les inviter, alors que le roi d'Italie était excommunié, que plusieurs présidents d'Amérique du Sud étaient francs-maçons, que l'empereur d'Autriche devait tolérer un gouvernement anticlérical ? Après de longues discussions, auxquelles prirent part Antonelli et le pape lui-même, on se mit d'accord sur un compromis : on n'inviterait pas expressément les princes à participer aux

---

(1) On prétendit officiellement à Rome qu'on avait eu l'intention de l'inviter, mais qu'on avait appris qu'il refuserait de venir (Antonelli à Schwarzenberg, 15 juillet 1868, dans T. GRANDERATH, *op. cit.*, t. I, p. 82). « Excuse diplomatique », note à bon droit Godet (*Revue du clergé français*, t. XXXVI, 1903, p. 378) ; voir les lettres de Doellinger à Schwarzenberg, dans T. GRANDERATH, *op. cit.*, t. I, p. 82, du cardinal Caterini au nonce à Vienne, dans E. CECCONI, *op. cit.*, t. IV, p. 704, et du nonce à Munich au cardinal Caterini, *ibid.*, t. II, p. 337. Sur le rôle de Reisach, voir le témoignage de Freppel dans *Germania*, Wissenschaft-Beilage, 1900, n° 45.
(2) Cité par F.-X. KRAUS dans *Deutsche Literaturzeitung*, 3 août 1901, col. 1954.
(3) Hefele à Schwarzenberg, 10 mai 1869, dans T. GRANDERATH, *op. cit.*, t. I, p. 84.
(4) MANSI, *A. C. C.*, t. 50, col. 193*-198*.

délibérations conciliaires, tout en choisissant pourtant les termes de la bulle qui les concernaient de façon à leur permettre une certaine coopération s'ils désiraient la fournir. Veuillot souligna bien la signification de cette décision :

> L'ordre sur lequel a vécu la société durant plus de dix siècles a cessé d'exister (...). Une autre ère commence. L'Église et l'État sont séparés de fait et tous deux le reconnaissent (...). C'en est fait, ce n'est pas un bien. L'État l'a voulu, non l'Église [1].

Lors des consultations préliminaires, plus d'un évêque avait suggéré de profiter du concile pour essayer une nouvelle fois de rentrer en rapports avec les chrétiens séparés. Après mûre délibération, on s'arrêta à la solution suivante : on adresserait une lettre « à tous les évêques des Églises de rite oriental qui ne sont pas en communion avec le Siège apostolique » pour les inviter à revenir à l'unité afin de pouvoir prendre part au concile ; un appel à l'union aux protestants et anglicans suivrait, mais quelques jours plus tard, pour bien montrer aux orientaux qu'on ne les traitait pas comme ceux qui n'ont pas d'évêques validement consacrés [2].

L'appel aux protestants fut en général considéré par les intéressés comme une provocation. Quelques pasteurs luthériens mis à part, seuls certains anglicans accueillirent l'invitation pontificale avec une certaine sympathie ; l'un de leurs évêques entra même en négociations avec un jésuite assez connu, de tendance libérale, le P. de Buck, mais l'attitude de Pusey, le grand homme du mouvement d'Oxford, rendit vite impossible la continuation de ces pourparlers [3]. Quant aux évêques orientaux, ils accueillirent la lettre pontificale par un silence méprisant ou, parfois, par une fin de non-recevoir catégorique. Les préjugés et les défiances contre Rome leur faisaient suspecter de « velléités impérialistes » les initiatives les plus désintéressées, mais il y avait eu également des maladresses du côté romain. Le terrain n'avait été aucunement préparé et les orthodoxes se plaignirent en outre de ce que la lettre d'invitation eût été divulguée par la presse avant d'être remise aux intéressés et de ce que la présentation de cette lettre n'eût pas revêtu la solennité désirable ; peut-être eut-on le tort aussi de la leur transmettre par des missionnaires latins [4].

LES REMOUS DE L'OPINION — Il suffit de mentionner pour mémoire, car son influence fut minime, la réaction de la franc-maçonnerie internationale, à l'annonce du concile ; sa principale manifestation extérieure fut l'organisation à Naples, en décembre 1869, par quelques libres-penseurs italiens, d'un « anti-concile », qui avorta assez piteusement [5].

---

(1) *L'Univers*, 11 juillet 1868.
(2) On trouve les lettres apostoliques *Arcano divinae Providentiae* du 8 septembre 1868 et *Jam vos omnes* du 13 septembre 1868 dans MANSI, *A. C. C.*, t. 50, col. 1255-1261.
(3) Cf. F. MOURRET, *Le concile du Vatican*, p. 124-126 (d'après des documents inédits).
(4) F. DE WYELS, *Le concile du Vatican et l'Union*, dans *Irenikon*, t. VI, 1929, p. 488-516, 655-686.
(5) *L'anticoncilio di Napoli del 1869 promosso e descritto dal già deputato G. Ricciardi*, Naples, 1870 ; *Enquête maçonnique à propos du convent extraordinaire du 8 décembre 1869*, Paris (1870).

Autrement importante fut l'agitation qui s'empara peu à peu du monde catholique et qui porta bientôt à son maximum l'opposition entre les courants d'opinion qui s'affrontaient depuis vingt ans : gallicans et libéraux, d'une part, ultramontains et intégristes, de l'autre.

La première annonce du concile avait en général été bien accueillie par les milieux libéraux, particulièrement en France : la plupart en attendaient une modernisation du droit canon, ainsi qu'une diminution de la toute-puissance du pape et de la Curie dans l'Église : « Du moins, on saura que si le pape est infaillible, il ne l'est pas en ce sens que le corps épiscopal ne participe pas à l'infaillibilité active », écrivait Mgr Ginouilhac [1], tandis que Maret se réjouissait de ce que « le parti absolutiste et antimoderne » allait être, « à la lumière du grand jour, réduit à sa faiblesse réelle » [2]. Cet optimisme, qui s'appuyait entre autres sur l'impression de modération donnée par beaucoup d'évêques lors de l'assemblée de 1867, semblait d'ailleurs justifié par les appréhensions des conservateurs, dont l'ambassadeur d'Autriche se faisait l'écho quand il avertissait son gouvernement que le concile allait « ouvrir une arène » aux champions des idées modernes dont la cause « défendue par de pareils avocats et sur un terrain aussi retentissant ne laisserait pas que de faire des prosélytes en dehors de l'assemblée » [3]. Aussi Mgr Dupanloup, dans une magnifique lettre pastorale inspirée par le comte de Falloux [4], saluait-il avec enthousiasme « cette grande tentative de l'Église catholique pour travailler à l'illumination et à la paix du monde » et il concluait en prévoyant que le concile serait « une aurore et non pas un couchant ».

Vers la fin de 1868 toutefois, les impressions se modifièrent. En dépit de quelques éléments défavorables, comme les dispositions incertaines de l'épiscopat austro-hongrois et la crainte de voir la participation espagnole réduite par suite de la révolution, les intransigeants reprirent espoir. Parallèlement, les libéraux commencèrent à s'inquiéter en présence de la « tactique » qu'ils croyaient pouvoir diagnostiquer : préparer le concile en secret, hors de tout débat contradictoire, puis faire accepter sans discussions « des propositions toutes faites » [5], dont le sens n'était rien moins que rassurant d'après les indications qui commençaient à filtrer.

Dans l'esprit de la majorité, *écrivait de Rome l'un des consulteurs*, le concile doit avoir un double but : confirmer le *Syllabus* et décréter l'infaillibilité du pape (...). C'est dans ce sens qu'opèrent les jésuites de la *Civiltà*, plusieurs du Collège romain (Franzelin, Schrader, Sanguinetti), tous les consulteurs français, deux ou trois allemands, et tout le parti des cardinaux et prélats qui tient le haut du pavé sous le pape actuel (...). La confiance que l'on a dans le succès de l'entreprise fait que le concile est vivement désiré de ceux-là même qui semblaient le redouter davantage il y a un an [6].

---

(1) Ginouilhac à Dupanloup, 12 mai 1867, *Arch. Nat.*, AB XIX, nº 526.
(2) Lettre citée dans G. BAZIN, *Vie de Mgr Maret*, t. II, p. 405.
(3) Dépêche de Hübner du 13 juillet 1867, dans S. JACINI, *Il tramonto del potere temporale*, p. 221.
(4) *Lettre sur le futur concile œcuménique*, 56 p. ; sur l'intervention de Falloux, cf. F. MOURRET, *op. cit.*, p. 46-47, en note.
(5) Freppel à Lagrange, 5 décembre 1868, dans E. TERRIEN, *Mgr Freppel*, t. I, p. 549.
(6) Freppel à Dupanloup, 17 février 1869, *Arch. Nat.*, AB XIX, nº 526.

L'ARTICLE DE LA « CIVILTA »   Les partisans des solutions extrêmes crurent
même le moment venu de frapper un grand
coup et, le 6 février 1869, l'officieuse *Civiltà cattolica* publiait — vraisemblablement avec l'assentiment du Saint-Siège [1] — une « Correspondance
de France » où se lisait entre autres :

> Nul n'ignore que les catholiques de France sont malheureusement divisés
> en deux partis : les uns simplement *catholiques*, les autres qui se disent *catholiques
> libéraux* (...). Les catholiques proprement dits croient que le concile sera fort
> court (...). Ils espèrent qu'il proclamera les doctrines du *Syllabus* en énonçant,
> au moyen de formules affirmatives et avec les développements nécessaires, les
> propositions qui y sont émises sous forme négative. Ils accueilleront avec joie
> la proclamation de l'infaillibilité dogmatique du souverain pontife. Personne
> ne trouve étonnant que, par un sentiment de juste réserve, Pie IX ne veuille
> pas prendre lui-même l'initiative d'une proposition qui semble se rapporter à
> lui directement, mais l'on espère que la manifestation unanime du Saint-
> Esprit, par la bouche des Pères du concile œcuménique, la définira par accla-
> mation [2].

Le retentissement de cet article fut immense, car la plupart des lecteurs
y virent un véritable manifeste des milieux romains. Non seulement la
crainte de voir étendre la portée du *Syllabus* se trouvait confirmée, mais,
ce qui était pire, on envisageait une définition de l'infaillibilité pontificale par acclamation, sans même laisser aux évêques le temps d'en préciser
les conditions et les limites. On a caractérisé à juste titre la publication
de cette « Correspondance de France » comme « une faute énorme » [3].
Elle allait, en effet, donner lieu à des polémiques publiques profondément
regrettables en faisant passer au premier plan de l'actualité la question
de l'infaillibilité qui avait jusqu'alors retenu beaucoup moins l'attention
que celle du *Syllabus*. Elle accentuait en outre l'impression que la Curie
elle-même poussait aux solutions extrêmes ; or ceci était inexact, Freppel,
tout prévenu qu'il était, l'avait noté dès son arrivée à Rome :

> S'il y a ici quantité de têtes exaltées, *surtout parmi les étrangers*, il ne manque
> pas de gens très sensés et fort peu disposés à pousser les choses à l'extrême. J'ai
> vu jusqu'ici près d'un tiers du Sacré Collège, et je vous assure que les cardinaux
> de Luca, Bizzarri, Bilio entre autres, tiennent un langage qui vous plairait beau-
> coup [4].

LES PREMIÈRES CONTROVERSES   Les réactions à l'article de *La Civiltà*
EN ALLEMAGNE   ne se firent pas attendre. C'est en
Allemagne qu'elles furent les plus vives.
Doellinger, qui travaillait depuis plusieurs mois à un grand ouvrage
sur la primauté pontificale, publia anonymement, avec l'aide de Huber,
dans l'*Allgemeine Zeitung* d'Augsbourg, cinq articles violents et partiaux

---

(1) Voir *Civ. catt.*, 1934, vol. IV, p. 628, par opposition à la version officielle donnée par
T. GRANDERATH, *op. cit.*, t. I, p. 215-218.
(2) *Civ. catt.*, sér. VII, vol. V, p. 349, 350 et 352 ; d'après le P. de Buck, l'auteur de l'article
serait l'abbé Darras (lettre citée dans F. MOURRET, *op. cit.*, p. 57).
(3) F. BRUGERETTE-É. AMANN, art. *Vatican*, dans *D. T. C.*, t. XV, col. 2542. Même des hommes
aussi peu suspects que Moufang, de Mayence, le confessaient. (Voir sa lettre à Ketteler, du 24 mars
1869, citée par U. RIED dans *Historisches Jahrbuch*, t. XLVII, 1927, p. 663.)
(4) Freppel à Maret, février 1869, dans E. TERRIEN, *Mgr Freppel*, t. I, p. 551. Même son de
cloche dans la lettre du 4 avril à Dupanloup citée par LAGRANGE, *Vie de Mgr Dupanloup*, t. III,
p. 125-126.

réunis ensuite sous le titre : *Le Pape et le Concile*, en un volume qui fit
grand bruit, même à l'étranger. Il y accusait les jésuites et le pape lui-
même de préparer une « révolution ecclésiastique » en prétendant imposer
comme dogme l'infaillibilité et la primauté de juridiction du pape alors
que, selon lui, l'histoire prouvait que les progrès de la papauté n'étaient
que le résultat d'une série d'usurpations datant du moyen âge, « une
tumeur qui défigure l'Église et la fait étouffer » [1]. On peut se faire une
idée des sentiments du vieux professeur à cette époque par ses confidences
au P. Hyacinthe Loyson lorsque celui-ci, de passage à Munich, lui avoua
son attrait vers le protestantisme :

> Des études profondes m'ont amené à reconnaître que l'état qui s'est déve-
> loppé dans l'Église depuis sa séparation d'avec les Églises orientales n'est pas
> selon Dieu. La papauté, par suite des fausses décrétales sous l'influence de
> Grégoire VII, a pris la forme d'un absolutisme que ma conscience chrétienne,
> et je dirai ma conscience historique, ne me permettent pas d'admettre. Toute-
> fois, on ne peut être chrétien en général : il faut appartenir à une Église posi-
> tive, et l'Église catholique en est encore la meilleure. D'ailleurs, il y a deux écoles
> dans l'Église catholique : l'école ultramontaine et l'école chrétienne (il m'a
> nommé alors le pieux évêque Sailer en Allemagne, Bossuet, Fénelon, Pascal,
> en France). Il faut rester fidèle aux institutions tout en luttant contre les abus.
> Lutter contre les tendances persistantes de la hiérarchie romaine [2].

Dans la préface de son ouvrage, Doellinger avait indiqué son but :
susciter contre le courant ultramontain « un mouvement d'opinion parmi
les hommes professant une foi positive », en particulier parmi les laïcs.
Effectivement, dans les mois qui suivirent, tandis qu'une agitation se
développait en Autriche, notamment autour du canoniste Schulte, allant
jusqu'à envisager la possibilité d'une Église nationale indépendante de
Rome et faisant une plus large part à l'élément laïque, on vit paraître
en Allemagne divers manifestes destinés à faire pression sur l'épiscopat
en indisposant l'opinion publique à l'égard du concile, notamment
un *Appel aux catholiques de Bade*, répandu également en Bavière, qui
opposait la « vieille Église catholique » à « l'Église absolutiste de Rome » [3].
Même dans des milieux beaucoup moins hostiles à l'autorité ecclésiastique,
et en dépit des réfutations de Janus par des théologiens de la valeur d'Hergenroether ou Scheeben, on se déclarait inquiet. Quelques hommes politiques, qui devaient compter parmi les plus zélés défenseurs de l'Église
lors du *Kulturkampf*, tels Reichensperger, Windthorst, Joerg, firent
parvenir, eux aussi, une adresse, mais non publique, à l'épiscopat allemand : tout en se soumettant d'avance à toutes les décisions du concile,
ils estimaient inopportune la définition de l'infaillibilité et insistaient
surtout sur la nécessité de tenir compte de la condition présente de la
société civile dans la détermination des rapports entre l'Église et l'État [4].
Et le nonce à Munich devait reconnaître que parmi les ecclésiastiques
« que l'on désigne sous le nom de romanistes et qui en effet méritent bien

(1) Janus, *Der Papst und das Konzil*, Leipzig, 1869, p. ix.
(2) Extrait du journal de Loyson à la date du 30 mai 1869, cité par A. Houtin, *Le P. Hyacinthe
dans l'Église romaine*, p. 292.
(3) Texte dans E. Cecconi, *Histoire du concile du Vatican*, t. III, p. 312-325.
(4) Texte *ibid.*, t. III, p. 341-344.

cette qualification », il y en avait fort peu qui « témoignaient désirer la promulgation de nouveaux dogmes »[1]. Heinrich lui-même, le théologien de Mayence, considérait la définition de l'infaillibilité comme inopportune.

Parmi les évêques, quelques-uns étaient de chauds partisans de cette définition, Senestrey, de Ratisbonne, par exemple, mais la plupart se rendaient compte du danger qu'il y avait à vouloir brusquer les choses à présent que les esprits étaient excités à un tel point. Aussi, lorsqu'ils se réunirent en septembre à Fulda pour arrêter leur ligne de conduite, la majorité, après des discussions animées, auxquelles Ketteler et Hefele prirent une part importante, se déclara pour l'inopportunité d'une définition de l'infaillibilité. On n'en fit toutefois pas mention dans la lettre pastorale collective, qui fut rédigée sur un ton apaisant, mais quatorze évêques sur vingt adressèrent au pape une lettre secrète où ils lui faisaient part de leurs appréhensions[2]. Ce mémoire, assez mal reçu par Pie IX, fit à Rome une grosse impression.

LES POLÉMIQUES EN FRANCE — En France non plus, on le pense bien, l'article de la *Civiltà* n'était pas passé inaperçu. Une ardente polémique s'engagea entre *L'Univers* et la presse anticléricale et c'est pour tenter de l'apaiser que Dupanloup publia dans *Le Français*, les 18 et 19 mars, deux articles anonymes, dont l'auteur fut d'ailleurs aisément reconnu : il y protestait contre le manifeste et déclarait « inopportune » la définition de l'infaillibilité, sans d'ailleurs critiquer la doctrine elle-même, comme l'avait fait Doellinger.

Deux ouvrages publiés au cours des mois suivants vinrent encore activer la controverse. Mgr Dechamps, archevêque de Malines, qui, comme beaucoup de Belges, était à la fois ultramontain dans la question des droits du pape et libéral en ce qui concernait les relations entre l'Église et la société moderne, publia en juin une brochure sur l'*Infaillibilité et le concile général*, où il expliquait aux « gens du monde » qu'il ne s'agissait nullement de « déifier » la personne et *tous* les actes du pape, mais de proclamer l'autorité infaillible de son enseignement *solennel* ; il ajoutait qu'à son avis le moment était venu de définir cette vérité puisqu'elle était publiquement contestée.

Dans le camp opposé, Mgr Maret préparait depuis des années un gros travail destiné à défendre scientifiquement le point de vue gallican modéré qui était le sien, beaucoup plus nuancé que celui de Doellinger. Il en avait communiqué les épreuves à de nombreux évêques et théologiens amis et beaucoup d'entre eux le pressaient de publier le livre avant l'ouverture du concile. C'était une erreur, car il allait fournir aux adversaires du gallicanisme une raison supplémentaire d'insister sur la nécessité pour le concile de prendre position, comme d'autres évêques, aussi hostiles à l'ultramontanisme mais plus perspicaces, comme Lavigerie ou Meignan,

---

(1) Rapport de Mgr Meglia du 17 mars 1869, *ibid.*, t. II, p. 430-431.
(2) Les protocoles des réunions sont publiés dans *Coll. Lac.*, t. VII, col. 1188-1191 ; il faut les compléter par J. FRIEDRICH, *Gesch. des Vat. Conc.*, t. II, p. 174 et 190-191 (d'après des indications fournies par Hefele et l'archevêque de Munich) et O. PFUELF, *Ketteler* t. III, p. 22-28 (d'après les papiers de Ketteler). La lettre pastorale et la lettre à Pie IX, dans *Coll. lac.*, t. VII, col. 1191-1197.

par exemple, essayèrent vainement de le lui faire comprendre [1]. L'ouvrage parut en septembre sous le titre : *Du concile général et de la paix religieuse*. Partant de la constatation que la souveraineté dans l'Église n'est pas le fait d'une monarchie absolue, mais d'une monarchie *efficacement* tempérée d'aristocratie, Maret en concluait qu'elle ne pouvait résider dans le pape seul, agissant sans le concours du corps épiscopal : le pape y a une part prépondérante, mais non pas unique ; en conséquence, l'infaillibilité, attribut de la souveraineté spirituelle, ne pouvait pas être « personnelle et séparée », mais exigeait obligatoirement le consentement de l'épiscopat, que seul d'ailleurs le pape avait un moyen certain de provoquer, en convoquant un concile général. On devine l'effet du livre dans les milieux ultramontains. Si les bruits de mise à l'*Index* furent démentis, le nonce ne cacha pas le mécontentement de Rome et plusieurs évêques, notamment Mgr Pie, dans une homélie à son clergé qui fut très remarquée, désavouèrent leur collègue, en regrettant à juste titre qu'il n'ait pas réservé ses considérations à ceux-là seuls qui étaient appelés à trancher éventuellement la question au concile. Quant à *L'Univers*, il en prit prétexte pour renforcer encore ses attaques contre les adversaires d'une définition, quels qu'ils fussent.

Dans le groupe du *Correspondant*, les préventions contre une définition de l'infaillibilité étaient moindres qu'on ne le dit parfois : « Un concile ne peut la proclamer sans la définir et il ne peut la définir sans la limiter » constatait Falloux [2]. Même Montalembert, qui fulminait contre la « transformation si prompte et si complète de la France catholique en une basse-cour de l'anticaméra du Vatican » [3], expliquait : « Ce n'est pas l'infaillibilité du pape en matière de foi qui me répugne, à moi ; ce n'est que son omnipotence sur les questions d'un autre genre, politique, scientifique, etc., que certains fanatiques cherchent à ériger derrière son infaillibilité dogmatique [4]. » Quoi qu'il en soit, les dirigeants de la revue avaient jugé plus sage de se tenir en dehors de la mêlée, mais devant l'attitude toujours plus violente du journal de Veuillot, ils estimèrent nécessaire de prendre position. Dans le numéro du 10 octobre 1869 parut un article écrit au nom de la rédaction par Albert de Broglie sous l'inspiration de Dupanloup. Après avoir expliqué qu'il était impossible que la convocation des « États généraux » de l'Église aboutisse à la proclamation d'une monarchie despotique, on y exprimait l'espoir que le concile, au lieu de consacrer les propositions du *Syllabus*, qui, mal comprises, avaient excité les incroyants contre l'Église, s'attacherait plutôt à les expliquer ; et qu'il écarterait comme inopportune la définition de l'infaillibilité, entre autres pour le motif que celle-ci, une fois proclamée, « s'appliquerait à l'œuvre des papes antérieurs, même à des actes que n'admet plus le droit public moderne » : les *Dictatus papae* de Grégoire VII, la

---

(1) On complétera les lettres publiées par G. Bazin, *Vie de Mgr Maret*, t. III, chap. iii à v, par celles que publie H. Boissonot, *Le cardinal Meignan*, p. 257-258 en note et p. 278, n. 2.
(2) Falloux à Montalembert, 1868, cité dans E. Lecanuet, *Montalembert*, t. III, p. 438.
(3) Montalembert à Doellinger, 7 novembre 1869, dans A. Houtin, *Autour d'un prêtre marié*, p. 241-242.
(4) Témoignage de Mme de Montalembert, dans E. Lecanuet, *op. cit.*, t. III, p. 471.

bulle *Unam Sanctam*, la déposition par le pape de certains princes, tels étaient en effet les points qui préoccupaient surtout les catholiques français, tandis que les théologiens allemands agitaient plutôt les cas d'Honorius ou de Libère.

L'INTERVENTION DE DUPANLOUP   Tandis que toute la presse religieuse de France, d'Italie, d'Allemagne et d'Angleterre discutait ce manifeste des catholiques libéraux, Dupanloup jugea qu'il fallait faire un pas de plus et intervenir personnellement. Il y était poussé par diverses raisons. D'abord l'idée d'une définition brusquée, et donc vague, de l'infaillibilité, semblait prendre corps, car ce n'était plus seulement Veuillot qui raillait *Le Correspondant* de demander « qu'on discute et que le Saint-Esprit prenne le temps de se former une opinion »[1], mais l'évêque de Nîmes qui affirmait dans un mandement : « Il n'en coûte pas plus à l'Esprit Saint de préserver l'Église d'erreur dans le feu d'une acclamation que dans les conclusions d'un débat [2]. » D'autres prônaient une conception de l'infaillibilité qui paraissait inacceptable, ainsi Manning qui venait de proclamer le pape infaillible *apart from the bishops*, ce qui pouvait s'entendre soit « sans le concours des évêques », soit « en opposition possible avec le corps épiscopal » ; l'expression était pour le moins équivoque [3]. Puis il y avait les remous causés par l'apostasie du P. Hyacinthe Loyson, le célèbre conférencier de Notre Dame [4] : tandis que *L'Univers* triomphait devant ce qui lui semblait l'aboutissement normal de sympathies libérales, les libres-penseurs dénonçaient l'absolutisme de l'Église qui obligeait les nobles esprits à s'éloigner d'elle. Dupanloup était en outre excité par Acton, par Doellinger, à qui il avait rendu visite au début de septembre, par Freppel, qui lui envoyait de Rome des avis alarmants, écrivant entre autres :

> Le concile vient trop tôt ou trop tard. Trop tard, car il arrive vers la fin du pontificat d'un vieillard fatigué et découragé (...) qui voit toutes choses à travers les malheurs qu'il a essuyés : pour lui tout ce qui se passe dans le monde moderne est et doit être l'abomination de la désolation. Trop tôt, car il est évident que la situation de l'Europe n'est pas définitive (...). Je ne puis m'empêcher de dire que la *Compagnie* assume en ce moment une terrible responsabilité par les imprudences de quelques-uns de ses membres... [5].

C'est sous l'empire de ces sentiments que, après bien des hésitations et malgré l'avis contraire de plusieurs de ses conseillers, Dupanloup fit paraître, le 11 novembre 1869, des *Observations sur la controverse soulevée relativement à la définition de l'infaillibilité au futur concile*, dans lesquelles il rassemblait toutes les raisons de considérer comme *inopportune* la définition en question. La brochure, envoyée à tous les évêques, produisit une sensation considérable, non seulement à cause

(1) *L'Univers*, 7 novembre 1869 (voir aussi le n° du 21 novembre 1869).
(2) Cité dans F. LAGRANGE, *Vie de Mgr Dupanloup*, t. III, p. 137.
(3) Texte dans E. MANNING, *Petri Privilegium. Three pastoral letters*, Londres, 1871. Un premier projet, bien plus extrémiste, avait été amendé à la suite des observations de Mgr Ullathorne et d'autres. (C. BUTLER, *Life and Times of Bishop Ullathorne*, t. II, p. 51.)
(4) Voir les détails et les documents dans A. HOUTIN, *Le P. Hyacinthe dans l'Église romaine*, Paris, 1920 ; *Le P. Hyacinthe, réformateur catholique*, Paris, 1922.
(5) Freppel à Lagrange, 15 juin 1869, dans E. TERRIEN, *Mgr Freppel*, t. I, p. 561-564.

du nom de son auteur et de la dépendance manifeste à l'égard d'une autre brochure, bien plus radicale, attribuée à Doellinger, mais surtout parce que c'était la première fois qu'un *évêque* présentait *publiquement* la définition comme inopportune. A Rome, où les prélats commençaient à affluer, plusieurs, hostiles pourtant à la définition, regrettèrent cet appel à l'opinion publique et se montrèrent froissés « d'une conduite qui semblait vouloir engager la leur »[1]. Pour comble de malheur, Veuillot ayant pris l'évêque vertement à partie et discuté ses arguments avec sa verve coutumière[2], le prélat, perdant tout sang-froid, riposta par un véhément *Avertissement à M. Veuillot* qui fit encore plus mauvaise impression. Le duc de Broglie caractérisait très justement la situation en disant à son fils : « Il a tort ; l'évêque prend position avant la réunion, il arrivera avec une apparence de parti-pris qui lui ôtera de l'autorité[3]. » Et M. Icard constatait l'exactitude de ce pronostic dès son arrivée à Rome : « Quant à l'évêque d'Orléans, c'est une idée générale qu'il s'est compromis et qu'il a perdu la position qu'il aurait eue au concile[4]. »

L'ATTITUDE DES GOUVERNEMENTS Tandis que l'opinion religieuse s'agitait, les gouvernements appréhendaient les décisions éventuelles du concile relatives au mariage civil, à l'enseignement laïque, aux libertés constitutionnelles. Ils s'inquiétaient des « idées archaïques dans lesquelles s'enlisaient en politique certains infaillibilistes »[5] et craignaient qu'on ne veuille définir la doctrine du « pouvoir indirect » de l'Église sur le pouvoir civil, voire les conceptions beaucoup plus extrêmes de la bulle *Unam Sanctam* ; en entendant exalter la Chrétienté médiévale, ils évoquaient l'humiliation de l'empereur Henri IV à Canossa ou le spectre de Charlemagne imposant par les armes le baptême aux Saxons. Tous ceux qui redoutaient le triomphe, au concile, du parti ultramontain s'employaient à attiser ces méfiances et ces appréhensions, Maret, Darboy et leurs amis auprès du gouvernement impérial, Doellinger auprès du président du conseil bavarois, le prince de Hohenlohe[6].

Ce dernier, à la suite de la publication du fameux article de *La Civiltà cattolica*, adressa, le 9 avril, à tous les gouvernements européens une circulaire, rédigée par le professeur de Munich, afin de les mettre en garde contre les projets du prochain concile, qui devaient aboutir à une sujétion totale du pouvoir civil au pape, et de les inviter à se réunir en une conférence internationale pour se concerter et élever à l'avance une protestation commune. Mais son initiative rencontra peu de succès[7].

---

(1) Dépêche du ministre de Belgique, 27 novembre 1869 (*Arch. Min. A. É. Bruxelles*, Saint-Siège, t. XIII, 2) ; cf. H. ICARD, *Journal de mon voyage à Rome*, p. 1 et 9-10.
(2) *L'Univers*, 18 novembre 1869.
(3) A. DE BROGLIE, *Mémoires*, t. I, p. 356.
(4) H. ICARD, *Journal*, p. 25 (5 décembre). Il « s'est noyé », disait avec joie Mgr Baillès. (*Ibid.*, p. 10.)
(5) G. GOYAU, *L'Allemagne religieuse*, t. IV, p. 332.
(6) On peut suivre ces tractations dans le t. III de la *Vie de Mgr Maret* de G. Bazin et dans les *Denkwürdigkeiten* du prince de Hohenlohe, édit. F. CURTIUS, t. I, Berlin, 1907.
(7) Outre T. GRANDERATH, *op. cit.*, t. I, p. 425-462, et É. OLLIVIER, *L'Église et l'État...*, t. I, chap. V, voir quelques pages très claires de J. ACTON dans *North British Review*, t. XIV, 1870, p. 191-197.

L'Espagne, le Portugal, la Belgique étaient décidés à garder la plus stricte neutralité. En Italie, le gouvernement Menabrea et surtout Minghetti désiraient produire une bonne impression sur les évêques en vue de préparer un rapprochement et ils déclarèrent qu'ils entendaient laisser pleine liberté au concile ; ce n'est qu'en novembre que les dispositions devaient changer, avec le nouveau gouvernement anticlérical. Bismarck, qui ne se souciait pas de mécontenter les catholiques prussiens à la veille d'une guerre qu'il voyait venir avec la France, déclara qu'une intervention dans les affaires de l'Église catholique se rattachait à un état de choses à jamais disparu. Enfin et surtout l'Autriche jugea le projet chimérique et juste bon à pousser Rome à bout et à indisposer les évêques qui comptaient résister. Autriche et Prusse étaient toutefois décidées à suivre de près les événements de Rome et à intervenir énergiquement si le concile empiétait sur les droits de l'État ; il semble qu'on en tint un certain compte à Rome.

Quant à la France, dont l'attitude avait une importance spéciale, puisque la sécurité du concile dépendait de la présence de ses troupes, elle se contenta d'une réponse évasive, car son gouvernement était indécis sur la ligne de conduite à suivre. On envisagea un moment d'envoyer un ambassadeur spécial au concile. Finalement, on décida de ne pas intervenir. Tel était en effet le sens des instructions envoyées le 19 octobre 1869 à l'ambassadeur à Rome [1], comme le note judicieusement Émile Ollivier, qui avait brillamment défendu cette politique à la Chambre : « Toute politique qui n'envoyait pas un ambassadeur extraordinaire au concile en mettant dans son portefeuille l'ordre de retirer nos troupes du territoire romain si l'on méprisait nos conseils était, de quelque nom qu'on la couvrît, une politique d'abstention [2]. »

Bref, on peut dire que jamais encore concile ne s'était réuni avec une certitude aussi grande de ne subir aucune pression de la part des pouvoirs civils tant qu'il resterait sur le terrain religieux.

§ 2. — Les premiers débats conciliaires [3].

LA QUESTION DU RÈGLEMENT Les controverses qui agitaient l'Allemagne, la France et l'Angleterre depuis plusieurs mois étaient l'indice d'un malaise qui n'existait pas seulement dans l'opinion publique, mais parmi les évêques, bien que, à part Dupanloup, ceux-ci eussent évité de le manifester publiquement. Beaucoup, même parmi ceux qui n'avaient aucune sympathie pour les idées libérales ou gallicanes, arrivaient à Rome avec certaines appréhensions : plus d'un craignait que l'on ne veuille aller trop vite et trop loin, en définis-

---

(1) Texte dans É. OLLIVIER, *op. cit.*, t. I, p. 519-530. Sur la politique du gouvernement français, outre Ollivier, voir J. MAURAIN, *La politique ecclésiastique du Second Empire*, p. 884-890.

(2) É. OLLIVIER, *op. cit.*, t. I, p. 530.

(3) BIBLIOGRAPHIE. — Aux sources et travaux mentionnés en tête du chapitre, p. 311, n. 1, ajouter A. VACANT, *Études théologiques sur les constitutions du concile du Vatican*, Paris, 1895, 2 vol. ; L. ORBAN, *Theologia Güntheriana et Concilium Vaticanum* (*Analecta Gregoriana*, XXVIII et L), Rome, 1942-1950, 2 vol., et, pour le chapitre III en particulier, R. AUBERT, *Le problème de l'acte de foi*, Louvain, 1945, p. 131-219.

sant par acclamation le *Syllabus* ou l'infaillibilité, sans se préoccuper d'en préciser le sens et la portée ; d'autres avaient l'impression qu'on leur demandait moins de venir délibérer que confirmer le travail des théologiens romains et jouer au fond un simple rôle d'apparat dans un scénario fixé d'avance par la Curie. Ces craintes et ces impressions étaient exagérées, mais elles parurent confirmées dès avant l'ouverture du concile lorsque, le 2 décembre, le règlement, contenu dans la bulle *Mulliplices*[1], fut communiqué aux évêques déjà arrivés à Rome.

Dans le but d'éviter les excès d'indépendance et surtout des discussions interminables, Pie IX avait décidé de rompre avec les usages des conciles antérieurs, qui avaient établi eux-mêmes leur règlement, et de fixer d'autorité la procédure à suivre. Or, ce règlement imposé d'en haut avait un caractère très centralisateur et restreignait la liberté d'initiative des Pères. Seul le pape pouvait proposer une question au concile, les évêques ne pouvant que présenter des vœux à une commission *de Poslulalis*, nommée par le pape, qui lui ferait rapport. Les projets de décret, élaborés en secret par des théologiens désignés eux aussi par le Saint-Siège, ne seraient remis aux Pères que quelques jours avant l'examen. Celui-ci devait avoir lieu en *Congrégations générales*, où l'on n'avait pas le droit d'intervenir spontanément dans la discussion et, s'il y avait lieu de modifier profondément le projet, il était renvoyé, non pas, comme à Trente, à une commission de spécialistes désignée par les Pères pour cet objet précis, mais à l'une des quatre *dépulations permanentes* de vingt-quatre membres (de la foi, de la discipline, des religieux et des rites orientaux). Les projets une fois amendés et réexaminés en congrégation générale devaient ensuite être approuvés en session solennelle et promulgués par le pape. Tous les débats étaient soumis au secret pontifical.

Comparé à celui de Trente, ce règlement rendait effectivement sensibles les progrès de l'ultramontanisme au cours de l'époque moderne. Il était cependant, pour l'essentiel, l'œuvre d'un théologien qui était loin d'appartenir au parti ultramontain, Hefele de Tubingue. Celui-ci avait été mû par la préoccupation de permettre un travail rapide et il s'était souvenu des débats aigres et prolongés auxquels la détermination des questions de personnel et de procédure avait donné lieu à Trente. Mais un certain nombre de Pères estimèrent que leur liberté était par trop limitée et, tout en estimant préférable de ne pas discuter le principe même d'un « règlement octroyé », ils s'empressèrent de suggérer des amendements[2], proposant notamment de s'inspirer davantage des méthodes de travail parlementaire en organisant des bureaux et des sous-commissions. Pie IX ne voulut pas modifier le texte primitif, mais il fit savoir officieusement que les évêques pourraient travailler par groupes en dehors des séances officielles et que, tout en désirant « mettre de l'ordre dès le commencement », il entendait bien que les Pères fussent parfaitement libres d'exposer leurs désirs et leurs opinions.

---

(1) Texte dans Mansi, *A. C. C.*, t. 50, col. 215*-222*.
(2) Une centaine d'évêques signèrent des pétitions, remises le 12 décembre et le 2 janvier d'autres préférèrent intervenir oralement auprès des présidents.

L'OUVERTURE DU CONCILE Le 8 décembre 1869, dans le transept droit de la basilique Saint-Pierre aménagé en salle des séances, au cours d'une cérémonie de près de 7 heures présidée par le pape et à laquelle assistaient vingt mille pèlerins étrangers, lecture fut donnée du décret d'ouverture du concile. Environ 700 Pères étaient présents [1], venus de tous les coins de l'univers, soit 70 pour cent de l'épiscopat catholique. L'Europe était abondamment représentée : plus de 200 Italiens, cardinaux, évêques et religieux, 70 archevêques et évêques de France, 40 d'Autriche-Hongrie, 36 d'Espagne, 19 d'Irlande, 18 d'Allemagne, 12 d'Angleterre, 19 de pays moins importants ; mais il y avait aussi près de 50 prélats de rite oriental, 40 évêques des États-Unis, 9 du Canada, 30 de l'Amérique latine et près d'une centaine d'évêques missionnaires. Pour la première fois, on entendait dans un concile œcuménique appeler des noms comme New-York, San Francisco, Toronto, Manille, Buenos-Ayres, Melbourne, Wellington, et l'on remarquait l'importance prise par l'Empire britannique — plus de 120 évêques — alors qu'à Trente il n'y avait que quatre évêques de langue anglaise.

Chacun de ces groupes nationaux avait ses caractères propres : les Italiens et les Espagnols se distinguèrent vite par leur habileté à manier les concepts scolastiques ; les Allemands par leur solide connaissance de l'antiquité chrétienne ; les Américains par leur tournure d'esprit démocratique ; les Belges par leurs tendances conciliatrices ; certains Austro-Hongrois par leurs allures de grands seigneurs, arrivant au concile en grand équipage, escortés de hussards, tandis que les évêques pauvres, logés aux frais du Saint-Père, circulaient à pied, même par temps de pluie. Les Français, le groupe le plus nombreux après les Italiens, tranchaient par leur brillant et en tiraient avantage en dehors de l'enceinte du concile, « dans les salons et les chaires de vérité (...) : ce terrain leur étant plus familier, leur assure aussi dans la société une faveur que leurs collègues de l'épiscopat n'atteignent pas » [2] ; la chose ne manquait pas d'importance, étant donné le rôle joué dans l'agitation extra-conciliaire par les membres de l'aristocratie italienne et européenne en séjour à Rome, et notamment par les grandes dames infaillibilistes ou anti-infaillibilistes que les Romains avaient surnommées les « Mères de l'Église » et Louis Veuillot, plus irrévérencieusement, les « commères du concile ».

En dépit de l'œcuménicité indiscutable de l'assemblée, la place tenue par les Italiens constituait une anomalie qui fut souvent relevée. Elle était le résultat de circonstances historiques qui avaient multiplié les diocèses dans l'Italie méridionale et centrale et donné aux Italiens une place prépondérante à la Curie. Non seulement les prélats italiens formaient à eux seuls 40 pour cent de l'épiscopat européen, mais les deux tiers des consulteurs et tous les secrétaires étaient Italiens, de même que

---

(1) Sur la difficulté de déterminer le nombre exact, cf. J. Fessler, *Das vatikanisches Konzilium*, 2e édit., Vienne, 1871, p. 13 et suiv. Au total, 744 Pères parurent au concile, dont 49 cardinaux, 10 patriarches, 127 archevêques, 529 évêques, 22 abbés, 26 généraux d'ordres ; 270 absents avaient été retenus par l'âge ou la maladie ou encore, comme les évêques de Russie, empêchés par leur gouvernement.

(2) Dépêche du ministre de Belgique, 19 janvier 1870, *Arch. Min. A. É. Bruxelles*, Saint-Siège, t. XIII, 2.

tous les présidents de commissions ; sur 48 personnes à remplir des charges auprès du concile, cinq seulement étaient étrangères. Conscient de cet inconvénient, le pape avait désigné le savant canoniste et historien autrichien Fessler comme secrétaire du concile et le cardinal de Reisach comme premier président. Malheureusement, ce dernier mourut quelques jours après l'ouverture du concile et les présidences furent dès lors toutes exercées également par des Italiens, relativements modérés du reste : le cardinal de Angelis, prélat énergique, qui avait été emprisonné à deux reprises, en 1849 par la République romaine et en 1860 par le gouvernement italien, « ce qui ne lui avait pas donné le goût des idées libérales »[1] ; le cardinal de Luca, esprit large et porté à la conciliation ; le cardinal Bizzarri, très en vue à la Curie par sa charge de préfet de la Congrégation des évêques et réguliers ; le cardinal Bilio, homme d'études, étranger aux intrigues et plutôt modéré malgré le rôle joué dans la composition du *Syllabus* ; enfin le bouillant cardinal Capalti, qui n'était pas non plus un extrémiste, mais dont les interventions faites sur un ton brusque et roide contribuèrent plus d'une fois à mettre de l'électricité dans l'air.

LA CONSTITUTION DES GROUPES   A la répartition nationale se superposa très vite une division idéologique. « Nous avons déjà notre droite et notre gauche », écrivait dès le 8 décembre l'évêque de Nancy[2], et cette division, qui paraissait d'abord surtout sensible dans l'épiscopat français, ne tarda pas à se généraliser. Il y avait d'une part les évêques qui souhaitaient que le concile définisse l'infaillibilité du pape et ne redoutaient pas de voir rappeler les principes qui devaient régir la société chrétienne idéale ; d'autre part, ceux qui croyaient que, par de pareils projets, « l'Église et la société (étaient) menacées, l'une dans sa constitution intime, l'autre dans ses institutions les plus chères et ses aspirations les plus légitimes »[3] ; entre les deux, un groupe important de modérés qui, se rendant compte qu'il était vain d'essayer d'écarter certaines questions du concile, se bornaient à demander qu'on évitât de « casser les carreaux » à propos des rapports entre l'Église et l'État, et que l'on ne définît l'infaillibilité qu'en en précisant bien les limites et en ajoutant aussi peu que possible à l'enseignement des théologiens anciens. C'est parmi ces derniers qu'il fut à plusieurs reprises question de constituer un « tiers-parti » : au début, des infaillibilistes modérés, comme l'autrichien Fessler et surtout l'archevêque de Baltimore Spalding, firent quelques tentatives pour l'organiser, mais il se composa finalement surtout d'un groupe de Français réunis autour du cardinal de Bonnechose, un ancien magistrat qui avait su, comme archevêque de Rouen, rester en bons termes avec l'empereur aussi bien qu'avec l'opposition cléricale et qui à Rome espéra

(1) É. OLLIVIER, *L'Église et l'État au concile du Vatican*, t. II, p. 16.
(2) Mgr Foulon à l'abbé Tapie, 8 décembre 1869, dans *Les Lettres*, mai 1928, p. 23.
(3) Ces lignes de Mgr de Las Cases (lettre à son frère, 11 mai 1870, *Arch. Min. A. É. Paris*, Rome, 1046, f⁰ 350) expriment bien la double préoccupation de la plupart des membres de la minorité. Il ne faut jamais oublier, pour comprendre l'atmosphère du concile du Vatican, que l'ombre du *Syllabus* planait tout autant que celle de l'infaillibilité.

vainement pendant plusieurs mois arriver à trouver un terrain d'entente entre les deux partis extrêmes [1].

LA MAJORITÉ   Le premier, que l'on désigna bientôt comme la « majorité », comprenait d'ailleurs aussi de nombreux modérés qui n'avaient jamais partagé les outrances du néo-ultramontanisme. C'était, par exemple, le cas de celui qui apparut à plusieurs reprises comme le chef réel du groupe infaillibiliste, Mgr Dechamps, archevêque de Malines, dont on connaît les sympathies pour le libéralisme catholique et qui, après s'être beaucoup dépensé pour que l'infaillibilité fût mise à l'ordre du jour, se montra prêt à de larges concessions quant aux termes de la définition. De même Mgr Ullathorne, l'un des plus écoutés parmi les évêques anglais, qui fit toujours montre de sentiments conciliateurs, ou le général des jésuites, le P. Beckx, qui avait du reste comme théologien le P. de Buck, un ami de Dupanloup [2] ; de même encore Mgr Pie, farouchement antilibéral mais qui, dans la question de l'infaillibilité, se montra plus réservé qu'on ne le dit souvent.

Le noyau le plus bruyant de la majorité était formé par quelques évêques français très ardents, tels Berteaud, Plantier ou Mabile, auxquels se rattachait Mermillod de Genève. Et les deux agents les plus actifs du groupe étaient des extrémistes : Senestrey, de Ratisbonne, un ancien *Germaniker*, et surtout Manning, qui sut mettre au service de son zèle pour la foi un talent diplomatique qui confinait parfois à l'esprit d'intrigue, « usant de sa langue aussi activement que Dupanloup de sa plume » [3] afin de gagner des votes pour une définition qu'il s'était engagé par vœu à promouvoir inconditionnellement et qu'il avait commencé à préparer méthodiquement plusieurs mois avant l'ouverture du concile [4].

La force de la majorité était constituée par les évêques des contrées traditionnellement catholiques : Italie, Espagne, Irlande, Amérique latine, mais on y trouvait également des prélats en contact avec le protestantisme, comme les Hollandais, les Suisses, beaucoup d'Anglais, Mgr Martin de Paderborn, ou aux prises avec les idées libérales, comme les Belges ; il y avait également quelques orientaux, formés à la Propagande. La plupart de ces évêques, ralliés depuis longtemps aux positions de

---

(1) Voir J.-L. SPALDING, *Life of archbishop Spalding*, chap. xxx et surtout R. AUBERT, *Documents concernant le Tiers Parti au concile du Vatican*, dans *Festschrift Karl Adam* (cité *supra*, p. 311, n. 1).

(2) Les adversaires du concile, Friedrich notamment, ont beaucoup exagéré le rôle des jésuites. Les tendances extrémistes du groupe italien de *La Civiltà* ou du P. Schrader n'étaient pas partagées par l'ensemble de la Compagnie, même en Italie : c'est même un jésuite et de surcroît préfet des études au Collège germanique, le P. Quarella, qui conseilla Ketteler dans son opposition à la définition et rédigea la *Quaestio* diffusée par celui-ci (cf. *infra*, p. 341). Voir d'ailleurs les notations recueillies par F. MOURRET, *Le concile du Vatican*, p. 67, 192, 221-222, 276, n. 2, corroborées par l'ambassadeur de France : « Le R. P. Beckx lui-même s'avoue effrayé des emportements auxquels on cède au Vatican. » (*Arch. Min. A. É. Paris*, Rome, 1046, fo 165 vo.)

(3) E. PURCELL, *Life of Manning*, t. II, p. 430. L'extrémisme de Manning déplaisait à la plupart des évêques anglais, même infaillibilistes ; il est souvent relevé dans ses lettres par Ullathorne qui le traitait de « fanatique ».

(4) Sur le vœu fait de concert avec Senestrey le 28 juin 1867, voir le récit de Manning lui-même dans E. PURCELL, *Life of Manning*, t. II, p. 420 ; sur les préparatifs plus immédiats, voir entre autres la lettre de Moufang à Ketteler du 29 janvier 1869 dans O. PFUELF, *Bischof Ketteler*, t. III, p. 13.

l'ultramontanisme modéré et venant de contrées où les controverses doctrinales étaient beaucoup moins vives qu'en France ou en Allemagne, ne voyaient guère d'inconvénients à ce que le concile reconnaisse solennellement l'infaillibilité pontificale, qui était, du moins en pratique, admise par l'ensemble de leurs fidèles et de leur clergé. Habitués à considérer cette doctrine comme traditionnelle, beaucoup d'entre eux, sans en avoir une conception aussi extensive que Manning et sans approuver toujours toutes les mesures centralisatrices de la Curie, trouvaient même normal de profiter de l'occasion du concile pour mettre un terme définitif à des controverses qui recommençaient à se développer dans certains milieux.

LA MINORITÉ Le second parti, désigné sous le nom d' « opposition » ou, plus souvent, de « minorité », présentait moins d'homogénéité. Un premier groupe était formé par la plupart des évêques allemands et la quasi-totalité des austro-hongrois. Il était présidé par l'archevêque de Prague, cardinal de Schwarzenberg, grand protecteur des *Deutsche Theologen*, dont le théologien, Friedrich, était un élève de Doellinger, à défaut du vieux maître lui-même qui s'y était refusé ; et par le savant cardinal-archevêque de Vienne, Rauscher, dont le prestige dépassait le groupe germanique et que Schulte considère à juste titre comme « le directeur spirituel de la minorité », car il exerça à plusieurs reprises sur celle-ci une influence modératrice inspirée par le respect dont il ne se départit jamais à l'égard du Saint-Siège, malgré les doutes que son immense érudition patristique lui inspirait à l'égard de la doctrine de l'infaillibilité personnelle du souverain pontife.

Parmi les membres les plus actifs de la minorité germanique, on comptait encore le primat de Hongrie, Simor, qui croyait à l'infaillibilité, mais craignait que sa définition n'envenimât les rapports, jusque-là très favorables, de l'Église et de l'État en Hongrie ; les Allemands Hefele et Ketteler, le premier combattant l'infaillibilité au nom de l'histoire ecclésiastique, le second partisan d'un épiscopalisme modéré conforme à son idéal corporatif médiéval ; le Hongrois Haynald, un libéral lié depuis longtemps avec Dupanloup ; enfin et surtout « l'agitateur » du groupe, le Croate Strossmayer, très opposé à la centralisation romaine dans laquelle il voyait le principal obstacle au retour des Slaves orthodoxes à l'Église catholique et qui ne craignait pas d'exposer ses critiques sans ménagement, au grand scandale de la majorité [1]. Ce ne fut pas l'une des moindres surprises des débuts du concile que de voir se ranger ainsi dans l'opposition des hommes comme les archevêques de Vienne et de Cologne et l'évêque de Mayence, qu'on s'était habitué à considérer comme les protecteurs des jésuites et des agents zélés de l'influence romaine en pays germanique.

On s'étonna beaucoup moins de voir se constituer un groupe d'opposants français. Présidé, sans beaucoup de mordant, par le cardinal Mathieu, animé par la fougue du bouillant évêque d'Orléans, mais dirigé en fait, avec

---

(1) Comme il s'opposait aux Hongrois en politique, il se rapprochait volontiers du groupe français. Sur cette figure marquante du concile, cf. *infra*, p. 409-410.

circonspection et perspicacité, par Darboy [1], il comptait un bon tiers des évêques français. Leurs motifs étaient assez divers. Quelques-uns justifiaient l'appréciation sévère du ministre de Belgique à Rome : « Ils se considèrent plutôt comme des fonctionnaires de l'empire que comme des dignitaires ecclésiastiques [2]. » Mais la plupart agissaient pour des raisons de conscience très nobles, que leurs adversaires n'ont souvent pas voulu reconnaître [3]. Certains, venus du libéralisme catholique, admettaient l'infaillibilité, mais redoutaient de voir le concile heurter l'opinion par des décisions peu nuancées dans le genre du *Syllabus* et consommer ainsi la rupture entre l'Église et la société moderne. Tout en partageant ces craintes, d'autres, plus nombreux qu'on ne l'a dit, hésitaient en outre à reconnaître comme révélée l'infaillibilité *personnelle* du souverain pontife, indépendamment d'une ratification par l'épiscopat ; leur état d'esprit est exactement caractérisé par ces mots de l'un d'eux :

> Qu'on n'aille donc pas dire de moi que j'étais seulement *inopportuniste* comme on l'a dit de certains de mes vénérables collègues de la minorité, qui n'auraient certainement pas accepté pendant leur vie cette interprétation (...). Élevé dans les doctrines anti-infaillibilistes, pendant le concile j'imposai, non sans douleur, silence à mon cœur, qui souffrait de ne point être sur ce point d'accord avec S. S. Pie IX, pour laisser parler ma conscience ; elle me faisait un devoir rigoureux de défendre ce que je croyais alors être la vérité [4].

Minorité germanique et minorité française étaient mues en fait par des motifs assez semblables : crainte de voir compromises les relations de l'Église avec les États modernes par des décisions qui renforceraient l'autorité de documents comme la bulle *Unam Sanctam* ; mentalité théologique trop centrée sur les sources et peu sensible à l'évolution du dogme ; compréhension erronée de la doctrine classique de l'infaillibilité personnelle par suite des exagérations des néo-ultramontains dont certains allaient jusqu'à opposer nettement le pape à l'Église et à envisager le cas où le pape demeurerait seul du côté de la vérité ; enfin et peut-être surtout mpression de voir l'autorité divine de l'épiscopat en danger — ce qu'un évêque anglais exprimait en disant que les évêques, arrivés au concile comme princes de l'Église, risquaient d'en repartir « comme satrapes d'un autocrate » et ce qui pouvait sembler justifié par le fait qu'il n'était pas fait mention de l'épiscopat dans le schéma sur l'Église [5] et par des paroles malheureuses comme celles d'un prélat romain disant à l'évêque de

---

(1) L'évêque d'Autun avouait qu'après avoir eu des préventions contre Darboy pendant les trois premiers mois, il avait changé d'avis « depuis qu'il l'avait vu de plus près et qu'il avait remarqué la mesure, la prudence de ce prélat et son sincère amour pour l'Église » (H. Icard, *Journal*, p. 361) ; et Mgr de Las Cases écrivait à l'archevêque, le 4 juillet 1870 : « Votre puissance a été non pas dans l'imposition de votre volonté, mais dans une direction que vous nous donniez en faisant comme si vous suiviez celle que nous vous donnions. » (*Archives archevêché Paris*, fonds Darboy, Casier 14 (11).)

(2) Dépêche du 25 mai 1870. (*Arch. Min. A. É. Bruxelles*, Saint-Siège, t. XIII, 2.)

(3) Cette remarque vaut tout autant pour les opposants des autres groupes et également pour les membres de la majorité. Les historiens sont trop tentés de s'arrêter aux excès par lesquels s'extériorisa souvent, sous l'effet de passions humaines inévitables, un zèle dont les racines étaient profondément religieuses (voir à ce propos la lettre du P. Whitty à Newman, du 15 mai 1870, dans C. Butler, *The Vatican Council*, t. II, p. 31).

(4) Testament de Mgr de Las Cases, dans *Coll. Lac.*, t. VII, col. 1002, n. 1.

(5) Mgr Pie lui-même reconnaissait que le schéma était « répréhensible » par suite de cette « lacune fâcheuse ». (H. Icard, *Journal*, p. 188.)

Metz : « Je suis bien aise que les évêques viennent à Rome pour voir que le pape est tout et que les évêques ne sont rien [1]. » Pourtant, si les raisons profondes de l'opposition étaient partout les mêmes, l'accent était placé différemment chez les uns et chez les autres. Dans le groupe germanique, l'opposition s'appuyait avant tout sur la science théologique, représentée par des hommes comme Rauscher et Hefele. Dans le groupe français, plutôt sur l'hostilité aux tendances absolutistes ; c'était l'avis non seulement de Mgr Gay : « Croyez bien qu'on redoute moins encore l'infaillibilité déclarée du souverain pontife que la condamnation des idées libérales » [2], mais aussi du comte Daru, l'ami de Dulanloup : « Au fond, le débat, quelle que soit la forme qu'il pourra prendre, revient à cette question : l'absolutisme triomphera-t-il à Rome [3] ? » Par suite de cette différence de préoccupations et aussi de leur chauvinisme, les deux groupes avaient relativement peu de contact entre eux, et leur caractère national très accusé les empêchait de devenir des points de cristallisation pour les évêques d'autres pays qui sympathisaient avec leur point de vue.

Il se constitua toutefois quelques groupes mineurs, gravitant autour du groupe français : quelques évêques d'Italie du Nord, dirigés par Mgr Nazari, archevêque de Milan : leur l'opposition s'expliquait par la formation joséphiste reçue dans les Facultés de théologie de Turin ou de Milan et par leurs sympathies pour le mouvement national italien ; une partie des évêques américains, que leur esprit libéral rapprochait des évêques de France et d'Italie du Nord et dont plusieurs, anciens élèves des Sulpiciens ou même Français d'origine, n'avaient aucune sympathie pour l'ultramontanisme extrême : leur représentant le plus marquant était Mgr Kenrick, archevêque de Saint-Louis, d'idées nettement gallicanes [4] ; enfin, beaucoup d'évêques orientaux, très émus par la récente bulle *Reversurus* [5] et redoutant de voir le concile accentuer encore la latinisation de leurs Églises : leur ignorance du latin, leur manque de culture théologique, l'influence du gouvernement français sur les chrétiens de l'empire turc, en faisaient des instruments faciles à manœuvrer.

C'est pour essayer d'unifier ces différents noyaux que, sur l'initiative de Mgr Haynald, très actif et habitué aux méthodes parlementaires, fut constitué avec l'appui enthousiaste de Dupanloup, qui devint vite le polémiste attitré de la minorité, un « comité international », présidé par le cardinal Rauscher. Mais l'action effective de ce comité ne répondit pas à l'attente de ses fondateurs : lord Acton regrettait qu'il n'ait constitué « que le cerveau et non pas la volonté de l'opposition » [6].

(1) H. ICARD, *Journal*, p. 256-257.
(2) Lettre du 16 mars 1870 dans *Correspondance de Mgr Gay*, t. II, p. 129-130.
(3) Lettre à Banneville, du 1er février 1870, *Arch. Min. A. É. Paris*, Rome, 1045, f° 257.
(4) Voir l'*Oratio habenda non habita*, qu'il comptait prononcer au concile, dans J. FRIEDRICH, *Documenta ad concilium vaticanum spectantia*, t. I, p. 187-246, ainsi que J. GIBBONS, *Retrospect of Fifty Years*, t. I ,p. 32.
(5) Cf. *infra*, p. 417-419.
(6) J. ACTON dans *North British Review*, t. LIII, 1870, p. 214.

ÉLECTION DE LA DÉPUTATION DE LA FOI

Si le sentiment de malaise qui régnait dès l'ouverture du concile se concrétisa si vite en une opposition organisée, la re sponsabilité en revient en partie à certains malentendus et surtout à une manœuvre très regrettable des dirigeants du groupe infaillibiliste.

Deux actes pontificaux, qui furent communiqués aux Pères les 10 et 14 décembre, froissèrent un certain nombre d'entre eux. Dans la constitution apostolique réglant l'élection pontificale en cas de vacance du Saint-Siège durant le concile, ils virent, bien à tort, une marque de défiance à leur égard. Et dans la bulle *Apostolicae Sedis*, qui remaniait profondément la législation des censures sans avoir pris l'avis de l'épiscopat dont la présence à Rome était imminente, ils virent, surtout à la suite des commentaires d'une certaine presse, un nouvel indice de la volonté du pape d'agir en maître absolu.

Par ailleurs, dès les premières séances, il était apparu que la salle du concile, « bien choisie et parfaitement organisée pour que les fidèles pussent jouir des splendides cérémonies observées dans les sessions publiques, manquait de toutes les conditions nécessaires pour être un lieu de discussion »[1] à cause de sa mauvaise acoustique. Il n'en fallut pas plus pour renforcer l'impression que, dans l'idée des organisateurs, les séances seraient de pur apparat, tout étant déjà réglé d'avance dans les commissions secrètes. Bien plus, cela sembla confirmer que les décisions auraient lieu par acclamation et plusieurs Pères, y compris Rauscher, prirent très au sérieux, durant les premières semaines, les bruits qui couraient avec insistance d'une définition par surprise de l'infaillibilité du pape[2].

Mais ce qui indisposa le plus, ce fut la manière dont fut élue la commission dogmatique, dite *députation de la foi*[3]. Deux sortes de « comités électoraux » s'étaient constitués, l'un autour de Dupanloup, l'autre autour de Manning. Or, ce dernier, au cours des réunions qui se tenaient chez Mgr Dechamps, insista pour qu'on exclue de la députation dont relèverait la question de l'infaillibilité tous les Pères opposés à la définition de cette vérité. Pareille intransigeance fut désapprouvée par Mgr Pie, par le cardinal Corsi, par Mgr Valerga, un familier de Pie IX, et par le pape lui-même, qui avait eu soin de nommer dans la commission *de Postulatis* trois anti-infaillibilistes et le cardinal de Bonnechose. Manning fut inflexible : « Les hérétiques, répondit-il, viennent à un concile pour être entendus et condamnés, non pour prendre part à la formulation de la doctrine. » Une liste rédigée par ses soins et approuvée par l'un des présidents fut alors distribuée parmi les Pères dont beaucoup, n'ayant encore aucune opinion personnelle sur les membres de l'assemblée, votèrent de confiance, le 14 décembre, pour les 24 candidats ainsi présentés. Cette

(1) Mgr COLLET, *Souvenirs du concile du Vatican* (inédits).
(2) Voir H. ICARD, *Journal*, p. 62 et 66 ; J. FRIEDRICH, *Tagebuch*, p. 13, 41, 45 ; dépêche de l'ambassadeur d'Autriche, copie aux *Arch. Min. A. É. Paris*, Rome, 1045, f° 60 ; et les *Souvenirs* de Mgr Collet.
(3) On peut suivre les faits, d'après le journal de Senestrey (*Coll. Lac.*, t. VII, col. 1646), dans T. GRANDERATH, *op. cit.*, t. II, p. 83-95 (apologétique) et C. BUTLER, *The Vatican Council*, t. I, p. 167-175 (sévère).

façon de procéder, qui excluait de la députation de la foi des hommes aussi
compétents que Rauscher, Hefele ou Ginouilhac, était une grave erreur de
jugement : les membres de la minorité, dont la liste était d'ailleurs tout
aussi unilatérale, furent exaspérés de cette partialité et, à l'extérieur
du concile, beaucoup eurent l'impression que les élections n'étaient qu'un
camouflage et que tout était réglé d'avance par la Curie. Dès ce moment
on commença, en divers milieux, à mettre en doute la liberté réelle du
concile et à prétendre que tout serait réglé à volonté grâce aux Italiens et
aux vicaires apostoliques que l'on affecta de considérer comme une
simple masse de manœuvre aux ordres du Saint-Siège.

LES PREMIERS DÉBATS   Il y eut pourtant une réelle détente lorsque,
après trois semaines de formalités — « j'observe
la façon dont on s'y prend à Rome pour perdre son temps », murmurait
Mgr Darboy [1] — on aborda enfin, le 28 décembre, l'examen du premier
projet de constitution, contre les multiples erreurs dérivant du rationalisme moderne, rédigé par les Pères Franzelin et Schrader. Le schéma,
distribué aux Pères à leur arrivée à Rome, avait été fort mal accueilli par
la majorité d'entre eux : la pensée était parfois obscure, le ton agressif
et il était trop affirmatif sur des points librement controversés. Mais
on lui reprochait surtout d'être une œuvre de professeurs, discutant de
problèmes techniques en un langage scolastique, plutôt qu'un exposé
pastoral destiné à éclairer les simples fidèles. C'est ce que dirent, sans
beaucoup de ménagement, dès le premier jour de la discussion, non seulement Rauscher ou Kenrick, mais plusieurs évêques italiens connus pour
leur adhésion totale aux doctrines romaines, et l'archevêque canadien
Connoly clôtura la journée en déclarant qu'il n'y avait même pas moyen
de retoucher le schéma pour l'améliorer : il ne reste qu'à « l'enterrer
avec honneur » [2]. Les critiques se poursuivirent tout aussi vives les jours
suivants, venant aussi bien des infaillibilistes que de la minorité, au grand
dépit du cardinal de Angelis qui gémissait : « Le concile semble ne se
composer que d'une gauche ! [3] »

Sans être encore tout à fait rassurés, les Pères se sentirent immédiatement plus à l'aise : les discussions seraient plus sérieuses que beaucoup ne l'avaient prédit et les évêques auraient leur mot à dire, même si
ce mot n'était pas élogieux ; le spectre d'une définition par acclamation
était écarté. Par contre les organisateurs du concile étaient consternés,
mais Pie IX, tout en avouant que « la physionomie des premiers débats
lui causait quelque émotion », fit savoir qu'il ne voulait rien faire qui
pût laisser soupçonner que les délibérations n'étaient pas entièrement
libres [4]. Le 10 janvier, après six séances, les présidents annoncèrent que le
schéma serait renvoyé à la députation de la foi pour être refondu et qu'on
aborderait, en attendant, les schémas sur la discipline ecclésiastique.

(1) Mgr Foulon à l'abbé Tapié, 22 décembre 1869, dans *Les Lettres*, juin 1928, p. 192.
(2) MANSI, *A. C. C.*, t. 50, col. 135.
(3) Cité dans C. WOLFSGRUBER, *Kardinal Schwarzenberg*, t. III, p. 236.
(4) Dépêche de Banneville d'après une audience de Mgr Lavigerie (*Arch. Min. A. É. Paris*,
Rome, 1045, f° 10).

PÉTITIONS POUR ET CONTRE L'INFAILLIBILITÉ

En dehors des congrégations générales, qui ne se tenaient que le matin, et encore à certains jours seulement, divers Pères déployaient une grande activité en vue de déterminer des courants d'opinion parmi les membres de l'assemblée, à propos des problèmes qui leur paraissaient importants. Une question surtout faisait l'objet de ces démarches, celle qui, tout en n'étant mentionnée ni dans la bulle de convocation ni dans le programme du concile, était à l'ordre du jour bien avant l'ouverture : la définition de l'infaillibilité du pape. A Rome même, on s'en était d'ailleurs occupé plus qu'on ne voulait bien le dire et la commission préparatoire, à l'instigation de Manning et du P. Schrader notamment, avait rédigé un projet qui pourrait éventuellement être inséré dans le schéma *de Ecclesia* ; il avait toutefois été décidé de ne pas le proposer spontanément et d'en laisser l'initiative aux évêques.

Dès le 25 décembre, Mgr Dechamps remettait à la commission des postulats une proposition de définition. Elle fut appuyée, quinze jours plus tard, par les autres évêques de Belgique, mais dans l'entre-temps, un mouvement beaucoup plus important s'était amorcé. Senestrey, après s'être concerté avec plusieurs Pères et avoir préparé un projet de pétition, provoqua une réunion, le 23 décembre, à la *Villa Caserta*, chez Dechamps, avec Manning, Martin et quelques autres. Le texte, définitivement arrêté le 28, fut mis en circulation le 30, d'abord confidentiellement puis ouvertement, au moment même où, profitant d'une série traditionnelle de sermons à l'occasion de l'Épiphanie, Mermillod, Pie et Berteaud, ainsi que l'abbé Combalot et un jésuite italien, prêchaient avec grand succès sur les prérogatives du souverain pontife. Malgré l'abstention d'un certain nombre de partisans de l'infaillibilité, qui trouvaient la formule trop absolue, la pétition recueillit dans le courant de janvier environ 380 signatures, et quelques autres pétitions analogues en recueillirent encore une centaine.

Les adversaires de l'infaillibilité, alertés dès les premiers jours de janvier, s'empressèrent de contre-attaquer. On envisagea d'abord une adresse commune, que Rauscher et Darboy auraient préparée, mettant en lumière les difficultés que soulèverait une définition et demandant qu'on en restât au *statu quo*, mais les évêques préférèrent finalement rédiger leurs demandes séparément, par groupes nationaux : Allemagne et Autriche-Hongrie, France, Amérique du Nord et Grande-Bretagne, Italie, Orientaux. Les cinq documents recueillirent 136 signatures et les animateurs de l'opposition furent très satisfaits de ce résultat. Ils n'étaient évidemment qu'une minorité, personne n'en avait jamais douté, mais qui représentait 20 pour cent du concile et s'imposait à l'attention par la notoriété scientifique de plusieurs de ses membres et par l'importance des diocèses représentés : pratiquement toute l'Autriche-Hongrie, dont l'influence restait considérable à Rome, une partie notable de l'épiscopat français, tous les grands sièges d'Allemagne, plusieurs archevêchés importants d'Amérique et d'Italie, trois patriarches orientaux. Dans ces conditions, ne pouvait-on espérer que Pie IX suivrait l'exemple de Pie IV, qui avait donné pour instruction à ses légats à Trente de retirer les propositions qui soulèveraient des discussions irritantes ? Dans cet espoir et pour impressionner davantage encore les milieux romains, Dupanloup estima utile d'intensifier la polémique à l'extérieur du concile, excitant notamment ses amis de Broglie et Cochin à donner au *Français* un ton « beaucoup plus belliqueux »[1] et « conjurant » le P. Gratry de ne pas hésiter à donner le plus de publicité possible à la réponse qu'il avait préparée à l'opuscule de Dechamps [2].

LES QUESTIONS DISCIPLINAIRES Tandis que se développait de la sorte autour de l'infaillibilité une agitation croissante à Rome et au dehors et que se succédaient les « brochures retentissantes »[3] et les articles sensationnels, le concile poursuivait ses travaux. En attendant que le schéma dogmatique eût été refondu, il avait abordé l'examen des questions de discipline ecclésiastique.

On se souvient que l'adaptation de celle-ci avait été considérée par beaucoup, lors de l'annonce du concile, comme la tâche principale de la future assemblée. Vingt-huit schémas avaient été préparés par la commission pour la discipline et dix-huit autres par celle des religieux, mais quatre seulement furent distribués aux Pères et discutés, sans que d'ailleurs on ait eu le temps d'en sanctionner aucun.

Les premiers schémas examinés furent ceux sur « les évêques, les vicaires généraux et les synodes » et sur « la vacance des sièges épiscopaux ». Ils ne furent pas reçus avec plus d'enthousiasme que le schéma dogmatique. Les Pères critiquèrent surtout la tendance des canonistes romains à penser avant tout aux petits diocèses italiens et à tenir trop peu compte des situations très différentes qui se rencontraient ailleurs. Bien que le cardinal de Luca ait prétendu avec humeur voir dans ces critiques un « esprit d'opposition »[4], beaucoup d'entre elles n'avaient rien à voir avec les dispositions des orateurs relativement à l'infaillibilité. Il faut pourtant reconnaître que la division des Pères en deux groupes se fit sentir à plus d'une reprise au cours de la discussion : « Tandis que les uns paraissaient toujours craindre qu'on empiétât sur les prérogatives de la papauté, les autres nourrissaient les mêmes appréhensions au sujet des droits des évêques [5]. » C'est ainsi que Schwarzenberg et Strossmayer suggérèrent de s'occuper d'abord de la réforme de la Curie et des Congrégations romaines et demandèrent pourquoi l'on ne parlait que des devoirs des évêques et jamais de leurs droits ou de leur place dans l'Église. L'archevêque de Cologne reprocha au projet sa tendance à une centralisation excessive, ce qui fit sensation, étant donnés ses sentiments bien connus d'attachement au Saint-Siège. Le patriarche chaldéen, Mgr Audo, souligna de son côté le danger de vouloir appliquer aux Églises d'Orient, si atta-

(1) Voir la lettre d'A. de Broglie à Dupanloup, du 7 janvier 1870, *Arch. Nat.*, AB XIX, n° 526.
(2) Voir A. CHAUVIN, *Le P. Gratry*, Paris, 1911, p. 430, n. 1. Sur l'intervention de Gratry, cf. *infra*, p. 344-345.
(3) F. MOURRET, *op. cit.*, p. 248.
(4) H. ICARD, *Journal*, p. 136.
(5) A. BOULENGER, *Histoire générale de l'Église*, vol. IX, 1e P., p. 428.

chées à leurs vieilles coutumes, la même discipline qu'en Occident et proposa que la réforme de ces Églises fût laissée à des synodes nationaux ; comme il se fit, quelques jours plus tard, vivement rappeler à l'ordre par le pape pour une affaire qui n'avait aucun rapport avec le concile [1], certains prétendirent bien à tort qu'il avait été victime de sa franchise, et l'incident accrut la tension entre la minorité et la cour romaine.

Du 25 janvier au 22 février furent examinés le schéma sur « la vie des clercs » et celui prévoyant le remplacement des catéchismes diocésains par un catéchisme universel. Dupanloup, entre autres, protesta contre ce projet, qui finit néanmoins par être adopté à une forte majorité.

Au cours de ces cinq semaines d'interminables discours, les Pères donnèrent des signes de plus en plus fréquents de lassitude et d'impatience. Il y avait bien de temps à autre un intermède de nature à dérider l'auguste assemblée, par exemple lorsqu'un évêque sicilien justifia l'obligation du port de la soutane en expliquant que, d'après Isaïe, l'Éternel portait dans le ciel une robe traînante. Mais les évêques, considérant qu'ils avaient abandonné leurs diocèses depuis trois mois sans qu'aucun résultat positif n'ait encore été atteint, avaient chaque jour davantage l'impression qu'on était en train de s'enliser. Et la mauvaise humeur causée par « l'expansion assoupissante » [2] de discours interminables était encore accrue par l'effort pénible d'attention qu'exigeait l'acoustique, demeurée lamentable en dépit de quelques modifications apportées les premiers jours.

LA MODIFICATION DU RÈGLEMENT Tout le monde était d'accord pour estimer que les choses ne pouvaient plus continuer de la sorte. Certains en profitèrent pour lancer l'idée d'une prorogation du concile, qui leur apparaissait comme une manière élégante de renvoyer dos à dos partisans et adversaires de l'infaillibilité. Mais une prorogation dans de pareilles conditions, alors qu'on n'avait encore abouti à rien, serait apparue comme un désaveu trop cinglant pour les responsables de la préparation et de l'organisation du concile. Aussi s'arrêta-t-on bientôt à une double décision : adapter le règlement à une situation qui se révélait bien différente de celle qui avait été prévue, à savoir un concile bref où les schémas n'auraient obtenu que l'admiration des Pères ; et transformer la salle de réunion, de manière à améliorer enfin l'acoustique.

La réalisation de ce dernier point demanda près d'un mois, du 22 février au 18 mars, temps pendant lequel les congrégations générales furent suspendues, mais le résultat fut aussi satisfaisant qu'il était possible de l'espérer dans un local aussi ingrat que la basilique Saint-Pierre.

Quant à l'amélioration du règlement, la question était à l'ordre du jour depuis le début et de nombreuses suggestions avaient déjà été faites de droite et de gauche aux présidents : la tâche de ceux-ci était délicate, car on les sollicitait d'un côté d'augmenter la liberté de mouvement des

---

(1) Voir quelques détails *infra*, p. 420-421, et surtout dans C. KOROLEVSKY, art. *Audo*, dans *D. H. G. E.*, t. V, col. 340-341, qui rectifie F. MOURRET, *op. cit.*, p. 237 et suiv.
(2) É. OLLIVIER, *op. cit.*, t. II, p. 44.

Pères et, de l'autre, d'accélérer la marche des travaux. C'est cette dernière préoccupation qui finit par l'emporter dans le nouveau règlement qui fut communiqué au concile, au nom du pape, le 22 février. Quatre modifications surtout sont à relever : au lieu de se borner à des critiques générales, les Pères devraient désormais proposer une rédaction écrite des amendements qu'ils désiraient voir introduire ; le vote sur ceux-ci aurait lieu par assis et levés (ce qui ferait gagner chaque fois une heure et demie sur le vote par *Placet* individuel) ; les présidents étaient autorisés à mettre aux voix la clôture anticipée de la discussion d'une question si celle-ci était demandée par dix Pères au moins ; enfin la majorité des voix suffirait pour qu'une constitution fût adoptée.

Dans l'ensemble, ce règlement n'était pas mauvais [1] et la première réaction de beaucoup de Pères fut favorable [2]. Mais les chefs de la minorité se montrèrent très émus des deux dernières mesures. L'abandon du principe de la nécessité d'une unanimité morale des votants, qu'ils considéraient, avec une certaine exagération, comme le seul traditionnel [3], était évidemment pour eux une grosse déception ; en outre, ils craignaient que les pouvoirs accordés aux présidents ne fussent utilisés pour étrangler la discussion [4]. Aussi les protestations ne se firent-elles pas attendre, l'une rédigée par Dupanloup avec l'approbation du Comité international [5] et signée par 30 évêques français, auxquels s'adjoignirent 20 étrangers, une autre, similaire, signée par 23 évêques germaniques, une troisième, signée par 14 autres évêques du groupe germanique, plus radicale sur la question de l'unanimité morale, sous l'influence de Ketteler et surtout d'Hefele, qui pendant les semaines suivantes continua à se montrer d'une intransigeance absolue sur ce point. Certains proposèrent même de refuser de continuer à participer au concile si l'on ne tenait pas compte de leurs objections, mais la plupart, satisfaits de constater à l'occasion de ce mouvement de protestation que la minorité restait aussi solidement unie qu'en janvier, estimèrent préférable d'éviter une décision aussi excessive, qui n'aurait rallié qu'un petit nombre de Pères.

REPRISE DES DÉBATS
La tempête soulevée parmi les évêques de la minorité par le nouveau règlement, et aussi par la distribution, le 6 mars, d'un projet de définition de l'infaillibilité,

---

(1) Voir notamment l'appréciation d'É. OLLIVIER, *op. cit.*, t. II, p. 81-85.

(2) Lettre de Ullathorne du 22 février 1870, dans C. BUTLER, *op. cit.*, t. I, p. 248-249.

(3) Voir les observations nuancées de C. BUTLER, *op. cit.*, t. I, p. 251 et d'É. OLLIVIER, *op. cit.*, t. II, p. 74-76.

(4) Rauscher, plus modéré et qui refusa de protester, estimait que si le danger existait, le règlement bien appliqué pourrait cependant être utile (*Tagebuch* de Mayer, cité dans C. WOLFSGRUBER, *Kardinal Schwarzenberg*, t. III, p. 246-247). Effectivement, tous les historiens impartiaux doivent reconnaître qu'il n'y eut jamais de clôture prononcée par surprise et que, d'une façon plus générale, en dépit de l'une ou l'autre pression regrettable, la conclusion du journal de M. Icard est exacte : « Le concile du Vatican, sans avoir eu une liberté pleine et parfaite, en a eu incontestablement assez pour la valeur de ses actes. Il y a eu liberté de la parole et liberté morale des votes. » (P. 392.) Sur cette question de la liberté du concile, voir, de préférence aux critiques par trop tendancieuses de J. Friedrich, celles plus mesurées du protestant Mirbt dans *Historische Zeitung*, t. CI, 1908, p. 574 et suiv. ; et, de préférence à l'apologie de Granderath, les observations d'É. OLLIVIER, *op. cit.*, t. II, p. 67-74 (voir aussi *ibid.*, p. 43-44), qui conclut : « Il est certain que la discussion a été libre autant qu'elle le fut jamais dans aucune assemblée humaine. »

(5) On peut suivre son élaboration dans les procès-verbaux de la minorité française, séances des 25, 26, 28 février et 1er mars 1870.

commençait à peine à s'apaiser lorsque les congrégations générales reprirent, pour examiner le projet remanié de constitution dogmatique.

L'activité n'avait pas chômé en commission : le lendemain même du jour où le schéma avait été renvoyé à la députation de la foi, celle-ci avait invité Franzelin à venir défendre son œuvre. Bien que l'éminent théologien ait tenté d'en justifier le caractère technique par le fait que les principaux adversaires de la doctrine catholique se rencontraient à présent dans les universités, la députation avait estimé qu'une refonte totale s'imposait. Trois membres en furent chargés : Pie, dont le théologien, Charles Gay, eut un rôle qu'on mentionne rarement [1] ; Dechamps, qui profita de l'occasion pour mettre encore mieux en relief ses idées sur la « méthode de la Providence », déjà exploitées d'ailleurs dans le schéma primitif [2] ; Martin enfin, qui eut la part principale et se fit aider par son ancien compagnon d'études, le P. Kleutgen. Le nouveau projet fut prêt à la fin de février ; très abrégé, il ne contenait plus que neuf chapitres, d'une rédaction beaucoup moins abstraite. La députation décida de laisser provisoirement de côté la partie consacrée à la dogmatique spéciale et de commencer par examiner les quatre premiers chapitres, qui formaient un tout, relatif à la théologie fondamentale.

Ce schéma, mis au point, revint devant l'assemblée le 18 mars. La nouvelle rédaction rencontra cette fois beaucoup de faveur et la discussion, qui se poursuivit jusqu'au 6 avril, ne porta plus que sur des amendements de détail ou sur des points particuliers comme l'ontologisme, dont Mgr Pecci, le futur Léon XIII, demanda vainement la condamnation explicite, ou le traditionalisme, que certains Pères auraient au contraire voulu voir condamner de façon moins radicale. Les orateurs, au nombre de cent, présentèrent de nombreuses observations judicieuses dont la commission tint souvent compte ; il est toutefois regrettable que des remarques comme celles de Mgr Meignan sur la question biblique, ou de Mgr Ginouilhac sur la liberté qu'il faudrait savoir laisser à la science, n'aient guère retenu l'attention d'une assemblée dont la grande majorité des membres étaient peu au fait du mouvement scientifique contemporain.

Le calme des discussions ne fut troublé qu'une seule fois, le 22 mars. Strossmayer ayant regretté la manière peu irénique dont il était parlé des protestants et signalé la bonne foi et l'esprit très chrétien de beaucoup d'entre eux, l'assemblée commença à murmurer et il se fit rappeler à l'ordre par le cardinal Capalti qui lui fit remarquer qu'il était question du protestantisme et non des protestants ; quand il poursuivit en se plaignant de la modification du règlement et de l'abandon de la règle de l'unanimité morale, de nombreux Pères s'indignèrent bruyamment :

« Et ce sont ces gens-là qui ne veulent pas de l'infaillibilité du pape ! Est-il donc infaillible, celui-là ? » D'autres : « C'est Lucifer. Anathème ! anathème ! »

(1) *Correspondance de Mgr Gay*, t. II, p. 120-121, 126.
(2) Le cardinal Bilio lui avait écrit : « En arrivant ici, vous pourrez constater que nous vous avons pillé. » (Dans H. Saintrain, *Vie du cardinal Dechamps*, Tournai, 1884, p. 86.) Cf. R. Kremer, *L'apologétique du cardinal Dechamps. Ses sources et son influence au concile du Vatican*, dans *Revue des sciences philosophiques et théologiques*, t. XIX, 1930, p. 679-702.

Ou bien : « C'est un autre Luther. Qu'on le chasse ! » Tous enfin criaient :
« Descendez ! Descendez ! » [1]

Granderath lui-même estime qu'« on aurait pu attendre des évêques
plus de calme et de dignité » [2], mais il faut noter que ce fut là la séance
la plus agitée de tout le concile et que les polémistes qui ont exploité
cet incident le présentent bien à tort comme caractéristique des rapports
habituels entre la majorité et la minorité ; l'atmosphère de l'assemblée,
même pendant la période où la discussion porta directement sur l'infaillibilité, resta presque toujours digne, souvent même courtoise.

Comme l'examen par la députation des 281 amendements se poursuivait parallèlement aux discussions, le vote sur l'ensemble du projet
retouché put avoir lieu dès le 12 avril. Quelques amendements furent
encore proposés à cette occasion par les 83 Pères qui émirent un vote
conditionnel : *Placet juxta modum* [3]. Quelques Pères, conduits par Strossmayer, envisagèrent un moment d'émettre le dernier jour un vote négatif
parce que la députation refusait de faire disparaître l'exhortation finale
où ils entrevoyaient une allusion à l'infaillibilité pontificale et surtout
parce qu'ils craignaient, en votant *Placet*, de paraître approuver le règlement tant critiqué, mais les cardinaux Rauscher et Schwarzenberg
réussirent à faire comprendre la maladresse d'une telle attitude qui
donnerait l'impression d'une opposition systématique. Le 24 avril, à la
IIIe session publique, les 667 membres présents votèrent à l'unanimité
la première constitution du concile.

LA CONSTITUTION « DEI FILIUS »   La constitution *Dei Filius* (ainsi
                                    désignée d'après les premiers mots
du texte) opposait au panthéisme, au matérialisme et au rationalisme
moderne un exposé dense et lumineux de la doctrine catholique sur Dieu,
la révélation et la foi. Le premier chapitre proclamait l'existence d'un
Dieu personnel, libre, créateur de toutes choses et absolument indépendant du monde matériel et spirituel qu'il a créé. Le second chapitre
enseignait que certaines vérités religieuses, comme l'existence de Dieu
notamment, « peuvent être connues avec certitude par la lumière naturelle
de la raison humaine » ; ceci portait contre l'athéisme, mais aussi contre
le fidéisme et le traditionalisme, et il y avait quelque chose de paradoxal,
en un siècle où la foi chrétienne était mise en accusation par le rationalisme, à voir un concile se faire le défenseur de la raison ; le chapitre
ajoutait d'ailleurs que, pour d'autres vérités, la révélation divine est
indispensable et précisait la doctrine du concile de Trente, d'après laquelle
cette révélation nous est transmise par des traditions orales remontant
aux apôtres et par les Écritures inspirées. Le troisième chapitre, tout en

---

(1) MANSI, *A. C. C.*, t. 51, col. 77. Strossmayer sortait en fait du sujet discuté en critiquant le
règlement et le lendemain Mgr Meignan put regretter à son tour les expressions trop dures à l'égard
des protestants, sans être interrompu ; l'évêque de Sirmium avait aisément des allures provocantes.
(2) T. GRANDERATH, *op. cit.*, t. II, 2, p. 57.
(3) Le souci que l'on eut de faire droit aux suggestions justifiées des Pères apparaît entre autres
dans l'adoption *in extremis* des observations d'Ullathorne concernant la formule *Sancta Romana
catholica Ecclesia*. (Cf. D. LATHOUD, *Catholica or Romana catholica*, dans *Unitas*, édit. franç.,
t. I, 1948, p. 33-36.)

rappelant le caractère raisonnable de la foi catholique (contre l'illumi-
nisme de certains protestants), la montrait avant tout comme une adhésion libre et un don de la grâce divine (contre l'hermésianisme) ; il montrait
aussi, conformément à la doctrine de Dechamps, comment l'Église,
gardienne du dépôt de la foi, porte en elle-même la garantie de son origine
divine. Enfin le quatrième chapitre délimitait les domaines respectifs
de la foi et de la raison et rappelait qu'un désaccord apparent entre la
science et la religion ne peut venir que d'une erreur sur la doctrine
prônée par celle-ci ou d'une idée fausse sur les conclusions de celle-là.

Cet exposé doctrinal était complété par dix-huit canons qui frappaient
d'anathème les erreurs opposées considérées comme hérétiques [1].

§ 3. — La définition des prérogatives pontificales.

VAINS ESSAIS DE TRANSACTION   Le mouvement de pétition en faveur
de l'infaillibilité, déclenché à la fin de
décembre, avait achevé de mettre à l'ordre du jour cette question brûlante. Mais plus il apparaissait qu'il serait impossible de l'éviter, comme
beaucoup l'auraient désiré, plus aussi augmentait, chez la plupart des
Pères, le désir de la régler d'une manière si modérée que l'on puisse finalement obtenir l'accord de la minorité et éviter ainsi d'étaler en pleine
lumière, au grand scandale des fidèles, le désaccord de l'épiscopat.

C'était notamment le point de vue des évêques français du tiers-parti,
groupés autour du cardinal de Bonnechose : « Les esprits modérés de l'épiscopat, expliquait l'un d'entre eux, devraient employer leurs efforts à
mitiger les termes de la définition, à lui enlever ce qu'on pourrait y mettre
d'outré, à la rendre telle enfin que Bossuet pourrait la signer [2]. »

Sans aller aussi loin, beaucoup parmi ceux qu'on considérait en bloc
comme faisant partie de la majorité faisaient preuve de dispositions
conciliantes et se désolidarisaient des vues extrémistes de Manning.
Certains suggéraient une définition indirecte condamnant par exemple
la prétention d'en appeler du pape à un concile ou de se contenter d'une
obéissance externe à l'autorité du pape ; Mgr Spalding, de Baltimore,
se dépensait beaucoup dans ce sens. D'autres proposaient d'éviter la
forme d'une définition dogmatique proprement dite, avec anathème ;
ainsi Mgr Pie, qui avait du reste refusé de signer le postulat demandant
la définition. L'ambassadeur de France relève que « les efforts pour
arriver à une transaction sous la forme d'une adresse ou d'une déclaration se multiplient. On s'attache, dans les rédactions que l'on essaie,
à faire disparaître ce mot d'infaillibilité et le mot personnel » [3].

(1) Il avait été décidé, dès la fin de 1868, que les canons seraient réservés aux hérésies n'ayant
point encore été condamnées comme telles par l'Église, tandis que dans les chapitres, on relèverait
encore d'autres erreurs. (MANSI, *A. C. C.*, t. 49, col. 654-655.) Cette règle fut strictement observée ;
voir par exemple les explications de Mgr Simor, le 18 mars 1870. (*Ibid.*, t. 51, col. 45.) Il y a donc
dans les chapitres des affirmations qui ne sont pas imposées à notre croyance, mais qui font
connaître de manière plus détaillée l'enseignement positif de l'Église contredit par les erreurs
condamnées dans les canons.

(2) Conversation de Lavigerie avec Ollivier, en février 1870, rapportée par ce dernier dans
L'Église et l'État au concile du Vatican, t. II, p. 96-97.

(3) Dépêche du 4 février 1870, *Arch. Min. A. É. Paris*, Rome, 1045, f°s 262-263.

Dès l'arrivée d'Hefele à Rome, au milieu de janvier, Fessler était venu le sonder pour voir s'il n'accepterait pas de prendre la tête d'un tiers-parti parmi les Austro-Allemands [1]. Le refus très net de l'ancien professeur de Tubingue n'avait pas découragé le secrétaire du concile qui, quelques jours plus tard, assurait à l'ambassadeur d'Autriche que les dirigeants viseraient « à faire prévaloir la plus grande modération » et lui laissait entrevoir que « dans la grande majorité proprement dite, un parti spécial, pénétré de la nécessité de cette modération, était sur le point de se former » [2]. De son côté, le supérieur de Saint-Sulpice, M. Icard, à qui on avait longtemps reproché à Rome ses doctrines trop peu ultramontaines, prêchait le calme aux évêques français de la minorité et s'efforçait de préparer les voies à un rapprochement en dissipant les équivoques et en travaillant lui aussi à la mise au point de formules nuancées [3].

Même, à la Curie, quoi qu'aient prétendu les adversaires du concile, l'atmosphère était loin d'être à l'intransigeance. Les Italiens apprécient volontiers les arrangements et les prélats romains en particulier ne goûtent guère les solutions violentes. Il n'y avait pas qu'Antonelli et les « politiques » à souhaiter de ne pas laisser allumer les brandons de discorde en un moment aussi difficile pour l'Église et pour l'État pontifical. Dès qu'on avait pris conscience de la force de l'opposition, le crédit des *zelanti* avait baissé. On estimait certes que les choses étaient allées trop loin pour qu'on ne fasse pas quelque chose, mais Ullathorne pouvait écrire à Newman, au début de février, que les milieux romains officiels désiraient une formule sur laquelle puissent s'accorder les modérés de tous les partis [4] et l'évêque de Nevers se réjouissait, quelques jours plus tard, de voir le pape « incliner visiblement de plus en plus vers la modération » [5].

LE TEXTE ADDITIONNEL SUR L'INFAILLIBILITÉ — Pendant ce temps, la commission compétente avait été saisie des postulats déposés les 28 et 29 janvier. Un second schéma de constitution dogmatique, traitant de l'Église et du pape, avait déjà été distribué aux Pères le 21 janvier. Il n'y était pas fait mention du privilège de l'infaillibilité, mais seulement de la primauté pontificale en général. La question se posait à présent de savoir s'il ne convenait pas de compléter ce schéma sans tarder, conformément aux désirs de la majorité. La chose fut examinée le 9 février et tranchée par l'affirmative, non seulement par les membres de la majorité, mais par Bonnechose et par deux sur trois des anti-infaillibilistes [6]. Le pape, après trois semaines d'hésitation, donna son accord le 1er mars et, cinq jours plus tard, on remettait aux Pères un texte assez bref énonçant le dogme de l'infaillibilité et se présentant comme un *addendum* au chapitre XI du schéma *de Ecclesia* consacré à la primauté.

---

(1) J. FRIEDRICH, *Tagebuch*, p. 111-112.
(2) Dépêche du 22 janvier 1870, copie aux *Arch. Min. A. É. Paris*, Rome, 1045, fo 149.
(3) Voir F. MOURRET, *op. cit.*, p. 140, 141, 220-222, 274-281.
(4) Ullathorne à Newman, 4 février 1870, dans C. BUTLER, *op. cit.*, t. I, p. 214-217.
(5) Mgr Forcade à un directeur du ministère des Cultes, 15 février 1870, *Arch. Nat.*, F19 1940.
(6) MANSI, *A. C. C.*, t. 51, col. 687-696.

L'émoi fut grand, car le texte distribué n'était autre que le projet préparé par les consulteurs bien avant l'ouverture du concile et ne portait donc aucune trace des récentes dispositions à la modération : cela donnait l'impression d'une « victoire de l'opinion extrême », d'autant plus qu'au même moment, le bruit recommençait à courir d'une éventuelle définition par acclamation et que plusieurs Pères appuyaient par des postulats les invitations de *L'Univers* à procéder à un examen immédiat du chapitre sur l'infaillibilité, séparé du schéma *de Ecclesia*, avant même que ne fût achevé l'examen de la première constitution dogmatique [1]. Après quelques jours toutefois, les modérés du tiers-parti, qui avaient d'abord été fort déçus, se rendirent compte qu'on n'était guère disposé à donner satisfaction à cette « précipitation passionnée » et ils reçurent l'assurance que le texte distribué n'était qu'une base de discussion qui restait susceptible d'être précisée et amendée. Ils reprirent dès lors leurs tractations avec un nouveau courage [2].

Malheureusement, du côté anti-infaillibiliste, le raidissement s'accentuait. Il y avait bien eu, en février, des signes de détente, une partie des Pères, émus par les attaques dont les prérogatives du pape étaient l'objet dans une certaine presse, ayant paru incliner vers une définition modérée de l'infaillibilité, qui éclairerait les fidèles sans entériner les vues radicales du néo-ultramontanisme. Mais, déjà mal impressionnés par la modification du règlement, la distribution du projet de définition dans les conditions qu'on a vues mit le comble à leur mécontentement. Du coup la minorité se trouva ressoudée autour de ses chefs, Schwarzenberg, Strossmayer, Dupanloup, Hefele, Ketteler, Darboy, qui, eux, n'avaient jamais envisagé de céder, et, tandis que le Comité international s'empressait de proposer la constitution d'une commission spéciale composée de six membres élus par la députation *de fide* et six désignés par les signataires du *Contrapostulatum* [3], on se mit à organiser la résistance à l'intérieur et même à l'extérieur du concile.

LA PROPAGANDE PARMI LES PÈRES — Les adversaires de l'infaillibilité, tout comme ses partisans, du reste, se rendaient compte de la nécessité d'agir pour gagner à leurs vues les hésitants, forcément nombreux dans une assemblée dont les 700 membres n'étaient pas tous de fortes personnalités, ni des compétences théologiques.

Les uns travaillaient surtout par contacts personnels, tels Strossmayer, toujours excité, ou Darboy, flegmatique et caustique, ou Moriarty, qui

---

(1) T. GRANDERATH, *op. cit.*, t. II, 1, p. 367-368 ; C. BUTLER, *op. cit.*, t. II, p. 33-34 ; voir *L'Univers* des 9 et 20 février 1870 et les postulats reproduits dans MANSI, *A. C. C.*, t. 51, col. 703-711.

(2) Sur cette évolution des sentiments du Tiers-Parti, voir R. AUBERT, *Documents concernant le Tiers-Parti...*, dans *Festschrift K. Adam* (cf. *supra*, p. 311, n. 1) ; sur la persistance des dispositions conciliatrices parmi les dirigeants du concile, voir le *Journal* de M. Icard cité par F. MOURRET, *op. cit.*, p. 278-279 et les lettres d'Ullathorne des 4 et 19 avril dans C. BUTLER, *op. cit.*, t. II, p. 35-36.

(3) Communication de Darboy à la minorité française, le 8 mars. (D'après les *Procès-verbaux* inédits.) Le cardinal de Angelis répondit, le 14, aux délégués que l'idée d'une commission mixte était contraire au règlement, « mais que rien ne lui paraissait mettre obstacle à ce que la discussion contradictoire dont il s'agit eût lieu officieusement ». (*Ibid.*, séance du 15 mars.) De fait, des contacts de ce genre eurent lieu à diverses reprises, sur lesquels nous restons insuffisamment renseignés.

écrivait à Newman : « Je suis la voix qui crie, non dans le désert ni dans l'assemblée, mais dans les rues et les salons »[1] ; et, dans l'autre camp, Dechamps, servi par sa formation théologique et ses qualités d'homme du monde, Mac Closkey, le sage archevêque de New-York, dont les avis étaient très écoutés, ou dom Pitra, l'un des rares savants que la majorité pouvait opposer à ses adversaires. C'était surtout le cas pour Manning, dont l'activité débordante indisposait certains cardinaux romains, qui ne comprenaient pas l'intensité de la foi religieuse qui inspirait ces allures rappelant un peu trop les mœurs parlementaires.

D'autres préféraient une méthode plus didactique et s'appliquèrent à défendre leur point de vue dans des mémoires imprimés à l'intention des Pères. Outre un certain nombre de brochures italiennes, on peut signaler, en faveur de l'infaillibilité, un travail de l'évêque arménien Azarian, sur l'appoint apporté à ce dogme par la tradition de son Église, un autre dû à Freppel, qui avait renoncé à ses vues inopportunistes peu avant sa nomination à l'évêché d'Angers, ainsi qu'un mémoire érudit de Fessler à propos de la prétendue nécessité de l'unanimité morale dans les conciles.

Du côté des adversaires de l'infaillibilité, quatre brochures surtout retinrent l'attention[2] : l'une, répandue par le cardinal de Schwarzenberg, mais rédigée par le prémontré Mayer, auquel il faisait généralement appel pour ses interventions théologiques ; une autre par le cardinal Rauscher, qui concluait d'une étude des témoignages patristiques et médiévaux que le pape ne peut définir une doctrine qu'avec le consentement de l'Église universelle ; une troisième due au P. Quarella et que Ketteler fit diffuser, où était exposée la thèse de Bossuet, à savoir que le magistère infaillible n'appartient au pape que s'il juge avec la collaboration et l'approbation des évêques, lesquels possèdent une part réelle dans le pouvoir donné par Dieu à l'Église ; dans la quatrième, Hefele examinait avec sa compétence habituelle un point particulier, mais qui joua dans la controverse un rôle considérable, la question du pape Honorius, qui semblait effectivement constituer une objection irréfutable contre la thèse des néo-ultramontains, lesquels considéraient *toute* décision du pape comme infaillible[3]. Comme les réponses à ces brochures ne tardèrent pas, les Pères se trouvèrent bientôt en possession de tout un dossier, avant même que la discussion ait commencé officiellement dans l'assemblée.

Quoi qu'en aient pensé certains, pendant le concile et après, l'activité déployée par quelques Pères pour gagner à leurs vues les autres membres de l'assemblée était parfaitement légitime et même utile, car elle permettait de juger en pleine connaissance de cause. Et surtout, il importe de ne jamais oublier que, des deux côtés, à l'exception de quelques personnages de second plan, ceux qui se démenaient

---

(1) Lettre du 3 février 1870 dans C. Butler, *op. cit.*, t. II, p. 29.
(2) Elles sont longuement analysées par T. Granderath, *op. cit.*, t. III, 1, p. 20-37.
(3) Il en va autrement si l'on considère, en accord avec ce qui fut finalement défini, que seules les décisions *ex cathedra* sont infaillibles. Sur la polémique autour de la question d'Honorius au moment du concile, voir la dissertation un peu partiale mais bien documentée de W. Plannet, *Die Honoriusfrage auf dem Vaticanischen Konzil*, Marbourg, 1912.

ainsi étaient animés par leur zèle pour les intérêts de l'Église et pour l'intégrité de la foi. Tandis que les uns craignaient que l'insistance toujours plus grande sur l'aspect autoritaire de l'Église en ce siècle de liberté n'éloignât d'elle de nombreuses âmes de bonne volonté, les autres étaient convaincus que la proclamation de l'infaillibilité pontificale concourrait à la réaction nécessaire contre le libre-examen, le grand mal du XIXe siècle. Et tandis que ceux-ci étaient convaincus que l'infaillibilité du pape était une vérité de foi qu'on n'était plus libre de discuter et qu'il importait donc d'affirmer solennellement à l'encontre de toutes les formes, même mitigées, de gallicanisme, les premiers se demandaient si l'on pouvait présenter aux fidèles comme faisant partie du dépôt révélé une doctrine, vraie peut-être, mais mal attestée par les anciens écrivains ecclésiastiques, et surtout ils croyaient déceler chez les *infaillibilistes* une tendance à « effacer les évêques et absorber leurs droits dans la juridiction suprême du pape »[1] au risque de « modifier ainsi la constitution divine de l'Église »[2].

On a déjà signalé ce dernier aspect, auquel on ne prête généralement pas suffisamment attention et qui est pourtant capital. Comme c'est si souvent le cas, chacun des deux groupes avait une conscience très nette d'un aspect de la vérité totale, mais, en y insistant, tendait, surtout dans la minorité, à minimiser un autre aspect, important lui aussi. Exaspérés par les outrances ou les maladresses de certains ultramontains, les membres de la minorité redoutaient qu'on en vînt à considérer les évêques comme de simples délégués du pape auprès de leurs fidèles, alors que le Christ les a institués pour en être les pasteurs ordinaires et qu'ils sont les authentiques successeurs des apôtres, héritiers notamment de leur pouvoir doctrinal. Mais cette préoccupation les rendait exagérément méfiants à l'égard de tout ce qui tendait à consolider et surtout à étendre l'exercice des privilèges et prérogatives du souverain pontife et portait certains d'entre eux à accorder quelque apparence de vérité aux thèses du gallicanisme modéré. De leur côté, les ultramontains, très lucides quand il s'agissait de discerner ce que le gallicanisme avait d'inacceptable, estimaient urgent de définir le rôle très particulier que la tradition avait de plus en plus nettement reconnu en fait au successeur du chef du Collège apostolique. Mais la volonté de réagir contre l'épiscopalisme gallican avait entraîné certains à méconnaître l'importance réelle de l'épiscopat et le caractère complexe de la constitution de l'Église. Par leur fait, c'était tout le courant ultramontain qui se trouvait discrédité aux yeux des défenseurs des droits légitimes de l'épiscopat, alors qu'il eût fallu distinguer entre les néo-ultramontains, comme Manning, et les tenants de l'ultramontanisme traditionnel, qui formaient la partie la plus nombreuse de l'assemblée.

Malheureusement, parallèlement à cette polémique intraconciliaire, qui demeura toujours digne et le plus souvent courtoise[3], se développait

---

(1) H. ICARD, *Journal*, p. 197.
(2) Mgr COLLET, *Souvenirs du concile du Vatican* (inédits).
(3) Ullathorne revient souvent dans ses lettres sur la courtoisie des rapports entre Pères d'opinion différente. Mourret toutefois (*op. cit.*, p. 279-280) rend un son de cloche différent, mais il

au dehors une agitation beaucoup plus discutable que plusieurs Pères, surtout dans la minorité, eurent le tort de vouloir utiliser pour faire pression sur le concile.

LA PROPAGANDE ULTRAMONTAINE EN DEHORS DU CONCILE

Les attaques de Janus et de Mgr Maret contre les thèses fondamentales de l'ultramontanisme, puis les critiques de Mgr Dupanloup, moins radicales mais de nature à frapper davantage le grand public, enfin l'organisation, qui n'était plus un secret, d'une opposition épiscopale au sein du concile, avaient profondément indigné tous ceux qui attendaient avec impatience la définition de l'infaillibilité pontificale comme le couronnement d'un demi-siècle de lutte victorieuse contre les restes du gallicanisme et du joséphisme. Les excitations, souvent dépourvues de mesure, de *La Civiltà cattolica*, de *L'Unità*, du *Tablet* ou de *L'Univers*, contribuèrent à nourrir cette indignation, mais celle-ci n'avait rien d'un phénomène provoqué artificiellement. Si ces journaux connurent un tel succès, c'est qu'ils correspondaient à la réaction spontanée de leurs lecteurs, peu au fait des difficultés théologiques, mais profondément convaincus du rôle providentiel de la papauté et, au surplus, admirateurs enthousiastes de Pie IX et de son œuvre de restauration intégrale des principes chrétiens. Une souscription ouverte en novembre par *L'Univers* pour concourir aux dépenses du concile fut l'occasion d'un véritable plébiscite, de nombreux donateurs accompagnant leur envoi de professions de leurs sentiments, reproduites par le journal à longueur de colonnes [1]. Puis, au cours des mois suivants, se développa, dans le clergé de France et d'Italie, un mouvement de pétitions, soutenu par la presse, en faveur de la proclamation du dogme, mouvement qui prit plus d'une fois l'allure d'une protestation des prêtres d'un diocèse contre leur évêque anti-infaillibiliste. Il y avait quelque chose de dangereux et somme toute d'assez peu catholique dans cette prétention des fidèles et des prêtres à dicter leur devoir à ceux qui sont les « juges de la foi », mais les champions de l'infaillibilité dans l'épiscopat ne s'en alarmèrent guère, soucieux seulement de l'appoint précieux qu'apportait à leur cause Veuillot et tous ceux qui se groupaient derrière lui : « Il est des nôtres et nous sommes siens », reconnaissait candidement l'un d'eux [2].

L'AGITATION EXTRACONCILIAIRE EN FRANCE

Bien plus regrettables d'ailleurs que les exagérations de la presse ultramontaine furent les tentatives inconsidérées de certains anti-infaillibilistes pour agiter l'opinion. Mgr Moriarty, membre décidé de la minorité, écrivait très sagement : « S'il plaît au Saint-Esprit et au concile que la définition ait quand même lieu, il serait dan-

---

s'agit uniquement des évêques français entre eux, plus bouillants et profondément divisés sur d'autres terrains encore.

(1) Elles se succédèrent pendant 184 numéros ; quelques-unes sont reproduites dans *Coll. Lac.*, t. VII, col. 1469-1471.

(2) H. ICARD, *Journal*, p. 70.

gereux que l'opinion publique se trouvât prévenue contre elle [1]. » Malheureusement, tous à Rome n'en jugeaient pas ainsi. Mgr Clifford, par exemple, ayant eu connaissance d'une lettre privée où Newman exprimait ses appréhensions sur les conséquences éventuelles d'une définition, la fit publier dans la presse au début d'avril [2], ce qui provoqua évidemment quelques remous. Mais c'est surtout en France et en Allemagne que des manœuvres de ce genre, beaucoup plus graves d'ailleurs, furent à déplorer.

En France, une grosse part de la responsabilité de l'agitation revient sans conteste à Dupanloup qui, pendant toute la durée du concile, se révéla particulièrement excité [3]. Il avait commis la faute d'en appeler à l'opinion publique dès avant le concile, et plus le concile avançait, plus il accentuait cette politique, entrant à nouveau dans la lice en publiant au début de mars une lettre ouverte à Mgr Dechamps, mais surtout s'efforçant de déterminer en France et ailleurs un courant de protestations qu'il encourageait en envoyant à ses amis, aidé par de nombreux secrétaires, lettres, circulaires ou projets d'articles. Il n'est pas certain qu'il fut l'inspirateur de l'article *La situation des choses à Rome*, paru dans *Le Moniteur universel* du 14 février, qui fut regardé à l'époque comme le manifeste des évêques de la minorité [4], mais c'est lui qui dès les premières semaines avait par ses instances contribué à provoquer l'intervention du P. Gratry. Encouragé par l'évêque d'Orléans et négligeant les conseils de prudence que lui prodiguaient ses amis, l'ardent oratorien, qui s'inquiétait des difficultés supplémentaires qu'une définition de l'infaillibilité entraînerait pour l'apostolat chrétien, entreprit de montrer, en exploitant notamment la fameuse question d'Honorius, que

dans l'histoire de l'esprit humain, il n'est pas une question théologique, philosophique, historique ou autre qui ait été aussi déshonorée par le mensonge, la mauvaise foi, le travail des faussaires, si totalement gangrénée par la fraude.

Philosophe et mathématicien, il n'avait pour ce faire aucune préparation théologique ni historique, mais son talent littéraire était incontestable et c'est beaucoup en France ; aussi ses *Lettres publiques à Mgr Dechamps* furent-elles l'événement qui passionna l'opinion parisienne au début de 1870. Les réponses ne tardèrent pas : l'évêque de Strasbourg, dans un mandement auquel s'associèrent de nombreux évêques, désavoua les brochures de Gratry en termes sévères ; l'archevêque de Malines publia coup sur coup quatre lettres d'une haute inspiration où alternaient réfutations et exhortations ; et dom Guéranger, qui venait précisément de publier une réponse très remarquée à Mgr Maret : *De la monarchie pontificale*, s'empressa de la compléter par une réponse à Gratry : *Défense*

---

(1) Moriarty à Newman, 20 février 1870, dans C. Butler, *op. cit.*, t. II, p. 30.
(2) Texte reproduit dans *Coll. Lac.*, t. VII, col. 1513-1514 ; voir sur l'incident W. Ward, *The Life of cardinal Newman*, t. II, p. 287-293 et C. Butler, *Life and Times of B. Ullathorne*, t. II, p. 61-63.
(3) H. Icard, *Journal*, p. 136, 275-276, 328.
(4) Le texte de l'article est reproduit par J. Friedrich, *Documenta spectantia...*, t. I, p. 132 et suiv. Cf. T. Granderath, *op. cit.*, t. II, 2, p. 218 et suiv. ; J. Friedrich, *Tagebuch*, p. 195, désigne plutôt Darboy comme l'inspirateur principal. L'auteur était l'abbé Hogan, sulpicien (Mgr Foulon à l'abbé Tapie, 23 mai 1870, dans *Les Lettres*, juillet 1928, p. 316). « Malheureusement, le fond de tout cet article est *très vrai* », notait M. Icard. (*Journal*, p. 211.)

de l'Église romaine, dans laquelle il n'eut pas de peine à écraser de son érudition bénédictine le pauvre « oiseau bleu » que Veuillot se chargea pour sa part de ridiculiser en termes plus badins [1].

Le bruit de cette polémique s'apaisait à peine que Montalembert livrait à la publicité une lettre dans laquelle il ne se bornait pas à saluer « le prêtre éloquent et intrépide » qui, à la suite du « grand et généreux évêque d'Orléans », avait eu le courage « de se mettre en travers du torrent d'adulation, d'imposture et de servilité où nous risquons d'être engloutis », mais où il prenait à parti directement

ces théologiens laïcs de l'absolutisme — lisez : Veuillot — qui ont commencé par faire litière de toutes nos libertés... devant Napoléon III, pour venir ensuite immoler la justice et la vérité, la raison et l'histoire, en holocauste à l'idole qu'ils se sont érigée au Vatican [2].

Le retentissement de cette lettre malheureuse fut encore accru par un double incident. La lettre était à peine publiée que Montalembert s'éteignait, à l'aube du 13 mars ; or, son beau-frère, Mgr de Mérode, ayant commandé un service solennel à l'*Ara Coeli*, Pie IX, encore sous le coup de son indignation, interdit la cérémonie [3]. Il avait en outre adressé la veille à dom Guéranger un bref de félicitation qui éclata comme « un coup de foudre » [4], car il apparaissait comme un blâme non déguisé à l'adresse de la minorité et de tous ceux qui « poussaient l'impudence jusqu'à désigner du nom de *parti ultramontain* l'ensemble de la famille catholique qui ne pense pas comme eux » [5].

ET EN ALLEMAGNE   A vrai dire, à part un bref succès de curiosité, ces controverses n'eurent qu'une action limitée sur la masse de l'opinion française, qui était ou trop indifférente au point de vue religieux ou trop ancrée déjà dans ses convictions ultramontaines. Il n'en alla pas de même dans les pays germaniques, où même des gens d'une orthodoxie à l'abri de tout soupçon et qui s'étaient jusque-là tenus soigneusement à l'écart de la controverse antiromaine, comme le patrologue Reithmayr, commençaient à s'agiter [6]. Témoin du trouble des esprits, Janssen, qui était loin d'être un excité, avertissait ses amis romains qu'il s'agissait d'autre chose que d'« une poignée de savants » et, en leur signalant qu'on commençait à parler d'une « communauté vieille-catholique », il prévoyait comme possible « une grande catastrophe ecclésiastique » [7].

Doellinger avait beau jeu d'attiser cette agitation. Il avait longtemps préféré sauvegarder ce qui lui restait de prestige auprès de l'opinion catholique en agissant sous le voile de l'anonymat, mais lorsqu'il fut

---

(1) Sur la polémique Gratry-Dechamps-Guéranger, voir A. CHAUVIN, *Le P. Gratry*, Paris, 1911 chap. XXIII ; T. GRANDERATH, *op. cit.*, t. II, 2, p. 191-217 ; dom DELATTE, *Dom Guéranger*, t. II p. 363-367 et une page brillante d'É. OLLIVIER, *op. cit.*, t. II, p. 49-52.
(2) Le texte de cette lettre à M. Lallemand, publié le 7 mars 1870 dans la *Gazette de France*, se trouve reproduit dans É. LECANUET, *Montalembert*, t. III, p. 466-469.
(3) L'incident est rapporté avec des détails précis et peu connus dans une lettre de Mgr Foulon du 23 mars 1870, dans *Les Lettres*, juin 1928, p. 200-204.
(4) Mgr Pie à Guéranger, 31 mars 1870, dans dom DELATTE, *Dom Guéranger*, t. II, p. 371.
(5) Le texte de ce bref du 12 mars 1870 se trouve dans *Coll. Lac.*, t. VII, col. 1537-1538.
(6) A.-M. WEISS, *Erinnerungen*, dans *H.-P. Bl.*, 1908, vol. I, p. 297-299.
(7) *Janssens Briefe*, édit. PASTOR, t. I, p. 387-395.

question du postulat en faveur de l'infaillibilité, il estima le moment venu de jeter le poids de ce prestige dans la bataille et, le 21 janvier, il publia sous son nom un article virulent contre la définition projetée. L'article, applaudi bruyamment par les catholiques ultralibéraux ainsi que par plusieurs universités, fut approuvé en sourdine par beaucoup de ceux qui redoutaient de voir le concile canoniser les théories extrêmes de Ward et de Veuillot. Mais les évêques allemands de la minorité, plus avisés que Dupanloup, se rendaient compte que ces attaques sans nuances, loin de servir leur cause, la compromettaient au contraire ; ils étaient d'ailleurs inquiets du radicalisme du professeur de Munich et craignaient que, par-delà l'infaillibilité du pape, ce ne fût l'autorité de tout le magistère ecclésiastique qui sortît ébranlée de ces discussions. Aussi évitèrent-ils de se solidariser avec l'agitation universitaire, tout en s'abstenant pour la plupart de la désavouer officiellement, comme les y conviaient certaines autorités romaines [1] ; Ketteler, suivi peu après par l'archevêque de Cologne, fut le seul à faire une mise au point qui montrait « que l'on peut être à la fois adversaire de la définition et adversaire des idées de Doellinger, ce que, à Rome, on ne peut ou ne veut pas comprendre » [2]. Indifférent à ces réserves épiscopales, Doellinger publia peu après un nouvel article, cette fois contre le règlement modifié, qui fit autant de sensation que le premier.

LES « *LETTRES ROMAINES* » ET
LA QUESTION DU SECRET
En réalité, la polémique de Doellinger allait beaucoup plus loin encore. Il était le principal inspirateur des *Lettres romaines* qui paraissaient depuis le 17 décembre dans l'*Allgemeine Zeitung* sous le pseudonyme de Quirinus [3]. C'était une chronique du concile qui, par une présentation tendancieuse des hommes et des événements, s'efforçait de discréditer par avance toutes les décisions que pourrait prendre l'assemblée, en faisant apparaître celle-ci comme s'apparentant davantage au brigandage d'Éphèse qu'à un véritable concile œcuménique. Le crédit qu'un certain nombre de catholiques sincères attachaient à ces lettres permet de mesurer l'exaspération des passions, mais il faut ajouter que leur succès fut facilité par la maladroite tactique de silence sur les événements de l'assemblée, que les organisateurs avaient voulu imposer à tous les participants. Ces lettres furent en effet, pendant les premiers mois, la seule relation un peu circonstanciée de ce qui se passait dans l'enceinte du concile et, si tout y était présenté à travers un prisme outrageusement déformant, on sentait bien qu'il y avait à la base des renseignements de première main.

(1) Celles-ci, du reste, n'attachaient qu'assez peu d'importance à l'agitation universitaire en Allemagne (cf. *Tagebuch* de Mayer dans C. Wolfsgruber, *Kardinal Schwarzenberg*, t. III, p. 249) et se préoccupaient beaucoup plus de l'opposition menée par Dupanloup (lettre d'Ullathorne du 20 mars 1870, dans C. Butler, *op. cit.*, t. I, p. 288).

(2) Comtesse von Hahn-Hahn à Moufang, 20 février 1870, dans *Hist. Jahrbuch*, t. XLVII, 1927, p. 671.

(3) Sur les informateurs de Doellinger, cf. J. Friedrich, dans *Revue internationale de théologie*, t. XI, 1903, p. 621-628 ; sur la valeur de ces lettres, très partiales mais contenant nombre de traits exacts, voir C. Mirbt, dans *Historische Zeitschrift*, t. CI, 1908, p. 535-536 et F. Vigener, *Drei Gestalten aus dem modernen Katholizismus*, p. 171, à nuancer par C. Butler, *op. cit.*, t. I, p. 256 et suiv. L'*Allgemeine Zeitung* gagna 10.000 nouveaux abonnés pendant le concile.

Effectivement, sans parler des fuites de documents conciliaires, grâce à la complicité de certains imprimeurs, dont profitait notamment l'ambassadeur de France, il y avait des indiscrétions de la part de quelques membres de la minorité qui ne s'estimaient pas tenus par un règlement qui leur avait été « imposé » et ne se faisaient pas faute de communiquer soit à des diplomates soit à des amis influents tous les renseignements qu'ils croyaient de nature à favoriser la campagne anti-infaillibiliste. Doellinger était de la sorte fort bien documenté par les lettres de Friedrich, théologien du cardinal de Hohenlohe, par les rapports de son jeune ami Acton [1], rédigés avec beaucoup de perspicacité, et par les dépêches des ambassadeurs de Bavière et de Prusse, que lui communiquait le président du conseil bavarois. Comme il n'y avait guère de mise au point possible de la part de la presse modérée ou ultramontaine, il avait dès lors beau jeu d'influencer l'opinion publique par un exposé qui dosait habilement les détails irréfutables et les traits caricaturaux.

On finit par se rendre compte à Rome de l'erreur commise : peu à peu, les Pères de la majorité commencèrent à leur tour à donner des renseignements à des journalistes sûrs, afin de les mettre en mesure de riposter et, bientôt, « l'intérieur du concile n'eut plus de secrets pour Louis Veuillot » [2], ce qui ne veut pas dire que ses exposés fussent toujours parfaitement impartiaux. Le pape lui-même autorisa explicitement certains évêques à faire connaître plus ou moins celles des affaires du concile qu'ils croiraient utile de divulguer [3]. Le cas le mieux connu est celui de Manning : pour contrecarrer les menées excitatrices d'Acton auprès du premier ministre d'Angleterre, Gladstone, l'archevêque de Westminster obtint de pouvoir parler ouvertement à son ami Odo Russel, l'informateur à Rome du ministre des Affaires étrangères, lequel trouva dans ses renseignements de quoi justifier sa politique de non-intervention [4].

L'ACTION DIPLOMATIQUE Les efforts de plusieurs gouvernements européens pour essayer de faire pression sur le concile sont un aspect moins bruyant mais non moins dangereux de l'agitation extra-conciliaire. On se souvient qu'en dépit des suggestions du prince de Hohenlohe, les pouvoirs civils avaient préféré se cantonner, avant l'ouverture du concile, dans un état d' « expectative bougonne » [5], mais ils suivaient de près la marche des événements, prêts à intervenir plus activement si l'Assemblée abordait des problèmes susceptibles

---

(1) Acton, qui séjourna à Rome pendant presque toute la durée du concile, y joua un rôle considérable, bien que caché, au service de la minorité. Lié d'amitié avec un bon nombre des évêques allemands et français qui composaient celle-ci, il les excitait dans leur opposition en les fournissant de textes et d'arguments historiques et les poussait à agir auprès de leurs gouvernements respectifs pour que ceux-ci interviennent en vue d'empêcher la définition ; lui-même s'efforça d'obtenir de Gladstone une intervention du gouvernement anglais (une partie des lettres hebdomadaires d'Acton à Gladstone a été publiée par FIGGIS et LAURENCE, *Selections from the Correspondence of the first Lord Acton*, Londres, 1917).

(2) E. VEUILLOT, *Louis Veuillot*, t. IV, p. 146.

(3) T. GRANDERATH, *op. cit.*, t. II, 2, p. 78.

(4) Voir E. PURCELL, *Life of Manning*, t. II, p. 433-447 et 455.

(5) E. JARRY, *L'Église contemporaine*, Paris, 1936, t. II, p. 20.

d'entraîner des répercussions politiques. Or, le schéma *De Ecclesia* remis aux Pères le 21 janvier, et dont les ambassadeurs de France et d'Autriche eurent connaissance même avant sa communication indiscrète à la presse[1], contenait un chapitre sur les pouvoirs de l'Église en matière civile, un chapitre sur la souveraineté temporelle du pape et trois chapitres sur les rapports de l'Église et de l'État : il n'y avait là rien de neuf par rapport à ce que les théologiens catholiques enseignaient depuis longtemps, mais l'excitation des esprits était telle que l'opinion libérale dans toute l'Europe y vit une reprise aggravée des thèses du *Syllabus*.

Le gouvernement autrichien fit aussitôt prévenir Antonelli qu'il allait se trouver forcé de « se départir de l'attitude d'abstention qu'il avait si strictement observée jusqu'ici »[2]. Quant au ministre français des Affaires étrangères, Daru, qui partageait les appréhensions de ses amis catholiques libéraux, et qui savait que plusieurs évêques de la minorité désiraient voir le gouvernement impérial intervenir pour engager le pape et le concile à plus de modération, il rédigea, le 20 février, sans consulter ses collègues, une note chargeant l'ambassadeur de protester avec force contre le schéma et, en invoquant l'article 16 du concordat, de demander la communication de tous les documents conciliaires et l'admission d'un ambassadeur spécial qui soutiendrait devant l'assemblée les réclamations de la France. Mais d'autres ministres, Ollivier en particulier, jugèrent cette démarche inopportune et obligèrent Daru à adoucir fortement le ton de sa note. Antonelli ayant répondu, le 19 mars, par une fin de non-recevoir, le conseil des ministres, tout en renonçant à l'idée d'un ambassadeur extraordinaire, décida de renouveler ses observations dans un *memorandum* qui serait remis au pape avec prière de le soumettre au concile. Ce *memorandum*, dont le texte fut publié au début d'avril, déclencha dans toute l'Europe une grande agitation et la plupart des chancelleries s'empressèrent de l'appuyer ; le représentant de la Prusse en particulier, von Arnim, qui brûlait du désir d'intervenir, le fit sur un ton bien plus violent que celui du *memorandum* lui-même.

Cette agitation inquiéta fort Antonelli et une partie de la Curie, mais Pie IX resta imperturbable. L'évolution de la politique française devait lui donner raison : des divergences de vues entre l'empereur et son ministre des Affaires étrangères sur la question du plébiscite amenèrent celui-ci à démissionner et, le 18 avril, parvenait à Rome une dépêche conçue en ces termes : « Daru se retire, Ollivier remplace, concile libre. » Effectivement, Ollivier se hâta de faire savoir aux autres gouvernements son intention de n'intervenir en rien dans les affaires du concile et, malgré de nouvelles sollicitations de certains évêques, dont Darboy[3], il s'en tint jusqu'à la fin à cette politique de totale abstention.

(1) Dès le 21 janvier, Banneville transmettait à son gouvernement de longs extraits et notamment le texte des canons. (*Arch. Min. A. É. Paris*, Rome, 1045, f⁰ 176.) Or, ce n'est que les 10 et 12 février que la presse bavaroise donna un aperçu du schéma et le texte des 21 canons.

(2) Beust à Trauttmannsdorff, 10 février 1870. (Copie à *Arch. Min. A. É. Paris*, Rome, 1045, f⁰ˢ 276-283.)

(3) Ces interventions des évêques français auprès du pouvoir civil afin de faire pression sur le concile ont été l'objet de nombreuses polémiques et tout n'est pas encore tout à fait clair. Il est certain que Darboy suggéra à plusieurs reprises à l'empereur d'intervenir (voir ses lettres des

ANTICIPATION DE LA QUESTION DE L'INFAILLIBILITÉ

Tandis que se développait en marge du concile l'agitation de l'opinion publique et des chancelleries, l'assemblée poursuivait ses travaux, mais il apparut vite qu'au rythme où l'on allait, le chapitre XI de la constitution sur l'Église, consacré aux prérogatives du pape, n'arriverait pas en discussion avant le printemps suivant. Or, comme l'avaient craint dès le début les évêques allemands de la minorité, et comme ne l'avaient compris ni Maret ni surtout Dupanloup, les manifestations contre l'infaillibilité avaient achevé de convaincre la majorité des évêques de la nécessité d'une prompte définition de cette vérité et, dans le courant de mars, huit pétitions demandèrent que le fameux chapitre fût abordé par anticipation aussitôt que serait terminé l'examen de la première partie de la constitution *De fide*. Certains à Rome voyaient un autre avantage à cette solution : elle permettrait d'« encommissionner » les délicates questions concernant les relations entre l'Église et l'État qui risquaient, plus encore que celle de l'infaillibilité, de susciter des difficultés sérieuses avec les gouvernements.

La minorité s'empressa évidemment de protester, appuyée du reste par un certain nombre d'évêques italiens, dont le cardinal Pecci, qui, bien que partisans de l'infaillibilité et de sa définition, trouvaient peu sage d'exaspérer l'opposition en intervertissant l'ordre normal des discussions [1]. Cette prudence était partagée par les présidents du concile, que Manning et Senestrey essayèrent vainement, au cours de la semaine sainte, de faire changer d'avis. Ils décidèrent alors de faire, avec neuf ou dix *zelanti*, une démarche personnelle auprès du pape. Ils retirèrent de cette visite, qui eut lieu le 19 avril, l'impression que des instructions seraient données le soir même aux présidents, mais leur déception fut grande quand ils constatèrent, les jours suivants, qu'on continuait imperturbablement à préparer en commission la suite du schéma dogmatique traitant d'erreurs relatives à la Trinité. N'y tenant plus, ils firent remettre au pape, le 23 avril, une pétition signée par environ cent Pères, demandant la mise à l'étude immédiate du schéma *De Summo Pontifice*. Mais Pie IX était l'objet de pressantes interventions en sens contraire de la part du

26 janvier, 2 et 21 mai dans *Coll. Lac.*, t. VII, col. 1551, 1567 et 1568) ; que Maret écrivit à l'empereur et à Ollivier (cf. G. BAZIN, *Vie de Mgr Maret*, t. III, p. 165-166) ; que Dupanloup intervint indirectement auprès de Daru et peut-être auprès d'Ollivier (cf. F. MOURRET, *op. cit.*, p. 255-258 ; *Études*, t. LVI, 1892, p. 463-464 ; *Le Correspondant*, déc. 1892, p. 939 ; *R. H. E.*, t. XVIII, 1922, p. 373-375 ; J. FRIEDRICH, *Tagebuch*, p. 186) ; que d'autres évêques n'hésitèrent même pas à suggérer des solutions radicales (voir notamment la copie d'une lettre très forte d'un évêque, dont le nom manque malheureusement, du 2 mai 1870, aux *Arch. Min. A. É. Paris*, Rome, 1046, f⁰ˢ 288-291). Il faut toutefois tenir compte de la remarque du chargé d'affaires français qui écrivait, le 23 mars : « Il est à remarquer que parmi ceux qui souhaitent de nous voir menacer le gouvernement pontifical du rappel de nos troupes, il n'en est peut-être pas un seul qui voie là autre chose qu'un simple moyen d'intimidation propre à ébranler le pape ou, si Sa Sainteté persiste dans ses desseins, à provoquer dans son entourage par la terreur une pression assez forte pour avoir finalement raison de la faiblesse de l'auguste vieillard. » (*Arch. Min. A. É. Paris*, Rome, 1046, f⁰ˢ 134 v⁰-135 r⁰). On connaît du reste la réponse de Dupanloup à Banneville : « Je lui disais que les résistances que l'on pourrait opposer à nos conseils nous conduiraient inévitablement au rappel des troupes et à l'abandon de Rome. Alors, m'a-t-il répondu, nous oublierions tout le reste et nous serions les premiers à nous ranger autour du pape », et l'ambassadeur d'ajouter : « Je n'en doutais pas. » (*Ibid.*, Rome, 1045, f⁰ 367.)

(1) Tel était aussi l'avis de deux infaillibilistes français notoires : Pie et Régnier, qui intervinrent dans ce sens à la députation de la foi (dépêche de Banneville du 4 mai 1870, *Arch. Min. A. É. Paris*, Rome, 1046, f⁰ 308).

cardinal Bilio, qui avait eu plusieurs conférences avec les principaux
membres du Comité international [1] ; de la part de l'ambassadeur de France,
qui faisait remarquer qu'une interversion des travaux « accréditerait
le sentiment que le concile n'avait été convoqué que pour définir l'infaillibilité du souverain pontife » [2] ; de la part de Dupanloup, qui, le 23 avril,
lui adressait une lettre angoissée, le suppliant de consulter « quelques
évêques de chaque nation, des plus expérimentés, des plus désintéressés »
sur cette question qui « met en ce moment l'Europe en feu » et le conjurant
de ne pas donner l'impression de « vouloir emporter l'affaire d'assaut » [3].

Après avoir longuement pesé le pour et le contre, Pie IX trancha
enfin dans le sens désiré par Manning et ses amis. L'assemblée en fut
avertie le 29 avril, mais dès le 27 la députation de la foi avait abordé,
avec l'aide de Franzelin et de Schrader, la mise au point du schéma.
On tomba facilement d'accord pour atténuer ce qu'il y avait d'étrange
à traiter le chapitre XI avant les autres en le développant de manière
à en faire une petite constitution en quatre chapitres : l'institution divine
de la primauté, sa transmission de Pierre aux pontifes romains, la nature
de la primauté, celle de l'infaillibilité. Par contre, deux courants se firent
jour concernant la formule de définition de l'infaillibilité. La formule
primitive était assez vague et dès lors extensible : le pape y était déclaré
infaillible lorsqu'il définissait « ce qui en matière de foi ou de mœurs
doit être admis par l'Église entière » [4] ; il pouvait donc s'agir non seulement de vérités à croire de foi divine, mais également de doctrines théologiques ou de faits dogmatiques, comme la condamnation d'un ouvrage
ou la canonisation d'un saint. Dans le but de se concilier les adversaires
de la définition, en réduisant la portée de celle-ci à ce qui était le moins
discutable, Mgr Martin suggéra de parler de « ce qui doit être *cru de foi
divine* » ; le cardinal Bilio se rallia volontiers à cette proposition, approuvé
par les membres les plus modérés de la députation, tels Dechamps et Spalding, et au grand mécontentement de Manning et Senestrey, qui tenaient
fort à la formule primitive, il fit passer la nouvelle formule, un peu remaniée par Schrader et Franzelin, dans le texte à soumettre aux Pères [5].
Celui-ci leur fut remis le 9 mai, le lendemain du jour où 71 membres
de la minorité, avec les cardinaux Schwarzenberg et Mathieu à leur
tête, avaient déposé une énergique protestation, rédigée par Ketteler,
insistant sur les graves inconvénients qu'il y avait à parler des prérogatives du pape avant d'avoir mis au clair la doctrine sur l'Église dans
son ensemble (ce que le cardinal Bilio avait précisément été amené à

(1) Dépêche de Banneville du 21 avril 1870 (*Arch. Min. A. É. Paris*, Rome, 1046, f⁰ 247) : « Assez
mal accueilli d'abord par le pape, le cardinal Bilio a été rappelé le lendemain matin par le Saint-
Père qui, après s'être plaint du trouble où la conversation de la veille l'avait jeté, a dit au cardinal
qu'il appréciait ses raisons et s'y rendait. »

(2) Télégramme de Banneville du 22 avril. (*Ibid.*, f⁰ 250 v⁰).

(3) Texte dans F. MOURRET, *op. cit.*, p. 270-272. Le pape lui répondit le 2 mai par une lettre
le mettant paternellement en garde contre les illusions du sens propre, dans MANSI, *A. C. C.*, t. 52,
col. 1-2.

(4) « *Quid in rebus fidei et morum ab universa Ecclesia tenendum vel rejiciendum sit.* » (MANSI,
A. C. C., t. 53, col. 243.)

(5) « *Quid in rebus fidei et morum ab universa Ecclesia tamquam de fide tenendum vel tamquam
fidei contrarium rejiciendum sit.* » (MANSI, *A. C. C.* , t. 52, col. 7.)

faire observer quelques jours auparavant en commission et ce qui allait apparaître manifeste au cours des débats).

LES DÉBATS   La discussion en congrégations générales s'ouvrit le 13 mai par un rapport modéré de Mgr Pie qui avait été choisi intentionnellement à cause de la réserve où il s'était tenu chaque fois qu'il s'était agi de mettre la question à l'ordre du jour [1]. Il souligna spécialement que, si certains avaient parlé de l'infaillibilité « personnelle et séparée », il n'était pas question d'attribuer le privilège de l'infaillibilité au pape en tant que personne privée, ni d'opposer le pape à l'Église comme si la tête pouvait vivre indépendamment du corps [2].

Le débat général sur le schéma considéré dans son ensemble se ramena en fait à une discussion sur l'opportunité de la définition, qui prit à certains moments une allure assez passionnée : « Plusieurs orateurs, écrivait l'évêque de Nancy le 23 mai, me font l'effet de parler les poings fermés ou le doigt sur la détente d'un revolver [3]. » Sur les conseils de Rauscher, la minorité s'était préparée depuis le milieu d'avril, en répartissant entre ses membres les critiques à faire, selon un plan rédigé par Hefele [4]. Les uns, comme Hefele lui-même, Maret ou Strossmayer, mirent en relief les difficultés théologiques et historiques que soulevait la doctrine elle-même ; les autres insistèrent surtout sur les inconvénients d'une définition dans les circonstances actuelles. On remarqua particulièrement le discours de Darboy ; il insista surtout sur l'apparente inutilité de la définition :

Je le dis en gémissant : l'Église s'en va de partout (...). Si le monde rejette la vérité quand elle lui est présentée par le corps entier de l'Église enseignante, combien plus ne la rejettera-t-il pas lorsqu'elle lui sera présentée par un docteur infaillible de la veille [5] !

Les orateurs de la majorité se placèrent également sur le double terrain de la justification de la doctrine et de l'opportunité de sa définition. Ils se montrèrent plus d'une fois inférieurs à leurs adversaires en matière de théologie positive, mais plusieurs de leurs exposés ne manquaient pas de solidité et ils se présentaient d'ailleurs surtout comme des témoignages de la foi de l'Église. La question de l'opportunité fut défendue d'une manière particulièrement brillante par Manning et Dechamps. Le premier, dont le long discours fut suivi par tous les Pères avec un intense intérêt, invoqua son expérience de converti, et le second s'appliqua à dissiper les malentendus soulevés autour de la notion d'infaillibilité personnelle, séparée et absolue, et à montrer que ce qu'on demandait au concile de définir, ce n'était pas l'ensemble des positions théologiques du néo-ultramontanisme, mais uniquement la doctrine traditionnelle de saint Thomas et de Bellarmin, qui ne méritait guère la qualification de « nouveau dogme ».

(1) Cf. Mgr BAUNARD, *Histoire du cardinal Pie*, t. II, p. 400-401.
(2) Texte de son discours dans MANSI, *A. C. C.*, t. 52, col. 29-37.
(3) Lettre citée dans *Les Lettres*, juillet 1928, p. 316.
(4) *Tagebuch* de Mayer dans C. WOLFSGRUBER, *Kardinal Schwarzenberg*, t. III, p. 253.
(5) MANSI, *A. C. C.*, t. 52, col. 161..

Au bout de quinze jours, on avait entendu sur le sujet 65 discours, dont 26 provenant de la minorité. La répétition des mêmes arguments devenait fastidieuse et il restait 40 orateurs inscrits. Aussi, le 2 juin, les présidents furent-ils saisis d'une requête signée par 150 Pères, demandant la clôture de la discussion générale, conformément au règlement du 20 février, et la proposition, mise aux voix le lendemain, fut acceptée par la grande majorité des présents, au grand dépit des intransigeants de la minorité [1]. Plusieurs parmi eux espéraient toujours éviter la définition en prolongeant les débats jusqu'à une éventuelle prorogation du concile. Quelques-uns suggérèrent alors d'abandonner celui-ci ; d'autres, dont Dupanloup, de cesser toute intervention active dans les débats ; mais les dirigeants du groupe austro-allemand, Rauscher et Hefele notamment, s'y opposèrent en faisant ressortir les inconvénients de ces propositions extrêmes [2].

LA DISCUSSION DU CHAPITRE III La discussion spéciale sur les détails du texte commença le 6 juin. Deux séances suffirent pour examiner le prologue et les deux premiers chapitres. Les retouches proposées furent surtout d'ordre rédactionnel, mais on notera l'introduction d'un développement rappelant opportunément que l'épiscopat était, lui aussi, tout comme le primat de Pierre, d'institution divine. Le débat sur le chapitre III, concernant la primauté de juridiction, se prolongea au contraire pendant cinq séances, du 9 au 14 juin, au cours desquelles 33 orateurs prirent la parole.

Bien que cette question de la primauté fût au moins aussi délicate que celle de l'infaillibilité, aucune opposition de principe ne se manifesta, mais plusieurs membres de la minorité formulèrent à nouveau la crainte, qui s'était déjà exprimée au cours de l'examen des schémas disciplinaires, de voir le pape et la Curie intervenir de trop près dans la vie des Églises locales sans tenir suffisamment compte du rôle réservé par le Christ à l'épiscopat. On s'en prit spécialement aux épithètes utilisées dans le schéma pour caractériser la puissance pontificale : *episcopalis, ordinaria, immediata*. Le terme « *ordinaire* », appliqué à la juridiction pontificale dans les Églises locales, ne risquait-il pas de faire oublier que les interventions romaines doivent rester exceptionnelles ? Plusieurs uniates exprimèrent la crainte que les termes employés ne soient un obstacle décisif au retour des Orientaux à l'unité et certains d'entre eux, excités par les leaders de la minorité, croyaient même qu'on envisageait de supprimer les privilèges patriarcaux, alors qu'il était uniquement question d'affirmer que ce n'étaient que des privilèges. Les orateurs de la majorité, le cardinal Pitra et Mgr Freppel en particulier, mirent les choses au point en précisant le sens technique des termes employés, sans réussir du reste

---

(1) Cette clôture anticipée du débat était raisonnable : tout avait été dit. Malheureusement, quelques jours auparavant, le pape avait assuré formellement à l'évêque d'Amiens qu'elle n'aurait pas lieu (Procès-verbaux de la minorité française, séance du 3 juin) et surtout elle fut prononcée juste après le discours de Mgr Maret, dont la surdité avait donné lieu à quelques incidents (voir le *Journal* de Mgr Ramadié, dans G. Bazin, *Vie de Mgr Maret*, t. III, p. 191-198) : l'exaspération de la minorité en fut encore accrue.

(2) A. Hagen dans *T. Q. S.*, t. CXXIII, 1942, p. 246-247.

à expliquer clairement comment le pouvoir pontifical se concilie avec l'exercice normal de la juridiction épiscopale. La députation de la foi tint compte en tout cas de certaines observations et, sur les quatre amendements qu'elle accepta, trois provenaient de la minorité.

LE CHAPITRE IV   On aborda le 15 juin la dernière étape, la discussion du chapitre IV. Pendant trois semaines, tandis que l'été romain devenait toujours plus accablant, on subit des discours bourrés de déjà dit. Quelques-uns cependant tranchèrent sur la monotonie générale : celui de Rauscher, le premier jour, qui suggéra comme un terrain d'entente possible entre la minorité et la majorité la formule proposée au XVe siècle par saint Antonin de Florence [1] ; celui de l'évêque de Calvo, un défenseur de l'infaillibilité, qui se révéla un orateur de grande classe ; celui de Ketteler, qui insista, conformément à ses vues sur la constitution corporative de l'Église, sur le recours nécessaire par le pape à ses conseillers naturels, les évêques ; celui surtout du cardinal dominicain Guidi qui, avec la proposition d'une nouvelle formule par le cardinal Cullen, fit du 18 juin la journée à sensation.

L'INTERVENTION DE GUIDI   Guidi, théologien renommé de la majorité, soucieux d'apaiser les objections relatives à l'infaillibilité « personnelle » et « séparée », proposa d'une part de ne plus parler de « l'infaillibilité du pontife romain », mais plutôt de « l'infaillibilité de ses définitions doctrinales », l'assistance divine n'affectant pas la personne du pape, mais seulement certains de ses actes ; et d'autre part, de bien expliquer que si le pape n'a pas besoin de l'approbation des évêques, il n'agit cependant pas indépendamment de l'épiscopat : en effet, le charisme de l'infaillibilité ne jouant que pour autant que le pape enseigne la doctrine traditionnelle de l'Église, un sérieux examen de la Tradition doit nécessairement précéder toute définition et le moyen le plus normal d'y procéder consiste à faire une enquête auprès des évêques, témoins naturels de la foi de leurs Églises. Ces vues mesurées, qui s'écartaient des thèses outrées du néo-ultramontanisme, furent une agréable surprise pour la minorité où l'on n'avait que trop tendance, on l'a déjà vu, à attribuer celles-ci indistinctement à la majorité dans son ensemble. L'évêque de Nancy écrivait à un ami :

Cette journée de samedi fut toute à l'espérance et, de tous ceux que je vis, l'évêque d'Orléans me parut espérer le plus (...). Tous, au moins parmi ceux que j'ai vus, se félicitaient d'avance et disaient qu'enfin le « pont de la concorde » était jeté entre les deux fractions de l'assemblée et que tout finirait bien après avoir commencé fort mal [2].

Mais les propositions de Guidi furent interprétées par certains extré-

(1) Saint Antonin de Florence distinguait entre le pape agissant *motu proprio, singularis* (de son propre mouvement et en son nom personnel) et le pape *utens consilio et requirens adjutorium universalis Ecclesiae*, ne lui reconnaissant l'infaillibilité que dans ce second cas. La proposition de Rauscher fut reprise par Ketteler, Ginouilhac, Maret, David et Meignan ; plusieurs d'entre eux l'interprétaient certainement dans un sens plus ou moins gallican.
(2) Mgr Foulon à l'abbé Tapie, 29 juin 1870, dans *Les Lettres*, juillet 1928, p. 321-322.

mistes de droite et de gauche comme une concession aux vues épiscopalistes d'une partie de la minorité. Ceci explique la réaction de certains membres de la majorité, comme Mgr Jacobini qui déclara que « l'orateur méritait d'être condamné à huit jours d'exercices spirituels »[1], et surtout la « scène » bien connue que Pie IX, informé de façon tendancieuse, fit le soir même au pauvre cardinal, scène doublement regrettable, parce que le pape donnait par là l'impression de vouloir faire pression sur les membres du concile[2] et parce qu'il semble bien avoir prononcé, sous le coup de l'emportement, une phrase maladroite, colportée dès le lendemain sous une forme encore plus agressive : *La Tradizione son'io !* La Tradition, c'est moi[3] !

Au bout de quinze jours, tout avait été dit et redit, et il restait encore plusieurs dizaines d'inscrits. L'accablement causé par la chaleur était tel que les opposants, qui avaient perdu l'espoir d'obtenir un ajournement des débats, jugèrent qu'il valait mieux mettre un terme à ces flots d'éloquence inutile. Les pourparlers décisifs s'amorcèrent le 1er juillet, conduits par Manning et Haynald et, le 4, le président annonça la clôture de la discussion et le renvoi à la députation des amendements proposés, en vue de mettre au point la formule définitive[4].

A LA RECHERCHE D'UNE NOUVELLE FORMULE A la vérité, la députation n'avait pas attendu ce jour pour examiner la question. Depuis plusieurs semaines, une partie de ses membres, encouragés par « plusieurs cardinaux et autres personnages importants du concile »[5], cherchaient un terrain d'entente qui permît d'éviter de faire éclater au grand jour les divisions de l'assemblée. Tandis que les discours se succédaient aux assemblées générales, les tractations se multipliaient « dans les coulisses »[6] en vue d'aboutir à une formule susceptible de rallier un accord à peu près général.

Les discussions avaient en effet mieux fait voir aux partisans de l'infaillibilité la complexité de la question et la nécessité d'être nuancé, et Hefele signalait qu'une partie d'entre eux envisageait maintenant un

---

(1) Procès-verbaux de la minorité française, séance du 18 juin.
(2) Il est difficile, dans l'état actuel de notre documentation, de déterminer dans quelle mesure Pie IX intervint réellement dans les débats et chercha à faire pression sur l'assemblée. Il avait au début déclaré qu'il entendait demeurer en dehors des débats et, pendant les premières semaines, il évita effectivement de prendre position. Mais bientôt les manœuvres des adversaires de la définition et surtout les intempérances de langage de certains d'entre eux l'exaspérèrent et il ne cacha plus ses sentiments de désapprobation à leur égard et le désir qu'il avait de voir l'infaillibilité reconnue par le concile, multipliant les amabilités et les brefs louangeurs envers les plus fervents soutiens de la majorité (voir notamment la lettre d'Ullathorne du 1er avril, dans C. BUTLER, *op. cit.*, t. II, p. 35 : « Le pape a complètement modifié sa manière de faire lors de notre arrivée, quand il affirmait sa complète neutralité vis-à-vis du concile »). Cette attitude contribua à aigrir les membres de l'opposition, qui trouvaient qu'elle manquait de *fair-play* tant que la décision finale n'était pas intervenue.
(3) Lettre de Mgr Foulon du 23 juin 1870, dans *Les Lettres*, juillet 1928, p. 322 ; dépêche de Banneville du 22 juin 1870, *Arch. Min. A. É. Paris*, Rome, 1047, fos 59-61 ; *Journal* de Dupanloup, 18 et 19 juin 1870, dans F. MOURRET, *op. cit.*, p. 299, n. 1 ; cf. J. FRIEDRICH, *Gesch. des Vatik. Conc.*, t. III, 2, p. 16-23 (voir à ce sujet C. MIRBT, dans *Historische Zeitschrift*, t. CI, 1908, p. 552-553).
(4) Voir notamment les extraits du *Tagebuch* cité par O. PFUELF, *Bischof Ketteler*, t. III, p. 98-99.
(5) Procès-verbaux de la minorité française, séance du 24 mai.
(6) Lettre d'Ullathorne du 25 mai 1870, dans C. BUTLER, *op. cit.*, t. II, p. 68.

mezzo termine [1]. De leur côté, les opposants avaient pu constater que la foi de l'Église sur ce point était plus générale et plus ferme qu'ils ne le supposaient ; en outre, l'ardeur même des discussions avait détruit le grand argument des inopportunistes : *Quod inopportunum dixerunt, necessarium fecerunt*, observait-on à juste titre. Déjà, à la fin de mai, à la suite de conférences entre Dechamps, Senestrey, Simor, Dupanloup et quelques évêques français, l'archevêque de Malines avait cru arriver à un accord [2]. Aussitôt après l'intervention de Guidi, il fit une nouvelle tentative, auprès de Darboy et de Ketteler cette fois, leur proposant de prendre ce discours comme base d'une entente [3]. Il savait que d'autres membres de la députation, notamment Mgr Martin, de Paderborn, travaillaient dans le même sens.

A la vérité, cependant, l'accord était moins facile qu'il ne semblait à première vue. En effet, si l'opposition se résignait à présent à l'idée d'une définition et si beaucoup de ses membres avaient renoncé à la thèse gallicane exigeant un accord *subséquent* de l'épiscopat pour qu'on reconnaisse une définition du pape comme infaillible, la plupart craignaient de lire entre les lignes des formules qu'on leur proposait « la doctrine que le pape est l'Église » [4] et ils désiraient en conséquence qu'il soit claire-ment indiqué que celui-ci ne peut, en définissant une vérité, faire abstrac-tion de la foi de l'Église, dont les évêques sont les témoins particulièrement autorisés. Et quand les infaillibilistes modérés de la députation offraient de faire en plein concile des déclarations précisant que tel était bien le sens du texte proposé, ils ripostaient : « Votre schéma n'est pas franc, ayez le courage de mettre dans les canons ce que vous nous offrez de dire » [5], et ils suggéraient de reprendre une formule comme celle de saint Antonin, où il était fait mention de « l'aide et du conseil de l'Église uni-verselle ». Mais c'était précisément ce que ne voulaient pas ceux qui désiraient extirper les dernières traces du gallicanisme et qui craignaient de laisser la porte ouverte à l'idée d'une intervention *nécessaire* de l'épis-copat. Il y avait en outre le mécontentement des ultramontains de la tendance Manning, qui s'agitaient pour qu'on en revienne à une for-mule qui ne limiterait pas l'infaillibilité du pape aux vérités révélées.

C'est dans ces conditions que le cardinal Bilio, qui se dépensait beau-coup en vue d'éviter une scission définitive, avait suggéré au cardinal Cullen de présenter à l'assemblée une formule nouvelle, rédigée avec l'aide de Franzelin et de Kleutgen, qui ne restreignait plus explicitement l'infaillibilité aux vérités de foi divine, mais ne parlait cependant que des *définitions doctrinales*, ce qui écartait notamment les interventions dans le domaine politique, qui inquiétaient tant les gouvernements [6]. Cette formule, sur laquelle Manning et son groupe marquèrent leur accord, paraissait à première vue acceptable à Ketteler, qui notait avec

---

(1) Hefele à von Gessler, 23 mai 1870, dans *T. Q. S.*, t. CXXIII, 1942, p. 246.
(2) Lettre de Ullathorne du 25 mai 1870, dans C. BUTLER, *op. cit.*, t. II, p. 68.
(3) F. MOURRET, *op. cit.*, p. 299-300 (d'après les procès-verbaux de la minorité).
(4) Rauscher à Darboy, 12 juin 1870, *Archives archevêché Paris*, fonds Darboy, casier 14 (11).
(5) Dépêche de Banneville du 25 mai 1870, *Arch. Min. A. É. Paris*, Rome, 1046, f° 393.
(6) Texte dans MANSI, *A. C. C.*, t. 52, col. 751-752 ; voir à ce sujet le *Diarium* de Senestrey dans *Coll. lac.*, t. VII, col. 1701.

satisfaction qu'elle ne présentait pas le pape comme « séparé » et « indépendant » de l'Église dans l'exercice de son infaillibilité [1].

Comme on s'en doute, parmi la centaine de formules proposées par les Pères, la députation retint celle de Cullen, légèrement amendée par l'heureuse addition de l'expression technique : *ex cathedra*. Mais elle la compléta par un long préambule historique, dû à Mgr Martin, qui marquait davantage la convenance qu'il y avait à ce que le pape prît conseil et s'inspirât de l'attitude de l'Église et de l'épiscopat ; en outre, conformément à la suggestion de Guidi, on intitula le chapitre : *De Romani Pontificis infallibili magisterio*.

ULTIMES DISCUSSIONS Le 11 juillet, Mgr Gasser, au nom de la députation, expliqua à l'assemblée la portée des modifications introduites [2]. Ce commentaire autorisé, qui dura près de quatre heures, est d'une importance capitale pour saisir les nuances du texte conciliaire. Il montre à quel point on avait tenu compte des désirs légitimes de la minorité et combien on était loin de la définition extensible à l'excès dont avaient rêvé certains au début du concile. Il mettait en relief les passages qui avaient été ajoutés pour sauvegarder les droits des évêques et l'union intime du pape et de l'Église ; il soulignait les conditions précises exigées pour qu'un décret pontifical fût infaillible ; enfin, en ce qui concerne l'objet de l'infaillibilité, il faisait observer que le texte se bornait à affirmer l'exacte coïncidence entre l'infaillibilité du pape et celle de l'Église ; dès lors, puisqu'il était hérétique de nier l'infaillibilité de l'Église en matière de vérités révélées, il le serait désormais de nier celle du pape sur ces mêmes objets ; mais par contre, puisque les théologiens n'étaient pas d'accord pour considérer comme un dogme l'infaillibilité de l'Église en matière de « faits dogmatiques », il ne serait pas hérétique de dénier celle-ci au pape. Les évêques français de l'opposition estimaient, après avoir entendu le discours de Gasser, qu'il avait « exposé la question dans un sens susceptible de satisfaire la minorité du concile » [3], à condition toutefois que l'on modifiât dans le sens de ce commentaire oral la formule de définition, en y disant explicitement que pour exercer son magistère infaillible le pape devait s'appuyer sur la foi ou sur la tradition de l'Église.

L'entente espérée allait-elle enfin se réaliser ? Tandis que Dechamps s'efforçait de réduire les dernières objections de Ketteler [4], Martin et Gasser, appuyés par quelques Italiens, laissaient entrevoir, à propos du fameux *consensus* de l'Église, une formule transactionnelle [5]. Pourtant, lorsque, le 13 juillet, les Pères durent se prononcer nommément sur

(1) Voir la minutieuse analyse qu'il en fit, dans *Historisches Jahrbuch*, t. XLVII, 1927, p. 715-721.
(2) Voir le texte du discours dans MANSI, *A. C. C.*, t. 52, col. 1204-1230.
(3) Procès-verbaux de la minorité française, séance du 11 juillet.
(4) Voir les deux lettres qu'il lui adressa le 7 juillet (dans RAICH, *Briefe an und von Ketteler*, p. 548 et suiv.) et le 12 juillet (dans *Revue des sciences philosophiques et théologiques*, t. XXIV, 1935, p. 296-299).
(5) Voir les lettres de Hefele à Friedrich des 7 et 9 juillet 1870, dans J. FRIEDRICH, *Tagebuch*, p. 403-405.

l'ensemble du projet, non seulement une cinquantaine d'entre eux, présents à Rome, s'abstinrent de paraître, mais sur les 601 votants, il y eut 88 *non placet* et 62 *placet juxta modum* (quelques-uns provenant d'ailleurs d'infaillibilistes qui regrettaient qu'on ne fût pas allé assez loin). Un quart de l'assemblée marquait donc son désaccord, y compris trois cardinaux, deux patriarches, les archevêques de sièges aussi importants que Paris, Lyon, Milan, Cologne, Munich, Halifax, Saint-Louis, de nombreux évêques, dont plusieurs de renommée mondiale, comme Dupanloup, ou considérés jusque-là comme des champions des idées romaines, tel Ketteler. « On était donc bien loin de cette fameuse unanimité morale, qui n'était sans doute pas une condition pour la validité des décisions conciliaires, mais dont il était à tous égards bien souhaitable que l'on se rapprochât [1]. »

Pourquoi la minorité, dont une bonne partie des membres ne demandait au fond que des concessions minimes, s'était-elle finalement, après quelques hésitations, refusée à une attitude plus conciliante ? Faut-il supposer qu'à la suite d'un télégramme de Paris, faisant prévoir comme imminente la guerre et ses conséquences [2], les opposants crurent de nouveau possible de voir ajourner la décision définitive ? Il est plus probable que beaucoup avaient été péniblement impressionnés par un incident qui s'était produit le 5 juillet : la modification *in extremis*, sous la pression de Manning, du canon annexé au chapitre III, afin de viser plus explicitement Maret [3], n'était pas seulement une irrégularité de procédure, c'était « une grave erreur de jugement de la part de la députation » [4], qui permettait de mettre en doute la sincérité de ses efforts conciliateurs et indiquait que l'influence des intransigeants, soutenus en sous-main par Pie IX [5], restait prépondérante. Mais la raison principale fut l'opinion qu'un vote massif contre le schéma proposé augmenterait les chances de le voir amendé en dernière minute.

Effectivement, aussitôt après le vote, les tractations reprirent, notamment entre Ketteler et Mgr Sforza [6], et Ginouilhac fit savoir au cardinal de Luca que « l'infaillibilité dans l'Église ayant été divinement accordée *cumulative* à son magistère, il ne paraît pas possible de souscrire à une définition de l'infaillibilité du pape sans le concours des évêques, au

---

(1) É. AMANN, art. *Vatican*, dans *D. T. C.*, t. XV, col. 2575.
(2) D'après le témoignage de Mgr Pie, cité par Mgr BAUNARD, *Histoire du cardinal Pie*, t. II, p. 411.
(3) L'incident est raconté en détail dans T. GRANDERATH, *op. cit.*, t. III, 1, p. 387-391, mais en dissimulant l'irrégularité. Sous prétexte de reprendre un amendement de l'évêque de Chartres, qui atténuait le texte du canon, la députation introduisait un nouveau membre de phrase qui en aggravait la portée en déclarant hérétiques ceux qui diraient que dans l'exercice de la juridiction suprême, le pape avait « *tantum potiores partes, non vero totam plenitudinem hujus supremae potestatis* ». (Déjà au début de juin, Mgr Dechamps avait proposé une insertion de ce genre, puis avait retiré son amendement dans des circonstances assez confuses, voir le *Journal* de Mgr Ramadié en date des 10, 11 et 12 juin, dans G. BAZIN, *Vie de Mgr Maret*, t. III, p. 200-202.) Haynald, puis Darboy, firent observer que ce passage contenant des éléments nouveaux, il fallait d'abord le soumettre à la discussion de l'assemblée ; on dut leur donner raison.
(4) C. BUTLER, *op. cit.*, t. II, p. 87.
(5) Devant les protestations véhémentes de la minorité, la députation de la foi avait décidé de retirer l'addition controversée (MANSI, *A. C. C.*, t. 53, col. 273) ; mais Pie IX, averti, fit prévenir son président qu'il désirait le maintien de l'addition. (MANSI, *A. C. C.*, t. 52, col. 1119-1120 en note.)
(6) Cf. U. RIED dans *Historisches Jahrbuch*, t. XLVII, 1927, p. 724.

moins indiqué dans la formule » [1] ; enfin le 15 au soir, après une vaine
tentative auprès du cardinal Bilio, Darboy, mandaté par le Comité
international, se rendit chez le pape à la tête d'une délégation comprenant
le primat de Hongrie, les archevêques de Lyon et de Munich et les évêques
de Mayence et de Dijon, afin de demander la suppression des mots controversés dans le canon du chapitre III et l'insertion, dans la définition du
chapitre IV, parmi les conditions de l'infaillibilité pontificale, de quelques
mots impliquant une certaine participation de l'épiscopat [2]. Dupanloup,
de son côté, écrivait au pape pour le conjurer d'accepter cette proposition [3] et Rauscher, le 17 juillet, fit une dernière tentative auprès de
Pie IX, le suppliant « de vouloir bien examiner les choses par lui-même,
en faisant ressortir le danger qu'il y aurait à définir comme de foi, sous
peine d'anathème, ce qui ne serait qu'opinion d'école » [4].

Mais les dirigeants du concile, loin de se laisser impressionner, décidèrent au contraire, puisque la minorité se montrait quand même intrai-
table, de faire droit au dernier moment à la demande de ceux qui souhaitaient voir souligner plus nettement encore la condamnation du gallicanisme, en ajoutant à la phrase : *Romani Pontificis definitiones esse ex
sese irreformabiles*, les mots : *non autem ex consensu Ecclesiae*, qui furent
acceptés par assis et levés le 16 juillet. Ils firent en outre désavouer par
l'assemblée unanime deux brochures françaises récentes, qui mettaient en
doute la liberté réelle du concile [5], ce qui était un moyen indirect de faire
reconnaître officiellement celle-ci.

LE VOTE FINAL   La minorité, grâce à laquelle la définition avait été
sensiblement nuancée et précisée, n'avait pas réussi
à écarter celle-ci. Quelle attitude allait-elle prendre ? Quelques-uns
estimèrent que les concessions obtenues étaient suffisantes et décidèrent
de renoncer à une opposition désormais inutile. Plusieurs trouvaient
au contraire qu'il fallait répondre *non placet* en séance publique pour
souligner l'absence d'unanimité morale. D'autres enfin, tout en refusant
d'accepter une formule qui ne leur donnait pas satisfaction, appréhendaient, comme Mgr Mathieu le fit remarquer à la réunion de la minorité
française, qu'un vote négatif n'apparût comme une bravade à l'égard
du souverain pontife et ne donnât lieu à des manifestations hostiles « partant des tribunes où seraient réunis tous les prêtres exaltés » [6]. Dupanloup
de son côté insista sur le danger qu'il y avait de scandaliser les fidèles

---

(1) Procès-verbaux de la minorité française, séance du 14 juillet.
(2) Voir le texte de la pétition remise au pape dans Mansi, *A. C. C.*, t. 52, col. 1322.
(3) Quelques heures plus tard, dans une nouvelle lettre, il faisait une nouvelle suggestion :
que le pape, après avoir recueilli les suffrages de la majorité à la session solennelle, déclare que
« par prudence et modération apostolique » il préférait attendre que le calme revînt dans les esprits
avant de confirmer le vote conciliaire. (Texte des deux lettres dans Mansi, *A. C. C.*, t. 52, col. 1321
et 1323.)
(4) Procès-verbaux de la minorité française, séance du 17 juillet.
(5) L'une de ces brochures, *Les derniers jours du concile*, était une véritable diatribe ; l'autre,
Ce qui se passe au concile, publiée en mai, était, quoique partiale, moins violente et constitue peut-
être l'acte d'accusation le plus sérieux qui ait été rédigé. Sur la liberté du concile, cf. *supra*, p. 335,
n. 4.
(6) Procès-verbaux de la minorité française, séance du 16 juillet.

en votant publiquement *non placet* [1]. En dépit de l'opposition de certains intraitables, comme Haynald [2], l'avis prévalut donc qu'il valait mieux s'abstenir et quitter Rome sur-le-champ. Une lettre, conçue en termes respectueux, avertit le pape de cette décision et des motifs qui l'inspiraient. Cinquante-cinq évêques la signèrent, sans compter six autres qui écrivirent dans le même sens en leur nom personnel.

De la sorte, lorsque le 18 juillet au matin, au milieu d'un orage épouvantable, lecture eut été donnée du texte définitif de la constitution *Pastor Aeternus*, à part deux prélats qui n'avaient pas été prévenus de la décision prise la veille, les 535 Pères présents l'approuvèrent unanimement et, lorsque le pape eut ratifié leur vote, ses paroles furent accueillies par une immense acclamation, qui se répandit vite à travers la basilique jusque sur le parvis. Les deux évêques qui avaient répondu *non placet* vinrent alors déclarer qu'ils adhéraient à leur tour au nouveau dogme et s'unirent au *Te Deum* qui clôtura la cérémonie.

§ 4. — Après le concile [3].

LA PRISE DE ROME ET LA PROROGATION DU CONCILE

Sous l'effet combiné des chaleurs estivales et de la déclaration de guerre franco-allemande, la plupart des Pères quittèrent Rome sitôt après la session solennelle du 18 juillet. Le pape, qui les avait, deux jours auparavant, autorisés à s'absenter jusqu'au 11 novembre, souhaitait cependant que le concile poursuive ses travaux au ralenti dans l'intervalle [4], mais les événements politiques en décidèrent autrement. Profitant du retrait de la brigade française qui protégeait le petit État pontifical, le gouvernement italien annonça, le 29 août, son intention de revoir la situation et de ne laisser au pape pleine juridiction et souveraineté que sur la Cité léonine. La défaite de la France et le renversement de l'empire précipitèrent les événements. Assurés que ni la France ni l'Autriche (cependant instamment sollicitée par le Saint-Siège) n'interviendraient, les Italiens, après l'échec d'une dernière tentative pour amener le pape à un accord à l'amiable, se décidèrent à attaquer.

L'armée pontificale se retira sur Civita-Vecchia, qui capitula le 15 septembre, puis sur Rome. Pie IX avait décidé que le général Kanzler capitulerait sitôt que les premiers coups de canon auraient manifesté la vio-

---

(1) Témoignage de Mgr Dupont des Loges recueilli par M. Branchereau dans une note du *Journal intime de Mgr Dupanloup*, Paris, 1902, p. 310-311.

(2) O. Russel à Manning, 17 juillet 1870, dans E. PURCELL, *Life of Manning*, t. II, p. 447. La décision de s'abstenir fut prise par 33 voix contre 22. (Lettre de Mgr Foulon du 18 juillet 1870, dans *Les Lettres*, juillet 1928, p. 328.)

(3) BIBLIOGRAPHIE. — Sur les origines du *vieux-catholicisme* : J.-F. VON SCHULTE, *Der Altkatholicismus*, Giessen, 1887 (nombreux documents) ; C.-B. Moss, *The Old Catholic Movement*, Londres, 1948 ; P. GESCHWIND, *Geschichte der Entstehung der christ-katholischen Kirche der Schweiz*, Berne, 1904-1910, 2 vol. (ces trois ouvrages sont à utiliser avec réserves) ; voir aussi les biographies de Doellinger, toutes deux partiales et à compléter l'une par l'autre, de J. FRIEDRICH (t. III, Munich, 1901) et de E. MICHAEL, S. J. (Innsbruck, 1894).

(4) Le 26 juillet, on avait distribué aux évêques qui demeuraient sur place un projet de constitution sur les missions, ainsi que le texte retravaillé des schémas concernant les évêques, et deux séances eurent encore lieu sur ce sujet le 23 août et le 1er septembre, en présence respectivement de 127 et de 104 Pères.

lence qui lui était faite. En fait, il y eut une brève résistance, mais dès
10 heures du matin, le 20 septembre, le général Cadorna pénétrait dans la
ville par la brèche pratiquée dans la muraille près de la *porta Pia*. Un
accord, signé dans l'après-midi, spécifiait que la Cité léonine ne serait pas
occupée, mais Antonelli lui-même demanda, le 25, à Cadorna d'y établir,
« comme dans le reste de Rome, des postes de police et un service régulier
d'administration militaire ». Les motifs de cette décision restent obscurs :
trahison intéressée, comme l'ont pensé certains ? crainte sincère de voir la
sécurité du Saint-Père compromise par les bandes d'anticléricaux qui
venaient manifester jusque sur la place Saint-Pierre ? ou peut-être simplement cette « politique du pire » déjà conseillée en 1848 par l'astucieux
cardinal qui espérait ainsi émouvoir davantage l'Europe catholique ?

Lorsque, le 9 octobre, Rome et les provinces adjacentes eurent été
annexées au royaume d'Italie à la suite d'un plébiscite [1], Pie IX estima
qu'en dépit des assurances prodiguées par le gouvernement, la liberté
du concile n'était plus assurée et, le 20 octobre, il déclara celui-ci prorogé
sine die. La proposition, avancée par quelques évêques, de le poursuivre à
Malines, au cours de l'hiver, ne rencontra aucun écho [2].

A la vérité, le concile n'avait fait que commencer : 51 schémas, dont 28
de nature disciplinaire, restaient encore à voter, et la plupart n'avaient
même pas été distribués aux Pères. A comparer l'ampleur du programme et
surtout des espérances que son annonce avait fait naître, avec les résultats
immédiats, le concile apparut comme un échec aux yeux de beaucoup de
contemporains, surtout de ceux qui estimaient sans grande utilité la
définition de l'infaillibilité pontificale. Avec le recul du temps, cependant,
on s'aperçoit mieux des conséquences importantes de l'intense mouvement
d'esprit provoqué par le concile. Les travaux des commissions de la discipline ecclésiastique, des religieux, des Églises orientales et des missions,
de même que les nombreuses suggestions remises sur ces sujets par les
Pères à la commission *De postulatis* furent loin d'être perdus : ils devaient
être en effet abondamment utilisés, quelques dizaines d'années plus
tard, par les rédacteurs du Code de droit canon [3]. Contrairement aux
apparences, le concile du Vatican a donc exercé, bien qu'à retardement,
une influence considérable sur l'adaptation de la discipline ecclésiastique.
On trouverait de même des traces sensibles des thèses élaborées par la
commission des affaires politico-ecclésiastiques dans les grandes encycliques sociales et politiques de Léon XIII. La constitution *Dei Filius*,
de son côté, a exercé sur l'enseignement théologique une profonde
influence, en particulier en ce qui concerne la question brûlante des rapports entre la raison et la foi : les traités *De religione revelata* et *De fide*

---

(1) Rome approuva l'annexion par 40.785 *oui* contre 46 *non* ; la campagne environnante par
133.681 *oui* contre 1.507 *non*. On notera toutefois que la population romaine, qui appréciait l'aspect
patriarcal du gouvernement de Pie IX, n'avait pas attendu l'entrée des Italiens avec impatience
et exaltation. (Cf. A.-M. GHISALBERTI, *A Roma, estate 1870. Dai dispacci di un diplomatico olandese*,
dans *Camiccia rossa*, t. XVII, 1941, p. 245-253.)

(2) La proposition, approuvée par Mgr Dechamps, et appuyée par Mgr Manning, fut adressée
par Mgr Spalding au cardinal Barnabo le 22 octobre. (Voir l'analyse de sa lettre dans T. GRANDE-
RATH, *op. cit.*, t. III, 2, p. 186-188.)

(3) Voir F. CIMETIER, *Les sources du droit ecclésiastique*, Paris (1930), p. 139-146.

ne peuvent plus être les mêmes depuis 1870 qu'auparavant. Enfin la consécration solennelle du mouvement ultramontain qui se développait depuis un demi-siècle, si elle n'a pas amené dans le gouvernement de l'Église la révolution qu'avaient fait prévoir les opposants, a cependant entraîné un renforcement de l'action directe du Saint-Siège dans les diocèses et accentué la centralisation pontificale. Par ailleurs, il faut reconnaître avec Mgr Amann que « en circonscrivant d'une manière plus précise la prérogative de l'infaillibilité doctrinale, à peu près unanimement reconnue dans la pratique avant 1870, le concile a écarté les idées exagérées et même fausses que certains théologiens avaient mises en circulation » [1]. C'est du reste ce caractère mesuré et nuancé de la définition, qui donnait au fond satisfaction aux légitimes exigences de la quasi-totalité des évêques, qui explique que tous les opposants s'y soumirent assez vite, une fois éloignés du feu de la lutte.

LA SOUMISSION DE L'ÉPISCOPAT L'adhésion au nouveau dogme ne posait pas de problème à ceux qui ne s'y étaient opposés que parce qu'ils en jugeaient la proclamation inopportune. Ceux qui avaient marqué des réserves quant au fond même de la doctrine, à cause de tendances plus ou moins gallicanes, ne tardèrent pas non plus à se soumettre, bien que quelques-uns aient quitté Rome décidés à ne pas donner leur adhésion au décret d'un concile qui leur paraissait avoir manqué de la liberté requise.

Cette soumission fut particulièrement rapide en ce qui concerne les évêques français. Outre la pression exercée sur beaucoup d'entre eux par l'ultramontanisme ardent de leur clergé et de la grande masse de leurs fidèles, il faut tenir compte de la guerre et de l'invasion qui les plaçaient soudain devant de tout nouveaux problèmes. Darboy le relevait avec sa coutumière impertinence : « Il ne s'agit plus guère en ce moment du concile ; la France est aux prises avec des difficultés qui relèguent au second plan les sacristains et leurs discussions byzantines [2]. » Mais il y avait aussi des raisons plus profondes. En vertu même du principe gallican, une définition promulguée par le pape et ratifiée par l'ensemble de l'épiscopat devait être considérée comme infaillible. Même si la procédure suivie par la concile avait été illégale, comme certains le pensaient, il fallait du moins reconnaître que l'affirmation de l'infaillibilité pontificale avait été ratifiée à Rome même par une fraction notable de l'épiscopat et que, en y ajoutant les adhésions individuelles qui s'étaient succédé depuis lors, on allait rapidement vers l'unanimité morale. D'ailleurs les nuances introduites dans la définition leur paraissaient, à l'examen, rejoindre l'essentiel de leurs préoccupations et écarter la thèse excessive des néo-ultramontains contre laquelle ils s'étaient insurgés. C'est ce que relevait par exemple Mgr Maret dans une note personnelle :

Le but du parti extrême a été de rendre le pape entièrement et absolument indépendant de l'épiscopat : les discours prononcés, les brochures et les ouvrages

---

(1) Art. *Vatican*, dans *D. T. C.*, t. XV, col. 2583.
(2) Mgr Darboy à Mgr Isoard, 1er septembre 1870, dans A. Bouzoud, *Mgr Isoard*, p. 118.

publiés en font preuve. Ce but n'a point été atteint par le décret, Dieu l'a voulu
ainsi. Un résultat est acquis : c'est que l'infaillibilité du pape ne vient pas de
l'épiscopat comme de sa source ; mais il reste également certain que le concours
et l'assentiment antécédent ou concomitant de l'épiscopat est une *condition*
essentielle de cette infaillibilité [1].

Des raisons analogues facilitèrent la soumission de l'épiscopat allemand. L'évêque d'Augsbourg, qui avait longtemps appuyé Doellinger,
n'écrivait-il pas au cardinal de Schwarzenberg que le décret réduisait
l'infaillibilité dans des limites si étroites qu'il devrait être considéré comme
une victoire de la minorité plutôt que de la majorité [2] ; et plus d'un
évêque, arrivé au concile avec de sérieuses objections doctrinales contre les
thèses ultramontaines, même modérées, avait fini par ne plus s'y opposer
que pour des raisons d'opportunité, impressionné par l'unanimité de la
croyance sur ce point dans une partie importante de l'Église, manifestée
au cours des débats [3]. A l'instigation de Mgr Haynald, les membres de
la minorité germanique s'étaient toutefois mis d'accord pour ne prendre
aucune décision individuellement, sans en référer d'abord aux cardinaux
Rauscher et Schwarzenberg, qui avaient dirigé la résistance pendant le
concile. Très vite cependant beaucoup d'évêques se rendirent compte
qu'il n'était guère possible de continuer à louvoyer sans mettre en danger
aux yeux des fidèles le principe même de l'autorité ecclésiastique. La
dislocation de la minorité fut encore accélérée par le fait que Rauscher,
dont l'influence eût pu être très grande, tomba gravement malade au
milieu d'août [4].

C'est ainsi qu'en Allemagne, où les évêques s'inquiétaient fort de
l'agitation qui régnait dans les milieux théologiques soumis à l'influence
de Doellinger, l'archevêque de Cologne provoqua dès la fin d'août une
réunion épiscopale à Fulda où, sans plus tenir compte des Autrichiens,
on mit au point une lettre pastorale affirmant que la validité des décisions
conciliaires ne pouvait être contestée. La publication de cette lettre,
accueillie avec soulagement par Pie IX et les milieux romains, fut une
grosse déception pour ceux qui avaient escompté une résistance massive
de l'épiscopat germanique. Même des prélats comme l'archevêque de
Munich, Scherr, qui avaient laissé entendre qu'ils n'exigeraient pas l'adhésion de leur clergé au nouveau dogme, s'y étaient ralliés et n'allaient pas
tarder à agir en conséquence.

Seuls cinq évêques s'étaient abstenus de signer le mandement de
Fulda, mais ils refusèrent d'appuyer ouvertement l'opposition au concile
qui s'organisait autour de Doellinger et, au cours des mois qui suivirent,
ils s'inclinèrent l'un après l'autre. Le dernier à le faire fut Hefele, qui ne
publia les décrets du concile dans son diocèse de Rottenbourg que le
10 avril 1871. Revenu de Rome très aigri contre Pie IX, le *perturbator*

---

(1) Reproduit dans G. BAZIN, *Vie de Mgr Maret*, t. III, p. 219.
(2) Mgr Dinkel à Schwarzenberg, 15 novembre 1870, dans T. GRANDERATH, *op. cit.*, t. III, 2,
p. 195.
(3) C'est ce que laissait par exemple clairement entendre l'archevêque de Cologne dans sa lettre
pastorale du 10 septembre 1870. (Voir *Archiv für katholischen Kirchenrecht*, t. XXIV, 1870, p. 109.)
(4) Il fut entre la vie et la mort pendant plusieurs semaines et ne reparut pour la première fois
en public que le 30 mai 1871. (C. WOLFSGRUBER, *J. O. Rauscher*, p. 96-97.)

Ecclesiae, comme il l'appelait, et contre la Curie — « Je croyais servir l'Église catholique et je servais la caricature qu'en a faite le jésuitisme » — le savant historien avait traversé une crise intérieure extrêmement pénible, se croyant obligé de choisir entre la fidélité à ses convictions scientifiques et sa soumission à l'Église : « Reconnaître comme divinement révélé ce qui n'est pas vrai en soi, le fasse qui peut, moi je ne le puis », écrivait-il en septembre à Doellinger [1]. Mais en dépit des espoirs que les chefs de file de la « théologie allemande » avaient placés en lui, son sens catholique finit par l'emporter, au fur et à mesure qu'il se rendait mieux compte de la portée exacte de la définition conciliaire et des rapports qu'elle maintenait malgré tout entre les déclarations du pape et l'enseignement de l'Église universelle [2].

Dans l'empire des Habsbourg, la soumission des évêques fut plus lente, beaucoup estimant qu'il valait mieux ne pas brusquer les choses si l'on voulait éviter un schisme. C'est ce que conseillait Schwarzenberg, qui eut au cours de l'automne et de l'hiver une correspondance intense avec les prélats de la minorité. *Kommt Zeit, kommt Rat*, laissons le temps arranger les choses, puisque le concile n'est quand même pas terminé, recommandait-il, le 4 septembre, dans une lettre où il décrivait ainsi la situation :

> J'ai trouvé, comme je m'y attendais, une grande excitation dans mon diocèse. Quelques personnes pieuses et quelques personnages haut placés qui voient dans la doctrine de l'infaillibilité un renforcement du principe d'autorité, se réjouissaient et s'attendaient à une rapide promulgation du décret. Mais la grande majorité est très troublée ; la quasi-totalité du clergé de langue allemande et slave, tous les hommes prudents, intelligents ou savants, sont abattus et attendent les pires conséquences de la promulgation. Je crois pouvoir dire que la majorité de mes administrés ne se soumettront pas avec vraie foi à la définition [3].

Le nonce à Vienne s'appliqua de son mieux à obtenir au plus vite une adhésion claire des évêques récalcitrants, tandis qu'à Rome on commençait à ne pas renouveler les indults ou les dispenses usuelles, en matière de mariage par exemple, à ceux qui n'avaient pas encore fait parvenir leur soumission. Peu à peu, en effet, l'émotion du grand public s'apaisait et la prétention du gouvernement de réintroduire le *placet* impérial pour la publication des décrets avait déplu aux évêques, qui furent par ailleurs impressionnés par le ralliement très net du cardinal Rauscher et par le commentaire modéré que donnait de la définition Mgr Fessler, l'ancien secrétaire du concile. Le premier estimait qu'on ne pouvait supposer qu'un concile régulier se fût trompé et s'il reconnaissait que toutes les difficultés n'étaient pas résolues, il assignait comme tâche à la science allemande de s'appliquer à les dissiper [4]. Quant au second, à l'encontre de Manning, qui tentait à nouveau d'étendre l'objet de l'infail-

---

(1) Hefele à Doellinger, 14 septembre 1870, dans J.-F. v. SCHULTE, *Der Altkatholicismus*, p. 223.
(2) On peut suivre l'évolution de cette crise dans l'article bien documenté de A. HAGEN, *Die Unterwerfung des Bischofs Hefele unter das Vatikanum*, dans *T. Q. S.*, t. CXXIV, 1943, p. 1-40.
(3) Schwarzenberg aux évêques de la minorité, 4 septembre 1870, dans C. WOLFSGRUBER, *Kardinal Schwarzenberg*, t. III, p. 259-260.
(4) Voir notamment sa lettre du 2 décembre 1870 au comité vieux-catholique de Bonn, dans J.-F. VON SCHULTE, *Der Altkatholicismus*, p. 237-241.

libilité pontificale en prenant le mot *definit* au sens large [1], il insistait sur la signification technique et restrictive de l'expression, comme il devait le répéter dans sa brochure sur *La vraie et la fausse infaillibilité des papes*, dont on sut bientôt que, formellement approuvée par Pie IX, elle pouvait être considérée comme un commentaire officieux.

Au cours des mois de décembre et de janvier, tous les évêques autrichiens se soumirent publiquement. Ceux de Hongrie le firent pendant les premiers mois de 1871, à l'exception des deux plus ardents opposants : Haynald ne s'inclina complètement qu'en octobre, après plusieurs avertissements du nonce et d'Antonelli ; quant à Strossmayer, qui conserva longtemps des relations assez cordiales avec Doellinger et d'autres théologiens en révolte contre le concile [2], il ne s'attaqua jamais ouvertement à celui-ci [3], mais il attendit jusqu'en décembre 1872 avant de s'y rallier sans aucune ambiguïté.

L'ADHÉSION DES FIDÈLES   En dehors des pays de langue allemande, le grand public n'avait guère attaché aux controverses autour de l'infaillibilité qu'un intérêt de curiosité et il n'y avait pas à s'attendre à de fortes réactions. En France, la grande majorité du clergé et des fidèles pieux accueillirent avec joie le nouveau dogme. Seuls certains catholiques libéraux manifestèrent leur mauvaise humeur, mais les défections ouvertes furent extrêmement rares. La seule marquante fut celle d'un correspondant de Doellinger, assez connu dans le monde intellectuel, l'abbé Michaud, vicaire à la Madeleine, qui rejoignit Hyacinthe Loyson en Suisse, mais la tentative de ce dernier, lors de son passage à Paris au début de 1872, pour organiser un noyau de résistance au concile autour d'une petite revue, *L'Espérance de Rome*, n'eut aucun succès. Quant au P. Gratry, qui hésita pendant de longues semaines, il finit par s'incliner, à la grande joie de Mgr Dechamps, qui avait multiplié les avances afin de dissiper l'amertume de son ancien ami.

Beaucoup se demandaient avec curiosité ce qu'allait faire lord Acton, le disciple de prédilection de Doellinger, le collaborateur des *Lettres romaines*. Mais il y avait en réalité chez le chef de file du libéralisme catholique anglais, en dépit de ses attaques passionnées contre l'ultramontanisme, un profond attachement à l'Église romaine. S'il continuait à manifester son aversion pour le système ultramontain et ne rompit jamais avec Doellinger [4], il se refusa à suivre son vieux maître dans sa révolte et lorsque, en 1874, à la suite d'une intervention dans le *Times* assez désobligeante pour la politique romaine, son évêque lui demanda s'il adhérait vraiment au concile du Vatican, sa réponse fut sans équivoque :

---

(1) Dans la longue lettre pastorale publiée en octobre 1870 sous le titre *The Vatican Council and his definitions*.

(2) Voir les lettres publiées par J.-F. von Schulte, *Der Altkatholicismus*, p. 251-263.

(3) En privé, par contre, il ne cachait pas son opinion ; le 10 juin 1871, il communiquait à Doellinger la copie d'une lettre à Mgr Dupanloup dans laquelle il écrivait : « Je ne puis reconnaître en aucune manière la légitimité du concile du Vatican, ni la validité de ses décisions. » (Cité dans J.-F. von Schulte, *op. cit.*, p. 259 ; cf. p. 255-258.)

(4) Voir entre autres *Letters of lord Acton to Mary daughter of the R. Hon. W. E. Gladstone*, Londres, 1904, p. LV, 141 et suiv. et 185 et suiv.

J'ai accepté les décrets comme les ont acceptés les évêques qui sont mes guides. Ces décrets sont une loi pour moi, comme ceux de Trente, en vertu, non d'aucune interprétation privée, mais de l'autorité dont ils émanent. Les difficultés de leur conciliation avec la Tradition, qui semblent si fortes à d'autres, ne me troublent pas, moi laïque, dont l'affaire n'est pas d'expliquer les questions théologiques et qui laisse cela à de meilleurs que moi [1].

Ce n'est vraiment que dans les pays germaniques qu'un nombre assez considérable de fidèles et des théologiens de marque s'obstinèrent dans leur opposition et préférèrent rompre avec l'Église plutôt que de s'incliner.

LE SCHISME VIEUX-CATHOLIQUE   Tous les gens au courant de la situation religieuse dans les pays germaniques n'avaient cessé de prédire que la proclamation de l'infaillibilité provoquerait inévitablement un schisme. Comme prévu, ce fut parmi les professeurs d'Université que le mouvement débuta. Lorsque, à son retour de Rome, l'archevêque de Munich, tout en avouant sa déception, invita les professeurs de la Faculté de théologie à recommencer comme lui à « travailler pour la Sainte Église », Doellinger répliqua sèchement : « Oui, pour l'*ancienne* Église ! — Il n'y a qu'une Église, reprit l'archevêque, il n'y en a pas de nouvelle ou d'ancienne. — On en a *fait* une nouvelle ! » répliqua le vieux professeur [2]. Dès la fin de juillet, avec l'aide du remuant professeur de droit canon de Prague, von Schulte, il provoquait la réunion à Nuremberg, pour les 25 et 26 août, de représentants des diverses universités, afin de mettre au point une déclaration rejetant les décrets relatifs à l'infaillibilité pontificale. D'anciens günthériens impénitents, comme Knoodt, s'y retrouvèrent avec des tenants de l'école de Munich et avec les chanoines Loewe et Mayer de Prague, envoyés comme observateurs par le cardinal de Schwarzenberg ; par contre, il n'y avait personne de Tubingue, où le sens catholique était trop profond pour qu'on acceptât de s'engager dans la voie de la révolte et où, par ailleurs, l'attitude temporisatrice d'Hefele, dont la Faculté dépendait, facilita les choses.

Quelles étaient les chances de voir le mouvement prendre de l'extension ? Plusieurs se montraient excessivement inquiets, l'historien Janssen par exemple, qui, en mai 1871, écrivait encore : « Si Strossmayer devait prendre la direction des vieux-catholiques, il faudrait s'attendre à un schisme comme l'histoire n'en connaît pas de semblable [3]. » A la vérité cependant, la situation était moins grave. Si des résistances étaient à prévoir parmi les prêtres d'un certain âge, hostiles à l'ultramontanisme et influencés par la conception hermésienne d'une foi basée exclusivement sur la science, le jeune clergé n'était guère touché. Et, parmi les laïcs, la plupart étaient soit des catholiques convaincus, prêts d'avance à accepter tout ce qui venait de Rome, soit des indifférents qui s'inquiétaient peu d'avoir à croire un dogme de plus ou de moins ; pour beaucoup du reste, leur fidélité catholique sera stimulée par le déclenchement du

(1) Lettre du 10 décembre 1874, dans A. Gasquet, *Lord Acton and his Circle*, p. 364. (Voir sur cet incident les cinq lettres publiées *ibid.*, p. 360-370 et H. Paul, *Introductory Memoir*, aux *Letters to Mary Gladstone*, p. xlvi-lv.)
(2) J. Friedrich, *Ignace von Doellinger*, t. III, p. 547.
(3) *Janssens Briefe*, édit. Pastor, t. I, p. 416.

Kulturkampf, qui devait purifier l'atmosphère interne de l'Église. Le vrai danger venait des milieux de fonctionnaires et d'anciens universitaires de l'Allemagne du Sud et de Bohême, qui souhaitaient une Église indépendante de Rome et étroitement inféodée à l'État. Mais ils durent bientôt renoncer à l'espoir qu'ils avaient eu pendant les premières semaines de voir une partie de l'épiscopat appuyer la résistance des professeurs de théologie et se mettre à la tête du mouvement des « vieux-catholiques »[1]. Ni Schwarzenberg, qui s'inquiétait entre autres des répercussions d'un schisme sur le mouvement séparatiste tchèque, ni Hefele, ni même Strossmayer, ne se prêtèrent à leurs vues et les plus avisés comprirent dès lors que, sans évêque, le mouvement n'avait plus de sens et se limiterait à la résistance de quelques intellectuels isolés[2].

Tandis que les laïcs s'interrogeaient, les évêques commençaient à exiger des professeurs de théologie l'acceptation explicite des décrets du Vatican. A Breslau, Reinkens avait déjà été suspendu de ses fonctions en août, les autres professeurs le furent en novembre. L'archevêque de Munich réussit à obtenir la soumission de quelques-uns de ses professeurs, mais, malgré plusieurs prolongations de délai, Doellinger resta intransigeant : « Comme chrétien, comme théologien, comme historien, comme citoyen, je ne puis accepter cette doctrine », répondit-il à son évêque[3] et, le 23 avril 1871, ce dernier dut se résoudre à prononcer contre lui l'excommunication solennelle, malgré la pénible impression qu'une telle mesure devait causer dans une partie de l'Allemagne catholique.

Un mois plus tard, le 28 mai, Doellinger réunissait chez lui les chefs de la résistance et il fut décidé de tenir à l'automne un congrès, qui eut lieu à la fin de septembre sous la présidence de Schulte. Tout en précisant les bases doctrinales du mouvement : la profession de foi de Pie IV, la primauté du pape « telle qu'elle était conçue par les Pères » et une part plus grande aux laïcs dans la direction de l'Église, on prit des mesures pour pourvoir à la vie ecclésiastique. Doellinger, qui n'avait pas envisagé autre chose qu'une protestation du genre prophétique contre la déviation de l'Église catholique, s'opposa à l'organisation de communautés schismatiques qui aboutirait à « dresser autel contre autel », mais la majorité, estimant qu'on ne pouvait laisser sans sacrements les quelques dizaines de milliers de fidèles qui avaient refusé de reconnaître le concile, décida de créer des paroisses partout où la chose serait possible et de faire appel à des évêques étrangers — on songeait à des évêques jansénistes du schisme d'Utrecht — en attendant d'être en mesure d'avoir sa propre hiérarchie. Le Congrès de Munich marquait de la sorte un tournant décisif : malgré Doellinger, les vieux-catholiques cessaient d'être

---

(1) Au mois d'août, d'autres évêques que Schwarzenberg avaient semblé approuver le projet de réunion à Nuremberg, l'archevêque de Munich, par exemple, l'évêque d'Augsbourg, Hefele. (Voir certains documents dans J. FRIEDRICH, *Ignace von Doellinger*, t. III, p. 549-551 et J.-F. VON SCHULTE, *op. cit., passim.*)

(2) Tel était notamment l'avis du prince de Hohenlohe, qui aurait pourtant vu avec satisfaction la constitution d'une Église nationale. (*Denkwürdigkeiten*, t. II, p. 52, en date du 30 avril 1871.) En fait, au moment de sa plus grande extension, vers 1875, le mouvement vieux-catholique compta en Allemagne et en Autriche 52.000 adhérents, groupés en 122 paroisses ; il faut y ajouter 46 paroisses en Suisse, groupant 73.000 fidèles.

(3) Doellinger à Mgr von Scherr, 29 mars 1871, dans J. FRIEDRICH, *op. cit.*, t. III, p. 571.

des catholiques en révolte pour devenir une nouvelle communauté schismatique, dont le développement ne relève plus désormais de l'histoire de l'Église et qui allait vite évoluer dans un sens progressiste et réformiste assez radical, bien différent de la réaction conservatrice qui était apparemment à son origine [1].

Ce caractère révolutionnaire du mouvement, où les tendances libérales l'emportèrent rapidement sur le souci de fidélité à l'antiquité chrétienne, apparut tout spécialement dans la branche suisse, où l'on renonça même au nom de *vieux-catholicisme* pour prendre celui d'*Église chrétienne catholique*. L'histoire de celle-ci est du reste assez particulière : la direction en fut prise par des laïcs, non par des théologiens comme en Allemagne, car les raisons de l'opposition au concile du Vatican y étaient bien plus politiques qu'historiques ; et le mouvement, tout en gardant un contact suivi avec Munich, interféra de près avec le *Kulturkampf* suisse, dont il sera question plus loin.

(1) La chose est bien mise en lumière par C.-B. Moss, *The Old Catholic Movement*, p. 256, 261, 271, 276.

CHAPITRE XI

L'ÉGLISE EN EUROPE AU LENDEMAIN
DU CONCILE DU VATICAN

§ 1. — L'Église en Italie [1].

LA LOI DES GARANTIES ET LA POLITIQUE RELIGIEUSE DU GOUVERNEMENT

Désireux de rassurer les puissances catholiques, les hommes d'État italiens s'empressèrent, au lendemain de l'occupation de Rome, de régler par une loi les deux questions, devenues intimement connexes, de la situation à faire au souverain pontife et des relations entre l'Église et l'État en Italie [2]. Votée le 13 mai 1871, la loi pour « les garanties de l'indépendance du souverain pontife et du libre exercice de l'autorité spirituelle du Saint-Siège » cherchait à assurer au pape « l'indépendance, la liberté et les dehors d'une souveraineté toute spirituelle » [3], mais, n'ayant ni portée internationale ni caractère d'irrévocabilité, elle mettait en fait le Saint-Siège à la merci des gouvernements italiens. C'est pourquoi Pie IX la repoussa solennellement, dès le 15 mai, par l'encyclique *Ubi nos* [4] et refusa les indemnités qu'on lui offrait.

En dépit de certaines mesures favorables, la façon dont la loi entendait régler la question religieuse italienne n'était pas non plus de nature à donner satisfaction à l'Église, car non seulement elle était inspirée par l'idéal libéral de la séparation de l'Église et de l'État, que les autorités ecclésiastiques n'étaient encore nullement disposées à accepter, mais elle se ressentait en outre, aux dépens de la logique d'ailleurs, et malgré les vues personnelles du premier ministre Lanza, des tendances régaliennes qui avaient souvent inspiré depuis quinze ans la politique anticléricale italienne.

---

(1) BIBLIOGRAPHIE. — Outre les ouvrages généraux de S. JACINI et de A. DELLA TORRE, signalés p. 9, ceux sur la Question romaine, cités p. 72, n. 1 et celui de F. OLGIATI, cité p. 97, n. 6, consulter sur le mouvement catholique : E. VERCESI, *Il movimento cattolico in Italia. 1870-1922*, Florence, 1923, chap. II et A. VIAN, *Il conte Paganuzzi (1841-1923)*, Rome, 1950.

(2) On peut en suivre l'élaboration dans S. JACINI, *La politica ecclesiastica italiana*, p. 390-497. (Cf. aussi F. RUESENBERG, *Entstehung und Bedeutung des italienischen Garantiegesetzes*, 1924.) La mise au point du texte définitif fut l'œuvre du jurisconsulte Bonghi (sur lequel on peut voir le jugement sévère de A.-C. JEMOLO, *Chiesa e Stato...*, p. 328-339). Certaines retouches furent demandées par le gouvernement français, notamment la suppression de l'article qui privait le pape de la jouissance des musées du Vatican.

(3) G. MOLLAT, *La Question romaine*, p. 367. L'art. 1 déclarait la personne du souverain pontife sacrée et inviolable ; les honneurs souverains lui étaient dus dans tout le royaume ; on lui reconnaissait le droit d'entretenir des nonces auprès des gouvernements étrangers et à ces derniers la possibilité de maintenir à Rome des ambassadeurs ; on garantissait au pape une rente annuelle de 3.225.000 lires, ainsi que la jouissance des palais du Vatican, du Latran et de Castel-Gandolfo, qui étaient déclarés inaccessibles à toute autorité italienne, de même que l'enceinte d'un conclave ou d'un concile œcuménique ; enfin la libre correspondance du pape avec l'épiscopat et le monde catholique était assurée.

(4) Texte dans H. BASTGEN, *Die Römische Frage*, t. II, p. 684.

D'ailleurs, dans les mois et les années qui suivirent, quel que fût le désir du gouvernement et surtout du roi de donner satisfaction à l'Église — et une fois de plus les interventions de don Bosco permirent de régler certaines choses à l'amiable — les éléments de gauche et les sectaires de la franc-maçonnerie ne négligèrent aucune occasion de donner libre cours à leurs sentiments antireligieux, alimentant par la presse ou des manifestations publiques « une campagne féroce et haineuse contre Pie IX, ses conseillers, la Curie romaine et les prêtres en général »[1] et obligeant les ministres à de nouvelles mesures vexatoires : confiscation du Collège romain[2] et de la plupart des couvents de Rome, y compris les maisons généralices des grands ordres ; obligation du service militaire pour les clercs ; abolition du serment religieux devant les tribunaux ; interdiction des processions et des pèlerinages.

L'INTRANSIGEANCE DU SAINT-SIÈGE — A en croire le comte d'Harcourt, Pie IX, au lendemain de la chute de Rome, ne se serait pas refusé à faire preuve de réalisme : « Tout ce que je demande, lui aurait-il déclaré en avril 1871, c'est un petit coin de terre où je serais le maître. Si l'on m'offrait de me rendre mes États, je refuserais, mais tant que je n'aurai pas ce petit coin de terre, je ne pourrai exercer dans leur plénitude mes fonctions spirituelles[3]. » Il continua en tout cas à entretenir avec le roi une correspondance assez cordiale, n'hésitant pas à lui écrire qu'il priait Dieu pour lui et *pour l'Italie* ; et, sans l'opposition d'Antonelli, il ne se serait pas refusé, au début, à recevoir un émissaire privé[4].

Sur la question de principe toutefois, le pape n'entendait pas capituler, ses paroles au comte d'Harcourt le montrent. Et l'atmosphère de sectarisme à laquelle les ministres italiens se virent obligés de céder n'était évidemment pas pour l'encourager dans la voie de la conciliation, pas plus que les manifestations des catholiques français et autrichiens en faveur de la restauration du pouvoir temporel, ou l'exaltation toujours plus enthousiaste de la papauté lors des grands pèlerinages internationaux organisés pour venir acclamer « le prisonnier du Vatican » ou à l'occasion des fêtes de ses vingt-cinq années de pontificat, en 1871, du huitième centenaire de Grégoire VII, en 1873, du septième centenaire de la victoire de Legnano, en 1876[5].

La plus grande partie de son entourage excellait du reste à l'entretenir dans l'espoir d'un « miracle » auquel il n'avait que trop tendance à croire lui-même et, pendant des années, sans se laisser décourager par aucune déception, le vieux pontife allait « dénoncer sans relâche

---

(1) F. HAYWARD, *Pie IX et son temps*, p. 400.
(2) Celui-ci ne disparut pourtant pas, car il trouva asile au Collège germanique, administré également par les jésuites et protégé par le pavillon allemand. Le 4 décembre 1873, Pie IX lui concéda le titre de *Pontificia Università Gregoriana del Collegio Romano* et, en août 1876, il y adjoignit aux Facultés de philosophie et de théologie une Faculté de droit canon, par suite de la sécularisation de l'Université pontificale de la *Sapienza*.
(3) *Archives diplomatiques*, t. II, 1874, p. 224.
(4) A. MONTI, *Vittorio Emanuele II*, Milan, 1941, p. 390-394.
(5) Voir sur ces fêtes grandioses J. SCHMIDLIN, *Papstgeschichte...*, t. II, p. 296-298.

les Attila et les Achab modernes, refaisant presque chaque jour le même discours, avec la même verve et la même vigueur, avec une fécondité inépuisable et un zèle toujours nouveau. Le triomphe de l'Église, la confusion de ses ennemis, il l'a attendu de chaque pays, de chaque événement, de chaque complication européenne »[1].

Si encore Pie IX s'était borné à affirmer, comme il le disait déjà en 1868 à Mgr Gay : « Le Saint-Siège a vu d'autres tempêtes ; des papes ont été chassés et s'ils ne sont pas rentrés eux-mêmes à Rome, leurs successeurs y sont revenus triomphants[2]. » Malheureusement, plus les années passaient et plus s'accusait chez lui la tendance à identifier les malheurs de l'Église avec les progrès des formes de gouvernement inspirées des principes libéraux, sans se rendre compte du danger qu'il y avait à présenter avec une telle insistance des réalités politiques qui s'avéraient dans le sens de l'histoire, comme l'incarnation nécessaire des idées antichrétiennes et comme associées inévitablement au triomphe d'une philosophie hostile à l'Église. Les porte-parole autorisés s'employaient bien à rassurer les gouvernements en rappelant la distinction entre la thèse et l'hypothèse, mais l'impression sur les masses était déplorable, car elles croyaient devoir souhaiter le triomphe de l'idéologie antichrétienne pour voir aboutir leur idéal politique et social.

L'ATTITUDE DES CATHOLIQUES   Tandis que le pape se durcissait dans son intransigeance, les catholiques italiens réagissaient de manière assez diverse. Pie IX réussit à garder fidèle à sa cause l'ensemble de l'aristocratie[3], mais la bourgeoisie se détourna de plus en plus d'une Église qui semblait exiger d'elle des sentiments antipatriotiques. A vrai dire, les apostasies formelles furent rares et les violences anticléricales demeurèrent le fait d'une minorité de sectaires et d'énergumènes, mais, notait en 1878 le P. Curci, avec quelque exagération d'ailleurs, « la plupart des laïques laissent de côté toutes les choses de la religion et ils attendent pour s'en occuper le moment suprême où l'on envoie chercher un prêtre »[4].

Ce n'était pas le cas pour tous. Il y avait d'une part le groupe, assez réduit à la vérité, de ceux qu'on devait appeler *cattolici transigenti*[5]. Ils croyaient possible de concilier leur foi catholique avec leurs sentiments italiens et même avec une participation effective à la vie publique, contrairement au mot d'ordre lancé par l'abbé Margotti et approuvé par le Vatican : *Ne eletti ne elettori.* Ils prenaient leur inspiration auprès de

---

(1) E. Lecanuet, *Les dernières années du pontificat de Pie IX*, p. 524. On peut suivre au jour le jour la pensée de Pie IX depuis 1870 dans les 556 discours sténographiés et publiés par S. De Franciscis, *Discorsi del S. Pont. Pio IX*, Rome, 1872-1878, 4 vol.

(2) *Correspondance de Mgr Gay*, t. II, p. 82.

(3) On a souvent mal interprété l'intérêt porté par Pie IX à l'aristocratie : « Ce n'est pas par une vaine passion héraldique que le pape flatte le patriciat et attire les patriciens à la cour. Il s'agit de retenir le plus possible à l'Église une classe qui garde encore une influence considérable : l'influence de la richesse territoriale, l'influence des traditions, l'influence du caractère. » (L. Teste, *Préface au Conclave*, p. 201.)

(4) C. Curci, *Il moderno dissidio tra la Chiesa e l'Italia*, Florence, 1878, p. 144 et suiv.

(5) Voir F. Fonzi, *Testimonianze e documenti. I « cattolici transigenti » italiani dell'ultimo ottocento*, dans *Convivium*, 1949, p. 955-972.

laïcs comme Tommaseo ou Conti, dont les convictions religieuses étaient parfois un peu floues, mais ils savaient que plus d'un prêtre inclinait comme eux vers la conciliation ; non seulement ces ecclésiastiques « réformistes » évoluent aux frontières de l'orthodoxie, dont le plus en vue était Mgr Puecher-Passavalli [1], mais également d'excellents prêtres comme le P. Tosti, du Mont-Cassin, le chanoine Audisio, qui devait publier en 1876 une défense de la formule : L'Église libre dans l'État libre [2], et même le P. Curci [3], l'ancien champion de l'intégrisme, que l'évolution des événements avait complètement retourné [4].

Il y avait d'autre part ceux, encore assez nombreux malgré tout, qui, comme la grande masse du clergé, suivaient les mots d'ordre de L'Unità, le pendant italien de L'Univers. C'étaient, pour la plupart, des « bien pensants » avec les qualités de fidélité mais aussi toute l'étroitesse d'esprit que le mot implique ; ils s'organisent de plus en plus en une société repliée sur elle-même, vivant en marge de la vie nationale, évitant autant que possible les contacts avec ceux qui ne pensent pas comme elle et se bornant à les anathématiser à la suite d'une presse dont le ton se fait toujours plus violent. On comptait toutefois parmi eux quelques militants entreprenants, qui allaient chercher à prolonger l'œuvre amorcée quelques années auparavant par l'Association de la Jeunesse catholique et jeter les bases d'une action catholique adulte.

L'ŒUVRE DES CONGRÈS ET L'ACTION CATHOLIQUE
L'idée de grouper les efforts catholiques était dans l'air. En 1869 déjà, un groupe de catholiques florentins avait lancé l'idée ; à Rome, la Société des intérêts catholiques, fondée en décembre 1870 par le P. Curci et le prince Chigi, et surtout la Société promotrice des bonnes œuvres, fondée en 1872, tendaient à devenir le noyau d'une fédération des œuvres, Federazione Piana, dont le principe fut à plusieurs reprises approuvé par Pie IX. Très vite, on estima que des congrès catholiques à l'image de ceux d'Allemagne ou de Belgique faciliteraient ces tentatives d'organisation sur le plan national.

Le conseil supérieur de la Jeunesse catholique prit la chose en main et réussit à réunir le premier congrès en juin 1874 à Venise [5]. L'esprit dans lequel on entendait travailler fut clairement précisé dans les deux discours

(1) Voir BEGEY et FAVERO, S. E. Mgr l'arcivescovo L. Puecher-Passavalli. Ricordi e lettere, Milan, 1911. On peut encore mentionner l'ancien capucin André d'Altagène (P. Panzani) et l'ancien barnabite Gavazzi. (Cf. A. HOUTIN, Le P. Hyacinthe, réformateur catholique, p. 70-73.)

(2) Della societa politica e religiosa rispetto al secolo XIX, Florence, 1876 ; l'ouvrage fut mis à l'Index.

(3) Il présenta en 1875 un mémoire au pape pour l'engager à transiger avec l'Italie. Le pape l'accueillit très mal et, deux ans plus tard, il fut obligé de quitter la Compagnie de Jésus. En 1878, il publia Il moderno dissidio tra la Chiesa e l'Italia, un livre d'une lucidité remarquable, analysant impitoyablement la situation du catholicisme en Italie et qui, lu à la lumière des Accords du Latran, apparaît le plus souvent d'un grand bon sens. Le P. Curci devait par la suite écrire d'autres livres, d'idées plus avancées et de ton plus violent, mais il finit par se soumettre totalement.

(4) Don Bosco lui-même était loin d'être un intransigeant. Et le saint cardinal Sforza, archevêque de Naples, fut le premier à conseiller la lutte électorale au clergé et aux fidèles, en dépit du mot d'ordre officiel ; il soumit ses raisons à Pie IX qui lui demanda de se limiter pour commencer à la politique municipale.

(5) Voir les Atti del primo congresso cattolico italiano, Bologne, 1874. Sur les difficultés de tout genre qui en retardèrent la convocation pendant deux ans, voir entre autres A. SASSOLI-TOMBA, Il primo congresso cattolico italiano, Bologne, 1873.

du professeur D'Ondes Reggio et de J. Saccheti. Le premier, l'une des
figures centrales du mouvement catholique italien au XIXe siècle, déclara
qu'on entendait être catholique « tout court », sans épithète d'aucune
sorte, y compris celle de libéral ; mais le second ajouta qu'il était temps
de ne plus se borner à attendre le « miracle » prédit depuis quinze ans
par la presse catholique et qu'il fallait désormais s'organiser et agir :
« Prions Dieu que la révolution meure demain ; mais travaillons ensuite
comme si elle devait vivre toujours ! »

L'année suivante, on se retrouvait à Florence et l'on décidait de fonder
une « Œuvre des congrès », distincte de la Jeunesse catholique, s'appuyant
sur des comités paroissiaux, sous la présidence des curés, à travers toute
l'Italie, et ayant pour but de coordonner et de promouvoir une floraison
d'œuvres sociales et religieuses. A la vérité, en dépit de beaucoup de
bonne volonté et de nombreux dévouements individuels, ces congrès
devaient rester longtemps des manifestations verbales. Ils étaient handi-
capés notamment par le fait que leur prise de position très nette dans la
Question romaine en écartait la plupart des laïcs cultivés et réalistes, au
point que le nombre des participants était inférieur à celui des congrès
de Malines, dans un pays six fois moins peuplé. La même raison empêcha
le développement d'une presse catholique vraiment convenable, cherchant à donner autre chose que des « nouvelles de sacristie », malgré
quelques efforts louables comme *Le Cittadino* de Gênes ou *L'Unione* de
Bologne, fondée en 1878, l'un des premiers journaux à s'occuper de questions sociales.

L'Association de la Jeunesse catholique, qui comptait déjà des cercles
dans 72 localités en 1874, devait elle aussi souffrir du même handicap.
En dehors de Rome, « qui est une ville privilégiée et peut offrir un beau
noyau de jeunes gens ouvertement catholiques », notait le P. Curci en
1878, « on trouve bien quelques douzaines de bons petits garçons qui
font une communion pour le Saint-Père ou qui quêtent pour le denier
de Saint-Pierre ; ce sont là des choses édifiantes et j'en suis édifié. Mais
de grâce, qu'on ne se moque pas de nous en affichant une grande organisation... La jeunesse a déserté et déserte continuellement l'Église » [1].
Cette mauvaise humeur avait cependant quelque chose d'excessif et,
sous la présidence active d'Acquaderni — Fani était mort prématurément en 1869 — un certain nombre de choses utiles furent réalisées :
dans le domaine purement religieux, l'essor des catéchismes et des congrégations mariales et une participation plus vivante à la liturgie ; dans
le domaine caritatif, le développement des conférences de Saint-Vincent-
de-Paul [2] et des écoles populaires du soir ; dans le domaine culturel,
l'organisation de conférences, de cercles d'études, de salles de lecture
et de bibliothèques ambulantes. C'est aussi à l'initiative de la Jeunesse
catholique que fut fondée en 1875 la *Ligue Daniel O'Connel* en faveur
de l'enseignement libre, qui dès l'année suivante put remettre au Parlement une pétition couverte de 30.000 signatures. En 1877, sous la pression

---

(1) C. CURCI, *Il moderno dissidio...*, p. 146-147.
(2) A. COZAZZI, *Ozanam e le conferenze di San Vincenzo, 1833-1933*, Turin, 1933.

de certains membres qui inclinaient vers la « conciliation », il fut même question d'orienter l'association vers l'action parlementaire, mais Pie IX s'empressa d'intervenir pour éviter ce glissement vers la politique, contraire aux statuts primitifs [1].

§ 2. — L'Église de France et les débuts de la Troisième République [2].

LA PARENTHÈSE DE LA COMMUNE — Au point de vue religieux comme à bien d'autres égards, les premières années de la Troisième République ne seront que la continuation du Second Empire.

Le clergé avait été fort inquiet en constatant qu'à l'exception du général Trochu, il n'y avait aucun catholique dans le gouvernement provisoire, et en voyant s'amorcer, à Paris et dans quelques villes du midi, des mesures contre les religieux ou l'enseignement chrétien. Mais les hommes de la Défense nationale, y compris Gambetta, eurent la sagesse de ne pas céder à cette poussée d'anticléricalisme qui risquait de diviser le pays devant l'ennemi et qui apparaissait particulièrement odieuse au moment où le clergé de France tout entier faisait magnifiquement son devoir au point de faire dire à Bismarck : « Nous n'avons trouvé debout que le clergé. » Puis tandis que les angoisses de la guerre ramenaient au pied des autels de nombreux Français que leurs plaisirs ou leurs affaires en avaient tenus éloignés, le pays envoya à l'Assemblée nationale, le 8 février 1871, une majorité de catholiques pratiquants, décidés à restaurer le pays sous le signe de l'*ordre moral* et de la religion ancestrale.

L'hostilité des républicains contre l'Église, à laquelle ils ne pardonnaient pas le soutien qu'elle avait apporté au régime impérial, se manifesta cependant au grand jour pendant les quelques semaines de l'insurrection parisienne. Le culte put continuer à peu près librement, mais la Commune, « considérant en fait que le clergé a été le complice des crimes

(1) Voir sa lettre au Conseil supérieur de la Jeunesse catholique dans F. OLGIATI, *La storia dell'Azione Cattolica*, p. 54-56.

(2) BIBLIOGRAPHIE. — I. SOURCES. — Outre les journaux catholiques (notamment *L'Univers* et *Le Français*) et les mandements d'évêques ainsi que les correspondances et souvenirs signalés p. 6, auxquels on peut ajouter le *Journal* de l'abbé Frémont (édit. A. SIEGFRIED, t. I, Paris, 1933), H. ANTOINE, *Trente ans de ministère paroissial à Paris (1870-1900)*, Paris, 1905, et M. DU CAMP, *La charité privée à Paris*, Paris, 1885, on trouvera de nombreux renseignements statistiques et autres dans E. KELLER, *Les congrégations religieuses en France : leurs œuvres et leurs services*, Paris, 1878, et A. DE MELUN, *Manuel des œuvres*, Paris, 1878. Pour la discussion des nombreuses lois à portée religieuse, voir le *Journal officiel*.

II. TRAVAUX. — E. LECANUET, *L'Église de France sous la IIIᵉ République. I. Les dernières années du pontificat de Pie IX*, Paris, 1907 ; J. BRUGERETTE, *Le prêtre français et la société contemporaine*, t. II. *Vers la Séparation*, Paris, 1935, chap. I à IV ; Ch. POUTHAS, *L'Église et les questions religieuses en France de 1848 à 1878* (Cours de Sorbonne), Paris, 1945, Vᵉ partie.

Pour les questions politico-religieuses, voir aussi W. GURIAN, *Die politischen und sozialen Ideen des französischen Katholizismus 1789-1914*, München-Gladbach, 1929, chap. XI et XII (nuancé), et A. DEBIDOUR, *L'Église catholique et l'État sous la IIIᵉ République*, t. I, Paris, 1906 (très tendancieux). On ne peut pas négliger les pages, souvent perspicaces, d'H. TAINE, *Les origines de la France contemporaine. Le Régime moderne*, t. III, chap. I et II.

En outre, les ouvrages généraux et les biographies, spécialement celles de Mgr Dupanloup, des cardinaux Pie et Guibert, de Mgr de Ségur (cf. p. 8), ainsi que celles de Keller (par GAUTHEROT, Paris, 1922), de Chesnelong (par M. DE MARCEY, 3 vol., Paris, 1903) et de Kolb-Bernard (par Mgr BAUNARD, Paris, 1889).

de la monarchie contre la liberté »[1], prononça la séparation de l'Église et de l'État, ainsi que la confiscation des biens des congrégations. Bien plus, il faut se garder d'identifier l'ensemble des communards, braves gens du peuple de Paris pour la plupart, avec les énergumènes qui déshonorèrent l'insurrection ; les rancœurs populaires étaient telles que certains agitateurs purent se faire acclamer en déclarant que « le seul moyen de purifier la société était de brûler les églises en mettant les prêtres dedans ». De fait, les pillages d'églises et de couvents[2] ainsi que les arrestations d'ecclésiastiques se multiplièrent jusqu'à la semaine tragique au cours de laquelle vingt-quatre d'entre eux, y compris l'archevêque de Paris, Mgr Darboy, furent fusillés[3].

Mais ce débordement antireligieux, d'ailleurs localisé, ne fut qu'une parenthèse et la répression impitoyable de l'insurrection y mit fin dès le mois de mai. Il contribua même à accentuer le réveil religieux chez tous ceux qui prêtèrent une oreille favorable aux mandements épiscopaux ou aux innombrables sermons où les massacres et les destructions de la Commune, suivant de près la défaite militaire, étaient présentés comme un châtiment divin ou du moins comme la conséquence logique de la baisse du sens chrétien dans la société.

LE RÉVEIL RELIGIEUX   Tandis que l'Assemblée nationale vote des prières publiques, se développent à travers le pays, souvent avec la participation des maires, de la magistrature et de l'armée, des manifestations expiatoires accompagnées d'un mouvement de piété spectaculaire qui ira en s'amplifiant jusqu'en 1873 et dont l'aboutissement sera la « consécration de la France pénitente au Sacré-Cœur de Jésus », lue à Paray-le-Monial, le 29 juin de cette année, par Gabriel de Belcastel au nom de cent cinquante de ses collègues de l'Assemblée. Des milliers de pèlerins, le chapelet au cou ou l'image du Sacré-Cœur sur la poitrine, se pressent dans une atmosphère fiévreuse vers les sanctuaires les plus célèbres, Lourdes, La Salette, Chartres, Paray surtout, et dans les rues ou les gares de Bourgogne, on n'entend plus que le cantique célèbre :

> Sauvez Rome et la France
> Au nom du Sacré-Cœur !

qui devient « le chant de guerre de l'ultramontanisme, une sorte de *Marseillaise* catholique »[4].

Qu'y a-t-il derrière ce déploiement de piété « tapageuse et conquérante »[5] ? Incontestablement un réel réveil du sentiment religieux, qui se traduit par une fidélité plus grande au devoir dominical et par une

---

(1) *Journal officiel de la Commune*, 2 avril 1871, p. 133.
(2) Cf. P. FONTOULIEU, *Les églises de Paris sous la Commune*, Paris, 1873.
(3) Soit un tiers des 74 otages exécutés, ce qui témoigne de la violence des sentiments antireligieux. Mais on tiendra compte de cette observation de Dansette : « Quant à la responsabilité des exécutions, il faut se souvenir, pour l'apprécier avec équité, qu'aucune d'entre elles ne fut commise avant le début de la semaine sanglante, au cours de laquelle l'armée de Versailles passa par les armes sans jugement au moins quinze mille communards. » (*Histoire religieuse de la France contemporaine*, t. I, p. 440.)
(4) P. DE QUIRIELLE, dans *Ann. E. Sc. P.*, t. VI, 1891, p. 144.
(5) A. DANSETTE, *op. cit.*, t. I, p. 443.

meilleure observance des commandements ; pour de nombreuses âmes, ces grandioses cérémonies ont été l'aliment qui a vivifié en elles la ferveur et les a fait flamber d'une ardeur nouvelle pour l'action catholique.

Mais on ne peut se dissimuler non plus le caractère superficiel, dans bien des cas, de cette exaltation religieuse, ni surtout le fait qu'en réalité « le mouvement n'entraîne qu'une minorité : c'était, si l'on osait parler ainsi, comme une clientèle que l'on retrouvait toujours la même, autour de tous les sanctuaires thaumaturgiques »[1]. Le monde paysan demeure méfiant à l'égard d'une Église qu'il soupçonne de préparer, avec la restauration de la monarchie, le rétablissement des dîmes de l'ancien régime. L'apostasie du monde savant n'est nullement enrayée, bien moins encore celle du peuple travailleur, plein de rancœur au souvenir de la sauvage répression de la Commune — près de 22.000 exécutions sommaires — par une Assemblée en majorité catholique. Ce n'est guère que dans la bourgeoisie, petite bourgeoisie provinciale, bourgeoisie aisée des grandes villes, où s'affermit la conviction que l'ordre social ne peut subsister sans le Christ, que s'accentue le retour à la foi amorcé une vingtaine d'années plus tôt, avec le danger de voir le catholicisme devenir, comme le lui reprochera Péguy, « une espèce de religion supérieure pour classes supérieures de la société, une misérable sorte de religion distinguée pour gens censément distingués ».

LA PROSPÉRITÉ DE L'ÉGLISE   Sur le moment toutefois, l'Église semble tirer profit de cette évolution. Formée de gentilshommes campagnards et de grands bourgeois conservateurs, l'Assemblée se montre très bienveillante à son égard, de même que le gouvernement.

Dans le choix des évêques, source fréquente de frottements entre Rome et Paris sous le Second Empire, tous les ministres des cultes, et Jules Simon plus qu'aucun autre, témoignent d'une parfaite déférence aux désirs du Saint-Siège, et comme les circonstances font qu'au cours de ces huit années plus de cinquante évêques doivent être remplacés, les derniers centres de résistance possible à l'ultramontanisme disparaissent ; l'influence conjuguée du nonce et de Mgr Pie écarte de même presque tous ceux qui manifestent des sympathies libérales.

La majorité de l'Assemblée, tout en prenant soin, au grand déplaisir des intégristes, de présenter le catholicisme comme une force sociale utile et respectable, plutôt que comme la religion officielle de l'État, se prête volontiers au désir de l'Église de voir pénétrer les institutions d'esprit chrétien. Elle s'associe — trop discrètement aux yeux de certains — au courant de ferveur expiatoire envers le Sacré-Cœur en déclarant d'utilité publique la construction de « l'église du vœu national » à Montmartre ; elle attribue divers avantages aux clercs et majore en plusieurs circonstances l'influence du clergé dans les grands services de l'État : armée, assistance publique, enseignement surtout.

L'Église, qui bénéficiait déjà, depuis la loi Falloux, d'une situation

---

(1) J. BRUGERETTE, *Le prêtre français et la société contemporaine*, t. II, p. 12.

privilégiée au Conseil supérieur de l'Instruction publique et de la liberté dans les enseignements primaire et secondaire, complète sa victoire de 1850 en obtenant encore la liberté de l'enseignement supérieur. Les partisans du monopole universitaire résistent cette fois avec acharnement ; ils tentent au moins de réserver à l'État la collation des grades, mais finalement, grâce à une transaction, les catholiques obtiennent satisfaction et la loi est votée, le 12 juillet 1875, par 316 voix contre 266. Bien que *L'Univers* lui-même reconnaisse que « les héros de cette campagne furent l'évêque d'Orléans, MM. Chesnelong et Lucien Brun », certains intransigeants, comme en 1850, accusent Dupanloup d'avoir trahi l'Église, mais Pie IX le remercie, dans un bref à vrai dire assez symptomatique :

Quoiqu'il répugne aux éternelles lois de la justice et même de la saine raison de mettre sur le même rang le vrai et le faux, cependant, comme l'iniquité des temps a transféré à l'erreur un droit qui, de sa nature, n'appartient qu'à la vérité... nous estimons, Vénérable Frère, que vos efforts pour tirer du poison communiqué à la société civile un antidote ont été tout à fait habiles et opportuns [1].

Pressentant que la gauche profiterait de la première occasion pour reprendre cet avantage obtenu de justesse, les évêques se hâtent d'ouvrir des Facultés catholiques dans l'espoir que les adversaires hésiteraient davantage si la loi était déjà en application. La générosité des fidèles leur fournit immédiatement par millions les ressources nécessaires, surtout à Lille et à Paris, où le cardinal Guibert installe la nouvelle université à l'École des Carmes et trouve dans l'abbé d'Hulst un collaborateur de choix [2].

Parallèlement, l'Église consolide et même étend ses positions dans les domaines primaire et secondaire. Tandis que les instituteurs prêtent toujours au clergé un concours efficace pour l'enseignement du catéchisme, les frères et surtout les « bonnes sœurs » continuent à diriger beaucoup d'écoles communales et prolongent leur influence en développant les patronages. De nombreuses écoles libres complètent l'enseignement congréganiste et la majorité de l'Assemblée écarte en 1872 le danger éventuel que pourrait constituer pour elles le projet de Jules Simon d'un enseignement primaire obligatoire, que les évêques considèrent comme le prélude d'une instruction gratuite et laïque.

Quant aux collèges secondaires libres, en constante augmentation, ils sont 309 en 1876, groupant 46.816 élèves, auxquels on peut ajouter les 23.000 des petits séminaires, contre 78.913 instruits par l'université et 31.000 dans les établissements libres laïques. Ces collèges, dont 91 sont dirigés par des congrégations, forment en 1872 l'*Alliance des Maisons d'Éducation chrétienne*, qui va fonder une revue et tenir des assemblées annuelles. Bien que les licenciés y demeurent rares parmi les professeurs, la qualité de leur enseignement s'améliore : en 1876, 14 pour cent des candidats reçus à Polytechnique et 33 pour cent de ceux admis à Saint-Cyr en sortent, ce qui commence à faire crier à l'invasion du cléricalisme à l'armée.

---

(1) Cité dans E. LECANUET, *L'Église de France sous la IIIe République*, t. I, p. 506.
(2) Cf. A. BAUDRILLART, *Vie de Mgr d'Hulst*, t. I, Paris, 1912, chap. x à xii.

L'extension de l'influence de l'Église dans la vie publique est facilitée par l'accroissement continu du personnel ecclésiastique. Certes, la baisse de la foi dans les campagnes commence à se traduire dans le chiffre des vocations sacerdotales : le grand séminaire de Reims, qui avait en moyenne 100 élèves avant 1870, n'en a plus que 55 en 1878, celui de Nîmes constate une baisse de plus de moitié, et pour l'ensemble du pays, on passera de 12.166 élèves en 1876 à 8.400 en 1880, tandis que les ordinations diminuent régulièrement à partir de 1875 ; c'est *Le grand péril de l'Église de France*, dénoncé par l'abbé Bougaud dans un livre retentissant. Mais le danger ne se matérialise pas encore par suite de la faible mortalité et le nombre de prêtres en exercice continue à augmenter [1] : en 1876, on en compte 1 pour 654 habitants contre 1 pour 730 en 1870. Et les vocations religieuses ne cessent de croître, au point que M. Pouthas considère le développement des congrégations comme le trait caractéristique de l'Église des débuts de la République : 30.287 religieux et 127.753 religieuses en 1877 [2], soit 158.040 congréganistes — 1 par 250 habitants —, qui desservent 3.086 écoles de garçons, 16.478 écoles de filles et 4.416 autres établissements. Les jésuites à eux seuls sont 1.840, répartis en 66 maisons, et leur influence est considérable : tout en ne dédaignant ni les réunions de servantes, ni les patronages de petits ramoneurs, ils dirigent dans leurs résidences un grand nombre de familles distinguées et surtout ils élèvent dans leurs 29 collèges les enfants de la haute société, ils ont la haute main sur les dames du Sacré-Cœur et bien d'autres congrégations enseignantes, ils prêchent les retraites des frères des Écoles chrétiennes, bref, ils contribuent pour une bonne part à former l'état d'esprit du clergé et des catholiques français.

VITALITÉ ET FAIBLESSES DU CATHOLICISME FRANÇAIS — Nulle part en Europe, il ne se forme autant de frères et de sœurs voués, leur vie durant, à l'enseignement populaire, autant de servantes et de serviteurs volontaires des pauvres, des malades, des infirmes et des enfants. Nulle part surtout il ne se forme autant de missionnaires : la France a pris la place que l'Espagne avait tenue pendant trois siècles et, en 1900, 70 pour cent des missionnaires catholiques seront français. Tout autant que ce grand nombre de vocations religieuses dans toutes les classes de la société, le développement continu des œuvres suffirait à prouver que cette Église extérieurement puissante est tout autre chose qu'une façade et qu'elle contient encore de fortes réserves de sève chrétienne.

Développement des œuvres de prière et de dévotion : *Ligue du Sacré-Cœur*, adoration nocturne, communion réparatrice, *Apostolat de la prière*, œuvre des pèlerinages à Lourdes, etc. Développement des œuvres missionnaires et des œuvres de propagande : journaux, brochures, bibliothèques militaires. Mais surtout développement considérable des œuvres

---

(1) A première vue, il semble avoir baissé, puisqu'il n'y a plus que 55.369 prêtres en 1876 au lieu de 56.295 en 1869, mais il faut tenir compte de la perte de l'Alsace et de la Lorraine.
(2) A titre de comparaison, il y en avait 25.000 et 37.000 en 1789.

de charité. Si le drame de la Commune n'a pas suffi à éveiller le sens social
de la bourgeoisie française, du moins a-t-il attiré son attention sur la
misère populaire. Ce sont surtout les œuvres destinées aux enfants (crèches,
orphelinats, écoles d'apprentissage) ou aux vieillards et aux malades
pauvres qui se multiplient, mais on ne peut négliger le beau départ de
l'Œuvre des cercles d'ouvriers, dont il sera question ailleurs : les erreurs
de conception, qui expliquent l'échec de cette première grande tentative
française de catholicisme social, n'enlèvent rien à l'abnégation généreuse
de ses promoteurs, ainsi qu'à leur sens profond de leurs responsabilités
chrétiennes.

Il est pourtant regrettable que ces fondations si variées, qui témoignent
de la ferveur de l'élite catholique, demeurent toujours trop isolées les
unes des autres, en dépit du progrès que constituent la fondation, en 1871,
de « l'Union des œuvres catholiques » avec ses congrès annuels, et surtout
celle, en 1873, des « Assemblées générales des comités catholiques de
France », sorties de l'initiative d'un médecin parisien, le Dr Frédault,
et dont Chesnelong, avec sa foi profonde, son cœur chaud et son esprit
modéré, sera l'âme pendant des années.

Chose plus grave encore, ce manque de coordination, on doit le déplorer
dans l'épiscopat lui-même. Alors que les évêques allemands se retrouvent
chaque année à Fulda et que les évêques belges ou hollandais sont en
communication constante, sous la direction de leur métropolitain, les
évêques français vivent isolés de leurs collègues et n'ont pratiquement
aucune occasion de concerter une action commune. Ils sont en outre
souvent opposés les uns aux autres par leurs attaches politiques ou par
des rivalités personnelles.

C'est du reste l'opinion catholique dans son ensemble qui se trouve
divisée. La vieille opposition entre les catholiques libéraux et les intransigeants devient plus acharnée encore que sous l'Empire, car les seconds
sont scandalisés par la politique religieuse modérée — lâche et pusillanime,
disent-ils — menée par leurs adversaires à l'Assemblée nationale ; ils
peuvent en outre se prévaloir des désaveux toujours plus nets que le
vieux Pie IX ne perd pas une occasion d'adresser au libéralisme catholique
dénoncé comme « un mal plus redoutable que la Révolution, plus redoutable que la Commune avec ses hommes échappés de l'enfer »[1]. Il faut du
reste reconnaître que si les idées politiques des « catholiques avant tout »
apparaissent avec le recul du temps singulièrement chimériques et maladroites, ils voyaient par contre fort juste quand ils dénonçaient le
caractère assez peu chrétien de la politique de l'ordre moral, prônée par
les catholiques libéraux sous la direction d'Albert de Broglie, et lui reprochaient de se borner à une simple « défense matérielle » de la société, inspirée surtout par un réflexe conservateur[2]. A la vérité, les libéraux, dont les
leaders sont des vieillards que nul parmi les jeunes ne se prépare à relayer,

---

(1) Discours adressé par Pie IX, le 18 juin 1871, à la députation française lors des fêtes du
25e anniversaire de son élévation au souverain pontificat.
(2) Cf. E. Beau de Loménie, Les responsabilités des dynasties bourgeoises, t. I, Paris, 1947 ;
D. Halévy, La fin des notables, Paris, 1930 et La République des ducs, Paris, 1937 ; R. Dreyfus,
De Monsieur Thiers à Marcel Proust, Paris, 1939.

ne sont guère plus lucides que leurs adversaires intégristes : ils ne se rendent pas compte du déplacement rapide des centres d'intérêts par suite de l'évolution sociale qui s'accomplit sous leurs yeux, et ne comprennent pas que ce que les masses reprochent désormais surtout à l'Église, c'est d'avoir partie liée avec les autorités sociales dont elles prétendent abattre la puissance.

Tandis que, sous la conduite de Mgr Pie, de Louis Veuillot, de Mgr de Ségur, du P. d'Alzon, des jésuites des *Études*, le clergé français s'absorbe dans des controverses bruyantes et stériles sur le régime politique idéal, il continue à négliger les véritables problèmes de l'heure : les problèmes intellectuels qui sont à la base de la religion, la formation des militants laïques capables de servir efficacement l'Église dans une société qui va se séculariser toujours davantage, les questions sociales.

« Ni les prédicateurs, ni les conférenciers, ni les catéchistes, ne sont en état de parler avec compétence des questions qui préoccupent aujourd'hui les hommes instruits et troublent les consciences », constatera Mgr Meignan dans un mémoire adressé à Léon XIII en 1881. Comment pourrait-il d'ailleurs en être autrement : « Les grands et les petits séminaires ne possèdent que rarement des professeurs capables de réfuter sérieusement les erreurs allemandes devenues françaises [1]. » On en est réduit à des manuels sulpiciens en mauvais latin, soi-disant *ad mentem Sancti Thomae*, le *Compendium Philosophiae* de Manier et les *Institutiones theologicae* de Bonal. Et les ouvrages qui s'offrent à ceux qui voudraient aller au delà du manuel ne valent guère mieux. Certes, on possède désormais quelques bons livres de spiritualité, *La vie surnaturelle* de l'abbé de Broglie, *La vie et les vertus chrétiennes* de Mgr Gay, et les carêmes du P. Monsabré, encore que manquant d'originalité, sont un exposé dogmatique estimable. Mais ces ouvrages s'adressent aux convaincus. L'Église n'a rien de sérieux à offrir pour sa défense : les cinq volumes de Bougaud, *Le christianisme et les temps présents*, éveillent plus d'émotions que d'idées ; en philosophie, la stérilité est complète ; en exégèse, on en est encore à rééditer les *Saints Livres vengés* de Glaire, et que valent les premières publications de Vigouroux en face de l'*Histoire des origines chrétiennes* de Renan, qui paraît de 1872 à 1878 ? En histoire ecclésiastique, il y a bien quelques travaux érudits et Mgr Baunard publie ses premières biographies, *Saint Jean*, *Saint Ambroise*, mais le lamentable Darras est encore présenté comme « le vrai livre de l'espérance nationale » [2]. La responsabilité de cette situation incombe pour une bonne part aux évêques, qui continuent, comme sous le Second Empire, à penser avec le cardinal de Bonald : « Des savants ? Que voulez-vous que j'en fasse ? » Il est caractéristique de constater que, lors de la fondation des universités catholiques, ils se préoccupent bien plus de former des avocats ou des médecins chrétiens que des prêtres savants.

Peut-on du moins dire que cette formation chrétienne du laïcat soit couronnée de succès ou simplement perspicace ? L'immense dévouement,

(1) Cité dans H. BOISSONOT, *Le cardinal Meignan*, p. 344-345.
(2) *L'Univers*, 22 juillet 1873.

l'immense bonne volonté qui sont à la base des collèges catholiques n'empêchent pas que dès 1873 la *Revue du Monde catholique* ne dénonce la faiblesse de l'instruction religieuse, où l'on se borne trop souvent à faire apprendre passivement et réciter machinalement quelque manuel sans attache avec la vie. On commence à voir se réaliser certaines des appréhensions formulées jadis par Dupanloup au lendemain de la loi Falloux : « Je crains beaucoup de choses : que nos collèges ne deviennent des lieux de refuge pour les enfants gâtés des grands bourgeois ; que la manie de la truelle ne mène le clergé à des dépenses inutiles ; que la routine des pratiques religieuses ne dégoûte l'enfant de l'église au lieu de l'y habituer [1]. » Et si, malgré tout, une partie de cette jeunesse reste fidèle, force est bien de constater avec l'abbé Brugerette que l'éducation à base d'obéissance, sans assez d'initiative, qu'elle reçoit, la prépare mal au rôle qu'on voudrait lui voir jouer : « Aucun champion de premier plan, vraiment digne d'être opposé à leurs adversaires, ne se lèvera parmi les anciens élèves des jésuites, pour la défense de leurs maîtres comme pour celle de l'Église. Les avocats de leur cause feront sans doute entendre des protestations éloquentes, ils ne seront pas des hommes de combat, d'action sociale, d'organisation électorale [2]. »

La plupart de ces « anciens élèves » forment du reste une caste fermée, qui ne fréquente que les siens et qui vit sans se douter de l'énorme évolution qui s'opère dans les couches profondes de la démocratie. Elle perd le contact avec une partie importante de la nation, et c'est malheureusement ce qui se produit également toujours plus pour le clergé lui-même.

L'IMPOPULARITÉ CROISSANTE DU CLERGÉ

Désormais préparé uniformément dans les séminaires diocésains par des prêtres vertueux qui le forment à une piété minutieuse et à la maîtrise de soi, jamais le clergé français n'a été si exemplaire ni si fervent et il s'acquitte avec conscience de toutes ses obligations pastorales. Aussi est-il généralement respecté, mais il n'est pourtant pas aimé.

Cette impopularité s'explique pour une part par des causes inhérentes à son ministère, qui répugnent tout particulièrement au Français moyen et que Taine a relevées dans un passage souvent cité [3] : au confessionnal et du haut de la chaire, il admoneste ses paroissiens et entend les régenter jusque dans le détail de leur vie familiale et privée ; parmi les observances qu'il prescrit, beaucoup paraissent insipides ou désagréables : maigres, jeûnes, assistance passive à de longs offices en latin, dont on ne comprend plus le sens symbolique, enfin et surtout la confession à échéance fixe. Il faut tenir compte également de ce que, pourvu de ressources modestes [4] mais suffisantes, convenablement logé,

---

(1) Lettre à Mme de L., citée dans *L'Éclair*, du 19 mai 1904.
(2) J. Brugerette, *Le prêtre français et la société contemporaine*, t. II, p. 63.
(3) H. Taine, *Le régime moderne*, t. III, Paris, 1907, p. 181-182.
(4) Les vicaires ont un traitement de 450 fr. par an, les curés-desservants de 900 fr. ; il faut y ajouter 200 à 300 fr. d'honoraires de messe et le casuel, qui souvent n'atteint pas 100 fr. par an. (Cf. E. Lecanuet, *op. cit.*, p. 295.)

ayant une table confortable, le prêtre apparaît à beaucoup comme un petit bourgeois qui a la vie facile, surtout dans les paroisses rurales où le catéchisme, quelques courtes séances de confession et, de temps à autre, un baptême ou un enterrement, lui laissent de nombreuses heures de liberté qu'il passe à des occupations considérées comme des divertissements dans d'autres professions : lecture du journal, jardinage, visites, rencontres avec les confrères.

Il y a toutefois autre chose encore. La formation intellectuelle très superficielle reçue au séminaire a fait trop souvent du prêtre un primaire, à l'esprit exagérément dogmatique, et la lecture quotidienne de *L'Univers* ne fait que renforcer ce manque d'esprit critique ; de là, un « ton impérieux et dominateur »[1] quand il réfute les adversaires de la religion, et qu'il conserve même quand il est descendu de la chaire. Enfin l'impopularité du prêtre s'explique encore par ses idées absolutistes et rétrogrades. « Le peuple a le sentiment que le clergé demande au passé son idéal politique et social [2]. » Les admirateurs des découvertes scientifiques constatent que beaucoup de prêtres se défient des progrès modernes ; les ouvriers, engagés dans le conflit qui se développe entre le capital et le travail, s'irritent d'entendre à leur cri de : Justice ! le prêtre répondre : Résignation ! Et tous ceux pour qui la libération du peuple est symbolisée par la République ne pardonnent pas au clergé d'attendre « comme un Messie »[3] le comte de Chambord, qui doit restaurer dans la société le règne du Christ et rendre ses États au pape détrôné.

Ces deux questions, en effet, question romaine et question monarchique, contribuent à élargir le fossé entre l'Église et le peuple français.

LES RÉPERCUSSIONS DE LA QUESTION ROMAINE

On a parfois accusé les catholiques français de s'être davantage intéressés à la restauration du pouvoir temporel du pape qu'au relèvement de leur patrie. L'affirmation est injuste, mais il est certain que la situation de Pie IX les préoccupait fort. Thiers, qui s'en rendait compte, multiplia les témoignages de déférence envers le souverain pontife, mais les extrémistes auraient voulu qu'il prît l'initiative d'une vaste campagne diplomatique qui obligerait l'Italie à rendre Rome au pape. Dès 1871, à l'instigation de *L'Univers* [4], des pétitions sommant le gouvernement d'intervenir circulèrent dans les diocèses avec l'appui d'une grande partie du clergé et certains évêques, qui trouvaient ces démarches inopportunes, se laissèrent entraîner par les autres de crainte de paraître manquer de zèle.

Ces manifestations tapageuses, que le Saint-Siège lui-même essaya de

---

(1) J. BRUGERETTE, *op. cit.*, t. II, p. 54.
(2) E. LECANUET, *op. cit.*, p. 297.
(3) Mgr Meignan à Léon XIII, Mémoire de 1881 dans II. BOISSONOT, *Le cardinal Meignan*, p. 344.
(4) L'extrait suivant donne une idée du ton : « La XXVe année de Pie IX, captif mais invincible, est une merveille qui en annonce une autre, celle de sa délivrance par le *Roi très chrétien* qui, du même coup, ramènera, tambour battant, l'usurpateur de l'Italie à sa principauté subalpine et effacera sur la Prusse écrasée le traité de Westphalie et sa monstrueuse erreur. Cet oracle est plus sûr que celui de Calchas. » (*L'Univers*, 13 juillet 1871.)

modérer, atteignirent leur paroxysme en 1873 et rebondirent en 1877, contribuant à amener la crise du 16 mai. Elles étaient inspirées par un sentiment généreux, mais elles ne pouvaient avoir aucune influence pratique et furent triplement néfastes : pour le pays, en poussant l'Italie dans les bras de l'Allemagne ; pour l'Église, en la faisant accuser injustement par ses adversaires de pousser à la guerre ; pour les catholiques, en accentuant leurs divisions et en discréditant davantage encore aux yeux de Rome et des ultramontains les partisans de Dupanloup, que ses conseils de prudence firent traiter par Veuillot de « Ponce Pilate », et du duc de Broglie, acculé à prendre à l'égard de *L'Univers*, pour atténuer la tension diplomatique, une mesure qu'il jugea lui-même « peu héroïque ».

LES CATHOLIQUES ET LES TENTATIVES DE RESTAURATION MONARCHIQUE

Pour comble de malheur, les catholiques, que la propagande présente en bloc comme des fanatiques dangereux pour la paix internationale, apparaissent au même moment aux yeux des masses comme des réactionnaires décidés à faire table rase des conquêtes de la Révolution.

A côté de Pie IX, le comte de Chambord, « qui réunit la loyauté d'Henri IV à la vertu de saint Louis » [1], personnifie pour le clergé l'ordre chrétien dont il attend impatiemment le triomphe. La plupart des catholiques partagent ces sentiments monarchistes ; ils suivent avec enthousiasme les tractations qui préparent le retour sur le trône de l'exilé de Frohsdorf [2], et les grands pèlerinages de 1873 prennent de ce fait une allure non équivoque de manifestation royaliste.

Mais tandis que les sympathies de nombreux laïques vont à la monarchie constitutionnelle et qu'ils souhaitent trouver dans le futur roi « un Louis-Philippe moins voltairien » [3], d'autres, et notamment la majorité du clergé, reportent sur Henri V l'espoir qu'ils avaient mis jadis en Napoléon III et que celui-ci avait amèrement déçu. Il est rare d'entrer dans un presbytère de campagne « sans y voir, à côté et sur le même rang, les images de Notre-Seigneur Jésus-Christ, de la Sainte Vierge et du comte de Chambord » [4]. Or, il est indéniable que les idées prônées dans *L'Univers* ou *Les Études* répondent aux sentiments intimes du prince, qui a grandi en exil dans « l'entourage archaïque » de Charles X et qui unit dans une même vénération les enseignements de Pie IX et les doctrines de de Maistre et de Bonald — la fameuse question du drapeau blanc n'étant qu'un symbole de l'opposition foncière aux principes de 1789 de celui qui rêve d'être le Roi Très Chrétien d'une France de la Contre-Révolution. Excités par Veuillot, de nombreux catholiques attendent de lui qu'il

(1) R. P. Ramière, S. J., dans *Les Études*, septembre 1875, p. 399.
(2) Sur ces tractations, où plusieurs évêques intervinrent directement, l'exposé de Lecanuet (suivi par Brugerette), qui représente le point de vue des catholiques libéraux, est à compléter par celui du collaborateur de Veuillot A. Loth, *L'échec de la Restauration en 1873*, Paris, 1910 et par celui, plus récent, de G. Hanotaux, *Histoire de la fondation de la III⁰ République*, t. I : *Le gouvernement de M. Thiers*, Paris, 1925.
(3) W. Gurian, *Die pol. und soz. Ideen des franz. Katholizismus*, p. 245.
(4) L. Chaine, *Les catholiques français et leurs difficultés actuelles*, p. 96.

rétablisse l'ancienne alliance du trône et de l'autel avec, en fin de compte, une certaine supériorité de la puissance spirituelle, et aussi qu'il mette un terme au funeste régime issu de 1789, qui prétend donner les mêmes droits à l'erreur et à la vérité. Cette école, qui brandit à tout propos le *Syllabus*, interprété dans un sens maximaliste, caractérise son idéal d'un mot : *Contre-Révolution*, que le gros public, objecte très justement Falloux, « traduit toujours par *ancien régime*, dans la plus mauvaise acception du terme ».

Les catholiques libéraux essaient bien de faire observer que ces idées extrêmes sont seulement celles d'une école et que tous les fidèles ne pensent pas comme Mgr Pie ; mais les hommes de gauche leur répondent, comme Challemel-Lacour en pleine Chambre :

Mais vous êtes désavoués ! Votre esprit de transaction est traité à Rome de complaisance coupable, de faiblesse irrémédiable ! Voilà vingt ans qu'on prépare votre condamnation ! Elle est aujourd'hui partout, dans les conciles, dans les livres orthodoxes, dans tout ce qui nous arrive de Rome [1].

On leur répond encore qu'ils sont d'ailleurs monarchistes eux aussi et que la monarchie, pour le moment, c'est le comte de Chambord. Effectivement, Dupanloup est intervenu à plusieurs reprises pour faciliter son retour, le cardinal Mathieu aussi ; on apprendra même plus tard que Lavigerie n'a pas hésité à conseiller au prince un coup d'État.

Ces sentiments monarchistes de la quasi-totalité des catholiques français se comprennent facilement dans les circonstances concrètes de l'époque. L'idée républicaine était incarnée dans un parti qui, surtout dans son aile marchante, s'affichait violemment anticlérical et même antireligieux, et il serait injuste de trop insister sur la responsabilité de certains catholiques de la génération précédente dans cette situation. Depuis des années, la plupart des républicains militants avaient partie liée avec les loges maçonniques ; Victor Hugo, dans ses *Châtiments*, avait flagellé l'Église aussi durement que l'Empire ; avant lui, Michelet, Quinet surtout, l'avaient combattue avec acharnement ; Proudhon dénonçait l'antagonisme existant entre la religion et l'idée de justice. On comprend dès lors que les catholiques devaient considérer l'avénement des républicains au pouvoir comme tout autre chose qu'une simple transformation politique, comme le triomphe d'une *Wellanschauung* foncièrement antichrétienne. Et si les républicains modérés comme Jules Simon ou Jules Favre se bornaient à souhaiter la séparation de l'Église et de l'État, les plus avancés étaient de vrais jacobins, qui menaçaient de dépouiller l'Église de ses libertés légitimes ; c'est Mérimée, peu suspect de cléricalisme, qui l'observait : « Si jamais ils sont au pouvoir, ils procéderont comme l'Église en ses plus sombres jours, car ils sont une église eux-mêmes, ces fanatiques d'anticléricalisme [2]. » On comprend dans

_____

(1) *Journal officiel*, 20 juillet 1876.
(2) Cité par Mme Adam, *Mes sentiments et nos idées avant 1870*, Paris, 1905, p. 15. Voir, à titre d'exemple, la correspondance d'Allain-Targé, *La République sous l'Empire. Lettres (1866-1870)*, Paris, 1939.

ces conditions qu'aux yeux de beaucoup de catholiques, la République apparaisse « sous les traits de la Terreur, rajeunis par la Commune » [1].

LA VICTOIRE DES RÉPUBLICAINS Malheureusement, au même moment, la République apparaît au contraire aux nouvelles couches sociales comme le symbole du progrès et la condition même de leur complète émancipation. Les publicistes bruyants et les prédicateurs, qu'ils soient ou non sous l'influence des « révélations » de la Salette, présentent l'humiliation de la défaite comme le signe de la colère divine contre la France qui a abandonné le pape après s'être adonnée à une existence toujours plus jouisseuse. Les partis de gauche et les militants de la Libre Pensée leur répondent que la France a en réalité payé les inepties de la politique impériale, les concessions à un cléricalisme vieillot, l'intervention en Italie, les entraves à la liberté de pensée, etc. Les maladresses ou l'incompréhension de nombreux catholiques, « illuminés par les visions mystiques ou clos par les préjugés de salon » [2], leur font la partie belle pour dénoncer le clergé comme un facteur d'obscurantisme, un fauteur de guerre, l'allié des puissances du passé, et pour promettre au peuple que la République, éclairée par la science, va achever de les libérer de toutes les tyrannies, celle du château, celle du curé, celle de l'enfer.

L'évolution politique est dès lors facile à prévoir. Au traditionnel esprit de résistance des Français contre tout ce qui ressemble à la domination politique de l'Église, à ce que Gambetta appelle « le gouvernement des curés », s'ajoute à présent l'hostilité de tous ceux qui, attachés jusqu'alors à l'Église par des habitudes sociales plutôt que par conviction, n'hésitent pas à se retourner contre elle dès qu'on la leur présente comme un obstacle à leur émancipation. Aussi les élections de 1876 font-elles entrer à la Chambre 360 républicains contre 130 conservateurs, et cette victoire est confirmée par celle de 1877, après l'échec du 16 mai. Le rêve d'une monarchie chrétienne s'effondre. Une nouvelle époque va s'ouvrir pour l'Église de France, où, face aux attaques de la République « anticléricale et laïque », elle va peu à peu découvrir sa faiblesse réelle en dépit des progrès apparents des trente années précédentes.

§ 3. — Le Kulturkampf [3].

LES CATHOLIQUES ALLEMANDS ET LE NOUVEL EMPIRE Les visées anticléricales et même antichrétiennes qui étaient allées en s'accentuant dans les différents États allemands depuis 1860 s'étaient affichées au grand jour au cours des polémiques

(1) A. Dansette, op. cit., t. I, p. 451.
(2) Ibid., t. I, p. 458.
(3) Bibliographie. — I. Sources. — H. v. Kremer-Auenrode, Aktenstücke zur Geschichte des Verhältnisses von Staat und Kirche im XIX. Jh., t. III et IV, Leipzig, 1877-1880 ; N. Siegfried (= V. Cathrein, S. J.), Aktenstücke betreffende den preussischen Kulturkampf, Fribourg-en-Br., 1882 ; L. Bergstraesser, Der politische Katholizismus. Dokumente zu seiner Entwicklung, t. II (1870-1914), Munich, 1923.
O. v. Bismarck, Politische Reden, édit. H.-Kohl, t. IV à VII, Stuttgart, 1895 et suiv. et Gedanken und Erinnerungen (trad. franç. Jaegle), t. I-II, Stuttgart, 1898 ; Cl. zu Hohenlohe, Denkwür-

auxquelles avait donné lieu le concile du Vatican. Les catholiques prussiens, encouragés bientôt par Ketteler qui gagna à leurs idées les cercles ecclésiastiques, estimèrent que le meilleur moyen de parer au danger était d'en revenir, en la perfectionnant, à la tactique qui leur avait si bien réussi dix ans auparavant. Ils décidèrent de constituer un parti qui ne serait plus strictement confessionnel, mais ouvert à tous ceux qui désiraient défendre les traditions chrétiennes et la liberté religieuse contre l'intransigeance laïcisante des nationaux-libéraux. Ce parti du *Centre*, qui ne visait pas seulement la défense religieuse, mais se présentait avec un programme démocratique et social très complet, défendu par un journal bien fait, la *Germania*, prit immédiatement une place importante dans la vie politique allemande, grâce à des chefs de valeur, politiques avisés et catholiques convaincus : les frères Reichensperger, deux vétérans de l'ancien Parlement prussien ; Mallinckrodt, le « Caton du Centre », qui avait assidûment fréquenté les réunions de Soest ; et surtout Windthorst, têtu parfois, mais jouteur parlementaire et manœuvrier de première force [1].

Ce nouveau parti, dont les allures populaires mécontentèrent certains aristocrates catholiques du Sud et de l'Est, était doublement antipathique aux nationaux-libéraux : ces bourgeois anticléricaux, représentants de la finance et de l'industrie, dénonçaient à la fois son orientation sociale progressiste et son attachement à une conception du monde et de l'homme qu'ils jugeaient incompatible avec les progrès de l'esprit humain.

Quant à Bismarck, protestant sincère qui croyait en Dieu et au Christ, il ne partageait pas leur antagonisme philosophique, mais il considérait comme un grand danger la fondation du Centre, dont les tendances particularistes pouvaient mettre en échec sa politique centralisatrice et qui sur-

---

digkeiten, édit. F. CURTIUS, t. II, Stuttgart, 1905 ; J. BACHEM, *Lose Blätter aus meinem Leben*, Fribourg-en-Br., 1911 ; extraits de la correspondance de Mallinckrodt publiés par O. PFUELF dans *Stimmen aus Maria Laach*, t. LXXXII et LXXXIII, 1912 ; E.-L. v. GERLACH, *Aufzeichnungen aus seinem Lebem*, Schwerin, 1903, 2 vol. ; J.-F. v. SCHULTE, *Lebenserinnerungen*, t. II et III, Giessen, 1908 (fort tendancieux). Enfin les *Protocole der Fuldaer Versammlungen*.

II. TRAVAUX. — Aux ouvrages classiques de J.-B. KISSLING, *Geschichte der Kulturkampf im Deutschen Reiche*, Fribourg-en-Br., 1911-1916, 3 vol. et G. GOYAU, *Bismarck et l'Église*, t. I et II, Paris, 1911 (point de vue catholique ; voir à leur sujet A. SCHNUTGEN, dans *Hochland*, t. I, 1917-1918, p. 641-657), ajoutent les ouvrages plus récents, utilisant des documents nouveaux, et conçus d'un point de vue moins confessionnel, de A. WAHL, *Vom Bismarck der 70ger Jahre*, Tubingue, 1920, à mettre au point par P. SATTLER, *Bismarcks Entschluss zum Kulturkampf*, dans *Forschungen zur Brandenburgischen und Preussischen Geschichte*, t. LII, 1940, p. 66-101 ; W. REICHLE, *Zwischen Staat und Kirche. Das Leben und Wirken des preussischen Kulturministers H. v. Mühler*, Berlin, 1938 ; E. FOERSTER, *Adalbert Falk*, Gotha, 1927, et surtout R. RUHENSTROTH-BAUER, *Bismarck und Falk*, Heidelberg, 1944. Aperçu synthétique serein par H. BORNKAMM, *Die Staatsidee im Kulturkampf*, dans *Historische Zeitschrift*, t. CLXX, 1950, p. 41-72, 273-306.

En outre, les ouvrages de K. BACHEM et J.-B. KISSLING et les biographies de Ketteler, de Martin et de A. Reichensperger, signalées p. 9, ainsi que celles de Windthorst (par H. E. HUESGEN, Cologne, 1920), de Mallinckrodt (par F. SCHMIDT, Fribourg-en-Br., 1921), de P. Reichensperger (par F. SCHMIDT, Fribourg-en-Br., 1913) et de Lieber (par M. SPAHN, Gotha, 1906).

Sur l'offensive contre l'Église en Autriche : GAUTSCH v. FRANKENSTURM, *Die Konfessionnellen Gesetze vom 7 und 20 Mei 1874 mit Materialien und Anmerkungen*, Vienne, 1874 ; M. HUSSAREK, *Die Krise und die Lösung des Konkordats* (cité p. 147, n. 1) ; C. WOLFSGRUBER, *Kirchengeschichte Œsterreichs*, p. 94-102 et *Kardinal Schwarzenberg*, t. III (cités p. 134, n. 2).

Sur le « Kulturkampf » suisse : *Histoire de la persécution religieuse à Genève. Essai d'un schisme d'État*, Paris, 1878 (documents intéressants) ; Ch. WOESTE, *Histoire du Culturkampf en Suisse*, Bruxelles, 1887 (catholique) ; P. GSCHWIND, *Geschichte der Entstehung der christkatholischer Kirche in der Schweiz*, Berne, 1910 (vieux-catholique) ; E. DANCOURT, *Scènes et récits du Kulturkampf dans le canton de Berne*, Saint-Maurice, 1921.

(1) L'ouvrage fondamental sur le Centre est celui de K. BACHEM, *Vorgeschichte, Geschichte und Politik der deutschen Zentrumspartei*, Cologne, 1926-1932, 9 vol. Voir surtout le t. III (*1870-1880*).

tout, s'appuyant sur les masses « fanatisées par l'Église », risquait de constituer un État dans l'État et de faire passer les intérêts du catholicisme
avant ceux de l'Allemagne. Ce danger lui paraissait d'autant plus grand qu'il
suspectait la sincérité du patriotisme des catholiques. Depuis longtemps,
il reprochait au clergé des provinces orientales d'encourager la résistance
des Polonais à la germanisation [1]. Il savait d'autre part combien la nostalgie à l'égard de l'idéal « grand-allemand » subsistait en Westphalie et
en Rhénanie et quelle attitude boudeuse les « Romains du Sud » avaient
prise à l'égard du nouvel Empire : en Bavière, en effet, le clergé était allé
parfois jusqu'à faire craindre aux catholiques de « devenir prussiens et
luthériens » si la Prusse victorieuse réalisait l'unification de l'Allemagne
à son profit [2], et la tendance qu'avaient certains protestants à présenter
Sedan comme une victoire de Luther sur la France catholique [3] devait
fatalement accentuer le malaise. La malheureuse question romaine vint
encore envenimer les choses : alors que Bismarck estimait qu'il était de
l'intérêt de l'Empire de rechercher l'alliance de l'Italie nouvelle, le Centre
se faisait l'écho du désir des catholiques allemands de voir le gouvernement
appuyer les protestations de Pie IX dépouillé de ses États, et le chancelier
alla jusqu'à le suspecter de provoquer sciemment l'agitation religieuse
en Allemagne pour affaiblir celle-ci et l'obliger à la neutralité en face d'une
éventuelle action franco-autrichienne contre l'Italie.

Ainsi, les catholiques s'étaient unis pour faire face au danger, mais
leur union même attirait contre eux de nouvelles méfiances. Bismarck, qui
en voulait moins à l'Église qu'au Centre, chercha à obtenir la dissolution du
nouveau parti en faisant pression sur les autorités ecclésiastiques et
c'est pour impressionner celles-ci qu'il se décida d'abord à soutenir les
menées des vieux-catholiques, puis qu'il accepta, en partie à contre-cœur,
d'approuver le programme de politique religieuse prôné par les nationaux-
libéraux. Le réalisateur de ce programme devait être le nouveau ministre
des Cultes, nommé au début de 1872, Adalbert Falk, théoricien rigide,
obstiné dans sa conception des droits tout-puissants de l'État en matière
ecclésiastique et qui avait en outre puisé dans la famille de pasteur
dont il était issu une antipathie profonde à l'égard du catholicisme.

PREMIÈRES MESURES
ANTICATHOLIQUES

Le changement d'attitude à l'égard des catholiques commença à se traduire dans les faits au
cours de l'été 1871, lorsqu'il devint clair que le
Vatican refusait de désavouer le Centre. Le 29 juin, Bismarck, après
plusieurs semaines d'hésitations, confirmait à Mgr Krementz que le
gouvernement prenait le parti des professeurs vieux-catholiques et interdisait aux évêques de les relever de leurs fonctions. Quelques jours plus
tard, la « division catholique » du ministère des Cultes, qui depuis trente

---

(1) Sur les difficultés entre l'Église et le gouvernement prussien en Silésie, voir B. v. SELCHOW,
Der Kampf um das Posener Erzbistum 1865. Graf Ledochovski und Oberpräsident v. Horr. Ein
Vorspiel zum Kulturkampf, Marbourg, 1923 (tendancieux). Cf. J. SCHMIDLIN, dans Theologische
Revue, XXIII, 1924, col. 403-404 et H. GRISAR dans Stimmen der Zeit., t. CVIII, 1925, p. 463-464.
(2) Marquard Barth à Miquel, 22 août 1870, dans Deutsche Revue, t. II, 1872, p. 62.
(3) Plusieurs témoignages dans G. GOYAU, Bismarck et l'Église, t. I, p. 70-73.

ans servait de tampon entre l'État et l'Église, était supprimée sous le prétexte que son chef Kraetzig favorisait les menées polonaises. Enfin, en août, Bismarck décidait brusquement de laïciser les écoles normales et l'inspection scolaire en Alsace-Lorraine, pour réagir contre l'influence d'un clergé accusé de rester trop attaché à la France [1].

En même temps, les violences de presse augmentaient contre les ultramontains. Les évêques, très inquiets, s'adressèrent en septembre à l'empereur, mais sa réponse ne fit que confirmer leurs craintes, car elle semblait étendre à l'épiscopat les griefs du gouvernement contre le Centre et surtout elle laissait entrevoir l'intention qu'avait l'État prussien de régler d'autorité et unilatéralement, sans accord préalable avec le Saint-Siège, un certain nombre de questions relatives à l'Église catholique. Bientôt après, en dépit d'une dernière tentative de Ketteler auprès de l'empereur et du chancelier, la guerre s'allumait sur le terrain parlementaire, sur l'initiative du ministre des Cultes de Bavière : le 16 décembre, le *Reichstag* [2] votait une loi, surnommée « Paragraphe de la Chaire », menaçant de peines de prison les prédicateurs qui critiqueraient un acte du gouvernement, fût-ce l'appui qu'il apportait au schisme vieux-catholique.

Plus grave fut la loi scolaire que Falk fit voter en mars 1872 par le *Landtag* prussien et qui enlevait aux curés ou aux pasteurs l'inspection des écoles primaires pour la confier à des agents de l'État. Pour Bismarck, il s'agissait de mettre fin à l'action antigermanique des curés de Silésie et il fut blessé au vif par l'opposition de Windthorst à cette loi qui lui paraissait une mesure indispensable de protection contre le polonisme abhorré. Mais le Centre avait tout de suite aperçu, comme certains protestants croyants du reste, que les nouvelles mesures scolaires inauguraient la politique de laïcisation réclamée depuis des années par les libéraux ; c'est ce qui devait apparaître au grand jour pendant la discussion qui s'évada vite du terrain national pour se transformer en un vaste débat pour ou contre l'Église. Falk projetait du reste d'autres réformes scolaires, destinées à desserrer encore davantage le lien, demeuré très ferme jusque-là, entre les Églises et l'école : celle-ci ne devait plus avoir comme objet « l'éducation chrétienne pour la vie », mais « l'acquisition de connaissances et d'aptitudes ». Et une circulaire du 15 juin 1872 excluait de l'enseignement public tous les ordres religieux, atteignant de la sorte près de 900 maisons.

Du plan politique, la lutte passait progressivement sur le terrain religieux. Bismarck s'y engagea d'autant plus aisément qu'il était au même moment exaspéré par l'attitude de Pie IX. Ce dernier s'était ouvertement solidarisé avec *La Correspondance de Genève*, une revue ultramontaine dont les rédacteurs attaquaient fréquemment le gouvernement allemand coupable de mener une politique en opposition avec le *Syllabus* et surtout de ne rien faire pour aider à replacer le pape sur son trône, et le chancelier avait beau jeu, surtout au lendemain de la défini-

---

(1) Voir à ce propos J. Régamey, *L'Alsace au lendemain de la conquête*, Paris, 1911 et F. Klein, *L'évêque de Metz. Vie de Mgr Dupont des Loges*, Paris, 1899.

(2) On se souviendra dans ce qui suit de la distinction entre le *Reichstag*, ou Diète réunissant les députés de tous les États de l'Empire, et le *Landtag* ou assemblée délibérante de chaque État.

tion de la primauté du pape, de dénoncer les évêques allemands comme les agents d'un système hiérarchique dont le chef avait pris position contre l'Allemagne ; ceux-ci avaient beau répondre qu'ils ne devaient en toute équité être jugés que d'après leurs paroles et leurs actes, ils « subissaient certaines solidarités onéreuses, où leur patriotisme même sentait un péril pour leur Église et que la déférence pour le Saint-Siège leur défendait cependant de décliner trop hautement » [1]. Pie IX multipliait d'ailleurs les protestations publiques et l'allocution consistoriale du 23 décembre 1872, où il dénonçait l'impudence des dirigeants allemands qui faisaient un crime de préférer aux lois de l'Empire celles de Dieu et de l'Église, acheva de pousser à bout Bismarck, qui rompit les relations diplomatiques avec le Vatican.

D'autres États allemands n'avaient pas tardé à prendre modèle sur la Prusse en matière d'anticléricalisme scolaire, en invoquant comme elle l'attachement des congrégations à « l'infaillibilisme », savamment présenté par les coryphées du libéralisme comme un redoutable danger pour l'État. C'est pour mieux extirper ce danger que le *Reichstag* vota, le 4 juillet 1872, avec l'appui enthousiaste de l'ancien premier ministre catholique de Bavière, le prince de Hohenlohe, une loi bannissant de l'Empire les jésuites, auxquels furent ensuite assimilées diverses congrégations soi-disant « affiliées », Dames du Sacré-Cœur, Rédemptoristes ou Lazaristes [2].

LES LOIS DE MAI   Ce n'étaient pourtant encore là que des mesures préliminaires. Bismarck désirait régler par un ensemble de mesures législatives la situation de l'Église de manière à l'empêcher d'intervenir efficacement dans la vie publique et de nuire à l'intérêt national. Il se serait volontiers rallié à la solution de la séparation totale de l'Église et de l'État, prônée par certains libéraux, mais Falk estimait que depuis vingt-cinq ans l'autonomie dont l'Église de Prusse avait joui grâce à la Constitution lui avait été trop favorable et qu'il fallait en revenir au principe luthérien de l'autorité souveraine de l'État sur l'Église, d'autant plus que le concile du Vatican avait accentué la dépendance des catholiques allemands à l'égard d'un « pouvoir étranger à la nation alle-mande et incapable de tout sentiment national ». Aussi élabora-t-il, en s'inspirant des suggestions des canonistes protestants les plus en vue, un ensemble de lois qu'on a pu comparer à la *Constitution civile du clergé* : elles s'appliquaient aussi bien à l'institution luthérienne qu'à l'Église romaine, mais atteignaient surtout cette dernière puisque, en prétendant soumettre toute la vie ecclésiastique au contrôle de l'État, elles bouleversaient l'organisation de l'Église sans se soucier de l'accord du Saint-Siège, ni même parfois des exigences de la doctrine catholique.

Bismarck se rallia aux projets de Falk qui, solennellement désavoués par l'épiscopat dès le 30 janvier, firent l'objet de débats passionnés au parlement prussien. L'opposition énergique menée par le Centre fut

---

(1) G. Goyau, *Bismarck et l'Église*, t. I, p. 278.
(2) Cf. B. Duhr, *Das Jesuitengesetz, sein Abbau und seine Aufhebung*, Fribourg-en-Br., 1919 (nombreux documents).

appuyée par certains protestants conservateurs, comme le vieux président de Gerlach, qui décelaient dans ce que le matérialiste athée Virchow avait dénommé un « combat pour la civilisation » (*Kulturkampf*), une lutte entre ceux qui croient au royaume de Dieu et ceux qui prônent l'autonomie complète de l'homme [1]. Mais la plupart des conservateurs suivirent Bismarck, qui expliqua que les mesures avaient une portée exclusivement politique et visaient non pas l'Église, mais uniquement « les courants souterrains qui aspiraient à la domination cléricale temporelle », tandis que le président du conseil prussien affirmait une fois de plus : « La faute en est au concile. »

Les lois furent finalement votées du 11 au 14 mai 1873, après que l'on eût modifié les articles 15 et 18 de la Constitution de 1850, qui accordaient aux Églises pleine liberté de s'organiser comme il leur convenait. La première réglait la formation du clergé : elle imposait à tous les candidats, prêtres ou pasteurs, un « examen de culture » et obligeait les séminaristes à faire trois années d'études dans une université allemande [2]. Il était en outre interdit aux évêques de faire aucune nomination ecclésiastique sans l'accord de l'*Oberpräsident* ou préfet de la province. Les autres lois restreignaient sensiblement le pouvoir de juridiction des évêques, leur interdisant notamment les excommunications publiques ou la déposition des curés, tandis qu'elles permettaient au gouvernement de déposer les prêtres jugés indésirables pour le bien de l'État ; elles établissaient en outre une *Cour royale de justice* devant laquelle les ecclésiastiques pouvaient en appeler des décisions de leurs supérieurs.

LA RÉSISTANCE CATHOLIQUE Au début de 1872, un attaché à la nonciature de Munich avait émis certaines craintes : « Il me semble que Geissel manque. Mgr Melchers, si éminent, si pieux, si apostolique, n'est pas un chef ; cela n'est pas donné à tout le monde [3]. » L'archevêque de Cologne devait, à l'heure du danger, se montrer à la hauteur de sa tâche, mais ce fut surtout l'énergique Ketteler, dont le diocèse comportait trois villages prussiens, qui stimula la résistance de ses collègues. Certains se demandaient si l'on ne pourrait pas accepter une partie des lois nouvelles puisque dans d'autres États, en Bavière ou au Wurtemberg par exemple, l'Église s'était accommodée de mesures analogues. Mais il s'agissait là de concessions faites par le Saint-Siège, tandis que le gouvernement prussien prétendait s'arroger ces droits de sa propre autorité. Ketteler jugea qu'il fallait repousser les lois en bloc et ce fut également l'avis de Rome, qui laissa du reste les évêques libres de fixer en commun les détails de l'attitude à prendre.

Devant le refus des évêques de soumettre aux autorités civiles les nominations ecclésiastiques, le gouvernement appliqua aussitôt les mesures

---

(1) G. Kraemer, *Die Stellung des Präsidents L. v. Gerlach zum politischen Katholizismus*, Breslau, 1931.

(2) Comme au même moment le gouvernement encourageait la résistance des professeurs vieux-catholiques, la mesure avait une portée particulièrement grave. Elle s'opposait d'autre part indirectement à l'envoi de séminaristes à Rome.

(3) Lettre du 23 mars 1872 dans O. Pfuelf, *Bischof Ketteler*, t. III, p. 174-175.

prévues : amendes, saisie des biens épiscopaux, et mise en prison des évêques insolvables. La première victime fut l'archevêque de Posnan, Mgr Ledochowski, particulièrement honni pour avoir refusé de faire enseigner le catéchisme en allemand aux petits Polonais.

Le clergé fut unanime à seconder l'épiscopat dans sa résistance et à affronter, lui aussi, les amendes et la prison, démentant les prévisions pessimistes de ceux qui avaient considéré la réserve de certains prêtres devant les progrès de l'ultramontanisme comme une preuve d'absence de sens du devoir. Même en Bavière, où la situation privilégiée du catholicisme avait assoupi le zèle du clergé, les énergies se trouvèrent galvanisées par l'exemple prussien, bien que le *Kulturkampf* y ait sévi avec beaucoup moins de rigueur qu'en Prusse [1].

Mais ce furent surtout les fidèles qui furent admirables. Il y eut quelques hésitations dans les hautes classes, parmi les intellectuels hostiles à l'ultramontanisme et aux jésuites ; certains prétendirent se désolidariser du « catholicisme politique » représenté par le Centre et s'affichèrent comme les tenants d'un « catholicisme religieux » qui, malheureusement, chez plusieurs, méritait plutôt l'épithète de « catholicisme d'État » que leur décochait la majorité de leurs coreligionnaires. Ceux-ci, paysans de l'Eifel ou de la vallée rhénane, artisans et ouvriers des grandes villes industrielles de l'Ouest, encadrés depuis plus de dix ans dans des associations vivantes grâce auxquelles ils étaient restés en contact étroit avec leur Église, se méfiaient spontanément des hobereaux conservateurs ou des représentants libéraux de la bourgeoisie capitaliste ; ils appréciaient le programme démocratique et social défendu par le Centre, mais ils avaient aussi compris la portée religieuse du conflit et la menace qui pesait sur leur foi. D'ailleurs, même dans la bourgeoisie, en présence des attaques contre Rome et des efforts tentés par Bismarck pour détourner l'Église du centre de la catholicité, beaucoup se découvraient des sentiments de fidélité insoupçonnée envers le Saint-Siège et, appréciant à ses fruits le travail qui s'était poursuivi depuis une génération, faisaient à présent bloc autour de leurs évêques et des chefs du mouvement catholique. Le brave Janssen notait avec une joyeuse surprise les « progrès réconfortants du sens de l'Église » [2] parmi les catholiques que quelques mois de *Kulturkampf* avaient suffi à détourner de la tentation qu'aurait pu constituer pour certains d'entre eux le schisme vieux-catholique.

RENFORCEMENT DU KULTURKAMPF La magnifique résistance des catholiques allemands faisait l'admiration du monde entier. Le pape allait bientôt créer le cardinal Ledochowski encore incarcéré et il aurait volontiers multiplié les protestations publi-

(1) Après avoir été l'un des instigateurs du *Kulturkampf*, le ministre Lutz, qui était plutôt un joséphiste de la nuance Montgelas qu'un véritable adversaire de l'Église, se tint de plus en plus sur la réserve, d'autant plus que l'influence du Centre en Bavière alla en s'amplifiant. (Cf. F. v. RUMMEL, *Das Ministerium Lutz und seine Gegner 1871 bis 1882*, Munich, 1935.)

Dans les autres États de l'Allemagne du Sud, la situation fut assez variée : le Wurtemberg fut à peine touché par le *Kulturkampf* et la loi de 1862 continua à assurer la paix religieuse, tandis que le grand-duché de Hesse et surtout celui de Bade marchèrent à fond dans la même ligne que Bismarck.

(2) Lettre du 8 août 1873, dans *Janssens Briefe*, édit. PASTOR, t. I, p. 438.

ques si les évêques n'avaient préféré qu'on évitât de jeter de l'huile sur le feu. L'épiscopat de France ou de Belgique envoyait à celui de Prusse des adresses qui mettaient le chancelier de fer hors de lui au point de provoquer des incidents diplomatiques [1].

Le gouvernement prussien espérait venir à bout de la résistance par l'intimidation, mais, sur les mesures concrètes à prendre, Bismarck et Falk différaient d'avis. Ce dernier voulait notamment profiter de la situation pour appliquer intégralement son programme de laïcisation en introduisant dans la législation le mariage civil obligatoire. Bismarck, tout en doutant de l'efficacité de la méthode préconisée par son ministre des cultes, la prit à son compte afin de ne pas perdre l'appui des libéraux [2], et, en mai 1874, de nouvelles lois renforcèrent les sanctions. Elles permettaient de destituer de leurs fonctions et de condamner à l'exil les évêques récalcitrants et confiaient la collation des cures, dans les diocèses vacants, à l'assemblée des paroissiens ; elles menaçaient d'amendes ou de prison les prêtres qui continueraient à obéir à leur ancien évêque. L'année suivante, d'autres lois supprimaient le traitement des membres du clergé paroissial qui refuseraient de se conformer à la nouvelle législation ecclésiastique, expulsaient de Prusse tous les ordres religieux, sauf les hospitaliers, remettaient aux mains des laïcs l'administration des biens d'Église et cherchaient à faciliter la mainmise sur les paroisses par les vieux-catholiques.

Bientôt, la plupart des sièges épiscopaux furent vacants. Mais les évêques avaient obtenu du Saint-Siège de pouvoir désigner des prêtres, connus seulement de quelques initiés, qui détiendraient les pouvoirs épiscopaux et, de Hollande, de Belgique, d'Autriche ou de Rome, où ils étaient exilés, ils continuèrent à diriger secrètement leurs diocèses. La vie de ceux-ci était profondément désorganisée : de nombreuses paroisses — près d'un quart à certains moments [3] — étaient sans pasteur, avec interdiction à aucun prêtre de dire la messe ou d'y administrer les sacrements, et certains libéraux commençaient à trouver que ces persécutions, qui aboutissaient à priver des mourants des secours de leur religion, n'avaient plus grand'chose à voir avec le libéralisme. Souvent d'ailleurs, des prêtres déguisés réussissaient à échapper à la surveillance des gendarmes et à exercer en secret leur ministère avec la complicité de leurs paroissiens.

VERS LA DÉTENTE Les évêques songeaient d'autant moins à céder que leur résistance était soutenue par celle du peuple catholique tout entier. On a généralement fait davantage l'histoire législative et parlementaire du *Kulturkampf* que celle de son application dans la

---

(1) Voir pour la France E. LECANUET, *L'Église de France sous la III* République, t. I : *Les dernières années de Pie IX*, p. 169-177 ; F. DE LANNOY, *Un incident germano-belge à propos du Kulturkampf*, Schaerbeek, 1938.

(2) L'apologie de Falk par Foerster, qui prétend rejeter sur Bismarck seul le renforcement de la persécution, ne paraît pas défendable, surtout après les plus récentes publications. Il est encore moins exact d'affirmer, comme certains catholiques, que le chancelier envisageait sérieusement la destruction de l'Église catholique en Allemagne.

(3) En 1880, il y en avait 1.100 sur 4.600, d'après le ministre des cultes (discours du 26 janvier 1881, cité par W. ONCKEN, *Das Zeitalter des Kaisers Wilhelm I*, t. I, Ratisbonne, 1890, p. 731).

vie quotidienne de l'Église [1] ; il serait cependant bien intéressant de
montrer dans le détail comment les rigueurs de la persécution achevèrent
de renforcer la cohésion des catholiques allemands tenus en haleine
par une presse qui se développait rapidement [2]. Ils prirent notamment
conscience de leur force véritable en face du protestantisme désorienté
par l'offensive antireligieuse et ce n'est pas un hasard si, par exemple,
le chiffre annuel des cotisations au *Bonifatiusverein*, chargé du soutien
des paroisses catholiques dans les régions protestantes, marqua pendant
le *Kulturkampf* un progrès notable sur les recettes de 1869. Et tandis que
le programme social des assemblées catholiques se précisait d'année en
année, un groupe d'intellectuels, dirigés par Georges von Hertling et
Julius Bachem, fondait, en 1877, l'année du centenaire du grand lutteur
munichois, une société destinée à encourager les jeunes savants catholiques, la *Goerresgesellschaft*.

Les observateurs perspicaces pouvaient dès lors prévoir que l'Église
sortirait de la lutte rajeunie, plus forte et plus disciplinée que jamais,
ayant appris, même dans le Sud, à se montrer plus indépendante de
l'État et à se tourner vers Rome avec plus de confiance. « Je crois, écrivait
Janssen en 1875, que depuis plusieurs siècles, la Sainte Église ne s'est
pas si bien portée en Allemagne qu'à présent [3]. » Protestants et libéraux
n'en revenaient pas et si certains idéologues sectaires désiraient malgré
tout continuer l'expérience, Bismarck, qui avait bien promis en commençant qu'il « n'irait jamais à Canossa », était trop réaliste pour ne
pas s'incliner devant les faits, surtout en présence du danger nouveau
que représentait la montée régulière du socialisme. Lorsque, au début de
1878, un pape à l'esprit conciliateur succéda au belliqueux Pie IX, toutes
les conditions se trouvèrent réunies pour un apaisement qui n'allait pas
tarder.

*L'OFFENSIVE LIBÉRALE CONTRE
L'ÉGLISE EN AUTRICHE*

Comme en Allemagne, le concile du
Vatican fut exploité en Autriche par
les libéraux, qui réclamaient depuis
plus de dix ans l'abrogation du concordat de 1855. Dès le 25 juillet 1870,
le gouvernement le dénonçait sous prétexte que « le pape infaillible
n'était plus le pape avec lequel l'Autriche avait conclu un concordat »,
et quelques jours plus tard, l'empereur chargeait le ministre des cultes
de préparer les lois nécessaires pour régler sur de nouvelles bases les rapports entre l'Église et l'État « conformément à la constitution et en tenant
compte des indications de l'histoire » [4].

La grave maladie qui contraignit le cardinal Rauscher, à son retour
de Rome, à garder la chambre pendant plusieurs mois retarda la résis-

---

(1) On aimerait disposer dans ce sens de nombreux documents dans le genre des souvenirs du
haut fonctionnaire L. FICKER, *Der Kulturkampf in Munster*, édit. O. HELLINGHAUS, Munster,
1928, qui révèlent l'héroïsme et l'ingéniosité de la résistance catholique, en même temps que certaines exagérations inévitables.
(2) Sous l'impulsion des *leaders* du Centre, le nombre des quotidiens catholiques passa de 126
en 1871 à 221 en 1881. (Cf. H.-J. REIBER, *Die katholische deutsche Tagespresse unter dem Einfluss
des Kulturkampfes*, Görlitz, 1936.)
(3) Janssen à Alberding Thijm, 21 mars 1875, dans *Janssens Briefe*, t. II, p. 18.
(4) Voir les textes dans M. HUSSAREK, *op. cit., passim*.

tance de l'épiscopat. Les libéraux étaient d'ailleurs décidés à aller de l'avant coûte que coûte, sans tenir compte ni des interventions réitérées de l'archevêque auprès de l'empereur, ni des protestations publiques des évêques qui insistèrent à plusieurs reprises sur le fait qu'en s'en prenant à l'Église, on détruisait la seule grande force antirévolutionnaire dans le pays. Il faut du reste reconnaître que, en dépit des efforts de certains curés pour organiser parmi les membres des associations catholiques un mouvement de pétition en vue de sauvegarder le caractère confessionnel de l'école publique, la réaction épiscopale ne trouva guère d'écho dans le peuple catholique, peu habitué à collaborer activement avec ses évêques pour la défense de sa foi ; à aucun moment, il ne fut question de voir se constituer, comme en Allemagne, un puissant parti issu de la masse des fidèles pour appuyer les revendications de l'Église sur le plan parlementaire [1].

Après avoir pris diverses mesures vexatoires, en faveur des vieux-catholiques et contre les jésuites notamment, puis avoir, en 1873, modifié le caractère exclusivement catholique des universités, le gouvernement déposa, en janvier 1874, quatre projets de lois qui devaient régler en dehors de toute intervention romaine la vie ecclésiastique autrichienne. Ils étaient moins radicaux que la législation allemande correspondante, mais l'esprit dans lequel ils étaient conçus apparaissait clairement dans l'exposé des motifs où l'on pouvait lire entre autres que « la conception d'après laquelle l'Église dans son domaine jouirait d'une souveraineté égale à celle de l'État dans le sien est moins que jamais acceptable à l'heure actuelle ». Les trois premières lois furent votées au mois de mai et ratifiées aussitôt par l'empereur, en dépit d'une ultime démarche de l'épiscopat. Elles soumettaient les nominations ecclésiastiques et l'exercice du culte à une étroite surveillance gouvernementale, réglaient la question de la propriété ecclésiastique et achevaient d'accorder l'égalité des droits aux différents cultes.

Étant donné la confiance qu'il avait longtemps mise dans l'Autriche « officiellement catholique », ce fut une dure déception pour Pie IX qui avait déjà ressenti avec amertume des mesures telles que la rupture unilatérale du concordat par une nation qu'il avait toujours considérée comme le boulevard contre les principes révolutionnaires, le refus de François-Joseph de prendre l'initiative d'une démarche auprès des États européens en vue de garantir l'indépendance du Saint-Siège après l'occupation de Rome, et l'envoi d'un ambassadeur autrichien auprès de Victor-Emmanuel, ce qui constituait à ses yeux un affront pour le pape. Aussi protesta-t-il cette fois solennellement, en stigmatisant « ces gouvernements catholiques qui dépassent les gouvernements protestants dans la honteuse carrière de l'oppression religieuse ».

L'épiscopat autrichien prit toutefois une attitude moins belliqueuse que le pape, bien que les évêques du Tyrol et certains militants, comme

---

(1) L'organisation même de la représentation parlementaire en Autriche rendait du reste la chose beaucoup plus difficile qu'en Allemagne. (Cf. ZANATTA, *I tempi e gli uomini che prepararono la « Rerum Novarum »*, Milan, 1931, p. 29-30.)

l'ancien ministre Thun, eussent souhaité qu'on imitât l'attitude radicale des évêques allemands et qu'on essayât de dresser les foules catholiques contre un gouvernement violateur des droits de l'Église [1]. Mais Rauscher, suivi par les évêques de tendance gouvernementale, ne voulut pas pousser la résistance jusqu'à la rupture et essaya au contraire de s'accommoder de la situation nouvelle : au lieu de protester contre les projets avant même qu'ils fussent votés, il préféra, sans se compromettre personnellement, en faire discuter les modalités avec le ministère des cultes par son évêque auxiliaire, Mgr Kutschker, afin de les rendre aussi acceptables que possible à l'Église [2]. Des motifs assez divers l'inclinaient au compromis. Il avait d'une part le souci de ne pas affaiblir l'État par des discordes intérieures et de continuer à prêter l'appui de l'Église à la politique centralisatrice du gouvernement de Vienne dans les régions allogènes. Il sentait d'autre part qu'au fond le gouvernement désirait simplement arriver à un régime analogue à celui du concordat français et était prêt, moyennant cela, à continuer à favoriser l'Église comme une institution officielle ; or, il lui semblait qu'il fallait à tout prix éviter la séparation totale entre l'Église et l'État, étant donné le caractère très traditionaliste et assez peu vivant du catholicisme autrichien [3]. Schwarzenberg, qui trouvait un peu faible l'attitude de Rauscher, pensait d'ailleurs de même, la plupart des évêques autrichiens ne concevant pas la vie ecclésiastique autrement que dans une entente étroite avec le gouvernement. Rome finit d'ailleurs par accepter le point de vue de l'archevêque de Vienne [4] et l'Autriche évita de la sorte la guerre religieuse : si les fidèles ne connurent pas le stimulant qu'avait été la persécution pour leurs frères allemands, du moins l'édifice extérieur de l'Église ne fut-il guère secoué.

La quatrième loi, déposée en janvier 1874, visait la suppression des couvents ; elle ne vint en discussion qu'au début de 1876, au lendemain de la mort de Rauscher. Schwarzenberg prit la tête de l'opposition épiscopale et l'empereur refusa finalement de signer la loi, offrant au contraire asile à certains moines expulsés d'Allemagne par le *Kulturkampf*, aux bénédictins de Beuron notamment, qui devaient par la suite fonder sur le sol autrichien les célèbres abbayes d'Emmaüs et de Seckau. L'année suivante, le nouvel archevêque de Vienne, Kutschker, réussit à faire écarter un projet de loi qui tendait à supprimer de la législation civile les empêchements canoniques de mariage. Toutefois, l'abandon progressif de la politique anticléricale des gouvernements précédents n'entraîna

---

(1) Outre C. WOLFSGRUBER, *Kard. Schwarzenberg*, t. III, p. 397-426, 453-458 et 580-581, voir surtout VON OER, *Fürstbischof J. Zwerger*, Graz, 1897, p. 275 et suiv.

(2) Voir entre autres la lettre de Rauscher à Schwarzenberg du 19 février 1874 dans C. WOLFSGRUBER, *O. Rauscher*, p. 211-212 : « Il a obtenu la suppression des mesures les plus mauvaises. » Schwarzenberg, qui n'avait rien d'un intransigeant, jugeait toutefois que Kutschker manquait d'énergie et s'inclinait trop facilement devant les désirs du gouvernement ou de la cour.

(3) Voir les observations éclairantes de M. PFLIEGLER, *Zum Verständnis des oesterreichischen Katholizismus*, dans *Schönere Zukunft*, t. VI, 1930, p. 151-152.

(4) Pie IX avait tout d'abord répondu à Mgr Gasser qui se plaignait de la passivité de Rauscher et demandait au pape de lui donner « *une piccola spinta* », qu'il interviendrait : « *non solamente una piccola spinta daro, ma una spinta molto forte, un spintone* » (C. WOLFSGRUBER, *Kardinal Schwarzenberg*, t. III, p. 432-433), et il avait adressé, le 7 mars 1874, une encyclique très nette à l'épiscopat autrichien. Mais en décembre, le nonce adressa à l'archevêque de Vienne une lettre qui fut une déception pour les partisans de la résistance à outrance.

pas un simple retour à la situation antérieure et les évêques se heurtèrent notamment à une fin de non-recevoir lorsque, en 1877, à la suite
du *Katholikentag*, ils dénoncèrent la politique scolaire antichrétienne
des dernières années comme préparant une génération mûre pour la
révolution.

LE KULTURKAMPF SUISSE   La proclamation de l'infaillibilité du pape
et l'agitation qui s'ensuivit en Suisse dans
certains milieux de catholiques ultralibéraux, qui aspiraient depuis
longtemps à une réforme du catholicisme dans un sens démocratique et
antiromain, fournit une excellente occasion d'intervention aux radicaux,
dont beaucoup rêvaient depuis longtemps de voir se constituer une Église
indépendante de Rome et étroitement soumise aux pouvoirs civils [1].
Mais, fait nouveau, le gouvernement fédéral, qui s'était abstenu jusqu'alors
d'appuyer la politique anticatholique menée par certains cantons, intervint à son tour dans la lutte contre l'Église.

C'est dans le diocèse de Bâle, qui englobait neuf cantons, que le mouvement vieux-catholique se développa surtout. L'évêque, Mgr Lachat [2],
ayant interdit les deux seuls prêtres qui refusaient d'accepter les décrets
du Vatican, les quelques dizaines de milliers de catholiques antiromains,
dont le concile avait porté l'effervescence à son comble, protestèrent
contre « l'allure de plus en plus despotique » que prenait l'Église romaine
et demandèrent aux autorités cantonales de protéger la « vraie » liberté
ecclésiastique. Celles-ci s'empressèrent évidemment d'intervenir, considérèrent l'évêque comme destitué de ses fonctions et en profitèrent même,
en plusieurs cantons, pour décider que dorénavant les curés seraient
élus par leurs paroissiens et soumis tous les six ans à réélection. Dans
le Jura bernois, où les catholiques protestaient contre ces mesures, le
gouvernement alla jusqu'à interdire aux prêtres fidèles à leur évêque de
continuer à administrer leurs paroisses ; leurs églises furent attribuées
à des prêtres vieux-catholiques venus d'Allemagne et les catholiques se
virent obligés, pendant plusieurs années, de célébrer leur culte dans des
granges ou des locaux de fortune.

Une situation analogue se présenta à la même époque dans le canton
de Genève. Le gouvernement, présidé par le franc-maçon Carteret, qui ne
pouvait supporter l'action de Mgr Mermillod, spécula sur le dépit que le
démembrement de son diocèse avait causé à l'évêque de Lausanne et
ordonna à Mermillod de cesser toute fonction épiscopale. Sur le refus de
celui-ci, il le fit arrêter et le bannit du pays en février 1873 [3]. Au mois de
mai suivant, il faisait voter une loi réorganisant l'Église catholique en
décidant notamment que les curés seraient nommés par l'assemblée
paroissiale à la majorité des voix. Le clergé et les fidèles, qui restaient en

(1) Voir sur ce courant libéral et antiromain et sur les tendances radicales vers une Église nationale, E. CAMPANA, *Il Concilio Vaticano*, vol. II, Lugano, 1926, p. 569-627.
(2) Cf. F. FOLLETÈTE, *L'évêque confesseur de Bâle, Mgr Lachat*, dans *Zeitschrift für schweizerische
Kirchengeschichte*, t. XIX, 1925, p. 19-38.
(3) Ce n'est qu'en 1884 qu'il put rentrer enfin dans son diocèse, lorsque grâce aux efforts de
Léon XIII une détente se produisit dans les relations entre le gouvernement suisse et l'Église
catholique.

contact étroit avec leur évêque, établi à la frontière française, refusèrent de s'incliner devant cet essai d'organisation d'une Église nationale. Le gouvernement s'empara alors des églises et fit appel à des prêtres français en révolte contre le concile du Vatican, à Hyacinthe Loyson notamment, qui fut élu, en octobre 1873, à la cure de Genève.

C'est dans cette atmosphère qu'eut lieu, en 1874, une revision de la constitution fédérale. Les articles dits « confessionnels » ne pouvaient que s'en ressentir fâcheusement : l'interdiction portée contre les jésuites fut aggravée [1] ; la fondation de nouveaux couvents ou ordres religieux fut défendue ; enfin l'interdiction d'établir de nouveaux évêchés sans l'approbation du pouvoir central fut confirmée. Sous l'effet d'un anti-cléricalisme rageur, les libéraux suisses emboîtaient de la sorte le pas aux libéraux d'Allemagne, sans prendre garde à la contradiction qui existait entre ces mesures et l'esprit de la constitution de leur pays.

§ 4. — L'Église catholique dans le reste de l'Europe occidentale [2].

MANNING ET LE DÉVELOPPEMENT DU CATHOLICISME EN ANGLETERRE

A son retour du concile du Vatican, Manning pouvait compter plus que jamais sur la faveur de Pie IX, qui le nommera cardinal en 1875, et son action devint prépondérante dans la vie catholique anglaise. Ses idées n'ont pas changé. Il poursuit envers et contre tous son rêve d'une Université catholique, qui n'aura d'ailleurs qu'une existence éphémère et sans gloire [3]. Il demeure l'apologiste du *Syllabus* et des droits de la papauté, l'adversaire implacable d'Acton et de Newman. Il cherche toujours, en tenant solidement en main la presse catholique [4], à imposer son point de vue et à laisser aux laïcs le moins d'initiative possible. Mais même ceux qui ne l'aiment pas sont bien obligés d'admirer ses beaux côtés, qui se manifestent de plus en plus : remarquables qualités d'homme d'action allant de pair avec un haut idéal de vie sacerdotale, un désintéressement total et un dévouement de tous les instants aux petits et aux pauvres.

C'est à partir de 1880 surtout que ses préoccupations sociales devien-

---

(1) « Cette interdiction peut s'étendre aussi par voie d'arrêté fédéral à d'autres ordres religieux dont l'action est dangereuse pour l'État ou trouble la paix entre les confessions. »

(2) BIBLIOGRAPHIE. — En ce qui concerne l'Angleterre : aux ouvrages signalés p. 9, ajouter M. WARD, *The Wilfrid Wards and the transition*, Londres, 1934.

En ce qui concerne la Belgique : aux travaux du P. de Moreau signalés p. 9, ajouter G. GUYOT DE MISHAEGEN, *Le parti catholique belge de 1830 à 1884*, Bruxelles, 1946, et H. SAINTRAIN, *Vie du cardinal Dechamps*, Tournai, 1884.

En ce qui concerne les Pays-Bas et l'Espagne, voir les ouvrages signalés p. 172, n. 1 et p. 179, n. 2.

(3) Manning, encouragé d'ailleurs par la Propagande, réexamina la question en 1871, d'abord avec Ullathorne et Clifford, puis avec les supérieurs de collèges et d'ordres religieux ; malgré l'attitude réservée de ses collègues et surtout le peu d'enthousiasme des laïcs qui, habitués au *self-government* des universités anglaises, ne comprenaient guère qu'une maison d'études supérieures fût placée sous la dépendance absolue et immédiate des évêques, il décida le concile provincial de 1873 à aller de l'avant et, en octobre 1884, une université fut ouverte à Kensington, dans le sud de Londres. La difficulté de trouver un recteur jouissant d'un prestige scientifique suffisant — Manning ne voulait évidemment pas entendre parler de Newman — acheva de rendre l'entreprise inviable.

(4) Son fidèle Ward dirigeait déjà la *Dublin Review* depuis 1863, et en 1868, il avait réussi à faire passer le *Tablet* d'une direction laïque indépendante à celle de son collaborateur dévoué et futur successeur Herbert Vaughan.

dront prépondérantes, mais il n'avait pas attendu cette date pour s'engager dans cette voie, poussé à la fois par la compassion pour les misères
de la classe populaire et par la conviction réfléchie que l'avenir de l'Église
était dans cette direction. Dès 1875, il était déjà dominé par l'idée qu'il
devait formuler en ces termes à la veille de sa mort :

> Dieu veuille que le peuple ne nous regarde jamais comme des *tories*, apparte-
> nant au parti qui fit obstacle à l'amélioration de sa condition, que nous ne lui
> paraissions pas les serviteurs de la ploutocratie au lieu d'être les guides et les
> protecteurs des pauvres [1].

Les interventions répétées du cardinal sur le terrain social et philanthropique eurent d'ailleurs un résultat très heureux dans l'immédiat.
Les sphères administratives s'habituèrent à considérer qu'il avait sa
place marquée dans toute commission où s'étudiait quelque problème de
ce genre et Manning vit ainsi se réaliser son désir le plus cher : « mettre
l'Église catholique en relations plus ouvertes avec le peuple et l'opinion
d'Angleterre » [2]. C'est dans ce même but qu'il n'hésita pas, malgré son
intransigeance doctrinale, à collaborer de plus en plus avec ses compatriotes de toute croyance, devenant même membre assidu de la Société
de Métaphysique, qui réunissait des anglicans, des protestants et même
des incroyants.

Cette réintégration du catholicisme anglais dans la vie nationale est
plus importante encore que sa lente progression numérique qui continue
sans défaillance et porte ses effectifs, vers la fin du pontificat de Pie IX,
à plus de deux millions d'âmes, encadrées par près de 2.000 prêtres.

LA HIÉRARCHIE RÉTABLIE
EN ÉCOSSE

L'influence de l'archevêque de Westminster
s'étend au delà des limites de l'Angleterre
jusqu'en Écosse, où il aide à mener à bien
l'épineuse question de la restauration d'une hiérarchie régulière [3]. Lorsque, en 1864, au cours du conflit qui avait dressé le clergé de Glasgow
contre le vicaire apostolique [4], l'idée avait été lancée par quelques prêtres
d'origine irlandaise, Wiseman, consulté, avait estimé la chose souhaitable, de même que Manning au début de son épiscopat, mais devant
l'opposition unanime des trois vicaires apostoliques, les autorités romaines
avaient estimé préférable d'ajourner le projet [5]. En 1877, une délégation
du clergé écossais vint prier Pie IX de reconsidérer la question et des
négociations s'engagèrent, auxquelles Manning prit une part importante.
Elles aboutirent après un an à la bulle du 4 mars 1878 qui établissait
pour l'Écosse deux archevêchés et quatre évêchés, groupant à cette date
257 prêtres et environ 380.000 fidèles, dont près des deux tiers concentrés
dans la région de Glasgow.

(1) Note autobiographique, 1890, dans E. PURCELL, *Life of Manning*, t. II, p. 637.
(2) Note de février 1887, *ibid.*, t. II, p. 678.
(3) Voir l'ouvrage de A. BELLESHEIM (cité p. 155, n. 1), p. 418-430.
(4) Cf. plus haut, p. 163.
(5) Lors de son passage à Rome, au début de 1869, Manning lui-même conseilla l'ajournement
(*Arch. Min. A. É. Paris*, Rome, n° 1043, f° 203 v°).

*LE CONCILE PLÉNIER
DE MAYNOOTH*

La catholique Irlande a aussi ses problèmes [1] : la question des écoles et des collèges universitaires, objets une fois de plus de discussions avec le gouvernement ; l'adaptation à la situation nouvelle créée par le « désétablissement » de l'Église anglicane en Irlande, proposé par Gladstone en 1869 et voté en 1870. C'est pour les mettre au point que le cardinal Cullen obtient de Rome en 1873 l'autorisation de convoquer un nouveau concile national qui se réunit à Maynooth à l'automne de 1875 [2].

*LES CATHOLIQUES BELGES FACE
AU RADICALISME LIBÉRAL*

Lors des congrès de Malines, Montalembert et Dupanloup avaient, avec le prestige de leur éloquence, formulé le programme autour duquel allaient se cristalliser les énergies des catholiques belges jusqu'à la naissance de la question ouvrière : accepter franchement les libertés constitutionnelles et, dans l'État libre, travailler à la grandeur d'une Église libre, agissante et conquérante, en s'efforçant spécialement de former l'âme de l'enfant par l'école chrétienne. Mais, comme on l'a vu, la plupart des participants n'avaient pas tardé à se convaincre que, au moment où la liberté de l'Église paraissait menacée par le radicalisme anticlérical des libéraux, « l'œuvre des œuvres était la fondation et l'organisation d'un parti qui assurerait toute son efficacité catholique à la constitution libérale » [3]. Si l'on voulait arriver à rénover la société par l'esprit religieux, il paraissait nécessaire de défendre d'abord l'Église sur le plan politique.

Habilement organisé autour des « cercles catholiques », transformés de sociétés d'agrément en associations électorales et groupés en une Fédération nationale, dirigé avec autorité par Adolphe Dechamps, évitant d'apparaître en dépendance trop étroite à l'égard du clergé ou de l'épiscopat, ce parti catholique (connu jusqu'en 1884 sous le nom de « parti conservateur ») remporta la victoire aux élections de 1870 et l'un de ses chefs, Jules Malou, prit la direction du gouvernement, qu'il conserva jusqu'en 1878.

Dès lors, bien que la propagande anticléricale n'en continuât pas moins activement [4], la politique de sécularisation progressive menée depuis quinze ans par les libéraux subit un temps d'arrêt. Mais si les œuvres, stimulées par les récents congrès, connurent un nouvel essor, l'action des catholiques fut cependant entravée par les dissensions provoquées par les « veuillotistes » qui, en dépit des lettres de Sterckx de 1864, continuaient à attaquer la constitution et ses « libertés de perdition ». Sous l'impulsion du journal *La Croix* et du professeur Charles

---

(1) Voir A. BELLESHEIM (ouvrage cité, p. 155, n. 1), p. 601-617 et 631-646.
(2) *Acta et Decreta Synodi plenariae episcoporum Hiberniae habitae apud Maynutiam anno 1875*, Dublin, 1877.
(3) A. MELOT, *Cinquante années de gouvernement parlementaire*, Louvain, 1935, p. 23.
(4) Celle-ci ne s'exerçait pas seulement sur le terrain scolaire. Elle se fit jour, entre autres, à l'occasion des manifestations de masse organisées sous la direction de Mgr Dechamps pour protester contre la spoliation des États pontificaux. Les étudiants des Universités de Gand, Bruxelles et Liège se livrèrent parfois à de véritables actes de sauvagerie ; ainsi, en 1875, au retour d'un pèlerinage à Oostacker, il y eut un mort et 169 blessés.

Périn de l'Université de Louvain, la controverse rebondit de plus belle après 1870, bien que le nouvel archevêque de Malines [1], frère d'Adolphe Dechamps et ami de Dupanloup, fût, comme son prédécesseur, un défenseur de la constitution. Il fallut attendre l'intervention de Léon XIII, en 1878, pour voir enfin cesser ces discussions stériles.

Mais, entre temps, elles avaient contribué à affaiblir le gouvernement catholique, qui fut renversé aux élections de 1878. Les libéraux, qui s'étaient déjà appliqués depuis des années, dans les communes où ils étaient les maîtres, à « corriger administrativement » la loi de 1842 sur l'enseignement primaire, en organisant en fait un enseignement neutre, souvent antireligieux, estimèrent le moment venu de supprimer officiellement cette loi déjà condamnée par les loges depuis 1854, par la *Ligue de l'Enseignement* depuis 1864 et par les 42 associations libérales du pays depuis 1870. Avec la « loi de malheur » de 1879, les catholiques belges allaient connaître la guerre scolaire, mais la réaction immédiate et victorieuse contre celle-ci devait révéler à quel point le sentiment catholique était encore profond dans de larges couches de la population, en dépit des progrès lents mais constants de la déchristianisation dans la bourgeoisie des villes, fortement travaillée par la franc-maçonnerie, et surtout dans les masses ouvrières, abandonnées à elles-mêmes dans les nouvelles agglomérations industrielles insuffisamment pourvues d'églises et de prêtres, et d'autant plus exposées à prêter l'oreille à la propagande irréligieuse des socialistes que les catholiques tardaient à s'engager dans la voie des réformes sociales.

LA RUPTURE DÉFINITIVE AVEC LE LIBÉRALISME AUX PAYS-BAS

Aux Pays-Bas, comme en Belgique, bien qu'avec un certain retard, à l'alliance entre catholiques et libéraux avait fait place une opposition qui allait en s'aggravant. La chute du pouvoir temporel du pape la mit en pleine lumière, lorsque le gouvernement Thorbecke, sans tenir compte de l'indignation des catholiques, décida au début de 1872 la suppression de l'ambassade près du Saint-Siège. Ce geste ne fit que rendre plus aiguë la lutte pour l'école, entamée depuis 1865 et que le mandement collectif des évêques du 23 juillet 1868 avait fait passer au premier plan des préoccupations catholiques.

Cette question scolaire était devenue peu à peu le pivot de la lutte des partis, et après la déclaration ministérielle de 1868 refusant d'amender la loi laïcisante de 1857, on se sépara, au Parlement et dans le pays, en partisans et en adversaires de l'école neutre : les uns, soutenus par l'Association des instituteurs néerlandais et par la Ligue de l'enseignement populaire, voulant le renforcement des tendances « laïques » ; les autres, au contraire, les « confessionnels » (*de Kerkelijken*), catholiques et protestants, proposant, à défaut de la suppression de la neutralité

---

(1) Sur Victor Dechamps, archevêque de Malines de 1867 à 1883, voir plus haut p. 221 ; sur son attitude vis-à-vis du catholicisme libéral et de la constitution belge, cf. H. SAINTRAIN, *Vie du cardinal Dechamps*, p. 224-238.

de l'enseignement officiel, l'égalité entre les écoles libres et les écoles publiques et l'octroi de subsides aux premières. Le ministère, présidé par un modéré depuis la mort de Thorbecke, essaya pendant plusieurs années de se dérober à cette épineuse question ; mais lorsque les élections de 1877 eurent donné aux libéraux une imposante majorité, les radicaux s'empressèrent de faire prévaloir leurs idées et firent voter une loi qui aggravait encore les dispositions de celle de 1857 et apparaissait pleine de menaces pour l'enseignement libre.

C'était une lourde défaite pour les défenseurs de l'école chrétienne. Elle fit mieux sentir aux catholiques la nécessité de combattre désormais non plus en ordre dispersé, mais en s'organisant en un parti politique. Dès 1877, le *Tijd* avait lancé l'idée en donnant comme exemple le « Centre » allemand ; au lendemain de la loi de 1878, un jeune prêtre, l'abbé Schaepman, allait la réaliser de main de maître [1]. Mais l'émotion suscitée par la loi de 1878 eut une autre conséquence tout aussi importante : elle acheva de rapprocher sur le terrain de la lutte politique les catholiques des protestants croyants, en mettant en pleine lumière que l'ère des rivalités confessionnelles était révolue et que la vraie bataille se livrait à présent entre la religion et l'athéisme. Depuis que le spiritualisme des libéraux de 1850 avait fait place à un rationalisme de plus en plus radical doublé d'un anticléricalisme décidé, deux conceptions de vie irréductibles opposaient les alliés d'hier, tandis qu'apparaissait au contraire le « sous-sol commun » [2], à savoir l'héritage du Christ, qui pouvait servir de base à une union entre calvinistes et catholiques. Cette « coalition chrétienne » dénoncée par les libéraux comme « l'alliance monstrueuse » (*het monsterverbond*) ne manquait pas d'être déconcertante au premier abord dans un pays où les guerres religieuses avaient été si violentes, et plus d'un catholique commença par la juger « indigne et déraisonnable ». Elle était pourtant dans la ligne de l'évolution : amorcée timidement depuis le début de la lutte scolaire et réalisée à la suite de la loi de 1878, sans atténuer du reste l'intransigeance doctrinale de part et d'autre, elle allait jouer un rôle considérable dans la vie publique néerlandaise au cours des décades suivantes et aboutir, en matière scolaire, à la législation la plus équitable et la plus satisfaisante de toute l'Europe occidentale.

L'ÉGLISE D'ESPAGNE DANS LA CRISE RÉVOLUTIONNAIRE La révolution de 1868, qui renversa la reine Isabelle et ramena les libéraux au pouvoir, apparut à juste titre comme étant « moins une révolution politique qu'une révolution antireligieuse » [3]. Les violences anticléricales reprirent aussitôt : fermeture d'églises, destruction de couvents, expulsion de religieux, tentatives pour supprimer l'enseignement chrétien dans les écoles. La nouvelle constitution votée

---

(1) Voir J. Witlox, *De Katholieke Staatspartij in haar oorsprong en ontwikkeling geschetst*, t. I, Bois-le-Duc, 1919.
(2) L'expression est du célèbre théologien calviniste Abraham Kuyper (*Volharden bij het ideaal*, Utrecht, 1901, p. 6).
(3) Déclaration du député républicain Carrido, citée par J.-A. Brandt, *Toward the New Spain*, Londres, 1933, p. 433.

par les Cortès en 1869, qui établissait le mariage civil, diminuait les subsides financiers à l'Église et soumettait celle-ci à diverses mesures restrictives, décrétait aussi la liberté des cultes, au grand scandale du clergé espagnol. Celui-ci refusa de prêter serment à cette constitution qui, comme le disait spirituellement *Le Journal de Paris*, proclamait la liberté de toutes les Églises à l'exception de la seule Église que connussent les Espagnols [1].

Les rapports entre l'Église et l'État ne firent qu'empirer pendant les trois ans du règne d'Amédée de Savoie, le second fils du roi d'Italie, et surtout pendant l'éphémère République de 1873. Comme, d'autre part, certains députés de gauche firent aux Cortès des déclarations inspirées d'un rationalisme sectaire et antireligieux [2], le libéralisme n'apparut plus seulement aux catholiques comme un mouvement hostile à l'influence de l'Église dans la société, mais comme un athéisme militant avec lequel aucun compromis n'était possible. Ce ne fut plus seulement la défense des positions de l'Église, mais la défense de la religion elle-même qui poussa désormais les catholiques à appuyer de toutes leurs forces le parti conservateur.

Ce dernier l'emporta de nouveau en 1874 et le régent Canovas s'empressa de renouer les relations avec le Saint-Siège et de remettre en vigueur le concordat. La liberté des cultes fut toutefois maintenue dans la constitution de 1876. En revanche, la liberté d'association reconnue par la même constitution rendait caduques les anciennes restrictions concernant les ordres religieux qui rentrèrent en masse à partir de 1877. Leur influence, jointe au calme dont l'Église allait jouir pendant une longue période, permit un certain réveil de la vie intellectuelle catholique, qui avait connu pendant des années une profonde décadence. Malheureusement, par un cercle vicieux presque fatal, l'hostilité des radicaux contre l'Église, que l'alliance de celle-ci avec les partis de droite entretenait, ne fit que renforcer l'intolérance des catholiques à l'égard des idéologies libérale et socialiste, dans lesquelles ils ne virent que l'aspect antireligieux sans discerner la valeur positive de certaines de leurs revendications. Le catholicisme espagnol continua donc à s'isoler toujours davantage des aspirations de son époque et acheva de perdre son influence sur beaucoup d'esprits cultivés qui ne virent plus en lui qu'une force de réaction en marge du progrès [3].

AU PORTUGAL Les péripéties de la lutte menée contre l'Église furent moins violentes au Portugal, mais les difficultés entre le gouvernement et l'Église persistèrent ; surtout, on vit s'accentuer l'orientation antireligieuse de la bourgeoisie qui se manifesta entre autres, en 1871, par une série de conférences antichrétiennes au casino de Lisbonne et, en 1874, par la nomination, très discutée, de Renan comme membre correspondant de l'Académie royale des Sciences.

---

(1) C'est de cette époque que date la première pénétration du protestantisme en Espagne, car les quelques essais de propagande secrète en 1830 n'avaient abouti à aucun résultat.
(2) Voir un aperçu de ces débats dans J.-A. BRANDT, *op. cit.*, p. 132-135.
(3) Cf. R. GARCIA Y GARCIA, *Los intellectuales y la Iglesia*, Madrid, 1934.

CHAPITRE XII

L'ÉGLISE A L'EST DE L'EUROPE ET DANS LE PROCHE-ORIENT DE 1846 A 1878

§ 1. — L'Église catholique aux frontières de l'orthodoxie [1].

LA GRANDE PITIÉ DE L'ÉGLISE CATHOLIQUE EN RUSSIE Au milieu du XIXe siècle, il subsistait plus d'un million de catholiques dans les provinces occidentales et méridionales de l'Empire russe, Pologne non comprise, mais, loin de Rome, presque séparé d'elle, ce malheureux troupeau dépérissait rapidement. Dans les hautes classes, où l'influence de Voltaire se perpétuait, l'indifférence religieuse prévalait souvent et le fait que presque toutes les carrières se fermaient aux catholiques invitait à l'apostasie. Les classes inférieures, fondues dans la grande masse orthodoxe, étaient ravagées par les mariages mixtes, tandis que les lois interdisant aux orthodoxes de changer de confession étaient appliquées avec rigueur. L'enseignement catholique était inexistant, les séminaires à peine organisés et de plus obligés d'utiliser des professeurs orthodoxes pour l'enseignement de l'histoire et de la littérature russe. Les paroisses, trop peu nombreuses, étaient dépourvues de ressources, le clergé insuffisant, et ce qui subsistait des anciens ordres religieux végétait misérablement, presque sans contact avec les supérieurs généraux résidant à l'étranger.

L'accord conclu en 1847 entre le Saint-Siège et la Russie, au terme d'une longue et pénible négociation [2], permettait bien d'espérer des jours meilleurs, surtout sous la direction du nouveau chef de la hiérarchie, Mgr Holowinski, dont « la noble et attachante figure fait oublier tous les autres métropolitains de Mohilev au XIXe siècle, et console de la lâcheté des uns et de l'incapacité des autres » [3]. Mais la situation était telle qu'un redressement était bien difficile, d'autant plus que le gouverne-

---

(1) BIBLIOGRAPHIE. — I. SOURCES. — Sur l'Église en Russie et en Pologne, outre les sources indiquées dans l'ouvrage de Boudou cité ci-dessous, on trouvera de nombreux documents dans l'*Esposizione documentata sulle costanti cure del S. P. Pio IX a riparo dei mali che soffre la Chiesa Cattolica nei Dominii di Russia e Polonia*, Rome, 1866 (trad. franç., souvent fautive, par P. LESCŒUR, *L'Église de Pologne. Exposé avec pièces à l'appui...*, Paris, 1868).

Sur les uniates de Galicie, M. MALINOWSKI, *Die Kirchen- und Staatssatzungen bezüglich des griechisch-katholischen Ritus der Ruthenen in Galizien*, Lemberg, 1861.

II. TRAVAUX. — Sur l'Église en Russie et en Pologne : A. BOUDOU, *Le Saint-Siège et la Russie*, t. II, Paris, 1925 (fondamental ; donne toute la bibliographie antérieure) ; P. LESCŒUR, *L'Église catholique en Pologne sous le gouvernement russe*, t. II, Paris, 1876.

Sur les uniates de l'Empire d'Autriche : A. KORCZOK, *Die griechisch-katholische Kirche in Galizien*, Leipzig, 1921.

Sur le catholicisme dans les Balkans : J. FRIEDRICH, *Die christlichen Balkanstaaten in Vergangenheit und Gegenwart*, Munich, 1916.

(2) Cf. plus haut, chap. I, p. 22.

(3) A. BOUDOU, *Le Saint-Siège et la Russie*, t. II, p. 9.

ment fit tout ce qu'il put pour l'entraver en multipliant les tracasseries administratives et en s'efforçant, en dépit des protestations romaines réitérées, de faire nommer aux hautes charges des ecclésiastiques sans énergie, toujours prêts à céder devant ses exigences, comme le pusillanime successeur d'Holowinski, Venceslas Zylinski, ou le lamentable Staniewski, et leur mauvais génie, le dominicain Stacewicz [1], quand il ne s'agissait pas de prêtres tarés et déjà à demi apostats, prêts, pour satisfaire leur ambition, à trahir les intérêts catholiques, tels les trois chanoines qui, à partir de 1866, dans les diocèses de Wilna et de Minsk, appuyèrent les efforts du gouvernement en vue d'imposer dans les églises l'usage de la langue russe pour toutes les prières et cérémonies paraliturgiques [2].

Cette attitude gouvernementale, franchement hostile à l'Église, était d'ailleurs inévitable. Pour les Russes, qui désignaient le catholicisme sous le nom de « foi polonaise », être Russe c'était être orthodoxe et tant que le catholicisme subsistait dans les provinces occidentales, celles-ci demeuraient à leurs yeux comme un corps étranger dans l'Empire.

EN POLOGNE   En Pologne, par contre, le gouvernement tsariste ne cherchait pas à supprimer le catholicisme latin. On se rendait compte que c'eût été impossible, étant donné l'attachement profond des masses paysannes à la foi de leurs ancêtres, ainsi qu'aux processions somptueuses, aux chants et aux pratiques héritées des anciens âges. Mais, prêt à tolérer l'Église catholique en Pologne, le gouvernement entendait qu'elle fût, comme c'était le cas pour l'Église orthodoxe, à l'entière disposition de l'État. S'il n'envisagea guère la solution, proposée par quelques utopistes, d'une « Église catholique slave » ou Église nationale polonaise officiellement séparée du Saint-Siège, il chercha inlassablement à réduire au minimum les interventions romaines en Pologne, à la fois parce qu'il estimait que c'était à l'État qu'il appartenait de régler les conditions d'exercice de la vie religieuse à l'intérieur de ses frontières, et parce qu'il soupçonnait le Vatican, bien à tort du reste, d'encourager la résistance nationale contre le régime russe.

Ce dernier point constituait en effet la question délicate. Profitant habilement du caractère peu éclairé de la foi de nombreux fidèles, aidés par certains prêtres éloquents, pleins d'enthousiasme juvénile mais sans expérience ni mesure, faisant pression sur le reste du clergé grâce à des procédés d'intimidation, les agitateurs nationalistes avaient réussi à organiser leur mouvement sous une forme religieuse, autant pour le

---

(1) Stacewicz, qui était recteur de l'Académie ecclésiastique catholique de Saint-Pétersbourg, déclarait en 1869 à un prélat belge « que quant au dogme catholique, il donnerait mille fois sa vie pour le défendre, mais que les affaires de discipline et d'administration regardaient avant tout le gouvernement ; qu'il aimait les libertés gallicanes ; qu'il ne voulait pas s'exposer à aller en Sibérie pour Rome ». (Cité par A. BOUDOU, *op. cit.*, t. II, p. 336.)

(2) Cette affaire, qui paraît à première vue anodine, était en réalité fort grave, car elle apparaissait comme le prélude d'une campagne ultérieure destinée à conduire par étapes vers l'Église orthodoxe les catholiques de ces régions. Rien qu'au cours des années 1865-1866, 50.000 catholiques de rite latin passèrent de la sorte à l'orthodoxie. Tant que ces causes subsistèrent (c'est-à-dire jusqu'en 1905), le Saint-Siège interdit formellement toute modification des usages existants. (Cf. A. BOUDOU, *op. cit.*, t. II, chap. x.)

rendre moins vulnérable que pour utiliser une puissance morale considérable dans le pays. A Rome, on se rendait compte de cette situation trouble et l'on était loin d'approuver cette confusion regrettable du politique et du religieux, qui devait apparaître au grand jour à l'occasion de l'insurrection de 1863 : « Les Polonais, disait Pie IX, cherchent avant tout la Pologne et non pas le règne de Dieu ; voilà pourquoi ils n'ont pas la Pologne [1]. » Mais il hésitait à intervenir ouvertement à cause du danger, signalé de divers côtés, de pousser hors de l'Église une partie des nationalistes dépités, qui, se recrutant surtout dans les villes et dans l'aristocratie, étaient loin d'avoir la foi patriarcale des paysans.

Dans ces conditions, on avait beau jeu à Saint-Pétersbourg d'identifier avec la révolte des patriotes exaspérés par une domination étrangère, le mécontentement des fidèles revendiquant les droits de leur Église foulés aux pieds par un gouvernement à tendances régaliennes. Dans l'espoir de faire pression sur le clergé et de l'amener à soutenir la politique russe, les représentants du tsar multipliaient les mesures vexatoires : création, en 1862, d'une Commission des cultes, qui dépossédait pratiquement les évêques de leur juridiction au mépris du droit canon et de l'accord de 1847 ; suppression, en 1864, de nombreux couvents (dont beaucoup, il faut le reconnaître, étaient en pleine décadence [2]) ; confiscation, en 1865, d'une partie des biens d'Église ; arrestation et déportation de nombreux prêtres et évêques, y compris l'archevêque de Varsovie, Mgr Felinski ; interdictions fréquentes de processions et de pèlerinages ; surveillance par la police des sermons et même des confessions [3].

Tout comme en Russie, les ruines causées par la tyrannie officielle s'accumulaient d'année en année et l'on en venait à se dire qu'une persécution ouverte eût peut-être été préférable, car l'usure de la vigueur morale des catholiques eût sans doute été moindre. Certes, un bon nombre supportait l'épreuve avec résignation, « et cette capacité d'endurer n'était point chose si méprisable » [4]. Mais la résistance manquait de mordant parce que le peuple catholique n'avait pas de chefs qu'on puisse comparer, même de loin, à ceux qui avaient encadré jadis les Irlandais ou qui allaient diriger la lutte des catholiques allemands contre le *Kulturkampf*. Comment s'en étonner d'ailleurs dans un État policier sans institutions libres, où les catholiques ne disposaient ni de journaux indépendants ni d'aucun moyen de se faire entendre ? De plus, bien des sièges épiscopaux étaient vacants et les autres avaient trop souvent à leur tête des hommes insuffisants, « fatigués de conduire à la défaite

---

(1) Cité d'après Br. Zaleski par A. Boudou, *op. cit.*, t. II, p. 136.
(2) D'après les renseignements recueillis par Mgr Chigi lors de sa mission de 1856, il fallait surtout regretter le trop grand nombre de petits couvents insuffisamment peuplés, les nombreuses entorses à la règle de la clôture et au vœu de pauvreté, le manque de directeurs spirituels et de maîtres des novices suffisamment formés, et l'admission de candidats sans vocation, désireux d'échapper au service militaire. (A. Boudou, *op. cit.*, t. II, p. 221.)
(3) Dans certaines provinces, les prêtres ne pouvaient pas entendre les confessions sans un gendarme planté auprès du confessionnal afin de vérifier si, parmi les pénitents, ne se trouvaient pas quelque orthodoxe converti ou d'anciens uniates qui auraient refusé d'accepter le retour à l'Église russe.
(4) A. Boudou, *op. cit.*, t. II, p. 475.

leurs troupes démoralisées »[1], et d'autant plus disposés à accepter quelques entorses au droit canon pour ne pas offusquer le tsar que leur éducation ecclésiastique avait été souvent imprégnée de joséphisme et de fébronianisme. La situation était sombre et l'avenir apparaissait plus sombre encore.

VAINES PROTESTATIONS ROMAINES Dès qu'il était apparu que les autorités russes ne respectaient pas l'accord de 1847, le Saint-Siège avait tenté d'intervenir, saisissant toutes les occasions pour protester auprès de l'ambassadeur russe à Rome. Espérant qu'il serait plus aisé de régler sur place nombre de difficultés pratiques, on chercha à plusieurs reprises à obtenir l'établissement d'une nonciature à Saint-Pétersbourg. En 1862, les négociations furent sur le point d'aboutir et le titulaire, Mgr Berardi, était même déjà désigné, quand les exigences russes firent abandonner, peut-être un peu vite, ce projet dont on avait espéré beaucoup.

Bientôt, par suite des représailles en Pologne, la situation s'aggrava sensiblement et Pie IX, auquel Antonelli avait longtemps conseillé la patience, ne put plus contenir son indignation. Il l'exprima publiquement dans une allocution improvisée, prononcée le 24 avril 1864 au Collège de la Propagande [2]. Si la rupture des relations diplomatiques qui faillit s'en suivre fut retardée grâce à l'intervention de l'Autriche, l'arrogance du chargé d'affaires russe Meyendorff la rendit inévitable à la fin de 1865. Dès lors, les événements se précipitèrent. Le 29 octobre 1866, dans une nouvelle allocution, soigneusement préparée cette fois par une commission de dix cardinaux et accompagnée d'un *Exposé* abondamment documenté, destiné à la presse, le pape condamna en plein consistoire, en les déclarant nuls, tous les décrets impériaux contraires aux droits de l'Église et du Saint-Siège [3]. Le 4 décembre, le gouvernement russe ripostait en abrogeant officiellement le concordat. Quelques semaines plus tard, un autre oukase, daté du 22 mai 1867, décrétait que désormais toute communication des évêques avec Rome devrait passer par un organisme sous contrôle officiel, le Collège ecclésiastique de Saint-Pétersbourg. Bien que cette mesure ne s'accordât guère avec le droit canon, la plupart des évêques estimèrent qu'on pouvait s'en accommoder, vu la gravité de la situation ; c'était entre autres l'avis de Mgr Lubienski, l'un des prélats les plus dévoués au Saint-Siège pendant cette période troublée [4]. Mais à Rome on n'en jugea pas ainsi : le 17 octobre, l'encyclique *Levate* [5] condamnait l'oukase du 22 mai comme attentatoire à la constitution divine de l'Église, et deux ans plus tard, le pape excommunia nommément Mgr Staniewski, administrateur de Mohilew, pour avoir refusé de tenir compte de cette

---

(1) A. BOUDOU, *op. cit.*, t. II, p. 449.
(2) Texte dans P. LESCŒUR, *L'Église catholique en Pologne*, t. II, p. 150-151.
(3) Texte dans l'*Esposizione documentata...*, p. 303-306.
(4) Sur Mgr Lubienski, qui joignait à une éducation très soignée une piété sincère et un zèle ardent, mais qui se laissait parfois emporter par son caractère impressionnable ou par son imagination, voir R. LUBIENSKI, *Vie de Mgr Constant Irénée comte Lubienski*, Roulers, 1898 (panégyrique).
(5) Texte dans *Acta Pii IX*, t. IV, p. 371-378.

condamnation et des avertissements qu'il avait reçus à ce sujet.

A partir de 1870, le gouvernement russe manifesta à plusieurs reprises son désir de se rapprocher de Rome, et pendant quelques années il y eut une détente. Mais un *memorandum* sur les griefs catholiques, rédigé en 1877 sur l'initiative du nouveau secrétaire d'État, le cardinal Simeoni, vint à nouveau tout gâter et Pie IX se préparait à protester une fois de plus à la face du monde quand il mourut, le 7 février 1878.

On pourrait se demander si la Curie ne manqua pas parfois de souplesse en ne comprenant pas qu'il était difficile de demander au gouvernement du tsar d'envisager les problèmes du point de vue des exigences canoniques d'une Église qui n'était dans l'Empire qu'une petite minorité à peine tolérée. Mais on doit bien reconnaître que l'hypocrisie et la mauvaise foi dont firent si souvent preuve à l'égard des catholiques les bureaucrates russes de tout rang et les plus hautes autorités gouvernementales elles-mêmes justifiaient amplement la méfiance de Rome. On y était d'autant plus en droit de se demander si la Russie ne méditait pas d'éliminer purement et simplement le catholicisme latin qu'au même moment elle achevait froidement, après une savante préparation, de liquider ce qui restait encore en Pologne du catholicisme de rite oriental.

LA FIN DE L'ÉGLISE UNIATE EN TERRITOIRE RUSSE   Après l'élimination des uniates de Russie en 1839, un diocèse de rite oriental, celui de Chelm, avait en effet subsisté à l'intérieur du Royaume de Pologne. Très vite, les autorités tsaristes souhaitèrent le russifier à son tour. Elles y furent aidées par certains membres du clergé, éblouis par le prestige de l'orthodoxie, et surtout par un groupe de prêtres, venus de Galicie, dont la haine antipolonaise l'emportait sur leurs sentiments catholiques.

Pour préparer le terrain, les fauteurs de schisme, sous la direction de Joseph Wojcicki et surtout de Marcel Popiel, entreprirent, à partir de 1865, d'« épurer » le rite ruthène des nombreux latinismes qui s'y étaient introduits depuis l'union de Brest-Litovsk, en particulier lors du concile de Zamosc. L'initiative était en soi légitime, mais, entreprise dans de telles circonstances et par des hommes aussi suspects, elle apparaissait comme une manœuvre cousue de fil blanc : il s'agissait, en supprimant tout ce qui donnait à l'Église unie sa physionomie propre, bien distincte de l'Église orthodoxe, de faciliter le passage définitif de la première à la seconde. Aussi le Saint-Siège mit-il immédiatement son *veto* et, lorsque, en 1873, Marcel Popiel, sur les injonctions d'un comité russe, décréta l'abolition radicale de tout ce qui distinguait encore les uniates des schismatiques, Pie IX réaffirma la position romaine dans l'encyclique adressée le 13 mai 1874 à tout l'épiscopat ruthène : « Personne n'a le droit, sans l'avis préalable du Saint-Siège, de faire dans la liturgie même les plus légers changements [1]. »

C'est en vain que les catholiques fidèles à Rome essayèrent de résister.

_____

(1) Texte dans *Coll. lac.*, t. II, col. 483-484 ; voir à ce sujet C. KOROLEVSKY, *L'Uniatisme*, Amay, 1927, p. 53-54.

Popiel, qui avait été chargé par le gouvernement de l'administration du diocèse, n'hésita pas à faire appel à la police impériale pour imposer ses projets. Puis il organisa une campagne de pétitions et finalement, au début de 1875, il signa au nom du haut clergé de Chelm l'acte d'apostasie sollicitant de la bienveillance de l'empereur « la réunion des Grecs-Unis à la sainte Église Orthodoxe d'Orient, qui fut l'Église de nos Pères »[1]. Ce que le P. Martynov devait appeler « le brigandage de Chelm »[2] était consommé : 204 prêtres, soit les deux tiers du clergé, suivirent Popiel et 250.000 fidèles furent agrégés en bloc à l'Église russe.

LES UNIATES DE GALICIE ET DE TRANSYLVANIE   S'il n'avait tenu qu'au tsar, l'Église catholique de rite slave aurait disparu totalement. Mais il y avait encore plus de deux millions de catholiques de rite oriental dans l'empire des Habsbourg.

Un premier groupe était constitué par les Ruthènes ou Ukrainiens de Galicie, qui avaient, depuis 1807, un siège métropolitain à Lemberg, avec un évêché suffragant à Przelisl. Des efforts très méritoires pour remédier à la situation assez chétive de cette Église furent tentés par Mgr Lewicki, archevêque de Lemberg de 1816 à 1858, mais ils n'aboutirent que partiellement[3]. C'est qu'il n'y avait pas de classe dirigeante, toute l'aristocratie ruthène ayant jadis abandonné le rite oriental pour jouir des avantages réservés aux Polonais, au point que l'on avait pris l'habitude d'opposer le « foi paysanne » du serf ruthène à la « foi du seigneur » du grand propriétaire de rite latin. Quant au clergé, sa culture était plus qu'élémentaire, quand il n'avait pas été déformé par l'éducation d'inspiration joséphiste reçue au séminaire central grec de Vienne. De plus, il manquait souvent de zèle et d'esprit de sacrifice et dépensait une bonne partie de son activité dans la lutte politique antipolonaise. Le passage sous régime autrichien n'avait en effet pas supprimé le traditionnel antagonisme entre Polonais et Ukrainiens et, depuis 1848 surtout, ceux-ci avaient mieux pris conscience de leur entité nationale et de la situation humiliée où les Polonais les avaient réduits. Par réaction, le jeune clergé ruthène s'habituait à regarder vers la Russie orthodoxe, encouragé d'ailleurs en sous-main par les agents tsaristes, et il suivait avec sympathie les entreprises de Michel Popiel dans le diocèse de Chelm en vue d'expurger le rite ruthène des infiltrations latines.

Dans cette atmosphère, les controverses séculaires entre les clergés catholiques des deux rites ne pouvaient que s'envenimer. Sous la pression du gouvernement autrichien et avec les encouragements du nonce Viale Prela, les évêques galiciens élaborèrent un projet d'accord, très raisonnable, qui fut soumis au Saint-Siège en 1853. Malheureusement, celui-ci ne trancha pas immédiatement et la situation ne tarda pas à se tendre

(1) Cf. P. LESCŒUR, *L'Église catholique en Pologne*, t. II, p. 397-407.
(2) Dans les *Études*, t. VII, 1875, p. 943-956. On trouvera les détails de l'affaire dans A. BOUDOU, *op. cit.*, t. II, p. 105-114, 263-274, 399-447.
(3) Voir la pastorale de l'archevêque Jachimovicz, du 25 mai 1862, dans *Archiv für Kirchenrecht*, t. IX, 1863, p. 200 et suiv. En reconnaissance de ses mérites, Mgr Lewicki fut promu en 1856 au cardinalat, dignité qui n'avait plus été conférée depuis le XVIe siècle à un prélat de rite oriental.

à nouveau, les deux partis s'accusant mutuellement de prosélytisme, surtout à l'occasion des mariages mixtes. Ce n'est qu'en 1863, et sur l'intervention de la Propagande, qu'on arriva enfin à un accord qui s'inspirait du projet de 1853 [1].

Le second groupe d'uniates comprenait les Roumains de Transylvanie, longtemps soumis à des évêques latins, mais pour qui Pie IX avait créé, en 1853, une province autonome, avec Fogaras (ou Alba Julia) comme métropole, et trois diocèses suffragants. Il fut un moment question à Rome, vers 1856, de mettre au point un concordat avec ces uniates, qui aurait pu servir de modèle lors d'arrangements futurs avec d'autres Églises orientales unies ou à unir [2], mais l'idée n'aboutit pas. Toutefois, bien que devant vivre dans une situation humiliée par rapport au clergé latin, représentant l'élément dominateur hongrois, les uniates de Transylvanie surent s'organiser assez vite : en 1862, un de leurs évêques, Mgr Papp-Szilàgyi, publia l'un des seuls manuels de droit canon oriental composés au XIXᵉ siècle, et en 1872 se tint à Fogaras un concile provincial qui est un « modèle du genre » [3].

En 1860, le nonce à Vienne profita du passage de dom Pitra, l'un des très rares spécialistes des choses orientales, pour lui faire étudier la question de la réforme des monastères basiliens, dont quelques-uns subsistaient encore en Galicie. Le bénédictin français s'intéressa de près à la chose : considérant que l'union entre Rome et l'Orient ne pourrait être mieux préparée que par la réforme de la vie monastique, si profondément liée à la vie chrétienne dans l'esprit des Russes, il estimait que l'Autriche, « l'unique puissance catholique slave », avait à cet égard une mission exceptionnelle. « Les Slaves doivent rester Slaves, expliquait-il au ministre des cultes von Thun, et nous ne pouvons les latiniser ; et nous ne pouvons slavifier des sociétés religieuses jésuites ou lazaristes (…). Force est bien de s'adresser aux basiliens [4]. » Mais la chute de Thun, qui avait encouragé le projet, puis les vicissitudes de la politique autrichienne obligèrent à remettre à plus tard la réforme envisagée, et ce n'est que sous Léon XIII, en 1882, à un moment où il ne restait plus que 54 religieux, qu'elle fut enfin réalisée, sous la direction des jésuites cette fois [5].

LE CATHOLICISME EN SERBIE   Les Slaves des Balkans proprement dits relevaient de trois États : l'Empire ottoman, la monarchie austro-hongroise et, entre les deux, la principauté indépendante de Serbie. Il sera question plus bas de la situation dans les provinces turques de Bosnie et d'Herzégovine, où les catholiques étaient près de 200.000. En Serbie, par contre, ils étaient à peine quelques milliers, et le fanatisme orthodoxe les empêchait souvent de pratiquer leur culte publiquement ; du temps où il était nonce à Vienne, Mgr Viale

---

(1) Texte dans *Coll. lac.*, t. II, col. 561-566.
(2) A. BATTANDIER, *Le cardinal Pitra*, Paris, 1896, p. 319.
(3) C. DE CLERCQ, *Les Églises unies d'Orient*, Paris, 1934, p. 89.
(4) Cité dans A. BATTANDIER, *Le cardinal Pitra*, p. 404.
(5) Ces tentatives de réformes sont étudiées par P. M. KAROVEC, *Velika reforma cina Sv. Vasilija v., 1882 g.*, Lvow, 1933, 2 vol.

Prela avait fait le voyage de Belgrade pour essayer d'obtenir un régime un peu plus satisfaisant, mais en vain.

LE CATHOLICISME DANS LE BALKAN HONGROIS La situation était plus favorable dans les provinces de Croatie et de Slavonie, qui dépendaient administrativement du royaume de Hongrie. Les catholiques, qui représentaient 70 pour cent de la population, y dépassaient très largement le million. Une province ecclésiastique avait été constituée en 1852, avec pour métropole Agram (ou Zagreb). Ce dernier diocèse, presque entièrement catholique, comptait 800.000 fidèles, mais ses 625 prêtres ne suffisaient pas à occuper toutes les paroisses. L'un des trois diocèses suffragants, celui de Diakovar, devait connaître une certaine célébrité, grâce à Joseph Georges Strossmayer [1], qui en fut évêque de 1849 à 1905 et dont l'action sur le catholicisme dans les Balkans fut des plus profondes.

L'ACTION DE STROSSMAYER Pieux et zélé, se dépensant sans compter pour le bien spirituel de son diocèse, Strossmayer était aussi un ardent patriote. Nommé évêque de Diakovar grâce à l'influence du leader croate Jellacic, qui avait dirigé la lutte contre les Hongrois en 1848, on a pu dire de lui qu'« il interpréta jusqu'au génie les aspirations de 1848 » [2], lesquelles, en pays slaves, n'impliquaient pas seulement la liberté politique, mais surtout l'émancipation nationale. Tout en restant loyal envers la dynastie, il devint peu à peu, surtout à partir de 1867, le chef moral de l'opposition croate contre l'oligarchie magyare, mais il eut l'intelligence de ne pas restreindre son action au terrain politique. Il s'appliqua surtout à promouvoir la cause slave sur le plan culturel, faisant de son palais un fastueux lieu de rendez-vous pour tous les slavophiles, catholiques, orthodoxes ou protestants, utilisant une bonne part de ses énormes revenus à subventionner journaux et publications sur la littérature, l'histoire ou le folklore slaves, et à doter sa patrie de deux institutions essentielles, une Académie, fondée en 1867, et une Université, inaugurée à Agram en 1874. Cette attitude lui valut auprès de tous les Slaves du Sud, orthodoxes aussi bien que catholiques, une popularité inégalée : son portrait se trouvait dans toutes les chaumières de l'Illyrie et il se faisait acclamer dans les rues de Belgrade, lui évêque catholique, quand il y venait confirmer les quelques fidèles relevant de sa juridiction.

Ce prestige, il chercha à l'utiliser en faveur de l'autre idéal de sa vie, qui était d'ailleurs pour lui connexe au premier : l'union des Églises.

---

(1) Sur Mgr Strossmayer (1815-1905), voir L. LÉGER, *Un évêque slave*, dans *Le Monde slave-*Paris, 1897, p. 117-136 ; Ch. LOISEAU, *Strossmayer*, dans *Le Correspondant*, t. CCXIX, 1905, p. 251, 271 ; A. FORTESCUE, *A Slav Bishop : J. G. Strossmayer*, dans *Dublin Review*, t. CLXIII, 1918, p. 234-257 ; A. WENZELIDES, *Il mecenate croato Josip Strossmayer*, dans *Europa Orientale*, 1930 ; N. LALIC, *Les idées de Strossmayer*, dans *Le Monde slave*, 1929, p. 442-450 ; S. JURINITCH, dans *Sbornik za narodni umotvorenia, nauka i knynina*, t. XXII, 1907, p. 1-71, a publié de nombreuses lettres intéressant l'histoire du catholicisme parmi les Slaves. — Cf. art. *Strossmayer*, dans *D. T. C.*, t, XIV (1941), col. 2630-2635.

(2) Ch. LOISEAU, *art. cit.*, p. 251.

Profondément attaché à l'Église catholique, tout en admirant la beauté du rite oriental et en abordant le problème du retour à l'unité chrétienne avec une mentalité irénique rare à cette époque, « il estimait que le monde slave, en général, ne s'approprierait pas sans danger la civilisation matérielle de l'Occident si un aliment plus substantiel que l'orthodoxie n'était point assuré en même temps à son mysticisme »[1]. C'est pour rendre plus aisé le rapprochement qu'il encouragea la liturgie romaine en vieux slavon, encore pratiquée par une centaine de milliers de catholiques de Croatie et de Dalmatie, et surtout qu'il se montra partisan, à l'époque du concile du Vatican et après, d'une politique de large décentralisation dans l'Église.

§ 2. — L'Église catholique dans l'Empire ottoman [2].

RÉPARTITION DES CATHOLIQUES — Ce n'est qu'à partir du traité de Berlin de 1878 que l'Empire ottoman commencera à se disloquer vraiment : jusqu'à cette date, il comporte encore en Europe d'importants territoires. Or, dans ce vaste empire, où l'armée, l'administration et toutes les positions-clefs sont aux mains des musulmans, près d'un tiers des trente millions d'habitants sont chrétiens, et l'on compte environ un million de catholiques, avec plus de 80 sièges épiscopaux.

Parmi ces catholiques, cinq à six cent mille sont de rite oriental, répartis en six patriarcats ; il en sera question à l'article suivant. Quant aux fidèles de rite latin, ils se répartissent comme suit aux environs de 1850 [3] : 70.000 « levantins », descendants des occidentaux venus à l'époque des Croisades et de l'expansion vénitienne, installés à Alexandrie, en Palestine, le long des côtes de l'Asie Mineure, dans les Iles et à Constantinople ; 15.000 en Bulgarie et en Valachie, dans une situation fort pénible, dirigés par des rédemptoristes autrichiens ; 60.000 Magyars constituant la préfecture apostolique de Moldavie, avec centre à Jassy ; une dizaine de milliers de Serbes, en butte aux vexations de leurs congénères orthodoxes ; 75.000 Albanais, dont le nombre diminue d'année en année par suite de leur émigration vers l'Italie ; 35.000 fidèles en Herzégovine ; enfin et surtout 125.000 Croates en Bosnie [4]. Ces derniers

(1) Ch. Loiseau, *art. cit.*, p. 264.
(2) Bibliographie. — I. Sources. — Silbernagel, *Verfassung und gegenwärtiger Stand sämtlicher Kirchen des Orients*, Landshut, 1865 ; *Papers relating to the condition of Christians in Turkey*, Londres, 1861 (rapports consulaires réunis par H. Buwler) ; L. Farley, *The massacres in Syria*, Londres, 1861.
  II. Travaux. — E. von Muelinen, *Die lateinische Kirche im türkischen Reiche*, Berlin, 1903 ; P. Piolet, *Les missions catholiques françaises au XIXᵉ siècle*. T. I, *Missions d'Orient*, Paris, s. d. ; K. Luebeck, *Die Katholische Orientmission in ihrer Entwicklung dargestellt*, Cologne, 1917 ; M. Jullien, *La nouvelle mission de la compagnie de Jésus en Syrie (1831-1895)*, Tours, 1898, 2 vol. ; G. Goyau, *Le protectorat de la France sur les chrétiens de l'Empire ottoman*, Paris, 1895 ; F.-W. Hasluck, *Christianity and Islam under the Sultans*, Oxford, 1929, 2 vol.
  Sur la question des Lieux Saints : B. Collin, *Les Lieux Saints*, Paris, 1948.
(3) Ces données sont empruntées à la 1ʳᵉ édition du *Kirchenlexikon* de Weltzer et Welte, Fribourg-en-Br., 1854 et suiv., art. *Turkei*.
(4) La Bosnie et l'Herzégovine avaient été séparées en 1847 pour constituer deux vicariats apostoliques, à la suite de violents conflits de juridiction. (Cf. B. Rupcic, *Entstehung der Franziskanerpfarreien in Bosnien und Herzegowina und ihre Entwicklung b. z. Jahre 1878*, Breslau, 1937

représentent 20 pour cent de la population, dont ils forment la partie la plus pauvre et la plus arriérée, le commerce et l'artisanat étant aux mains d'orthodoxes de race serbe ; et le clergé local étant quasi inexistant, les paroisses sont administrées par les franciscains, dont les couvents constituent les seuls centres de culture catholique en ces régions [1]. Détail caractéristique des difficultés d'organisation que rencontre l'Église dans ces régions balkaniques, même lorsqu'elle groupe des masses catholiques assez nombreuses : alors que l'Albanie possède six évêchés, datant du moyen âge, dont les évêques peuvent se réunir à deux reprises en synode, en 1863 et en 1868 [2], la Bosnie, au contraire, où se trouvent concentrés la moitié des catholiques de la Turquie d'Europe, ne forme qu'un vicariat apostolique et devra attendre l'occupation autrichienne pour avoir une hiérarchie régulière.

AMÉLIORATION DE LA SITUATION Bien que les sultans aient ordinairement entretenu de courtoises relations avec les souverains pontifes, la situation des catholiques de l'Empire turc était celle des chrétiens en général : méprisés par les musulmans, en butte à de nombreuses vexations de la part des autorités locales, victimes parfois de massacres, occasionnés par des agitations politiques ou des haines de race plutôt que par des raisons purement religieuses. Le *Hatti-Shérif* de Ghulané, en 1839, qui concédait aux chrétiens la pleine liberté religieuse, n'avait guère été appliqué. La situation s'améliora par contre à la suite de la guerre de Crimée, car, désireuses d'enlever à la Russie tout prétexte de réclamer un protectorat sur les chrétiens de Turquie, les grandes puissances occidentales exigèrent le fameux *Hatti-Houmayoum* de 1856, mentionné à l'article 9 du traité de Paris, par lequel le sultan octroyait à tous ses sujets chrétiens les droits civils et politiques qui leur avaient été refusés jusqu'alors et accordait aux patriarches et évêques le libre exercice de leur autorité spirituelle.

Ce décret supprimait également la peine de mort qui frappait toute conversion au christianisme, mais la cohésion de l'Islam et les préjugés religieux demeurèrent les plus forts. Les seules conversions possibles provenaient donc des communautés schismatiques, mais, durant cette période, elles furent assez peu nombreuses, à part dans le groupe arménien ; on ne vit plus, comme sous les prédécesseurs de Pie IX, de retour d'évêques à la tête de communautés relativement importantes. Aussi, si l'on fait abstraction des étrangers, attirés de plus en plus nombreux dans les ports et les grandes villes d'Orient par le développement du commerce international, on ne peut guère enregistrer de différence

p. 129-142.) De 1846 à 1878, le chiffre des catholiques s'accrut sensiblement par suite de la natalité : on comptait, en 1878, 210.000 catholiques, répartis en 96 paroisses et 23 chapellenies. (*Ibid.*, p. 101-102.)

(1) Ces couvents passent de 4 à 10 de 1850 à 1878. A cette dernière date, 54 écoles en dépendent. Avant la fondation, en 1857, avec l'aide de Strossmayer, d'un collège philosophico-théologique à Diakovar, les jeunes moines devaient être envoyés pour leur formation en Italie ou en Hongrie, ce qui nuisait à l'unité d'esprit. (B. Rupcic, *op. cit.*, p. 162-163.)

(2) Décrets dans *Ius Pontificium de Propaganda Fide*, édit. R. de Martinis, t. VI, 1, p. 129 ; t. VI, 2, p. 24.

notable, au point de vue quantitatif, dans la population catholique de l'Empire ottoman entre 1846 et 1878.

L'amélioration n'en fut pas moins sensible au cours de ces trente années. En effet, la guerre de Crimée ne valut pas seulement aux catholiques un statut légal adouci : en même temps qu'elle ouvrait aux influences occidentales un monde replié sur lui-même depuis des siècles, elle attira davantage l'attention des chrétiens d'Occident vers l'Orient.

L'APPUI DE L'OCCIDENT   Cette attention de l'Occident devait s'exprimer parfois sous des formes maladroites, soit qu'elle apparût trop manifestement liée à des intérêts politiques et soulignât davantage encore le caractère allogène des minorités catholiques, soit qu'elle ne tînt pas assez compte de l'originalité propre du christianisme oriental et confondît trop facilement la pureté du catholicisme avec les usages traditionnels de l'Église latine. Elle eut du moins l'avantage d'apporter aux catholiques de l'Empire turc une protection réelle, ainsi qu'un appui matériel et intellectuel, dont ils avaient le plus grand besoin. En outre, en les faisant sortir de leur trop long isolement, elle stimula leur zèle et leur donna l'occasion de redresser un certain nombre d'abus : car l'interruption de communications régulières avec Rome n'avait pas seulement eu pour conséquence de fréquentes vacances dans les diocèses de rite latin, mais aussi le développement de l'ignorance et de l'indiscipline, jusque dans le clergé.

L'Occident n'avait pourtant jamais cessé d'être présent grâce aux missionnaires italiens, dont l'importance s'explique par l'origine italienne de beaucoup de « levantins ». Mais il semble qu'un grand nombre d'entre eux manquaient des qualités requises pour infuser à ces Églises en décadence un sang nouveau. « On reçoit le rebut des diocèses, gémissait l'archevêque de Smyrne, Mgr Spaccapietra. Tout est bon, dit-on, pour les missions, et avec ce principe, il est facile de prévoir les conséquences [1]. »

C'est en fait d'Autriche et de France que vinrent les concours les plus utiles aux catholiques d'Orient. Déjà, en 1830, à la suite d'une violente persécution excitée contre les Arméniens catholiques par le patriarche schismatique, la pression conjuguée de ces deux puissances avait obligé le sultan à émanciper les catholiques de rite oriental de la juridiction civile des patriarches non catholiques de leur rite respectif, dont ils dépendaient jusqu'alors ; c'est notamment à partir de ce firman d'émancipation que les prêtres catholiques de rite grec purent — chose qui leur avait été interdite jusqu'alors et qui revêtait une importance considérable en Orient — porter la coiffure cylindrique caractéristique du clergé byzantin.

L'influence autrichienne se faisait principalement sentir dans les Balkans : le gouvernement intervenait à l'occasion pour faire respecter les droits des catholiques ; bon nombre de sièges épiscopaux et la plupart des paroisses étaient administrés par des religieux en provenance d'Au-

---

(1) Mgr Spaccapietra à M. Girard, de Grenoble, 8 mai 1868, *Arch. Nat.*, AB XIX, n° 524.

triche, et les quelques prêtres autochtones avaient généralement fait leurs études dans les séminaires de Hongrie.

INFLUENCE FRANÇAISE
DANS LE LEVANT

C'était au contraire dans le Levant que se faisait surtout sentir l'action de la France, tant celle du gouvernement que celle des missionnaires [1]. L'antique protectorat de la France sur les chrétiens du Levant avait survécu à la crise révolutionnaire et de nombreux catholiques de Syrie n'hésitaient pas à se nommer *Frangis* pour désigner à la fois leur religion et le fait qu'ils n'étaient pas de race turque. L'accroissement de prestige consécutif à la guerre de Crimée, puis l'intervention des troupes françaises au Liban en 1860, à la suite du massacre de plusieurs milliers de chrétiens par les Druzes, avaient permis à la France de réaffirmer avec plus d'insistance son rôle traditionnel en Orient. Et comme la tentative amorcée au début du pontificat de Pie IX pour arriver à une entente directe du Saint-Siège avec le sultan [2] n'avait pratiquement donné aucun résultat, on était heureux à Rome de continuer à profiter de ce protectorat, quitte à regretter parfois que certaines initiatives du gouvernement de Napoléon III en Orient ne respectassent pas mieux les stipulations du droit canon, par exemple dans le cas du lycée français de Galata, auquel le représentant pontifical reprocha vivement de mêler les élèves catholiques aux non-catholiques [3].

Sous la protection du drapeau tricolore, l'activité des missionnaires français, lazaristes, frères des Écoles chrétiennes, filles de la Charité, sœurs de Saint-Joseph, etc., s'intensifia, surtout après 1860. Les écoles et les établissements hospitaliers se développèrent, aidés par une œuvre spécifiquement française, l'*Œuvre des Écoles d'Orient*. Fondée en 1856, à l'occasion de la guerre de Crimée, par quelques laïcs militants groupés autour du baron Cauchy, cette œuvre allait rapidement jouer un rôle de premier plan grâce à l'esprit d'entreprise de son premier directeur, l'abbé Lavigerie [4].

Parmi les œuvres françaises en Orient, l'action des jésuites en Syrie doit être spécialement mentionnée. Très intelligemment conçue, elle s'appliqua notamment à protéger la classe cultivée contre la propagande protestante, active depuis le milieu du siècle. Les Églises melkite et même maronite n'avaient ni les hommes, ni la culture, ni les ressources pour résister. Les jésuites y parvinrent grâce à la fondation d'une imprimerie moderne, puis, en 1871, d'un grand journal catholique, *Al-Bashir*,

---

(1) Surtout, mais pas exclusivement : il faut spécialement mentionner l'action exercée par les assomptionnistes du P. d'Alzon en Bulgarie, où ils vinrent s'installer en 1863 à la demande expresse de Pie IX.

(2) Voir chap. I, p. 21.

(3) Voir la note adressée de Paris le 1er juin 1869 à l'ambassadeur français à Rome. (*Arch. Min. A. É. Paris*, Rome, 1043, fº 393 et suiv.) Le représentant pontifical à Constantinople, Mgr Testa, était un homme intransigeant et même étroit d'esprit ; il voulait obliger aussi les Frères des Écoles chrétiennes établis en Turquie à renvoyer les élèves non catholiques de certaines de leurs écoles. (*Ibid.*, Rome, 1043, fº 384.)

(4) Cf. Mgr BAUNARD, *Le cardinal Lavigerie*, t. I, Paris, 1896, chap. III et IV ; E. BELUZE, *Vie de Mgr Dauphin (1806-1882) avec une lettre de S. É. le cardinal Lavigerie sur le commencement de l'œuvre des Écoles d'Orient*, Paris, 1886 ; H. DE LACOMBE, *Note sur l'œuvre d'Orient*, Paris, 1906.

enfin d'un collège moderne, qui deviendra, en 1881, l'Université Saint-Joseph. Féconde par son rayonnement, l'œuvre des jésuites de Beyrouth mérite toutefois un reproche. Contrairement aux dominicains, qui s'efforçaient d'aider les Églises syrienne et chaldéenne à remettre en valeur leurs antiques traditions liturgiques et spirituelles, ils se bornèrent à diffuser des livres de prières d'inspiration purement latine et des livres de piété qui n'étaient que la traduction d'ouvrages modernes occidentaux ou des vies de saints occidentaux. « C'est là un défaut que le souci de ne favoriser aucun rite au dépens des autres veut excuser, mais qui tient surtout à une mentalité qui est celle de presque tous les missionnaires latins d'Orient [1]. »

L'ACTION DES DÉLÉGUÉS APOSTOLIQUES — Plus les années passent et plus le Vatican s'efforce de suivre de près et de diriger cette action missionnaire multiforme et la vie propre des diverses communautés catholiques de l'Empire ottoman. Cette action centralisatrice s'exerce surtout grâce aux délégués apostoliques. Pour la partie septentrionale de l'Empire, il faudra attendre jusqu'en 1868 avant de voir le vicaire patriarcal de Constantinople chargé de cette fonction, mais il y avait depuis longtemps des délégués en Égypte, en Mésopotamie et en Syrie. L'agent le plus actif de cette présence romaine en Orient fut Mgr Valerga, missionnaire à Mossoul depuis 1841, devenu en 1847 patriarche latin de Jérusalem, et délégué apostolique en Syrie pendant de longues années, prélat d'une énergie parfois un peu cassante, et volontiers porté à pousser à la latinisation, mais, tout compte fait, « un des meilleurs connaisseurs de l'Orient asiatique à cette époque » [2].

LA QUESTION DES LIEUX SAINTS — Un coin de l'Empire ottoman intéressait tout spécialement les catholiques : la Palestine. Ils y avaient eu longtemps une influence prépondérante, grâce au protectorat français, mais, au cours du XVIIIe siècle, les orthodoxes avaient réussi à se faire attribuer par les autorités turques un certain nombre de sanctuaires auxquels étaient attachés des souvenirs évangéliques, et il était à craindre que ces usurpations ne se poursuivissent par suite du prestige croissant de la Russie en Orient. En 1847, au moment où Pie IX rétablissait avec l'accord du sultan le patriarcat latin de Jérusalem, un grave incident éclata. La France en profita pour remettre en question tout le problème des Lieux Saints et exiger l'abandon par les Grecs des sanctuaires qu'ils avaient usurpés sur les Latins depuis cent ans. Les revendications françaises se firent surtout énergiques lorsque, grâce à l'influence de Montalembert, le marquis de La Valette, très dévoué au Saint-Siège, eut été nommé ambassadeur à Constantinople [3]. La Turquie, coincée entre la Russie et la France, essaya d'abord

---

(1) C. KOROLEVSKY, art. *Beyrouth*, dans *D. H. G. E.*, t. VIII, col. 1329-1330.
(2) ID., art. *Bahouth*, dans *D. H. G. E.*, t. VI, col. 231.
(3) Le cardinal Antonelli remercia Montalembert de cette nomination comme d'un service rendu à toute la Chrétienté. (E. LECANUET, *Montalembert*, t. III, p. 12, n. 1.)

de s'en tirer par des faux-fuyants, puis par des demi-mesures : le firman
du 9 février 1852 notamment n'accordait aux Latins que quelques petits
avantages et maintenait au fond le *statu quo*, c'est-à-dire l'équilibre
existant de fait entre les différentes confessions qui se partageaient les
Lieux Saints. Les discussions continuèrent donc et elles furent partiellement l'occasion de la guerre de Crimée, mais la France ne profita pas
de sa victoire pour faire modifier le *statu quo*, qui resta en vigueur jusqu'à
la fin du XIXe siècle [1].

La présence catholique en Terre Sainte avait été assurée, depuis leur
fondation, par les franciscains. Ils conservèrent la place importante qu'ils
occupaient, même après l'établissement d'un patriarche à Jérusalem,
mais ce dernier, soucieux d'accroître l'influence catholique et désireux
de multiplier les conversions parmi les orthodoxes, fit appel à diverses
autres congrégations religieuses, surtout enseignantes. Il faut faire une
place spéciale aux sœurs de Notre-Dame de Sion, fondées par le P. Théo-
dore Ratisbonne et puissamment soutenues par son frère Alphonse,
le célèbre converti, qui vint s'établir en Palestine en 1855 et qui devait
y déployer, pendant trente ans, un zèle apostolique infatigable [2].

§ 3. — Les patriarcats orientaux [3].

ROME ET LES ÉGLISES UNIATES   Jusqu'au milieu du XIXe siècle, l'indifférence de l'Occident pour les choses
de l'Orient avait été quasi totale. Cette situation commença à se modifier
par suite notamment de l'espoir de voir les remous politiques qui se
produisaient dans les Balkans et le Proche-Orient provoquer d'importants

---

(1) On était très mécontent à Rome de ce que le gouvernement de Napoléon III n'eût pas mis
plus d'énergie à défendre les revendications catholiques. En 1869 encore, l'ambassadeur de France,
après un entretien avec le cardinal Barnabo, écrivait : « L'un de ses griefs principaux est à Jérusalem. » (*Arch. Min. A. É. Paris*, Rome, 1043, fo 432.)

(2) Cf. *Le P. Ratisbonne et Notre-Dame de Sion*, Paris, 1928.

(3) BIBLIOGRAPHIE. — I. SOURCES. — *Ius Pontificium de Propaganda fide*, édit. R. DE MARTINIS,
Rome, 1886-1895, 6 vol.

Sur la bulle *Reversurus* et les discussions subséquentes : MANSI, *S. Conciliorum amplissima collectio*, t. 40, col. 745-1132.

Sur les débuts de l'union bulgare, MANSI, *ibid.*, t. 45, col. 1-176.

II. TRAVAUX. — C. DE CLERCQ, *Les Églises unies d'Orient*, Paris, 1934 et *Conciles des Orientaux
catholiques* (t. XI de l'*Histoire des conciles* de HEFELE-LECLERCQ), Ie Partie, Paris, 1949 ; D.
ATTWATER, *The Catholic Eastern Churches*, Milwaukee, 1935 ; K. LUEBECK, *op. cit.*, p. 410, n. 2.

Sur la bulle *Reversurus* et la crise du patriarcat arménien : F. TOURNEBIZE, art. *Arménie*, dans
D. H. G. E., t. IV, col. 290 et suiv., en particulier col. 338-342 ; Th. LAMY, *La question arménienne*, dans *Revue catholique*, t. XII, 1874 et t. XIII, 1875.

Sur le patriarcat chaldéen et la question malabare : J. TFINKDGI, *L'Église chaldéenne catholique
autrefois et aujourd'hui*, Paris, 1913 ; B. M. GOORMACHTIGH, *Histoire de la mission dominicaine en
Mésopotamie*, dans *Analecta S. Ordinis Fratrum Praedicatorum*, t. II, 1895-96, et t. III, 1897-98 ;
J.-C. PANJIKARAN, *Christianity in Malabar*, Rome, 1926, et surtout C. KOROLEVSKY, art. *Audo*,
dans *D. H. G. E.*, t. V, col. 317-356.

Sur le patriarcat syrien : MAMARBASHI, *Les Syriens catholiques et leur patriarche Samhiri*, Paris,
1855 ; C. KOROLEVSKY, art. *Arqous*, dans *D. H. G. E.*, t. IV, col. 676-681.

Sur le patriarcat melkite : C. KOROLEVSKI, *Histoire des patriarcats melkites depuis le schisme
monophysite jusqu'à nos jours*, t. II, Rome, 1910 ; K. LUEBECK, *Patriarch Maximos III Mazloum*,
Aix-la-Chapelle, 1920.

Sur le patriarcat maronite : CHURCHILL, *The Druses and Maronites under the Turkisch Rule
from 1840 to 1860*, Londres, 1862 ; P. DIB, art. *Maronites*, dans *D. T. C.*, t. IX, col. 1-142 (utilisant une importante bibliographie en arabe).

Sur la question bulgare : C. ARMANET, *Le mouvement des Bulgares vers Rome en 1860*, dans *Échos
d'Orient*, t. XII, 1909 et t. XIII, 1910.

retours d'orthodoxes à l'Église romaine [1], et si ces espoirs furent finalement déçus, le Saint-Siège n'en continua pas moins à s'intéresser désormais de beaucoup plus près à la consolidation et à la réorganisation des
minorités uniates qui « doublaient » les églises schismatiques d'Orient.

Une mesure importante dans ce sens fut la décision prise par Pie IX,
en janvier 1862, de diviser la Congrégation de la Propagande en deux
sections dont l'une, ayant ses membres spéciaux et un secrétaire particulier, serait consacrée aux affaires du rite oriental [2]. Jusqu'en 1874, elle
eut comme préfet le cardinal Barnabo, un homme de grand mérite, à
l'esprit vif et pénétrant. Énergique et droit, il se montra plus d'une fois
exaspéré par les tergiversations ou les réticences de certains prélats
orientaux qui lui donnaient l'impression de vouloir éluder leur devoir
par des compromis ; mais il avait toutefois nettement conscience de
la nécessité de respecter autant que possible les traditions légitimes
des vénérables Églises d'Orient. Il déclarait à dom Pitra :

> Il faut se tenir dans un milieu équitable, qui sauve les principes et ne froisse
> ni les usages légitimes des Unis, ni les préjugés respectables des insoumis schis-
> matiques. Les uns se plaignent qu'en belles paroles, on promet la conservation
> des rites et de la discipline, mais qu'en réalité on réduit peu à peu tout ce qu'ils
> ont au rite latin. Les autres nous reprochent d'ignorer leur discipline et d'inno-
> ver sur des choses consacrées par les conciles et les saints Pères... Il faut remon-
> ter aux sources et interroger les canons grecs [3].

On a pourtant fait grief aux autorités romaines d'avoir poursuivi
pendant le pontificat de Pie IX une politique systématique de latinisation. La réalité est plus nuancée. Il y a certes des faits incontestables,
les uns qui s'expliquent par les circonstances, comme le bref du 13 mai
1874 au métropolite de Lemberg, désavouant la réaction du jeune clergé
ruthène contre le caractère hybride du rite uniate [4], d'autres plus discutables, comme le refus opposé à plusieurs reprises aux chrétiens du Malabare d'avoir des évêques de leur rite, ou l'approbation donnée à certaines
rééditions d'ouvrages liturgiques farcis de latinismes. Mais il faut tenir
compte que, le plus souvent, c'est l'ignorance des traditions orientales
qui amena la Propagande à « des latinisations de fait, qui n'étaient certes
pas voulues positivement et contre lesquelles on protestait tout en les
commettant sans s'en rendre compte » [5]. Pie IX lui-même était bien
décidé à respecter les liturgies orientales et son encyclique du 6 avril
1862 [6], affirmant la légitime diversité des rites, tranchait sur l'opinion
courante, défendue encore d'un ton catégorique par dom Guéranger,
d'après laquelle l'unité de l'Église devait avoir pour corollaire normal
l'uniformité liturgique [7].

---

(1) Sur ces espoirs et les initiatives auxquelles ils donnèrent lieu (notamment la mission de dom
Pitra en Russie), voir *infra*, p. 481-482.

(2) Voir la bulle *Romani Pontifices* du 6 janvier 1862 dans *Coll. lac.*, t. II, col. 557-558.

(3) A. BATTANDIER, *Le cardinal J.-B. Pitra*, Paris, 1896, p. 319.

(4) Des sympathies nationalistes pour la Russie orthodoxe encouragées en sous main par le
gouvernement tsariste se mêlaient de manière assez trouble à ce mouvement de rénovation liturgique, qui se présentait en fait comme le prélude à un glissement vers le schisme. (Cf. *supra*, p. 406.)

(5) C. KOROLEVSKY, art. *Audo*, dans *D. H. G. E.*, t. V, col. 341.

(6) Texte dans *Coll. lac.*, t. II, col. 558 et suiv.

(7) Voir les passages du t. III de ses *Institutions liturgiques* (1851), cités par H. CHIRAT dans

Il est par contre indéniable que, sous l'influence des idées centralisatrices régnantes, il y avait à Rome une nette tendance à faire beaucoup moins de cas de la discipline orientale. Ne fut-il pas question, dans les commissions préparatoires au concile du Vatican, de la supprimer purement et simplement ! Sans aller jusque-là, plusieurs estimaient que, vu la nécessité de réformer sans tarder les communautés orientales, où le relâchement de la discipline avait fait de grands ravages, le plus simple était, puisque la législation des Orientaux était encore si mal connue, de leur appliquer en attendant la discipline latine, quelles que fussent leurs répugnances. Il est certain que, sur ce point, l'attitude du pape fut moins réservée que celle de la Propagande et que, porté, comme beaucoup d'Occidentaux, à voir une tendance au schisme dans l'attachement des Orientaux à leurs antiques privilèges, il entendait bien arriver à changer graduellement l'ancienne discipline. La mesure la plus grave en ce domaine fut la décision de modifier les règles relatives à l'élection des évêques. Rendue publique par la bulle *Reversurus*, du 12 juillet 1867, elle allait agiter l'Orient plusieurs années durant et y soulever une profonde méfiance contre Rome.

LA BULLE « REVERSURUS » Depuis que les autorités turques avaient émancipé les catholiques orientaux de la juridiction civile des patriarches schismatiques, les élections épiscopales avaient pris une portée temporelle qu'elles n'avaient pas eue dans les siècles antérieurs, et il était à craindre qu'elles ne subissent une pression de la part de laïcs influents, plus préoccupés de leurs intérêts matériels que du bien spirituel de leur Église. Des mesures étaient donc nécessaires pour obvier à un danger très réel, mais Pie IX voulut en profiter pour renforcer l'influence romaine dans le choix des évêques, conformément à ce qui s'était passé en Occident depuis le moyen âge. L'occasion lui en fut fournie par le transfert à Constantinople du siège patriarcal arménien.

Les Arméniens de l'Empire turc se trouvaient surtout concentrés en Cilicie et en Syrie, et leur patriarche résidait au Liban, mais il existait en outre à Constantinople, depuis 1830, un archevêque-primat. A la mort du patriarche Pierre VIII, en 1866, les évêques assemblés sous la présidence du délégué apostolique de Syrie, Mgr Valerga, élurent d'une voix unanime pour lui succéder l'archevêque de Constantinople, Mgr Hassoun, ancien élève de la Propagande et très dévoué au Saint-Siège. L'année suivante, Pie IX confirmait la fusion des deux juridictions supra-épiscopales et transférait le siège patriarcal dans la capitale, où le nombre des fidèles atteignait les 25.000. Mais, dans le même document, la bulle *Reversurus*, il déterminait également les règles à suivre à l'avenir pour le choix des évêques et du patriarche, en étendant à tout le patriarcat un mode d'élection analogue à celui qu'il avait déjà introduit en 1853 pour la région de Constantinople, avec l'assentiment de Mgr Hassoun,

---

La Maison-Dieu, n° 7, p. 138-140. Par contre, dans un bref mémoire rédigé en 1865, un autre liturgiste, dom Gréa, devait se prononcer nettement contre ceux qui préconisaient l'abolition des rites traditionnels des communautés chrétiennes d'Orient et la constitution en ces régions d'un clergé indigène de rite latin. (*Ibid.*, p. 140-141.)

par l'instruction *Licet episcopalis*. Les laïcs et le clergé inférieur étaient absolument exclus, non seulement de l'élection patriarcale, mais même de la présentation d'une liste préparatoire aux élections épiscopales ; le pape se réservait le droit de nommer les évêques en dehors des candidats choisis par le patriarche ; enfin, contrairement à l'usage suivi jusqu'alors, avant d'avoir reçu la confirmation en consistoire, le patriarche élu ne pouvait être intronisé ni procéder à aucun des cinq actes majeurs de juridiction ; il devait en outre accomplir tous les cinq ans la visite *ad limina*.

Ces stipulations rompaient avec une tradition remontant aux premiers âges du christianisme et plusieurs d'entre elles étaient en contradiction avec une décision de la Propagande datant de 1827, qui laissait au patriarche élu la faculté de commencer à exercer sa juridiction conformément à l'ancien usage oriental, même avant d'avoir été confirmé par le Saint-Siège. L'interdiction des cinq actes majeurs et l'obligation de la visite *ad limina* étaient une extension à l'Église arménienne de dispositions du droit canonique occidental. Mais le point le plus grave était la suppression radicale de l'ancienne autonomie des élections épiscopales.

On a souvent reproché à Mgr Hassoun d'avoir trop aisément sacrifié à cette occasion les antiques privilèges patriarcaux. En réalité, dans le projet qu'il avait soumis à Rome et que la Propagande avait d'abord approuvé, l'intervention du Saint-Siège se bornait à prescrire au délégué apostolique, toutes les fois que cela serait jugé nécessaire, d'assister au synode électoral avec des facultés pouvant aller jusqu'à suspendre l'élection ou à interdire le choix d'un candidat. Cette méthode sauvegardait l'autorité du Saint-Siège tout en conservant l'autonomie traditionnelle, mais Pie IX hésita, consulta Mgr Valerga et l'évêque arménien Araquelean et finalement trancha la question dans le sens du choix définitif des évêques par Rome et de l'extension aux Arméniens de la prohibition des cinq actes majeurs.

L'émotion produite dans la communauté arménienne par la bulle *Reversurus* gagna bientôt les autres Églises unies quand on apprit que le pape avait l'intention d'étendre ces mesures à tous les patriarcats. Cette seconde décision était plus grave encore, car le patriarcat arménien n'étant pas un des patriarcats dits « majeurs », il pouvait paraître normal qu'on ne lui accordât pas les mêmes privilèges qu'à ceux-ci ; mais Pie IX, à l'encontre de l'avis qui prévalait à la Propagande, semble avoir embrassé l'opinion de certains canonistes pour qui les patriarches orientaux modernes sont à assimiler aux patriarches dits mineurs de l'Église latine. Il était logique dans ces conditions de supprimer dans les autres patriarcats ce qu'on avait refusé à Mgr Hassoun. Le pape avait incontestablement le droit d'agir ainsi, puisque les droits patriarcaux ne sont que des privilèges de droit ecclésiastique, mais il n'en supprimait pas moins d'un trait de plume et sans ménager les susceptibilités « toute une législation que Benoît XIV n'avait pas hésité à sanctionner *in forma specifica* dans le synode maronite de 1736 »[1]. Il sera intéressant

(1) C. KOROLEVSKY, art. *Arqous*, dans *D. H. G. E.*, t. IV, col. 679. Voir sur toute cette affaire ID., art. *Audo*, dans *D. H. G. E.*, t. V, col. 337-338 et 345.

de déterminer un jour, à l'aide des pièces d'archives, si cette erreur de tactique doit être attribuée exclusivement à Pie IX ou si elle doit être partagée par les deux prélats qu'il consulta.

LE SCHISME ARMÉNIEN La communauté arménienne, qui comportait vers 1870 environ 120.000 fidèles en Turquie [1], avait connu pendant les vingt premières années du pontificat de Pie IX une période de calme et de prospérité, spécialement dans les régions environnant Constantinople, où les conversions s'étaient multipliées. Le zèle entreprenant de Mgr Hassoun avait su mener à bien plusieurs réalisations importantes : fondations, en 1852, des sœurs arméniennes de l'Immaculée-Conception pour l'éducation des filles, construction d'une cinquantaine d'églises ou d'écoles. Mais les événements consécutifs à la bulle *Reversurus* allaient troubler profondément cette Église pendant plus de dix ans.

Après deux ans d'accalmie, les protestations contre le nouveau règlement avaient repris en juillet 1869, lors du synode national, et tandis que Mgr Hassoun se trouvait à Rome pour le concile du Vatican, son vicaire patriarcal prit le parti des mécontents ; bientôt, en dépit des menaces du délégué apostolique, un parti dissident se constitua, groupant quatre évêques, quelques prêtres et un certain nombre de religieux Antonins, dont le supérieur, Mgr Gazandjian, fut l'un des meneurs les plus ardents de l'opposition [2]. Peu soucieux de l'excommunication prononcée contre eux et se sentant soutenus par le gouvernement français, qui était heureux d'exploiter tous les incidents pouvant entraver le concile, les quatre évêques, dont plusieurs avaient fui Rome dans des conditions quelque peu mélodramatiques, prononcèrent la déchéance de Mgr Hassoun et élirent à sa place Jacques Bahdiarian ; puis ils réussirent à se faire reconnaître par le gouvernement turc comme les seuls véritables Arméniens catholiques, s'approprièrent un grand nombre d'églises et d'écoles et obtinrent même le bannissement de Mgr Hassoun. Les catholiques demeurés fidèles finirent pourtant par obtenir, grâce à l'appui du nouvel ambassadeur de France, le marquis de Vogüé, d'être considérés par les autorités ottomanes comme une communauté distincte des néo-schismatiques, et Mgr Hassoun put reprendre ses fonctions jusqu'à ce que Léon XIII, en 1880, le fasse venir à Rome comme cardinal de Curie et le remplace par un homme plus souple, Mgr Azarian, afin de hâter l'extinction du schisme.

LES DIFFICULTÉS DU PATRIARCAT CHALDÉEN ET LA QUESTION MALABARE L'agitation consécutive à la bulle *Reversurus*, qui avait gagné l'Église chaldéenne de Mésopotamie, contribua, avec la question de la juridiction sur le Malabar, à y fomenter également un schisme où le patriarche lui-même manqua d'être entraîné.

(1) Il y en avait plusieurs dizaines de milliers d'autres en Géorgie et en Pologne et un certain nombre en Autriche où les célèbres religieux méchitaristes possédaient plusieurs couvents, avec leur centre à Vienne.

(2) Voir F. TOURNEBIZE, art. *Antonins arméniens*, dans *D. H. G. E.*, t. III, col. 868-870.

La situation de cette Église, qui devait son origine à des nestoriens réconciliés avec Rome au cours des XVIIe et XVIIIe siècles, et comptait environ 40.000 fidèles vers 1850, était fort misérable : aucune école, *a fortiori* aucun centre de formation cléricale, à l'exception de l'unique monastère situé à Rabban Hormizd ; en l'absence d'imprimerie, tous les ouvrages, liturgiques ou autres, devaient être transcrits à la main comme au moyen âge ; le droit canonique était pratiquement inexistant. Bref « tout était à faire » lorsque Joseph Audo fut élu patriarche en 1847. Très pieux, très zélé, il avait réussi à ramener à l'Église vingt villages nestoriens, mais sa formation était plus qu'élémentaire et son ignorance du latin et de l'italien le mettait dans ses rapports avec Rome à la merci de conseillers peu scrupuleux, surtout une fois que ses relations avec la mission dominicaine de Mossoul eurent commencé à s'aigrir.

Les premières difficultés se produisirent vers 1860. Il y avait sur la côte du Malabar, à la pointe sud de l'Inde, un groupe de catholiques d'origine nestorienne, placés sous la juridiction d'un vicaire apostolique latin, qui souhaitaient depuis longtemps avoir un évêque de leur rite, et comme, jusqu'au XVIe siècle, le Malabar avait reçu ses évêques du patriarche chaldéen de Mésopotamie, quelques prêtres plus turbulents décidèrent de s'adresser directement à Audo. Celui-ci, incapable de saisir la complexité de la question, prit l'affaire à cœur et se déclara prêt à consacrer un évêque pour le Malabar. Il en référa toutefois à la Propagande qui, par deux fois, lui répondit qu'il n'avait pas à intervenir aux Indes où il n'avait aucune juridiction. La délégation malabare, déçue, menaça alors d'aller demander son évêque au catholicos nestorien et, soucieux de prévenir ce malheur, excités d'ailleurs par l'attitude du délégué apostolique, qui outrepassa ses pouvoirs, les évêques chaldéens décidèrent, lors d'une réunion secrète, de consacrer un évêque titulaire qu'on enverrait au Malabar sans juridiction, afin de reconnaître la situation. Le Saint-Siège, alerté, manda aussitôt Audo à Rome et celui-ci fut obligé de reconnaître qu'il avait méconnu l'autorité du délégué et dut désavouer publiquement sa conduite après une audience où Pie IX se montra particulièrement sévère. « Le crut-il de mauvaise foi ? Toujours est-il qu'il lui fit de vifs reproches, ce qui était assez dans son caractère primesautier et impulsif. On en a d'autres exemples. Le vieux patriarche, éclatant en sanglots, demanda pardon et absolution »[1], mais on garda contre lui, à Rome, une certaine méfiance, tandis que lui-même revint chez lui humilié et très monté contre les missionnaires dominicains, qu'il rendait responsables de tout.

Le différend devait rebondir dix ans plus tard, mais, dans l'entre-temps, les rapports entre Audo et Rome allaient se tendre encore davantage à la suite de l'extension de la bulle *Reversurus* à la Chaldée par la constitution *Cum ecclesiastica disciplina* du 31 août 1869. Audo, consulté quelques mois auparavant, avait répondu qu'il n'y voyait pas d'inconvénient, mais ensuite, pendant son séjour à Rome lors du concile, il se rétracta et refusa même de consacrer deux évêques désignés par le pape conformément à la nouvelle constitution. On sait comment cette résis-

---

(1) C. KOROLEVSKY, art. *Audo*, dans *D. H. G. E.*, t. V, col. 334.

tance fut l'occasion d'une audience dramatique où, une fois de plus, Mgr Valerga semble avoir poussé Pie IX à l'intransigeance et Audo, encore excité par certains membres de la minorité, quitta Rome ulcéré, convaincu que le pape visait à supprimer tous les privilèges patriarcaux, et il fut l'un des derniers à donner son adhésion aux décisions du concile, après avoir ouvertement appuyé, à plusieurs reprises, les néo-schismatiques arméniens.

A la suite de ces divers événements, l'épiscopat chaldéen se trouva partagé en deux clans, dont l'un devenait de plus en plus hostile à Rome et, par deux fois, en 1874 et 1875, Audo procéda à des consécrations épiscopales sans se soucier des nouvelles stipulations romaines ni d'un rappel à l'ordre de la Propagande. Bien plus, ayant été sollicité à nouveau par les mécontents du Malabar, il leur envoya un évêque dont l'orthodoxie était pour le moins douteuse, Élie Mellous, lequel, en se prévalant d'un faux bref pontifical, procéda à une série d'ordinations, généralement assez malheureuses, et réussit à arracher aux catholiques un certain nombre d'églises [1].

A Rome, on était prêt à montrer de l'indulgence pour un vieillard de 87 ans, qu'on savait être le jouet de mauvais conseillers, mais on tenait à ce que le dernier mot restât à l'autorité et, le 1er septembre 1876, l'encyclique *Quae in patriarchatu*, adressée à tout le patriarcat chaldéen, annonçait que si Audo ne s'était pas soumis dans les quarante jours, il serait frappé d'excommunication majeure et suspens de toute juridiction patriarcale. Tout l'Orient avait les yeux fixés sur le vieux patriarche de Babylone. Heureusement, celui-ci, dont « la bonne foi paraît avoir été sincère, quelles qu'aient pu être les conséquences de son entêtement et de sa versatilité » [2], se rendit compte de la voie où l'avait entraîné un clan de révoltés et, le 1er mars 1877, il adressa à Pie IX l'expression de sa soumission totale.

LE PATRIARCAT SYRIEN   La réconciliation avec Rome, à la fin du XVIIIe siècle, du patriarche jacobite Michel Garweh n'avait pas été suivie d'un retour en corps de l'Église monophysite syrienne, mais elle avait du moins abouti à la constitution d'un nouveau patriarcat uniate qui groupait, à l'avénement de Pie IX, une vingtaine de milliers de fidèles et dix évêques, nés jacobites pour la plupart, car les adhésions au catholicisme s'étaient multipliées surtout à la suite de la conversion, en 1827, de l'archevêque de Jérusalem et de son vicaire général Samhiri. Ce dernier, homme plein de zèle et d'énergie, qui peut être considéré comme l'un des fondateurs du patriarcat syrien catholique, devint lui-même patriarche en 1854 [3] et fit beaucoup, au

---

(1) Il ne devait quitter les Indes qu'en 1882 et le schisme dit « mellusien » continua après son départ sous la direction d'Antoine Thondanatta, qui était depuis 1856 à l'origine de toute l'agitation.

(2) C. KOROLESVKI, art. *Audo*, col. 350. Le même auteur remarque : « Le schisme arménien n'aurait probablement pas été évité, car il y avait trop d'éléments troubles dans cette communauté ; mais avec un peu de doigté, celui des chaldéens aurait pu l'être. » (*Ibid.*, col. 345.)

(3) Au cours du synode de Sajida-el-Scharfé, sur lequel on peut voir P. BACEL, *Le premier synode syrien de Charfé*, dans *Échos d'Orient*, t. XIV, 1911, p. 293 et suiv.

cours de ses dix ans de gouvernement, pour intéresser l'Europe catholique au sort de sa pauvre Église. Après sa mort, le synode qui se tint à Alep en mai 1866, pour choisir son successeur, s'efforça, avec l'aide du délégué apostolique de Mésopotamie, de mettre un peu d'uniformité dans la discipline canonique et liturgique, demeurée jusqu'alors fort vague, mais sans arriver encore à des résultats définitifs.

Le successeur de Samhiri, Mgr Arqous, évêque de Diarbekir, était loin d'avoir l'envergure de son prédécesseur. Il ne semble pas avoir manifesté beaucoup de zèle ni d'énergie dans son administration et son attitude en face des remous causés par la bulle *Reversurus* fut caractéristique de son manque de personnalité. Se rendant bien compte qu'il n'aurait pu l'appliquer sans susciter des troubles dans son Église, il n'eut le courage ni d'affronter ceux-ci, ni de faire au pape les représentations nécessaires. Aussi, lorsque la nouvelle constitution romaine lui fut communiquée, avec invitation d'avoir à s'y conformer à l'avenir, lors des élections épiscopales, il commença par prétendre qu'il ne l'avait pas reçue, puis, lorsqu'on lui en eut transmis un duplicata, sans répondre directement, il envoya sa démission à Rome en invoquant sa mauvaise santé, et renouvela son offre à l'occasion du concile du Vatican. Pie IX l'ayant refusée, il resta patriarche jusqu'à sa mort en 1874, mais il continua à ignorer pratiquement la bulle et adopta à l'égard des décisions du concile une attitude pour le moins ambiguë.

LE PATRIARCAT MELKITE   D'origine beaucoup plus ancienne que le patriarcat syrien, l'Église melkite unie comptait, au milieu du XIXe siècle, environ 50.000 fidèles et près de 200 prêtres, religieux pour la plupart. Elle était dirigée, depuis 1833, par Maxime III Mazloum, prélat intelligent et relativement instruit pour l'époque, qui savait unir la souplesse à l'énergie. Sincèrement attaché à la foi catholique, il avait, comme beaucoup de melkites du XIXe siècle, une propension marquée pour les principes gallicans, et il resta toujours partisan des doctrines du synode de Qarqafé de 1806, que Rome s'était vu obligée de condamner[1]. En 1849, il réunit à Jérusalem un concile[2] dont les canons, soigneusement préparés par le patriarche au cours des années précédentes, formaient un code de discipline ecclésiastique assez complet et auraient pu servir de base à une réforme de l'Église melkite, qui, comme toutes les communautés orientales de l'époque, en avait bien besoin. Mais il ne réussit pas à obtenir l'approbation de la Propagande, car l'influence du concile de Qarqafé était trop visible en plusieurs points, comme le montrèrent Mgr Valerga et le P. Van Everbroeck, qui avait déjà provoqué la condamnation de 1835. Le Saint-Siège était d'autant plus mécontent que Mazloum avait tardé plus d'un an avant d'envoyer les canons à Rome et qu'il les avait mis immédiatement en vigueur sans attendre l'approbation. Les discussions se prolongèrent et s'envenimèrent

---

(1) Sur ce concile et sa condamnation par le bref *Melchitarum catholicorum synodus* du 16 septembre 1835, voir C. DE CLERCQ, *Conciles des Orientaux catholiques*, vol. I, p. 337-360.
(2) Cf. C. DE CLERCQ, *op. cit.*, vol. I, p. 390-414.

au point que Pie IX envisageait d'interdire l'exercice de toute juridiction au patriarche qui, convoqué à Rome, s'était dérobé, lorsqu'il mourut, en 1855.

Le choix de son successeur s'annonçait difficile, étant données les divisions de l'épiscopat. La Propagande, par une mesure « hardie mais nécessaire »[1], attribua au délégué apostolique des pouvoirs spéciaux sur le synode électoral et parvint de la sorte à faire élire le saint évêque Clément Bahouth[2], très dévoué à Rome où il avait fait ses études et qui s'était jusque-là soigneusement tenu à l'écart des luttes de parti. Aussitôt élu, celui-ci renonça spontanément aux canons du concile de Jérusalem, conformément aux suggestions de la Propagande, qui préférait éviter une condamnation, peu honorable pour les évêques qui les avaient souscrits. Il fut convenu qu'une nouvelle assemblée serait convoquée peu après, sous la présidence du délégué apostolique, mais, en fait, il fallut attendre jusqu'en 1909, avant qu'un concile ne se réunisse à nouveau.

Rome comptait beaucoup sur Bahouth pour introduire des réformes, car si la communauté melkite s'était accrue de 20.000 âmes, depuis l'élection de Mazloum, il restait beaucoup à faire pour la restauration de la discipline, notamment dans les couvents, où la vie intérieure s'épuisait en intrigues et en vaines querelles causées par l'esprit de clocher et le manque de formation sérieuse. Malheureusement l'heure n'était guère favorable, car l'Église syrienne allait bientôt être profondément secouée par les tragiques massacres du printemps de 1860 où les Druses firent presque autant de victimes parmi les melkites que parmi les maronites ; et de plus, si les intentions du nouveau patriarche étaient excellentes, c'était plus un ascète qu'un homme de gouvernement et son manque de doigté mit son Église à deux doigts du schisme. Invité par Rome à imiter les Églises syriennes et chaldéennes, qui avaient, vingt ans auparavant, adopté le calendrier grégorien pour se distinguer des orthodoxes, il brusqua les choses et imposa celui-ci d'autorité en janvier 1857, sans se préoccuper de préparer d'abord les esprits comme plusieurs évêques l'avaient suggéré. Le métropolite de Beyrouth, Agapios Riachi, qui était jaloux d'avoir été écarté du patriarcat par Rome, en profita pour organiser la résistance et, après avoir gagné trois autres évêques à sa cause, il essaya de constituer, avec l'appui intéressé de représentants de l'Église russe, une communauté dissidente qui prit le nom de *sarqiîn* (les « orientaux », par opposition aux *gharbiîn* ou occidentaux). « Ce que voulaient les *sarqiîn*, ce n'était pas se faire orthodoxes, mais bien former une nation à part, catholique si l'on veut, mais d'un catholicisme spécial et tout gallican, ne reconnaissant à Rome qu'une primauté d'honneur, niant toute obéissance au patriarche Bahouth et, comme signe de ralliement, gardant l'ancien calendrier[3]. » Tout finit toutefois par rentrer dans l'ordre, grâce à l'attitude compréhensive du successeur de Bahouth.

(1) C. KOROLEVSKY, art. *Antioche*, dans *D. H. G. E.*, t. III, col. 657.
(2) Sur Mgr Clément Bahouth, (1799-1882), mort en odeur de sainteté sous l'habit d'un moine convers, voir l'art. *Bahouth* dans *D. H. G. E.*, t. VI, col. 229-236.
(3) C. KOROLEVSKY, art. *Bahouth*, dans *D. H. G. E.*, t. VI, col. 234.

Ce dernier, en effet, qui avait déjà offert sa démission dès le début des incidents, renouvela sa proposition en avril 1864 et le pape, qui l'avait refusée en 1858, l'accepta cette fois, à condition que l'on fût assuré que le nouveau patriarche serait l'évêque d'Acre, Mgr Grégoire Yûssof, qui devait se révéler comme le patriarche le plus remarquable du xixe siècle. « Son grand mérite aura été d'avoir su, durant 33 ans de patriarcat, avec un clergé généralement très inférieur à sa tâche, maintenir l'ordre et la discipline et même promouvoir le retour d'un certain nombre d'orthodoxes »[1], au point qu'à sa mort l'Église melkite dépassait les 100.000 fidèles. Tout en encourageant les œuvres d'enseignement pour lutter contre le prosélytisme protestant, il chercha à obvier aux inconvénients que présentaient certaines écoles tenues par des missionnaires latins : c'est ainsi qu'à Damas, le refus opposé par les lazaristes à leurs élèves melkites de fréquenter l'église de leur rite l'amena à fonder, en 1874, le collège Saint-Jean-Damascène. Un de ses principaux soucis fut d'améliorer la formation du clergé local. Mazloum s'était déjà efforcé de préparer un clergé séculier célibataire. Yûssof commença par rouvrir en 1866 le séminaire d'Aîn-Traz, puis, de concert avec Mgr Lavigerie, il mit au point le projet, inspiré de principes diamétralement opposés à ceux suivis jusqu'alors par les missionnaires, d'un séminaire basé sur le respect absolu du rite et des usages nationaux, tout en donnant aux élèves une solide formation spirituelle et l'instruction ecclésiastique européenne ; et l'idée une fois réalisée par la fondation du séminaire Sainte-Anne de Jérusalem, confié aux Pères Blancs, il soutint celui-ci de toute son influence.

Par contre, malgré l'utilité qu'il y aurait eu à réunir un concile pour reprendre l'œuvre de réorganisation tentée lors du synode de Jérusalem, Yûssof ne s'y décida jamais, « vu qu'il ne pouvait le faire sans l'aide de Rome et qu'il redoutait d'être obligé d'y recourir »[2]. En effet, bien qu'il fût ancien élève du collège de la Propagande, il demeura toute sa vie préoccupé d'éviter tout ce qui pourrait heurter les tendances particularistes de son Église, d'autant plus que la bulle Reversurus et le manque de nuance des néo-ultramontains lors du concile du Vatican avaient renforcé sa persuasion que Rome visait la destruction des antiques privilèges patriarcaux. Son attitude pendant le concile s'explique par ces craintes, de même que le peu de solennité avec lequel il promulga la constitution Pastor aeternus. Rome eut la sagesse, étant donné l'état des esprits, de ne pas exiger davantage.

LE PATRIARCAT MARONITE   L'histoire du patriarcat maronite, le plus important des patriarcats orientaux, puisqu'il comptait plus de 250.000 fidèles, est beaucoup plus brève. En effet, si cette Église fut fort agitée par les événements extérieurs — révolte des paysans du Liban contre la féodalité des cheiks en 1858, terribles massacres druses en 1860 — elle n'eut guère à souffrir des dissensions

---

(1) Id., art. *Antioche*, dans *D. H. G. E.*, t. III, col. 663.
(2) *Ibid.*

intérieures qui troublèrent les autres patriarcats pendant le pontificat de Pie IX. Elle le doit en partie à son attachement traditionnel au Saint-Siège, qui lui permit de voir d'un œil plus serein la politique romaine de réforme et de centralisation, peut-être aussi au fait que les rapports étroits et anciens avec la France, grande protectrice du Liban, favorisaient la bonne entente avec les missionnaires latins.

De 1854 à 1890, l'Église maronite fut dirigée par l'énergique et perspicace patriarche Pierre Paul Mas'ad, qui joignait à de grandes qualités de gouvernement une culture intellectuelle remarquable. Il inaugura son pontificat par la préparation d'un concile national, qui eut lieu à Bèkorki en 1856, sous la présidence du délégué apostolique ; le patriarche y avait convoqué, outre les évêques, les supérieurs des trois ordres religieux maronites [1], les recteurs des missions latines et quelques notables. Mais, bien que le pape eût écrit au patriarche pour le féliciter de l'œuvre accomplie, la Propagande ne confirma jamais officiellement les actes de l'assemblée, et ceux-ci restèrent en conséquence lettre morte.

Très dévoué au pape, Mgr Mas'ad fut le premier patriarche maronite depuis le XIIIe siècle à faire sa visite *ad limina*, mais lorsqu'il apprit, de la bouche même de Pie IX, que les dispositions de la bulle *Reversurus* seraient bientôt étendues à son Église, il n'hésita pas à faire remarquer, en termes respectueux mais très nets, combien cette mesure était en contradiction avec les assurances solennelles données jadis par Benoît XIV.

VAINES TENTATIVES D'UNIATISME EN BULGARIE ET EN GRÈCE

On put croire un moment, aux environs de 1860, que, parmi les populations orthodoxes des Balkans, allait se produire un mouvement de retour à l'Église romaine analogue à celui qui avait eu lieu antérieurement dans le Levant et donnant naissance, lui aussi, à de nouvelles Églises uniates. C'est en Bulgarie que ces espoirs, bientôt démentis, parurent le plus près de se réaliser.

Depuis 1830, s'était développé parmi les Bulgares un double mouvement d'émancipation intellectuelle à l'égard des Grecs et d'indépendance politique à l'égard des Turcs. Or, ils s'étaient vite rendu compte de l'appoint qu'apporterait à ces aspirations la constitution d'une Église bulgare autonome : celle-ci obligerait les Turcs à reconnaître leur entité ethnique, conformément au principe que tout groupement religieux officiellement reconnu obtenait le titre de « nation » ; et des évêques nationaux mettraient fin à la politique systématique d'hellénisation poursuivie depuis un demi-siècle par les autorités orthodoxes de Constantinople.

Pendant plus de vingt ans, le Phanar repoussa systématiquement ces revendications et c'est dans ces conditions que, en 1859, un parti qui se trouvait sous l'influence du comité polonais de Paris commença à défendre l'idée que les Bulgares ne pouvaient attendre leur émancipation

---

(1) Ceux-ci comptaient, en 1852, 800 religieux prêtres et un millier de frères lais. Il y avait, en outre, dans l'Église maronite un millier de prêtres séculiers. (*Kirchenlexikon* de WETZER et WELTE, 2e édit., Fribourg-en-Br., 1893, art. *Maroniten*, t. VIII, col. 899.)

religieuse que du pape, et à préconiser dès lors l'union avec Rome, à
condition de pouvoir conserver la liturgie en slavon. Un certain nombre de
conversions se produisirent bientôt parmi les Bulgares de Constantinople,
pour des raisons où la politique jouait un rôle au moins aussi grand que
les convictions religieuses ; ils entrèrent en contact avec le délégué apostolique et Pie IX répondit à leurs suppliques, le 21 janvier 1861, en leur
garantissant le maintien de leurs coutumes religieuses. Les interventions
de l'ambassadeur de Russie, que la chose inquiétait beaucoup, ralen-
tirent un moment le mouvement, mais celui-ci reprit un peu plus tard
en Thrace et en Macédoine et, pour se donner plus de consistance, la
jeune communauté demanda à recevoir un évêque. Pie IX accueillit
paternellement le vieil higoumène Joseph Sokolski qui avait été choisi
dans ce but et il le consacra lui-même, le 14 avril 1861, en lui conférant
le titre d'archevêque et de vicaire apostolique des Bulgares unis.

Ce geste fit monter en quelques jours le nombre des conversions à
60.000, mais les déconvenues n'allaient pas tarder par suite du manque
d'expérience du nouvel évêque, qui savait à peine lire et écrire, et la
Russie, qui désirait que l'émancipation religieuse des Bulgares se fît
à son avantage, s'empressa de profiter de cette situation. Dans le courant
de l'été, un de ses agents secrets réussit à entraîner en Russie Sokolski
dont « on n'a jamais su de façon certaine s'il avait été la victime ou le
complice de son enlèvement »[1]. Cette disparition arrêta net le mouvement
vers Rome : quelques ecclésiastiques donnèrent le signal de la défection,
le nombre des fidèles retomba rapidement à 4.000 et bientôt tous les
prêtres repassèrent à l'orthodoxie. Pie IX commença par confier le
petit noyau de catholiques restants aux assomptionnistes du P. d'Alzon,
qui s'adjoignit quelques religieux polonais, et ce n'est qu'après plus de
quatre ans qu'il nomma un nouvel évêque bulgare, Raphaël Popoff.
Mais l'occasion était passée, car l'organisation d'un exarchat orthodoxe
bulgare, indépendant de Constantinople, fit bientôt disparaître aux
yeux de l'immense majorité des fidèles l'intérêt qu'on avait cru voir
à une union avec Rome.

Tandis que ces événements se déroulaient en Bulgarie, quelques conversions se produisaient également dans l'Église grecque. Bien que la famille
royale de Grèce, d'origine bavaroise, fût catholique, les Grecs identifiaient
fanatiquement la religion orthodoxe avec leur nationalité et les quelques
milliers de catholiques que comptait le pays étaient des descendants
de commerçants italiens, desservis par des prêtres latins. Toutefois,
vers 1856, un prêtre latin de Syrie, Jean Marengo, parvint à convertir
quelques schismatiques et lorsque, en novembre 1861, un archevêque
se convertit à son tour, certains crurent y voir les prodromes d'un vaste
mouvement vers l'unité. Il n'en était rien cependant et les véritables
débuts de l'Église grecque unie ne datent que de 1882, lorsque débuta,
en Thrace, l'action d'Isaïe Papadopoulos.

---

(1) R. JANIN, art. *Bulgarie*, dans *D. H. G. E.*, t. VIII, col. 1185.

L'ÉGLISE CATHOLIQUE EN AMÉRIQUE
DE 1846 A 1878

§ 1. — L'Église catholique aux États-Unis [1].

L'ÉGLISE DES ÉTATS-UNIS A L'AVÉNEMENT DE PIE IX

Les « Pères de l'Église américaine » avaient eu à faire face, pendant les premières décades du siècle, à des difficultés de tout genre. Les unes, inévitables, provenaient de l'extension démesurée du champ à évangéliser, sans cesse accru par le déplacement de la « frontière » vers l'Ouest, et de l'insuffisance du clergé, qui ne comptait encore en 1842 que 482 prêtres, dont beaucoup étaient des hommes que leur évêque avait laissés s'expatrier sans regret.

D'autres difficultés furent provoquées par des circonstances malheureuses. De nombreux prêtres de la première génération, les plus vertueux et les plus instruits en tout cas, étaient des Français qui avaient fui la Révolution, et c'est parmi eux qu'avaient été choisis jusque vers 1825 la plupart des évêques. La chose n'allait pas sans inconvénients dans un pays qui évoluait rapidement dans un sens assez différent de la vieille Europe et où l'élément anglo-saxon et irlandais prenait chaque jour plus d'importance. Sans parler de l'imperfection avec laquelle ces évêques maniaient souvent la langue anglaise, l'esprit entreprenant et volontiers turbulent de certains prêtres d'origine irlandaise souffrait de la modération, qu'ils traitaient d'inertie, de prélats formés dans l'esprit sulpicien. De plus, la prépondérance française dans la direction de l'Église des États-Unis risquait de faire apparaître le catholicisme comme une religion d'étrangers. Mgr England, un prélat irlandais dont un contemporain disait qu'il fut « le premier à rendre la religion catholique respectable aux yeux du public américain » [2], réussit à conjurer la crise et, à sa mort, en 1842, l'épiscopat américain commençait à prendre cette allure

(1) BIBLIOGRAPHIE. — I. SOURCES. — D. SHEARER, *Pontificia Americana. A documentary, History of the Catholic Church in the United States (1784-1884)*, Washington, 1933 ; R. GUILDAY, *The National Pastorals of the American Hierarchy. 1792-1910*, Washington, 1923 ; *Relazione completa rimessa da mons. Bedini all'Em. Seg. Card. Prefetto dello stato di quelle vaste regioni nell'anno 1853*, Rome, 1854 ; *Coll. lac.*, t. III, Fribourg-en-Br., 1884 (conciles provinciaux jusqu'en 1869, conciles pléniers de Baltimore de 1852 et 1866). Nombreux renseignements statistiques et autres dans les *Sedliers' Catholic Directory*, publiés annuellement à New-York, depuis 1832.

II. TRAVAUX. — Bonne synthèse par Th. MAYNARD, *Histoire du catholicisme américain*, trad. franç., Paris, 1948. L'ouvrage fondamental de J.-G. SHEA, *History of the Catholic Church in the United States*, 4 vol., New-York, 1886-1902, s'arrête à l'année 1866. On le complétera par P. GUILDAY, *A History of the Councils of Baltimore (1791-1884)*, New-York, 1932 ; G. SHAUGHNESSY, *Has the Immigrant kept the Faith ? A Study of Immigration and Catholic Growth in the United States*, New-York, 1925 ; G. HERBERMANN, *The Sulpicians in the United States*, New-York, 1916 ; Th. ROEMER, *The Ludwig-Missionsverein and the Church in the United States (1838-1918)*, Washington, 1941 ; J.-B. CODE, *Dictionary of the American Hierarchy*, New-York, 1940.

(2) Sur John England (1786-1842), voir l'excellent ouvrage de P. GUILDAY, *The Life and Times of J. England, first Bishop of Charleston*, New-York, 1927, 2 vol.

irlandaise qui allait le caractériser à partir du milieu du XIXe siècle.

Une autre difficulté provenait de l'esprit d'indépendance de beaucoup de laïques, arrivés en Amérique avec des notions outrées sur la liberté et qui prétendaient s'immiscer dans la direction des paroisses. C'est ainsi que les administrateurs ou *trustees*, chargés par la loi du temporel des églises, n'étaient pas seulement imbus d'idées protestantes sur l'administration des biens ecclésiastiques, mais avaient tendance à empiéter sur les pouvoirs spirituels des curés, voire des évêques. Ce fut un autre prélat d'origine irlandaise, John Hughes, évêque puis premier archevêque de New-York de 1842 à 1866 [1], qui réussit à mettre au point une solution satisfaisante à cette épineuse question du *Trusteism*. Grâce à son énergie et à son influence sur le public, il obtint pour l'État de New-York une loi donnant toute liberté à l'autorité ecclésiastique, tout en maintenant un contrôle modéré de l'administration des fabriques par l'élément laïque, et progressivement les autres États imitèrent celui de New-York [2].

C'est vers 1840, au moment où l'Église des États-Unis achevait de surmonter sa crise de croissance, que la population commença soudain à s'accroître dans des proportions dépassant toute prévision. Le nombre des catholiques, qui avait déjà progressé de 318.000 à 663.000 entre 1830 et 1840, augmenta de près d'un million au cours de la décade suivante, par suite de l'arrivée massive d'émigrants irlandais fuyant la grande famine qui sévissait dans leur pays, et de réfugiés allemands qui s'expatrièrent au moment des troubles de 1848.

RÉORGANISATION DES CADRES ECCLÉSIASTIQUES

Cette augmentation brusque de la population catholique, combinée avec le mouvement qui poussait les colons toujours plus à l'Ouest, posa aux autorités ecclésiastiques une série de problèmes, d'autant plus que précisément à ce moment, de 1845 à 1848, les États-Unis achevaient de se constituer, en englobant au Nord-Ouest l'Orégon et l'Idaho et en s'emparant au Sud du Texas, du Nouveau-Mexique et de la Californie.

Dans les immenses territoires du *Far-West*, tout était à créer et, sur la proposition de l'entreprenant vicaire apostolique, Mgr Blanchet, la Propagande organisa, un peu prématurément, dès 1846, une province ecclésiastique nouvelle, avec Oregon-City pour métropole.

Dans les anciennes provinces mexicaines, tout était à recommencer : l'attitude hostile d'un gouvernement dominé par les francs-maçons, jointe à l'insouciance d'un clergé ignorant et souvent corrompu, avaient fait tomber ces régions théoriquement catholiques à un degré de langueur tel que le nouvel évêque de Santa-Fé, Mgr Lamy, dut commencer par priver tous les prêtres de son diocèse de leurs facultés sacerdotales et par faire appel, pour remplacer ceux qui s'avéraient incorrigibles, à des prêtres venant de France.

---

(1) Sur John Hughes (1797-1864), l'un des plus remarquables parmi les bâtisseurs de l'Église des États-Unis, voir H.-A. BRANN, *Life of Mgr Hughes*, New-York, 1892.

(2) Cf. P.-J. DIGNAN, *A History of the legal Incorporation of catholic Property in the U. S. (1784-1932)*, Washington, 1933.

Dans les États du Centre, encore peu peuplés et où l'on ne rencontrait guère que de petits villages, la situation était restée longtemps celle des pays de mission. Les prêtres étant rares ne pouvaient passer que de loin en loin pour la célébration de la messe et les confessions. Aussi des laïcs étaient-ils souvent désignés pour administrer le baptême ou lire les prières le dimanche ; ils y ajoutaient parfois une petite homélie dont la rigueur théologique n'était d'ordinaire pas la qualité maîtresse, ou certaines pratiques pour le moins étranges, telle cette petite colonie de Slaves où les services étaient agrémentés de coups de grosse caisse et de salves de fusils. Malgré tout, l'influence de ces laïcs fut utile pour maintenir la foi vivante, mais il importait que les autorités reprissent le plus vite possible les choses en main.

Dans les États de l'Est enfin, où se concentraient surtout les émigrants, les difficultés, pour être d'un autre genre, n'en étaient pas moins grandes. Les prêtres, pour la plupart zélés et entreprenants, ne suffisaient plus à la tâche en face de l'accroissement rapide du nombre des catholiques, et la diversité des langues parlées par les émigrants achevait de compliquer les choses [1]. Heureusement, les évêques américains apparaissent dès lors doués au plus haut degré de l'énergie, de l'initiative et de l'esprit pratique qui caractérise les émigrants des Iles britanniques, et d'autre part l'Europe répond avec générosité à leurs appels : tandis que l'afflux des prêtres irlandais, grâce auxquels l'effectif du clergé passe en cinq ans de 737 à 1.303, permet de fonder de nouvelles paroisses, l'arrivée régulière de religieux et religieuses, Français ou Belges pour la plupart, ainsi que les contributions financières des sociétés missionnaires de France, d'Allemagne et d'Autriche, rendent possible la multiplication des écoles et la création dans les ports d'œuvres de charité destinées à accueillir et à soutenir les immigrants.

Cet apport massif d'éléments européens, qui venait enrichir le catholicisme américain, présentait cependant un danger : les nouveaux venus amenaient avec eux leurs coutumes et leurs traditions séculaires très diverses et étaient même parfois tentés de constituer des groupes nationaux distincts. La préoccupation de fondre en un seul corps ces éléments variés, en uniformisant la discipline ecclésiastique, fut l'une des raisons qui poussèrent la hiérarchie à solliciter de Rome l'autorisation de réunir un concile national. La chose était d'autant plus souhaitable que le nombre des diocèses ne cessait d'augmenter : il atteignait déjà 21 en 1846, date à laquelle Baltimore cessa d'être la seule et unique métropole ecclésiastique [2]. Retardé par la maladie de Mgr Eccleston, dont la mort

---

(1) Une lettre du 3 juillet 1847 de la congrégation de la Propagande recommandait de nommer des évêques et des prêtres connaissant la langue de leurs fidèles et attirait spécialement l'attention sur les fidèles de langue allemande : les immigrés de cette catégorie passent en effet de 152.000 en 1840 à 434.000 en 1850. (Cf. D. SHEARER, *Pontificia Americana*, p. 246-249.) L'arrivée de nombreux prêtres allemands et de jésuites expulsés de Suisse en 1847 vint faciliter la solution de ce problème.

(2) On créa successivement les nouvelles provinces d'Oregon-City en 1846, de Saint-Louis en 1847, de New-York, de Cincinatti et de la Nouvelle-Orléans en 1850. Après que San-Francisco fût devenu également métropole, en 1852, à la demande du concile, ces cadres ne se modifièrent plus pendant une quinzaine d'années.

en avril 1851 fut une perte sensible pour la jeune Église américaine, qu'il avait dirigée pendant dix-sept ans, le premier concile plénier de Baltimore réunit enfin, du 8 au 20 mai 1852, sous la présidence du nouvel archevêque, Francis Patrick Kenrick [1], cinq archevêques, vingt-cinq évêques, un abbé trappiste, le supérieur des sulpiciens, très influents depuis l'époque de Mgr Carroll, et les autres supérieurs religieux. C'était la plus importante assemblée ecclésiastique qu'on eût encore vue sur le continent américain. Elle étendit à tous les diocèses la législation canonique qui avait été mise au point par le concile provincial de Baltimore de 1849, attira l'attention sur l'importance des écoles catholiques et régla de manière satisfaisante l'irritante question de l'intervention des laïcs dans l'administration des biens ecclésiastiques.

LA CONSOLIDATION   En 1850, à la veille du premier concile plénier, les catholiques étaient 1.600.000, soit un peu plus de 5 pour cent de la population. Quinze ans plus tard, au moment du second concile, ils seront près de quatre millions, grâce au jeu combiné de la fécondité naturelle et de l'émigration, irlandaise, allemande et canadienne. Il faut toutefois noter que ces catholiques de plus en plus nombreux ne se répartissent pas également sur l'ensemble du territoire : ce sont surtout les États de New-York, Pensylvanie, Ohio, Kentucky, Illinois, Wisconsin, Iowa et Minnesota qui bénéficient de l'émigration catholique, de même que le Texas et, dans une mesure moindre, la côte californienne. La grande majorité de ces catholiques sont de pauvres gens, petits fermiers, et surtout ouvriers, journaliers ou domestiques ; mais, encadrés par un clergé ingénieux et plein de zèle, ils réussissent, à force d'économies, à couvrir le pays d'églises et d'écoles.

Malheureusement, le recrutement sacerdotal continue à rester très déficient malgré les efforts de la hiérarchie pour développer les maisons de formation [2]. La raison devrait en être cherchée, d'après la Pastorale collective de 1866, dans le fait que trop de parents catholiques orientent leurs enfants vers la poursuite de la richesse : cette tentation matérialiste avait déjà été relevée en 1858 par Pie IX qui reprochait aux catholiques américains d'être trop absorbés par les biens de ce monde [3]. Force est, dans ces conditions, de continuer à compter avant tout sur les prêtres venant d'Europe ; en 1857, sur l'initiative de Mgr Spalding, un collège est même ouvert à Louvain pour recruter et former des séminaristes se destinant à l'apostolat aux États-Unis [4]. On arrive de la sorte à tripler

---

(1) Sur F.-P. Kenrick, né à Dublin en 1797, évêque de Philadelphie puis, de 1851 à sa mort en 1863, archevêque de Baltimore, voir O' CONNOR, *Archbishop Kenrick and his work*, Philadelphie, 1867. Son frère Richard (1806-1893) devint en 1840 coadjuteur, puis archevêque de Saint-Louis.

(2) En 1854, il y avait déjà 34 séminaires diocésains. Le Collège américain de Rome date de 1859. En 1848, après plusieurs essais infructueux, avait été ouvert, dans le diocèse de Baltimore, sous la direction des sulpiciens, le premier petit séminaire. (Cf. J.-R. FOLEY, *St Charle's College*, dans *The Catholic World*, t. CLXVIII, 1948, p. 61-67.)

(3) W. ELLIOTT, *Vie du P. Hecker*, trad. franç., Paris, 1897, p. 248-249 et 259.

(4) Cf. J. VAN DER HEYDEN, *The Louvain American College*, Louvain, 1909. Le fondateur et premier recteur du collège fut un Belge, Pierre Kindekens, ancien vicaire général de Détroit. La grande majorité des étudiants (près de 800 prêtres en 50 ans) était composée de Belges, d'Allemands, d'Irlandais et de Hollandais.

en vingt ans le nombre des prêtres, qui passe de 1.320 en 1852 à 2.270 en 1866 et à 3.780 en 1870.

Les anciens colons, d'origine anglaise et puritaine, n'avaient pas vu sans un profond mécontentement la Nouvelle-Angleterre littéralement submergée par les émigrés irlandais, d'autant plus qu'à l'antagonisme religieux venait s'ajouter chez ces « vieux américains » la crainte de voir leurs traditions nationales et leur *standing* de vie compromis par l'arrivée massive de cette main-d'œuvre étrangère à bon marché en quête de travail. Sous le nom de *Native Americanism* et aux cris de « l'Amérique aux Américains ! » ou de « *No Popery !* », une campagne d'agitation s'était développée au début des années quarante. Apaisées pendant quelques temps, les violences anticatholiques reprirent, sous le nouveau nom de *Know-nothingism*, au lendemain du concile de Baltimore, pour atteindre leur point culminant entre 1853 et 1855 [1] : des églises et des couvents furent incendiés, des ecclésiastiques molestés, des manifestations organisées contre l'envoyé du pape, Mgr Bedini, qui venait se rendre compte de la possibilité d'établir une délégation apostolique et qui fut obligé de se retirer devant la tempête après un court séjour [2]. La loi vexatoire, portée en 1854 dans le Massachusetts sur l'inspection des couvents de religieuses, et le *Bloody Monday* du 5 août 1855 à Louisville, montrèrent jusqu'où pouvait aller le fanatisme.

La crainte d'une « agression catholique » était d'autant plus ridicule que, provenant de pays où les catholiques avaient été considérés pendant des siècles en citoyens de seconde zone, et étant au surplus, pour la plupart, de condition sociale modeste, les immigrants n'avaient que trop tendance, au contraire, à se replier sur eux-mêmes et à s'abstenir de tout prosélytisme. C'est parmi les convertis d'origine américaine, touchés entre autres par le mouvement d'Oxford, que naquit le désir d'adopter une attitude moins passive. L'un des premiers à s'engager dans cette voie fut l'essayiste Oreste Brownson, un vrai Yankee, volontiers excentrique et tempêtant à l'occasion contre la « domination irlandaise » dans l'Église, un autodidacte arrogant et entêté, mais, au demeurant, « l'esprit le plus remarquable qu'ait produit le catholicisme américain » [3] ; clamant à temps et à contretemps ce qu'il estimait être la vérité, il justifiait bien la réflexion d'un de ses amis, apprenant que l'évêque de Boston lui avait conseillé de ne pas cacher sa lumière sous le boisseau : « Autant supplier un taureau de ne pas se conduire comme un agneau [4] ! » Dans la revue qu'il avait fondée en 1844, l'année même de sa conversion, et qu'il dirigea jusqu'à sa mort en 1875, la *Brownson's Quarterly*, il ne se contenta pas de prendre à partie la mentalité matérialiste de beaucoup de ses contemporains ou les tendances au latitudinarisme de certains catholiques, mais il s'engagea avec sa brutalité coutumière dans la controverse anti-

---

(1) Cf. R.-A. BILLINGTON, *The Protestant Crusade*, New-York, 1938.

(2) A l'antipapisme protestant vint s'ajouter dans ce dernier cas l'hostilité d'émigrés italiens opposés au pouvoir temporel et à la politique antilibérale de Pie IX.

(3) T. MAYNARD, *Histoire du catholicisme américain*, p. 423. Voir sa biographie par son fils H.-F. BROWNSON, *Brownson's Life*, Détroit, 1898-1900, 2 vol.

(4) Cité dans Th. MAYNARD, *op. cit.*, p. 207.

protestante. Un peu plus tard, en 1859, un autre converti au tempérament entreprenant, Isaac Hecker [1], fondait, avec quelques anciens
rédemptoristes mal à l'aise devant la mentalité trop européenne de leurs
supérieurs, la congrégation des paulistes, destinée à travailler spécialement à la conversion des protestants ; malgré le nombre restreint de
ses membres, elle allait rapidement occuper une place importante dans
la vie catholique des États-Unis.

LA GUERRE DE SÉCESSION   La guerre civile qui, de 1861 à 1865, dressa
les États du Sud contre le reste de l'Union
vint mettre à l'épreuve la solidité de la jeune Église américaine [2]. Celle-ci
n'avait jamais pris nettement position dans la controverse sur l'esclavage.
Dans le Sud, les autorités ecclésiastiques ne s'y montraient pas hostiles,
pourvu qu'il fût humanisé, et même dans le Nord, les catholiques avaient
été irrités par l'alliance entre les abolitionnistes et le mouvement nati-
viste ; les Irlandais, de surcroît, ne voyaient nulle objection à ce que
les nègres fussent tenus à leur place, ce qui explique la participation
de certains d'entre eux aux émeutes qui, en 1863, faillirent livrer New-
York aux troupes sudistes. Cette attitude attira évidemment de nouvelles
critiques contre l'Église. Pourtant, dans l'ensemble, celle-ci sortit de la
guerre avec un prestige accru : la charité agissante des prêtres et des
religieuses, dont le dévouement ne faisait pas de distinction entre sudistes
et nordistes, leur assura le respect unanime de la nation, et l'on admira
comment, seule entre les différents corps religieux, l'Église catholique
sut demeurer unie bien que ses fidèles eussent pris parfois très nettement
position pour un camp ou pour l'autre [3].

Mais si le bénéfice moral fut grand, les pertes furent sévères. L'Église
eut à la fois à souffrir de la destruction ou de l'occupation par les troupes
de nombreux bâtiments ecclésiastiques et de l'ébranlement moral provoqué par cinq années de lutte à outrance et de misères de tout genre.
Dans le Sud, la guerre eut des conséquences particulièrement graves.
La ruine de la vieille aristocratie terrienne en ces régions ne priva pas
seulement le clergé de bienfaiteurs qui s'étaient toujours montrés très
généreux, mais elle fit disparaître le seul élément cultivé de l'Église
des États-Unis. Par ailleurs, les nègres catholiques avaient jusqu'alors
dépendu entièrement de cette aristocratie et leurs anciens maîtres étant
désormais incapables d'entretenir encore des prêtres, les affranchis
catholiques, au moment où ils auraient eu davantage besoin d'aide

(1) Voir sa biographie par Elliott, citée p. 430, n. 3. Né à New York en 1819, converti en
1844 en même temps que Brownson, rédemptoriste de 1845 à 1857, il fut encouragé par Pie IX
lui-même à fonder la « Société missionnaire de l'apôtre saint Paul » qu'il dirigea jusqu'à sa mort
en 1888.
(2) Cf. R.-J. Murphy, *The Catholic Church in the United States during the Civil War Period*,
dans *Record of the American Catholic Historical Society*, t. XXXIX, 1928, p. 271-346.
(3) Certains évêques mirent même leur influence au service de leurs gouvernements respectifs.
Mgr Hughes, le patriote archevêque de New-York, accepta en 1861 de remplir en faveur du gouvernement de Washington une mission non officielle en Angleterre et en France, et contribua beaucoup à l'apaisement des émeutes de 1863. De son côté, l'évêque Lynch, de Charleston, essaya,
mais en vain, d'obtenir du pape qu'il reconnût le gouvernement sudiste. (Cf. L.-F. Stock, *Catholic
Participation in the Diplomacy of the Southern Confederacy*, dans *Catholic Historical Review*, t. XVI,
1930, p. 1-18, où il est également question de la mission de l'abbé Bannon en Irlande.)

spirituelle, se trouvèrent abandonnés à eux-mêmes, sans amarres morales
ou religieuses, tandis que les nègres protestants, ayant des pasteurs de
leur race, souffraient beaucoup moins de la nouvelle situation. En outre,
alors que les relations entre catholiques blancs et noirs étaient cordiales
avant la guerre, et que maîtres et esclaves s'agenouillaient souvent
côte à côte à la Sainte Table, l'amertume qui suivit la guerre amena la
séparation des races même à l'Église, ce qui rendit les contacts encore
plus difficiles. Les noirs, représentant 10 pour cent de la population des
États-Unis, furent pratiquement perdus pour l'Église, malgré quelques
efforts tentés par des religieux étrangers, notamment les Pères de Saint-
Joseph de Mill-Hill, envoyés d'Angleterre en 1871 par le futur cardinal
Vaughan, à la demande du pape lui-même [1].

LE CONCILE DE 1866
ET LA QUESTION SCOLAIRE
Si la pénible question noire resta sans solution [2], dans tous les autres domaines les
catholiques s'empressèrent, sitôt la guerre
finie, de relever les ruines, sous l'active direction d'évêques tels que Mgr
Vérot, l'un des grands artisans de la « reconstruction » dans le Sud, dont
les enfantillages au concile du Vatican ne peuvent faire oublier l'œuvre
remarquable accomplie en cette période troublée [3], ou Mgr Spalding,
archevêque de Baltimore de 1864 à 1872, un Américain de vieille souche,
éminemment représentatif des hommes qui organisèrent l'Église des
États-Unis, unissant en lui une foi solide et une tendre dévotion, les
qualités de l'homme d'étude et une grande capacité administrative,
le zèle pour l'Église et le souci de la prospérité nationale [4].

L'une des premières préoccupations fut de réunir un nouveau concile
plénier, qui se tint à Baltimore en octobre 1866 et dont l'œuvre fut encore
plus décisive que celle du précédent. A côté des problèmes que continuait
à soulever l'augmentation du nombre des catholiques et de l'attitude
à prendre à l'égard des sociétés secrètes, un des éléments caractéristiques de la vie américaine à cette époque [5], on s'y occupa surtout de la
question scolaire.

La constitution ayant garanti la liberté de religion, le ministre de
l'Instruction publique Horace Mann avait d'abord tenté d'organiser

(1) Voir l'ouvrage de leur supérieur le P. SLATTERY, *Our Africa*, Baltimore, 1875.
(2) On pourrait adresser aux catholiques américains un reproche analogue en ce qui concerne
les Indiens : l'évangélisation de ceux-ci fut surtout l'œuvre de religieux étrangers, parmi lesquels
le jésuite belge De Smet et une française, la mère Duchesne, des Dames du Sacré-Cœur, méritent
une mention spéciale.
(3) Venu de France en 1830 comme professeur de séminaire, il était devenu en 1857 vicaire apostolique de Floride et, en 1861, évêque de Savannah. Il ne fut pas seulement un zélé pasteur d'âmes,
mais favorisa avec beaucoup d'intelligence le développement économique de la Floride. (Cf.
O' CONNEL, *Catholicity in the Carolinas and Georgia*, New-York, 1878.)
(4) Cf. J.-L. SPALDING, *Life of Mgr Spalding*, New-York, 1873.
(5) A côté des sociétés secrètes irlandaises, affiliées au mouvement *Fenian*, d'autres groupements
à but philanthropique prirent une extension considérable (tels les *Knights of Pythias*, fondés en
1864). Bien qu'elles n'eussent pas, comme la franc-maçonnerie en Europe, de visées révolutionnaires ou antichrétiennes, elles présentaient un réel danger d'indifférentisme. Les évêques étaient
divisés sur l'attitude à prendre, et le décret du concile de Baltimore, qui tenta d'établir des règles
uniformes, s'avéra à l'usage peu pratique, entre autres parce qu'il eut pour effet de multiplier les
recours à Rome où l'on ne se rendait pas exactement compte de la situation américaine (voir l'ouvrage très documenté de F. McDONALD, *The Catholic Church and the Secret Societies in the U. S.*,
New-York, 1946).

dans les écoles publiques un enseignement religieux non confessionnel ; mais comme la chose s'était avérée difficile en pratique, l'enseignement officiel était devenu assez vite presque laïque. Dès lors, les catholiques jugèrent qu'il leur fallait organiser eux-mêmes, quel qu'en fût le prix, leur propre enseignement et l'archevêque Hughes lança le *leitmotiv*, bientôt adopté par la majorité de l'épiscopat : « L'école avant l'église ! » A l'exemple de ce qui se faisait en Irlande, et malgré la lourde charge que cet effort impliquait, les écoles paroissiales se multiplièrent, surtout après que le concile provincial de Cincinnati de 1858 et le concile plénier de Baltimore de 1866 en eurent fait une grave obligation de conscience pour les curés. Les invitations pressantes et répétées de la hiérarchie aux familles catholiques pour qu'elles évitent dans la mesure du possible d'envoyer leurs enfants dans les écoles officielles furent ratifiées par une déclaration de la congrégation de la Propagande du 24 novembre 1875, qui donnait des directives analogues à celles en vigueur en Angleterre et en Irlande [1].

NOUVEAUX PROGRÈS   Le développement de l'enseignement catholique, limité en fait, malheureusement, jusqu'à la fin du XIXe siècle, au seul enseignement primaire, fut facilité par l'expansion des congrégations religieuses, qui s'accéléra encore après 1870, lorsque le *Kulturkampf* poussa vers les États-Unis de nombreuses religieuses en provenance d'Allemagne. Cette augmentation du nombre des religieux favorisa également l'efflorescence de nombreuses œuvres, anciennes et nouvelles : établissements charitables ou hospitaliers, cercles de toute nature, œuvres de presse, comme la *Catholic Publication Society* des paulistes du P. Hecker.

En même temps qu'il s'affermissait de la sorte, le catholicisme américain continuait à progresser quantitativement : le grand *rush* des pays méridionaux, qui commença vers 1870, amena des centaines de milliers de nouveaux catholiques, et il fallut bientôt ajouter encore huit diocèses aux quatorze nouveaux qui avaient été créés au lendemain du concile.

LE CATHOLICISME AMÉRICAIN   En 1878, le nombre des fidèles approchait
A LA MORT DE PIE IX   des six millions, chiffre d'autant plus remarquable qu'il ne s'agit pas de catholiques « théoriques », comme c'est trop souvent le cas en Europe : aux États-Unis, tant parmi les hommes que parmi les femmes, le catholique ayant abandonné la pratique religieuse reste une exception. Mais la répartition des fidèles sur l'ensemble du territoire est très inégale : alors qu'un Américain sur huit est catholique vers 1875, la proportion n'est que de un sur vingt-cinq — et parfois bien moindre encore [2] — dans les États du Sud, d'où le bon marché de la main-d'œuvre d'origine servile écarte les immigrants, principale source du catholicisme américain.

---

(1) D. SHEARER, *Pontificia Americana*, p. 367 et suiv. L'ouvrage fondamental sur la question scolaire est celui de J.-A. BURNS, *The Growth and Developpement of the Catholic School System in the United States*, New-York, 1912.
(2) Ainsi la Caroline du Nord ne comptait pas mille catholiques sur un million d'habitants en 1868. (A. WILL, *Vie du cardinal Gibbons*, Paris, 1925, p. 41.)

Plus remarquable encore est l'inégale répartition entre villes et campagnes, et le caractère essentiellement urbain du catholicisme aux États-Unis. Cette situation, apparemment paradoxale pour un Européen, s'explique aisément : les Irlandais n'aimaient pas la campagne et surtout la plupart des immigrants étaient trop pauvres pour entreprendre la colonisation agricole et devaient chercher à s'employer immédiatement comme ouvriers ou comme domestiques. La hiérarchie les encouragea d'ailleurs à demeurer en ville. En effet, les fermes, aux États-Unis, sont isolées au milieu de vastes territoires, au lieu d'être concentrées en villages autour d'un clocher ; les protestants peuvent aisément y vivre en lisant leur bible et en recevant de loin en loin la visite du pasteur ; mais il n'en allait pas de même pour les catholiques et le danger était encore accru par le fait que beaucoup de ces gens très simples, qui avaient jusqu'alors été rattachés à l'Église avant tout par leur entourage ou des traditions locales, semblaient incapables de résister à l'influence d'un milieu protestant ou indifférent. On comprend dès lors l'attitude des évêques du XIXᵉ siècle, et le résultat positif de leur manière de faire fut que peu d'immigrants perdirent la foi [1].

Il y avait toutefois une contrepartie, et pour assurer une solution immédiate, on risquait de compromettre l'avenir, pour le jour où le flot de l'immigration se ralentirait, les taux de natalité étant plus bas en ville qu'à la campagne. Il y eut bien quelques essais de fondations de colonies agricoles catholiques : déjà pendant la première moitié du siècle, certains prêtres étrangers avaient poussé à la fondation dans l'Ouest de villages réservés aux catholiques de telle ou telle nationalité [2] ; l'idée fut reprise en 1870 par trois évêques qui fondèrent l'*Irish Catholic Colonization Society* et achetèrent entre autres au gouvernement d'immenses bandes de terrain le long de certains chemins de fer, mais ces tentatives demeurèrent isolées.

Le fait que la source principale du catholicisme aux États-Unis fut le courant d'immigration n'a pas été sans quelques inconvénients. D'abord, si réelle qu'elle soit, la piété américaine conservera longtemps de ses origines irlandaises ou italiennes quelque chose d'un peu extérieur et mécanique. Ensuite, les prêtres venus d'Irlande ont souvent gardé l'habitude de traiter leurs paroissiens comme des enfants mineurs, et s'il y a quelque chose de beau dans la confiance de leurs ouailles, qui s'abandonnent aveuglément à eux non seulement en matière spirituelle, mais même pour leurs affaires temporelles, le peu d'initiative laissée aux laïcs dans le domaine religieux est regrettable.

Le catholicisme américain, concentré dans les ports et les villes industrielles du Nord-Est, présente une allure nettement prolétarienne. Ceci explique en partie le petit nombre des conversions. « Dans un pays où la différence des rangs et des races est observée avec un soin infiniment plus jaloux qu'on ne l'imagine en Europe, la peur de se déclasser a détourné

---

(1) Il est courant d'affirmer le contraire, mais les calculs minutieux de G. SHAUGHNESSY, *Has the Immigrant kept the faith ?* New-York, 1925, ont scientifiquement prouvé qu'il n'en était rien.
(2) Voir M. G. KELLY, *Catholic Immigrant Colonization Projects in the U. S., 1815-1860*, New-York, 1939.

de l'Église des Irlandais et des servantes plus d'un Américain distingué[1]. »
Mais ce peuple indigent sait donner avec générosité, et la cathédrale
Saint-Patrick de New-York, par exemple, construite avec les sous des
gagne-petit, sera, à la fin du siècle, le plus riche édifice religieux de la
ville. Chose étonnante, si le catholicisme aux États-Unis apparaît centré
sociologiquement sur les milieux populaires et non sur la bourgeoisie,
comme c'est de plus en plus le cas en Europe, à cette époque, le clergé et
surtout la hiérarchie — à quelques exceptions près — considèrent pourtant d'un œil méfiant les tentatives d'organisation ouvrière et les revendications sociales[2] : la raison en est que les idées sociales étaient d'ordinaire importées d'Europe par des révolutionnaires anticléricaux, dont
les tendances radicales paraissaient dangereuses pour l'Église.

Le fait que la grande majorité des fidèles appartenaient aux classes
économiquement inférieures, sans culture intellectuelle, joint à la tendance du clergé américain d'organiser des communautés étroitement
refermées sur elles-mêmes, qui n'aient besoin, du berceau à la tombe, de
recourir à aucune institution protestante, empêcha longtemps les catholiques d'exercer dans le pays une influence proportionnelle à leur nombre.
Pourtant, peu à peu, l'ancien ostracisme disparaissait et, vers 1850, le président Lincoln avait même émis le vœu de voir un prélat américain élevé
au cardinalat. A Antonelli, qui trouvait la suggestion ridicule, Pie IX
aurait répondu, paraît-il : « Je suis, parmi les successeurs de l'Apôtre,
le seul qui ait foulé le sol de cette grande Amérique et c'est à moi
qu'appartient l'honneur de créer des cardinaux américains[3]. » Les circonstances retardèrent un peu sa décision, mais, le 15 mars 1875, il conférait
la barette à l'archevêque de New-York, Mgr Mc Closkey, soulignant
ainsi, de manière spectaculaire, la place que commençait à prendre dans
l'Église universelle la jeune Église américaine, qui avait su appliquer
de manière si heureuse, en dehors de toute idéologie, la célèbre formule
de l'Église libre dans l'État libre.

§ 2. — L'Église catholique au Canada[4].

DÉVELOPPEMENT DE L'ÉGLISE   La situation de l'Église dans les pos-
sessions anglaises de l'Amérique du Nord
pendant la première moitié du xixᵉ siècle avait été plus difficile qu'aux
États-Unis, mais heureusement la politique tendant à assujettir l'élément
catholique à l'élément protestant avait pris fin à la suite de l'Acte d'Union
de 1840 et le gouverneur, lord Elgin, qui avait toujours fait preuve

(1) Cte DE MEAUX, *L'Église catholique aux États-Unis*, Paris, 1893, p. 62.
(2) H.-J. BROWNE, *The Catholic Church and the Knights of Labor*, Washington, 1949, chap. I.
(3) D'après L. TESTE, *Préface au Conclave*, Paris, 1878, p. 248.
(4) BIBLIOGRAPHIE. — Il n'existe pas encore d'ouvrage d'ensemble sur l'histoire de l'Église
au Canada pendant le xixᵉ siècle. On trouvera des renseignements succincts dans GEORGES DE
QUÉBEC, O. F. M. Cap., *L'Église catholique au Canada. Précis historique et statistique*, Montréal,
1944, ainsi que dans les articles *Canada* de la *Catholic Encyclopedia*, t. III, p. 234-242 et du *D. H.
G. E.*, t. XI, col. 675-698 ; et des exposés partiels dans TÉTU, *Les évêques de Québec*, Québec, 1889
(notices biographiques) ; ALEXIS, *Histoire de la province ecclésiastique d'Ottawa*, Ottawa, 1897;
MORICE, *Histoire de la province ecclésiastique d'Ottawa dans l'Ouest canadien*, Saint-Boniface, 1921-
1923, 4 vol. Le *Dictionnaire général du Canada*, publié par L. LEJEUNE (Ottawa, 1931, 2 vol.),
est très précieux.

à l'égard des catholiques de dispositions libérales, obtint, en 1851, de la reine Victoria, une reconnaissance explicite de la liberté religieuse. Dès lors, le catholicisme canadien ne cessera plus de faire des progrès réguliers, moins spectaculaires toutefois que dans la grande république voisine, car le pays attire beaucoup moins les immigrants européens.

A l'avénement de Pie IX, en 1846, le Canada ne comptait qu'une province ecclésiastique, divisée en huit diocèses, dont le plus occidental était celui de Toronto. Vers cette époque, des missionnaires en provenance des États-Unis commencèrent à remonter vers le Nord le long du Pacifique et, en 1847, l'un d'entre eux, l'abbé Demers, fut sacré évêque et se fixa dans la région de Vancouver, mais, avant 1890, il n'y eut pratiquement pas de colons européens dans cette région. Dans les régions du Centre et du Nord, également, les colons restèrent très rares jusqu'à la fin du siècle, mais dès avant 1850, les missionnaires, en particulier les oblats de Marie Immaculée, y pénétrèrent en vue de convertir les Indiens et les Esquimaux (le premier baptême d'Esquimau semble avoir eu lieu en 1860).

Le long de la frontière américaine, par contre, les colons avaient commencé à pousser vers l'Ouest et, dès 1820, un évêque d'origine française, Mgr Provencher, s'était établi aux environs de la future ville de Winnipeg, dans la région de la Rivière Rouge qui, en 1846, comptait un peu plus de 5.000 habitants, dont près de 3.000 catholiques, métis franco-canadiens pour la plupart. Lorsque, en 1869, la Confédération canadienne, constituée deux ans plus tôt, racheta à la Compagnie de la Baie d'Hudson les riches plaines du Canada central, ces métis, sous la direction de Louis Riel, tentèrent de s'opposer par les armes à cette annexion. Comme ils étaient catholiques et ouvertement soutenus par leur clergé, leur attitude aurait pu avoir des conséquences dommageables pour l'Église [1]. Heureusement, le gouvernement central se montra compréhensif, grâce à l'habile médiation de Mgr Taché, un homme remarquable qui avait succédé en 1853 à Mgr Provencher et qui devait, au cours d'un épiscopat de plus de quarante ans, devenir le grand organisateur de l'Église de « l'Ouest », réussissant à encadrer et à protéger les immigrants catholiques, dans des conditions difficiles, analogues à celles qui avaient été longtemps celles de la « frontière » aux États-Unis. Ses efforts se matérialisèrent dans la création, en 1871, d'une nouvelle province ecclésiastique divisée en quatre diocèses, lesquels du reste ne comptaient que quelques milliers de fidèles, la population de ces immenses territoires restant très clairsemée jusqu'à la fin du siècle.

Les forces vives du catholicisme canadien restent donc concentrées pendant toute cette période dans la zone maritime et la région du Saint-Laurent. Et même là, il ne faut pas se faire d'illusions : la population totale du pays, aux alentours de 1848, ne dépasse pas les deux millions d'habitants et elle est composée en bonne partie de pionniers, de défricheurs, de trappeurs et de marins, souvent obligés de lutter contre le

---

(1) Sur ces événements qui ont été à tort considérés par des historiens protestants de langue anglaise comme une rébellion contre le gouvernement légitime, voir l'exposé serein de MORICE, *op. cit.*, t. II, p. 237-320.

froid et les bêtes sauvages ; en 1867, à un moment où les États-Unis sont en pleine expansion déjà, le Canada ne possède encore que quatre villes de plus de quarante mille habitants et une vingtaine à peine dépassant les trois mille. Toutefois, en dépit de cette situation peu favorable et malgré le double handicap que constitue la nette prépondérance protestante parmi les immigrés et le départ de nombreux catholiques de la province de Québec vers les plaines plus hospitalières du Nord des États-Unis, le catholicisme se développe régulièrement. Deux facteurs y contribuent : l'immigration d'Irlandais, puis, plus tard, de Polonais, assez vite assimilés par l'élément britannique, ce qui amène la constitution, dans la région du Haut-Saint-Laurent et des Grands Lacs, d'un groupe non négligeable de catholiques canadiens de langue anglaise, rappelant par plus d'un trait le catholicisme américain, avec lequel les contacts iront en se multipliant ; et surtout l'exceptionnel coefficient de natalité des franco-canadiens, chez qui les familles de douze et quinze enfants ne sont pas rares. Aussi, vers 1880, les catholiques approchent des deux millions, sur une population totale de 4.800.000 habitants.

Ces progrès se traduisent par des remaniements des circonscriptions ecclésiastiques. A Terre-Neuve (qui, par la suite, préférera ne pas entrer dans la Confédération), deux nouveaux diocèses furent constitués, en 1847 à Saint-Jean, en 1856 au Havre-de-Grâce et, en 1870, un vicariat apostolique à Saint-Georges. Les catholiques, d'origine surtout irlandaise, représentaient près de 70 pour cent de la population de l'île en 1844, mais l'arrivée d'immigrants anglicans et protestants fera tomber progressivement cette proportion à 35 pour cent [1]. Dans les provinces maritimes de la côte orientale (Nouveau-Brunschwig et Nouvelle-Écosse), on constitua, en 1852, avec les quatre diocèses déjà existants, la nouvelle province ecclésiastique de Halifax, où la vie catholique bénéficia notamment, à partir de 1864, de la forte impulsion du P. Lefebvre, « l'apôtre des Acadiens », l'animateur du collège Saint-Joseph de Memrancook [2]. Dans l'Ontario, la population catholique passa en quinze ans de 78.000 à 258.000 fidèles, grâce en partie à l'afflux de nombreux franco-canadiens en quête de terre ; trois nouveaux diocèses y furent érigés de 1847 à 1859 et, en 1870, Toronto devint à son tour métropole.

De la sorte, les limites de la province ecclésiastique de Québec furent progressivement ramenées à celles de la province civile, mais celle-ci, qui demeura presque intégralement française et catholique, comptait, en 1878, plus d'un million de fidèles au lieu de 600.000 en 1846, et le nombre des diocèses y passa de deux à sept. Ce résultat fut dû entre autres aux efforts du clergé pour orienter vers le défrichement de nouvelles terres, dans la province même, l'excédent de population tenté de s'expatrier au risque de perdre la foi ancestrale : l'abbé Lebelle, qui fonda à lui seul plus de quarante nouvelles paroisses, est le type le plus caractéristique de ces « prêtres colonisateurs » que les évêques encourageaient volontiers [3].

(1) M. F. HOWLEY, *Ecclesiastical History of Newfoundland*, Boston, 1888.
(2) Cf. POIRIER, *Le P. Lefebvre et l'Acadie*, Montréal, 1898.
(3) Cf. LABELLE, *Mémoire sur la colonisation des terres incultes du Bas-Canada*, Québec, 1867 ; *La colonisation dans la vallée d'Ottawa, au nord de Montréal*, 1880.

L'ENSEIGNEMENT CATHOLIQUE Comme dans tous les pays de religion mixte, l'école, et spécialement l'école primaire, fut souvent au Canada un sujet de friction. Dès 1863 pourtant, des compromis assez satisfaisants pour les catholiques étaient intervenus et lorsque, en 1867, l'acte constitutif du Dominion canadien laissa à chaque province le soin d'adopter le système d'éducation de son choix, il leur prescrivit de respecter les droits acquis au moment de l'union par les divers groupes de la population jouissant d'écoles séparées. Parfois cependant, dans les provinces où les protestants étaient en majorité, ceux-ci cherchèrent à priver les catholiques du bénéfice des subsides officiels : ainsi dans le Nouveau-Brunschwig, en 1871, mais la résistance des catholiques fut telle que, dès 1874, ils obtinrent un *modus vivendi* convenable.

Dans l'Ontario, où les efforts de l'Église anglicane pour conserver le contrôle général sur l'enseignement échouèrent devant la politique de sécularisation vigoureusement menée par Ryerson, surintendant de l'Éducation pour le Haut-Canada, de 1844 à 1866, les catholiques purent développer leurs écoles d'une manière satisfaisante sous l'impulsion de Mgr Guigues et de ses oblats de Marie-Immaculée ; ils réussirent même à mettre sur pied un enseignement supérieur confessionnel : en 1866, l'autorisation de conférer les grades académiques était accordée au *Regiopolis College* de Kingstone et, après quelques difficultés, au collège d'expression française d'Ottawa, fondé en 1848 par les Pères oblats, dirigé depuis 1853 avec une compétence exceptionnelle par le P. Tabaret, et appelé à devenir l'une des forteresses du catholicisme franco-ontarien.

La situation était évidemment beaucoup plus facile dans la province de Québec où les catholiques, formant 85 pour cent de la population, étaient les maîtres absolus et où l'Église put s'assurer, directement ou indirectement, la direction de tout l'enseignement, du primaire au supérieur. Tandis qu'un catholique d'élite, le docteur Meilleur, surintendant de l'Éducation à partir de 1842, organisait l'enseignement primaire officiel, les collèges secondaires étaient tous dirigés par des prêtres ou des religieux et, en 1854, cet édifice scolaire fut couronné par l'inauguration, à Québec, en présence de lord Elgin, du Corps législatif et des évêques, de la première université catholique canadienne, dont la fondation, décidée par le concile provincial de 1850, avait été aisément autorisée par la reine Victoria : l'Université Laval [1], dont une succursale devait être ouverte à Montréal en 1876.

L'EFFLORESCENCE DE LA VIE RELIGIEUSE Le rapide développement des œuvres d'enseignement fut rendu possible par l'efflorescence des congrégations religieuses [2], systématiquement favorisée par l'entreprenant Mgr Bourget, évêque de Montréal de 1840 à 1876. A peine installé, celui-ci avait fait venir de France les dames du Sacré-Cœur et les sœurs du Bon Pasteur, les jésuites et

(1) Nom du premier évêque et fondateur du séminaire de Québec ; c'est en effet autour du séminaire et de son supérieur, l'abbé Courtauld, que la nouvelle université se constitua.
(2) Voir les deux ouvrages *Au service de l'Église*, Montréal, 1924 (notices sur les ordres religieux) et *Sur les pas de Marthe et Marie*, sous la direction du P. ARCHAMBAULT, Montréal, 1929 (notices sur les congrégations féminines).

surtout les oblats de Marie-Immaculée, destinés à prendre une place de plus en plus prépondérante dans le catholicisme canadien ; en 1847, il introduisit les clercs de Saint-Viateur et les pères de Sainte-Croix, qui seront vite plus florissants sur le sol américain qu'ils ne l'étaient en Europe. C'est toute la vie catholique, non seulement dans le Bas-Canada, mais au delà, qui bénéficia de ses intelligentes initiatives et bientôt, pour répondre aux besoins multiples d'une Église en pleine expansion, il fonda lui-même plusieurs congrégations proprement canadiennes, qui se répandront par la suite à travers toute l'Amérique : sœurs des Saints-Noms de Jésus et Marie et sœurs de Sainte-Anne, pour l'éducation des jeunes filles ; sœurs de la Providence et sœurs de la Miséricorde, consacrées au soin des malades et des pauvres. Contrairement à ce qui fut longtemps le cas aux États-Unis, le recrutement de ces diverses congrégations, masculines et féminines, comme le recrutement sacerdotal d'ailleurs, est très abondant, grâce à la ferveur et à la générosité morale des franco-canadiens, qui tranche nettement sur l'ambiance matérialiste de l'époque.

LE CATHOLICISME FRANCO-CANADIEN Ce groupe franco-canadien, qui ne compte qu'un bon million d'âmes aux environs de 1870, constitue dans l'Église universelle un cas très typique de société qui demeure, en plein XIXe siècle, intégralement dominée par les principes catholiques, non pas légalement mais réellement. En effet, l'idée dominante du clergé, depuis la fin du XVIIIe siècle, avait été de maintenir les canadiens aussi français que possible afin de les maintenir catholiques en les isolant des Anglo-Saxons protestants ; or, les descendants des colons français se rendirent compte que, dans un pays où l'élite laïque faisait encore totalement défaut, les mesures prises par le clergé pour préserver leurs traditions ancestrales constituaient en fait la meilleure des armatures pour la sauvegarde de leur personnalité ethnique, et ils acceptèrent dès lors aisément de se voir enveloppés dans un réseau serré d'influences ecclésiastiques s'étendant à tous les domaines de la vie privée, familiale, sociale, politique. De la sorte, bien que, officiellement, les protestants eussent conservé une liberté totale, l'atmosphère du milieu canadien français était devenue telle que la vie sociale était moralement impraticable à ceux qui ne voulaient pas accepter, extérieurement du moins, les conditions imposées par l'Église.

Ce « cordon sanitaire », primitivement destiné à servir de sauvegarde contre le protestantisme anglais, fut dans la suite utilisé par l'Église pour résister à l'infiltration de la libre-pensée française et, grâce à son influence omniprésente, elle réussit à éviter presque totalement la pénétration des idées modernes et à maintenir l'ensemble des fidèles dans la docilité la plus complète à l'égard de leurs prêtres, en ville aussi bien qu'à la campagne. Dans ces conditions, bien qu'aucune loi ne restreigne la liberté de la presse, il suffit que les autorités ecclésiastiques désavouent un journal du haut de la chaire pour que son tirage diminue rapidement et beaucoup de journaux préfèrent dès lors se soumettre spontanément à une sorte de censure préalable.

Ce régime, où l'Église, avec un minimum de privilèges officiels [1], jouit d'une autorité morale à peu près incontestée parce qu'elle a su se rendre indispensable, ne va pas toutefois sans de réels inconvénients, car on aboutit vite au cléricalisme. Celui-ci se manifeste lors des élections, où les interventions du clergé sont fréquentes et patentes, si patentes que Rome jugera prudent d'inviter les évêques à plus de discrétion [2] ; mais il se manifeste en bien d'autres circonstances encore : ayant été amené pour le salut de la race à assumer des fonctions excessives, le clergé n'entend plus partager, même lorsque les conditions redeviennent plus normales ; il continue à ne tolérer aucune action laïque indépendante de lui ; s'il en naît une, il s'y glisse et s'y impose, ou s'arrange pour l'empêcher de s'exercer efficacement. L'exemple des bibliothèques publiques que tentèrent d'organiser sous le nom d'*Institut canadien* de jeunes catholiques de tendance libérale est caractéristique : devant le succès de l'entreprise, l'Église leur opposa immédiatement des *Instituts nationaux* et Mgr Bourget refusa en 1863 d'accepter la transaction qu'on lui proposait, à savoir que l'autorité épiscopale désignerait les mauvais livres à retirer de la circulation ; c'est la suppression pure et simple qu'il désirait et qu'il finit par obtenir. On comprend aisément qu'il n'est dès lors pas question de voir s'épanouir au Canada une culture catholique qui ne soit pas strictement conformiste.

L'Église franco-canadienne au XIXe siècle, très influencée par les traditions héritées de la « vieille France », se caractérise encore par une tendance un peu jansénisante qui donna naissance à un « catholicisme plus volontariste que sage, plus raisonneur qu'intelligent, plus moralisateur que nourrissant, plus satisfait de supprimer que d'incorporer, plus traditionaliste qu'accueillant, plus routinier qu'avide de valeurs humaines à sanctifier » [3]. Mais sur un point important, le clergé canadien rompt avec les traditions de la France du XVIIIe siècle : les anciennes sympathies gallicanes ne tardent pas à faire place à une attitude de totale soumission à l'égard du Saint-Siège et l'on juge sévèrement les allures plus réservées, parfois plus indépendantes à l'égard de Rome, de certains évêques des États-Unis. L'ultramontanisme est d'autant plus décidé que Rome apparaît sous Pie IX comme la citadelle de la résistance aux idées libérales abhorrées par la hiérarchie ; celle-ci, en 1875, opposera à la formule *L'Église libre dans l'État libre* l'affirmation d'une thèse qui, en dépit de la séparation légale, est aussi en fait l'hypothèse dans le Québec : « L'Église n'est pas seulement indépendante de la société civile, elle lui est supérieure par son étendue et par sa fin... Ce n'est pas l'Église qui est dans l'État, c'est l'État qui est dans l'Église [4]. »

---

(1) Car on ne peut pas parler d'un véritable régime de séparation pour la province de Québec comme l'a bien montré A. SIEGFRIED dans son livre perspicace, encore qu'un peu tendancieux, *Le Canada. Les deux races*, Paris, 1906, p. 12-13.

(2) Voir différents exemples dans l'ouvrage, assez partial, de Ch. LINDSEY, *Rome in Canada. The Ultramontane Struggle for supremacy over the Civil Authority*, Toronto, 1877, et dans celui de J. WILLISON, *Sir Wilfrid Laurier and the liberal party*, t. I, Toronto, 1926.

(3) *L'Action nationale* (Montréal), mars 1950, p. 178.

(4) Lettre pastorale collective de l'épiscopat de la province de Québec du 22 septembre 1875.

§ 3. — L'Église catholique en Amérique latine [1].

TRAITS GÉNÉRAUX   Jusqu'à la fin du XIXe siècle, la situation du catholicisme tant au Brésil que dans les républiques issues du démembrement de l'empire colonial espagnol présente un lamentable contraste avec la partie septentrionale du continent américain.

A l'exception des tribus indiennes, peu nombreuses, restées attachées au paganisme de leurs ancêtres, la quasi-totalité de la population y est nominalement catholique et beaucoup de constitutions reconnaissent même le catholicisme comme religion d'État. Mais la réalité est profondément déconcertante. La grande masse des métis et des descendants des colonisateurs semble plus superstitieuse que chrétienne [2] et l'appoint des immigrants, catholiques pour la plupart cependant, n'apporte pas comme dans l'Amérique septentrionale un sang nouveau à ces Églises en décadence, car les nouveaux venus se laissent vite gagner par la tiédeur générale et la nonchalance à l'égard de la pratique religieuse.

La responsabilité tient en partie à des conditions de vie qui ne favorisent guère la trempe des caractères, mais elle revient pour une large part aux insuffisances du clergé. Insuffisance numérique d'abord, à laquelle les évêques ne cherchent guère à remédier en faisant appel, comme leurs confrères des États-Unis, à des prêtres venant d'Europe. Des catégories entières de chrétiens se trouvent dès lors dans l'impossibilité matérielle de recevoir l'enseignement de la doctrine chrétienne, de participer au culte et de s'approcher des sacrements. Et cette disette de secours spirituels réguliers est particulièrement néfaste dans une société en perpétuel renouvellement, que ne soutiennent pas de solides traditions morales ou sociales.

Ces prêtres trop peu nombreux manquent en outre de dynamisme et de zèle et, d'une manière assez générale, mènent une vie très relâchée. Le clergé régulier surtout est en pleine décadence depuis la révolution, qui a bouleversé ses conditions de vie, mais si le clergé séculier lui est généralement quelque peu supérieur, il est loin d'être à la hauteur de sa tâche. Le prêtre « isolé de ses confrères, vivant à six ou sept journées de marche d'un autre prêtre, habitant seul au milieu de populations de mœurs plutôt faciles, court trop de dangers et a trop peu de secours pour ne pas se laisser entraîner sur la pente des passions » [3]. On ne faisait d'ailleurs

---

(1) BIBLIOGRAPHIE. — I. SOURCES. — F.-J. HERNAEZ, *Coleccion de Bulas, Breves y otros Documentos relativos a la Iglesia de America y Filipinas*, Bruxelles, 1878, 2 vol.

II. TRAVAUX. — V.-L. TAPIÉ, *Histoire de l'Amérique latine au XIXe siècle*, Paris, 1945 ; J.-T. BERTRAND, *Histoire de l'Amérique espagnole*, t. II, Paris, 1929 ; J. CORREDOR DE LA TORRE, *L'Église romaine dans l'Amérique latine*, Paris, 1910 ; J.-C. ZURETTI, *Historia eclesiastica argentina*, Buenos-Aires, 1945 ; E. BADARO, *L'Église du Brésil pendant l'Empire et pendant la République*, Rome, 1895 ; *La provincia eclesiastica de Chile*, Barcelone, 1891 ; J.-P. RESTREPO, *La Iglesia y el Estado en Colombia*, Londres, 1885 ; M. TALAVERA Y GARCIA, *Apuntes de historia eclesiastica de Venezuela*, Caracas, 1929 ; M. CUEVAS, *Historia de la Iglesia en Mexico*, t. V, Tlalpam, 1928 ; J. BRAVO-UGARTE, *Diocesis y obispos de la Iglesia mexicana, 1519-1939*, Mexico, 1941 ; R. PEREZ, *La Compañia de Jesùs restaurada en la Rep. Argentina, y Chile, Uruguay y Brazil*, Barcelone, 1931.

(2) Ceci vaut *à fortiori* pour les nègres amenés jadis par la traite au Brésil ou dans les Antilles, dont un bon nombre sont devenus catholiques tout en conservant pas mal de leurs anciennes pratiques fétichistes.

(3) Art. *Amérique latine*, dans *D. T. C.*, t. I, col. 1100.

aucune difficulté à Rome pour reconnaître les déficiences du clergé sud-américain tant dans sa vie privée que dans ses rapports avec les autorités civiles, où il fait parfois montre de prétentions exagérées. C'est ce que notait en 1870 un diplomate au sortir d'une conversation avec Antonelli : « Il s'est étendu avec moi volontiers sur les graves reproches qu'on peut adresser trop souvent au clergé de l'Amérique du Sud et plus particulièrement encore, a-t-il ajouté, au clergé brésilien, dont l'attitude répréhensible fixe la sérieuse attention du Saint-Siège [1]. »

La plupart des États de l'Amérique latine sont en outre le théâtre de poussées périodiques d'anticléricalisme. Tantôt elles sont le fait de gouvernants imbus de principes régaliens hérités des siècles antérieurs ou des idées « laïques » venues d'Europe, qui entrent en conflit avec un clergé préoccupé à juste titre de sauvegarder l'indépendance et les droits de l'Église, mais trop attaché parfois à des privilèges qui apparaissent de plus en plus anachroniques. Tantôt, elles sont provoquées par « une lutte que l'Europe connaît également, mais à un degré moins aigu et surtout avec infiniment plus de nuances intermédiaires, entre l'idéologie catholique favorable au gouvernement autoritaire d'une élite, et l'idéologie libérale, nourrie par les loges maçonniques, hostile au principe d'autorité sous toutes ses formes, poussant jusqu'aux extrêmes conséquences les droits personnels de l'individu » [2]. Parfois enfin, ces conflits, envenimés par un rationalisme sectaire, se transforment en une véritable lutte contre la foi chrétienne elle-même que commence à rejeter une partie de la bourgeoisie cultivée très influencée, au Brésil et au Mexique notamment, par le positivisme de Comte [3].

LA SOLLICITUDE DE PIE IX Cette triste situation préoccupait d'autant plus Pie IX qu'il avait conservé de son voyage en Amérique du Sud un intérêt très vif pour ces régions lointaines et, pour tenter d'y remédier, il multiplia les initiatives tout au long de son pontificat au point de se voir désigner parfois comme « le » pape de l'Amérique latine [4] : consolidation de la hiérarchie et multiplication des diocèses, afin de faciliter les contacts des pasteurs avec leurs fidèles ; signature de conventions et concordats en profitant des périodes où les hommes au pouvoir étaient favorables à l'Église [5] ; envoi de délégués apostoliques afin de resserrer les liens du clergé local avec Rome et d'améliorer les relations avec les gouvernements ; érection à Rome, en 1858, d'un séminaire confié aux jésuites, le *Collegio Pio latino americano*, destiné aux clercs de l'Amérique latine fréquentant les universités romaines.

Cette sollicitude ne put toutefois porter tous ses fruits par suite des

---

(1) Dépêche de l'ambassadeur de France du 18 janvier 1870, *Arch. Min. A. É. Paris*, Rome, 1045, f⁰ 119.
(2) V.-L. Tapié, *Histoire de l'Amérique latine au XIXᵉ siècle*, p. 120.
(3) Pour le Mexique, voir C. Zea, *El positivismo en Mexico*, Mexico, 1943 ; *Apogeo y decadencia del Positivismo en Mexico*, Mexico, 1944.
(4) Cf. P. de Leturia, dans *Gregorio XVI. Miscellanea Commemorativa*, t. II, Rome, 1948, p. 352.
(5) Concordat avec le Guatémala et Costa-Rica en 1853 ; avec Haïti en 1860 ; avec le Venezuela et le Honduras en 1861 ; avec San Salvador en 1862 ; avec le Nicaragua et l'Équateur en 1863 ; avec le Pérou en 1874.

vicissitudes politiques. En outre, les années qui suivirent 1870 virent un peu partout une recrudescence de l'anticléricalisme, car les éléments libéraux s'inquiétèrent comme en Europe du progrès des idées ultra-montaines mis en lumière à l'occasion du concile du Vatican : le renfor-cement de l'autorité pontificale et l'affirmation plus nette des droits de l'Église dans la société leur paraissait d'autant plus redoutable qu'ils ne pouvaient pas opposer comme en Europe au prestige du clergé sur les masses l'indépendance d'un État fort ; ils ne voyaient dès lors de défense efficace que dans des mesures d'exception qui n'étaient parfois que vexatoires, mais qui prenaient facilement une allure de persécution.

LES AVATARS DE L'ÉGLISE MEXICAINE — L'excessive richesse et les mœurs relâchées du clergé et des moines au Mexique ren-daient indispensable une réforme dont Pie IX chargea Mgr Nunguia, « une des plus hautes personnalités religieuses du Mexique contemporain » [1]. Mais, tandis que celui-ci essayait vaine-ment de venir à bout des résistances opposées par les bénéficiaires des abus, la situation politique évolua et le pouvoir passa en 1855 aux démocrates, qui s'empressèrent de supprimer la plupart des privilèges du clergé, de confisquer les propriétés ecclésiastiques qui n'étaient pas directement exploitées par les curés pour les besoins du culte ou les œuvres charitables, puis de disperser les ordres religieux, d'enlever l'enseignement à la surveillance de l'Église et de décréter le mariage civil. L'âme de cette politique anticléricale était Benito Juarez, un Indien tenace et astucieux, jurisconsulte pénétré des principes régaliens qui avaient inspiré la constitution civile du clergé ; il ne renia jamais formellement son catholicisme d'origine, mais, la foi dût-elle pâtir de la guerre menée contre les hommes d'Église, il entendait abattre la puis-sance d'un clergé qui depuis longtemps se mêlait à la politique « parfois de la manière la plus compromettante » [2].

Le clergé, soutenu par une partie de la population qui criait au sacri-lège, tenta de faire front contre les mesures anticléricales systématisées dans les « lois de Réforme » de 1859, en appuyant les généraux conser-vateurs, puis, lorsque ceux-ci eurent été définitivement vaincus après trois ans de guerre civile, en soutenant l'intervention française qui aboutit en 1864 à l'avénement de l'empereur Maximilien. La hiérarchie, dont une partie des membres avait été exilée, put dès lors être réorganisée et la situation religieuse s'améliora, mais le couple impérial ne tarda pas à décevoir les catholiques les plus ardents. Les propos de l'impératrice Charlotte, empreints des idées libérales de son père le roi des Belges, déconcertèrent ceux qui s'attendaient à trouver en cette princesse catho-lique une « Mère de l'Église » conforme à leurs vues intégristes. Quant au clergé, il fut fort mécontent de voir le nouveau gouvernement ratifier les ventes de biens d'Église et proclamer la liberté des cultes, dans l'espoir de rallier les libéraux modérés. Encouragé à l'intransigeance par le nonce

(1) J.-T. BERTRAND, *Histoire de l'Amérique espagnole*, t. II, p. 204.
(2) *Ibid.*, p. 203.

Meglia, qui refusait d'envisager un concordat analogue au concordat français de 1801, l'archevêque de Mexico, Mgr Labastida, prit la tête de la résistance au nom du droit canon.

Mais la chute de Maximilien, en 1867, ramena au pouvoir les jacobins, plus montés que jamais contre l'Église et bien décidés à réduire le clergé au rang de parias. Non seulement les lois de Réforme furent immédiatement renforcées, mais en 1874, après la mort de Juarez, son ami et successeur Sebastien Lerdo de Tejada, qui avait l'âme encore plus sectaire, décida de les incorporer à la constitution et dès lors le mariage civil, la laïcité de l'enseignement à tous les degrés, la séparation de l'Église et de l'État et la nationalisation des biens d'Église, devenus articles de la Loi fondamentale mexicaine, furent considérés comme intangibles. A la vérité cependant, l'attachement d'une grande partie de la population au catholicisme, pourvu qu'il ne fût pas trop exigeant, et la grande générosité des fidèles rendirent la situation de fait moins tragique qu'elle ne le semblait légalement.

L'AMÉRIQUE CENTRALE ET LES ANTILLES   Le Mexique fut, de tous les États de l'Amérique espagnole, celui où le conflit avec l'Église, compliqué du reste d'une rancœur raciale des Indiens contre les Espagnols, prit les proportions les plus graves et se révéla le plus durable. Ailleurs, l'alternance fréquente des partis au pouvoir eut pour conséquence une politique religieuse tour à tour pleine de déférence à l'endroit de l'autorité ecclésiastique, puis sectaire et persécutrice. Tel fut le cas notamment dans les petites républiques de l'Amérique centrale. Rome réussit à conclure entre 1853 et 1863 une série de concordats satisfaisants, mais après 1870 la situation évolua avec le retour au pouvoir des libéraux au Guatémala et par suite des intrigues des États-Unis au Nicaragua et des protestants anglais à Costa-Rica.

Par contre, les îles des Antilles réservèrent au souverain pontife des satisfactions plus durables. A Cuba, la seule possession que l'Espagne eût conservée en Amérique, l'Église jouit jusqu'à la fin du siècle de la protection gouvernementale et, si sa situation matérielle était plus brillante que la situation religieuse proprement dite, elle se ressentit cependant favorablement des efforts de saint Antoine Marie Claret, nommé archevêque en 1850. Vers le même temps, le président de la république nègre de Haïti, Fabre Geffrard, qui estimait qu'il n'y avait de véritable civilisation que française et catholique, conclut en 1860 un concordat suivi, en 1861, par la création de cinq diocèses confiés à des missionnaires français ; il en résulta d'heureux changements dans la conduite et la culture intellectuelle d'un clergé jusque-là très inférieur à sa tâche.

L'EMPIRE BRÉSILIEN   Au Brésil, l'État le plus peuplé et le plus étendu de l'Amérique du Sud, l'Église connut une tranquillité relative sous le long règne de l'empereur philosophe Pedro II (1831-1889), libéral de tendance, mais croyant respectueux de la religion catholique. La situation était toutefois loin d'être satisfaisante. L'alliance officielle de l'Église et de l'État, qui rendait d'indéniables services, se trouvait « grevée de servitudes abusives et stérilisantes » [1] par suite de la tradition de Pombal, aggravée par le libéralisme régalien du XIXᵉ siècle. Au congrès de Malines de 1864, le délégué brésilien s'en plaignait :

> Chez nous aussi, il y a dans le gouvernement de ces hommes nourris des pré-jugés de la vieille Europe, qui, tout en parlant beaucoup de liberté, ne peuvent se faire à l'idée d'accorder cette liberté à l'Église [2].

Aussi dans un mémoire publié en 1900 à l'occasion du quatrième centenaire de la découverte du Brésil, le P. Julio Maria pouvait considérer l'empire comme une période de décadence pour le catholicisme :

> Rationalistes, matérialistes et sceptiques, tels étaient en majorité les hommes qui dirigeaient la société brésilienne pendant les dernières années de l'Empire. Au point de vue social, vraiment prodigieux fut le nombre d'hôpitaux, d'asiles, d'orphelinats, mais, exception faite de ce développement des œuvres de charité, qui révèle le cœur du peuple brésilien, le catholicisme ne prit durant la monarchie aucun essor ni même aucune initiative, en dehors des actes individuels de la foi et des cérémonies du culte [3].

Il faut surtout déplorer la profonde décadence du clergé, plus profonde que dans le reste de l'Amérique latine. Les anciens ordres, bénédictins, carmes et franciscains, soumis à la tutelle de l'État et soustraits à tout contrôle effectif des autorités religieuses, étaient « victimes d'abus intérieurs qui rendaient inévitable leur disparition prochaine à moins d'une réforme radicale » [4] ; les légistes brésiliens leur interdirent du reste de recevoir encore des novices, afin d'accélérer l'extinction de ces communautés qui avaient cessé, à leur avis, de rendre les services correspondant à leur institution primitive, et dans l'espoir de voir l'État entrer bientôt en possession de leur immense patrimoine destiné à tomber en déshérence. Par contre les congrégations religieuses étrangères, particulièrement celles vouées à l'enseignement et aux œuvres de charité, accueillies volontiers, purent se développer paisiblement et travailler à la formation progressive d'une élite catholique dont le pays avait le plus grand besoin.

La situation du clergé séculier, encore moins nombreux au Brésil qu'ailleurs, était du reste déplorable à tout point de vue, au témoignage des évêques eux-mêmes : « L'honneur de Dieu et l'honneur de l'Église réclament une réforme aussi prompte que radicale » [5], gémissait en 1865 l'évêque de Para, qui se plaignait un peu plus tard de ce qu'on trouvait dans certains diocèses des séminaristes qui ne s'étaient plus confessés depuis quatre ans. L'archevêque de Rio de Janeiro confiait de son côté : « Les prêtres âgés de plus de 35 ans sont presque tous dans un état moral qui ne donne pas d'espérance ; plusieurs ne disent pas de bréviaire, un grand nombre ont femme et enfants publiquement [6]. » La racine du mal

---

(1) Cité par J. SERRANO, art. Brésil, dans D. H. G. E., t. X, col. 578-579.
(2) Communication de M. Macedo y Maia dans Assemblée générale des catholiques en Belgique, IIᵉ Session, 1864, t. I, p. 393.
(3) Memoria historica (1900), cité dans D. H. G. E., art. Brésil, t. X, col. 578-579.
(4) Y. DE LA BRIÈRE, Au Brésil, Bruges-Paris, 1929, p. 62.
(5) Lettre à Dupanloup du 10 mars 1865, Arch. S. Sulp., fonds Dupanloup.
(6) Paroles recueillies par M. Icard à Rome en décembre 1869. (Journal de mon voyage à Rome, p. 37, Arch. S. Sulp., fonds Icard.)

se trouvait pour une part dans les séminaires et certains prélats firent appel à des religieux européens pour y remédier, mais ils n'avaient pas toujours leur liberté de mouvement. Vers 1860, le gouvernement avait envisagé de soumettre à son contrôle les auteurs employés pour l'enseignement et, en 1870, un évêque se trouvait en conflit avec les autorités civiles pour avoir révoqué deux professeurs de son grand séminaire affiliés à la franc-maçonnerie [1].

Cette question de la franc-maçonnerie fut l'occasion en 1872 d'un conflit assez aigu. La maçonnerie était alors au Brésil ce qu'elle avait été en Europe au XVIIIe siècle, un foyer d'éclectisme intellectuel et d'idées « progressistes » au sens le plus vague du terme, et beaucoup ne voyaient pas d'incompatibilité entre leur foi catholique et leur adhésion à la secte. Mais les loges maçonniques, où l'influence des idées de Pombal restait vivante, se trouvaient être aussi des foyers de régalisme et, au lendemain du concile du Vatican, quelques prêtres ultramontains commencèrent à réagir. Ils furent appuyés par le jeune évêque réformateur d'Olinda, Mgr Vital de Oliveira, et par l'évêque de Para, Mgr Antonio de Macedo Costa, un orateur distingué, qui avait nourri au contact de ses relations françaises, de Dupanloup entre autres, son désir de libérer l'Église de l'emprise du pouvoir séculier. En 1872, ces deux évêques obligèrent leurs diocésains à quitter la franc-maçonnie et prirent des sanctions contre les récalcitrants en dépit des protestations des milieux « régaliens », où l'on arguait du fait que les documents pontificaux condamnant la franc-maçonnerie n'avaient pas reçu le *placet* gouvernemental. Approuvé par tous ceux qui s'inquiétaient du progrès des idées ultramontaines, le président du Conseil, Rio Branco, qui était lui-même maçon et catholique pratiquant, fit prononcer par le Conseil d'État l'annulation des décisions épiscopales, et comme les prélats ripostaient en déniant au pouvoir civil le droit d'intervenir dans les questions religieuses, il les fit emprisonner, au grand scandale de la presse catholique mondiale qui salua en eux des martyrs de la lutte pour l'indépendance de l'Église [2]. Les autres évêques n'appuyèrent leurs deux collègues qu'avec beaucoup de tiédeur, mais cette affaire fut le point de départ d'un mouvement d'opinion chez les plus énergiques des catholiques brésiliens pour revendiquer, contre l'intrusion abusive de l'autorité civile, les franchises spirituelles de l'Église.

EN ARGENTINE   Dans la grande république du Rio de la Plata, où l'Église avait eu beaucoup à souffrir sous la dictature de Rozas, la chute de celui-ci, en 1852, ouvrit des perspectives meilleures. Les négociations entamées en vue d'un concordat n'aboutirent pas, mais du moins la constitution de 1855 contint plusieurs clauses très favorables et le président Mitre favorisa ouvertement le catholicisme. Toute-

(1) Lettre de Rome de Mgr Boudinet, du 1er décembre 1869, dans *Le Dimanche* (*Semaine religieuse d'Amiens*), 10 mai 1925, p. 316.
(2) Cf. M.-C. THORNTON, *The Church and freemasonry in Brazil, 1872-1875. A study in regalism*, Washington, 1948 ; LOUIS DE GONZAGUE, *Mgr Vital* (*Ant. G. de Oliveira*), *frère mineur capucin*, Paris, 1912.

fois, à part quelques chrétiens d'élite comme José Manuel Estrada, le « Louis Veuillot argentin » et le plus éloquent défenseur des droits des catholiques, l'ensemble de la population, peu soutenue d'ailleurs par un clergé très libre d'allures, continua à faire preuve d'une profonde indifférence religieuse. Les congrégations européennes toutefois, qui purent se développer librement, eurent une excellente influence, en particulier dans le domaine de l'enseignement, mais leur action n'atteignit guère que la classe riche.

Comme dans la plupart des autres États américains, l'anticléricalisme connut un renouveau après 1870. La presse libérale, soutenue par les loges, prit notamment prétexte d'une lettre pastorale de Mgr Aneiros, archevêque de Buenos-Aires, pour déclencher une violente campagne contre l'Église, qui atteignit son point culminant en 1880, sous la présidence du général Roca.

AU CHILI   La situation du catholicisme au Chili, sans être excellente, se présente dans de meilleures conditions. Diego Portales, ministre quasi inamovible du parti conservateur pendant plus d'un quart de siècle, sans être personnellement un chrétien exemplaire, voyait dans la religion la garantie de l'ordre public et s'appliqua à renforcer l'influence du catholicisme dans les écoles publiques, favorisant l'influence des congrégations venues d'Europe comme les jésuites ou les dames du Sacré-Cœur. L'arrivée au pouvoir des libéraux en 1871, coïncidant avec un afflux d'émigrés protestants, eut pour conséquence un assouplissement de l'ancienne législation dans un sens favorable à la tolérance religieuse, et certaines mesures de laïcisation telles que la suppression de la juridiction ecclésiastique ou l'introduction du mariage civil. Les catholiques s'inquiétèrent et organisèrent pour se défendre un parti catholique d'inspiration ultramontaine, mais ils n'eurent pourtant guère à souffrir des gouvernements libéraux dont le principal inspirateur, le positiviste Victoriano Lastarria, estimait que partout où la religion catholique était une institution nationale, il importait de la maintenir et de la favoriser.

A ces circonstances politiques favorables s'ajoute l'action d'un clergé de qualité bien supérieure au reste de l'Amérique latine, recruté dans les meilleures familles chiliennes et formé dans des collèges et des séminaires bien tenus, tel le grand séminaire de Santiago, fondé par Mgr Valdivieso, bientôt célèbre dans tout le continent sud-américain.

LES ÉTATS ANDINS DU NORD   Les catholiques connurent par contre de mauvais jours dans plusieurs des républiques de l'Ouest et du Nord. Si au Pérou le projet de loi visant à restreindre les privilèges et les libertés de l'Église dut être retiré devant les protestations de l'opinion publique et grâce en particulier à l'influence des dames de la haute société, en Colombie, le général Mosquera, qui gouverna de fait le pays durant plus de vingt ans, déchaîna contre les privilèges de l'Église une guerre acharnée qui s'inspirait de l'exemple mexicain : après avoir nationalisé les biens ecclésiastiques, supprimé la dîme, confisqué les couvents, il proclama la liberté absolue des cultes bien qu'il n'y eût pas un dissident dans le pays et n'hésita pas à laisser expulser du territoire son frère, l'archevêque de Bogota, qui avait protesté contre le décret abolissant le for ecclésiastique. Les jésuites, considérés comme les soutiens du parti antifédéraliste et absolutiste, furent particulièrement en butte à la haine du président.

Le Venezuela connut d'abord une période assez favorable du point de vue religieux et le méritant archevêque de Caracas, Mgr Guevara y Lira, parvint même à faire conclure en 1861 un concordat avec Rome. Mais la situation se modifia du tout au tout avec l'arrivée au pouvoir du libéral Guzman-Blanco, adversaire fanatique de l'Église, qui s'apparentait par bien des traits avec ceux qui à la même époque entamaient la lutte pour l'école laïque dans la France républicaine ou déclenchaient le *Kulturkampf* en Allemagne et, à l'instar de ces derniers, tandis qu'il expulsait les prélats légitimes, il essaya d'organiser une Église nationale totalement dépendante de l'État, dont les fidèles éliraient eux-mêmes leurs curés et leurs évêques.

L'ÉQUATEUR SOUS GARCIA MORENO — Les déboires répétés de l'Église, du nord au sud de l'Amérique latine, furent une source fréquente de douleur pour Pie IX, qui protesta à plusieurs reprises avec indignation contre une politique anticléricale qui l'indignait d'autant plus qu'elle était le fait de pays officiellement catholiques. Il eut toutefois une compensation pendant les quinze années où Garcia Moreno [1] présida aux destinées du petit État d'Équateur, en s'attachant à faire de son pays une république chrétienne idéale.

Personnalité exceptionnelle, idéaliste et homme d'action tout ensemble, Garcia Moreno tenait de ses ancêtres l'esprit de domination et l'intransigeance de la foi catholique qui caractérisent les Castillans. Mais ce serait une erreur de le considérer comme un réactionnaire se bornant à lutter avec les souvenirs du passé contre la montée de l'esprit moderne et à tenter de prolonger sans changement la situation acquise par l'Église au cours de l'ancien régime. Il savait que l'Église équatorienne était faible et corrompue et il se mit directement à l'école de Rome, maîtresse de vérité religieuse, morale et sociale. Avec une audace qui ignorait les compromissions auxquelles se sentaient obligés les catholiques européens, cet Américain avait décidé de faire passer intégralement dans les faits la doctrine exposée dans les encycliques de Pie IX et spécialement dans le *Syllabus*. De là le caractère du concordat qu'il signa en 1862, où le gouvernement affranchissait le clergé de toute tutelle laïque, renonçant aux droits de *Patronato*, de *placet* et d'*exequatur*, que tous les États sud-américains avaient jalousement conservés de l'héritage colonial ; l'État y acceptait en outre que l'*Index* eût force de loi, même dans les universités officielles, et qu'aucune société interdite par l'Église ne fût admise dans

(1) Sur Gabriel Garcia Moreno (1821-1875), voir A. BERTHE, *Garcia Moreno*, Paris, 1888, 2 vol. et R. PATTEE, *G. Garcia Moreno y el Ecuador de su tiempo*, Quito, 1941.

le pays, s'engageant par contre à favoriser les congrégations religieuses, lesquelles, parce que dépendant directement de Rome, apparaissaient à Moreno comme le meilleur gage de régénération.

En 1859, il publia une nouvelle constitution où les « extravagants privilèges accordés à l'Église » [1] — seuls les catholiques pouvaient être citoyens de l'Équateur — paraissaient un défi à l'opinion contemporaine, et qui frappe par « cette allure hiérarchique, constructive et autoritaire, traduisant les incompatibilités d'un ordre catholique avec les relâchements du pouvoir, l'amenuisement des contraintes disciplinaires, l'éparpillement des responsabilités, le nivellement des individus, tout ce qu'à l'extrême peut impliquer une constitution fédéraliste, démocrate ou populaire » [2].

En même temps, au nom des mêmes principes catholiques, parce que la volonté de Dieu implique que l'on mette en œuvre les biens que sa Providence a concédés à l'homme, le président se préoccupait d'améliorer les conditions de vie dans l'État, de développer la prospérité économique, d'assurer par le travail le relèvement des pauvres et l'adoucissement des misères.

Bref, le régime de Garcia Moreno, qui n'hésita pas à consacrer officiellement son pays au Sacré-Cœur en 1873, « offre à l'histoire le type d'un État catholique, non point par fidélité au passé, mais par l'efficacité d'une foi vivante, animant un programme cohérent et doctrinal » [3]. Mais on comprend à quel point le gouvernement de cet autocrate, d'une sincérité et d'une probité absolues, mais quelque peu excessif, dut paraître étouffant à la bourgeoisie libérale qui refusait d'humilier la société devant l'Église. Sous la conduite du polémiste Montalvo, elle s'acharna à abattre la « forteresse confessionnelle » [4] dressée par Moreno et, désespérant d'y parvenir par les voies légales, elle fit assassiner en 1875 le président qui venait d'être réélu pour un nouveau terme.

(1) W. ROBERTSON, *History of the Latin American Nations*, p. 516.
(2) V.-L. TAPIÉ, *Histoire de l'Amérique latine au XIXe siècle*, p. 125.
(3) *Ibid.*, p. 126.
(4) *Ibid.*, p. 227.

CHAPITRE XIV

LA VIE CATHOLIQUE SOUS LE PONTIFICAT DE PIE IX [1]

§ 1. — Le clergé diocésain et l'évolution de l'apostolat.

UN CLERGÉ PLUS DISCIPLINÉ Tandis que la centralisation romaine triomphe définitivement des particularismes locaux, une évolution parallèle se poursuit à l'intérieur des diocèses, plus lente en Europe centrale et méridionale, où l'ancien régime a partiellement survécu à la Révolution française, plus radicale en Europe occidentale.

Les chapitres perdent beaucoup de leur ancienne indépendance et de leur importance. Leurs membres, choisis désormais par l'évêque lui-même parmi les fonctionnaires ecclésiastiques, ne sont plus que des sous-ordres qui ne songent guère à entrer en conflit avec leur supérieur et l'obligation canonique de requérir leur approbation pour certains actes devient dans ces conditions une pure formalité. De plus en plus d'ailleurs, les tâches qui revenaient jadis aux chanoines sont à présent confiées à des secrétaires, qui deviennent, avec les vicaires généraux, les vrais collaborateurs de l'évêque de l'époque contemporaine.

L'autorité de l'évêque sur ses curés est de son côté fortement accrue par la généralisation du principe français de l'amovibilité des desservants, que Geissel, par exemple, réussit à faire admettre par le gouvernement prussien ; sur ce point toutefois, beaucoup d'évêques d'Autriche, de Bavière, d'Italie, voient encore leur liberté d'action entravée par le principe ancien de l'inamovibilité des charges, par l'obligation de conférer de nombreuses cures au concours, voire parfois par les droits de certains « patrons » laïques ou ecclésiastiques.

Ce n'est pas seulement en matière de nominations que les évêques contrôlent de plus près leur clergé. Ces prêtres qui, surtout en Europe occidentale, ne sont plus attirés par l'appât de gros bénéfices et qui songent uniquement à restaurer la vie catholique ébranlée, se rendent compte que la complexité croissante des problèmes exige une modification de la pratique épiscopale et une part beaucoup plus grande d'organisation et d'administration que sous l'ancien régime. Sibour, Bonnechose, Dupanloup en France, Geissel ou Ketteler en Allemagne, Manning en Angleterre, Sterckx en Belgique, ne sont que les plus marquants de cette génération d'évêques qui avaient subi l'influence des conceptions napoléoniennes

---

(1) BIBLIOGRAPHIE. — Excellente synthèse par M.-H. VICAIRE, dans le chapitre : *Rayonnement spirituel de l'Église (XIXe-XXe siècle)*, de l'*Histoire de l'Église illustrée*, sous la direction de G. DE PLINVAL et R. PITTET, t. II, Paris, 1948, p. 261-324.

d'une action ecclésiastique centralisée et presque fonctionnarisée et qui, avec une conscience très nette de leur autorité pastorale, cherchaient à diriger et à systématiser autant que possible l'action apostolique de leurs prêtres. C'est l'époque où l'abbé Combalot, témoin de cette évolution, proposait, dans une de ces boutades dont il était coutumier, de modifier la formule de la consécration épiscopale : « Au lieu de dire : *Accipe baculum pastorale*, il faudra dire : *Accipe calamum administrativum ut possis scribere, scribere, scribere usque in sempiternum et ultra* [1]. »

UN CLERGÉ PLUS PIEUX   Si les évêques renforcent leur autorité sur leurs prêtres, ce n'est pas seulement en vue d'une meilleure administration diocésaine, c'est avant tout par souci de développer chez eux le zèle et l'esprit sacerdotal : ils généralisent la pratique des retraites périodiques ; en beaucoup d'endroits, ils organisent sous une forme ou sous une autre les récollections mensuelles ; ils prescrivent ou conseillent divers exercices de piété réguliers et notamment la méditation quotidienne. Ces préoccupations épiscopales répondent d'ailleurs aux aspirations d'un nombre croissant de prêtres dont la vocation a une origine plus purement religieuse que par le passé et qui souscrivent volontiers au mot d'un des leurs, le P. Chevrier : « Le prêtre est un homme mangé. » Le relèvement sensible du niveau spirituel du clergé est l'un des aspects les moins spectaculaires mais les plus importants de l'histoire de l'Église pendant la seconde moitié du XIXe siècle.

Quelques évêques essaient même de préconiser la vie commune du clergé : Ketteler à Mayence, Mgr Régnier et Dupanloup en France [2]. Ces efforts n'aboutissent pas, mais ne sont pas entièrement perdus. En effet, s'inspirant d'un règlement élaboré au XVIe siècle par Balthazar Holzhauser pour un Institut de clercs séculiers et sur lequel l'évêque d'Orléans avait attiré l'attention [3], M. Lebeurier fonde en 1862 dans le diocèse de Coutances l'*Union apostolique des prêtres séculiers*, appelée à jouer un rôle si bienfaisant dans le clergé, français d'abord, étranger ensuite. Dans le même esprit de soutien spirituel aux prêtres diocésains isolés, l'abbé Chaumont fondera quelques années plus tard la *Société des prêtres de Saint-François de Sales*, et, en 1868, à Vienne, l'abbé Muller mettra sur pied l'*Associatio perseverantiae sacerdotalis*. On voit également se constituer des groupements de prêtres qui, à l'imitation des oblats de Saint-Ambroise, institués par saint Charles Borromée, s'engagent plus strictement, sous la direction d'un supérieur, au service de l'évêque et à la pratique de la perfection sacerdotale : Manning, à peine converti, organise à Londres les oblats de Saint-Charles, tandis qu'en France, aux oblats de Marie-Immaculée, fondés en 1842 par Mgr de Mazenod et qui se répandent dans seize diocèses, viennent s'ajouter les oblats de Saint-Hilaire à Poitiers, les oblats du Sacré-Cœur à Saint-Quentin,

---

(1) A. RICARD, *L'abbé Combalot*, p. 67.
(2) Les conciles provinciaux y reviennent à plusieurs reprises (cf. *Coll. lac.*, t. IV, col. 758, 898, 984, 1044, 1179).
(3) En faisant publier en français par un de ses prêtres, l'abbé Gaduel, une vie de Balthasar Holzhauser (Orléans, 1861).

les oblats de Saint-François de Sales à Troyes et dans six autres diocèses, les prêtres du Prado à Lyon. On pourrait aussi rattacher à ce mouvement les efforts, sans grands résultats au moment même, de ce précurseur de génie que fut dom Gréa, le « théologien de l'Église particulière » [1].

UN CLERGÉ PLUS ZÉLÉ   Les différences restent grandes, évidemment, entre un vicaire de banlieue industrielle de Rhénanie, un curé de campagne français, un bénéficier d'Autriche-Hongrie, un ecclésiastique espagnol ou un curé d'émigrants aux États-Unis. Pourtant, la formation qu'ils reçoivent au séminaire, par lequel ils doivent désormais tous passer, s'inspire de plus en plus des mêmes principes, ainsi que les directives qu'ils reçoivent de leurs évêques, et, peu à peu, par delà les nuances qu'expliquent les traditions nationales variées et les circonstances de temps et de lieu, on voit se dégager le type classique du curé du XIXe siècle. Le prêtre mondain faisant le bel esprit dans les salons, l'érudit à qui son bénéfice assure avec l'aisance des loisirs studieux, l'ecclésiastique campagnard aux mœurs faciles que seul son habit semblait parfois distinguer de la masse de ses paroissiens, sont des cas de plus en plus exceptionnels après 1850, surtout en Europe occidentale.

Curés et vicaires ont désormais pris conscience des responsabilités qui pèsent sur leurs épaules et se comportent en véritables pasteurs. Certes, à la campagne, ils continuent souvent à vérifier la célèbre définition de Taine : « factionnaire fidèle dans sa guérite, patient, attentif au mot d'ordre, montant correctement sa faction solitaire et monotone ». Leur activité en effet consiste essentiellement à célébrer la messe, à prêcher la doctrine chrétienne le dimanche, à expliquer le catéchisme aux enfants qui se préparent à la première communion et à administrer les sacrements. Pourtant, de plus en plus, du moins dans les villes, ils se montrent préoccupés de développer par les moyens les plus variés la glorification de Dieu et l'éducation des chrétiens dans le cadre de la paroisse, qui constitue le champ d'action par excellence du prêtre de l'époque contemporaine. Dans un monde toujours plus individualiste, le sens communautaire de l'ancienne Chrétienté s'est « réfugié maintenant à l'ombre des clochers » [2]. Il se manifeste, avec une maladresse qui fait parfois sourire mais qui ne doit pas nous dissimuler ses aspects positifs, tantôt par l'éclosion ou la renaissance d'innombrables confréries, dont certaines, comme les *Enfants de Marie*, finiront par s'étendre à la catholicité des deux hémisphères, tantôt par l'organisation de ces cérémonies caractéristiques que sont les processions de la Fête-Dieu et surtout la communion solennelle, particulièrement importante, car « elle organise le système de l'éducation religieuse et met une liaison profonde, que le folklore vient encore renforcer, entre la paroisse, l'école et la famille » [3].

L'action paroissiale est effectivement complétée par l'école : école

---

(1) Cf. F. VERNET, *Dom Gréa (1828-1917)*, Paris, 1938.
(2) M.-H. VICAIRE, dans *Histoire de l'Église illustrée* de G. DE PLINVAL, t. II, p. 273.
(3) *Ibid.*, p. 272.

publique inspirée et surveillée par le prêtre, si possible ; école libre, établie à côté et parfois en concurrence de l'école officielle, dans le cas contraire, de plus en plus fréquent. La question scolaire avait déjà commencé à se poser avec acuité dès le pontificat de Grégoire XVI, mais plus l'instruction se généralise, et plus les autorités ecclésiastiques estiment que l'éducation dans un esprit positivement catholique est une nécessité primordiale qui justifie tous les efforts et tous les sacrifices. On peut aujourd'hui se demander si l'importance attribuée par les catholiques du xixᵉ siècle à la création d'écoles libres n'a pas contribué parfois à les enfermer dans une sorte de ghetto spirituel, avec tous les inconvénients d'ordre apostolique et même défensif que la chose comporte. Mais on ne peut oublier que, dans de nombreux cas, ils y furent quasi acculés par l'impossibilité de trouver dans l'école publique le minimum d'atmosphère respirable pour la foi de leurs enfants. L'existence même de ces écoles libres a du reste plus d'une fois obligé par la concurence les écoles officielles à une « neutralité » moins agressive. Et l'on doit en tout cas admirer la somme de générosité et de dévouement que représente la reconstitution, après la destruction quasi totale de la Révolution française, de cette multitude d'établissements d'enseignement supérieur, moyen, primaire ou professionnel qui s'étend progressivement sur l'Europe occidentale et les États-Unis.

UNE IMAGINATION PASTORALE LIMITÉE — Les trésors de zèle et de dévouement que dépensent avec toujours plus d'enthousiasme tant de milliers de prêtres sur le triple terrain de la paroisse, de l'école et des œuvres, suscitent l'admiration et dénotent une Église bien vivante. Pourtant, avec le recul du temps, on commence à mieux voir ce qui manque à ces pasteurs, curés, vicaires généraux ou évêques, pour que leur œuvre soit complète. A part quelques rares exceptions, ils ne semblent pas avoir pressenti la nécessité de repenser les méthodes pastorales pour tenir compte des formidables changements qui se produisaient autour d'eux. De plus, même en Allemagne, les autorités responsables n'ont guère pris conscience de la révolution intellectuelle qui s'accomplissait sous leurs yeux et des problèmes nouveaux qui allaient se poser à la religion en face d'exigences scientifiques nouvelles. Le courant ultramontain, quels qu'aient été par ailleurs ses mérites, a certainement exercé sur ce point une influence néfaste en amenant une partie notable du clergé à renchérir encore sur l'attitude systématiquement antimoderne qui prévalut de plus en plus au Vatican pendant la seconde moitié du pontificat de Pie IX ; il en résulta chez beaucoup une tendance, au fond peu chrétienne, à encourager le milieu « bien pensant » à se replier sur lui-même et à ne rien faire pour diminuer le fossé qui le séparait des incroyants, considérés *a priori* comme des gens de mauvaise foi, qu'il est inutile d'espérer convertir [1].

D'un autre côté, si l'on a affirmé un peu vite que le clergé avait tout

---

(1) Voir sur ce point les observations pertinentes de W. WARD, *W. G. Ward and the catholic Revival*, Londres, 1912, p. 121-123.

ignoré de la question sociale avant le pontificat de Léon XIII, il est certain qu'il n'a guère soupçonné les conséquences qui allaient résulter pour l'Église du développement de la civilisation industrielle. Il est à la rigueur excusable de n'avoir pas prévu, un demi-siècle à l'avance, l'avénement du prolétariat et la substitution progressive de son influence à celle des anciennes classes dirigeantes. On comprend moins le manque d'initiatives vraiment neuves en face du problème constitué par l'extension des « villes tentaculaires » [1]. Tandis que des masses populaires toujours plus nombreuses abandonnent leurs paisibles paroisses de campagne pour venir s'entasser dans les quartiers suburbains, la seule réaction des autorités ecclésiastiques est de construire quelques églises nouvelles [2], sans s'apercevoir qu'il faudrait avant tout mettre au point de nouvelles méthodes d'apostolat permettant de remettre ces masses déracinées en contact effectif avec l'Église. Le prolétariat industriel va se développer de la sorte en dehors de l'Église et « sa détresse matérielle s'accompagne d'un véritable abandon spirituel, prélude et principal facteur de sa déchristianisation » [3]. A cet égard, l'époque de Pie IX, si féconde à tant d'égards au point de vue spirituel, porte une lourde part dans « le grand scandale du XIXe siècle ».

On a souvent dit que l'erreur fondamentale du clergé de cette époque avait été de dépenser beaucoup de zèle et de charité pour atteindre des individus, mais d'avoir trop négligé d'exercer une influence sur les idées et les institutions. C'est très vrai, bien que, présentée en ces termes, la critique ne soit pas tout à fait exacte. En effet, si le clergé, et notamment le haut clergé, apparaît si soucieux de rester en bons termes avec les gouvernements et s'oppose à la séparation entre l'Église et l'État, c'est précisément dans l'espoir de résister à la laïcisation des institutions et d'obtenir de l'État le contrôle de la diffusion des idées contraires aux principes catholiques. La véritable erreur a plutôt été de ne pas voir que cette action directe de l'Église sur les pouvoirs publics allait devenir de plus en plus illusoire dans une société qui entendait s'affranchir de tout « cléricalisme », et surtout que, dans le monde nouveau qui s'élaborait, bien plus encore que les institutions officielles, c'étaient les « milieux de vie » qui exerceraient toujours davantage leur influence sur les masses et que c'étaient dès lors ceux-ci qu'il aurait fallu avant tout imprégner d'esprit chrétien.

Est-ce à dire que rien ne fut fait en ce sens ? Ce serait exagéré. De divers côtés notamment, on se rendit compte de l'importance qu'il y avait à développer les œuvres de presse si l'on voulait agir sur l'opinion. Sans parler de l'action, le plus souvent bienfaisante malgré quelques

---

(1) Il n'y avait, au début du XIXe siècle, que 20 villes de plus de 100.000 habitants et seules Paris et Londres dépassaient les 500.000. Il y en aura 149 à la fin du siècle (dont 19 de plus de 500.000), groupant 47 millions d'habitants contre 5 en 1801.

(2) Et encore cet effort apparaît-il singulièrement limité. C'est que, dans le système moderne, où le temporel du culte dépend en bonne partie des pouvoirs publics, la construction d'une église et surtout l'érection d'une nouvelle paroisse pose des problèmes administratifs complexes, d'autant plus que les États, administrés souvent par des bourgeois libéraux, ne subventionnent que chichement l'expansion de l'Église, sauf pendant les périodes de réaction antirévolutionnaire.

(3) M.-H. VICAIRE, dans *Histoire de l'Église illustrée* de G. DE PLINVAL, t. II, p. 274.

étroitesses, exercée par les journaux catholiques dans de nombreux pays, qu'il suffise de rappeler les heureux résultats de l'action des assomptionnistes en France, de la société Saint-Étienne pour la diffusion des bons livres en Hongrie, d'un don Bosco en Italie, d'un chanoine Alban Stolz en Allemagne, et tout spécialement, dans ce dernier pays, du *Borromaeusverein*.

Les *Vereine*, dont nous avons dit le remarquable essor en pays germaniques et que l'étranger commence à imiter, apparaissent du reste comme une première amorce de ce qui deviendra l'Action catholique : groupements de laïcs, dirigés en bonne partie par des laïcs et visant à faire pénétrer les principes chrétiens dans les divers secteurs de l'activité profane. Toutefois, en dehors de l'Allemagne, on hésite encore le plus souvent à faire directement appel aux laïcs et, pour aider le clergé dans les tâches toujours plus variées et toujours plus complexes où l'on comprend que l'Église doit être présente, on fait avant tout appel aux congrégations religieuses dont le xixe siècle voit un développement sans précédent, qui suffirait à lui seul à prouver de quelles réserves spirituelles cachées dispose cette Église d'apparence parfois si bourgeoise.

§ 2. — Le développement des ordres et des congrégations [1].

LA RÉORGANISATION DE LA VIE RELIGIEUSE

La restauration des ordres religieux, si durement éprouvés par la double crise de l'*Aufklärung* et de la Révolution, avait déjà atteint des résultats sensibles au cours des trente années qui vont de la chute de Napoléon à l'avénement de Pie IX. Mais le pontificat de ce dernier devait être décisif pour donner à la vie religieuse sa physionomie contemporaine : augmentation considérable de l'effectif total, favorisée par la multiplication des congrégations nouvelles ; centralisation toujours plus poussée, qui renforce l'unité à l'intérieur des ordres eux-mêmes et accentue leur dépendance à l'égard des Congrégations romaines ; ferveur plus grande et retour au respect des règles anciennes, en réaction contre le relâchement qui s'était introduit au cours de l'époque moderne.

Un pape aussi pieux que Pie IX et aussi préoccupé de la vitalité interne de l'Église devait prendre ce dernier point particulièrement à cœur. A peine élu, il établissait, en septembre 1846, une commission cardinalice chargée de promouvoir la restauration de la vie religieuse dans les pays où elle avait été bouleversée et, l'année suivante, dans l'encyclique *Ubi primum arcano*, il invitait les supérieurs à veiller à l'intégrité de leurs ordres, dont la collaboration était si nécessaire à l'Église pour la bonne exécution de sa mission spirituelle. Pendant son séjour à Gaëte, témoin de la situation lamentable des rédemptoristes dans le royaume de Naples, où se trouvait alors leur maison généralice, il appuya les efforts pour

(1) BIBLIOGRAPHIE. — M. HEIMBUCHER, *Die Orden und Kongregationen der katholischen Kirche*, t. III, Paderborn, 1908 ; BRAUNSBERGER, *Rückblick auf das katholische Ordenswesen im XIX. Jh.*, Fribourg-en-Br., 1901 ; Ch. TYCK, *Notices historiques sur les congrégations et communautés religieuses du XIXe siècle*, Louvain, 1892.

transférer celle-ci à Rome et « dénapolitaniser » la congrégation, selon l'expression du cardinal préfet des évêques et réguliers [1], et il se réserva la nomination du recteur majeur qui aurait normalement dû être élu par un chapitre général. De même, préoccupé de remédier à la décadence où était tombé l'ordre dominicain, il décida, le 1er décembre 1850, de surseoir à l'élection d'un maître général et désigna lui-même un vicaire général de l'ordre ; son choix se porta, au grand mécontentement des pères italiens, sur un des compagnons de Lacordaire, le P. Jandel, qui prit aussitôt des mesures en vue de rétablir l'ancienne observance. La même année 1850, il appela dom Casaretto comme abbé à Subiaco dans un but de réforme et, deux ans plus tard, c'est sur ses invitations pressantes que la congrégation cassinienne décida d'en revenir à l'observation exacte de la vie commune et de la clôture, ainsi qu'à l'exercice de l'oraison.

Les efforts de réforme provoqués ou encouragés par Pie IX aboutirent assez vite à d'heureux résultats dans les ordres centralisés, mais ils se heurtèrent à une grande force d'inertie dans les abbayes, restées très indépendantes, de l'Europe centrale et méridionale. Dans les pays du Sud, la vie religieuse fut en outre durement éprouvée par les révolutions espagnoles et par l'extension à toute l'Italie, après 1860, des mesures piémontaises de sécularisation. Dans les pays d'Europe occidentale par contre, ainsi qu'en Amérique du Nord, les progrès furent vraiment remarquables, tant par le nombre des membres que par la qualité de la vie religieuse, et les congrégations y devinrent un facteur essentiel de l'efflorescence des œuvres et de l'intensité de la vie spirituelle.

L'IMPORTANCE CROISSANTE DES JÉSUITES

Nul ordre ne devait exercer une influence plus profonde sur la vie de l'Église pendant la seconde moitié du XIXe siècle que celui des jésuites [2]. Après la période un peu difficile qui avait suivi le rétablissement de l'ordre en 1815, il avait progressé rapidement et comptait déjà plus de 4.500 membres à l'avénement de Pie IX. Les premières années du nouveau pontificat furent assez pénibles : situation tendue en France, expulsion de Suisse, fermeture de la plupart des maisons en Italie et même à Rome, à la suite de la campagne déclenchée par Gioberti, puis en Autriche, lors de la révolution de mars 1848 ; à quoi s'ajoutait la froideur témoignée par le nouveau pape qu'indisposait la collusion des jésuites italiens avec la réaction « grégorienne ». Mais on a vu comment les désillusions de Pie IX en 1848 le rapprochèrent des jésuites et comment, à partir de ce moment, leur rôle alla grandissant à la Curie romaine [3],

---

(1) Lettre du P. Dechamps du 26 nov. 1849, *Archives de la province belge des Rédemptoristes*, cl. II. Sur les difficiles tractations qui devaient aboutir au décret du 8 octobre 1853 inaugurant la « période romaine » de la congrégation et au premier chapitre général de 1855, voir M. DE MEULEMEESTER, *Histoire sommaire de la congrégation du T. S. Rédempteur*, Bruxelles, 1950, p. 140-154.

(2) Aperçu d'ensemble dans L. DE JONGE, *De Orde der Jezuieten*, t. III, Wassenaar, 1931. Voir en particulier, pour la France, J. BURNICHON, *La Compagnie de Jésus en France*, t. III (*1845-1860*) et IV (*1860-1880*), Paris, 1919-1922 ; pour l'Espagne, L. FRIAS, *Historia de la Compañia de Jesus en su asistencia moderna de España*, t. II, Madrid, 1923 ; pour l'Italie, A. MONTI, *La compania di Gesu nel territorio della Provincia Torinese*, t. V, Chieri, 1920 ; pour l'Allemagne, B. DUHR, *Aktenstücke zur Geschichte der Jesuiten-Missionen in Deutschland 1848-1872*, Fribourg-en-Br., 1903.

(3) Voir *supra*, p. 286.

surtout lorsque, en 1853, le P. Beckx eut succédé à la tête de la Compagnie au P. Roothaan, avec qui le pape ne s'entendait qu'à moitié.

Le P. Beckx [1], que ses adversaires ont souvent présenté comme un fanatique, était en réalité un esprit modéré, modeste et très pieux. Sous sa direction, la Cpomagnie vit prospérer ses entreprises et ses effectifs doubler en une quinzaine d'années : quand il se retira en 1884, la Compagnie de Jésus comptait 11.480 membres répartis en 19 provinces, contre 5.209 en 10 provinces en 1853, en dépit des difficultés croissantes occasionnées depuis 1860 par le gouvernement italien (en 1873, le général fut même obligé de quitter Rome) et malgré les déboires consécutifs à la révolution espagnole de 1869 et au *Kulturkampf* allemand. Bien plus importante encore que ce développement numérique est l'influence sans cesse accrue prise par les jésuites tant dans le domaine des sciences ecclésiastiques ou dans la direction des Congrégations romaines, que dans la vie des Églises particulières où ils agissent à la fois sur les masses par l'organisation de missions paroissiales et d'œuvres populaires, et sur les classes dirigeantes par leurs collèges, leurs prédications et leurs nombreux contacts personnels. Leur action est particulièrement sensible dans les progrès du mouvement ultramontain et dans l'orientation de la dévotion des fidèles ou de la spiritualité sacerdotale : si la piété du XIXe siècle devient de plus en plus individualiste, c'est en partie à eux qu'on le doit, mais en revanche ils ont beaucoup contribué à répandre le culte du Sacré-Cœur, la pratique de l'oraison mentale (selon la méthode de saint Ignace interprétée par le P. Roothaan) et le goût des retraites fermées.

ANCIENS ORDRES ET CONGRÉGATIONS NOUVELLES

Des grands ordres de l'ancien régime, celui des jésuites est le seul à se retrouver en pleine prospérité dès le milieu du XIXe siècle. L'ordre de saint Dominique [2], par exemple, qui comptait comme celui de saint Ignace environ 4.500 membres vers 1845, n'en compte plus que 3.341 trente ans plus tard : les progrès notables en France, grâce au prestige de Lacordaire (il y aura, en 1872, 304 religieux contre 13 en 1844), en Allemagne, en Autriche ou en Angleterre ne compensent pas les pertes subies en Pologne, en Italie, en Espagne, au Portugal et en Amérique du Sud, et il est caractéristique de constater que la renaissance thomiste sera l'œuvre des jésuites bien plus que celle des dominicains ; toutefois, la réorganisation de l'ordre se poursuivit au cours du long généralat du P. Jandel [3], qui se préoccupa aussi du renouveau des études, et si certains crurent déceler avec regret certaines influences ignatiennes dans ses projets d'adaptation des constitutions, son œuvre apparaît au total très prometteuse pour l'avenir.

---

(1) Sur le P. Beckx (1795-1887), voir A.-M. VERSTRAETEN, *Leven van den H. E. P. Petrus Beckx*, Anvers, 1889. Né en Belgique dans un milieu très simple, il avait passé toute sa vie religieuse en Allemagne puis en Autriche et avait dirigé cette dernière province avec beaucoup d'habileté.

(2) A.-M. WALZ, *Compendium historiae Ordinis Praedicatorum*, 2e édit., Rome, 1948, p. 517 et suiv.

(3) Vicaire général de l'ordre depuis 1850 et nommé par le pape maître général en 1855, il fut confirmé dans cette charge pour 12 ans par le chapitre général de 1862 (Pie IX avait préféré ne pas rétablir le généralat à vie). Il mourut en décembre 1872. (Cf. H. M. CORMIER, *Vie du Rme P. Jandel*, Paris, 1890.)

La restauration des franciscains [1] fut plus lente. S'ils purent reprendre en 1856 la tradition interrompue depuis longtemps des chapitres généraux et mettre à jour leurs statuts, la question du retour à l'unité des trois branches (observantins ou réformés, récollets et alcantarins), envisagée dès 1862, n'aboutira que sous Léon XIII et les effectifs continuent à fondre dans les pays méridionaux où l'ordre avait largement essaimé [2]. Il faut surtout regretter que cet ordre, qui devait par sa tradition propre être essentiellement populaire, n'ait guère réussi à s'adapter à la transformation industrielle de l'Europe et à s'intégrer efficacement aux masses prolétariennes qui s'accumulent dans les grandes villes.

Les ordres monastiques avaient été plus ébranlés que les ordres mendiants par le joséphisme ou la crise révolutionnaire et il leur était en outre particulièrement difficile de retrouver leur place exacte dans la société nouvelle du XIXe siècle. Ils souffraient souvent de voir la prépondérance très nette que prenaient les jésuites ou l'intérêt croissant porté par Rome aux congrégations modernes plus soumises au Saint-Siège et orientées vers un apostolat plus actif : ils devaient relever avec amertume que dans la « commission des religieux » préparatoire au concile du Vatican, à part un franciscain, il n'y avait aucun représentant des ordres médiévaux. D'ailleurs, de bon gré ou à contre-cœur, il leur fallut bien s'aligner quelque peu sur les usages des instituts religieux modernes. Les bénédictins eux-mêmes [3], si fiers pourtant de leur indépendance, n'y échappèrent pas : les réformes introduites en 1853 dans la congrégation cassinienne, les nouveaux statuts de la congrégation de Subiaco, élaborés en 1867, en portent des traces très nettes, et Pie IX obtint qu'à partir de 1859 il y ait un noviciat commun pour les abbayes de la congrégation anglaise. Dans ces abbayes anglaises, du reste, la direction de grands collèges et le ministère paroissial occupaient beaucoup plus de place que la vie monastique proprement dite. Il en était de même en Europe centrale, où, de plus, la discipline laissait souvent à désirer [4]. En dépit de ces déficiences, plus le siècle avance et plus l'ordre de saint Benoît prospère, en même temps qu'il reprend conscience de sa mission propre dans l'Église. C'est avant tout au rayonnement de Solesmes qu'il le doit : l'histoire se doit de signaler les faiblesses de dom Guéranger, mais « les petits côtés de son existence, les défauts de son caractère, les fautes de son administration disparaissent devant la grandeur des résultats obtenus » [5]. C'est vraiment son esprit qui anime, à l'insu même de ceux qui sont tentés

---

(1) H. HOLZAPFEL, *Manuale historiae ordinis Fratrum Minorum*, Fribourg-en-Br., 1909, p. 331 et suiv.

(2) Alors qu'ils avaient dépassé les 70.000 vers la fin du XVIIIe siècle, les franciscains n'atteignaient plus que 23.000 membres en 1862 et 15.000 en 1882 ; à cette date, les provinces avaient été ramenées à 30 alors que, en 1862, on en comptait encore, théoriquement du moins, 103, vestiges de la splendeur passée.

(3) Ph. SCHMITZ, *Histoire de l'ordre de saint Benoît*, t. IV, Maredsous, 1948, p. 175 et suiv.

(4) C'était surtout le cas en Autriche où le vœu de pauvreté était très mal observé ; la grande visite canonique, qui se poursuivit de 1852 à 1859 sous la direction des cardinaux Schwarzenberg et Scitowsky, ne changea pas grand'chose à la situation. (C. WOLFSGRUBER, *Kardinal Schwarzenberg*, t. II, p. 197, 222-281.)

(5) Ph. SCHMITZ, *Histoire de l'ordre de saint Benoît*, t. IV, p. 177. L'ouvrage essentiel sur l'œuvre de dom Guéranger reste celui de dom DELATTE, *Dom Guéranger, abbé de Solesmes*, Paris, 1909, 2 vol.

de le nier, le mouvement de retour vers les grandes traditions du passé qui s'affirme toujours plus nettement dans les diverses congrégations bénédictines ; c'est à son école que viennent notamment se perfectionner les fondateurs de la congrégation allemande de Beuron, dont l'influence devait être prépondérante dans le renouveau liturgique contemporain [1].

A côté des ordres anciens qui reprennent vie, à côté des congrégations plus récentes qui se développent rapidement, tel l'Institut des frères des Écoles chrétiennes, qui, sous la direction avisée du frère Philippe, rayonne dans le monde entier [2], des congrégations et communautés nouvelles de tout genre surgissent de tous côtés, aux États-Unis ou au Canada aussi bien qu'en Europe, avec plus d'intensité en France que partout ailleurs. Le mouvement avait commencé avant Pie IX, mais il s'amplifie sous son pontificat, encouragé par le cardinal Bizzarri, qui préside avec énergie la Congrégation des évêques et réguliers de 1854 à 1877.

Parmi ces congrégations nouvelles, quelques-unes se consacrent au culte, ainsi les pères du Saint-Sacrement, fondés en 1856 par Julien Eymard, les sœurs de l'Adoration Réparatrice, fondées au moment de la révolution de 1848 par une ancienne servante, ou les religieuses de Marie Réparatrice fondées en 1857 par une noble dame belge, la baronne d'Hoogvorst. Mais la plupart se vouent à un apostolat plus actif : missions, œuvres de presse, soulagement des pauvres, service des malades, relèvement des malheureuses, enseignement surtout, et particulièrement l'enseignement des filles, car c'est un des principes fondamentaux du clergé de cette époque qu'il faut préparer des foyers chrétiens en donnant à la jeunesse féminine une solide éducation chrétienne. C'est dans ce dernier secteur que l'on constate une prolifération de minuscules congrégations, diocésaines et parfois presque paroissiales. Faut-il l'attribuer à l'individualisme des temps modernes ou, comme le pensait le cardinal de Reisach [3], à une certaine « anarchie ecclésiastique » et au désir d'être maître chez soi ? Peut-être, mais on peut y voir aussi un signe des préoccupations pastorales d'évêques et de curés beaucoup plus soucieux que dans les siècles antérieurs d'assurer à leurs paroisses le maximum d'efficacité ; et on doit y voir certainement la preuve de l'intensité de ferveur religieuse diffuse dans la masse chrétienne de ce temps et qui jaillit spontanément au contact des besoins multiples de la vie journalière de l'Église.

L'APPARITION DES « INSTITUTS SÉCULIERS » Depuis des siècles, l'influence des ordres religieux se prolongeait dans le monde des laïcs pieux grâce aux tiers-ordres, dont le plus connu est celui dirigé par les franciscains [4]. Mais dans la seconde

(1) La fondation de Beuron date de 1863. L'abbaye essaimera rapidement, même à l'étranger ; dès 1872, elle fondait en Belgique le monastère de Maredsous. Cf. O. WOLFF, *Beuron,* Fribourg-en-Br., 1923 ; en outre, sur dom Maur Wolter (1825-1890), le vrai fondateur de la congrégation, surnommé « le dom Guéranger de l'Allemagne », voir *Maurus Wolter. Dem Gründer Beurons zum 100. Geburtstag,* Beuron, 1925.

(2) L'effectif total de l'Institut passe au cours du généralat du frère Philippe (1838-1874) de 2.300 religieux, en 313 communautés, à plus de 10.000 répartis en un bon millier d'établissements ; 276 fondations sont réalisées en dehors de France, mais le recrutement demeure pendant toute cette période très national.

(3) *Janssens Briefe,* édit. PASTOR, t. I, p. 248-249.

(4) Fort ralentie par la crise révolutionnaire, leur action reprend au cours de la décade 1860-

moitié du XIXᵉ siècle commence à apparaître une nouvelle forme de vie religieuse aux confins de la vie laïque, des sociétés de personnes vivant la vie religieuse tout en restant dans le monde. L'une des premières réalisations fut la fondation par Mme Carré de Malmberg, assistée par l'abbé Chaumont, des filles de Saint-François-de-Sales [1]. En ce domaine comme en tant d'autres, la France est à l'avant-garde.

§ 3. — Les formes de dévotion et l'évolution de la spiritualité [2].

VERS UNE PIÉTÉ PLUS CHAUDE ET PLUS EXTÉRIEURE

Le vrai triomphe de l'ultramontanisme, a-t-on pu écrire, ne fut pas tant la proclamation de l'infaillibilité du pape que « la transformation intérieure du catholicisme » dans les pays du nord des Alpes. Il n'a guère fallu plus d'une génération pour que, malgré la résistance parfois très vive des partisans de l'ancien ordre de choses, une piété de type italien, plus indulgente, plus superficielle parfois mais aussi plus humaine et plus populaire, faisant davantage sa part au sentiment et au besoin d'extériorisation, mais basée sur une fréquentation plus grande des sacrements et sur la multiplication d'exercices précis, se substitue au rigorisme janséniste et à la piété solide et profonde mais austère et peu démonstrative que prônaient les disciples de Sailer en Allemagne, les anciens du collège d'Ushaw en Angleterre, Saint-Sulpice en France. Désormais, la dévotion s'oriente de plus en plus vers le Christ miséricordieux montrant son cœur « qui a tant aimé les hommes », vers Jésus « prisonnier d'amour dans le tabernacle », vers Marie sous les traits plus sensibles de Notre-Dame de Lourdes, vers un certain nombre de saints particulièrement populaires, saint Antoine, saint Joseph, que Pie IX proclame en 1870 patron de l'Église universelle [3].

Divers facteurs intervinrent pour provoquer cette transformation, à commencer par l'enthousiasme romantique pour tout ce qui rappelait le moyen âge : dévotion mariale, culte des saints, vénération des reliques, processions, pèlerinages et autres manifestations publiques de la foi. Mais il faut faire une large part aussi aux influences délibérées : celle des jésuites, avec leur théologie optimiste et leur souci d'organiser de manière systématique la dévotion des masses ; celle aussi des prêtres de for-

1870 en France, où les capucins commencent à publier en septembre 1861 les *Annales franciscaines*, la plus ancienne revue franciscaine, puis en Italie, en Espagne, en Angleterre, en Allemagne, à partir de 1870. (Cf. FRÉDEGAND D'ANVERS, *Le Tiers-Ordre de saint François d'Assise*, Paris, 1923, chap. VII.)

(1) Voir Mgr LAVEILLE, *Mme Carré de Malmberg (1829-1891)*, Tours-Paris, 1917 ; *L'abbé Henri Chaumont (1838-1896)*, 1919.

(2) BIBLIOGRAPHIE. — On trouvera de nombreuses indications dans M. NÉDONCELLE, *Les leçons spirituelles du XIXᵉ siècle*, Paris, 1936. Voir aussi K. KEMPF, *Die Heiligkeit der Kirche im XIX.Jh.*, Einsiedeln, 1928.

Sur le culte du Sacré-Cœur : J. BAINVEL, *La dévotion au Cœur de Jésus. Doctrine. Histoire*, Paris, 1921, IIIᵉ part., chap. VII à IX, et NILLES, *De Rationibus festorum S. S. Cordis Jesu et purissimi Cordis Mariae*, Innsbruck, 1885, 2 vol.

Sur la dévotion mariale : SAINT-JOHN, *L'épopée mariale en France au XIXᵉ siècle*, trad. franç., Paris, 1904.

Sur la littérature spirituelle : H. POURRAT, *Histoire de la spiritualité*, t. IV, Paris, 1930, chap. XVIII à XX.

(3) Pour satisfaire à une demande des Pères du concile du Vatican. Déjà en 1847, Pie IX avait étendu à toute l'Église la fête du patronage de saint Joseph.

mation romaine, en particulier d'hommes comme le P. Faber [1], qui s'attache à vulgariser les dévotions populaires qui l'avaient charmé en Italie et à faire connaître des recueils comme les *Gloires de Marie* de saint Alphonse de Ligori, ou l'abbé Gaume [2], dont les innombrables publications, notamment l'*Horloge de la Passion*, adaptation d'un autre ouvrage de saint Alphonse, remplacèrent dans maintes bibliothèques cléricales les œuvres de Bossuet et de Fénelon dont s'était nourri le clergé des générations précédentes.

Ce n'est pas un hasard si le nom de saint Alphonse de Ligori vient de revenir par deux fois : son influence fut considérable dans l'évolution que nous évoquons. Influence de sa théologie morale moins rigoriste, que le jeune clergé accueille avec une ferveur croissante malgré les réserves des professeurs de la vieille école [3], que les Congrégations romaines recommandent avec insistance [4], que Pie IX ratifie solennellement en proclamant le saint prélat docteur de l'Église le 23 mars 1871 à la demande de près de 600 évêques [5]. Influence aussi de la spiritualité antijanséniste de ce « saint François de Sales italien », tout imprégnée de confiance en la divine miséricorde et de tendre dévotion envers la Vierge et l'Eucharistie.

Cette orientation nouvelle de la piété, que les partisans des usages anciens devaient dénoncer comme l'un des principaux méfaits de l'ultramontanisme, n'alla pas sans certains inconvénients : elle n'évita pas toujours les excès de sentimentalité ni même certaines tendances à la superstition ; l'insistance mise sur les dévotions particulières rétrécit encore les perspectives de nombreux fidèles, qui avaient déjà perdu le contact avec la Bible et la liturgie ; la préoccupation de « gagner des indulgences » l'emporta trop souvent sur le souci de transformer ses dispositions profondes [6] ; et l'insistance mise sur la répétition fréquente des pratiques extérieures aboutit parfois à une piété moins personnelle et trop mécanique. Le caractère plus populaire et, il faut bien l'avouer, parfois un peu puéril, pris par la dévotion catholique, contribua à éloigner de l'Église un certain nombre d'intellectuels et de libéraux incapables de faire le partage entre l'accessoire et l'essentiel. Pourtant, cette évolution fut heureuse et bienfaisante, car, sous des formes un peu maladroites, elle exprimait la saine réaction du sentiment chrétien contre le christianisme édulcoré, confinant presque au déisme du siècle précédent.

(1) Sur Frédéric William Faber (1814-1863), converti en 1845 et organisateur avec Newman de l'Oratoire anglais, voir J.-E. Bowden, *The Life and Letters of F.-W. Faber*, Baltimore, 1869.
(2) Voir la liste complète des ouvrages de Gaume (1802-1879) dans *D. T. C.*, t. VI, col. 1169-1171.
(3) Certains allaient jusqu'à traiter publiquement saint Alphonse de « farceur », ainsi au grand séminaire de Chartres vers 1845. (Cf. É. Sévrin, dans *R. H. É. F.*, t. XXV, 1939, p. 340.)
(4) Voir par exemple les réponses du Saint-Office, 18 juillet 1860 ; de la Sacrée Congrégation du Concile, 29 août 1860 (« *consulat theologos, praesertim S. Alphonsum a Ligorio* ») ; de la Sacrée Pénitencerie, 10 décembre 1860.
(5) Voir les *Acta doctoratus*, Rome, 1870 et *Il concilio vaticano e il titolo di Dottore della Chiesa decretato a S. Alph. M. de Ligorio*, dans *Civ. catt.*, 8e série, vol. III, 1871, p. 285-297.
(6) Pie IX encouragea beaucoup le renouveau de faveur accordé aux indulgences et multiplia notamment les occasions de gagner l'indulgence du Jubilé : à son avénement, lors de son retour de Gaëte, à l'occasion de la définition de l'Immaculée-Conception, après son voyage de 1857 à travers ses États ; pour le 10e anniversaire de l'Immaculée-Conception ; en 1869 pour son jubilé sacerdotal, en 1871, pour ses 25 années de pontificat, sans parler de l'année sainte normale.

LA REDÉCOUVERTE DU CHRIST La piété allait désormais se centrer à nouveau sur la Crèche, sur la Croix, sur l'Eucharistie. Les appels des jésuites et de leurs émules en faveur de la confession et de la communion fréquente ramenèrent l'attention sur le caractère essentiellement sacramentaire de la vie catholique. Et surtout, on redécouvrit la réalité centrale du christianisme, le Christ, vrai Dieu et vrai homme, Incarnation de l'amour de Dieu invitant chaque homme à l'aimer en retour, personne actuellement vivante et agissante dans la vie des chrétiens. Il suffit de parcourir les sermons, les catéchismes ou les lettres pastorales de la première moitié du siècle pour se rendre compte à quel point ces thèmes, qui nous paraissent aujourd'hui banals, s'étaient estompés chez beaucoup pour qui le surnaturel était devenu un mot plutôt qu'une vie, et chez qui la majesté abstraite de Dieu avait pris une importance plus grande que le Christ au point de parler de l'Eucharistie comme de « Dieu demeurant parmi nous » ou du « bon Dieu descendant sur l'autel ».

La diffusion des soi-disant révélations de Catherine Emmerich, publiées par Brentano et vulgarisées par le rédemptoriste Schmöger [1] ; la reprise, en 1844, de l'ostension de la Sainte Tunique de Trèves ; les efforts du P. Faber pour diffuser le culte du Précieux Sang et les formes italiennes de la dévotion à l'Eucharistie ; l'action de M. Dupont, « le saint homme de Tours », en faveur du culte de la sainte Face ; les appels émouvants de Lacordaire pour qui « la grande affaire, c'est d'aimer Jésus-Christ » et qui y revient sans cesse dans ses sermons et ses lettres de direction ; bien d'autres influences analogues, redonnèrent en quelques années à la piété catholique une orientation christocentrique dont le développement de la dévotion eucharistique et le succès extraordinaire du culte du Sacré-Cœur furent les expressions les plus marquantes.

LA DÉVOTION EUCHARISTIQUE La piété eucharistique se manifeste sur deux plans : la communion fréquente et l'adoration du Saint-Sacrement. La réaction contre la sévérité janséniste relativement à la communion s'affirme toujours davantage, bien qu'avec une timidité qui nous étonne aujourd'hui. Beaucoup de théologiens et de directeurs spirituels, jusqu'à la fin du siècle, considéreront la communion bihebdomadaire comme un maximum : en 1870 encore, au Collège américain ou au Collège germanique de Rome, la plupart des séminaristes ne communient qu'une fois par semaine [2] et au séminaire de Mayence, seuls quelques-uns ajoutent une communion, parfois deux, à celle du dimanche [3]. Pourtant, dès le milieu du siècle, la tendance à encourager

---

(1) La *Douloureuse passion de Notre-Seigneur* avait été publiée en 1833. La *Vie de la Sainte Vierge* parut en 1852 et la *Vie de Notre-Seigneur* de 1858 à 1860. Le succès de ces ouvrages fut très grand et ils furent aussitôt traduits en diverses langues. Sur le problème d'authenticité, voir W. Huempfner, *Cl. Brentanos Glaubwürdigkeit in seinem Emmerick-Aufzeichnungen*, Wurzbourg, 1923, qui conclut que Brentano « a mystifié le public catholique en falsifiant les communications que la religieuse lui avait faites ».

(2) Renseignements recueillis par M. Icard lors du concile du Vatican. (*Journal de mon voyage à Rome*, p. 268, 271 et 338, *Arch. S. Sulp.*, fonds Icard.)

(3) *Ibid.*, p. 314.

la communion fréquente se précise. Le prêtre poète Guido Gezelle la préconise en Westflandre dès 1850 ; Dupanloup réédite en 1855 une lettre de Fénelon favorable à la communion quotidienne, dont 100.000 exemplaires s'écoulent en peu de temps ; Mgr de Ségur s'en fait l'ardent et inlassable propagandiste et y engage les jeunes ouvriers dont il s'occupe avec prédilection. Mais c'est d'Italie surtout que viennent les encouragements. Joseph Frassinetti, dans son *Banquet de l'Amour divin* (1867), fait l'apologie de la communion fréquente au nom de l'antiquité chrétienne. Don Bosco se déclare en outre partisan de la communion précoce : « Quand un enfant sait distinguer entre le pain ordinaire et le pain eucharistique, quand il a une instruction suffisante, il ne faut pas s'occuper de son âge. » Dans le sud d'ailleurs, la chose était déjà courante : en 1870, M. Icard apprenait, au cours d'un séjour à Naples, que « chaque parent peut avec l'avis du confesseur faire faire la première communion en dehors du catéchisme aux enfants de 7 à 8 ans. Le curé n'a rien à y voir, c'est l'affaire du confesseur »[1]. Et si, à Rome, il n'est pas encore question de mesures générales en vue d'abaisser l'âge de la première communion, les autorités commencent pourtant à réagir : en 1851, la Congrégation du concile corrige un chapitre du concile provincial de Rouen où on interdisait d'admettre les enfants à la communion avant douze ans ; et en 1866, une lettre d'Antonelli aux évêques français réprouve vivement la coutume de différer la première communion jusqu'à un âge tardif et fixe.

C'est d'Italie aussi que vinrent les différentes formes d'adoration du Saint-Sacrement qui prirent tant d'extension au cours de la seconde moitié du XIXe siècle. L'adoration perpétuelle, officiellement recommandée par Pie IX en 1851, fut propagée en Angleterre par les deux convertis Faber et Dalgairns ; elle fut introduite au Canada par Mgr Bourget et pénétra aux États-Unis au cours de la décade 1850-1860, tandis qu'en France, où elle existait déjà dans deux diocèses à l'avènement de Pie IX, vingt diocèses l'adoptèrent entre 1849 et 1860, et trente-sept autres au cours des quinze années suivantes. Quant à la pratique romaine de l'adoration nocturne, le carme Hermann Cohen, juif converti, célèbre prédicateur et musicien de talent, auteur de mélodies touchantes en l'honneur de l'Eucharistie, l'introduisit en Allemagne puis en France, en 1848, avec l'aide de l'abbé de la Bouillerie ; trente ans plus tard, elle existera dans vingt diocèses et sera particulièrement florissante dans le nord grâce aux efforts de Philibert Vrau.

LA DÉVOTION AU SACRÉ-CŒUR   L'adoration nocturne met en avant un aspect particulièrement cher à la piété du XIXe siècle : l'union au Christ souffrant. Le développement du culte du Sacré-Cœur allait lui permettre de s'épanouir sous ses différentes formes : compassion douloureuse devant la victime pitoyable du Calvaire, telle que le moyen âge l'avait pratiquée ; compensation pour les trahisons et les outrages des pécheurs, en esprit d'amour et de réparation, conformément au message de Marguerite-Marie ; aspiration apostolique enfin à

« compléter ce qui manque aux souffrances du Christ » en prenant sur soi, à son imitation, les fautes des hommes et leurs conséquences, trait propre à l'époque contemporaine. Ce siècle bourgeois, individualiste et positif, a compris mieux que beaucoup d'autres l'influence que peut avoir une âme chrétienne lorsqu'elle a découvert ses responsabilités dans le salut du monde, et jamais époque ne vit pareille floraison d'âmes ou d'instituts se vouant à l'apostolat du sacrifice [1].

C'est surtout à partir du pontificat de Pie IX que le XIXe siècle mérite d'être appelé, comme le proposait Mgr d'Hulst, le « siècle du Sacré-Cœur ». Activement propagée par les jésuites, la dévotion est accueillie avec enthousiasme par les fidèles qui apprécient en elle, avec son caractère de tendresse exaltée et de mysticisme réaliste, le meilleur moyen de protester contre les tendances rationalistes et jouisseuses de l'époque. En France, un élément spécial vient encore contribuer à son succès pour en faire la dévotion française par excellence : les légitimistes, dont l'influence est si importante dans la vie catholique, et les nombreux prêtres qui conservent le culte du « roi martyr », n'oublient pas que Louis XVI avait, pendant sa captivité au Temple, pris l'engagement de consacrer la France au Sacré-Cœur, ni que des Vendéens s'étaient battus avec le divin emblème cousu sur leur poitrine ; ces souvenirs coloreront d'une teinte particulière les cérémonies toujours plus grandioses qui se dérouleront à Paray-le-Monial puis, à partir de 1876, à Montmartre, où dès la première année le sanctuaire voit défiler 3 cardinaux, 26 évêques, 140.760 pèlerins [2].

Les théologiens, à la suite de Perrone, qui semble avoir été le premier à le faire, commencent à introduire la question du Sacré-Cœur dans les traités *de Verbo Incarnato*, tandis que de nombreux auteurs spirituels s'appliquent à montrer qu'il s'agit de bien plus que d'une simple dévotion de sentiment ou même d'un dogme isolé : « C'est la synthèse de toute la doctrine catholique, résumée dans l'amour de Jésus-Christ pour nous et de nous pour Jésus-Christ [3]. » Pie IX, qui s'était fait inscrire dans la garde d'honneur du Cœur de Jésus, se plaît à encourager le mouvement. Un des premiers actes de son pontificat avait été de proclamer Marguerite-Marie vénérable ; en 1856, il se décide à répondre à un vœu déjà maintes fois exprimé et, à la demande des évêques français réunis à Paris pour le baptême du prince impérial, il étend enfin la fête du Sacré-Cœur à l'Église universelle. En 1864, la béatification de Marguerite-Marie lui fournit l'occasion d'affirmer nettement que le Christ a choisi l'humble visitandine pour nous révéler par elle son amour et nous pousser à y répondre en l'honorant sous le symbole du cœur.

Le message de Marguerite-Marie comportait un aspect social. Demeuré longtemps secondaire, il fut mis en relief à partir du milieu du siècle par

---

(1) Voir R. Plus, *La folie de la croix*, Toulouse, 1927, IIIe Partie.
(2) Pendant les dix premières années, les dons pour la basilique dépasseront 12 millions de francs-or, et les longues listes de souscription (dans le *Bulletin du Vœu National*) attestent le caractère populaire de cette dévotion où l'élan simplement chrétien l'emporte de beaucoup sur les préoccupations monarchistes : « 3.000 fr., toutes les économies d'une vieille domestique depuis 14 ans », « 125 fr. économisés sous par sous par un pauvre nègre du Basutoland ». (Cf. E. Lecanuet, *L'Église de France sous la IIIe République*, t. I, p. 380.)
(3) Mgr Baunard, *Un siècle de l'Église de France*, p. 200.

les milieux ultramontains, qui s'efforcèrent de faire reconnaître par l'univers entier l'absolue souveraineté du Sacré-Cœur et le devoir de travailler à son « règne social ». C'est dans cette perspective que le musicien Vervoitte, inspecteur des maîtrises subventionnées à la fin du Second Empire, composa le motet célèbre « *Christus vincit, Christus regnat, Christus imperat* », et que le P. Ramière, de la Compagnie de Jésus, conçut l'*Apostolat de la Prière*, « ligue de prière en union avec le Cœur de Jésus » fondée en 1861, dont les membres prient et communient à la même intention proposée chaque mois avec l'approbation du souverain pontife. C'est aussi dans cette perspective, à laquelle se mêlaient chez certains d'incontestables sympathies pour la théocratie, que s'orienta de plus en plus le mouvement de consécration au Sacré-Cœur ; après les consécrations des individus, des familles, des congrégations religieuses, des diocèses, on préconisa la consécration solennelle des États. A la veille du concile du Vatican, Mgr Dechamps consacrait la Belgique au Sacré-Cœur. En juin 1873, dans l'attente fiévreuse d'une imminente restauration monarchique dont on attendait le salut religieux du pays, des délégués de tous les diocèses de France et, pour finir, un groupe de représentants de l'Assemblée nationale en firent autant pour leur patrie, en attendant qu'Henri V puisse le faire officiellement comme venait de le faire pour l'Équateur le saint président Garcia Moreno. Peu après, l'archevêque de Toulouse écrivit à tous les évêques de la catholicité pour les inviter à renouveler la demande d'une consécration de l'univers au Sacré-Cœur, qui avait déjà été présentée à Pie IX à la fin du concile du Vatican, signée par presque tous les évêques et supérieurs d'ordres et par plus d'un million de fidèles. 525 évêques envoyèrent leur adhésion et, en avril 1875, le P. Ramière, qui avait été l'âme du mouvement, pouvait offrir au pape leur pétition. Pie IX, toutefois, préféra ne pas brusquer les choses et se contenta de faire envoyer par la S. Congrégation des Rites une formule de consécration approuvée par lui, qu'il encourageait à réciter publiquement le 16 juin 1875, second centenaire de la grande apparition. La réponse à cet appel dans les deux hémisphères constitua l'une des plus grandes solennités qu'ait encore jamais vues le monde catholique.

LA DÉVOTION MARIALE ET LES APPARITIONS   La redécouverte du Christ au XIXᵉ siècle s'accompagna, comme il était naturel, d'un nouveau développement de piété envers sa mère : nouvelle floraison de congrégations mariales [1], remise en honneur des lieux de pèlerinage en l'honneur de Marie [2], succès du « rosaire vivant » de Pauline Jaricot, généralisation progressive des exercices du mois de mai, dont la pratique sera à peu près générale à la fin du siècle. La proclamation du dogme de l'Immaculée-Conception en 1854 [3] par Pie IX, résultat elle-même d'un vaste courant de piété dont les pétitions

(1) Cf. P. LOEFFLER, *Die marianischen Kongregationen in ihren Wesen und ihrer Geschichte* hrgb. v. G. HARRASSEN, Fribourg-en-Br., 1924.
(2) Notamment, pour la France, sous l'influence de l'ouvrage du curé de St-Sulpice, M. HAMON, *Notre-Dame de France. Histoire du culte de la sainte Vierge en France*, 7 vol. (1861-1866).
(3) Cf. *supra*, p. 278-280.

épiscopales furent l'expression, contribua à augmenter encore la dévotion des fidèles, en même temps qu'elle orientait l'intérêt des théologiens vers l'étude plus approfondie des privilèges de Marie.

Dieu encouragea ce mouvement de dévotion par diverses apparitions qui firent une grosse impression sur les fidèles et contribuèrent à faire de la France la terre mariale. La Vierge, qui était apparue en 1830 à Paris à Catherine Labouré et en 1836 au curé de Notre-Dame des Victoires, l'abbé Desgenettes, apparut, le 19 septembre 1846, sur le plateau de La Salette, à deux jeunes pâtres savoyards, Maximin Giraud et Mélanie Calvat ; sous les traits d'une « belle dame » en pleurs, elle les chargea de communiquer « à tout son peuple » ses plaintes concernant la sanctification du dimanche, le respect du nom de Dieu et le précepte de l'abstinence [1]. Comme quelques miracles se produisirent, les fidèles s'habituèrent à venir en pèlerinage au lieu de l'apparition, surtout lorsque, après une enquête minutieuse, l'évêque de Grenoble eut conclu, en 1851, à la réalité des faits [2], et la renommée de la « sainte montagne » dépassa bientôt les frontières de la France [3].

Elle sera toutefois éclipsée, quelques années plus tard, par un autre pèlerinage, appelé à une célébrité sans précédent. Du 11 février au 16 juillet 1858, la Vierge apparut dix-huit fois dans une grotte voisine de Lourdes à une fillette de 14 ans, Bernadette Soubirous. Là aussi, des miracles ne tardèrent pas à se produire et finirent par convaincre les autorités ecclésiastiques, très réservées au début, de la véracité de la voyante : le 18 janvier 1862, l'évêque de Tarbes se prononçait favorablement [4]. Deux ans plus tard, on installa dans la grotte, où les fidèles affluaient toujours, la célèbre statue de l'Immaculée sculptée par Fabisch sur les indications de Bernadette. L'ouvrage d'Henri Lasserre, *Notre-Dame de Lourdes* (1869), relatant les deux cents premiers miracles, et la dérobade des médecins rationalistes devant les défis que leur adressaient des croyants enthousiastes, comme M. E. Artus, contribuèrent à attirer l'attention des foules. L'exaltation religieuse des années d'après-guerre coïncidant avec le développement des voies ferrées permit aux assomptionnistes, qui obtinrent des tarifs spéciaux des compagnies de chemin de fer, d'organiser des pèlerinages, qui prirent une ampleur jamais atteinte ; en 1872, on en compte 149 avec 119.000 pèlerins, en 1873, 183 avec 140.000 pèlerins ; de 1870 à 1878, 661.000 pèlerins conduits par 282 évêques de toutes langues et de toutes régions visitent la célèbre basilique

---

(1) Voir J. BERTRAND, *La Salette, documents et bibliographie*, Paris, 1889 et L. CARLIER, *Histoire de l'apparition de La Salette*, Paris, 1904. La Vierge aurait en plus de son message public révélé un secret qui fut en 1851 envoyé à Pie IX sous pli scellé. Mélanie publia par la suite ce « secret » sous une forme sans doute déformée ; il importe absolument de « distinguer deux Mélanies : la petite voyante de 1846 et la visionnaire de 1878 ». (J. VERDUNOY, *La Salette*, Paris, 1906, p. 104.)

(2) Ce jugement fut confirmé en 1854 par le nouvel évêque, Mgr Ginouilhac, en dépit de l'opposition de quelques prêtres et du scepticisme du cardinal de Bonald, archevêque de Lyon. (Cf. L. BASSETTE, *Les origines du culte de Notre-Dame de La Salette*, dans *Vie spirituelle*, t. LXXV, 1946, p. 216-225.)

(3) En 1854, l'évêque de Birmingham, Mgr Ullathorne, écrivit un livre qui connut 6 éditions en Angleterre et fut traduit en allemand : *The Holy Mountain of La Salette*.

(4) Voir J.-M. CROS, *Histoire de Notre-Dame de Lourdes*, Paris, 1925-1926, 3 vol. (nombreux documents) ; H. PETITOT, *Histoire des apparitions de N.-D. de Lourdes*, Paris, 1935 et *Sainte Bernadette*, Paris, 1940.

qui devient « le rendez-vous de la piété française et bientôt de la piété mondiale »[1]. Pie IX, qui avait affirmé la «lumineuse évidence» de l'apparition et placé son image dans son oratoire, accorda à la statue de Notre-Dame de Lourdes les honneurs d'un couronnement solennel auquel le nonce procéda solennellement, le 3 juillet 1876, en présence de 34 évêques, pendant que 3.000 prêtres et 100.000 fidèles faisaient retentir les Pyrénées de leurs acclamations : Vive l'Immaculée-Conception ! Vive Pie IX ! Vive la France catholique ! [2]

LE GOUT DU MERVEILLEUX — Les faits miraculeux de Lourdes ne sont qu'un cas privilégié. On dirait que Dieu prend plaisir à réagir contre le rationalisme positiviste de l'époque en multipliant les interventions du surnaturel : autour du curé d'Ars et de don Bosco se produisent des miracles qui ne le cèdent en rien à ceux des temps anciens que les ouvrages d'Ozanam et de Montalembert ont remis en honneur ; et dans la vie de nombreux héros plus humbles de la sainteté, pauvres religieuses, curés héroïques, fondateurs d'hôpitaux ou d'orphelinats, on perçoit ce même témoignage sensible d'une action surnaturelle[3]. Les canonisations, que Pie IX aime entourer d'un lustre peu ordinaire et dont le nombre frappe les contemporains après la réserve des papes antérieurs, contribuent à attirer l'attention des fidèles sur le caractère permanent dans l'Église des merveilles de la grâce divine.

Que certains passent d'un extrême à l'autre et, pour réagir contre le scepticisme du siècle précédent, se montrent d'une crédulité excessive, la chose est inévitable. La faveur que rencontrent sous le Second Empire, même auprès de certains prêtres, les excentricités d'un Vintras, est révélatrice à cet égard[4]. Sans aller aussi loin, trop de personnes pieuses et d'ecclésiastiques fervents se mettent à considérer comme un manque de foi la moindre réserve à l'égard des faits merveilleux du présent ou du passé. Non seulement on se jette avec une curiosité malsaine, contre laquelle certains évêques comme Dupanloup estiment nécessaire de réagir, sur toute une littérature colportant sans la moindre critique toute sorte de prophéties, de miracles ou de récits de stigmatisation ; mais on assiste à un regain des vieilles légendes hagiographiques du moyen âge. Elles étaient restées vivaces en Italie et surtout en Espagne. Le P. Faber s'emploie à les faire revivre en Angleterre. Mais c'est surtout en France, dans les milieux ultramontains influencés par le traditionalisme, que le phénomène se développe. Au moment même où les bollandistes se reconstituent pour reprendre leur œuvre de vérité, et où les historiens catholiques allemands formés aux méthodes universitaires entreprennent de récrire l'histoire ecclésiastique à l'aide de documents strictement authentiques, des auteurs pieux comme l'abbé Gaume, au nom du principe qu'il faut accepter toute croyance traditionnelle du moment qu'elle

(1) A. Dansette, Histoire religieuse de la France contemporaine, t. I, p. 443.
(2) Cf. E. Lecanuet, L'Église de France sous la IIIᵉ République, t. I, p. 382.
(3) Cf. M.-H. Vicaire, dans Histoire de l'Église illustrée, de G. de Plinval, t. II, p. 278.
(4) Voir l'article objectif d'A. Amann, dans D. T. C., t. XV, col. 3055-3062.

favorise la dévotion, se font les ardents défenseurs de tous les récits légendaires aussi invraisemblables qu'émouvants dont Mabillon et les mauristes avaient fait justice un siècle plus tôt [1].

SPIRITUALITÉ D'ACTION ET APOSTOLAT — Ces excès de crédulité sont un des points faibles de la piété de cette époque. Il en est d'autres, notamment un certain formalisme et un certain sentimentalisme. Mais les éléments positifs sont appréciables. Ce n'est pas un mince mérite d'avoir rendu au Christ la place centrale qui lui revient, d'avoir osé affirmer bien haut la réalité du surnaturel, d'avoir retrouvé le sens de la croix, d'avoir mieux compris la nécessité de s'alimenter à la source vive des sacrements. Par ailleurs, on est stupéfait de constater à quel point les chrétiens de ce temps, qui devaient mettre si longtemps à comprendre la justice sociale, ont su pratiquer la charité sous toutes ses formes, y compris les plus pénibles. Et même si l'on fait des réserves sur les méthodes, on doit admirer les ressources de dévouement qu'ils surent consacrer à la vaste entreprise de restauration religieuse à laquelle leur siècle s'était attelé avec un tel optimisme. L'ardeur apostolique de cette époque, dont les grands saints furent des curés, des missionnaires et des religieux éducateurs, fait pardonner beaucoup de choses et invite à juger avec plus de compréhension son intransigeance à l'égard de tout ce qui pouvait entraver le rayonnement de la vérité.

VIE INTÉRIEURE ET ORAISON MÉTHODIQUE — Ces chrétiens actifs, dont la plupart préfèrent l'ascèse à la contemplation, savent pourtant être aussi des hommes de prière, en aimant d'ailleurs donner à leur prière un sens apostolique. A partir de 1850, la conviction s'implante chez les meilleurs prêtres de la nécessité de faire régulièrement oraison et même les laïcs, pour lesquels on commence à organiser des retraites fermées, découvrent dans la méditation un moyen de sanctification indispensable pour réagir contre l'atmosphère de naturalisme qui les environne.

C'est la méditation discursive, sous la forme schématisée préconisée par le P. Roothaan dans son *De ratione meditandi*, qui a les faveurs des chrétiens de ce temps. La chose se comprend aisément si l'on songe que beaucoup de retraites sacerdotales et la plupart des retraites pour gens du monde sont prêchées par des jésuites et que la réaction qui devait ramener la Compagnie vers les méthodes authentiques de saint Ignace n'a pas encore commencé. Quelques foyers de résistance à cette prépondérance de la « spiritualité jésuite » apparaissent cependant au cours des années soixante : dom Guéranger commence à attirer l'attention sur la méthode d'oraison bénédictine [2], et il rencontre l'approbation du P. Faber

---

(1) Voir notamment, à propos de la controverse sur l'apostolicité des Églises de Gaule, *supra*, p. 215.
(2) Dans la préface de l'*Enchiridion benedictinum*, publié en 1862. Cf. dom DELATTE, *Dom Guéranger*, t. II, p. 241-242.

et de Mgr Gay, lequel va lui-même bientôt contribuer par ses propres écrits à un élargissement des perspectives en matière de spiritualité.

LE RENOUVEAU DE LA LITTÉRATURE SPIRITUELLE   La vitalité spirituelle du XIXe siècle n'a trouvé que difficilement des modes d'expression artistiques ou littéraires satisfaisants, spécialement en France. Les fadeurs de « l'art Saint-Sulpice » ont leur pendant dans le sentimentalisme superficiel de ces « affreux petits livres de piété » dénoncés par Ernest Hello : dépourvus de doctrine solide, on les croirait écrits pour aider les incroyants « à se persuader que la faiblesse, la médiocrité, la niaiserie sont les attributs nécessaires de la parole catholique » [1]. Leur succès est d'ailleurs en baisse : alors que, pendant la première moitié du siècle, les libraires catholiques avaient écoulé sans difficulté une masse d'opuscules pieux, les collections religieuses qu'on essaie de lancer au cours du Second Empire se vendent plutôt mal [2].

Pourtant, même dans ce domaine de la littérature spirituelle, un renouveau est amorcé dont les pontificats suivants recueilleront tous les fruits. Hello lui-même [3] devait contribuer, par ses traductions si prenantes des *Visions et Révélations* d'Angèle de Foligno (1868) et des *Œuvres choisies* de Ruysbroeck (1869), à ramener aux grands maîtres de la spiritualité, tandis qu'en Allemagne, le grand ouvrage de Goerres sur la mystique chrétienne avait été le point de départ d'une foule de travaux, soit d'érudition, soit de mystique spéculative : il faut spécialement relever les premières œuvres de Scheeben, *Nature et Grâce* (1861) et *Les splendeurs de la divine grâce* (1863), où l'on retrouve en même temps les horizons des Pères grecs révélés au jeune professeur de Cologne par son maître Franzelin.

L'Angleterre produit également des œuvres originales et nourrissantes dues à des convertis qui ont conservé de leur formation anglicane un sens biblique et patristique beaucoup plus prononcé que chez la plupart des catholiques du continent : Newman, dont les sermons dans leur sobre retenue révèlent une âme d'une ferveur peu commune ; Manning, qui remet en lumière l'action du Saint-Esprit dans la direction de l'Église et dans la sanctification des âmes [4] ; Dalgairns, qui cherche à révéler à l'Angleterre puritaine « les côtés attrayants de la religion du Christ » ; Faber enfin, que le succès de son premier ouvrage, *Tout pour Jésus* (1853), encourage à continuer et qui, travaillant jusqu'à seize heures par jour et sans rature, fournit en quelques années une œuvre, trop abondante sans doute, mais qui, rapidement traduite, contribue à diffuser sur le continent comme dans les pays de langue anglaise une spiritualité s'inspirant à la fois de l'école italienne de saint Alphonse et des grands oratoriens du XVIIe siècle.

(1) Cité dans P. POURRAT, *op. cit.*, t. IV, p. 636.
(2) M. NÉDONCELLE, *Les leçons spirituelles du XIXe siècle*, p. 24.
(3) Sur Hello (1823-1885), l'un des très rares représentants, au cours du Second Empire, d'une culture catholique un peu originale, voir J. SERRE, *Ernest Hello. L'Homme, le Penseur, l'Écrivain*, Paris, s. d.
(4) *The temporal Mission of the Holy Ghost* (1865) ; *The internal Mission of the Holy Ghost* (1875). Son ouvrage le plus célèbre, *Le Sacerdoce éternel*, date de la période suivante.

En France enfin, quelques œuvres importantes tranchent sur la banalité de la production courante et en reviennent aux grandes traditions de saint François de Sales et de Bérulle. Dès 1854, M. Hamon, curé de Saint-Sulpice, publie une vie de l'évêque de Genève appelée à un succès durable et, après 1870, l'abbé Chaumont contribue à remettre en honneur la spiritualité salésienne par ses livres et par la direction des deux sociétés de prêtres et de dames du monde mises sous le patronage du saint évêque. C'est aussi l'esprit de saint François de Sales qu'on retrouve en maintes pages des opuscules pieux de Mgr de Ségur [1], directeur de conscience apprécié, ultramontain militant, qui fut mêlé de près à toutes les grandes initiatives religieuses de Paris pendant le troisième quart du XIXe siècle. Mais chez ce dernier apparaît également l'influence de l'école française du XVIIe siècle dont la tradition s'était conservée, bien qu'édulcorée, dans les séminaires sulpiciens ainsi que chez les Pères du Saint-Esprit grâce à leur second fondateur, le vénérable Libermann, dont l'action personnelle, jusqu'à sa mort en 1852, fut très grande sur le clergé.

Celui qui a fait le plus toutefois pour ramener les âmes aux grandes perspectives bérulliennes, c'est Mgr Gay [2], prédicateur et directeur très goûté, dont le grand ouvrage *De la vie et des vertus chrétiennes*, publié en 1874, devait faire « l'auteur spirituel français le plus classique du XIXe siècle » [3] : la spiritualité paulinienne et johannique de notre incorporation au Christ, tel est le fond de ce livre austère dont 10.000 exemplaires s'enlèvent en dix-huit mois à la grande surprise de l'auteur et de ses conseillers, qui ne se doutaient pas que leurs contemporains étaient déjà préparés à ce point à se nourrir de ces grandes vérités.

§ 4. — La liturgie et l'art sacré [4].

LES DÉBUTS DU MOUVEMENT LITURGIQUE — L'époque de Pie IX est marquée par une prépondérance très nette de ce qu'on a coutume de nommer la « piété moderne », sous la double forme des dévotions particulières et de l'oraison méthodique. Parallèlement pourtant, se produit une lente renaissance de l'esprit liturgique qui prépare une évolution de la spiritualité dont les effets apparaîtront au XXe siècle.

Il s'en faut du reste de beaucoup que cette renaissance soit générale.

(1) Sur Louis-Gaston de Ségur (1820-1881), fils de la comtesse de Ségur, née Rostopchine, si connue par ses livres pour enfants, voir H. CHAUMONT, *Mgr de Ségur, directeur des âmes*, Paris, 1884, 2 vol. ; *Œuvres* en 10 vol., Paris, 1876 et suiv.

(2) Sur Charles Gay (1815-1892), converti par Lacordaire et le P. de Ravignan, vicaire général puis évêque auxiliaire de Mgr Pie, voir B. DU BOISROUVRAY, *Mgr Gay, évêque d'Anthédon*, Tours, 1922, 2 vol.

(3) P. POURRAT, *La Spiritualité chrétienne*, t. IV, p. 619.

(4) BIBLIOGRAPHIE. — Sur le mouvement liturgique : outre la biographie de dom Guéranger, voir O. ROUSSEAU, *Histoire du mouvement liturgique*, Paris, 1945 ; P. W. TRAPP, *Vorgeschichte und Ursprung der liturgischen Bewegung*, Ratisbonne, 1940.

Sur la musique sacrée : R. AIGRAIN, *La musique religieuse*, Paris, 1929 ; P. WAGNER, *Der Kampf gegen die Editio Vaticana*, 1907 ; N. ROUSSEAU, *L'école grégorienne de Solesmes*, Rome-Tournai, 1910.

Sur l'art religieux : A. MICHEL, *Histoire de l'Art*, t. VIII, 2e et 3e vol., Paris, 1925 ; A. REICHENSPERGER, *L'art gothique au XIXe siècle*, trad. franç., Bruxelles, 1867 ; J. KREITMAIER, *Beuroner Kunst*, Fribourg-en-Br., 1914.

L'Italie, jusque vers 1900, ne s'intéresse à la liturgie que sous l'aspect de la restauration du chant grégorien et la commission désignée en 1856 pour la réforme du bréviaire n'aboutit pratiquement à rien [1]. L'Espagne, devenue « le pays classique des dévotions particulières et innombrables », gardera longtemps encore « un visage tout à fait aliturgique » [2]. En Angleterre aussi, le catholique moyen manque totalement d'intérêt pour la liturgie en dépit de l'exemple anglican et des efforts de Wiseman, qui aurait aimé voir le sens des tractariens pour l'antiquité chrétienne et leur profonde estime de l'office divin se développer dans le catholicisme renaissant. L'indifférence à l'égard de la piété traditionnelle de l'Église est patente jusque dans l'ordre bénédictin, où l'influence des méthodes ignatiennes avait modifié l'atmosphère spirituelle des monastères, en Allemagne notamment [3], et où l'exécution de l'office choral trahissait parfois une absence totale de sens liturgique, témoin ces observations sur l'abbaye romaine de Saint-Paul-hors-les-Murs en 1870 :

> Le chant des Vêpres doit durer près de trois heures et ne sont qu'une sorte d'Académie de musique. Des chantres à la tribune, des solos et des chœurs, voix affectées, voix de théâtre, tout le monde presque assis et tournant le dos à l'autel pour jouir de la musique. C'est un spectacle plus profane que religieux [4].

Le renouveau fut vraiment l'œuvre personnelle de dom Guéranger, dont le laborieux effort [5] se poursuit tout au long du pontificat de Pie IX. La méfiance romaine fait place progressivement aux encouragements et, en 1875, à l'occasion de la mort de l'abbé de Solesmes, un bref pontifical fera l'éloge de ses mérites et accordera, en témoignage des services rendus par lui à l'Église et à la liturgie, qu'il y ait toujours un bénédictin parmi les membres de la Sacrée Congrégation des Rites [6]. Si, dans le domaine des études scientifiques, le travail inauguré par les *Institutions liturgiques* (qui s'achèvent en 1851 et dont le premier volume est traduit en allemand en 1854) ne portera vraiment ses fruits que vers la fin du siècle, par contre, le succès des volumes de *L'Année liturgique* [7] s'affirme d'année en année. Beaucoup de lecteurs s'étaient au début sentis dépaysés, mais le journal maçonnique qui écrivait : « Voilà un ouvrage qui fera autant de mal que les contes de Voltaire ont fait de bien » [8] voyait juste en prévoyant l'action profonde qu'elle finirait par exercer sur le renouveau de la piété catholique.

C'est en Allemagne que cette œuvre, née en France, devait d'abord porter ses fruits. En 1864, dom Maur Wolter, le restaurateur de Beuron,

---

(1) Les changements proposés par la bulle du 9 juillet 1868 sont en effet très superficiels. (Cf. P. BATIFFOL, *Histoire du bréviaire romain*, Paris, 1911, p. 425-426.)

(2) M.-L. CARRERA, *Le mouvement liturgique en Espagne*, dans *Questions liturgiques et paroissiales*, t. XV, 1930, p. 334.

(3) *La Maison-Dieu*, n° 7 (1946), p. 143.

(4) H. ICARD, *Journal de mon voyage à Rome*, p. 145. (*Arch. S. Sulp.*, fonds Icard.)

(5) On a pu en suivre les débuts dans le volume précédent, p. 508-509, où sont signalés les points faibles de l'œuvre liturgique de l'abbé de Solesmes ; sur la valeur permanente de celle-ci, voir les excellentes pages d'O. ROUSSEAU, *op. cit.*, p. 7-14 et 45-53.

(6) On trouvera ce bref du 19 mars 1875, de même que ceux à Mgr Pie, du 29 mars, et à l'évêque d'Angers, du 18 avril, en tête du premier volume de la 2e édit. des *Institutions liturgiques*.

(7) Neuf volumes paraissent de 1841 à 1866. Les cinq derniers (le temps après la Pentecôte) seront l'œuvre des disciples de Guéranger, en particulier de dom Lucien Fromage. *L'Année liturgique* fut traduite en allemand à partir de 1875, puis en italien et en anglais.

(8) Cité par O. ROUSSEAU, *op. cit.*, p. 52.

traduisait les célèbres *Exercices spiriluels* de sainte Gertrude, fondés sur la vie liturgique annuelle et journalière et, dans une longue introduction où l'on sent vibrer l'enthousiasme d'un disciple [1], il faisait part au public allemand du mouvement suscité en France par dom Guéranger, dont il exposait les principes ; l'année suivante, il publiait un petit opuscule, *Le Plain-Chant el la Lilurgie*, qui citait des pages entières de l'introduction à *L'Année lilurgique* et constituait un résumé des principales idées de l'abbé de Solesmes sur la richesse insurpassable de la nourriture spirituelle contenue dans l'Office divin. Cet opuscule eut une grande répercussion et, avec le *Gerlrudenbuch*, devait puissamment concourir à orienter les catholiques allemands vers la piété liturgique qui, en France, restera pendant plusieurs dizaines d'années encore le lot d'une petite élite.

LE TRIOMPHE DE LA LITURGIE ROMAINE
Sur un point toutefois, l'œuvre de dom Guéranger est couronnée en France d'un plein succès : sa campagne pour le retour à la liturgie romaine, appuyée par tous ceux qui militent en faveur de l'ultramontanisme et positivement encouragée par Rome à partir de 1850, aboutit en une quinzaine d'années à la disparition totale des anciennes liturgies gallicanes contaminées par le jansénisme. Seul le diocèse de Lyon parvint à conserver quelque chose de ses vénérables coutumes, après un conflit assez vif qui opposa un moment à Pie IX son vieil archevêque et la quasitotalité du clergé [2]. Le succès du mouvement unificateur fut si total que l'abbé de Solesmes ne réussit même pas à obtenir pour sa congrégation le « propre » qu'il avait préparé [3].

Si le particularisme liturgique s'était surtout développé en France sous l'influence du gallicanisme, de nombreuses particularités locales moins importantes, pour le cérémonial des évêques ou pour l'office canonial par exemple, subsistaient cependant souvent dans les autres pays également. Malgré la mauvaise humeur du vieux clergé, qui tenait à ses traditions, les évêques s'appliquèrent à les faire disparaître pour répondre au désir du Saint-Siège, qui souhaitait voir les usages romains adoptés uniformément dans toute la chrétienté latine de manière à mieux manifester sensiblement l'unité de l'Église.

L'hostilité de Rome envers tout ce qui évoquait les particularismes d'autrefois devint bientôt telle qu'en 1863 la Congrégation des Rites, à l'instigation de Mgr Carazza, maître des cérémonies de la chapelle papale, fut sur le point d'interdire absolument les chasubles gothiques. Pie IX ne permit toutefois pas que le rapport très violent, qui accumulait tous les arguments imaginables contre les ornements s'écartant du type romain, fût distribué aux consulteurs et il se contenta de faire adresser une circulaire

(1) Sur les rapports entre Solesmes et Beuron, voir le recueil collectif *Maurus Wolter, dem Grün der Beurons*, Beuron, 1925, et surtout la correspondance entre Wolter et Guéranger, publiée par D. ZAEHRINGER, dans *Benediktinische Monatschrift*, t. XIX, 1937, p. 234-241.
(2) Cf. J. MAURAIN, *La politique ecclésiastique du Second Empire*, p. 695-699.
(3) Il dut se contenter d'une transaction, en 1856. (Cf. dom DELATTE, *Dom Guéranger*, t. II, p. 118-128.)

aux évêques de France, d'Allemagne, de Belgique et d'Angleterre, pour leur demander des explications [1] ; cette désapprobation mitigée suffit pour que certains évêques s'empressent de donner des instructions à leur clergé en conséquence [2].

LA MUSIQUE SACRÉE   Dom Guéranger avait pressenti que la réforme liturgique entraînerait un relèvement de tous les arts religieux. La chose était particulièrement nécessaire en ce milieu du XIXᵉ siècle où le chant et l'ornementation des églises apparaissaient trop souvent, selon le mot de Huysmans, comme « une revanche du diable ». Les hommes de goût et les hommes religieux se préoccupaient surtout de la décadence de la musique sacrée dont la réforme s'imposait d'urgence, mais « il fallut laisser mourir tout doucement une ou deux générations de chanteurs et d'auditeurs pour qu'une amélioration sensible puisse se faire sentir » [3] ; au lendemain de la mort de Pie IX, il restait encore beaucoup à faire, comme le montre un décret de la Congrégation des Rites de 1884 interdisant

de jouer dans l'église toute phrase, si minime soit-elle, ou toute réminiscence des œuvres théâtrales, de morceaux de danses de tout genre comme polkas, valses, mazurkas, menuets, rondels, scottisch (...), chansons populaires érotiques ou bouffonnes, romances, etc., et d'utiliser les instruments de musique trop bruyants comme tambours, grosses caisses, cymbales et semblables, comme les instruments de saltimbanques [4].

Le renouveau se manifesta dans le triple domaine de la musique instrumentale, de la musique polyphonique et du chant grégorien. La musique d'orgue se régénéra en revenant à l'esprit de Bach. La Belgique y eut une part importante avec Nicolas-Jacques Lemmens (1823-1881) et César Franck (1822-1890), lequel appartient du reste plus encore à la France par sa carrière et ses principaux élèves. Le premier, dont l'*École d'orgue* devint bientôt classique et qui devait fonder en 1879 l'École de musique religieuse de Malines sous le patronage de l'épiscopat belge, eut une grosse influence, mais qui n'égale pas celle de César Franck. L'œuvre de celui-ci, d'une science et d'un charme qui rappellent Bach et Beethoven, manifeste la foi profonde de son auteur et contient une profondeur de tendresse et d'émotion pieuse qui ne se retrouve que chez les primitifs.

En matière polyphonique, les Italiens demeurent loin de Palestrina et leurs productions évoquent davantage le *bel canto* que la musique d'église ; Veuillot lui-même, malgré son admiration pour tout ce qui venait de Rome, devait bien l'avouer [5]. Les compositions françaises, pour leur

---

(1) Lettre du 21 août 1863, reproduite dans Mgr Barbier de Montault, *Le costume et les usages ecclésiastiques selon la tradition romaine*, t. II, p. 26, qui donne tous les détails utiles sur l'histoire de la lettre. (Voir le texte du rapport de Mgr Corazza dans *Analecta Juris Pontificii*, sér. XXVIII, col. 867-892 et 965-1016.)

(2) Par exemple le cardinal Sterckx, de Malines. (Cf. mandement du 17 octobre 1863 dans *Collectio Epistolarum Pastoralium*, t. III, Malines, 1870, p. 652-654.)

(3) O. Rousseau, *Histoire du mouvement liturgique*, p. 154.

(4) Cité *ibid.*, p. 155.

(5) L. Veuillot, *Les Parfums de Rome*, t. II, p. 69.

part, ne sont le plus souvent que « de la musique mondaine, d'une piété toute en surface, faite bien plus pour plaire que pour édifier » [1] ; seul Gounod (1818-1893) tranche un peu sur la médiocrité générale, non certes dans son « fâcheux *Ave Maria* » [2], mais dans quelques pages de ses *oratorios, Rédemption* et *Mors et Vita*. De même en dépit de leur indéniable puissance artistique, les œuvres de Liszt, notamment la fameuse *Messe de Gran* et la *Messe hongroise du couronnement*, écrite en 1867 sur des thèmes nationaux, négligeaient trop le véritable esprit de la liturgie pour concourir efficacement à la restauration de la musique religieuse. Les compositions d'Anton Brueckner (1827-1896), l'organiste de la cour de Vienne, marquent déjà un net progrès à ce point de vue, mais c'est surtout en Allemagne où, depuis 1860, les conciles provinciaux s'occupèrent avec insistance de la réforme de la musique sacrée [3], que se produisit le vrai retour à l'esprit des grands maîtres de l'époque classique. Il avait été préparé par le chanoine Proske (1794-1861) qui, après avoir laborieusement rassemblé une collection unique des chefs-d'œuvre de la polyphonie ancienne, réussit, avec l'aide de Mettenleiter, à faire apprécier le style palestrinien par un nombre croissant d'auditeurs. Mais à sa mort, son action restait encore limitée à la seule ville de Ratisbonne. C'est au prêtre bavarois F.-X. Witte (1834-1883) qu'était réservé le mérite d'étendre son œuvre à l'Allemagne entière. En novembre 1865, il lançait sa brochure-programme : *L'état de la musique dans l'Église catholique en Bavière* ; l'année suivante, il fondait ses *Reformblätter* et, en 1868, lors du congrès catholiques de Bamberg, il réussit enfin à mettre sur pied la société à laquelle il songeait depuis plusieurs années, le *Caecilienverein*, dont les ramifications s'étendirent vite dans tous les pays de langue allemande. Sous l'égide de cette association, une équipe de compositeurs entreprit de constituer un répertoire moderne dans l'esprit ancien, où les œuvres de réelle valeur sont l'exception, mais qui eut l'avantage de « fournir des chants faciles et toujours corrects aux groupes choraux très répandus en Allemagne et en Suisse, jusque dans les plus petites églises » [4].

LA RESTAURATION DU CHANT GRÉGORIEN — Plus importants encore par leurs conséquences furent le renouveau d'estime pour le plain-chant et le travail accompli en vue de retrouver la pureté primitive des mélodies grégoriennes. L'influence de dom Guéranger fut, ici encore, prépondérante. Alors que tous les monastères qui avaient survécu à la Révolution continuaient à chanter des messes et des motets à plusieurs voix, il avait réintroduit à Solesmes l'usage exclusif de l'antique chant romain, et la multiplication des rééditions du Graduel ou de l'Antiphonaire à partir de 1848 (Malines, 1848 ; Reims-Cambrai, 1851 ; Rennes, 1853 ; Dijon, 1855 ; Digne, 1858) indiquent que ce renou-

(1) K. WEINMANN, *La musique d'Église*, trad. franç., Paris, 1912, p. 178.
(2) R. AIGRAIN, *La musique religieuse*, p. 207.
(3) F. ROMITA, *Jus musicae liturgicae*, Rome, 1936, p. 108 et suiv.
(4) R. AIGRAIN, *op. cit.*, p. 207.

veau ne resta pas confiné à son monastère. Les moines de Beuron, en
particulier dom Benoît Sauter qui avait été novice à Solesmes, intéressèrent à cette renaissance le public catholique d'Allemagne, où la pratique
du grégorien s'était beaucoup plus perdue qu'en France, et le mouvement,
encouragé par l'épiscopat, bénéficia bientôt de l'appui du *Caecilienverein*.

Mais il ne suffisait pas d'en revenir à l'usage du plain-chant, il fallait
dégager celui-ci des nombreuses altérations qui s'y étaient infiltrées
au cours des siècles. Depuis que le musicologue Fétis, directeur du
conservatoire de Bruxelles, avait attiré l'attention sur ces altérations
et que le jésuite belge Lambillotte avait publié en fac-similé le fameux
manuscrit de Saint-Gall, qu'il prenait pour l'antiphonaire même de saint
Grégoire, la question faisait l'objet de discussions passionnées où souvent
« l'ardeur des convictions tint lieu de compétence et d'où les préoccupations d'amour-propre et d'intérêt ne furent pas toujours exclues »[1]. La
clarté devait toutefois se faire progressivement. Déjà, alors que les éditions courantes se contentaient de réimprimer celles du XVIe ou du
XVIIe siècle, l'édition de Reims-Cambrai s'inspirait d'un important manuscrit découvert à Montpellier en 1847 ; en Allemagne, des travaux, sérieux
pour l'époque, de Schubiger et de Schlecht préparaient la voie à une
édition plus correcte encore, d'après les manuscrits anciens, dont les
éditions de Trèves en 1863 et de Cologne en 1865 donnaient un avant-
goût. Mais c'est à Solesmes que se faisait le travail le plus scientifique :
encouragé par dom Guéranger, dom Jausions parcourut pendant des
années les bibliothèques, en quête de manuscrits à notation ancienne, et
élabora peu à peu la théorie que son collaborateur dom Pothier devait
exposer en 1881 dans son célèbre ouvrage : *Les mélodies grégoriennes*.

Une malencontreuse décision devait toutefois retarder cette renaissance de quelques années. En 1869, la Congrégation des Rites, soucieuse
de l'unification du chant liturgique, chargea l'éditeur Pustet de Ratisbonne d'entreprendre une édition officielle. Or, le texte choisi pour le
Graduel, qui parut en 1871, fut celui, très déformé, de l'édition du
XVIe siècle dite « médicéenne », que l'édition de Malines de 1848 avait remis
en honneur et que beaucoup croyaient être le résultat d'une revision due
à Palestrina. Conseillé par certains dirigeants du *Caecilienverein* et par des
consulteurs dont la compétence scientifique n'égalait pas la bonne volonté,
Pie IX, dans des brefs très laudatifs, ne se borna pas à recommander le
texte comme ayant les préférences de Rome, mais il le présenta comme
« le chant grégorien authentique (*genuinus*)... qui peut être considéré
comme le plus conforme, d'après la tradition, à celui dont saint Grégoire
le Grand avait doté la liturgie sacrée »[2]. Comme l'éditeur avait reçu pour
une durée de trente ans le monopole des éditions liturgiques, l'introduction d'un texte établi sur des bases scientifiques subit un certain retard,
mais on peut dire que, dès la mort de Pie IX, les bases de la grande réforme
de Pie X étaient posées.

---

(1) H. LECLERCQ, art. *Lambillotte*, dans *D. A. C. L.*, t. VIII, col. 1076.
(2) Brefs du 20 janvier et du 14 août 1871.

L'ARCHITECTURE RELIGIEUSE On a rarement restauré ou construit autant d'églises que pendant la seconde moitié du XIXᵉ siècle, mais rarement aussi les architectes ont fait preuve d'un pareil manque de goût et d'originalité.

Le meilleur de leur œuvre est encore la continuation de la vaste entreprise de restauration des chefs-d'œuvre de l'art médiéval. Celle-ci se poursuit en Allemagne sous l'égide du « cercle de Trèves » constitué autour de Pierre Reichensperger : les travaux d'achèvement du dôme de Cologne durent jusqu'en 1880, tandis que Ketteler lance de vibrants appels pour la restauration des dômes de Worms et de Mayence et que Beda Weber en fait autant pour celui de Francfort. Le mouvement gagne la France, où il aura pour principal artisan Viollet-le-Duc (1813-1879) qui, après s'être révélé à Vézelay, sera chargé par la *Commission des Monuments historiques* de restaurer la Sainte-Chapelle, Notre-Dame de Paris, les cathédrales d'Amiens, de Chartres, de Reims, l'abbatiale Saint-Denis. Soutenu par la faveur de Napoléon III, il s'acquitta de sa tâche avec une science incontestable, bien qu'on ait pu lui reprocher à juste titre d'avoir retouché ces vénérables monuments d'après ses idées personnelles en les refaisant « non pas tels qu'ils étaient, mais tels qu'ils devaient être ».

En matière de constructions neuves, par contre, le bilan est tout à fait décevant. Guère d'originalité jusqu'en 1885 : tout au plus peut-on signaler, outre l'apparition du fer, qui n'avait pas encore été admis dans l'architecture religieuse et que Baltard emploie pour la première fois à Saint-Augustin de Paris, achevé en 1869, deux préoccupations nouvelles, surtout à Paris : les églises sont situées dans une perspective monumentale et leur plan vise moins à faciliter la prière du peuple chrétien et les cérémonies ordinaires du culte qu'à fournir un « décor » aux grands mariages et aux premières communions somptueuses. Quelques architectes construisent, en s'inspirant de la tradition du XVIIIᵉ siècle, des églises « clinquantes » et théâtrales comme celle de la *Trinité*, inaugurée en 1867 ; d'autres font du néo-roman, ou du néo-byzantin ; la plupart préfèrent le néo-gothique, dont les formules stéréotypées et sans art triomphent partout, non seulement en Allemagne, où la passion exclusive pour l'art gothique apparaît quelque peu intolérante aux yeux des Romains [1], mais également en France, où en vingt ans la campagne entreprise par Montalembert, Rio et Viollet-le-Duc parvient à changer le goût du public, et jusqu'en Amérique ou dans les pays de mission.

LA PEINTURE RELIGIEUSE La peinture religieuse ne se renouvelle pas davantage. L'art italien se meurt, on en trouve une preuve entre beaucoup dans cette « salle affreuse que Pie IX crut élever à la gloire de l'Immaculée-Conception » [2] dans le palais du Vatican. Dans les pays germaniques, l'influence d'Overbeck, qui vit jusqu'en 1869, se prolonge dans une atmosphère émouvante de piété, mais

(1) Cf. *Civ. catt.*, ser. III, vol. X, 1858, p. 123.
(2) N. MAURICE-DENIS, *Romée*, 2ᵉ édit., Paris, 1948, p. 82.

sans originalité artistique, avec Steinle et les bons imagiers de l'école de Dusseldorf, Deger, Schadow, Ittenbach, ainsi qu'avec Fuehrich à Vienne, dont le chemin de croix peint en 1846 pour l'église Saint-Jean du Prater devait être reproduit dans le monde entier. C'est aussi à l'école des Nazaréens que se rattache par ses origines l'école de Beuron, fondée en 1870 par Pierre Lenz, devenu frère Didier, qui sortait de l'Académie de Munich : intéressante à ses débuts, cette école, qui prétendait faire revivre les traditions de l'art hiératique des Égyptiens, devait assez vite sombrer dans la géométrie et l'esprit de système.

L'influence des Nazaréens allemands unie à celle des Préraphaëlites anglais se retrouve en France chez certains peintres mystiques comme Ary Scheffer (1795-1858) ou Hippolyte Flandrin (1809-1864). A part eux, on a le choix entre de belles œuvres, mais que leur titre seul distingue du genre profane [1] (tels l'*Héliodore* peint par Delacroix à Saint-Sulpice en 1861, ou ces innombrables Madeleine éplorées dont seule l'auréole relève du genre religieux), et de fades scènes évangéliques sans valeur plastique, qui se veulent émouvantes et « vont bientôt rivaliser avec la littérature militaire de Meisonnier » [2], Bouguereau et *tutti quanti*. Puvis de Chavannes pourtant, tout à la fin de la période considérée, apporte un accent nouveau ; mais il n'est guère apprécié du catholique moyen et ceux qui ont la responsabilité des décorations d'église ne font pas appel à lui.

§ 5. — Premières tentatives unionistes [3].

LE SOUCI DE L'UNITÉ CHRÉTIENNE — L'expansion missionnaire, qui est l'une des grandes caractéristiques du pontificat de Pie IX, et l'affermissement croissant du pouvoir pontifical, devaient inévitablement aviver chez les catholiques le désir de voir revenir à l'unité romaine les millions de chrétiens orthodoxes ou protestants séparés du siège de Pierre. Ceux qui suivaient d'un peu près le mouvement du monde et se souvenaient des conversions sensationnelles de la première moitié du siècle en Allemagne, en Angleterre et même, dans une mesure moindre, en Russie, croyaient trouver des raisons d'espérer dans la persistance du mouvement d'Oxford et la crise engendrée au sein du luthéranisme allemand par les progrès du protestantisme libéral, de même que dans les transformations sociales et politiques qui se produisaient dans l'Empire des Tsars, dans les Balkans, dans le Proche-Orient, mettant à nouveau en contact avec l'Occident ces vastes régions où la situation religieuse était restée figée depuis des siècles.

C'est un peu partout qu'à partir de 1850 s'affirment les préoccupations unionistes : en Angleterre, où à l'action qu'exerçaient déjà certains

---

(1) Mais où s'arrête l'art profane ? Le pittoresque P. Martin écrivait à peu près en ce temps : « Je vois de l'art chrétien partout où la pensée chrétienne maîtrise un style quelconque. »

(2) *Les Études*, t. CCLXVIII, 1951, p. 75.

(3) BIBLIOGRAPHIE. — Quelques renseignements dans C. QUENET, *L'unité de l'Église*, chap. I, (dans l'encyclopédie *Tu es Petrus*, Paris, 1934, p. 874 et suiv.) et dans A. PAUL, *L'unité chrétienne. Schismes et rapprochements*, Paris, 1930, IIIᵉ Part., chap. I, III et IV.

Sur l'attitude de Rome à l'égard des Églises orientales : G. SMIT, *Roma e l'Oriente christiano. L'azione dei Papi per l'unità della Chiesa*, Rome, 1944.

convertis de 1830 comme Ignace Spencer, vient s'ajouter l'influence de
Newman, qui a vraiment ressenti en chrétien le drame théologique de la
séparation ; dans les Balkans, où Strossmayer, dont l'œuvre a déjà été
évoquée plus haut [1], souffre du même tourment de l'unité ; en Allemagne,
où Doellinger suit avec intérêt l'évolution de la Réforme, où Léopold
Schmid élabore un programme, à vrai dire assez aventureux, de réconciliation du catholicisme et du protestantisme [2], où le baron von Haxthou-
sen engage une correspondance avec le métropolite Philarète de Moscou
après avoir fondé en 1857, avec le patronage des évêques de Munster
et de Paderborn, une ligue de prières pour la conversion de la Russie,
le *Petrus-Verein* [3], tandis que le publiciste Riess envisage, avec l'approbation d'un théologien jésuite, une association analogue où les catholiques
appliqueraient aux protestants le fruit de leurs communions [4]. Même en
France, où manquait pourtant le stimulant d'un contact constant avec
des non-catholiques, les esprits éclairés commencent à attirer l'attention
sur l'importance qu'il y aurait à « mettre un terme à ces divisions funestes
qui affaiblissent le christianisme et l'empêchent d'accomplir ses destinées
dans le monde » [5].

L'ATTITUDE DU SAINT-SIÈGE   Le centre de la catholicité ne pouvait
rester indifférent à ce mouvement, mais,
comme il est normal, on y envisageait les choses beaucoup plus froidement.

En ce qui concerne les chrétientés issues de la Réforme, comme on ne
voyait pas qu'on pût espérer autre chose que des conversions individuelles,
on s'appliqua surtout à renforcer les centres d'attraction que constituaient les noyaux catholiques en ces régions — rétablissement de la
hiérarchie en Angleterre et aux Pays-Bas, pénétration en Scandinavie,
consolidation de la vie ecclésiastique en Allemagne, multiplication des
diocèses aux États-Unis — et à encourager la tendance des théologiens,
des jésuites en particulier, à en revenir, après une éclipse de deux siècles,
à la tradition des controversistes de la Contre-Réforme.

Ces mesures, qui fortifiaient la situation du catholicisme en pays
protestant, élargissaient par ailleurs le fossé séparant les deux confessions. Du côté de l'orthodoxie également, le Saint-Siège fut amené, sur
certains points, à raidir son attitude. Se basant sur le fait que l'union de
Florence n'avait jamais été dénoncée officiellement, on avait longtemps
fermé les yeux à Rome sur les cas, assez fréquents en Grèce et dans le
Proche-Orient, de *communicatio in sacris* entre catholiques et orthodoxes.
Cette attitude conciliante se modifia à partir du milieu du siècle pour un
double motif : non seulement la multiplication des contacts, par suite des
communications plus faciles, faisait apparaître dans toute son évidence

(1) Cf. *supra*, p. 409-410.
(2) *Der Geist des Katholizismus oder Grundlegung der christlichen Irenik*, Giessen, 1848-1850,
2 vol. Cf. B. Schroeder et F. Schwarz, *L. Schmids Leben und Denken*, Leipzig, 1871.
(3) A. Boudou, *Le Saint-Siège et la Russie*, t. II, p. 98-99.
(4) Florian Riess à Doellinger, 18 novembre 1854, dans J. Friedrich, *Ign. v. Doellinger*, t. III,
p. 153.
(5) Mémoire remis par Mgr Maret aux évêques de France en 1862, cité dans G. Bazin, *Vie de
Mgr Maret*, t. II, p. 263. Maret revint encore sur la question dans le mémoire adressé au pape à
l'occasion du *Syllabus* (*ibid.*, p. 344).

la réalité du schisme, mais surtout la Russie prenait une attitude de plus en plus persécutrice à l'égard des catholiques et on jugea dès lors que « la participation d'un catholique à des cérémonies orthodoxes dans ces conditions ferait scandale, ce qui n'était pas le cas jadis »[1].

Toutefois, s'ils estimaient de leur devoir d'accuser les distances, le pape et plusieurs de ses conseillers n'en continuaient pas moins à désirer sincèrement l'union et estimaient que, du côté grec et slave, certaines possibilités existaient qu'on ne pouvait négliger. Pie IX ne se borna pas à prescrire des prières dans ce but ou à encourager des entreprises comme la *Société chrétienne orientale*, fondée par J. G. Pitzipios en 1853, et qui devait finir plutôt mal ; il rouvrit également la série, interrompue depuis longtemps, des initiatives pontificales à visée unioniste.

ROME ET L'ORIENT   La première, plutôt malencontreuse il est vrai, remonte aux débuts de son pontificat. Le 6 janvier 1848, Pie IX adressait aux autorités orthodoxes grecques l'encyclique *In suprema Petri sede*, pour « les exhorter et supplier de retourner dans la communion du siège de Pierre »[2]. L'idée en avait été suggérée par quelques notabilités romaines qui, vers la fin du règne de Grégoire XVI, avaient constitué autour du cardinal préfet de la Propagande un comité destiné à promouvoir le retour à l'unité des schismatiques orientaux. Malheureusement, ce document, que rien n'avait préparé et qui n'était pas exempt de maladresses, fit plus de mal que de bien : les orthodoxes y virent une provocation et il réveilla bien involontairement « après un armistice de quatre siècles » la polémique entre Grecs et Latins ; au mois de mai suivant, en effet, quatre patriarches et vingt-neuf archevêques réunis à Constantinople ripostèrent hargneusement en stigmatisant les innovations latines, les prétentions papales et la propagande des missionnaires latins parmi les orthodoxes[3].

Des mesures positives plus heureuses devaient être prises une dizaine d'années plus tard sous l'influence du cardinal de Reisach, l'un des rares intellectuels de la Curie, que ses relations avec l'Autriche poussaient à s'intéresser à la question slave et qui avait compris qu'il importait avant tout d'avoir à Rome quelques spécialistes des choses orientales. Presque tout était à faire dans ce domaine. En 1858, admirant chez Mgr Malou, évêque de Bruges, la seule bibliothèque catholique où l'on pût à l'époque trouver rassemblés les principaux ouvrages concernant la controverse avec les orthodoxes, dom Pitra écrivait mélancoliquement : « N'est-il point surprenant qu'on ait comme oublié les Grecs, tandis qu'un immense travail a amoncelé les livres contre la Réforme ?[4] » L'ignorance était particulièrement grande en ce qui concernait la liturgie

(1) Mémoire remis par dom Pitra au cardinal de Reisach en 1860, reproduit dans A. Battandier, *Le cardinal Pitra*, p. 435-439. Peut-être des conseillers d'un tempérament moins intransigeant auraient-ils pu suggérer des solutions plus souples, évitant de rompre tout à fait les ponts comme ce fut le cas à partir de cette époque.
(2) Texte et traduction dans *Irenikon*, t. VI, 1929, p. 666-686.
(3) Texte dans Mansi, *Collectio Conciliorum*, t. XL, col. 377-418. Cf. T. Popescu, *Enciclica Patriarhilor ortodocsi dela 1848. Studiu introductiv, text si traducere*, Bucarest, 1935 (résumé en français).
(4) A. Battandier, *Le cardinal Pitra*, p. 361 ; cf. p. 400.

et le droit canon byzantins ; les quelques travaux de valeur, publiés par les érudits des siècles antérieurs, les Assemani, Thomas de Jésus, Renaudot, Jean Morin, restaient ignorés dans la poussière des bibliothèques et les rares hommes — canonistes de la Propagande ou missionnaires — qui étaient amenés par leurs fonctions à s'intéresser aux chrétientés de rite oriental donnaient facilement dans l'erreur déjà stigmatisée par Benoît XIV : blâmer ou condamner en bloc tout ce qui ne se faisait pas comme chez eux.

Dès qu'il se rendit compte de la compétence que possédait en la matière dom Pitra, alors moine de Solesmes, Reisach suggéra de le mander à Rome et il eut facilement gain de cause, car le moment était bien choisi, comme le pape lui-même l'expliqua au savant bénédictin :

En ce moment, une des portions de l'Église tend les bras vers moi. Ce sont les Grecs unis de l'empire d'Autriche. Je viens de leur donner un archevêque, de créer une province de Fogaras dont le titulaire m'écrivait, hier encore, pour me presser de lever les difficultés de sa position. Je n'ai pu lui répondre qu'en termes généraux qui ne peuvent longtemps suffire [1].

La première idée avait été d'attacher Pitra à une commission chargée des affaires slaves et orientales qui pourrait commencer par régulariser la situation de la Transylvanie. Mais il obtint du pape de pouvoir d'abord rassembler une documentation aussi large que possible sur les sources du droit canonique oriental et il entreprit dans ce but un voyage d'étude en Russie. Soigneusement préparé par des recherches dans les bibliothèques d'Italie, puis, en Belgique, chez Mgr Malou et les bollandistes, il put mener à bonne fin, au cours des années 1859 et 1860, à Saint-Pétersbourg et à Moscou, puis à Prague et chez les méchitaristes de Vienne, « la plus vaste enquête qui eût jamais été entreprise sur l'histoire et les monuments du droit des Grecs » [2]. Il avait en outre eu l'occasion, grâce à son habit bénédictin, de prendre un contact étroit avec les monastères russes et, dès son retour, il adressait au cardinal Antonelli un mémoire extrêmement perspicace attirant l'attention sur le fait que, s'il était nécessaire de pouvoir réfuter théologiquement les erreurs des orthodoxes, il était capital de tenir également compte de « deux choses importantes, indispensables avec les Slaves : la liturgie et la vie monastique », et il proposait en conséquence de porter tous les efforts sur la restauration de ce qui restait des monastères basiliens dans les régions frontières de l'Empire d'Autriche [3].

Lorsque dom Pitra rentra à Rome, en 1861, les problèmes orientaux étaient plus que jamais à l'ordre du jour. A son départ, le pape lui avait confié :

Nous sommes à la veille de grands événements. Le seul affranchissement des serfs est un très grand événement, même pour la religion. Cela entraîne en Russie toutes sortes de mouvements où il semble que Dieu veuille faire sa part. Il faut donc aviser à seconder les vues de la Providence et nous préparer [4].

(1) A. BATTANDIER, *Le cardinal Pitra*, p. 317.
(2) F. CABROL, *Histoire du cardinal Pitra*, p. 235. Voir sur ce voyage, cf. F. CABROL, *op. cit.*, chap. xv et A. BATTANDIER, *op. cit.*, chap. xvi.
(3) Cité dans A. BATTANDIER, *Le cardinal Pitra*, p. 410 et suiv., en particulier p. 412 et 426.
(4) F. CABROL, *Histoire du cardinal Pitra*, p. 223.

Puis il y avait eu les massacres de Syrie de 1860, qui avaient attiré l'attention vers le Levant, et le désir d'avoir à Rome un homme connaissant ces régions ne fut probablement pas étranger à la désignation comme auditeur de rote pour la France de l'abbé Lavigerie, avec lequel le pape s'était longuement entretenu à son retour du Liban [1]. A présent, c'était vers les Balkans que se portaient les regards et l'on s'attendait à Rome à d'importants mouvements de conversion dans ces régions, ainsi que le montre une circulaire distribuée à l'occasion de la réunion des évêques de 1862 :

> Les retours à l'unité, autrefois rares et isolés, se multiplient désormais et ne se bornent plus à de simples particuliers, mais embrassent des populations entières. C'est parmi les Arméniens que cet heureux mouvement s'est fait sentir tout d'abord et qu'il a pris les plus vastes proportions (...). Il y a deux ans à peine, les Bulgares se sont ébranlés à leur tour et le mouvement, commencé à Constantinople, s'est rapidement propagé dans les provinces (...). Enfin ce ne sont plus seulement les races arméniennes et slaves qui subissent l'influence de ce mouvement, l'ébranlement se fait sentir parmi la race grecque elle-même. Un archevêque grec, Mgr Melethios, est rentré, le 21 novembre dernier, dans le sein de l'Église catholique, et déjà, quoique sans ressources, il voit se former autour de lui un noyau de grecs convertis... [2].

S'abandonnant à ces prévisions optimistes, Pie IX, qui avait tenu à consacrer lui-même le premier évêque bulgare uni, chargeait dom Pitra de préparer une nouvelle encyclique aux orthodoxes [3], mais on a vu plus haut que la réalité devait décevoir ces espérances. Toutefois, une mesure prise au moment où tous les espoirs semblaient permis, la constitution, en 1862, d'une section de la Congrégation de la Propagande « pour les affaires du rite oriental », devait avoir une influence durable et heureuse, en provoquant la constitution du noyau de spécialistes rêvé par le cardinal de Reisach. Non seulement chacun des cardinaux attachés à la nouvelle section fut affecté à un rite déterminé et obligé ainsi à acquérir une certaine compétence, mais quelques-uns des meilleurs orientalistes catholiques du temps furent désignés pour les assister : Erculei et Scapaticci, de la Bibliothèque Vaticane, et les Allemands Franzelin, Theiner, Laemmer et Zingerle. Pie IX éteignit à cette occasion la commission instituée par Clément XI pour la correction des livres orientaux et confia au nouvel organisme le soin de reviser la littérature ecclésiastique des Églises non latines, quel qu'en fût l'objet : traductions de la Bible, recueils canoniques ou liturgiques, traités de catéchèse. Les travaux dont fut chargé Pitra après son retour à Rome — revision des textes liturgiques byzantins, publication des résultats de ses recherches sur le droit canonique oriental [4] — s'inscrivent dans la même ligne. En même temps, ce dernier ne manqua pas une occasion d'attirer l'attention des théologiens sur la

---

(1) Mgr BAUNARD, *Le cardinal Lavigerie*, t. I, p. 89 et 107.
(2) Un exemplaire de cette circulaire se trouve aux *Arch. Malines*, fonds Sterckx, dossier Voyages à Rome. (Sur les faits mentionnés, cf. *supra*, p. 425-426.) On profita de la présence des évêques pour organiser à Sant'Andrea della Valle une messe solennelle par le patriarche arménien de Constantinople au cours de laquelle Mgr Dupanloup prêcha sur le retour des Églises d'Orient.
(3) Pitra à Guéranger, 24 mai 1862, dans F. CABROL, *op. cit.*, p. 243.
(4) Le premier travail, qui l'occupa pendant de nombreuses années, ne fut mené à bonne fin qu'en 1886. Il publia par contre de 1864 à 1868 deux gros volumes sous le titre : *Juris ecclesiastici Graecorum historia et monumenta*.

pauvreté des ouvrages traitant de la question cruciale de la procession du Saint-Esprit et de les inviter à en reprendre l'examen en remontant à la tradition patristique, mais il dut attendre quinze ans avant de voir son vœu partiellement satisfait [1].

LE GROUPE DES CONVERTIS
RUSSES DE PARIS

Pendant des siècles, la question orthodoxe avait été un problème essentiellement grec, mais plus le XIXe siècle avançait et plus il devenait évident que l'accession du slavisme à un rôle de premier plan sous l'égide de la Russie allait dominer désormais la question des rapports avec les chrétientés orientales. Ce fait donne une importance spéciale au petit groupe d'intellectuels russes convertis qui s'était constitué à Paris autour de Mme Swetchine et qui peut être considéré comme l'un des premiers foyers de l'unionisme contemporain [2].

Dans ce milieu, qui était en relations étroites avec les traditionalistes français, on avait compris, à la suite de Joseph de Maistre et de Bautain [3], que d'éventuels accords officiels entre les confessions séparées n'auraient de chance d'aboutir que s'ils sanctionnaient un rapprochement préalable des esprits et des cœurs. Leur mentalité est bien exprimée par ces lignes d'un intime du groupe :

> Lorsqu'on traite une question aussi grave que la réconciliation de l'Église russe avec l'Église catholique, je ne crois pas qu'on puisse y mettre trop de charité, trop de procédés. Je ne puis m'empêcher de croire que si l'on avait dépensé, pour se rapprocher, autant d'encre qu'on en a employé pour se séparer davantage de jour en jour, il y a longtemps que l'union serait faite [4].

La mort de Mme Swetchine en 1857 provoqua la dislocation de son cercle, mais son action fut prolongée par son parent, le prince Gagarine, converti en 1842 et entré depuis lors dans la Compagnie de Jésus [5]. L'ouvrage de ce dernier, *La Russie sera-t-elle catholique ?* publié en 1856 et traduit aussitôt en plusieurs langues, avait constitué un petit événement et, la même année, il fondait, avec l'aide d'un confrère français, Ch. Daniel, et deux compatriotes, les P. Martinov et Balabine, une revue destinée à éclairer les catholiques français sur les questions religieuses orientales, les *Études de théologie, de philosophie et d'histoire*. La décision prise peu après par le comité de rédaction et approuvée par le père général de transformer la revue en périodique d'intérêt général fut pour lui une grosse désillusion, mais il ne se découragea pas. Tout en poursuivant ses publications, il s'appliqua à l'*Œuvre des SS. Cyrille et Méthode*, qu'il

---

(1) En 1876, Franzelin donna en appendice à son *De Trinitate* une brève étude sur la question (*Examen doctrinae Macarii Bulgakov...*) et, deux ans plus tard, L. Vincenzi publia un *De processione Spiritus Sancti ex Patre Filioque adversus Graecos thesis dogmatica*.
(2) J'utilise ici des notes du P. Olivier Rousseau, qui prépare un travail sur l'unionisme au XIXe siècle. Voir aussi M. J. ROUET DE JOURNEL, *Mme Swetchine et les conversions russes*, dans *Études*, t. CXCI, 1927, p. 183 et suiv. et 321 et suiv.
(3) Cf. M. JUGIE, *Joseph de Maistre et l'orthodoxie gréco-russe*, Paris, 1922 ; E. BAUDIN, *L'union des Églises d'Orient et d'Occident d'après une correspondance inédite de Bautain*, dans *Revue des Sciences religieuses*, t. II, 1922, p. 393-410.
(4) I. GAGARINE, dans *L'Univers*, 20 janvier 1857.
(5) G. REMMERS, *Een Herenigingsapostel uit de vorige eeuw, Ivan S. Gagarin*, dans *Het christelijk Oosten en Hereniging*, t. II, 1949, p. 91-105 ; M.-J. ROUET DE JOURNEL, *L'œuvre des saints Cyrille et Méthode et la Bibliothèque slave*, dans *Revue des Études slaves*, t. III, 1923, p. 90 et suiv.

avait fondée en 1858 dans le double but de faire mieux connaître l'orthodoxie aux catholiques et le catholicisme aux orthodoxes, travaillant
notamment à rassembler à Paris une importante collection d'ouvrages
sur l'histoire et la théologie russe, qui est à l'origine de la *Bibliothèque
slave* de la rue de Sèvres.

Bien qu'il ait vécu davantage en Italie qu'en France, on peut rattacher
à ce groupe le P. Schouvalof. C'est lui qui devait éveiller la vocation
unioniste du barnabite Tondini, le futur collaborateur de Strossmayer, et
qui l'engagea à fonder une association de prières pour l'union à Rome des
Églises d'Orient. Pie IX approuva cette initiative par un bref du 2 septembre 1872 et, lorsqu'un peu plus tard, la branche anglaise de l'association proposa d'y ajouter la prière pour les Anglais dissidents, cette
nouvelle disposition fut encore approuvée par le pape à la demande du
P. Tondini.

*LES CATHOLIQUES ANGLAIS
ET LA « CORPORATE REUNION »*

Pie IX avait déjà eu l'occasion, une
vingtaine d'années auparavant, de témoigner l'intérêt qu'il portait aux chrétiens
séparés d'Angleterre lorsque le passionniste Ignace Spencer, un ancien
anglican converti, lui avait demandé, lors d'une audience, que le mot
hérétique ne fût plus employé dans les prières pour la conversion de l'Angleterre, étant donné qu'il ne s'était jamais senti coupable d'hérésie avant
sa conversion et qu'il ne pensait pas que ses compatriotes pris en bloc
fussent davantage coupables de ce péché ; après un premier moment de
surprise, le pape avait cédé et donné ordre à la Propagande de remplacer
le mot : *hérétiques* par celui d'*acatholiques* [1].

Le Saint-Siège se montra par contre moins favorable à une autre
initiative d'origine anglaise. Un groupe de *clergymen* de la nuance *High
Church* avait fondé en 1857 une société visant à grouper catholiques et
anglicans en vue de préparer le retour en corps de l'Église d'Angleterre à
l'unité avec Rome, l'*Association for the Promotion of the Union of Chris-
tendom* (A.P.U.C.). Du côté catholique, le converti Ambrose Phillips
s'enthousiasma pour l'entreprise, mais la plupart la considéraient comme
chimérique, voire comme dangereuse, Manning surtout qui devait expliquer par la suite, dans *L'Angleterre et le Christianisme* (1867), qu'à son
avis, entrer en pourparlers avec une confession non catholique équivalait
à nier l'infaillibilité de l'unique vraie Église ; seules des conversions individuelles lui paraissaient admissibles. Les évêques anglais finirent par s'inquiéter et, en avril 1864, adressèrent à Rome un mémorandum défavorable
à la participation des catholiques à cette association mixte. Le Saint-
Siège répondit en condamnant sévèrement ce qui lui paraissait un excès
de libéralisme religieux. Les principaux membres anglicans de l'A.P.U.C.
essayèrent vainement d'obtenir que le décret soit modifié. Wiseman
serait peut-être intervenu dans ce sens, mais Manning l'en dissuada et,
devenu archevêque, insista à Rome pour qu'on se montrât intransigeant :

---

(1) DE MADAUNE, *Ignace Spencer et la renaissance catholique en Angleterre*, Paris, 1875.

le 8 novembre 1865, un nouveau décret, un peu moins sec dans la forme, confirmait le précédent [1].

ACTON ET DOELLINGER Peu d'hommes furent aussi préoccupés que John Acton du problème de la réunion des confessions chrétiennes. Mais il lui paraissait vain d'escompter un mouvement de conversion généralisé tant que ne serait pas intervenue une certaine réforme de l'Église catholique elle-même, qui, sans rien changer à sa structure essentielle, remédierait aux déficiences qui expliquaient pour une bonne part les préventions des non-catholiques. Aussi s'appliqua-t-il surtout, au risque de heurter ses coreligionnaires, à prêcher un catholicisme large, spirituel, dégagé de toute exagération populaire et de tout ce qui risquait d'entretenir le malaise entre catholiques et anglicans. Ce faisant, il estimait travailler à la cause de l'union plus efficacement que s'il avait composé des ouvrages d'apologétique dont il savait bien qu'ils n'auraient guère de prise sur l'âme anglaise [2].

C'est également dans cette direction que s'engagea de plus en plus le savant professeur de Munich, Ignace Doellinger. Sans doute, à l'époque où il avait composé ses études sur les origines de la Réforme, envisageait-il la question sous un angle avant tout polémique, mais déjà son ouvrage sur *L'Église et les Églises*, en 1860, rendait un autre son, bien qu'il comportât encore certains jugements sévères ; il faisait dès lors sienne l'idée du cardinal Diepenbrock que les catholiques devaient « supporter la scission religieuse en esprit de pénitence pour les fautes communes » et « accepter une sérieuse correction de tout ce qui paraîtrait nuisible ». Ces idées ne firent que s'affirmer à mesure que se développaient ses contacts avec des savants protestants et, lors de la controverse soulevée par l'*Eirenicon* de Pusey, il écrivit à celui-ci dans un sens très compréhensif [3]. On a vu toutefois comment le souci de réagir contre l'évolution contemporaine du catholicisme, à laquelle il reprochait de renforcer les obstacles à un rapprochement des Églises, l'amena finalement à sortir de la sienne, au grand détriment de ce qu'il y avait de fondé dans certaines de ses revendications.

LE CONCILE DU VATICAN Le concile du Vatican ne constitua certes pas, dans l'immédiat, un facteur d'apaisement. On n'en doit pas moins le signaler ici à un double titre.

D'abord, l'annonce du concile fournit l'occasion de certains contacts, sans lendemain d'ailleurs, en Angleterre et en Allemagne. Un évêque écossais, Mgr Forbes, au nom d'un groupe de ritualistes très attirés par Rome, entra en rapports avec un bollandiste belge, le P. de Buck, auquel le Saint-Office enjoignit après quelque temps de rompre les pourparlers [4]. D'autre part, quelques pasteurs luthériens s'adressèrent à Mgr Martin,

---

(1) Cf. E. PURCELL, *Life and letters of A. Phillips de Lisle*, t. I, chap. v (où l'on trouvera, p. 386 et suiv., le texte des deux décrets) et C. BUTLER, *Life and Times of B. Ullathorne*, t. I, chap. XII.
(2) A. DE LILIENFELD, dans *Irenikon*, t. XIV, 1937, p. 378-379.
(3) Voir ses lettres du 30 mai 1866 à Pusey et à Oxenham dans J. FRIEDRICH, *Ign. v. Doellinger*, t. III, p. 412-415.
(4) F. MOURRET, *Le concile du Vatican*, p. 124-126 (d'après des documents inédits).

de Paderborn ; celui-ci, qui avait à cœur depuis des années le retour des protestants, insista à plusieurs reprises au cours du concile pour qu'on fît au moins un geste [1].

Mais il faut surtout tenir compte de l'intérêt unioniste de certaines thèses ecclésiologiques d'inspiration plus patristique que Franzelin eut l'occasion de remettre en lumière à propos du schéma *De Ecclesia* ; et aussi de l'importante documentation liturgique et canonique qui fut réunie par la commission chargée des questions orientales et qui devait être abondamment utilisée sous Léon XIII, quelques années plus tard.

§ 6. — Les débuts du catholicisme social [2].

AUX ORIGINES DU CATHOLICISME SOCIAL

Les catholiques avaient assez vite pris conscience des problèmes politico-religieux posés par la Révolution française. Ils réalisèrent beaucoup plus lentement que, parallèlement, une seconde révolution, d'une tout autre nature, ébranlait profondément la société traditionnelle : le développement du machinisme avec, pour conséquence, l'apparition d'une classe nouvelle, et la « misère imméritée » de ce prolétariat industriel dont l'importance numérique croissait d'année en année.

Dès 1830, des théoriciens et des militants, tels Robert Owen et les chartistes en Angleterre, Saint-Simon, Fourier ou Proudhon en France, avaient dénoncé les injustices du capitalisme libéral et ébauché la résistance ouvrière. Dès 1847, Marx et Engels avaient élaboré dans le *Manifeste communiste* la charte du socialisme scientifique et, au cours des quinze années suivantes, en dépit de la réaction bourgeoise consécutive à la crise de 1848, Marx allait, avec un véritable génie révolutionnaire, amener la classe ouvrière à s'organiser efficacement et réussir à mettre sur pied, en 1864, la première Internationale des Travailleurs.

Mais tandis que le mouvement ouvrier s'organisait de la sorte et unissait dans une commune espérance les masses révoltées par la misère et l'injustice, beaucoup de catholiques, jusqu'à la fin du siècle, moins par égoïsme d'ailleurs que par incompréhension des problèmes nouveaux posés par la révolution industrielle, se refusèrent à envisager la nécessité de ce qu'on nomme aujourd'hui des « réformes de structure » et à s'associer aux efforts tentés pour modifier la condition ouvrière. Le caractère

---

(1) Voir les lettres des pasteurs dans *Coll. lac.*, t. VII, col. 1137-1144. Sur l'action unioniste de Mgr Martin, cf. G. GOYAU, *L'Allemagne religieuse*, t. III, p. 315-316 ; sur son insistance pendant le concile, cf. J. FRIEDRICH, *Tagebuch während des Vat. Concils*, p. 137, 150-151, 177.

(2) BIBLIOGRAPHIE. — Une histoire d'ensemble du catholicisme social manque encore. On trouvera des notices utiles dans R. KOTHEN, *La pensée et l'action sociale des catholiques (1789-1944)*, Louvain, 1945.

Pour la France, voir les deux excellents ouvrages de J.-B. DUROSELLE, *Les débuts du catholicisme social en France, 1822-1870*, Paris, 1951, et de H. ROLLET, *L'action sociale des catholiques en France (1871-1901)*, Paris, 1947.

Pour l'Allemagne, A. FRANZ, *Das soziale Katholizismus in Deutschland b. z. Tode Kettelers (1877)*, München-Gladbach, 1914 ; M. MOENNIG, *Die Stellung der deutschen katholischen Sozialpolitiker des XIX. Jh. zur Staatsintervention in der sozialen Frage*, Munster, 1927 ; T. BRAUER, *Der deutsche Katholizismus in die soziale Entwicklung des kapitalischen Zeitalters*, dans *Archiv für Rechts- und Wirtschaftsphilosophie*, t. XXIV, 1930, p. 209-254 ; voir aussi V. CRAMER, *Bücherkunde zur Geschichte der Katholischen Bewegung im XIX. Jh.*, Cologne, 1914, p. 29-62.

lamentable de celle-ci était reconnu par un nombre croissant d'entre eux, mais ils ne voyaient d'autre solution à proposer que la charité privée ou les œuvres de bienfaisance. Pourtant, une minorité plus clairvoyante s'ouvrit assez vite à de véritables préoccupations sociales en se rendant compte que la question ouvrière posait un problème non seulement de charité mais de justice. C'est surtout en Allemagne que se développèrent des préoccupations de ce genre, bientôt suivies de réalisations pratiques, et c'est dans ce pays qu'il faut situer l'origine du mouvement social catholique qui devait trouver, en 1891, sa première expression officielle dans l'encyclique *Rerum Novarum*.

Pareilles préoccupations, toutefois, ne demeurèrent pas totalement étrangères aux catholiques français, contrairement à ce qu'on a trop souvent affirmé, et les travaux de M. Duroselle nous ont révélé, bien avant l'intervention d'Albert de Mun, de La Tour du Pin ou d'Harmel, « un grouillement inattendu de personnages et un foisonnement d'œuvres, dont l'insuccès habituel ne décourage généralement pas les initiatives ultérieures » [1].

Il faut ajouter que Pie IX lui-même, bien qu'il se soit beaucoup plus préoccupé des répercussions du libéralisme dans l'ordre politique et doctrinal que de ses néfastes conséquences sociales, n'ignorait pas cet aspect du problème, auquel il s'était intéressé dès avant son avénement [2]. On oublie trop souvent que, dans l'encyclique *Quanta cura*, il avait déjà donné « une première ébauche des enseignements sociaux que Léon XIII devait synthétiser » [3], en dénonçant à la fois l'illusion du socialisme, qui prétend remplacer la Providence par l'État, et le caractère matérialiste et païen du libéralisme économique, qui exclut la morale des relations entre le capital et le travail. Ces avertissements ne passèrent pas tout à fait inaperçus [4].

L'ÉCHEC DE LA PREMIÈRE DÉMOCRATIE CHRÉTIENNE EN FRANCE

Sous la Monarchie de Juillet, deux tendances étaient apparues parmi les catholiques s'intéressant au problème social : l'une née dans les milieux légitimistes, assez timide ; l'autre, plus démocratique, assez proche des mouvements socialistes romantiques. La révolution de 1848 joua en faveur de cette dernière pendant les premières semaines d'euphorie, mais bientôt le courant démocrate chrétien se trouva englobé dans la réaction générale contre tout ce qui tenait de près ou de loin au socialisme [5]. Ce qui en subsistait après la chute de *L'Ère nouvelle* fut frappé à mort par le coup d'État du 2 dé-

---

(1) B. Mirkine-Guetzevitch et M. Prelot, dans la préface de J.-B. Duroselle, *op. cit.*, p. XI.
(2) Le cardinal Mastai possédait dans sa bibliothèque une « abondante littérature » sur les problèmes sociaux et notamment sur l'amélioration de la situation du prolétariat. (L. Sandri, *La biblioteca privata di Pio IX*, dans *Rass. St. R.*, t. XXV, 1938, p. 1431.)
(3) E. Jarry, *L'Église contemporaine*, t. II, p. 19.
(4) L'un de ceux qui dégagea le mieux de l'encyclique les bases d'une action catholique à orientation sociale fut le député alsacien E. Keller, dans le chapitre XVII (p. 279-301) de son ouvrage : *L'encyclique du 8 décembre 1864 et les principes de 1789* (Paris, 1865).
(5) Voir plus haut, p. 44-50. On trouvera un exposé détaillé de cette évolution dans J.-B. Duroselle, *op. cit.*, IIe partie : *L'attitude sociale des catholiques sous la Seconde République* (p. 291-490).

cembre qui disloqua le parti républicain dans son ensemble et si les républicains anticléricaux se ranimèrent à partir de 1859, il n'en alla pas de même de la démocratie chrétienne, qui n'avait pas dans le pays, ni surtout dans le clergé, de racines assez profondes. Elle mena dès lors « une vie languissante, marquée seulement par la publication de quelques ouvrages sans retentissement »[1], et il n'est même pas sûr qu'il y ait une filiation entre elle et la deuxième démocratie chrétienne de la fin du siècle. Pendant le Second Empire et les débuts de la Troisième République, il n'y aura plus désormais pour représenter le catholicisme social que des conservateurs, partisans convaincus des méthodes paternalistes et acquis pour la plupart à la doctrine de la Contre-Révolution.

LES PREMIÈRES RÉALISATIONS SOCIALES DES CATHOLIQUES ALLEMANDS — En Allemagne au contraire, bien que l'évolution industrielle se fût amorcée plus tard qu'en France, les catholiques furent très vite sensibles à l'ébranlement des structures sociales anciennes par le capitalisme libéral. Ils étaient de mentalité encore plus conservatrice que les catholiques français et attachaient au moins autant d'importance que ceux-ci au maintien de « l'ordre ». Seulement, à la différence de la France, où trop de catholiques n'avaient en vue que l'ordre extérieur et la soumission des travailleurs à l'état de choses existant, l'ordre que les catholiques allemands cherchaient à restaurer était l'ordre traditionnel, l'organicisme social de l'ancien régime, fort peu démocratique certes, mais qui avait au moins l'avantage de protéger les petites gens contre une exploitation sans frein par les détenteurs de la richesse.

A vrai dire, si les catholiques allemands ne manifestèrent guère de velléité de s'associer avec la bourgeoisie libérale pour constituer un grand parti de l'Ordre destiné à résister aux revendications populaires, il faut reconnaître que, longtemps, beaucoup d'entre eux s'inquiétèrent surtout de la crise de l'artisanat et de l'agriculture, plutôt que des misères du prolétariat industriel : le relèvement du petit métier, l'organisation des milieux ruraux, telles étaient encore les préoccupations dominantes des participants des rencontres de Soest vers 1865 et de la plupart des députés catholiques lors de la constitution du parti du Centre en 1871. Toutefois, depuis plus de trente ans déjà, la situation lamentable des ouvriers avait également retenu l'attention. Sans parler de Baader, qui avait souligné, bien avant Marx, une série d'aspects dont on attribue souvent la découverte à ce dernier, mais qui demeura toujours aux frontières du catholicisme, Buss, le grand militant badois, avait, dès 1837, dénoncé la misère des classes populaires et les dangers de l'industrialisation non contrôlée. Tout en réclamant, en opposition avec les principes sacro-saints du libéralisme économique et contrairement à l'avis de nombreux catholiques militants tels que Lennig, Heinrich ou Joerg, une intervention de l'État pour régler la situation des travailleurs, il avait

---

(1) J.-B. Duroselle, *op. cit.*, p. 493. Cf. p. 657-672.

proclamé le devoir, pour l'Église, de prendre leur défense, puisqu'ils étaient privés d'une représentation officielle dans la société. Parallèlement, sous l'influence des idées de Sailer, on avait vu se substituer au type de prêtre issu de l'*Aufklärung*, très cultivé mais coupé du peuple, des pasteurs en contact étroit avec la masse et conscients de leurs responsabilités sociales [1].

C'est grâce à des militants laïcs du genre de Buss et à ces prêtres qui avaient pris pour devise de « rapprocher l'Église du peuple afin de rapprocher le peuple de l'Église » qu'on vit dès 1848 les assemblées annuelles des catholiques allemands mettre à l'avant-plan, à côté de la défense des droits de l'Église et de la propagation des principes chrétiens, la préoccupation du relèvement populaire. La question ouvrière forma même le thème principal du congrès de Francfort, en 1863 : le curé Thissen y dénonça un régime qui « bien loin d'estimer l'homme dans le travailleur, ne le considère que comme une machine et le traite même plus mal qu'une machine » [2] et, après avoir entendu des exposés théoriques et quelques témoignages concrets, les participants conclurent en « recommandant instamment aux catholiques de s'occuper de l'étude de la grande question sociale qui, certainement, ne peut être amenée vers une solution convenable qu'à la lumière du christianisme ». Détail caractéristique, qui montre l'unanimité du catholicisme allemand sur ce point, tandis que le chanoine Heinrich de Mayence faisait voter cette résolution, Doellinger, son grand adversaire sur le terrain théologique, déposait quelques jours plus tard, à l'assemblée des savants catholiques de Munich, une motion « invitant le clergé à s'occuper plus à fond de la question sociale » [3].

Il y avait longtemps du reste que le catholicisme social en Allemagne avait dépassé le stade des résolutions de congrès. C'est en 1846 qu'un ancien apprenti cordonnier devenu prêtre, l'abbé Kolping, avait, sur la suggestion du maître d'école d'Elberfeld, mis sur pied le premier *Gesellenverein* [4]. Grâce à des collaborateurs d'élite et à la participation active des apprentis eux-mêmes, à qui Kolping, comme dans la J.O.C. moderne, laissait une large initiative, l'œuvre, bénie par le cardinal Geissel, avait rapidement prospéré et, de la Rhénanie, poussé des rameaux à travers toute l'Allemagne et jusqu'en Autriche et en Suisse. Toute une série d'autres œuvres et d'activités s'étaient encore développées en Rhénanie, qui devaient aboutir un jour à la fondation du *Volksverein* et au mouvement de München-Gladbach : au souci de préserver les âmes et de secourir les misères, qui fut longtemps prédominant, s'ajoutait peu à peu celui d'organiser la profession et de fournir à l'action ouvrière une solide base d'action en vue d'aboutir à une modification du régime du travail.

(1) Cf. F. Schnabel, *Deutsche Geschichte im XIX. Jh.*, t. IV : *Die religiösen Kräfte*, p. 202-207.
(2) Cité par J. May, *Geschichte der Generalversammlung der Katholiken Deutschlands*, p. 154-155. Sur le congrès de Francfort et son influence, voir J.-B. Kissling, *Geschichte der Katholikentage*, t. I, p. 395 et 437-440.
(3) *Verhandlungen der Versammlung katholischer Gelehrten in München*, Ratisbonne, 1863, p. 81.
(4) Cf. S.-G. Schaeffer et J. Dahl, *Adolf Kolping. Sein Leben und sein Werk*, 6e édit., Cologne, 1947 ; *Hundert Jahre Kolpingsfamilie Köln 1849-1949*, sous la direction de D. Weber, Cologne, 1949.

L'ŒUVRE DE KETTELER   Dans cette évolution où, de plus en plus, le problème ouvrier apparaissait comme une question de réformes de structure plutôt que comme une question de secours, quel fut le rôle de Mgr Ketteler [1] ? On l'a parfois exagéré et surtout mal compris en présentant l'évêque de Mayence comme le pionnier de la démocratie chrétienne et comme l'initiateur du florissant mouvement d'œuvres sociales de l'Allemagne contemporaine. Or, beaucoup de ces œuvres, originaires de la région de Cologne, sont nées en dehors de son initiative, et, par ailleurs, si Ketteler, lorsqu'il s'engagea dans la voie des réalisations pratiques, s'inspira parfois de près des doctrines socialistes, il n'avait guère de sympathies pour la démocratie : quand il s'élevait contre l'oppression dont les économiquement faibles étaient les victimes dans le régime social de son temps, cet aristocrate westphalien songeait avant tout à un retour à la société corporativement organisée telle que l'avait connue le Saint-Empire germanique du moyen âge [2]. L'influence de Ketteler sur le catholicisme social contemporain n'en reste pas moins considérable.

Dans l'ordre idéologique d'abord, où se situe sa véritable originalité. En condensant, en 1864, dans *La question ouvrière et le christianisme*, les résultats de quinze années de réflexion, il n'entendait pas seulement suggérer certaines réformes concrètes, mais surtout montrer que la solution du problème ouvrier ne se concevait qu'en fonction d'une conception générale de l'État et de la société, en opposition directe tant avec l'individualisme libéral qu'avec le totalitarisme de l'État centralisateur moderne. Il prenait vigoureusement position contre les solutions prônées par la bourgeoisie capitaliste ou par le socialisme étatiste et exaltait, sous l'influence du romantisme catholique qui avait marqué sa jeunesse, une conception de la société présentée comme un organisme vivant, animé par l'unité de foi et fortement hiérarchisé, où les métiers seraient organisés en vue de la bonne marche de l'ensemble de l'État réduit au rôle d'une simple fonction de ce grand corps. Ce faisant, il apparaissait comme le premier théoricien de cet organicisme social à base corporative qui devait constituer pendant plus d'un demi-siècle le fond de la doctrine sociale catholique.

Même dans l'ordre pratique, où il fut moins original, on ne peut minimiser l'appoint que représenta pour la cause ouvrière l'intervention de l'énergique prélat qui mit au service de celle-ci tout son prestige épiscopal. Se rendant bien compte qu'il ne fallait se faire aucune illusion sur le temps que demanderait une réforme sociale aussi profonde que celle qu'il préconisait, et soucieux de parer, en attendant, au plus pressé, en améliorant la situation des travailleurs dans le cadre du régime existant, il reconnut l'utilité qu'il pouvait y avoir à faire appel à l'intervention de

---

(1) Sur la personnalité de Ketteler, voir plus haut, p. 142-144. Ajouter à la bibliographie T. BRAUER, *Ketteler, der deutsche Bischof und soziale Reformer*, Hambourg, 1927 et R. DE GIRARD, *Ketteler et la question ouvrière*, Berne, 1896, ainsi que G. GOYAU, *Ketteler*, Paris, 1908 (extraits de ses œuvres sur les problèmes politiques et sociaux).

(2) Ce point a été bien mis en lumière par F. VIGENER, *Ketteler*, Munich, 1924, qui minimise trop, par ailleurs, le rôle social de l'évêque de Mayence. (Voir la judicieuse mise au point de M. SPAHN dans *Hochland*, t. XXII, 1925, p. 144-146.)

l'État, dont il s'était d'abord méfié, afin de contrebalancer la puissance du patronat et d'obtenir certains résultats concrets tels que l'augmentation des salaires, la diminution des heures de travail, le repos dominical, l'interdiction du travail des enfants dans les fabriques. Comme Ketteler s'attachait à montrer même du haut de la chaire l'indéniable équité, au nom de la religion, de ces revendications ouvrières, un vicaire d'Aix-la-Chapelle lui écrivait un jour : « Sur d'autres lèvres que les vôtres, nos bourgeois catholiques n'auraient pu supporter de telles vérités [1]. »

En 1869, Ketteler prépara pour les évêques allemands réunis à Fulda un rapport détaillé leur demandant d'appuyer un programme de réformes précises : participation aux bénéfices, mesures en faveur de la mère de famille, intervention de l'État pour limiter les heures de travail, etc., et proposant en outre de désigner dans chaque diocèse un prêtre ou un laïc chargé d'étudier l'état de la classe ouvrière.

Les dernières années de Ketteler l'amenèrent à préciser encore le rôle de l'État dans la solution de la question ouvrière. Dans son livre, publié en 1873, *Les catholiques dans l'Empire d'Allemagne*, qui devait servir de programme au parti du Centre, il engageait celui-ci à s'orienter nettement dans la voie d'un programme social très concret en attendant la réorganisation corporative du monde du travail. Et il avait la satisfaction de constater que, tandis que les ouvriers protestants passaient en masse au parti socialiste, dont ils adoptaient bientôt les doctrines irréligieuses, les travailleurs catholiques restaient fidèles à l'Église et apportaient à celle-ci un appui efficace dans sa résistance au *Kulturkampf*.

Ketteler, dont Léon XIII devait dire qu'il avait été son « grand prédécesseur », mourut en 1877, mais son esprit ne disparut pas avec lui et l'on peut attribuer en bonne partie à son impulsion les premières lois sociales, très progressistes pour l'époque, qui furent votées avec l'appui du Centre au Reichstag à partir de 1879 et qu'un auteur récent n'hésitait pas à qualifier de « législation sociale à direction catholique » [2].

VOGELSANG ET LE CATHOLICISME SOCIAL EN AUTRICHE

Les idées lancées par Ketteler et ses collaborateurs, parmi lesquels il faut au moins mentionner le chanoine Moufang, rencontrèrent un succès tout particulier en Autriche, grâce au baron Charles von Vogelsang [3]. Ce haut fonctionnaire prussien, de vieille noblesse mecklembourgeoise, s'était intéressé de plus en plus aux problèmes sociaux sous l'influence de l'école de Mayence, qui l'avait converti jadis au catholicisme. Il s'était rallié aux revendications ouvrières de celle-ci — il en ajoutait même une nouvelle : l'assurance contre les principaux risques — mais il cherchait en même temps à envisager le problème social dans son ensemble, y compris la question paysanne, qui commençait alors à préoccuper l'Europe centrale.

(1) G. GOYAU, *Ketteler*, p. XLI-XLII.
(2) H. SOMERVILLE, *Studies in the catholic social Movement*, Londres, 1933.
(3) Cf. W. KLOPP, *Leben und Wirken des Sozialpolitikers Karel von Vogelsang*, Vienne, 1930 ; A. LESOWSKY, *K. V. Vogelsang. Zeitwichtige Gedanken aus seine Schriften*, Vienne, 1927.

Retiré à Vienne depuis 1864, il devint après 1870 collaborateur de *Das Vaterland*, l'organe de l'aristocratie fédéraliste et agrarienne, très hostile au gouvernement de la haute bourgeoisie capitaliste et libérale. A la suite de son maître Ketteler, il mit en lumière la fonction sociale de la propriété et le vice fondamental du régime capitaliste basé sur la recherche exclusive du profit ; plus que lui, il insista sur la nécessité d'organiser les professions, non pas, comme le voulaient les socialistes, dans un sens étatiste, mais à la manière des corporations médiévales.

Assez vite, ces idées, auxquelles s'était rallié tout un groupe de jeunes catholiques autrichiens, furent répandues à l'étranger grâce à une revue internationale, *La Correspondance de Genève*, dirigée par l'un d'entre eux, le comte von Blome, et Vogelsang, après la mort de Ketteler, fut pendant tout un temps considéré par beaucoup de catholiques comme le doctrinaire du catholicisme social.

CATHOLICISME SOCIAL ET PATERNALISME SOUS LE SECOND EMPIRE   En comparaison avec ce qui se passait dans les pays germaniques, la pensée et l'action sociale des catholiques français pendant le troisième quart du XIX⁰ siècle apparaissent singulièrement décevantes. Certes, un nombre croissant d'entre eux prennent conscience des obligations sociales impliquées par la foi chrétienne, mais leurs tentatives rencontrent encore une grande indifférence ; elles se heurtent même parfois à l'hostilité, à celle de Veuillot notamment, dont l'influence est alors si grande dans l'Église de France ; en tout cas, elles ne sont guère encouragées par le clergé, d'origine avant tout paysanne, ni par la hiérarchie. Beaucoup de prêtres et d'évêques, dont l'intelligence est rendue rigide par l'âge ou que les tâches administratives détournent des vues lointaines, sont incapables de percevoir les conséquences de l'alliance de fait du conservatisme social et de l'Église, qui paraît au premier abord la seule alternative acceptable d'un dilemme. Ils sont portés à confondre les tendances sociales avec le socialisme, dont ils connaissent surtout le mot de Proudhon : « La propriété, c'est le vol ! » ou les théories subversives de Fourier en matière de morale familiale. Il faut d'ailleurs tenir compte, pour juger équitablement cette attitude timorée, de l'allure que prenaient volontiers les revendications socialistes, « les volontés de violence pour la transformation sociale qu'on rencontre chez les novateurs, le naturalisme au nom duquel ils la promouvaient ordinairement, et surtout la fréquente hostilité directe dont ils témoignaient à l'égard de l'idée religieuse pour des motifs idéologiques, qui évidemment dépassent le problème social ou même politique et lui sont antérieurs » [1].

Rien d'étonnant, dans ces conditions, si le résultat des efforts de ceux qui, sous la conduite d'Armand de Melun et d'Augustin Cochin, se préoccupent d'améliorer la condition ouvrière, apparaît fort limité. On est d'abord frappé par leur manque de doctrine. Les catholiques sociaux français n'ont qu'une connaissance sommaire de la science économique,

---

(1) P. DROULERS, dans *Bulletin de littérature ecclésiastique*, 1950, p. 112.

pourtant en plein essor à cette époque, et leurs nombreux écrits témoignent tout autant de l'imprécision et de l'inefficacité de leur pensée que de leur bonne volonté. Ils n'ont eu ni leur Marx ni leur Proudhon ; la théorie dont ils s'inspirent le plus est celle de Le Play, dont le caractère scientifique laisse à désirer et qui, combinée avec une interprétation étroite du *Syllabus*, contribue à les orienter vers la doctrine « contre-révolutionnaire », hostile aux « droits de l'homme » et à l'égalitarisme démocratique.

Leur action apparaît aussi hésitante que leur pensée. Un certain nombre de chefs d'entreprises s'engagent bien dans la voie du « patronat social » en créant des institutions visant au bien-être de leurs ouvriers, mais la plupart des réalisations de ce genre semblent moins le fruit d'un désintéressement réel, comme c'est le cas chez un Cochin, que du souci de maintenir l'ordre au prix de quelques concessions.

Par ailleurs, si Melun réussit partiellement à orienter la *Société d'Économie charitable* vers des activités proprement sociales, celle-ci restera jusqu'à la fin tiraillée entre trois solutions : la vieille méthode du « patronage » des œuvres ouvrières par les classes dirigeantes, l'appui donné aux ouvriers qui aspirent à la liberté syndicale, et le développement du mutuellisme. Dans ce dernier domaine, certains résultats sont acquis, du moins en province, mais les sociétés catholiques de secours mutuels ne réussissent pas à entamer sérieusement la masse ouvrière.

Quant aux prêtres, trop rares, qui se vouent à l'apostolat populaire, ils cherchent avant tout à « ramener la classe ouvrière au catholicisme en formant les esprits plus malléables des jeunes ouvriers »[1] et leur attention se porte bien plus sur la formation religieuse ou la « récréation honnête » que sur l'éducation populaire ou l'amélioration matérielle du sort des jeunes travailleurs. C'est ainsi que l'attitude intransigeante de l'abbé Timon-David, si méritant par ailleurs, contribua à faire échouer la réalisation d'un « compagnonnage chrétien » analogue à celui organisé par Kolping. A ce point de vue, Maurice Maignen[2], qui avait pris contact avec le mouvement allemand, apparaît comme un précurseur, tout comme lorsqu'il se décide à accorder aux jeunes travailleurs une plus grande autonomie dans la direction des cercles qu'il avait fondés.

Mais cette façon de voir reste une exception et, dans l'ensemble, « nulle part, sauf chez quelques écrivains peu connus et dont l'œuvre passe inaperçue, on ne voit poindre l'idée que c'est en confiant aux ouvriers catholiques la responsabilité des œuvres sociales catholiques que l'on éviterait la déchristianisation et la lutte des classes et que l'on aboutirait à une réelle amélioration sociale »[3]. Or, c'est à ce moment que le mouvement ouvrier évolue précisément en France comme dans toute l'Europe

---

(1) J.-B. DUROSELLE, *op. cit.*, p. 418. On notera la remarque pertinente du même auteur : « Le demi-échec qui marque cet immense effort ne provient-il pas du caractère artificiel qu'il présentait ? Travailler les esprits sans songer à réformer les conditions sociales en luttant pour la justice, pour l'amélioration de la vie matérielle, aboutissait soit à déraciner les persévérants de leur milieu, soit à l'abandon des œuvres par les jeunes gens dès qu'ils se sentaient solidaires de ce milieu. » (Dans *R. H. E. Fr.*, t. XXXIV, 1948, p. 61.)

(2) Cf. Ch. MAIGNEN, *Maurice Maignen, directeur du Cercle Montparnasse*, Luçon, 1927, 2 vol.

(3) J.-B. DUROSELLE, *op. cit.*, p. 496.

dans un sens de plus en plus opposé au paternalisme ; rien d'étonnant si les ouvriers prennent une attitude réticente à l'égard des initiatives des catholiques sociaux et si celles-ci ne réussissent pas à toucher la véritable élite ouvrière.

ALBERT DE MUN ET L'ŒUVRE DES CERCLES — C'est en définitive le même jugement qui doit être porté sur l'*Œuvre des Cercles catholiques d'ouvriers*, fondée à la Noël 1871 par un jeune officier, Albert de Mun, qui avait pris contact avec l'œuvre de Ketteler pendant sa captivité à Aix-la-Chapelle, et découvert sa « vocation sociale » au moment de la Commune. En étroite union avec Maurice Maignen, il avait entrepris de multiplier à travers la France des cercles analogues à celui que ce dernier dirigeait à Montparnasse, en adjoignant à chacun d'eux un comité protecteur recruté dans la classe dirigeante, le tout étant unifié par un comité général résidant à Paris.

Sous l'impulsion de son infatigable secrétaire général, qui était « doué d'une magnifique prestance et du plus beau talent oratoire de son temps avec Jaurès »[1], l'*Œuvre des Cercles* connut d'abord un rapide essor, en dépit de la réticence du clergé, qui, par étroitesse de vue, s'inquiétait de voir ces foyers catholiques se constituer en dehors de la paroisse et sous une direction laïque. Mais si elle compta, en 1878, jusqu'à 8.000 membres protecteurs, provenant surtout des milieux légitimistes[2], cet engouement ne devait pas durer et surtout les 35.000 travailleurs qu'elle groupait à cette date demeuraient en marge du véritable milieu ouvrier et n'étaient trop souvent que « des employés de librairies cléricales, des bedeaux en rupture de hallebarde, des sacristains retraités, des concierges de communautés, des garçons de bureau des œuvres »[3]. Les ouvriers se méfiaient de l'orientation très paternaliste donnée à l'œuvre par la majorité de ses dirigeants, malgré les efforts du fondateur, et surtout ils étaient heurtés par l'idéologie antirévolutionnaire de l'*Œuvre*, qui devait la faire apparaître comme une entreprise réactionnaire, bien que, en réalité, les préoccupations d'Albert de Mun et de son groupe les situent très en avance non seulement sur les orléanistes, mais même sur la plupart des chefs républicains d'alors.

HARMEL ET LA TOUR DU PIN — Malgré cet échec apparent, l'*Œuvre des Cercles* devait exercer sur le développement du mouvement social chrétien en France une influence durable en assurant une large publicité aux réalisations concrètes de Léon Harmel et au programme doctrinal de René de La Tour du Pin.

(1) H. ROLLET, *Les étapes du catholicisme social*, Paris, 1949, p. 28. Sur Albert de Mun, outre le volume de souvenirs, *Ma vocation sociale*, Paris, 1911, voir H. FONTANILLE, *L'œuvre sociale d'Albert de Mun*, Paris, 1926.

(2) Ce qu'on savait des dispositions du comte de Chambord en matière sociale (il avait des idées assez proches de celles de Vogelsang) contribua à orienter vers l'*Œuvre des Cercles* la société légitimiste ; les milieux orléanistes et républicains étaient au contraire favorables au libéralisme économique.

(3) E. BARBIER, *Histoire du catholicisme libéral et du catholicisme social en France*, Bordeaux, 1923.

C'est en 1874, à l'occasion du congrès de l'*Union des Œuvres ouvrières catholiques*, que l'*Œuvre* s'était assuré la collaboration d'Harmel [1], un précurseur qui, fidèle à son principe : « le bien de l'ouvrier avec lui, jamais sans lui, à plus forte raison jamais malgré lui », avait réalisé dans son usine du Val-des-Bois une véritable association des ouvriers à la gestion des œuvres sociales de l'entreprise : il peut être considéré comme le pionnier de la démocratie chrétienne des dernières années du siècle.

Quant à La Tour du Pin, qui, comme penseur, « apparaît comme la figure dominant de très haut toutes les autres dans l'histoire du catholicisme social après 1870 » [2], il avait été aux côtés d'Albert de Mun dès la première heure, mais son influence doctrinale devait surtout se faire sentir à partir de la fondation, le 30 janvier 1878, du *Conseil des Études* annexé à l'*Œuvre des Cercles*, qui allait orienter définitivement celle-ci vers des recherches d'économie sociale assez éloignées de la pensée initiale de ses fondateurs. Rapprochant ses réflexions personnelles des idées de Vogelsang, La Tour du Pin commença dès lors à élaborer un plan complet de restauration d'une société chrétienne à base corporative, fondée sur la proclamation de la royauté sociale de Jésus-Christ.

LE LENT ÉVEIL DES CATHOLIQUES BELGES AUX PROBLÈMES SOCIAUX

Les influences doctrinales dominantes et les circonstances politiques maintinrent les catholiques belges pendant de longues années dans une attitude assez réservée en matière sociale.

Le grand théoricien de ce que La Tour du Pin appelait en 1878 « l'école belge » fut Charles Périn [3], brillant professeur d'économie politique à Louvain de 1845 à 1881, et dont l'ouvrage fondamental, *De la richesse dans les sociétés chrétiennes* (1861), fut traduit dans la plupart des langues européennes. S'il prône l'association ouvrière, le retour aux corporations et un régime du travail plus humain, il confie l'exécution de ces vœux à la seule initiative privée et surtout à la responsabilité des patrons chrétiens. En effet, cet ultramontain, qui dénonce avec énergie les abus lamentables dont la classe ouvrière est victime et enseigne que les lois morales doivent dominer l'économie, reste partisan du libéralisme économique, refuse à l'État toute intervention dans le domaine du travail et cherche la solution du problème social dans le seul progrès de la moralité et de l'esprit chrétien chez les patrons et les ouvriers.

Ces idées sociales, d'inspiration libérale et paternaliste, rencontrèrent beaucoup de faveur parmi les catholiques. Ceux-ci, en effet, basaient précisément toute la défense de leurs droits religieux sur l'appel à la liberté et craignaient de favoriser les tendances étatistes de leurs adversaires ; en outre, contre la politique anticléricale de ces derniers, ils avaient besoin de l'appui du roi Léopold Ier et ils sentaient que, pour

---

(1) Voir G. GUITTON, *Léon Harmel*, Paris, 1927, 2 vol.
(2) H. ROLLET, *L'action sociale des catholiques en France*, p. 65. Voir à son sujet Ch. BAUSSAN, *La Tour du Pin*, Paris, 1931.
(3) A défaut d'une étude d'ensemble qui manque encore, voir R. KOTHEN, *La pensée et l'action sociale des catholiques*, p. 139-148.

l'avoir, ils devaient apparaître comme les défenseurs des intérêts conservateurs : « La démocratie chrétienne viendra plus tard, sinon trop tard ; le roi lui a fait perdre son temps [1]. »

On comprend dans ces conditions l'accueil plutôt froid que les congrès de Malines réservèrent à ceux qui, à la suite de Ducpétiaux [2], auraient voulu que l'on donnât au mouvement catholique, comme en Allemagne, une allure sociale progressiste et favorable à une législation ouvrière. Si le congrès de 1867 fut l'occasion de la constitution d'une *Fédération des sociétés ouvrières catholiques* et proclama dans ses conclusions que « l'étude des relations des chefs d'industrie avec leurs ouvriers appelle impérieusement l'attention des catholiques », par contre, quand il s'agit de préconiser une solution, il se borna à une « résolution incolore » [3] où l'on lisait entre autres, à propos du désordre social, dont on prenait de plus en plus conscience : « La cause principale en est dans l'oubli des devoirs que la religion catholique impose tant aux ouvriers qu'aux patrons : le remède principal est dans le retour à la pratique de leurs devoirs [4]. »

VERS UNE ORGANISATION INTERNATIONALE DU CATHOLICISME SOCIAL

La participation d'un certain nombre d'étrangers aux congrès de Malines aurait pu faire de ceux-ci le point de départ d'une sorte d'internationale du catholicisme social théorique, mais les idées n'étaient pas encore mûres et il faudra attendre les débuts de l'*Union de Fribourg*, en 1884, pour voir s'organiser cette internationale catholique. Elle se préparait pourtant depuis la fin du pontificat de Pie IX.

La question sociale passait en effet partout à l'ordre du jour. Même en Italie, où les effets de la révolution industrielle ne se firent sentir que tardivement, on voit le marquis Sassoli-Tomba parler en faveur des syndicats ouvriers au congrès catholique de Bergame de 1877, où il évoque l'exemple de Léon Harmel [5]. Cette même année, au moment où disparaissait Ketteler, qui avait donné la preuve que la hiérarchie ne craignait pas de prendre sur ce point une position en flèche, Manning prononçait sa célèbre conférence de Leeds sur *Les droits et la dignité du Travail* et s'engageait résolument dans la voie qui sera désormais la sienne pendant les quinze dernières années de sa vie, le soutien actif de toutes les initiatives tendant à protéger les ouvriers contre les abus du régime capitaliste.

(1) A. SIMON, *L'Église catholique et les débuts de la Belgique indépendante*, p. 75.
(2) Ducpétiaux fut un précurseur social. Il s'était beaucoup intéressé à la misère ouvrière. (Voir notamment : *Le paupérisme en Belgique*, 1842 ; *De la condition physique et morale des jeunes ouvriers et des moyens de l'améliorer*, 1843 ; *Enquête sur la condition des classes ouvrières et sur le travail des enfants en Belgique*, 1855.) Après avoir préconisé jusqu'en 1848 des solutions proches du socialisme utopique de Louis Blanc, il s'orienta ensuite vers des solutions plus personnalistes tout en restant partisan d'une certaine intervention de l'État .(Cf. E. RUBBENS, *Edouard Ducpétiaux, sa vie, son œuvre*, t. II, Louvain, 1934.)
(3) Cf. *Actes du IVe Congrès catholique de Malines* (1891), t. I, p. 20-21.
(4) *Assemblée générale des catholiques en Belgique*. IIIe Session, Bruxelles, 1868, 1re partie, p. 327.
(5) *Atti e documenti del IVº Congresso cattolico italiano tenuosi in Bergamo*, Bologne, 1877, p. 81 et suiv.

L'utilité de rapprocher les bonnes volontés et de confronter les expériences et les doctrines s'imposait de plus en plus. C'est en 1877, semble-t-il, qu'eut lieu la première rencontre internationale du catholicisme social, à l'initiative d'un prélat acquis aux idées de Ketteler, Mgr Mermillod [1], l'évêque exilé de Genève, qui dénonçait depuis des années l'injustice de la situation sociale. Il invita La Tour du Pin à venir le voir à Ferney et lui fit rencontrer quelques disciples de Vogelsang, les princes Aloys et Alfred de Liechtenstein, ainsi que le comte von Blome. Quelques-uns des principaux participants de la future *Union* de Fribourg se trouvaient ainsi réunis, et, tant par Mermillod que par Blome, le fondateur de la très ultramontaine *Correspondance de Genève*, le contact avec les milieux romains était établi d'emblée.

§ 7. — Le bilan d'un pontificat.

LE CRÉPUSCULE Les dernières années de Pie IX présentent paradoxalement le double caractère d'une apothéose et d'une mélancolique liquidation. Auprès des foules catholiques, auprès du clergé ultramontain — et c'est maintenant l'immense majorité du du clergé — la popularité du vieux pontife ne fait que croître avec ses malheurs. Ses noces d'or épiscopales, célébrées le 3 juin 1877, furent l'occasion d'un triomphe sans précédent, qui permit de mesurer une fois de plus à quel point le prestige de la papauté avait grandi depuis trente ans et combien les forces chrétiennes restaient vivantes dans l'ancien et le nouveau monde.

Pourtant, malgré les acclamations des pèlerins toujours plus enthousiastes, malgré les adresses de fidélité toujours plus nombreuses, qui affluent des quatre coins du monde à l'occasion de chaque réunion catholique de quelque importance, l'atmosphère à Rome devenait de plus en plus lourde. Les déceptions avaient succédé aux déceptions : l'un après l'autre, les gouvernements sur lesquels on avait cru pouvoir compter s'étaient dérobés ou même devenaient ouvertement hostiles ; le monde intellectuel de son côté se détournait nettement de l'Église ; et les idées qui avaient inspiré la politique ecclésiastique depuis l'époque de Grégoire XVI n'étaient plus prises à partie seulement par des esprits avancés comme Montalembert, Doellinger, Passaglia ou Hyacinthe Loyson, mais par des collaborateurs aussi fidèles que le P. Curci, l'animateur de *La Civiltà cattolica*.

Au milieu de ces désastres, les grands collaborateurs du règne disparaissaient les uns après les autres : en 1874, le cardinal Barnabo et Mgr de Mérode, en 1876 le cardinal Antonelli [2]. D'autres se survivaient, mais il devenait de plus en plus évident que leur temps était passé. Man-

---

(1) Cf. C. MASSARD, *L'œuvre sociale du cardinal Mermillod*, Louvain, 1914. Sur les rencontres de **Ferney**, voir H. ROLLET, *L'action sociale des catholiques français*, p. 108-109.

(2) Après 1870, l'activité politique d'Antonelli fut évidemment beaucoup plus effacée. Mais ses qualités financières, qui l'avaient fait remarquer au début de sa carrière, trouvèrent à s'employer utilement et il eut le mérite d'organiser sur des bases saines la situation matérielle du Saint-Siège dans sa nouvelle position. (Cf. L. TESTE, *Préface au conclave*, p. 63-65.)

ning, qui avait été longtemps l'un des plus zélés défenseurs de la Curie et de tout ce qui s'y décidait, ne cachait pas sa déception à son retour de Rome en décembre 1876. Il notait avec désenchantement :

> Quelle impression de stagnation ! Six ans ont passé depuis 1870 et l'organisation de la curie a périclité d'année en année !
> Il semble y avoir à Rome une carence d'hommes jeunes et d'avenir. Ceux qui étaient dans la force de l'âge sont maintenant trop vieux pour prendre en main de nouvelles tâches. Le Saint-Siège me paraît en ce moment bien bas en ce qui concerne les conseillers capables et les hommes d'action [1].

Pie IX, qui sans perdre son entrain coutumier, se sentait pourtant de plus en plus isolé, avait bien conscience d'assister à la fin d'une époque. Le cardinal Ferrata nous a transmis une réflexion qu'il fit peu avant sa mort à Mgr Czacki et qui suffirait à montrer qu'il n'était pas au fond aussi aveugle qu'il pouvait y paraître à première vue :

> Mon successeur devra s'inspirer de mon attachement à l'Église et de mon désir de faire le bien ; quant au reste, tout a changé autour de moi ; mon système et ma politique ont fait leur temps, mais je suis trop vieux pour changer d'orientation : ce sera l'œuvre de mon successeur [2].

C'est d'ailleurs sous l'empire de cette préoccupation, semble-t-il, qu'il avait fait choix, à la surprise générale, pour remplacer Antonelli, du vieux cardinal Simeoni, excellent homme sans doute, mais manifestement dépourvu du minimum de qualités requises pour s'acquitter des fonctions de secrétaire d'État, surtout dans des circonstances aussi délicates :

> On prétend, expliquait le ministre de Belgique, que Sa Sainteté aurait dit : J'ai choisi Simeoni ; ce ne sera pas pour longtemps et je laisse ainsi toute liberté au conclave et à mon successeur en prenant un cardinal qui n'est indiqué ni pour la papauté ni pour des fonctions politiques [3].

LA MORT DE PIE IX   Toujours ardent, Pie IX, qui était entré dans sa quatre-vingt-sixième année, vieillissait visiblement. En 1877, son entourage l'avait amené à mettre définitivement au point les règles à suivre pour l'élection de son successeur [4], tandis que le gouvernement italien se préoccupait de la manière dont seraient réglées les funérailles du pape. Contrairement à toute attente, celui-ci eut encore la surprise avant de mourir de voir disparaître son rival, le roi Victor-Emmanuel, qu'une brève maladie emporta, le 9 janvier 1878, mais, quatre semaines plus tard, il s'alitait à son tour, au moment où l'on commençait à préparer pour l'été suivant la célébration solennelle de ce fait unique dans l'histoire : un pontificat ayant atteint les 32 ans qu'avait duré l'épiscopat de saint Pierre à Antioche et à Rome. En quelques heures, il s'avéra que tout espoir était perdu et, le 7 février, à cinq heures

---

(1) Notes personnelles citées par E. PURCELL, *Life of Manning*, t. II, p. 575 (cf. p. 572).
(2) D. FERRATA, *Mémoires*, t. I, Rome, 1920, p. 32-33.
(3) Dépêche du 15 novembre 1876, *Arch. Min. A. É. Brux.*, Saint-Siège, t. XV.
(4) Déjà deux bulles, en août 1871 et en septembre 1874, avaient réglé l'essentiel en veillant notamment à exclure du conclave toute intervention séculière. La bulle d'octobre 1877 ne fit que les compléter. (Cf. à ce sujet EISLER, *Das Veto der katholischen Staten bei der Papstwahl*, Vienne, 1907, p. 248 et suiv.)

quarante de l'après-midi, Pie IX s'éteignait doucement. Ce fut une consternation générale, à Rome d'abord, où la population restait très attachée à son ancien souverain, puis dans le monde chrétien tout entier, où l'on oublia pour un temps les critiques que certains avaient cru devoir adresser à la politique religieuse suivie depuis trente ans, pour ne plus voir que les mérites de celui que d'aucuns étaient déjà prêts à canoniser et dont les autres saluaient du moins la ferveur et les vertus, l'indomptable force d'âme au service de la vérité et l'inlassable dévouement à la cause de l'Église.

LE PASSIF DE LA SUCCESSION   Pie IX était mort. Les problèmes qu'il avait laissés sans solution ne disparaissaient pas avec lui. Avec trois quarts de siècle de recul, l'historien discerne mieux encore que les contemporains les points faibles et les lacunes de ce long pontificat, si fécond pour l'Église à tant de points de vue.

Un point frappe tout d'abord. Pie IX laisse en mourant l'Église plus forte intérieurement, mais isolée devant l'hostilité générale des gouvernements et de l'opinion publique. Les succès du catholicisme au cours de son pontificat ont été pour une large part le résultat des progrès du courant ultramontain et de l'alliance de l'Église avec les régimes anti-révolutionnaires. Mais le triomphe même de l'ultramontanisme déclenche la réaction des gouvernements, mécontents de voir le clergé local s'affranchir de leur emprise et suivre toujours plus docilement les mots d'ordre de Rome. Et l'opinion démocratique, dont l'influence s'affirme d'année en année dans la vie publique et la direction des États, ne pardonne pas à l'Église l'appui que celle-ci n'a cessé d'apporter depuis 1848 aux partis conservateurs et qui la fait apparaître comme irrémédiablement solidaire de l'ancien régime et des forces de réaction. Ces heurts avec des gouvernements habitués depuis des siècles à s'immiscer beaucoup trop dans les affaires de l'Église et ce conflit avec les partis de gauche étaient pour une part inévitables, car tant que l'Église ne renoncera pas à sa « vocation terrestre », — renonciation impossible, du reste, — elle suscitera des réactions anticléricales. Pourtant, lorsqu'on constate que quelques années, quelques mois parfois, suffiront à l'habile Léon XIII pour amener une détente notable dans la plupart des cas, on ne peut s'empêcher de penser que bien des crises auraient pu être amorties ou même évitées en freinant quelque peu les progrès irréversibles de la centralisation romaine, et en prenant à l'égard de la transformation des institutions sous l'influence du libéralisme une attitude plus conciliante et moins exclusivement doctrinaire.

Pie IX, très maladroitement conseillé par son entourage, n'a pas réussi à adapter l'Église à la profonde évolution politique qui transforme du tout au tout l'organisation de la société civile au cours du xixe siècle. Il ne s'est pas assez rendu compte par ailleurs de l'urgente nécessité de s'adapter à une autre évolution, la transformation progressive de l'ancienne économie agricole en un monde industrialisé et la prise de conscience de sa misère mais aussi de sa force par un prolétariat urbain dont l'importance numérique croît d'année en année. Heureusement, des catholiques et des évêques ont commencé à aborder le problème, en Allemagne, puis en Autriche, en France, en Angleterre. Mais il est regrettable que le Saint-Siège, trop absorbé par la lutte contre le libéralisme doctrinal et politique et contre les dernières traces du gallicanisme et du joséphisme, n'ait encore donné aucune directive quelque peu précise ni sur le plan des principes, ni sur celui de l'organisation pastorale. Du moins n'a-t-il freiné aucune initiative.

Sur le plan intellectuel par contre, non seulement Pie IX n'a pas réussi à donner les impulsions nécessaires, mais, peu averti personnellement de cet aspect des choses, il a de plus en plus abandonné la direction et le contrôle de la vie scientifique dans l'Église à trop d'esprits étroits, qui, effrayés des progrès du rationalisme et du positivisme, ont cru qu'il suffisait d'anathématiser les courants doctrinaux nouveaux incompatibles à première vue avec la foi chrétienne et de se raidir dans les positions considérées comme traditionnelles, sans même se demander s'il ne convenait pas de reprendre à la base certaines questions. Cette attitude bien intentionnée mais peu intelligente ne fait que reculer le problème et bientôt il ne sera plus possible d'éviter le conflit qui va éclater ouvertement entre un enseignement ecclésiastique figé et les résultats du développement des sciences naturelles et historiques, en particulier sur le terrain biblique. « C'est la grande lacune du pontificat de Pie IX de n'avoir pas vu l'immensité du danger et de n'avoir pas opposé une doctrine à ses adversaires [1]. » Sans doute, le néo-thomisme est né, mais ce n'est encore qu'un espoir et l'idée maîtresse du cardinal Mercier : repenser la philosophie de saint Thomas en fonction de la philosophie et de la science modernes, n'a encore été qu'entrevue par Kleutgen. Sans doute encore un effort méritoire a-t-il été réalisé par les savants catholiques allemands pour répondre aux exigences de la critique historique, mais ce travail s'accomplit sous l'œil de plus en plus méfiant d'une partie de la hiérarchie et surtout des autorités romaines qui n'ont pas su discerner ce qu'il y avait de légitime dans une œuvre un peu trop influencée par le rationalisme ambiant. Cette attitude timorée a fait perdre un temps précieux et les directives plus ouvertes qui vont prévaloir sous Léon XIII viendront trop tard pour pouvoir rattraper entièrement le temps perdu : les vraies racines de la crise moderniste se situent sous le pontificat de Pie IX.

L'ŒUVRE POSITIVE DE PIE IX   Ces lacunes sont graves, et pourtant, aux yeux de l'historien de l'Église, le bilan du pontificat apparaît nettement favorable.

L'Église, d'abord, s'est développée et affermie extérieurement. L'expansion missionnaire, brillamment amorcée sous Grégoire XVI, s'est poursuivie tout au long de ces trente-deux années parallèlement à l'expansion coloniale européenne, sous l'impulsion centralisatrice du Vatican [2].

_____

(1) Ch. Pouthas, *Le pontificat de Pie IX* (cours de Sorbonne), p. 299.
(2) Il manque à ce volume un chapitre essentiel sur l'expansion missionnaire durant le ponti-

Grâce aux efforts conjugués de la Congrégation de la Propagande et de missionnaires toujours plus nombreux, parmi lesquels les Français occupent de loin la première place, l'Église catholique se trouve, à la mort de Pie IX, implantée dans les cinq parties du monde. Au moment où la disparition de l'État pontifical élimine la papauté de l'échiquier diplomatique européen, l'Église commence à apparaître, par suite de sa présence universelle, comme « une grande puissance mondiale dont toute politique doit tenir compte » [1]. Indépendamment du mouvement missionnaire, l'immigration catholique a donné naissance à de nouvelles Églises, pleines d'avenir, en Angleterre, au Canada, et surtout aux États-Unis. Bref, de 1846 à 1878, Pie IX a érigé 206 nouveaux diocèses et vicariats apostoliques. Pendant ce temps, d'anciennes Églises qui vivaient depuis la Réforme dans des conditions précaires ont été réorganisées ; c'est le cas aux Pays-Bas et surtout en Allemagne, où le *Kulturkampf* ne fait que mettre en relief à quel degré de vitalité cette Église a su parvenir en quelques années en s'appuyant toujours davantage sur le Saint-Siège.

En même temps qu'elle s'est étendue quantitativement, l'Église catholique s'est en effet davantage resserrée autour du pape. Il a été longuement question au cours de ce volume des progrès constants de l'unification et de la centralisation romaine, qui constituent l'un des phénomènes les plus marquants du pontificat de Pie IX. Le concile du Vatican, tout en purifiant le mouvement ultramontain de ce qu'il avait parfois d'outré, a marqué la défaite définitive des tendances particularistes dans l'Église : depuis lors, « il peut y avoir encore des oppositions gouvernementales, il n'y a plus de gallicanisme ecclésiastique » [2]. Ces progrès de l'influence romaine, qui n'ont pas été sans susciter d'amers regrets chez ceux qui avaient connu les avantages du pluralisme, ne tardent pas à produire leurs heureux effets dans les Églises affaiblies par les traditions régaliennes de l'ancien régime qui commencent lentement à se régénérer.

Incapable de prendre la direction des efforts d'adaptation de l'enseignement catholique au mouvement intellectuel contemporain, Pie IX a toutefois exercé un rôle doctrinal important, souvent ingrat, mais indispensable et fécond. Bien qu'il ait été amené à définir deux nouveaux dogmes, l'Immaculée-Conception et l'infaillibilité pontificale, son œuvre doctrinale ne présente guère d'originalité et s'inscrit en parfaite continuité avec celle de Grégoire XVI, mais elle est beaucoup plus systématique et plus ample que cette dernière et amène à sa pleine réalisation le travail collectif et diffus des théologiens de la génération précédente. Inaugurée dès le début du pontificat avec l'encyclique *Qui pluribus*, elle s'est poursuivie sans trêve, le pape profitant de toute circonstance locale ou occasionnelle pour rappeler les principes qui doivent commander la restauration chrétienne de la société ; elle aboutit à la synthèse de l'encyclique *Quanta cura* en 1864 et aurait dû être couronnée par le vaste ensemble

ficat de Pie IX : conformément au plan de la collection, tout ce qui concerne les missions au xixᵉ siècle sera traité au t. XXIII.

(1) H. Marc-Bonnet, *La papauté contemporaine*, Paris, 1946, p. 51.
(2) Ch. Pouthas, *Le pontificat de Pie IX*, p. 298.

des décrets du concile du Vatican. On a souvent été frappé par l'aspect négatif de cette œuvre qui apparaît essentiellement comme une condamnation sans cesse répétée du libéralisme sous toutes ses formes : rationalisme ou tendance de l'esprit humain à s'affranchir de l'autorité de la révélation et du magistère doctrinal ; indifférentisme moral et religieux ou tendance à rejeter les normes de la moralité et les exigences de la vérité au nom des droits de l'individu ; laïcisme ou rejet de l'influence de l'Église dans la vie des sociétés ; gallicanisme, dans lequel on voyait de plus en plus à Rome une tendance à concevoir l'organisation de l'Église à l'image des gouvernements parlementaires et à réduire l'autorité divine du pape au profit des pouvoirs subordonnés. Mais derrière cette œuvre de condamnation, il y a une affirmation positive toujours sous-jacente : le véritable rapport de la créature à Dieu et la réalité de l'ordre surnaturel, qui conditionnent la vision catholique de l'homme et de la société civile et religieuse. Le pontificat de Pie IX, caractérisé dans les institutions ecclésiastiques par la liquidation définitive du gallicanisme et du joséphisme de l'ancien régime, marque dans l'ordre de la pensée un courageux effort pour éliminer les dernières traces du déisme naturaliste qui avait caractérisé la pensée chrétienne pendant la période de l'*Aufklärung* et pour centrer à nouveau celle-ci sur les données fondamentales de la Révélation : les mystères du Verbe Incarné, de l'Église, de la grâce et des sacrements.

Ce vaste effort de restauration doctrinale a son pendant dans un effort d'approfondissement de la vie chrétienne qui est sans doute le résultat le plus notable et le principal mérite de ce long pontificat : l'Église en sort sensiblement plus « religieuse ». Pie IX inaugure la lignée des papes contemporains, qui ont eu à cœur de rester sur la chaire de saint Pierre avant tout prêtres et pasteurs d'âmes : au delà de la sauvegarde intégrale du message doctrinal et de la défense des droits de l'Église, il a conscience d'être également responsable devant Dieu de la vie chrétienne des fidèles et de la sanctification des prêtres. Bien des choses ont changé dans le monde et dans l'Église entre 1846 et 1878. Aucune peut-être n'a changé davantage que la qualité de la vie catholique moyenne. Au prix du sacrifice de certaines traditions précieuses qui avaient fait la valeur de l'ancien clergé de France ou de l'école de Sailer, mais dont la qualité même paraissait devoir les réserver à une petite élite, un vaste courant de dévotion populaire et de spiritualité sacerdotale est allé en s'épanouissant toujours davantage ; on lui a souvent reproché d'être superficiel et beaucoup trop extérieur, mais l'efflorescence des œuvres et l'énorme développement des congrégations religieuses sont là pour démentir ce jugement simpliste. Or, si bien des éléments locaux ainsi que l'action de la Compagnie de Jésus ont joué dans cette évolution un rôle indiscutable, Pie IX lui-même y a contribué pour une large part. D'abord, parce qu'il est apparu personnellement comme un exemple pour le mouvement de dévotion : on peut sourire de cette « sorte d'hagiographie naïve ou intéressée qui entoure sa personne d'une auréole »[1] et qui prend parfois les allures d'une

---

(1) Ch. Pouthas, *Le pontificat de Pie IX*, p. 297.

adulation assez déplaisante ; mais on ne peut nier son influence sur la renaissance spirituelle du XIXᵉ siècle. Ensuite, et surtout, parce que, homme de décision et d'autorité, il a consacré une bonne part de son temps et de ses efforts à activer, parfois même à bousculer, la lente évolution qui s'était amorcée dès le lendemain de la grande crise révolutionnaire. Et c'est précisément parce qu'il croyait indispensable à la réussite de cette œuvre de restauration chrétienne une attitude intransigeante dans les domaines pratique et doctrinal, qu'il se força, malgré ses tendances personnelles à la conciliation et à l'apaisement, à répéter sans cesse, parfois avec un manque regrettable de nuances, les grands principes dont il était réservé au génie de Léon XIII de formuler les applications à la fois parfaitement orthodoxes et pourtant adaptées au monde nouveau qui était né avec Jean-Marie Mastai à la fin du XVIIIᵉ siècle.

TABLE DES MATIÈRES

Imprimé en France par l'Imprimerie André Brulliard, Saint-Dizier (Haute-Marne).
Numéro d'impression : 1952-6. — Dépôt légal : N° 1338.